# Marine Carotenoids

# Marine Carotenoids

Editors

**Tatsuya Sugawara**
**Takashi Maoka**

MDPI • Basel • Beijing • Wuhan • Barcelona • Belgrade • Manchester • Tokyo • Cluj • Tianjin

*Editors*
Tatsuya Sugawara
Division of Applied Biosciences,
Graduate School of Agriculture,
Kyoto University
Japan

Takashi Maoka
Research Institute for
Production Development
Japan

*Editorial Office*
MDPI
St. Alban-Anlage 66
4052 Basel, Switzerland

This is a reprint of articles from the Topical Collection published online in the open access journal *Marine Drugs* (ISSN 1660-3397) (available at: https://www.mdpi.com/journal/marinedrugs/special_issues/marine_carotenoids_collection).

For citation purposes, cite each article independently as indicated on the article page online and as indicated below:

LastName, A.A.; LastName, B.B.; LastName, C.C. Article Title. *Journal Name* **Year**, *Article Number*, Page Range.

**ISBN 978-3-03943-190-8 (Hbk)**
**ISBN 978-3-03943-191-5 (PDF)**

# Contents

# About the Editors

**Tatsuya Sugawara** received an MS in Agriculture from Tohoku University in 1993. Tatsuya was a research worker for NOF CORPORATION from 1993 to 1997, before receiving a PhD in Agriculture from Tohoku University in 2000. Tatsuya was a Postdoctoral Fellow of the National Food Research Institute from 2000 to 2002 and a Postdoctoral Fellow of the National Institute of Health and Nutrition from 2002 to 2004. Finally, Tatsuya worked as an Associate Professor at Kyoto University from 2004 to 2013, before becoming a Full Professor in 2013.

**Takashi Maoka** was awarded an MS from Kyoto Pharmaceutical University in 1981. Takashi then began as a research assistant in the Department of Natural Product Chemistry of Kyoto Pharmaceutical University and graduated with a PhD in 1991. Takashi became Chief Researcher of Food Chemistry in Research Institute for Production Development in 1993 and Director of Research Institute for Production Development in 2017.

# Preface to "Marine Carotenoids"

Carotenoids represent a large group of isoprenoid structures with many different structural characteristics and biological activities. They are the most important pigments of those occurring in nature and are responsible for the various colors of different fruits, vegetables, and plant parts. Marine carotenoids are responsible for the color of many fish, shellfish, and algae. However, while there are many available papers and reviews on carotenoids of terrestrial origin, there has been relatively little research on the impact of marine carotenoids on human health.

Research on the potential beneficial effects of marine carotenoids has been particularly focused on astaxanthin and fucoxanthin, as they are major marine carotenoids. Both carotenoids show strong antioxidant activity, which is attributed to quenching singlet oxygen and scavenging free radicals. The potential role of the carotenoids as dietary antioxidants has been suggested to be one of the main mechanisms for their preventive effects against cancer, inflammation, etc. However, it is difficult to explain their biological activities based only antioxidant activity. Other mechanisms of action that are independent to their antioxidant properties are also likely to play a role. The mechanisms should be based on the regulatory effects of marine carotenoids on particular bio-molecules. This activity of carotenoids is responsible for the characteristic chemical structures, which differ depending on the length of the polyene, nature of the end group, and the substituents they contain.

This Topical Collection of *Marine Drugs* is dedicated to marine carotenoids, and will be focused on the benefits of carotenoids for human beings. For a better understanding of the physiological effects of marine carotenoids, this collection will include recent developments in the presence, analysis, chemistry, and biochemistry of marine carotenoids. We are pleased to serve as Editors for this collection and sincerely hope it will encourage other scientists to work in the exciting field of marine carotenoids.

**Tatsuya Sugawara, Takashi Maoka**
*Editors*

marine drugs

MDPI

*Review*

# Astaxanthin: A Potential Therapeutic Agent in Cardiovascular Disease

**Robert G. Fassett [1,2,*] and Jeff S. Coombes [2]**

1 Renal Research Royal Brisbane and Women's Hospital and The University of Queensland School of Medicine, Level 9 Ned Hanlon Building, Butterfield Street, Brisbane, Queensland 4029, Australia

2 School of Human Movement Studies, The University of Queensland, St. Lucia, Brisbane, Queensland 4072, Australia; E-Mail: jcoombes@uq.edu.au

* Author to whom correspondence should be addressed; E-Mail: r.fassett@uq.edu.au; Tel.: +61-419399571; Fax: +61-736368572.

*Received: 7 February 2011; in revised form: 14 March 2011 / Accepted: 18 March 2011 / Published: 21 March 2011*

**Abstract:** Astaxanthin is a xanthophyll carotenoid present in microalgae, fungi, complex plants, seafood, flamingos and quail. It is an antioxidant with anti-inflammatory properties and as such has potential as a therapeutic agent in atherosclerotic cardiovascular disease. Synthetic forms of astaxanthin have been manufactured. The safety, bioavailability and effects of astaxanthin on oxidative stress and inflammation that have relevance to the pathophysiology of atherosclerotic cardiovascular disease, have been assessed in a small number of clinical studies. No adverse events have been reported and there is evidence of a reduction in biomarkers of oxidative stress and inflammation with astaxanthin administration. Experimental studies in several species using an ischaemia-reperfusion myocardial model demonstrated that astaxanthin protects the myocardium when administered both orally or intravenously prior to the induction of the ischaemic event. At this stage we do not know whether astaxanthin is of benefit when administered after a cardiovascular event and no clinical cardiovascular studies in humans have been completed and/or reported. Cardiovascular clinical trials are warranted based on the physicochemical and antioxidant properties, the safety profile and preliminary experimental cardiovascular studies of astaxanthin.

**Keywords:** antioxidants; xanthophyll carotenoid; inflammation; *Haematococcus pluvialis*; oxidative stress

## 1. Introduction

Astaxanthin is a xanthophyll carotenoid of predominantly marine origin, with potent antioxidant and anti-inflammatory effects demonstrated in both experimental and human studies. Oxidative stress and inflammation are common pathophysiological features of atherosclerotic cardiovascular disease hence astaxanthin may have a potential therapeutic role in this condition. This review will summarise the available evidence suggesting astaxanthin may be of therapeutic value in cardiovascular disease.

## 2. Oxidative Stress and Inflammation

Oxidative stress and inflammation are established non-traditional risk factors for atherosclerosis associated cardiovascular morbidity and mortality [1]. Dietary antioxidants can reduce the oxidation of lipids and proteins and have the potential to protect from the development of arterial stiffening and atherosclerosis [2–4]. Cross-sectional and prospective observational studies have demonstrated an association between the intake of dietary antioxidants and/or their plasma levels and a reduction of cardiovascular events [5–10]. This supports the theory that oxidative stress is a pathophysiological process involved in atherosclerotic vascular damage. Also, a reduced dietary antioxidant intake is associated with oxidative stress and inflammation [11]. Newer more potent dietary antioxidants such as astaxanthin have yet to be studied in this setting. Studies that have assessed the intake of β-carotene or dietary β-carotene supplementation have shown higher β-carotene consumption is associated with a reduction in cardiovascular disease [6,12–17]. Other than a few studies [18–20], cardiovascular intervention trials using antioxidants have not demonstrated benefits [21–23]. This may be because study participants did not have oxidative stress and/or the antioxidants used were insufficiently potent. In addition, it is becoming recognized that there is communication between oxidative stress and inflammatory processes leading to the additional hypothesis that antioxidants may be able to modify both deleterious events. Further research is needed studying antioxidants with different biological actions in patients with demonstrated oxidative stress.

## 3. Carotenoids

Carotenoids are ubiquitous, and found in high concentrations in plants, algae and microorganisms. Humans and other animals cannot synthesize them and therefore are required to source them in their diet [24]. Carotenoids are classified, according to their chemical structure, into carotenes and xanthophylls. The carotene carotenoids include β-carotene and lycopene and the xanthophyll carotenoids include lutein, canthaxanthin, zeaxanthin, violaxanthin, capsorubin and astaxanthin [25,26].

The effects of carotenoids vary dependent on how they interact with cell membranes [25]. The effects of astaxanthin, zeaxanthin, lutein, β-carotene and lycopene on lipid peroxidation have been assessed using a polyunsaturated fatty acid enriched membrane model [25,27]. Non-polar carotene carotenoids such as lycopene and β-carotene caused membrane disorder and lipid peroxidation in contrast to the polar xanthophyll carotenoid astaxanthin, which preserved membrane structure [27]. Contrasting effects of different carotenoids may be responsible for the differing biological effects seen in clinical studies. For instance, in some studies the non-polar carotenoid, β-carotene has been shown to have no benefit on cardiovascular disease [28–32] and in fact it may be pro-oxidant at high doses [33]. In contrast, the polar carotenoid astaxanthin has protective effects on the cardiovascular system demonstrated in animal studies. However, this has not been studied in human clinical trials [34–36]. β-carotene at physiological levels may act in differing ways when ultraviolet A light A (UVA) acts on keratinocytes including vitamin A-independent pathways [37]. Astaxanthin, canthaxanthin and β-carotene had differential effects on UVA-induced oxidative damage [38]. In addition, carotenoids may also alter the immune response [39] and transcription [40].

## 4. Astaxanthin

Astaxanthin contains two oxygenated groups on each ring structure (see Figure 1), which is responsible for its enhanced antioxidant features [41]. It is found in living organisms particularly in the marine environment where it is present in microalgae, plankton, krill and seafood. It gives salmon, trout, and crustaceans such as shrimp and lobster their distinctive reddish coloration [42]. It is also present in yeast, fungi, complex plants and the feathers of some birds including flamingos and quail [42]. In 1987, the United States Food and Drug Administration approved astaxanthin as a feed additive for use in the aquaculture industry and in 1999 it was approved for use as a dietary supplement (nutraceutical) [41]. The microalgae *Haematococcus pluvialis* produces the astaxanthin isomer (3*S*, 3*S*′), which is the same as the form found in wild salmon. Synthesis of astaxanthin is not possible in humans and it cannot be converted to vitamin A, which means excess intake will not cause hypervitaminosis A toxicity [43,44]. Astaxanthin and canthaxanthin are scavengers of free radicals, chain-breaking antioxidants and potent quenchers of reactive oxygen and nitrogen species including singlet oxygen, single and two electron oxidants [45–47]. They (astaxanthin and canthaxanthin) have terminal carbonyl groups that are conjugated to a polyene backbone [26] and are more potent antioxidants and scavengers of free radicals than carotene carotenoids such as β-carotene [47,48]. For these reasons dietary supplementation with astaxanthin has the potential to provide antioxidant protection of cells and from atherosclerotic cardiovascular disease [49].

**Figure 1.** Molecular structure of astaxanthin.

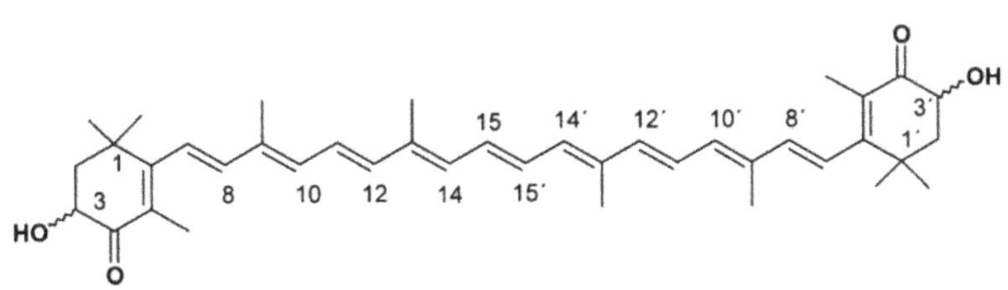

## 5. Astaxanthin Formulations

### *5.1. Astaxanthin of Marine Origin*

Astaxanthin used in nutritional supplements is usually a mixture of configurational isomers produced by *Haematococcus pluvialis,* a unicellular microalga [50]. Astaxanthin can be produced in its natural forms on a large scale [51]. The initial production of astaxanthin from *Haematococcus pluvialis* uses closed culture systems followed by a 5–7 day, "reddening" cycle, conducted in open culture ponds. At each production stage, the cultures are closely monitored by microscopic examination to ensure they remain free of contamination. After the reddening cycle, *Haematococcus pluvialis* cultures are harvested, washed and dried. The final step for the production of astaxanthin is extraction of dried *Haematococcus pluvialis* biomass using supercritical carbon dioxide to produce a purified oleoresin, which is free of any contamination. Other sources used for the commercial production of astaxanthin include cultures of *Euphausia pacifica* (Pacific krill), *Euphausia superba* (Antarctic krill), *Pandalus borealis* (shrimp) and *Xanthophyllomyces dendrorhous,* formerly *Phaffia rhodozyma* (yeast). Astaxanthin from natural sources varies considerably from one organism to another. For instance, the astaxanthin found in seafood will depend on the stereoisomer ingested. Astaxanthin produced by *haematococcus pluvialis,* consists of the (3-*S*,3′-*S*) stereoisomer which is most commonly used in aquaculture. It is therefore the form most commonly consumed by humans.

### *5.2. Synthetic Astaxanthin*

There are three stereoisomers of astaxanthin; (3-*R*,3′-*R*), (3-*R*,3′-*S*) and (3-*S*,3′-*S*). Disodium disuccinate astaxanthin (DDA) is a synthetic astaxanthin containing a mixture of all three stereoisomers, in the proportions 1:2:1. DDA was manufactured by Cardax Pharmaceuticals and used in animal studies investigating the myocardial ischemia-reperfusion injury models [34–36,52–54]. This form of astaxanthin was touted to have better aqueous solubility, unlike other carotenoids, and this enabled both oral and intravenous administration. DDA is no longer available but the same company now produces a second synthetic astaxanthin compound; Heptax/XanCor, CDX-085. The company claims that it is developed for thrombotic protection, triglyceride reduction, metabolic syndrome, and inflammatory liver disease. In addition, it has increased water dispersibility and enhanced bioavailability compared to natural astaxanthin and DDA. The synthetic forms are metabolized via hydrolysis in the intestine yielding free astaxanthin for intestinal absorption. CDX-085 has been used in one study, discussed below [55].

It is not yet clear which form of astaxanthin should be administered in clinical studies, the natural form from the marine environment or a synthetic form. As the proportions of stereoisomers, vary between these different forms of astaxanthin they may not be therapeutically equivalent [56]. Thus synthetic astaxanthin could result in different outcomes when assessed clinically [57].

## 6. Astaxanthin-Experimental Studies

Experimental studies undertaken with astaxanthin specifically relevant to the cardiovascular system are summarised in Table 1. Astaxanthin attenuates mediators of oxidative stress and inflammation and has shown beneficial effects in non-cardiovascular models of disease [58–69]. In addition, astaxanthin has decreased markers of lipid peroxidation [70], inflammation [61,62,67,68] and thrombosis [55].

**Table 1.** Animal studies investigating the cardiovascular effects of astaxanthin.

| Study | Model | Dosage | Duration/timing of supplementation | Effects of astaxanthin |
|---|---|---|---|---|
| Lauver *et al.* 2008 [34] | Dog with occlusive carotid artery thrombus | DDA 10, 30, or 50 mg/kg/body weight IV | 30 min after occlusion | - Reduced incidence of secondary thrombosis |
| Aoi *et al.* 2003 [63] | C57BL/6 mice | Diet supplemented with astaxanthin 0.02% weight/weight and food intake recorded | 3 weeks | - Attenuation of exercise increased 4-hydroxy-2-nonenal-modified protein and 8-hydroxy-2′-deoxyguanosine in cardiac and gastrocnemius muscle<br>- Attenuation of exercise increases in creatine kinase and myeloperoxidase activity in cardiac and gastrocnemius muscle<br>- Astaxanthin accumulated in cardiac and gastrocnemius muscle |
| Gross and Lockwood 2004 [35] | Myocardial infarct model Sprague-Dawley rats | DDA 25/50/75 mg/kg body weight intravenously daily | 4 days prior to myocardial infarction | - Myocardial infarct size significantly reduced |
| Hussein *et al.* 2005 [71] | Stroke prone Spontaneously hypertensive rats | Astaxanthin 50 mg/kg body weight/day | 5 weeks | - Significant blood pressure reduction<br>- Delayed incidence of stroke |
| Lauver *et al.* 2005 [52] | Rabbit model of myocardial ischemia/reperfusion | DDA 50 mg/kg body weight/day intravenously | 5 days | - Significant reduction in complement activation<br>- Significant reduction in myocardial infarct size |
| Gross *et al.* 2005 [54] | Canine model of myocardial ischemia/reperfusion | DDA 50 mg/kg body weight/day intravenously | 2 h or daily for four days | - Significant reduction in myocardial infarct size<br>- Two of three dogs treated for four days had 100% cardiac protection |

**Table 1.** *Cont.*

| | | | | |
|---|---|---|---|---|
| Gross *et al.* 2006 [36] | Sprague-Dawley rats Left anterior descending coronary artery occlusion/reperfusion | DDA 125 or 500 mg/kg body weight/day orally | 7 days | - Astaxanthin loading of myocardium indicating good bioavailability<br>- Trends in lowering of lipid peroxidation products<br>- Significant reduction in myocardial infarct size |
| Hussein *et al.* 2006 [72] | Spontaneously hypertensive rats | Astaxanthin 5% in olive oil (5 mg/kg/day orally) | 7 days | - Significant reduction in nitric oxide end products<br>- Significant reduction in elastin bands in aorta<br>- Significant reduction in wall/lumen arterial ratio in coronary arteries |
| Aoi *et al.* 2008 [73] | ICR mice | Astaxanthin 0.02% w/w | 4 weeks | Astaxanthin increased fat utilization during exercise and prolonged exercise<br>Astaxanthin prevented increase in hexanoyl-lysine modification of CPT I with exercise |
| Nakao *et al.* 2010 [74] | BALC/c mice | Astaxanthin 0, 0.02, 0.08% orally/day | 8 weeks | - No change in blood glutathione concentration<br>- No change in lymphocyte mitochondrial membrane potential<br>- Higher myocardial mitochondrial membrane potential and contractility index |
| Khan *et al.* 2010 [55] | C57BL/6 mice<br><br>Human umbilical vein endothelial cells and platelets from Wistar-Kyoto rats | CDX-085 500 mg/kg body weight/day | 14 days | - CDX-085 administered orally to C57BL/6 mice was associated with presence of free astaxanthin in the plasma, heart, liver and platelets<br>- Mice fed astaxanthin had significantly increased basal arterial blood flow and delay in occlusive thrombosis after endothelial injury<br>- Human umbilical vein endothelial cells and platelets from Wistar-Kyoto rats treated with free astaxanthin has significantly increased release of nitric oxide and decreased peroxynitrite levels |

### 6.1. Cardiovascular Studies

A series of experiments have been conducted to assess the efficacy of DDA in protecting the myocardium using the myocardial ischemia-reperfusion model in rats, rabbits and dogs [35,36,53,54]. Prior treatment for four-days with intravenous DDA using doses of 25, 50 and 75 mg/kg body weight in Sprague-Dawley rats significantly reduced myocardial infarct size [35]. The degree of cardiac protection correlated with the dose of DDA administered. In a study in rabbits using a myocardial ischaemia-reperfusion model prior intravenous treatment with 50 mg/kg/day of DDA for four days resulted in a significant decrease in the size of the myocardial infarction and an improvement in myocardial salvage [52]. Animals treated with DDA had an attenuation of inflammation and complement activation suggesting there was a reduction in tissue inflammation [52]. In another study using a dog model intravenous DDA was administered daily for four-days prior to occlusion of the left anterior descending coronary artery or two hours prior to coronary artery occlusion [54]. After an hour of coronary occlusion and three hours of reperfusion there was a significant reduction in myocardial infarct size in the dogs treated with DDA. In the four-day treatment group, two out of three dogs had complete cardiac protection [54]. In a rat study, the effects of seven days of pre-treatment with oral DDA, 125 and 500 mg/kg/day on the concentrations of free astaxanthin in myocardial tissue [36]. The astaxanthin concentration in the myocardium was 400 nM after oral DDA at a dose of 125 mg/kg/day for seven-days and it was 1634 nM after 500 mg/kg/day. There was also a reduction of multiple lipid peroxidation products. The doses of DDA used in these experiments were quite high and at this stage it is not known whether such doses would be safe to use in humans.

The effects of astaxanthin on blood pressure (BP) were assessed in spontaneously hypertensive rats (SHR). There was a significant reduction in BP after 14-days of oral astaxanthin administration whereas this did not occur in normotensive Wistar Kyoto rats [71]. Astaxanthin administered orally for five-weeks in stroke prone SHR also resulted in a significant BP reduction [71]. Oral astaxanthin also enhanced nitric oxide induced vascular relaxation in the rat aortas [71] In experiments in SHR, oral astaxanthin significantly decreased nitric oxide end products indicating that it may be exerting its BP effects via this pathway [72]. Studies using the SHR aorta and coronary arteries demonstrated that astaxanthin reduced the wall/lumen ratio in coronary arteries and decreased elastin bands in the aorta [72]. This suggests that astaxanthin may beneficially mediate atherosclerotic CVD processes.

Recently, a series of two experiments were reported in the one article, one using the synthetic astaxanthin (CDX-085) and the other using free astaxanthin [55]. CDX-085 administered orally to C57BL/6 mice resulted in the presence of free astaxanthin in the plasma, heart, liver and platelets. Mice that were fed astaxanthin had significantly increased basal arterial blood flow and a delay in occlusive thrombosis after endothelial injury. Also, in an *in vitro* study, human umbilical vein endothelial cells and platelets isolated from Wistar-Kyoto rats that were treated with free astaxanthin has significantly increased nitric oxide release and a decrease in peroxynitrite levels [55]. The authors concluded the results support the potential of astaxanthin as a potential therapy to prevent thrombosis associated with cardiovascular disease.

Astaxanthin administered to C57BL/6 mice resulted in a reduction in exercise-induced increases in the oxidative stress biomarkers 8-hydroxy-2′-deoxyguanosine and 4-hydroxy-2-nonenal-modified protein in both cardiac and gastrocnemius muscle [63]. Increases in myeloperoxidase and creatinine kinase activity in cardiac and gastrocnemius muscle were also reduced by astaxanthin. After three-weeks of astaxanthin supplementation there was evidence of accumulation of astaxanthin in gastrocnemius and cardiac muscle. Astaxanthin given to female BALB/c mice for eight-weeks resulted in a dose dependent increase in plasma astaxanthin but no significant changes in blood glutathione or change in lymphocyte mitochondrial membrane potential and cardiac contractility index measured on echocardiography. The mice that were fed 0.08% astaxanthin in the diet had higher cardiac mitochondrial membrane potential and contractility index compared with control animals [74]. This suggests dietary astaxanthin provides cardiac protection. Astaxanthin administered for four weeks to eight week old ICR mice resulted in increased exercised induced fat utilization and prevention of increased hexanoyl-lysine modification of carnitine palmitoyltransferase I (CTP I) [73]. In a canine carotid artery thrombosis model, administration of DDA resulted in a dose-dependent reduction in carotid artery re-thrombosis and a reduction of re-thrombosis after thrombolysis but there was no effect on hemostasis [34].

### *6.2. Diabetes Studies*

Diabetes mellitus and its associated nephropathy is a common cause of chronic kidney disease and is complicated by accelerated atherosclerotic cardiovascular disease [75]. In studies involving diabetic db/db mice, supplementation with astaxanthin produced a reduction in the levels of blood glucose [60]. In the kidney there was significantly decreased relative mesangial area in the animals supplemented with astaxanthin. Also proteinuria and urinary excretion of 8-OHdG were attenuated. Mice supplemented with astaxanthin had less glomerular 8-OHdG immunoreactive cells [60]. Hyperglycemia induced reactive oxygen species production, activation of transcription factors, and cytokine expression and production by normal human mesangial cells was suppressed significantly by astaxanthin [66].

## 7. Astaxanthin Studies in Humans

Although no cardiovascular outcomes studies using astaxanthin have been reported in humans there have been clinical studies that have investigated the effects of astaxanthin in human health and other diseases (Table 2). The majority of these have been conducted in healthy participants who volunteered to assess astaxanthin dose, bioavailability, safety and oxidative stress, which are all potentially relevant to the cardiovascular system. Studies have also been conducted in other medical conditions such as reflux oesophagitis, where measurements of oxidative stress and/or inflammation have been included.

### *7.1. Dosing*

Human clinical studies have used oral astaxanthin in a dose that ranges from 4 mg up to 100 mg/day, given from a one off dose up to durations of one-year (Table 2).

**Table 2.** Clinical studies investigating the safety, bioavailability and effects of astaxanthin on oxidative stress.

| Study | Study population ($n$ = subject numbers) | Dosage | Study design | Duration of supplementation | Effects of astaxanthin |
|---|---|---|---|---|---|
| Iwamoto *et al.* 2000 [70] | Volunteers ($n$ = 24) | Different doses: 1.8, 3.6, 14.4, 21.6 mg/day | Open labelled | 2 weeks | - Reduction of LDL oxidation |
| Osterlie *et al.* 2000 [76] | Middle aged male volunteers ($n$ = 3) | 100 mg | Open labelled | Single dose | - Astaxanthin taken up by VLDL chylomicrons |
| Mercke Odeberg *et al.* 2003 [77] | Healthy male volunteers ($n$ = 32) | 40 mg | Open labelled parallel | Single dose | - Enhanced bioavailability with lipid based formulation |
| Spiller *et al.* 2003 [78] | Healthy adults ($n$ = 35) | 6 mg/day (3 × 2 mg tablets/day) | Randomised, double blind, placebo controlled | 8 weeks | - Demonstrated safety assessed by measures of blood pressure and biochemistry |
| Coral-Hinostroza *et al.* 2005 [79] | Healthy adult males ($n$ = 3) | 10 mg and 100 mg | Open labelled | Single dose or 4 weeks | - $C_{max}$ 0.28 mg/L at 11.5 h at high dose and 0.08 mg/L at low dose<br>- Elimination half life 52 ± 40 h<br>- *z*-isomer selectively absorbed |
| Karppi *et al.* 2007 [80] | Healthy non-smoking Finnish males ($n$ = 40) | 8 mg/day | Randomised, double blind, placebo controlled | 12 weeks | - Intestinal absorption adequate with capsules<br>- Reduced levels of plasma 12 and 15 hydroxy fatty acids<br>- Decreased oxidation of fatty acids |

**Table 2.** *Cont.*

| | | | | | |
|---|---|---|---|---|---|
| Parisi *et al.* 2008 [81] | Non-advanced age related macular degeneration ($n = 27$) | 4 mg/day | Randomised controlled trial open labelled no placebo | 12 months | - Improved central retinal dysfunction in age related macular degeneration when administered with other antioxidants |
| Miyawaki *et al.* 2008 [82] | Healthy males ($n = 20$) | 6 mg/day | Single blind, placebo controlled | 10 days | - Decreased whole blood transit time (improved blood rheology) |
| Rufer *et al.* 2008 [83] | Healthy males ($n = 28$) | 5 µg/g salmon flesh (wild *vs.* aquacultured ) | Randomised, double blind, placebo controlled | 4 weeks | - Bioavailability initially better with ingestion of aquacultured salmon but equivalent at day 28. Isomer (3*S*, 3′*S*) greater in plasma compared with isomer proportion in salmon flesh |
| Park *et al.* 2010 [84] | Healthy females ($n = 14$) | 0, 2, 8 mg/day | Randomised, double blind, placebo controlled | 8 weeks | - Decreased plasma 8-hydroxy-2′-deoxyguanosine after week four in those taking astaxanthin.<br>- Lower CRP after week four in those taking 2 mg/day astaxanthin |

### 7.2. Bioavailability

Astaxanthin bioavailability from the marine environment was assessed in a randomised double blind trial in 28 volunteers [83]. Participants were given either 250 g of wild salmon or aquaculture salmon (5 μg/g) to eat. Wild salmon ingest astaxanthin naturally from krill whereas aquacultured salmon acquire it from fish that are fed astaxanthin that might be derived from a synthetic source. Plasma levels of astaxanthin were higher at 3, 6, 10 and 14 days during ingestion of the aquacultured compared with the wild salmon. Plasma levels of the (3-*S*, 3′-S) isomer of astaxanthin appeared at higher levels than its proportionate level in the flesh of the salmon. This suggests that isomers of astaxanthin might have different bioavailability. The plasma isomers of astaxanthin have also been studied after ingestion of single oral dose of 10mg and also 100 mg over four-weeks. Astaxanthin plasma elimination half-life was 52 (SD 40) h and there was a non-linear dose response and selective absorption of z-isomers [79].

### 7.3. Safety

The safety of astaxanthin administered orally was assessed in a double-blind, randomised placebo-controlled trial undertaken in healthy adults [78]. Volunteers took either 6 mg/day of astaxanthin or placebo for eight-weeks. BP and biochemistry measured after four and eight weeks of therapy revealed no significant differences in these parameters between treatment and placebo groups and these did not differ from baseline. The authors concluded that healthy adults could safely consume 6 mg/day of astaxanthin derived from a *Haematococcus pluvialis* algal extract. Measuring whole blood transit time in 20 healthy males was used to assess the effects of astaxanthin on blood rheology in humans. Six milligrams of oral astaxanthin per day for ten days improved blood rheology as evidenced by decreased whole blood transit time [82]. Escalating concentrations of astaxanthin were tested *in vitro* with blood taken from volunteers, 8 of whom were taking asprin and 12 who were not [85]. Even supra-therapeutic concentrations of astaxanthin had no adverse effects on indices of platelet, coagulation and fibrinolytic function. These results support the safety profile of astaxanthin for future clinical trials. No significant side effects have been reported so far in published human studies in which astaxanthin was administered to humans.

### 7.4. Oxidative Stress and Inflammation

Oral supplementation with astaxanthin in studies in healthy human volunteers and patients with reflux oesophagitis demonstrated a significant reduction in oxidative stress, hyperlipidemia and biomarkers of inflammation [70,80,86]. In a study involving 24 healthy volunteers who ingested astaxanthin in doses from 1.8 to 21.6 mg/day for two weeks, LDL lag time, as a measure of susceptibility of LDL to oxidation, was significantly greater in astaxanthin treated participants indicating inhibition of the oxidation of LDL [70]. Plasma levels of 12- and 15-hydroxy fatty acids were significantly reduced in 40 healthy non-smoking Finnish males given astaxanthin [80] suggesting astaxanthin decreased the oxidation of fatty acids [80]. The effects of dietary astaxanthin in doses of

0, 2 or 8 mg/day, over 8 weeks, on oxidative stress and inflammation were investigated in a double blind study in 14 healthy females [84]. Although these participants did not have oxidative stress or inflammation those taking 2 mg/day had lower CRP at week eight. There was also a decrease in DNA damage measured using plasma 8-hydroxy-2′-deoxyguanosine after week four in those taking astaxanthin. Astaxanthin therefore appears safe, bioavailable when given orally and is suitable for further investigation in humans.

## 8. Clinical Trial Using Astaxanthin

A double-blind randomised placebo-controlled clinical trial (Xanthin study) is currently being conducted to assess the effects of astaxanthin 8mg orally day on oxidative stress, inflammation and vascular function in patients that have received a kidney transplant [87]. Patients in the study undertake measurements of surrogate markers of cardiovascular disease including aortic pulse wave velocity, augmentation index, brachial forearm reactivity and carotid artery intima-media thickness. Depending on the results from this pilot study a large randomised controlled trial assessing major cardiovascular outcomes such as myocardial infarction and death may be warranted.

## 9. Conclusions

Experimental evidence suggests astaxanthin may have protective effects on cardiovascular disease when administered prior to an induced ischemia-reperfusion event. In addition, there is evidence that astaxanthin may decrease oxidative stress and inflammation which are known accompaniments of cardiovascular disease. At this stage we do not know whether astaxanthin has any therapeutic value in human cardiovascular disease either in a preventative capacity or when administered after a cardiovascular insult. It has been proposed that astaxanthin may provide cardiovascular protection through reducing oxidative stress, which is one of the non-traditional risk factors for the development of atherosclerotic cardiovascular disease. The role of oxidative stress in cardiovascular disease is supported by evidence from observational studies that have found associations between antioxidant intake, oxidative stress and cardiovascular outcomes. Despite this, clinical intervention studies using antioxidants including vitamin E, β-carotene and vitamin C, have not proved successful [22,23]. These intervention studies may have failed because of flawed design where patients were not included based on the presence of oxidative stress. Hence, many participants may not have been in a state of oxidative stress and able to benefit from antioxidant therapy. Also, in those participants where oxidative stress may have existed there was no way of assessing whether the therapy adequately corrected this. Thus, the antioxidants used such as vitamin E, β-carotene and vitamin C may not have been effective because insufficient doses were used or an inadequate length of therapy followed to correct the oxidative stress. Some antioxidants such as β-carotene may be pro-oxidant at higher doses, which could have confounded study results.

Astaxanthin is a potent antioxidant and based on its physicochemical properties and the results of preliminary experimental studies in ischaemia-reperfusion models of cardiovascular disease, it warrants consideration for testing in human clinical trials. There have been no safety concerns noted so far in human clinical studies where astaxanthin has been administered. As astaxanthin is a potent antioxidant and is associated with membrane preservation, it may protect against oxidative stress and inflammation and provide cardiovascular benefits.

## Acknowledgements

Cyanotech the manufacturer of BioAstin, a proprietary brand of astaxanthin, is providing financial support and astaxanthin capsules and placebo for a clinical trial being conducted by the authors. The sponsor has no role in the study itself.

## References

1. Dzau, V.J.; Antman, E.M.; Black, H.R.; Hayes, D.L.; Manson, J.E.; Plutzky, J.; Popma, J.J.; Stevenson, W. The cardiovascular disease continuum validated: Clinical evidence of improved patient outcomes. Part II: Clinical trial evidence (acute coronary syndromes through renal disease) and future directions. *Circulation* **2006**, *114*, 2871–2891.
2. Ellingsen, I.; Seljeflot, I.; Arnesen, H.; Tonstad, S. Vitamin C consumption is associated with less progression in carotid intima media thickness in elderly men: A 3-year intervention study. *Nutr. Metab. Cardiovasc. Dis.* **2009**, *19*, 8–14.
3. Carty, J.L.; Bevan, R.; Waller, H.; Mistry, N.; Cooke, M.; Lunec, J.; Griffiths, H.R. The effects of vitamin C supplementation on protein oxidation in healthy volunteers. *Biochem. Biophys. Res. Commun.* **2000**, *273*, 729–735.
4. Carpenter, K.L.; Kirkpatrick, P.J.; Weissberg, P.L.; Challis, I.R.; Dennis, I.F.; Freeman, M.A.; Mitchinson, M.J. Oral alpha-tocopherol supplementation inhibits lipid oxidation in established human atherosclerotic lesions. *Free Radic. Res.* **2003**, *37*, 1235–1244.
5. Stampfer, M.J.; Hennekens, C.H.; Manson, J.E.; Colditz, G.A.; Rosner, B.; Willett, W.C. Vitamin E consumption and the risk of coronary disease in women. *N. Engl. J. Med.* **1993**, *328*, 1444–1449.
6. Rimm, E.B.; Stampfer, M.J.; Ascherio, A.; Giovannucci, E.; Colditz, G.A.; Willett, W.C. Vitamin E consumption and the risk of coronary heart disease in men. *N. Engl. J. Med.* **1993**, *328*, 1450–1456.
7. Gey, K.F.; Puska, P. Plasma vitamins E and A inversely correlated to mortality from ischemic heart disease in cross-cultural epidemiology. *Ann. N. Y. Acad. Sci.* **1989**, *570*, 268–282.
8. Willcox, B.J.; Curb, J.D.; Rodriguez, B.L. Antioxidants in cardiovascular health and disease: Key lessons from epidemiologic studies. *Am. J. Cardiol.* **2008**, *101*, 75D–86D.
9. Frei, B. Cardiovascular disease and nutrient antioxidants: Role of low-density lipoprotein oxidation. *Crit. Rev. Food Sci. Nutr.* **1995**, *35*, 83–98.
10. Steinberg, D. Antioxidants in the prevention of human atherosclerosis. Summary of the proceedings of a National Heart, Lung, and Blood Institute Workshop: September 5–6, 1991, Bethesda, Maryland. *Circulation* **1992**, *85*, 2337–2344.
11. Helmersson, J.; Arnlov, J.; Larsson, A.; Basu, S. Low dietary intake of beta-carotene, alpha-tocopherol and ascorbic acid is associated with increased inflammatory and oxidative stress status in a Swedish cohort. *Br. J. Nutr.* **2009**, *101*, 1775–1782.

12. Osganian, S.K.; Stampfer, M.J.; Rimm, E.; Spiegelman, D.; Manson, J.E.; Willett, W.C. Dietary carotenoids and risk of coronary artery disease in women. *Am. J. Clin. Nutr.* **2003**, *77*, 1390–1399.
13. Ford, E.S.; Giles, W.H. Serum vitamins, carotenoids, and angina pectoris: Findings from the National Health and Nutrition Examination Survey III. *Ann. Epidemiol.* **2000**, *10*, 106–116.
14. Klipstein-Grobusch, K.; Geleijnse, J.M.; den Breeijen, J.H.; Boeing, H.; Hofman, A.; Grobbee, D.E.; Witteman, J.C. Dietary antioxidants and risk of myocardial infarction in the elderly: The Rotterdam Study. *Am. J. Clin. Nutr.* **1999**, *69*, 261–266.
15. Gaziano, J.M.; Manson, J.E.; Branch, L.G.; Colditz, G.A.; Willett, W.C.; Buring, J.E. A prospective study of consumption of carotenoids in fruits and vegetables and decreased cardiovascular mortality in the elderly. *Ann. Epidemiol.* **1995**, *5*, 255–260.
16. Morris, D.L.; Kritchevsky, S.B.; Davis, C.E. Serum carotenoids and coronary heart disease. The Lipid Research Clinics Coronary Primary Prevention Trial and Follow-up Study. *JAMA* **1994**, *272*, 1439–1441.
17. Knekt, P.; Reunanen, A.; Jarvinen, R.; Seppanen, R.; Heliovaara, M.; Aromaa, A. Antioxidant vitamin intake and coronary mortality in a longitudinal population study. *Am. J. Epidemiol.* **1994**, *139*, 1180–1189.
18. Stephens, N.G.; Parsons, A.; Schofield, P.M.; Kelly, F.; Cheeseman, K.; Mitchinson, M.J. Randomised controlled trial of vitamin E in patients with coronary disease: Cambridge Heart Antioxidant Study (CHAOS). *Lancet* **1996**, *347*, 781–786.
19. Tepel, M.; van der Giet, M.; Statz, M.; Jankowski, J.; Zidek, W. The antioxidant acetylcysteine reduces cardiovascular events in patients with end-stage renal failure: A randomized, controlled trial. *Circulation* **2003**, *107*, 992–995.
20. Boaz, M.; Smetana, S.; Weinstein, T.; Matas, Z.; Gafter, U.; Iaina, A.; Knecht, A.; Weissgarten, Y.; Brunner, D.; Fainaru, M.; Green, M.S. Secondary prevention with antioxidants of cardiovascular disease in endstage renal disease (SPACE): Randomised placebo-controlled trial. *Lancet* **2000**, *356*, 1213–1218.
21. Steinhubl, S.R. Why have antioxidants failed in clinical trials? *Am. J. Cardiol.* **2008**, *101*, 14D–19D.
22. Heart Protection Study Collaborative Group. MRC/BHF Heart Protection Study of antioxidant vitamin supplementation in 20,536 high-risk individuals: A randomised placebo-controlled trial. *Lancet* **2002**, *360*, 23–33.
23. Yusuf, S.; Dagenais, G.; Pogue, J.; Bosch, J.; Sleight, P. Vitamin E supplementation and cardiovascular events in high-risk patients. The Heart Outcomes Prevention Evaluation Study Investigators. *N. Engl. J. Med.* **2000**, *342*, 154–160.
24. Sandmann, G. Carotenoid biosynthesis in microorganisms and plants. *Eur. J. Biochem.* **1994**, *223*, 7–24.
25. McNulty, H.; Jacob, R.F.; Mason, R.P. Biologic activity of carotenoids related to distinct membrane physicochemical interactions. *Am. J. Cardiol.* **2008**, *101*, 20D–29D.
26. Jackson, H.; Braun, C.L.; Ernst, H. The chemistry of novel xanthophyll carotenoids. *Am. J. Cardiol.* **2008**, *101*, 50D–57D.
27. McNulty, H.P.; Byun, J.; Lockwood, S.F.; Jacob, R.F.; Mason, R.P. Differential effects of carotenoids on lipid peroxidation due to membrane interactions: X-ray diffraction analysis. *Biochim. Biophys. Acta* **2007**, *1768*, 167–174.
28. Brown, B.G.; Zhao, X.Q.; Chait, A.; Fisher, L.D.; Cheung, M.C.; Morse, J.S.; Dowdy, A.A.; Marino, E.K.; Bolson, E.L.; Alaupovic, P.; Frohlich, J.; Albers, J.J. Simvastatin and niacin, antioxidant vitamins, or the combination for the prevention of coronary disease. *N. Engl. J. Med.* **2001**, *345*, 1583–1592.

29. The Alpha-Tocopherol, Beta Carotene Cancer Prevention Study Group. The effect of vitamin E and beta carotene on the incidence of lung cancer and other cancers in male smokers. *N. Engl. J. Med.* **1994**, *330*, 1029–1035.
30. Omenn, G.S.; Goodman, G.E.; Thornquist, M.D.; Balmes, J.; Cullen, M.R.; Glass, A.; Keogh, J.P.; Meyskens, F.L.; Valanis, B.; Williams, J.H.; Barnhart, S.; Hammar, S. Effects of a combination of beta carotene and vitamin A on lung cancer and cardiovascular disease. *N. Engl. J. Med.* **1996**, *334*, 1150–1155.
31. Lee, I.M.; Cook, N.R.; Manson, J.E.; Buring, J.E.; Hennekens, C.H. Beta-carotene supplementation and incidence of cancer and cardiovascular disease: The Women's Health Study. *J. Natl. Cancer Inst.* **1999**, *91*, 2102–2106.
32. Hennekens, C.H.; Buring, J.E.; Manson, J.E.; Stampfer, M.; Rosner, B.; Cook, N.R.; Belanger, C.; LaMotte, F.; Gaziano, J.M.; Ridker, P.M.; Willett, W.; Peto, R. Lack of effect of long-term supplementation with beta carotene on the incidence of malignant neoplasms and cardiovascular disease. *N. Engl. J. Med.* **1996**, *334*, 1145–1149.
33. Burton, G.W.; Ingold, K.U. beta-Carotene: An unusual type of lipid antioxidant. *Science* **1984**, *224*, 569–573.
34. Lauver, D.A.; Driscoll, E.M.; Lucchesi, B.R. Disodium disuccinate astaxanthin prevents carotid artery rethrombosis and *ex vivo* platelet activation. *Pharmacology* **2008**, *82*, 67–73.
35. Gross, G.J.; Lockwood, S.F. Cardioprotection and myocardial salvage by a disodium disuccinate astaxanthin derivative (Cardax). *Life Sci.* **2004**, *75*, 215–224.
36. Gross, G.J.; Hazen, S.L.; Lockwood, S.F. Seven day oral supplementation with Cardax (disodium disuccinate astaxanthin) provides significant cardioprotection and reduces oxidative stress in rats. *Mol. Cell. Biochem.* **2006**, *283*, 23–30.
37. Wertz, K.; Hunziker, P.B.; Seifert, N.; Riss, G.; Neeb, M.; Steiner, G.; Hunziker, W.; Goralczyk, R. beta-Carotene interferes with ultraviolet light A-induced gene expression by multiple pathways. *J. Invest. Dermatol.* **2005**, *124*, 428–434.
38. Camera, E.; Mastrofrancesco, A.; Fabbri, C.; Daubrawa, F.; Picardo, M.; Sies, H.; Stahl, W. Astaxanthin, canthaxanthin and beta-carotene differently affect UVA-induced oxidative damage and expression of oxidative stress-responsive enzymes. *Exp. Dermatol.* **2009**, *18*, 222–231.
39. Camara, B.; Bouvier, F. Oxidative remodeling of plastid carotenoids. *Arch. Biochem. Biophys.* **2004**, *430*, 16–21.
40. Sharoni, Y.; Agbaria, R.; Amir, H.; Ben-Dor, A.; Bobilev, I.; Doubi, N.; Giat, Y.; Hirsh, K.; Izumchenko, G.; Khanin, M.; Kirilov, E.; Krimer, R.; Nahum, A.; Steiner, M.; Walfisch, Y.; Walfisch, S.; Zango, G.; Danilenko, M.; Levy, J. Modulation of transcriptional activity by antioxidant carotenoids. *Mol. Asp. Med.* **2003**, *24*, 371–384.
41. Guerin, M.; Huntley, M.E.; Olaizola, M. Haematococcus astaxanthin: Applications for human health and nutrition. *Trends Biotechnol.* **2003**, *21*, 210–216.
42. Hussein, G.; Sankawa, U.; Goto, H.; Matsumoto, K.; Watanabe, H. Astaxanthin, a carotenoid with potential in human health and nutrition. *J. Nat. Prod.* **2006**, *69*, 443–449.
43. Schweigert, F. *Metabolism of Carotenoids in Mammals;* Birkhauser Verlag: Basel, Switzerland, 1998.

44. Jyonouchi, H.; Sun, S.; Tomita, Y.; Gross, M.D. Astaxanthin, a carotenoid without vitamin A activity, augments antibody responses in cultures including T-helper cell clones and suboptimal doses of antigen. *J. Nutr.* **1995**, *125*, 2483–2492.
45. Shimidzu, N. Carotenoids as singlet oxygen quenchers in marine organisms. *Fish. Sci.* **1996**, *62*, 134–137.
46. Miki, W. Biological functions and activities of animal carotenoids. *Pure Appl. Chem.* **1991**, *63*, 141–146.
47. Krinsky, N.I. Antioxidant functions of carotenoids. *Free Radic. Biol. Med.* **1989**, *7*, 617–635.
48. Beutner, S.; Bloedorn, B.; Frixel, S.; Blanco, I.H.; Hoffman, T.; Martin, H.D.; Mayer, B.; Noach, P.; Rack, C.; Schmidt, M.; *et al.* Quantitative assessment of antioxidant properties of natural colorants and phytochemicals; carotenoids. flavonoids, phenols and indigoids: The role of β-carotene in antioxidant functions. *J. Sci. Food Agric.* **2001**, *81*, 559–568.
49. Pashkow, F.J.; Watumull, D.G.; Campbell, C.L. Astaxanthin: A novel potential treatment for oxidative stress and inflammation in cardiovascular disease. *Am. J. Cardiol.* **2008**, *101*, 58D–68D.
50. Kobayashi, M.; Kakizono, T.; Nishio, N.; Nagai, S.; Kurimura, Y.; Tsuji, Y. Antioxidant role of astaxanthin in the green alga Haematococcus pluvialis. *Appl. Microbiol. Biotechnol.* **1997**, *48*, 351–356.
51. Ernst, H. Recent advances in industrial carotenoid synthesis. *Pure Appl. Chem.* **2002**, *74*, 2213–2226.
52. Lauver, D.A.; Lockwood, S.F.; Lucchesi, B.R. Disodium Disuccinate Astaxanthin (Cardax) attenuates complement activation and reduces myocardial injury following ischemia/reperfusion. *J. Pharmacol. Exp. Ther.* **2005**, *314*, 686–692.
53. Lockwood, S.F.; Gross, G.J. Disodium disuccinate astaxanthin (Cardax): Antioxidant and antiinflammatory cardioprotection. *Cardiovasc. Drug Rev.* **2005**, *23*, 199–216.
54. Gross, G.J.; Lockwood, S.F. Acute and chronic administration of disodium disuccinate astaxanthin (Cardax) produces marked cardioprotection in dog hearts. *Mol. Cell. Biochem.* **2005**, *272*, 221–227.
55. Khan, S.K.; Malinski, T.; Mason, R.P.; Kubant, R.; Jacob, R.F.; Fujioka, K.; Denstaedt, S.J.; King, T.J.; Jackson, H.L.; Hieber, A.D.; Lockwood, S.F.; Goodin, T.H.; Pashkow, F.J.; Bodary, P.F. Novel astaxanthin prodrug (CDX-085) attenuates thrombosis in a mouse model. *Thromb. Res.* **2010**, *126*, 299–305.
56. Shargel, L.; Yu, A. *Applied Biopharmaceutics and Pharmacokinetics*; Appleton-Lange: Stamford, CT, USA, 1999.
57. Rishton, G.M. Natural products as a robust source of new drugs and drug leads: Past successes and present day issues. *Am. J. Cardiol.* **2008**, *101*, 43D–49D.
58. Kang, J.O.; Kim, S.J.; Kim, H. Effect of astaxanthin on the hepatotoxicity, lipid peroxidation and antioxidative enzymes in the liver of CCl4-treated rats. *Methods Find. Exp. Clin. Pharmacol.* **2001**, *23*, 79–84.
59. Kamath, B.S.; Srikanta, B.M.; Dharmesh, S.M.; Sarada, R.; Ravishankar, G.A. Ulcer preventive and antioxidative properties of astaxanthin from Haematococcus pluvialis. *Eur. J. Pharmacol.* **2008**, *590*, 387–395.
60. Naito, Y.; Uchiyama, K.; Aoi, W.; Hasegawa, G.; Nakamura, N.; Yoshida, N.; Maoka, T.; Takahashi, J.; Yoshikawa, T. Prevention of diabetic nephropathy by treatment with astaxanthin in diabetic db/db mice. *Biofactors* **2004**, *20*, 49–59.
61. Ohgami, K.; Shiratori, K.; Kotake, S.; Nishida, T.; Mizuki, N.; Yazawa, K.; Ohno, S. Effects of astaxanthin on lipopolysaccharide-induced inflammation *in vitro* and *in vivo*. *Invest. Ophthalmol. Vis. Sci.* **2003**, *44*, 2694–2701.
62. Lee, S.J.; Bai, S.K.; Lee, K.S.; Namkoong, S.; Na, H.J.; Ha, K.S.; Han, J.A.; Yim, S.V.; Chang, K.; Kwon, Y.G.; Lee, S.K.; Kim, Y.M. Astaxanthin inhibits nitric oxide production and inflammatory gene expression by suppressing I(kappa)B kinase-dependent NF-kappaB activation. *Mol. Cells* **2003**, *16*, 97–105.

63. Aoi, W.; Naito, Y.; Sakuma, K.; Kuchide, M.; Tokuda, H.; Maoka, T.; Toyokuni, S.; Oka, S.; Yasuhara, M.; Yoshikawa, T. Astaxanthin limits exercise-induced skeletal and cardiac muscle damage in mice. *Antioxid. Redox Signal.* **2003**, *5*, 139–144.
64. Uchiyama, K.; Naito, Y.; Hasegawa, G.; Nakamura, N.; Takahashi, J.; Yoshikawa, T. Astaxanthin protects beta-cells against glucose toxicity in diabetic db/db mice. *Redox Rep.* **2002**, *7*, 290–293.
65. Nakajima, Y.; Inokuchi, Y.; Shimazawa, M.; Otsubo, K.; Ishibashi, T.; Hara, H. Astaxanthin, a dietary carotenoid, protects retinal cells against oxidative stress *in vitro* and in mice *in vivo*. *J. Pharm. Pharmacol.* **2008**, *60*, 1365–1374.
66. Manabe, E.; Handa, O.; Naito, Y.; Mizushima, K.; Akagiri, S.; Adachi, S.; Takagi, T.; Kokura, S.; Maoka, T.; Yoshikawa, T. Astaxanthin protects mesangial cells from hyperglycemia-induced oxidative signaling. *J. Cell. Biochem.* **2008**, *103*, 1925–1937.
67. Nakano, M.; Onodera, A.; Saito, E.; Tanabe, M.; Yajima, K.; Takahashi, J.; Nguyen, V.C. Effect of astaxanthin in combination with alpha-tocopherol or ascorbic acid against oxidative damage in diabetic ODS rats. *J. Nutr. Sci. Vitaminol. (Tokyo)* **2008**, *54*, 329–334.
68. Choi, S.K.; Park, Y.S.; Choi, D.K.; Chang, H.I. Effects of astaxanthin on the production of NO and the expression of COX-2 and iNOS in LPS-stimulated BV2 microglial cells. *J. Microbiol. Biotechnol.* **2008**, *18*, 1990–1996.
69. Liu, X.; Shibata, T.; Hisaka, S.; Osawa, T. Astaxanthin inhibits reactive oxygen species-mediated cellular toxicity in dopaminergic SH-SY5Y cells via mitochondria-targeted protective mechanism. *Brain Res.* **2009**, *1254*, 18–27.
70. Iwamoto, T.; Hosoda, K.; Hirano, R.; Kurata, H.; Matsumoto, A.; Miki, W.; Kamiyama, M.; Itakura, H.; Yamamoto, S.; Kondo, K. Inhibition of low-density lipoprotein oxidation by astaxanthin. *J. Atheroscler. Thromb.* **2000**, *7*, 216–222.
71. Hussein, G.; Nakamura, M.; Zhao, Q.; Iguchi, T.; Goto, H.; Sankawa, U.; Watanabe, H. Antihypertensive and neuroprotective effects of astaxanthin in experimental animals. *Biol. Pharm. Bull.* **2005**, *28*, 47–52.
72. Hussein, G.; Goto, H.; Oda, S.; Sankawa, U.; Matsumoto, K.; Watanabe, H. Antihypertensive potential and mechanism of action of astaxanthin: III. Antioxidant and histopathological effects in spontaneously hypertensive rats. *Biol. Pharm. Bull.* **2006**, *29*, 684–688.
73. Aoi, W.; Naito, Y.; Takanami, Y.; Ishii, T.; Kawai, Y.; Akagiri, S.; Kato, Y.; Osawa, T.; Yoshikawa, T. Astaxanthin improves muscle lipid metabolism in exercise via inhibitory effect of oxidative CPT I modification. *Biochem. Biophys. Res. Commun.* **2008**, *366*, 892–897.
74. Nakao, R.; Nelson, O.L.; Park, J.S.; Mathison, B.D.; Thompson, P.A.; Chew, B.P. Effect of astaxanthin supplementation on inflammation and cardiac function in BALB/c mice. *Anticancer Res.* **2010**, *30*, 2721–2725.
75. Yamamoto, R.; Kanazawa, A.; Shimizu, T.; Hirose, T.; Tanaka, Y.; Kawamori, R.; Watada, H. Association between atherosclerosis and newly classified chronic kidney disease stage for Japanese patients with type 2 diabetes. *Diabetes Res. Clin. Pract.* **2009**, *84*, 39–45.
76. Osterlie, M.; Bjerkeng, B.; Liaaen-Jensen, S. Plasma appearance and distribution of astaxanthin *E*/*Z* and *R*/*S* isomers in plasma lipoproteins of men after single dose administration of astaxanthin. *J. Nutr. Biochem.* **2000**, *11*, 482–490.

77. Mercke Odeberg, J.; Lignell, A.; Pettersson, A.; Hoglund, P. Oral bioavailability of the antioxidant astaxanthin in humans is enhanced by incorporation of lipid based formulations. *Eur. J. Pharm. Sci.* **2003**, *19*, 299–304.
78. Spiller, G.A.; Dewell, A. Safety of an astaxanthin-rich Haematococcus pluvialis algal extract: A randomized clinical trial. *J. Med. Food* **2003**, *6*, 51–56.
79. Coral-Hinostroza, G.N.; Ytrestoyl, T.; Ruyter, B.; Bjerkeng, B. Plasma appearance of unesterified astaxanthin geometrical *E*/*Z* and optical *R*/*S* isomers in men given single doses of a mixture of optical 3 and 3′*R*/*S* isomers of astaxanthin fatty acyl diesters. *Comp. Biochem. Physiol. C Toxicol. Pharmacol.* **2004**, *139*, 99–110.
80. Karppi, J.; Rissanen, T.H.; Nyyssonen, K.; Kaikkonen, J.; Olsson, A.G.; Voutilainen, S.; Salonen, J.T. Effects of astaxanthin supplementation on lipid peroxidation. *Int. J. Vitam. Nutr. Res.* **2007**, *77*, 3–11.
81. Parisi, V.; Tedeschi, M.; Gallinaro, G.; Varano, M.; Saviano, S.; Piermarocchi, S. Carotenoids and antioxidants in age-related maculopathy italian study: Multifocal electroretinogram modifications after 1 year. *Ophthalmology* **2008**, *115*, 324–333.e2.
82. Miyawaki, H.; Takahashi, J.; Tsukahara, H.; Takehara, I. Effects of astaxanthin on human blood rheology. *J. Clin. Biochem. Nutr.* **2008**, *43*, 69–74.
83. Rufer, C.E.; Moeseneder, J.; Briviba, K.; Rechkemmer, G.; Bub, A. Bioavailability of astaxanthin stereoisomers from wild (*Oncorhynchus* spp.) and aquacultured (*Salmo salar*) salmon in healthy men: A randomised, double-blind study. *Br. J. Nutr.* **2008**, *99*, 1048–1054.
84. Park, J.S.; Chyun, J.H.; Kim, Y.K.; Line, L.L.; Chew, B.P. Astaxanthin decreased oxidative stress and inflammation and enhanced immune response in humans. *Nutr. Metab. (Lond.)* **2010**, *7*, 18.
85. Serebruany, V.; Malinin, A.; Goodin, T.; Pashkow, F. The *in vitro* effects of Xancor, a synthetic astaxanthine derivative, on hemostatic biomarkers in aspirin-naive and aspirin-treated subjects with multiple risk factors for vascular disease. *Am. J. Ther.* **2010**, *17*, 125–132.
86. Andersen, L.P.; Holck, S.; Kupcinskas, L.; Kiudelis, G.; Jonaitis, L.; Janciauskas, D.; Permin, H.; Wadstrom, T. Gastric inflammatory markers and interleukins in patients with functional dyspepsia treated with astaxanthin. *FEMS Immunol. Med. Microbiol.* **2007**, *50*, 244–248.
87. Fassett, R.G.; Healy, H.; Driver, R.; Robertson, I.K.; Geraghty, D.P.; Sharman, J.E.; Coombes, J.S. Astaxanthin *vs.* placebo on arterial stiffness, oxidative stress and inflammation in renal transplant patients (Xanthin): A randomised controlled trial. *BMC Nephrol.* **2008**, *9*, 17.

*marine drugs*

MDPI

*Review*

# Marine Carotenoids and Cardiovascular Risk Markers

**Graziano Riccioni [1,2,*], Nicolantonio D'Orazio [2], Sara Franceschelli [3] and Lorenza Speranza [3]**

[1] Cardiology Unit, San Camillo de Lellis Hospital, via Isonzo, Manfredonia, Foggia, 71043, Italy

[2] Human Nutrition, Department of Biomedical Science, via Dei vestini, University G. D'Annunzio, Chieti, 66013, Italy; E-Mail: ndorazio@unich.it

[3] Department of Human Movement Sciences, via Dei Vestini, University G. D'Annunzio, Chieti, 66013, Italy; E-Mails: s.francheschelli@unich.it (S.F.); l.speranza@unich.it (L.S.)

* Author to whom correspondence should be addressed; E-Mail: griccioni@hotmail.com;
Tel.: +39-0882-227022; Fax: +39-0882-227022.

*Received: 17 February 2011; in revised form: 26 May 2011 / Accepted: 15 June 2011 / Published: 27 June 2011*

**Abstract:** Marine carotenoids are important bioactive compounds with physiological activities related to prevention of degenerative diseases found principally in plants, with potential antioxidant biological properties deriving from their chemical structure and interaction with biological membranes. They are substances with very special and remarkable properties that no other groups of substances possess and that form the basis of their many, varied functions and actions in all kinds of living organisms. The potential beneficial effects of marine carotenoids have been studied particularly in astaxanthin and fucoxanthin as they are the major marine carotenoids. Both these two carotenoids show strong antioxidant activity attributed to quenching singlet oxygen and scavenging free radicals. The potential role of these carotenoids as dietary anti-oxidants has been suggested to be one of the main mechanisms for their preventive effects against cancer and inflammatory diseases. The aim of this short review is to examine the published studies concerning the use of the two marine carotenoids, astaxanthin and fucoxanthin, in the prevention of cardiovascular diseases.

**Keywords:** marine carotenoids; astaxanthin; fucoxantin; atherosclerosis; cardiovascular disease; free radicals; oxidized LDL

## 1. Introduction

Cardiovascular disease (CVD), especially coronary heart disease (CHD) and stroke, is the leading killer in Western Society and its prevalence is increasing dramatically in developing nations [1].

Atherothrombotic disease is the consequence of conventional risk factors such as smoking, hypertension, hyperlipidemia, insulin resistance and diabetes, and obesity. Novel risk factors include highly sensitive C-reactive protein (hs-CRP) and other markers of inflammation, homocysteine,
and lipoprotein (a) [2].

Along with genetic factors and age, lifestyle and diet are also considered important risk factors. Reduction in dietary consumption of animal fat, cholesterol, and sodium should be the mainstay of population-wide CHD prevention [3]. Dietary interventions should be the initial step in the treatment of CVD. In particular, a group of phytochemical substances in carotenoids which is responsible for the color of food play an important role in the prevention of human diseases and the maintenance of good health [4].

## 2. Oxidative Stress and Antioxidants

Oxidative modification of low density lipoprotein cholesterol (LDL-C) play an important role in the initiation and progression of the atherosclerotic process, a *continuum pathophysiological process* that includes oxidative stress, endothelial dysfunction, inflammatory process, and vascular remodelling [5].

Specifically, CVD are associated with increased production of reactive oxygen species (ROS) and a compromised endogenous anti-oxidant defense system. Oxidative stress is tightly regulated by a balance between production and removal of ROS [6], which are natural by-products of metabolism with important roles in cell signaling. However, excessive levels of ROS can be toxic to cells, *i.e.*, whenever the expression of anti-oxidant enzymes, including superoxide dismutases (SODs), heme oxygenase-1 (HO-1), NAD(P)H quinine oxidoreductase-1 (NQO-1), catalase and thioredoxin are not sufficient to control ROS and minimize ROS-induced damage [7]. A compromised anti-oxidant defense system can lead to excessive oxidative stress and ultimately result in cell damage [8].

Numerous studies indicate that increased oxidative stress may be involved in the pathogenesis of CVD. Several animal models suggest that when endogenous anti-oxidant systems are overwhelmed, exogenous antioxidant supplementation can be used for preventive and/or therapeutic intervention of oxidative cardiovascular disorders [9]. In particular, SODs, catalase and glutathione peroxidase ($GSH_{Px}$) are endogenous natural antioxidants present within human cells. In addition, antioxidants such as vitamin E, vitamin C, polyphenols and carotenoids are available from foods [10].

Current dietary guidelines to combat chronic diseases, including cancer and CHD, recommend increased intake of plant foods, including fruits and vegetables, which are rich sources of antioxidants [11]. The role of such dietary antioxidants in disease prevention has received much attention recently and appears to have a wide range of antiatherogenic properties [12,13]. These observations may explain the epidemiological data indicating that diets rich in fruits and vegetables are associated with a reduced risk of numerous chronic diseases [14].

Carotenoids are ubiquitous in nature and present in plants, algae and microorganisms. However, humans and other animals are unable to manufacture carotenoids and hence require these in their diet. There are two class types of carotenoids based on their chemical composition: carotenes and xanthopylls [15]. Astaxanthin and fucoxanthin are major marine carotenoids. Both these carotenoids show strong antioxidant activity which is attributed to quenching singlet oxygen and scavenging free radicals (FRs) [16].

## 3. Astaxanthin

### 3.1. Chemical Structure and Mechanism of Action

Astaxanthin is a xanthophyll carotenoid which contains two additional oxygenated groups on each ring structure compared with other carotenoids, resulting in enhanced antioxidant properties, approved in the 1999 as a dietary supplement by Food and Drug Amninistration. This compound occurs naturally in a wide variety of living organisms including microalgae (*Haematococcus pluvialis, Chlorella zofingiensis*, and *Chlorococcum* sp.), fungi (red yeast *Phaffia rhodozyma*), complex plants, seafood, and some birds such as flamingos and quail; it is reddish-coloured, and gives salmon, shrimp and lobster their distinctive colouration [17]. The microalga *H. pluvialis* has the highest capacity to accumulate astaxanthin up to 4–5% of cell dry weight. Astaxanthin has been attributed with extraordinary potential for protecting the organism against a wide range of diseases, and has considerable potential and promising applications in the prevention and treatment of various diseases, such as cancers, chronic inflammatory diseases, metabolic syndrome, diabetes, diabetic nephropathy, CVD, gastrointestinal and liver diseases, and neurodegenerative diseases [18].

Astaxanthin cannot be manufactured in animals or converted to vitamin A and therefore must be consumed in the diet. Xanthophyll carotenoids such as astaxanthin and canthaxanthin have antioxidant activity, are free radical scavengers, potent quenchers of ROS and nitrogen oxygen species (NOS), and chain-breaking antioxidants. They are superior antioxidants and scavengers of free radicals (FRs) compared to the carotenoids such as β-carotene [19].

Available forms of astaxanthin are represented by natural forms on an industrial scale of
production [20,21]. The disodium disuccinate astaxanthin (DDA), a synthetic form of astaxanthin, overcame the limitations of careteniods related to their poor aqueous solubility and enabled investigation of this agent in the animal models of myocardial ischaemia and reperfusion using both intravenous and oral routes of administration [22]. DDA has been shown to be very effective in animal cardiovascular studies administered both intravenously and orally [23–25].

### 3.2. *Astaxanthin and Cardiovascular Disease: Experimental Studies*

Astaxanthin has undergone investigation in a large number of experimental studies related to the cardiovascular and cerebrovascular systems. Only few studies have investigated the potential benefits of astaxanthin in human health and disease and most of these have been performed in healthy volunteers to assess dosing [26], bioavailability [27,28], safety [29], oxidative stress, and inflammation [26,30].

Studies conducted in healthy human volunteers have found significant reductions in oxidative stress, hyperlipidemia and inflammatory markers after astaxanthin oral supplementation. In particular, Cicero *et al.* [31] have demonstrated that food supplements for four weeks containing a combination of natural products such as berberine, policosanol, red yeast extracts, folic acid and astaxanthin could be a useful support to diet and life style changes to correct dyslipidemias and to reduce cardiovascular (CV) risk in 40 subjects with moderate mixed dyslipidemias. In this study total cholesterol (TC),
LDL-C, high density lipoprotein cholesterol (HDL-C), non HDL-C, Apo-B, Apo-A, Lp(a) and triglycerides (TG) were measured before and at the end of treatments. Berberine and combinations of natural products (policosanol, red yeast extracts, folic acid and astaxanthin) significantly reduced TC (respectively by 16% and 20%), LDL-C (by 20% and 25%), Apo-B (by 15% and 29%) and TG (by 22% and 26%), and increased HDL-C (by 6.6% and 5.1%). Even Yoshida *et al.* [32] demonstrated in a randomized, placebo-controlled human study (61 non-obese subjects aged 20–65 years) that astaxanthin consumption (0, 6, 12, 18 mg/day for 12 weeks) ameliorates TG and HDL-C in correlation with increased adiponectin in humans. In this study, before and after tests, body mass index (BMI) and LDL-C were unaffected at all doses, however, TG decreased, while HDL-C increased significantly. Multiple comparison tests showed that 12 and 18 mg/day doses significantly reduced TG, and 6 and 12 mg doses significantly increased HDL-C. Serum adiponectin was increased by astaxanthin (12 and 18 mg/day), and changes of adiponectin correlated positively with HDL-C changes independent of age and BMI. Iwamoto *et al.* [26] demonstrated a significant inhibition of LDL-C oxidation in 24 healthy volunteers who took doses

of astaxanthin (1.8, 3.6, 14.4 and 21.6 mg/day for 2 weeks). Even Karppi *et al.* [30] in a 12-week randomized double-blind study involving 40 healthy non-smoking Finnish males assessed a significant plasma reduction levels of 12- and 15-hydroxy fatty acids in those taking astaxanthin (8 mg/day) suggesting an important reduced fatty acid oxidation due to astaxanthin.

## 4. Fucoxanthin

### 4.1. Chemical Structure and Mechanism of Action

Fucoxanthin is a naturally occurring brown pigment that belongs to the class of non-provitamin A carotenoids, a class of 40-carbon organic molecules that consist of two groups: xanthophylls if their structure contains oxygen, and carotenes if there is no oxygen in their chemical formula. Fucoxanthin is a xanthophyll whose distinct structure includes an unusual allenic bond, epoxide group and conjugated carbonyl group in polyene chain [33] with antioxidant properties [34]. The difference, however is that fucoxanthin acts as an antioxidant under anoxic conditions whereas other carotenoids have practically no quenching abilities. Most tissues under physiological conditions have low oxygen presence. Furthermore, the typical antioxidants are usually proton donors (ascorbic acid, $\alpha$-tocopherol, glutathione). Fucoxanthin, on the other hand, donates electron as a part of its free-radical quenching function. A combination of these distinct properties is very rarely found among naturally occurring food-derived compounds [35,36]. During normal metabolism the body produces heat. Fucoxanthin increases the amount of energy released as heat in fat tissue, a process also called thermogenesis. A published study reports that fucoxanthin affects multiple enzymes involved in fat metabolism causing an increase in the production of energy from fat [37].

### 4.2. Fucoxantin and Cardiovascular Disease: Experimental Studies (Human and Animal)

Experiments on stroke-prone spontaneously hypertensive rats (SHRSP) show the possible protective role of fucoxanthin in CVD. Thirty-three male SHRSP rats, 5 weeks of age, were purchased from the Disease Model Cooperative Research Association (Kyoto, Japan). Animals were divided into three groups: (1) kaolin group, which was given a normal diet [5% (w/w) kaolin, a non-nutrient material]; (2) Wakame (Undaria Pinnatifida) group [normal diet containing 5% (w/w) Wakame powder]; and (3) cellulose group [normal diet containing 5% (w/w) cellulose]. In this study, Wakame delayed the incidence of stroke signs and increased the life span of SHRSP. Wakame did not attenuate the development of hypertension in SHRSP [38].

Thrombosis is a major complication of coronary atherosclerosis that can lead to myocardial infarction. Docosahexaenoic acid (DHA) inhibits the synthesis of thromboxane $A_2$ ($TxA_2$) from arachidonic acid (AA) in platelets [39]. In addition, DHA enhances the production of prostacyclin, a prostaglandin that produces vasodilation and less sticky platelets [40]. Also epidemiologic and clinical trials demonstrated that fish oil such as eicosapentaenoic acid (EPA) and DHA, decreased LDL-C, TG and increased HDL-C concentrations. DHA content in fish oil fed to experimental animals inhibits the development of atherosclerosis, so the fucoxanthin may have an potential role in the modulation and prevention of human diseases, particularly to reducing the incidence of CVD [41,42].

### *4.3. Fucoxantin and Metabolic Syndrome: Experimental Studies (Human and Animal)*

Among marine carotenoids, attention has been paid towards fucoxanthin in recent years, that is actually used for the treatment of methabolic syndrome and obesity [43], two important risk factors of CVD. Fucoxanthin has a unique structure including an allenic bond and 5,6-monoepoxide in
the molecole, is a major carotenoid found in edible sceaweeds such as *Undaria pinnatifida, Hijikia fusiformis* and *Sargassum fulvelum* [44]. This molecule is under study for possible application in the fight against overweight and obesity since it promotes the reduction of abdominal fat. In this regard it is interesting to note an increased concentration of fat in the abdomen statistically correlated with an increased risk of CVD. A study conducted by Maeda *et al.* [43] in male Wister rats and female KK-Ay mice under different experimental diets (soybean oil, Undaria lipids, Undaria glycolipid fraction, crude fucoxanthin and purified fucoxanthin fed to different concentration according to their groups) for 4 weeks shows that Undaria lipids (containing 9.6% fucoxanthin) reduced significantly the weights of abdominal white adipose tissue (WAT) of both rats and mice. Body weights of mice fed Undaria lipid was significantly lower than that of controls [45]. Animal studies by one group of researchers suggest that fucoxanthin might prevent the growth of fat tissue and reduce abdominal fat, has a beneficial effects in stroke prevention, reduction of inflammation, and slowing the growth of various cancer cell type [38,45]. Woo *et al.* [37] in a recent study shows as, the fucoxanthin facilitates youthful energy metabolism by activating a special cellular mitochondrial protein called UCP-1, which induces the thermogenesis. This is important because obesity markedly increases the risk of CVD. Fucoxanthin has been found to reduce blood glucose in animals with diabetes and in normal mice that are fed high fat diets [46]. It appears that fucoxanthin is capable of upregulating glucose transporter, mRNA expression of L6 myotubes which are responsible for glucose transport in adult muscle
tissue [47]. An interesting, extra, metabolic benefit of fucoxanthin administration in rodents is the promotion of the synthesis of DHA in the liver [43].

Because the metabolic syndrome is a collection of risk factors that substantially increase the chances of damage in the CVS, which can lead to a heart attack or stroke, the importance of the fucoxanthin in the regular metabolic syndrome would be very important to prevent CV damage. Clinical research also indicated that the metabolic boost from taking fucoxanthin did not stimulate the central nervous system, meaning it did not cause the jitters or lost sleep like caffeine, nicotine, or thyroid hormones. Only one study has currently been conducted in humans which has evaluated the effectiveness of fucoxanthin supplementation for weight loss. This study reports that the supplement, Xanthigen, which contains 300 mg pomegranate seed oil and 300 mg brown marine algae fucoxanthin significantly increased weight loss and reduced body and liver fats content in obese women treated for 16 weeks [48].

Fucoxanthin proved safe with no side effects, and even provided other health benefits, including improved cardiovascular health, reduction of inflammation (a major cause of heart disease), healthy cholesterol and TG levels, improvements in blood pressure levels, and healthy liver function [49–51].

## 5. Conclusion

Oxidative stress and inflammation play an important role in the pathophysiology of many chronic diseases including CVD [52]. The xanthophyll carotenoid dietary supplement, astaxanthin, has demonstrated to be a potential antioxidant and anti-inflammatory therapeutic agent in models of CVD. There have been human clinical studies using astaxanthin to assess its safety, bioavailability and clinical aspects relevant to oxidative stress and inflammation in CVS. There were no adverse effects reported. These demonstrated reduced markers of oxidative stress and inflammation and improved blood rheology. Astaxanthin has great potential as a potent antioxidant to be tested in human clinical trials based on theoretical grounds related to its physicochemical properties and on the basis of exciting preliminary experimental studies in CV models. Although its use in human clinical studies has been limited, so far no safety concerns have arisen [53]. We predict that because of its greater antioxidant potency and membrane preservation, astaxanthin will reduce measures of oxidative stress and inflammation and provide vascular benefits [54].

The versatile effects of fucoxanthin on intermediate metabolism make this carotenoid of great potential value in the prevention or management of the metabolic syndrome and obesity. The animal experiments with fucoxanthin stimulated researchers to recommend human clinical trials with fucoxanthin. As a carotenoid, fucoxanthin is a powerful antioxidant that protects cells from FRs damage. Future clinical studies and trials will help determine the efficacy of these marine carotenoids (asthaxantin and fucoxanthin) on vascular structure, function, oxidative stress and inflammation in a variety of patients at risk of, or with established CVD. These may lead to large interventional trials assessing CV morbidity and mortality.

## References


1. World Health Organization. *Diet, Nutrition, and the Prevention of Chronic Disease*; WHO Technical Report Series 916; WHO: Geneva, Switzerland, 2003.
2. Neaton, J.; Wentworth, D. Serum cholesterol, blood pressure, cigarette smoking, and death from coronary heart disease. Overall findings and difference by age for 316,099 white man. Multiple Risk Factor Intervention Trial Research Group. *Arch. Intern. Med.* **1992**, *152*, 56–64.
3. Riccioni, G. Carotenoids and cardiovascular disease. *Curr. Atheroscler. Rep.* **2009**, *11*, 434–439.
4. Riccioni, G.; D'Orazio, N.; Speranza, L.; Di Ilio, E.; Glade, M.; Bucciarelli, V.; Scotti, L.; Martini, F.; Pennelli, A.; Bucciarelli, T. Carotenoids and asymptomatic carotid atherosclerosis. *J. Biol. Regul. Homeost. Agents* **2010**, *24*, 447–452.
5. Gori, T.; Nzel, T.M. Oxidative stress and endothelial dysfunction: Therapeutic implications. *Ann. Med.* **2011**, *43*, 259–272.
6. Lee, S.; Park, Y.; Zuidema, M.Y.; Hannink, M.; Zhang, C. Effects of interventions on oxidative stress and inflammation of cardiovascular diseases. *World J. Cardiol.* **2011**, *3*, 18–24.
7. Gao, L.; Mann, G.E. Vascular NAD(P)H oxidase activation in diabetes: A double-edged sword in redox signalling. *Cardiovasc. Res.* **2009**, *82*, 9–20.
8. Rasmussen, H.H.; Hamilton, E.J.; Liu, C.C.; Figtree, G.A. Reversible oxidative modification: Implications for cardiovascular physiology and pathophysiology. *Trends Cardiovasc. Med.* **2010**, *20*, 85–90.
9. Yeh, C.T.; Ching, L.C.; Yen, G.C. Inducing gene expression of cardiac antioxidant enzymes by dietary phenolic acids in rats. *J. Nutr. Biochem.* **2009**, *20*, 163–171.
10. Khansari, N.; Shakiba, Y.; Mahmoudi, M. Chronic inflammation and oxidative stress as a major cause of age-related diseases and cancer. *Recent Pat. Inflamm. Allergy Drug Discov.* **2009**, *3*, 73–80.
11. *Dietary Guidelines for Americans*, 5th ed.; Home and Garden Bulletin no. 232; US Department of Agriculture, US Department of Health and Human Services: Washington, DC, USA, 2000. Available online: http://www.nal.usda.gov/fnic/dga (accessed on 13 March 2011).
12. Sies, H.; Stahl, W. Vitamins E and C, beta-carotene, and other carotenoids as antioxidants. *Am. J. Clin. Nutr.* **1995**, *62*, 1315–1321.
13. Lichtenstein, A.H. Nutrient supplements and cardiovascular disease: A heartbreaking story. *J. Lipid Res.* **2009**, *50*, 429–433.
14. Gaziano, J.M.; Manson, J.E.; Branch L.G.; Colditz, G.A.; Willett, W.C.; Buring, J.E. A prospective study of consumption of carotenoids in fruits and vegetables and decreased cardiovascular mortality in the elderly. *Ann. Epidemiol.* **1995**, *5*, 255–260.

15. Jackson, H.; Braun, C.L.; Ernst, H. The chemistry of novel xanthophyll carotenoids. *Am. J. Cardiol.* **2008**, *101*, 50–57.
16. Miyashita, K. Function of marine carotenoids. *Forum Nutr.* **2009**, *61*, 136–146.
17. Paterson, E.; Gordon, M.H.; Niwat, C.; George, T.W.; Parr, L.; Waroonphan, S.; Lovegrove, J.A. Supplementation with fruit and vegetable soups and beverages increas plasma carotenoid concentrations but does not alter marker of oxidative stress or cardiovascular risk factors. *J. Nutr.* **2006**, *136*, 2849–2855.
18. Yuan, J.P.; Peng, J.; Yin, K.; Wang, J.H. Potential health-promoting effects of astaxanthin: A high-value carotenoid mostly from microalgae. *Mol. Nutr. Food Res.* **2011**, *55*, 150–165.
19. Pashkow, F.J.; Watumull, D.G.; Campbell, C.L. Astaxanthin: A novel potential treatment for oxidative stress and inflammation in cardiovascular disease. *Am. J. Cardiol.* **2008**, *101*, 58–68.
20. Ernst, H. Recent advances in industrial carotenoid synthesis. *Pure Appl. Chem.* **2002**, *74*, 2213–2226.
21. Montanti, J.; Nghiem, N.P.; Johnston, D.B. Production of astaxanthin from cellulosic biomass sugars by mutants of the yeast *Phaffia rhodozyma*. *Appl. Biochem. Biotechnol.* **2011**, *164*, 655–665.
22. Lockwood, S.F.; Gross, G.J. Disodium disuccinate astaxanthin (Cardax): Antioxidant and antiinflammatory cardioprotection. *Cardiovasc. Drug Rev.* **2005**, *23*, 199–216.
23. Gross, G.J.; Lockwood, S.F. Cardioprotection and myocardial salvage by a disodium disuccinate astaxanthin derivative (Cardax). *Life Sci.* **2004**, *5*, 215–224.
24. Gross, G.J.; Hazen, S.L.; Lockwood, S.F. Seven day oral supplementation with Cardax (disodium disuccinate astaxanthin) provides significant cardioprotection and reduces oxidative stress in rats. *Mol. Cell Biochem.* **2006**, *283*, 23–30.
25. Lauver, D.A.; Lockwood, S.F.; Lucchesi, B.R. Disodium disuccinate astaxanthin (Cardax) attenuates complement activation and reduces myocardial injury following ischemia/reperfusion. *J. Pharmacol. Exp. Ther.* **2005**, *314*, 686–692.
26. Iwamoto, T.; Hosoda, K.; Hirano, R. Inhibition of low-density lipoprotein oxidation by astaxanthin. *J. Atheroscler. Thromb.* **2000**, *7*, 216–222.
27. Rufer, C.E., Moeseneder, J.; Briviba, K. Bioavailability of astaxanthin stereoisomers from wild (*Oncorhynchus* spp.) and aquacultured (*Salmo salar*) salmon in healthy men: A randomised, double-blind study. *Br. J. Nutr.* **2008**, *99*, 1048–1054.
28. Coral-Hinostroza, G.N.; Ytrestoyl, T.; Ruyter, T.; Bjerkeng, B. Plasma appearance of unesterified astaxanthin geometrical E/Z and optical R/S isomers in men given single doses of a mixture of optical 3 and 3′R/S isomers of astaxanthin fatty acyl diesters. *Comp. Biochem. Physiol. C Toxicol. Pharmacol.* **2004**, *139*, 99–110.

29. Spiller, G.A.; Dewell, A. Safety of an astaxanthin-rich *Haematococcus pluvialis* algal extract: A randomized clinical trial. *J. Med. Food* **2003**, *6*, 51–56.
30. Karppi, J.; Rissanen, T.H.; Nyyssonen, K. Effects of astaxanthin supplementation on lipid peroxidation. *Int. J. Vitam. Nutr. Res.* **2007**, *77*, 3–11.
31. Cicero, A.F.; Rovati, L.C.; Setnikar, I. Eulipidemic effects of berberine administered alone or in combination with other natural cholesterol-lowering agents. A single-blind clinical investigation. *Arzneimittel-Forschung* **2007**, *57*, 26–30.
32. Yoshida, H.; Yanai, H.; Ito, K.; Tomono, Y.; Koikeda, T.; Tsukahara, H.; Tada, N. Administration of natural astaxanthin increases serum HDL-cholesterol and adiponectin in subjects with mild hyperlipidemia. *Atherosclerosis* **2010**, *209*, 520–523.
33. Mercadante, A.Z.; Egeland, E.S. Carotenoids with a C40 skeleton. In *Carotenoids—Handbook*; Britton, G., Liaaen-Jensen, S., Pfander, H., Eds.; Birkhauser: Basel, Switzerland, 2004; p. 563.
34. Hu, T.; Liu, D.; Chen, Y.; Wu, J.; Wang, S. Antioxidant activity of sulfated polysaccharide fractions extracted from *Undaria pinnitafida in vitro*. *Int. J. Biol. Macromol.* **2010**, *46*, 193–198.
35. Nomura, T.; Kikuchi, M.; Kubodera, A.; Kawakami, Y. Proton-donative antioxidant activity of fucoxanthin with 1,1-diphenyl-2-picrylhydrazyl (DPPH). *Biochem. Mol. Biol. Int.* **1997**, *42*, 361–370.
36. Yan, X.; Chuda, Y.; Suzuki, M.; Nagata, T. Fucoxanthin as the major antioxidant in *Hijikia fusiformis*, a common edible seaweed. *Biosci. Biotechnol. Biochem.* **1999**, *63*, 605–607.
37. Woo, M.N.; Jeon, S.M.; Shin, Y.C.; Lee, M.K.; Kang, M.A.; Choi, M.S. Anti-obese property of fucoxanthin is partly mediated by altering lipid-regulating enzymes and uncoupling proteins of visceral adipose tissue in mice. *Mol. Nutr. Food Res.* **2009**, *53*, 1603–1611.
38. Ikeda, K.; Kitamura, A.; Machida, H.; Watanabe, M.; Negishi, H.; Hiraoka, J.; Nakano, T. Effect of *Undaria pinnatifida* (Wakame) on the development of cerebrovascular diseases in stroke-prone spontaneously hypertensive rats. *Clin. Exp. Pharmacol. Physiol.* **2003**, *30*, 44–48.
39. Connor, W.E. Importance of n-3 fatty acids in health and disease. *Am. J. Clin. Nutr.* **2000**, *71*, 171–175.
40. Adan, Y.; Shibata, K.; Sato, M.; Ikeda, I.; Imaizumi, K. Effects of docosahexaenoic and eicosapentaenoic acid on lipid metabolism, eicosanoid production, platelet aggregation and atherosclerosis in hypercholesterolemic rats. *Biosci. Biotechnol. Biochem.* **1999**, *63*, 111–119.

41. Wang, S.; Wu, D.; Matthan, N.R.; Lamon-Fava, S.; Lecker, J.L.; Lichtenstein, A.H. Reduction in dietary omega-6 polyunsaturated fatty acids: Eicosapentaenoic acid plus docosahexaenoic acid ratio minimizes atherosclerotic lesion formation and inflammatory response in the LDL receptor null mouse. *Atherosclerosis* **2009**, *204*, 147–155.
42. Egert, S.; Kannenberg, F.; Somoza, V.; Erbersdobler, H.F.; Wahrburg, U. Dietary alpha-linolenic acid, EPA, and DHA have differential effects on LDL fatty acid composition but similar effects on serum lipid profiles in normolipidemic humans. *J. Nutr.* **2009**, *139*, 861–868.
43. Maeda, H.; Tsukui, T.; Sashima, T.; Hosokawa, M.; Miyashita, K. Seaweed carotenoid, fucoxanthin, as a multi-functional nutrient. *Asia Pac. J. Clin. Nutr.* **2008**, *1*, 196–199.
44. Maeda, H.; Hosokawa, M.; Sashima, T.; Miyashita, K. Dietary combination of fucoxanthin and fish oil attenuates the weight gain of white adipose tissue and decreases blood glucose in obese/diabetic KK-Ay mice. *J. Agric. Food Chem.* **2007**, *55*, 7701–7706.
45. Tsukui, T.; Konno, K.; Hosokawa, M.; Maeda, H.; Sashima, T.; Miyashita, K. Fucoxanthin and fucoxanthinol enhance the amount of docosahexaenoic acid in the liver of KKAy obese/diabetic mice. *J. Agric. Food Chem.* **2007**, *55*, 5025–5029.
46. Park, H.J.; Lee, M.K.; Park, Y.B.; Shin, Y.C.; Choi, M.S. Beneficial effects of *Undaria pinnatifida* ethanol extract on diet-induced-insulin resistance in C57BL/6J mice. *Food Chem. Toxicol.* **2010**, *13*, 357–363.
47. Hosokawa, M.; Miyashita, T.; Nishikawa, S.; Emi, S.; Tsukui, T.; Beppu, F.; Okada, T.; Miyashita, K. Fucoxanthin regulates adipocytokine mRNA expression in white adipose tissue of diabetic/obese KK-Ay mice. *Arch. Biochem. Biophys.* **2010**, *504*, 17–25.
48. Abidov, M.; Ramazanov, Z.; Seifulla, R.; Grachev, S. The effects of Xanthigen in the weight management of obese premenopausal women with non-alcoholic fatty liver disease and normal liver fat. *Diabetes Obes. Metab.* **2010**, *12*, 72–81.
49. Kim, K.N.; Heo, S.J.; Kang, S.M.; Ahn, G.; Jeon, Y.J. Fucoxanthin induces apoptosis in human leukemia HL-60 cells through a ROS-mediated Bcl-xL pathway. *Toxicol. Vitro* **2010**, *24*, 1648–1654.
50. Jeon, S.M.; Kim, H.J.; Woo, M.N.; Lee, M.K.; Shin, Y.C.; Park, Y.B.; Choi, M.S. Fucoxanthin-rich seaweed extract suppresses body weight gain and improves lipid metabolism in high-fat-fed C57BL/6J mice. *Biotechnol. J.* **2010**, *5*, 961–969.
51. Shiratori, K.; Okgami, K.; Ilieva, I.; Jin, X.H.; Koyama, Y.; Miyashita, K.; Yoshida, K.; Kase, S.; Ohno, S. Effects of fucoxanthin on lipopolysaccharide-induced inflammation *in vitro* and *in vivo*. *Exp. Eye Res.* **2005**, *81*, 422–428.
52. Granger, D.N.; Rodrigues, S.F.; Yildirim, A.; Senchenkova, E.Y. Microvascular responses to cardiovascular risk factors. *Microcirculation* **2010**, *17*, 192–205.

53. Pashkow, F.J.; Watumull, D.G.; Campbell, C.L. Astaxanthin: A novel potential treatment for oxidative stress and inflammation in cardiovascular disease. *Am. J. Cardiol.* **2008**, *101*, 58–68.
54. Fassett, R.G.; Healy, H.; Driver, R.; Robertson, I.K.; Geraghty, D.P.; Sharman, J.E.; Coombes, J.S. Astaxanthin *vs.* placebo on arterial stiffness, oxidative stress andinflammation in renal transplant patients (Xanthin): A randomised controlledtrial. *BMC Nephrol.* **2008**, *18*, 9–17.

*marine drugs*

*Article*

# Accumulation of Astaxanthin by a New *Haematococcus pluvialis* Strain BM1 from the White Sea Coastal Rocks (Russia)

**Konstantin Chekanov [1], Elena Lobakova [1], Irina Selyakh [1], Larisa Semenova [1], Roman Sidorov [2] and Alexei Solovchenko [1,2,*]**

1 Biological Faculty of Lomonosov Moscow State University, 1/12 Leninskie Gori, Moscow 119234, Russia; E-Mails: chekanov@mail.bio.msu.ru (K.C.); elena.lobakova@mail.ru (E.L.); i-savelyev@mail.ru (I.S.), semelar@mail.ru (L.S.)

2 Timiryazev Institute of Plant Physiology, Russian Academy of Sciences, 35, Botanicheskaya str., Moscow 127276, Russia; E-Mail: roman.sidorov@mail.ru

* Author to whom correspondence should be addressed; E-Mail: solovchenko@mail.bio.msu.ru; Tel.: +7-495-939-2587; Fax: +7-495-939-4309.

*Received: 12 June 2014; in revised form: 17 July 2014 / Accepted: 4 August 2014 / Published: 11 August 2014*

**Abstract:** We report on a novel arctic strain BM1 of a carotenogenic chlorophyte from a coastal habitat with harsh environmental conditions (wide variations in solar irradiance, temperature, salinity and nutrient availability) identified as *Haematococcus pluvialis* Flotow. Increased (25‰) salinity exerted no adverse effect on the growth of the green BM1 cells. Under stressful conditions (high light, nitrogen and phosphorus deprivation), green vegetative cells of *H. pluvialis* BM1 grown in BG11 medium formed non-motile palmelloid cells and, eventually, hematocysts capable of a massive accumulation of the keto-carotenoid astaxanthin with a high nutraceutical and therapeutic potential. Routinely, astaxanthin was accumulated at the level of 4% of the cell dry weight (DW), reaching, under prolonged stress, 5.5% DW. Astaxanthin was predominantly accumulated in the form of mono- and diesters of fatty acids from C16 and C18 families. The palmelloids and hematocysts were characterized by the formation of red-colored cytoplasmic lipid droplets, increasingly large in size and number. The lipid droplets tended to merge and occupied almost the entire volume of the cell at the advanced stages of stress-induced carotenogenesis. The potential application of the new strain for the production of astaxanthin is discussed in comparison with the *H. pluvialis* strains currently employed in microalgal biotechnology.

**Keywords:** astaxanthin; carotenogenesis; fatty acids; green microalgae

---

## 1. Introduction

The red ketocarotenoid Astaxanthin (Ast) is a potent antioxidant exerting a plethora of health-beneficial effects in human and animal organisms. It is of high demand as an ingredient of cosmetic, medical and dietary formulations [1,2] as well as quality feed for aquaculture. In particular, the red color of the crustacean shells and salmon meat is due to the presence of Ast; the only source of Ast in animals is their diet [3]. At present, most of the feed Ast is chemically synthesized although the synthetic pigment, unlike natural Ast, is a racemic mixture containing a substantial proportion of the stereoisomers lacking the biological activity [3].

The richest natural source of Ast is the chlorophyte *Haematococcus pluvialis* Flotow [4] that accumulates the pigment in an amount of up to 3%–6% of cell dry weight (DW) under unfavorable environmental conditions [5]. Essentially a freshwater alga, *H. pluvialis* survives in small rain pools under extremely volatile conditions such as extreme temperatures, low nutrient availability and solar irradiance [6] mainly in form of Ast-rich non-motile coccoid cells with an exceptional tolerance of the adverse conditions [7–10]. The massive accumulation of Ast in *H. pluvialis* depends on and is accompanied by the enhanced biosynthesis of neutral lipids, mainly triacylglycerols (TAG) [11]. The reason for this is that Ast is deposited in cytoplasmic lipid droplets (LD) comprised by TAG, mainly in the form of FA esters. Accordingly, *H. pluvialis* can also be a source of valuable FA e.g., oleic acid [5,12].

In spite of its high bioavailability and numerous beneficial effects, natural Ast from microalgae hardly can compete with its synthetic analog due to high production cost and limited productivity of the commercial *H. pluvialis* strains [2,3]. Obviously, at least a two-fold increase in the Ast productivity of current strains (which is at the level of ca. 3% DW) is necessary for natural Ast to outcompete the synthetic pigment [13]. Moreover, mass cultivation of *H. pluvialis* is highly demanding of fresh water, which may not be available at the site of cultivation. Therefore bioprospecting of more stress-tolerant *H. pluvialis* strains is important to reduce the costs of the Ast-enriched biomass production e.g., by the use of brackish water. We paid close attention to White Sea coastal rocks characterized by a particularly harsh environment expecting to obtain microalgal isolates naturally adapted to the adverse conditions. In the present work, we obtained a detailed characteristic of a novel *H. pluvialis* strain from an arctic sea and estimated its suitability for Ast production.

## 2. Results and Discussion

### 2.1. The Habitat of the New Strain

The carotenogenic microalga designated as BM1 described in this paper was originally discovered as reddish crusts on the northern slope of black granite-gneiss coastal rocks on Kost'yan island (66°29′47″ N, 33°24′22″ E), White Sea (Figure 1). This habitat is characterized by harsh environmental conditions even during the warm season. Thus, during polar days (March to August), the northern slopes of the cliffs are constantly illuminated by sun. As a result, the water filling the rock baths inhabited by the microalga is considerably warmer than the seawater (Supplementary Table S1). Sharp fluctuations of salinity are also typical of this habitat due to enhanced evaporation from sun-heated rocks, especially during windy weather, and regular inflow of seawater from high tides or fresh water from rain and melting snow.

**Figure 1.** (**a**) Coastal rocks at Kost'yan Island, White Sea; (**b**) The red crust formed by the astaxanthin-rich hematocysts in a drying rock bath.

(**a**) (**b**)

Until the end of June, the microalga dwelled in the baths mainly as motile biflagellate green zoospores or coccoid non-motile cells (Figure 2b). In July, the microalga was represented mainly by large (up to 80 μm in diameter) bright red-colored coccoid resting cells (Supplementary Figure S2a) referred to below as "red" cells. By the end of July, the baths usually dried out and the "cells" formed the reddish crusts on the rock surface (Figure 1). In addition to the microalgal cells, thin (1–3 μm in diameter) filamentous heterocyst-lacking cyanobacteria (III subsection, presumably Oscillatoriaceae) were encountered in the samples. The surface of the "red" cells was often decorated by numerous bacterial rod-like cells attached by their apical ends (not shown).

To the best of our knowledge, the literature on isolation of *H. pluvialis* from arctic regions is scarce and limited to the isolates from freshwater habitats. A cold-tolerant strain of *H. pluvialis* capable of growth at a low (4–10 °C) temperature was recently isolated from a freshwater lagoon in Blomstrandhalvøya Island (Svalbard) [14]. In contrast to BM1, this strain was incapable of sustained growth at temperatures higher than 15 °C since these conditions triggered the formation of the red-colored cysts in the latter. Notably, the enrichment culture of the freshwater arctic strain also contained filamentous heterocyst-lacking cyanobacteria from the III subsection.

## 2.2. *The Cell Morphology and Ultrastructure*

Several morphological types of cells were found in BM1 cultures. The first cell type was represented by motile spherical zoospores (15–20 μm in diameter with a mucous sheath) with two isokont anterior flagellae and a discoid eyespot near the cytoplasmic membrane at the cell anterior (Figure 2a). The zoospores featured a cup-shaped chloroplast occupying almost the entire cell volume. Another type was comprised by non-motile coccoid cells (20–40 μm in diameter; Figure 2b). The cells contained a centrally located spherical nucleus. Palmelloid clusters of two to eight cells were formed occasionally. Under adverse environmental conditions, the coccoid cells increased in size (up to 80 μm) and accumulated red-colored spherical inclusions in the cytoplasm, which tended to cluster around the nucleus (Figure 2c). The red inclusions gradually occupied the entire cytoplasm volume resulting in the formation of resting "red" cells (Supplementary Figure S2a).

**Figure 2.** Life cycle stages of BM1 isolate (**a**) Zoospore; (**b**) Coccoid green cell; (**c**) Coccoid green cell with red-colored lipid droplets; (**d**) Sporangium; (**e**) Putative isogamous sexual process.

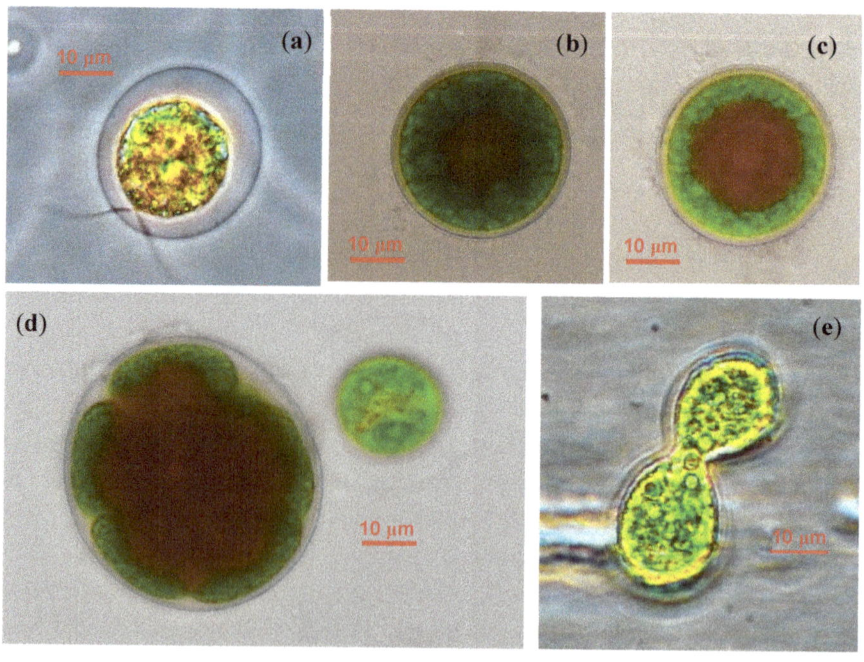

The isolated microalgae propagated mainly via asexual reproduction forming sporangia containing two to eight autospores (Figure 2d). After shedding the sporangium wall, the newly formed cells remained attached to each other for a long time; as a result, the culture tended to accumulate four-cell clusters. Under non-optimal cultivation conditions, e.g., in a highly diluted culture, small, pear-shaped fast-moving biflagellate cells similar to gametes as described by Triki *et al.* [15] were encountered. Occasionally, these cells underwent conjugation resembling isogamous sexual process (Figure 2e). Dead cells with transparent content were also present in the culture (not shown). One could conclude that the life cycle and cell morphology recorded in the BM1 isolate as well as the ability to accumulate the red pigment in the "red" cells are consistent with the characteristic traits of *Haematococcus pluvialis* Flotow [15–17].

To characterize the newly isolated microalga, we investigated its cell ultrastructure. It should be noted that ultrastructural studies of *H. pluvialis* are generally more difficult in comparison to most of green microalgae, mainly due to the presence of tough cell walls complicating the chemical fixation, embedding and preparation of ultrathin sections [18]. Indeed, we found that the thick cell wall of BM1 was, like that of *H. pluvialis* aplanospores, extremely resistant to mechanical disruption and chemical agents and presented difficulties for electron microscopy. Nevertheless, both transmission (Figure 3a,c,d) and scanning (Figure 3b) electron micrographs of "green" and "red" BM1 cells were obtained.

**Figure 3.** Electron micrographs of *H. pluvialis* BM1: (**a**) transmission electron microscopy (TEM) of a "green" cell; (**b**) scanning electron microscopy (SEM) of enrichment culture comprised of different cell types; (**c**) TEM of a "red" cell; (**d**) Pyrenoid structure typical of BM1 cells. CW—cell wall; LD—lipid droplets; SG—starch grains; T—thylakoids. Note the absence of LD in the "green" cells (**a**) and their presence in the "red" cells (**c**).

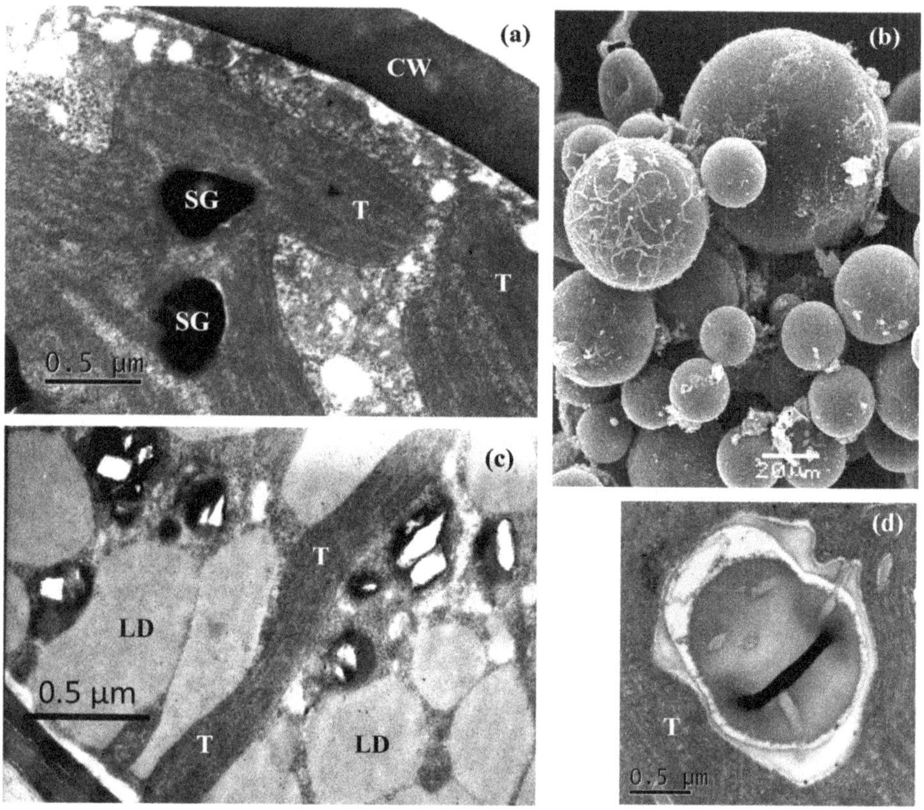


As was shown by electron microscopy, the BM1 cells at different stages of life cycle were spherical and 18–59 µm in diameter (Figure 3b). Green flagellated and palmelloid cells possessed a thick (0.64–0.8 µm) extracellular matrix, which was thinner in the palmelloid and "red" cells. All kinds of palmelloid cells formed (up to 0.4 µm) the thick cell wall. The chloroplast contained two to ten pyrenoids and a few starch grains (Figure 3a,d). In the resting "red" cells, a pronounced decrease in the thylakoid volume and number was recorded; large merging lipid droplets were also present, which (Figure 3c) eventually occupied almost the entire cell volume.

## *2.3. Molecular Identification*

In order to reveal the taxonomic identity of the BM1 isolate we obtained a partial sequence of its 18S rRNA gene (GenBank ID JQ867352.1). The homology search using the Basic Logical Alignment Search Tool (BLAST) showed the maximum (99%–100%) identity of the sequence with respective sequences of a number of known *H. pluvialis* strains (Figure 4). The phylogenetic analysis of the BM1 isolate showed that it belongs to *Haematococcus pluvialis* Flotow, the single species in the genus *Haematococcus* nested in the Chlorogonium clade, Volvocales, Chlorophyceae which is consistent with the generally accepted results of molecular identification of *H. pluvialis* [19]. Basing on the findings described above, the isolate BM1 was designated as *H. pluvialis* BM1.

**Figure 4.** Phylogenetic relationships of BM1 isolate as revealed by 18S rRNA gene sequence. The optimal tree is shown. The percentage of replicate trees in which the associated taxa clustered together in the bootstrap test (1000 replicates) are shown under the branches for maximum likelihood/neighboring-joining (ML/NJ) method, respectively. All positions containing gaps and missing data were eliminated from the dataset. There were a total of 782 positions and 25 taxa in the final dataset. Phylogeny analysis was conducted in PhyML 3.0 and BioNJ. The tree was rendered using TreeDyn 198.3 software (GEMI Bioinformatics, Montpellier, France).

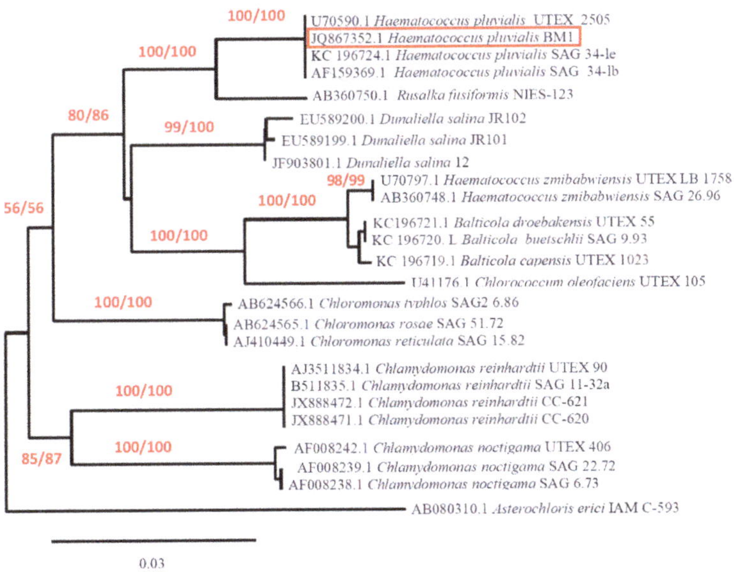

## 2.4. Growth and Carotenogenesis

### 2.4.1. Biomass Accumulation and Pigment Composition

In order to evaluate the potential of *H. pluvialis* BM1 for biotechnology the isolate was cultivated in a closed bubble-column photobioreactor as described in the "Experimental" Section (see also Supplementary Figure S2d). The "green" cell cultures reached a maximum cell density of 1.6 × $10^6$ $mL^{-1}$ (37 mg·$mL^{-1}$ Chl, 6 mg·$mL^{-1}$ Car, ca. 1.0 g·$L^{-1}$ of cell dry weight, DW) in 5–7 days corresponding to the specific growth rate, $\mu = 0.095$ $day^{-1}$ at the exponential phase (Supplementary Figure S2). At the exponential growth phase, the culture was comprised, to a considerable extent, by dividing "green" cells (Figure 2). The cell suspension was bright green in color due to low (ca. 0.18 ± 0.01) Car/Chl ratio (by weight). The Car at this growth phase were represented exclusively by primary carotenes and xanthophylls, there was no detectable presence of Ast (see Section 2.4.3 below and Figure 6). After 10 day of cultivation, accumulation of astaxanthin was detected and the Car/Chl ratio gradually increased, apparently, due to nitrate depletion in the medium.

### 2.4.2. Stress-Induced Astaxanthin Accumulation

To induce the massive accumulation of Car, the "green" cells of *H. pluvialis* BM1 were transferred to the stressful conditions mimicking, to a certain extent, the nutrient deficiency and excessive solar irradiation in their natural habitat. Specifically, the cells were washed, resuspended in distilled water, and exposed to irradiance one order of magnitude higher than that optimal to the "green cells" (see the "Experimental" section). Under these conditions, most of the cells displayed a rapid induction of Ast biosynthesis and turned into non-motile "red" cells (Supplementary Figure S2a).

The induction of carotenogenesis occurred in the background of a sharp decline of Chl content. As a result, the shape of the absorption spectra of extracts from the "red" cells was governed by Ast absorption (Supplementary Figure S2b). Notably, the cells of *H. pluvialis* BM1 even after prolonged stress exposure always retained a certain amount of Chl; only dead colorless cells possessed no detectable Chl content. On the whole, the dynamics of stress-induced Car accumulation displayed by BM1 was similar to that recorded in known *H. pluvialis* strains [20].

High performance liquid chromatography (HPLC) analysis (see Section 2.4.4 below) confirmed that nearly 99% of total Car in the "red" cells were represented by Ast reaching 5.0%–5.5% DW by the 6th day of stress (corresponding to a Car/Chl of 13.0 ± 0.1). After the 6th day of stress exposition, the Ast content declined sharply (Supplementary Figure S2c). This process was manifested by a massive appearance of bleached cells.

### 2.4.3. Salinity Effects on the Growth of the "Green Cells"

The abrupt changes of salinity characteristic of the habitat of BM1 (see Section 2.1) suggest an increased ability to acclimate to this factor in the microalga under investigation. To obtain a preliminary estimation of BM1 salinity tolerance, we cultivated the microalga under salinity similar to that of the rock bath water (25‰), which is typically below the White Sea water salinity (29‰) because of dilution with rainwater.

It was found that the increase in salinity *per se* did not trigger a decline in Chl accumulation by the culture (Figure 5a) or an increase in Car accumulation over Chl (Figure 5b) typical of the stress-induced carotenogenic response. During first 5–7 days, the kinetics of growth on the saline medium did not differ significantly from that on the medium lacking NaCl (see Section 2.4.1 above and Supplementary Figure S1). Only a limited accumulation of Ast was detected in the cultures grown at 25‰ NaCl (insert in Figure 5c).

**Figure 5.** Effects of 25‰ NaCl on (**a**) chlorophyll accumulation; (**b**) carotenoid-to-chlorophyll ratio; and (**c**) normalized absorption spectra of the cell dimethyl sulfoxide (DMSO) extracts of *H. pluvialis* BM1 "green" cell culture. The spectra for (*1*) initial culture (Day 0) as well as those recorded after one day (*2*, *2′*) and five days (*3*, *3′*) of cultivation in the medium containing (*2′*, *3′*) or lacking (*2*, *3*) NaCl are shown. Insert: different absorbance spectra of the extract spectra presented in the panel (**c**). Note a positive peak in the green region indicative of a limited accumulation of astaxanthin by the fifth day of cultivation in the presence of NaCl.

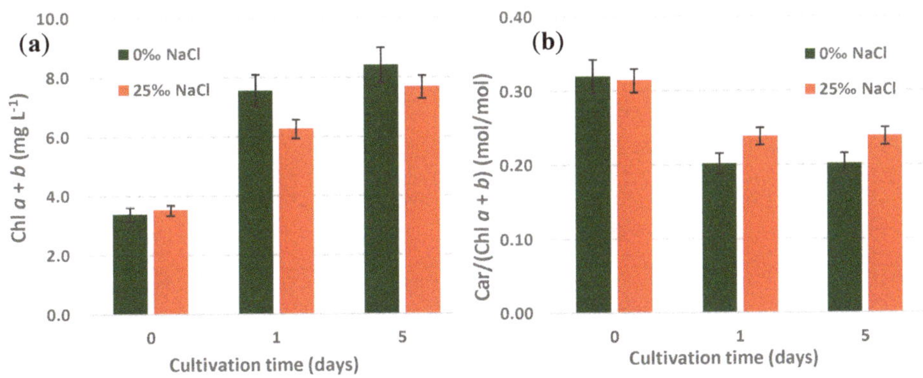

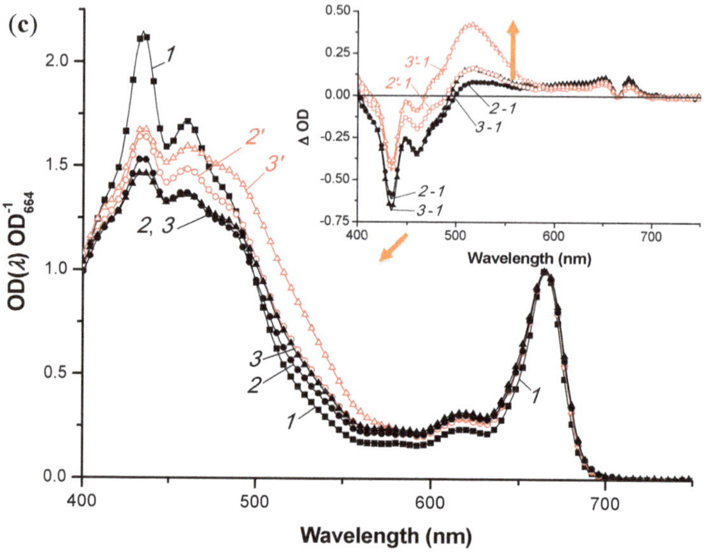


**Figure 6.** Pigment composition of *H. pluvialis* BM1 cells at different cultivation stages (**a**) The "green" cells (high performance liquid chromatography (HPLC)); (**b**) The "red" cells (thin-layer chromatography (TLC) + HPLC).

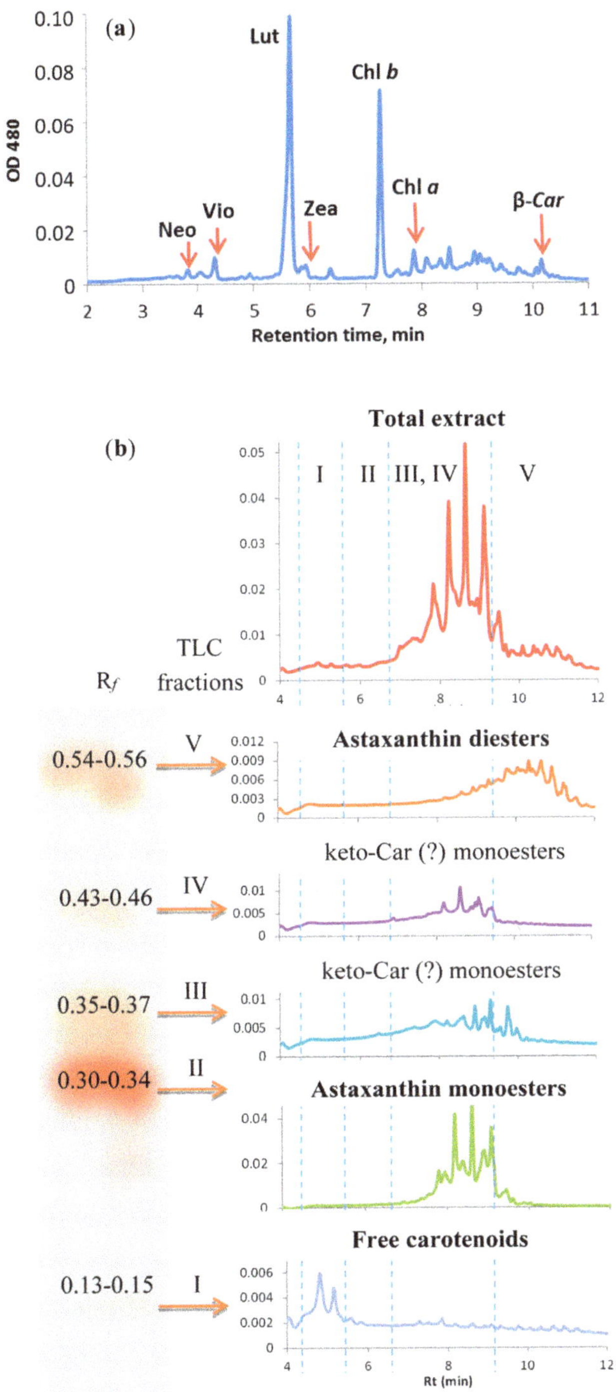


At the same time, it is known that the addition of 0.8% NaCl (8‰) to the medium normally causes a cessation of growth of *H. pluvialis* [7,20]. In view of these facts, the new strain *H. pluvialis* BM1 seems to have a higher tolerance to salinity stress although a more detailed investigation of the limits and possible side effects of its salinity tolerance is necessary. Nevertheless, this finding may be important for the biotechnology of Ast production in the areas with a limited supply of fresh water suggesting the possibility of cultivation of "green" cells of *H. pluvialis* BM1 in brackish water.

### 2.4.4. Stress-Induced Changes in Pigment and Fatty Acid Composition

Under conditions conducive to rapid growth of the culture, the pigment composition of *H. pluvialis* BM1 was typical of green algae thylakoid membranes [21] including Chl *a* and *b* as well as primary Car; only trace quantities of Ast esters were detected (Figure 6a).

Thin-layer chromatographical (TLC) separation of the extracts from the "red" cells yielded five pigment fractions (Fractions I–V, Figure 6b). The absorption spectra of all fractions resembled those of pure Ast ($\lambda_{max}$ = 490). Incubation in air at room temperature for 10–15 min resulted in the long-wave shift of the maximum to 494 nm typical for astacene, an oxidation product of Ast [22]. The fast conversion of the pigment from the Fraction I to astacene suggests that the Fraction I contained non-esterified (free) Ast. The bulk (ca. 70%) of the Ast in the "red" cells was found in the Fractions II ($R_f$ = 0.30–0.34) and III ($R_f$ = 0.35–0.37). The Fractions IV ($R_f$ = 0.43–0.46) and V ($R_f$ = 0.54–0.56) were substantially less abundant. Free Ast (the Fraction I) comprised less than 3.5% of the total Ast.

**Table 1.** Fatty acid composition (*mas.-%*) of esterified carotenoid fractions of *H. pluvialis* BM1 "red" cells.

| Fatty Acid | TLC Fraction | | | |
|---|---|---|---|---|
| | **II** [(a)] | **III** [(b)] | **IV** [(c)] | **V** [(d)] |
| 14:0 | 7.8 | 4.9 | 5.0 | - |
| 16:0 | **41.1** | **55.3** | **50.4** | - |
| 7–16:1 | 5.9 | 6.5 | 4.4 | 2.3 |
| 7,10–16:2 | -[*] | - | - | **12.0** |
| 18:0 | 5.7 | 7.9 | 6.7 | 1.7 |
| 9–18:1 | **16.3** | **17.3** | **17.7** | **23.8** |
| 11–18:1 | 1.1 | - | - | 1.3 |
| 9,12–18:2 | **10.2** | **7.4** | **9.4** | **15.8** |
| 6,9,12–18:3 | 1.0 | - | - | 0.8 |
| 9,12,15–18:3 | 5.9 | - | - | 5.0 |
| 13–22:1 | 2.2 | - | 4.1 | 0.5 |
| **UI** [**] | **0.694** | **0.387** | **0.450** | **1.156** |

[*] not detected; [**] unsaturation index [(a)] 6,9,12,15–18:4, 20:0, 22:0, 24:0 also were present; the concentration of each was 0.6%–0.8%; [(b)] also contained 0.5% of 20:0 FA; [(c)] 12:0, 20:0 and 22:0 were present, the concentration each was 0.7%–0.9%; [(d)] also contained 7,10,13–16:3—1.9%, 4,7,10,13–16:3—2.3% and 20:0, 22:0, 24:0, the concentration of each was 0.1%–0.2%.

The HPLC-diode-array detector (DAD) analysis of the Fraction I as obtained by TLC confirmed the presence of free Ast. The HPLC elution profile of the Fractions II and III pigments contained eight major peaks ($R_t$ 8–10 min). One could suggest that these peaks correspond to individual molecular species of Ast monoesters. The Fraction V harbored a large number of nonpolar compounds containing Ast chromophore, most probably Ast fatty acid diesters.

Notably, accumulation of Ast in *H. pluvialis* BM1 cells took place along with a significant increase in neutral lipids in cytoplasmic lipid droplets as evidenced by vital staining of the "red" cells with Nile Red. The gas chromatography-mass spectrometry (GC-MS) fatty acid analysis (Table 1) demonstrated that the fatty acid profile of the Ast monoesters was dominated by palmitic (16:0), oleic (Δ9–18:1), and linoleic (Δ9,12–18:2) acids; it was similar to that of the known *H. pluvialis* strains [11,12,23]. Interestingly, The FA from the diester fraction (V) possessed a nearly two-fold higher unsaturation index in comparison with those from the monoester fractions (II–IV). Taking into account the strong differences in FA composition of the mono- and diesters of Ast in *H. pluvialis* BM1, one may speculate that (i) their FA originate from different pools and (ii) the enzymes responsible for biosynthesis of the different classes of Ast esters possess a different substrate specificity.

## 3. Experimental Section

### 3.1. BM1 Isolation and Obtaining of Its Culture

Samples were collected from red-colored bath tiles filled with semi-saline water found in the supralittoral zone of the Kost'yan Island (66°29′47″ N, 33°24′22″ E) in the White Sea. Water samples and red plaque (Figure 1b) scrapings from the surface of parched "bath-tile" mud cracks were collected in June–August of 2011 and 2012. The samples were sealed in sterile polypropylene bags. Within three hours, the precipitate of the water samples was transferred to two mL of BG-11 medium [24]. The plaque scrapings were transferred in five mL of BG-11 in sterile 15 mL glass vials. Samples were incubated in 15 mL sterile glass vials at 20 °C under daylight for one month. Algological pure cultures were obtained at Pertsov White Sea Biological Station of Moscow State University. Single red-colored cells and sporangia were placed in five mL of BG11 in sterile tubes using a micromanipulator and incubated under a white fluorescent lamp (40 $\mu E \cdot m^{-2} \cdot s^{-1}$) at room temperature for one month. The resulting cultures were inoculated onto Petri dishes with solidified BG11 medium (containing 2% agarose) using a dilution method and incubated under the same conditions until green-colored colonies became apparent (2–3 weeks). The colonies were transferred to the tubes containing sterile BG-11 medium and incubated under the same conditions.

### *3.2. Culture Handling and Maintaining*

The "green cells" of the microalgal isolate BM1 were cultivated in 400 mL of BG11 medium in 600-mL glass columns (6.6 cm internal diameter, 1.5 L·volume). The cultures were grown under continuous photosynthetically active radiation (PAR) illumination of moderate (40 $\mu E \cdot m^{-2} \cdot s^{-1}$) intensity as measured with a LiCor 850 quantum sensor (LiCor, Lincoln, NE, USA) in the center of an empty column by white light emitting diode source. For the induction of carotenogenesis, the cells were incubated in distilled water at high (480 $\mu E \cdot m^{-2} \cdot s^{-1}$) PAR intensity. The cultures were continuously bubbled with air (1 $v \cdot v^{-1} \cdot min^{-1}$); the temperature was maintained at 27 °C.

To obtain a preliminary estimation of salinity tolerance of BM1, the cells were cultured in 250-mL flasks containing 80 mL of BG-11 medium supplemented with 25 $g \cdot L^{-1}$ NaCl (BG-11 lacking NaCl was used as the control) in a shaker incubator at 100 RPM, 27 °C and 40 $\mu E \cdot m^{-2} \cdot s^{-1}$ PAR.

Cell dry weight (DW) was measured according to [20]. Cells were counted in a hemocytometer.

### *3.3. Molecular Identification*

To destroy the tough cell wall, the 1.5-mg cell samples collected for DNA extraction were subjected to three cycles of freezing in liquid nitrogen and subsequent thawing at room temperature. The samples were incubated for one hour in 300 μL of citrate-phosphate buffer (pH 5.0) containing 10 $mg \cdot mL^{-1}$ cellulase (Fermentas, Vilnius, Lithuania), 10 $mg \cdot mL^{-1}$ pectinase (Fermentas, Vilnius, Lithuania) and 1mM EDTA at 37 °C for one hour. The samples were incubated with 2% sodium dodecyl sulfate for one hour at 40 °C. Next 400 μL of 1M NaCl were added and allowed to stand overnight for protein salting. Then standard phenol-chloroform extraction was performed [25]. The DNA sample purity was evaluated by electrophoresis in 1.5% agarose gel with ethidium bromide.

For amplification of the 18S rRNA gene fragment, the following primers were designed using Gene Runner 4.0.9.68: 5′-tggctcattaaatcagttatag-3′, 5′-ccaagaatttcacctctgaca-3′. Polymerase chain reaction (PCR) was performed on a Bio-Rad DNA engine (PTC 200, Hercules, CA, USA) using the amplification profile of 94 °C for 3 min initial denaturation, 94 °C for 20 s, 60 °C for 25 s, 72 °C for 35 s, 30 cycles; final elongation—72 °C for 5 min. The PCR mixture contained 10 ng of the algae genomic DNA in 25 μL of 1× PCR Buffer for Tersus (Evrogen, Moscow, Russia) containing 200 μM of each dNTP, 0.2 μM of each primer and 0.5 μL of 50× Tersus Taq polymerase (Evrogen, Moscow, Russia). The PCR products were purified using a Cleanup Standard PCR purification kit (Evrogen, Moscow, Russia) and sequenced on ABI Prizm 3730 (Applied Biosystems, Life Technologies, Grand Island, NY, USA) in both directions.

Sequences were searched against the NCBI GeneBank (nucleotide collection nr/nt database) using BLAST. For the data analysis, 24 sequences from GenBank were selected. Multiple alignment of the sequences were conducted using ClustalW2 online tool [26]. There were a total of 782 positions and 25 taxa in the final dataset. The following parameters of pair-wise alignment were used Alignment type: slow, DNA weight matrix: IUB, gap open: 100, gap extinction: 10.0. The parameters of multiple alignments are given below. DNA weight matrix: IUB, gap open: 100, gap extinction: 10.0, gap distances: 10, no end gaps, iteration type: "tree", number of iterations: 10. Phylogenetic trees for multiple alignments were designed with using maximum likelihood (ML) method [27] in PhyML 3.0 [28] and neighbor-joining method (NJ) [29] in BioNJ [30,31].The best of nearest neighbor interchange (NNI) and subtree pruning and regrafting (SPR) were used for tree improvement in PhyML 3.0. HKY85 model of DNA evolution [32] were used. For other parameters of the analysis, the default values were left. Trees were rendered using program TreeDyn 198.3 [33]. The accuracy of the tree topology was tested using bootstrap analysis [34] with 1000 replicates.

### 3.4. *Pigment Analysis*

Routinely, the 0.5–1.0 mL aliquots of cells suspension were extracted with dimethyl sulfoxide (DMSO). The cells were pelleted by centrifugation (5 min, 12,000 rpm), the supernatant was removed and the cells were incubated with 1.5 mL of DMSO at 75 °C for 15 min, and the cells were removed by centrifugation; then the extraction was repeated. Chlorophyll (Chl) *a*, Chl *b* and total carotenoids (Car) in the extract were assayed spectrophotometrically [35] with an Agilent Cary 300 (Agilent, Santa Clara, CA, USA) spectrophotometer in standard 1-cm cuvettes.

For the fatty acid (FA) and pigment assay, the extraction by Folch was used [36]. In this case Chl *a, b,* and total Car contents were determined in lower (chloroform) fraction [37]. The pigments were initially separated by TLC on silica gel plates (*Sulifol*, Kavalier, Prague, Czech Republic). The hexane:benzene:chloroform (5:5:2, by volume) and hexane:chloroform:benzene (10:20:1, by vol.) were used for separation of "green" and "red" cell extracts, respectively. The spots obtained after separation of the "red" cell extracts were gently scrapped from the plate by scalpel and eluted with chloroform.

Both the total pigment extracts and eluted fractions were subjected to HPLC analysis using an Alliance 2995 separation module (Waters, Milford, MD, USA) equipped with a 150 × 4.5-mm Prontosil RP C-18 column (Knauer, Berlin, Germany) maintained at 25°C and a Waters e2695 DAD detector (Waters, Milford, MD, USA). The gradient elution of pigments was achieved at a flow rate of 1 mL·min$^{-1}$ using (A) acetonitrile, (B) water, and (C) ethyl acetate mixtures (vol. %): for the "red" cell extracts, 98:2:0 (2 min), 40:0:60 (10 min), 0:0:100 C (2 min) followed by 6-min re-equilibration of the column; for the "green" cell extracts: 98:2:0% B (5 min); 48.5:

1.2:50, 0:0: 100 (3 min) followed by 6-min re-equilibration. Eluted component spectra were monitored in the range 400–700 nm. Free pigments were identified and quantified using authentic standards (Sigma, St. Louis, MS, USA). Astaxanthin esters were identified by their chromatographic mobility and quantified as free Ast.

### 3.5. Fatty Acid Analysis

Heptadecanoic acid (17:0) was added as an internal standard to the chloroform phase of the extracts obtained as specified above and the samples were transmethylated according to [38]. Methyl esters were extracted with *n*-hexane and immediately subjected to GC-MS analysis with an Agilent 7890 gas chromatograph equipped with DB-23 capillary column (Ser. No. US8897617H, 60 m × 0.25 mm, containing a grafted (50% cyanopropyl)-methylpolysiloxane polar liquid phase as a 0.25 μm-thick film) coupled with Agilent 5975C mass-selective detector (Agilent, Santa Clara, CA, USA). The fatty acid methyl esters (FAME) were separated under the following conditions: carrier gas (helium) pressure in the injector, 191 kPa; operational gas pressure in the column at 1 mL/min, carrier gas flow linear velocity in the column, 18 cm/s; sample volume, 1 μL; flow split ratio, 1:1; evaporator temperature, 260 °C. The oven temperature program was as follows: from 130 to 170 °C at 6.5 °C/min, to 215 °C at 2.75 °C/min (25 min at this temperature), to 240 °C at 40 °C/min, and 50 min at 240 °C, operational temperature of the mass selective detector, 240 °C, energy of the ionization, 70 eV. Identification of FA was done according to the retention times of standards (Sigma, St. Louis, MS, USA) and by characteristic mass spectra. The unsaturation index (UI) of FA mixtures was calculated as follows: $UI = \Sigma p_i \times e_i/100$, where $p_i$ is the percentage and $e_i$—the double bond number of *i*-th FA [38].

### 3.6. Microscopy

#### 3.6.1. Light and Luminescent Microscopy

The cells from the cultures were studied by bright field and fluorescence microscopy on Eclipse 90i (Nikon, Tokyo, Japan) motorized photomicroscope. For neutral lipid express assay, the cells were vitally stained with Nile Red fluorescent stain [39].

#### 3.6.2. Transmission Electron Microscopy

To prepare the samples for TEM, the cell suspension was washed twice with 0.1 M cacodylate buffer (pH 7.2) and fixed for transmission electron microscopic investigation according to a modified method suitable for the cells featuring a tough cell wall [40]. Ultrathin sections were prepared using an LKB 4800 ultramicrotome (LKB Produkter, Broma, Sweden), contrasted with lead citrate [41], and analyzed with a Hitachi HU-11F electron microscope (Hitachi Ltd., Tokyo, Japan) at an accelerating voltage of 80 kV.

### 3.5.3. Scanning Electron Microscopy

For SEM sample preparation, the cells fixed and dehydrated in ethanol were transferred in absolute acetone and dried at a critical point in the HCP-2 dryer (Hitachi, Hitachi, Japan), coated with gold and palladium on by ion-sputtering IB Ion Coater (Eiko, Ibaraki, Japan) and examined in a JSM-6380LA (JEOL, Tokyo, Japan) microscope at an accelerating voltage of 15 kV.

## 4. Conclusions

The present work describes, to the best of our knowledge, the first strain of *H. pluvialis* isolated from an arctic seashore habitat characterized with moderate and highly variable salinity levels [42]. Indeed, the current compendia of White Sea algoflora do not include species from the genus *Haematococcus* other than freshwater representatives [43] which are not yet confirmed rigorously. The new strain was capable of massive accumulation of Ast as a secondary Car, mainly in the form of esters of the FA from the C16 and C18 families. Depending on the cultivation conditions, it turned out to be able to accumulate Ast in the amounts of 3%–5.5% of DW, which compares favorably with most of the currently known strains [2]. Our preliminary studies have shown that *H. pluvilais* BM1 seems to have an elevated salt tolerance, the extent of which, however, remains to be elucidated. Nevertheless, given its growth and biosynthetic capacity and potential salt tolerance, the new BM1 strain is a promising organism for biotechnological production of valuable carotenoids using microalgae. In particular, it merits close attention as a potential producer of Ast potentially capable of growth in brackish water.

## Acknowledgments

The dedicated technical assistance of Ekaterina Lobanova and Ksenia Klokova is greatly appreciated.

## Author Contributions

K.C. performed the cultivation experiments, carried out pigment analysis, molecular identification and wrote the manuscript; E.L. isolated the strain and supervised the cultivation; I.S. and L.S. supervised the culture purification and electron microscopy; R.S. supervised the fatty acid analysis; A.S. conceived the experiments and wrote the manuscript.

## Conflicts of Interest

The authors declare no conflict of interest.

## References

1. Guerin, M.; Huntley, M.; Olaizola, M. *Haematococcus* astaxanthin: Applications for human health and nutrition. *Trends Biotechnol.* **2003**, *21*, 210–216.
2. Han, D.; Li, Y.; Hu, Q. Biology and commercial aspects of *Haematococcus pluvialis*.
In *Handbook of Microalgal Culture: Applied Phycology and Biotechnology*, 2nd ed.; Richmond, A., Hu, Q., Eds.; Blackwell: Hoboken, NJ, USA, 2013; pp. 388–405.
3. Lorenz, R.T.; Cysewski, G.R. Commercial potential for *Haematococcus* microalgae as a natural source of astaxanthin. *Trends Biotechnol.* **2000**, *18*, 160–167.
4. Sussela, M.; Toppo, K. *Haematococcus pluvialis*–A green alga, richest natural source of astaxanthin. *Curr. Sci.* **2006**, *90*, 1602–1603.
5. Lemoine, Y.; Schoefs, B. Secondary ketocarotenoid astaxanthin biosynthesis in algae: A multifunctional response to stress. *Photosynth. Res.* **2010**, *106*, 155–177.
6. Droop, M. Carotenogenesis in *Haematococcus pluvialis*. *Nature* **1955**, *175*, 42.
7. Boussiba, S. Carotenogenesis in the green alga *Haematococcus pluvialis*: Cellular physiology and stress response. *Physiol. Plant.* **2000**, *108*, 111–117.
8. Solovchenko, A. Pigment composition, optical properties, and resistance to photodamage of the microalga *Haematococcus pluvialis* cultivated under high light. *Russ. J. Plant Physiol.* **2011**, *58*, 9–17.
9. Solovchenko, A. Physiology and adaptive significance of secondary carotenogenesis in green microalgae. *Russ. J. Plant Physiol.* **2013**, *60*, 1–13.
10. Solovchenko, A. Physiological role of neutral lipid accumulation in eukaryotic microalgae under stresses. *Russ. J. Plant Physiol.* **2012**, *59*, 167–176.
11. Zhekisheva, M.; Zarka, A.; Khozin-Goldberg, I.; Cohen, Z.; Boussiba, S. Inhibition of astaxanthin synthesis under high irradiance does not abolish triacylglycerol accumulation in the green alga *Haematococcus pluvialis* (Chlorophyceae). *J. Phycol.* **2005**, *41*, 819–826.
12. Zhekisheva, M.; Boussiba, S.; Khozin-Goldberg, I.; Zarka, A.; Cohen, Z. Accumulation of oleic acid in *Haematococcus pluvialis* (Chlorophyceae) under nitrogen starvation or high light is correlated with that of astaxanthin esters. *J. Phycol.* **2002**, *38*, 325–331.
13. Del Campo, J.; García-González, M.; Guerrero, M. Outdoor cultivation of microalgae for carotenoid production: Current state and perspectives. *Appl. Microbiol. Biotechnol.* **2007**, *74*, 1163–1174.
14. Klochkova, T.A.; Kwak, M.S.; Han, J.W.; Motomura, T.; Nagasato, C.; Kim, G.H. Cold-tolerant strain of *Haematococcus pluvialis* (Haematococcaceae, Chlorophyta) from Blomstrandhalvøya (Svalbard). *Algae* **2013**, *28*, 185–192.
15. Triki, A.; Maillard, P.; Gudin, C. Gametogenesis in *Haematococcus pluvialis* Flotow (Volvocales, Chlorophyta). *Phycologia* **1997**, *36*, 190–194.

16. Kobayashi, M.; Kurimura, Y.; Kakizono, T.; Nishio, N.; Tsuji, Y. Morphological changes in the life cycle of the green alga *Haematococcus pluvialis*. *J. Ferment. Bioeng.* **1997**, *84*, 94–97.
17. Elliot, A. Morphology and life history of *Haematococcus pluvialis*. *Arch. Protistenk* **1934**, *82*, 250–272.
18. Hagen, C.; Siegmund, S.; Braune, W. Ultrastructural and chemical changes in the cell wall of *Haematococcus pluvialis* (Volvocales, Chlorophyta) during aplanospore formation. *Eur. J. Phycol.* **2002**, *37*, 217–226.
19. Buchheim, M.A.; Sutherland, D.M.; Buchheim, J.A.; Wolf, M. The blood alga: Phylogeny of *Haematococcus* (Chlorophyceae) inferred from ribosomal RNA gene sequence data. *Eur. J. Phycol.* **2013**, *48*, 318–329.
20. Boussiba, S.; Vonshak, A. Astaxanthin accumulation in the green alga *Haematococcus pluvialis*. *Plant Cell Physiol.* **1991**, *32*, 1077–1082.
21. Lohr, M.; Im, C.-S.; Grossman, A.R. Genome-Based Examination of Chlorophyll and Carotenoid Biosynthesis in *Chlamydomonas reinhardtii*. *Plant Physiol.* **2005**, *138*, 490–515.
22. Cooper, R.D.; Davis, J.B.; Leftwick, A.P.; Price, C.; Weedon, B.C. Carotenoids and related compounds. Part XXXII. Synthesis of astaxanthin, phoenicoxanthin, hydroxyechinenone, and the corresponding diosphenols. *J. Chem. Soc. Perkin Trans. 1* **1975**, 2195–2204.
23. Damiani, M.C.; Popovich, C.A.; Constenla, D.; Leonardi, P.I. Lipid analysis in *Haematococcus pluvialis* to assess its potential use as a biodiesel feedstock. *Bioresour. Technol.* **2010**, *101*, 3801–3807.
24. Stanier, R.; Kunisawa, R.; Mandel, M.; Cohen-Bazire, G. Purification and properties of unicellular blue-green algae (order Chroococcales). *Microbiol. Mol. Biol. Rev.* **1971**, *35*, 171–205.
25. Sambrook, J.; Russell, D.W. Purification of nucleic acids by extraction with phenol: Chloroform. *Cold Spring Harb. Protoc.* **2006**, *2006*, doi:10.1101/pdb.prot4455.
26. Larkin, M.; Blackshields, G.; Brown, N.; Chenna, R.; McGettigan, P.A.; McWilliam, H.; Valentin, F.; Wallace, I.M.; Wilm, A.; Lopez, R. Clustal W and Clustal X version 2.0. *Bioinformatics* **2007**, *23*, 2947–2948.
27. Aldrich, J. RA Fisher and the making of maximum likelihood 1912–1922. *Stat. Sci.* **1997**, *12*, 162–176.
28. Guindon, S.; Dufayard, J.-F.; Lefort, V.; Anisimova, M.; Hordijk, W.; Gascuel, O. New algorithms and methods to estimate maximum-likelihood phylogenies: Assessing the performance of PhyML 3.0. *Syst. Biol.* **2010**, *59*, 307–321.
29. Saitou, N.; Nei, M. The neighbor-joining method: A new method for reconstructing phylogenetic trees. *Mol. Biol. Evol.* **1987**, *4*, 406–425.
30. Gascuel, O. BIONJ: An improved version of the NJ algorithm based on a simple model of sequence data. *Mol. Biol. Evol.* **1997**, *14*, 685–695.

31. Dereeper, A.; Audic, S.; Claverie, J.-M.; Blanc, G. BLAST-EXPLORER helps you building datasets for phylogenetic analysis. *BMC Evol. Biol.* **2010**, *10*, 8.
32. Hasegawa, M.; Kishino, H.; Yano, T. Dating of the human-ape splitting by a molecular clock of mitochondrial DNA. *J. Mol. Evol.* **1985**, *22*, 160–174.
33. Chevenet, F.; Brun, C.; Bañuls, A.-L.; Jacq, B.; Christen, R. TreeDyn: Towards dynamic graphics and annotations for analyses of trees. *BMC Bioinfor.* **2006**, *7*, 439.
34. Felsenstein, J. Confidence limits on phylogenies: An approach using the bootstrap. *Evolution* **1985**, *39*, 783–791.
35. Solovchenko, A.; Merzlyak, M.; Khozin-Goldberg, I.; Cohen, Z.; Boussiba, S. Coordinated carotenoid and lipid syntheses induced in *Parietochloris incisa* (Chlorophyta, Trebouxiophyceae) mutant deficient in Δ5 desaturase by nitrogen starvation and high light. *J. Phycol.* **2010**, *46*, 763–772.
36. Folch, J.; Lees, M.; Sloane-Stanley, G. A simple method for the isolation and purification of total lipids from animal tissues. *J. Biol. Chem.* **1957**, *226*, 497–509.
37. Wellburn, A. The spectral determination of chlorophyll *a* and chlorophyll *b*, as well as total carotenoids, using various solvents with spectrophotometers of different resolution. *J. Plant Physiol.* **1994**, *144*, 307–313.
38. Kates, M. *Techniques of Lipidology: Isolation, Analysis and Identification of Lipids*, 2nd ed.; Elsevier: Amsterdam, The Netherlands, 1986; p. 464.
39. Pick, U.; Rachutin-Zalogin, T. Kinetic anomalies in the interactions of Nile red with microalgae. *J. Microbiol. Methods* **2012**, *88*, 189–196.
40. Rainina, E.; Zubatov, A.; Buchwalow, I.; Luzikov, V. A cytochemical study of the localization of acid phosphatase in *Saccharomyces cerevisiae* at different growth phases. *Histochem. J.* **1979**, *11*, 299–310.
41. Reynolds, E. The use of lead citrate at high pH as an electron-opaque stain in electron microscopy. *J. Cell Biol.* **1963**, *17*, 208.
42. Guiry, M. AlgaeBase. Available online: http://www.algaebase.org (accessed on 31 May 2014).
43. Tchesunov, A.; Kaliakina, N.; Bubnova, E. *Katalog Bioti Belomorskoi Biologicheskoi Stantsii MGU (Catalog of the Biota of White Sea Biological Research Station Of MSU)*; KMK: Moscow, Russia, 2008; Volume 2008, p. 384.

marine drugs

MDPI

Article

# Production of the Marine Carotenoid Astaxanthin by Metabolically Engineered *Corynebacterium glutamicum*

**Nadja A. Henke, Sabine A. E. Heider †, Petra Peters-Wendisch and Volker F. Wendisch ***

Genetics of Prokaryotes, Faculty of Biology & CeBiTec, Bielefeld University, Bielefeld D-33615, Germany; n.henke@uni-bielefeld.de (N.A.H.); saeheider@aol.com (S.A.E.H.); petra.peters-wendisch@uni-bielefeld.de (P.P.-W.)

* Correspondence: volker.wendisch@uni-bielefeld.de; Tel.: +49-521-106-5611

† Current Address: GSK Vaccines S.r.l., Siena 53100, Italy.

Academic Editor: Tatsuya Sugawara

Received: 16 May 2016; Accepted: 24 June 2016; Published: 30 June 2016

**Abstract:** Astaxanthin, a red C40 carotenoid, is one of the most abundant marine carotenoids. It is currently used as a food and feed additive in a hundred-ton scale and is furthermore an attractive component for pharmaceutical and cosmetic applications with antioxidant activities. *Corynebacterium glutamicum*, which naturally synthesizes the yellow C50 carotenoid decaprenoxanthin, is an industrially relevant microorganism used in the million-ton amino acid production. In this work, engineering of a genome-reduced *C. glutamicum* with optimized precursor supply for astaxanthin production is described. This involved expression of heterologous genes encoding for lycopene cyclase CrtY, β-carotene ketolase CrtW, and hydroxylase CrtZ. For balanced expression of *crtW* and *crtZ* their translation initiation rates were varied in a systematic approach using different ribosome binding sites, spacing, and translational start codons. Furthermore, β-carotene ketolases and hydroxylases from different marine bacteria were tested with regard to efficient astaxanthin production in *C. glutamicum*. In shaking flasks, the *C. glutamicum* strains developed here overproduced astaxanthin with volumetric productivities up to 0.4 mg$\cdot$ $L^{-1}\cdot h^{-1}$ which are competitive with current algae-based production. Since *C. glutamicum* can grow to high cell densities of up to 100 g cell dry weight (CDW)$\cdot$ $L^{-1}$, the recombinant strains developed here are a starting point for astaxanthin production by *C. glutamicum*.

**Keywords:** astaxanthin production; carotenoids; genome-reduced *Corynebacterium glutamicum*; systematic approach; metabolic engineering

## 1. Introduction

Carotenoids are natural pigments with yellow-to-red coloring properties, found ubiquitously in plants, algae, fungi, and bacteria. These pigments form a subfamily of the large and diverse group of terpenoids with more than 55,000 different structures. Terpenoids are natural secondary metabolites composed of isoprene units, which typically exhibit flavoring, fragrance and coloring properties. Carotenoids and their derivatives have become more and more important for the health care industry due to their beneficial effects on human and animal health and their possible pharmaceutical, medical, and nutraceutical applications. For example, carotenoids are suggested to have beneficial effects on the human immune system and to protect against degenerative diseases and cancer [1–3]. Astaxanthin is a marine, red, cyclic C40 carotenoid and the third most important carotenoid on the global market after β-carotene and lutein, with a predicted sales volume of 670 metric tons valued at 1.1 billion US$ in 2020 [4]. Currently, astaxanthin is primarily used as a food and beverage colorant, in animal feed and in nutraceuticals [5] with e.g., an annual demand of 130 tons for coloration of poultry, salmon, lobster and fish [6]. Astaxanthin shows the strongest hitherto demonstrated anti-oxidant effect due to its keto and hydroxy groups at 4,4'- and 3,3'-beta-ionone ring positions, respectively. Those functional groups result in a more polar nature of astaxanthin and explain its unique antioxidative properties [7]. Furthermore, astaxanthin can be esterified which leads to increased stability [8]. Therefore, the demand for astaxanthin is particularly rising in the health sector [5]. Astaxanthin has been described to promote skin health and to have potential anti-aging effect [9]. Moreover, it alleviates the fatigue, inflammation, and aging of the eye [10–12]. Astaxanthin has a positive effect on blood rheology and potential antihypertensive properties, which makes it interesting for therapy of cardiovascular diseases [13,14]. Its wide potential for the reduction of inflammation also promotes the immune system functions [15]. In addition, astaxanthin was reported to have a positive impact on muscle recovery when used as a nutritional supplement [16].

Although the chemical synthesis of astaxanthin from petrochemical precursors is so far more cost-efficient and therefore dominates the market [17], consumer demand for naturally produced carotenoids is increasing [18]. Synthetic astaxanthin is often a mixture of *R*- and *S*-enantiomers and, thus, inferior to natural-based astaxanthin [19] and not suitable as a neutraceutical supplement without further complex and cost-intensive purification steps before application. Consequently, the demand for an efficient, environmentally friendly production of natural astaxanthin, and carotenoids in general, by microbial hosts is on the rise [20–22].

*C. glutamicum* is a Gram-positive soil bacterium with a long biotechnological history: its relevance goes back to the 1950s when it was first discovered as a natural glutamate producer [23]. Over centuries it has been used for the million-ton scale production of different amino acids for the feed and food industry. Moreover, its potential for biotechnological application has been further exploited [24]: besides amino acids, e.g., diamines [25], alcohols [26], and terpenoids [27,28] can be produced by engineered *C. glutamicum*. This bacterium has the ability to grow aerobically on a variety of carbon sources like glucose, fructose, sucrose, mannitol, propionate, and acetate [29,30]. In addition, it has been engineered to grow with alternative carbon sources such as glycerol [31], pentoses [32], amino sugars [33,34], β-glucans [35], and starch [36]. *C. glutamicum* is pigmented due to synthesis of the C50 carotenoid decaprenoxanthin and its glucosides. Its potential to produce carotenoids has been explored over recent years [28,37–39]. The carotenogenic pathway of *C. glutamicum* was identified [40] and several metabolic engineering strategies were applied to convert this biotechnologically established bacterium into a carotenoid producer [41,42].

In order to enable C40 carotenoid production by *C. glutamicum*, the conversion of lycopene to decaprenoxanthin needs to be prevented by deletion of the genes encoding lycopene elongase and ε-cyclase. As consequence of deletion of the lycopene elongase encoding gene *crtEb*, the cells exhibited a slight red color due to accumulation of the intermediate lycopene [37]. Additional overexpression of the endogenous genes *crtE*, *crtB*, and *crtI* in *C. glutamicum* Δ*crtEb* intensified the red phenotype as conversion of GGPP to the red chromophore lycopene was improved. Thereby, the lycopene content could be increased 80 fold with $2.4 \pm 0.3$ mg·(g·CDW)$^{-1}$ and showed for the first time enhanced C40 carotenoid production in *C. glutamicum* [37]. Heterologous expression of *crtY* from *Pantoea ananatis* ($crtY_{Pa}$) in a lycopene accumulating platform strain led to the production of the orange pigment β-carotene. Zeaxanthin was accumulated when *crtZ* from *P. ananatis* ($crtZ_{Pa}$) was expressed in addition [38]. Furthermore, carotenoid biosynthesis was improved by enhancing the precursor supply, which was accomplished by overexpression of the *dxs* gene encoding the enzyme for the initial condensation of pyruvate and GAP in the MEP-pathway [42].

In this study production of the marine carotenoid astaxanthin by *C. glutamicum* was developed using a β-carotene producing strain (Figure 1). Two strategies were followed: (i) the implementation of a combinatorial gene assembly for $crtW_{Ba}$ and $crtZ_{Pa}$ to optimize the ratio of enzyme quantities (ketolase and hydroxylase) by variation of translation initiation rates (TIR) based on different ribosome binding sites, spacing lengths, and translation start codons and (ii) the use of alternative *crtW* and *crtZ* genes from marine and non-marine prokaryotes in a two-vector system in order to find enzymes with higher activities or affinities for the intermediates of the astaxanthin biosynthesis pathway (Figure 1). Combined expression of *crtW* and *crtZ* from the marine bacterium *Fulvimarina pelagi* yielded a *C. glutamicum* strain producing astaxanthin as the major carotenoid in shaking flasks with productivities of up to 0.35 mg· $L^{-1}$· $h^{-1}$.

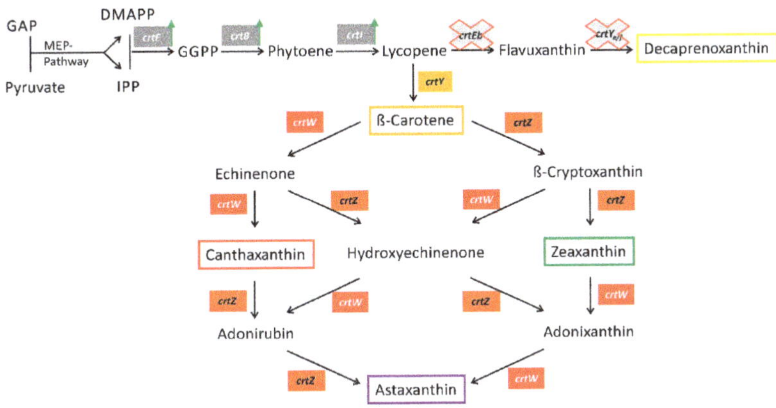


**Figure 1.** Scheme of C40 cyclic carotenoid biosynthesis in recombinant *C. glutamicum*. The biosynthesis of C40 cyclic carotenoids derived from precursor molecules dimethylallyl pyrophosphate (DMAPP) and isopentenyl pyrophosphate (IPP) is illustrated. Genes are shown next to the reaction catalyzed by the encoded enzyme (*crtE*: Prenyl transferase, *crtB*: Phytoene synthase, *crtI*: Phytoene desaturase, *crtEb*: Lycopene elongase, *crtYe/f*: C45/50 carotenoid ε-cyclase, *crtY*: Lycopene β-cyclase, *crtZ*: β-Carotene hydroxylase (3,3'-beta-ionone ring hydroxylase), *crtW*: β-Carotene ketolase (4,4'-beta-ionone ring ketolase). Endogenous genes are shown in grey boxes and their overexpression indicated by green arrows. Heterologous genes are highlighted in colored boxes.

## 2. Results

### 2.1. Construction of a β-Carotene Producing *C. glutamicum* Base Strain

*C. glutamicum* was metabolically engineered for plasmid-independent lycopene overproduction (Table 1). Chromosomal integration of the synthetic operon *crtEBI* under the control of the endogenous promoter of the gene coding for the translational elongation factor ($P_{tuf}$) in the $crtY_eY_fEb$ deletion mutant of *C. glutamicum* MB001 (LYC3) [37] was performed in order to improve the expression of prenyltransferase CrtE, phytoene synthase CrtB and phytoene desaturase CrtI encoding genes. Thereby, the flux from the precursor molecules IPP and DMAPP to lycopene was enhanced and an 8-fold higher lycopene titer resulted for strain LYC4. When *dxs*, encoding the first enzyme of the MEP-pathway, was additionally overexpressed by chromosomal exchange of its natural promoter by the strong promoter $P_{tuf}$, the lycopene titer was further improved by 34% and the respective strain LYC5 produced 0.43 ± 0.02 mg· (g· CDW)$^{-1}$ (Table 1).

**Table 1.** Lycopene production by plasmid-free recombinant *C. glutamicum* strains. Cells were grown in glucose CGXII minimal medium for 24 h. Means and standard deviations of three replicates are given.

| Name | Strain | Lycopene (mg· (g· CDW)$^{-1}$) |
|---|---|---|
| LYC3 | $crtY_eY_fEb$ deletion mutant of *C. glutamicum* MB001 | 0.04 ± 0.01 |
| LYC3-$P_{tuf}$*dxs* | LYC3::$P_{tuf}$*dxs* | 0.09 ± 0.01 |
| LYC4 | LYC3::$P_{tuf}$*crtEBI* | 0.32 ± 0.01 |
| LYC5 | LYC4::$P_{tuf}$*dxs* | 0.43 ± 0.02 |

Strain LYC5 was converted to a β-carotene producing strain (Table 2) by heterologous expression of the lycopene β-cyclase gene *crtY* from *P. ananatis*. Plasmid-borne expression of *crtY* under the control of the isopropyl β-D-1-thiogalactopyranoside (IPTG) inducible *tac* promoter (pEKEx3_$crtY_{Pa}$) allowed for β-carotene production. Constitutive expression of *crtY* under control of the $P_{tuf}$ promoter from the newly constructed expression and shuttle vector pSH1 resulted in a comparable production titer. Similarly, a β-carotene titer of 6.5 mg·g$^{-1}$ was achieved by BETA3, a strain having $crtY_{Pa}$ under the control of $P_{tuf}$ integrated into the genome of *C. glutamicum* strain LYC5 (Table 2).

**Table 2.** β-Carotene production in recombinant *C. glutamicum* strains. Cells were grown in glucose CGXII minimal medium for 24 h induced by 1 mM isopropyl β-D-1-thiogalactopyranoside (IPTG). Means and standard deviations of three replicates are given.

| Name | Strain | β-Carotene (mg·(g·CDW)$^{-1}$) |
|---|---|---|
| BETA1 | LYC5 (pEXEx3_$crtY_{Pa}$) | 5.2 ± 1.0 |
| BETA2 | LYC5 (pSH1_$crtY_{Pa}$) | 5.9 ± 0.8 |
| BETA3 | LYC5::$P_{tuf}$_ $crtY_{Pa}$ | 6.5 ± 1.3 |

### *2.2. Design of the Combinatorial Gene Assembly and Library Construction for Engineering Astaxanthin Production in C. glutamicum*

Metabolic flux in a synthetic pathway may require well-adjusted activities of the enzymes involved. Prediction of the flux from gene expression is rather difficult, hence, a combinatorial gene assembly was used to screen for balanced expression of the β-carotene ketolase and β-carotene hydroxylase encoding genes with respect to astaxanthin production. Since *crtY* from *P. ananatis* has previously been expressed successfully in *C. glutamicum* for production of β-carotene, the β-carotene hydroxylase *crtZ* gene from this organism was chosen. However, *P. ananatis* lacks β-carotene ketolase, and therefore the β-carotene ketolase gene *crtW* from *Brevundimonas aurantiaca* was used, which on the contrary lacks a *crtZ* gene. *CrtW* from *B. aurantiaca* and *crtZ* from *P. ananatis* were combined in an artificial operon under the control of the constitutive $P_{tuf}$ promoter in the vector pSH1. Gene expression was varied by combining different ribosome-binding sites (RBS) and start codons separated by spacers of different lengths (Figure 2). The theoretical translation initiation rates were calculated using the RBS calculator [43] and ranged from 14 to 33,626 for *crtW* and from 40 to 30,731 for *crtZ*. A library of combinatorially assembled *crtW* and *crtY* genes was generated and the constructed library of pSH1_$crtW_{Ba}$_$crtZ_{Pa}$ plasmids was used to transform the β-carotene accumulating strain *C. glutamicum* BETA1.

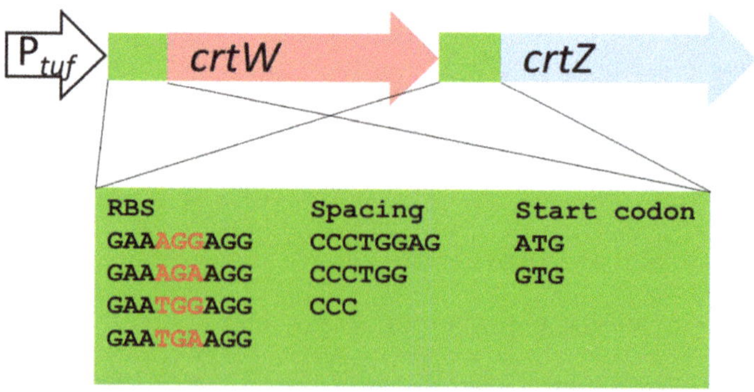


**Figure 2.** Combinatorial gene assembly for varied translation initiation of β-carotene ketolase and hydroxylase genes. Combinations of different RBS sequences (differences given in red letters), translation start codons (ATG/GTG) and spacers (3, 6 or 8 bp in length) between them are highlighted in a green box.

For each gene four different RBS, three different spacer lengths, and two different translational start codons were chosen. These were introduced by the forward primers and equimolar mixture of these primers and one reverse primer by PCR. The resulting DNA products were gel-extracted and combined by cloning via Gibson Assembly [44] in pSH1. Thus, theoretically 24 different constructs per gene resulted (Figure 2). With this approach 576 different combinations of *crtW* and *crtZ* genes are theoretically possible and the event of creating a specific combination of the two genes follows the Poisson distribution [45] with a probability of 1/576 (Equation (1)). To cover with 99% probability that a single specific combination is present at least once in the library, approximately 2650 clones are required (Equation (1)). The necessary number of transformants for creating a library with each of the 576 combinations can be calculated employing the path rules [45]. For creating a library that includes each of the 576 specific combinations at least once with a 99% probability, approximately 6315 transformants are required (Equation (2)). Preliminary experiments showed that correct assembling of an insert with the restricted vector via Gibson assembly occurs in about 90% of the events. Consequently, the number of transformants had to be corrected by multiplication by 1.11, thus, a minimum of 7000 transformants had to be screened.

$$p_{\lambda}(k) = \frac{\lambda^k}{k!} * e^{-\lambda} \tag{1}$$

Equation (1): Poisson distribution. $\lambda = n * p$; $n$: library size; $p$: probability of one specific gene assembly of *crtW* and *crtZ*, $k$: number of one specific gene assembly in library with size $n$.

$$p_{all}(k \geqslant 1) = \left(1 - e^{-\lambda}\right)^N \tag{2}$$

Equation (2): Path rules. $\lambda = n * p$; $n$: library size; $p$: probability of one specific gene assembly of *crtW* and *crtZ*, $k$: number of one specific gene assembly in library with size $n$; $N$: number of possible gene assemblies.

Around 8000 transformants were visually color-screened on plates and 46 colonies with different colors ranging from yellow to red were selected for further analysis. The plasmid DNA was isolated and sequenced to identify the sequences (RBS, spacer, translational start codon) of *crtW* and *crtZ*. The set of 46 transformants represented 20 of the 24 possible variants for *crtW* and 19 of 24 variants of *crtZ*. Furthermore, three plasmids harbored only the *crtW* gene and two plasmids harbored only the *crtZ* gene.

### 2.3. Combinatorial Engineering Covered Vastly Different Astaxanthin, β-Carotene, Zeaxanthin and Canthaxanthin Titers

To evaluate which of the gene combinations was best in terms of astaxanthin production, the 46 selected transformants referred to as COMB strains, were characterized with respect to carotenoid production. After growth in CGXII minimal medium with 100 mM glucose, appropriate antibiotics and 1 mM IPTG in a Biolector micro fermentation system (Figure 3), carotenoids were quantified by HPLC using standards for β-carotene, canthaxanthin, zeaxanthin, and astaxanthin.

As expected, the parental strain BETA1 (Figure 3) produced β-carotene (6.7 mg·(g·CDW)$^{-1}$), but no further carotenoids. The 46 COMB strains could be categorized in six groups according to their carotenoid production profiles (group I: only lycopene, group II: only β-carotene, group III: β-carotene and zeaxanthin, group IV: β-carotene, zeaxanthin and astaxanthin, groupV: β-carotene and canthaxanthin, group VI: β-carotene, canthaxanthin and astaxanthin; Figure 4). For all COMB strains, the TIRs were calculated with the RBS calculator tool [41], which takes (amongst others) the free binding energy of the RBS and the 16S rRNA into consideration as well as the free energy of secondary structures of the mRNA itself.

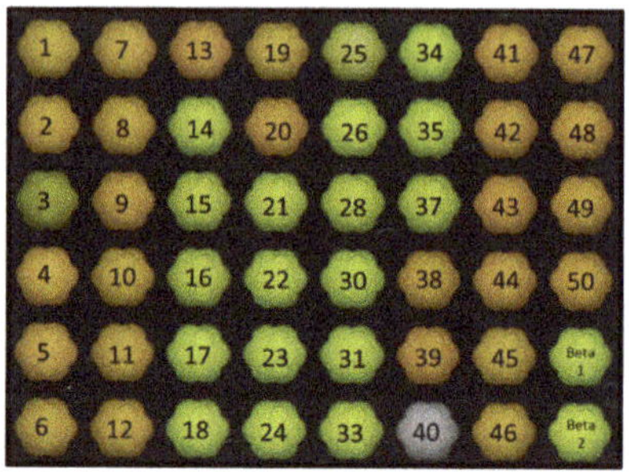


**Figure 3.** COMB strains expressing *crtW* from *B. aurantiaca* and *crtZ* from *P. ananatis* with varied translation initiation signals after growth in the Biolector micro fermentation system. Color phenotypes of 46 different COMB strains and the parental strain BETA1 (bottom right) after 24 h of cultivation.

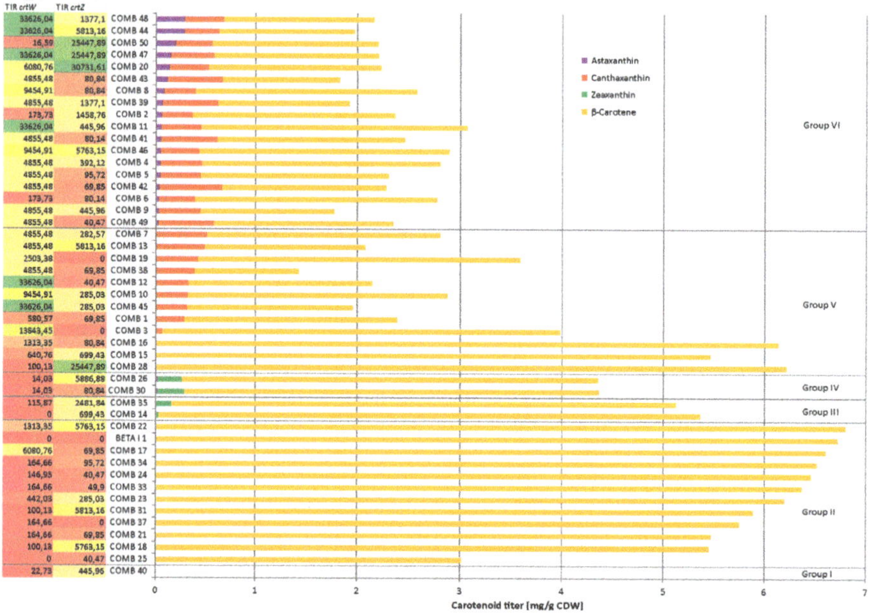


**Figure 4.** Carotenoid profiles and calculated translational initiation rates (TIRs) for *C. glutamicum* strains expressing *crtW* from *B. aurantiaca* and *crtZ* from *P. ananatis* with varied translation initiation signal. TIRs were calculated by applying the RBS calculator tool [46] on the mRNA sequence. TIRs were classified as follows: TIRs <200: low; 200 < TIRs < 2000: medium; TIRs >2000: high. Production of β-carotene, zeaxanthin, canthaxanthin and astaxanthin was determined after 24 h of cultivation in CGXII + 100 mM glucose in Biolector micro fermenter.

COMB 40 (Figure 4) produced none of the cyclic carotenoids, but about as much lycopene (0.39 mg· (g· CDW)$^{-1}$) as LYC5 (0.5 $\pm$ 0.1 mg· (g· CDW)$^{-1}$), the parental strain of BETA1. Sequencing of pEKEx3_*crtY*$_{Pa}$ isolated from COMB 40 revealed a deletion of 11 base pairs in the coding region of *crtY*, hence, β-carotene production was not possible in this strain. By contrast, the other 45 strains produced β-carotene with a titer of at least 1 mg· (g· CDW)$^{-1}$ (Figure 4). For about 24% of the strains, β-carotene was the only cyclic carotenoid being produced. In these cases, the calculated TIRs of *crtW* and/or *crtZ* were rather low (less than 200 for at least one gene; Figure 4). Zeaxanthin, one of the intermediates in the pathway towards astaxanthin, was detected in only four strains (COMB 14, COMB 26, COMB 30, and COMB 35) and these strains exhibited very diverse TIRs for *crtZ* (from 81 to 5887) (Figure 4). The highest production of zeaxanthin was detected in group IV for strain COMB 30 with 0.3 mg· (g· CDW)$^{-1}$, although this strain possessed a low TIR for *crtZ*. The highest titers of canthaxanthin and astaxanthin were observed among the strains of the large group VI (39%) and these strains co-produced β-carotene along with canthaxanthin and astaxanthin (Figure 4). The intermediate canthaxanthin was detected in 30 strains with strain COMB 42 showing the highest titer for canthaxanthin (0.6 mg· (g· CDW)$^{-1}$; Figure 4). In average, the TIR for *crtW* of these strains was high (10,299). Astaxanthin was identified in 20 of the 46 strains, but only two strains, COMB 44 and COMB 48, exhibited reasonably high astaxanthin yields (approximately 0.3 mg· (g· CDW)$^{-1}$). These two strains exhibited high *crtW* TIRs (33,626) and medium to high *crtZ* TIRs (5813 and 1377) (Figure 4).

In general, it was found that the higher the TIR of *crtW* the higher was the astaxanthin production, with three exceptions, COMB 11, COMB 12, and COMB 45 (Figure 4). In the latter three strains, however, the TIRs for *crtZ* were 3 to 145-fold lower than in the best astaxanthin producing strains COMB 44 and COMB 48. For strains COMB 44 and COMB 48, a spacing length of six base pairs, the RBS sequence GAAAGGAGG, and the translation start codon ATG was found for *crtW*. The *crtZ* gene in COMB 44 showed the consensus RBS sequence, a spacer length of eight base pairs and ATG as translational start codon. The *crtZ* gene variant of COMB 48 had a slightly lower TIR and possessed the RBS sequence GAAAGAAGG, six base pairs of spacing and ATG as start codon.

Three strains (COMB 37, COMB 3 and COMB 19) did not express *crtZ* due to an incorrect gene assembly. Strains COMB 3 and COMB 19 accumulated canthaxanthin besides β-carotene, while COMB 37, which also showed a low TIR for *crtW*, only accumulated β-carotene (Figure 4). Strains COMB 25 and COMB 14 did not express *crtW* due to an incorrect gene assembly. While COMB 25 only produced β-carotene probably because of a very low TIR for *crtZ* (Figure 4), strain COMB 14 produced zeaxanthin besides β-carotene.

Taken together, widely varied carotenoid production was represented by the library, but none of the combinations tested yielded high astaxanthin product levels.

### 2.4. In Silico Analysis of the Carotenogenic Genes crtZ and crtW from Marine and Non-Marine Bacteria for Heterologous Expression in C. glutamicum

In the above described experiments the bacteria *B. aurantiaca* and *P. ananatis* were chosen as sources for *crtW* and *crtZ*, respectively, although not producing astaxanthin themselves. *B. aurantiaca* lacks *crtZ*, but possesses the *crtG* gene coding for a 2,2'-beta-ionone ring hydroxylase and produces canthaxanthin and 2-hydroxycanthaxanthin as main carotenoids. *P. ananatis* lacks *crtW* and produces glycosylated zeaxanthin involving CrtZ. Thus, on the basis of available genome sequences, reported carotenoid production and biological diversity, four alternative prokaryotic natural carotenoid producers were selected as donors for *crtW* and *crtZ*. Since *crt* genes of *Brevundimonas* species were reported to show a high potential for heterologous carotenoid productions [47] two different *Brevundimonas* strains were selected: *Brevundimonas vesicularis*, a non-marine bacterium suggested to be a suitable gene donor for astaxanthin production [47,48], and as alternative *Brevundimonas bacteroides* [49]. The marine bacterium *Fulvimarina pelagi* was chosen due to its promising brownish-yellow color as a consequence of carotenoid accumulation [50] and the evolutionary distance to *Brevundimonas*. In addition, the red-pigmented marine bacterium *Sphingomonas astaxanthinifaciens* was selected since experimental evidence that astaxanthin is the major carotenoid produced by this bacterium has been reported [51,52].

The organization of carotenogenic gene clusters of the considered donors *B. aurantiaca*, *B. bacteroides*, *B. vesicularis*, *F. pelagi*, *P. ananatis*, and *S. astaxanthinifaciens* was analyzed on the basis of the partly available genome sequences/contigs or scaffolds in GenBank: the carotenoid gene cluster of *B. bacteroides*, an orange-red pigmented bacterium, comprises *crtW* and *crtZ* as well as the gene *idi* encoding the IPP isomerase of the MEP-pathway and several other genes encoding for enzymes of the astaxanthin biosynthesis pathway, however, a *crtE* gene is not present in its genome. The genome of *B. vesicularis* DC263, a red-pigmented soil bacterium, possesses a large carotenoid gene cluster with 10 coding sequences, eight of which encode enzymes for the biosynthesis pathway of astaxanthin or the terpenoid precursors IPP and DMAPP. In addition, a second hydroxylase CrtG is encoded, which is responsible for further hydroxylation of astaxanthin leading to 2-hydroxyastaxanthin. Carotenogenic genes of *F. pelagi*, a Mn(II)-oxidizing bacterium [53], are found in at least four different loci of the genome. Genes encoding for enzymes of the astaxanthin biosynthesis and glycosylation as well as enzymes for the spirilloxanthin biosynthesis (CrtC, CrtD, CrtF) are present. Furthermore, two genes coding for an ABC-transporter and a MFS-transporter are located next to the carotenogenic genes *crtZ* and *crtY*. *S. astaxanthinifaciens*, producing astaxanthin and its glycosides, has at least two carotenoid gene clusters in its genome also including farnesyl pyrophosphate synthase. Moreover, a gene encoding a putative carotenoid transporter is located in this cluster.

### *2.5. High Astaxanthin Production by C. glutamicum Strains Expressing crtW and crtZ from F. pelagi*

β-Carotene ketolase and hydroxylase genes (*crtW* and *crtZ*, respectively) from *B. aurantiaca*, *B. bacteroides*, *B. vesicularis*, *F. pelagi*, *P. ananatis*, and *S. astaxanthinifaciens* were expressed in the plasmid-free β-carotene overproducing *C. glutamicum* strain BETA4. The affinities of the β-carotene ketolases and hydroxylases for the various substrates and intermediates of the branched astaxanthin biosynthesis pathway may vary and it is conceivable that astaxanthin production proceeds e.g., only via canthaxanthin or only via zeaxanthin. However, also various other routes via hydroxyechinenone are possible (Figure 1). Thus, in a first step only either *crtW* or *crtZ* was expressed in the parental strain BETA4 that produced ~12 mg· (g· CDW)$^{-1}$ β-carotene with a productivity of ~3.4 mg· L$^{-1}$· h$^{-1}$ in 24 h of cultivation and a growth rate of 0.32 ± 0.01 h$^{-1}$. Zeaxanthin accumulated (0.52 and 1.1 mg· (g· CDW)$^{-1}$, respectively) when *crtZ* from *P. ananatis* or *F. pelagi* were expressed (data not shown). Canthaxanthin accumulated (0.34 to 1.0 mg· (g· CDW)$^{-1}$) when *crtW* from *S. astaxanthinifaciens*, *F. pelagi* or *B. aurantiaca* were expressed, while only traces were observed as consequence of expression of *crtW* from *B. bacteroides* or *B. vesicularis* (data not shown). Thus, *crtW* from *S. astaxanthinifaciens*, *F. pelagi* or *B. aurantiaca* and *crtZ* from *P. ananatis* or *F. pelagi* appeared suitable for further analysis.

Subsequently, combinations of *crtW* from *S. astaxanthinifaciens* and *B. aurantiaca* with *crtZ* from *F. pelagi* were co-expressed in strain BETA4 using the two expression vectors pSH1 and pEC-XT99A. In addition, the *crtW* and *crtZ* genes from species known to synthesize astaxanthin (*B. bacteroides*, *B. vesicularis*, *F. pelagi*, and *S. astaxanthinifaciens*) were co-expressed in BETA4. Carotenoids of these strains were extracted and analyzed in the stationary growth phase 24 h after inoculation. Transformants harboring the genes from *B. vesicularis* grew poorly and were not analyzed further.

Co-expression of *crtW* and *crtZ* from *B. bacteroides* and *crtW* and *crtZ* from *S. astaxanthinifaciens* genes led to less than 0.1 mg· (g· CDW)$^{-1}$ of astaxanthin (Table 3). Strains expressing *crtZ* from *F. pelagi* in combination with *crtW* from *S. astaxanthinifaciens*, *B. aurantiaca*, and *F. pelagi* produced about 0.7, 1.7 and 1.6 mg· (g· CDW)$^{-1}$ of astaxanthin, respectively (Table 3). In all strains except for BETA4(pSH1_$crtW_{Fp}$)(pEC-XT_$crtZ_{Fp}$), β-carotene was the major carotenoid (~2 mg· (g· CDW)$^{-1}$). *C. glutamicum* BETA4 co-expressing *crtW* and *crtZ* from *F. pelagi* (subsequently named ASTA1) produced astaxanthin as main carotenoid (1.6 ± 0.3 mg· (g· CDW)$^{-1}$) and accumulated little canthaxanthin and β-carotene (0.1 ± 0.1 and 0.3 ± 0.1 mg· (g· CDW)$^{-1}$, respectively) as side products (Table 3). This strain produced astaxanthin with a volumetric productivity of 0.4 ± 0.1 mg· L$^{-1}$· h$^{-1}$ in shaking flasks with a growth rate of 0.29 ± 0.05 h$^{-1}$.

**Table 3.** Astaxanthin, canthaxanthin, and β-carotene production by strains overexpressing various combinations of *crtW* and *crtZ* genes. Titers, productivities, and final ODs are given as means and standard deviations ($n$ = 3) after 24 h of cultivation in CGXII + 100 mM glucose. *B.a.*: *Brevundimonas aurantiaca*; *B.b.*: *Brevundimonas bacteroides*; *F.p.*: *Fulvimarina pelagi*; *S.a.*: *Sphingomonas astaxanthinifaciens*.

| Strain Growth | | Carotenoid Titer (mg·g$^{-1}$·CDW) | | | Volumetric Productivity (mg·L$^{-1}$·h$^{-1}$) | | |
|---|---|---|---|---|---|---|---|
| BETA4 Transformed with | Final $OD_{600\,nm}$ | Astaxanthin | Canthaxanthin | β-Carotene | Astaxanthin | Canthaxanthin | β-Carotene |
| - | 28 ± 1 | <0.1 | <0.1 | 11.7 ± 2.0 | <0.1 | <0.1 | 3.4 ± 0.5 |
| (pSH1_*crtW*$_{Bb}$) (pEC-XT_*crtZ*$_{Bb}$) | 21 ± 1 | <0.1 | <0.1 | 4.9 ± 0.4 | <0.1 | <0.1 | 1.1 ± 0.1 |
| (pSH1_*crtW*$_{Sa}$) (pEC-XT_*crtZ*$_{Sa}$) | 22 ± 2 | < 0.1 | 0.3 ± 0.1 | 3.3 ± 0.5 | <0.1 | <0.1 | 0.8 ± 0.1 |
| (pSH1_*crtW*$_{Sa}$) (pEC-XT_*crtZ*$_{Fp}$) | 24 ± 1 | 0.7 ± 0.3 | 0.2 ± 0.1 | 1.8 ± 0.1 | 0.2 ± 0.1 | <0.1 | 0.5 ± 0.1 |
| (pSH1_*crtW*$_{Ba}$) (pEC-XT_*crtZ*$_{Fp}$) | 22 ± 1 | 1.7 ± 0.3 | 0.1 ± 0.1 | 2.0 ± 0.5 | 0.4 ± 0.1 | <0.1 | 0.4 ± 0.2 |
| (pSH1_*crtW*$_{Fp}$) (pEC-XT_*crtZ*$_{Fp}$) = ASTA1 | 23 ± 1 | 1.6 ± 0.3 | 0.1 ± 0.1 | 0.3 ± 0.1 | 0.4 ± 0.1 | <0.1 | 0.1 ± 0.1 |

## 3. Discussion

In this study, *Corynebacterium glutamicum* was engineered for the production of the marine carotenoid astaxanthin. *C. glutamicum* grows fast to high cell densities [54] and, thus, is suitable for production of carotenoids and other compounds that are stored within the cell. Here, *C. glutamicum* was shown to produce β-carotene to about 12 mg·(g·CDW)$^{-1}$ within 24 h at a volumetric productivity of about 3.4 mg·L$^{-1}$·h$^{-1}$. Growth and production of carotenoids by *C. glutamicum* is monophasic and strains BETA4 and ASTA1 showed growth rates of 0.32 ± 0.01 h$^{-1}$ and 0.29 ± 0.05 h$^{-1}$, respectively. This is in contrast to biphasic growth/production of carotenoids e.g., by the alga *Haematococcus pluvialis* [55]. As a consequence, the volumetric productivity for β-carotene exceeds that reported for the industrially used microalga *Dunaliella bardawil* [56] or the yeast *Saccharomyces cerevisiae* [57] by about a factor of three.

Combined expression of the genes coding for β-carotene ketolase and hydroxylase from microorganisms that do not synthesize astaxanthin (*B. aurantiaca* and *P. ananatis*) in a β-carotene producing *C. glutamicum* led to astaxanthin production. However, astaxanthin was not the main carotenoid being produced. Since a balanced expression of the β-carotene ketolase and hydroxylase genes are essential for an efficient astaxanthin production [48,58] we assumed that the activities of the respective enzymes in the tested recombinants were not matched. Therefore, translation initiation rates of the respective genes, *crtW* and *crtZ*, were varied in a combinatorial approach. However, a strict correlation between TIR and production titers was not observed. As tendencies, the lower the TIRs of both *crtW* and *crtZ* the lower were the canthaxanthin and astaxanthin titers, and the higher the TIR of *crtW* the higher were astaxanthin titers (Figure 4).

In *E. coli* astaxanthin biosynthesis from β-carotene was reported to proceed more efficiently via zeaxanthin rather than canthaxanthin since ketolated intermediates did not accumulate [48,58]. Both ketolase and hydroxylase compete for their substrates and accept β-carotene as well as canthaxanthin and zeaxanthin, respectively, as substrates [59,60]. Independently induced expression of *crtZ* from *P. ananatis* and *crtW148* of *Nostoc puntiforme* PC73102 revealed that hydroxylation occurred fast with β-carotene, echinenone, adonirubin, and canthaxanthin [58]. In their system, CrtW148 was identified as the limiting step in conversion of zeaxanthin to astaxanthin [58]. Expression of *crtZ* from *P. ananatis* in β-carotene producing *C. glutamicum* also yielded zeaxanthin [38] as did expression of *crtZ* from *F. pelagi* in this study (data not shown). Varying expression levels of *crtW*$_{Ba}$ and *crtZ*$_{Pa}$ led to accumulation of zeaxanthin only if TIR for *crtW*$_{Ba}$ was low (Figure 4). On the other hand, canthaxanthin accumulated as intermediate typically if TIR of *crtW*$_{Ba}$ was medium to high (Figure 4). Canthaxanthin accumulation may be explained best by the assumption that β-carotene ketolase CrtW from *B. aurantiaca* did not accept the non-natural substrate zeaxanthin well. It is likely that astaxanthin production by this approach was not only limited by an imperfect match between expression levels of the β-carotene ketolase and hydroxylase genes, but rather by imperfect compatibility of the substrate spectra of the chosen β-carotene ketolase and hydroxylase enzymes.

Consequently, *crtW* and *crtZ* genes from marine and non-marine bacteria known to synthesize astaxanthin were examined in the second approach. Astaxanthin was produced in combinations of CrtZ from the marine bacterium *F. pelagi* and CrtW from either *F. pelagi*, *S. astaxanthinifaciens* or *B. aurantiaca*. *F. pelagi* was isolated from ocean surface water, an aerated environment at least transiently exposed to high solar radiation [45]. It is hypothesized that carotenoids play an important role as antioxidants for survival of *F. pelagi* under these conditions [50]. Analysis of the codon usage of *crtW* and *crtZ* from *F. pelagi* revealed a good fit to the codon usage of *C. glutamicum*, which is in compliance with the achieved astaxanthin titers of the recombinants. Co-expression of *crtW* from *B. aurantiaca* and *crtZ* from *F. pelagi* led to comparable astaxanthin titers, but considerable β-carotene amounts accumulated as side-product (Table 3), co-expression of *crtW* and *crtZ* from *F. pelagi*, instead, yielded astaxanthin as major carotenoid (80%; Table 3).

As compared to β-carotene production of about 12 mg· (g· CDW)$^{-1}$ by the parent strain BETA4, the astaxanthin titers were at least seven fold lower (Table 3). Thus, conversion of β-carotene to astaxanthin is incomplete; however, other carotenoids besides canthaxanthin and residual β-carotene did not accumulate to significant titers (data not shown and Table 3). The partial conversion of β-carotene to astaxanthin may, thus, indicate that astaxanthin and/or intermediate(s) of its biosynthesis are inhibitory. This is in line with our finding that overexpression of only *crtW* from *F. pelagi* resulted in 0.5 mg· (g· CDW)$^{-1}$ canthaxanthin and 1.7 mg· (g· CDW)$^{-1}$ remaining β-carotene. Similarly, overexpression of only *crtZ* yielded 1.1 mg· (g· CDW)$^{-1}$ zeaxanthin and 5.6 mg· (g· CDW)$^{-1}$ β-carotene remained. Similarly, heterologous expression of *crtW148* and *crtZ* in the β-carotene-producing *E. coli* strain reduced the overall formation of carotenoids, indicating that the formation of the carotenoid precursors were affected [58].

High product purities and titers are beneficial for downstream processing. The astaxanthin producing *C. glutamicum* strain overexpressing *crtW* and *crtZ* from *F. pelagi* accumulated astaxanthin (about 1.6 mg· (g· CDW)$^{-1}$) as major (about 80%) carotenoid. The fact that little β-carotene and canthaxanthin accumulated (about 0.3 and 0.1 mg· (g· CDW)$^{-1}$, respectively) may be an important advantage for downstream processing. Nevertheless, higher product purities can be obtained by algae with 95% of total carotenoids being astaxanthin [58]. Purification of astaxanthin from the cell walls of algae and red yeasts is challenging since algae like *H. pluvialis* accumulate astaxanthin in response to stress and heavily walled cysts are formed in the red stage [55]. Extraction of carotenoids from microalgae does not only require the removal of chlorophyll [61], but also efficient cell breakage technology [55]. Ethoxyquin or other antioxidants are added to the cells in order to minimize oxidation of the carotenoids during drying and cracking [58]. Because of laborious and time-consuming extraction processes of astaxanthin from algal systems, its production by a prokaryotic host, *Escherichia coli*, has emerged for substitution [62]. It has to be noted that *H. pluvialis* produces esterified astaxanthin, which is more stable than the free form astaxanthin as it does not cross react with proteins and e.g., lipoproteins [8], and which is incorporated easier by marine animals [63]. But hydrolysis of the ester narrows the bioavailability of astaxanthin e.g., to salmon [64]. The rigid cell walls of the red yeast *X. dendrorhous* also requires cell breakage prior to astaxanthin extraction [65,66]. In contrast to that, a simple methanol-acetone extraction was sufficient to recover astaxanthin from *C. glutamicum* cells at lab scale.

The volumetric productivities of up to about 0.4 mg· L$^{-1}$· h$^{-1}$ obtained in simple shaking flask cultures by the recombinant *C. glutamicum* strains compare favorably with those reported for the commercially used production hosts such as the green microalgae *H. pluvialis* [55,67] and the red yeast *Xanthophyllomyces dendrorhous* (formerly *Pfaffia rhodozyma*) [6,68] under similar conditions as well as recombinant *E. coli* [58]. Under optimal conditions, astaxanthin titers obtained e.g., with

*H. pluvialis* are very high (up to about 40 mg· (g· CDW)$^{-1}$), but slow growth, biphasic growth (green stage) and production (red stage) properties and the low final biomass concentrations reduce the maximal volumetric productivity [55]. After the non-productive green phase (about 4 days), the volumetric productivity for astaxanthin in the red stage is about 1 mg· L$^{-1}$· h$^{-1}$ and can be maintained for extended periods [55]. Although astaxanthin product titers from red yeasts such as *X. dendrorhous* are generally lower than from algae [69], higher growth rates and easier cultivation conditions argue in favor of these yeasts [70]. After optimization of a glucose-based fed-batch process a volumetric productivity of about 5 mg· L$^{-1}$· h$^{-1}$ was achieved [65,71]. Can it be envisioned that comparably high volumetric productivities can be obtained using the recombinant *C. glutamicum* strains described here? In pressurized high-cell-density fed-batch cultivations *C. glutamicum* grows to biomass concentrations of about 220 g· CDW· L$^{-1}$ within 24 h [54]. If this growth could be achieved with the *C. glutamicum* strains accumulating astaxanthin to titers of about 1.6 mg· (g· CDW)$^{-1}$, theoretically volumetric productivities of about 14 mg· L$^{-1}$· h$^{-1}$ may be achieved. Future work focused on process intensification, however, needs to be performed in order to evaluate if scale-up to such high astaxanthin volumetric productivities can be realized with *C. glutamicum*.

## 4. Materials and Methods

### 4.1. Bacterial Strains, Media and Growth Conditions

The strains and plasmids used in this work are listed in Table 4. *C. glutamicum* ATCC 13032 was used as wild type (WT), for metabolic engineering the prophage-cured *C. glutamicum* MB001 [72] was used as platform strain. Pre-cultivation of *C. glutamicum* strains was performed in LB medium or LB with 50 mM glucose. For cultivation in CGXII medium [73], pre-cultivated cells were washed once with CGXII medium without carbon source and inoculated to an initial $OD_{600}$ of 1. Glucose was added as carbon and energy source to a concentration of 100 mM. Standard cultivations of *C. glutamicum* were performed at 30 °C in a volume of 50 mL in 500 mL flasks with two baffles shaking at 120 rpm. The $OD_{600}$ was measured in dilutions using a Shimadzu UV-1202 spectrophotometer (Duisburg, Germany). Alternatively, cultivations were performed in 1 mL volume in micro-titerplates at 1100 rpm at 30 °C using Biolector® micro fermentation system (m2p-labs GmbH, Baesweiler, Germany). For cloning, *E. coli* DH5α was used as host and cultivated in LB medium at 37 °C. When appropriate, kanamycin, tetracycline or spectinomycin was added to concentrations of 25, 5, and 100 µg· mL$^{-1}$, respectively. Gene expression was induced by addition of 1 mM IPTG, at inoculation of the main culture.

**Table 4.** Strains and plasmids used in this study.

| Strain; Plasmid | Relevant Characteristics | Reference |
|---|---|---|
| | ***C. glutamicum*** **Strains** | |
| WT | Wild type, ATCC 13032 | [74] |
| MB001 | prophage cured, genome reduced ATCC 13032 | [72] |
| LYC3 | *crtY$_e$Y$_f$Eb* deletion mutant of *C. glutamicum* MB001 | [42] |
| LYC4 | LYC3 derivative with an artificial operon containing *crtE, crtB*, and *crtI* under control of the $P_{tuf}$ promoter integrated into the chromosome | this work |
| LYC5 | LYC4 derivative with *dxs* under control of the $P_{tuf}$ promoter integrated into the chromosome | this work |
| BETA1 | LYC5 derivative (pEKEx3_ *crtY$_{Pa}$*) | this work |
| BETA2 | LYC5 derivative (pSH1_ *crtY$_{Pa}$*) | this work |
| BETA3 | LYC5 derivative with *crtY$_{Pa}$* under control of the $P_{tuf}$ promoter integrated into the chromosome | this work |
| BETA4 | *cg0725* deletion mutant of *C. glutamicum* BETA3 | this work |
| ASTA1 | *C. glutamicum* BETA4 carrying pSH1_*crtW1$_{Fp}$* and pEC-XT_*crtZ$_{Fp}$* | this work |

**Table 4.** *Cont.*

| Strain; Plasmid | Relevant Characteristics | Reference |
|---|---|---|
| | **Other Strains** | |
| *E. coli* DH5α | F- *thi-1 endA1 hsdr17*(r-, m-) *supE44 ΔlacU169* (Φ80*lacZ*ΔM15) *recA1 gyrA96* | [75] |
| *Pantoea ananatis* | Wild type, ATCC 33244, DSM 17873, Z96081 | [76] |
| *Brevundimonas aurantiaca* | Wild type, ATCC 15266, DSM 4731, NR028889 | [77] |
| *Brevundimonas bacteroides* | Wild type, ATCC 15254, DSM 4726, AJ227782 | [49] |
| *Brevundimonas vesicularis* | Wild type, ATCC 11426, DSM 7226, LN681560 | [78] |
| *Fulvimarina pelagi* | Wild type, ATCC BAA-666, DSM 15513, AY178860 | [50] |
| *Sphingomonas astaxanthinifaciens* | Wild type, NBRC 102146, DSM 22298, AB277583 | [52] |
| | **Plasmids** | |
| pEC-XT99A (pEC-XT) | $Tet^R$, $P_{trc}lacI^q$, pGA1 $oriV_{Cg}$, *C. glutamicum*/*E. coli* expression shuttle vector | [79] |
| pEC-XT_$crtZ_{Bb}$ | pEC-XT derivative for IPTG-inducible expression of *crtZ* from *B. bacteroides* containing an artificial ribosome binding site | this work |
| pEC-XT_$crtZ_{Bv}$ | pEC-XT derivative for IPTG-inducible expression of *crtZ* from *B. vesicularis* containing an artificial ribosome binding site | this work |
| pEC-XT_$crtZ_{Fp}$ | pEC-XT derivative for IPTG-inducible expression of *crtZ* from *F. pelagi* containing an artificial ribosome binding site | this work |
| pEC-XT_$crtZ_{Sa}$ | pEC-XT derivative for IPTG-inducible expression of *crtZ* from *S. astaxanthinifaciens* containing an artificial ribosome binding site | this work |
| pEKEx3 | $Spec^R$, $P_{tac}lacI^q$, pBL1 $oriV_{Cg}$, *C. glutamicum*/*E. coli* expression shuttle vector | [80] |
| pEKEx3_$crtY_{Pa}$ | pEKEx3 derivative for IPTG-inducible expression of *crtY* from *P. ananatis* containing an artificial ribosome binding site | this work |
| pVWEx1 | $Km^R$, $P_{tac}lacI^q$, pHM519 $oriV_{Cg}$, *C. glutamicum*/*E. coli* expression shuttle vector | [81] |
| pSH1 | $Km^R$, $P_{tuf}$, pHM519 $oriV_{Cg}$, *C. glutamicum*/*E. coli* expression shuttle vector | this work |
| pSH1_$crtY_{Pa}$ | pSH1 derivative for constitutive expression of *crtY* from *P. ananatis* containing an artificial ribosome binding site | this work |
| pSH1_$crtW_{Ba}$_$crtZ_{Pa}$ | pSH1 derivative for constitutive expression of *crtW* from *B. aurantiaca* and *crtZ* from *P. ananatis* containing artificial ribosome binding sites | this work |
| pSH1_$crtW_{Ba}$ | pSH1 derivative for constitutive expression of *crtW* from *B. aurantiaca* containing an artificial ribosome binding site | this work |
| pSH1_$crtW_{Bb}$ | pSH1 derivative for constitutive expression of *crtW* from *B. bacteroides* containing an artificial ribosome binding site | this work |
| pSH1_$crtW1_{Bv}$ | pSH1 derivative for constitutive expression of *crtW* from *B. vesicularis* containing an artificial ribosome binding site | this work |
| pSH1_$crtW2_{Bv}$ | pSH1 derivative for constitutive expression of *crtW* from *B. vesicularis* containing an artificial ribosome binding site | this work |
| pSH1_$crtW1_{Fp}$ | pSH1 derivative for constitutive expression of *crtW* from *F. pelagi* containing an artificial ribosome binding site | this work |
| pSH1_$crtW2_{Fp}$ | pSH1 derivative for constitutive expression of *crtW* from *F. pelagi* containing an artificial ribosome binding site | this work |
| pSH1_$crtW3_{Fp}$ | pSH1 derivative for constitutive expression of *crtW* from *F. pelagi* containing an artificial ribosome binding site | this work |
| pSH1_$crtW_{Sa}$ | pSH1 derivative for constitutive expression of *crtW* from *S. astaxanthinifaciens* containing an artificial ribosome binding site | this work |
| pK19*mobsacB* | $Km^R$; *E. coli*/*C. glutamicum* shuttle vector for construction of insertion and deletion mutants in *C. glutamicum* (pK18 $oriV_{Ec}$ *sacB lacZα*) | [82] |
| pK19*mobsacB-cg0725* | pK19*mobsacB* with a *cg0725* deletion construct | - |
| pK19*mobsacB*-$P_{tuf}$-*dxs* | pK19*mobsacB* derivative with a *tuf* promoter region (200 bp upstream of the coding sequence of the *tuf* gene(cg0587) construct for the promoter exchange of *dxs* | [42] |
| pK19*mobsacB*-Int*crtEBI* | pK19*mobsacB* derivative containing the artificial operon *crtE_crtBI* under the control of the $P_{tuf}$ promoter with an addition ribosome binding site in front of *crtB* for integration in the cgp2 cured region of *C. glutamicum* MB001 | this work |
| pVWEx1-*crtEBI* | pVWEx1 derivative for IPTG-inducible expression of *crtE*, *crtB* and *crtI* from *C. glutamicum* containing artificial ribosome binding sites in front of *crtE* and *crtBI* | [38] |
| pK19*mobsacB*-Int*crtY* | pK19*mobsacB* derivative containing *crtY* of *Pantoea ananatis* under the control of the $P_{tuf}$ promoter for integration in the cgp1 cured region of *C. glutamicum* MB001 | this work |

### 4.2. Recombinant DNA Work

Plasmids were constructed in *E. coli* DH5α from PCR-generated fragments (All-in HiFi, highQu, Kraichtal, Germany) and isolated with the Plasmid GeneJET Miniprep kit (Thermo Fisher Scientific, Schwerte, Germany). Oligonucleotides used in this study were obtained from Metabion (Planegg/Steinkirchen, Germany) and are listed in Table 5. Standard reactions like restriction, and PCR were performed as described previously [83]. Besides cloning by restriction, Gibson assembly was applied for the construction of plasmids [44]. If applicable, PCR products were purified using the PCR clean-up and gel extraction kit (Macherey-Nagel, Düren, Germany). For transformation of *E. coli* DH5α, the RbCl method was used and *C. glutamicum* was transformed via electroporation [84] at 2.5 kV, 200 Ω, and 25 μF. All cloned DNA fragments were confirmed by sequencing.

**Table 5.** Oligonucleotides used in this study.

| Oligonucleotide | Sequence (5'→3') |
|---|---|
| N1 | *CATGCCTGCAGGTCGACTCTAGAG***GAAAGGAGGCCCTTCAG**ATGACCGCCGCCGTCGCCGAG |
| N2 | *CATGCCTGCAGGTCGACTCTAGAG***GAAAGGAGGCCCTTCAG**GTGACCGCCGCCGTCGCCGAG |
| N3 | *CATGCCTGCAGGTCGACTCTAGAG***GAAAGGAGGCCCTTC**ATGACCGCCGCCGTCGCCGAG |
| N4 | *CATGCCTGCAGGTCGACTCTAGAG***GAAAGGAGGCCCTT**CGTGACCGCCGCCGTCGCCGAG |
| N5 | *CATGCCTGCAGGTCGACTCTAGAG***GAAAGGAGGCCC**ATGACCGCCGCCGTCGCCGAG |
| N6 | *CATGCCTGCAGGTCGACTCTAGAG***GAAAGGAGGCCC**GTGACCGCCGCCGTCGCCGAG |
| N7 | *CATGCCTGCAGGTCGACTCTAGAG***GAAAGAAGGCCCTTCAG**ATGACCGCCGCCGTCGCCGAG |
| N8 | *CATGCCTGCAGGTCGACTCTAGAG***GAAAGAAGGCCCTTCAG**GTGACCGCCGCCGTCGCCGAG |
| N9 | *CATGCCTGCAGGTCGACTCTAGAG***GAAAGAAGGCCCTTC**ATGACCGCCGCCGTCGCCGAG |
| N10 | *CATGCCTGCAGGTCGACTCTAGAG***GAAAGAAGGCCCTTC**GTGACCGCCGCCGTCGCCGAG |
| N11 | *CATGCCTGCAGGTCGACTCTAGAG***GAAAGAAGGCCC**ATGACCGCCGCCGTCGCCGAG |
| N12 | *CATGCCTGCAGGTCGACTCTAGAG***GAAAGAAGGCCC**GTGACCGCCGCCGTCGCCGAG |
| N13 | *CATGCCTGCAGGTCGACTCTAGAG***GAATGGAGGCCCTTCAG**ATGACCGCCGCCGTCGCCGAG |
| N14 | *CATGCCTGCAGGTCGACTCTAGAG***GAATGGAGGCCCTTCAG**GTGACCGCCGCCGTCGCCGAG |
| N15 | *CATGCCTGCAGGTCGACTCTAGAG***GAATGGAGGCCCTTC**ATGACCGCCGCCGTCGCCGAG |
| N16 | *CATGCCTGCAGGTCGACTCTAGAG***GAATGGAGGCCCTTC**GTGACCGCCGCCGTCGCCGAG |
| N17 | *CATGCCTGCAGGTCGACTCTAGAG***GAATGGAGGCCC**ATGACCGCCGCCGTCGCCGAG |
| N18 | *CATGCCTGCAGGTCGACTCTAGAG***GAATGGAGGCCC**GTGACCGCCGCCGTCGCCGAG |
| N19 | *CATGCCTGCAGGTCGACTCTAGAG***GAATGAAGGCCCTTCAG**ATGACCGCCGCCGTCGCCGAG |
| N20 | *CATGCCTGCAGGTCGACTCTAGAG***GAATGAAGGCCCTTCAG**GTGACCGCCGCCGTCGCCGAG |
| N21 | *CATGCCTGCAGGTCGACTCTAGAG***GAATGAAGGCCCTTC**ATGACCGCCGCCGTCGCCGAG |
| N22 | *CATGCCTGCAGGTCGACTCTAGAG***GAATGAAGGCCCTTC**GTGACCGCCGCCGTCGCCGAG |
| N23 | *CATGCCTGCAGGTCGACTCTAGAG***GAATGAAGGCCC**ATGACCGCCGCCGTCGCCGAG |
| N24 | *CATGCCTGCAGGTCGACTCTAGAG***GAATGAAGGCCC**GTGACCGCCGCCGTCGCCGAG |
| N25 | *AACTGCCACACGAAC***GAAAGGAGGCCCTTCAG**ATGTTGTGGATTTGGAATGCCCTGATC |
| N26 | *AACTGCCACACGAAC***GAAAGGAGGCCCTTCAG**GTGTTGTGGATTTGGAATGCCCTGATC |
| N27 | *AACTGCCACACGAAC***GAAAGGAGGCCCTTC**ATGTTGTGGATTTGGAATGCCCTGATC |
| N28 | *AACTGCCACACGAAC***GAAAGGAGGCCCTTC**GTGTTGTGGATTTGGAATGCCCTGATC |
| N29 | *AACTGCCACACGAAC***GAAAGGAGGCCC**ATGTTGTGGATTTGGAATGCCCTGATC |
| N30 | *AACTGCCACACGAAC***GAAAGGAGGCCC**GTGTTGTGGATTTGGAATGCCCTGATC |
| N31 | *AACTGCCACACGAAC***GAAAGAAGGCCCTTCAG**ATGTTGTGGATTTGGAATGCCCTGATC |
| N32 | *AACTGCCACACGAAC***GAAAGAAGGCCCTTCAG**GTGTTGTGGATTTGGAATGCCCTGATC |
| N33 | *AACTGCCACACGAAC***GAAAGAAGGCCCTTC**ATGTTGTGGATTTGGAATGCCCTGATC |
| N34 | *AACTGCCACACGAAC***GAAAGAAGGCCCTTC**GTGTTGTGGATTTGGAATGCCCTGATC |
| N35 | *AACTGCCACACGAAC***GAAAGAAGGCCC**ATGTTGTGGATTTGGAATGCCCTGATC |
| N36 | *AACTGCCACACGAAC***GAAAGAAGGCCC**GTGTTGTGGATTTGGAATGCCCTGATC |
| N37 | *AACTGCCACACGAAC***GAATGGAGGCCCTTCAG**ATGTTGTGGATTTGGAATGCCCTGATC |
| N38 | *AACTGCCACACGAAC***GAATGGAGGCCCTTCAG**GTGTTGTGGATTTGGAATGCCCTGATC |
| N39 | *AACTGCCACACGAAC***GAATGGAGGCCCTTC**ATGTTGTGGATTTGGAATGCCCTGATC |
| N40 | *AACTGCCACACGAAC***GAATGGAGGCCCTTC**GTGTTGTGGATTTGGAATGCCCTGATC |
| N41 | *AACTGCCACACGAAC***GAATGGAGGCCC**ATGTTGTGGATTTGGAATGCCCTGATC |
| N42 | *AACTGCCACACGAAC***GAATGGAGGCCC**GTGTTGTGGATTTGGAATGCCCTGATC |
| N43 | *AACTGCCACACGAAC***GAATGAAGGCCCTTCAG**ATGTTGTGGATTTGGAATGCCCTGATC |
| N44 | *AACTGCCACACGAAC***GAATGAAGGCCCTTCAG**GTGTTGTGGATTTGGAATGCCCTGATC |
| N45 | *AACTGCCACACGAAC***GAATGAAGGCCCTTC**ATGTTGTGGATTTGGAATGCCCTGATC |
| N46 | *AACTGCCACACGAAC***GAATGAAGGCCCTTC**GTGTTGTGGATTTGGAATGCCCTGATC |
| N47 | *AACTGCCACACGAAC***GAATGAAGGCCC**ATGTTGTGGATTTGGAATGCCCTGATC |
| N48 | *AACTGCCACACGAAC***GAATGAAGGCCC**GTGTTGTGGATTTGGAATGCCCTGATC |
| N49 | *GTTCGTGTGGCAGTT*TTAGTGGTGGTGGTGGTGGTGAGACTCGCCGCGCCACAGACG |
| N50 | *ATTCGAGCTCGGTACCCGGGGATC*TTACTTTTCGAACTGTGGGTGGGACCACTTCCCGGATGCGGGCTC |
| BaW1 | *CATGCCTGCAGGTCGACTCTAGAG***GAAAGGAGGCCCTTCAG**ATGACCGCCGCCGTCGCCGAG |
| BaW2 | *CGGTACCCGGGGATC*TCAAGACTCGCCGCGCCAC |
| BbW1 | *CATGCCTGCAGGTCGACTCTAGAG***GAAAGGAGGCCCTTCAG**ATGACGCGGGAACGCCA |
| BbW2 | *ATTCGAGCTCGGTACCCGGGGATC*TTAGAGACGTTCGCTACGC |

**Table 5.** *Cont.*

| Oligonucleotide | Sequence (5'→3') |
|---|---|
| BbZ1 | *ATGGAATTCGAGCTCGGTACCCGGG***GAAAGGAGGCCCTTCAG**ATGACGATCGTCTGGTTCAC |
| BbZ2 | *GCATGCCTGCAGGTCGACTCTAGAGGATC*TTACTCGGCCGGGATGTCC |
| BvW1 | *CATGCCTGCAGGTCGACTCTAGAG***GAAAGGAGGCCCTTCAG**ATGGGGCAAGCGAACAG |
| BvW2 | *CATGCCTGCAGGTCGACTCTAGAG***GAAAGGAGGCCCTTCAG**ATGCGGCAAGCGAACAG |
| BvW3 | *ATTCGAGCTCGGTACCCGGGGATC*CTAGCTGAACAAACTCCACCAG |
| BvZ1 | *ATGGAATTCGAGCTCGGTACCCGGG***GAAAGGAGGCCCTTCAG**ATGTCCTGGCCGACGATG |
| BvZ2 | *GCATGCCTGCAGGTCGACTCTAGAGGATC*TTAGGCGCCGTTGCTGGAT |
| FpW1 | *CATGCCTGCAGGTCGACTCTAGAG***GAAAGGAGGCCCTTCAG**ATGACCCTCAGCCCAACCTC |
| FpW2 | *CATGCCTGCAGGTCGACTCTAGAG***GAAAGGAGGCCCTTCAG**ATGAGACCCTACCAAACGACG |
| FpW3 | *CATGCCTGCAGGTCGACTCTAGAG***GAAAGGAGGCCCTTCAG**ATGCATGGTTCGCTGGC |
| FpW4 | *ATTCGAGCTCGGTACCCGGGGATC*TTAGGACTGGCGAGTATGCG |
| FpZ1 | *ATGGAATTCGAGCTCGGTACCCGGG***GAAAGGAGGCCCTTCAG**ATGACGATCTGGACTCTCTACTAC |
| FpZ2 | *GCATGCCTGCAGGTCGACTCTAGAGGATC*TTACCGAACCGGCGCGT |
| SaW1 | *CATGCCTGCAGGTCGACTCTAGAG***GAAAGGAGGCCCTTCAG**ATGGCCCCCATGCTCAGTG |
| SaW2 | *ATTCGAGCTCGGTACCCGGGGATC*TTAGGCGGGAAGCGCAAG |
| SaZ1 | *ATGGAATTCGAGCTCGGTACCCGGG***GAAAGGAGGCCCTTCAG**ATGTCCTGGCCTGCCG |
| SaZ2 | *GCATGCCTGCAGGTCGACTCTAGAGGATC*TTAGGCCCGCTCCTCGTG |
| pSH1 fw | ACCGGCTCCAGATTTATCAG |
| pVWEx/ pSH1 rv | ATCTTCTCTCATCCGCCA |
| pEC-XT fw | AATACGCAAACCGCCTCTCC |
| pEC-XT rv | TACTGCCGCCAGGCAAATTC |
| pV_$P_{tuf}$-fw | *CGGAATCTTGCACGCCCTT*GGCCGTTACCCTGCGAATG |
| pV_$P_{tuf}$-rv | *CTGCAGGCATGCAAGCTTT*GTATGTCCTCCTGGACTTC |
| pV1-fw | *GAAGTCCAGGAGGACATACA*AAGCTTGCATGCCTGCAG |
| pV6962-rv | *CATTCGCAGGGTAACGGCCA*AGGGCGTGCAAGATTCCG |
| *cg0725*-A | *GCAGGTCGACTCTAGAGGATCCCC*GCGCGAAGATTTGATGGG |
| *cg0725*-B | *CCCATCCACCCCGGGTAAACA*TTCCTGCATATTCAGCATAGTAATC |
| *cg0725*-C | *TGTTTACCCGGGGTGGATGGG*TCCCTTAATAATGCACCATGGC |
| *cg0725*-D | *CCAGTGAATTCGAGCTCGGTACCC*CTTGTCACCACAGCACTACT |
| *cg0725*-E | GCGCGAAGATTTGATGGG |
| *cg0725*-F | ACTTGTCACCACAGCACTAC |
| *crtY*-Int1 | GCAGGTCGACTCTAGAGGATCCCCCAGTGAAGGATCGGTGCG |
| *crtY*-Int2 | CATTCGCAGGGTAACGGCCACCTATCTGCTGGCCGGTG |
| *crtY*-Int3 | CACCGGCCAGCAGATAGGTGGCCGTTACCCTGCGAATG |
| *crtY*-Int4 | CAGATCATAATGCGGTTGCATTGTATGTCCTCCTGGACTTC |
| *crtY*-Int5 | GAAGTCCAGGAGGACATACAATGCAACCGCATTATGATCTG |
| *crtY*-Int6 | TCTTACTACTTGCGCTAGGTACAGTTAACGATGAGTCGTCATAATGG |
| *crtY*-Int7 | CCATTATGACGACTCATCGTTAACTGTACCTAGCGCAAGTAGTAAGA |
| *crtY*-Int8 | CCAGTGAATTCGAGCTCGGTACCCCTGCTCATCCTTCAACAACGT |
| cgp1-E | GTGGTGCTCGAGAACATAAG |
| cgp1-F | CGGTCACCCGTAACAATCAG |
| *crtEBI*-Int1 | *GCAGGTCGACTCTAGAGGATCCCC*GTGCTTCGCATCGTCTATGTC |
| *crtEBI*-Int2 | CATTCGCAGGGTAACGGCCAATAGTTGGGGGAATTTATAAGGATTTG |
| *crtEBI*-Int3 | CAAATCCTTATAAATTCCCCCAACTATTGGCCGTTACCCTGCGAATG |
| *crtEBI*-Int4 | GATTGTCATGCCATTGTCCATTGTATGTCCTCCTGGACTTC |
| *crtEBI*-Int5 | GAAGTCCAGGAGGACATACAATGGACAATGGCATGACAATC |
| *crtEBI*-Int6 | CTAATGGACGGTGAAGTATCATTTATGTTAATGATCGTATGAGGTCTTTTGAG |
| *crtEBI*-Int7 | CTCAAAAGACCTCATACGATCATTAACATAAATGATACTTCACCGTCCATTAG |
| *crtEBI*-Int8 | *CCAGTGAATTCGAGCTCGGTACCC*CGCCGTATGTAACAAGATTTG |
| Cgp2-E | TCGCACCATCTACGACAACC |
| Cgp2-F | CTACGAAGCTGACGCCGAAG |

Sequence in bold: artificial ribosome binding site; sequence underlined: tag site; sequence in italics: linker sequence for hybridization.

### *4.3. Construction of Expression Vector pSH1*

The plasmid pSH1 was constructed based on the expression vector pVWEx1 [81]. The backbone of pSH1 was amplified from pVWEx1 omitting the *lacIq* and $P_{tac}$ region using the oligonucleotides pV1-fw and pV6962-rv (Table 5) with All-in HiFi polymerase (highQu, Kraichtal, Germany). The promoter of the *C. glutamicum tuf* gene (cg0587) was amplified using the primers pV_$P_{tuf}$-fw and pV_$P_{tuf}$-rv (Table 5). Both fragments were assembled using the Gibson method [44]. Vector sequence was confirmed via sequencing to exclude mutations.

### *4.4. Deletion and Exchenge Mutagenesis in the Genome of* C. glutamicum

For targeted deletion of *cg0725*, which encodes a transcriptional regulator and is part of the carotenogenesis gene cluster of *C. glutamicum*, the suicide vector pK19*mobsacB* was used [82]. Genomic regions flanking *cg0725* were amplified from genomic DNA of *C. glutamicum* WT using primer pairs *cg0725*-A/B and *cg0725*-C/D (Table 5), respectively. Subsequently the purified PCR products were linked by crossover PCR using the primer pair *cg0725*-A/D (Table 5). The resulting amplificate was cloned into pK19*mobsacB* resulting in the construction of deletion vector pK19*mobsacB*-*cg0725* (Table 4). Deletion of *cg0725* via two-step homologous recombination as well as the selection for the first and second recombination events were carried out as described previously [85]. Successful removal of *cg0725* was verified by PCR analysis of the constructed mutant using primer pair *cg0725*-E/F (Table 5).

The integration of the synthetic operon *crtEBI* and the lycopene cyclase gene of *Pantoea ananatis crtY*, respectively, was conducted by using the suicide vector pK19*mobsacB* [82]. Operon *crtEBI* consists of the carotenogenic genes *crtE* (cg0723), *crtB* (cg0721) and *crtI* (cg0720) and was amplified from the expression vector pVWEx1-*crtEBI* [38] using the oligonucleotides *crtEBI*-Int5 and *crtEBI*-Int6. The $P_{tuf}$ promoter region was amplified using the oligonucleotides *crtEBI*-Int3/4 or *crtY*-Int3/4, respectively. Genomic regions flanking the selected insertion region were amplified from genomic DNA of *C. glutamicum* MB001 using primer pairs *crtEBI*-Int1/2 and *crtEBI*-Int7/8 for integration in the cgp2 cured region in the case of the *crtEBI* operon, or *crtY*-Int1/2 and *crtY*-Int7/8 for integration of *crtY* in the CGP1 cured region (Table 5), respectively. *CrtY* was amplified from genomic DNA of *P. ananatis* using the primer pair *crtY*-Int5/6. The purified PCR products were directly combined together with the plasmid by Gibson assembly [44]. The final assembly of the insert with linearized pK19*mobsacB* led to the construction of the respective integration vectors pK19*mobsacB*-Int*crtEBI* and pK19*mobsacB*-Int*crtY* (Table 4). The following integration of the operon by two-step homologous recombination was performed according to the deletion of genes. The integration in the cgp1 or cgp2 region was verified by PCR using the primers cgp1-E/F and cgp2-E/F, respectively.

The plasmid pK19*mobsacB*-$P_{tuf}$-*dxs* was constructed to replace the native *dxs* promoter with the *tuf* promoter region from *C. glutamicum* WT as described earlier [42]. The promoter exchange was verified by PCR using the primers *dxs E* and 33, and sequencing of the PCR product.

### 4.5. Combinatorial Gene Assembly, Library Construction and Overexpression of Carotenogenic Genes

The combinatorial assembly of genes $crtW_{Ba}$ and $crtZ_{Pa}$ was performed with Gibson Assembly [44]. The *crtW* gene was amplified from the genomic DNA of *Brevundimonas aurantiaca* in a one-pot-PCR containing an equimolar mixture of forward primers (N1-N24) and a reverse primer (N49) (Table 5). The *crtZ* gene was amplified from the genomic DNA of *Pantoea ananatis* in a one-pot-PCR containing an equimolar mixture of forward primers (N25-N48) and a reverse primer (N50) (Table 5). PCR products of both genes were gel-extracted (Macherey-Nagel) and cloned in *Bam*HI-restricted pSH1 applying Gibson Assembly. The transformation of *E. coli* DH5α was done as described above. 1/10 of the transformed cells were plated on selective agar plates for colony number calculation while 9/10 of the transformants were grown in selective liquid medium for plasmid isolation. Isolated plasmids were used for transformation of *C. glutamicum* strain BETA1.

Plasmids harboring a carotenogenic gene (general abbreviation *crt*), pECXT99A_*crt*, pEKEX3_*crt* or pVWEx1_*crt* allowed an IPTG-inducible overexpression of *crt*. The vector pSH1 expressed *crt* constitutively under the *tuf*-promoter ($P_{tuf}$). The plasmids were constructed on the basis of pECXT99A [79] pEKEx3 [80] or pVWEx1 [81], respectively. Amplification of *crt* was achieved by polymerase chain reaction (PCR) from genomic DNA of *C. glutamicum* ATCC 13032, *P. ananatis*, *B. aurantiaca*, *B. bacteroides*, *B. vesicularis*, *F. pelagi*, and *S. astaxanthinifaciens*, respectively, that was prepared as described [86] using the respective primers (Table 5). The amplified products were cloned into the *Bam*HI restricted pEC-XT99A, pEKEx3, pVWEx1 or pSH1 plasmid DNA by Gibson assembly.

### 4.6. Extraction and Quantification of Carotenoids

For extracting carotenoids from *C. glutamicum*, 1 mL of the culture was harvested by centrifugation for 7 min at 14,000 rpm. Carotenoid pigments were extracted with 800 μL methanol:acetone (7:3) containing 0.05% BHT at 60 °C for 15 min with careful vortexing every 5 min. Cell debris was spun down for 7 min at 14,000 rpm and the supernatant was used for high performance liquid chromatography (HPLC) analysis. For HPLC the Agilent 1200 series system (Agilent Technologies

Sales & Services GmbH & Co. KG, Waldbronn, Germany) was used. The UV/visible (Vis) spectrum was recorded with a diode array detector (DAD). The quantification of carotenoids was performed by the integration of the extracted wavelength chromatogram at $\lambda_{max}$ 470 nm for every maximum and by the analysis of the appropriate UV/Vis profiles. Standard calibration curves were generated with lycopene (Sigma-Aldrich), β-carotene (Sigma-Aldrich), canthaxanthin (Sigma-Aldrich), zeaxanthin (Sigma-Aldrich) and astaxanthin (Sigma-Aldrich) to quantify carotenoid titers. All standards were dissolved in chloroform according to their solubility and diluted in methanol:acetone (7:3) containing 0.05% BHT.

As column system, a precolumn (10 × 4 mm MultoHigh 100 RP18-5, CS Chromatographie Service GmbH, Langerwehe, Germany) and a main column (ProntoSIL 200–5 C30, 250 × 4 mm, CS Chromatographie Service GmbH) were used. The HPLC protocol ensured a gradient elution for 10 min and a mobile phase composition of (A) methanol and (B) methanol/methyl tert-butyl ether/ethyl acetate (5:4:1) starting from 10% to 100% of eluent B, followed by 20 min of isocratic elution with 100% B. After that, the eluent composition was set back to 10% B for 3 min. The injection volume was 100 μL and the flow rate was kept constant at 1.4 mL/min.

**Acknowledgments:** This work was supported in part by EU-FP7 project PROMYSE (289540). We acknowledge support for the Article Processing Charge by the Deutsche Forschungsgemeinschaft and the Open Access Publication Fund of Bielefeld University.

**Author Contributions:** All authors planned and designed the experiments. N.A.H. and S.A.E.H. performed the experiments and analyzed data. P.P.W. and V.F.W. analyzed data. N.A.H., S.A.E.H. and P.P.W. drafted the manuscript. V.F.W. coordinated the study and finalized the manuscript. All authors read and approved the final manuscript.

**Conflicts of Interest:** The authors declare no conflict of interest. The founding sponsors had no role in the design of the study; in the collection, analyses, or interpretation of data; in the writing of the manuscript, and in the decision to publish the results.

## Abbreviations

The following abbreviations are used in this manuscript:

| | |
|---|---|
| TIR | Translation Initiation Rate |
| MDPI | Multidisciplinary Digital Publishing Institute |
| DOAJ | Directory of open access journals |
| TLA | Three letter acronym |
| LD | Linear dichroism |

## References

1. Cooper, D.A.; Eldridge, A.L.; Peters, J.C. Dietary carotenoids and certain cancers, heart disease, and age-related macular degeneration: A review of recent research. *Nutr. Rev.* **1999**, *57*, 201–214. [CrossRef] [PubMed]
2. Krinsky, N.I.; Johnson, E.J. Carotenoid actions and their relation to health and disease. *Mol. Asp. Med.* **2005**, *26*, 459–516. [CrossRef] [PubMed]
3. Ye, V.M.; Bhatia, S.K. Pathway engineering strategies for production of beneficial carotenoids in microbial hosts. *Biotechnol. Lett.* **2012**, *34*, 1405–1414. [CrossRef] [PubMed]
4. BBC Research. The Global Market for Carotenoids. Available online: http://www.bccresearch.com/market-research/food-and-beverage/carotenoids-global-market-fod025d.html (accessed on 13 June 2016).
5. Belviranli, M.; Okudan, N. Well-Known Antioxidants and Newcomers in Sport Nutrition: Coenzyme Q10, Quercetin, Resveratrol, Pterostilbene, Pycnogenol and Astaxanthin. In *Antioxidants in Sport Nutrition*; Lamprecht, M., Ed.; Taylor & Francis Group: Boca Raton, FL, USA, 2015.
6. Gassel, S.; Schewe, H.; Schmidt, I.; Schrader, J.; Sandmann, G. Multiple improvement of astaxanthin biosynthesis in *Xanthophyllomyces dendrorhous* by a combination of conventional mutagenesis and metabolic pathway engineering. *Biotechnol. Lett.* **2013**, *35*, 565–569. [CrossRef] [PubMed]

7. Liu, X.; Osawa, T. *Cis* astaxanthin and especially 9-*cis* astaxanthin exhibits a higher antioxidant activity in vitro compared to the all-*trans* isomer. *Biochem. Biophys. Res. Commun.* **2007**, *357*, 187–193. [CrossRef] [PubMed]
8. Hussein, G.; Nakamura, M.; Zhao, Q.; Iguchi, T.; Goto, H.; Sankawa, U.; Watanabe, H. Antihypertensive and neuroprotective effects of astaxanthin in experimental animals. *Biol. Pharm. Bull.* **2005**, *28*, 47–52. [CrossRef] [PubMed]
9. Giuffrida, D.; Sutthiwong, N.; Dugo, P.; Donato, P.; Cacciola, F.; Girard-Valenciennes, E.; le Mao, Y.; Monnet, C.; Fouillaud, M.; Caro, Y.; et al. Characterisation of the C50 carotenoids produced by strains of the cheese-ripening bacterium *Arthrobacter arilaitensis*. *Int. Dairy J.* **2016**, *55*, 10–16. [CrossRef]
10. Chen, C.W.; Hsu, S.H.; Lin, M.T.; Hsu, Y.H. Mass production of C50 carotenoids by *Haloferax mediterranei* in using extruded rice bran and starch under optimal conductivity of brined medium. *Bioprocess Biosyst. Eng.* **2015**, *38*, 2361–2367. [CrossRef] [PubMed]
11. Furubayashi, M.; Ikezumi, M.; Takaichi, S.; Maoka, T.; Hemmi, H.; Ogawa, T.; Saito, K.; Tobias, A.V.; Umeno, D. A highly selective biosynthetic pathway to non-natural C50 carotenoids assembled from moderately selective enzymes. *Nat. Commun.* **2015**, *6*, 7534. [CrossRef] [PubMed]
12. Umeno, D.; Arnold, F.H. A C35 carotenoid biosynthetic pathway. *Appl. Environ. Microbiol.* **2003**, *69*, 3573–3579. [CrossRef] [PubMed]
13. Takaichi, S.; Sandmann, G.; Schnurr, G.; Satomi, Y.; Suzuki, A.; Misawa, N. The carotenoid 7,8-dihydro-y end group can be cyclized by the lycopene cyclases from the bacterium *Erwinia uredovora* and the higher plant *Capsicum annuum*. *Eur. J. Biochem.* **1996**, *241*, 291–296. [CrossRef] [PubMed]
14. Fassett, R.G.; Coombes, J.S. Astaxanthin in cardiovascular health and disease. *Molecules* **2012**, *17*, 2030–2048. [CrossRef] [PubMed]
15. Ohgami, K.; Shiratori, K.; Kotake, S.; Nishida, T.; Mizuki, N.; Yazawa, K.; Ohno, S. Effects of astaxanthin on lipopolysaccharide-induced inflammation in vitro and in vivo. *Investig. Ophthalmol. Vis. Sci.* **2003**, *44*, 2694–2701. [CrossRef]
16. Baralic, I.; Andjelkovic, M.; Djordjevic, B.; Dikic, N.; Radivojevic, N.; Suzin-Zivkovic, V.; Radojevic-Skodric, S.; Pejic, S. Effect of Astaxanthin Supplementation on Salivary IgA, Oxidative Stress, and Inflammation in Young Soccer Players. *Evid. Based. Complement. Altern. Med.* **2015**, *2015*, 783761. [CrossRef] [PubMed]
17. Li, J.; Zhu, D.; Niu, J.; Shen, S.; Wang, G. An economic assessment of astaxanthin production by large scale cultivation of *Haematococcus pluvialis*. *Biotechnol. Adv.* **2011**, *29*, 568–574. [CrossRef] [PubMed]
18. Zhu, F.; Zhong, X.; Hu, M.; Lu, L.; Deng, Z.; Liu, T. In vitro Reconstitution of Mevalonate Pathway and targeted engineering of farnesene overproduction in *Escherichia coli*. *Biotechnol. Bioeng.* **2014**, *111*, 1396–1405. [CrossRef] [PubMed]
19. Capelli, B.; Bagchi, D.; Cysewski, G. Synthetic astaxanthin is significantly inferior to algal-based astaxanthin as an antioxidant and may not be suitable as a human nutraceutical supplement. *Nutrafoods* **2013**, *12*, 145–152. [CrossRef]
20. Lee, P.C.; Schmidt-Dannert, C. Metabolic engineering towards biotechnological production of carotenoids in microorganisms. *Appl. Microbiol. Biotechnol.* **2002**, *60*, 1–11. [PubMed]
21. George, B.; Synnove, L.J.; Hanspeter, P. *Carotenoids Handbook*; Birkhauser Verlag: Basel, Switzerland, 2004.
22. Cutzu, R.; Coi, A.; Rosso, F.; Bardi, L.; Ciani, M.; Budroni, M.; Zara, G.; Zara, S.; Mannazzu, I. From crude glycerol to carotenoids by using a *Rhodotorula glutinis* mutant. *World J. Microbiol. Biotechnol.* **2013**, *29*, 1009–1017. [CrossRef] [PubMed]
23. Kinoshita, S.; Udaka, S.; Shimono, M. Studies on the amino acid fermentation. Production of *L*-glutamic acid by various microorganisms. *J. Gen. Appl. Microbiol.* **1957**, *3*, 193–205. [CrossRef]
24. Zahoor, A.; Otten, A.; Wendisch, V.F. Metabolic engineering of *Corynebacterium glutamicum* for glycolate production. *J. Biotechnol.* **2014**, *192*, 366–375. [CrossRef] [PubMed]
25. Schneider, J.; Wendisch, V.F. Putrescine production by engineered *Corynebacterium glutamicum*. *Appl. Microbiol. Biotechnol.* **2010**, *88*, 859–868. [CrossRef] [PubMed]
26. Blombach, B.; Eikmanns, B.J. Current knowledge on isobutanol production with *Escherichia coli*, *Bacillus subtilis* and *Corynebacterium glutamicum*. *Bioeng. Bugs* **2011**, *2*, 346–350. [CrossRef] [PubMed]
27. Frohwitter, J.; Heider, S.A.; Peters-Wendisch, P.; Beekwilder, J.; Wendisch, V.F. Production of the sesquiterpene (+)-valencene by metabolically engineered *Corynebacterium glutamicum*. *J. Biotechnol.* **2014**, *191*, 205–213. [CrossRef] [PubMed]

28. Heider, S.A.; Peters-Wendisch, P.; Wendisch, V.F.; Beekwilder, J.; Brautaset, T. Metabolic engineering for the microbial production of carotenoids and related products with a focus on the rare C50 carotenoids. *Appl. Microbiol. Biotechnol.* **2014**, *98*, 4355–4368. [CrossRef] [PubMed]
29. Kinoshita, S.; Tanaka, K. Glutamic acid. In *The Microbial Production of Amino Acids*; Yamada, K., Kinoshita, S., Tsunoda, T., Aida, K., Eds.; Halsted Press: New York, NY, USA, 1972; pp. 263–324.
30. Blombach, B.; Seibold, G.M. Carbohydrate metabolism in *Corynebacterium glutamicum* and applications for the metabolic engineering of *L*-lysine production strains. *Appl. Microbiol. Biotechnol.* **2010**, *86*, 1313–1322. [CrossRef] [PubMed]
31. Meiswinkel, T.M.; Rittmann, D.; Lindner, S.N.; Wendisch, V.F. Crude glycerol-based production of amino acids and putrescine by *Corynebacterium glutamicum*. *Bioresour. Technol.* **2013**, *145*, 254–258. [CrossRef] [PubMed]
32. Gopinath, V.; Meiswinkel, T.M.; Wendisch, V.F.; Nampoothiri, K.M. Amino acid production from rice straw and wheat bran hydrolysates by recombinant pentose-utilizing *Corynebacterium glutamicum*. *Appl. Microbiol. Biotechnol.* **2011**, *92*, 985–996. [CrossRef] [PubMed]
33. Uhde, A.; Youn, J.W.; Maeda, T.; Clermont, L.; Matano, C.; Kramer, R.; Wendisch, V.F.; Seibold, G.M.; Marin, K. Glucosamine as carbon source for amino acid-producing *Corynebacterium glutamicum*. *Appl. Microbiol. Biotechnol.* **2013**, *97*, 1679–1687. [CrossRef] [PubMed]
34. Matano, C.; Uhde, A.; Youn, J.W.; Maeda, T.; Clermont, L.; Marin, K.; Kramer, R.; Wendisch, V.F.; Seibold, G.M. Engineering of *Corynebacterium glutamicum* for growth and *L*-lysine and lycopene production from *N*-acetyl-glucosamine. *Appl. Microbiol. Biotechnol.* **2014**, *98*, 5633–5643. [CrossRef] [PubMed]
35. Tsuchidate, T.; Tateno, T.; Okai, N.; Tanaka, T.; Ogino, C.; Kondo, A. Glutamate production from beta-glucan using Endoglucanase-secreting *Corynebacterium glutamicum*. *Appl. Microbiol. Biotechnol.* **2011**, *90*, 895–901. [CrossRef] [PubMed]
36. Seibold, G.; Auchter, M.; Berens, S.; Kalinowski, J.; Eikmanns, B.J. Utilization of soluble starch by a recombinant *Corynebacterium glutamicum* strain: Growth and lysine production. *J. Biotechnol.* **2006**, *124*, 381–391. [CrossRef] [PubMed]
37. Heider, S.A.; Peters-Wendisch, P.; Wendisch, V.F. Carotenoid biosynthesis and overproduction in *Corynebacterium glutamicum*. *BMC Microbiol.* **2012**, *12*, 198. [CrossRef] [PubMed]
38. Heider, S.A.; Peters-Wendisch, P.; Netzer, R.; Stafnes, M.; Brautaset, T.; Wendisch, V.F. Production and glucosylation of C50 and C40 carotenoids by metabolically engineered *Corynebacterium glutamicum*. *Appl. Microbiol. Biotechnol.* **2014**, *98*, 1223–1235. [CrossRef] [PubMed]
39. Heider, S.A.; Wendisch, V.F. Engineering microbial cell factories: Metabolic engineering of *Corynebacterium glutamicum* with a focus on non-natural products. *Biotechnol. J.* **2015**, *10*, 1170–1184. [CrossRef] [PubMed]
40. Krubasik, P.; Kobayashi, M.; Sandmann, G. Expression and functional analysis of a gene cluster involved in the synthesis of decaprenoxanthin reveals the mechanisms for C50 carotenoid formation. *Eur. J. Biochem.* **2001**, *268*, 3702–3708. [CrossRef] [PubMed]
41. Heider, S.A.; Peters-Wendisch, P.; Beekwilder, J.; Wendisch, V.F. IdsA is the major geranylgeranyl pyrophosphate synthase involved in carotenogenesis in *Corynebacterium glutamicum*. *FEBS J.* **2014**, *281*, 4906–4920. [CrossRef] [PubMed]
42. Heider, S.A.; Wolf, N.; Hofemeier, A.; Peters-Wendisch, P.; Wendisch, V.F. Optimization of the IPP precursor supply for the production of lycopene, decaprenoxanthin and astaxanthin by *Corynebacterium glutamicum*. *Front. Bioeng. Biotechnol.* **2014**, *2*, 28. [CrossRef] [PubMed]
43. Farasat, I.; Kushwaha, M.; Collens, J.; Easterbrook, M.; Guido, M.; Salis, H.M. Efficient search, mapping, and optimization of multi-protein genetic systems in diverse bacteria. *Mol. Syst. Biol.* **2014**, *10*, 731. [CrossRef] [PubMed]
44. Gibson, D.G.; Young, L.; Chuang, R.Y.; Venter, J.C.; Hutchison, C.A.; Smith, H.O. Enzymatic assembly of DNA molecules up to several hundred kilobases. *Nat. Methods* **2009**, *6*, 343–345. [CrossRef] [PubMed]
45. Takano, H.; Agari, Y.; Hagiwara, K.; Watanabe, R.; Yamazaki, R.; Beppu, T.; Shinkai, A.; Ueda, K. LdrP, a cAMP receptor protein/FNR family transcriptional regulator, serves as a positive regulator for the light-inducible gene cluster in the megaplasmid of *Thermus thermophilus*. *Microbiology* **2014**, *160*, 2650–2660. [CrossRef] [PubMed]

46. Takano, H.; Mise, K.; Hagiwara, K.; Hirata, N.; Watanabe, S.; Toriyabe, M.; Shiratori-Takano, H.; Ueda, K. The role and function of LitR, an $AdoB_{12}$-bound light-sensitive regulator of *Bacillus megaterium* QM B1551, in the regulation of carotenoid production. *J. Bacteriol.* **2015**, *197*, 2301–2315. [CrossRef] [PubMed]
47. Tao, L.; Rouviere, P.E.; Cheng, Q. A carotenoid synthesis gene cluster from a non-marine *Brevundimonas* that synthesizes hydroxylated astaxanthin. *Gene* **2006**, *379*, 101–108. [CrossRef] [PubMed]
48. Scaife, M.A.; Ma, C.A.; Ninlayarn, T.; Wright, P.C.; Armenta, R.E. Comparative analysis of beta-carotene hydroxylase genes for astaxanthin biosynthesis. *J. Nat. Prod.* **2012**, *75*, 1117–1124. [CrossRef] [PubMed]
49. Poindexter, J.S. Biological Properties and Classification of the *Caulobacter* Group. *Bacteriol. Rev.* **1964**, *28*, 231–295. [PubMed]
50. Cho, J.C.; Giovannoni, S.J. Fulvimarina pelagi gen. nov., sp. nov., a marine bacterium that forms a deep evolutionary lineage of descent in the order "*Rhizobiales*". *Int. J. Syst. Evol. Microbiol.* **2003**, *53*, 1853–1859. [PubMed]
51. Asker, D.; Amano, S.; Morita, K.; Tamura, K.; Sakuda, S.; Kikuchi, N.; Furihata, K.; Matsufuji, H.; Beppu, T.; Ueda, K. Astaxanthin dirhamnoside, a new astaxanthin derivative produced by a radio-tolerant bacterium, *Sphingomonas astaxanthinifaciens*. *J. Antibiot.* **2009**, *62*, 397–399. [CrossRef] [PubMed]
52. Asker, D.; Beppu, T.; Ueda, K. *Sphingomonas astaxanthinifaciens* sp. nov., a novel astaxanthin-producing bacterium of the family Sphingomonadaceae isolated from Misasa, Tottori, Japan. *FEMS Microbiol. Lett.* **2007**, *273*, 140–148. [CrossRef] [PubMed]
53. Kang, I.; Oh, H.M.; Lim, S.I.; Ferriera, S.; Giovannoni, S.J.; Cho, J.C. Genome sequence of *Fulvimarina pelagi* $HTCC2506^T$, a Mn(II)-oxidizing alphaproteobacterium possessing an aerobic anoxygenic photosynthetic gene cluster and Xanthorhodopsin. *J. Bacteriol.* **2010**, *192*, 4798–4799. [CrossRef] [PubMed]
54. Knoll, A.; Bartsch, S.; Husemann, B.; Engel, P.; Schroer, K.; Ribeiro, B.; Stockmann, C.; Seletzky, J.; Buchs, J. High cell density cultivation of recombinant yeasts and bacteria under non-pressurized and pressurized conditions in stirred tank bioreactors. *J. Biotechnol.* **2007**, *132*, 167–179. [CrossRef] [PubMed]
55. Aflalo, C.; Meshulam, Y.; Zarka, A.; Boussiba, S. On the relative efficiency of two- vs. one-stage production of astaxanthin by the green alga *Haematococcus pluvialis*. *Biotechnol. Bioeng.* **2007**, *98*, 300–305. [PubMed]
56. Landry, A.P.; Ding, H. Redox Control of Human Mitochondrial Outer Membrane Protein MitoNEET [2Fe-2S] Clusters by Biological Thiols and Hydrogen Peroxide. *J. Biol. Chem.* **2014**, *289*, 4307–4315. [CrossRef] [PubMed]
57. Xie, W.; Liu, M.; Lv, X.; Lu, W.; Gu, J.; Yu, H. Construction of a controllable beta-carotene biosynthetic pathway by decentralized assembly Strategy in *Saccharomyces Cerevisiae*. *Biotechnol. Bioeng.* **2014**, *111*, 125–133. [CrossRef] [PubMed]
58. Lemuth, K.; Steuer, K.; Albermann, C. Engineering of a plasmid-free *Escherichia coli* strain for improvedin vivo biosynthesis of astaxanthin. *Microb. Cell Fact.* **2011**, *10*, 29. [CrossRef] [PubMed]
59. Steiger, S.; Sandmann, G. Cloning of two carotenoid ketolase genes from *Nostoc punctiforme* for the heterologous production of canthaxanthin and astaxanthin. *Biotechnol. Lett.* **2004**, *26*, 813–817. [CrossRef] [PubMed]
60. Fraser, P.D.; Miura, Y.; Misawa, N. In vitro characterization of astaxanthin biosynthetic enzymes. *J. Biol. Chem.* **1997**, *272*, 6128–6135. [CrossRef] [PubMed]
61. Dong, S.; Huang, Y.; Zhang, R.; Wang, S.; Liu, Y. Four different methods comparison for extraction of astaxanthin from green alga *Haematococcus pluvialis*. *Sci. World J.* **2014**, *2014*, 694305. [CrossRef] [PubMed]
62. Scaife, M.A.; Burja, A.M.; Wright, P.C. Characterization of cyanobacterial beta-carotene ketolase and hydroxylase genes in *Escherichia coli*, and their application for astaxanthin biosynthesis. *Biotechnol. Bioeng.* **2009**, *103*, 944–955. [CrossRef] [PubMed]
63. Tejera, N.; Cejas, J.R.; Rodriguez, C.; Bjerkeng, B.; Jerez, S.; Bolanos, A.; Lorenzo, A. Pigmentation, carotenoids, lipid peroxides and lipid composition of skin of red porgy (*Pagrus pagrus*) fed diets supplemented with different astaxanthin sources. *Aquaculture* **2007**, *270*, 218–230. [CrossRef]
64. Rodriguez-Saiz, M.; de la Fuente, J.L.; Barredo, J.L. *Xanthophyllomyces dendrorhous* for the industrial production of astaxanthin. *Appl. Microbiol. Biotechnol.* **2010**, *88*, 645–658. [CrossRef] [PubMed]
65. Schmidt, I.; Schewe, H.; Gassel, S.; Jin, C.; Buckingham, J.; Humbelin, M.; Sandmann, G.; Schrader, J. Biotechnological production of astaxanthin with *Phaffia rhodozyma/Xanthophyllomyces dendrorhous*. *Appl. Microbiol. Biotechnol.* **2011**, *89*, 555–571. [CrossRef] [PubMed]

66. Ni, H.; Chen, Q.H.; He, G.Q.; Wu, G.B.; Yang, Y.F. Optimization of acidic extraction of astaxanthin from *Phaffia rhodozyma*. *J. Zhejiang Univ. Sci. B* **2008**, *9*, 51–59. [CrossRef] [PubMed]
67. Monnet, C.; Loux, V.; Gibrat, J.F.; Spinnler, E.; Barbe, V.; Vacherie, B.; Gavory, F.; Gourbeyre, E.; Siguier, P.; Chandler, M.; et al. The *Arthrobacter arilaitensis* Re117 genome sequence reveals its genetic adaptation to the surface of cheese. *PLoS ONE* **2010**, *5*, e15489. [CrossRef] [PubMed]
68. Brown, N.L.; Stoyanov, J.V.; Kidd, S.P.; Hobman, J.L. The MerR family of transcriptional regulators. *FEMS Microbiol. Rev.* **2003**, *27*, 145–163. [CrossRef]
69. Lorenz, R.T.; Cysewski, G.R. Commercial potential for *Haematococcus* microalgae as a natural source of astaxanthin. *Trends Biotechnol.* **2000**, *18*, 160–167. [CrossRef]
70. Bumbak, F.; Cook, S.; Zachleder, V.; Hauser, S.; Kovar, K. Best practices in heterotrophic high-cell-density microalgal processes: Achievements, potential and possible limitations. *Appl. Microbiol. Biotechnol.* **2011**, *91*, 31–46. [CrossRef] [PubMed]
71. Jacobson, G.; Jolly, S.; Sedmak, J.; Skatrud, T.; Wasileski, J. Astaxanthin over-Producing Strains of Phaffia Rhodozyma, Methods for their Cultivation, and their Use in Animal Feeds. U.S. Patent 6,015,684, 12 August 1999.
72. Baumgart, M.; Unthan, S.; Ruckert, C.; Sivalingam, J.; Grunberger, A.; Kalinowski, J.; Bott, M.; Noack, S.; Frunzke, J. Construction of a prophage-free variant of *Corynebacterium glutamicum* ATCC 13032 for use as a platform strain for basic research and industrial biotechnology. *Appl. Environ. Microbiol.* **2013**, *79*, 6006–6015. [CrossRef] [PubMed]
73. Eggeling, L.; Reyes, O. Experiments. In *Handbook of Corynebacterium Glutamicum*; Eggeling, L., Bott, M., Eds.; CRC Press: Boca Raton, FL, USA, 2005; pp. 3535–3566.
74. Abe, S.; Takayarna, K.; Kinoshita, S. Taxonomical studies on glutamic acid producing bacteria. *J. Gen. Appl. Microbiol.* **1967**, *13*, 279–301. [CrossRef]
75. Hanahan, D. Studies on transformation of *Escherichia coli* with plasmids. *J. Mol. Biol.* **1983**, *166*, 557–580. [CrossRef]
76. Misawa, N.; Nakagawa, M.; Kobayashi, K.; Yamano, S.; Izawa, Y.; Nakamura, K.; Harashima, K. Elucidation of the Erwinia uredovora carotenoid biosynthetic pathway by functional analysis of gene products expressed in *Escherichia coli*. *J. Bacteriol.* **1990**, *172*, 6704–6712. [PubMed]
77. Abraham, W.R.; Strömpl, C.; Meyer, H.; Lindholst, S.; Moore, E.R.; Christ, R.; Vancanneyt, M.; Tindall, B.J.; Bennasar, A.; Smit, J.; et al. Phylogeny and polyphasic taxonomy of Caulobacter species. Proposal of Maricaulis gen. nov. with Maricaulis maris (Poindexter) comb. nov. as the type species, and emended description of the genera Brevundimonas and Caulobacter. *Int. J. Syst. Bacteriol.* **1999**, *49*, 1053–1073. [CrossRef] [PubMed]
78. Busing, K.H.; Doll, W.; Freytag, K. Bacterial flora of the medicinal leech. *Arch. Mikrobiol.* **1953**, *19*, 52–86. [PubMed]
79. Kirchner, O.; Tauch, A. Tools for genetic engineering in the amino acid-producing bacterium *Corynebacterium glutamicum*. *J. Biotechnol.* **2003**, *104*, 287–299. [CrossRef]
80. Stansen, C.; Uy, D.; Delaunay, S.; Eggeling, L.; Goergen, J.L.; Wendisch, V.F. Characterization of a *Corynebacterium glutamicum* lactate utilization operon induced during temperature-triggered glutamate production. *Appl. Environ. Microbiol.* **2005**, *71*, 5920–5928. [CrossRef] [PubMed]
81. Peters-Wendisch, P.G.; Schiel, B.; Wendisch, V.F.; Katsoulidis, E.; Mockel, B.; Sahm, H.; Eikmanns, B.J. Pyruvate carboxylase is a major bottleneck for glutamate and lysine production by *Corynebacterium glutamicum*. *J. Mol. Microbiol. Biotechnol.* **2001**, *3*, 295–300. [PubMed]
82. Schäfer, A.; Tauch, A.; Jäger, W.; Kalinowski, J.; Thierbach, G.; Puhler, A. Small mobilizable multi-purpose cloning vectors derived from the *Escherichia coli* plasmids pK18 and pK19: Selection of defined deletions in the chromosome of *Corynebacterium glutamicum*. *Gene* **1994**, *145*, 69–73. [CrossRef]
83. Sambrook, J.; Russell, D. *Molecular Cloning: A Laboratory Manual*, 3rd ed.; Cold Spring Harbor Laboratoy Press: New York, NY, USA, 2001.
84. Van der Rest, M.E.; Lange, C.; Molenaar, D. A heat shock following electroporation induces highly efficient transformation of *Corynebacterium glutamicum* with xenogeneic plasmid DNA. *Appl. Microbiol. Biotechnol.* **1999**, *52*, 541–545. [CrossRef] [PubMed]

85. Eggeling, L., Bott, M., Eds.; *Handbook of Corynebacterium Glutamicum*; CRC Press: Boca Raton, FL, USA, 2005.
86. Eikmanns, B.J.; Rittmann, D.; Sahm, H. Cloning, sequence analysis, expression, and inactivation of the *Corynebacterium glutamicum* icd gene encoding isocitrate dehydrogenase and biochemical characterization of the enzyme. *J. Bacteriol.* **1995**, *177*, 774–782. [PubMed]

*marine drugs*

MDPI

*Article*

# Metabolic Engineering of *Escherichia coli* for Producing Astaxanthin as the Predominant Carotenoid

**Qian Lu †, Yi-Fan Bu † and Jian-Zhong Liu ***

Institute of Synthetic Biology, Biomedical Center, Guangdong Province Key Laboratory for Aquatic Economic Animals and South China Sea Bio-Resource Exploitation, School of Life Sciences, Sun Yat-sen University, Guangzhou 510275, China; luqian5@mail2.sysu.edu.cn (Q.L.); lssljz@outlook.com (Y.-F.B.)

* Correspondence: lssljz@mail.sysu.edu.cn; Tel.: +86-20-8411-0115; Fax: +86-20-8403-6461

† These authors contributed equally to this work.

Received: 4 August 2017; Accepted: 20 September 2017; Published: 22 September 2017

**Abstract:** Astaxanthin is a carotenoid of significant commercial value due to its superior antioxidant potential and wide applications in the aquaculture, food, cosmetic and pharmaceutical industries. A higher ratio of astaxanthin to the total carotenoids is required for efficient astaxanthin production. β-Carotene ketolase and hydroxylase play important roles in astaxanthin production. We first compared the conversion efficiency to astaxanthin in several β-carotene ketolases from *Brevundimonas* sp. SD212, *Sphingomonas* sp. DC18, *Paracoccus* sp. PC1, *P.* sp. N81106 and *Chlamydomonas reinhardtii* with the recombinant *Escherichia coli* cells that synthesize zeaxanthin due to the presence of the *Pantoea ananatis crtEBIYZ*. The *B.* sp. SD212 *crtW* and *P. ananatis crtZ* genes are the best combination for astaxanthin production. After balancing the activities of β-carotene ketolase and hydroxylase, an *E. coli* ASTA-1 that carries neither a plasmid nor an antibiotic marker was constructed to produce astaxanthin as the predominant carotenoid (96.6%) with a specific content of 7.4 ± 0.3 mg/g DCW without an addition of inducer.

**Keywords:** astaxanthin; *Escherichia coli*; metabolic engineering; β-carotene ketolase; β-carotene hydroxylase

## 1. Introduction

Astaxanthin is a carotenoid of significant commercial value due to its superior antioxidative, anti-inflammatory and anticancer features [1]. It has wide applications in the aquaculture, food, cosmetic and pharmaceutical industries. Currently, commercial astaxanthin is mainly synthesized chemically or extracted from natural producers such as the green algae *Haematococcus pluvialis* or the red yeast *Xanthophyllomyces dendrorhous*. Considering the limited productivity of astaxanthin via extraction and the biosafety issues of chemical synthesis, microbial production of astaxanthin via metabolic engineering has become an attractive alternative [2,3].

In recent years, *Escherichia coli* [4], *Saccharomyces cerevisiae* [5,6] and *Corynebacterium glutamicum* [7] have been used as a host strain for astaxanthin production by the introduction of the astaxanthin biosynthesis pathway (Figure 1) into these non-carotenogenic microorganisms. Metabolic engineering *E. coli* for astaxanthin production has been widely reported in recent years. It has been demonstrated that the pathway from β-carotene to astaxanthin is a crucial step in astaxanthin synthesis [8]. The pathway requires two enzymes, β-carotene ketolase CrtW and β-carotene hydroxylase CrtZ. It has been shown that many bacterial CrtWs and CrtZs are bifunctional, with respect to their substrate [9,10]. They can accept β-carotene as well as its hydroxylated or ketolated products as a substrate, resulting in the formation of eight carotenoid intermediates which affect astaxanthin conversion as measured by the percentage of astaxanthin produced relative to the total carotenoid content (Figure 1). The astaxanthin ratio affects the production costs. To increase the astaxanthin ratio, many bacterial CrtWs and CrtZs have been identified and characterized [4,8,11–17]. However, the ratio reported in the above papers was lower than 90%. Thus, to increase the astaxanthin ratio, we first compared the conversion efficiency to astaxanthin in several CrtWs, which had a higher efficiency for astaxanthin production reported in the literature, with recombinant *E. coli* cells that synthesize zeaxanthin. Then, balancing the expressions of the two enzymes was carried out to obtain a plasmid-free *E. coli*, which produced astaxanthin of 7.4 ± 0.3 mg/g dry cell weight (DCW) with the astaxanthin ratio of 96.6% without the addition of an inducer.

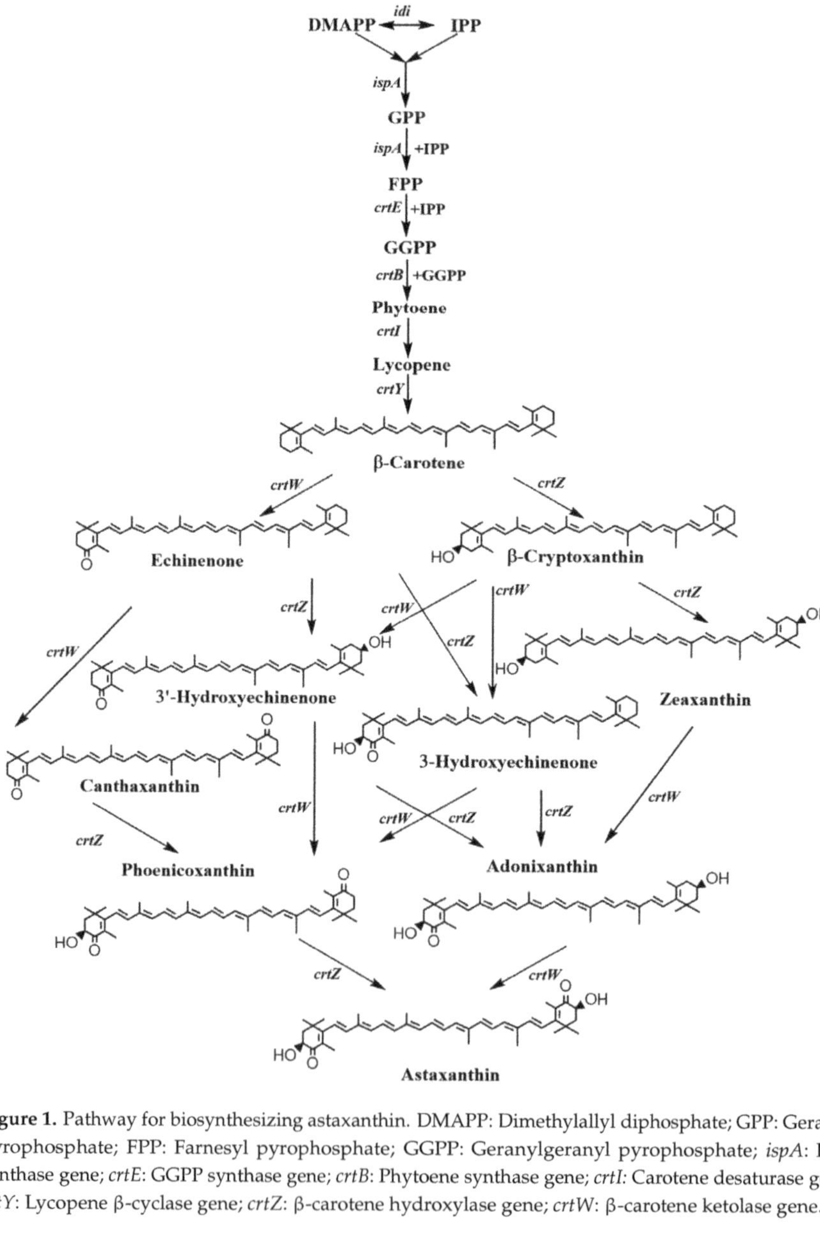


**Figure 1.** Pathway for biosynthesizing astaxanthin. DMAPP: Dimethylallyl diphosphate; GPP: Geranyl pyrophosphate; FPP: Farnesyl pyrophosphate; GGPP: Geranylgeranyl pyrophosphate; *ispA*: FPP synthase gene; *crtE*: GGPP synthase gene; *crtB*: Phytoene synthase gene; *crtI:* Carotene desaturase gene; *crtY*: Lycopene β-cyclase gene; *crtZ*: β-carotene hydroxylase gene; *crtW*: β-carotene ketolase gene.

## 2. Results and Discussion

### 2.1. Screening of β-Carotene Ketolase

It has been demonstrated that astaxanthin biosynthesis proceeds from β-carotene through hydroxylation by CrtZ and then ketolation by CrtW [14]. Moreover, our previous study has proven that *Pantoea ananatis crtZ* is superior to that of *P. agglomerans* or *H. pluvialis* for zeaxanthin production [18]. Thus, we first compared the catalytic efficiency for ketolating zeaxanthin to astaxanthin by different CrtWs. We selected four β-carotene ketolases with higher efficiencies for astaxanthin production reported in the literature as candidates. The four ketolases were *Brevundimonas* sp. SD212 CrtW [11,12], *Sphingomonas* sp. DC18 $CrtW^{F213L/R203W}$ [8], *Paracoccus* sp. N81106 $CrtW^{L175W}$ [16] and *Chlamydomonas reinhardtii* β-carotene ketolase (Bkt) [17]. The plasmids containing the β-carotene ketolase gene were transferred into an engineered zeaxanthin-producing strain *E. coli* ZEAX [19]. One copy of *P. ananatis crtZ* under the control of the P37 promoter was integrated into the chromosome of the β-Carotene producing strain *E. coli* BETA-1 [18]. Table 1 presents the results of astaxanthin production by the different engineered *E. coli*. Among the four β-carotene ketolase genes, the strain harboring *B.* sp. SD212 *crtW* produced a higher level of astaxanthin (2.7 ± 0.1 mg/g DCW), indicating that *B.* sp. SD212 *crtW* and *P. ananatis crtZ* genes are the best combinations for astaxanthin production. Misawa's group also demonstrated that *B.* sp. SD212 *crtW* and *P. ananatis crtZ* genes are a combination of the most promising gene candidates for astaxanthin production [10–12]. Then we assembled two genes into one plasmid to increase the dose of the gene using BglBrick assembly technology and investigated its effect on the combination of different genes on astaxanthin production. As shown in Table 1, increasing the dose of the gene indeed enhanced astaxanthin production. *E. coli* ZEAX (pZS-2$crtW_{Bsp}$) produced 4.6 ± 0.1 mg/g DCW of astaxanthin.

**Table 1.** Effect of the overexpression of different β-carotene ketolase genes on astaxanthin production in *Escherichia coli* ZEAX.

| Plasmid | $OD_{600}$ * | Astaxanthin Concentration, mg/L | Astaxanthin Content, mg/gDCW |
|---|---|---|---|
| | | Single gene | |
| pZS-$crtW_{Bsp}$ | 9.55 ± 0.16 | 8.1 ± 0.1 | 2.7 ± 0.1 |
| pZS-$crtW_{Psp}$ | 8.05 ± 0.47 | 1.3 ± 0.1 | 0.5 ± 0.1 |
| pZS-$crtW_{Ssp}$ | 8.61 ± 0.21 | 0.8 ± 0.1 | 0.3 ± 0.1 |
| pZS-*bkt* | 11.07 ± 0.20 | 5.0 ± 0.2 | 1.4 ± 0.1 |
| | | Double genes | |
| pZS-2$crtW_{Bsp}$ | 19.67 ± 0.33 | 28.8 ± 0.2 | 4.6 ± 0.1 |
| pZS-2*bkt* | 22.15 ± 0.33 | 20.9 ± 1.6 | 3.5 ± 0.1 |
| pZS-2$crtW_{Psp}$ | 16.82 ± 0.56 | 7.0 ± 0.1 | 1.3 ± 0.1 |
| pZS-2$crtW_{Ssp}$ | 14.91 ± 0.31 | 1.1 ± 0.6 | 0.2 ± 0.1 |
| | | Mixed genes | |
| pZS-$crtW_{Bsp}crtW_{Psp}$ | 19.93 ± 0.38 | 12.6 ± 0.6 | 2.0 ± 0.1 |
| pZS-$crtW_{Bsp}crtW_{Ssp}$ | 21.73 ± 0.19 | 24.6 ± 0.4 | 3.5 ± 0.1 |
| pZS-$crtW_{Bsp}bkt$ | 19.9 ± 1.27 | 20.9 ± 1.6 | 3.3 ± 0.1 |
| pZS-$crtW_{Psp}crtW_{Ssp}$ | 16.6 ± 0.17 | 8.5 ± 0.2 | 1.6 ± 0.2 |
| pZS-$crtW_{Psp}bkt$ | 20.01 ± 0.12 | 10.5 ± 1.6 | 1.6 ± 0.1 |
| pZS-$crtW_{Ssp}bkt$ | 21.57 ± 0.38 | 11.2 ± 0.7 | 1.6 ± 0.1 |

* The $OD_{600}$ value was expressed as cell growth.

It has been shown that ketolase activity on zeaxanthin is the limiting step of astaxanthin biosynthesis in a bacterial and plant system [4,17]. To increase the astaxanthin ratio and produce efficiency, many bacterial CrtWs have been characterized and compared. It has been reported that the CrtW enzyme from *B.* sp. SD212 had a higher efficiency for converting zeaxanthin to astaxanthin than that from *P.* sp. PC1 and *P.* sp. N81106 [11]. Of the three β-carotene ketolase enzymes from *H. pluvialis*, *Chlorella zofingien* and *C. reinhardtii*, *C. reinhardtii* β-carotene ketolase had the highest activity for the conversion of zeaxanthin to astaxanthin [17]. Among *Rhodococcus erythropolis* PR4 CrtO, *Synechosistis* sp. PCC6803 CrtO and *B.* sp. SD212 CrtW, only *B.* sp. SD212 CrtW could synthesize astaxanthin from zeaxanthin [12]. Comparative analysis of the CrtO and CrtW revealed that CrtW was more efficient for the conversion of carotene to canthaxanthin than CrtO [14]. The conversion efficiency of *Gloeobacter violaceus* PCC 7421, *Anabaena* (also known as *Nostoc*) sp. PCC 7120 and *Nostoc punctiforme* PCC 73102 CrtW was compared in engineered *E. coli* [13]. The results demonstrated that the CrtW from *A.* sp. PCC 7120 as well as *N. punctiforme* PCC 73102 (CrtW148) can convert not only β-carotene but also zeaxanthin into canthaxanthin and astaxanthin, respectively [13].

Protein engineering of CrtW has been successfully used to improve astaxanthin production in recombinant *E. coli* cells that synthesize zeaxanthin. To improve *S.* sp. DC18 CrtW activity in hydroxylated carotenoids for astaxanthin production, *S.* sp. DC18 CrtW was evolved to obtain the R203W/F213L double mutant that yielded the highest improvement for astaxanthin production [8]. The strain harboring the double mutant produced astaxanthin as the predominant carotenoid (88%) [8]. By using random mutagenesis, *P.* sp. N81106 *crtW* mutants were generated [16]. The zeaxanthin producer *E. coli* harboring the $crtW^{L175}$ mutant produced 78% of astaxanthin in the total carotenoid [16].

### 2.2. Balancing the Activities of β-Carotene Ketolase and Hydroxylase

We analyzed the accumulated carotenoids in *E. coli* ZEAX (pZS-2$crtW_{Bsp}$) as shown in Figure 2A. The engineered strain produced 51.9% astaxanthin, 13.4% phoenicoxanthin and 30.4% canthaxanthin. From the biosynthetic pathway as shown in Figure 1, canthaxanthin and phoenicoxanthin are the intermediates of the pathway through first ketolation and then hydroxylation. Their accumulation indicates that the expression level of the hydroxylase gene *crtZ* may be low in this strain. Thus, we expressed pBAD-*crtZ* in *E. coli* ZEAX (pZS-2$crtW_{Bsp}$) to verify our hypothesis. As shown in Figure 2B, co-expressing pBAD-*crtZ* with pZS-2$crtW_{Bsp}$ in *E. coli* ZEAX indeed increased the astaxanthin ratio to the total carotenoid content from 51.9% to 87.5%. We also co-overexpressed pBAD-$crtW_{\mathrm{Bsp}}$ with pZS-2$crtW_{Bsp}$ in *E. coli* ZEAX. As shown in Figure 2C, the co-overexpression decreased the astaxanthin ratio from 51.9% to 46.4% and increased the canthaxanthin and phoenicoxanthin ratio. This stands in contrast to the study by Lemuth et al. [4], who found that increasing the ketolase activity or decreasing the hydroxylase activity would be necessary for astaxanthin production. Our results demonstrated that increasing CrtZ activity would be necessary for producing astaxanthin as the predominant carotenoid. Thus, to increase the expression level of *crtZ*, we integrated two copies of the *crtZ* into the chromosome of the β-carotene producer *E. coli* BETA to generate the zeaxanthin producer *E. coli* ZEAX-4. The recombinant *E. coli* ZEAX-4 harboring pZS-2$crtW_{Bsp}$ produced 88.6% astaxanthin, 3.9% phoenicoxanthin and 3.0% canthaxanthin (Figure 2D).

To reduce the metabolic burden and to avoid antibiotic markers resulting from the plasmid, we integrated *B.* sp. SD212 *crtW* into the chromosome of the zeaxanthin producer *E. coli* ZEAX-4 to generate an astaxanthin producer *E. coli* ASTA. The resulting strain *E. coli* ASTA produced 92.6% astaxanthin (Figure 2E). However, lycopene was also detected in *E. coli* ASTA. We guess the phenomenon may be due to the lower expression level of the *crtY*. Thus, we integrated another copy of the *crtY* into the chromosome of *E. coli* ASTA to obtain *E. coli* ASTA-1. This integration enhanced the astaxanthin ratio to 96.6% (Figure 2F). *E. coli* ASTA-1 produced astaxanthin as the predominant carotenoid (96.6%) with a specific content of 7.4 ± 0.3 mg/g DCW (Figure 2F).

It is supposed that β-carotene hydroxylase and ketolase compete for their substrate and that only a balanced expression of these two enzymes might result in a complete conversion of β-carotene to astaxanthin [4,14,20,21]. To allow a variable expression of *crtZ* compared to the tac-promoter controlled *N. punctiforme* PCC 73102 *crtW148* and *P. ananatis crtEBIY*, *P. ananatis crtZ* was expressed under the control of the rhamnose-promoter [4]. The engineered strain *E. coli* BW-ASTA produced astaxanthin as the predominant carotenoid (95%) at a concentration of 1.4 mg/g DCW in minimal medium with glucose and Isopropyl β-D-thiogalactoside (IPTG) [4]. The *E. coli* strain with the pTrc*CrtW*-pBAD*CrtZ* dual expression systems had an increased selectivity for astaxanthin production (1.99 mg/g DCW, about 90%) [14]. Our study also suggests that appropriate activities of β-carotene hydroxylase and ketolase are important for astaxanthin production.

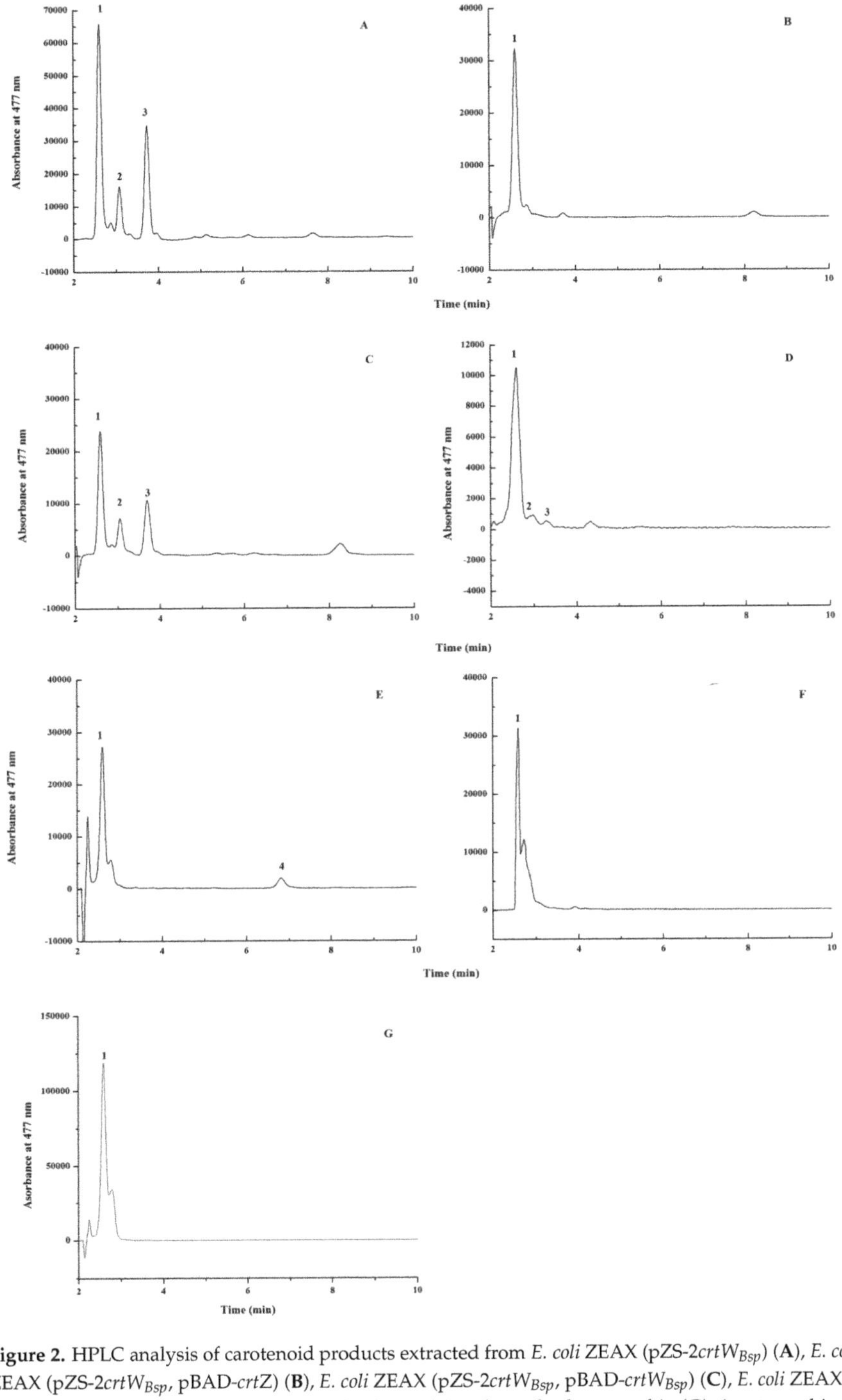


**Figure 2.** HPLC analysis of carotenoid products extracted from *E. coli* ZEAX (pZS-2*crtW*$_{Bsp}$) (**A**), *E. coli* ZEAX (pZS-2*crtW*$_{Bsp}$, pBAD-*crtZ*) (**B**), *E. coli* ZEAX (pZS-2*crtW*$_{Bsp}$, pBAD-*crtW*$_{Bsp}$) (**C**), *E. coli* ZEAX-4 (pZS-2*crtW*$_{Bsp}$) (**D**), *E. coli* ASTA (**E**), *E. coli* ASTA-1 (**F**) and standard astaxanthin (**G**). 1. astaxanthin; 2. Phoenicoxanthin; 3. Canthaxanthin; 4. Lycopene.

Astaxanthin production by microorganisms is summarized in Table 2. Although Lemuth et al. first engineered a plasmid-free *E. coli* strain for astaxanthin production, the strain produced astaxanthin of 1.4 mg/g DCW with an astaxanthin ratio of 95% only with IPTG induction [4]. In this study, we engineered a plasmid-free *E. coli* for astaxanthin production, which reached 7.4 ± 0.3 mg/g DCW with the astaxanthin ratio of 96.6% without the addition of an inducer. From Table 2, we can see that the astaxanthin ratio obtained in this study is the highest value. However, the astaxanthin yield obtained in this study is slightly lower than that (8.64 mg/g DCW) in *E. coli* reported by Ma et al., which is the highest astaxanthin yield reported to date [22]. In their study, the upper mevalonate (MEV) pathway operon from *S. cerevisiae*, the lower MEV pathway operon from *S. cerevisiae*, plus *E. coli idi* and the optimized astaxanthin biosynthetic pathway genes were expressed on three different plasmids [22]. The optimized astaxanthin biosynthetic pathway genes contain *P. ananatis crtEBI* under the control of the $P_{T7}$ promoter, *P. agglomerans crtY* and *crtZ*, *B. sp.* SD212 *crtW* and *E. coli idi* under the control of the $P_{T7}$ promoter [22]. Thus, the introduction of an MEV pathway in our strain *E. coli* ASTA-1 may further increase astaxanthin production.

**Table 2.** Astaxanthin production by different microorganisms.

| Strain | Astaxanthin Yield | Astaxanthin Ratio (%) | Reference |
|---|---|---|---|
| *E. coli* | 5.8 mg/g DCW | N.D. * | [23] |
| *E. coli* | 8.64 mg/g DCW | N.D. | [22] |
| *E. coli* | 1.4 mg/g DCW | 95 | [4] |
| *E. coli* | 1.99 mg/g DCW | 90 | [14] |
| *E. coli* | 7.4 ± 0.3 mg/g DCW | 96.6 | This study |
| *S. cerevisiae* | 4.7 mg/g DCW | N.D. | [5] |
| *S. cerevisiae* | 8.10 mg/g DCW | N.D. | [6] |
| *C. glutamicum* | 0.4 mg/L/h | N.D. | [7] |

* N.D. = not determined.

## 3. Materials and Methods

### 3.1. Strains, Plasmids and Primers

Strains, plasmids and primers used in this study were listed in Table 3.

**Table 3.** Strains and plasmids used in this study.

| Name | Description | Reference/Sources |
|---|---|---|
| | Strain | |
| *E. coli* BETA-1 | β-Carotene producing strain | [18] |
| *E. coli* ZEAX | Zeaxanthin producing strain, one copy of *Pantoea ananatis crtZ* under the control of the P37 promoter was integrated into *E. coli* BETA-1 chromosome | [19] |
| *E. coli* ZEAX-4 | Zeaxanthin producing strain, two copies of *P. ananatis crtZ* under the control of the P37 promoter was integrated into *E. coli* BETA-1 chromosome | This study |
| *E. coli* ASTA | Astaxanthin producer, one of *B.* sp. SD212 *crtW* under the control of the P37 promoter was integrated into *E. coli* ZEAX-4 chromosome | This study |
| *E. coli* ASTA-1 | Astaxanthin producer, another copy of *P. ananatis crtY* under the control of the P37 promoter was integrated into *E. coli* ASTA-1 chromosome | This study |
| | Plasmid | |
| pZSABP | Constitute expression vector, pSC101 *ori*, P37 promoter, $Amp^r$, BglBrick, ePathBrick containing four isocaudamer (*Avr*II, *Nhe*I, *Spe*I and *Xba*I) | [18] |
| pBAD33 | Expression vector, $P_{BAD}$, p15A *ori*, $Cm^r$ | [24] |
| pZS-$crtW_{Bsp}$ | pZSABP containing *Brevundimonas* sp. SD212 *crtW* under the control of the P37 promoter | This study |
| pZS-$crtW_{Psp}$ | pZSABP containing *Paracoccus* sp. N81106 $crtW^{L175W}$ under the control of the P37 promoter | This study |

**Table 3.** *Cont.*

| Name | Description | Reference/Sources |
|---|---|---|
| pZS-*crtW*$_{Ssp}$ | pZSABP containing *Sphingomonas* sp. DC18 *crtW*$^{F213L/R203W}$ under the control of the P37 promoter | This study |
| pZS-*bkt* | pZSABP containing *Chlamydomonas reinhardtii bkt* under the control of the P37 promoter | This study |
| pZS-2*crtW*$_{Bsp}$ | pZSABP containing two copies of *B.s* sp. SD212 *crtW* under the control of the P37 promoter | This study |
| pZS-2*bkt* | pZSABP containing two copies of *C. reinhardtii bkt* under the control of the P37 promoter | This study |
| pZS-2*crtW*$_{Psp}$ | pZSABP containing two copies of *P.* sp. N81106 *crtW*$^{L175W}$ under the control of the P37 promoter | This study |
| pZS-2*crtW*$_{Ssp}$ | pZSABP containing two copies of *S.* sp. DC18 *crtW*$^{F213L/R203W}$ under the control of the P37 promoter | This study |
| pZS-*crtW*$_{Bsp}$*crtW*$_{Psp}$ | pZSABP containing *B.s* sp. SD212 *crtW* under the control of the P37 promoter and *P.* sp. N81106 *crtW*$^{L175W}$ under the control of the P37 promoter | This study |
| pZS-*crtW*$_{Bsp}$*crtW*$_{Ssp}$ | pZSABP containing *B.s* sp. SD212 *crtW* under the control of the P37 promoter and *S.* sp. DC18 *crtW*$^{F213L/R203W}$ under the control of the P37 promoter | This study |
| pZS-*crtW*$_{Bsp}$*bkt* | pZSABP containing *B.s* sp. SD212 *crtW* under the control of the P37 promoter and *C. reinhardtii bkt* under the control of the P37 promoter | This study |
| pZS-*crtW*$_{Psp}$*crtW*$_{Ssp}$ | pZSABP containing *P.* sp. N81106 *crtW*$^{L175W}$ under the control of the P37 promoter and *S.* sp. DC18 *crtW*$^{F213L/R203W}$ under the control of the P37 promoter | This study |
| pZS-*crtW*$_{Psp}$*bkt* | pZSABP containing *P.* sp. N81106 *crtW*$^{L175W}$ under the control of the P37 promoter and *C. reinhardtii bkt* under the control of the P37 promoter | This study |
| pZS-*crtW*$_{Ssp}$*bkt* | pZSABP containing *S.* sp. DC18 *crtW*$^{F213L/R203W}$ under the control of the P37 promoter and *C. reinhardtii bkt* under the control of the P37 promoter | This study |
| pBAD-*crtZ* | pBAD33 containing *P. ananatis crtZ* | This study |
| pBAD-*crtW*$_{Bsp}$ | pBAD33 containing *B.s* sp. SD212 *crtW* | This study |

### 3.2. Genetic Methods

After codon optimization for *E. coli* codon usage by using the 31C method reported by Boël et al. [25], *B.* sp. SD212 *crtW*, *S.* sp. DC18 *crtW*$^{F213L/R203W}$, *P.* sp. N81106 *crtW*$^{L175W}$ and *C. reinhardtii bkt* genes were synthesized by Suzhou GENEWIZ, Inc. (Suzhou, China) and ligated into pUC57. The gene fragment was then digested and inserted into the *Nhe*I/*Kpn*I sites of pZSBP [18] to obtain pZS-*crtW*$_{BSP}$, pZS-*crtW*$_{SSP}$, pZS-*crtW*$_{PSP}$ and pZS-*bkt*, respectively. The BglBrick standard assembling method was used to assemble the above any two genes into a plasmid.

### 3.3. Astaxanthin Production in Shake Flasks

A single colony was inoculated into 5 mL of Luria-Bertani (LB) medium supplemented with 5 g/L KAc in a falcon tube which was incubated overnight at 37 °C. The overnight seed culture was then inoculated into 50 mL of the Super Broth with ammonium and sucrose (SBMSN) medium with an initial $OD_{600}$ of 0.1. The SBMSN medium (pH 7.0) contained 5 g/L sucrose, 12 g/L peptone, 24 g/L yeast extract, 1.7 g/L $KH_2PO_4$, 11.42 g/L $K_2HPO_4$, 1 g/L $MgCl_2 \cdot 6H_2O$, 1.42 g/L ammonium oxalate, and 2 g/L Tween-80. The cultures were incubated at 37 °C for 48 h in a rotary shaking incubator set to 150 rpm. Cell growth was measured according to the $OD_{600}$ and converted into DCW (g/L) using a standard curve.

### 3.4. Extraction and Quantification of Carotenoids

Cells were extracted with acetone to isolate carotenoids as described previously [9]. *E. coli* cultures (250 μL) were harvested by centrifugation at 12,000 rpm for 5 min. The cell pellet was washed with water and extracted with 1 mL of acetone at 55 °C for 15 min with intermittent vortexing. The acetone supernatant after centrifugation was transferred to a new tube. Carotenoids were analyzed by HPLC

(Shimadzu HPLC system, Model LC-20A, Shimadzu, Japan) using an Inertsil ODS-SP column (5 μm, 4.6 × 150 mm, GL Sciences Inc., Tokyo, Japan). The mobile phase of acetonitrile-methanol (65:35 $v/v$) at a flow rate of 1 mL/min was used. The absorbance of carotenoids at 477 nm was detected using a photodiode array detector (SPD-M20A). Carotenoid compounds were identified on the basis of their retention times relative to standard compounds (Sigma-Aldrich, St. Louis, MO, USA). Astaxanthin was quantified by comparing the integrated peak areas with that of authentic standards. The contents of total carotenoids were approximated via application of the astaxanthin curve.

### 3.5. Statistical Analysis

All experiments were performed in triplicate, and the data are presented as the mean of the three experiments ± standard deviation. Tukey's test was carried out for the statistical analysis using the OriginPro (version 7.5) package. Statistical significance was defined as $p < 0.05$.

**Acknowledgments:** We are grateful to the National Natural Science Foundation of China (Grant No. 21276289, J1310025), the Natural Science Foundation of Guangdong Province (No. 2015A030311036), the Project of the Scientific and Technical Program of Guangdong Province (No. 2015A010107004) and the Project of the Scientific and Technical Program of Guangzhou (No. 201607010028) for their financial support.

**Author Contributions:** Q.L. and Y.-F.B. performed the experiments. They contributed equally to this work. J.-Z.L. directed the project and wrote the paper.

**Conflicts of Interest:** The authors declare no conflicts of interest.

## Abbreviations

| | |
|---|---|
| *B.* sp. | *Brevundimonas* sp. |
| *S.* sp. | *Sphingomonas* sp. |
| *P.* sp. | *Paracoccus* sp. |
| *P. ananatis* | *Pantoea ananatis* |
| *H. pluvialis* | *Haematococcus pluvialis* |
| *P. agglomerans* | *Pantoea agglomerans* |
| *C. reinhardtii* | *Chlamydomonas reinhardtii* |
| *A.* sp. | *Anabaena* sp. |
| *N. punctiforme* | *Nostoc punctiforme* |
| MEV | mevalonate |

## References


1. Hussein, G.; Sankawa, U.; Goto, H.; Matsumoto, K.; Watanabe, H. Astaxanthin, a carotenoid with potential in human health and nutrition. *J. Nat. Prod.* **2006**, *69*, 443–449. [CrossRef] [PubMed]
2. Lee, P.C.; Schmidt-Dannert, C. Metabolic engineering towards biotechnological production of carotenoids in microorganisms. *Appl. Microbiol. Biotechnol.* **2002**, *60*, 1–11. [PubMed]
3. Ye, V.M.; Bhatia, S.K. Pathway engineering strategies for production of beneficial carotenoids in microbial hosts. *Biotechnol. Lett.* **2012**, *34*, 1405–1414. [CrossRef] [PubMed]
4. Lemuth, K.; Steuer, K.; Albermann, C. Engineering of a plasmid-free *Escherichia coli* strain for improved in vivo biosynthesis of astaxanthin. *Microb. Cell Fact.* **2011**, *10*, 29. [CrossRef] [PubMed]
5. Zhou, P.P.; Ye, L.D.; Xie, W.P.; Lv, X.M.; Yu, H.W. Highly efficient biosynthesis of astaxanthin in *Saccharomyces cerevisiae* by integration and tuning of algal *crtZ* and *bkt*. *Appl. Microbiol. Biotechnol.* **2015**, *99*, 8419–8428. [CrossRef] [PubMed]
6. Zhou, P.P.; Xie, W.P.; Li, A.P.; Wang, F.; Yao, Z.; Bian, Q.; Zhu, Y.Q.; Yu, H.W.; Ye, L.D. Alleviation of metabolic bottleneck by combinatorial engineering enhanced astaxanthin synthesis in *Saccharomyces cerevisiae*. *Enzyme Microb. Technol.* **2017**, *100*, 28–36. [CrossRef] [PubMed]
7. Henke, N.A.; Heider, S.A.E.; Peters-Wendisch, P.; Wendisch, V.F. Production of the marine carotenoid astaxanthin by metabolically engineered *Corynebacterium glutamicum*. *Mar. Drugs* **2016**, *14*, 124. [CrossRef] [PubMed]

8. Tao, L.A.; Wilczek, J.; Odom, J.M.; Cheng, Q.O. Engineering a β-carotene ketolase for astaxanthin production. *Metab. Eng.* **2006**, *8*, 523–531. [CrossRef] [PubMed]
9. Martin, J.F.; Gudina, E.; Barredo, J.L. Conversion of β-carotene into astaxanthin: Two separate enzymes or a bifunctional hydroxylase-ketolase protein? *Microb. Cell Fact.* **2008**, *7*, 3. [CrossRef] [PubMed]
10. Misawa, N. Carotenoid beta-ring hydroxylase and ketolase from marine bacteria-promiscuous enzymes for synthesizing functional xanthophylls. *Mar. Drugs* **2011**, *9*, 757–771. [CrossRef] [PubMed]
11. Choi, S.K.; Nishida, Y.; Matsuda, S.; Adachi, K.; Kasai, H.; Peng, X.; Komemushi, S.; Miki, W.; Misawa, N. Characterization of beta-carotene ketolases, CrtW, from marine bacteria by complementation analysis in *Escherichia coli*. *Mar. Biotechnol.* **2005**, *7*, 515–522. [CrossRef] [PubMed]
12. Choi, S.K.; Harada, H.; Matsuda, S.; Misawa, N. Characterization of two beta-carotene ketolases, CrtO and CrtW, by complementation analysis in *Escherichia coli*. *Appl. Microbiol. Biotechnol.* **2007**, *75*, 1335–1341. [CrossRef] [PubMed]
13. Makino, T.; Harada, H.; Ikenaga, H.; Matsuda, S.; Takaichi, S.; Shindo, K.; Sandmann, G.; Ogata, T.; Misawa, N. Characterization of cyanobacterial carotenoid ketolase CrtW and hydroxylase CrtR by complementation analysis in *Escherichia coli*. *Plant Cell Physiol.* **2008**, *49*, 1867–1878. [CrossRef] [PubMed]
14. Scaife, M.A.; Burja, A.M.; Wright, P.C. Characterization of cyanobacterial β-carotene ketolase and hydroxylase Genes in *Escherichia coli*, and their application for astaxanthin biosynthesis. *Biotechnol. Bioeng.* **2009**, *103*, 944–955. [CrossRef] [PubMed]
15. Scaife, M.A.; Ma, C.A.; Ninlayarn, T.; Wright, P.C.; Armenta, R.E. Comparative analysis of beta-carotene hydroxylase genes for astaxanthin biosynthesis. *J. Nat. Prod.* **2012**, *75*, 1117–1124. [CrossRef] [PubMed]
16. Ye, R.W.; Stead, K.J.; Yao, H.; He, H.X. Mutational and functional analysis of the beta-carotene ketolase involved in the production of canthaxanthin and astaxanthin. *Appl. Environ. Microb.* **2006**, *72*, 5829–5837. [CrossRef] [PubMed]
17. Zhong, Y.J.; Huang, J.C.; Liu, J.; Li, Y.; Jiang, Y.; Xu, Z.F.; Sandmann, G.; Chen, F. Functional characterization of various algal carotenoid ketolases reveals that ketolating zeaxanthin efficiently is essential for high production of astaxanthin in transgenic *Arabidopsis*. *J. Exp. Bot.* **2011**, *62*, 3659–3669. [CrossRef] [PubMed]
18. Li, X.R.; Tian, G.Q.; Shen, H.J.; Liu, J.Z. Metabolic engineering of *Escherichia coli* to produce zeaxanthin. *J. Ind. Microbiol. Biotechnol.* **2015**, *42*, 627–636. [CrossRef] [PubMed]
19. Shen, H.J.; Cheng, B.Y.; Zhang, Y.M.; Tang, L.; Li, Z.; Bu, Y.F.; Li, X.R.; Tian, G.Q.; Liu, J.Z. Dynamic control of the mevalonate pathway expression for improved zeaxanthin production in *Escherichia coli* and comparative proteome analysis. *Metab. Eng.* **2016**, *38*, 180–190. [CrossRef] [PubMed]
20. Fraser, P.D.; Miura, Y.; Misawa, N. In vitro characterization of astaxanthin biosynthetic enzymes. *J. Biol. Chem.* **1997**, *272*, 6128–6135. [CrossRef] [PubMed]
21. Fraser, P.D.; Shimada, H.; Misawa, N. Enzymic confirmation of reactions involved in routes to astaxanthin formation, elucidated using a direct substrate in vitro assay. *Eur. J. Biochem.* **1998**, *252*, 229–236. [CrossRef] [PubMed]
22. Ma, T.; Zhou, Y.J.; Li, X.W.; Zhu, F.Y.; Cheng, Y.B.; Liu, Y.; Deng, Z.X.; Liu, T.G. Genome mining of astaxanthin biosynthetic genes from *Sphingomonas* sp. ATCC 55669 for heterologous overproduction in *Escherichia coli*. *Biotechnol. J.* **2016**, *11*, 228–237. [CrossRef] [PubMed]
23. Zelcbuch, L.; Antonovsky, N.; Bar-Even, A.; Levin-Karp, A.; Barenholz, U.; Dayagi, M.; Liebermeister, W.; Flamholz, A.; Noor, E.; Amram, S.; et al. Spanning high-dimensional expression space using ribosome-binding site combinatorics. *Nucleic Acids Res.* **2013**, *41*. [CrossRef] [PubMed]
24. Guzman, L.M.; Belin, D.; Carson, M.J.; Beckwith, J. Tight regulation, modulation, and high-level expression by vectors containing the arabinose PBAD promoter. *J. Bacteriol.* **1995**, *177*, 4121–4130. [CrossRef] [PubMed]
25. Boel, G.; Letso, R.; Neely, H.; Price, W.N.; Wong, K.H.; Su, M.; Luff, J.D.; Valecha, M.; Everett, J.K.; Acton, T.B.; et al. Codon influence on protein expression in *E. coli* correlates with mRNA levels. *Nature* **2016**, *529*, 358–363. [CrossRef] [PubMed]

marine drugs

MDPI

*Article*

# Astaxanthin Prevents Human Papillomavirus L1 Protein Binding in Human Sperm Membranes

**Gabriella Donà [1], Alessandra Andrisani [2], Elena Tibaldi [1], Anna Maria Brunati [1], Chiara Sabbadin [3], Decio Armanini [3], Guido Ambrosini [2], Eugenio Ragazzi [4] and Luciana Bordin [1,*]**

1 Department of Molecular Medicine-Biological Chemistry, University of Padova, 35131 Padova, Italy; gabriella.dona@unipd.it (G.D.); elena.tibaldi@unipd.it (E.T.); annamaria.brunati@unipd.it (A.M.B.)
2 Department of Women's and Chilren's Health, University of Padova, 35131 Padova, Italy; alessandra.andrisani@unipd.it (A.A.); guido.ambrosini@unipd.it (G.A.)
3 Department of Medicine-Endocrinology, University of Padova, 35131 Padova, Italy; ChiaraSabbadin@libero.it (C.S.); decio.armanini@unipd.it (D.A.)
4 Department of Pharmaceutical and Pharmacological Sciences, University of Padova, 35131 Padova, Italy; eugenio.ragazzi@unipd.it
* Correspondence: luciana.bordin@unipd.it; Tel.: +39-049-827-6113; Fax: +39-049-807-3310

Received: 15 September 2018; Accepted: 1 November 2018; Published: 2 November 2018

**Abstract:** Astaxanthin (Asta), red pigment of the carotenoid family, is known for its anti-oxidant, anti-cancer, anti-diabetic, and anti-inflammatory properties. In this study, we evaluated the effects of Asta on isolated human sperm in the presence of human papillomavirus (HPV) 16 capsid protein, L1. Sperm, purified by gradient separation, were treated with HPV16-L1 in both a dose and time-dependent manner in the absence or presence of 30 min-Asta pre-incubation. Effects of HPV16-L1 alone after Asta pre-incubation were evaluated by rafts (CTB) and Lyn dislocation, Tyr-phosphorylation (Tyr-P) of the head, percentages of acrosome-reacted cells (ARC) and endogenous reactive oxygen species (ROS) generation. Sperm membranes were also analyzed for the HPV16-L1 content. Results show that HPV16-L1 drastically reduced membrane rearrangement with percentage of sperm showing head CTB and Lyn displacement decreasing from 72% to 15.8%, and from 63.1% to 13.9%, respectively. Accordingly, both Tyr-P of the head and ARC decreased from 68.4% to 10.2%, and from 65.7% to 14.6%, respectively. Asta pre-incubation prevented this drop and restored values of the percentage of ARC up to 40.8%. No alteration was found in either the ROS generation curve or sperm motility. In conclusion, Asta is able to preserve sperm by reducing the amount of HPV16-L1 bound onto membranes.

**Keywords:** human papillomavirus 16 (HPV16); astaxanthin (Asta); acrosome reaction; cholera toxin subunit B (CTB); L1 protein

---

## 1. Introduction

HPV (Human Papilloma Virus) is responsible for the 5.2% (3% in women and 2% in men) of cancers in the world, with prevalence in cervical, ano-genital, head, and neck cancers [1]. Oncogenic strains of HPV DNA were demonstrated in almost all cervical malignancies in women [2], and, on a global scale, also in 5% of men [3], in the form of penile cancer (45%) with the HPV 16 member mostly responsible (60%) followed by the HPV 18 accounting for 13% of the cases [4].

Besides sexual inter-infections between sexual partners leading to the wider virus propagation, it has been recently shown that HPV DNA can cause a detrimental effect on early embryo development and clinical reproductive outcomes, since HPV DNA can be harbored inside the blastocyst stage by spermatozoa carrying HPV virions, which represent the viral DNA included inside the L1–L2 capsid [5–7].

HPV infection is considered a cause of male infertility or subfertility [8], even if the mechanism of the reduced sperm motility and DNA degradation is still debated [9]. HPV is a non-enveloped double stranded DNA virus with a genome of 8kb pairs encoding two protein types: (i) the "Late proteins" L1 and L2 which are the structural components of viral capsid and are involved in the packaging of the virus; and (ii) the "Early proteins" E1,2,4,5,6,7 which regulate the replication of viral DNA. The early proteins are expressed throughout the life cycle of the virus, whereas the late proteins are expressed only during the initial stages of infection [10]. During the uncoating (process needed for the releasing of the viral genomes into the host nuclei), the protective capsids undergo sequential structural changes. HPV16 L1 binds primarily to heparan sulfate proteoglycans (HSPGs) on the host cell, in particular with glycosaminoglycan (GAG) chains, Upon HSPG binding, capsid goes through conformational changes that are required for a secondary binding event which, in turn, triggers the uptake of the virus through a still unknown receptor [10]. One of these conformational changes leads to the exposition of the L2 protein on the capsid surface followed by a furin convertase dependent cleavage of the L2 N-terminus. Only upon L2 cleavage can the virus capsids be transferred from the membrane to a secondary receptor on the cell surface, and, hence, be endocytosed via receptors [10].

L1, the main constituent of the capsid envelope, spontaneously self-assembles into virus-like particles (VLPs), that are an icosahedral structure composed of 72 pentamers of L1 and an unknown number of the L2 minor coat proteins. Due to its ability of self-assembling in VLPs also in the absence of the L2 protein, L1 has been considered for vaccine development against HPV infections, since the L1-based VLPs share the same immunogen properties of native HPV, but lack genome and other proteins [10,11].

Astaxanthin (Asta) is mainly produced by the microalgae *Haematococcus pluvialis* in the presence of stressing conditions including, deficiency of nitrogen, high salinity, and high temperature [12].

For its molecular structure, Asta belongs to the carotenoid family, with an extended nonpolar zone in the middle, which is made up of a series of carbon–carbon double bond termed "conjugated atoms" and two polar regions at either ends. This nonpolar–polar structure allows Asta to fit precisely into the polar–nonpolar-polar area of the cell membrane [13,14] with a suitable capacity for neutralizing free radicals, which is 65 times more powerful than vitamin C, 54 times stronger than β-carotene, and 100 times more effective than α-tocopherol [14,15]. Asta has a wide range of applications in the food, feed, cosmetic, aquaculture, nutraceutical, and pharmaceutical industries. In the last decades, Asta was described to have anti-inflammatory and pain relieving activity, due to its ability of blocking different biochemical factors involved in pain [14]. More specifically, Asta inhibits cyclooxygenase 2 (COX2) enzyme activities, which are related to many diseases, such as osteoarthritis, rheumatoid arthritis, dysmenorrhea, and acute pain [14]. Asta and *H. pluvialis* extracts prevents the development of human colon cancer cells by blunting the progression of the cell cycle, ameliorating apoptosis, and suppressing the expression of inflammatory cytokines (e.g., NF-kβ, TNF-a and IL-1β) [15–18]. In HCT116 and HT29 cells, Asta induced the expression of the negative regulators of the cell cycle [19].

Recent evidence has demonstrated that Asta treatment improves human sperm capacitation by inducing the membrane rafts relocation [20,21]. During the process of capacitation sperm undergoes a series of transformations, including reactive oxygen species (ROS) production, membrane micro-domains rafts translocation, Lyn displacement and activation, and sperm head Tyr-phosphorylation (Tyr-P) [20–24]. This transformation is needed to let sperm undergo the acrosome reaction, with the release of lytic enzymes responsible for sperm-oocyte fusion occurring.

The aim of this study was to evaluate the effect of HPV capsid protein L1 on human sperm capacitation, and the effect of Asta in L1 binding to plasma sperm membranes.

## 2. Results and Discussion

### 2.1. L1 Treatment: Effect on Sperm L1 Location

Aliquots of sperm, isolated and pooled as described in Methods, were analyzed for L1 protein binding and distribution with immunocytochemistry, immediately ($T_0$), or after incubation in capacitating conditions, in the absence (C), or presence of 1 (L1 1), 10 (L1 10) or 13 (L1 13) µg/mL of L1 protein (Figure 1a). The expected absence of anti-L1 fluorescence in $T_0$ and C confirmed that samples were free from any HPV pre-existing infection. Conversely, L1-capacitated sperm fluorescence indicated that L1 bound to cells in a concentration-dependent way, with a percentage of marked cells at 1, 10 and 13 µg/mL reaching 55.3% ± 3.6%, 98.1% ± 1.6% and 99.1% ± 0.7% of cells compared to control 0%, respectively ($p < 0.0001$) (Figure 1b). Due to its ability to bind to all sperm, a concentration of 10 µg/mL was chosen as suitable concentration to further investigate the time-dependent effect.

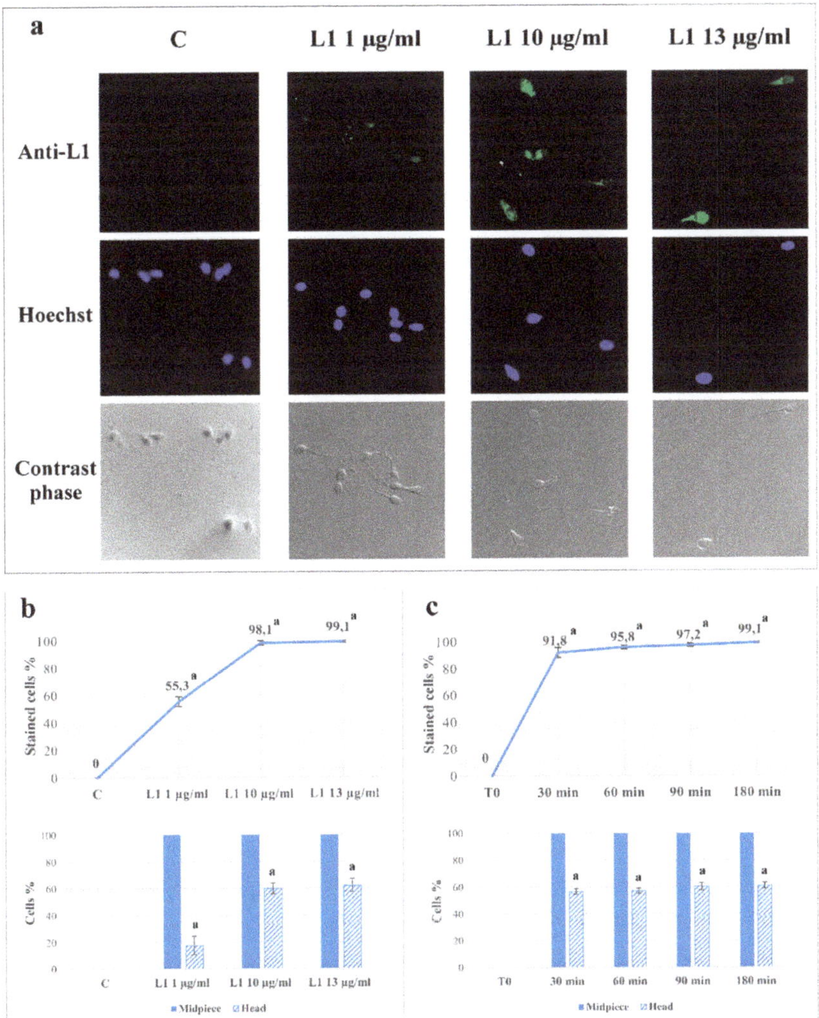


Figure 1. Before ($T_0$) or after ($T_1$) 180 min of incubation in capacitating conditions in absence (C) or presence of L1 (1, 10 or 13 µg/mL), sperm were analyzed for L1 presence by immunofluorescence

cytochemistry as described in Methods. Similarly, time-dependent evaluation was carried out with L1 (10 μg/mL) for 30, 60, 90, 180 min. (**a**) Immunofluorescence analysis was performed with anti-L1 antibody (green) and Hoechst (blue) was used to visualize nuclei. Corresponding phase-contrast images for each condition were shown. (**b**) Dose dependent evaluation of protein L1 localization in C, L1 1 μg/mL, L1 10 μg/mL, and L1 13 μg/mL samples. The graph in the upper side shows the percentage of cell marked by L1 protein, whereas the histogram at the bottom shows L1 distribution between the mid-piece (filled blue) and the head (striped). (**c**) Time dependent evaluation of protein L1 localization in sample L1 10 μg/mL at different incubation times (30, 60, 90, 180 min). The number of sperm stained with anti-L1 antibody in any part of the cells (stained) is expressed as % of the total number of cells analyzed, whereas the number of cells showing L1 in the head or mid-piece is expressed as means ± SD% of marked cells. The figure is representative of 12 separate experiments conducted in triplicate ([a]: $p < 0.0001$ comparing each sample against C values, by using Dunnett's test, following a significant one-way ANOVA).

Experiments with L1 10 μg/mL were carried out for 30, 60, 90, 180 min of incubation and sperm fluorescence evaluated. Interestingly, already after 30 min of L1 incubation, 91.8% ± 3.7% of sperm presented L1 related fluorescence ($p < 0.0001$ compared to control), which was progressively more marked by prolonging incubation time up to 3 h.

A qualitative analysis was also performed to evaluate the main site of protein binding (bottom panels of b and c). In the dose-dependent experiments, almost 100% of the marked cells showed fluorescence at the mid-piece, eligible as the principal binding point right after the first 30 min of incubation with 10 μg/mL, HPV L1 (Figure 1c). Interestingly, with 1 μg/mL only 20% of the marked cells showed L1 binding also in the head, whereas the maximal effect was observed with 10 μg/mL (60%). The L1 distribution between mid-piece and head did not change with incubation time (Figure 1, panels b and c).

### 2.2. L1 Treatment: Effect on Sperm Motility

Aliquots were then assessed for motility by computer-assisted sperm analysis (CASA) (Table 1). Samples were incubated in the presence of increasing L1 concentration (1, 10, and 13 μg/mL as representing the concentrations able to infect 100% of cells), after a pre-incubation of 30 min in the presence or absence of 2 μM Asta [20,21]. The increase of values of all parameters (except motility) is evident with all treatments compared to $T_0$, meaning that capacitation had occurred leading to hyper-motility. However, when values from treated cells were compared to C (capacitated without Asta and/or L1), there was no significant difference, also for samples pre-incubated alone or with Asta and L1.

**Table 1.** Sperm motility and kinematic parameters observed in different samples. Motility and kinematic parameters of spermatozoon were evaluated with computer-assisted sperm analysis (CASA) at $T_0$ (before starting incubation) and after 180 min of incubation in capacitating conditions in absence (C) or presence of L1, Asta or both. Motility = progressive and non-progressive motility (%); VSL = straight-line velocity (μm/s); VAP = average path velocity (μm/s); ALH = amplitude of lateral head displacement (μm).

| | **Motility (%)** | **VSL (μm/s)** | **VAP (μm/s)** | **ALH (μm)** |
|---|---|---|---|---|
| $T_0$ | 68 ± 9 | 58.4 ± 8.9 | 54.0 ± 6.7 | 3.1 ± 0.5 |
| C | 75 ± 11 | 77.8 ± 13.9 [a] | 67.6 ± 11.0 [a] | 4.9 ± 0.9 [a] |
| L1 1 μg/mL | 74 ± 10 | 76.9 ± 9.4 [a] | 67.3 ± 9.0 [a] | 4.7 ± 0.6 [a] |
| L1 10 μg/mL | 73 ± 8 | 76.3 ± 8.8 [a] | 67.5 ± 7.2 [a] | 4.8 ± 0.7 [a] |
| L1 13 μg/mL | 73 ± 8 | 76.4 ± 8.9 [a] | 67.4 ± 8.1 [a] | 4.8 ± 0.6 [a] |
| Asta | 75 ± 9 | 78.2 ± 13.6 [a] | 67.9 ± 9.3 [a] | 5.0 ± 0.6 [a] |
| Asta+L1 10 | 74 ± 6 | 77.7 ± 11.3 [a] | 67.0 ± 7.5 [a] | 4.9 ± 0.7 [a] |

[a]: $p < 0.001$, comparing each parameter under different treatment against $T_0$, by using Dunnett's test, following a significant one-way ANOVA; no significant difference was observed comparing each treatment against C. Values are expressed as the mean ± SD.

These results suggest that capacitated-related hyper-motility was not affected by Asta, by L1, or by both Asta and L1.

### 2.3. *Effect of L1 and Asta Alone or in Association on Sperm Capacitation*

We previously showed that capacitation of sperm is linked to membrane rearrangement with lipid rafts relocation to the apical part of the head [20]. This rearrangement allows the gathering of the src kinase family member, Lyn, which activates and increases the Tyr-phosphorylation of the sperm head proteins [21]. To evidence the effects of L1 on sperm capacitation parameters, aliquots of sperm treated with L1 (10 μg/mL), after a pre-incubation of 30 min in the presence or absence of Asta, as described above, were analyzed for their shifted rafts (CTB), Lyn location, Tyr-P, and acrosome reaction (ACR).

L1 incubation affected rafts relocation (Figure 2, CTB), with the sample showing only 20% of sperm with rafts relocated on the head (sample L1 compared to C). The corresponding quantification is in Table 2 (21.3% ± 4.2% compared to 72.0% ± 3.6% for L1 10 μg/mL and C, respectively, $p < 0.001$).

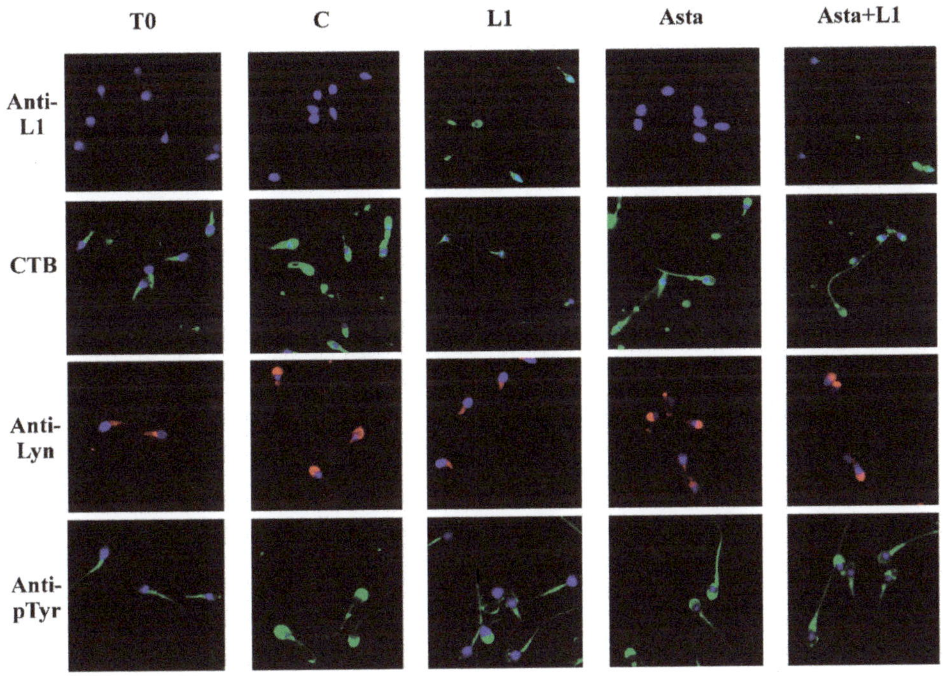


**Figure 2.** Localization of L1 protein, membrane rafts, Lyn and Tyr-P protein in human sperm during capacitation in absence or presence of L1 and/or Asta. Sperm cells, at $T_0$ or incubated in capacitating conditions for 180 min in absence (C) or presence of L1 10 μg/mL, Asta 2 μM or both, were analyzed for L1 localization (green), CTB labelling (green), Lyn (red) and Tyr-P (green) localization by immunofluorescence cytochemistry as described in Methods. Hoechst (blue) was used to visualize nuclei and images were merged. The figure is representative of 12 separate experiments.

**Table 2.** Sperm biochemical parameters observed in different samples. Sperm biochemical parameters were evaluated with fluorescence microscopy at $T_0$ (before starting incubation) and after 180 min of incubation in capacitating conditions in absence (C) or presence of L1 (L1 1, 10 or 13 µg/mL), Asta or both (Asta+L1 10 µg/mL) as described in Methods.

| | L1 (%) | | CTB (%) | Lyn (%) | Tyr-P (%) | ARC (%) | NVC (%) |
|---|---|---|---|---|---|---|---|
| | Tot. Stained | Head | Head | Head | Head | | |
| $T_0$ | ND | ND | 13.4 ± 1.7 | 9.2 ± 1.2 | 15.3 ± 2.1 | 8.1 ± 1.3 | 7.1 ± 1.2 |
| C | ND | ND | 72.0 ± 3.6 | 63.1 ± 2.5 | 68.4 ± 2.9 | 65.7 ± 4.2 | 10.1 ± 0.9 |
| L1 1 µg/mL | 55.3 ± 3.6 [a] | 17.6 ± 7.0 [a] | 25.5 ± 3.5 [a] | 20.5 ± 2.7 [a] | 19.7 ± 2.3 [a] | 20.9 ± 3.2 [a] | 10.3 ± 1.1 |
| L1 10 µg/mL | 98.1 ± 1.6 [a] | 59.9 ± 3.9 [a] | 21.3 ± 4.2 [a] | 16.3 ± 3.5 [a] | 11.3 ± 1.7 [a] | 19.4 ± 2.8 [a] | 10.6 ± 1.3 |
| L1 13 µg/mL | 99.1 ± 0.7 [a] | 62.1 ± 4.8 [a] | 15.8 ± 3.5 [a] | 13.9 ± 2.2 [a] | 10.2 ± 1.3 [a] | 14.6 ± 3.3 [a] | 11.4 ± 1.0 [b] |
| Asta | ND | ND | 72.9 ± 2.8 | 65.4 ± 2.8 | 71.3 ± 1.8 [b] | 67.9 ± 3.5 | 9.3 ± 1.5 |
| Asta+L1 10 | 45.5 ± 4.2 [a,c] | 43.8 ± 9.0 [a,c] | 50.2 ± 2.7 [a,c] | 43.2 ± 3.0 [a,c] | 43.6 ± 3.9 [a,c] | 40.8 ± 2.6 [a,c] | 9.6 ± 1.0 |

[a]: $p < 0.001$, [b]: $p < 0.05$, comparing each sample against C by using Dunnett's test, following a significant one-way ANOVA; [c]: $p < 0.001$, comparing Asta+L1 10 sample to L1 10 µg/mL by using Dunnett's test, following a significant one-way ANOVA. ND: not detectable, and assumed as zero for calculation. Values are expressed as the mean ± SD.

The lack of translocation causes the consequent decrease of Lyn gathered on the apical part of the head (Figure 2, Anti-Lyn) (Table 2, 16.3% ± 3.5% compared to 63.1% ± 2.5% for L1 10 µg/mL and C, respectively, $p < 0.001$), followed by the net reduction of the percentage of cells showing Tyr-P of the head (Figure 2, Anti-P-Tyr) (Table 2).

In these conditions only a reduced percentage of cells underwent the acrosome reaction compared to the C sample (19.4% ± 2.8% of acrosome-reacted cells (ARC) in the L1 10 µg/mL sample compared to 65.7% ± 4.2% in C sample, $p < 0.001$) (Table 2).

When Asta was added to the capacitating medium, the values of capacitation only slightly improved, as previously shown [21], since sperm were already in optimal conditions and did not need membrane raft translocation (CTB), Lyn gathering, and Tyr-P. Asta also prevented L1 binding to the membranes of sperm (Figure 2 Anti-L1, compare L1 with Asta+L1 panels) with a net reduction of protein of about 52% (98.1% ± 1.6% of sperm infected by L1 compared to 45.5 ± 4.2% in the presence of Asta, $p < 0.001$, Table 2). In the same way, also Lyn relocation and the following Tyr-P of the head were partially restored. Our results show that Asta can recover about 50% of the L1-induced alterations, with cell showing rafts relocation switching from 21.3% ± 4.2% to 50.2% ± 2.7% ($p < 0.001$), Lyn relocation from 16.3% ± 3.5% to 43.2% ± 3.0% ($p < 0.001$) and Tyr-P of the head from 11.3% ± 1.7% to 43.6% ± 3.9% ($p < 0.001$) (Table 2, comparing line L1 10 µg/mL to Asta+L1 10). Consequently, also the L1-induced reduction of the ARC percentage was greatly restored by the presence of Asta (Table 2), shifting from 19.4% ± 2.8% to 40.8% ± 2.6% ($p < 0.001$), which, also if still far from the 65.7% ± 4.2% of the C sample, accounts for twice the value of the L1-treated sample (L1 10 µg/mL: 19.4% ± 2.8%).

### 2.4. Effect of Asta on L1 Binding to Sperm Membranes

Samples were also analyzed for the L1 protein binding to the subcellular fractions. After incubation in the absence or presence of Asta, L1, or both, aliquots from each sample were sonicated, sub-cellular fractions were separated on gradient and further centrifuged, as described in Methods. Membranes, cytosol, heads, and flagella were subjected to Western blotting and revealed with ant-L1 antibodies.

L1 was detected on membranes from L1-treated sperm, (Figure 3, lanes L1 and Asta+L1), but not in $T_0$, C and Asta samples did not show any response, as expected. The sample treated with Asta+L1 showed a reduction in the L1 content, with Asta reducing the amount of L1 by about 30%, compared to L1 sample (206 ± 17 vs. 145 ± 14 ng of L1 protein/30 × $10^6$ cells $p < 0.001$).

When analyzed in other compartments, such as heads and tails, L1 detection was practically negligible (Figure S1), thus indicating that the main site of L1 binding was on the membrane (Figure S2)

Compared to data in Figure 1 and Table 2, showing that Asta decreased the percentage of marked cells from 98.1% ± 1.6% to 45.5% ± 4.2% ($p < 0.001$) (Table 2), data from the Western blotting of membranes confirmed the net reduction of the amount of L1 in the presence of Asta. The decrease seems not to account for the higher reduction found with immunofluorescence. We must consider that, when evaluated with immunocytochemistry, we account for the number of cells presenting L1-related fluorescence, without considering the amount of the protein effectively linked. By observing the L1 distribution between mid-piece and head, it is reasonable that the binding of L1 starts from the mid-piece and gradually spreads over the whole cell, thus increasing the concentration of L1 for each cell.

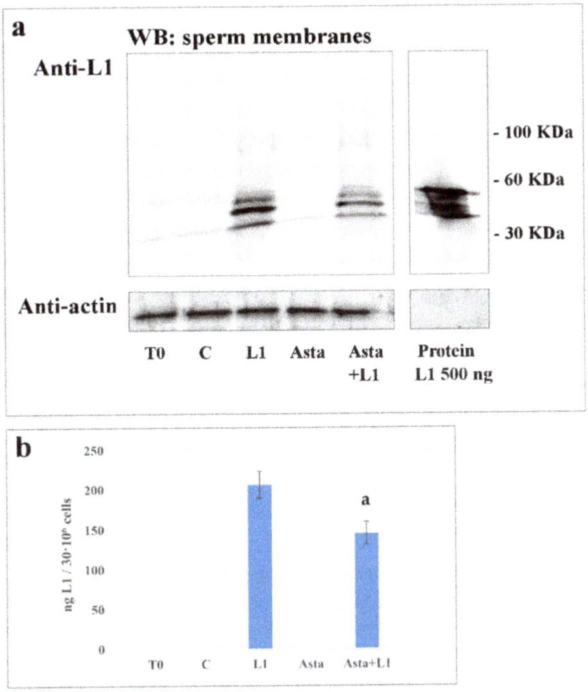


**Figure 3.** Detection of protein L1 in membrane of human sperm during capacitation in absence or presence of L1 and/or Asta. Western blot analysis (panel **a**) of plasma membrane (obtained as described in Methods) of sperm cells, at $T_0$ or incubated in capacitating conditions for 180 min in absence (C) or presence of L1 10 µg/µL, Asta 2 µM or Asta+ L1. Membranes of different samples were analyzed by SDS-PAGE, transferred to nitrocellulose and immuno-revealed with anti-L1 antibody and then with anti-β actin as loading control. Bands were densitometrically analyzed (panel **b**) and the amount of bound L1 was expressed as ng calculated by the ratio between samples and the bands of L1 protein (500 ng). a: $p < 0.001$ comparing Asta+L1 sample to L1. Values are expressed as the mean ± SD. The figure is representative of seven separate experiments conducted in triplicate. When analyzed in other compartments, L1 detection was practically negligible (Figure S1), thus indicating that the main site of L1 binding was on the membrane.

### 2.5. *Effect of Asta and L1 on ROS Production*

When the effects of the different treatments were assayed on the ROS generation curve (Figure 4), samples were incubated in the absence (C), or presence of Asta, or L1, or both (Asta+L1), and analyzed for ROS generation in a luminometer for 180 min with luminol as luminescent source.

All samples showed similar curves of ROS generation compared to the control (C), thus suggesting that none of the treatments affected $H_2O_2$ formation in sperm.

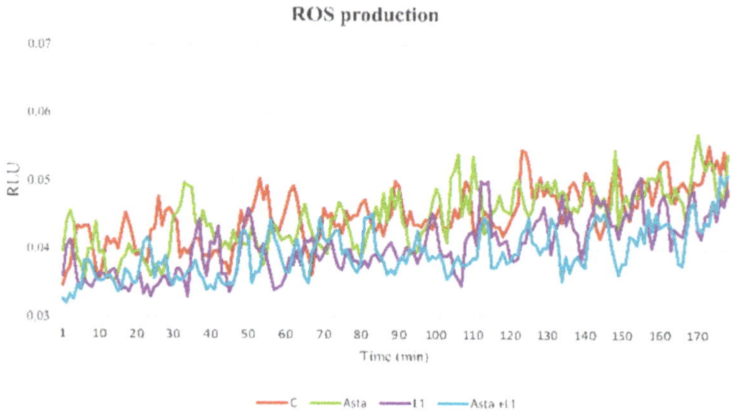


**Figure 4.** Reactive oxygen species (ROS) generation curves of sperm samples and effects of Asta, L1, or both. Sperm cells from 3 volunteers for each experiment were collected to form a pool with a sufficient number of cells. Sperm were incubated for up to 180 min in capacitating conditions in the absence (C) or presence of Asta (2 μM), L1 (10 μg/mL), or both (Asta+L1). Luminol chemiluminescence was monitored during sperm capacitation. Results are expressed as moving averages of Relative Luminescence Units (RLU)/30 sec for $2 \times 10^6$ cells. Figure is representative of $n = 7$ separate experiments. Detection was carried out in triplicate.

To initiate a successful infection, HPV must bind to heparan sulfate proteoglycan (HSPG) by exposing L1 from its capside structure and through L1 form an endocytic complex [25], that is also responsible for virus internalization [26].

It has been recently confirmed that also in human sperm, GAGs mediate the viral binding to the cell surface, in particular the GAG syndecan-1 (Synd-1) seems to interact with the viral capsid protein L1 of HPV16 [27,28]. Synd-1 belongs to the heparan sulfate GAGs (HSGAGs) family which regulates cell proliferation and cell-matrix and cell–cell adhesion by modulating the ligand-dependent activation of GAG receptors at the cell surface, interacting with components of the actin-based cytoskeleton with intracellular domain [29]. A direct interaction between the extracellular portion of Synd-1 and HPV-L1 is involved in L1 binding to sperm membrane, as confirmed by heparinase-mediated abolition of the co-localization of Synd-1 and HPV-1 [25,26].

In the present study, the L1 binding would act as an anchor complex, preventing further rafts translocation and relocation to the sperm head. This hypothesis is consistent with the results of Chen et al. [30] who demonstrated that Synd-1, clustered upon L1 ligand binding, induced recruitment and binding to cortactin, a protein involved in the membrane-cytoskeleton formation and modulation through actin–cortactin net interaction. From these considerations, we postulate that the probable binding site in human sperm membrane may be included in raft domains, thus being regulated and regulating, in turn, further L1 binding/inclusion and raft relocation. Asta, binding to plasma membranes, probably alters raft composition by sterically filling the loci dedicated to the interaction between external viral L1 and internal Synd.1. The possible mechanism underlying Asta-related induction of capacitation would rely on its ability to insert into the lipid bilayer, disengaging rafts from the blocks that maintain the membrane in non-capacitated form. Consistently, Asta would also release the protein links mediated by L1 and cytoskeleton proteins, both allowing/facilitating the correct engagement of proteins in rafts, such as in the case of actin polymerization [31], and the correct relocation of enzymes, such as Lyn, to the acrosome region.

## 3. Experimental Section

### 3.1. Chemicals

Recombinant HPV16 L1 protein was purchased from Abcam (Cambridge, UK). Anti- HPV16 L1 mouse monoclonal, goat anti-mouse and anti-rabbit IgG- fluorochrome fluorescein isothiocyanate (FITC) conjugate antibodies were purchased from Santa Cruz Biotechnology (Heidelberg, Germany). Anti-P-Tyr mouse monoclonal and anti-Lyn rabbit polyclonal antibody were obtained by Upstate (Becton Dickinson Italia SpA, Milan, Italy) and Millipore (Temecula, CA, USA), respectively. Density gradient (Pure Sperm 40/80) and pure sperm wash buffer (PSW) were obtained from Nidacon International AB (Göteborg, Sweden). Asta was supplied by FERpharma s.r.l. (Milan, Italy). 12-myristate-13-acetate phorbol ester (PMA) was purchased from Calbiochem (Nottingham, UK) and all other reagents from Sigma-Aldrich (Milan, Italy).

### 3.2. Semen Collection and Analysis

Forty healthy male donors (age range: 25–43 years, average age: 34.6 years) were enrolled at the Centre of Assisted Reproduction-U.O.C. Obstetrics and Gynecology Clinic–Padua, Italy. After 3 days of abstinence, semen samples were collected by masturbation in a sterile container and then assessed for sperm parameters. All sperm samples used in this study were normal in terms of sperm count, motility, morphology, volume, and pH, according to the World Health Organization criteria [32]. All samples presenting any kind of contamination were discarded. This study was approved by the Ethics Committee for Research and Clinical Trials of our University (Code: 4400-AO18), and all recruited donors gave their informed written consent and provided detailed lifestyle histories.

### 3.3. Sample Preparation

After semen analysis, samples were laid on a discontinuous gradient (Pure Sperm 40%/80%) and centrifuged at 500× *g* for 30 min at room temperature. The seminal plasma and sperm from the 40% gradient interface were discarded, and the sperm cells from the bottom pellet (80% gradient) were gathered. After gradient separation, sperm samples were washed with PSW and collected three by three in a single pool (stock sample) to obtain a sufficient number of cells to perform all tests. Stock samples (concentration adjusted to $80 \times 10^6$ sperm cells/mL in PSW) were divided in aliquots, analyzed immediately ($T_0$), or incubated for up to 180 min in capacitating conditions, in the absence (C) or presence of Asta 2 μM (Asta) from stock solutions of 100 mM dissolved in dimethyl sulphoxide (DMSO), L1 1, 10 or 13 μg/mL (L1), or together with 30 min-Asta pre-incubation (Asta+L1 10).

### 3.4. Computer Assisted Sperm Analysis (CASA)

Sperm motility and hyperactivation were analyzed using a computer-assisted sperm analyzer (CASA, Ab scientific, London, UK). For each sample, the following parameters were evaluated: the percentage of motile spermatozoa and VAP (average path velocity), VSL (straight-line velocity), and ALH (amplitude of lateral head displacement) to determine the percentage of hyper-activated (HA) cells [33]. All measurements were performed at 37 °C. A minimum of 100 cells and 5 fields were analyzed for each aliquot.

### 3.5. Anti-L1, Anti-P-Tyr and Anti-Lyn Evaluations with Confocal Microscopy

Aliquots of sperm ($15 \times 10^6$ cells) from each sample were accurately washed with phosphate buffer saline (PBS) containing vanadate 1 mM and protease inhibitor cocktail, fixed with 2% (*w*/*v*) paraformaldehyde and incubated overnight at 4 °C on slides pre-coated with poly-L-lysine [20–22]. Slides were rinsed twice with PBS and sperm cells were permeabilized with 0.2% (*v*/*v*) Triton X-100 for 15 min at 4 °C and then incubated with anti-L1, anti-P-Tyr, or anti-Lyn antibodies for 1 h at 37 °C in a humid chamber. Slides were washed with PBS, stained with anti-mouse or anti-rabbit IgG-FITC conjugate for 1 h at 37 °C in a humid chamber and then rinsed with PBS and mounted. Staining without primary antibody was used as negative control. Fluorescence was detected with the UltraView LCI confocal system (Perkin Elmer, Waltham, MA, USA).

### 3.6. Evaluation of Membrane Rafts

GM1 membrane raft marker was visualized in live human spermatozoa by staining with the cholera toxin subunit B (CTB)-FITC [22,34]. For this purpose, suspensions of cells ($15 \times 10^6$ cells) from each sample were mixed with an equal volume of CTB (50 μg/mL) and incubated for 15 min at 37 °C. The sperm cells were then washed twice in PBS before being fixed in 2% paraformaldehyde for 30 min, mounted on poly-L-lysine coated glass microscope slides, and viewed using the confocal microscope as described above. For each treatment, at least 200 cells were counted.

### 3.7. Evaluation of Acrosome Reaction

Acrosome status was monitored with acrosome-specific FITC-labeled peanut (*Arachis hypogaea*) agglutinin (FITC-PNA) in conjunction with DNA-specific fluorochrome propidium iodide (PI) as a viability test [23,24]. Briefly, in order to induce AR, aliquots ($15 \times 10^6$ cells) of each sample were incubated for 30 min at 37 °C, in the presence of 10 μM $Ca^{2+}$ ionophore A23187. Samples containing DMSO, but not ionophore, were used as control. After incubation sperm cells were centrifuged, resuspended in PBS, and treated for 10 min at room temperature with 12 μM PI. Sperm was washed with PBS, fixed with 2% (*w*/*v*) paraformaldehyde and incubated overnight at 4 °C on poly-L-lysine-treated slides. Permeabilized sperm cells, as described above, were stained with 1 mg FITC-PNA/mL for 15 min at 37 °C in the dark, washed and mounted. At least 200 cells were evaluated for each sample, and fluorescence was detected as described above. Only sperm cells showing evenly distributed fluorescence over the acrosomal region were considered acrosome-intact.

### 3.8. Protein L1 Distribution

To determine whether and how much L1 was present in the plasma membrane, head or flagellum, intact spermatozoa ($30 \times 10^6$ cells) from each treatment were accurately washed in PBS-VI (containing vanadate 1 mM and protease inhibitor cocktail), resuspended in 200 μL of the same PBS-VI, and sonicated 3 times (30 s followed by a 10-s rest period each) on ice. Heads and flagellar fragments were then separated by a 15-min centrifugation (700× *g*) at 4 °C through a 75% Percoll layer in PSW. Flagellar fragments were recovered at the surface of the Percoll layer while the heads were found in the pellet. The purity of each fraction was assessed by microscopy prior to proceeding to analysis. The supernatant was centrifuged for 10 min (10,000× *g*, 4 °C) and the resulting supernatant was further centrifuged (100,000× *g*) to separate the membrane from the cytosol [21]. Each resulting fraction (membranes M, cytosol C, head H and flagella F) was diluted with PBS to the initial volume except for membranes, which were resuspended in 30 μL of PBS-VI. The presence of L1 protein was investigated by Western blotting and immuno-revealed with anti-L1 antibody.

### 3.9. Western Blotting

Cell fractions (membrane, heads and tails) or corresponding cytosol (Figure S1) were solubilized by adding SDS and β-mercaptoethanol (2% final concentration), boiled at 100 °C for 5 min, and subjected to SDS/PAGE (10% polyacrylamide gels). Proteins were electrotransferred to a nitrocellulose membrane and immunorevealed with anti-L1 antibodies. Loading control was performed with anti-actin antibodies [35].

Densitometric analysis. Bands corresponding to L1 protein bands were counted by ImageJ software 1.48v.

### 3.10. ROS Enhanced Chemiluminescence (ECL)

Production of ROS was measured by the chemiluminescence assay method with luminol (5-amino-2,3-dihydro-1,4-phthalazinedione) as probe [23,24]. Briefly, 2 μL of 25 mM luminol and 4 μL of 10 mg/mL horseradish peroxidase, both prepared in DMSO, were added to 200 μL of a sperm suspension at a concentration of $10 \times 10^6$ cells/mL. ROS levels were determined by a luminometer (Fluoroskan Ascent FL, Labsystems, Helsinki, Finland) in the integrated mode for 180 min at 37 °C. Results are expressed as Relative Luminescence Units (RLU) per $2 \times 10^6$ sperm cells. Lastly, 2 μL of a 10 mM N-formylmethionyl-leucyl-phenylalanine (FMLP) stock was added and, after a further 10 min of incubation, 4 μL of a 1 nM stock solution of PMA was added, to exclude leukocyte contamination. Only samples with negative response to FLMP and PMA were processed.

### 3.11. Statistical Analysis

Results are expressed as means ± SD. Comparisons among multiple groups were obtained with ANOVA followed by Dunnett's test. Statistical significance was set at $p < 0.05$ (two-tailed). Statistical analysis were performed with JMP® 13 software (SAS Institute, Cary, NC, USA).

## 4. Conclusions

Recent evidence substantially strengthens the possibility that HPV is responsible not only for HPV-associated cancers, but also for idiopathic infertility due to its detrimental effect on sperm parameters [36–38] and the potential stage arrest of embryo development [39]. In this study we showed that L1, at 10 μg/mL, binds to quit 100% of sperm membrane, locating at the sides of mid-piece and head. This protein, blocks rafts translocation, Lyn displacement, and Tyr-P of the head, thus resulting in a dramatic decrease of the percentage of ARC. Asta treatment prevented L1 binding to the membranes reducing by more than 50% the presence of L1, thus protecting cells from HPV L1 binding. Further studies are needed to better clarify the mechanism of HPV L1 entry, and Asta could be considered for its potential antiviral effect.

**Supplementary Materials:** The following are available online at http://www.mdpi.com/1660-3397/16/11/427/s1, Figure S1: Detection of protein L1 in different fractions of human sperm capacitated in presence of L1, Figure S2: Detection of protein L1 in cytosol of human sperm during capacitation in absence or presence of L1 and/or Asta.

**Author Contributions:** Conceptualization, L.B.; Data curation, G.D., A.A., E.T., A.M.B. and G.A.; Formal analysis, E.R. and L.B.; Funding acquisition, D.A.; Investigation, G.D., A.A., C.S. and L.B.; Methodology, G.D., D.A., E.R. and L.B.; Project administration, L.B.; Resources, E.T., A.M.B. and D.A.; Supervision, D.A., G.A. and L.B.; Validation, E.R.; Visualization, G.D. and E.R.; Writing—original draft, G.D., A.A., E.T., A.M.B., C.S., D.A., G.A., E.R. and L.B.; Writing—review and editing, G.D., A.A., C.S., D.A., G.A., E.R. and L.B.

**Funding:** This research received no external funding.

**Acknowledgments:** This work was supported by the Italian Ministero dell'Università e della Ricerca Scientifica e Tecnologica (MURST).

**Conflicts of Interest:** The authors declare no conflict of interest.

## References

1. Kumar, S.; Biswas, M.; Jose, T. HPV vaccine: Current status and future directions. *Med. J. Armed. Forces India* **2015**, *71*, 171–177. [CrossRef] [PubMed]
2. Walboomers, J.M.; Meijer, C.J. Do HPV-negative cervical carcinomas exist. *J. Pathol.* **1997**, *181*, 253–254. [CrossRef]
3. Zur Hausen, H. Papillomaviruses in the causation of human cancers—A brief historical account. *Virology* **2009**, *384*, 260–265. [CrossRef] [PubMed]
4. De Vuyst, H.; Clifford, G.M.; Nascimento, M.C.; Madeleine, M.M.; Franceschi, S. Prevalence and type distribution of human papillomavirus in carcinoma and intraepithelial neoplasia of the vulva, vagina and anus: A meta-analysis. *Int. J. Cancer* **2009**, *124*, 1626–1636. [CrossRef] [PubMed]

5. Cabrera, M.; Chan, P.J.; Kalugdan, T.H.; King, A. Transfection of the inner cell mass and lack of a unique DNA sequence affecting the uptake of exogenous DNA by sperm as shown by dideoxy sequencing analogues. *J. Assist. Reprod. Genet.* **1997**, *14*, 120–124. [CrossRef] [PubMed]
6. Chan, P.J.; Seraj, I.M.; Kalugdan, T.H.; King, A. Blastocysts exhibit preferential uptake of DNA fragments from the E6-E7 conserved region of the human papillomavirus. *Gynecol. Oncol.* **1995**, *58*, 194–197. [CrossRef] [PubMed]
7. Depuydt, C.E.; Beert, J.; Bosmans, E.; Salembier, G. Human Papillomavirus (HPV) virion induced cancer and subfertility, two sides of the same coin. *Facts Views Vis. ObGyn* **2016**, *8*, 211–222. [PubMed]
8. Foresta, C.; Noventa, M.; De Toni, L.; Gizzo, S.; Garolla, A. HPV-DNA sperm infection and infertility: From a systematic literature review to a possible clinical management proposal. *Andrology* **2015**, *3*, 163–173. [CrossRef] [PubMed]
9. Pereira, N.; Kucharczyk, K.M.; Estes, J.L.; Gerber, R.S.; Lekovich, J.P.; Elias, R.T.; Spandorfer, S.D. Human Papillomavirus Infection, Infertility, and Assisted Reproductive Outcomes. *J. Pathog.* **2015**, *2015*, 578423. [CrossRef] [PubMed]
10. Aksoy, P.; Gottschalk, E.Y.; Meneses, P.I. HPV entry into cells. *Mutat. Res. Rev. Mutat. Res.* **2017**, *772*, 13–22. [CrossRef] [PubMed]
11. Raff, A.B.; Woodham, A.W.; Raff, L.M.; Skeate, J.G.; Yan, L.; Da Silva, D.M.; Schelhaas, M.; Kast, W.M. The evolving field of human papillomavirus receptor research: A review of binding and entry. *J. Virol.* **2013**, *87*, 6062–6072. [CrossRef] [PubMed]
12. Boussiba, S.; Bing, W.; Yuan, J.P.; Zarka, A.; Chen, F. Changes in pigments profile in the green alga Haeamtococcus pluvialis exposed to environmental stresses. *Biotechnol. Lett.* **1999**, *21*, 601–604. [CrossRef]
13. McNulty, H.P.; Byun, J.; Lockwood, S.F.; Jacob, R.F.; Mason, R.P. Differential effects of carotenoids on lipid peroxidation due to membrane interactions: X-ray diffraction analysis. *Biochim. Biophys. Acta Biomembr.* **2007**, *1768*, 167–174. [CrossRef] [PubMed]
14. Lee, S.J.; Bai, S.K.; Lee, K.S.; Namkoong, S.; Na, H.J.; Ha, K.S.; Han, J.A.; Yim, S.V.; Chang, K.; Kwon, Y.G.; et al. Astaxanthin inhibits nitric oxide production and inflammatory gene expression by suppressing IκB kinase-dependentNF-κB activation. *Mol. Cells* **2003**, *16*, 97–105. [PubMed]
15. Ambati, R.R.; Phang, S.M.; Ravi, S.; Aswathanarayana, R.G. Astaxanthin: Sources, extraction, stability, biological activities and its commercial applications—A review. *Mar. Drugs* **2014**, *12*, 128–152. [CrossRef] [PubMed]
16. Hosokawa, M.; Yasui, Y. Chemopreventive effects of astaxanthin on inflammatory bowel disease and inflammation-related colon carcinogenesis. In *Carotenoids and Vitamin A in Translational Medicine*; Sommerburg, O., Siems, W., Kraemer, K., Eds.; CRC Press: Boca Raton, FL, USA, 2013; pp. 289–304.
17. Kochi, T.; Shimizu, M.; Sumi, T.; Kubota, M.; Shirakami, Y.; Tanaka, T.; Moriwaki, H. Inhibitory effects of astaxanthin on azoxymethaneinduced colonic preneoplastic lesions in C57/BL/KsJdb/db mice. *BMC Gastroenterol.* **2014**, *14*, 212. [CrossRef] [PubMed]
18. Palozza, P.; Torelli, C.; Boninsegna, A.; Simone, R.; Catalano, A.; Mele, M.C.; Picci, N. Growth-inhibitory effects of the astaxanthinrich alga Haematococcus pluvialis in human colon cancer cells. *Cancer Lett.* **2009**, *283*, 108–117. [CrossRef] [PubMed]
19. Liu, X.; Song, M.; Gao, Z.; Cai, X.; Dixon, W.; Chen, X.; Cao, Y.; Xiao, H. Stereoisomers of Astaxanthin Inhibit Human Colon Cancer Cell Growth by Inducing G2/M Cell Cycle Arrest and Apoptosis. *J. Agric. Food Chem.* **2016**, *64*, 7750–7759. [CrossRef] [PubMed]
20. Donà, G.; Kožuh, I.; Brunati, A.M.; Andrisani, A.; Ambrosini, G.; Bonanni, G.; Ragazzi, E.; Armanini, D.; Clari, G.; Bordin, L. Effect of astaxanthin on human sperm capacitation. *Mar. Drugs* **2013**, *11*, 1909–1919. [CrossRef] [PubMed]
21. Andrisani, A.; Donà, G.; Tibaldi, E.; Brunati, A.M.; Sabbadin, C.; Armanini, D.; Alvisi, G.; Gizzo, S.; Ambrosini, G.; Ragazzi, E.; et al. Astaxanthin improves human sperm capacitation by inducing Lyn displacement and activation. *Mar. Drugs* **2015**, *13*, 5533–5551. [CrossRef] [PubMed]
22. Andrisani, A.; Donà, G.; Ambrosini, G.; Bonanni, G.; Bragadin, M.; Cosmi, E.; Clari, G.; Armanini, D.; Bordin, L. Effect of various commercial buffers on sperm viability and capacitation. *Syst. Biol. Reprod. Med.* **2014**, *60*, 239–244. [CrossRef] [PubMed]

23. Donà, G.; Fiore, C.; Tibaldi, E.; Frezzato, F.; Andrisani, A.; Ambrosini, G.; Fiorentin, D.; Armanini, D.; Bordin, L.; Clari, G. Endogenous reactive oxygen species content and modulation of tyrosine phosphorylation during sperm capacitation. *Int. J. Androl.* **2011**, *34*, 411–419. [CrossRef] [PubMed]
24. Donà, G.; Fiore, C.; Andrisani, A.; Ambrosini, G.; Brunati, A.; Ragazzi, E.; Armanini, D.; Bordin, L.; Clari, G. Evaluation of correct endogenous reactive oxygen species content for human sperm capacitation and involvement of the NADPH oxidase system. *Hum. Reprod.* **2011**, *26*, 3264–3273. [CrossRef] [PubMed]
25. Surviladze, Z.; Dziduszko, A.; Ozbun, M.A. Essential roles for soluble virion-associated heparan sulfonated proteoglycans and growth factors in human papillomavirus infections. *PLoS Pathog.* **2012**, *8*, e1002519. [CrossRef] [PubMed]
26. Lipovsky, A.; Popa, A.; Pimienta, G.; Wyler, M.; Bhan, A.; Kuruvilla, L.; Guie, M.A.; Poffenberger, A.C.; Nelson, C.D.; Atwood, W.J.; et al. Genome-wide siRNA screen identifies the retromer as a cellular entry factor for human papillomavirus. *Proc. Natl. Acad. Sci. USA* **2013**, *110*, 7452–7457. [CrossRef] [PubMed]
27. Foresta, C.; Ferlin, A.; Bertoldo, A.; Patassini, C.; Zuccarello, D.; Garolla, A. Human papilloma virus in the sperm cryobank: An emerging problem? *Int. J. Androl.* **2011**, *34*, 242–246. [CrossRef] [PubMed]
28. Foresta, C.; Patassini, C.; Bertoldo, A.; Menegazzo, M.; Francavilla, F.; Barzon, L.; Ferlin, A. Mechanism of human papillomavirus binding to human spermatozoa and fertilizing ability of infected spermatozoa. *PLoS ONE* **2011**, *6*, e15036. [CrossRef] [PubMed]
29. Carey, D.J. Syndecans: Multifunctional cell-surface co-receptors. *Biochem. J.* **1997**, *327*, 1–16. [CrossRef] [PubMed]
30. Chen, K.; Williams, K.J. Molecular mediators for raft-dependent endocytosis of syndecan-1, a highly conserved, multifunctional receptor. *J. Biol. Chem.* **2013**, *288*, 13988–13999. [CrossRef] [PubMed]
31. Lee, J.S.; Kwon, W.S.; Rahman, M.S.; Yoon, S.J.; Park, Y.J.; Pang, M.G. Actin-related protein 2/3 complex-based actin polymerization is critical for male fertility. *Andrology* **2015**, *3*, 937–946. [CrossRef] [PubMed]
32. World Health Organization. *WHO Laboratory Manual for the Examination and Processing of Human Semen*; Cambridge University Press: Cambridge, UK, 2010.
33. Mortimer, S.T.; Swan, M.A.; Mortimer, D. Effect of seminal plasma on capacitation and hyperactivation in human spermatozoa. *Hum. Reprod.* **1998**, *13*, 2139–2146. [CrossRef] [PubMed]
34. Nixon, B.; Mitchell, L.A.; Anderson, A.L.; McLaughlin, E.A.; O'bryan, M.K.; Aitken, R.J. Proteomic and functional analysis of human sperm detergent resistant membranes. *J. Cell. Physiol.* **2011**, *226*, 2651–2665. [CrossRef] [PubMed]
35. Liu, D.Y.; Clarke, G.N.; Baker, H.W. Exposure of actin on the surface of the human sperm head during in vitro culture relates to sperm morphology, capacitation and zona binding. *Hum. Reprod.* **2005**, *20*, 999–1005. [CrossRef] [PubMed]
36. Lai, Y.M.; Lee, J.F.; Huang, H.Y.; Soong, Y.K.; Yang, F.P.; Pao, C.C. The effect of human papillomavirus infection on sperm cell motility. *Fertil. Steril.* **1997**, *67*, 1152–1155. [CrossRef]
37. Rintala, M.A.; Grenman, S.E.; Pollanen, P.P.; Suominen, J.J.; Syrjänen, S.M. Detection of high risk HPV DNA in semen and its association with the quality of semen. *Int. J. STD AIDS* **2004**, *15*, 740–743. [CrossRef] [PubMed]
38. Yang, Y.; Jia, C.W.; Ma, Y.M.; Zhou, L.Y.; Wang, S.Y. Correlation between HPV sperm infection and male infertility. *Asian J. Androl.* **2013**, *15*, 529–532. [CrossRef] [PubMed]
39. Henneberg, A.A.; Patton, W.C.; Jacobson, J.D.; Chan, P.J. Human papilloma virus DNA exposure and embryo survival is stage-specific. *J. Assist. Reprod. Genet.* **2006**, *23*, 255–259. [CrossRef] [PubMed]

marine drugs

Article

# Astaxanthin Attenuates Environmental Tobacco Smoke-Induced Cognitive Deficits: A Critical Role of p38 MAPK

**Xia Yang †, An-Lei Guo †, Yi-Peng Pang, Xiao-Jing Cheng, Ting Xu, Xin-Rui Li, Jiao Liu, Yu-Yun Zhang and Yi Liu ***

Jiangsu Key Laboratory of New Drug Research and Clinical Pharmacy, Xuzhou Medical University, Xuzhou 221004, China; XiaYangco@163.com (X.Y.); gal303827537@163.com (A.-L.G.); xzmu_pyp@163.com (Y.-P.P.); 15771378196@163.com (X.-J.C.); xutingt@126.com (T.X.); 15266427365@163.com (X.-R.L.); 15298370510@163.com (J.L.); 15094359563@163.com (Y.-Y.Z.)

* Correspondence: liuyi@xzhmu.edu.cn; Tel.: +86-0516-83262084

† These authors equally contributed to this work.

Received: 7 November 2018; Accepted: 19 December 2018; Published: 3 January 2019

**Abstract:** Increasing evidence indicates that environmental tobacco smoke (ETS) impairs cognitive function and induces oxidative stress in the brain. Recently, astaxanthin (ATX), a marine bioactive compound, has been reported to ameliorate cognitive deficits. However, the underlying pathogenesis remains unclear. In this study, ATX administration (40 mg/kg and 80 mg/kg, oral gavage) and cigarette smoking were carried out once a day for 10 weeks to investigate whether the p38 MAPK is involved in cognitive function in response to ATX treatment in the cortex and hippocampus of ETS mice. Results indicated that ATX administration improved spatial learning and memory of ETS mice ($p < 0.05$ or $p < 0.01$). Furthermore, exposure to ATX prevented the increases in the protein levels of the p38mitogen-activated protein kinase (p38 MAPK; $p < 0.05$ or $p < 0.01$) and nuclear factor-kappa B (NF-κB p65; $p < 0.05$ or $p < 0.01$), reversed the decreases in the mRNA and protein levels of synapsin I (SYN) and postsynaptic density protein 95 (PSD-95) (all $p < 0.05$ or $p < 0.01$). Moreover, ATX significantly down-regulated the increased levels of pro-inflammatory cytokines including interleukin-6 (IL-6) and tumor necrosis factor (TNF-$\alpha$) (all $p < 0.05$ or $p < 0.01$). Meanwhile, the increased level of malondialdehyde (MDA) and the decreased activities of superoxide dismutase (SOD), glutathione (GSH), and catalase (CAT) were suppressed after exposure to ATX (all $p < 0.05$ or $p < 0.01$). Also, the results of the molecular docking study of ATX into the p38 MAPK binding site revealed that its mechanism was possibly similar to that of PH797804, a p38 MAPK inhibitor. Therefore, our results indicated that the ATX might be a critical agent in protecting the brain against neuroinflammation, synaptic plasticity impairment, and oxidative stress in the cortex and hippocampus of ETS mice.

**Keywords:** astaxanthin; cigarette smoke exposure; p38 MAPK; antioxidant inflammatory; synaptic-associated plasticity

## 1. Introduction

Environmental tobacco smoke (ETS), the combination of the side-stream smoke emitted from the burning end of a tobacco product and the mainstream smoke exhaled by the smoker, contains more than 6000 chemicals that are harmful to human body and may lead to many serious health problems, such as cognitive impairment and dementia [1,2]. For instance, compared with nonsmokers, smokers are reported to have remarkably decreased prefrontal attention network activity, and such a deficit is related with the length of smoking time [3]. Moreover, pregnant women exposed to tobacco smoke may present fetal neurobehavioral damages [4].

Although the pathogenesis of cognitive impairments due to tobacco smoke exposure has not been completely understood, several factors have been implicated such as oxidative stress and inflammation. For instance, long-term exposure to tobacco smoke led to oxidative stress [5]. Oxidative stress, which is mainly attributable to excessive generation of reactive oxygen species (ROS), mediates the activation of the mitogen-activated protein kinases (MAPKs) MAPK signaling cascades, especially the p38 MAPK pathway. As an important member of the MAPK family, p38 MAPK has been demonstrated to play a key role in nuclear factor – kappa B (NF-Kb) activation and pro-inflammatory expression [6]. NF-κB, one of the ubiquitous transcriptional factors, is the main medium that leads to the enlargement of inflammatory responses and then promotes the expression of proinflammatory cytokines like TNF-α and IL-6 [7,8]. Furthermore, researches demonstrate that the tobacco smoking induced oxidative stress and inflammation are involved in brain dysfunction [9]. Meanwhile, excessive ROS and inflammatory cytokines can impair hippocampal structure and function on learning and memory-related synaptic plasticity and neurogenesis [10]. Therefore, we hypothesized that attenuation of oxidative stress and inflammation might reverse the cognitive impairment induced by ETS.

Astaxanthin (ATX), a naturally occurring red carotenoid pigment, is abundant in red yeast *Phaffia rhodozyma*, green algae *Haematococcus pluvialis* and many kinds of marine organisms such as salmon and lobsters [11,12]. ATX has hydrophobic polyunsaturated polar structure on both ends of the conjugated olefins structure that facilitate its precise positioning within cell membranes and circulating lipoproteins, before exhibiting potent antioxidant functions as a powerful scavenger of oxygen free radicals so as to decrease oxidative stress and lipid peroxidation [13,14]. Recent studies revealed that ATX can relieve ischemia-related injury in brain tissue by suppressing oxidative stress, glutamate release, and anti-apoptosis [15]. Furthermore, some researches find that ATX can exert neuroprotective effects by weakening neuroinflammation [16]. More excitingly, ATX can attenuate subarachnoid hemorrhage induced neuroinflammation in rats and improve hippocampal plasticity and cognitive functions in male C57BL/6J mice [17]. However, the protective effects of ATX against ETS-induced cognitive decline have not been investigated. Therefore, the current work was designed to evaluate whether ATX can alleviate ETS-induced cognitive decline, and investigate the mechanisms involved.

## 2. Results

### 2.1. Effects of ATX Treatment on Exposure to ETS Induced Cognitive Decline

In order to investigate whether ATX could improve the cognitive impairments induced by ETS, we evaluated the learning and memory by the Morris water maze (MWM) test. The trained mice in all groups showed a decrease in mean escape latency during the learning trials (Figure 1A), and from the second day to the fourth day, an apparent elevation appeared in transfer latency in the ETS groups compared with the control group (all $p < 0.05$; Figure 1A). On the fifth day, the escape latency significantly elevated in ETS group compared with the control group ($p < 0.05$; Figure 1A). Meanwhile, ATX treatment (40 mg/kg and 80 mg/kg) significantly inhibited the elevation of escape latency in the ETS mice (all $p < 0.05$; Figure 1A). Administration with ATX (80 mg/kg) alone exhibited no visible difference in the escape latency compared with the control mice ($p > 0.05$; Figure 1A).

The probe trial was performed on the fifth day. In the ETS mice, the percentage of time spent in the target quadrant ($p < 0.05$; Figure 1C) and the number of crossings of the platform area ($p < 0.05$; Figure 1D) appeared to decrease in comparison with the control mice, while the decrease of the percentage of time spent in the target quadrant ($p < 0.05$; Figure 1C) and the number of crossings of the platform area ($p < 0.05$; Figure 1D) were prevented by ATX (40 mg/kg and 80 mg/kg) treatment. ATX (80 mg/kg) alone treated mice presented no visible differences in comparison with the control mice ($p > 0.05$; Figure 1C,D), suggesting that ATX itself had no influence on the learning and memory in the control group. The swimming speed exhibited similar performance among the five groups during the five-days MWM test (all $p > 0.05$, Figure 1B), which indicates that the differences in escape latency, the number of crossings, and the time spent in the target quadrant do not affect the movement defects.

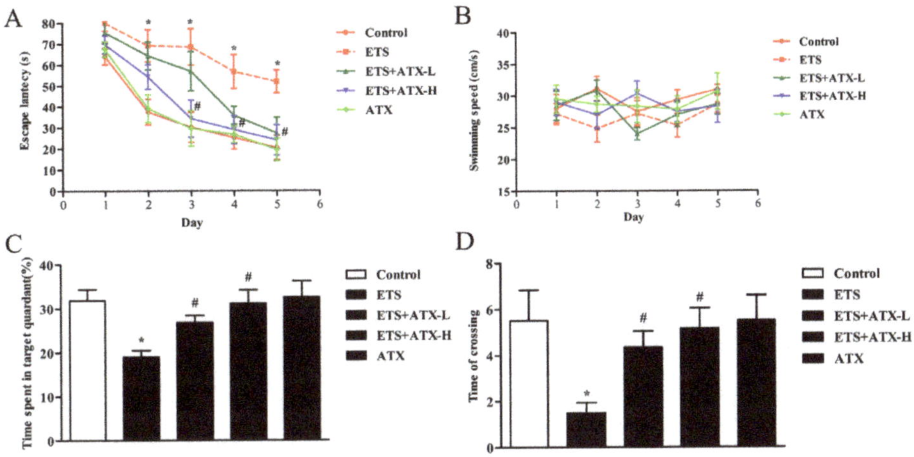


**Figure 1.** Effects of chronic astaxanthin (ATX) treatment on environment tobacco smoke (ETS) induced cognitive decline ($n$ = 12). (**A**) Escape latency appeared during the training and the probe sessions. Data are reported as mean ± SE. (* $p < 0.05$) versus Group Control at the corresponding days; # $p < 0.05$ versus Group ETS at the corresponding days). (**B**) The swimming speed among the four groups during the five-day period. Data are reported as mean ± SE ($p > 0.05$). (**C**) The percentage of time spent in the target quadrant during the probe trial. Data are reported as mean ± SE. (* $p < 0.05$ versus Group Control; # $p < 0.05$ versus Group ETS). (**D**) The number of crossings of the platform area. Data are reported as mean ± SE (* $p < 0.05$ versus Group Control; # $p < 0.05$ versus Group ETS).

### 2.2. Effects of ATX Treatment on Exposure to ETS Induced Parameters of Oxidative Stress in the Mouse Brain

The MDA levels, SOD activities, CAT activities and GSH levels were detected to investigate whether ATX have the effects on the ETS exposed brain antioxidant system. According to the results described, the ETS mice presented a remarkable increase in the MDA levels ($p < 0.01$, Figure 2A) and a striking decrease in the SOD activities ($p < 0.01$, Figure 2B), CAT activities ($p < 0.01$, Figure 2C) and GSH levels ($p < 0.01$, Figure 2D) in the hippocampus and prefrontal cortex in comparison with the control group. Administration with ATX (40 mg/kg and 80 mg/kg) inhibited the ETS caused elevation of MDA levels ($p < 0.05$ or $p < 0.01$; Figure 2A) and prevented the ETS caused decrease of SOD activities ($p < 0.05$ or $p < 0.01$; Figure 2B), CAT activities ($p < 0.05$ or $p < 0.01$; Figure 2C) and GSH levels ($p < 0.05$ or $p < 0.01$; Figure 2D) in the hippocampus and prefrontal cortex. ATX (80 mg/kg) alone treatment presented no difference in these parameters of oxidative stress in comparison with the control mice in the hippocampus and prefrontal cortex ($p > 0.05$; Figure 2). These results indicated that chronic exposure to ETS caused oxidative stress in mice, and ATX treatment could attenuate the ETS caused oxidative stress.

### 2.3. Effects of ATX Treatment on Exposure to ETS-Induced Inflammation in the Hippocampus and Prefrontal Cortex

Inflammatory response is closely linked to the pathogenesis of cognitive disorder, which damages hippocampal synaptic plasticity by increasing the levels of pro-inflammatory cytokines. Thus, the effects of ATX on ETS induced alteration of the levels of inflammatory factors (such as TNF-α and IL-6) in the brain were tested by ELISA. The levels of TNF-α ($p < 0.01$; Figure 3A) as well as IL-6 ($p < 0.01$; Figure 3B) were found to be increased remarkably in the hippocampus and cortex in ETS mice in comparison with control mice, while ATX (40 mg/kg and 80 mg/kg) administration attenuated the ETS induced increase in the levels of TNF-α ($p < 0.05$ or $p < 0.01$; Figure 3A) and IL-6 ($p < 0.05$ or $p < 0.01$; Figure 3B) in the hippocampus and cortex. ATX (80 mg/kg) treatment alone did not change inflammation levels in the hippocampus and cortex in comparison with the control group ($p > 0.05$;

Figure 3A and B). These results inferred that ETS caused inflammatory response, and ATX could inhibit ETS caused inflammatory response.

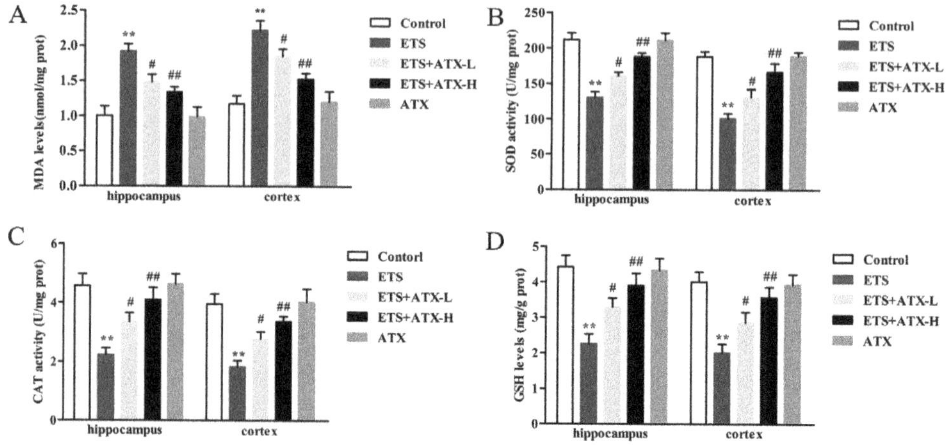


**Figure 2.** Effects of ATX on parameters of oxidative stress in the mouse brain ($n = 12$). (**A**) The level of MDA. Data are reported as mean ± SE. (** $p < 0.01$ versus Group Control; # $p < 0.05$ versus Group ETS; ## $p < 0.01$ versus Group ETS). (**B**) The activity of SOD. Data are reported as mean ± SE. (** $p < 0.01$ versus Group Control; # $p < 0.05$ versus Group ETS; ## $p < 0.01$ versus Group ETS). (**C**) The activity of CAT (** $p < 0.01$ versus Group Control; # $p < 0.05$ versus Group ETS; ## $p < 0.01$ versus Group ETS). (**D**) The level of GSH. Data are reported as mean ± SE. (** $p < 0.01$ versus Group Control; # $p < 0.05$ versus Group ETS; ## $p < 0.01$ versus Group ETS).

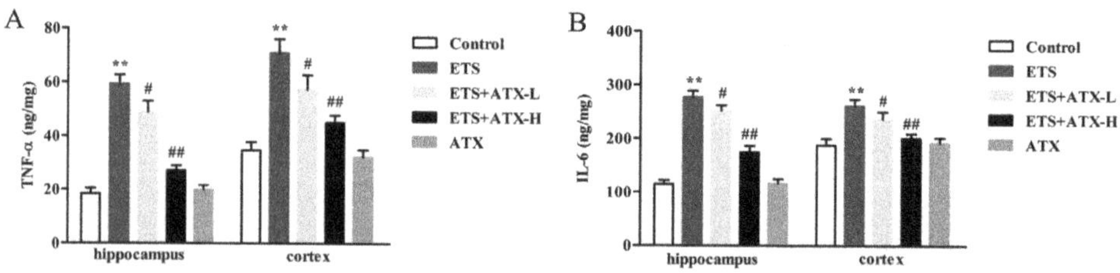


**Figure 3.** Effects of ATX on inflammation in the hippocampus and cortex ($n = 12$). (**A**) The levels of TNF-α. Data are reported as mean ± SE. (** $p < 0.01$ versus Group Control; # $p < 0.05$ versus Group ETS; ## $p < 0.01$ versus Group ETS). (**B**) The levels of IL-6. Data are reported as mean ± SE. (** $p < 0.01$ versus Group Control; # $p < 0.05$ versus Group ETS; ## $p < 0.01$ versus Group ETS).

### 2.4. *Effects of ATX on the Expressions of NF-κB p65 in the Hippocampus and Prefrontal Cortex*

NF-κB p65 expression was carried out to study the potential mechanisms of the neuroprotective changes in ATX treatment of ETS caused cognitive impairment. As shown in Figure 4, there was an obvious enhancement in NF-κB p65 levels ($p < 0.01$) in the hippocampus and cerebral cortex in ETS mice compared with control mice and this enhancement was repressed by ATX (40 mg/kg and 80 mg/kg) treatment ($p < 0.05$ or $p < 0.01$). Treatment with ATX (80 mg/kg) alone exhibited no difference compared to the control mice ($p > 0.05$). These results suggested that ETS enhanced the levels of the NF-kB, and ATX administration could prevent the enhancement of the levels of the NF-κB p65.

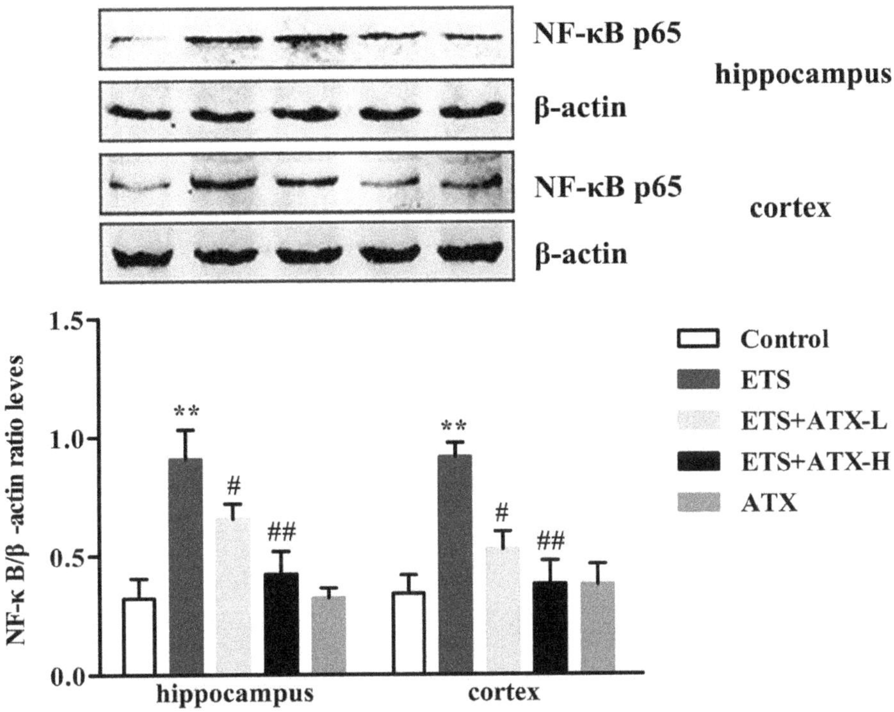


**Figure 4.** Effects of ATX on the expressions of NF-κB p65 in the cortex and hippocampus ($n$ = 12). The levels of NF-κB p65 in the hippocampus and cortex of mice. Data are reported as mean ± SE. (** $p < 0.01$ versus Group Control; # $p < 0.05$ versus Group ETS; ## $p < 0.01$ versus Group ETS).

### 2.5. Effects of ATX on the Expressions of p38 MAPK and p- p38MAPK in the Hippocampus and Prefrontal Cortex of ETS Mice

The protein expression of total-p38 MAPK and p- p38 MAPK in the hippocampus and prefrontal cortex were tested by Western blot and the results are shown in Figure 5. In the ETS mice, the levels of phosphorylated p38 MAPK were remarkably increased in the hippocampus and cerebral cortex in comparison with the control mice ($p < 0.01$), while the increased levels of phosphorylated p38 MAPK were prevented in the ATX (40 mg/kg and 80 mg/kg) mice ($p < 0.05$ or $p < 0.01$). ATX (80 mg/kg) alone groups exhibited no obvious difference compared to the control group ($p > 0.05$). The levels of the total-p38 MAPK exerted no obvious differences among all groups in the hippocampus and cerebral cortex ($p > 0.05$). These results indicated that ETS caused the excessive activation of p38 MAPK, and ATX could inhibit the ETS caused the activation of p38 MAPK.

### 2.6. Effects of ATX on the Expression of SYN mRNA and PSD-95 mRNA in the Mouse Brain of ETS mice

SYN and PSD-95 are two major synaptic associated proteins that can directly or indirectly affect cognitive function [18]. Accordingly, reverse transcriptase-PCR (RT-PCR) was used to estimate the levels of SYN and PSD-95 mRNA. In the ETS mice, the expression of SYN mRNA ($p < 0.01$; Figure 6A) and PSD-95 mRNA ($p < 0.01$; Figure 6B) were markedly down-regulated in the hippocampus and cortex compared with the control mice, while this down-regulation of the SYN mRNA ($p < 0.05$ or $p < 0.01$; Figure 6A) and PSD-95 mRNA ($p < 0.05$ or $p < 0.01$; Figure 6B) levels were elevated by ATX (40 mg/kg and 80 mg/kg) administration. ATX (80 mg/kg) alone had no influence on the expression of SYN mRNA ($p > 0.05$; Figure 6A) and PSD-95 mRNA ($p > 0.05$; Figure 6B) in comparison with the control mice. These results inferred that ETS led to a reduction of the SYN mRNA and PSD-95 mRNA, and ATX could reverse this change.

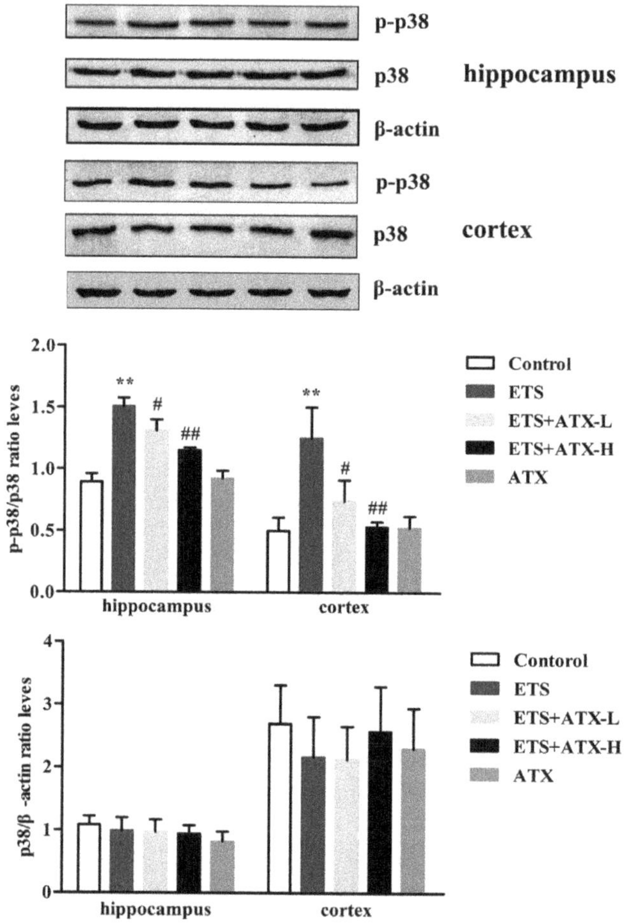


**Figure 5.** Effects of ATX on the expressions of p38 and p-p38 in the cortex and hippocampus ($n$ = 12). The levels of p-p38 in the hippocampus and cortex of mice. Data are reported as mean ± SE. (** $p < 0.01$ versus Group Control; # $p < 0.05$ versus Group ETS; ## $p < 0.01$ versus Group ETS).

## 2.7. *Effects of ATX on the Expression of Synaptic Proteins in the Mouse Brain*

In order to detect whether ATX could protect synaptic plasticity from ETS impairment Western blot was used to examine the expression of SYN and PSD-95 proteins in the hippocampus and cortex. In the ETS mice, SYN ($p < 0.01$; Figure 7A) and PSD-95 ($p < 0.01$; Figure 7B) were noticeably reduced in the hippocampus and cortex compared with the control mice, while treatment with ATX (40 mg/kg and 80 mg/kg) inhibited this reduction of both SYP ($p < 0.05$ or $p < 0.01$) and PSD-95 ($p < 0.05$ or $p < 0.01$) expressions in ETS mice. Administration with ATX (80 mg/kg) alone presented no difference in SYP and PSD-95 expressions compared with the control mice ($p > 0.05$). These results inferred that ATX could prevent ETS induced changes of the SYN and PSD-95 protein expressions.

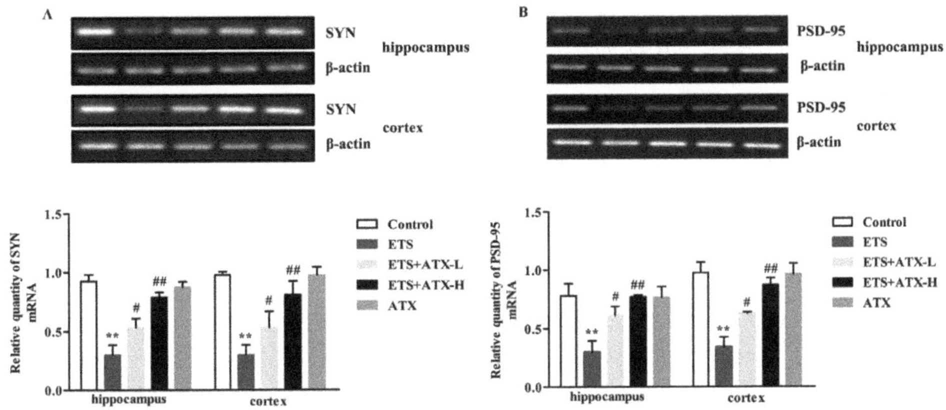


**Figure 6.** Effects of ATX on the expression of SYN mRNA and PSD-95 mRNA in the mouse brain ($n$ = 12). (**A**) The levels of SYN mRNA. Data are reported as mean ± SE. (** $p < 0.01$ versus Group Control; # $p < 0.05$ versus Group ETS; ## $p < 0.01$ versus Group ETS). (**B**) The levels of PSD-95 mRNA. Data are reported as mean ± SE. (** $p < 0.01$ versus Group Control; # $p < 0.05$ versus Group ETS; ## $p < 0.01$ versus Group ETS).

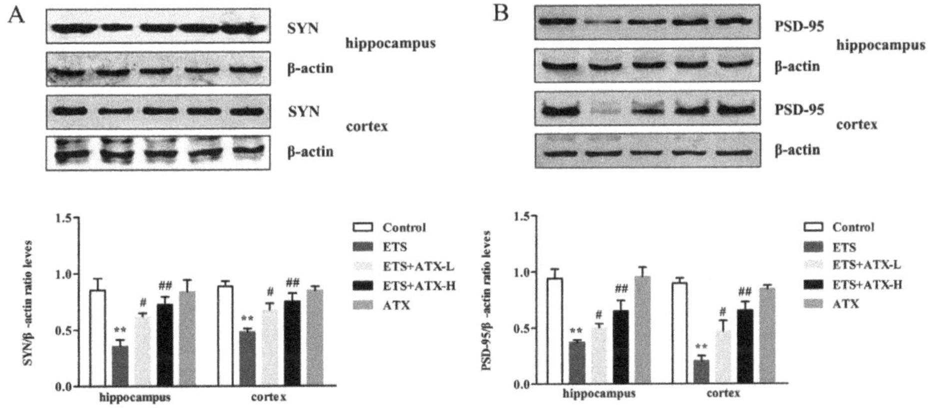


**Figure 7.** Effects of ATX on the expression of synaptic proteins in the mouse brain ($n$ = 12). (**A**) The levels of SYN in the hippocampus and cortex of mice. Data are reported as mean ± SE. (** $p < 0.01$ versus Group Control; # $p < 0.05$ versus Group ETS; ## $p < 0.01$ versus Group ETS). (**B**) The levels of PSD-95 in the hippocampus and cortex of mice. Data are reported as mean ± SE. (** $p < 0.01$ versus Group Control; # $p < 0.05$ versus Group ETS; ## $p < 0.01$ versus Group ETS).

### *2.8. Effects of ATX on the Structure and Morphology of the Hippocampal Neurons*

Microphotographies of the cerebral cortex and the hippocampal CA1 subfield in each group are shown in Figure 8. In the ETS group, no obvious differences were observed in the neurons in the cerebral cortex in comparison with the control group. Meanwhile, no obvious differences were found in intact neuron counts in the cerebral cortex among the groups (Figure 8B). In contrast, in the hippocampal CA1 subfield of ETS mice, the neurons appeared in a noticeably wrinkled, irregular pattern, and a weak staining effect, and most Nissl bodies were lost, which inferred that extensively they were injured or dead (Figure 8A). Also, a significant decrease in the number of surviving neurons was observed in the ETS mice as compared to the control mice, while ATX (40 mg/kg and 80 mg/kg) treatment remarkably attenuated this decrease in ETS mice. Additionally, the ATX (80 mg/kg) group alone and the control group showed no difference in the number of surviving neurons, which revealed that ATX itself had no effect on the neurons in different areas of brain.

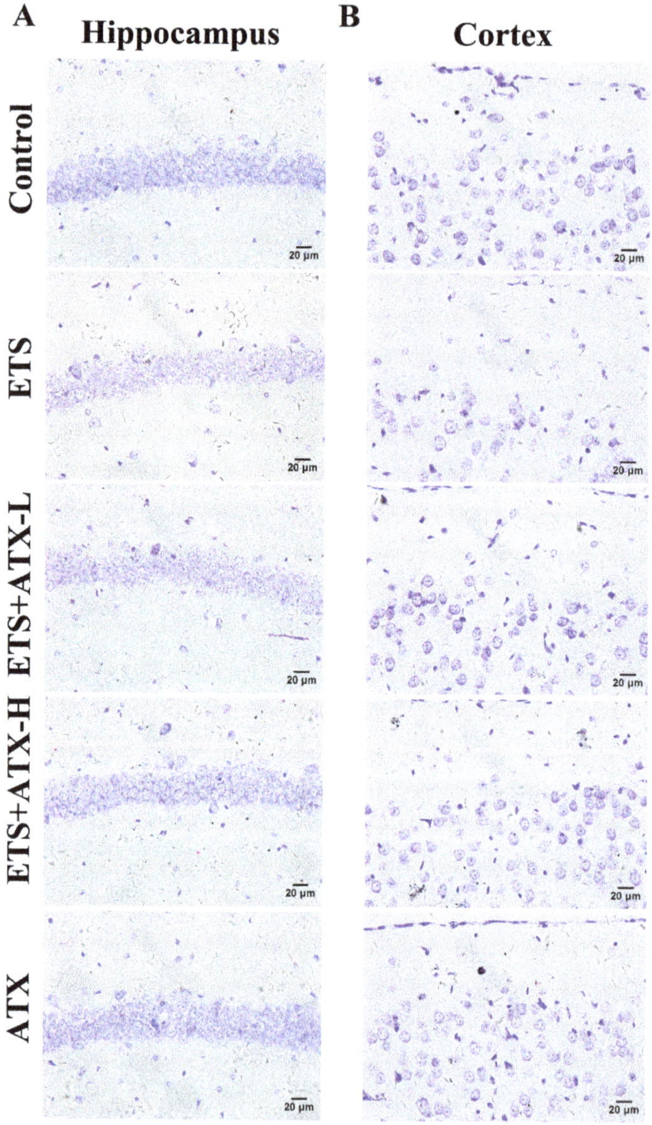


**Figure 8.** Effects of ATX on the structure and morphology of the hippocampal neurons ($n$ = 12). Nissl stained neurons in the hippocampal CA1 subfield (**A**) and cortex (**B**). Bar = 20 μm.

## 2.9. Molecular Docking Studies

The interactions between ATX and human p38 and the interaction between human p38 alpha and p38 inhibitor PH797804 are shown as Figure 9. The p38 alpha and ATX docking pocket was formed by the residues of Glu-71, Leu-167, Phe-169, Leu-171, Thr-175, Arg-49, Leu-108, Met-109, Thr-106 and Leu-104, and a hydrogen bond was formed to Glu-71 from helix C. Meanwhile, another two hydrogen bonds were formed between the side chain of Arg-49 and the opposite end of ATX. p38 alpha and p38 inhibitor PH797804 docking pocking were formed by a hydrogen bond of Gly-110. The docking result demonstrated that ATX occupied the active site of the p38 and generated an interaction with surrounding amino acids like p38 inhibitor PH797804.

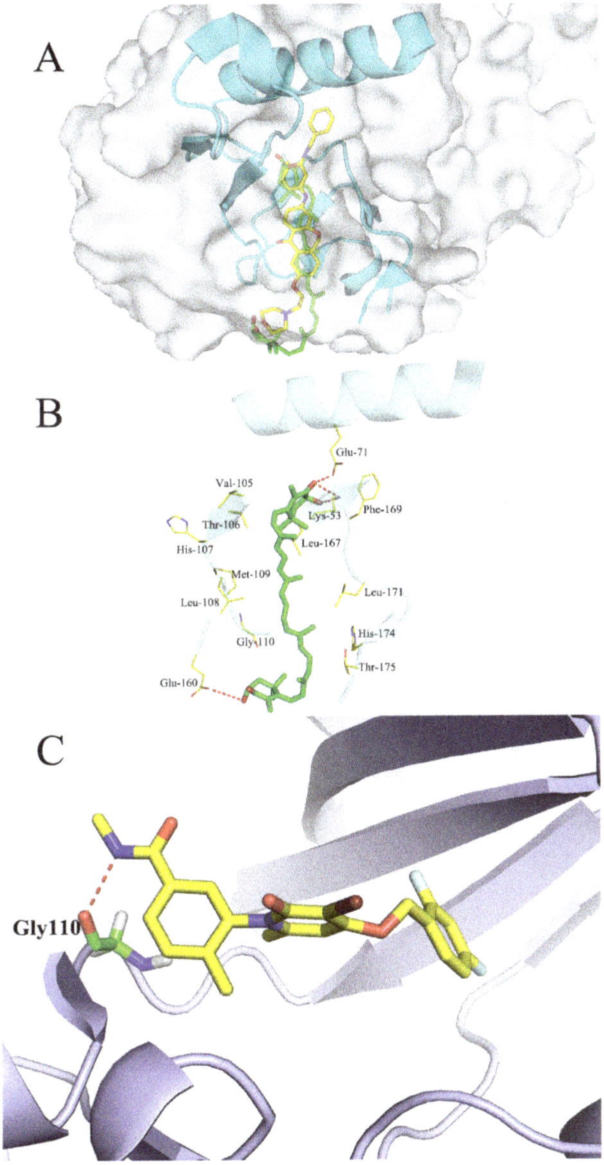


**Figure 9.** Molecular docking model for ATX (green and stick) with the active site of the ATP pocket of p38 alpha and the p38 alpha with the p38 inhibitor PH797804. (**A**) The whole picture of the molecular docking model for ATX with the p38 alpha. (**B**) Molecular docking model for ATX with the p38 alpha. Highlighting the hydrogen bonds (red dashed lines) coordination between the oxygen atoms in ATX and residues Glu-71 and Arg-49. (**C**) Molecular docking model for p38 MAPK with p38 MAPK inhibitor PH797804 binding pocket. The red color bonds indicate hydrogen bonds.

## 3. Discussion

According to the present research, chronic ATX administration reversed the ETS-induced cognitive deficits of mice. Notably, we found that ATX administration normalized the oxidative stress markers, decreased the levels of inflammatory cytokines, phospho-p38 MAPK, and NF-κB p65 proteins in the hippocampus and prefrontal cortex. In addition, the levels of SYN and PSD-95 were increased in the prefrontal cortex and hippocampus of ATX-treated mice. What is more important is that p38 MAPK may be the key factor in the reduction of cognitive deficits.

Previous studies have reported that the capacity for learning and memory were impaired by cigarette smoke exposure [19,20]. However, the mechanism of ETS-induced cognitive impairment remains unclear. ROS are closely related with neuroinflammation and synaptic plasticity impairment [21]. Chronic cigarette smoke exposure induced an excessive ROS generation followed by the loss of the dynamic balance between ROS generation and elimination [16,22]. As a marine bioactive compound, ATX is reported to have antioxidant and anti-inflammation properties [16]. The present data indicated that ETS induced impairment in learning and memory function was improved by ATX treatment. To our knowledge, it is the first report that ETS-induced cognitive deficits can be improved by ATX treatment and the p38 MAPK may be the key factor in the reduction of cognitive deficits.

Generally, the MWM test is widely applied to measure the spatial learning of rodents [23]. Some researches demonstrate that the performance in the MWM test is usually related with both neurotransmitter systems and drug effects [24]. Several studies have confirmed that long term exposure to tobacco smoke could cause cognitive deficits [19,25]. Importantly, ATX can enhance cognitive function and attenuate depression-like behavior. So, we used MWM to observe the ETS-induced cognitive deficits and explore the therapeutic effect of ATX. The result of MWM indicated that the mice exposed to ETS showed enhanced escape latency and reduced time spent in the target quadrant (revealing an impairment of spatial learning and memory), which is consistent with published results [26]. Above all, long-term administration with low or high doses of ATX markedly reversed these behavioral changes, suggesting that ATX is the potential to protect ETS-induced cognition damage.

What is well recognized is that oxidative damage plays a crucial role in many brain dysfunction diseases [27]. Importantly, the brain is particularly vulnerable to oxidative stress because of a relatively high production rate of ROS without commensurate levels of antioxidative defense [28]. Tobacco smoke contains a large number of ROS which can permeate the blood brain barrier and mobilize the antioxidant defenses [29]. In the current research, we found an elevation of MDA, and a reduction of GSH, SOD, and CAT activities in the cerebral cortex and hippocampus of ETS mice, which is consistent with published results [30]. It has been proved that several flavonoids have strong antioxidant properties and improve memory and learning [31]. Moreover, treatment with ATX could decrease the MDA level and increase the SOD level in aging rats [32]. Our results showed that, the MDA level was suppressed, but GSH content, SOD, and CAT activity were raised when chronic administration with ATX in the hippocampus and prefrontal cortex of ETS mice. Consequently, these results support the hypothesis that ATX can inhibit the chronic ETS-induced pro-oxidant–antioxidant disequilibrium contributing to cognition improvement.

p38 MAPK as a stress-activated kinase, is sensitive to various exogenous and endogenous stimulations, and highly responded to oxidative stress and proinflammatory cytokines [33]. In addition, recent studies have found that the activation of the p38 MAPK signaling pathway is closely related with neuronal death or apoptosis, which may be the main reason of cognitive dysfunction [34]. In the current work, we found that the phosphorylation level of p38 MAPK was remarkably raised in the hippocampus and prefrontal cortex of ETS mice. And the chronic ATX administration attenuated the p38 MAPK phosphorylation level. Thus, we speculate that the cognition impairment of ETS mice may contribute to oxidative stress and the activation of p38 MAPK, where the activation of p38 MAPK may be more important.

It is well established that inflammation and oxidative stress are intricately interrelated. Oxidative stress is considered to be a crucial factor in regulating proinflammatory signaling pathways [35]. Long-term exposed to ETS induced oxidative stress and the activation of NF-κB followed by the release of the pro-inflammatory [36]. In addition, many studies confirm that the activation of NF-κB and the release of inflammation cytokines play a key part in the cognitive dysfunction that may explain cognitive decline [37]. In the current research, we also detected that the mice exposed to tobacco smoke showed up-regulated levels of NF-κB p65 and TNF-α and IL-6. However, chronic treatment with ATX remarkably suppresses the expression of NF-κB p65 and attenuates the excessive release of TNF-α and IL-6 [38].

The alterations of structural plasticity of dendrites and spines in the hippocampus and prefrontal cortex were found as a result of cognition deficits [39]. Research indicates that morphological alterations in the brain development of mice exposed to smoke may disrupt neural prediction [40]. Generally, synaptic plasticity is associated with the synapse related proteins, including presynaptic SYN and postsynaptic PSD-95 [41]. In our research, the reduction in the expression of synaptic proteins was observed in ETS-exposed mice, which may result in cognitive impairment. However, the reduction in both SYN and PSD-95 levels in ETS exposure mice was remarkably overturned, by chronic administration with ATX. Both SYN and PSD-95 were regulated by the inflammatory response caused by p38 MAPK and NF-κB p65. Therefore, these neurochemical findings imply that the neuroprotective response of ATX is attributable to reducing the phosphorylation level of p38 MAPK and relieving inflammatory responses. Thus, cognitive impairment in ETS-exposed mice can be improved by increasing the level of plastic-related proteins (SYN and PSD-95).

In conclusion, these findings manifest that ATX exerted protective effects on the cognition decline caused by ETS in mice. These improvements in the behaviors and neurochemicals implied that supplementation with ATX-enriched food may be an effective novel therapy and provide a hopeful mitigation to chronic ETS-induced cognition decline. Administration of ATX reduced oxidative stress and inflammatory responses, as well as enhanced the synapse-related proteins in the hippocampus and prefrontal cortex of ETS mice, and p38 MAPK plays an important role in the protection process. Therefore, our results provide ideals for further studies on the anti-inflammatory or antioxidant aspects of ATX and ATX derivatives in CNS related diseases in the future.

## 4. Materials and Methods

### 4.1. Reagents

ATX (97% purity) was purchased from Xi'an Fengzu Biotechnology Co., Ltd (Shaanxi, China) and dissolved in olive oil (1 mL/kg) immediately before use. MDA, SOD, GSH, CAT, and BCA assay kits were obtained from Nanjing Jiancheng Biotechnology Co., Ltd (Nanjing, China). Antibodies against phospho-p38 MAPK ($T^{180}/Y^{182}$), p38 MAPK, SYN, PSD-95 and NF-κB p65 were from Cell Signaling Technology Inc., (Danvers, MA USA) and β-actin was from ZSGB-BIO, Beijing, China. All other reagents were from Sigma-Aldrich (St. Louis, MO, USA) unless otherwise indicated.

### 4.2. Animals

Adult male Kunming mice weighing between 18 and 22 g were purchased from the Laboratory Animal Center, Xuzhou Medical University. The whole experimental schedule was depicted in Figure 10. The mice were housed with a 12 h light/dark cycle and free access to food and water under controlled temperatures 22 ± 2 °C and humidity 50 ± 10%. The animals were sacrificed within 24 h after the final test. All animal experiments in the current study were conducted in accordance with the Animal Ethics Committee, Xuzhou Medical University, China, and followed the National Institutes of Health Guidelines for the Care and Use of Laboratory Animals (Ethical approval number: XZMC2014-AN-39).

The mice were randomly divided into five groups according to their corresponding treatments ($n$ = 12): (1) an ETS group: mice were exposed to ETS once a day for 2 h with an interval of 10 min between each cigarette, using 8 cigarettes per day for 10 consecutive weeks; (2) an ETS+ATX-L group: mice were exposed to ETS once a day for 2 h with an interval of 10 min between each cigarette, using 8 cigarettes followed by treatment with a low dose of ATX (40 mg/kg) once a day for 10 consecutive weeks; (3) an ETS+ATX-H group: mice were exposed to ETS once a day for 2 h with an interval of 10 min between each cigarette, using 8 cigarettes followed by treatment with a high dose of ATX (80 mg/kg) per day for 10 consecutive weeks; (4) an ATX group: mice were treated with 80 mg/kg ATX alone per day for 10 consecutive weeks; (5) a control group: under normal conditions with an equal volume of olive oil as ATX treatment once a day for 10 consecutive weeks. ATX was dissolved in olive oil before administration. Either ATX or the equal volume of olive oil was administered by oral gavage.

### 4.3. Smoke Generation

In the current study, smoke was generated according to previous descriptions [42]. Each cigarette contains 10 mg tar, 0.8 mg nicotine, and 10 mg carbon monoxide. After the mice were placed within a chamber (56.4 cm × 38.5 cm × 37.1 cm), four cigarettes (Jiangsu Tobacco Industrial Co. Ltd., China) were lit at one time and the chamber was shut down immediately, leaving a small hole (371 mm × 40 mm) in both ends for ventilation, and the cigarettes were burned up within 15 min. In order to keep an air flow inside the chamber, the smoke generated within the chamber was pumped by a noiseless extractor fan. The diluted side-stream smoke exposed to mice was adopted to imitate the ETS experienced for non-smokers.

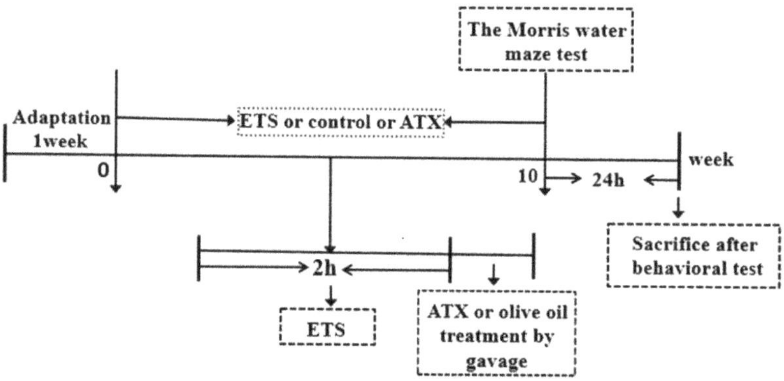


**Figure 10.** Schematic figure of the treatment protocol ($n$ = 12).

### 4.4. Morris Water Maze (MWM)

The Morris water maze test was performed according to previous descriptions [43,44]. Mice were trained in a black circular pool (120 cm in diameter and 60 cm in height) filled with water (20–22 °C). The pool was divided into four quadrants with a clear 10 cm diameter escape platform hidden 1.5 cm beneath the surface in one of the quadrants. Training trials were conducted in the first consecutive four days, and the escape latency was recorded according to the time spent to reach the hidden platform. Then a probe trial was performed on the fifth day and the hidden platform was removed. The total time spent in each target quadrant was recorded.

### 4.5. Measurement of Oxidative Stress

After the behavioral assessments, the mice were sacrificed. The hippocampus and prefrontal cortex were dissected and homogenized (1:9 $w/v$) with cold normal saline (4 °C) to prepare 10%

cerebral homogenate in an ice bath. The homogenized tissue was centrifuged at 4000 rpm at 4 °C for 10 min and the supernatant was collected for the following tests.

#### 4.5.1. Determination of Lipid Peroxidation

The MDA level was measured by supernatants reacted with thiobarbituric (TBA) to form thiobarbituric acid reactive substances using a commercial kit (Nanjing Jiancheng Bioengineering Institute, Nanjing, China) [45] and the absorption was determined at the wavelength of 532 nm.

#### 4.5.2. Determination of SOD Activity

The activity of SOD was assayed according to the method previously described [46]. Xanthine reacts with xanthine oxidase to produce superoxide radicals which then react with nitro-blue tetrazolium (NTB) to form a colored formazan dye. The amount of formazan generated was determined by the absorption at the wavelength of 550 nm. One unit of enzyme was defined as the amount of enzyme required at an inhibition rate of 50%. Enzyme specific activity was expressed in units per milligram protein.

#### 4.5.3. Determination of CAT Activity

The activity of CAT was assayed based on the method previously described [47]. Briefly, 0.1 mL of supernatant of tissues in hippocampus and cortex was added to 1.91 mL of 50 mmol/L phosphate buffer (pH 7.0). Then 1 mL freshly prepared 30 mmol/L $H_2O_2$ was added to start the reaction. The decrease in $H_2O_2$ content was determined by the absorption at the wavelength of 240 nm.

#### 4.5.4. Determination of GSH

The concentration of GSH was assayed according to a previous method [48]. In brief, 160 μL of supernatant of tissues in hippocampus and cortex was added to 2 mL of Ellman's reagent (5, 5′-dithiobis [2-nitrobenzoic acid] 10 mM, $NaHCO_3$ 15 mM). The mixture was incubated at room temperature for 5 min and the absorption was measured at the wavelength of 412 nm.

### 4.6. *Enzyme-linked Immunosorbent Assay (ELISA)*

The frozen brain cortex and hippocampal tissues were homogenized in ice-cold normal saline and centrifuged at 12,000 rpm at 4 °C for 5 min. The supernatants were then collected, and the total protein concentration was assayed using Micro BCA procedures (Beyotime Institute of Biotechnology, Shanghai, China). On the basis of the manufacturer's instructions, enzyme-linked immunosorbent assay (ELISA) kits (Immuno-Biological Laboratories Co., Ltd., Japan) were used to quantify TNF-α and IL-6 in the supernatants.

### 4.7. *Western Blotting*

The frozen cerebral cortex and hippocampus tissues were homogenized in ice-cold extraction buffer (20 mM Tris-HCl buffer, pH 7.6, 150 mM NaCl, 2 mM EDTA·2Na, 50 mM sodium fluoride, 1 mM sodium vanadate, 1% Nonidet P-40, 1% sodium deoxycholate, 0.1% SDS, 1 mg/mL aprotinin, and 1 mg/mL leupeptin). The resultant homogenates were centrifuged at 10000× *g* for 10 min at 4 °C to obtain the final supernatants. Nuclear and cytoplasmic extracts for Western blot analysis were extracted using a nuclear/cytoplasmic isolation kit (Beyotime Institute of Biotechnology, Shanghai, China). Pierce BCA Protein Assay Kit (ibid.) was used to determine protein concentrations. Equal amounts of protein (20 μg) for each sample were separated by SDS–PAGE and transferred onto nitrocellulose membranes. And 5% skim milk powder in Tris-buffered saline containing 0.05% (v/v) Tween 20 (TBST) was used to block the membranes at 25 °C for 2 h, before incubation with the primary antibodies to NF-κB p65 (1:1000), p38 MAPK (1:1000), phospho- p38 MAPK (1:1000), SYN (1:1000), and PSD-95 (1:2000) and β-actin (1:1000) at 4 °C overnight. Then, the membranes were washed three times every 15 min with

TBST and then incubated with the secondary horseradish peroxidase-linked anti-rabbit (1:1000) or anti-mouse (1:1000) antibodies (ZSGB-BIO, Beijing, China) at 37 °C for 1 hour. Bands were scanned, and the density was analyzed by the Quantity One analysis software (Bio-Rad Laboratories, Hercules, CA, USA). All quantitative analyses were performed based on our former researches [49].

### 4.8. Reverse Transcriptase-PCR (RT-PCR)

The assay was performed based on previous researches [44,48]. The total RNA was extracted using trizol reagent. A High Capacity RNA-to-cDNA kit was applied to synthesize cDNA. The sequences of the forward and reverse primers for SYN, PSD-95 and the housekeeping gene β-actin (Sangon Biotech Co. Ltd., Shanghai, China) are shown in Table 1. Electrophoresis on a 1% agarose gel was used to separate amplified products followed by photography for visualization under a UA trans-illuminator. In order to verify reproducibility, duplicate reaction was performed. The values obtained for the target gene expression were normalized to β-actin and quantified relative to the expression in the control samples. The products were analyzed with densitometry using the Quantity One 1-D analysis software (Bio-Rad, Hercules, CA, USA).

**Table 1.** Sequences and annealing temperatures of the oligo primers used in this study.

| Target mRNA Sequences | Primer Sequence | Annealing Tm (°C) |
|---|---|---|
| β-actin | 5′ ATGGTCACGCACGATTTCCC 3′<br>5′ GAGACCTTCAACACCCCAGC 3′ | 59 |
| SYN | 5′-TCTTCCTGCAGAACAAGTACC-3′<br>5′-CCTTGCATGTGTTCCCTGTCTG-3′ | 200 |
| PSD-95 | 5′- CCCAGACATCACAACCTCAT -3′<br>5′- ACACCATTGACCGACAGGAT -3′ | 324 |

### 4.9. Histological Analysis

After the behavioral test, mice were immediately anesthetized with sodium pentobarbital (50 mg/g, i.p. injection) and then perfused with ice-cold normal saline followed by 4% paraformaldehyde via the left ventricle. The whole brain was removed and fixed in 4% paraformaldehyde, then in 15% cane sugar for 24 h, followed by dehydration in 30% cane sugar for 12 h. For histological analysis with Nissl's staining, all specimens were frozen and cut into consecutive coronal sections (30 μm in thickness). The number of intact cells in the cerebral cortex and hippocampal CA1 subfield were counted by an investigator blinded to sample identity, and the average value from adjacent two sections was used for each animal. Data were represented as cells per $mm^2$. The histological analysis was performed as previous research described [50].

### 4.10. Molecular Docking Studies

In order to investigate the possible binding modes of ATX with human p38 alpha and human p38 alpha with p38 inhibitor PH797804, a molecular docking study was carried out using the Sybyl v7.1 program package (Tripos International, St. Louis, MO). The three-dimensional structure of human p38 and p38 inhibitor PH797804 alpha were taken from the Protein Data Bank (PDB ID: 4l8m; http://www.rcsb.org/), hydrogen atoms were added to the crystallographic structures and all the water were removed subsequently. The energy of human p38 and p38 inhibitor PH797804 alpha were minimized, before ATX had been docked into the active site of the ATP pocket of p38 alpha.

### 4.11. Statistical Analysis

All values are expressed as the mean ± SEM and analyzed by SPSS v16.0 (SPSS, Inc., Chicago, IL, USA). Differences between the groups were assessed by the one-way ANOVA and the Turkey's test. Significant differences were represented as * $p < 0.05$.

**Author Contributions:** X.Y. and Y.L. designed the experiments. X.Y. and A.-L.G. wrote the manuscript. X.Y., A.-L.G., Y.-P.P., X.-J.C., T.X., X.-R.L., J.L. and Y.-Y.Z. performed experiments and analyzed data.

**Funding:** This research was funded by the Open Research Project of Jiangsu Key Laboratory of Immunity and Metabolism (JSKIM201703), the Technology Innovation Project for the Special Social Development of Public Health in Xuzhou City (KC16SW164), the Open Program of Key Laboratory of Nuclear Medicine, and Ministry of Health and Jiangsu Key Laboratory of Molecular Nuclear Medicine (KF201503).

**Acknowledgments:** We especially thank editor Hao Guo for conscientious guidance and modification on this article.

**Conflicts of Interest:** The authors declare no conflict of interest.

## References

1. Anstey, K.J.; Sanden, C.V.; Salim, A.; O'Kearney, R. Smoking as a Risk Factor for Dementia and Cognitive Decline: A Meta-Analysis of Prospective Studies. *Am. J. Epidemiol.* **2008**, *166*, 367–378. [CrossRef] [PubMed]
2. Chen, R.; Hu, Z.; Orton, S.; Chen, R.L.; Wei, L. Association of passive smoking with cognitive impairment in nonsmoking older adults: a systematic literature review and a new study of Chinese cohort. *J. Geriatr. Psychiatry Neurol.* **2013**, *3*, 199–208. [CrossRef] [PubMed]
3. Hall, B.J.; Cauley, M.; Burke, D.A.; Kiany, A.; Slotkin, T.A.; Levin, E.D. Cognitive and Behavioral Impairments Evoked by Low Level Exposure to Tobacco Smoke Components: Comparison with Nicotine Alone. *Toxicol. Sci.* **2016**, *151*, 236–244. [CrossRef] [PubMed]
4. Fitzpatrick, C.; Barnett, T.A.; Pagani, L.S. Parental bad habits breed bad behaviors in youth: exposure to gestational smoke and child impulsivity. *Int. J. Psychophysiol.* **2014**, *93*, 17–21. [CrossRef] [PubMed]
5. Renda, T.; Baraldo, S.; Pelaia, G.; Bazzan, E.; Turato, G.; Papi, A.; Maestrelli, P.; Maselli, R.; Vatrella, A.; Fabbri, L.M. Increased activation of p38 MAPK in COPD. *Eur. Respir. J.* **2008**, *31*, 62–69. [CrossRef]
6. Lin, X.; Zhang, S.; Huang, R.; Wei, L.; Tan, S.; Liang, C.; Lv, S.; Chen, Y.; Liang, S.; Tian, Y.; et al. Protective effect of madecassoside against cognitive impairment induced by D-galactose in mice. *Pharmacol. Biochem. Behav.* **2014**, *124*, 434–442. [CrossRef]
7. Xie, C.; Kang, J.; Ferguson, M.E.; Nagarajan, S.; Badger, T.M.; Wu, X. Blueberries reduce pro-inflammatory cytokine TNF-α and IL-6 production in mouse macrophages by inhibiting NF-κB activation and the MAPK pathway. *Mol. Nutr. Food Res.* **2011**, *55*, 1587–1591. [CrossRef]
8. Feng, D.; Ling, W.H.; Duan, R.D. Lycopene suppresses LPS-induced NO and IL-6 production by inhibiting the activation of ERK, p38MAPK, and NF-kappaB in macrophages. *Inflamm. Res.* **2010**, *59*, 115–121. [CrossRef]
9. Kaisar, M.A.; Sivandzade, F.; Bhalerao, A.; Cucullo, L. Conventional and electronic cigarettes dysregulate the expression of iron transporters and detoxifying enzymes at the brain vascular endothelium: In vivo evidence of a gender-specific cellular response to chronic cigarette smoke exposure. *Neurosci. Lett.* **2018**, *682*, 1–9. [CrossRef]
10. Fernando, B.T.; José, F.O.; Roberto, D.P. Reactive Oxygen Species: Physiological and Physiopathological Effects on Synaptic Plasticity. *J. Exp. Neurosci.* **2016**, *10*, 23–48. [CrossRef]
11. Yeh, P.T.; Huang, H.W.; Yang, C.M.; Yang, W.S.; Yang, C.H. Astaxanthin Inhibits Expression of Retinal Oxidative Stress and Inflammatory Mediators in Streptozotocin-Induced Diabetic Rats. *PLoS ONE* **2016**, *11*, e0146438. [CrossRef] [PubMed]
12. Shah, M.M.R.; Liang, Y.; Cheng, J.J.; Daroch, M. Astaxanthin-Producing Green MicroalgaHaematococcus pluvialis: From Single Cell to High Value Commercial Products. *Front. Plant Sci.* **2016**, *7*, 531–559. [CrossRef] [PubMed]
13. Huang, X.G.; Zhang, W.; Sun, J.X.; Bai, W.B. Research progress of the structure, function, extraction and analysis of astaxanthin. *Sci. Tech. Food Ind.* **2009**, *30*, 355–359.
14. Al-Amin, M.M.; Reza, H.M.; Saadi, H.M.; Mahmud, W.; Ibrahim, A.A.; Alam, M.M.; Kabir, N.; Saifullah, A.R.M.; Tropa, S.T.; Quddus, A.H.M.R. Astaxanthin ameliorates aluminum chloride-induced spatial memory impairment and neuronal oxidative stress in mice. *E. J. Pharmacol.* **2016**, *777*, 60–69. [CrossRef] [PubMed]
15. Dae-Hee, L.; Lee, Y.J.; Han, K.K. Neuroprotective Effects of Astaxanthin in Oxygen-Glucose Deprivation in SH-SY5Y Cells and Global Cerebral Ischemia in Rat. *J. Clin. Biochem. Nutr.* **2010**, *47*, 121–129. [CrossRef]

16. Tingting, Y.; Yan, Z.; Xia, Z.; Xiaotong, L. Astaxanthin Inhibits Acetaldehyde-Induced Cytotoxicity in SH-SY5Y Cells by Modulating Akt/CREB and p38MAPK/ERK Signaling Pathways. *Mar. Drugs* **2016**, *14*, 56. [CrossRef]
17. Zhang, X.S.; Zhang, X.; Wu, Q.; Li, W.; Wang, C.X.; Xie, G.B.; Zhou, X.M.; Shi, J.X.; Zhou, M.L. Astaxanthin offers neuroprotection and reduces neuroinflammation in experimental subarachnoid hemorrhage. *J. Surg. Res.* **2014**, *192*, 206–213. [CrossRef]
18. Wu, D.M.; Lu, J.; Zheng, Y.L.; Zhou, Z.; Shan, Q.; Ma, D.F. Purple sweet potato color repairs d-galactose-induced spatial learning and memory impairment by regulating the expression of synaptic proteins. *Neurobiol. Learn. Memory* **2008**, *90*, 19–27. [CrossRef]
19. Neal, R.E.; Jagadapillai, R.; Jing, C.; Webb, C.; Stocke, K.; Greene, R.M.; Pisano, M.M. Developmental clgarette smoke exposure II: Hippocampus proteome and metabolome profiles in adult offspring. *Reprod. Toxicol.* **2016**, *65*, 436–447. [CrossRef]
20. Polanska, K.; Krol, A.; Merecz-Kot, D.; Ligocka, D.; Mikolajewska, K.; Mirabella, F.; Chiarotti, F.; Calamandrei, G.; Hanke, W. Environmental Tobacco Smoke Exposure during Pregnancy and Child Neurodevelopment. *Int. J. Environ. Res. Public Health* **2017**, *14*, 796. [CrossRef]
21. Oswald, M.C.W.; Brooks, P.S.; Zwart, M.F.; Mukherjee, A.; West, R.J.H.; Giachello, C.N.G.; Morarach, K.; Baines, R.A.; Sweeney, S.T.; Landgraf, M. Reactive oxygen species regulate activity-dependent neuronal plasticity in Drosophila. *eLife* **2018**, *7*, e39393. [CrossRef] [PubMed]
22. Ma, J.; Chen, X.; Xin, G.; Li, X. Chronic exposure to the ionic liquid [C8mim]Br induces inflammation in silver carp spleen: Involvement of oxidative stress-mediated p38MAPK/NF-kappaB signalling and microRNAs. *Fish Shellfish Immunol.* **2018**, *84*, 627–638. [CrossRef] [PubMed]
23. Jing, Z.; Yu, C.; Xuan, Z.; Chen, H.; Dong, J.; Lu, W.; Song, Z.; Wei, Z. Porphyromonas gingivalis lipopolysaccharide induces cognitive dysfunction, mediated by neuronal inflammation via activation of the TLR4 signaling pathway in C57BL/6 mice. *J. Neuroinflamm.* **2018**, *15*, 37–51. [CrossRef]
24. Du, C.N.; Min, A.Y.; Kim, H.J.; Shin, S.K.; Yu, H.N.; Sohn, E.J.; Ahn, C.W.; Jung, S.U.; Park, S.H.; Kim, M.R. Deer Bone Extract Prevents Against Scopolamine-Induced Memory Impairment in Mice. *J. Med. Food* **2015**, *18*, 157–165. [CrossRef] [PubMed]
25. Chen, R.; Clifford, A.; Lang, L.; Anstey, K.J. Is exposure to secondhand smoke associated with cognitive parameters of children and adolescents?- a systematic literature review. *Ann. Epidemiol.* **2013**, *23*, 652–661. [CrossRef] [PubMed]
26. Amoskroohs, R.M.; Williams, M.T.; Braun, A.A.; Graham, D.L.; Webb, C.L.; Birtles, T.S.; Greene, R.M.; Vorhees, C.V.; Pisano, M.M. Neurobehavioral phenotype of C57BL/6J mice prenatally and neonatally exposed to cigarette smoke. *Neurotoxicol. Teratol.* **2013**, *35*, 34–45. [CrossRef] [PubMed]
27. Swomley, A.M.; Butterfield, D.A. Oxidative stress in Alzheimer disease and mild cognitive impairment: evidence from human data provided by redox proteomics. *Arch. Toxicol.* **2015**, *89*, 1669–1680. [CrossRef]
28. Shi, Y.; Pulliam, D.A.; Liu, Y.; Hamilton, R.T.; Jernigan, A.L.; Bhattacharya, A.; Sloane, L.B.; Qi, W.; Chaudhuri, A.; Buffenstein, R. Reduced mitochondrial ROS, enhanced antioxidant defense, and distinct age-related changes in oxidative damage in muscles of long-lived Peromyscus leucopus. *Am. J. Physiol. Regul. Integr. Comp. Physiol.* **2013**, *304*, R343–R355. [CrossRef]
29. Seo, S.B.; Choe, E.S.; Kim, K.S.; Shim, S.M. The effect of tobacco smoke exposure on generation of reactive oxygen species and cellular membrane damage using a co-culture model system of blood-brain barrier with astrocytes. *Toxicol. Ind. Health* **2017**, *33*, 530–536. [CrossRef]
30. Ambrożewicz, E.; Augustyniak, A.; Gęgotek, A.; Bielawska, K.; Skrzydlewska, E. Black-currant protection against oxidative stress formation. *J.Toxicol. Environ. Health A* **2013**, *76*, 1293–1306. [CrossRef]
31. George, A.; Ng, C.; O'Callaghan, M.; Jensen, G.; Wong, H. In vitro and ex-vivo cellular antioxidant protection and cognitive enhancing effects of an extract of Polygonum minus Huds (Lineminus™) demonstrated in a Barnes Maze animal model for memory and learning. *BMC Complement. Altern. Med.* **2014**, *14*, 161–171. [CrossRef] [PubMed]
32. Wu, W.; Wang, X.; Xiang, Q.; Meng, X.; Peng, Y.; Du, N.; Liu, Z.; Sun, Q.; Wang, C.; Liu, X. Astaxanthin alleviates brain aging in rats by attenuating oxidative stress and increasing BDNF levels. *Food Funct.* **2013**, *5*, 158–166. [CrossRef] [PubMed]

33. Arab, H.H.; Salama, S.A.; Maghrabi, I.A. Camel Milk Ameliorates 5-Fluorouracil-Induced Renal Injury in Rats: Targeting MAPKs, NF-κB and PI3K/Akt/eNOS Pathways. *Cell. Physiol. Biochem.* **2018**, *46*, 1628–1642. [CrossRef] [PubMed]
34. Khan, A.Q.; Khan, R.; Qamar, W.; Lateef, A.; Rehman, M.U.; Tahir, M.; Ali, F.; Hamiza, O.O.; Hasan, S.K.; Sultana, S. Geraniol attenuates 12-O-tetradecanoylphorbol-13-acetate (TPA)-induced oxidative stress and inflammation in mouse skin: Possible role of p38 MAP Kinase and NF-κB. *Exp. Mol. Pathol.* **2013**, *94*, 419–429. [CrossRef] [PubMed]
35. Hahn, W.S.; Kuzmicic, J.; Burrill, J.S.; Donoghue, M.A.; Foncea, R.; Jensen, M.D.; Lavandero, S.; Arriaga, E.A.; Bernlohr, D.A. Proinflammatory cytokines differentially regulate adipocyte mitochondrial metabolism, oxidative stress, and dynamics. *Am. J. Physiol. Endocrinol. Metab.* **2014**, *306*, E1033–E1045. [CrossRef]
36. Niu, J.; Wang, K.; Kolattukudy, P.E. Cerium oxide nanoparticles inhibit oxidative stress and nuclear factor-κB activation in H9c2 cardiomyocytes exposed to cigarette smoke extract. *J. Pharmacol. Exp. Ther.* **2011**, *338*, 53–61. [CrossRef]
37. Ding, Y.; Shi, C.; Chen, L.; Ma, P.; Li, K.; Jin, J.; Zhang, Q.; Li, A. Effects of andrographolide on postoperative cognitive dysfunction and the association with NF-κB/MAPK pathway. *Oncol. Lett.* **2017**, *14*, 7367–7373. [CrossRef]
38. Karunakaran, S.; Ravindranath, V. Activation of p38 MAPK in the substantia nigra leads to nuclear translocation of NF-kappaB in MPTP-treated mice: implication in Parkinson's disease. *J. Neurochem.* **2010**, *109*, 1791–1799. [CrossRef]
39. Kasai, H.; Fukuda, M.S.; Hayashi, T.A.; Noguchi, J. Structural dynamics of dendritic spines in memory and cognition. *Trends Neurosci.* **2010**, *33*, 121–129. [CrossRef]
40. Jauniaux, E.; Burton, G. Morphological and biological effects of maternal exposure to tobacco smoke on the feto-placental unit. *Early Hum. Dev.* **2007**, *83*, 699–706. [CrossRef]
41. Ma, J.; Zhang, Z.; Kang, L.; Geng, D.; Wang, Y.; Wang, M.; Cui, H. Repetitive transcranial magnetic stimulation (rTMS) influences spatial cognition and modulates hippocampal structural synaptic plasticity in aging mice. *Exp. Gerontol.* **2014**, *58*, 256–268. [CrossRef] [PubMed]
42. Jaques, J.A.; Doleski, P.H.; Castilhos, L.G.; Da, R.M.; Souza, V.C.; Carvalho, F.B.; Marisco, P.; Thorstenberg, M.L.; Rezer, J.F.; Ruchel, J.B. Free and nanoencapsulated curcumin prevents cigarette smoke-induced cognitive impairment and redox imbalance. *Neurobiol. Learn. Memory* **2013**, *100*, 98–107. [CrossRef] [PubMed]
43. Tian, X.; Sun, L.; Gou, L.; Ling, X.; Feng, Y.; Wang, L.; Yin, X.; Liu, Y. Protective effect of l-theanine on chronic restraint stress-induced cognitive impairments in mice. *Brain Res.* **2013**, *1503*, 24–32. [CrossRef] [PubMed]
44. Liu, Y.; Fu, X.; Lan, N.; Li, S.; Zhang, J.; Wang, S.; Li, C.; Shang, Y.; Huang, T.; Zhang, L. Luteolin protects against high fat diet-induced cognitive deficits in obesity mice. *Behav. Brain Res.* **2014**, *267*, 178–188. [CrossRef] [PubMed]
45. Liu, Y.; Lan, N.; Ren, J.; Wu, Y.; Wang, S.; Huang, X.; Yu, Y. Orientin improves depression-like behavior and BDNF in chronic stressed mice. *Mol. Nutr. Food Res.* **2015**, *59*, 1130–1142. [CrossRef] [PubMed]
46. Liu, Y.; Tian, X.; Gou, L.; Fu, X.; Li, S.; Lan, N.; Yin, X. Protective effect of l -citrulline against ethanol-induced gastric ulcer in rats. *Environ. Toxicol. Pharmacol.* **2012**, *34*, 280–287. [CrossRef] [PubMed]
47. Aebi, H. Catalase in vitro. *Methods Enzymol.* **1984**, *105*, 121–126.
48. Wang, S.; Yu, Y.; Feng, Y.; Zou, F.; Zhang, X.; Huang, J.; Zhang, Y.; Zheng, X.; Huang, X.F.; Zhu, Y. Protective effect of the orientin on noise-induced cognitive impairments in mice. *Behav. Brain Res.* **2016**, *296*, 290–300. [CrossRef]
49. Liu, Y.; Zhang, Y.; Zheng, X.; Fang, T.; Yang, X.; Luo, X.; Guo, A.; Newell, K.A.; Huang, X.F.; Yu, Y. Galantamine improves cognition, hippocampal inflammation, and synaptic plasticity impairments induced by lipopolysaccharide in mice. *J. Neuroinflamm.* **2018**, *15*, 112–127. [CrossRef]
50. Liu, Y.; Tian, X.; Gou, L.; Sun, L.; Ling, X.; Yin, X. Luteolin attenuates diabetes-associated cognitive decline in rats. *Brain Res. Bull.* **2013**, *94*, 23–29. [CrossRef]

*marine drugs*

MDPI

*Article*

# Astaxanthin Suppresses Cigarette Smoke-Induced Emphysema through Nrf2 Activation in Mice

**Hiroaki Kubo, Kazuhisa Asai *, Kazuya Kojima, Arata Sugitani, Yohkoh Kyomoto, Atsuko Okamoto, Kazuhiro Yamada, Naoki Ijiri, Tetsuya Watanabe, Kazuto Hirata and Tomoya Kawaguchi**

Department of Respiratory Medicine, Graduate School of Medicine, Osaka City University, Osaka 545-8585, Japan; khiro@med.osaka-cu.ac.jp (H.K.); kazuya-kojima@maia.eonet.ne.jp (K.K.); arata.sugitani@gmail.com (A.S.); kitoyou79@yahoo.co.jp (Y.K.); m1169499@med.osaka-cu.ac.jp (A.O.); kazuhironishiyamato@yahoo.co.jp (K.Y.); naru_hodou@yahoo.co.jp (N.I.); tetsuyaw5353@gmail.com (T.W.); kazutoh@msic.med.osaka-cu.ac.jp (K.H.); kawaguchi.tomoya@med.osaka-cu.ac.jp (T.K.)

* Correspondence: kazuasai@med.osaka-cu.ac.jp; Tel.: +81-6-6645-3916

Received: 31 October 2019; Accepted: 26 November 2019; Published: 28 November 2019

**Abstract:** Oxidative stress plays an important role in the pathogenesis of chronic obstructive pulmonary disease (COPD). The activation of nuclear factor erythroid 2-related factor 2 (Nrf2) is a key cellular defense mechanism against oxidative stress. Recent studies have shown that astaxanthin protects against oxidative stress via Nrf2. In this study, we investigated the emphysema suppression effect of astaxanthin via Nrf2 in mice. Mice were divided into four groups: control, smoking, astaxanthin, and astaxanthin + smoking. The mice in the smoking and astaxanthin + smoking groups were exposed to cigarette smoke for 12 weeks, and the mice in the astaxanthin and astaxanthin + smoking groups were fed a diet containing astaxanthin. Significantly increased expression levels of Nrf2 and its target gene, heme oxygenase-1 (HO-1), were found in the lung homogenates of astaxanthin-fed mice. The number of inflammatory cells in the bronchoalveolar lavage fluid (BALF) was significantly decreased, and emphysema was significantly suppressed. In conclusion, astaxanthin protects against oxidative stress via Nrf2 and ameliorates cigarette smoke-induced emphysema. Therapy with astaxanthin directed toward activating the Nrf2 pathway has the potential to be a novel preventive and therapeutic strategy for COPD.

**Keywords:** chronic obstructive pulmonary disease; oxidative stress; astaxanthin; nuclear factor erythroid 2-related factor 2; heme oxygenase-1

## 1. Introduction

Chronic obstructive pulmonary disease (COPD) is caused by the prolonged inhalation of noxious gases, primarily cigarette smoke [1]. Cigarette smoke contains many harmful substances such as oxidants [2]. It has been hypothesized that the etiology of COPD stems from an oxidant–antioxidant imbalance and a protease–antiprotease imbalance. Oxidative stress is an important factor in COPD pathogenesis [3]. Therefore, antioxidant treatment has recently attracted attention in COPD research [4]. Nuclear factor erythroid 2-related factor 2 (Nrf2) is a transcription factor that regulates antioxidant capacity [5]. Nrf2 translocates to the nucleus of the cell and binds to the antioxidant response element (ARE) in response to oxidative stress. Subsequently, Nrf2 initiates the transcription of antioxidant genes and the expression of corresponding proteins. The activation of the Nrf2–ARE signaling pathway is known to be a primary mechanism in the defense against oxidative stress [6]. It has been reported that Nrf2-deficient mice are highly susceptible to cigarette smoke-induced lung injury [7,8]. In addition, the overexpression of Nrf2 was reported to protect against cigarette smoke-induced cell apoptosis [9,10]. These reports suggest that Nrf2 activation protects against the oxidative stress seen in cigarette smoke-induced emphysema.

Astaxanthin is a xanthophyll carotenoid that is widely distributed throughout the world, particularly in marine environments. This compound has potent antioxidant activity, which has been shown to be greater than that of other carotenoids and vitamin E [11,12]. In addition, several studies have reported that astaxanthin activates the Nrf2–ARE signaling pathway as the mechanism for exerting its antioxidant effects [13–17]. However, the suppression of cigarette smoke-induced emphysema by astaxanthin via its antioxidant activity has not yet been reported.

Based on these reports, we hypothesize that astaxanthin enhances Nrf2 expression in the lungs, attenuates oxidative stress, and ameliorates cigarette smoke-induced emphysema. To address this hypothesis, we examined the Nrf2–ARE signaling pathway and the emphysema suppression effect by administering astaxanthin in a murine model of COPD.

## 2. Results

### 2.1. Body Weight Changes in Mice

One mouse from the control group and one mouse from the astaxanthin group were excluded from the analysis due to missing or sample collection failure. Two weeks after starting cigarette smoke exposure, one mouse in the astaxanthin + smoking group died of unknown causes after cigarette smoke exposure.

After the 12-week experimental period, the mice in the smoking group showed lower rates of weight gain than those in the control or astaxanthin groups. Although the mice in the astaxanthin + smoking group had higher rates of weight gain than those in the smoking group, the weight gain was less than that of the mice in the control and astaxanthin groups (Figure 1).

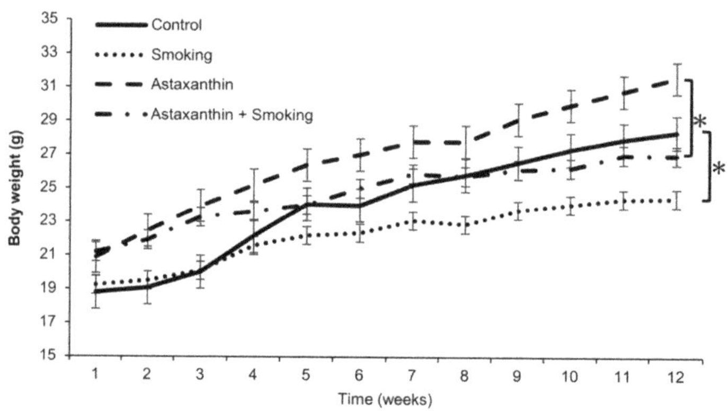


**Figure 1.** Body weight changes in each group. In both standard diet and astaxanthin-fed groups, smoking exposure significantly decreased weight gain. Although it did not reach statistical significance, the astaxanthin + smoking group gained more weight than the smoking group. Values represent the means ± SD. * $p < 0.05$.

### 2.2. *Nrf2 and HO-1 Expression Levels Were Increased in Astaxanthin-Fed Mice*

The Nrf2 mRNA expression levels (as evaluated by real-time PCR) in lung homogenates were significantly increased in the mice in the astaxanthin and astaxanthin + smoking groups compared to those in the control and smoking groups ($p < 0.05$; Figure 2a). No significant difference was observed in Nrf2 mRNA expression levels between the mice in the astaxanthin and astaxanthin + smoking groups. In addition, no significant difference was observed in Nrf2 mRNA expression levels between the mice in the control and smoking groups.

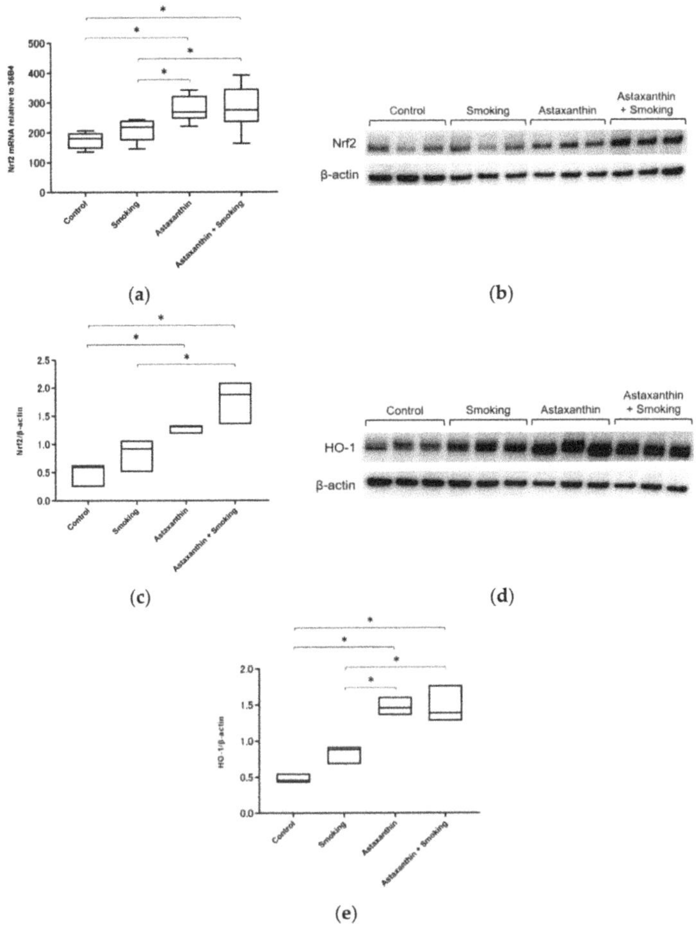


**Figure 2.** Nrf2 and HO-1 expression in the astaxanthin group was significantly increased compared to that in the control group. Similarly, Nrf2 and HO-1 expression in the astaxanthin + smoking group was significantly increased compared to that in the smoking group. Nrf2 mRNA expression in lung homogenates (**a**). Western blot analysis of Nrf2 in lung homogenates (**b**). The blots were normalized to β-actin and measured by densitometry (**c**). Western blot analysis of HO-1 in lung homogenates (**d**). The blots were normalized to β-actin and measured by densitometry (**e**). * $p < 0.05$.

Nrf2 protein expression levels were assessed by Western blot analysis. Nrf2 protein expression was increased in the mice in the astaxanthin group compared to the control group and in the astaxanthin + smoking groups compared to the smoking group ($p < 0.05$; Figure 2b,c). No significant difference in Nrf2 protein expression was observed between the mice in the control and smoking groups. Similarly, no significant difference in Nrf2 protein expression was observed between the mice in the astaxanthin and the astaxanthin + smoking groups.

To evaluate the Nrf2–ARE signaling pathway, heme oxygenase-1 (HO-1), which is regulated by Nrf2, was also assessed by Western blot analysis. HO-1 protein expression was also increased in the mice in the astaxanthin and astaxanthin + smoking groups compared to that in the control and smoking groups ($p < 0.05$; Figure 2d,e). No significant difference in HO-1 protein expression was observed between the mice in the control and smoking groups. Similarly, no significant difference in HO-1 protein expression was observed between the astaxanthin and astaxanthin + smoking groups.

### 2.3. *Astaxanthin Ameliorated Inflammatory Cell Increase in BALF of Cigarette Smoke-Induced COPD*

A representative image of the bronchoalveolar lavage fluid (BALF) from each group is shown in Figure 3a. To examine the influence of cigarette smoke exposure on BALF and the changes induced by astaxanthin, we enumerated the cell populations and evaluated the number of cells in the BALF. No significant differences in total cell count, the number of macrophages, the number of neutrophils, or the number of lymphocytes in the BALF of mice were observed in the control and astaxanthin groups (Figure 3b–e). The number of neutrophils was significantly higher in the BALF of mice in the smoking and astaxanthin + smoking groups compared to the control and astaxanthin groups due to the effects of smoking exposure (Figure 3d). Total cell count and the number of macrophages and lymphocytes were significantly higher in the BALF of mice in the smoking group compared to the control and astaxanthin groups (Figure 3b,c,e). Total cell count and the number of macrophages and neutrophils were significantly lower in the BALF of mice in the astaxanthin + smoking group compared to the smoking group ($p < 0.05$; Figure 3b–d). No significant difference was observed in the number of lymphocytes in the BALF of mice in the smoking and astaxanthin + smoking groups (Figure 3e).

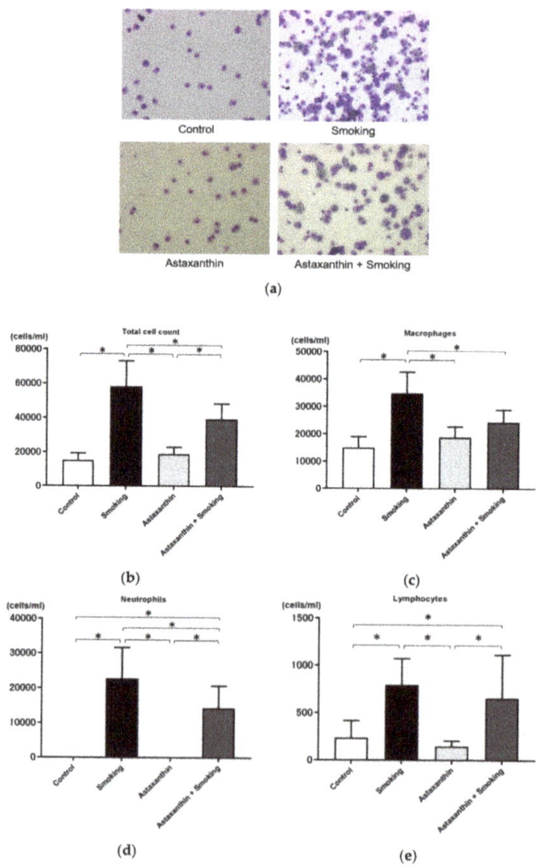


**Figure 3.** Total cell count and the number of macrophages and neutrophils in the bronchoalveolar lavage fluid (BALF) were significantly lower in the BALF of mice in the astaxanthin + smoking group than in the smoking group, but the number of lymphocytes was not attenuated. Representative images of the BALF from each group are shown at 200× magnification (**a**). Number of total cells (**b**), number of macrophages (**c**), number of neutrophils (**d**), and number of lymphocytes in the BALF (**e**). Values represent the means ± SD. * $p < 0.05$.

### 2.4. Astaxanthin Ameliorated Cigarette Smoke-Induced Emphysema

A representative histologic image of the lung from each group is shown in Figure 4a. The exposure to cigarette smoke for 12 weeks resulted in the development of pulmonary emphysema in the mice in the smoking group. The lung tissues from the astaxanthin + smoking group showed lower alveolar destruction than the lung tissues of the mice in the smoking group. Mean linear intercept (MLI) was significantly larger in the mice in the smoking group than in the control group, and MLI for the mice in the astaxanthin + smoking group was significantly lower than that reported for the mice in the smoking group ($p < 0.05$; Figure 4b). No significant difference was observed in MLI between the mice in the control, astaxanthin, and astaxanthin + smoking groups. Moreover, the destructive index was significantly larger in the mice in the smoking group than in the control or astaxanthin groups, and the destructive index was significantly lower in the mice in the astaxanthin + smoking group than in the smoking group ($p < 0.05$; Figure 4c). No significant difference was observed in the destructive index between the mice in the control, astaxanthin, and astaxanthin + smoking groups.

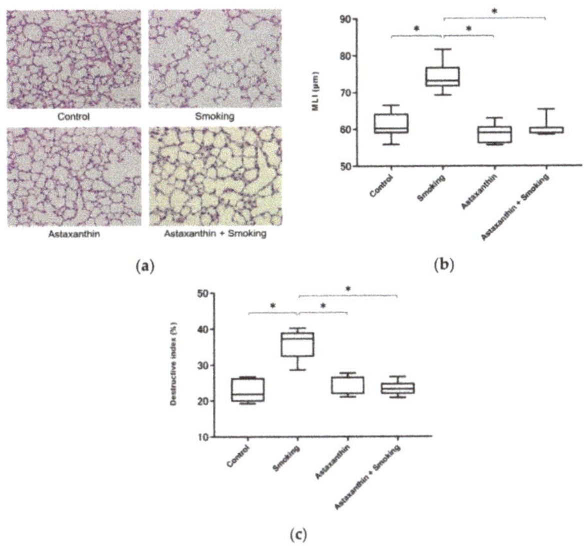


**Figure 4.** Mean linear intercept (MLI) and destructive index were significantly larger in the smoking group than in the control group. MLI and destructive index were significantly smaller in the astaxanthin + smoking group than in the smoking group. No significant difference was observed in MLI and destructive index between the control, astaxanthin, and astaxanthin + smoking groups. Representative histologic image of lung sections from each group stained with hematoxylin-eosin. Destructed alveolar lesions are indicated by arrows (**a**). MLI data (**b**). Destructive index data (**c**). * $p < 0.05$.

## 3. Discussion

In this study, we showed that astaxanthin increased Nrf2 and HO-1 expression in lung tissue and suppressed cigarette smoke-induced emphysema in mice. Our results indicate that the ingestion of astaxanthin suppresses cigarette smoke-induced inflammatory cell infiltration in the BALF and emphysema by activating the Nrf2–ARE signaling pathway in the lungs in a murine model of COPD.

COPD is the third leading cause of death in the world [18]. However, current therapies for COPD provide only limited benefit and fail to halt progression. Therefore, the development of new prevention and treatment strategies for COPD is necessary. Cigarette smoke is the primary cause of COPD, and it contains many oxidants [2]. An insufficient antioxidant capacity is related to COPD pathogenesis [19]. An excess of oxidants has been reported to induce emphysema through epithelial cell apoptosis [20].

In recent decades, oxidative stress has been recognized as a key factor responsible for the pathogenesis of COPD [21].

Previously, we reported that N-acetylcysteine significantly suppressed cigarette smoke extract-induced apoptosis of airway epithelial cells [22]. This result suggests that antioxidants such as N-acetylcysteine may suppress cigarette smoke-induced apoptosis and emphysema in models of COPD. Epidemiologic evidence also supports the potential beneficial effects of an antioxidant-rich diet on pulmonary function and COPD risk [23]. Antioxidant therapy or supplemental treatment with an external antioxidant to neutralize excess oxidants may have great therapeutic potential in COPD [24].

Nrf2 is a transcription factor involved in the regulation of various antioxidants. In response to oxidative stress, Nrf2 translocates to the nucleus and binds the ARE of target genes involved in an antioxidant response. Subsequently, Nrf2 initiates the transcription and expression of antioxidant proteins. Then, antioxidant proteins induced by Nrf2, such as HO-1, protect against oxidative stress [25]. Nrf2 is expressed in various organs including the lung. Nrf2-deficient mice show reduced activity of antioxidant enzymes, are susceptible to cigarette smoke, and develop severe lung emphysema [7,8]. Moreover, increased Nrf2 activation was shown to attenuate the oxidative stress of cigarette smoke and protect cells from apoptosis induced by oxidative stress [9,10]. We previously showed that Nrf2 expression was significantly reduced in the airway epithelial cells of COPD patients [22]. In addition, other studies indicate the relationship of Nrf2 polymorphisms and airflow limitations in smokers [26,27]. Recently, we reported that a polymorphism of the Nrf2 gene contributed to the progression of lung emphysema in smokers [28]. From these findings, Nrf2 is considered to be prominently involved in the pathogenesis of COPD.

To our knowledge, there is no report that astaxanthin is related to the prevention of COPD. Astaxanthin, a carotenoid xanthophyll, is a natural reddish-orange pigment widely present in nature. Astaxanthin is especially abundant in marine organisms such as shrimp, crab, salmon, krill, and algae. Since ancient times, these crustaceans and fishes have been eaten by humans. Astaxanthin ingestion is safe, and pure astaxanthin was approved as a dietary supplement by the Food and Drug Administration in the United States in 1999 [29]. Recently, the technology for mass-producing astaxanthin by culturing *Haematococcus pluvialis* was developed, and it has become simple to obtain a large amount of astaxanthin. In fact, astaxanthin is widely used in cosmetics because it has been reported to protect the skin from ultraviolet rays and help maintain healthy skin [30]. Astaxanthin has attracted attention due to its strong antioxidant properties, and there have been many reports focusing on its antioxidant activity. Astaxanthin has been shown to protect various cells from oxidative stress in vitro [31–34] and to protect the brain, eyes, salivary glands, skeletal muscle, liver, kidney, and lungs from oxidative stress in vivo [12–17,35–39]. These results indicate that astaxanthin is distributed throughout the body and has systemic effects. Moreover, previous studies have reported that astaxanthin enhances Nrf2 expression in various organs including the lungs [13–17]. Additionally, some studies have investigated the pathway of Nrf2 activation by astaxanthin. Astaxanthin facilitates the dissociation and nuclear translocation of Nrf2 through activation of the PI3K/Akt and ERK signaling pathways [40,41].

In our study, Nrf2 expression in the lungs was slightly higher in the smoking group than in the control group; however, no significant difference was observed. Cigarette smoke-induced oxidants were potentially stronger than the protective effect of the antioxidants in the smoking group, which may have caused emphysema. In contrast, Nrf2 expression was significantly increased in the astaxanthin + smoking group compared to the smoking group. Therefore, the antioxidants may have exerted stronger effects than the cigarette smoke-induced oxidants and suppressed the development of emphysema in the astaxanthin + smoking group.

Previous studies have reported that emphysema was suppressed by administering antioxidant substances to mice. Hydrogen has been found to be a strong antioxidant, and administration of hydrogen-rich water was reported to attenuate cigarette smoke-induced lung damage and reduce the MLI in senescence marker protein-30 knockout mice [42]. 2-Cyano-3,12-dioxooleana-1,9-dien-28-oic acid (CDDO) has also been reported to have an Nrf2 activation effect. CDDO-imidazolide administered during a period of cigarette smoke inhalation was shown to suppress pulmonary emphysema via Nrf2 in mice [43]. These reports support our results.

We showed that astaxanthin inhibited cigarette smoke-induced inflammatory cell infiltration in BALF. Although Nrf2 suppresses inflammation as a secondary consequence of its antioxidant effect, astaxanthin has been also reported to directly suppress inflammation [44,45]. The suppression of inflammatory cell infiltration in BALF may also be related to this property of astaxanthin.

Oxidative stress caused by cigarette smoke has been reported to persist long after smoking cessation [46]. Prolonged oxidative stress is a primary factor in the enhancement of both airway and systemic inflammation in COPD patients and is known to play an important role in the development of COPD and its comorbidities [47]. Therefore, it may be possible to suppress persistent oxidative stress and inflammation by the ingestion of astaxanthin even after smoking cessation; it may also be possible to treat COPD and its comorbidities with a single therapeutic agent. Ingestion of astaxanthin has been proven to be safe, it is widely used in beauty products, and mass production methods have been established. Therefore, astaxanthin may have the potential to serve as a therapeutic agent or a supplement for COPD in the near future.

This study has some limitations. First, the concentrations of astaxanthin in the blood of mice were not determined and bioavailability is unknown. Second, the concentration of astaxanthin (0.02% w/w) in the diet was taken from a previous study [48]. In addition, 50 mg/kg of astaxanthin was reported to be effective in mice [37]. Therefore, we decided to use the diet to contain 0.02% (w/w) astaxanthin. However, the optimal effective concentration of astaxanthin is unknown. Further research is needed to clarify these points.

## 4. Materials and Methods

### 4.1. Experimental Animals

C57BL/6 mice (male, four weeks old, 18–20 g) were obtained from Japan SLC (Shizuoka, Japan) and kept under pathogen-free conditions. The mice were maintained at a controlled temperature of 23 °C ± 2 °C under a 12:12 h light–dark cycle with free access to water. The mice were divided into four groups as follows: (1) control (n = 8), (2) smoking (n = 8), (3) astaxanthin (n = 8), (4) astaxanthin + smoking (n = 8). All mice were acclimatized to the environment for one week. The mice in the astaxanthin and astaxanthin + smoking groups were fed a diet containing astaxanthin (FUJIFILM ASTAXANTHIN 10O; FUJIFILM Corporation, Tokyo, Japan). We prepared the diet to contain 0.02% (w/w) astaxanthin; the concentration of astaxanthin was measured by using high-performance liquid chromatography after enzymatic degradation of fatty acid ester form of astaxanthin to free form of astaxanthin. The actual concentration of the diet was determined to be 0.0158% (w/w). The mice in the control and smoking groups were fed a standard diet. All experimental protocols were approved by the Ethics Committee of the Institutional Animal Care and Use of Osaka City University Graduate School of Medicine (17023, 6/11/2017). Animal experiments were conducted in accordance with the Regulations on Animal Experiments in Osaka City University following the Guidelines for Proper Conduct of Animal Experiments in Japan.

### 4.2. Experimental Model of Cigarette Smoke-Induced COPD

The mice in the smoking and astaxanthin + smoking groups were exposed to cigarette smoking (18 cigarettes/day) for 60 min once daily, 5 days per week. Commercially available Peace® nonfilter cigarettes (2.3 mg nicotine and 28 mg tar/cigarette; Japan Tobacco, Tokyo, Japan) and a cigarette smoke generator model SG-300 for small animals (Shibata Scientific Technology, Tokyo, Japan) were used for the cigarette smoke exposure. The mice in the control and astaxanthin groups were exposed to fresh air. Cigarette smoke and fresh air exposure was performed for 12 weeks.

### 4.3. Treatments and Preparation for Evaluation

At the end of the 12-week experimental period, all mice were sacrificed under deep anesthesia. The mice were tracheotomized and cannulated, and bronchoalveolar lavage (BAL) was performed three times with 0.5 mL phosphate-buffered saline for sampling of the BALF. After BAL, the right lung of each mouse was carefully excised. The right lower lobe was instantly soaked in RNAlater (Invitrogen by Thermo Fisher Scientific, Waltham, MA, USA) for mRNA expression analysis. The other right lobe was immediately frozen in liquid nitrogen for protein expression analysis. The left lung was excised and immediately soaked in 10% formalin for further histological analysis.

### 4.4. *Nrf2* mRNA Expression Analysis

The right lower lobe was homogenized in RLT lysis buffer (Qiagen NV, Venlo, Netherlands). RNeasy mini kit (Qiagen NV, Venlo, Netherlands) was used for the extraction of total RNA. Complementary DNA (cDNA) was obtained by reverse transcription of the mRNA with the Ready-to-Go T-primed first-strand kit (GE Healthcare, Little Chalfont, UK). The cDNA was used in a real-time quantitative PCR reaction in an Applied Biosystems 7500 real-time PCR system (Thermo Fisher Scientific, Waltham, MA, USA) using TaqMan gene expression assays for Nrf2 (Mm00477784_m1). The housekeeping gene *36B4* (Mm00725448_s1) was used for the normalization of Nrf2 mRNA as previously described [49].

### 4.5. Western Blot Analysis

The right lung other than the lower lobe was used for Western blot analysis. Approximately 30 mg of the lung sample was soaked in 300 μL of radioimmunoprecipitation assay (RIPA) buffer (Beyotime Biotechnology, Shanghai, China) supplemented with the protease inhibitor phenylmethanesulfonyl fluoride (Beyotime Biotechnology, Shanghai, China) and Protease Inhibitor Cocktail (Cell Signaling Technology Japan, Tokyo, Japan). After the lung samples were homogenized in RIPA buffer, the samples were placed on ice for 5 min and centrifuged at 11,800× *g* and 4 °C for 4 min. The supernatant was collected, and the protein concentration was determined with the colorimetric bicinchoninic acid protein assay kit (Pierce, Waltham, MA, USA) according to the manufacturer's instructions. The supernatant was subjected to sodium dodecyl sulfate polyacrylamide gel electrophoresis with Mini-PROTEAN TGX Precast Protein Gels (4561023, Bio-Rad, Hercules, California, USA). Next, the separated bands on the gel were transferred onto polyvinylidene fluoride membranes. The membranes were then incubated with primary anti-Nrf2 antibody (1:500; ab137550, Abcam, Cambridge, UK), anti-HO-1 antibody (1:250; ab13248, Abcam), or anti-β-actin antibody (1:1000; ab8227, Abcam) at 4 °C overnight. The next day, the membranes were incubated with the corresponding secondary antibodies for 2 h at 25 °C. After washing the membranes three times, SuperSignal West Dura Extended Duration Substrate (Thermo Fisher Scientific, Waltham, MA, USA) was used for detection. Western blot signals were acquired with a Fuji LAS-4000 fluorescence imager (Fujifilm Corporation, Tokyo, Japan). The target protein levels were normalized to β-actin.

### 4.6. Bronchoalveolar Lavage Fluid Analysis

Each BALF sample was centrifuged at 1200× *g* and 4 °C for 10 min, and the supernatant was collected. The cell pellet was resuspended in 1 mL of phosphate-buffered saline and applied to cytospin columns in a Shandon Cytospin 3 centrifuge (Shandon Scientific Co., London, England), and the cytospin protocol was followed. The slides sprayed with the cells were stained with Diff-Quick (Sysmex, Kobe, Japan), and the enumeration of cells and the differential cell counts (macrophages, neutrophils, and lymphocytes) were performed in a blind manner.

### 4.7. Quantitative Evaluation of Lung Emphysema

The left lung was fixed with 10% formalin for 24–48 h at positive pressure (25 cm $H_2O$) and subjected to histological analysis. Three-micrometer-thick slices were stained with hematoxylin and eosin for the analysis of the level of airspace size in the lung. Emphysema was evaluated by determining the MLI as previously described [50]. Moreover, destruction was evaluated by determining the destructive index as previously described [51].

### 4.8. Statistical Analysis

Data are expressed as the mean ± standard deviation. For multiple-group comparisons, differences were evaluated using one-way ANOVA followed by Tukey's multiple comparison test. Statistical significance was considered at $p < 0.05$. All statistical analyses were performed using GraphPad Prism 7.04 (GraphPad Software, San Diego, CA, USA).

## 5. Conclusions

COPD is associated with an excessive oxidant burden; therefore, the rationale for exploring antioxidant therapies in COPD is clear. Astaxanthin increases Nrf2 and HO-1 expression in the lung and suppresses cigarette smoke-induced emphysema in mice. Therapy directed toward activating the Nrf2–ARE pathway, such as the use of astaxanthin, may be a novel preventive and therapeutic strategy for attenuating oxidative stress in the pathogenesis of COPD.

**Author Contributions:** H.K. performed the data collection, data and statistical analysis, interpretation of the results, and article drafting. K.A. designed this study and performed article drafting. K.K., A.S., Y.K., and A.O. performed the sample collection. K.Y., N.I., T.W., K.H., and T.K. interpreted the results and revised the manuscripts. All authors have read and approved the final manuscript.

**Funding:** This work was supported by research grants from JSPS KAKENHI Grant Number 19K08660 to Kazuhisa Asai.

**Acknowledgments:** Real-time PCR analysis and Western blot analysis were performed at the Research Support Platform of Osaka City University Graduate School of Medicine. The authors would like to thank FUJIFILM Corporation for providing astaxanthin.

**Conflicts of Interest:** The authors declare no conflict of interest.

## References

1. Hogg, J.C.; Timens, W. The pathology of chronic obstructive pulmonary disease. *Annu. Rev. Pathol.* **2009**, *4*, 435–459. [CrossRef]
2. Pryor, W.A.; Stone, K. Oxidants in cigarette smoke. Radicals, hydrogen peroxide, peroxynitrate, and peroxynitrite. *Ann. N. Y. Acad. Sci.* **1993**, *686*, 12–27. [CrossRef]
3. Sundar, I.K.; Yao, H.; Rahman, I. Oxidative stress and chromatin remodeling in chronic obstructive pulmonary disease and smoking-related diseases. *Antioxid. Redox Signal.* **2013**, *18*, 1956–1971. [CrossRef]
4. Vezina, F.A.; Cantin, A.M. Antioxidants and Chronic Obstructive Pulmonary Disease. *Chronic Obstr. Pulm. Dis.* **2018**, *5*, 277–288. [CrossRef]
5. Kensler, T.W.; Wakabayashi, N.; Biswal, S. Cell survival responses to environmental stresses via the Keap1-Nrf2-ARE pathway. *Annu. Rev. Pharmacol. Toxicol.* **2007**, *47*, 89–116. [CrossRef]
6. Nguyen, T.; Nioi, P.; Pickett, C.B. The Nrf2-antioxidant response element signaling pathway and its activation by oxidative stress. *J. Biol. Chem.* **2009**, *284*, 13291–13295. [CrossRef] [PubMed]
7. Iizuka, T.; Ishii, Y.; Itoh, K.; Kiwamoto, T.; Kimura, T.; Matsuno, Y.; Morishima, Y.; Hegab, A.E.; Homma, S.; Nomura, A.; et al. Nrf2-deficient mice are highly susceptible to cigarette smoke-induced emphysema. *Genes Cells* **2005**, *10*, 1113–1125. [CrossRef] [PubMed]
8. Rangasamy, T.; Cho, C.Y.; Thimmulappa, R.K.; Zhen, L.; Srisuma, S.S.; Kensler, T.W.; Yamamoto, M.; Petrache, I.; Tuder, R.M.; Biswal, S. Genetic ablation of Nrf2 enhances susceptibility to cigarette smoke-induced emphysema in mice. *J. Clin. Investig.* **2004**, *114*, 1248–1259. [CrossRef] [PubMed]

9. Huang, C.; Wang, J.J.; Ma, J.H.; Jin, C.; Yu, Q.; Zhang, S.X. Activation of the UPR protects against cigarette smoke-induced RPE apoptosis through up-regulation of Nrf2. *J. Biol. Chem.* **2015**, *290*, 5367–5380. [CrossRef]
10. Blake, D.J.; Singh, A.; Kombairaju, P.; Malhotra, D.; Mariani, T.J.; Tuder, R.M.; Gabrielson, E.; Biswal, S. Deletion of Keap1 in the lung attenuates acute cigarette smoke-induced oxidative stress and inflammation. *Am. J. Respir. Cell Mol. Biol.* **2010**, *42*, 524–536. [CrossRef]
11. Naguib, Y.M. Antioxidant activities of astaxanthin and related carotenoids. *J. Agric. Food Chem.* **2000**, *48*, 1150–1154. [CrossRef] [PubMed]
12. Ni, Y.; Nagashimada, M.; Zhuge, F.; Zhan, L.; Nagata, N.; Tsutsui, A.; Nakanuma, Y.; Kaneko, S.; Ota, T. Astaxanthin prevents and reverses diet-induced insulin resistance and steatohepatitis in mice: A comparison with vitamin E. *Sci. Rep.* **2015**, *5*, 17192. [CrossRef] [PubMed]
13. Xie, X.; Chen, Q.; Tao, J. Astaxanthin Promotes Nrf2/ARE Signaling to Inhibit HG-Induced Renal Fibrosis in GMCs. *Mar. Drugs* **2018**, *16*. [CrossRef]
14. Zhu, X.; Chen, Y.; Chen, Q.; Yang, H.; Xie, X. Astaxanthin Promotes Nrf2/ARE Signaling to Alleviate Renal Fibronectin and Collagen IV Accumulation in Diabetic Rats. *J. Diabetes Res.* **2018**, *2018*, 6730315. [CrossRef]
15. Wu, Q.; Zhang, X.S.; Wang, H.D.; Zhang, X.; Yu, Q.; Li, W.; Zhou, M.L.; Wang, X.L. Astaxanthin activates nuclear factor erythroid-related factor 2 and the antioxidant responsive element (Nrf2-ARE) pathway in the brain after subarachnoid hemorrhage in rats and attenuates early brain injury. *Mar. Drugs* **2014**, *12*, 6125–6141. [CrossRef]
16. Feng, Y.; Chu, A.; Luo, Q.; Wu, M.; Shi, X.; Chen, Y. The Protective Effect of Astaxanthin on Cognitive Function via Inhibition of Oxidative Stress and Inflammation in the Brains of Chronic T2DM Rats. *Front. Pharmacol.* **2018**, *9*, 748. [CrossRef]
17. Liu, N.; Zhang, W.; Luo, S.; Cao, J.; Peng, M.; Liu, Z. Astaxanthin suppresses cigarette smoke and lipopolysaccharide-induced airway inflammation through induction of heme oxygenase-1. *Cell. Mol. Biol.* **2019**, *65*, 94–99.
18. Quaderi, S.A.; Hurst, J.R. The unmet global burden of COPD. *Glob. Health Epidemiol. Genom.* **2018**, *3*, e4. [CrossRef]
19. Domej, W.; Oettl, K.; Renner, W. Oxidative stress and free radicals in COPD—Implications and relevance for treatment. *Int. J. Chron. Obs. Pulm. Dis.* **2014**, *9*, 1207–1224. [CrossRef]
20. Hoshino, Y.; Mio, T.; Nagai, S.; Miki, H.; Ito, I.; Izumi, T. Cytotoxic effects of cigarette smoke extract on an alveolar type II cell-derived cell line. *Am. J. Physiol. Lung Cell. Mol. Physiol.* **2001**, *281*, L509–L516. [CrossRef]
21. McGuinness, A.J.; Sapey, E. Oxidative Stress in COPD: Sources, Markers, and Potential Mechanisms. *J. Clin. Med.* **2017**, *6*. [CrossRef] [PubMed]
22. Yamada, K.; Asai, K.; Nagayasu, F.; Sato, K.; Ijiri, N.; Yoshii, N.; Imahashi, Y.; Watanabe, T.; Tochino, Y.; Kanazawa, H.; et al. Impaired nuclear factor erythroid 2-related factor 2 expression increases apoptosis of airway epithelial cells in patients with chronic obstructive pulmonary disease due to cigarette smoking. *BMC Pulm. Med.* **2016**, *16*, 27. [CrossRef] [PubMed]
23. Scoditti, E.; Massaro, M.; Garbarino, S.; Toraldo, D.M. Role of Diet in Chronic Obstructive Pulmonary Disease Prevention and Treatment. *Nutrients* **2019**, *11*, 1357. [CrossRef] [PubMed]
24. Bernardo, I.; Bozinovski, S.; Vlahos, R. Targeting oxidant-dependent mechanisms for the treatment of COPD and its comorbidities. *Pharmacol. Ther.* **2015**, *155*, 60–79. [CrossRef]
25. Loboda, A.; Damulewicz, M.; Pyza, E.; Jozkowicz, A.; Dulak, J. Role of Nrf2/HO-1 system in development, oxidative stress response and diseases: An evolutionarily conserved mechanism. *Cell. Mol. Life Sci.* **2016**, *73*, 3221–3247. [CrossRef]
26. Masuko, H.; Sakamoto, T.; Kaneko, Y.; Iijima, H.; Naito, T.; Noguchi, E.; Hirota, T.; Tamari, M.; Hizawa, N. An interaction between Nrf2 polymorphisms and smoking status affects annual decline in FEV1: A longitudinal retrospective cohort study. *BMC Med. Genet.* **2011**, *12*, 97. [CrossRef]
27. Siedlinski, M.; Postma, D.S.; Boer, J.M.; van der Steege, G.; Schouten, J.P.; Smit, H.A.; Boezen, H.M. Level and course of FEV1 in relation to polymorphisms in NFE2L2 and KEAP1 in the general population. *Respir. Res.* **2009**, *10*, 73. [CrossRef]
28. Sugitani, A.; Asai, K.; Watanabe, T.; Suzumura, T.; Kojima, K.; Kubo, H.; Sato, K.; Ijiri, N.; Yamada, K.; Kimura, T.; et al. A Polymorphism rs6726395 in Nrf2 Contributes to the Development of Emphysema-Associated Age in Smokers without COPD. *Lung* **2019**. [CrossRef]

29. Riccioni, G.; D'Orazio, N.; Franceschelli, S.; Speranza, L. Marine carotenoids and cardiovascular risk markers. *Mar. Drugs* **2011**, *9*, 1166–1175. [CrossRef]
30. Singh, K.N.; Patil, S.; Barkate, H. Protective effects of astaxanthin on skin: Recent scientific evidence, possible mechanisms, and potential indications. *J. Cosmet. Dermatol.* **2019**. [CrossRef]
31. O'Connor, I.; O'Brien, N. Modulation of UVA light-induced oxidative stress by beta-carotene, lutein and astaxanthin in cultured fibroblasts. *J. Dermatol. Sci.* **1998**, *16*, 226–230. [CrossRef]
32. Lee, D.H.; Kim, C.S.; Lee, Y.J. Astaxanthin protects against MPTP/MPP+-induced mitochondrial dysfunction and ROS production in vivo and in vitro. *Food Chem. Toxicol.* **2011**, *49*, 271–280. [CrossRef] [PubMed]
33. Ikeda, Y.; Tsuji, S.; Satoh, A.; Ishikura, M.; Shirasawa, T.; Shimizu, T. Protective effects of astaxanthin on 6-hydroxydopamine-induced apoptosis in human neuroblastoma SH-SY5Y cells. *J. Neurochem.* **2008**, *107*, 1730–1740. [CrossRef] [PubMed]
34. Regnier, P.; Bastias, J.; Rodriguez-Ruiz, V.; Caballero-Casero, N.; Caballo, C.; Sicilia, D.; Fuentes, A.; Maire, M.; Crepin, M.; Letourneur, D.; et al. Astaxanthin from Haematococcus pluvialis Prevents Oxidative Stress on Human Endothelial Cells without Toxicity. *Mar. Drugs* **2015**, *13*, 2857–2874. [CrossRef]
35. Aoi, W.; Naito, Y.; Sakuma, K.; Kuchide, M.; Tokuda, H.; Maoka, T.; Toyokuni, S.; Oka, S.; Yasuhara, M.; Yoshikawa, T. Astaxanthin limits exercise-induced skeletal and cardiac muscle damage in mice. *Antioxid. Redox Signal.* **2003**, *5*, 139–144. [CrossRef]
36. Kuraji, M.; Matsuno, T.; Satoh, T. Astaxanthin affects oxidative stress and hyposalivation in aging mice. *J. Clin. Biochem. Nutr.* **2016**, *59*, 79–85. [CrossRef]
37. Harada, F.; Morikawa, T.; Lennikov, A.; Mukwaya, A.; Schaupper, M.; Uehara, O.; Takai, R.; Yoshida, K.; Sato, J.; Horie, Y.; et al. Protective Effects of Oral Astaxanthin Nanopowder against Ultraviolet-Induced Photokeratitis in Mice. *Oxid. Med. Cell. Longev.* **2017**, *2017*, 1956104. [CrossRef]
38. Bi, J.; Cui, R.; Li, Z.; Liu, C.; Zhang, J. Astaxanthin alleviated acute lung injury by inhibiting oxidative/nitrative stress and the inflammatory response in mice. *Biomed. Pharm.* **2017**, *95*, 974–982. [CrossRef]
39. Song, X.; Wang, B.; Lin, S.; Jing, L.; Mao, C.; Xu, P.; Lv, C.; Liu, W.; Zuo, J. Astaxanthin inhibits apoptosis in alveolar epithelial cells type II in vivo and in vitro through the ROS-dependent mitochondrial signalling pathway. *J. Cell. Mol. Med.* **2014**, *18*, 2198–2212. [CrossRef]
40. Li, Z.; Dong, X.; Liu, H.; Chen, X.; Shi, H.; Fan, Y.; Hou, D.; Zhang, X. Astaxanthin protects ARPE-19 cells from oxidative stress via upregulation of Nrf2-regulated phase II enzymes through activation of PI3K/Akt. *Mol. Vis.* **2013**, *19*, 1656–1666.
41. Wang, H.Q.; Sun, X.B.; Xu, Y.X.; Zhao, H.; Zhu, Q.Y.; Zhu, C.Q. Astaxanthin upregulates heme oxygenase-1 expression through ERK1/2 pathway and its protective effect against beta-amyloid-induced cytotoxicity in SH-SY5Y cells. *Brain Res.* **2010**, *1360*, 159–167. [CrossRef] [PubMed]
42. Suzuki, Y.; Sato, T.; Sugimoto, M.; Baskoro, H.; Karasutani, K.; Mitsui, A.; Nurwidya, F.; Arano, N.; Kodama, Y.; Hirano, S.I.; et al. Hydrogen-rich pure water prevents cigarette smoke-induced pulmonary emphysema in SMP30 knockout mice. *Biochem. Biophys. Res. Commun.* **2017**, *492*, 74–81. [CrossRef] [PubMed]
43. Sussan, T.E.; Rangasamy, T.; Blake, D.J.; Malhotra, D.; El-Haddad, H.; Bedja, D.; Yates, M.S.; Kombairaju, P.; Yamamoto, M.; Liby, K.T.; et al. Targeting Nrf2 with the triterpenoid CDDO-imidazolide attenuates cigarette smoke-induced emphysema and cardiac dysfunction in mice. *Proc. Natl. Acad. Sci. USA* **2009**, *106*, 250–255. [CrossRef] [PubMed]
44. Park, J.H.; Yeo, I.J.; Han, J.H.; Suh, J.W.; Lee, H.P.; Hong, J.T. Anti-inflammatory effect of astaxanthin in phthalic anhydride-induced atopic dermatitis animal model. *Exp. Dermatol.* **2018**, *27*, 378–385. [CrossRef]
45. Ohgami, K.; Shiratori, K.; Kotake, S.; Nishida, T.; Mizuki, N.; Yazawa, K.; Ohno, S. Effects of astaxanthin on lipopolysaccharide-induced inflammation in vitro and in vivo. *Investig. Ophthalmol. Vis. Sci.* **2003**, *44*, 2694–2701. [CrossRef]
46. Louhelainen, N.; Rytila, P.; Haahtela, T.; Kinnula, V.L.; Djukanovic, R. Persistence of oxidant and protease burden in the airways after smoking cessation. *BMC Pulm. Med.* **2009**, *9*, 25. [CrossRef]
47. Cavailles, A.; Brinchault-Rabin, G.; Dixmier, A.; Goupil, F.; Gut-Gobert, C.; Marchand-Adam, S.; Meurice, J.C.; Morel, H.; Person-Tacnet, C.; Leroyer, C.; et al. Comorbidities of COPD. *Eur. Respir. Rev.* **2013**, *22*, 454–475. [CrossRef]
48. Liu, P.H.; Aoi, W.; Takami, M.; Terajima, H.; Tanimura, Y.; Naito, Y.; Itoh, Y.; Yoshikawa, T. The astaxanthin-induced improvement in lipid metabolism during exercise is mediated by a PGC-1alpha increase in skeletal muscle. *J. Clin. Biochem. Nutr.* **2014**, *54*, 86–89. [CrossRef]

49. Okamoto, A.; Nojiri, T.; Konishi, K.; Tokudome, T.; Miura, K.; Hosoda, H.; Hino, J.; Miyazato, M.; Kyomoto, Y.; Asai, K.; et al. Atrial natriuretic peptide protects against bleomycin-induced pulmonary fibrosis via vascular endothelial cells in mice: ANP for pulmonary fibrosis. *Respir. Res.* **2017**, *18*, 1. [CrossRef]
50. Thurlbeck, W.M. The internal surface area of nonemphysematous lungs. *Am. Rev. Respir. Dis.* **1967**, *95*, 765–773. [CrossRef]
51. Saetta, M.; Shiner, R.J.; Angus, G.E.; Kim, W.D.; Wang, N.S.; King, M.; Ghezzo, H.; Cosio, M.G. Destructive index: A measurement of lung parenchymal destruction in smokers. *Am. Rev. Respir. Dis.* **1985**, *131*, 764–769. [CrossRef] [PubMed]

*marine drugs*

*Article*

# Comparative Transcriptome Analyses Provide Potential Insights into the Molecular Mechanisms of Astaxanthin in the Protection against Alcoholic Liver Disease in Mice

**Huilin Liu [1], Huimin Liu [2,3], Lingyu Zhu [4], Ziqi Zhang [4], Xin Zheng [4], Jingsheng Liu [2,3,*] and Xueqi Fu [1,*]**

1 College of Life Science, Jilin University, Changchun 130012, China; ireneliuhl@163.com
2 College of Food Science and Engineering, Jilin Agricultural University, Changchun 130118, China; liuhuimin@jlau.edu.cn
3 National Engineering Laboratory for Wheat and Corn Deep Processing, Changchun 130118, China
4 College of Animal Science and Technology, Jilin Agricultural University, Changchun 130118, China; 13610728530@163.com (L.Z.); zzqknight@163.com (Z.Z.); zhengxin@jlau.edu.cn (X.Z.)
* Correspondence: liujs1007@vip.sina.com (J.L.); fxq@jlu.edu.cn (X.F.); Tel.: +86-136-3431-5137 (X.F.)

Received: 14 February 2019; Accepted: 15 March 2019; Published: 19 March 2019

**Abstract:** Alcoholic liver disease (ALD) is a major cause of chronic liver disease worldwide. It is a complex process, including a broad spectrum of hepatic lesions from fibrosis to cirrhosis. Our previous study suggested that astaxanthin (AST) could alleviate the hepatic inflammation and lipid dysmetabolism induced by ethanol administration. In this study, a total of 48 male C57BL/6J mice were divided into 4 groups: a Con group (fed with a Lieber–DeCarli liquid diet), an AST group (fed with a Lieber–DeCarli liquid diet and AST), an Et group (fed with an ethanol-containing Lieber–DeCarli liquid diet), and a EtAST group (fed with an ethanol-containing Lieber–DeCarli liquid diet and AST). Then, comparative hepatic transcriptome analysis among the groups was performed by Illumina RNA sequencing. Gene enrichment analysis was conducted to identify pathways affected by the differentially expressed genes. Changes of the top genes were verified by quantitative real-time PCR (qRT-PCR) and Western blot. A total of 514.95 ± 6.89, 546.02 ± 15.93, 576.06 ± 21.01, and 690.85 ± 54.14 million clean reads were obtained for the Con, AST, Et, and EtAST groups, respectively. Compared with the Et group, 1892 differentially expressed genes (DEGs) (including 351 upregulated and 1541 downregulated genes) were identified in the AST group, 1724 differentially expressed genes (including 233 upregulated and 1491 downregulated genes) were identified in the Con group, and 1718 DEGs (including 1380 upregulated and 338 downregulated genes) were identified in the EtAST group. The enrichment analyses revealed that the chemokine signaling, the antigen processing and presentation, the nucleotide-binding and oligomerization domain (NOD)-like receptor signaling, and the Toll-like receptor signaling pathways enriched the most differentially expressed genes. The findings of this study provide insights for the development of nutrition-related therapeutics for ALD.

**Keywords:** astaxanthin; comparative transcriptome analyses; alcoholic liver disease; bioinformatic analysis

---

## 1. Introduction

A recent report from the World Health Organization indicates that three million deaths every year result from the harmful use of alcohol (representing 5.3% of all global deaths) [1,2]. Alcohol abuse is one of the leading causes of more than 200 disease and injury conditions worldwide [3]. As the liver is the major organism for the metabolism of alcohol, long-term use and over-consumption of alcohol leads to alcoholic liver disease (ALD). ALD is a complex process including a broad clinical-histologic spectrum of hepatic lesions, ranging from simple fatty liver to liver injury, from steatosis to cirrhosis, and even hepatocellular carcinoma [3,4]. The pathogenesis of ALD has been well characterized, but no specific drugs and therapy were available to reverse this progress in humans. Emerging evidence showed that inflammation, oxidative stress, cell injury, and regeneration are predominant drivers for ALD [5,6]. Recent studies also showed that some intracellular signaling pathways and transcriptional factors are involved in the mechanisms of pathogenesis of ALD [7].

Astaxanthin (AST) is a significant xanthophyll carotenoid, mainly derived from marine organisms and algae; it cannot be synthesized in humans. AST is a lipid-soluble compound which cannot be converted to vitamin A in the human body [8]. Growing research indicates that AST provides many benefits to humans, such as antioxidant effects, anti-apoptosis effects, anti-inflammation effects, neuroprotective effects, cardiovascular disease prevention, and immune-modulation effects [9,10]. Notably, it is a potential protector against liver damage, such as liver fibrosis and non-alcoholic fatty liver disease. However, previously published studies are limited to the protective effect of AST on ALD [11,12]. In vivo studies have suggested that decreased AST serum levels of aspartate transaminase and alanine transaminase in the livers of the AST administrated group could alleviate the hepatic inflammation and lipid dysmetabolism induced by ethanol administration [12,13]. However, the molecular mechanisms of AST in the protective effect of ALD are still unclear. Hence, in the present study, a genome-wide comparison of the transcriptome with or without AST in ALD mice was done using RNA-sequencing (RNA-Seq) analysis. Then, differential protein expression was confirmed using immunoblot. This provides insight on how astaxanthin affects ALD and gives evidence on the molecular mechanism of AST in ALD protection.

## 2. Results

### 2.1. Overview of RNA-Sequencing Analysis

After removing the low-quality reads and quality control, a total of 514.95 ± 6.89, 546.02 ± 15.93, 576.06 ± 21.01, and 690.85 ± 54.14 million clean reads were obtained for the Con, AST, Et, and EtAST groups, respectively (see Table 1). The clean GC content of each group ranged from 47.78 to 49.2%, and the value of Q30 ranged from 92.11 to 93.56% (see Table S1). To evaluate the quality of the RNA-Seq data, the total clean reads were mapped to the reference genome. A high proportion of the clean reads were mapped to the mouse reference genome using Tophat2 (http://tophat.cbcb.umd.edu/); that is, 92.02% from Con, 91.85% from AST, 91.59% from Et, and 91.5% from EtAST (see Table 1). Through nucleotide basic local alignment search tool (BLAST) analysis, more than 97% of the reads of each group were mapped to known genes, and more than 95% of the reads were mapped to exons. Furthermore, principal component analysis revealed high correlations among biological replicates (Figure S1). Together, all the results indicated that the RNA-Seq data was reliable.

**Table 1.** Summary of RNA-sequencing data.

| Sample | Con | AST | Et | EtAST |
|---|---|---|---|---|
| Total reads ($\times 10^5$) | 514.95 ± 6.89 | 546.02 ± 15.93 | 576.06 ± 21.01 | 690.85 ± 54.14 |
| Total mapped reads ($\times 10^5$) | 473.87 ± 6.11 | 501.50 ± 14.26 | 527.61 ± 19.32 | 632.12 ± 49.06 |
| Mapped to reference genome % | 92.02 | 91.85 | 91.59 | 91.5 |
| Mapped to gene % | 97.47 | 97.25 | 97.04 | 97.64 |
| Mapped to exon % | 95.35 | 95.67 | 95.56 | 96.24 |
| Mapped to intergene % | 2.55 | 2.76 | 2.72 | 2.37 |

Con: mice fed with fed with a Lieber–DeCarli liquid diet, Astaxanthin (AST): mice fed with a Lieber–DeCarli liquid diet and astaxanthin, Et: mice fed with an ethanol-containing Lieber–DeCarli liquid diet, EtAST: mice fed with ethanol-containing Lieber–DeCarli liquid diet and astaxanthin.

### 2.2. Gene Annotation and Functional Analysis

The genes were aligned with public databases, such as the Gene Ontology (GO) database, the Kyoto Encyclopedia of Genes and Genomes (KEGG), and eggNOG (i.e., the Evolutionary Genealogy of Genes: Non-supervised Orthologous Groups). As shown in Table 2, most of the genes were annotated using the GO database (97.06%), followed by eggNOG (74.58%) and KEGG (61.52%).

**Table 2.** Functional annotation of transcriptome data in three public databases.

| Database | Annotated | Percent |
|---|---|---|
| GO | 21,430 | 97.06 |
| KEGG | 13,582 | 61.52 |
| eggNOG | 16,465 | 74.58 |
| Ensembl | 22,078 | 100 |

GO is an international standardized gene functional classification system. In total, there were 21,430 genes mapped in the GO database (Figure S2). The biological process group possessed more terms than the cellular component and molecular function groups. The highly enriched GO terms were in the cellular process (GO: 0009987), biological regulation (GO: 0065007), metabolic process (GO: 0008152), response stimulus (GO: 0050896), multicellular organismal process (GO: 0032501), and signaling (GO: 0023052) groups.

Furthermore, the genes were annotated and classified using the KEGG database. As shown in Figure S3, genes assigned to human diseases (2705) occupied the maximum proportion, followed by those assigned to signal transduction (1784) and cellular processes (1728).

### 2.3. Identification of Differentially Expressed Genes (DEGs)

Gene expression levels of Con, AST, Et, and EtAST were quantified and compared (Figure 1). The genes with a reads per kilobases per million (RPKM) ratio greater than twofold were defined as DEGs. As shown in Figure 1a, a total of 15,779, 15,740, 16,136, and 15,877 DEGs were identified in the Con, AST, Et, and EtAST groups, respectively. Among these DEGs, there were 168, 163, 351, and 201 DEGs uniquely expressed in Con, AST, Et, and EtAST, respectively. Moreover, 14,917 DEGs were commonly expressed in all the groups.

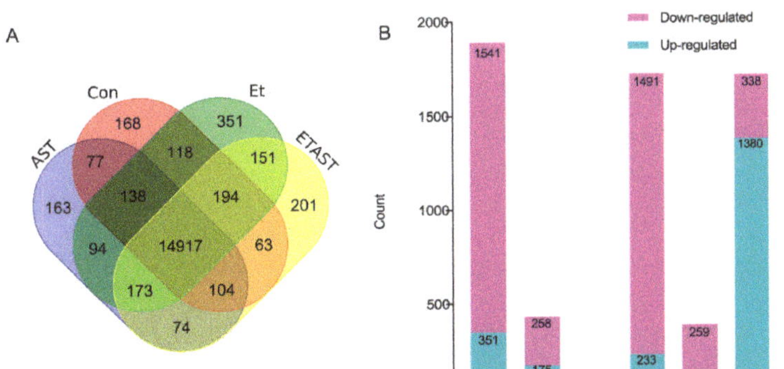


**Figure 1.** Statistical analysis of the gene expression detected by RNA-sequencing (RNA-Seq). (**A**) Venn diagram of gene counts expressed in the Con, AST, Et, and EtAST groups. (**B**) Number of total differentially expressed genes (DEGs) and down- or upregulated DEGs, respectively.

Significant DEGs, including upregulated or downregulated genes, were identified by DEGseq (Figure 1B). Compared with the Et group, 1892 DEGs, including 351 upregulated and 1541 downregulated genes, were identified in the AST group; 1724 DEGs, including 233 upregulated and 1491 downregulated genes, were identified in the Con group; and 1718 DEGs, including 1380 upregulated and 338 downregulated genes, were found in the EtAST group.

To confirm the gene expression level acquired by RNA-Seq, 16 genes were determined by quantitative real-time PCR (qRT-PCR). The results showed that the expression levels of the 16 genes were consistent with the RNA-Seq data (Figure 2), which indicated that the RNA-Seq data in the present study were reliable.

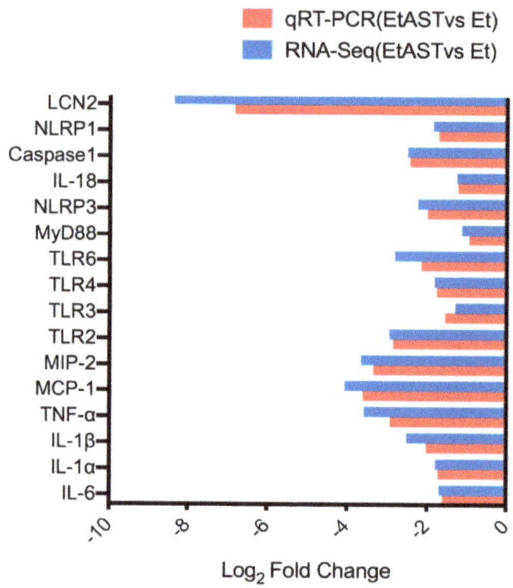


**Figure 2.** Quantitative real-time PCR (qRT-PCR) verification of RNA-sequencing results. The x-axis represents genes; the y-axis represents the logarithm of fold change; and the red column and blue column represents the qRT-PCR results and RNA-sequencing results, respectively.

## *2.4. KEGG Enrichment Analyses of DEGs*

To uncover the potential mechanisms of AST involved in the protective effect against ALD, we performed KEGG enrichment analyses using a path-finder software. In the previous study, we found that AST could influence the immune system to ameliorate liver injury [12]. Thus, to identify the specific biological pathways involved in the immune system, we conducted KEGG enrichment analyses on pairs of comparison groups (Con versus Et and Et versus EtAST). The top 10 ranked KEGG pathways for each comparison group are summarized in Tables 3 and 4.

As shown in Table 3, compared with Con mice, most of the DEGs related to the immune system were up-regulated in Et mice. It indicated that alcohol consumption disrupts the immune system in complex ways. However, AST administration could ameliorate these disruptions (Table 4). Between Con and Et mice, the chemokine signaling pathway (ko04062), the antigen processing and presentation pathway (ko04612), and the NOD-like receptor signaling pathway (ko04621) were the most significantly enriched pathways (Table 3). To Et and EtAST mice, the top 3 pathways were the natural killer cell mediated cytotoxicity pathway (ko04650), the NOD-like receptor signaling pathway (ko04621), and the chemokine signaling pathway (ko04062) (see Table 4). Together, the chemokine signaling pathway, the NOD-like receptor signaling pathway, and the Toll-like receptor signaling pathway were chosen for further validation.

**Table 3.** Statistics on the Kyoto Encyclopedia of Genes and Genomes (KEGG) pathway enrichment of DEGs between Con and Et.

| Pathway ID | Pathway | Con up | Et up | P Value | FDR |
|---|---|---|---|---|---|
| ko04062 | Chemokine signaling pathway | 1 | 46 | $9.13 \times 10^{-11}$ | $7.03 \times 10^{-9}$ |
| ko04612 | Antigen processing and presentation | 0 | 26 | $5.62 \times 10^{-9}$ | $2.49 \times 10^{-7}$ |
| ko04621 | NOD-like receptor signaling pathway | 0 | 19 | $1.71 \times 10^{-7}$ | $4.38 \times 10^{-6}$ |
| ko04650 | Natural killer cell mediated cytotoxicity | 0 | 32 | $3.38 \times 10^{-7}$ | $7.80 \times 10^{-6}$ |
| ko04672 | Intestinal immune network for IgA production | 0 | 15 | $1.47 \times 10^{-6}$ | $2.83 \times 10^{-5}$ |
| ko04610 | Complement and coagulation cascades | 0 | 23 | $2.04 \times 10^{-6}$ | $3.62 \times 10^{-5}$ |
| ko04611 | Platelet activation | 0 | 27 | $6.64 \times 10^{-6}$ | $1.10 \times 10^{-4}$ |
| ko04620 | Toll-like receptor signaling pathway | 1 | 21 | $6.48 \times 10^{-5}$ | $7.87 \times 10^{-4}$ |
| ko04623 | Cytosolic DNA-sensing pathway | 0 | 16 | $7.96 \times 10^{-5}$ | $9.20 \times 10^{-4}$ |
| ko04666 | Fc gamma R-mediated phagocytosis | 0 | 19 | $9.31 \times 10^{-5}$ | $9.90 \times 10^{-4}$ |

Con up: the DEGs which were up-regulated in control group, Et up: the DEGs which were up-regulated in ethanol group, FDR: false discovery rate.

**Table 4.** Statistics on the KEGG pathway enrichment of DEGs between EtAST and Et.

| Pathway ID | Pathway | Et up | EtAST up | P Value | FDR |
|---|---|---|---|---|---|
| ko04650 | Natural killer cell mediated cytotoxicity | 40 | 1 | $1.81 \times 10^{-12}$ | $1.33 \times 10^{-10}$ |
| ko04621 | NOD-like receptor signaling pathway | 22 | 0 | $7.34 \times 10^{-10}$ | $2.70 \times 10^{-8}$ |
| ko04062 | Chemokine signaling pathway | 43 | 1 | $3.38 \times 10^{-9}$ | $9.34 \times 10^{-8}$ |
| ko04612 | Antigen processing and presentation | 26 | 0 | $5.62 \times 10^{-9}$ | $1.38 \times 10^{-7}$ |
| ko04620 | Toll-like receptor signaling pathway | 28 | 1 | $8.00 \times 10^{-9}$ | $1.77 \times 10^{-7}$ |
| ko04610 | Complement and coagulation cascades | 24 | 0 | $5.28 \times 10^{-7}$ | $9.71 \times 10^{-6}$ |
| ko04672 | Intestinal immune network for IgA production | 15 | 0 | $1.47 \times 10^{-6}$ | $2.3 \times 10^{-5}$ |
| ko04666 | Fc gamma R-mediated phagocytosis | 22 | 0 | $2.07 \times 10^{-6}$ | $2.92 \times 10^{-5}$ |
| ko04611 | Platelet activation | 26 | 2 | $2.11 \times 10^{-6}$ | $2.92 \times 10^{-5}$ |
| ko04640 | Hematopoietic cell lineage | 19 | 2 | $1.18 \times 10^{-5}$ | $1.45 \times 10^{-4}$ |

Et up: the DEGs which were up-regulated in ethanol group, EtAST up: the DEGs which were up-regulated in ethanol plus astaxanthin group, FDR: false discovery rate.

### 2.5. qRT-PCR Validation of Differentially Expressed Genes

To verify the results of the transcriptome sequencing and further analyze the key gene expressions involved in astaxanthin regulating alcoholic liver disease, 13 representative genes were selected from the chemokine signaling pathway, the NOD-like receptor signaling pathway, and the Toll-like receptor signaling pathway, and were quantified by qRT-PCR.

As shown in Figure 3A, ethanol significantly upregulated the expression of certain genes, including Interleukin-1 alpha (IL-1$\alpha$), Interleukin-1 beta (IL-1$\beta$), PYD domains-containing protein (NLRP3), Caspase1, and Interleukin-18 (IL-18) in the Et group, compared with Con group. However, AST supplement in the EtAST group reversed this effect, and showed no significant difference compared with the Con group

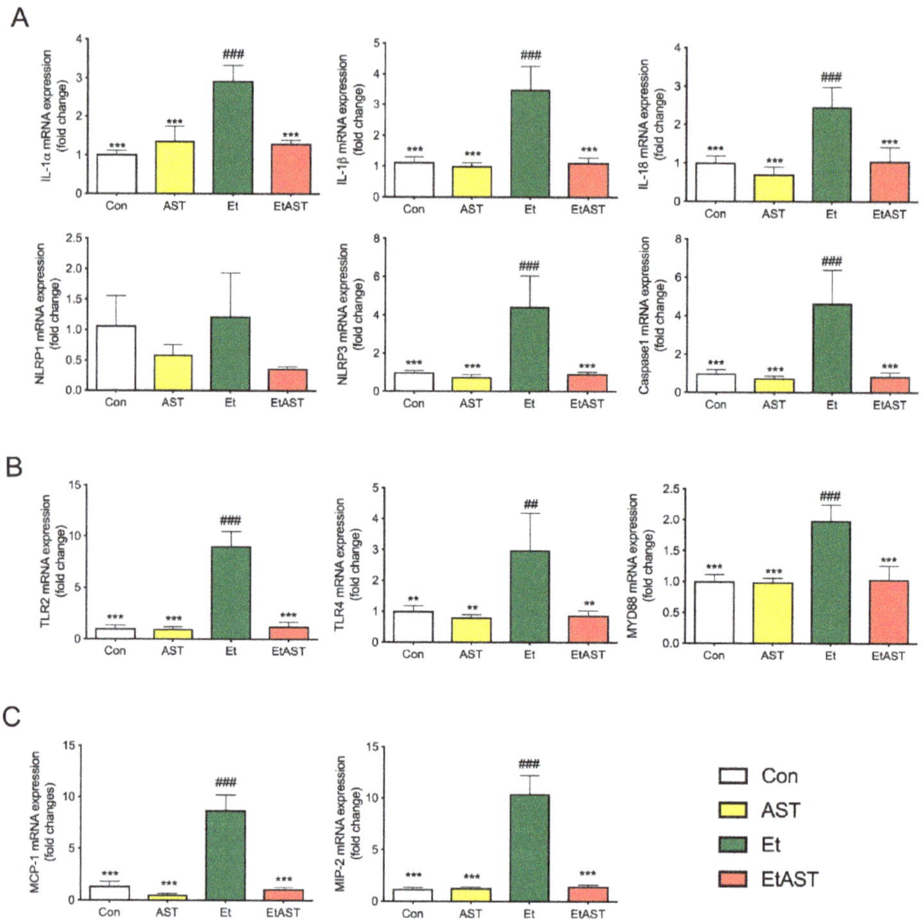


**Figure 3.** Hepatic mRNA expression levels of associated genes in (**A**) the NOD-like pathway, (**B**) Toll-like pathway, and (**C**) chemokines pathway in the Con, AST, Et, and EtAST groups. Relative mRNA expression levels were determined by real-time RT-PCR and normalized to 18s rRNA as an internal control. Data was represented as means ± SEM ($n = 6$; ** $p < 0.01$ versus Et; *** $p < 0.001$ versus Et; ## $p < 0.01$ versus Con; and ### $p < 0.001$ versus Con).

As shown in Figure 3B, the detection of the representative genes in the Toll-like receptor signal pathway—including Toll-like receptors 2, 3, 4, and 6 (TLR2, 3, 4, 6) and myeloid differential protein-88 (MyD88)—were significantly upregulated in the Et group, compared to the Con and AST groups, whereas an AST supplement in the EtAST group reversed this effect, which showed no difference compared to the Con group.

As shown in Figure 3C, compared with the Con group, ethanol significantly upregulated the representative genes from the chemokine signaling pathway, including the monocyte chemoattractant protein-1 (MCP-1) and macrophage inflammatory protein 2 (MIP-2). AST significantly downregulated the two genes, but showed no significant difference compared with the Con group. The qRT-PCR results showed a similar downregulated trend with the gene expression found through RNA-Seq, and the coincidence rate was more than 82%; therefore, the qRT-PCR expression validates the findings of RNA-Seq.

Overall, these results suggest that AST reversed the inflammation caused by ethanol through the regulated chemokine signaling pathway, the NOD-like receptor signaling pathway, and the Toll-like receptor signaling pathway.

*2.6. Western Blot Validation of Differentially Expressed Genes*

To further investigate the mechanism underlying the hepatoprotective effects of AST on alcohol-induced liver inflammation, we examined the protein expression levels of the Toll-like receptor and NOD-like receptor. Compared with the Et group, the protein levels of MYD88, TLR4, NLRP3, and IL-1β were significantly decreased in the Con and EtAST groups. However, there was no significant difference in the levels of MYD88, TLR4, and IL-1β in the AST group (Figure 4). It has been reported that the Toll-like receptor and the NOD-like receptor were relevant to the NF-κB and MAPK families. Next, the representative proteins—including JNK, p38, ERK 1/2, and p65—involved in these two families were detected. The phosphorylation levels of JNK, p38, ERK 1/2, and p65 were significantly increased in the Et group when compared with the Con group, and these proteins decreased in level after the AST supplement was administered, compared with the Et group (Figure 4). Taken together, these results suggest that AST has protective effects on alcoholic liver injury and causes an associated depression in the expression of p65, JNK, p38, and ERK1/2.

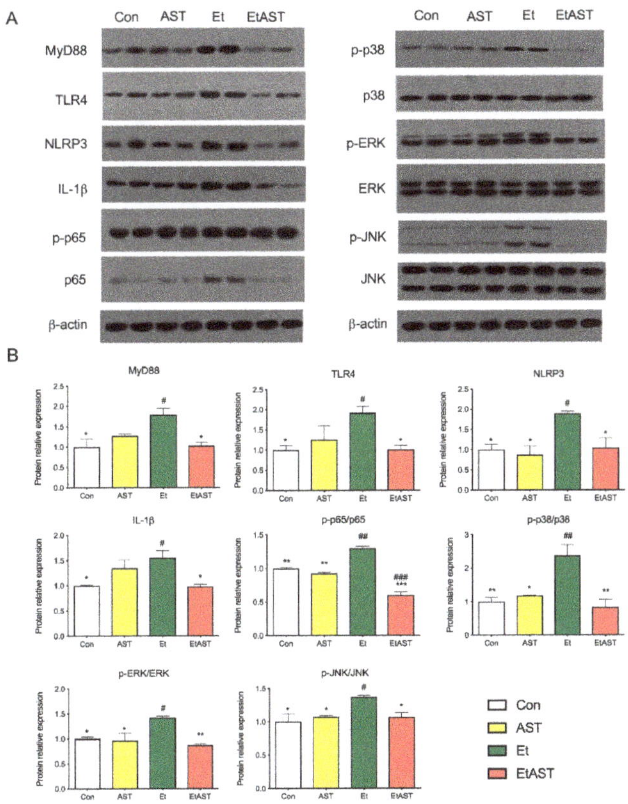


**Figure 4.** Hepatic protein expression levels of selected genes involved in the NOD-like pathway, Toll-like pathway, and chemokine pathway. The protein expression (**A**) and relative protein levels (**B**) were measured by western blot analysis. The relative protein levels were measured by Western blot analysis. Data was represented as means ± SD ($n = 6$). * $p < 0.05$ versus Et; ** $p < 0.01$ versus Et; # $p < 0.05$ versus Con; ## $p < 0.01$ versus Con; and ### $p < 0.001$ versus Con.

## 3. Discussion

AST is similar to β-carotene in molecular structure and possesses a strong antioxidative effect [10]. Recently, researchers have shown an increased interest in AST due to the demand in the promotion of human health [9]. Previous research has established that AST can relieve ischemia-related brain injury by suppressing oxidative stress [14], exerting neuroprotective effects by weakening neuroinflammation [15], and modulating the endogenous antioxidant defense system [16]. Moreover, AST is also a potential protector against liver damage [11]. It can inhibit liver fibrosis and lipid peroxidation [17]; inhibit liver tumorigenesis and inflammation [18]; attenuate hepatic ischemia reperfusion-induced apoptosis and autophagy [19,20]; and prevent ethanol-induced hepatic injury through the in vivo inhibition of oxidant and inflammatory responses [21]. Additionally, AST possesses anti-fibrogenic effects, through the TGFβ1–Smad3 signaling pathway in hepatic stellate cells [22]. However, most of the in vitro studies were performed on hepatocellular carcinoma cells and hepatic stellate cells. In term of the in vivo studies, nonalcoholic steatohepatitis mice, diabetic mice, and obese mice were the major mouse models. Thus far, very little attention has been paid to the molecular mechanisms of ALD nutrition prevention and protection. Our previous study indicated that AST administration significantly relieves inflammation, decreases lipid accumulation, and improves serum marker levels relating to ethanol-induced liver injury [12]. The present study was designed to determine possible molecular mechanisms involved in the protective effect of AST in ALD. We found that AST may prevent the progress of ALD through the chemokine signaling pathway, the NOD-like receptor signaling pathway, and the TLR signaling pathway (Figures 3 and 4).

Currently, alcohol abuse is one of the major drivers of chronic liver disease in Western countries [2]. Growing evidence suggests that TLRs play a vital role in the pathogenesis and progression of liver diseases, such as alcoholic liver disease (ALD), non-alcoholic fatty liver disease (NAFLD), and autoimmune liver disease [23]. TLRs belong to a family of pattern recognition receptors that recognize pathogen-associated molecular patterns (PAMPs) and damage-associated patterns (DAMPs). They play an important role in the initiation of the immune system and inflammation process [24]. Once activated, TLRs are expressed by liver-resident cells and trigger the production of cytokines and chemokines [23]. The mRNA expression levels of TLRs is deficit in healthy liver cells, and it seems that the TLR-signaling pathway is not activated [25]. As shown in Figure 3A, compared with Et mice, the mRNA expression levels of TLR2 and TLR4 were very low in normal mice. The protein level of TLR4 showed the same tendency (see Figure 4). Myeloid differentiation primary response 88 (MYD88) is the canonical adaptor for inflammatory signaling pathways downstream from members of the TLRs [26]. It is suggested that the TLR4 downstream signaling is regulated by the MyD88-independent pathway in ALD [26]. Both mRNA and the protein level of MyD88 were examined in the present study (see Figures 3 and 4). These results are in line with the previous study and provide further support for the hypothesis that TLRs participate in the progress of ALD, and could be a therapeutic target for the treatment of ALD.

Activation of the immune signaling pathways plays a critical role in the pathogenesis of ALD [27]. NOD-like receptors (NLRs) are intracellular innate immune sensors that could recognize pathogen-PAMPs and -DAMPs [28]. Recent studies suggest that NLRs are not only expressed and activated in innate immune cells, but also in parenchymal cells in the liver. NLRs can work with TLRs and regulate the inflammatory response [29]. The NLRs consist of two major subfamilies, NODs and NLRPs, containing 23 members in humans and 34 in mice [28]. NLRs are key mediators of the inflammasome, which are the major drivers of inflammation. There are several inflammasomes, such as NLRP1, NLRP3, and NLRC4. Thus far, the most characterized and investigated member is NLRP3 [30]. Recently, it was demonstrated that the NLRP3 inflammasome was involved in the development of chronic liver diseases, such as alcoholic steatohepatitis and NAFLD. Once activated, it upregulates the expression of caspase-1, then promotes the secretion of IL-1β and IL-18, which play key roles in the induction and progression of liver inflammation [31,32]. The results of this study indicate that AST administration can significantly downregulate the expression of NLRP3, but not NLRP1. Thus, the mRNA levels of downstream factors, such as caspase-1 and IL-18, were decreased. Compared with Et mice, AST treatment notably downregulated the hepatic protein expression of NLRP3 and IL-1β in EtAST mice (Figures 3 and 4). Taken together, these findings suggest that AST may protect against alcohol-induced liver injury through the NOD pathway.

As mentioned above, innate immune cells play a key role in the pathophysiology of ALD. Resident macrophages in the liver are activated by PAMPs and DAMPs. The expression of TLRs and NLRs are then stimulated. This, finally, triggers the production of cytokines and chemokines [33]. It was revealed that chemokines were important determinants in the pathogenesis of liver disease. They regulated the migration and activities of the resident cells in the liver [33,34]. Chemokines were initially discovered for their role in regulating cell trafficking. To date, more than 50 chemokine ligands and 19 receptors have been identified, several of which have been described as relevant in the progress of liver disease [35]. Monocyte chemoattractant protein-1 (MCP1) and macrophage inflammatory protein-2 (MIP2) were the most-studied chemokines involved in ALD. MCP1, also known as CCL2, regulates macrophage activation, proinflammatory responses, and hepatic steatosis in the liver [36]. Previous studies have indicated that the plasma and hepatic levels of MCP1 were elevated in ALD patients. Moreover, it was shown that the deficiency of MCP1 protects against alcoholic liver injury by inhibiting the production of proinflammatory cytokines in mice. All these findings suggested that MCP1 might be a potential therapeutic target in ALD [36–38]. In Figure 3, it is evident that there is a significant difference in the hepatic expression levels of MCP1 between Et and EtAST mice. MIP2, also named CXCL2, is mainly activated by Kupffer cells in liver injury. It accelerates liver inflammation by releasing various inflammatory mediators [39]. In vivo studies have indicated that plasma and hepatic MIP2 concentrations were increased in ALD mice [40]. In our study, the hepatic mRNA expression of EtAST mice was dramatically lower compared to Et mice (see Figure 3). According to this data, we can infer that AST could improve hepatic inflammation by inhibiting the expression of MCP1 and MIP2 in ALD mice or patients.

## 4. Materials and Methods

### 4.1. Animal Experimentation

Male C57BL/6J mice (20–24 g, six-weeks-old) were purchased from the Beijing Vital River Laboratory Animal Technology Co., Ltd (Beijing, China). Mice were housed individually in cages for a 12 h light/dark cycle at 23 ± 2 °C with optimum access to chow and water ad libitum. After one-week acclimation, a total of 48 mice were randomly divided into four groups: (1) the Con group ($n$ = 12), given a Lieber–DeCarli liquid diet (Table S1) for 12 weeks; (2) the AST group ($n$ = 12), given a Lieber–DeCarli liquid diet for the first two weeks, then a Lieber–DeCarli liquid diet with astaxanthin (AST, 50 mg/kg body weight) for another 10 weeks; (3) the Et group ($n$ = 12), given a Lieber–DeCarli liquid diet for the first two weeks, then an ethanol-containing Lieber–DeCarli liquid diet (i.e., 5% ethanol *v/v* accounted for 36% the total caloric intake) for another 10 weeks; and (4) the EtAST group ($n$ = 12), given a Lieber–DeCarli liquid diet for the first two weeks, then an ethanol-containing Lieber–DeCarli liquid diet plus astaxanthin (50 mg/kg body weight) for another 10 weeks. The AST was purchased from Sigma–Aldrich (St Louis, MO, USA. SML0982) and dissolved in corn oil for further use. The AST dose was done with reference to the previous research [41], while the ethanol uptake amount was increased for two weeks, and the final concentration was 5% (*v/v*). After the mice were sacrificed, the liver tissues were collected and frozen in liquid nitrogen overnight for RNA extraction. All experiment protocols were approved by the Institutional Animal Care and Use Committee at the Jilin Institute of Traditional Chinese Medicine (Approval Number SYXK (JI) 2015-0009).

### 4.2. RNA Sequencing

Total fresh liver tissues were suspended in TRIzol reagent (Invitrogen Life Technologies, Carlsbad, CA, USA) according to the manufacturer's protocol. The concentration and purity of RNA was determined using a NanoDrop 2000 microspectrophotometer (Thermo Fisher Scientific, Waltham, MA, USA).

The sequencing libraries were generated using the TruSeq RNA Sample Preparation Kit (Illumina, San Diego, CA, USA), which consists of an mRNA purification process that uses poly-T beads, mRNA fragmentation, reverse transcription, end repair, the addition of a single 'A' base, ligation of the adapters, and purification and enrichment with PCR. The library fragments were purified using the AMPure XP system (Beckman Coulter, Beverly, CA, USA) and quantified using the Agilent High Sensitivity DNA assay on a Bioanalyzer 2100 system (Agilent, Santa Clara, CA, USA). Sequencing was carried out using an Illumina Hiseq Xten platform (150 bp paired-end reads).

### 4.3. *Differential Expression Gene Analysis*

The gene expression level was quantified using RPKM and analyzed using the HTSeq software (Version 0.6.1p2, http://www-huber.embl.de/users/anders/HTSeq), using union as the counting model. A cut-off value of RPKM > 1 was used to define the gene expression. DEG analysis was performed using DESeq (Version 1.18.0), and the fold change and Fisher-test were used to choose differentially expressed genes [42]. The false discovery rate (FDR) criterion was introduced to adjust the $p$-values. In this study, the differential genes with the $p$-value $< 0.05$ and the false discovery rate (FDR) $< 0.02$ were considered to be statistically significant.

### 4.4. *The Enrichment Analyses of Differentially Expressed Genes*

Differentially expressed gene enrichment analyses were performed using KEGG databases. KEGG is a database resource dealing with genomes, biological pathways, diseases, drugs, and chemical substances. Here, KEGG enrichment analyses was conducted using the online Path-Finder software (http://www.genome.jp).

### 4.5. *Quantitative Real-Time PCR*

qRT-PCR was performed to validate the DEGs obtained from the RNA-Seq results. Specific primers of 16 candidate DEGs were designed using the Primer 5 software (Version 5.0, Premier biosoft, Palo Alto, CA, USA)., and were synthesized by Sheng Gong (Shanghai, China) (Table S2). Total RNA was extracted as previously described and reverse-transcribed using the PrimeScript™ RT Reagent Kit with a gDNA eraser (Takara, Tokyo, Japan). SYBR Green Mix (Takara, Tokyo, Japan) and a CFX96 Real Time PCR System (Bio-Red Laboratories, Hercules, CA, USA) were used to perform the qRT-PCR; following the manufacturer's protocol, the melting curves were as follows: 95 °C for 60 s, followed by 55 °C for 30 s, and then 95 °C for 30 s. For an endogenous reference gene, 18 s rRNA was used. The mRNA expression levels were calculated using the $2^{-\Delta\Delta Ct}$ method. The QPCR assays were performed as compliant with MIQE.

### 4.6. *Western Blotting*

Western blotting was performed to evaluate the protein expression levels of DEGs. The liver tissues were homogenized using a liquid nitrogen pre-cooled high-speed tissue homogenizer (Gering Scientific Instruments Ltd, Beijing, China) and lysed using 10 μM phenylmethanesulfonyl fluoride (PMSF, Beyotime Institute of Biotechnology, Jiangsu, China) and a 1% protease inhibitor cocktail (104 mM AEBSF, 80 μM Aprotinin, 4 mM Bestatin, 1.4 mM E-64, 2 mM Leupeptin, and 1.5 mM Pepstatin A; Sigma). The protein contents were analyzed with the BCA Protein Assay Kit (Vazyme Biotech Co., Ltd, Nanjing, China). Standard Western blotting was performed, and the blots were visualized via a chemiluminescent system using ImageQuant LAS 500 imaging instruments (GE Healthcare Life Sciences, Shanghai, China) and quantified using the Image J analyzer software. Antibodies against ERK 1/2, p-ERK1/2, p-NF-kB p65, MAPK p38, p-MAPK p38, JNK, and p-JNK were purchased from Cell Signaling Technology (Danver, MA, USA). Antibodies against TLR4, NLRP3, IL-1b, MYD88, and NF-kB p65 were obtained from Abcam (Cambridge, MA, USA).

### 4.7. Statistical Analysis

Statistical significance in the DEG analyses was performed using the R statistical package. Values of $p < 0.05$ were considered statistically significant.

The qRT-PCR and Western blot results were presented as means ± SEM and calculated using GraphPad Prism version 7.01 (GraphPad Software, Inc., La Jolla, CA, USA). Statistical significance was determined using the Tukey's multiple-comparison test. Values of $p < 0.05$ were considered statistically significant.

## 5. Conclusions

In this study, comparative hepatic transcriptome analyses were performed to elucidate the possible mechanisms of AST in the protection against ALD. Through RNA-Seq, a total of 22,078 genes were identified. Then, KEGG enrichment analyses were conducted to reveal potential signaling pathways related to the immune system. We found that most DEGs were enriched in the chemokine signaling pathway, the NOD-like receptor signaling pathway, and the Toll-like receptor signaling pathway. Furthermore, 13 genes associated with these three pathways were selected to identify the RNA-sequencing results. This is the first report, to our knowledge, describing comparative transcriptome analyses of AST in the protection against ALD. We found that AST may prevent the progress of ALD through the chemokine signaling pathway, the NOD-like receptor signaling pathway, and the TLR signaling pathway. Our findings provide important insights into the possible molecular mechanisms of AST in the nutritional intervention of ALD.

**Supplementary Materials:** The following are available online at http://www.mdpi.com/1660-3397/17/3/181/s1, Table S1: Statistics of RNA-Seq, Table S2: Sequences of primers used for real-time RT-PCR analyses, Figure S1: Principal Components Analysis. x-axis represents the first principal component, y-axis represents the second principal component; different shapes represent different groups; different colors represent the different samples, Figure S2: Histogram presentation of gene distribution in Gene Ontology (GO) functional classification. The x-axis represents level to GO terms; the left y-axis represents gene numbers in each GO term. Genes were further classified into sub-groups in biological process, cellular component, and molecular function, Figure S3: Histogram presentation of gene distribution in KEGG classification. The x-axis represents level to KEGG terms; the left y-axis represents gene numbers in each term. Genes were further classified into sub-groups in metabolism, signal transduction, human diseases, and cell process, Figure S4: Determination of qPCR melting curves and melting peaks for selected genes involved in NOD-like pathway, Toll-like pathway, and chemokines pathway. Left side: qPCR melting curves, the x-axis represents melting temperature; the left y-axis represents relative fluorescence intensity. Right side: qPCR melting peaks, the x-axis represents melting temperature and the y-axis is the relative fluorescence unit (RFU) rate of change over time (T) (-d(RFU)/dT).

**Author Contributions:** HL.L. and X.Z. conceived and designed the experiments; HL.L., L.Z., Z.Z., and performed the experiments and analyzed the data; J.L., X.F., HM.L., and X.Z. contributed reagents/materials/analysis tools; HM.L. and HL.L. wrote the paper.

**Funding:** The program was supported by the National Key R&D Program of China (2016YFD0400700,2016YFD0400702), and the National Natural Science Foundation of China (31672511).

**Acknowledgments:** Sequencing service was provided by Personal Biotechnology Co., Ltd. of Shanghai, China.

**Conflicts of Interest:** The authors declare no conflict of interest.

**Availability of Data and Materials:** The datasets generated and/or analyzed during the current study are available at NCBI project PRJNA524945 with accession number (SRR8689617-SRR8689631).

## References

1. Kharbanda, K.K.; Ronis, M.J.J.; Shearn, C.T.; Petersen, D.R.; Zakhari, S.; Warner, D.R.; Feldstein, A.E.; McClain, C.J.; Kirpich, I.A. Role of nutrition in alcoholic liver disease: Summary of the symposium at the esbra 2017 congress. *Biomolecules* **2018**, *8*, 16. [CrossRef]
2. Singal, A.K.; Bataller, R.; Ahn, J.; Kamath, P.S.; Shah, V.H. ACG clinical guideline: Alcoholic liver disease. *Am. J. Gastroenterol.* **2018**, *113*, 175–194. [CrossRef]

3. Kawaratani, H.; Moriya, K.; Namisaki, T.; Uejima, M.; Kitade, M.; Takeda, K.; Okura, Y.; Kaji, K.; Takaya, H.; Nishimura, N.; et al. Therapeutic strategies for alcoholic liver disease: Focusing on inflammation and fibrosis (review). *Int. J. Mol. Med.* **2017**, *40*, 263–270. [CrossRef]
4. Gao, B.; Bataller, R. Alcoholic liver disease: Pathogenesis and new therapeutic targets. *Gastroenterology* **2011**, *141*, 1572–1585. [CrossRef]
5. Louvet, A.; Mathurin, P. Alcoholic liver disease: Mechanisms of injury and targeted treatment. *Nat. Rev. Gastroenterol. Hepatol.* **2015**, *12*, 231–242. [PubMed]
6. Dasarathy, S. Nutrition and alcoholic liver disease: Effects of alcoholism on nutrition, effects of nutrition on alcoholic liver disease, and nutritional therapies for alcoholic liver disease. *Clin. Liver Dis.* **2016**, *20*, 535–550. [CrossRef] [PubMed]
7. Beier, J.I.; McClain, C.J. Mechanisms and cell signaling in alcoholic liver disease. *Biol. Chem.* **2010**, *391*, 1249–1264. [CrossRef] [PubMed]
8. Ambati, R.R.; Phang, S.M.; Ravi, S.; Aswathanarayana, R.G. Astaxanthin: Sources, extraction, stability, biological activities and its commercial applications—A review. *Mar. Drugs* **2014**, *12*, 128–152. [CrossRef]
9. Fakhri, S.; Abbaszadeh, F.; Dargahi, L.; Jorjani, M. Astaxanthin: A mechanistic review on its biological activities and health benefits. *Pharmacol. Res.* **2018**, *136*, 1–20.
10. Higuera-Ciapara, I.; Felix-Valenzuela, L.; Goycoolea, F.M. Astaxanthin: A review of its chemistry and applications. *Crit. Rev. Food Sci. Nutr.* **2006**, *46*, 185–196. [PubMed]
11. Chen, J.T.; Kotani, K. Astaxanthin as a potential protector of liver function: A review. *J. Clin. Med. Res.* **2016**, *8*, 701–704. [CrossRef]
12. Liu, H.; Liu, M.; Fu, X.; Zhang, Z.; Zhu, L.; Zheng, X.; Liu, J. Astaxanthin prevents alcoholic fatty liver disease by modulating mouse gut microbiota. *Nutrients* **2018**, *10*, 1298. [CrossRef] [PubMed]
13. Dong, L.Y.; Jin, J.; Lu, G.; Kang, X.L. Astaxanthin attenuates the apoptosis of retinal ganglion cells in db/db mice by inhibition of oxidative stress. *Mar. Drugs* **2013**, *11*, 960–974. [PubMed]
14. Wu, W.; Wang, X.; Xiang, Q.; Meng, X.; Peng, Y.; Du, N.; Liu, Z.; Sun, Q.; Wang, C.; Liu, X. Astaxanthin alleviates brain aging in rats by attenuating oxidative stress and increasing BDNF levels. *Food Funct.* **2014**, *5*, 158–166. [CrossRef]
15. Zhang, X.-S.; Zhang, X.; Wu, Q.; Li, W.; Wang, C.-X.; Xie, G.-B.; Zhou, X.-M.; Shi, J.-X.; Zhou, M.-L. Astaxanthin offers neuroprotection and reduces neuroinflammation in experimental subarachnoid hemorrhage. *J. Surg. Res.* **2014**, *192*, 206–213. [CrossRef]
16. Naguib, Y.M. Antioxidant activities of astaxanthin and related carotenoids. *J. Agric. Food Chem.* **2000**, *48*, 1150–1154.
17. Kim, B.; Farruggia, C.; Ku, C.S.; Pham, T.X.; Yang, Y.; Bae, M.; Wegner, C.J.; Farrell, N.J.; Harness, E.; Park, Y.-K. Astaxanthin inhibits inflammation and fibrosis in the liver and adipose tissue of mouse models of diet-induced obesity and nonalcoholic steatohepatitis. *J. Nutr. Biochem.* **2017**, *43*, 27–35. [CrossRef]
18. Ohno, T.; Shimizu, M.; Shirakami, Y.; Miyazaki, T.; Ideta, T.; Kochi, T.; Kubota, M.; Sakai, H.; Tanaka, T.; Moriwaki, H. Preventive effects of astaxanthin on diethylnitrosamine-induced liver tumorigenesis in C57/BL/KsJ-db/db obese mice. *Hepatol. Res.* **2016**, *46*, E201–E209. [CrossRef]
19. Li, J.; Dai, W.; Xia, Y.; Chen, K.; Li, S.; Liu, T.; Zhang, R.; Wang, J.; Lu, W.; Zhou, Y.; et al. Astaxanthin inhibits proliferation and induces apoptosis of human hepatocellular carcinoma cells via inhibition of nf-kappab p65 and wnt/beta-catenin in vitro. *Mar. Drugs* **2015**, *13*, 6064–6081. [CrossRef] [PubMed]
20. Li, J.; Wang, F.; Xia, Y.; Dai, W.; Chen, K.; Li, S.; Liu, T.; Zheng, Y.; Wang, J.; Lu, W.; et al. Astaxanthin pretreatment attenuates hepatic ischemia reperfusion-induced apoptosis and autophagy via the ros/mapk pathway in mice. *Mar. Drugs* **2015**, *13*, 3368–3387. [CrossRef]
21. Han, J.H.; Ju, J.H.; Lee, Y.S.; Park, J.H.; Yeo, I.J.; Park, M.H.; Roh, Y.S.; Han, S.B.; Hong, J.T. Astaxanthin alleviated ethanol-induced liver injury by inhibition of oxidative stress and inflammatory responses via blocking of stat3 activity. *Sci. Rep.* **2018**, *8*, 14090. [CrossRef]
22. Yang, Y.; Kim, B.; Park, Y.-K.; Koo, S.I.; Lee, J.-Y. Astaxanthin prevents tgfβ1-induced pro-fibrogenic gene expression by inhibiting smad3 activation in hepatic stellate cells. *Biochim. Biophys. Acta* **2015**, *1850*, 178–185. [PubMed]
23. Roh, Y.S.; Seki, E. Toll-like receptors in alcoholic liver disease, non-alcoholic steatohepatitis and carcinogenesis. *J. Gastroenterol. Hepatol.* **2013**, *28* (Suppl. 1), 38–42.
24. Kiziltas, S. Toll-like receptors in pathophysiology of liver diseases. *World J. Hepatol.* **2016**, *8*, 1354–1369.

25. Guo, J.; Friedman, S.L. Toll-like receptor 4 signaling in liver injury and hepatic fibrogenesis. *Fibrogenesis Tissue Repair* **2010**, *3*, 21. [CrossRef] [PubMed]
26. Soares, J.B.; Pimentel-Nunes, P.; Roncon-Albuquerque, R.; Leite-Moreira, A. The role of lipopolysaccharide/toll-like receptor 4 signaling in chronic liver diseases. *Hepatol. Int.* **2010**, *4*, 659–672. [CrossRef]
27. Del Campo, J.A.; Gallego, P.; Grande, L. Role of inflammatory response in liver diseases: Therapeutic strategies. *World J. Hepatol.* **2018**, *10*, 1. [CrossRef] [PubMed]
28. Chen, G.; Shaw, M.H.; Kim, Y.-G.; Nuñez, G. NOD-like receptors: Role in innate immunity and inflammatory disease. *Ann. Rev. Pathol.* **2009**, *4*, 365–398. [CrossRef] [PubMed]
29. Xu, T.; Du, Y.; Fang, X.-B.; Chen, H.; Zhou, D.-D.; Wang, Y.; Zhang, L. New insights into nod-like receptors (NLRs) in liver diseases. *Int. J. Physiol. Pathophysiol. Pharmacol.* **2018**, *10*, 1.
30. Szabo, G.; Csak, T. Inflammasomes in liver diseases. *J. Hepatol.* **2012**, *57*, 642–654.
31. Wu, X.; Dong, L.; Lin, X.; Li, J. Relevance of the NLRP3 inflammasome in the pathogenesis of chronic liver disease. *Front. Immunol.* **2017**, *8*, 1728. [CrossRef]
32. Benetti, E.; Chiazza, F.; Patel, N.S.; Collino, M. The NLRP3 Inflammasome as a novel player of the intercellular crosstalk in metabolic disorders. *Mediat. Inflamm.* **2013**, *2013*. [CrossRef]
33. Marra, F.; Tacke, F. Roles for chemokines in liver disease. *Gastroenterology* **2014**, *147*, 577–594. [CrossRef]
34. Keane, M.P.; Strieter, R.M. Chemokine signaling in inflammation. *Crit. Care Med.* **2000**, *28*, N13–N26. [CrossRef]
35. Wang, J.; Knaut, H. Chemokine signaling in development and disease. *Development* **2014**, *141*, 4199–4205. [PubMed]
36. Deshmane, S.L.; Kremlev, S.; Amini, S.; Sawaya, B.E. Monocyte chemoattractant protein-1 (MCP-1): An overview. *J. Interferon Cytokine Res.* **2009**, *29*, 313–326. [CrossRef] [PubMed]
37. Baeck, C.; Wehr, A.; Karlmark, K.R.; Heymann, F.; Vucur, M.; Gassler, N.; Huss, S.; Klussmann, S.; Eulberg, D.; Luedde, T. Pharmacological inhibition of the chemokine CCL2 (MCP-1) diminishes liver macrophage infiltration and steatohepatitis in chronic hepatic injury. *Gut* **2012**, *61*, 416–426. [CrossRef] [PubMed]
38. Mandrekar, P.; Ambade, A.; Lim, A.; Szabo, G.; Catalano, D. An essential role for monocyte chemoattractant protein-1 in alcoholic liver injury: Regulation of proinflammatory cytokines and hepatic steatosis in mice. *Hepatology* **2011**, *54*, 2185–2197. [CrossRef] [PubMed]
39. Qin, C.-C.; Liu, Y.-N.; Hu, Y.; Yang, Y.; Chen, Z. Macrophage inflammatory protein-2 as mediator of inflammation in acute liver injury. *World J. Gastroenterol.* **2017**, *23*, 3043. [CrossRef]
40. Bautista, A.P. Chronic alcohol intoxication induces hepatic injury through enhanced macrophage inflammatory protein-2 production and intercellular adhesion molecule-1 expression in the liver. *Hepatology* **1997**, *25*, 335–342. [CrossRef] [PubMed]
41. Zheng, D.; Li, Y.; He, L.; Tang, Y.; Li, X.; Shen, Q.; Yin, D.; Peng, Y. The protective effect of astaxanthin on fetal alcohol spectrum disorder in mice. *Neuropharmacology* **2014**, *84*, 13–18. [CrossRef] [PubMed]
42. Anders, S.; Huber, W. Differential expression analysis for sequence count data. *Genome Biol.* **2010**, *11*, R106. [CrossRef] [PubMed]

marine drugs

*Article*

# Sublingual Delivery of Astaxanthin through a Novel Ascorbyl Palmitate-Based Nanoemulsion: Preliminary Data

**Andrea Fratter [1], Damiano Biagi [1] and Arrigo F. G. Cicero [2,*]**

1 Research and Innovation Technology, Nutraceutical Department, Labomar Research, 31036 Istrana, Italy
2 Medical and Surgical Sciences Department, University of Bologna, 40138 Bologna, Italy
* Correspondence: arrigo.cicero@unibo.it; Tel.: +39-5-1214-2224

Received: 27 July 2019; Accepted: 23 August 2019; Published: 29 August 2019

**Abstract:** Astaxanthin is a carotenoid extracted from several seaweeds with ascertained therapeutic activity. With specific reference, astaxanthin is widely used in clinical practice to improve ocular tissue health and skin protection from UV ray damages. Despite its well-documented pleiotropic actions and demonstrated clinical efficacy, its bioavailability in humans is low and limited because of its hydrophobicity and poor dissolution in enteric fluids. Furthermore, astaxanthin is very unstable molecule and very sensitive to light exposure and thermal stress. Taken together, these pharmacological and chemical–physical features strongly limit pharmaceutical and nutraceutical development of astaxanthin-based products and as a consequence its full clinical usage. This work describes the preliminary in vitro investigation of sublingual absorption of astaxanthin through a novel ascorbyl palmitate (ASP) based nanoemulsion.

**Keywords:** astaxanthin; nanoemulsion; sublingual delivery; ascorbyl palmitate; Franz cell

## 1. Introduction

Astaxanthin (C40H52O4;6S)-6-Hydroxy-3 [(1E,3E,5E,7E,9E,11E,13E,15E,17E)-18-[(4S)-4-hydroxy-2,6,6-trimethyl-3-oxo-1-cyclohexenyl]-3,7,12,16-tetramethyloctadeca-1,3,5,7,9,11,13,15,17-nonaenyl]-2,4,4-trimethyl-1-cyclohex-2-enone) (Figure 1) is a lipid-soluble xanthophyll keto-carotenoid with molecular mass 596.841 g/mol [1].

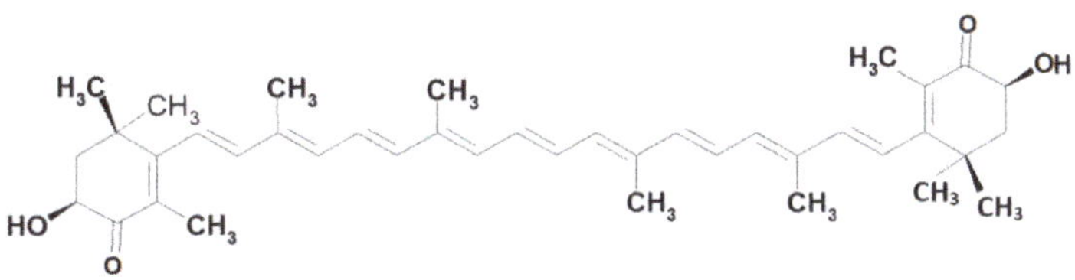


**Figure 1.** Structure of Astaxanthin.

It is the responsible of the red color of some crustaceous and fishes [2]. The most known pharmacological activity of astaxanthin is the antioxidant one [3] but, contrary to other carotenoids, [4] it seems to also exert direct anti-inflammatory activity and to activate Peroxisome Proliferator-Activated Receptors [5].

In (usually small and short-term) clinical trials, oral supplementation with astaxanthin not associated with other nutraceuticals has demonstrated to be protective against UV-induced skin deterioration and helps maintain healthy skin in healthy people [6], to improve liver parameters in climacteric women [7], to protect the vocal fold from injury and inflammation caused by vocal loading [8], to increase the choroidal blood flow velocity in healthy subjects [9], to reduce LDL-cholesterolemia and oxidative stress in overweight patients [10,11], to increase HDL-cholesterolemia and serum adiponectin levels in mildly dyslipidaemic subjects, [12] to prevent oxidative damage in smokers by suppressing lipid peroxidation and stimulating the activity of the antioxidant system [13], and to improve symptoms in patients affected by functional dyspepsia (especially if infected by *Helicobacter pylori*) [14].

In some trials, supplementation with astaxanthin was also shown to prevent and reduce oxidative stress in young soccer players, [15,16] but not in well-trained cyclists [17]. In a previous study, it improved performance in cyclists, [18] while in a recent study it does not augment fat use or improve endurance performance [19]. However, a recent randomized controlled clinical trial showed no effect of astaxanthin on arterial stiffness, oxidative stress, or inflammation in renal transplant recipients [20].

The apparent contrast between positive and neutral effects observed in clinical trials are mainly related to the different dosage used, but also to the largely different bioaccessibility of the tested pharmaceutical forms [21].

It is well ascertained, indeed, that astaxanthin is poorly bioavailable in humans [22] from the conventional pharmaceutical forms, particularly because of its high lipophilicity that precludes the overall enteric bioaccessibility and because it can be enhanced by modified lipids and surfactants capable of making it more hydro-dispersible [23].

Many attempts have been dedicated to projecting pharmaceutical forms with the aim of improving bioaccessibility and overall bioavailability of astaxanthin and in this frame, nanoemulsions seem to play a pivotal role according to numerous published papers [24–26].

Given the potential interest of developing more effective forms of astaxanthin supplements, the aim of our study was to evaluate, likely for the first time, a novel liquid nanoemulsion to promote astaxanthin sublingual delivery by means of an in vitro model assessing its permeation through porcine lingual specimens.

## 2. Materials and Methods

### 2.1. Materials

A Franz cells device was purchased from Copley Scientific (Nottingham, UK), surgical blades from Tekno Optik-Chirurgie GmbH (Tuttlingen, Germany), and scalpel handle from Moretti Spa (Cavriglia, Arezzo, Italy). Malvern Zetasizer Nano series (DLS device) was purchased from Malvern Panalytical (Mlavern, UK). A HPLC-DAD device was purchased from Perkin-Elmer (Series 200, diode array, Waltham, MA, USA). A mechanical stirrer (LG series) and heating plate (RC series) were purchased from Velp Scientifica (Usmate, MB, Italy). Astaxanthin (Astapure$^{TM}$ 10% titration in astaxanthin) was purchased from AlgaTech (New York, NY, USA), astaxanthin standard analytic (>97% from *Hematococcus pluvialis*) was purchased from Sigma-Aldrich (Milan, Italy), physiological solutions were purchased from BS Medital Spa (Grosotto, SO, Italy). Polysorbate 80 (Veremul T 80) was purchased from Veronelli SPA, Milan, Italy; ascorbyl palmitate was purchased from ACEF, Fiorenzuola, Piacenza, Italy; caprylic/capric triglycerides (Labrafac Lipophile WL 1349) were purchased from Gattefossè, Milan, Italy; deionized water was obtained from inverse osmosis industrial device.

### 2.2. Preparation of Porcine Sublingual Epithelium

For the experiments, fresh pork tongue was used. Pig's tongue was obtained from a 6-month-old pig, weighing around 80 kg. The tongue was withdrawn in a local slaughterhouse, transported to the laboratory under vacuum, and immediately used (within 2 h).

Initially, the tongue was placed in physiological solution for about ten minutes, then washed out with new physiologic solution three times, and then sectioned to get the epithelium specimens. The sublingual (ventral) epithelium was sectioned using a scalpel to separate it from underlying connective tissue (Figure 2).

Before clamping the epithelium into the Franz cell chamber (Figure 3), it was washed with PBS 10× pH 6.6–7.4 three times [27,28].

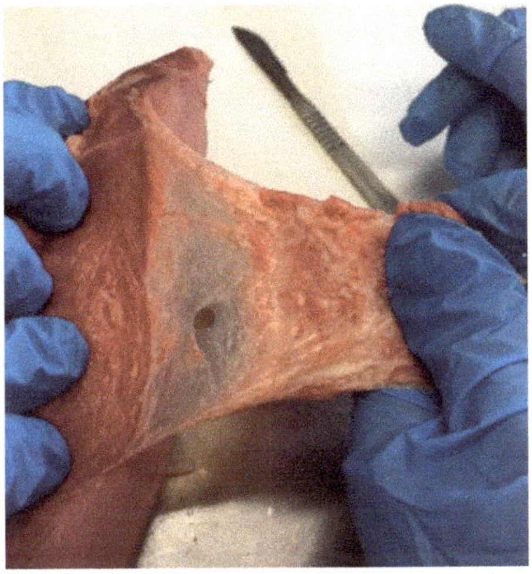

**Figure 2.** Sectioning of the epithelium from pig tongue.

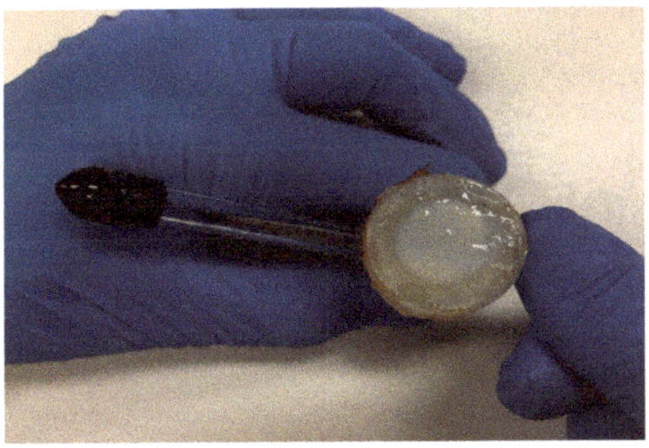

**Figure 3.** Epithelium specimen placed over the Franz cell chamber.

In order to ascertain the correct device assembly and the porcine lingual epithelium integrity, the donor compartment was filled with physiologic solution, to be sure that no liquid overcame the membrane reaching the receptor compartment, meaning that no lesion occurred in the membrane and that the cell was well assembled.

*2.3. Preparation of the Astaxanthin Containing Nanoemulsion*

Astaxanthin nanoemulsion was prepared using the components listed in Table 1.

**Table 1.** Nanoemulsion qualitative composition.

| Component | % *w/w* |
|---|---|
| Caprylic/Capric triglycerides | 4.2 |
| Polysorbate 80 | 2.6 |
| Ascorbyl Palmitate | 1.3 |
| Glycerine | 2.0 |
| Deionized water | Up to 100 g |
| Astapure™ | 0.15% w/2 Astaxanthin |

Emulsion was prepared incorporating Astapure™ (10% *w/w* in astaxanthin) in the oily phase (equal to 0.15% *w/w* theoretic value of astaxanthin on the total nanoemulsion) composed of caprylic/capric triglyceride, and polysorbate 80 (PS 80) and ascorbyl palmitate, as the main high hydrophilic–lipophilic balance emulsifying agent and co-emulsifying agent, respectively. The preparation of the oily phase and the further steps to achieve the nanoemulsion were carried out in a dark room to protect astaxanthin from UV rays. Both the oily and water phases were warmed up at 50 °C and the final emulsification process was carried out at this temperature, slowly pouring the water phase into the oily phase under high-speed mechanical stirring. Soon after, the system was cooled down by placing the beaker containing nanoemulsion, appearing as a perfectly clear system with an intense red color (Figure 4), in an ice-water bath, maintaining low-speed stirring until the room temperature was reached, according to low energy PIT method [29–31].

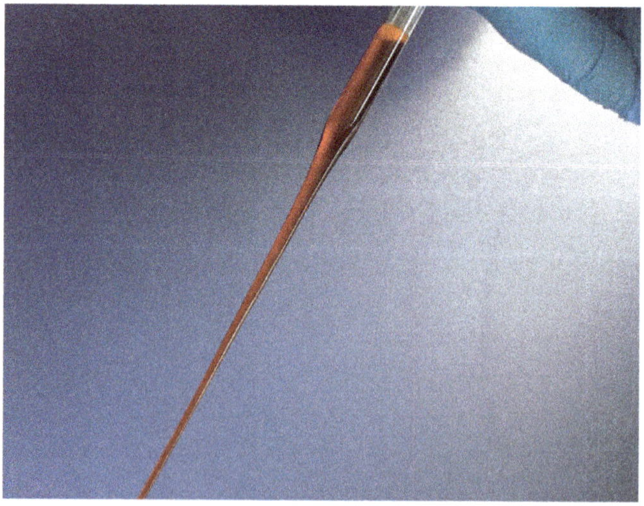

**Figure 4.** Astaxanthin containing nanoemulsion. The system appears perfectly clear with an intense red color conferred by the carotenoid entrapped.

## 2.4. Dimensional Characterization of Astaxanthin Containing Nanoemulsion

Samples of nanoemulsion containing astaxanthin were analyzed in triplicate with Dynamic Back Scattering device (DLS) to assess the average dimensional size of the oily droplets.

## 2.5. Permeation Experiments

The incisions were made starting from the ventral part of the swine tongue, with particular attention given to removing only the outermost layer which, once cut, is completely transparent.

Once the external epithelial tissue was sectioned, it was gently positioned in the appropriate space between the donor and the receiving chamber. A total amount of 1.26 g of the astaxanthin-containing nanoemulsion (equal to total 0.189 g of astaxanthin) was inserted from the upper apex, in the donor compartment of the Franz cell and placed at 37 °C under magnetic agitation (210 rpm). The acceptor chamber was filled up with degassed solution of PBS 10× (10 mL). The cells, before the start of the experiments, were allowed to equilibrate for 60 min in a water bath at 37 °C (according to Franz-Montan 2016) [26].

A stirrer was necessary to keep the system in a continuous flow to carry out solution withdrawals in which the active permeation is homogeneously dispersed. The samples were withdrawn after 15′, 30′, 60′, 2 h, 4 h. Samples (1 mL) were withdrawn from the receiving chamber for the HPLC analysis and the volume was replaced with the same amount of fresh buffer PBS, taking account of dilution effects.

The data obtained from the HPLC analysis relating to the title of astaxanthin in the receptor chamber were converted into mass per unit of surface (μg/cm$^2$) of the permeating membrane. The surface area of the porcine lingual epithelium inserted in the chamber was calculated, starting from the chamber diameter, according to Equation (1):

$$A = \frac{(\pi d^2)}{4} cm^2 \tag{1}$$

where A is the surface of the chamber and $d$ is the diameter of the chamber.

From the linear correlation obtained by relating the astaxanthin content in the donor chamber with time, the slope of the linear tract of the plot ($s$) was calculated: This data permitted the calculation of the apparent permeability coefficient ($P_e$) through the following relation (Equation (2)) derived from the first Fick's equation considering $C_d > C_a$:

$$P_e = \frac{dC_a}{dt} \times \frac{1}{A} \times \frac{V_a}{C_d} \tag{2}$$

where $dC_a/dt$ is the slope ($s$) of the linear correlation between the change in concentration of permeated astaxanthin in the infinitesimal time change, A is the permeation surface, $C_d$ is the concentration on the donor compartment, and $V_a$ is the volume of the acceptor chamber [32]. Once the permeability coefficient ($P_e$) was calculated, according to Bortolotti F. et al. (2009), the flux at the steady state ($J_{ss}$) was also calculated through the following Equation (3):

$$J_{ss} = P_e \times C_d \tag{3}$$

where $J_{ss}$ is the flux at the steady state, $P_e$ is the permeability coefficient, and $C_d$ is the concentration of astaxanthin in the donor chamber [33].

Experiments were realized in triplicate ($n = 3$) and mean value of astaxanthin concentration (±SEM) permeated in the receptor liquid, at any time of withdrawal, was used to calculate the concentration of astaxanthin per cm$^2$ (±SEM), flux ($J_{ss}$) (±SEM), and apparent permeability coefficient ($P_e$) (±SEM). Statistical elaboration of the data collected was realized through software SPSS according to T-Student method ($p < 0.05$).

### 2.6. Titration of Astaxanthin in Raw Material, Nanoemulsion, and Permeation Specimens

Firstly, 100 mg of Astapure™ (10% astaxanthin containing oil, raw material employed to fabricate nanoemulsion) was solubilized in 100 mL of acetone. Then, 500 mg of nanoemulsion containing astaxanthin was weighted in analytical balance and then solubilized in 100 mL of acetone. After that, the samples were sonicated for 15 min and then centrifuged and placed in vials. The samples were analyzed through High-Performance Liquid Chromatography with Diode-Array Detection coupled with UV-analyzer (HPLC-DAD-UV), DAD scan range was from 200 to 800 nm, with stationary phase composed of YMC Carotenoid column 4.6 mm I.D.× 250 mm (C30 bonded silica, Particle size: 5 μm, usable pH range: 2.0–7.5): YMC Carotenoid stationary phase provides sufficient phase thickness to enhance interaction with long chained molecules, therefore, geometric and positional isomers of conjugated double bonding systems, typical of carotenoids and their esters, are recognized and resolved [34]. Mobile phase composed of methylterbutil ether (MTBE)/methanol 90:10 and methanol. The flux was set at 1.3 mL/min and the wave length set at 470 nm. The same method was applied to assess titration of astaxanthin from the permeation specimens. Thanks to this procedure, encapsulation efficiency $E_e$ of the fresh fabricated nanoemulsion was calculated according to Equation (4) [35]:

$$E_e\ (\%) = C_{Ast} \times 100 \tag{4}$$

where $E_e$ is the encapsulation efficiency and $C_{Ast}$ is the concentration of astaxanthin loaded in the nanoemulsion during fabrication (time 0).

The titration results were expressed as the mean and standard error of the mean (±SEM) for each variable studied.

## 3. Results and Discussion

### 3.1. Dimensional Characterization of Astaxanthin Containing Nanoemulsion

Figures 5–7 clearly show that astaxanthin containing nanoemulsion is characterized by a single, narrow, well-shaped pick, both for the measures by number and volume weighting. The average diameter of the oily droplets is around 20 nm (z-average d.nm). The PDI of 0.2 indicates a low poly-dispersion profile with a quite uniform dispersion of the droplets. Under physical point of view, considering the average dimension of the particles and according to the published papers (refer to the section Supplementary File), this system can be considered "border-line" between a nanoemulsion and a microemulsion. Nanoemulsion, indeed, is defined as a clear kinetically stable and thermodynamically unstable liquid system, with average particles size ranging from 100 to 200 nm and a microemulsion is defined as a clear, bicontinous, kinetically and thermodynamically stable liquid system with average particle sizes lower than 50 nm. Thanks to these evidences, it is possible to argue that this system can be particularly capable of enhancing superficial surface of contact between the active ingredient entrapped in the dispersed oily phase and epithelium, in this case, lingual epithelium. Since astaxanthin is a lipophilic molecule easily soluble in oils and insoluble in water, nano-encapsulation can be hugely advantageous for efficient delivery through the lingual epithelium.

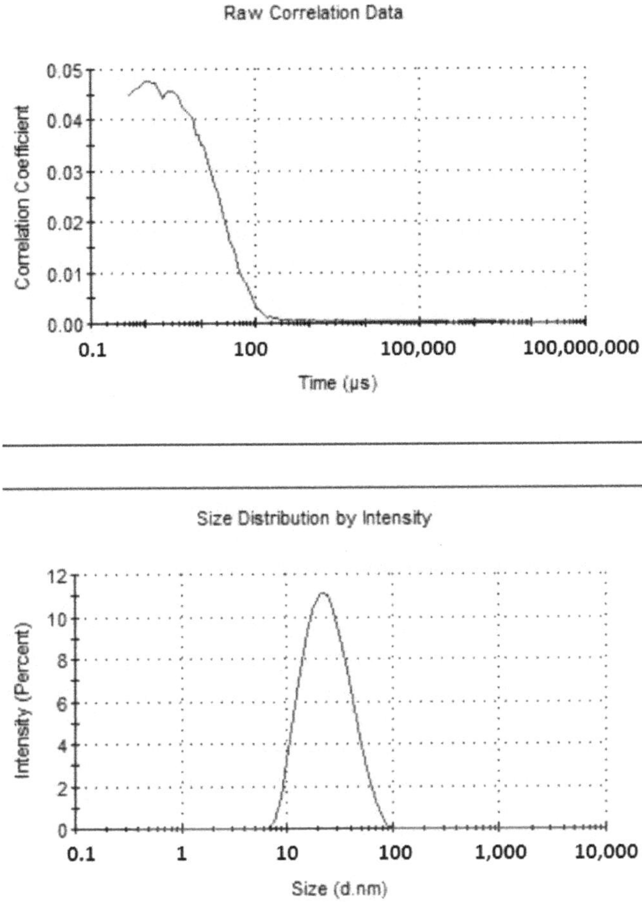


**Figure 5.** Graph plotting the correlation coefficient of astaxanthin nanoemulsion over time (**upper**) and size distribution by intensity (**lower**).

| | | | |
|---|---|---|---|
| **Sample Name:** | Nanoemulsion | | |
| **SOP Name:** | mansettings.nano | | |
| **File Name:** | Nanoemulsion.dts | **Dispersant Name:** | Water |
| **Record Number:** | 1215 | **Dispersant RI:** | 1.330 |
| **Material RI:** | 1.40 | **Viscosity (cP):** | 0.8872 |
| **Material Absorbtion:** | 0.001 | | |

| | | | |
|---|---|---|---|
| **Temperature (°C):** | 25.0 | **Duration Used (s):** | 50 |
| **Count Rate (kcps):** | 180.9 | **Measurement Position (mm):** | 4.65 |
| **Cell Description:** | Disposable sizing cuvette | **Attenuator:** | 7 |

| | | | Size (d.nm): | % Number: | St Dev (d.nm): |
|---|---|---|---|---|---|
| **Z-Average (d.nm):** 20.68 | | **Peak 1:** | 10.69 | 100.0 | 3.245 |
| **PdI:** 0.227 | | **Peak 2:** | 0.000 | 0.0 | 0.000 |
| **Intercept:** 0.218 | | **Peak 3:** | 0.000 | 0.0 | 0.000 |

**Result quality :** Good

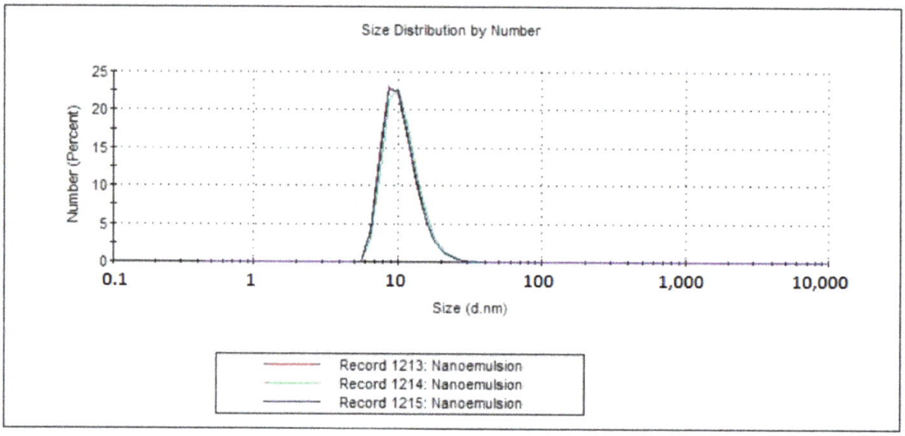


**Figure 6.** Graph plotting the size distribution of astaxanthin nanoemulsion by number.

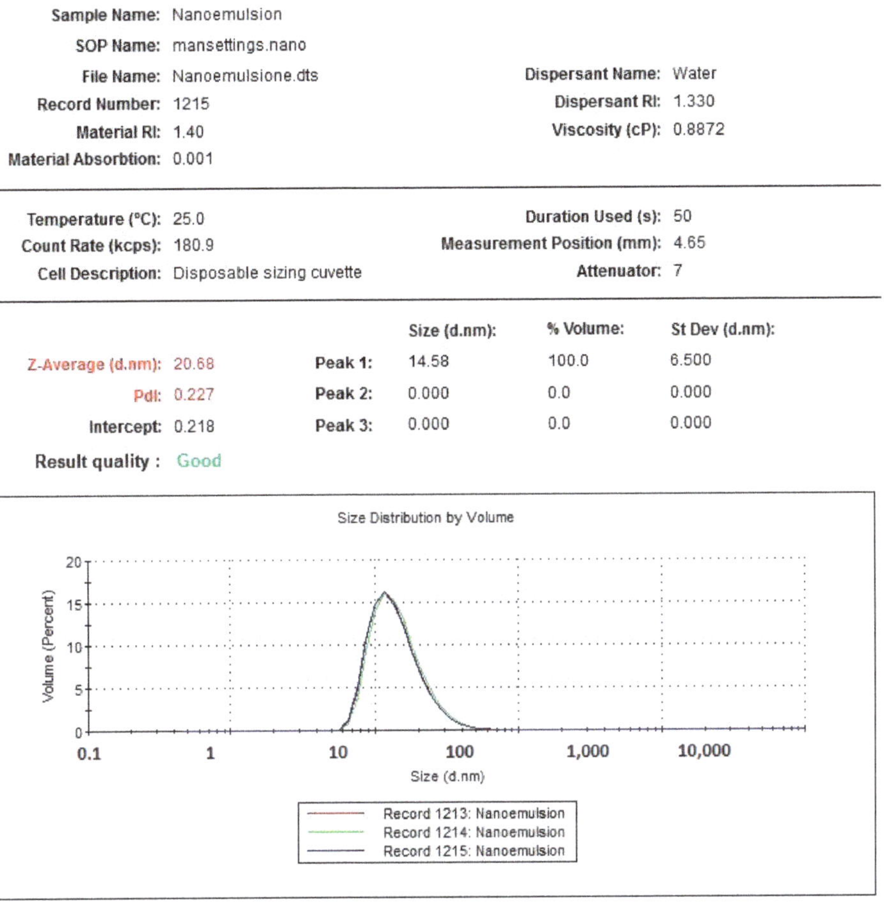


**Figure 7.** Graph plotting the size distribution of astaxanthin nanoemulsion by volume.

### *3.2. Titration of Astaxanthin in Raw Material and Nanoemulsion*

Figure 8 confirms the presence of astaxanthin in both Astapure$^{TM}$ raw material and nanoemulsion with the characteristic pick at 6, 7 min. The analysis of the graph also shows a group of picks (8, 5–12 min) indicating products derived from or analogues of astaxanthin characterized by the same UV spectrum with peculiar absorption pick at 470 nm. According to Ranga et al. (2009), the mentioned picks (8–12 min) recognized in the chromatogram obtained with HPLC-DAD are mainly ascribable to astaxanthin mono and diesters [36], present in Astapure$^{TM}$, an extract derived from *Haematococcus pluvialis* that is indeed characterized by the high content of astaxanthin in the form of fatty acid esters with the predominant presence of monoester, about 70% *w*/*w* [36], that very likely corresponds to the highest pick at about 9 min in chromatogram 1. As further confirmation, the analytical facts data sheet of Astapure$^{TM}$ reports the presence of natural astaxanthin complex in addition to other free carotenoids such as lutein and zeaxanthin. From the titration, it was confirmed that nanoemulsion contains 0.15% *w*/*w* of astaxanthin, according to the 1.5% *w*/*w* amount of Astapure$^{TM}$ titrated at 10% *w*/*w* of astaxanthin complex, loaded during the preparation of the nanoemulsion (Table 2).

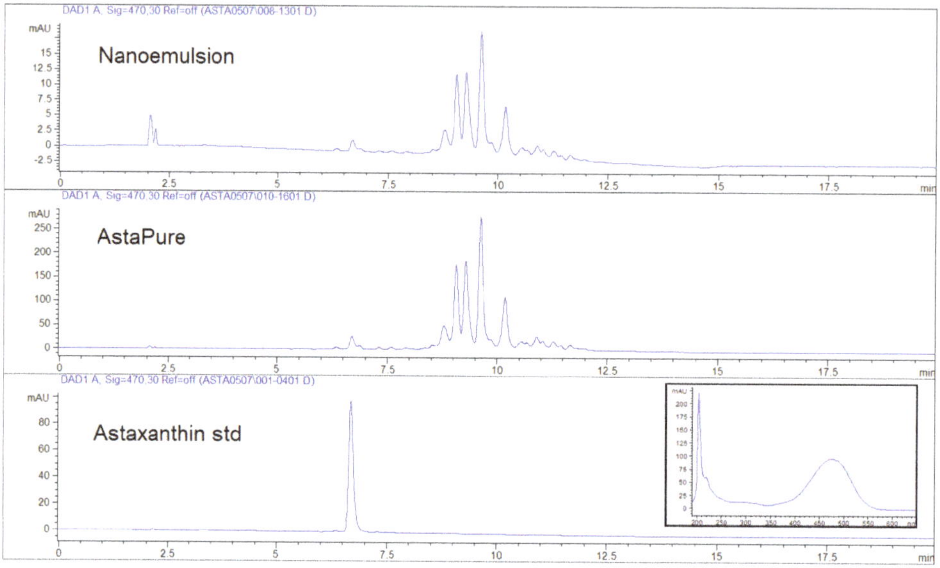


**Figure 8.** Titration of astaxanthin and correlated molecules from Astapure™ (raw material), nanoemulsion and astaxanthin (analytical standard).

**Table 2.** Titration of astaxanthin in Astapure™ raw material (theoretical 10% *w/w* astaxanthin) and nanoemulsion. Results are expressed as mean ± SEM.

| Sample | *w/w* as Total Astaxanthin |
|---|---|
| Astapure™ | 10.04 ± 0.09 |
| Nanoemulsion | 0.151 ± 0.006 |

This data, apart from the confirmation of the correct content of astaxanthin in the nanoemulsion, gives account of the stability of astaxanthin during nanoemulsion preparation: Astaxanthin is notoriously a very unstable molecule [37] and can be promptly and massively degraded when subjected to heat and UV exposure. The measures of dark room and low thermal energy chosen during the phases of preparation of the nanoemulsion allowed the full recovery of astaxanthin into the final system as the excellent data of encapsulation efficiency $E_e$ confirms (Table 3).

**Table 3.** $E_e$ of Astaxanthin in the nanoemulsion.

| **Astapure™ loaded** | 1.5 g/100 mL |
|---|---|
| $C_{Ast}$ in nanoemulsion | 0.15 gt/100 mL |
| Nanoemulsion $E_e$ (%) | 100 |

Moreover, it can be argued that astaxanthin can be stabilized throughout the nanoemulsion preparation process and over time by ascorbyl palmitate (ASP), that is notoriously a lipidic antioxidant that can protect isoprenoid structure of astaxanthin and prevent it from being oxidized.

### 3.3. Titration of Astaxanthin in Permeation Specimens

According to Figures 9 and 10 in which the permeation rate of astaxanthin and its derivatives is represented through the HPLC-DAD titration on the five permeates specimens, as reported in Sections 2.6 and 3.2, it is possible to recognize the peculiar astaxanthin pick in every sample. Increasing concentration of astaxanthin and its derivatives was registered in specimens over time (Table 4), indicating that astaxanthin and its derivatives accumulated in the permeated receptor liquid reaching the plateau at 2 h. From 1 h specimen to 2 h specimen, the concentration of astaxanthin and its derivatives shows a 21-fold increase (Table 4). Flux ($J_{ss}$) and Apparent Permeability ($P_e$) of astaxanthin from the nanoemulsion are equal to 6.27 ± 0.022 μm/cm$^2$/h and 41.55 ± 0.442 cm/h respectively; the total amount of astaxanthin permeated after 4 h is equal to 23.6% (25.1 ± 0.24 mcg/cm$^2$) of the total amount loaded in the donor compartment entrapped in the nanoemulsion (189 mcg). The amount of astaxanthin retained in the lingual tissue was not determined.

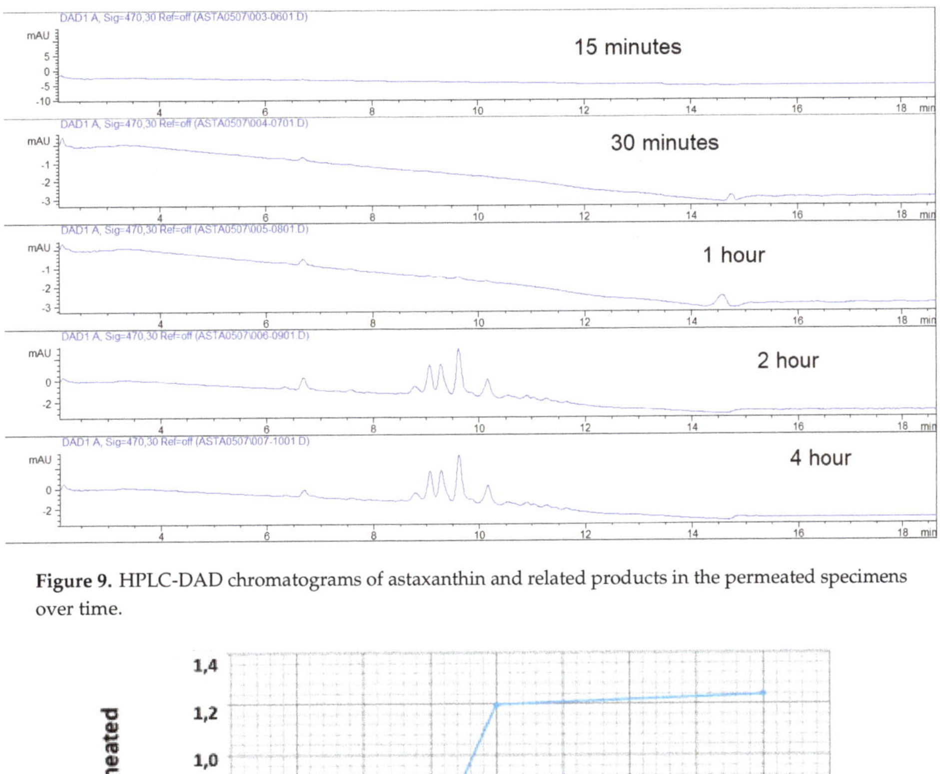


**Figure 9.** HPLC-DAD chromatograms of astaxanthin and related products in the permeated specimens over time.

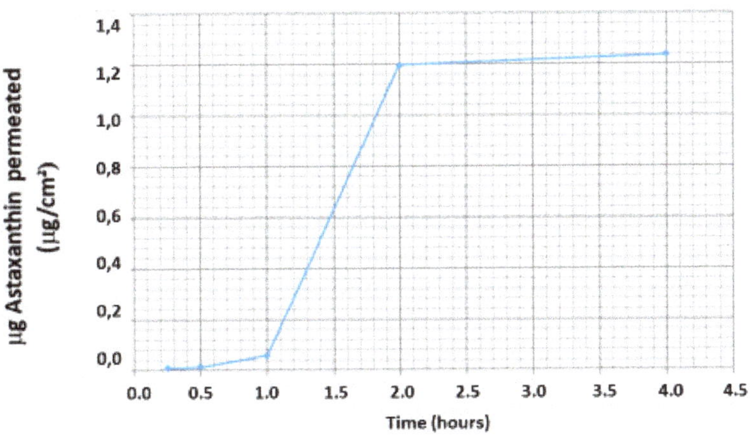


**Figure 10.** Graph of the permeation profile of astaxanthin (nanoemulsion) over time.

**Table 4.** Concentration of astaxanthin (nanoemulsion) permeated expressed as μg/mL and μg/cm$^2$. Results are expressed as mean ± SEM.

| Specimen | μg/mL Astaxanthin and Derivates | μg/cm$^2$ Astaxanthin and Derivates |
|---|---|---|
| Permeated 15 min | 0.019 ± 0.001 | 0.0108 ± 0.001 |
| Permeated 30 min | 0.025 ± 0.004 | 0.0142 ± 0.001 |
| Permeated 1 h | 0.102 ± 0.023 | 0.0577 ± 0.008 |
| Permeated 2 h | 2.116 ± 0.002 | 1.1980 ± 0.012 |
| Permeated 4 h | 2.184 ± 0.073 | 1.2365 ± 0.009 |

Comparing this data with that collected from Odeberg et al. [23] (Wagner–Nelson method), assessing astaxanthin bioavailability in humans from different formulations assumed orally containing 40 mg of the carotenoid with or without absorption enhancers (PS 80, Glycerol mono and dioleate, SPAN 80), it is possible to recognize comparable data of percentage of absorption over time (4 h) ranging from 4% (formulation without absorption enhancers) to 34% (PS 80 + SPAN 80). Despite this, the data are far from being completely comparable given the differences in experimental design and approach, model and the route of administration, the high hydrophilic–lipophilic balance surfactants such as PS 80, and the technological approaches capable of enhancing astaxanthin hydro-dispersibility that improve astaxanthin bioavailability seem to be confirmed.

## 4. Conclusions

Astaxanthin is a carotenoid that attracts the attention of clinicians for its well ascertained potential therapeutic activity, especially in the fields of ocular and skin health and cardiovascular disease prevention. Despite this clinical potential, astaxanthin bioavailability and stability are low and limited. Since astaxanthin is a very lipophilic molecule, the only way to deliver it in a water-based liquid system is to create an emulsion. According to this concept, liquid nanoemulsion seems to be a promising technical system for releasing astaxanthin through the lingual epithelium.

The data collected in this work show for the first time that astaxanthin can be delivered in a nanoemulsion through the lingual district, opening an interesting path to improving its poor bioavailability when assumed by the oral route. In particular, the nanoemulsion herewith described, is characterized by a good uniformity of dispersion, very low dimension of the oily droplets (around 20 nm), close to those of a microemulsion, and overall can be considered an effective and innovative pharmaceutical form for entrapping astaxanthin as the encapsulation efficiency data confirms. It is also important to emphasize that the peculiar association of surfacing agents employed to achieve this system, with specific reference to ASP, provides an overall good stability of astaxanthin during the process of nanoemulsion fabrication. This fact is of particular significance in consideration of the notorious instability of the molecule that prevents its full clinical usage. From the data collected regarding astaxanthin permeation behavior, it is possible to argue that astaxanthin reaches a pseudo-linear permeation trend during the second hour and then reaches a plateau, confirming that a quasi-steady state diffusion can be described. Taken together, these data confirm that a novel pharmaceutical form projected to deliver astaxanthin through the sublingual route has been achieved and characterized, and thanks to this peculiar form, this route can be considered, even though preliminarily, a potential effective alternative to enhance the bioavailability of this carotenoid, potentially improving its therapeutic potential. A confirmation of these preliminary data should be achieved with a pharmacological assessment of the kinetic profile of astaxanthin-containing nanoemulsion through the sublingual route in healthy human volunteers.

**Supplementary Materials:** The following are available online at http://www.mdpi.com/1660-3397/17/9/508/s1, Text S1 Thermodynamic topics of nanoemulsions and microemulsions; Text S2. Synergy between Ascorbyl Palmitate and Polysorbate in the production of nanoemulsion.

**Author Contributions:** Conceptualization, A.F. and A.F.G.C.; methodology, A.F.; formal analysis, A.F. and D.B.; investigation, A.F. and D.B.; data curation, A.F.; writing—original draft preparation, A.F. and A.F.G.C.; writing—review and editing, A.F. and A.F.G.C.

**Funding:** This research received no external funding. Nanoemulsion technology herewith described is intellectual property of Labomar SPA, Istrana, TV (Patent number: WO 2011/073726).

**Conflicts of Interest:** Andrea Fratter and Damiano Biagi are employers of Labomar SpA, the company owner of the patented nanoemulsion technology. Arrigo F.G. Cicero has no direct nor indirect conflict of interest in the publication of this paper.

## References

1. Fakhri, S.; Abbaszadeh, F.; Dargahi, L.; Jorjani, M. Astaxanthin: A mechanistic review on its biological activities and health benefits. *Pharmacol. Res.* **2018**, *136*, 1–20. [CrossRef] [PubMed]
2. Cicero, A.F.G.; Colletti, A. Effects of Carotenoids on Health: Are All the Same? Results from Clinical Trials. *Curr. Pharm. Des.* **2017**, *23*, 2422–2427. [CrossRef] [PubMed]
3. Kim, S.H.; Kim, H. Inhibitory Effect of Astaxanthin on Oxidative Stress-Induced Mitochondrial Dysfunction-A Mini-Review. *Nutrients* **2018**, *10*, 1137. [CrossRef] [PubMed]
4. Chuyen, H.V.; Eun, J.B. Marine carotenoids: Bioactivities and potential benefits to human health. *Crit. Rev. Food Sci. Nutr.* **2017**, *57*, 2600–2610. [CrossRef] [PubMed]
5. Choi, C.I. Astaxanthin as a Peroxisome Proliferator-Activated Receptor (PPAR) Modulator: Its Therapeutic Implications. *Mar. Drugs* **2019**, *17*, 242. [CrossRef] [PubMed]
6. Ito, N.; Seki, S.; Ueda, F. The Protective Role of Astaxanthin for UV-Induced Skin Deterioration in Healthy People-A Randomized, Double-Blind, Placebo-Controlled Trial. *Nutrients* **2018**, *10*, 817. [CrossRef] [PubMed]
7. Chen, J.T.; Kotani, K. Effects of Astaxanthin on Liver and Leukocyte Parameters in Healthy Climacteric Women: Preliminary Data. *J. Med. Food* **2017**, *20*, 724–725. [CrossRef]
8. Kaneko, M.; Kishimoto, Y.; Suzuki, R.; Kawai, Y.; Tateya, I.; Hirano, S. Protective Effect of Astaxanthin on Vocal Fold Injury and Inflammation Due to Vocal Loading: A Clinical Trial. *J. Voice* **2017**, *31*, 352–358. [CrossRef]
9. Saito, M.; Yoshida, K.; Saito, W.; Fujiya, A.; Ohgami, K.; Kitaichi, N.; Tsukahara, H.; Ishida, S.; Ohno, S. Astaxanthin increases choroidal blood flow velocity. *Graefe's Arch. Clin. Exp. Ophthalmol.* **2012**, *250*, 239–245. [CrossRef]
10. Choi, H.D.; Youn, Y.K.; Shin, W.G. Positive effects of astaxanthin on lipid profiles and oxidative stress in overweight subjects. *Plant Foods Hum. Nutr.* **2011**, *66*, 363–369. [CrossRef]
11. Choi, H.D.; Kim, J.H.; Chang, M.J.; Kyu-Youn, Y.; Shin, W.G. Effects of astaxanthin on oxidative stress in overweight and obese adults. *Phytother. Res.* **2011**, *25*, 1813–1818. [CrossRef] [PubMed]
12. Yoshida, H.; Yanai, H.; Ito, K.; Tomono, Y.; Koikeda, T.; Tsukahara, H.; Tada, N. Administration of natural astaxanthin increases serum HDL-cholesterol and adiponectin in subjects with mild hyperlipidemia. *Atherosclerosis* **2010**, *209*, 520–523. [CrossRef] [PubMed]
13. Kim, J.H.; Chang, M.J.; Choi, H.D.; Youn, Y.K.; Kim, J.T.; Oh, J.M.; Shin, W.G. Protective effects of Haematococcus astaxanthin on oxidative stress in healthy smokers. *J. Med. Food* **2011**, *14*, 1469–1475. [CrossRef] [PubMed]
14. Kupcinskas, L.; Lafolie, P.; Lignell, A.; Kiudelis, G.; Jonaitis, L.; Adamonis, K.; Andersen, L.P.; Wadström, T. Efficacy of the natural antioxidant astaxanthin in the treatment of functional dyspepsia in patients with or without Helicobacter pylori infection: A prospective, randomized, double blind, and placebo-controlled study. *Phytomedicine* **2008**, *15*, 391–399. [CrossRef] [PubMed]
15. Djordjevic, B.; Baralic, I.; Kotur-Stevuljevic, J.; Stefanovic, A.; Ivanisevic, J.; Radivojevic, N.; Andjelkovic, M.; Dikic, N. Effect of astaxanthin supplementation on muscle damage and oxidative stress markers in elite young soccer players. *J. Sports Med. Phys. Fitness* **2012**, *52*, 382–392.
16. Baralic, I.; Djordjevic, B.; Dikic, N.; Kotur-Stevuljevic, J.; Spasic, S.; Jelic-Ivanovic, Z.; Radivojevic, N.; Andjelkovic, M.; Pejic, S. Effect of astaxanthin supplementation on paraoxonase 1 activities and oxidative stress status in young soccer players. *Phytother. Res.* **2013**, *27*, 1536–1542. [CrossRef] [PubMed]
17. Klinkenberg, L.J.; Res, P.T.; Haenen, G.R.; Bast, A.; van Loon, L.J.; van Dieijen-Visser, M.P.; Meex, S.J. Effect of antioxidant supplementation on exercise-induced cardiac troponin release in cyclists: A randomized trial. *PLoS ONE* **2013**, *8*, e79280. [CrossRef]

18. Earnest, C.P.; Lupo, M.; White, K.M.; Church, T.S. Effect of astaxanthin on cycling time trial performance. *Int. J. Sports Med.* **2011**, *32*, 882–888. [CrossRef]
19. Res, P.T.; Cermak, N.M.; Stinkens, R.; Tollakson, T.J.; Haenen, G.R.; Bast, A.; Van Loon, L.J. Astaxanthin supplementation does not augment fat use or improve endurance performance. *Med. Sci. Sports Exerc.* **2013**, *45*, 1158–1165. [CrossRef]
20. Coombes, J.S.; Sharman, J.E.; Fassett, R.G. Astaxanthin has no effect on arterial stiffness, oxidative stress, or inflammation in renal transplant recipients: A randomized controlled trial (the XANTHIN trial). *Am. J. Clin. Nutr.* **2016**, *103*, 283–289. [CrossRef]
21. Viera, I.; Pérez-Gálvez, A.; Roca, M. Bioaccessibility of Marine Carotenoids. *Mar. Drugs.* **2018**, *16*, 397. [CrossRef] [PubMed]
22. Ambati, R.R.; Phang, S.M.; Ravi, S.; Aswathanarayana, R.G. Astaxanthin: Sources, extraction, stability, biological activities and its commercial applications—A review. *Mar. Drugs* **2014**, *12*, 128–152. [CrossRef]
23. Odeberg, J.M.; Lignell, A.; Pettersson, A.; Höglund, P. Oral bioavailability of the antioxidant astaxanthin in humans is enhanced by incorporation of lipid based formulations. *Eur. J. Pharm. Sci.* **2003**, *19*, 299–304. [CrossRef]
24. Yuan, Y.; Gao, Y.; Zhao, J.; Mao, L. Characterization and stability evaluation of β-carotene nanoemulsions prepared by high pressure homogenization under various emulsifying conditions. *Food Res. Int.* **2008**, *41*, 61–68. [CrossRef]
25. Redzuan, M.; Effendi, T.J.B.; Majeed, A.B.A. Development and stability evaluation of Astaxanthin nanoemulsion. *Asian J. Pharm. Clin. Res.* **2011**, *4*, 143–148.
26. Shanmugapriya, K.; Kim, H.; Kang, H.W. In vitro antitumor potential of astaxanthin nanoemulsion against cancer cells via mitochondrial mediated apoptosis. *Int. J. Pharm.* **2019**, *560*, 334–346. [CrossRef] [PubMed]
27. Franz-Montan, M.; Serpe, L.; Martinelli, C.C.; da Silva, C.B.; Santos, C.P.; Novaes, P.D.; Volpato, M.C.; de Paula, E.; Lopez, R.F.; Groppo, F.C. Evaluation of different pig oral mucosa sites as permeability barrier models for drug permeation studies. *Eur. J. Pharm. Sci.* **2016**, *81*, 52–59. [CrossRef]
28. Narang, N.; Sharma, J. Sublingual mucosa as a route for systemic drug delivery. *Int. J. Pharm. Pharm. Sci.* **2011**, *3*, 18–22.
29. McClements, D.J. Nanoemulsions versus microemulsions: Terminology, differences, and similarities. *Soft Matter* **2012**, *8*, 1719–1729. [CrossRef]
30. Kotta, S.; Khan, A.W.; Ansari, S.H.; Sharma, R.K.; Ali, J. Formulation of nanoemulsion: A comparison between phase inversion composition method and high-pressure homogenization method. *Drug Deliv.* **2015**, *22*, 455–466. [CrossRef]
31. Jintapattanakit, A. Preparation of nanoemulsions by phase inversion temperature (PIT) method. *Pharm. Sci. Asia* **2018**, *45*, 1–12. [CrossRef]
32. Hsu, H.H.; Kracht, J.K.; Harder, L.E.; Rudnik, K.; Lindner, G.; Schimek, K.; Marx, U.; Pörtner, R. A Method for Determination and Simulation of Permeability and Diffusion in a 3D Tissue Model in a Membrane Insert System for Multi-Well Plates. *J. Vis. Exp.* **2018**, *132*. [CrossRef] [PubMed]
33. Bortolotti, F.; Balducci, A.G.; Sonvico, F.; Russo, P.; Colombo, G. In vitro permeation of desmopressin across rabbit nasal mucosa from liquid nasal sprays: The enhancing effect of potassium sorbate. *Eur. J. Pharm. Sci.* **2009**, *37*, 36–42.
34. Sander, L.C.; Wise, S.A. Shape selectivity in reversed-phase liquid chromatography for the separation of planar and non-planar solutes. *J. Chromatogr.* **1993**, *656*, 335–351. [CrossRef]
35. Shen, X.; Fang, T.; Zheng, J.; Guo, M. Physicochemical Properties and Cellular Uptake of Astaxanthin-Loaded Emulsions. *Molecules* **2019**, *24*, 727. [CrossRef] [PubMed]
36. Ranga, R.; Sarada, A.R.; Baskaran, V.; Ravishankar, G.A. Identification of Carotenoids from Green Alga Haematococcus pluvialis by HPLC and LC-MS (APCI) and Their Antioxidant Properties. *J. Microbiol. Biotechnol.* **2009**, *19*, 1333–1341. [PubMed]
37. Bustos-Garza, C.; Yáñez-Fernández, J.; Barragán-Huerta, B.E. Thermal and pH stability of spray-dried encapsulated astaxanthin oleoresin from Haematococcus pluvialis using several encapsulation wall materials. *Food Res. Int.* **2013**, *54*, 641–649. [CrossRef]

*marine drugs*

*Review*

# Absorption and Metabolism of Xanthophylls

**Eiichi Kotake-Nara and Akihiko Nagao** *

National Food Research Institute, National Agriculture and Food Research Organization,
2-1-12 Kannondai, Tsukuba, Ibaraki 305-8642, Japan; E-Mail: ekotake@affrc.go.jp

* Author to whom correspondence should be addressed; E-Mail: nagao@affrc.go.jp;
Tel.: +81-29-838-8039; Fax: +81-29-838-7996.

*Received: 1 April 2011; in revised form: 3 June 2011 / Accepted: 7 June 2011 / Published: 10 June 2011*

**Abstract:** Dietary carotenoids, especially xanthophylls, have attracted significant attention because of their characteristic biological activities, including anti-allergic, anti-cancer, and anti-obese actions. Although no less than forty carotenoids are ingested under usual dietary habits, only six carotenoids and their metabolites have been found in human tissues, suggesting selectivity in the intestinal absorption of carotenoids. Recently, facilitated diffusion in addition to simple diffusion has been reported to mediate the intestinal absorption of carotenoids in mammals. The selective absorption of carotenoids may be caused by uptake to the intestinal epithelia by the facilitated diffusion and an unknown excretion to intestinal lumen. It is well known that β-carotene can be metabolized to vitamin A after intestinal absorption of carotenoids, but little is known about the metabolic transformation of non provitamin A xanthophylls. The enzymatic oxidation of the secondary hydroxyl group leading to keto-carotenoids would occur as a common pathway of xanthophyll metabolism in mammals. This paper reviews the absorption and metabolism of xanthophylls by introducing recent advances in this field.

**Keywords:** absorption; bioavailability; carotenoid; metabolism; xanthophyll

## 1. Introduction

Carotenoids, which are synthesized *de novo* by microorganisms and plants, accumulate in various biological tissues throughout the food chain. More than 700 carotenoids, including the metabolites in animals, are present in nature. Most of the carotenoids contain oxygen functions in the molecules, and these carotenoids are referred to as xanthophylls. In recent years, a great deal of attention has been focused on biological activities of dietary xanthophylls such as lutein, zeaxanthin, β-cryptoxanthin, capsanthin, astaxanthin, and fucoxanthin.

Lutein is one of the major xanthophylls present in green leafy vegetables. Lutein and zeaxanthin are known to selectively accumulate in the macula of the human retina. They have been thought to work as antioxidants [1,2] and as blue light filters [3] to protect the eyes from such oxidative stresses as cigarette smoking and sunlight exposure, which can lead to age-related macular degeneration and cataracts. β-Cryptoxanthin, a major xanthophyll in fruits such as papaya and mandarin orange, was reported to decrease the gene expression of interleukin-1β in mouse macrophage RAW264 cells [4], to promote osteoblastic differentiation of mouse MC3T3 cells [5], and to prevent the decrease of calcium content in the bone of ovariectomized rat [6]. Capsanthin, a major xanthophyll in paprika, was reported to increase high-density lipoprotein (HDL)-cholesterol in rat plasma [7].

Astaxanthin and fucoxanthin are abundant in green algae and brown algae, respectively. Numerous studies have reported that astaxanthin has the potential to prevent cancers, diabetes, and inflammatory and cardiovascular diseases [8,9]. Fucoxanthin has been shown to inhibit the growth of various cancer cell lines [10–14] and chemically induced mouse carcinogenesis [15]. Furthermore, the anti-allergic and anti-obese activities of fucoxanthin were recently shown in rodent mast cells [16] and in mice [17], respectively. Neoxanthin, which has a structure similar to that of fucoxanthin, is present in green leafy vegetables. These two xanthophylls have a 5,6-monoepoxide and an allenic bond as the common characteristic functional groups (Figure 1). We found that fucoxanthin and neoxanthin showed the highest inhibitory effect on the proliferation of human prostate cancer cells among the fifteen carotenoids tested [13].

**Figure 1.** Chemical structures of various epoxy xanthophylls. The geometrical configuration of neoxanthin in nature was recognized as 9′-*cis*.

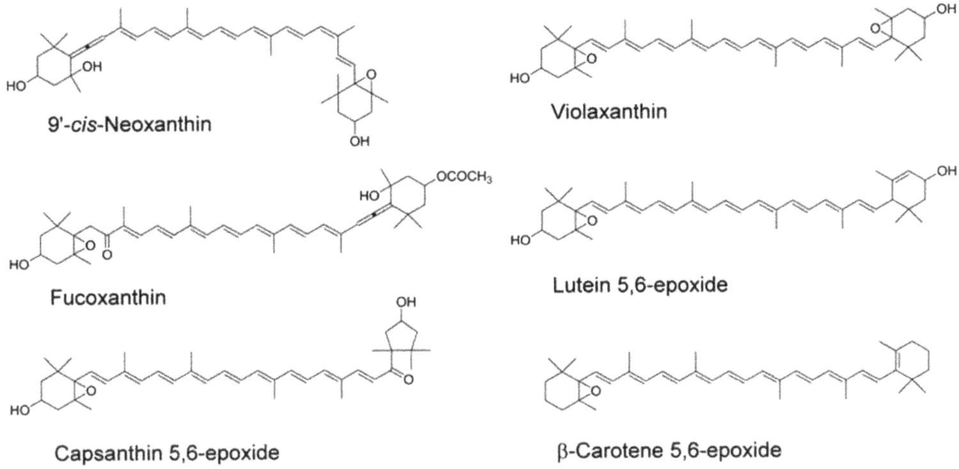


Thus, the characteristic biological activities of several xanthophylls have attracted a great deal of attention. Although no less than forty carotenoids are ingested from common foods, only six major carotenoids (β-carotene, α-carotene, lycopene, β-cryptoxanthin, lutein, and zeaxanthin), their proposed metabolites, and several acyclic carotenes such as phytoene, phytofluene, and ζ-carotene have been found in the plasma of human subjects under usual dietary habits [18,19]. For example, neoxanthin and violaxanthin are ingested together with lutein from green leafy vegetables, but the accumulation of the former two xanthophylls was not confirmed in human plasma [18]. Carotenoids are thought to be selectively absorbed in the human intestine. Moreover, carotenoid accumulations in the biological tissues are known to differ greatly among animal species [20]. However, the mechanisms underlying these phenomena have not been determined.

After intestinal absorption of dietary carotenoids, conversion from β-carotene to vitamin A is well known in animals. In fishes and birds, oxidative and reductive metabolisms of the end group in xanthophylls were also proposed by the identification of the metabolic products, but details as to the mechanism of their metabolic transformation are yet to be elucidated. In mammals, several proposed metabolites of xanthophylls have been detected in the tissues, but the metabolic pathway is still uncertain. It is necessary to reveal the carotenoid metabolism after intestinal absorption in order to elucidate not only the mechanism of the biological activities but also the exact bioavailability. Here, we describe the absorption and metabolism of xanthophylls in mammals.

## 2. Bioaccessibility of Carotenoids

Although xanthophylls have the potential to prevent various degenerative diseases as described above, the bioavailability of carotenoids is lower than that of other fatty components such as
α-tocopherol and triacylglycerols [21–24]. The major cause of the low bioavailability is the poor solubility of carotenoids in digestive fluid. Carotenoids must be solubilized in the digestive fluid via several steps before uptake by intestinal epithelial cells can occur [25]. First, carotenoids are released from the food matrix. In some types of food, the matrix interferes with the release of carotenoids. Carotenoids are hardly released from raw vegetables due to the solid structure of the cell walls, but processing and heat treatment of foods accelerate the release of carotenoids by destroying the structures [26]. The released carotenoids must be well dispersed in the gastrointestinal tract. However, the carotenoid dispersion is greatly limited in digestive fluid due to the high hydrophobicity of C40 isoprenoid carbon skeletons. In this step, dietary lipids facilitate the carotenoid dispersion. Carotenoids are dissolved into the dietary lipids and then dispersed as an emulsion in the digestive fluid. The digestion of the dietary lipids in the emulsion progresses with the aid of lipolytic enzymes and bile fluid, and finally the carotenoids are solubilized in the mixed micelle. The mixed micelle consisting of bile acids, phospholipids, cholesterol, fatty acids, and monoacylglycerols has a disk-like shape, in which the outside is surrounded by the bile acids [27]. Carotenoids solubilized in the mixed micelle are thought to become accessible to uptake by the intestinal epithelial cells. Thereby, the bioaccessibility is defined as the ratio of carotenoids solubilized in the mixed micelles to the total carotenoids ingested. The bioaccessibility, dependent on the food matrix, processing, cooking, and structures of carotenoids, is an important factor for bioavailability.

## 3. Intestinal Absorption of Xanthophylls

In addition to the bioaccessibility, carotenoid uptake by intestinal epithelial cells is also a critical factor for the carotenoid bioavailability. Only one part of the accessible carotenoid is taken up by the intestinal epithelial cells and secreted into lymph as chylomicrons for circulating in blood stream. After the chylomicrons are degraded by lipoprotein lipase, carotenoids in chylomicron remnants are taken up by the liver. The carotenoids are stored in liver or resecreted as very-low-density lipoprotein into the blood stream, and then delivered as low-density lipoprotein (LDL). Finally, carotenoids are taken up to tissues through the LDL receptor. Highly hydrophobic carotenoids such as β-carotene and lycopene are localized in the inner part of LDL, while less hydrophobic xanthophylls such as lutein and zeaxanthin are equally distributed to LDL and HDL, and localized in the outer surface area of the lipoprotein particles [28].

The intestinal absorption of carotenoids had been thought to be mediated by simple diffusion [29,30]. To characterize the human intestinal absorption of carotenoids, we compared the uptakes of various carotenoids by human intestinal Caco-2 cells [31]. The carotenoids solubilized at the same concentration in mixed micelles were incubated with the Caco-2 cells. The uptakes were correlated with their lipophilicity, suggesting that simple diffusion mediated the intestinal uptake of the carotenoids. The amounts of fucoxanthin and neoxanthin taken up by the cells were approximately 25% of that of lutein and were the lowest among the eleven carotenoids tested. These results indicated that fucoxanthin and neoxanthin were certainly absorbed in the Caco-2 cells, although the amounts were lower than that of lutein.

In addition to the experiments using Caco-2 cells, we were able to confirm the absorption of these two xanthophylls in mice [32–34]. The xanthophylls solubilized in mixed micelles were orally administrated to male ICR mice. Fucoxanthinol and amarouciaxanthin A derived from fucoxanthin were detected in plasma and the liver [32,33]. A similar result was also reported in rats fed with fucoxanthin [35]. Neoxanthin and neochromes (formed from neoxanthin by intragastric acidity) were detected in plasma and the liver [34]. The plasma concentrations in the mice 2 h after administration of four purified carotenoids (40 nmol) in the independent experiments under almost the same condition were as follows: 36 nM for β-carotene [36]; 10 nM for lutein [36]; 35 nM for neoxanthin (neoxanthin and neochromes) [34] and 50 nM for fucoxanthin (fucoxanthinol and amarouciaxanthin A) [33]. Neoxanthin and fucoxanthin were confirmed to be absorbed at a similar level to those of β-carotene and lutein, and no selectivity for carotenoids tested was found in mice.

In addition to rodents, it has been reported that fucoxanthin is absorbed in other animals such as tunicates [37,38], chicken [39], and aquatic insects [40]. However, fucoxanthin was not absorbed in freshwater fish [40]. East Asian people ingest fucoxanthin from foodstuffs such as sea squirt, sea urchins, mussel, and brown algae. However, no information on the absorption of fucoxanthin in humans has been available. Although neoxanthin and violaxanthin are ingested from green leafy vegetables under usual dietary habits, they were not found in human serum and milk by a detailed analysis of carotenoids [19]. Thus, it has been uncertain whether fucoxanthin and neoxanthin are absorbed in humans.

We reported for the first time the bioavailability of fucoxanthin from edible brown algae (wakame) and of neoxanthin and violaxanthin from spinach in humans [41]. After the daily intake of stir-fried wakame containing 6.1 mg fucoxanthin for 1 week, the concentrations of fucoxanthin and its metabolites in plasma were analyzed by HPLC. Fucoxanthin and amarouciaxanthin A were not detected. Fucoxanthinol was detected, but the plasma concentration was under the quantification limit (1.0 nM). Similar to the case of fucoxanthin, the plasma concentrations of neoxanthin and violaxanthin after the intake of stir-fried spinach were under the quantification limit.

On the other hand, both β-carotene and lutein, which were present with these epoxy xanthophylls in the same food matrix of spinach, were increased in the plasma [41], suggesting that little neoxanthin and violaxanthin in spinach were absorbed in humans. In contrast to the case of mice, selective absorption of carotenoids may occur in humans.

The low bioavailability of these epoxy xanthophylls may be caused by their low bioaccessibility from spinach and wakame. However, the bioaccessibility of neoxanthin (neoxanthin and neochromes) from spinach was comparable with that of lutein and was greater than that of β-carotene in *in vitro* digestion study [34]. Similarly, the bioaccessibility of fucoxanthin from wakame was sufficiently high [41]. These results suggested that the bioaccessibility was not a limiting factor of the bioavailability.

The absence of these epoxy xanthophylls in human plasma may be due to the rapid metabolism. However, the concentrations of these epoxy xanthophylls and their metabolites in the plasma were under the quantification limit even shortly after the intake of spinach and wakame [41], indicating that the rapid disappearance might not occur.

It is possible that the level of these epoxy xanthophylls in plasma were estimated to be low due to unknown metabolic transformation such as hydrolysis of epoxide or formation of conjugates by detoxification enzymes after the intestinal uptake. For instance, fucoxanthinol 3′-sulphate found in the egg yolk of hens fed with seaweed meal [39] might be formed from fucoxanthin in humans.

The dietary water-soluble fibers, alginates in wakame may be associated with the low bioavailability of fucoxanthin from wakame in humans, because dietary water-soluble fibers inhibited the β-carotene and lutein uptake by Caco-2 cells [42]. Thus, it is necessary to reveal the bioavailability of isolated carotenoid to avoid the influence of the food matrix.

There are several reports on the bioavailability of epoxy xanthophylls in the purified preparations and the oleoresins in human subjects. Oleoresin, which is extracted from plant materials, does not contain dietary fibers and any other polar substances. Capsanthin 5,6-epoxide and violaxanthin were not detected in chylomicron after ingestion of paprika oleoresin containing these epoxy xanthophylls [43]. However, 9-*cis* zeaxanthin, which was present at a lower amount than epoxy xanthophylls in paprika oleoresin, was found in chylomicron [43]. This result suggested that little capsanthin 5,6-epoxide and violaxanthin in paprika were absorbed in humans. Moreover, after a single oral dose of purified violaxanthin or lutein 5,6-epoxide suspended in corn oil, the two epoxy xanthophylls were not detected in the plasma [44]. In contrast, after an oral dose of purified β-carotene 5,6-epoxide (9.1 μmol) suspended in corn oil, the plasma concentration reached 2.29 μM [45]. Considering these experimental results with the oleoresin and purified xanthophylls, little epoxy xanthophylls that have higher polarity than β-carotene 5,6-epoxide would be absorbed by humans, consistent with the results of our human study using spinach and wakame. The chemical structures of these epoxy xanthophylls are shown in Figure 1.

To summarize the intestinal absorption of carotenoids, little of highly polar epoxy xanthophylls such as neoxanthin and violaxanthin were absorbed in humans independent of the food matrix. Fucoxanthin was absorbed in mice and several other animals, but not in humans and freshwater fishes. A selective absorption mechanism for carotenoids would be present in humans, but not in mice. Moreover, the selectivity in the intestinal absorption of carotenoids appears to differ among animal species.

## 4. Mechanisms of the Intestinal Absorption

The selective absorption for carotenoids in humans cannot be explained by the simple diffusion mechanism alone. On the other hand, recent studies have suggested that the carotenoid uptake is partly mediated by facilitated diffusion [46–53]. For example, the ratio of the uptake mediated by scavenger receptor class B type 1 (SR-B1) to the total uptake of carotenoids in Caco-2 cells was as follows:
50% for β-carotene; 20% for β-cryptoxanthin and 7% for lutein/zeaxanthin [53]. The efficiency of
β-carotene absorption was remarkably reduced in SR-B1 knockout mice [54]. The physiological relevance of SR-B1 as an mediator of intestinal uptake for provitamin A carotenoids was indicated
by the report that retinoic acid and the intestinal transcription factor ISX regulated expressions of
both SR-B1 and β-carotene-15,15′-oxygenase (BCO1), an enzyme responsible for vitamin A production [55]. The facilitated diffusion may cause the selective absorption of carotenoids in humans. However, even if SR-B1 does not mediate intestinal uptake of the highly polar epoxy xanthophylls, they can pass across membranes via the simple diffusion pathway. Thus, these absorption mechanisms could not account for the strict selectivity that was observed in humans. The strict selective absorption might occur if most parts of the highly polar epoxy xanthophylls taken up by intestinal epithelial cells were excreted back into intestinal lumen.

The ATP-binding cassette (ABC) transporters such as ABCG5 and ABCG8 are well known to mediate the excretion of dietary phytosterols [56,57]. Although phytosterols such as β-sitosterol and campesterol are ingested from vegetables, grains, and cooking oils, the serum concentrations of the phytosterols are much lower than that of cholesterol in mammals [56,57]. Interestingly, ABCG5 polymorphism was suggested to be associated with the lutein bioavailability from egg in human subjects [58]. ABCG5 may excrete lutein and highly polar epoxy xanthophylls to intestinal lumen.

**Figure 2.** Proposed mechanisms of selectivity in the intestinal absorption of the dietary carotenoids.

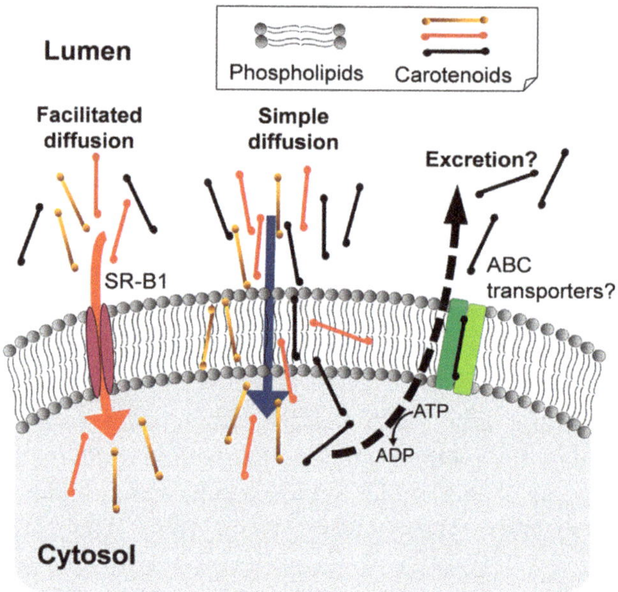


Multi-drug resistance 1 (MDR1, ABCB1) is well known as a major efflux pump for lipid-soluble compounds. As the affinity of substrates for MDR1 has been suggested to be related to their polarity [59], the highly polar xanthophylls may be excreted by MDR1. Carotenoids were evaluated for a substrate of MDR1 expressed in certain cancer cells. Neoxanthin and violaxanthin, compared with other carotenoids tested, showed higher affinity for transfected-human MDR1 in mouse lymphoma L1210 cells [60], but similar results were not found in several human breast and colon cancer cell lines [61,62]. Further study is required to confirm the involvement of MDR1 in the excretion of carotenoids in intestinal cells. Thus, the selectivity in the intestinal absorption of carotenoids in humans is likely to be caused by these proteins that mediate uptake and excretion (Figure 2). The specificity of these proteins would cause the differences in the intestinal absorption of carotenoids among animal species.

## 5. Metabolism of Xanthophylls in Mammals

It is necessary to explore the metabolism of carotenoids after intestinal absorption in order to elucidate the mechanism of their biological activities, and to achieve safe and effective applications to human subjects. Although β-carotene is known to be metabolized to vitamin A through action of BCO1, little is known about the metabolic transformation of non provitamin A xanthophylls in mammals.

Recently, we obtained evidence that the oxidative transformation of fucoxanthin and lutein to
keto-carotenoids occurred in mammals. Fucoxanthinol and amarouciaxanthin A were found in the plasma and liver of mice fed with fucoxanthin, whereas fucoxanthin itself was not detected [32,33]. Fucoxanthinol was hydrolyzed from fucoxanthin in the intestinal tract, circulated in the body, and
then oxidatively converted into amarouciaxanthin A (Figure 3). The conversion of fucoxanthinol into amarouciaxanthin A was also found to occur in human hepatoma HepG2 cells. Moreover, we found for the first time that the oxidative conversion was mediated in mouse liver microsomal fractions and required $NAD^+$ as a cofactor, demonstrating the metabolic conversion of the 3-hydroxyl end group in xanthophylls at the level of enzyme reaction in animals [33].

**Figure 3.** Proposed metabolic transformation of fucoxanthin.

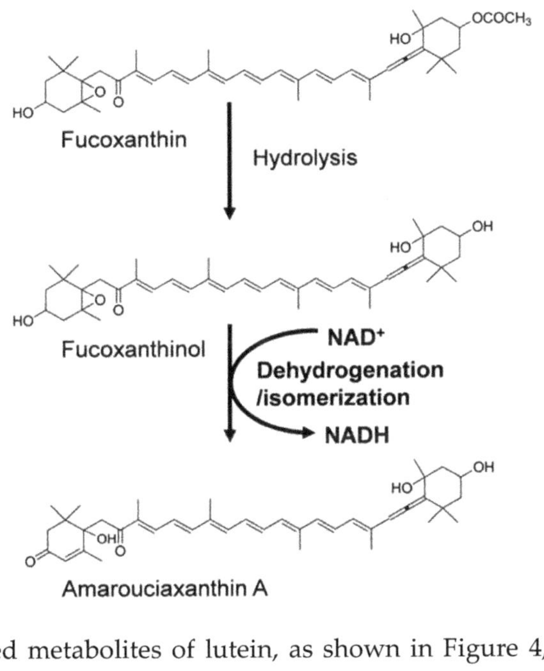


Several proposed metabolites of lutein, as shown in Figure 4, were previously known to be present in such human tissues as plasma, milk, liver, and retina [18,63–66]. Moreover, we found a remarkable accumulation of metabolites in mice fed with lutein [67]. 3′-Hydroxy-ε,ε-caroten-3-one and lutein were the predominant carotenoids in the plasma, liver, kidney, and adipose, accompanied by
ε,ε-carotene-3,3′-dione, indicating that mice actively convert lutein to keto-carotenoids by oxidizing the secondary hydroxyl group. However, 3-hydroxy-β,ε-caroten-3′-one (3′-oxolutein), the major metabolite of lutein in human plasma [67] and the retina [64], was not detected in the tissues of the mice.

**Figure 4.** Chemical structures of lutein and its metabolites.

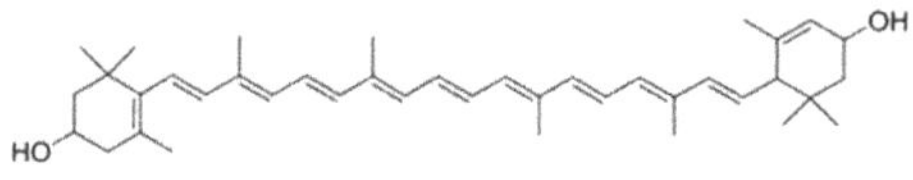

Lutein

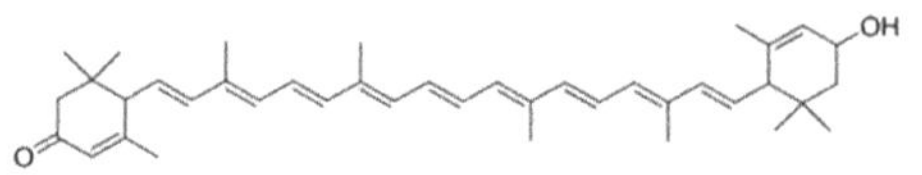

3'-Hydroxy- ε,ε-caroten-3-one

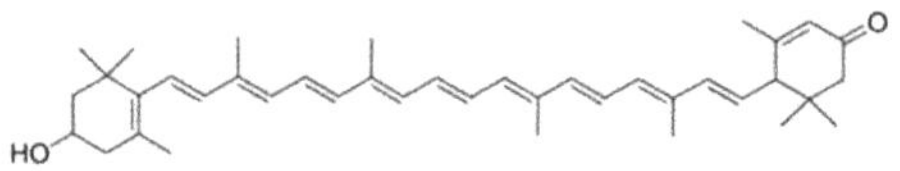

3-Hydroxy-β,ε-caroten-3'-one (3'-oxolutein)

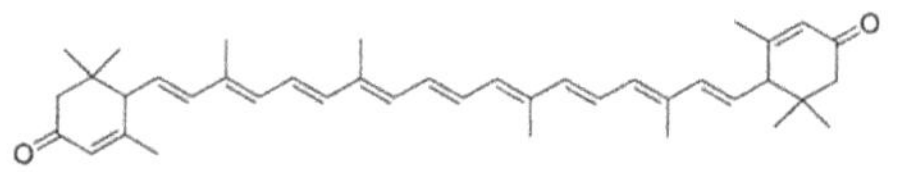

ε,ε-Carotene-3,3'-dione

These metabolites would be formed by the same enzyme that mediated the conversion of fucoxanthinol to amarouciaxanthin A. The combined level of the lutein metabolites in the liver of the mice was 72.4% of the total (intact lutein and the metabolites) [67]. This indicates that quantification of the metabolites is necessary to estimate the lutein bioavailability. Moreover, intact lutein and the metabolites may differ in their biological activities. Differences among lutein and its metabolites as antioxidants and blue light filters deserved further study.

Similar to the case of lutein in mice, the oxidative metabolism of the other xanthophylls was reported to occur in human subjects. After the ingestion of paprika juice containing capsanthin as a major xanthophyll, capsanthon in addition to capsanthin was found in the plasma [68]. Capsanthon may be formed from capsanthin by the oxidation of the 3′-hydroxyl group to the 3′-keto group. After an oral dose of 4,4′-dimethoxy-β-carotene in peanut oil, both 4-keto-β-carotene and canthaxanthin were found in the plasma [69]. These studies certainly indicate that humans have potential metabolic activity for the oxidation of secondary hydroxyl groups in various xanthophylls.

In human tissues, other metabolites of lutein were detected. 3′-Epilutein might be formed by a back reduction of 3′-oxolutein that was produced from lutein [64]. *meso*-Zeaxanthin, which is detected in the retina only, might be formed by double bond migration from lutein [64]. The dehydration products of lutein, 3-hydroxy-3′,4′-didehydro-β,γ-carotene and 3-hydroxy-2′,3′-didehydro-β,ε-carotene [19] were thought to be formed non-enzymatically under acidic conditions of stomach [70,71].

Recent studies have indicated the cleavage reaction of xanthophylls occurred in mammals. BCO1 catalyzes the central cleavage of provitamin A carotenoids, while β-carotene 9′,10′-oxygenase (BCO2) expressed *in vitro* can cleave a double bond at C-9′ and C-10′ of β-carotene, lycopene and xanthopylls [72–74]. Nonsense mutation of BCO2 was found to be associated with a yellow fat phenotype in sheep, in which xanthophylls were accumulated in adipose tissues [75]. The BCO2 knockout mice fed with lutein remarkably accumulated lutein metabolites, compared with the wild-type mice [76]. BCO2 might reduce the accumulation of xanthophylls by converting to smaller molecules, although the cleavage products and their further metabolites have not been detected in animal tissues yet. Thus, in addition to oxidation of secondary hydroxyl group in xanthophylls, the cleavage reaction of carbon skeleton by BCO2 would be also a major metabolic transformation of xanthophylls in mammals.

## 6. Conclusions

Various carotenoids, in particular, xanthophylls are ingested under usual dietary habits. However, carotenoids accumulated in human tissues are limited, suggesting selectivity in the intestinal absorption and different metabolic fates of carotenoids. The responses to the feeding of highly polar xanthophylls indicated that, for humans, intestinal absorption would be strictly selective in comparison with mice. The selectivity and its differences among animal species cannot be explained by simple diffusion mechanism alone. Instead, facilitated diffusion via SR-B1 and an unknown excretion to luminal side might cause the selectivity. After intestinal absorption of xanthophylls, the enzymatic oxidation of the secondary hydroxyl group leading to keto-carotenoids would occur as a common pathway of xanthophyll metabolism in mammals. We have no knowledge about the relation of these metabolites to the biological activities of parental xanthophylls. The potential biological activities of xanthophyll metabolites and their further metabolic fates warrant future studies with respect to the beneficial effects of xanthophylls on human health.

## References

1. Miller, N.J.; Sampson, J.; Candeias, L.P.; Bramley, P.M.; Rice-Evans, C.A. Antioxidant activities of carotenes and xanthophylls. *FEBS Lett.* **1996**, *384*, 240–242.
2. Di Mascio, P.; Kaiser, S.; Sies, H. Lycopene as the most efficient biological carotenoid singlet oxygen quencher. *Arch. Biochem. Biophys.* **1989**, *274*, 532–538.
3. Junghans, A.; Sies, H.; Stahl, W. Macular pigments lutein and zeaxanthin as blue light filters studied in liposomes. *Arch. Biochem. Biophys.* **2001**, *391*, 160–164.
4. Katsuura, S.; Imamura, T.; Bando, N.; Yamanishi, R. Beta-carotene and beta-cryptoxanthin but not lutein evoke redox and immune changes in RAW264 murine macrophages. *Mol. Nutr. Food Res.* **2009**, *53*, 1396–1405.
5. Yamaguchi, M.; Weitzmann, M.N. The bone anabolic carotenoid beta-cryptoxanthin enhances transforming growth factor-beta1-induced SMAD activation in MC3T3 preosteoblasts. *Int. J. Mol. Med.* **2009**, *24*, 671–675.
6. Uchiyama, S.; Yamaguchi, M. Oral administration of beta-cryptoxanthin prevents bone loss in ovariectomized rats. *Int. J. Mol. Med.* **2006**, *17*, 15–20.
7. Aizawa, K.; Inakuma, T. Dietary capsanthin, the main carotenoid in paprika (Capsicum annuum), alters plasma high-density lipoprotein-cholesterol levels and hepatic gene expression in rats. *Br. J. Nutr.* **2009**, *102*, 1760–1766.

8. Hussein, G.; Sankawa, U.; Goto, H.; Matsumoto, K.; Watanabe, H. Astaxanthin, a carotenoid with potential in human health and nutrition. *J. Nat. Prod.* **2006**, *69*, 443–449.
9. Yuan, J.P.; Peng, J.; Yin, K.; Wang, J.H. Potential health-promoting effects of astaxanthin:
A high-value carotenoid mostly from microalgae. *Mol. Nutr. Food Res.* **2011**, *55*, 150–165.
10. Okuzumi, J.; Nishino, H.; Murakoshi, M.; Iwashima, A.; Tanaka, Y.; Yamane, T.; Fujita, Y.; Takahashi, T. Inhibitory effects of fucoxanthin, a natural carotenoid, on n-myc expression and cell cycle progression in human malignant tumor cells. *Cancer Lett.* **1990**, *55*, 75–81.
11. Hosokawa, M.; Wanezaki, S.; Miyauchi, K.; Kurihara, H.; Kohno, H.; Kawabata, J.; Odashima, S.; Takahashi, K. Apoptosis-inducing effect of fucoxanthin on human leukemia cell line HL-60. *Food Sci. Technol. Res.* **1999**, *5*, 243–246.
12. Kotake-Nara, E.; Terasaki, M.; Nagao, A. Characterization of apoptosis induced by fucoxanthin in human promyelocytic leukemia cells. *Biosci. Biotechnol. Biochem.* **2005**, *69*, 224–227.
13. Kotake-Nara, E.; Kushiro, M.; Zhang, H.; Sugawara, T.; Miyashita, K.; Nagao, A. Carotenoids affect proliferation of human prostate cancer cells. *J. Nutr.* **2001**, *131*, 3303–3306.
14. Kotake-Nara, E.; Sugawara, T.; Nagao, A. Antiproliferative effect of neoxanthin and fucoxanthin on cultured cells. *Fish. Sci.* **2005**, *71*, 459–461.
15. Okuzumi, J.; Takahashi, T.; Yamane, T.; Kitao, Y.; Inagake, M.; Ohya, K.; Nishino, H.; Tanaka, Y. Inhibitory effects of fucoxanthin, a natural carotenoid, on *N*-ethyl-*N'*-nitro-*N*-nitrosoguanidine-induced mouse duodenal carcinogenesis. *Cancer Lett.* **1993**, *68*, 159–168.
16. Sakai, S.; Sugawara, T.; Matsubara, K.; Hirata, T. Inhibitory effect of carotenoids on the degranulation of mast cells via suppression of antigen-induced aggregation of high affinity IgE receptors. *J. Biol. Chem.* **2009**, *284*, 28172–28179.
17. Maeda, H.; Hosokawa, M.; Sashima, T.; Miyashita, K. Dietary combination of fucoxanthin and fish oil attenuates the weight gain of white adipose tissue and decreases blood glucose in obese/diabetic KK-Ay mice. *J. Agric. Food Chem.* **2007**, *55*, 7701–7706.
18. Khachik, F.; Beecher, G.R.; Goli, M.B.; Lusby, W.R. Separation, identification, and quantification of carotenoids in fruits, vegetables and human plasma by high performance liquid chromatography. *Pure Appl. Chem.* **1991**, *63*, 71–90.
19. Khachik, F.; Spangler, C.J.; Smith, J.C., Jr.; Canfield, L.M.; Steck, A.; Pfander, H. Identification, quantification, and relative concentrations of carotenoids and their metabolites in human milk and serum. *Anal. Chem.* **1997**, *69*, 1873–1881.
20. Slifka, K.A.; Bowen, P.E.; Stacewicz-Sapuntzakis, M.; Crissey, S.D. A survey of serum and dietary carotenoids in captive wild animals. *J. Nutr.* **1999**, *129*, 380–390.

21. Richelle, M.; Enslen, M.; Hager, C.; Groux, M.; Tavazzi, I.; Godin, J.P.; Berger, A.; Métairon, S.; Quaile, S.; Piguet-Welsch, C.; *et al.* Both free and esterified plant sterols reduce cholesterol absorption and the bioavailability of beta-carotene and alpha-tocopherol in normocholesterolemic humans. *Am. J. Clin. Nutr.* **2004**, *80*, 171–177.
22. Maiani, G.; Castón, M.J.; Catasta, G.; Toti, E.; Cambrodón, I.G.; Bysted, A.; Granado-Lorencio, F.; Olmedilla-Alonso, B.; Knuthsen, P.; Valoti, M.; *et al.* Carotenoids: actual knowledge on food sources, intakes, stability and bioavailability and their protective role in humans. *Mol. Nutr. Food Res.* **2009**, *53* (Suppl. 2), S194–S218.
23. Holst, B.; Williamson, G. Nutrients and phytochemicals: from bioavailability to bioefficacy beyond antioxidants. *Curr. Opin. Biotechnol.* **2008**, *19*, 73–82.
24. Zaripheh, S.; Erdman, J.W., Jr. Factors that influence the bioavailablity of xanthophylls. *J. Nutr.* **2002**, *132*, 531S–534S.
25. Yonekura, L.; Nagao, A. Intestinal absorption of dietary carotenoids. *Mol. Nutr. Food Res.* **2007**, *51*, 107–115.
26. Rock, C.L.; Lovalvo, J.L.; Emenhiser, C.; Ruffin, M.T.; Flatt, S.W.; Schwartz, S.J. Bioavailability of beta-carotene is lower in raw than in processed carrots and spinach in women. *J. Nutr.* **1998**, *128*, 913–916.
27. Small, D.M.; Penkett, S.A.; Chapman, D. Studies on simple and mixed bile salt micelles by nuclear magnetic resonance spectroscopy. *Biochim. Biophys. Acta.* **1969**, *176*, 178–189.
28. Yeum, K.J.; Russell, R.M. Carotenoid bioavailability and bioconversion. *Annu. Rev. Nutr.* **2002**, *22*, 483–504.
29. Hollander, D.; Ruble, P.E., Jr. Beta-carotene intestinal absorption: Bile, fatty acid, pH, and flow rate effects on transport. *Am. J. Physiol. Endocrinol. Metab.* **1978**, *235*, E686–E691.
30. Scita, G.; Aponte, G.W.; Wolf, G. Uptake and cleavage of β-carotene by cultures of rat small intestinal cells and human lung fibroblasts. *J. Nutr. Biochem.* **1992**, *3*, 118–123.
31. Sugawara, T.; Kushiro, M.; Zhang, H.; Nara, E.; Ono, H.; Nagao, A. Lysophosphatidylcholine enhances carotenoid uptake from mixed micelles by Caco-2 human intestinal cells. *J. Nutr.* **2001**, *131*, 2921–2927.
32. Sugawara, T.; Baskaran, V.; Tsuzuki, W.; Nagao, A. Brown algae fucoxanthin is hydrolyzed to fucoxanthinol during absorption by Caco-2 human intestinal cells and mice. *J. Nutr.* **2002**, *132*, 946–951.
33. Asai, A.; Sugawara, T.; Ono, H.; Nagao, A. Biotransformation of fucoxanthinol into amarouciaxanthin A in mice and HepG2 cells. *Drug Metab. Dispos.* **2004**, *32*, 205–211.
34. Asai, A.; Terasaki, M.; Nagao, A. An epoxide–furanoid rearrangement of spinach neoxanthin occurs in the gastrointestinal tract of mice and *in vitro*:

Formation and cytostatic activity of neochrome stereoisomers. *J. Nutr.* **2004**, *134*, 2237–2243.

35. Sangeetha, R.K.; Bhaskar, N.; Divakar, S.; Baskaran, V. Bioavailability and metabolism of fucoxanthin in rats: structural characterization of metabolites by LC-MS (APCI). *Mol. Cell. Biochem.* **2010**, *333*, 299–310.
36. Baskaran, V.; Sugawara, T.; Nagao, A. Phospholipids affect the intestinal absorption of carotenoids in mice. *Lipids* **2003**, *38*, 705–711.
37. Matsuno, T.; Ookubo, M. A new carotenoid, halocynthiaxanthin from the sea squirt, Halocynthia Roretzi. *Tetrahedron Lett.* **1981**, *22*, 4659–4660.
38. Matsuno, T.; Ookubo, M.; Komori, T. Carotenoids of tunicates, III. The structural elucidation of two new marine carotenoids, amarouciaxanthin A and B. *J. Nat. Prod.* **1985**, *48*, 606–613.
39. Strand, A.; Herstad, O.; Liaaen-Jensen, S. Fucoxanthin metabolites in egg yolks of laying hens. *Comp. Biochem. Physiol. A Mol. Integr. Physiol.* **1998**, *119*, 963–974.
40. Matsuno, T.; Ohkubo, M.; Toriiminami, Y.; Tsushima, M.; Sakaguchi, S.; Minami, T.; Maoka, T. Carotenoids in food chain between freshwater fish and aquatic insects. *Comp. Biochem. Physiol. A Mol. Integr. Physiol.* **1999**, *124*, 341–345.
41. Asai, A.; Yonekura, L.; Nagao, A. Low bioavailability of dietary epoxyxanthophylls in humans. *Br. J. Nutr.* **2008**, *100*, 273–277.
42. Yonekura, L.; Nagao, A. Soluble fibers inhibit carotenoid micellization *in vitro* and uptake by Caco-2 cells. *Biosci. Biotechnol. Biochem.* **2009**, *73*, 196–199.
43. Pérez-Gálvez, A.; Martin, H.D.; Sies, H.; Stahl, W. Incorporation of carotenoids from paprika oleoresin into human chylomicrons. *Br. J. Nutr.* **2003**, *89*, 787–793.
44. Barua, A.B.; Olson, J.A. Xanthophyll epoxides, unlike beta-carotene monoepoxides, are not detectibly absorbed by humans. *J. Nutr.* **2001**, *131*, 3212–3215.
45. Barua, A.B. Intestinal absorption of epoxy-beta-carotenes by humans. *Biochem. J.* **1999**, *339* (Pt. 2), 359–362.
46. Reboul, E.; Abou, L.; Mikail, C.; Ghiringhelli, O.; André, M.; Portugal, H.; Jourdheuil-Rahmani, D.; Amiot, M.J.; Lairon, D.; Borel, P. Lutein transport by Caco-2 TC-7 cells occurs partly by a facilitated process involving the scavenger receptor class B type I (SR-BI). *Biochem. J.* **2005**, *387* (Pt. 2), 455–461.
47. Kiefer, C.; Sumser, E.; Wernet, M.F.; Von Lintig, J. A Class B scavenger receptor mediates the cellular uptake of carotenoids in Drosophila. *Proc. Natl. Acad. Sci. USA* **2002**, *99*, 10581–10586.
48. Borel, P.; Moussa, M.; Reboul, E.; Lyan, B.; Defoort, C.; Vincent-Baudry, S.; Maillot, M.; Gastaldi, M.; Darmon, M.; Portugal, H.; *et al.* Human plasma levels of vitamin E and carotenoids are associated with genetic polymorphisms in genes involved in lipid metabolism. *J. Nutr.* **2007**, *137*, 2653–2659.
49. Moussa, M.; Landrier, J.F.; Reboul, E.; Ghiringhelli, O.; Coméra, C.; Collet, X.; Fröhlich, K.; Böhm, V.; Borel, P. Lycopene absorption in human intestinal cells

and in mice involves scavenger receptor class B type I but not Niemann-Pick C1-Like 1. *J. Nutr.* **2008**, *138*, 1432–1436.
50. O'Sullivan, L.; Aisling, S.A.; O'Brien, N.M. Investigation of beta-carotene and lutein transport in Caco-2 cells: carotenoid-carotenoid interactions and transport inhibition by ezetimibe. *Int. J. Vitam. Nutr. Res.* **2009**, *79*, 337–347.
51. During, A.; Hussain, M.M.; Morel, D.W.; Harrison, E.H. Carotenoid uptake and secretion by CaCo-2 cells: Beta-carotene isomer selectivity and carotenoid interactions. *J. Lipid Res.* **2002**, *43*, 1086–1095.
52. During, A.; Harrison, E.H. Mechanisms of provitamin A (carotenoid) and vitamin A (retinol) transport into and out of intestinal Caco-2 cells. *J. Lipid Res.* **2007**, *48*, 2283–2294.
53. During, A.; Dawson, H.D.; Harrison, E.H. Carotenoid transport is decreased and expression of the lipid transporters SR-BI, NPC1L1, and ABCA1 is downregulated in Caco-2 cells treated with ezetimibe. *J. Nutr.* **2005**, *135*, 2305–2312.
54. van Bennekum, A.; Werder, M.; Thuahnai, S.T.; Han, C.H.; Duong, P.; Williams, D.L.; Wettstein, P.; Schulthess, G.; Phillips, M.C.; Hauser, H. Class B scavenger receptor-mediated intestinal absorption of dietary beta-carotene and cholesterol. *Biochemistry* **2005**, *44*, 4517–4525.
55. Lobo, G.P.; Hessel, S.; Eichinger, A.; Noy, N.; Moise, A.R.; Wyss, A.; Palczewski, K.;
von Lintig, J. ISX is a retinoic acid-sensitive gatekeeper that controls intestinal beta, beta-carotene absorption and vitamin A production. *FASEB J.* **2010**, *24*, 1656–1666.
56. Fransen, H.P.; de Jong, N.; Wolfs, M.; Verhagen, H.; Verschuren, W.M.; Lütjohann, D.;
von Bergmann, K.; Plat, J.; Mensink, R.P. Customary use of plant sterol and plant stanol enriched margarine is associated with changes in serum plant sterol and stanol concentrations in humans.
*J. Nutr.* **2007**, *137*, 1301–1306.
57. Yu, L.; von Bergmann, K.; Lutjohann, D.; Hobbs, H.H.; Cohen, J.C. Selective sterol accumulation in ABCG5/ABCG8-deficient mice. *J. Lipid Res.* **2004**, *45*, 301–307.
58. Herron, K.L.; McGrane, M.M.; Waters, D.; Lofgren, I.E.; Clark, R.M.; Ordovas, J.M.;
Fernandez, M.L. The ABCG5 polymorphism contributes to individual responses to dietary cholesterol and carotenoids in Eggs. *J. Nutr.* **2006**, *136*, 1161–1165.
59. Shen, J.; Cross, S.T.; Tang-Liu, D.D.; Welty, D.F. Evaluation of an immortalized retinal endothelial cell line as an *in vitro* model for drug transport studies across the blood-retinal barrier. *Pharm. Res.* **2003**, *20*, 1357–1363.

60. Gyémánt, N.; Tanaka, M.; Molnár, P.; Deli, J.; Mándoky, L.; Molnár, J. Reversal of multidrug resistance of cancer cells *in vitro*: modification of drug resistance by selected carotenoids. *Anticancer Res.* **2006**, *26*, 367–374.
61. Molnár, J.; Gyémánt, N.; Mucsi, I.; Molnár, A.; Szabó, M.; Körtvélyesi, T.; Varga, A.; Molnár, P.; Tóth, G. Modulation of multidrug resistance and apoptosis of cancer cells by selected carotenoids. *In Vivo* **2004**, *18*, 237–244.
62. Ugocsai, K.; Varga, A.; Molnár, P.; Antus, S.; Molnár, J. Effects of selected flavonoids and carotenoids on drug accumulation and apoptosis induction in multidrug-resistant colon cancer cells expressing MDR1/LRP. *In Vivo* **2005**, *19*, 433–438.
63. Khachik, F.; Bernstein, P.S.; Garland, D.L. Identification of lutein and zeaxanthin oxidation products in human and monkey retinas. *Invest. Ophthalmol. Vis. Sci.* **1997**, *38*, 1802–1811.
64. Khachik, F.; de Moura, F.F.; Zhao, D.Y.; Aebischer, C.P.; Bernstein, P.S. Transformations of selected carotenoids in plasma, liver, and ocular tissues of humans and in nonprimate animal models. *Invest. Ophthalmol. Vis. Sci.* **2002**, *43*, 3383–3392.
65. Bhosale, P.; Bernstein, P.S. Quantitative measurement of 3′-oxolutein from human retina by normal-phase high-performance liquid chromatography coupled to atmospheric pressure chemical ionization mass spectrometry. *Anal. Biochem.* **2005**, *345*, 296–301.
66. Bhosale, P.; Zhao, D.Y.; Serban, B.; Bernstein, P.S. Identification of 3-methoxyzeaxanthin as a novel age-related carotenoid metabolite in the human macula. *Invest. Ophthalmol. Vis. Sci.* **2007**, *48*, 1435–1440.
67. Yonekura, L.; Kobayashi, M.; Terasaki, M.; Nagao, A. Keto-carotenoids are the major metabolites of dietary lutein and fucoxanthin in mouse tissues, *J. Nutr.* **2010**, *140*, 1824–1831.
68. Etoh, H.; Utsunomiya, Y.; Komori, A.; Murakami, Y.; Oshima, S.; Inakuma, T. Carotenoids in human blood plasma after ingesting paprika juice. *Biosci. Biotechnol. Biochem.* **2000**, *64*, 1096–1098.
69. Zeng, S.; Furr, H.C.; Olson, J.A. Metabolism of carotenoid analogs in humans. *Am. J. Clin. Nutr.* **1992**, *56*, 433–439.
70. Khachik, F. An efficient conversion of (3*R*,3′*R*,6′*R*)-lutein to (3*R*,3′*S*,6′*R*)-lutein (3′-epilutein) and (3*R*,3′*R*)-zeaxanthin. *J. Nat. Prod.* **2003**, *66*, 67–72.
71. Khachik, F.; Englert, G.; Beecher, G.R.; Smith, J.C., Jr. Isolation, structural elucidation, and partial synthesis of lutein dehydration products in extracts from human plasma. *J. Chromatogr. B Biomed. Appl.* **1995**, *670*, 219–233.
72. Kiefer, C.; Hessel, S.; Lampert, J.M.; Vogt, K.; Lederer, M.O.; Breithaupt, D.E.; von Lintig, J. Identification and characterization of a mammalian enzyme catalyzing the asymmetric oxidative cleavage of provitamin A. *J. Biol. Chem.* **2001**, *276*, 14110–14116.

73. Hu, K.Q.; Liu, C.; Ernst, H.; Krinsky, N.I.; Russell, R.M.; Wang, X.D. The biochemical characterization of ferret carotene-9′,10′-monooxygenase catalyzing cleavage of carotenoids *in vitro* and *in vivo*. *J. Biol. Chem.* **2006**, *281*, 19327–19338.
74. Mein, J.R.; Dolnikowski, G.G.; Ernst, H.; Russell, R.M.; Wang, X.D. Enzymatic formation of apo-carotenoids from the xanthophyll carotenoids lutein, zeaxanthin and β-cryptoxanthin by ferret carotene-9′,10′-monooxygenase. *Arch. Biochem. Biophys.* **2011**, *506*, 109–121.
75. Våge, D.I.; Boman, I.A. A nonsense mutation in the *beta-carotene oxygenase 2* (*BCO2*) gene is tightly associated with accumulation of carotenoids in adipose tissue in sheep (*Ovis aries*). *BMC Genet.* **2010**, *11*, 10.
76. Amengual, J.; Lobo, G.P.; Golczak, M.; Li, H.N.; Klimova, T.; Hoppel, C.L.; Wyss, A.; Palczewski, K.; von Lintig, J. A mitochondrial enzyme degrades carotenoids and protects against oxidative stress. *FASEB J.* **2011**, *25*, 948–959.

*Samples Availability:* Available from the authors.

*marine drugs*

MDPI

*Article*

# Fucoxanthin Enhances the Level of Reduced Glutathione via the Nrf2-Mediated Pathway in Human Keratinocytes

**Jian Zheng, Mei Jing Piao, Ki Cheon Kim, Cheng Wen Yao, Ji Won Cha and Jin Won Hyun ***

School of Medicine and Institute for Nuclear Science and Technology, Jeju National University,
Jeju 690-756, Korea; E-Mails: zhengjian0317@hotmail.com (J.Z.);
mjpiao@hanmail.net (M.J.P.);
svv771@hotmail.com (K.C.K.); vane1989923@hotmail.com (C.W.Y.);
cjw102700@hanmail.net (J.W.C.)

* Author to whom correspondence should be addressed; E-Mail:
jinwonh@jejunu.ac.kr;
Tel.: +82-2-64-754-3838; Fax: +82-2-64-702-2687.

*Received: 4 June 2014; in revised form: 1 July 2014 / Accepted: 4 July 2014 / Published: 10 July 2014*

**Abstract:** Fucoxanthin, a natural carotenoid, is abundant in seaweed with antioxidant properties. This study investigated the role of fucoxanthin in the induction of antioxidant enzymes involved in the synthesis of reduced glutathione (GSH), synthesized by glutamate-cysteine ligase catalytic subunit (GCLC) and glutathione synthetase (GSS), via Akt/nuclear factor-erythroid 2-related (Nrf2) pathway in human keratinocytes (HaCaT) and elucidated the underlying mechanism. Fucoxanthin treatment increased the mRNA and protein levels of GCLC and GSS in HaCaT cells. In addition, fucoxanthin treatment promoted the nuclear translocation and phosphorylation of Nrf2, a transcription factor for the genes encoding GCLC and GSS. Chromatin immune-precipitation and luciferase reporter gene assays revealed that fucoxanthin treatment increased the binding of Nrf2 to the antioxidant response element (ARE) sequence and transcriptional activity of Nrf2. Fucoxanthin treatment increased phosphorylation of Akt (active form), an up-regulator of Nrf2 and exposure to LY294002, a phosphoinositide 3-kinase (PI3K)/Akt inhibitor, suppressed the fucoxanthin-induced activation of Akt, Nrf2, resulting in decreased

GCLC and GSS expression. In accordance with the effects on GCLC and GSS expression, fucoxanthin induced the level of GSH. In addition, fucoxanthin treatment recovered the level of GSH reduced by ultraviolet B irradiation. Taken together, these findings suggest that fucoxanthin treatment augments cellular antioxidant defense by inducing Nrf2-driven expression of enzymes involved in GSH synthesis via PI3K/Akt signaling.

**Keywords:** fucoxanthin; NF-E2-related factor 2; oxidative stress; cytoprotection; PI3K/Akt; GCLC; GSS; GSH

---

## 1. Introduction

Oxidative stress is the most common cause of skin aging and can be effectively eliminated by the organism itself, pharmacological agents, and natural antioxidants. There are two theoretical methods to deal with these harmful stimuli, namely, early and delayed responses. Early responses rapidly remove reactive oxygen species (ROS) and free radicals via chemical reactions soon after their generation [1]. By contrast, delayed responses involve the expression of genes encoding antioxidant enzymes and proteins to reduce the generation of noxious substances [2]. Nuclear factor-erythroid
2-related factor (Nrf2) is often the central signaling switch that modulates the activation of phase II bio-transferase/antioxidant enzymes, which include glutamate-cysteine ligase catalytic subunit (GCLC) and glutathione synthetase (GSS) [3,4]. As an extremely important antioxidant, GSH, which is synthesized by GCLC and GSS [5,6], not only scavenges free radicals [7], but also maintains the redox-sensitive active sites of many enzymes from an oxidized form to a reduced form [8]. Therefore, the correct balance between reduced GSH and oxidized GSH is required for cellular homeostasis [9].

Genes that encode antioxidant enzymes, such as GCLC and GSS, contain an antioxidant responsive element (ARE) in their promoter region [3]. Transduction of the ARE sequence-containing genes encoding GCLC and GSS mainly occurs via activation of Nrf2 protein [10]. Nrf2 is a transcription factor that detects variation in oxidative stress within cells [11] and induces the transcription of its target genes by binding to the ARE in their promoters. The target genes of Nrf2 include many antioxidant and phase II detoxifying genes [12], including those encoding GCLC and GSS. The synthesis of GSH catalyzed by GCLC and GSS via up-regulation of Nrf2 is associated with protection of cells against oxidative stress [13].

Fucoxanthin is a major carotenoid found in edible brown seaweeds [14] and contains several functional groups, including an unusual allenic bond, a conjugated carbonyl group, and an acetyl group [15]. Fucoxanthin has antioxidant, anti-cancer, anti-obesity, anti-diabetic, and anti-photoaging activities [16–20]. We previously reported that fucoxanthin reduced levels of ROS, inhibited DNA damage, restored mitochondrial membrane potential, and suppressed apoptosis [21]. In the present study, we examined whether fucoxanthin could increase the level of GSH by inducing GCLC and GSS expression via the Akt/Nrf2 pathway.

## 2. Results

### 2.1. Fucoxanthin Increases Expression of GCLC and GSS at the mRNA and Protein Levels

HaCaT cells were treated with various concentrations of fucoxanthin for 12 h. The mRNA levels of GCLC and GSS in fucoxanthin-treated cells increased, demonstrating that 20 μM of fucoxanthin caused the maximal induction both in GCLC and GSS expression (Figure 1A). Next, HaCaT cells were treated with 20 μM fucoxanthin for various amounts of time. The mRNA levels of GCLC and GSS were highest at 12 h of treatment (Figure 1B). Furthermore, when HaCaT cells were treated with various concentrations of fucoxanthin for 12 h, the protein levels of GCLC and GSS increased, exhibiting that 20 μM of fucoxanthin induced maximal in GCLC and GSS expression (Figure 1C). Treatment with
20 μM of fucoxanthin for 12 h markedly increased the protein levels of GCLC and GSS (Figure 1D). These results indicate that fucoxanthin treatment increases expression of GCLC and GSS at both the mRNA and protein levels.

**Figure 1.** Effects of fucoxanthin treatment on the mRNA and protein expression of glutamate-cysteine ligase catalytic subunit (GCLC) and glutathione synthetase (GSS) in HaCaT cells. Cells were incubated with various concentrations of fucoxanthin (0–20 μM) for various amounts of time (0–24 h). The mRNA levels of GCLC and GSS were detected by reverse transcription-PCR (RT-PCR) following treatment (**A**) with various concentrations of fucoxanthin for 12 h; and (**B**) with 20 μM fucoxanthin for various amounts of time. The protein levels of GCLC and GSS were detected by Western blotting following treatment (**C**) with various concentrations of fucoxanthin for 12 h; and (**D**) with 20 μM fucoxanthin for various amounts of time. * and # indicates significantly different from control of GCLC and GSS, respectively ($p < 0.05$).

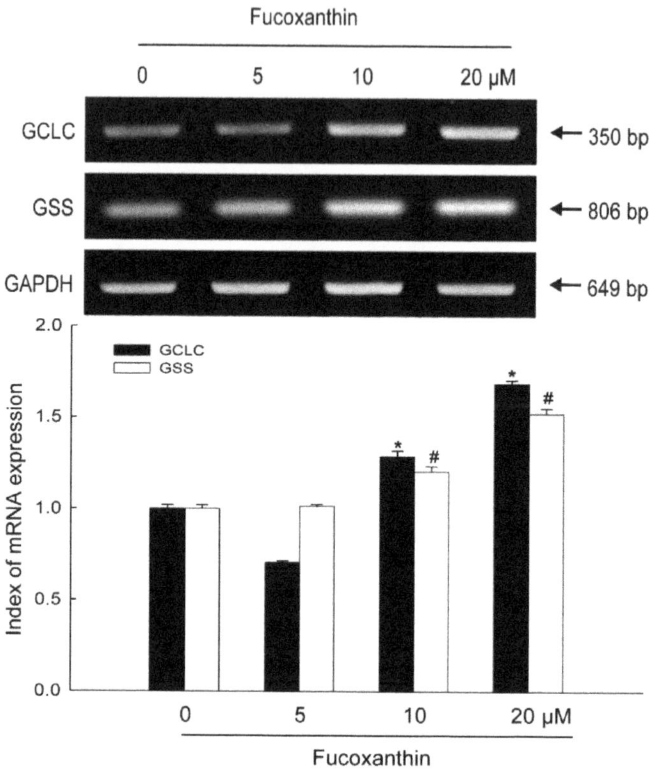
Fucoxanthin
0
5
10
20 μM
GCLC
GSS
GAPDH
350 bp
806 bp
649 bp
GCLC
GSS
Index of mRNA expression
2.0
1.5
1.0
0.5
0.0
*
#
0
5
10
20 μM
Fucoxanthin

(**A**)

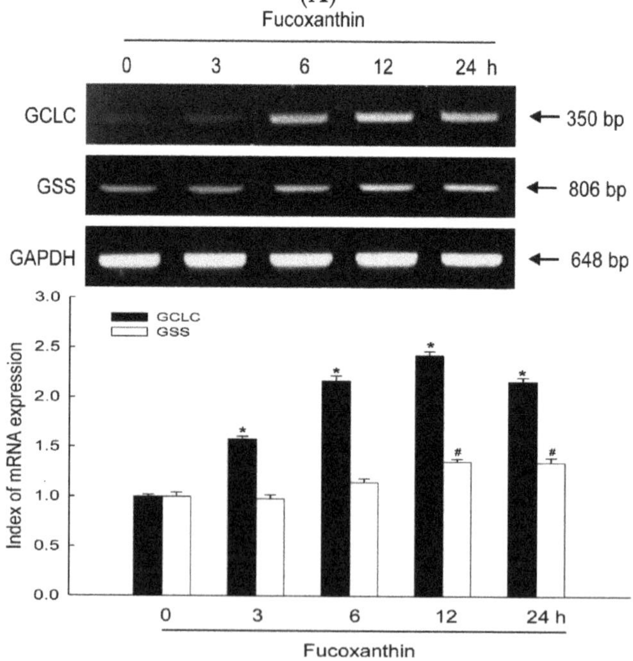
Fucoxanthin
0
3
6
12
24 h
GCLC
GSS
GAPDH
350 bp
806 bp
648 bp
GCLC
GSS
Index of mRNA expression
3.0
2.5
2.0
1.5
1.0
0.5
0.0
*
#
0
3
6
12
24 h
Fucoxanthin

(**B**)

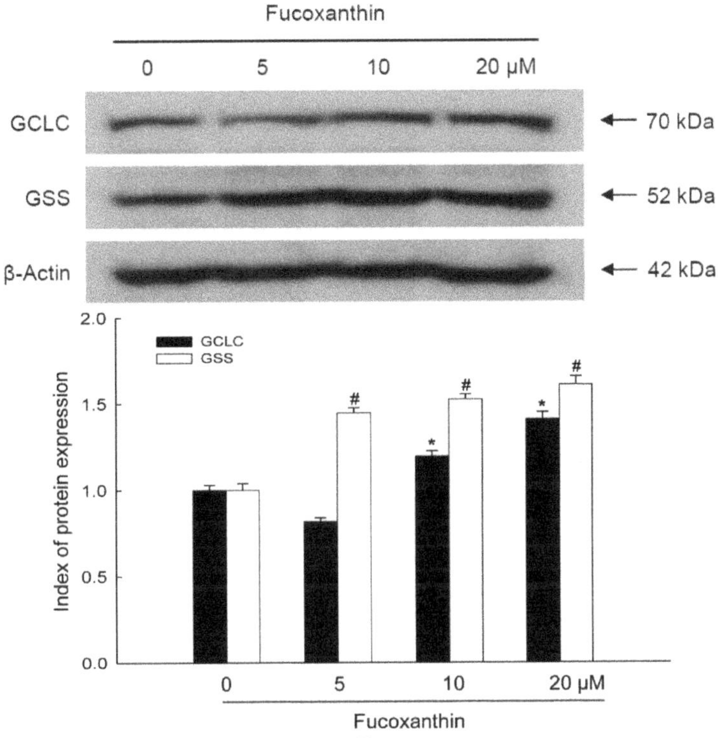
Fucoxanthin
0
5
10
20 μM
GCLC
GSS
β-Actin
70 kDa
52 kDa
42 kDa
GCLC
GSS
Index of protein expression
2.0
1.5
1.0
0.5
0.0
#
*
0
5
10
20 μM
Fucoxanthin

(C)

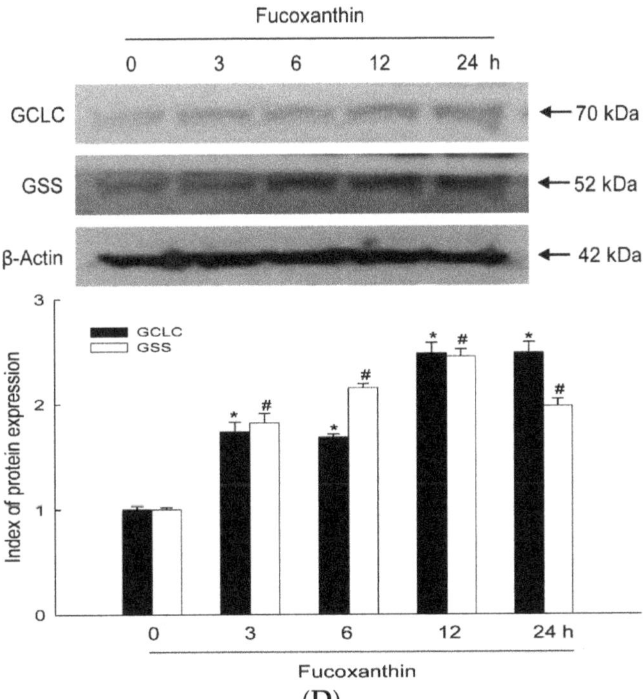
Fucoxanthin
0
3
6
12
24 h
GCLC
GSS
β-Actin
70 kDa
52 kDa
42 kDa
GCLC
GSS
Index of protein expression
3
2
1
0
*
#
0
3
6
12
24 h
Fucoxanthin

(D)

### 2.2. Fucoxanthin Induces Activation of Nrf2 and Enhances Binding of Nrf2 to the ARE in the Promoters of the GCLC and GSS Genes

The genes encoding GCLC and GSS have an ARE sequence in their promoter regions. Nrf2 is an important transcription factor that regulates ARE-driven expression of these genes [22]. We examined whether fucoxanthin treatment activated Nrf2, resulting in the up-regulation of these enzymes. Fucoxanthin treatment increased protein levels of Nrf2 and phospho Nrf2 (active form) (Figure 2A), and resulted in the translocation of Nrf2 protein from the cytosol into the nucleus (Figure 2B). Moreover, chromatin immune-precipitation (ChIP) analysis revealed that binding of Nrf2 to the ARE in the promoters of the genes encoding GCLC and GSS was markedly increased in fucoxanthin-treated cells, as determined by comparison to binding of histone H3 as the internal control (Figure 2C). To verify the functional relevance of Nrf2 binding to the ARE sequence of these two genes, a construct was used that contained a promoter containing an ARE sequence (bearing the consensus Nrf2-binding site) linked to a luciferase reporter gene. Fucoxanthin treatment increased the transcriptional activity of Nrf2 (Figure 2D). These results suggest that Nrf2 mediates fucoxanthin-induced transcription of GCLC and GSS.

**Figure 2.** Effects of fucoxanthin treatment on the expression, nuclear translocation, and antioxidant response element (ARE) sequence-binding activity of Nrf2. (**A**) Nuclear extracts were prepared from HaCaT cells following treatment with 20 μM fucoxanthin for the indicated amount of time. Western blotting of the nuclear lysates was performed using Nrf2 and phospho Nrf2 antibodies. * and # indicates significantly different from Nrf2 and phospho Nrf2 of control, respectively ($p < 0.05$); (**B**) An anti-Nrf2 antibody and a FITC-conjugated secondary antibody were used to detect Nrf2 localization (green) by using confocal microscopy. DAPI staining indicates the locations of nuclei (blue). The merged images show the nuclear localization of Nrf2 protein; (**C**) Nuclear extracts were prepared from HaCaT cells treated with 20 μM fucoxanthin for 6 h. A ChIP assay was performed to assess binding of Nrf2 to the ARE in the promoters of the genes encoding GCLC and GSS; (**D**) Transcriptional activity of Nrf2 in HaCaT cells following treatment with 20 μM fucoxanthin for 6 h was assessed by using luciferase reporter assay.
* Significantly different from control cells.

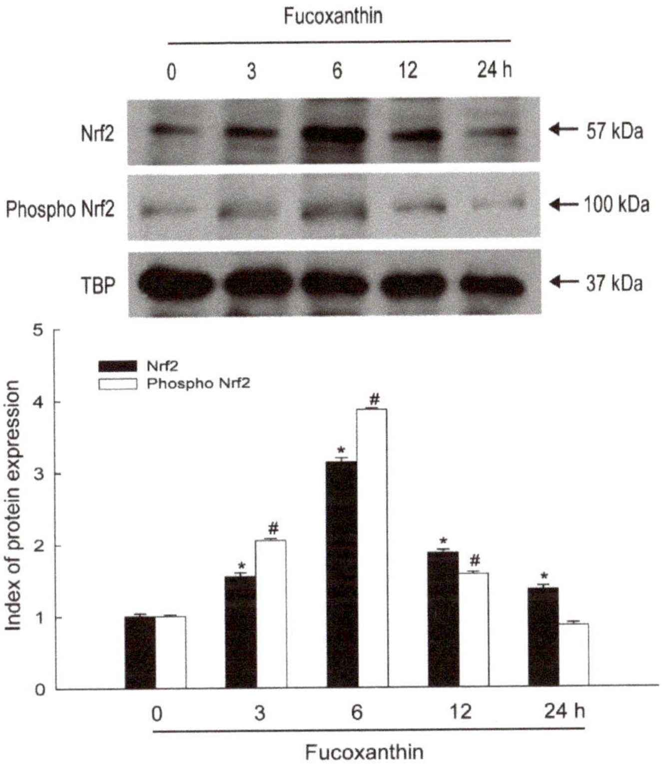
Fucoxanthin
0
3
6
12
24 h
Nrf2
57 kDa
Phospho Nrf2
100 kDa
TBP
37 kDa
Index of protein expression
Nrf2
Phospho Nrf2
0
1
2
3
4
5
0
3
6
12
24 h
Fucoxanthin

(**A**)

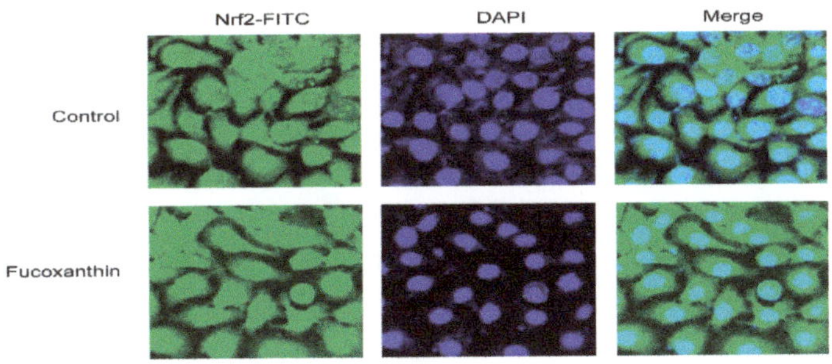
Nrf2-FITC
DAPI
Merge
Control
Fucoxanthin

(**B**)

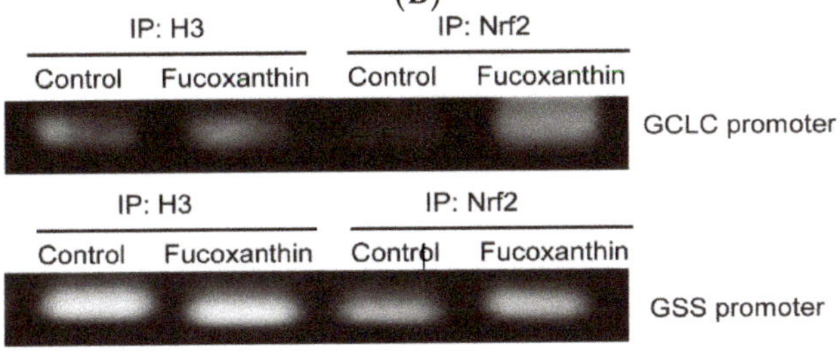
IP: H3
IP: Nrf2
Control
Fucoxanthin
Control
Fucoxanthin
GCLC promoter
IP: H3
IP: Nrf2
Control
Fucoxanthin
Control
Fucoxanthin
GSS promoter

(**C**)

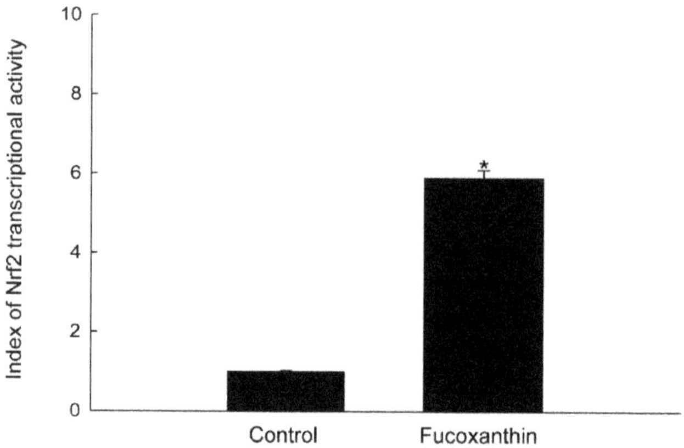


(D)

### 2.3. *Fucoxanthin Involves Nrf2-Driven GCLC and GSS via Phosphorylation of Akt*

To further elucidate the up-stream signaling pathway involved in fucoxanthin-mediated activation of Nrf2 and induction of GCLC and GSS, we examined activation of Akt, which is a signaling enzyme that is involved in the phosphorylation and nuclear translocation of Nrf2 [23]. Activation of Akt by fucoxanthin was assessed by performing Western blotting with an antibody against phosphorylated Akt. Fucoxanthin treatment increased phosphorylation of Akt (Figure 3A). A LY294002, phosphoinositide 3-kinase (PI3K)/Akt inhibitor, specifically represses the phosphorylation of Akt [24]. This inhibitor reduced the fucoxanthin-induced phospho Akt expression (Figure 3B). Furthermore, this inhibitor suppressed the fucoxanthin-induced Nrf2, GCLC and GSS expression (Figure 3C,D).

**Figure 3.** Effects of fucoxanthin treatment on Akt and its related protein. (**A**) Cells were incubated with 20 μM fucoxanthin for various amounts of time (0–12 h). Cell lysates were prepared and Western blotting was performed with anti-Akt and anti-phospho Akt antibodies. * indicates significantly different from control cells ($p < 0.05$); After treatment with LY294002, cell lysates were subjected to electrophoresis (**B**) with anti-Akt, anti-phospho Akt. * indicates significantly different from control cells ($p < 0.05$) and # significantly different from fucoxanthin-treated cells ($p < 0.05$); (**C**) with anti-Nrf2 and anti-phospho Nrf2. * and ** indicates significantly different from Nrf2 and phospho Nrf2 of control, respectively ($p < 0.05$), # and ## indicates significantly different from Nrf2 and phospho Nrf2 of fucoxanthin-treated cells, respectively ($p < 0.05$); (**D**) with anti-GCLC and anti-GSS antibodies. * and ** indicates significantly different from GCLC and GSS of control, respectively ($p < 0.05$), # and ## indicates significantly different from GCLC and GSS of fucoxanthin-treated cells, respectively ($p < 0.05$).

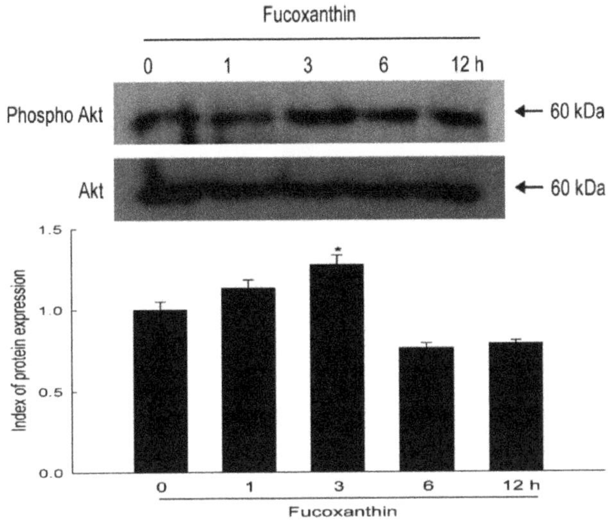
Fucoxanthin
0
1
3
6
12 h
Phospho Akt
60 kDa
Akt
60 kDa
Index of protein expression
1.5
1.0
0.5
0.0
*
0
1
3
6
12 h
Fucoxanthin

**(A)**

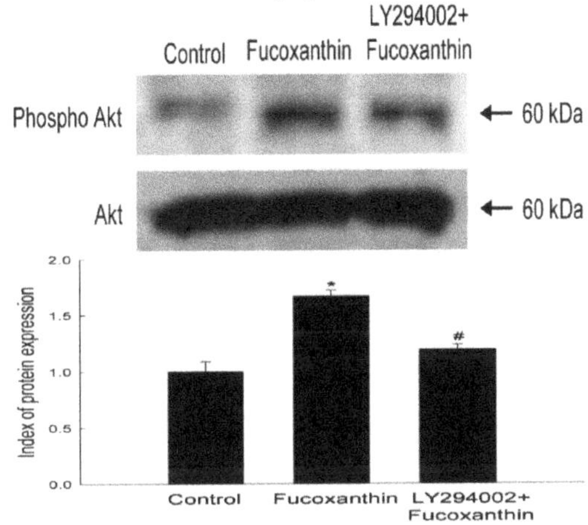
Control
Fucoxanthin
LY294002+ Fucoxanthin
Phospho Akt
60 kDa
Akt
60 kDa
Index of protein expression
2.0
1.5
1.0
0.5
0.0
*
#
Control
Fucoxanthin
LY294002+ Fucoxanthin

**(B)**

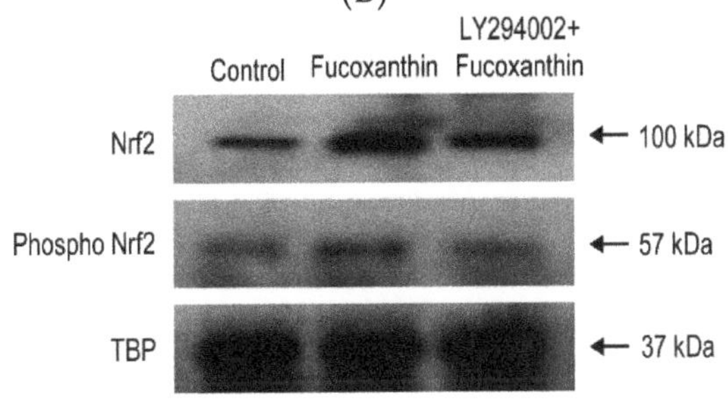
Control
Fucoxanthin
LY294002+ Fucoxanthin
Nrf2
100 kDa
Phospho Nrf2
57 kDa
TBP
37 kDa

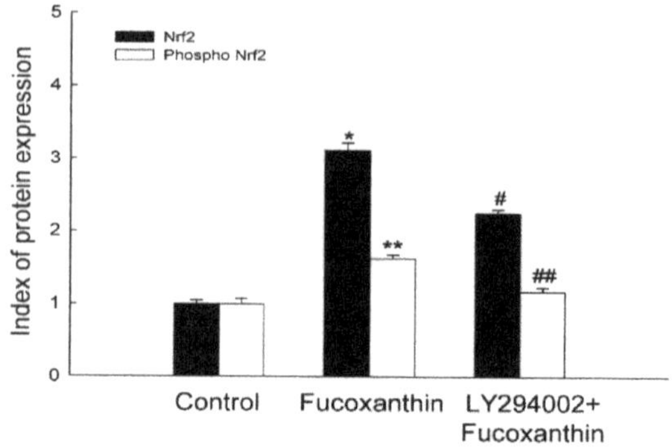


(C)

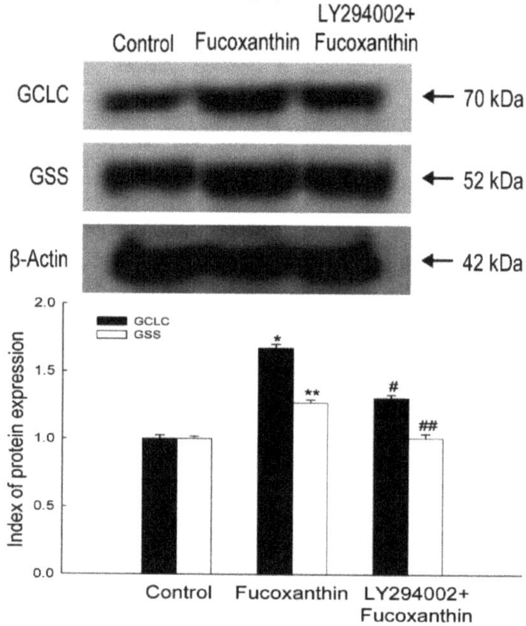


(D)

### 2.4. *Fucoxanthin Promotes the Synthesis of GSH Catalyzed by GCLC and GSS*

GSH is a tri-peptide formed via GCLC and GSS, and has powerful antioxidant effects against free radicals. GSH was detected by confocal microscopy using 7-amino-4-chloromethylcoumarin (CMAC), a dye that specifically labels GSH. The fluorescence intensity of CMAC, indicative of the level of GSH, was notably higher in fucoxanthin-treated cells than in control cells (Figure 4A). Consistently, fucoxanthin increased the concentration of GSH, as determined by a GSH detection kit (Figure 4B). To evaluate whether fucoxanthin induced GSH production to protect cells against ultraviolet B (UVB)-induced oxidative stress, cells were pretreated with fucoxanthin and then exposed to UVB irradiation. The level of GSH was reduced by UVB exposure, and this decrease was significantly restored in cells pretreated with fucoxanthin (Figure 4C,D). These data suggest that fucoxanthin partially recovers the reduction of GSH induced by UVB.

**Figure 4.** Effect of fucoxanthin on the level of reduced glutathione (GSH). The level of GSH was assessed in cells treated with 20 μM fucoxanthin for 12 h by (**A**) performing confocal microscopy after CMAC staining and (**B**) using a GSH detection kit. * indicates significantly different from control ($p < 0.05$). The level of GSH in UVB-treated HaCaT cells incubated for 12 h, with or without pretreatment with 20 μM fucoxanthin, was detected by (**C**) performing confocal microscopy after CMAC staining and (**D**) using a GSH detection kit. * indicates significantly different from control ($p < 0.05$) and # significantly different from UVB-irradiated cells ($p < 0.05$).

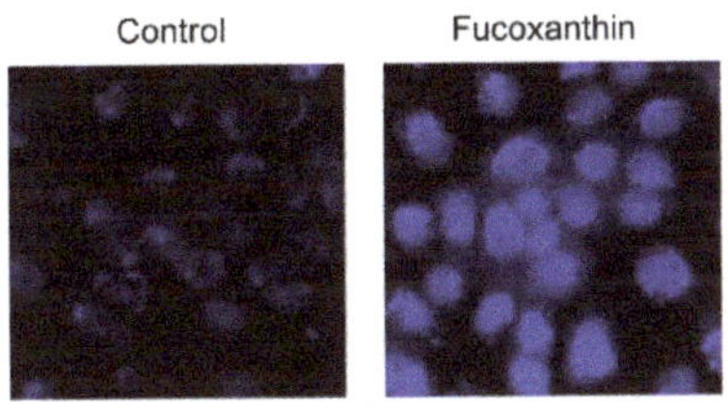


(**A**)

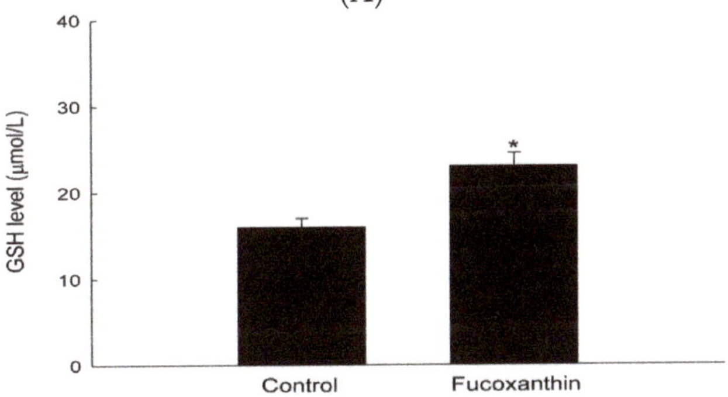


(**B**)

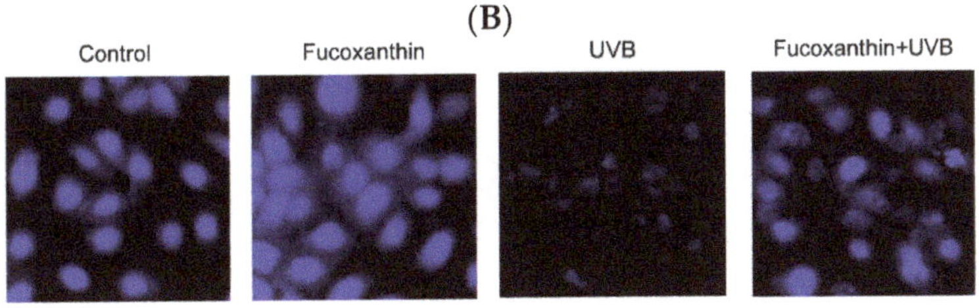


(**C**)

**Figure 4.** *Cont.*

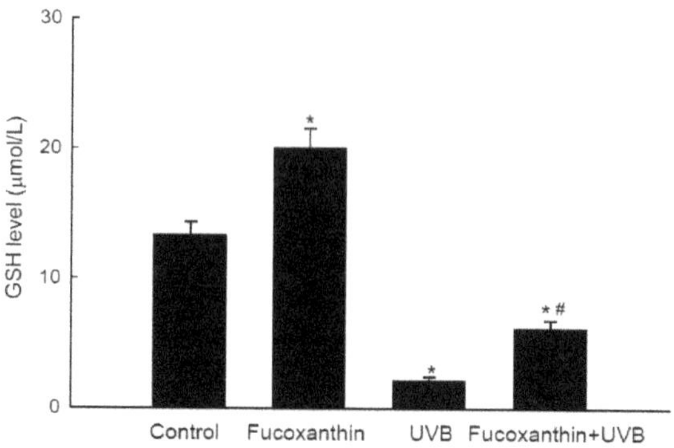


**(D)**

## 3. Materials and Methods

### 3.1. Materials

Anti-TATA-binding protein (TBP) and anti-phospho Nrf2 antibodies were purchased from Abcam, Inc. (Cambridge, MA, USA). Anti-Nrf2, anti-Akt, and anti-phospho Akt antibodies were purchased from Cell Signaling Technology (Beverly, MA, USA). Fucoxanthin, anti-GCLC and anti-GSS antibodies were purchased from Santa Cruz Biotechnology (Santa Cruz, CA, USA). [3-(4,5-Dimethylthiazol-2-yl)-2, 5-diphenyltetrazolium] bromide (MTT) and an anti-β-actin antibody were purchased from Sigma-Aldrich Chemical Company (St. Louis, MO, USA). Cell Tracker™ Blue CMAC was purchased from Molecular Probes (Eugene, OR, USA). LY294002 was provided by Calbiochem (San Diego, CA, USA). All other chemicals and reagents were of analytical grade.

### 3.2. Cell Culture

The human keratinocyte cell line HaCaT was supplied by Amore Pacific Company (Gyeonggi-do, Korea) and maintained at 37 °C in an incubator with a humidified atmosphere of 5% $CO_2$ and 95% air. Cells were grown in RPMI 1640 medium containing 10% fetal calf serum, streptomycin (100 μg/mL), and penicillin (100 units/mL).

### 3.3. Reverse Transcription-PCR (RT-PCR)

Total RNA was isolated from cells using the easy-BLUE™ total RNA extraction kit (iNtRON Biotechnology Inc., Seongnamsi, Korea). cDNA was amplified using 1 μL of reverse transcription reaction buffer, primers, dNTPs, and 0.5 U of Taq DNA polymerase in a final volume of 20 μL. The PCR conditions were initial denaturation at 94 °C for 5 min, followed by 26 cycles of 94 °C for 30 s, 63 °C for 45 s, and 72 °C for

1 min, and a final elongation step at 72 °C for 7 min. The following primers were used: human GCLC, forward (5′-AACCAAGCGCCATGCCGACC-3′) and reverse (5′-CCTCCTTCCGGCGTTTTCGC-3′); human GSS, forward (5′-GCCCCATTCACGCTCTTCCCC-3′) and reverse (5′-ATGCCCGGCCTGCTTAGCTC-3′); human GAPDH, forward (5′-TCAAGTGGGGCGATGCTGGC-3′) and reverse (5′-TGCCAGCCCCAGCGTCAAAG-3′). The amplified products were mixed with blue/orange 6 × loading dye, resolved by electrophoresis on a 1% agarose gel, stained with RedSafe™ nucleic acid staining solution (iNtRON Biotechnology Inc., Seongnamsi, Korea), and photographed under UV light using Image Quant™ TL analysis software (Amersham Biosciences, Uppsala, Sweden).

### 3.4. Western Blot Analysis

Cells were lysed on ice for 30 min in 100 μL lysis buffer (120 mM NaCl, 40 mM Tris (pH 8), and 0.1% NP-40) and centrifuged at 13,000× *g* for 15 min. Supernatants were collected and the protein concentrations were determined. Aliquots containing 40 μg of protein were boiled for 5 min and electrophoresed on 10% SDS-polyacrylamide gels. Proteins were transferred to nitrocellulose membranes, which were subsequently incubated with a primary antibody overnight at 4 °C. The membranes were further incubated with horseradish peroxidase-conjugated anti-immunoglobulin-G (Pierce, Rockford, IL, USA). Protein bands were detected using an enhanced chemiluminescence western blotting detection kit (Amersham Biosciences, Little Chalfont, Buckinghamshire, UK).

### 3.5. Immunocytochemistry

Cells at a density of $1.0 \times 10^5$ cells/mL were seeded into a 4-well chamber slide. After incubation for about 16 h, cells were exposed to 20 μM fucoxanthin for a further 6 h. Subsequently, cells were fixed with 1% paraformaldehyde for 30 min and then washed three times with phosphate-buffered saline (PBS) for 5 min each time. Cells were permeabilized with PBS containing 1% Triton X-100 for 30 min and then washed with PBS. Cells were blocked with PBS containing 5% bovine serum albumin for 1 h at 37 °C, and then treated with an anti-Nrf2 antibody diluted in blocking medium (1:125 dilution) overnight. To visualize the primary anti-Nrf2 antibody, cells were treated with a FITC-conjugated secondary antibody (1:125) for 2 h. After washing with PBS, stained cells were mounted onto microscope slides in mounting medium containing DAPI (Vector, Burlingame, CA, USA). Images were collected using the LSM 510 program on a Zeiss confocal microscope.

### 3.6. ChIP Assay

Cells were processed using the SimpleChIP® enzymatic chromatin IP kit (Cell signaling technology, Beverly, MA, USA) according to the manufacturer's instructions. Briefly, proteins were cross-linked to DNA by adding 1% formaldehyde to the culture dishes and rocking on a platform for 10 min at room temperature. The cross-linking was stopped by the addition of glycine solution. Cells were washed twice with ice-cold PBS, pelleted by centrifugation, and re-suspended in 1 mL cell lysis buffer containing protease inhibitors. Soluble chromatin was sheared by sonication and then centrifuged at 15,000× *g*. Diluted supernatants were pre-cleared and blocked with protein A/G agarose, and the sonicated chromatin-DNA complex was precipitated overnight with the antibodies of interest. Bound DNA was eluted by incubating the beads in elution buffer, purified, and amplified using primers flanking the Nrf2-binding site within the promoters of the genes encoding human GCLC and GSS. The oligonucleotide containing the transcription factor-binding site of the GCLC and GSS promoter was obtained from Bioneer (Seoul, Korea). The ChIP procedure was analyzed by PCR using human GCLC and GSS promoter-specific primers as follows: GCLC, forward (5′-ATCTCCACGGTCCAGGTT-3′) and reverse (5′-CTCCCTCACCCTATCCATTT-3′); GSS, forward (5′-CTGGGAATAACCAGACACCTA-3′) and reverse (5′-CAGGTTCAAGCAATTCTCCTG-3′). The cycle parameters were as follows: initial denaturation at 95 °C for 5 min, followed by 40 cycles of 95 °C for 30 s, 60 °C for 30 s, and 72 °C for 30 s, and a final extension at 72 °C for 7 min. The amplified products were resolved by electrophoresis on a 3% agarose gel, stained with RedSafe™ nucleic acid staining solution, and photographed under UV light using Image Quant™ TL analysis software.

### 3.7. Luciferase Reporter Assay

HaCaT cells were transiently transfected with 0.5 μg of the luciferase reporter and 0.2 μg of the ARE expression vector using Lipofectamine™ 2000 (Invitrogen Corporation, Carlsbad, CA, USA). Co-transfection with 0.02 μg of pRL-TK Renilla reniformis luciferase served as a normalizing control. Luciferase assays were performed using the dual luciferase assay system (Promega, Madison, WI, USA).

### 3.8. Detection of GSH

For image analysis of GSH, cells were seeded in four-well chamber slides at a density of
$1 \times 10^5$ cells/mL. Sixteen hours after plating, cells were treated with 20 μM fucoxanthin and then irradiated with UVB 1 h later. After 12 h, 10 μM of CMAC was added to each well, and samples were incubated for an additional 30 min at 37 °C. After washing with PBS, the stained cells were mounted onto a chamber slide in

mounting medium. Images were collected on a confocal microscope using the LSM 5 PASCAL software (Carl Zeiss, Jena, Germany). In addition, the GSH concentration was measured using a BIOXYTECH GSH-400 assay kit (Foster City, CA, USA).

### 3.9. Statistical Analysis

All measurements were performed in triplicate and all values are expressed as the mean ± of the standard error. The results were subjected to an analysis of variance followed by Tukey's test to analyze differences between means. In each case, a $p$ value less than 0.05 was considered statistically significant.

## 4. Conclusions

In this study, we demonstrated that fucoxanthin induced activation of nucleus Nrf2 via PI3K/Akt, which in turn activated the transcription of ARE-driven GCLC and GSS genes, leading the synthesis of antioxidant GSH. We recently reported that fucoxanthin did not show toxic effect on HaCaT cells at 20 μM and this concentration prevented cells against oxidative damage [21]. Therefore, we chose the same concentration to examine the effect of fucoxanthin on GSH induction in the present study.

GSH is an important biological antioxidant and has diverse functions in nutrient metabolism [25], gene expression, and DNA/protein synthesis [26], and particularly in eliminating oxidants [27]. GSH is formed from glutamate, cysteine, and glycine in a reaction that is catalyzed by two cytosolic enzymes, namely, GCLC and GSS [6]. The GSH is associated with the inhibition of tumor cell growth [28], prevention of apoptosis [29], and reduced inflammation [30]. In this study, fucoxanthin increased the mRNA and protein expression of GCLC and GSS (Figure 1). Nrf2, a major transcription factor of antioxidant enzymes, is tightly controlled by a master regulator Keap-1. Keap-1 has a high affinity for Nrf2 owing to its cysteine residues that form a covalent bond with Nrf2 [31]. Subsequently, Nrf2 signaling is switched off by Keap-1-mediated ubiquitination and degradation of Nrf2 [32]. However, the phospho form of Nrf2 can translocate into nucleus [33].

The nuclear Nrf2 recognizes the ARE sequence within the promoters of its target genes [34], and binding of Nrf2 to ARE sequences stimulates the transcription of genes that are involved in cellular defense. As shown in data, the phosphorylated Nrf2 expression was induced by fucoxanthin treatment (Figure 2A,B). The Nrf2 then binds to ARE sequence in GCLC and GSH promoters, which was assessed by ChIP assays (Figure 2C). The increased Nrf2 binding ability to ARE sequence in fucoxanthin-treated cells led to the increased transcriptional activity of Nrf2 (Figure 2D). These results indicated that fucoxanthin promotes release of Nrf2 from Keap-1 and subsequent translocation to nucleus, which induced the synthesis of GSH by enhancing expression of GSS and GCLC through interaction between Nrf2 and ARE sequence. It recently reported that fucoxanthin enhanced heme oxygenase-1 and

NAD(P)H:quinone oxidoreductase 1 expression via activation of Nrf2/ARE system [35]. The phosphorylation of Nrf2 plays a pivotal role in its nuclear accumulation, and this phosphorylation can occur via Akt pathways [36]. Akt is a classic signal-transducing protein which can activate the primary cellular defense mechanism Nrf2/ARE in skin cells [37]. In our study, fucoxanthin treatment increased the level of phosphorylated Akt, which is the active form of this kinase (Figure 3A) and could elevate the nuclear level of Nrf2 (Figure 2A). Furthermore, LY294002, a specific inhibitor of PI3K/Akt, significantly suppressed the active form of Akt (Figure 3A) which resulted in reduction of Nrf2 accumulation, following decreased protein expression of GSS and GCLC (Figure 3B–D). In addition, working in concert with the effects of fucoxanthin on GSS and GCLC expression, the amount of GSH was increased in fucoxanthin-treated cells, as indicated by the increased fluorescence intensity of CMAC (Figure 4A) and the increased concentration of GSH (Figure 4B). In our system, UVB irradiation, an inducer of oxidative stress, suppressed the GSH level. However, treatment with fucoxanthin prior to UVB damage partially mitigated the reduction in GSH levels (Figure 4C,D).

In conclusion, fucoxanthin substantially increased the mRNA and proteins levels of GCLC and GSS in human keratinocytes, and these effects were dependent on the nuclear translocation of Nrf2 following its phosphorylation by the protein kinase Akt. In addition, this study demonstrates that the Akt/Nrf2 pathway plays an essential role in the mechanism underlying the effects of fucoxanthin. Taken together, one of the major ways by which fucoxanthin treatment prevents or eliminates oxidative damage is to enhance the Akt/Nrf2/GSH-dependent antioxidant response.

## Acknowledgments

This work was supported by the National Research Foundation of Korea Grant funded by the Korean Government (MEST) (NRF-C1ABA001-2011-0021037).

## Author Contributions

Jian Zheng designed the research, performed the experiments, and wrote the manuscript. Mei Jing Piao and Ki Cheon Kim analyzed some data. Cheng Wen Yao and Ji Won Cha gave the advice on some experiments. Jin Won Hyun designed the research, analyzed and interpreted data, drafted the manuscript.

## Conflicts of Interest

The authors declare no conflict of interest.

## References


1. Halliwell, B. Biochemistry of oxidative stress. *Biochem. Soc. Trans.* **2007**, *35*, 1147–1150.
2. Surh, Y.J.; Kundu, J.K.; Na, H.K. Nrf2 as a master redox switch in turning on the cellular signaling involved in the induction of cytoprotective genes by some chemopreventive phytochemicals. *Planta Med.* **2008**, *74*, 1526–1539.
3. Wu, K.C.; Liu, J.J.; Klaassen, C.D. Nrf2 activation prevents cadmium-induced acute liver injury. *Toxicol. Appl. Pharmacol.* **2012**, *263*, 14–20.
4. Harvey, C.J.; Thimmulappa, R.K.; Singh, A.; Blake, D.J.; Ling, G.; Wakabayashi, N.; Fujii, J.; Myers, A.; Biswal, S. Nrf2-regulated glutathione recycling independent of biosynthesis is critical for cell survival during oxidative stress. *Free Radic. Biol. Med.* **2009**, *46*, 443–453.
5. Yang, Y.; Chen, Y.; Johansson, E.; Schneider, S.N.; Shertzer, H.G.; Nebert, D.W.; Dalton, T.P. Interaction between the catalytic and modifier subunits of glutamate-cysteine ligase.
*Biochem. Pharmacol.* **2007**, *74*, 372–381.
6. Wu, G.; Fang, Y.Z.; Yang, S.; Lupton, J.; Turner, N.D. Glutathione metabolism and its implications for health. *J. Nutr.* **2004**, *134*, 489–492.
7. Fang, Y.Z.; Yang, S.; Wu, G. Free radicals, antioxidants, and nutrition. *Nutrition* **2002**, *18*, 872–879.
8. Filomeni, G.; Rotilio, G.; Ciriolo, M.R. Cell signalling and the glutathione redox system. *Biochem. Pharmacol.* **2002**, *64*, 1057–1064.
9. Clément, M.V.; Ponton, A.; Pervaiz, S. Apoptosis induced by hydrogen peroxide is mediated by decreased superoxide anion concentration and reduction of intracellular milieu. *FEBS Lett.* **1998**, *440*, 13–18.
10. Pi, J.; Zhang, Q.; Woods, C.G.; Wong, V.; Collins, S.; Andersen, M.E. Activation of Nrf2-mediated oxidative stress response in macrophages by hypochlorous acid. *Toxicol. Appl. Pharmacol.* **2008**, *226*, 236–243.
11. Marrot, L.; Jones, C.; Perez, P.; Meunier, J.R. The significance of Nrf2 pathway in (photo)-oxidative stress response in melanocytes and keratinocytes of the human epidermis. *Pigment Cell Melanoma Res.* **2008**, *21*, 79–88.
12. Kensler, T.W.; Wakabayashi, N.; Biswal, S. Cell survival responses to environmental stresses via the Keap1-Nrf2-ARE pathway. *Annu. Rev. Pharmacol. Toxicol.* **2007**, *47*, 89–116.
13. Eggler, A.L.; Gay, K.A.; Mesecar, A.D. Molecular mechanisms of natural products in chemoprevention: Induction of cytoprotective enzymes by Nrf2. *Mol. Nutr. Food Res.* **2008**, *52*, 84–94.
14. Sugawara, T.; Baskaran, V.; Tsuzuki, W.; Nagao, A. Brown algae fucoxanthin is hydrolyzed to fucoxanthinol during absorption by Caco-2 human intestinal cells and mice. *J. Nutr.* **2002**, *132*, 946–951.

15. Yan, X.; Chuda, Y.; Suzuki, M.; Nagatat, T. Fucoxanthin as the Major Antioxidant in *Hifikia fusiformis*, a Common Edible Seaweed. *Biosci. Biotechnol. Biochem.* **1999**, *63*, 605–607.
16. Hu, T.; Liu, D.; Chen, Y.; Wu, J.; Wang, S. Antioxidant activity of sulfated polysaccharide fractions extracted from *Undaria pinnitafida in vitro*. *Int. J. Biol. Macromol.* **2010**, *46*, 193–198.
17. Das, S.K.; Hashimoto, T.; Kanazawa, K. Growth inhibition of human hepatic carcinoma HepG2 cells by fucoxanthin is associated with down-regulation of cyclin D. *Biochim. Biophys. Acta* **2008**, *1780*, 743–749.
18. Woo, M.N.; Jeon, S.M.; Shin, Y.C.; Lee, M.K.; Kang, M.; Choi, M.S. Anti-obese property of fucoxanthin is partly mediated by altering lipid-regulating enzymes and uncoupling proteins of visceral adipose tissue in mice. *Mol. Nutr. Food Res.* **2009**, *53*, 1603–1611.
19. Maeda, H.; Hosokawa, M.; Sashima, T.; Miyashita, K. Dietary combination of fucoxanthin and fish oil attenuates the weight gain of white adipose tissue and decreases blood glucose in obese/diabetic KK-Ay mice. *J. Agric. Food Chem.* **2007**, *55*, 7701–7706.
20. D'Orazio, N.; Gemello, E.; Gammone, M.A.; de Girolamo, M.; Ficoneri, C.; Riccioni, G. Fucoxantin: A treasure from the sea. *Mar. Drugs* **2012**, *10*, 604–616.
21. Zheng, J.; Piao, M.J.; Keum, Y.S.; Kim, H.S.; Hyun, J.W. Fucoxanthin protects cultured human keratinocytes against oxidative stress by blocking free radicals and inhibiting apoptosis. *Biomol. Ther.* **2013**, *21*, 270–276.
22. Nguyen, T.; Huang, H.C.; Pickett, C.B. Transcriptional regulation of the antioxidant response element Activation by Nrf2 and repression by MafK. *J. Biol. Chem.* **2000**, *275*, 15466–15473.
23. Nakaso, K.; Yano, H.; Fukuhara, Y.; Takeshima, T.; Wada-Isoe, K.; Nakashima, K. PI3K is a key molecule in the Nrf2-mediated regulation of antioxidative proteins by hemin in human neuroblastoma cells. *FEBS Lett.* **2003**, *546*, 181–184.
24. Vlahos, C.J.; Matter, W.F.; Hui, K.Y.; Brown, R.F. A specific inhibitor of phosphatidylinositol 3-kinase, 2-(4-morpholinyl)-8-phenyl-4H-1-benzopyran-4-one (LY294002). *J. Biol. Chem.* **1994**, *269*, 5241–5248.
25. Chen, J.S.; Huang, P.H.; Wang, C.H.; Lin, F.Y.; Tsai, H.Y.; Wu, T.C.; Lin, S.J.; Chen, J.W. Nrf-2 mediated heme oxygenase-1 expression, an antioxidant-independent mechanism, contributes to anti-atherogenesis and vascular protective effects of Ginkgo biloba extract. *Atherosclerosis* **2011**, *214*, 301–309.
26. Ogasawara, Y.; Takeda, Y.; Takayama, H.; Nishimoto, S.; Ichikawa, K.; Ueki, M.; Suzuki, T.; Ishii, K. Significance of the rapid increase in GSH levels in the protective response to cadmium exposure through phosphorylated Nrf2 signaling in Jurkat T-cells. *Free Radic. Biol. Med.* **2014**, *69*, 58–66.
27. Sies, H. Glutathione and its role in cellular functions. *Free Radic. Biol. Med.* **1999**, *27*, 916–921.

28. Kudugunti, S.K.; Vad, N.M.; Ekogbo, E.; Moridani, M.Y. Efficacy of caffeic acid phenethyl ester (CAPE) in skin B16-F0 melanoma tumor bearing C57BL/6 mice. *Investig. New Drugs* **2011**, *29*, 52–62.
29. Zhou, B.R.; Yin, H.B.; Xu, Y.; Wu, D.; Zhang, Z.H.; Yin, Z.Q.; Permatasari, F.; Luo, D. Baicalin protects human skin fibroblasts from ultraviolet A radiation-induced oxidative damage and apoptosis. *Free Radic. Res.* **2012**, *46*, 1458–1471.
30. Bolfa, P.; Vidrighinescu, R.; Petruta, A.; Dezmirean, D.; Stan, L.; Vlase, L.; Damiane, G.; Catoia, C.; Filip, A.; Clichici, S. Photoprotective effects of Romanian propolis on skin of mice exposed to UVB irradiation. *Food Chem. Toxicol.* **2013**, *62*, 329–342.
31. Cullinan, S.B.; Gordan, J.D.; Jin, J.; Harper, J.W.; Diehl, J.A. The Keap1-BTB protein is an adaptor that bridges Nrf2 to a Cul3-based E3 ligase: Oxidative stress sensing by a Cul3-Keap1 ligase. *Mol. Cell. Biol.* **2004**, *24*, 8477–8486.
32. McMahon, M.; Thomas, N.; Itoh, K.; Yamamoto, M.; Hayes, J.D. Redox-regulated turnover of Nrf2 is determined by at least two separate protein domains, the redox-sensitive Neh2 degron and the redox-insensitive Neh6 degron. *J. Biol. Chem.* **2004**, *279*, 31556–31567.
33. Apopa, P.L.; He, X.; Ma, Q. Phosphorylation of Nrf2 in the transcription activation domain by casein kinase 2 (CK2) is critical for the nuclear translocation and transcription activation function of Nrf2 in IMR-32 neuroblastoma cells. *J. Biochem. Mol. Toxicol.* **2008**, *22*, 63–76.
34. Jaiswal, A.K. Nrf2 signaling in coordinated activation of antioxidant gene expression. *Free Radic. Biol. Med.* **2004**, *36*, 1199–1207.
35. Liu, C.L.; Chiu, Y.T.; Hu, M.L. Fucoxanthin enhances HO-1 and NQO1 expression in murine hepatic BNL CL. 2 cells through activation of the Nrf2/ARE system partially by its pro-oxidant activity. *J. Agric. Food Chem.* **2011**, *59*, 11344–11351.
36. Martin, D.; Rojo, A.I.; Salinas, M.; Diaz, R.; Gallardo, G.; Alam, J.; Ruiz de Galarreta, C.M.; Cuadrado, A. Regulation of heme oxygenase-1 expression through the phosphatidylinositol 3-kinase/Akt pathway and the Nrf2 transcription factor in response to the antioxidant phytochemical carnosol. *J. Biol. Chem.* **2004**, *279*, 8919–8929.
37. Rodriguez, K.J.; Wong, H.K.; Oddos, T.; Southall, M.; Frei, B.; Kaur, S. A purified Feverfew extract protects from oxidative damage by inducing DNA repair in skin cells via a PI3-kinase-dependent Nrf2/ARE pathway. *J. Dermatol. Sci.* **2013**, *72*, 304–310.

*marine drugs*

*Article*

# Dietary Fucoxanthin Increases Metabolic Rate and Upregulated mRNA Expressions of the PGC-1alpha Network, Mitochondrial Biogenesis and Fusion Genes in White Adipose Tissues of Mice

**Meng-Ting Wu [1], Hong-Nong Chou [2] and Ching-jang Huang [1,*]**

[1] Department of Biochemical Science and Technology, National Taiwan University, No. 1, Sec. 4, Roosevelt Road, Taipei 10617, Taiwan; E-Mail: dreamstop@gmail.com

[2] Institute of Fisheries Science, National Taiwan University, No. 1, Sec. 4, Roosevelt Road, Taipei 10617, Taiwan; E-Mail: unijohn@ntu.edu.tw

* Author to whom correspondence should be addressed; E-Mail: cjjhuang@ntu.edu.tw;
Tel.: +886-2-3366-2276; Fax: +886-2-2362-1301.

*Received: 17 December 2013; in revised form: 22 January 2014 / Accepted: 23 January 2014 /*
*Published: 14 February 2014*

**Abstract:** The mechanism for how fucoxanthin (FX) suppressed adipose accumulation is unclear. We aim to investigate the effects of FX on metabolic rate and expressions of genes related to thermogenesis, mitochondria biogenesis and homeostasis. Using a 2 × 2 factorial design, four groups of mice were respectively fed a high sucrose (50% sucrose) or a high-fat diet (23% butter + 7% soybean oil) supplemented with or without 0.2% FX. FX significantly increased oxygen consumption and carbon dioxide production and reduced white adipose tissue (WAT) mass. The mRNA expressions of peroxisome proliferator-activated receptor (PPAR) $\gamma$ coactivator-1$\alpha$ (PGC-1$\alpha$), cell death-inducing DFFA-like effecter a (CIDEA), PPAR$\alpha$, PPAR$\gamma$, *estrogen*-related receptor $\alpha$ (ERR$\alpha$), β3-adrenergic receptor (β3-AR) and deiodinase 2 (Dio2) were significantly upregulated in inguinal WAT (iWAT) and epididymal WAT (eWAT) by FX. Mitochondrial biogenic genes, nuclear respiratory factor 1 (NRF1) and NRF2, were increased in eWAT by FX. Noticeably, FX upregulated genes of mitochondrial fusion, mitofusin 1 (Mfn1), Mfn2 and optic atrophy 1 (OPA1), but not mitochondrial fission, Fission 1, in both iWAT and eWAT. In conclusion, dietary FX enhanced the metabolic rate and lowered adipose mass irrespective of the diet. These were associated with upregulated genes of the PGC-1$\alpha$ network and mitochondrial fusion in eWAT and iWAT.

**Keywords:** fucoxanthin; adipose tissue; metabolic rate; PGC-1$\alpha$ network; mitochondrial biogenesis and fusion

---

## Abbreviations

DFFA, DNA fragmentation factor, alpha subunit; HF, High fat diet; HF + F, High fat diet supplemented with fucoxanthin; HS, High sucrose diet; HS + F, High sucrose diet supplemented with fucoxanthin; qRT, Quantitative real-time; RQ, Respiratory quotient.

## 1. Introduction

Obesity, defined as excess accumulation of adipose, is a worldwide endemic health problem. Obesity and its related disorders are associated with increased morbidity, mortality and healthcare costs [1]. Mitochondria play an important role in adipose biology [2]. Mitochondria dysfunction is linked to obesity [3] and type 2 diabetes [4,5]. Mitochondria biogenesis is thus considered a potential target for the intervention of insulin resistance in obesity and diabetes [6–8]. Mitochondrion is a dynamic organelle that is continuously remodeled by fusion and fission [9], which is important for bioenergetic adaptation to metabolic demand. Cells exposed to a nutrient-rich environment tend to maintain their mitochondria in a separated state, while cells under starvation tend to have mitochondria in a connected state [10]. Reduction in the mitochondrial network, but unaltered mitochondrial mass have been reported in skeletal muscle of obese Zucker rats [3] and type 2 diabetic patients [11]. Moreover, mRNA and the protein, Mfn2, an important protein located in the mitochondrial outer membrane and mediating mitochondrial fusion, were reduced in skeletal muscle of obese subjects compared to lean subjects [12].

Peroxisome proliferator-activated receptor $\gamma$ coactivator-1$\alpha$ (PGC-1$\alpha$) is a strong promoter of mitochondrial biogenesis and oxidative metabolism through nuclear respiratory factor, NRF1 and NRF2, ERR$\alpha$, PPAR$\gamma$ and PPAR$\alpha$ [13–15]. ERR$\alpha$ further activated the transcriptional activity of the Mfn2 promoter, and the effects were synergic with those of PGC-1$\alpha$ [16]. In addition, NRF1 and NRF2 themselves are able to stimulate mitochondrial transcription factor A (TFAM), a mitochondrial matrix protein essential for the replication and transcription of mitochondria DNA [17,18]. PGC-1$\alpha$ mRNA expression is reduced in subcutaneous fat of morbidly obese patients [19]. Ectopic expression of PGC-1$\alpha$ in white adipose tissue (WAT) increased brown adipocyte specific genes, including uncoupling protein 1 (UCP1) and mitochondrial activity [20].

Fucoxanthin (FX) is a major carotenoid in brown algae and has an unusual allenic structure [21]. FX has a suppressive effect on adipose accumulation in genetically diabetic KK-*A*$^{y}$ mice [22–24], Wistar rats [25,26] and diet-induced obese C57BL/6J mice [27–29]. FX feeding significantly increased fecal triglyceride and cholesterol excretion and β-oxidation in liver and epididymal WAT (eWAT) and also reduced fatty acid, cholesterol synthesis-related enzyme activity, serum and hepatic lipid accumulation [28,30,31]. Previous studies also showed that FX increased UCP1 mRNA expression in abdominal WAT of mice with reduced adipose mass [25,28,31]. PGC-1$\alpha$ expression in the skeletal muscles was upregulated in KK-*A*$^{y}$ mice fed FX [32].

UCP1 is a thermogenic mitochondria protein that was thought to express exclusively in brown adipose tissue (BAT). However, its expression in WAT has been confirmed in most recent studies, and these UCP1 expressing adipocytes in WAT were named "Beige/Brite" or "recruited (in contrast to the traditional "constituted") brown adipocytes [33,34]. Increases in the recruited brown adipocytes can enhance thermogenic activity and reduce adiposity in rodents [33,34] and humans [35]. As a dominant regulator of mitochondrial biogenesis and oxidative metabolism [15], PGC-1$\alpha$ regulates thermogenic activity by inducing the expression of UCP1 and key enzymes of the mitochondrial respiratory chain in both constituted and recruited brown adipocytes [33]. Mitochondrial biogenesis and dynamics regulate mitochondrial function, respiratory capacity, apoptosis and oxidative phosphorylation and, thus, impact energy balance [3,12,36]. Whether FX increased expressions of genes in the PGC-1$\alpha$ regulated pathways in WAT and metabolic rate has not been reported.

Here, we examined the suppressive effect of adipose accumulation of FX under a high sucrose or a high saturated fat diet using a 2 × 2 factorial design. Whole body $O_2$ consumption and $CO_2$ production were measured and expressions of PGC-1$\alpha$ regulated genes in WAT analyzed.

## 2. Results and Discussion

### 2.1. FX Decreased White Adipose Weight without Altering Food Intake

Four groups of mice were respectively fed the 4 test diets shown in Table 1. The results of two-way ANOVA on the four groups (Table 2) of mice showed that neither FX nor diet affected final body weight, body weight gain and energy efficiency ($p >$ 0.05). The high fat diet, but not FX, decreased food and energy intake ($p < 0.05$) (Table 2). FX significantly decreased relative weight of both abdominal, eWAT and retroperitoneal WAT (rWAT), and subcutaneous iWAT ($p < 0.001$). High fat diet decreased, but FX increased, the relative weight of BAT, ($p < 0.05$) (Table 3). Liver, kidney, spleen and heart relative weights were significantly increased by FX (Supplementary Table S1).

**Table 1.** Composition of the 4 test diets: HS, high sucrose diet; HS + F, high sucrose diet supplemented with fucoxanthin; HF, high fat diet; HF + F, high fat diet supplemented
with fucoxanthin. The composition of vitamin mix and mineral mix are in accordance with AIN-93 and AIN-93G, respectively [37].

| Ingredients of Diet (g/kg) | HS | HS + F | HF | HF + F |
|---|---|---|---|---|
| Corn starch | 129.5 | 129.5 | 209.35 | 209.35 |
| Sucrose | 500 | 500 | 100 | 100 |
| Butter | – | – | 230 | 228 |
| Soybean oil | 70 | 68 | 70 | 70 |
| Casein | 200 | 200 | 260 | 260 |
| Cellulose | 50 | 50 | 65 | 65 |
| AIN-93 vitamin mix | 10 | 10 | 13 | 13 |
| AIN-93G mineral mix | 35 | 35 | 45.5 | 45.5 |
| L-Cystine | 3 | 3 | 3.9 | 3.9 |
| Choline bitartrate | 2.5 | 2.5 | 3.25 | 3.25 |
| Fucoxanthin | – | 2 | – | 2 |

**Table 2.** Initial body weight, final body weight, body weight gain, food/energy intake and energy efficiency of C57BL/6J male mice fed test diets for five weeks. Values are the means ± SD ($n$ = 4). * denotes significant influence by either dietary factor at $p < 0.05$ analyzed by two-way ANOVA. Energy efficiency = grams of body weight gain/1000 kcal energy intake. FX, fucoxanthin.

| Dietary groups | Initial body weight | Final body weight | Body weight gain | Food intake | Energy intake | Energy efficiency |
|---|---|---|---|---|---|---|
| | g | g | g | g/day | kcal/day | |
| HS | 19.27 ± 1.06 | 24.19 ± 1.75 | 4.92 ± 0.76 | 3.12 ± 0.39 | 12.37 ± 1.53 | 11.42 ± 1.66 |
| HS + F | 19.24 ± 0.58 | 24.63 ± 1.03 | 5.39 ± 0.59 | 3.16 ± 0.20 | 12.45 ± 0.78 | 12.35 ± 0.92 |
| HF | 19.28 ± 0.69 | 24.27 ± 1.71 | 4.99 ± 1.31 | 2.29 ± 0.14 | 11.44 ± 0.72 | 12.45 ± 3.01 |
| HF + F | 19.21 ± 0.61 | 24.84 ± 1.26 | 5.63 ± 0.98 | 2.27 ± 0.06 | 11.28 ± 0.30 | 14.28 ± 2.67 |
| | | | *p* values | | | |
| Diet | 0.9769 | 0.8469 | 0.7474 | <0001 * | 0.0460 * | 0.2085 |
| FX | 0.9001 | 0.5050 | 0.2656 | 0.9331 | 0.9391 | 0.2381 |
| Diet * FX | 0.9717 | 0.9257 | 0.8628 | 0.7604 | 0.8075 | 0.6943 |

**Table 3.** Relative tissue weights (percent of body weight) of C57BL/6J male mice fed test diets for five weeks. Values are the means ± SD ($n$ = 4). * denotes significant influence by either dietary factor at $p < 0.05$ analyzed by two-way ANOVA. iWAT, inguinal white adipose tissue (WAT); eWAT, epididymal WAT; rWAT, retroperitoneal WAT; BAT, brown adipose tissue.

| Group | HS | HS+F | HF | HF + F | *p*-values | | |
|---|---|---|---|---|---|---|---|
| | | | | | Diet | FX | Diet * FX |
| iWAT | 0.56 ± 0.13 | 0.39 ± 0.03 | 0.63 ± 0.08 | 0.34 ± 0.08 | 0.8149 | 0.0002 * | 0.1599 |
| eWAT | 1.77 ± 0.48 | 1.10 ± 0.12 | 1.90 ± 0.42 | 1.21 ± 0.21 | 0.3944 | 0.0006 * | 0.9527 |
| rWAT | 0.40 ± 0.18 | 0.13 ± 0.04 | 0.44 ± 0.24 | 0.14 ± 0.06 | 0.7321 | 0.0003 * | 0.9982 |
| BAT | 0.29 ± 0.05 | 0.37 ± 0.04 | 0.26 ± 0.03 | 0.30 ± 0.02 | 0.0198 * | 0.0066 * | 0.3840 |

### 2.2. FX Enhanced Metabolic Rate

Mice were placed in respiratory chambers at week 3 for six days to continuously monitor $O_2$ consumption ($VO_2$), $CO_2$ production ($VCO_2$). These and calculated respiratory quotients (RQ) values throughout a day and respective area under the curve (AUC) values were shown in Figure 1. At seven out of 24 hour time points, high sucrose diet supplemented with fucoxanthin (HS + F) or high-fat diet (HF + F) groups had significantly higher $VO_2$ and $VCO_2$, respectively, compared to HS and HF groups (Figure 1A,B). Diet did not change the AUC of $VO_2$ (Figure 1A), but the high fat diet decreased that of the $VCO_2$ (Figure 1B). FX increased the AUC of both $VO_2$ and $VCO_2$ through the dark phase and all day without an interaction with the diet. Mice fed high fat diets had significantly lower AUC of RQ

($p$ < 0.05), indicating that these mice used more fat as their energy source. FX supplementation did not change RQ (Figure 1C), implying that FX might have enhanced the common aerobic metabolic pathway of glucose and fat metabolism. Indeed, FX was shown to enhance the utilization of glucose in skeletal muscle [32] and liver [29], as well as fatty acid β-oxidation in WAT [28] and liver [30].

**Figure 1.** $O_2$ consumption (**A**), $CO_2$ production (**B**), respiratory quotient (**C**) and their area under the curve (AUC) of test mice at week 3. Mice were individually placed in metabolic chambers with free access to water and their respective diet and monitored for six days. Values are means and error bars are SD ($n$ = 4). * denotes significant difference between groups HS and HS + F, $p < 0.05$; # denotes significant difference between groups HF and HF + F, $p < 0.05$, analyzed by the Student's $t$-test. AUCs of $O_2$ consumption ($VO_2$), $CO_2$ production ($VCO_2$) were analyzed by two-way ANOVA. The AUC of RQ was analyzed by the Wilcoxon rank-sum test.

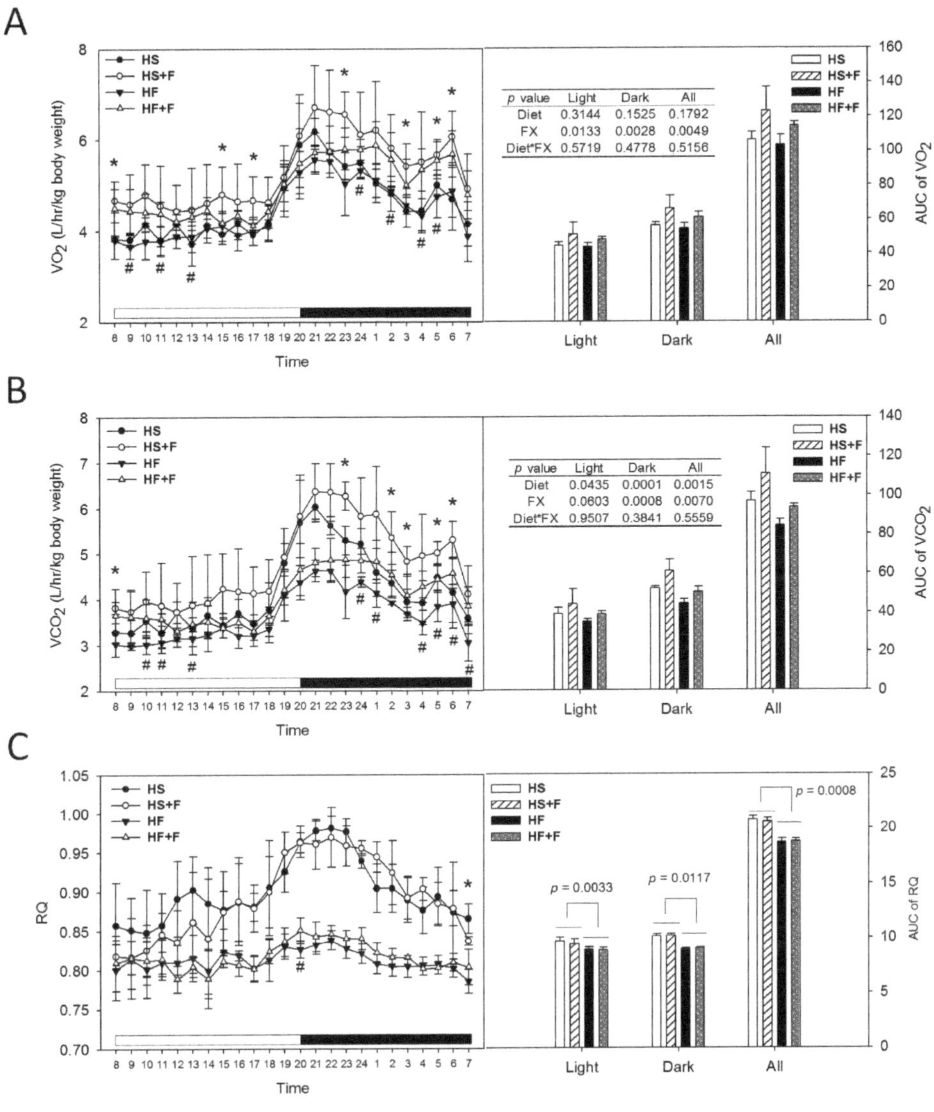


**Figure 2.** Thermogenic (**A**) and mitochondrial homeostasis-related (**B**) mRNA levels in brown adipose tissue of mice fed test diets for five weeks. Values are means, and error bars are SEM ($n$ = 4). * denotes significant effect by either dietary factor at $p < 0.05$ analyzed by two-way ANOVA. When an interaction ($p$ < 0.05) between diet and FX existed, the significance of differences among the HS, HS + F, HF and HF + F groups was further analyzed by Duncan's multiple range test. HS, HS + F, HF and HF + F: As indicated in Figure 1. Relative mRNA expression was measured by real-time qRT-PCR using β-actin as the internal control and normalized to group HS.

A

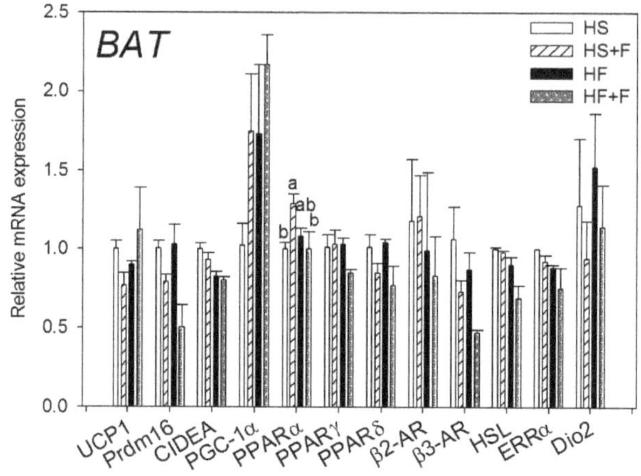


| *p* values | UCP1 | Prdm16 | CIDEA | PGC-1α | PPARα | PPARγ | PPARδ | β2-AR | β3-AR | HSL | ERRα | Dio2 |
|---|---|---|---|---|---|---|---|---|---|---|---|---|
| **Diet** | 0.4328 | 0.2144 | 0.0005 * | 0.0908 | 0.1677 | 0.2415 | 0.7560 | 0.4437 | 0.0375 * | 0.6750 | 0.0098 * | 0.5102 |
| **FX** | 0.4638 | 0.0029 * | 0.1679 | 0.0808 | 0.1791 | 0.2320 | 0.0156 * | 0.8537 | 0.0035 * | 0.6637 | 0.0912 | 0.2936 |
| **Diet * FX** | 0.1014 | 0.1443 | 0.5263 | 0.6426 | 0.0285 * | 0.1345 | 0.5130 | 0.7989 | 0.3115 | 0.9766 | 0.8158 | 0.9622 |

B

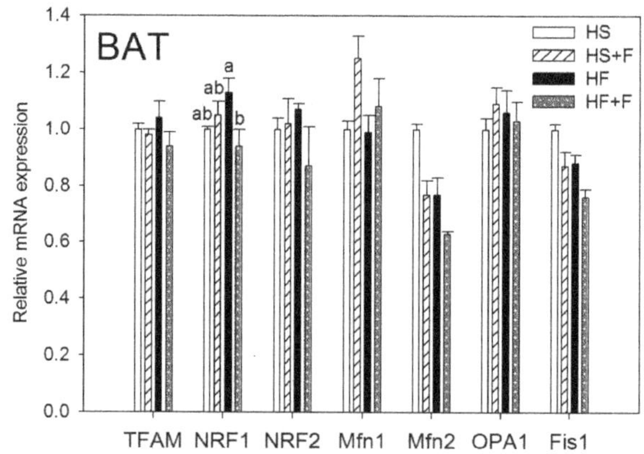


| *p* values | TFAM | NRF1 | NRF2 | Mfn1 | Mfn2 | OPA1 | Fis1 |
|---|---|---|---|---|---|---|---|
| **Diet** | 0.9681 | 0.8508 | 0.7589 | 0.2273 | 0.0006 * | 0.9604 | 0.0065 * |
| **FX** | 0.1807 | 0.1698 | 0.4806 | 0.0390 * | 0.0006 * | 0.6677 | 0.0028 * |
| **Diet * FX** | 0.3594 | 0.0253 * | 0.2438 | 0.2862 | 0.2619 | 0.3839 | 0.8775 |

### *2.3. Effect of FX on Thermogenic and Mitochondrial Homeostasis-Related Gene Expressions in BAT and Serum Hormone Concentration*

Both high fat diet and FX lowered β3-AR, Mfn2 and Fis1 mRNA levels in the BAT ($p < 0.05$). Mice fed high fat diets had significantly lower CIDEA and ERRα mRNA expression in the BAT ($p < 0.05$). FX also decreased Prdm16 and PPARδ but increased Mfn1 mRNA in the BAT. Neither diet nor FX changed PGC-1α UCP1, PPARα, PPARγ, β2-AR, HSL, Dio2, TFAM, NRF1, NRF2 and OPA1 mRNA expressions in

BAT. Among them, diet and FX had an interaction effect on the PPARα and NRF1 mRNA expression ($p < 0.05$). FX increased PPARα mRNA expression in the high sucrose diet-fed mice, but not in the high fat diet-fed mice. FX reduced NRF1 mRNA expression in the high fat diet-fed mice, but not in the high sucrose diet-fed mice (Figure 2).

In this study, BAT mass was significantly increased by FX as in some of previous reports [23–25], although other studies did not observe such an effect [31,38]. This discrepancy might be associated with the different mouse strain, gender and diet formula used. Although BAT mass was increased by FX and positively correlated to carbon dioxide production ($p = 0.02$) during the dark period (Supplementary Figure S1), we did not observe any thermogenic genes upregulated in BAT. β3-AR, Prdm16 and PPARδ mRNA were even downregulated by FX. To validate these results, we even conducted the Housekeeping Genes PCR array (Qiagen, Germantown, MD, USA) and employed three popular software packages (GeNorm, NormFinder and BestKeeper) to confirm the use of β-actin as the most stable housekeeping gene.

On the other hand, our serum hormone analysis showed that the norepinephrine concentration was decreased by FX (Table 4). Norepinephrine, the sympathetic neurotransmitter, is the main *in vivo* stimulator of the adrenergic signaling mediating the BAT thermogenic machinery. It is not known whether this low serum norepinephrine in FX-fed mice is related to the very minor change in the BAT gene expression irrespective of the enlarged mass observed in this study. Moreover, no significant correlation between BAT mass and $O_2$ consumption was observed (Supplementary Figure S1). Taken together, it is speculated that BAT contributes little, if any, to the FX enhanced $O_2$ consumption.

**Table 4.** Serum thyroid hormone, (nor)epinephrine and corticosterone concentrations of C57BL/6J male mice fed test diets for five weeks. Values are the means ± SD ($n$ = 4). * denotes a significant effect by either dietary factor at $p < 0.05$ analyzed by two-way ANOVA. T4, thyroxine; T3, triiodothyronine; NE, norepinephrine; E, epinephrine; Cort, corticosterone.

| | HS | HS + F | HF | HF + F | *p*-values | | |
|---|---|---|---|---|---|---|---|
| | | | | | Diet | FX | Diet * FX |
| T4, nM | 24.59 ± 3.69 | 20.73 ± 5.22 | 18.94 ± 2.88 | 17.55 ± 3.79 | 0.0467 * | 0.2117 | 0.5464 |
| T3, nM | 0.80 ± 0.21 | 0.87 ± 0.34 | 0.80 ± 0.14 | 0.82 ± 0.21 | 0.8297 | 0.6998 | 0.8432 |
| T4/T3 | 32.70 ± 11.11 | 27.42 ± 14.18 | 23.80 ± 1.92 | 21.62 ± 2.66 | 0.1346 | 0.4312 | 0.7415 |
| NE, nM | 92.26 ± 16.52 | 66.75 ± 11.69 | 105.72 ± 20.01 | 72.05 ± 10.17 | 0.2383 | 0.0021 * | 0.5989 |
| E, nM | 3.05 ± 0.75 | 3.52 ± 1.21 | 5.61 ± 3.09 | 4.98 ± 3.03 | 0.0949 | 0.9931 | 0.5768 |
| NE/E | 31.87 ± 10.69 | 20.62 ± 7.69 | 22.09 ± 7.96 | 17.53 ± 6.75 | 0.1512 | 0.0843 | 0.4413 |
| Cort, ng/mL | 71.11 ± 10.15 | 63.99 ± 20.76 | 55.55 ± 32.38 | 58.69 ± 23.35 | 0.3898 | 0.8384 | 0.6634 |

## 2.4. FX Induced Thermogenic-Related Gene Expressions in eWAT and iWAT

While the high fat diet decreased CIDEA mRNA expression, FX significantly increased mRNA expressions of CIDEA, PGC-1α, ERRα, PPARγ, β3-AR, Dio2, PPARα and HSL in eWAT (Figure 3A). The high fat diet decreased UCP1, CIDEA, Prdm16, ERRα, PPARγ, Dio2 and PPARα mRNA expression, but FX increased CIDEA, PGC-1α, ERRα, β3-AR, Dio2 and PPARα mRNA expression in iWAT (Figure 3B). FX did not significantly affect mRNA expressions of these genes in the abdominal rWAT (data not shown), except for upregulating PGC-1α.

Expressions of UCP1, Prdm16, CIDEA and PGC-1α in WAT are indicators for the "browning" of WAT, which could reduce the adverse effects of WAT and help to improve metabolic health [34,39]. In this study, we observed increases in the mRNA of CIDEA and PGC-1α, but not UCP1 and Prdm16 in eWAT and iWAT in mice fed the FX diets. The expression of UCP1 is transcriptionally regulated by the adrenergic signaling (β-AR, cAMP-dependent protein kinase A (PKA), *etc.*) coupled to PGC-1α and PPAR [40]. PGC-1α has been shown to be required for exercise-induced UCP1 expression in WAT [41]. Prdm16, by co-activating PGC-1α increased mitochondrial content, as well as enhanced uncoupled respiration. Prdm16 activates a robust brown fat phenotype, including the induction of PGC-1α, UCP1 and Dio2 [42–44]. In this study, although UCP1 mRNA in these WATs of FX-fed mice was slightly higher than that of the respective controls, the difference did not reach statistical significance, due to the low number of animals per group ($n = 4$), as well as the large individual variations. Like our study, Woo *et al.* did not observe increased UCP1 in eWAT of mice fed the 0.2% FX diet, although mice fed the 0.05% FX diet did show an increase of UCP1 in eWAT [31]. CIDEA has been shown to inhibit uncoupling activity of UCP1 when co-expressed in yeast and increases mitochondrial coupling by suppressing UCP1 expression [45]. Therefore, the unchanged UCP1 mRNA level in this study could also be related to the dose of FX and the elevated CIDEA mRNA expression in WAT. Induction of CIDEA has been used as a marker for the emergence of brown adipocyte-like cells in WAT [46]. As we found that only two of the four "beige" adipocyte-specific genes were upregulated, whether FX induces the whole program of WAT "browning" cannot be confirmed.

On the other hand, we observed that β3-AR, Dio2, PGC-1α, ERRα and PPARα mRNA in both eWAT and iWAT and HSL in eWAT were elevated by FX. The adrenergic signaling (β3-AR, PKA, *etc.*) coupled to PGC-1α and PPAR regulates UCP1 expression [40]. PGC-1α also controls mitochondria biogenesis and respiratory function through targeting multiple transcription factors, like NRF1, NRF2 and ERRα. β3-AR is expressed abundantly and predominantly in brown and white adipocytes. Treatment of mice with β3-AR agonists increases oxygen consumption [47]. Under β3-AR stimulation, PKA is activated and further turns on the transcription of Dio2 and PGC-1α and phosphorylation of HSL [48,49]. Dio2 catalyzes the conversion of T4 to T3 (the active form of thyroid hormone) in most peripheral tissues and is essential for the adrenergic receptor mediated thermogenesis in BAT [50]. T3 increases oxygen consumption, which is mediated through PGC-1$\alpha$ and NRF1 [51]. In this study, increased Dio2 mRNA in WATs is speculated to increase local T3 content and further stimulate PGC-1$\alpha$ mRNA expression and the metabolic rate of mice.

Fucoxanthinol and amarouciaxanthin A are two major metabolites of FX in mouse plasma and liver [52]. In mouse adipose tissue, amarouciaxanthin A is the major metabolite of FX [53]. *In vitro*, amarouciaxanthin A showed a higher activity in suppressing 3T3-L1 adipocyte differentiation than FX, fucoxanthinol and amarouciaxanthin B [54]. The suppression is associated with downregulations of PPAR$\gamma$ and CCAAT-enhancer-binding protein $\alpha$ (C/EBP$\alpha$) mRNA levels. In the present *in vivo* study, however, PPAR$\gamma$ mRNA expression in eWAT was upregulated by FX, implying that inhibition of adipocyte differentiation might have a minor role on the low WAT mass observed in our *in vivo* study.

**Figure 3.** Thermogenic gene expressions in epididymal (**A**) and inguinal (**B**) white adipose tissue. Values are means, and error bars are SEM ($n$ = 4). * denotes significant effect by either dietary factor at $p < 0.05$ analyzed by two-way ANOVA. HS, HS + F, HF and HF + F: as indicated in Figure 1. The measurement of relative mRNA expression is as indicated in Figure 2.

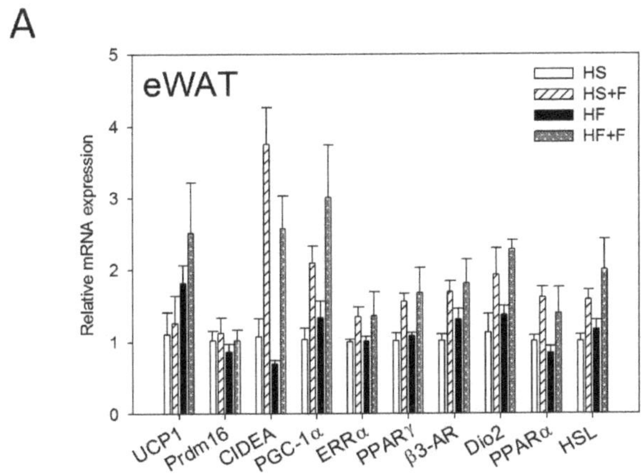


| $p$ values | UCP1 | Prdm16 | CIDEA | PGC-1$\alpha$ | ERR$\alpha$ | PPAR$\gamma$ | $\beta$3-AR | Dio2 | PPAR$\alpha$ | HSL |
|---|---|---|---|---|---|---|---|---|---|---|
| Diet | 0.0501 | 0.4106 | 0.0462 * | 0.1888 | 0.4433 | 0.6824 | 0.3115 | 0.2471 | 0.1837 | 0.2784 |
| FX | 0.4079 | 0.4087 | <0.0001 * | 0.0018 * | 0.0426 * | 0.0054 * | 0.0109 * | 0.0044 * | 0.0082 * | 0.0053 * |
| Diet * FX | 0.5842 | 0.8328 | 0.9069 | 0.8885 | 0.5496 | 0.7961 | 0.6489 | 0.8222 | 0.9477 | 0.9359 |

B

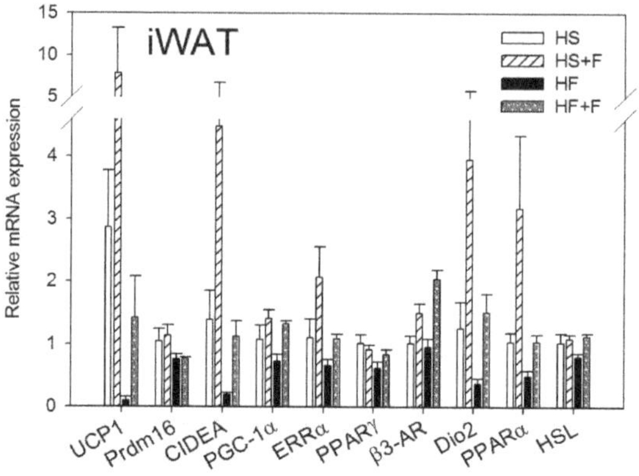


| *p* values | UCP1 | Prdm16 | CIDEA | PGC-1α | ERRα | PPARγ | β3-AR | Dio2 | PPARα | HSL |
|---|---|---|---|---|---|---|---|---|---|---|
| **Diet** | 0.0613 | 0.0160 * | 0.0057 * | 0.1721 | 0.0183 * | 0.0280 * | 0.1056 | 0.0397 * | 0.0016 * | 0.3091 |
| **FX** | 0.1014 | 0.6004 | 0.0015 * | 0.0080 * | 0.0092 * | 0.5649 | <0.0001 * | 0.0016 * | 0.0013 * | 0.0518 |
| **Diet * FX** | 0.1545 | 0.7362 | 0.2874 | 0.4150 | 0.7144 | 0.1186 | 0.0506 | 0.9451 | 0.6116 | 0.1610 |

### *2.5. FX Increased Mitochondrial Biogenesis and Fusion-Related Gene Expressions in eWAT and iWAT*

FX increased NRF1 and NRF2 mRNA expressions in eWAT, but not iWAT. Diet did not affect TFAM, NRF1 and NRF2 mRNA expressions in eWAT and iWAT (Figure 4). The high fat diet decreased Mfn1, Mfn2, OPA1 and Fis1 mRNA levels in iWAT, but not eWAT. In contrast, FX increased mitochondrial fusion-related genes, including Mfn1, Mfn2 and OPA1, but not fission-related gene Fis1 and mRNA expressions in both eWAT and iWAT (Figure 5). Mitochondria provide ~90% of the cellular energy supply and are also the most important organelle of metabolism. PGC-1α is a strong promoter of mitochondrial biogenesis and oxidative metabolism through NRF1, NRF2, ERRα, PPARγ and PPARα [13–15]. PGC-1α targets NRF1 and NRF2 directly or indirectly through ERRα in stimulating nuclear genes required for mitochondrial biogenesis and respiration function [14]. PGC-1α interacts with PPARα in the transcription control of genes encoding mitochondrial fatty acid oxidation enzymes [55]. PGC-1α acts as a partner of PPARγ in the induction of adaptive thermogenesis in brown fat and adipocyte differentiation [14]. The increased metabolic rate of mice fed the FX diets in this study coincides with increased PGC-1α networks. ERRα further activates the transcriptional activity of the Mfn2 promoter, and the effect was synergic with those of PGC-1α [16]. In addition to Mfn2, other mitochondrial fusion-related genes (Mfn1 and OPA1) were also increased by FX in this study. In contrast, mitochondrial fission-related gene Fis1 in WATs was not affected by FX. Mitochondrial fusion is frequently found in metabolically active cells [56], and elevated mitochondrial fusion genes agreed with enhanced $VO_2$ and $VCO_2$ and the decreased WAT mass of mice in the present study.

Among these genes, Mfn2 was shown to be stimulated under cold, adrenergic agonist or exercise treatment through the regulation of PGC-1$\alpha$ and ERR$\alpha$ [16,57]. Therefore, the elevated mitochondrial fusion genes by FX were in accordance with increased PGC-1$\alpha$ and ERR$\alpha$ mRNA in WATs.

### *2.6. Overall Discussion*

Using the 2 × 2 factorial design, we demonstrated that irrespective of the diet, FX increased both $O_2$ consumption and $CO_2$ production, indicating that FX enhanced metabolic rate. This increase in energy expenditure agreed with smaller adipose mass and increases of the PGC-1$\alpha$ network gene expressions in iWAT and eWAT. These included β3-AR, PGC-1$\alpha$, CIDEA, Dio2, PPAR$\alpha$ and genes regulating mitochondria biogenesis and fusion (ERR$\alpha$, NRF1, NRF2, Mfn1, Mfn2 and OPA1). We speculate that dietary FX might increase the energy expenditure through changes in mitochondria biogenesis, homeostasis and/or activity mediated by the PGC-1$\alpha$ network in eWAT and iWAT.

PGC-1$\alpha$ is a master regulator of energy metabolism that orchestrates cellular responses to various types of metabolic stress, such as fasting, cold temperature and physical exertion. The expression of PGC-1$\alpha$ is induced by such factors as β3-AR and Dio2. It in turn activates the expressions of transcription factors, including PPARs, NRF1/2, ERR$\alpha$ and Mfn2. In this study, FX upregulated PGC-1$\alpha$ and both of its upstream and downstream genes in iWAT and eWAT, implying that the "PGC-1$\alpha$ network" is upregulated. As this network is known to regulate energy metabolism, it coincides with our observation that FX increased energy expenditure.

**Figure 4.** Mitochondrial biogenic gene expressions in epididymal (**A**) and inguinal (**B**) white adipose tissue. Values are means, and error bars are SEM ($n$ = 4). * denotes significant effect by either dietary factor at $p < 0.05$ analyzed by two-way ANOVA. HS, HS + F, HF and HF + F: As indicated in Figure 1. The measurement of relative mRNA expression is as indicated in Figure 2.

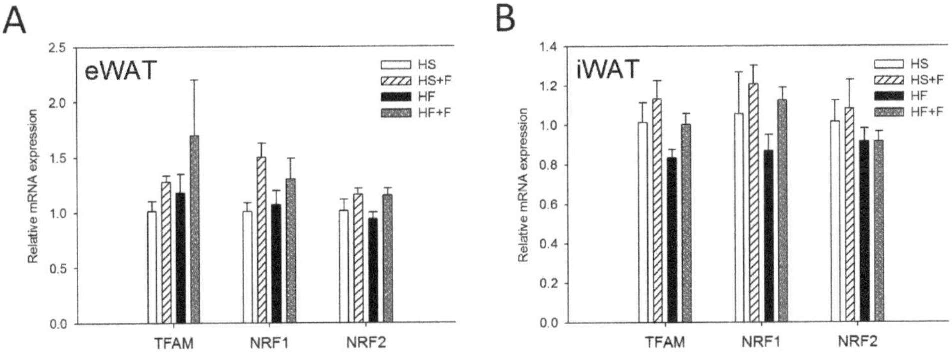


| *p* values | TFAM | NRF1 | NRF2 |
|---|---|---|---|

| *p* values | TFAM | NRF1 | NRF2 |
|---|---|---|---|

| Diet | 0.3522 | 0.6200 | 0.5750 |
|---|---|---|---|
| FX | 0.0886 | 0.0229 * | 0.0340 * |
| Diet * FX | 0.9868 | 0.3666 | 0.6506 |

| Diet | 0.0672 | 0.3140 | 0.2352 |
|---|---|---|---|
| FX | 0.0890 | 0.1362 | 0.7799 |
| Diet * FX | 0.7627 | 0.6892 | 0.8213 |

**Figure 5.** Mitochondrial homeostatic gene expressions in epididymal (**A**) and inguinal (**B**) white adipose tissue. Values are means, and error bars are SEM ($n$ = 4). * denotes significant effect by either dietary factor at $p < 0.05$ analyzed by two-way ANOVA. HS, HS + F, HF and HF + F: As indicated in Figure 1. The measurement of relative mRNA expression is as indicated in Figure 2.

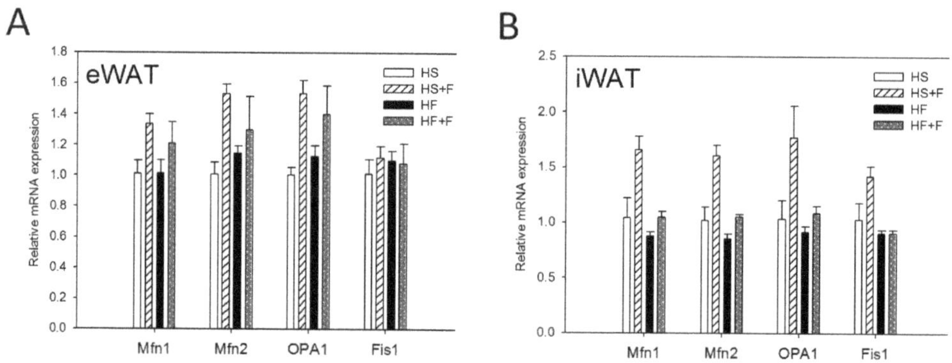


| $p$ values | Mfn1 | Mfn2 | OPA1 | Fis1 |
|---|---|---|---|---|
| Diet | 0.5359 | 0.6993 | 0.9566 | 0.7986 |
| FX | 0.0207 * | 0.0161 * | 0.0037 * | 0.6620 |
| Diet * FX | 0.5162 | 0.1534 | 0.2755 | 0.5300 |

| $p$ values | Mfn1 | Mfn2 | OPA1 | Fis1 |
|---|---|---|---|---|
| Diet | 0.0090 * | 0.0010 * | 0.0367 * | 0.0121 * |
| FX | 0.0040 * | 0.0003 * | 0.0070 * | 0.3346 |
| Diet * FX | 0.1166 | 0.0771 | 0.1624 | 0.2077 |

As we only measured the whole body $O_2$ consumption, it is difficult to specify the contribution of different tissues by FX. However, comparison of changes in the gene expression pattern in WATs and BAT prompt us to speculate that WAT might contribute more than BAT in promoting $O_2$ consumption.

Moreover, our data together with data of previous reports did not support a major role of BAT in the FX enhanced oxygen consumption. These include: (1) FX downregulated β3-AR, Mfn2, Fis1, Prdm16 and PPARδ mRNA levels in the BAT and did not change other thermogenic genes in this study; (2) some studies [31,38] found an anti-obesity effect of FX without BAT enlargement; (3) oxygen consumption did not correlate to BAT mass in our data (Supplementary Information, Figure S1); (4) FX lowered both serum NE (norepinephrine) and BAT β3-AR expression, which are key regulators of BAT adaptive thermogenesis. Lowered serum NE might be related to a decreased sympathetic activity or an increased NE degradation. It is not known why and how FX lowered serum NE and downregulated BAT β3-AR expression. Although it might not be applicable to FX, there has been a study showing that beta-carotene suppresses exhaustive exercise-induced plasma levels of adrenocorticotropic hormone, norepinephrine and epinephrine by inhibiting the secretion of corticotropin-releasing hormone [58]. Although NE concentration is very important for the activation of PGC-1alpha networks in whole body, factors other than NE, such as PPARs and thyroid receptor activation, *etc.*, can also lead to the activation of the PGC-1$\alpha$ network.

In this study, FX did not significantly increase UCP1 mRNA expression in iWAT, eWAT and BAT ($p > 0.05$). In contrast, expressions of CIDEA and the PGC-1$\alpha$ network in eWAT and iWAT were upregulated, although to different extents. These changes were not noted in liver and skeletal muscle. Moreover, differential extents of induction were also observed in other study. For example, mRNA expression levels of adipogenic marker genes (PPAR$\gamma$, aP2, adiponectin, C/EBP$\alpha$, FATP1, LPL and UCP2) and thermogenic genes and mitochondrial biogenesis genes (UCP1, PGC-1$\alpha$, NRF1, TFAM, Prdm16, CIDEA and Elovl3) induced by triiodothyronine (T3) were also to different extents [59].

## 3. Experimental Section

### 3.1. Preparation of FX

FX was isolated from the dried *Hincksia mitchellae* (Harvey) P. C. Silva brown algae. The brown algae, *Hincksia mitchellae* (Harvey) P. C. Silva, was originally collected from the Taiwan coast and characterized. The material used in this study was obtained by cultivation in enriched seawater medium (SWM-III) in the lab. The dried powder was extracted with acetone in a brown bottle for 24 h, filtered (No. 2, Whatman filter paper, Maidstone, Kent, UK) and the solvent removed, with a temperature lower than 35 °C. The crude extracts were separated by using a silica gel flash column chromatography (Geduran® Si 60, 0.040–0.063 mm, Merck, Darmstadt, Germany), eluted with ethyl acetate/*n*-hexane (4.5:5.5, v/v). The red-orange color fraction was collected and passed through a second silica gel column eluted with ethanol/acetone/*n*-hexane (1:19:80, v/v). After removing the solvent, the residue was re-dissolved in acetone/*n*-hexane (45:55, v/v). An equal volume of Millipore-Q water was added, and the mixture was standing at −20 °C for 4 h until a red precipitate was produced. The precipitate was filtered and dried in a freeze drier. The obtained solid was confirmed as FX by H-NMR and the purity checked by HPLC (PU-980 pump and UV-visible absorbance detector, Jasco, Japan and LiChrospher® 100 RP-18 column, 5 μm, Merck, Darmstadt, Germany). The mobile phase used was methanol/$H_2O$ (90:10, v/v) with a flow rate of 1 mL/min. The FX was detected at 450 nm and quantified from the peak area by using a standard curve with previous purified and identified FX (95%). The purity of FX prepared was >95% by this HPLC analysis.

### 3.2. Animals and Diets

The animal study was approved by the Institution Animal Care and Use Committee of National Taiwan University (No.NTU-98-EL-106). Three-week old male C57BL/6J mice were purchased from the National Laboratory Animal Center (Taipei, Taiwan) and housed individually in stainless steel wire cages in an animal room with a 12-hr light, 12-hr dark cycle (light period: 0800–2000) and constant temperature (22 ± 2 °C). Mice were fed a non-purified diet (LabDiet® 5001, PMI® Nutrition International Inc., Brentwood, MO, USA) for 1 week and switched to the high sucrose (HS, Table 1) diet for another week before the experiment. After the 2-week acclimation, mice were randomly assigned into four groups (4 mice/group) and respectively fed: HS (50% sucrose, 7% soybean oil), HF (high-fat diet, 23% butter plus 7% soybean oil), HS + F (HS diet supplemented with 0.2% FX) or HF + F (HF diet supplemented with 0.2% FX) test diets for 5 weeks. The composition of the 4 test diets was modified from AIN-93G [37] and our previous studies [60,61] (Table 1). The amounts of casein, cellulose, vitamin and mineral mixtures in the high-fat diets were adjusted to make the nutrient/energy ratios equivalent to those of the HS diet. Mice had free access to the diet and water. Body weight and food intake were recorded weekly.

### 3.3. Metabolic Rate Measurement

After feeding test diets for 2 weeks, mice were placed in the respiratory chamber individually (AccuScan Instruments, Inc., Columbus, OH, USA) with free access to their respective diet food and water. After a 24-h acclimation, mice were continuously monitored in the metabolic chambers for 6 days. Gas samples were collected every minute per mouse, and the data were averaged for each hour. Output parameters from the average of 6 days included the $O_2$ consumption ($VO_2$), $CO_2$ production ($VCO_2$) and respiratory quotients (RQ = $VO_2/VCO_2$).

### 3.4. Tissue and Serum Collection

After feeding test diets for 5 weeks, mice were deprived of diet for 12 h and killed by $CO_2$ asphyxiation. Blood was withdrawn through retro-orbital. Serum was obtained after centrifugation at 12,000 rpm for 10 min at 4 °C and stored at −80 °C. Liver, spleen, kidney, heart, lung, gastrocnemius muscle, retroperitoneal WAT (rWAT), eWAT, inguinal (iWAT) and intra-scapular BAT were excised, weighed and immediately frozen in liquid nitrogen and stored at −80 °C for mRNA analysis.

### 3.5. Housekeeping Genes PCR Array

rWAT and BAT total RNA (0.5 μg) of 2 mice per group were reversed transcribed using the RT² First Strand Kit and further analyzed by the Mouse Housekeeping Genes PCR array (Qiagen, Germantown, MD, USA), which analyzed 12 commonly used housekeeping genes in a 96-well plate. Briefly, 1 μL cDNA (1 ng/μL), 12.5 μL RT[2] SYBR Green Mastermix and 11.25 μL water were loaded into primer-precoated wells, and PCR was performed using a StepOnePlus Real-Time PCR System (Applied Biosystems, Foster City, CA, USA). Data were analyzed by three popular algorithms: GeNorm [62], NormFinder [63] and BestKeeper [64]. β-actin was shown to be the most stable and suitable reference gene in our study.

### *3.6. RNA Extraction and Quantitative Real-Time RT-PCR*

Total RNA of eWAT, rWAT, iWAT and BAT were isolated by an RNeasy® Plus Universal Mini Kit (*Qiagen*, Stockach, Germany), according to the manufacturer's instruction. Total RNA (2 μg) was reverse transcribed to cDNA using a High-Capacity cDNA Reverse Transcription kit (Applied Biosystems, Foster City, CA, USA). PCR was performed in a final volume of 25 μL containing 10 μL cDNA (1 ng/μL), 12.5 μL TaqMan® Gene Expression Master Mix, 1.25 μL probe/primer assay mix and 1.25 μL water. PCR primers and probes were purchased from Applied Biosystems: UCP1, CIDEA (cell death-inducing DFFA-like effecter a), Prdm16 (PR domain containing 16), PGC-1α, NRF1, NRF2, TFAM, ERRα, PPARα, PPARγ, HSL (hormone-sensitive lipase), β3-AR (β3-adrenergic receptor), Dio2 (deiodinase 2), Mfn1, Mfn2, OPA1 (optic atrophy 1), Fis1 (fission 1) and β-actin. These genes are associated with thermogenesis, mitochondrial biogenesis, browning of WAT and mitochondrial fusion and fission. The mRNA expression levels were determined by quantitative real-time RT-PCR using the StepOnePlus Real-Time PCR System (Applied Biosystems, Foster City, CA, USA). The expression levels were normalized to β-actin. Data were analyzed by StepOne Software (v2.2.2, Applied Biosystems, Foster City, CA, USA).

### *3.7. Serum Hormone Analysis*

Commercially available ELISA kits were used to measure serum thyroxine (T4) and triiodothyronine (T3) (Calbiotech, Spring Valley, CA, USA), norepinephrine and epinephrine (Labor Diagnostika Nord, Nordhorn, Germany), as well as corticosterone (Cayman chemical company, Ann Arbor, MI, USA).

### *3.8. Statistical Analysis*

Data are expressed as the means ± SD or SEM. Statistical analysis were performed using SAS 9.1 (SAS Institute Inc., Cary, NC, USA). For all statistical analyses, data were transformed logarithmically if the variances were non-homogeneous. In order to examine the effect of diet, FX and their interaction in this 2 × 2 factorial design, two-way ANOVAs were used. The effect is considered significant if $p < 0.05$. When an interaction ($p < 0.05$) existed between diet and FX, data were further analyzed by one-way ANOVA and Duncan's multiple range test. In the metabolic rate study, differences of $VO_2$, $VCO_2$ and RQ at each time point between HS and HS + F and between HF and HF + F were respectively analyzed by the Student's *t*-test. Areas under curve (AUCs) of $VO_2$ and $VCO_2$ were analyzed by two-way ANOVA. The AUC of RQ was analyzed by the Wilcoxon rank-sum test.

## 4. Conclusions

In conclusion, we demonstrated that dietary FX elevated the metabolic rate and reduced both subcutaneous and abdominal WAT mass irrespective of the diet used. These were associated with upregulated genes of the PGC-1$\alpha$ network, including those regulating mitochondrial biogenesis and fusion in iWAT and eWAT.

## Acknowledgments

This study was supported by a grant from the National Science Council of Taiwan (NSC 99-2313-B-002-013-MY3).

## Conflicts of Interest

The authors declare no conflict of interest.

## References

1. Malik, V.S.; Willett, W.C.; Hu, F.B. Global obesity: Trends, risk factors and policy implications. *Nat. Rev. Endocrinol.* **2013**, *9*, 13–27.
2. Medina-Gomez, G. Mitochondria and endocrine function of adipose tissue. *Best Pract. Res. Clin. Endocrinol. Metab.* **2012**, *26*, 791–804.
3. Bach, D.; Pich, S.; Soriano, F.X.; Vega, N.; Baumgartner, B.; Oriola, J.; Daugaard, J.R.; Lloberas, J.; Camps, M.; Zierath, J.R.; *et al.* Mitofusin-2 determines mitochondrial network architecture and mitochondrial metabolism. A novel regulatory mechanism altered in obesity. *J. Biol. Chem.* **2003**, *278*, 17190–17197.
4. Patti, M.E.; Corvera, S. The role of mitochondria in the pathogenesis of type 2 diabetes. *Endocr. Rev.* **2010**, *31*, 364–395.
5. Lowell, B.B.; Shulman, G.I. Mitochondrial dysfunction and type 2 diabetes. *Science* **2005**, *307*, 384–387.
6. Liu, J.; Shen, W.; Zhao, B.; Wang, Y.; Wertz, K.; Weber, P.; Zhang, P. Targeting mitochondrial biogenesis for preventing and treating insulin resistance in diabetes and obesity: Hope from natural mitochondrial nutrients. *Adv. Drug Deliv. Rev.* **2009**, *61*, 1343–1352.
7. Joseph, A.M.; Joanisse, D.R.; Baillot, R.G.; Hood, D.A. Mitochondrial dysregulation in the pathogenesis of diabetes: Potential for mitochondrial biogenesis-mediated interventions. *Exp. Diabetes Res.* **2012**, *2012*, 642038.
8. Kusminski, C.M.; Scherer, P.E. Mitochondrial dysfunction in white adipose tissue. *Trends Endocrinol. Metab.* **2012**, *23*, 435–443.
9. Youle, R.J.; van der Bliek, A.M. Mitochondrial fission, fusion, and stress. *Science* **2012**, *337*, 1062–1065.
10. Liesa, M.; Shirihai, O.S. Mitochondrial dynamics in the regulation of nutrient utilization and energy expenditure. *Cell Metab.* **2013**, *17*, 491–506.

11. Bach, D.; Naon, D.; Pich, S.; Soriano, F.X.; Vega, N.; Rieusset, J.; Laville, M.; Guillet, C.; Boirie, Y.; Wallberg-Henriksson, H.; *et al.* Expression of Mfn2, the Charcot-Marie-Tooth neuropathy type 2A gene, in human skeletal muscle: Effects of type 2 diabetes, obesity, weight loss, and the regulatory role of tumor necrosis factor alpha and interleukin-6. *Diabetes* **2005**, *54*, 2685–2693.
12. Zorzano, A.; Liesa, M.; Palacin, M. Role of mitochondrial dynamics proteins in the pathophysiology of obesity and type 2 diabetes. *Int. J. Biochem. Cell Biol.* **2009**, *41*, 1846–1854.
13. Wu, Z.; Puigserver, P.; Andersson, U.; Zhang, C.; Adelmant, G.; Mootha, V.; Troy, A.; Cinti, S.; Lowell, B.; Scarpulla, R.C.; *et al.* Mechanisms controlling mitochondrial biogenesis and respiration through the thermogenic coactivator PGC-1. *Cell* **1999**, *98*, 115–124.
14. Scarpulla, R.C. Metabolic control of mitochondrial biogenesis through the PGC-1 family regulatory network. *Biochimi. Biophys. Acta* **2011**, *1813*, 1269–1278.
15. Lin, J.; Handschin, C.; Spiegelman, B.M. Metabolic control through the PGC-1 family of transcription coactivators. *Cell Metab.* **2005**, *1*, 361–370.
16. Soriano, F.X.; Liesa, M.; Bach, D.; Chan, D.C.; Palacin, M.; Zorzano, A. Evidence for a mitochondrial regulatory pathway defined by peroxisome proliferator-activated receptor-gamma coactivator-1 alpha, estrogen-related receptor-alpha, and mitofusin 2. *Diabetes* **2006**, *55*, 1783–1791.
17. Clayton, D.A. Replication and transcription of vertebrate mitochondrial DNA. *Annu. Rev. Cell Biol.* **1991**, *7*, 453–478.
18. Virbasius, J.V.; Scarpulla, R.C. Activation of the human mitochondrial transcription factor A gene by nuclear respiratory factors: A potential regulatory link between nuclear and mitochondrial gene expression in organelle biogenesis. *Proc. Natl. Acad. Sci. USA* **1994**, *91*, 1309–1313.
19. Semple, R.K.; Crowley, V.C.; Sewter, C.P.; Laudes, M.; Christodoulides, C.; Considine, R.V.; Vidal-Puig, A.; O'Rahilly, S. Expression of the thermogenic nuclear hormone receptor coactivator PGC-1alpha is reduced in the adipose tissue of morbidly obese subjects. *Int. J. Obes. Relat. Metab. Disord.* **2004**, *28*, 176–179.
20. Tiraby, C.; Tavernier, G.; Lefort, C.; Larrouy, D.; Bouillaud, F.; Ricquier, D.; Langin, D. Acquirement of brown fat cell features by human white adipocytes. *J. Biol. Chem.* **2003**, *278*, 33370–33376.
21. Haugan, J.A.; Aakermann, T.; Liaaen-Jensen, S. Isolation of fucoxanthin and peridinin. *Methods Enzymol.* **1992**, *213*, 231–245.
22. Okada, T.; Mizuno, Y.; Sibayama, S.; Hosokawa, M.; Miyashita, K. Antiobesity Effects of Undaria Lipid Capsules Prepared with Scallop Phospholipids. *J. Food Sci.* **2011**, *76*, H2–H6.
23. Maeda, H.; Hosokawa, M.; Sashima, T.; Funayama, K.; Miyashita, K. Effect of medium-chain triacylglycerols on anti-obesity effect of fucoxanthin. *J. Oleo Sci.* **2007**, *56*, 615–621.

24. Maeda, H.; Hosokawa, M.; Sashima, T.; Miyashita, K. Dietary combination of fucoxanthin and fish oil attenuates the weight gain of white adipose tissue and decreases blood glucose in obese/diabetic KK-Ay mice. *J. Agric. Food Chem.* **2007**, *55*, 7701–7706.
25. Maeda, H.; Hosokawa, M.; Sashima, T.; Funayama, K.; Miyashita, K. Fucoxanthin from edible seaweed, Undaria pinnatifida, shows antiobesity effect through UCP1 expression in white adipose tissues. *Biochem. Biophys. Res. Commun.* **2005**, *332*, 392–397.
26. Maeda, H.; Tsukui, T.; Sashima, T.; Hosokawa, M.; Miyashita, K. Seaweed carotenoid, fucoxanthin, as a multi-functional nutrient. *Asia Pac. J. Clin. Nutr.* **2008**, *17*, 196–199.
27. Maeda, H.; Hosokawa, M.; Sashima, T.; Murakami-Funayama, K.; Miyashita, K. Anti-obesity and anti-diabetic effects of fucoxanthin on diet-induced obesity conditions in a murine model. *Mol. Med. Rep.* **2009**, *2*, 897–902.
28. Jeon, S.M.; Kim, H.J.; Woo, M.N.; Lee, M.K.; Shin, Y.C.; Park, Y.B.; Choi, M.S. Fucoxanthin-rich seaweed extract suppresses body weight gain and improves lipid metabolism in high-fat-fed C57BL/6J mice. *Biotechnol. J.* **2010**, *5*, 961–969.
29. Park, H.J.; Lee, M.K.; Park, Y.B.; Shin, Y.C.; Choi, M.S. Beneficial effects of Undaria pinnatifida ethanol extract on diet-induced-insulin resistance in C57BL/6J mice. *Food Chem. Toxicol.* **2011**, *49*, 727–733.
30. Woo, M.N.; Jeon, S.M.; Kim, H.J.; Lee, M.K.; Shin, S.K.; Shin, Y.C.; Park, Y.B.; Choi, M.S. Fucoxanthin supplementation improves plasma and hepatic lipid metabolism and blood glucose concentration in high-fat fed C57BL/6N mice. *Chemico-Biol. Interact.* **2010**, *186*, 316–322.
31. Woo, M.N.; Jeon, S.M.; Shin, Y.C.; Lee, M.K.; Kang, M.A.; Choi, M.S. Anti-obese property of fucoxanthin is partly mediated by altering lipid-regulating enzymes and uncoupling proteins of visceral adipose tissue in mice. *Mol. Nutr. Food Res.* **2009**, *53*, 1–9.
32. Nishikawa, S.; Hosokawa, M.; Miyashita, K. Fucoxanthin promotes translocation and induction of glucose transporter 4 in skeletal muscles of diabetic/obese KK-A(y) mice. *Phytomedicine* **2012**, *19*, 389–394.
33. Wu, J.; Cohen, P.; Spiegelman, B.M. Adaptive thermogenesis in adipocytes: Is beige the new brown? *Genes Dev.* **2013**, *27*, 234–250.
34. Bartelt, A.; Heeren, J. Adipose tissue browning and metabolic health. *Nature Rev. Endocrinol.* **2014**, *10*, 24–36.
35. Yoneshiro, T.; Aita, S.; Matsushita, M.; Kayahara, T.; Kameya, T.; Kawai, Y.; Iwanaga, T.; Saito, M. Recruited brown adipose tissue as an antiobesity agent in humans. *J. Clin. Investig.* **2013**, *123*, 3404–3408.
36. Chan, D.C. Fusion and fission: interlinked processes critical for mitochondrial health. *Annu. Rev. Genet.* **2012**, *46*, 265–287.
37. Reeves, P.G.; Nielsen, F.H.; Fahey, G.C., Jr. AIN-93 purified diets for laboratory rodents: Final report of the American Institute of Nutrition ad hoc writing

committee on the reformulation of the AIN-76A rodent diet. *J. Nutr.* **1993**, *123*, 1939–1951.

38. Hu, X.; Li, Y.; Li, C.; Fu, Y.; Cai, F.; Chen, Q.; Li, D. Combination of fucoxanthin and conjugated linoleic acid attenuates body weight gain and improves lipid metabolism in high-fat diet-induced obese rats. *Arch. Biochemi. Biophys.* **2012**, *519*, 59–65.
39. Shan, T.; Liang, X.; Bi, P.; Zhang, P.; Liu, W.; Kuang, S. Distinct populations of adipogenic and myogenic Myf5-lineage progenitors in white adipose tissues. *J. Lipid Res.* **2013**, *54*, 2214–2224.
40. Collins, S.; Yehuda-Shnaidman, E.; Wang, H. Positive and negative control of Ucp1 gene transcription and the role of beta-adrenergic signaling networks. *Int. J. Obes.* **2010**, *34*, S28–S33.
41. Ringholm, S.; Grunnet Knudsen, J.; Leick, L.; Lundgaard, A.; Munk Nielsen, M.; Pilegaard, H. PGC-1alpha is required for exercise- and exercise training-induced UCP1 upregulation in mouse white adipose tissue. *PloS One* **2013**, *8*, e64123.
42. Kajimura, S.; Seale, P.; Tomaru, T.; Erdjument-Bromage, H.; Cooper, M.P.; Ruas, J.L.; Chin, S.; Tempst, P.; Lazar, M.A.; Spiegelman, B.M. Regulation of the brown and white fat gene programs through a PRDM16/CtBP transcriptional complex. *Genes Dev.* **2008**, *22*, 1397–1409.
43. Seale, P.; Kajimura, S.; Yang, W.; Chin, S.; Rohas, L.M.; Uldry, M.; Tavernier, G.; Langin, D.; Spiegelman, B.M. Transcriptional control of brown fat determination by PRDM16. *Cell Metab.* **2007**, *6*, 38–54.
44. Seale, P.; Conroe, H.M.; Estall, J.; Kajimura, S.; Frontini, A.; Ishibashi, J.; Cohen, P.; Cinti, S.; Spiegelman, B.M. Prdm16 determines the thermogenic program of subcutaneous white adipose tissue in mice. *J. Clin. Investig.* **2011**, *121*, 96–105.
45. Zhou, Z.; Yon Toh, S.; Chen, Z.; Guo, K.; Ng, C.P.; Ponniah, S.; Lin, S.C.; Hong, W.; Li, P. Cidea-deficient mice have lean phenotype and are resistant to obesity. *Nature Genet.* **2003**, *35*, 49–56.
46. Meyer, C.W.; Willershauser, M.; Jastroch, M.; Rourke, B.C.; Fromme, T.; Oelkrug, R.; Heldmaier, G.; Klingenspor, M. Adaptive thermogenesis and thermal conductance in wild-type and UCP1-KO mice. *Am. J. Physiol.* **2010**, *299*, R1396–R1406.
47. Susulic, V.S.; Frederich, R.C.; Lawitts, J.; Tozzo, E.; Kahn, B.B.; Harper, M.E.; Himms-Hagen, J.; Flier, J.S.; Lowell, B.B. Targeted disruption of the beta 3-adrenergic receptor gene. *J. Biol. Chem.* **1995**, *270*, 29483–29492.
48. Lowell, B.B.; Spiegelman, B.M. Towards a molecular understanding of adaptive thermogenesis. *Nature* **2000**, *404*, 652–660.
49. Mottillo, E.P.; Bloch, A.E.; Leff, T.; Granneman, J.G. Lipolytic products activate peroxisome proliferator-activated receptor (PPAR) alpha and delta in brown

adipocytes to match fatty acid oxidation with supply. *J. Biol. Chem.* **2012**, *287*, 25038–25048.
50. de Jesus, L.A.; Carvalho, S.D.; Ribeiro, M.O.; Schneider, M.; Kim, S.W.; Harney, J.W.; Larsen, P.R.; Bianco, A.C. The type 2 iodothyronine deiodinase is essential for adaptive thermogenesis in brown adipose tissue. *J. Clin. Investig.* **2001**, *108*, 1379–1385.
51. Weitzel, J.M.; Iwen, K.A.; Seitz, H.J. Regulation of mitochondrial biogenesis by thyroid hormone. *Exp. Physiol.* **2003**, *88*, 121–128.
52. Asai, A.; Sugawara, T.; Ono, H.; Nagao, A. Biotransformation of fucoxanthinol into amarouciaxanthin A in mice and HepG2 cells: Formation and cytotoxicity of fucoxanthin metabolites. *Drug Metab. Dispos.* **2004**, *32*, 205–211.
53. Yonekura, L.; Kobayashi, M.; Terasaki, M.; Nagao, A. Keto-carotenoids are the major metabolites of dietary lutein and fucoxanthin in mouse tissues. *J. Nutr.* **2010**, *140*, 1824–1831.
54. Yim, M.J.; Hosokawa, M.; Mizushina, Y.; Yoshida, H.; Saito, Y.; Miyashita, K. Suppressive Effects of Amarouciaxanthin A on 3T3-L1 Adipocyte Differentiation through Down-regulation of PPARgamma and C/EBPalpha mRNA Expression. *J. Agric. Food Chem.* **2011**, *59*, 1646–1652.
55. Vega, R.B.; Huss, J.M.; Kelly, D.P. The coactivator PGC-1 cooperates with peroxisome proliferator-activated receptor alpha in transcriptional control of nuclear genes encoding mitochondrial fatty acid oxidation enzymes. *Mol. Cell. Biol.* **2000**, *20*, 1868–1876.
56. Westermann, B. Molecular machinery of mitochondrial fusion and fission. *J. Biol. Chem.* **2008**, *283*, 13501–13505.
57. Cartoni, R.; Leger, B.; Hock, M.B.; Praz, M.; Crettenand, A.; Pich, S.; Ziltener, J.L.; Luthi, F.; Deriaz, O.; Zorzano, A.; *et al.* Mitofusins 1/2 and ERRalpha expression are increased in human skeletal muscle after physical exercise. *J. Phys.* **2005**, *567*, 349–358.
58. Hasegawa, T. Anti-stress effect of beta-carotene. *Ann. N. Y. Acad. Sci.* **1993**, *691*, 281–283.
59. Lee, J.Y.; Takahashi, N.; Yasubuchi, M.; Kim, Y.I.; Hashizaki, H.; Kim, M.J.; Sakamoto, T.; Goto, T.; Kawada, T. Triiodothyronine induces UCP-1 expression and mitochondrial biogenesis in human adipocytes. *Am. J. Physiol. Cell Physiol.* **2012**, *302*, C463–C472.
60. Hsu, S.C.; Huang, C.J. Reduced fat mass in rats fed a high oleic acid-rich safflower oil diet is associated with changes in expression of hepatic PPARalpha and adipose SREBP-1c-regulated genes. *J. Nutr.* **2006**, *136*, 1779–1785.
61. Lu, K.N.; Hsu, C.; Chang, M.L.; Huang, C.J. Wild bitter gourd increased metabolic rate and upregulated genes related to mitochondria biogenesis and UCP-1 in mice. *J. Funct. Foods* **2013**, *5*, 668–678.
62. Vandesompele, J.; De Preter, K.; Pattyn, F.; Poppe, B.; Van Roy, N.; De Paepe, A.; Speleman, F. Accurate normalization of real-time quantitative RT-PCR data

by geometric averaging of multiple internal control genes. *Genome Biol.* **2002**, *3*, doi:10.1186/gb-2002-3-7-research0034.
63. Andersen, C.L.; Jensen, J.L.; Orntoft, T.F. Normalization of real-time quantitative reverse transcription-PCR data: a model-based variance estimation approach to identify genes suited for normalization, applied to bladder and colon cancer data sets. *Cancer Res.* **2004**, *64*, 5245–5250.
64. Pfaffl, M.W.; Tichopad, A.; Prgomet, C.; Neuvians, T.P. Determination of stable housekeeping genes, differentially regulated target genes and sample integrity: BestKeeper—Excel-based tool using pair-wise correlations. *Biotechnol. Lett.* **2004**, *26*, 509–515.

*marine drugs*

MDPI

*Article*

# Fucoxanthin, a Marine Carotenoid, Reverses Scopolamine-Induced Cognitive Impairments in Mice and Inhibits Acetylcholinesterase *in Vitro*

**Jiajia Lin [1], Ling Huang [1], Jie Yu [1], Siying Xiang [1], Jialing Wang [1], Jinrong Zhang [2], Xiaojun Yan [2], Wei Cui [1,*], Shan He [2,*] and Qinwen Wang [1,*]**

1 Ningbo Key Laboratory of Behavioral Neuroscience, Zhejiang Provincial Key Laboratory of Pathophysiology, School of Medicine, Ningbo University, Ningbo 315211, China; linjiajiamed@yahoo.com (J.L.); nbhl1226@yahoo.com (L.H.); xiao.tai.yang_jie@163.com (J.Y.); xiangsi.ying@163.com (S.X.); wjl58452@yahoo.com (J.W.)

2 School of Marine Sciences, Ningbo University, Ningbo 315211, China; zhangjinrong@nbu.edu.cn (J.Z.); Yanxiaojun@nbu.edu.cn (X.Y.)

* Correspondence: cuiwei@nbu.edu.cn (W.C.); heshan@nbu.edu.cn (S.H.); wangqinwen@nbu.edu.cn (Q.W.); Tel.: +86-574-8760-9589 (W.C.); +86-574-8760-0458 (S.H.); +86-574-8760-8922 (Q.W.)

Academic Editors: Tatsuya Sugawara and Takashi Maoka

Received: 7 January 2016; Accepted: 21 March 2016; Published: 25 March 2016

**Abstract:** Fucoxanthin, a natural carotenoid abundant in edible brown seaweeds, has been shown to possess anti-cancer, anti-oxidant, anti-obesity and anti-diabetic effects. In this study, we report for the first time that fucoxanthin effectively protects against scopolamine-induced cognitive impairments in mice. In addition, fucoxanthin significantly reversed the scopolamine-induced increase of acetylcholinesterase (AChE) activity and decreased both choline acetyltransferase activity and brain-derived neurotrophic factor (BDNF) expression. Using an *in vitro* AChE activity assay, we discovered that fucoxanthin directly inhibits AChE with an $IC_{50}$ value of 81.2 μM. Molecular docking analysis suggests that fucoxanthin likely interacts with the peripheral anionic site within AChE, which is in accordance with enzymatic activity results showing that fucoxanthin inhibits AChE in a non-competitive manner. Based on our current findings, we anticipate that fucoxanthin might exhibit great therapeutic efficacy for the treatment of Alzheimer's disease by acting on multiple targets, including inhibiting AChE and increasing BDNF expression.

**Keywords:** fucoxanthin; scopolamine; acetylcholinesterase; Alzheimer's disease; cognitive impairments

## 1. Introduction

Alzheimer's disease (AD) is one of the most common neurodegenerative disorders leading to cognitive impairments [1]. Although the etiology of this disease is not precisely known, many factors, including cholinergic system dysfunction, β-amyloid (Aβ) deposits and oxidative stress, have been considered to play important roles in the occurrence and development of AD [2]. Currently, the primary treatment for AD, which only provides symptomatic relief, is a cholinergic replacement therapy, represented by four FDA-approved acetylcholinesterase (AChE) inhibitors, donepezil, galantamine and rivastigmine [3].

Scopolamine is an antagonist of the muscarinic acetylcholine receptor. Many studies have shown that a high dose of scopolamine impairs short-term learning and memory in animal models and humans [4–6]. Therefore, scopolamine has become widely accepted as an experimental model of AD and is used to screen for anti-amnesic drugs [7,8]. Scopolamine induces cognitive impairment associated with an attenuation of cholinergic neurotransmission, as well as increased oxidative stress and inflammation in the brain [9,10]. Therefore, drugs that can effectively enhance cholinergic transmission and/or reduce oxidative stress and inflammation might reverse scopolamine-induced cognitive impairments [11].

Marine carotenoids are widely present in both plants and animals. Many marine carotenoids, including fucoxanthin, astaxanthin and lutein, are reported to produce anti-oxidant, anti-inflammatory, anti-obesity and anti-diabetic effects [12,13]. Fucoxanthin is the most abundant marine carotenoid and contributes more than 10% of the total production of carotenoids in nature [13]. Fucoxanthin contains unique functional groups, including an unusual allenic bond and a 5,6-monoepoxide structure [14]. Previous studies have shown that fucoxanthin exhibits various health benefits [15–17]. For example, fucoxanthin in diet significantly reduced weight gain in experimental animals by increasing fatty acid oxidation and heart production in white adipose tissue [18]. Moreover, rats that were fed with fucoxanthin displayed decreased levels of oxidative stress markers and increased activities of antioxidant enzymes [19].

Interestingly, a recent study has also shown that fucoxanthin can ameliorate Aβ-induced oxidative stress in microglia cells, suggesting that fucoxanthin might be useful for the treatment of AD [20]. In this study, we first evaluated the effects of fucoxanthin on scopolamine-induced cognitive impairments in mice. We further examined if fucoxanthin could directly inhibit AChE *in vitro*. Finally, we investigated the molecular basis of the interaction between fucoxanthin and AChE by molecular docking simulation.

## 2. Results

### *2.1. Fucoxanthin Does Not Affect Locomotor Activity in the 5-min Open-Field Test*

To examine whether fucoxanthin could reverse cognitive impairments in an animal model, we used the scopolamine-induced cognitive impairment mouse model and evaluated its effects with the open-field, novel object recognition (NOR) and Morris water maze tests. The schedule of animal experiments is shown in Figure 1. Drugs were given 30 min prior to the test each day. Locomotor activity was examined by testing the number of line crossings and rearings in the open-field test for 5 min. As shown in Figure 2, none of the treatments significantly altered the number of line crossings or rearings (for line crossing, one-way ANOVA, $F(5, 42) = 0.784$, $p > 0.05$; for rearing, $F(5, 42) = 1.443$, $p > 0.05$).

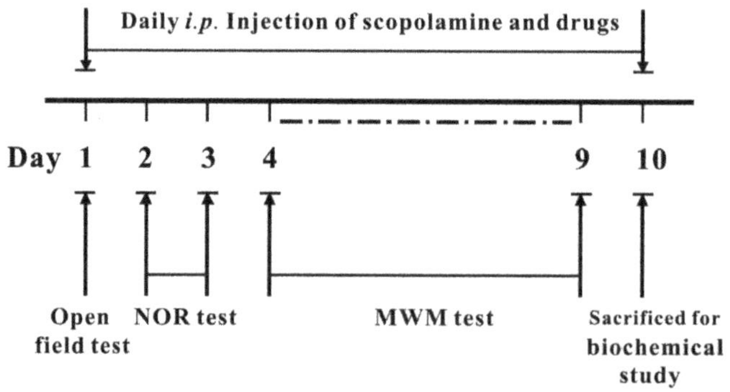


**Figure 1.** Experimental design and schedule of animal tests. NOR, novel object recognition; MWM: Morris water maze.

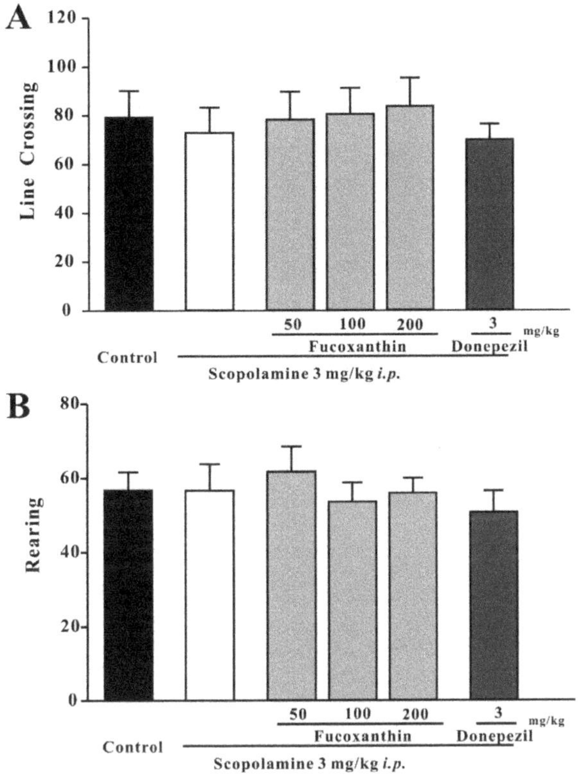


**Figure 2.** Treatments do not affect locomotor activity in the 5-min open field tests. The number of line crossings and rearings in the open field test are shown in (**A**) and (**B**), respectively. Data are expressed as the mean ± SEM ($n = 8$).

## 2.2. Fucoxanthin Reverses Scopolamine-Induced Recognition Impairment in the NOR Test

The NOR test was used to evaluate whether fucoxanthin could reverse scopolamine-induced recognition impairment. On the first day of the NOR test, mice were habituated to the experimental arena without any behaviorally-relevant stimulus. The training session was conducted one day after the habituation session. The exploration time of two identical objects was recorded in the training sessions 30 min after drug treatments. In this session, all groups exhibited similar total exploring times and recognition indexes for both objects (for total exploring times, one-way ANOVA, $F(5, 42) = 0.672$, $p > 0.05$; for the recognition index, $F(5, 42) = 1.513$, $p > 0.05$; Figure 3).

The retention session was conducted one day after the training session. Although the total exploration time was similar among groups (one-way ANOVA, $F(5, 42) = 0.434$, $p > 0.05$; Figure 4A), the recognition index for the novel object was significantly changed (one-way ANOVA, $F(5, 42) = 12.966$, $p < 0.01$; Figure 4B). The recognition index in the control mice was significantly higher than that in the scopolamine-induced mice (Tukey's test, $p < 0.01$). Moreover, treatment with fucoxanthin or donepezil significantly reversed the scopolamine-induced decrease of the recognition index (Tukey's test, $p < 0.01$).

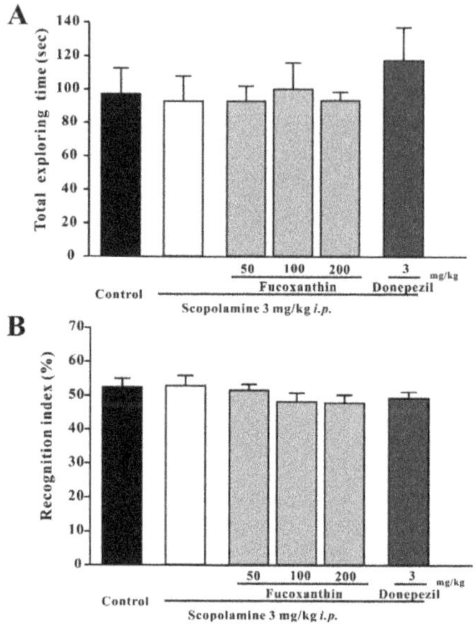


**Figure 3.** In the training session, all groups were found to possess similar total exploring times and recognition indexes for two identical objects. The total exploring times and recognition indexes in the training session are shown in (**A**) and (**B**), respectively. Data are expressed as the mean ± SEM ($n$ = 8).

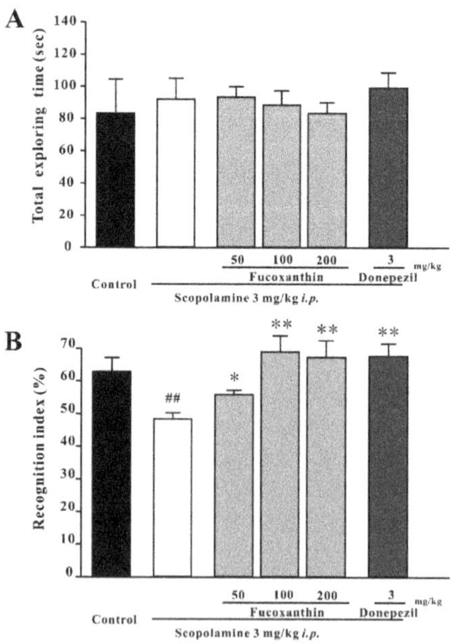


**Figure 4.** Fucoxanthin reverses scopolamine-induced recognition impairments in mice. The total exploring time and recognition index in the retention session are shown in (**A**) and (**B**), respectively. Data are expressed as the mean ± SEM ($n$ = 8; ## $p < 0.01$ *versus* the control group, * $p < 0.05$ and ** $p < 0.01$ *versus* the scopolamine-treated group (one-way ANOVA and Tukey's test)).

### 2.3. Fucoxanthin Reverses Scopolamine-Induced Spatial Learning and Memory Impairments in the Morris Water Maze Test

We further studied whether fucoxanthin could reverse scopolamine-induced spatial learning and memory impairments using the Morris water maze. Two-way repeated-measures ANOVA revealed significant changes in treatment effect (two-way ANOVA $F(5, 168) = 6.333$, $p < 0.01$) and time effect ($F(3, 168) = 9.211$, $p < 0.01$), but not for the treatment × time interaction ($F(15, 168) = 0.717$, $p > 0.05$; Figure 5A). The performance of mice in all groups improved throughout the training session, as indicated by the shortened escape latency during this period (Figure 5A). The scopolamine-treated group took a significantly longer time to find the platform on the last two days of training when compared to the control group, suggesting that scopolamine causes spatial learning impairments ($p < 0.01$; Figure 5). Fucoxanthin at medium and high doses significantly decreased the scopolamine-induced increase of the mean latency on the third and fourth day of the training session ($p < 0.05$; Figure 5). Under the same conditions, donepezil also reserved the scopolamine-elevated mean latency on the last two days of training ($p < 0.05$; Figure 5).

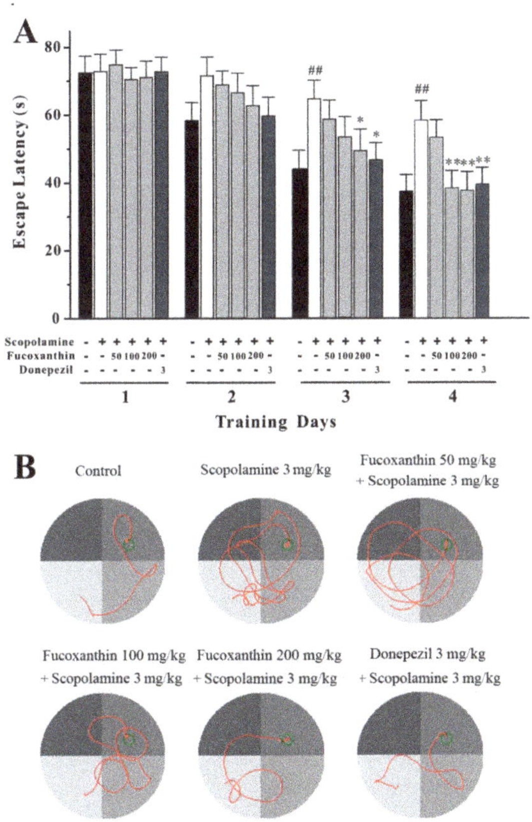


**Figure 5.** Fucoxanthin reverses scopolamine-induced spatial learning and memory deficits in the training session of the Morris water maze task. (**A**) Mean latencies to escape from the water onto the hidden platform in training trials. Each mouse was subjected to four trials per day for four consecutive days. Data are expressed as the mean ± SEM ($n = 8$); ## $p < 0.01$ *versus* the control group, * $p < 0.05$ and ** $p < 0.01$ *versus* the scopolamine-treated group (two-way repeated-measures ANOVA and LSD test). (**B**) Representative swimming-tracking paths of various groups as indicated in the training session.

In the probe trial, the time spent in the target quadrant was significantly different among groups (one-way ANOVA, $F(5, 42) = 13.415$, $p < 0.01$; Figure 6). Scopolamine significantly decreased the time spent in the target quadrant when compared to the control group (Tukey's test, $p < 0.01$; Figure 6). However, fucoxanthin (100 and 200 mg/kg) or donepezil caused a significant increase in the time spent in the target quadrant when compared to scopolamine alone (Tukey's test, $p < 0.01$; Figure 6).

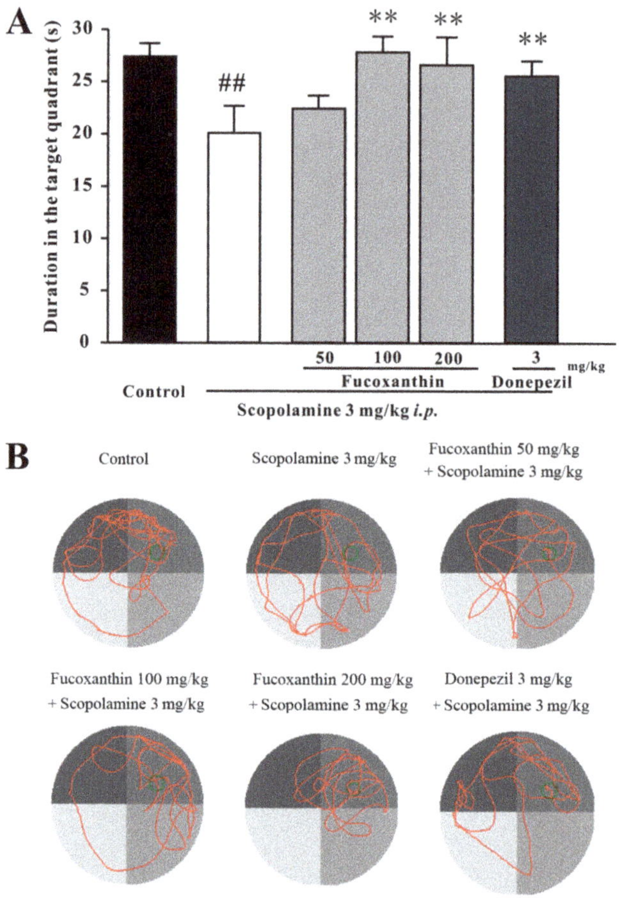


**Figure 6.** Fucoxanthin reverses scopolamine-induced spatial learning and memory deficits in the probe trial of the Morris water maze task. (**A**) The swimming time in the target quadrant (in which the platform had been placed during the training session) in the probe trial (swimming 90 s without platform). Data are expressed as the mean ± SEM ($n = 8$); ## $p < 0.01$ *versus* the control group, ** $p < 0.01$ *versus* the scopolamine-treated group (ANOVA and Tukey's test). (**B**) Representative swimming-tracking paths of various groups as indicated in the probe trials are demonstrated.

## *2.4. Fucoxanthin Increases the Expression of BDNF in the Hippocampus and Cortex of Mice*

Brain-derived neurotrophic factor (BDNF) levels in the brains of mice were detected by Western blot analysis. On the 10th day of treatment, the mice were sacrificed, and the hippocampus and cortex were dissected. BDNF levels of the control group were significantly higher than those of the scopolamine-treated group in both the hippocampus and cortex (Tukey's test, $p < 0.05$; Figure 7). Moreover, fucoxanthin significantly increased the expression of BDNF in the hippocampus and cortex (Tukey's test, $p < 0.05$; Figure 7). Under the same conditions, donepezil also reversed the scopolamine-induced decrease of BDNF expression (Tukey's test, $p < 0.05$; Figure 7).

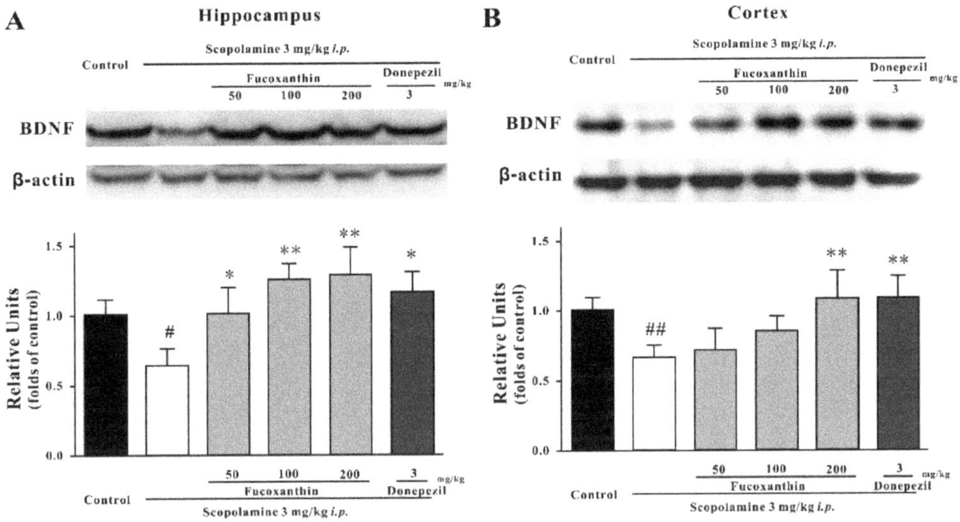


**Figure 7.** Fucoxanthin significantly reverses the scopolamine-induced decrease in the expression of BDNF in mice. BDNF expression in (**A**) the hippocampus and (**B**) the cortex of mice was detected by Western blotting. Data are expressed as the mean ± SEM ($n = 3$); # $p < 0.05$ and ## $p < 0.01$ *versus* the control group, * $p < 0.05$ and ** $p < 0.01$ *versus* the scopolamine-treated group (ANOVA and Tukey's test).

### 2.5. Fucoxanthin Increases ChAT Activity and Decreases AChE Activity in the Hippocampus and Cortex of Mice

Choline acetyltransferase (ChAT) and AChE activities were also measured on the 10th day of treatment. As shown in Table 1, compared to the control group, ChAT and AChE activities in both the hippocampus and the cortex were significantly changed in the scopolamine-induced groups (Tukey's test, $p < 0.05$; Tables 1 and 2). Notably, both fucoxanthin (100 and 200 mg/kg) and donepezil significantly reduced the scopolamine-induced increase of AChE activity in the hippocampus and cortex (Tukey's test, $p < 0.01$; Table 1). Moreover, fucoxanthin (100 and 200 mg/kg) and donepezil also significantly reversed the scopolamine-induced decrease of ChAT activity in the hippocampus and cortex (Tukey's test, $p < 0.05$; Table 2).

**Table 1.** Fucoxanthin decreases AChE activity in the hippocampus and cortex of scopolamine-treated mice.

| Brain Region | AChE Activity | | | | | |
|---|---|---|---|---|---|---|
| | Control | Scopolamine 3 mg/kg *i.p.* | | | | |
| | | Vehicle | Fucoxanthin (mg/kg) | | | Donepezil (mg/kg) |
| | | | 50 | 100 | 200 | 3 |
| Hippocampus | 60.3 ± 2.7 | 80.2 ± 2.3 ## | 76.2 ± 4.0 | 68.1 ± 3.2 ** | 67.6 ± 4.3 ** | 62.4 ± 2.6 ** |
| Cortex | 52.1 ± 2.8 | 71.2 ± 1.3 ## | 60.5 ± 3.0** | 62.4 ± 2.9 ** | 57.4 ± 3.2 ** | 61.6 ± 1.4 ** |

Data, expressed as the mean ± SEM ($n = 8$), are nanomoles of acetylcholine degraded/milligram protein/hour. ## $p < 0.01$ *versus* the control group, ** $p < 0.01$ *versus* the scopolamine-treated group (ANOVA and Tukey's test).

**Table 2.** Fucoxanthin increases ChAT activity in the hippocampus and cortex of scopolamine-treated mice.

| Brain Region | ChAT Activity | | | | | |
|---|---|---|---|---|---|---|
| | Control | Scopolamine 3 mg/kg *i.p.* | | | | |
| | | Vehicle | Fucoxanthin (mg/kg) | | | Donepezil (mg/kg) |
| | | | 50 | 100 | 200 | 3 |
| Hippocampus | 32.8 ± 1.1 | 27.6 ± 0.8 # | 30.0 ± 2.0 | 32.9 ± 1.6 * | 30.5 ± 1.8 * | 31.2 ± 1.8 * |
| Cortex | 30.0 ± 1.3 | 25.2±1.0 # | 27.7 ± 1.9 | 34.0 ± 2.4 * | 30.7 ± 1.5 * | 33.7 ± 2.8 * |

Data, expressed as the mean ± SEM ($n$ = 8), are nanomoles of acetylcholine formed/milligram protein/hour. # $p < 0.01$ *versus* the control group, * $p < 0.05$ *versus* the scopolamine-treated group (ANOVA and Tukey's test).

### 2.6. Fucoxanthin Directly Inhibits AChE in a Non-Competitive Manner in Vitro

Our *in vivo* results have shown that fucoxanthin could effectively reverse scopolamine-induced cognitive impairments. Because AChE inhibitors used for AD treatment were reported to have similar effects, we speculated that fucoxanthin may also directly inhibit AChE [11]. To test this, we used an *in vitro* AChE activity assay. In our study, fucoxanthin directly inhibits AChE with an $IC_{50}$ of 81.2 μM (Figure 8). To further explore the mode of AChE inhibition by fucoxanthin, three concentrations (25, 50 or 75 μM) of fucoxanthin were used in the *in vitro* AChE activity assay containing various concentrations (0.125–1 mM) of substrate. Analysis of these assays with Lineweaver–Burk plots shows that fucoxanthin acts as a non-competitive inhibitor of AChE, and the $K_i$ value of AChE inhibition by fucoxanthin is 42 μM (Figure 9).

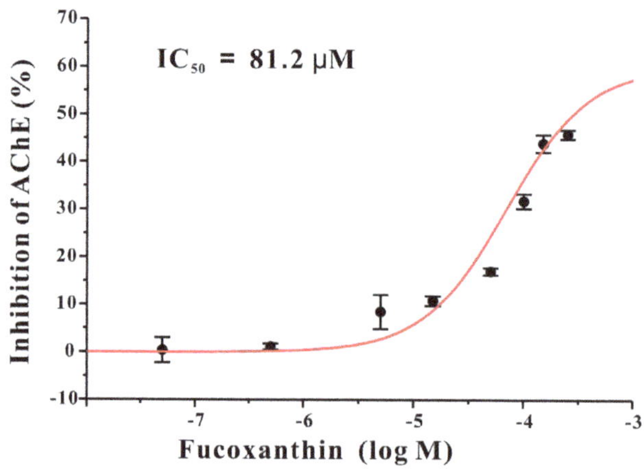


**Figure 8.** Fucoxanthin directly inhibits AChE enzyme activity in a concentration-dependent manner. The inhibitory effect of fucoxanthin on AChE is shown in the graph. The $IC_{50}$ value is also indicated in the graph. Data are expressed as the mean ± SEM ($n$ = 3).

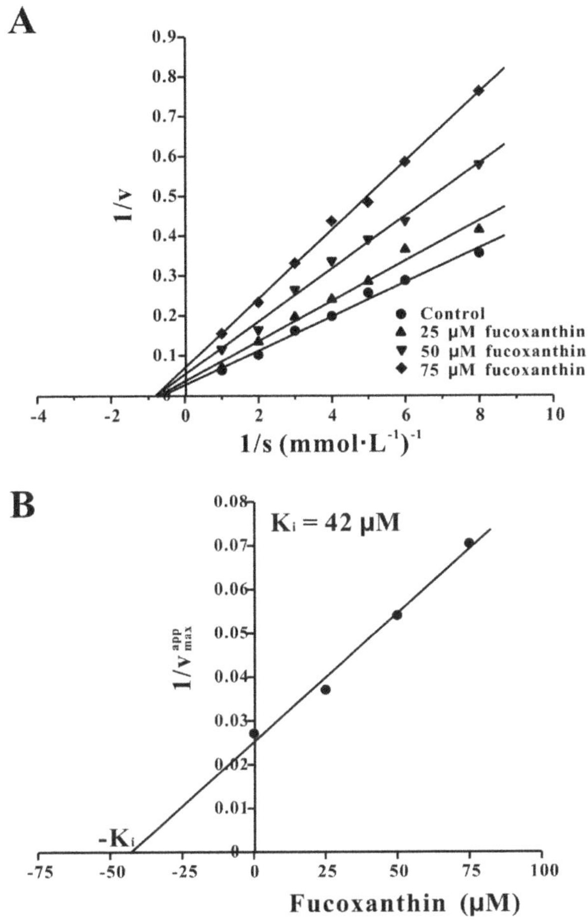


**Figure 9.** Fucoxanthin inhibits AChE by a non-competitive mechanism. (**A**) Kinetic analysis of AChE inhibition by fucoxanthin. The AChE enzyme was assayed in either the presence (25, 50 or 75 µM) or absence of fucoxanthin over a range of concentrations of acetylthiocholine iodide (0.125–1 mM). The ranges of 1/v *versus* 1/[S] were fitted by a Lineweaver-Burk plot. The data are expressed as the mean of three independent experiments. (**B**) The $K_i$ value of fucoxanthin in the inhibition of AChE.

### 2.7. Molecular Docking Analysis of the Interaction between Fucoxanthin and AChE

To gain insight into the molecular interaction between fucoxanthin and AChE, computational docking was performed. Fucoxanthin showed favorable interaction mainly with the peripheral anionic site (PAS) of AChE (Figure 10). The Surflex-Dock score is expressed as $-\log(K_d)$ (the unit of $K_d$ is M). The Surflex-Dock score, evaluated automatically by the software, was 4.74. Therefore, the value of $K_d$ could be easily calculated and equals $1.8 \times 10^{-5}$ M (18 µM), which is close to the experimental value determined by our *in vitro* activity assay. Moreover, the docking analysis suggests that fucoxanthin might form hydrogen bonds with Asp283 and Ser286, respectively, in the PAS of AChE.

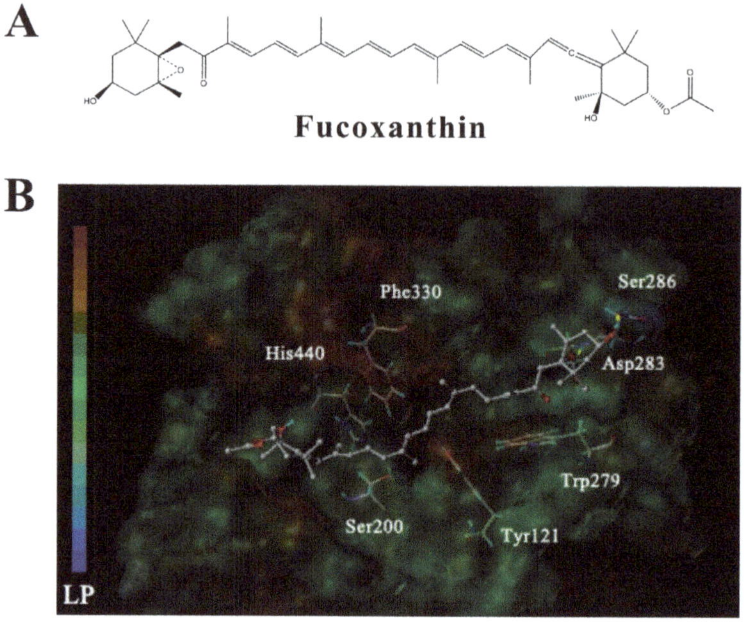


**Figure 10.** Molecular docking simulation of the interaction between AChE and fucoxanthin. (**A**) Chemical structure of fucoxanthin; (**B**) low-energy conformation of fucoxanthin mainly bound to the peripheral anionic site (PAS) of AChE generated by molecular docking. Fucoxanthin is depicted as a stick model showing carbon (white), oxygen (red) and hydrogen (green). Yellow line: hydrogen bond. LP: lipophilic potential.

## 3. Discussion

In the current study, we have reported for the first time that fucoxanthin, a carotenoid extracted from brown seaweeds, can effectively reverse scopolamine-induced impairments of cognition in mice, suggesting that fucoxanthin might be used in the treatment of AD. We have further demonstrated that fucoxanthin significantly increases ChAT activity and BDNF expression and decreases AChE activity in the brains of scopolamine-treated mice. Moreover, fucoxanthin directly inhibits AChE in a non-competitive manner *in vitro*, possibly via interacting with the PAS of AChE.

Fucoxanthin is a compound safe for use in humans and animals [21]. Previous studies have shown that repeated oral administration of fucoxanthin at 1000 mg/kg for 30 days in ICR mice resulted in no mortality, no abnormalities in gross appearance and no abnormal findings in histological observations [21,22]. Moreover, oral administration of fucoxanthin at 50–200 mg/kg exhibited anti-cancer, anti-obesity and anti-diabetic affects *in vivo* [15–17]. Therefore, we tested the effects of repeated oral administration of fucoxanthin at 50–200 mg/kg on cognitive impairments in our study.

Administration of scopolamine in animals produces short-term learning and memory deficits, which are considered to be characteristic of cholinergic dysfunction in AD [4–6]. Therefore, the scopolamine-induced amnesic model has been widely accepted as a pharmacological model of cognitive impairments useful for screening potential cognitive-enhancing agents. In our study, fucoxanthin effectively reversed scopolamine-induced recognition impairments in the NOR test and spatial learning and memory impairments in the Morris water maze test, suggesting that fucoxanthin has cognitive-enhancing potential. BDNF is a neurotrophin that facilitates synaptic transmission and neuronal plasticity [23]. Importantly, BDNF plays an essential role in the formation and storage of memory. The expression of BDNF is downregulated not only in the brains of AD patients, but also in amnesic animal models, including the scopolamine-induced cognitive impairment model [11]. In our study, fucoxanthin significantly reversed the reduction in BDNF expression caused by scopolamine treatment, suggesting that fucoxanthin might increase memory formation in animals by increasing BDNF levels.

Alternatively, decreased acetylcholine expression also generally results in a diminished ability to learn and form new memories. The concentration of acetylcholine in the brain is dynamically regulated by the activities of AChE and ChAT [24]. Therefore, we evaluated the activity of these two enzymes in the brains of our treated mice. In our study, scopolamine decreased ChAT activity and increased AChE activity in both the hippocampus and the cortex, which is in accordance with previous studies [9,10]. Interestingly, fucoxanthin significantly reversed the scopolamine-induced alterations of ChAT and AChE activities, suggesting that fucoxanthin might directly affect enzymes in the cholinergic system.

We further explored if fucoxanthin could directly inhibit AChE, a key enzyme in the cholinergic system. AChE hydrolyzes acetylcholine into acetic acid and choline, and AChE inhibitors prolong the action of acetylcholine synapses and enhance cholinergic neurotransmission [25]. Therefore, many AChE inhibitors are clinically used in the treatment of neurological diseases associated with cholinergic deficits, including AD in particular [26]. Our results show that fucoxanthin directly inhibits the activity of AChE *in vitro*. Moreover, Lineweaver–Burk plots suggest that fucoxanthin acts as a non-competitive inhibitor of AChE.

AChE possesses two important binding sites, the catalytic anion site (CAS) and the PAS [27]. The CAS and the PAS are responsible for catalytic activity and ligand binding, respectively. The PAS also serves as a binding site for non-competitive inhibitors [28]. By using molecular docking analysis, we investigated the interaction between fucoxanthin and AChE. In the simulated binding of fucoxanthin and AChE, the 5,6-monoepoxide structure of fucoxanthin formed two hydrogen bonds with Asp283 and Ser286 in the PAS of AChE. Therefore, our molecular docking analysis further supports our finding that fucoxanthin acts as a non-competitive inhibitor of AChE.

Why was the concentration of fucoxanthin (50–200 mg/kg) used in our study much higher than that of donepezil (3 mg/kg)? Donepezil is much more potent than fucoxanthin in inhibiting AChE *in vitro*. Therefore, it is rational to expect that a higher concentration of fucoxanthin than that of donepezil is needed to reverse AChE-involved scopolamine-induced cognitive impairments *in vivo*. Does fucoxanthin reverse scopolamine-induced cognitive impairments solely via the direct inhibition of AChE? Donepezil was reported to inhibit AChE with an $IC_{50}$ around 10 nM [29]. However, in our study, fucoxanthin inhibited AChE with an $IC_{50}$ of 81.2 μM. Interestingly, fucoxanthin at 100 mg/kg produced anti-cognitive impairment effects as strong as that produced by 3 mg/kg donepezil in our animal model. These results suggest that the beneficial effects of fucoxanthin in the scopolamine-induced cognitive impairments mouse model may not solely result from direct inhibition of AChE. We speculated that besides inhibition of AChE, fucoxanthin might act on additional targets that are involved in cognitive enhancement. A recent study has shown that fucoxanthin attenuates pro-inflammatory cytokine secretion, reduces reactive oxygen species formation and elevates anti-oxidative enzymes in Aβ-treated microglia cells [20]. Because oxidative stress and inflammation play important roles in scopolamine-induced cognitive impairments, it is rational to speculate that fucoxanthin might protect against cognitive impairments by reducing the scopolamine-induced increases in oxidative stress and inflammation. To further determine whether fucoxanthin protects against cognitive impairments via anti-oxidant and anti-inflammatory mechanisms, additional experiments, including analysis of pro-inflammatory cytokine production and anti-oxidative enzyme expression, are being undertaken in our lab. Moreover, it is agreed that several risk factors, including diabetes and obesity in particular, are related to AD [30]. Therefore, fucoxanthin with anti-diabetic and anti-obesity potential might also be helpful for the prevention of AD.

## 4. Materials and Methods

### 4.1. Chemicals and Reagents

Fucoxanthin was obtained from *Sargassum horneri*, a genus of brown seaweeds, by a purification method based on solvent extraction, low temperature concentration and ethanol precipitation. Briefly, fucoxanthin extraction was performed at 30 °C for 2 h with an ethanol to sample ratio of 4:1 mL/g. The extraction solution was then concentrated at 25 °C. The precipitation of lipid and chlorophylls further occurred when the concentration solution contained 63% ethanol. Fucoxanthin was then purified by precipitation when the solution reached an ethanol content of 40%. The purity of fucoxanthin was more than 80% by HPLC. Fucoxanthin was stored at −20 °C before use. Donepezil, scopolamine, dithiobisnitrobenzoic acid (DTNB), ethopropazine hydrochloride and acetylthiocholine iodide (ATCI) were purchased from Sigma-Aldrich (St. Louis, MO, USA).

### 4.2. Drug Treatment for Animal Study

Male Institute of Cancer Research (ICR) mice weighing 25–30 g were obtained from Zhejiang Academy of Medical Sciences. The ICR mouse is a strain of albino mice originating in Switzerland [31]. The animals were maintained on a 12-h light/dark cycle under controlled temperature (22 ± 2 °C) and humidity (50% ± 10%) and given standard diet and water. Animals were allowed to acclimatize for 3 days before the experiments. All procedures followed the National Institutes of Health (NIH) Guide for the Care and Use of Laboratory Animals (NIH Publications No. 80-23, revised 1996) and were approved by the Animal Care and Use Committee of Ningbo University (SYXK-2008-0110).

Fucoxanthin was dissolved in sterile saline containing 0.5% Tween-20. Donepezil and scopolamine were dissolved in sterile saline. Mice were randomly assigned into six groups of 8 animals each: control, 3 mg/kg scopolamine, 3 mg/kg scopolamine plus low (50 mg/kg), medium (100 mg/kg) and high (200 mg/kg) doses of fucoxanthin and 3 mg/kg scopolamine plus donepezil (3 mg/kg). Fucoxanthin was given by intragastric (*i.g.*) administration. Donepezil or scopolamine was given by intraperitoneal (*i.p.*) injection. All drugs were given 30 min prior to each trial once a day for 10 consecutive days. Mice were sacrificed for biochemical study on the 10th day. All animals received the last injection of drugs 30 min prior to being sacrificed.

### 4.3. Open-Field Test

To analyze the exploratory and locomotor activities, animals were placed in the left rear quadrant of a 50 × 50 × 39-cm open field with white plywood walls and a brown floor divided into 4 identical squares of equal dimensions (25 × 25 cm) [32]. The animals were placed one by one at the center of the box and allowed to explore it for 5 min. Hand-operated counters and stopwatches were used to score the number of line crossing with four paws and the number of rearings (number of times the animals stood on its hind legs), which were used as indicators of locomotor and exploratory activities, respectively. The person counting was blind to the drug status of the subjects. To avoid perturbation of the animals due to urine and feces, between two tests, the open field was cleaned with 70% ethanol solution and a dry cloth.

### 4.4. Novel Object Recognition Test

The NOR test was carried out in an open-field arena (30 × 30 × 30 cm) built with polyvinyl chloride plastic, plywood and transparent acrylic as described before [33,34]. The task consisted of three sessions: habituation, training and retention sessions that were carried out over a period of three consecutive days. On the first day, the animals were habituated to the experimental arena by allowing them to freely explore the arena for 5 min in the absence of any behaviorally-relevant stimulus. On the second day, the animals were allowed to explore two identical objects for 5 min. On the third day, one of the objects was changed to a novel one with a different shape and color, and the animals were allowed to explore the arena for 5 min. To avoid perturbation of the animals due to urine and feces, between two tests, the field was cleaned thoroughly with 70% ethanol solution and a dry cloth. Exploration was defined as sniffing or touching the objects with the nose and/or forepaws at a distance of less than 2 cm. Sitting on or turning around the objects closely was not considered exploratory behavior. The exploration was scored manually using a video camera positioned over the arena by an observer blind to testing conditions. Total exploring time is the amount of time spent exploring both objects. A recognition index, which is the ratio of the amount of time spent exploring either of the two objects (training session) or the novel object (retention session) over the total exploring time, was used to measure cognitive function.

### 4.5. Morris Water Maze Task

The Morris water maze task was carried out as described before [35]. The water maze apparatus consisted of a circular pool 110 cm in diameter, filled with water at 23 ± 2 °C to cover a platform. The platform always resided in the center of the northeast quadrant, except on the last day. Each mouse's swimming was monitored by a video camera linked to a computer-based image analyzer. Learning performance was tested for 4 consecutive days beginning on the 5th day after the first injection of scopolamine. Each mouse was trained to find the platform with four trials per day. In each trial, the time required to escape onto the hidden platform was recorded. On the 9th day, a probe trial was made by removing the platform and allowing the mice to swim for 90 s in search of it. Swimming time in each of the four quadrants in the pool was calculated. A persistent preference for the quadrant previously occupied by the platform was taken to indicate that the mice had acquired and remembered the spatial task.

### 4.6. Western Blot Analysis

Western blot analysis was performed as previously described [36]. Briefly, after decapitation, the brains of mice were removed quickly. Half of the brain tissue was analyzed by Western blot analysis, and the other half was analyzed by AChE and ChAT activity assays. For Western blot analysis, the hippocampi and cortex were dissected on ice. Brain tissue samples were homogenized at 4 °C for 1 min in lysis buffer (50 mM Tris-HCl, 150 mM NaCl, 1% Triton X-100, 1 mM EDTA, 1 mM phenylmethanesulfonyl fluoride, 0.1% sodium dodecyl sulfate, 1% sodium deoxycholate, 5 μg/mL leupeptin, 1 μg/mL aprotinin and 5 μg/mL pepstatin). After centrifugation at 16,000× *g* for 10 min, the protein concentration in the supernatant was determined by the Bradford assay. Samples (40 μg) were separated by SDS-PAGE and transferred to polyvinylidene fluoride membranes for 2 h at 100 V. The membranes were further blocked with 5% non-fat milk in PBST (0.1% Tween 20 in phosphate-buffered saline) for 2 h. The blots were incubated overnight at 4 °C with antibodies against BDNF (1:500, Santa Cruz Biotechnology, Santa Cruz, CA, USA) and β-actin (1:1000, Santa Cruz Biotechnology). After three washes with PBST, the membranes were further incubated with secondary antibody. Blots were developed by using an enhanced chemiluminescence plus kit (Amersham Bioscience, Aylesbury, UK) and then exposed to Kodak films. All data are the results from three independent experiments. Data were expressed as a ratio to the optical density (OD) values of control samples for statistical analyses.

### 4.7. Measurement of Choline ChAT Activity ex Vivo

Mouse brain was weighted, followed by the addition of a 10-times volume of lysis buffer (10 mM HEPES, pH 7.5, 1 mM EDTA, 1 mM EGTA, 150 mM NaCl and 0.5% Triton X-100). Homogenization was then performed by vortexing on ice for 15 min. The homogenates were clarified by centrifugation at 3000 rpm for 15 min at 4 °C. The protein concentration of the supernatant was then determined by the Bradford assay. The activity of ChAT from each sample was determined spectrophotometrically using the assay kit from Nanjing Jiancheng Bioengineering Institute (Nanjing, China). Briefly, the assay medium containing CoA-SH as a substrate was incubated at 37 °C for 20 min. The activity was determined by measuring the absorbance at 412 nm.

### 4.8. Measurement of AChE Activity ex Vivo

For the analysis of AChE activity *ex vivo*, the brains of mice were dissected on ice into the frontal cortex and hippocampus. Samples were weighted, and a 10-times volume of lysis buffer was added (10 mM HEPES, pH 7.5, 1 mM EDTA, 1 mM EGTA, 150 mM NaCl and 0.5% Triton X-100). The homogenization was done by vortexing on ice for 15 min. The homogenates were clarified by centrifugation for 15 min at 3000 rpm at 4 °C. The assay medium contained 0.1 M $Na_2HPO_4$ (pH 7.5), 10 mM DTNB and 1 mM ATCI. Samples was prepared to give 0.5 µg/mL and incubated with 0.1 mM ethopropazine hydrochloride for 5 min to inhibit BuChE activity. The reaction was undertaken at 37 °C for 15 min. The activity was determined by measuring the absorbance at 412 nm.

### 4.9. Measurement of AChE Activity in Vitro

The colorimetric method, which was modified for use in 96-well-plates with a final volume of 200 µL, was used for the detection of AChE activity *in vitro* [37]. Briefly, brains of rats collected immediately after decapitation were used as the source of AChE. Brain was weighted, and a 10-times volume of lysis buffer was added (10 mM HEPES, pH 7.5, 1 mM EDTA, 1 mM EGTA, 150 mM NaCl and 0.5% Triton X-100). The homogenization was done by vortexing on ice for 15 min. The homogenates were clarified by centrifugation for 15 min at 3000 rpm at 4 °C. The assay medium contained 0.1 M $Na_2HPO_4$ (pH 7.5), 10 mM DTNB and 1 mM ATCI. The brain lysate was incubated with 0.1 mM ethopropazine hydrochloride for 5 min to inhibit BuChE activity. Test compounds were added to the assay solution and pre-incubated at 37 °C with the enzyme for 15 min followed by the addition of substrate. The activity was determined by measuring the absorbance at 412 nm. The inhibitory curve and $IC_{50}$ were determined from a series of fucoxanthin concentrations.

### 4.10. Molecular Docking

Molecular docking analyses were accomplished by using the SYBYL (Tripos Inc., St. Louis, MO, USA) software and the programs embedded therein. The three-dimensional crystal structure of AChE complexed with donepezil was retrieved from the Protein Data Bank (PDB code: 1EVE) [38]. The three-dimensional structure of fucoxanthin was constructed using the standard geometric parameters of SYBYL and then optimized using the Powell method. The Surflex-Dock program, which uses an empirically-derived scoring function based on the binding affinities of protein-ligand complexes, was employed to perform docking analysis. As a flexible docking method, Surflex-Dock has been proven to be efficient in analyzing a variety of receptors [39]. The active site of AChE was defined relative to the coordinates of donepezil. During the simulations, the rotatable bonds of the ligands were defined, whereas the receptor was kept rigid. Standard parameters were used to estimate the binding affinity characterized by Surflex-Dock scores.

### 4.11. Data Analysis and Statistics

The data are expressed as the mean ± SEM. Statistical significance was determined by one-way ANOVA and Tukey's or Dunnett's test for *post hoc* multiple comparison, with the exception of mean escape latency, which was analyzed using two-way repeated-measures ANOVA followed by the LSD *post hoc* test. Differences were accepted as significant at $p < 0.05$.

## 5. Conclusions

In summary, we have found that fucoxanthin effectively reverses cognitive impairments in a scopolamine-induced mouse model. Moreover, fucoxanthin directly inhibits AChE by a non-competitive mechanism *in vitro*, possibly via interacting with the PAS of AChE. Based on these novel findings, we anticipate that fucoxanthin might provide significant therapeutic efficacy for the treatment of AD by acting on multiple targets, including inhibiting AChE and increasing BDNF expression in particular.

**Acknowledgments:** This work was supported by the Natural Science Foundation of Zhejiang Province (LY15H310007, LY15090010, LQ13B020004), the Applied Research Project on Nonprofit Technology of Zhejiang Province (2016C37110), the National Natural Science Foundation of China (81202510, 81471398, 41306134, 41406163, U1503223), the 863 Program of China (2013AA092902), the Ningbo international science and technology cooperation project (2014D10019), the Ningbo Marine Algae Biotechnology Team (2011B81007), the Scientific Research Foundation for the Returned Overseas Chinese Scholars, State Education Ministry, and the K. C. Wong Magna Fund at Ningbo University.

**Author Contributions:** Lin Jiajia, Huang Lin, Yu Jie and Xiang Siying and Wang Jialin performed the experiments. Cui Wei and Wang Qinwen designed the experiments. Zhang Jinrong, He Shan and Yan Xiaojun contributed materials.

**Conflicts of Interest:** The authors declare no conflict of interest.

## References

1. Kumar, A.; Singh, A.; Ekavali. A review on Alzheimer's disease pathophysiology and its management: An update. *Pharmacol. Rep.* **2015**, *67*, 195–203. [CrossRef] [PubMed]
2. Anand, R.; Gill, K.D.; Mahdi, A.A. Therapeutics of Alzheimer's disease: Past, present and future. *Neuropharmacology* **2014**, *76*, 27–50. [CrossRef] [PubMed]
3. Zemek, F.; Drtinova, L.; Nepovimova, E.; Sepsova, V.; Korabecny, J.; Klimes, J.; Kuca, K. Outcomes of Alzheimer's disease therapy with acetylcholinesterase inhibitors and memantine. *Expert Opin. Drug Saf.* **2014**, *13*, 759–774. [PubMed]
4. Min, A.Y.; Doo, C.N.; Son, E.J.; Sung, N.Y.; Lee, K.J.; Sok, D.E.; Kim, M.R. *N*-palmitoyl serotonin alleviates scopolamine-induced memory impairment via regulation of cholinergic and antioxidant systems, and expression of BDNF and p-CREB in mice. *Chem. Biol. Interact.* **2015**, *242*, 153–162. [CrossRef] [PubMed]
5. Singh, P.; Konar, A.; Kumar, A.; Srivas, S.; Thakur, M.K. Hippocampal chromatin-modifying enzymes are pivotal for scopolamine-induced synaptic plasticity gene expression changes and memory impairment. *J. Neurochem.* **2015**, *134*, 642–651. [CrossRef] [PubMed]
6. Araujo, J.A.; Studzinski, C.M.; Milgram, N.W. Further evidence for the cholinergic hypothesis of aging and dementia from the canine model of aging. *Prog. Neuropsychopharmacol. Biol. Psychiatry* **2005**, *29*, 411–422. [CrossRef] [PubMed]
7. Beatty, W.W.; Butters, N.; Janowsky, D.S. Patterns of memory failure after scopolamine treatment: Implications for cholinergic hypotheses of dementia. *Behav. Neural Biol.* **1986**, *45*, 196–211. [CrossRef]
8. Kopelman, M.D.; Corn, T.H. Cholinergic "blockade" as a model for cholinergic depletion. A comparison of the memory deficits with those of Alzheimer-type dementia and the alcoholic Korsakoff syndrome. *Brain* **1988**, *111*, 1079–1110. [CrossRef] [PubMed]
9. Kwon, S.H.; Lee, H.K.; Kim, J.A.; Hong, S.I.; Kim, H.C.; Jo, T.H.; Park, Y.I.; Lee, C.K.; Kim, Y.B.; Lee, S.Y.; *et al.* Neuroprotective effects of chlorogenic acid on scopolamine-induced amnesia via anti-acetylcholinesterase and anti-oxidative activities in mice. *Eur. J. Pharmacol.* **2010**, *649*, 210–217. [CrossRef] [PubMed]
10. Ghumatkar, P.J.; Patil, S.P.; Jain, P.D.; Tambe, R.M.; Sathaye, S. Nootropic, neuroprotective and neurotrophic effects of phloretin in scopolamine induced amnesia in mice. *Pharmacol. Biochem. Behav.* **2015**, *135*, 182–191. [CrossRef] [PubMed]
11. Wu, Y.; Luo, X.; Liu, X.; Liu, D.; Wang, X.; Guo, Z.; Zhu, L.; Tian, Q.; Yang, X.; Wang, J.Z. Intraperitoneal Administration of a Novel TAT-BDNF Peptide Ameliorates Cognitive Impairments via Modulating Multiple Pathways in Two Alzheimer's Rodent Models. *Sci. Rep.* **2015**, *5*, 15032. [CrossRef] [PubMed]
12. Miyashita, K. Function of marine carotenoids. *Forum Nutr.* **2009**, *61*, 136–146. [PubMed]
13. D'Orazio, N.; Gammone, M.A.; Gemello, E.; De Girolamo, M.; Cusenza, S.; Riccioni, G. Marine bioactives: Pharmacological properties and potential applications against inflammatory diseases. *Mar. Drugs* **2012**, *10*, 812–833. [CrossRef] [PubMed]
14. Yan, X.; Chuda, Y.; Suzuki, M.; Nagata, T. Fucoxanthin as the major antioxidant in *Hijikia fusiformis*, a common edible seaweed. *Biosci. Biotechnol. Biochem.* **1999**, *63*, 605–607. [CrossRef] [PubMed]
15. Das, S.K.; Hashimoto, T.; Kanazawa, K. Growth inhibition of human hepatic carcinoma HepG2 cells by fucoxanthin is associated with down-regulation of cyclin D. *Biochim. Biophys. Acta* **2008**, *1780*, 743–749. [CrossRef] [PubMed]

16. Woo, M.N.; Jeon, S.M.; Shin, Y.C.; Lee, M.K.; Kang, M.A.; Choi, M.S. Anti-obese property of fucoxanthin is partly mediated by altering lipid-regulating enzymes and uncoupling proteins of visceral adipose tissue in mice. *Mol. Nutr. Food Res.* **2009**, *53*, 1603–1611. [CrossRef] [PubMed]
17. Maeda, H.; Hosokawa, M.; Sashima, T.; Miyashita, K. Dietary combination of fucoxanthin and fish oil attenuates the weight gain of white adipose tissue and decreases blood glucose in obese/diabetic KK-Ay mice. *J. Agric. Food Chem.* **2007**, *55*, 7701–7706. [CrossRef] [PubMed]
18. Gammone, M.A.; D'Orazio, N. Anti-obesity activity of the marine carotenoid fucoxanthin. *Mar. Drugs* **2015**, *13*, 2196–2214. [CrossRef] [PubMed]
19. Ha, A.W.; Na, S.J.; Kim, W.K. Antioxidant effects of fucoxanthin rich powder in rats fed with high fat diet. *Nutr. Res. Pract.* **2013**, *7*, 475–480. [CrossRef] [PubMed]
20. Pangestuti, R.; Vo, T.S.; Ngo, D.H.; Kim, S.K. Fucoxanthin ameliorates inflammation and oxidative reponses in microglia. *J. Agric. Food Chem.* **2013**, *61*, 3876–3883. [CrossRef] [PubMed]
21. Beppu, F.; Niwano, Y.; Sato, E.; Kohno, M.; Tsukui, T.; Hosokawa, M.; Miyashita, K. *In vitro* and *in vivo* evaluation of mutagenicity of fucoxanthin (FX) and its metabolite fucoxanthinol (FXOH). *J. Toxicol. Sci.* **2009**, *34*, 693–698. [CrossRef] [PubMed]
22. Beppu, F.; Niwano, Y.; Tsukui, T.; Hosokawa, M.; Miyashita, K. Single and repeated oral dose toxicity study of fucoxanthin (FX), a marine carotenoid, in mice. *J. Toxicol. Sci.* **2009**, *34*, 501–510. [CrossRef] [PubMed]
23. Chen, J.; Li, C.R.; Yang, H.; Liu, J.; Zhang, T.; Jiao, S.S.; Wang, Y.J.; Xu, Z.Q. proBDNF Attenuates Hippocampal Neurogenesis and Induces Learning and Memory Deficits in Aged Mice. *Neurotox. Res.* **2015**, *29*, 47–53. [CrossRef] [PubMed]
24. Lee, B.; Shim, I.; Lee, H.; Hahm, D.H. Rehmannia glutinosa ameliorates scopolamine-induced learning and memory impairment in rats. *J. Microbiol. Biotechnol.* **2011**, *21*, 874–883. [CrossRef] [PubMed]
25. Siow, N.L.; Xie, H.Q.; Choi, R.C.; Tsim, K.W. ATP induces the post-synaptic gene expression in neuron-neuron synapses: Transcriptional regulation of AChE catalytic subunit. *Chem. Biol. Interact.* **2005**, *157–158*, 423–426. [CrossRef]
26. Anand, P.; Singh, B. A review on cholinesterase inhibitors for Alzheimer's disease. *Arch. Pharm. Res.* **2013**, *36*, 375–399. [CrossRef] [PubMed]
27. Jiang, H.; Zhang, X.J. Acetylcholinesterase and apoptosis. A novel perspective for an old enzyme. *FEBS J.* **2008**, *275*, 612–617. [CrossRef] [PubMed]
28. Tsim, K.; Soreq, H. Acetylcholinesterase: Old questions and new developments. *Front Mol. Neurosci.* **2012**, *5*, 101. [PubMed]
29. Ota, T.; Shinotoh, H.; Fukushi, K.; Kikuchi, T.; Sato, K.; Tanaka, N.; Shimada, H.; Hirano, S.; Miyoshi, M.; Arai, H.; *et al.* Estimation of plasma IC50 of donepezil for cerebral acetylcholinesterase inhibition in patients with Alzheimer disease using positron emission tomography. *Clin. Neuropharmacol.* **2010**, *33*, 74–78. [CrossRef] [PubMed]
30. Smith, A.D.; Yaffe, K. Dementia (including Alzheimer's disease) can be prevented: Statement supported by international experts. *J. Alzheimers Dis.* **2014**, *38*, 699–6703. [PubMed]
31. Chen, X.; Hu, J.; Jiang, L.; Xu, S.; Zheng, B.; Wang, C.; Zhang, J.; Wei, X.; Chang, L.; Wang, Q. Brilliant Blue G improves cognition in an animal model of Alzheimer's disease and inhibits amyloid-beta-induced loss of filopodia and dendrite spines in hippocampal neurons. *Neuroscience* **2014**, *279*, 94–101. [CrossRef] [PubMed]
32. Ketcha Wanda, G.J.; Djiogue, S.; Zemo Gamo, F.; Guemnang Ngitedem, S.; Njamen, D. Anxiolytic and sedative activities of aqueous leaf extract of *Dichrocephala integrifolia* (Asteraceae) in mice. *J. Ethnopharmacol.* **2015**, *176*, 494–498. [CrossRef] [PubMed]
33. Lu, Y.; Wang, C.; Xue, Z.; Li, C.; Zhang, J.; Zhao, X.; Liu, A.; Wang, Q.; Zhou, W. PI3K/AKT/mTOR signaling-mediated neuropeptide VGF in the hippocampus of mice is involved in the rapid onset antidepressant-like effects of GLYX-13. *Int. J. Neuropsychopharmacol.* **2015**, *18*, 1–12. [CrossRef] [PubMed]
34. Zhang, J.; Guo, J.; Zhao, X.; Chen, Z.; Wang, G.; Liu, A.; Wang, Q.; Zhou, W.; Xu, Y.; Wang, C. Phosphodiesterase-5 inhibitor sildenafil prevents neuroinflammation, lowers beta-amyloid levels and improves cognitive performance in APP/PS1 transgenic mice. *Behav. Brain Res.* **2013**, *250*, 230–237. [CrossRef] [PubMed]
35. Chang, L.; Cui, W.; Yang, Y.; Xu, S.; Zhou, W.; Fu, H.; Hu, S.; Mak, S.; Hu, J.; Wang, Q.; *et al.* Protection against beta-amyloid-induced synaptic and memory impairments via altering beta-amyloid assembly by bis(heptyl)-cognitin. *Sci. Rep.* **2015**, *5*, 10256. [CrossRef] [PubMed]

36. Cui, W.; Zhang, Z.; Li, W.; Mak, S.; Hu, S.; Zhang, H.; Yuan, S.; Rong, J.; Choi, T.C.; Lee, S.M.; Han, Y. Unexpected neuronal protection of SU5416 against 1-Methyl-4-phenylpyridinium ion-induced toxicity via inhibiting neuronal nitric oxide synthase. *PLoS ONE* **2012**, *7*, e46253. [CrossRef] [PubMed]
37. Mak, S.; Luk, W.W.; Cui, W.; Hu, S.; Tsim, K.W.; Han, Y. Synergistic inhibition on acetylcholinesterase by the combination of berberine and palmatine originally isolated from Chinese medicinal herbs. *J. Mol. Neurosci.* **2014**, *53*, 511–516. [CrossRef] [PubMed]
38. Kryger, G.; Silman, I.; Sussman, J.L. Structure of acetylcholinesterase complexed with E2020 (Aricept): Implications for the design of new anti-Alzheimer drugs. *Structure* **1999**, *7*, 297–307. [CrossRef]
39. Jain, A.N. Surflex: Fully automatic flexible molecular docking using a molecular similarity-based search engine. *J. Med. Chem.* **2003**, *46*, 499–511. [CrossRef] [PubMed]

*marine drugs*

*Article*

# Degradation of Fucoxanthin to Elucidate the Relationship between the Fucoxanthin Molecular Structure and Its Antiproliferative Effect on Caco-2 Cells

**Shiro Komba** *, **Eiichi Kotake-Nara** and **Wakako Tsuzuki**

Food Component Analysis Unit, Food Research Institute, National Agriculture and Food Research Organization, 2-1-12, Kannondai, Tsukuba, Ibaraki 305-8642, Japan; ekotake@affrc.go.jp (E.K.-N.); wakako@affrc.go.jp (W.T.)
* Correspondence: skomba@affrc.go.jp; Tel.: +81-29-838-7298

Received: 18 July 2018; Accepted: 2 August 2018; Published: 6 August 2018

**Abstract:** Fucoxanthin has an antiproliferative effect on cancer cells, but its detailed structure–activity correlation has not yet been elucidated. To elucidate this correlation, fucoxanthin was degraded by ozonolysis. The degraded compounds of fucoxanthin obtained by ozonolysis were purified by HPLC and analyzed by NMR. The polyene chain of fucoxanthin was cleaved by ozonolysis, and the fucoxanthin was divided into two types of cyclohexyl derivatives, one with a $\beta,\gamma$-epoxy ketone group and the other with an allenic bond. In order to elucidate the structure–activity correlation, Caco-2 cells (human colorectal carcinoma) were treated with fucoxanthin degradation compounds. It was found that the entire structure of fucoxanthin is not essential for its antiproliferative effect and that even a partial structure exerts this effect.

**Keywords:** fucoxanthin; ozonolysis; apo-fucoxanthinone; Caco-2; antiproliferative effect

## 1. Introduction

Fucoxanthin (**1**) is a xanthophyll-type carotenoid which is mainly distributed in brown algae [1–10]. It has been reported that it has anticarcinogenic [11–22], anti-obesity [23–25], anti-inflammatory [26], anti-angiogenic [27], and antioxidative [28] effects. We focused on the anticarcinogenic activity of fucoxanthin.

It is well known that fucoxanthin and its metabolite, fucoxanthinol, induce $G_1$ cell-cycle arrest and apoptosis in various cell lines and can inhibit cancer development in animal models [22]. However, its detailed structure–activity correlation has not been elucidated, particularly regarding what kind of molecular structure of fucoxanthin is responsible for this activity. Therefore, we decided to decompose a fucoxanthin molecule in order to elucidate the detailed structure–activity relationship and elucidate the mechanisms involved in its antiproliferative effect. By investigating the antiproliferative effect using each part of the degraded fucoxanthin, the structure contributing to the activity was able to be elucidated.

Fucoxanthin has a very complicated structure, with an allenic bond, a polyene chain, an acetyl, and a $\beta,\gamma$-epoxy ketone group. In addition, the two six-membered ring derivatives bound by the polyene chain are not symmetrical, with one having an allenic bond and the other having a $\beta,\gamma$-epoxy ketone group. In particular, we focused on the two six-membered ring derivatives that are connected by a polyene chain. We anticipated that one of the two might contribute to this activity and separated them by cutting the polyene chain. In regard to its anti-obesity activity, the allenic bond on fucoxanthin may be a key structure, according to Miyashita et al. [25]. Using these two six-membered ring derivatives, the antiproliferative effect on Caco-2 cells was examined. Although Caco-2 cells have been widely used as a model of intestinal absorption, we used them here as colon cancer cells, as we did in our previous study [29].

In order to chemically decompose fucoxanthin, which is sensitive to light, acid, and basic conditions, it was necessary to perform the reaction under mild, neutral conditions. Therefore, ozone oxidation was chosen, in which the reaction proceeds in a neutral condition at a relatively low temperature, and only gas is used as a reagent. Ozone oxidation is a reaction involving the oxidization of a double bond to decompose it into two aldehydes. It was predicted that fucoxanthin has a polyene chain, and that ozone easily oxidizes the double bond and decomposes it into two six-membered ring derivatives.

## 2. Results and Discussion

### 2.1. Chemistry

Fucoxanthin was purified from wakame (*Undaria pinnatifida*) (1.09 mg fucoxanthin/g dry wakame), as referenced in previous experiments [30]. The $^{1}$H-NMR and $^{13}$C-NMR spectra were identical to those previously reported. Purified fucoxanthin was dissolved in dichloromethane/methanol, and ozone gas was bubbled through the stirred mixture at 0 °C for 2 h (Figure 1). We then replaced ozone gas with nitrogen gas, and dimethyl sulfide was added as a reducing agent. Silica gel was added to the reaction mixture to separate the polar substances. The degradation products of fucoxanthin were purified by reverse phase HPLC by detecting the absorbance at both 215 nm and 280 nm (Figure 2). Though many peaks appeared on the HPLC chart, the main peaks were collected and analyzed by NMR. Peak A in Figure 2, which has a strong absorbance at 280 nm, was obtained at 1.87 mg (6.9% yield). On the other hand, peak B in Figure 2, which has a strong absorbance at 215 nm, was obtained at 7.85 mg (33% yield). From the ESI-Orbitrap-MS analysis and $^{1}$H-NMR, COSY-NMR, $^{13}$C-NMR, HSQC-NMR, and HMBC-NMR analyses, peak A in Figure 2 was identified as apo-13-fucoxanthinone (**2**) [4] (Table 1) and peak B in Figure 2 was identified as apo-9′-fucoxanthinone (**3**) [4] (Table 2). The NMR spectrum of apo-13-fucoxanthinone (**2**) (Figure S1), which has five methyl groups {$\delta$ = 0.94 (s, 3H, Me-15), 1.03 (s, 3H, Me-14), 1.21 (s, 3H, Me-16), 2.03 (s, 3H, Me-17), 2.35 (s, 3H, Me-18) ppm}, three methylene groups {$\delta$ = 1.34 and 1.50 (each m, 2H, H-2a and H-2b), 1.79 and 2.31 (each m, 2H, H-4a and H-4b), 2.61 (d, 1H, $J_{gem}$ 18.6 Hz, H-7a), 3.63 (d, 1H, $J_{gem}$ 19.0 Hz, H-7b) ppm}, three methine groups {$\delta$ = 6.46 (d, 1H, $J_{11,12}$ 15.4 Hz, H-12), 7.05 (d, 1H, $J_{10,11}$ 11.2 Hz, H-10), 7.47 (dd, 1H, $J_{10,11}$ 11.3 Hz, $J_{11,12}$ 15.4 Hz, H-11) ppm} and one oxymethine group {$\delta$ = 3.81 (m, 1H, H-3) ppm}, agreed well with the NMR data of the compound reported by Mori et al. [4]. In addition, the NMR spectrum of apo-9′-fucoxanthinone (**3**) (Figure S2), which has four methyl groups {$\delta$ = 1.16 (s, 3H, Me-10′), 1.43 (s, 6H, Me-11′ and Me-12′), 2.19 (s, 3H, Me-13′) ppm}, one acetyl group {$\delta$ = 2.05 (s, 3H, Ac) ppm}, two methylene groups {$\delta$ = 1.44 (1H, H-2′a), 1.53 (dd, 1H, $J$ 11.4 Hz, $J$ 12.9 Hz, H-4′a), 2.02 (dd, 1H, $J$ 2.2 Hz, $J$ 4.3 Hz, H-2′b), 2.33 (ddd, 1H, $J$ 2.2 Hz, $J$ 4.3 Hz, $J$ 12.9 Hz, H-4′b) ppm}, one methine group {$\delta$ = 5.87 (s, 1H, H-8) ppm}, and one oxymethine group {$\delta$ = 5.39 (tt, 1H, $J$ 4.2 Hz, $J$ 11.5 Hz, H-3′) ppm} also agreed well with the NMR data of the compound reported by Mori et al. [4]. Since these two types of degradation product have already been found in edible brown algae cultivated in deep seawater by Mori et al. [4], there is a possibility that this experiment artificially reproduced the decomposition of fucoxanthin which occurs in nature. Furthermore, based on the discovery made by Mori et al. and the results from this experiment, positions 13 and 9′ of fucoxanthin are easily oxidatively cleavable sites in the polyene chain of fucoxanthin. As a result, we succeeded in separating the two six-membered rings.

### 2.2. Biology

These obtained compounds were evaluated for their antiproliferative activity using Caco-2 cells. After 72 h of cultivation, cell viability was evaluated by 3-(4,5-dimethylthiazol-2-yl)-2,5-diphenyl tetrazolium bromide (MTT) assay [31] (Figure 3). As a result, it was found that each of the degraded compounds inhibited the proliferation of Caco-2 cells in a concentration-dependent manner. In addition, apo-9′-fucoxanthinone (**3**) was found to inhibit proliferation more strongly than apo-13-fucoxanthinone (**2**). However, its activity was weaker than that of the original fucoxanthin (**1**). Both structures may be necessary to exert a powerful effect like that of fucoxanthin. The synergistic effect of these two degraded compounds will be discussed in the future. Although the activity of apo-13-fucoxanthinone (**2**) has not been studied extensively, the activity of apo-9′-fucoxanthinone (**3**) has recently drawn attention. For example, anti-inflammatory effects [32–34] and hair growth effects [35] have been reported. Detailed examination of the fucoxanthin when decomposed will lead to the discovery of new activity of the degradation products. Furthermore, this study suggests that not only fucoxanthin itself, but also the degradation products, may exert effects as a mechanism of fucoxanthin's activity in humans.

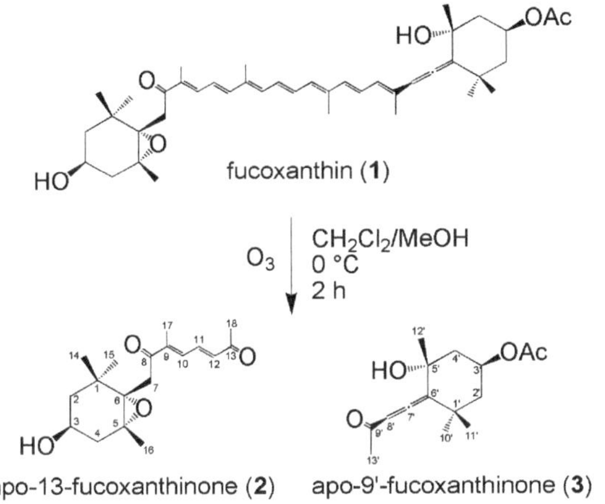


**Figure 1.** Synthesis of two six-membered ring derivatives (apo-13-fucoxanhinone (**2**) and apo-9′-fucoxanthinone (**3**)).

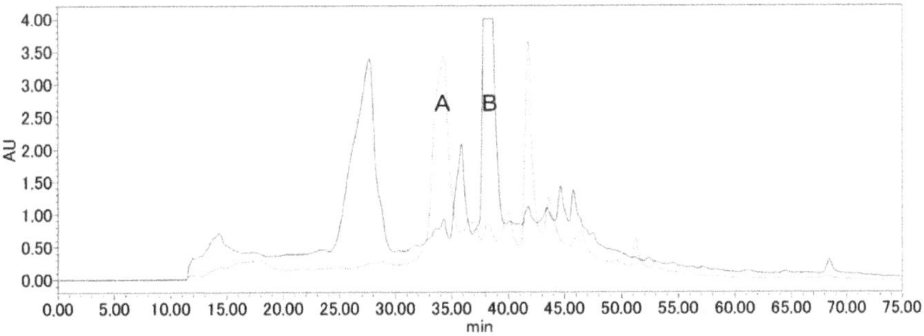


**Figure 2.** HPLC profile of fucoxanthin degradation compounds. Column: Mightysil RP-18 GP 250-20 column (20 × 250 mm, 5 μm; Kanto Chemical, Tokyo, Japan); solvent A: water; solvent B: EtOH. The solvent profile was as follows (linear gradient between each step): initial conditions: A/B = 75:25, 1 mL/min; 1 min: A/B = 75:25, 5 mL/min; 10 min: A/B = 75:25, 5 mL/min; 60 min: A/B = 0:100, 5 mL/min. A straight line indicates an absorbance of 215 nm. The dashed line indicates an absorbance of 280 nm. Peak A: apo-13-fucoxanthinone (**2**) (34.3 min); peak B: apo-9′-fucoxanthinone (**3**) (38.6 min).

**Table 1.** $^{1}$H-NMR and $^{13}$C-NMR chemical shift assignments ($\delta$ = ppm, $CDCl_3$, reference TMS = 0.00 ppm) for peak A in Figure 2, apo-13-fucoxanthinone (**2**).

| **Comp. 2** | $^{1}$H | $^{13}$C |
|---|---|---|
| C-1 | - | 35.1 |
| $CH_2$-2 | 1.34, 1.50 | 47.2 |
| CH-3 | 3.81 | 64.2 |
| $CH_2$-4 | 1.79, 2.31 | 41.6 |
| C-5 | - | 66.1 |
| C-6 | - | 66.7 |
| $CH_2$-7 | 2.61, 3.63 | 41.4 |
| C-8 | - | 198.1 |
| C-9 | - | 143.1 |
| CH-10 | 7.05 | 134.5 |
| CH-11 | 7.47 | 136.7 |
| CH-12 | 6.46 | 135.6 |
| C-13 | - | 197.7 |
| $CH_3$-14 | 1.03 | 24.9 |
| $CH_3$-15 | 0.94 | 28.1 |
| $CH_3$-16 | 1.21 | 21.0 |
| $CH_3$-17 | 2.03 | 12.5 |
| $CH_3$-18 | 2.35 | 28.2 |

**Table 2.** $^{1}$H-NMR and $^{13}$C-NMR chemical shift assignments ($\delta$ = ppm, $CDCl_3$, reference TMS = 0.00 ppm) for peak B in Figure 2, apo-9′-fucoxanthinone (**3**).

| **Comp. 3** | $^{1}$H | $^{13}$C |
|---|---|---|
| C-1′ | - | 36.0 |
| $CH_2$-2′ | 1.44, 2.02 | 45.00 |
| CH-3′ | 5.39 | 67.4 |
| $CH_2$-4′ | 1.53, 2.33 | 45.04 |
| C-5′ | - | 72.0 |
| C-6′ | - | 118.5 |
| C-7′ | - | 209.5 |
| CH-8′ | 5.87 | 100.9 |
| C-9′ | - | 198.0 |
| $CH_3$-10′ | 1.16 | 31.6 |
| $CH_3$-11′ | 1.43 | 28.9 |
| $CH_3$-12′ | 1.43 | 30.8 |
| $CH_3$-13′ | 2.19 | 26.4 |
| $CH_3$-Ac | 2.05 | 21.3 |
| C-Ac | - | 170.4 |

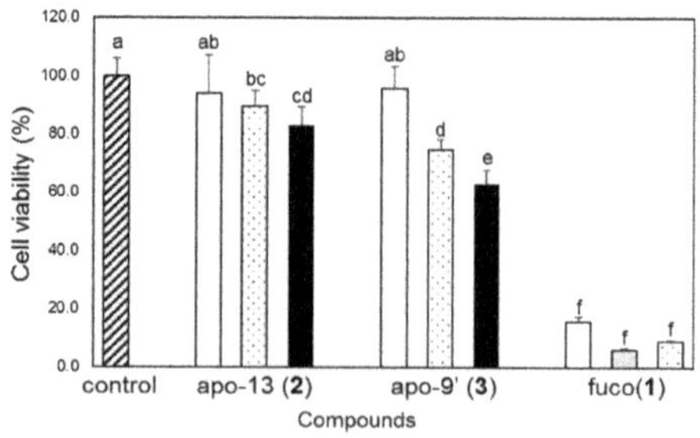


**Figure 3.** Concentration-dependent antiproliferative effect of apo-13-fucoxanthinone (apo-13, **2**), apo-9′-fucoxanthinone (apo-9′, **3**), and fucoxanthin (fuco, **1**) on Caco-2 cells. Caco-2 cells were cultured with apo-13 (**2**), apo-9′ (**3**) and fuco (**1**) in supplemented medium for 72 h. Cell viability was estimated by MTT assay and is expressed relative to the control cells treated with the vehicle alone. The pattern of each bar shows the concentration, as follows: 0 μM: hatched bar, 10 μM: open bar, 20 μM: shaded bar, 30 μM: dotted bar, 50 μM: solid bar. The data represent the mean ± standard deviation of three wells. Replicate experiments demonstrated similar trends. Values not sharing common alphabets (from a to f) were shown to be significantly different with the Tukey–Kramer test ($p < 0.05$).

## 3. Materials and Methods

### 3.1. Reagents and Conditions

$^1$H NMR and $^{13}$C NMR spectra were obtained in $CDCl_3$ on a Bruker BioSpin spectrometer (AV 400, Bruker Corporation, Madison, MA, USA). Chemical shifts are given in ppm and referenced to $Me_4Si$ (δ 0.00). The following abbreviations are used for the characterization of NMR signals: s = singlet, d = doublet, t = triplet, m = multiplet. The ozone generator was made by combining the ozone-generating electrode ZC-60-MM (Silver Seiko Ltd., Tokyo, Japan) with a non-noise s500 air pump (approximately 1 L/min) (Japan Pet Design Co., Ltd., Tokyo, Japan). ESI-Orbitrap-MS spectra were recorded on a Thermo Fisher Scientific instrument (VELOS PRO, Thermo Fisher Scientific Inc., Waltham, MA, USA). The optical rotations were determined in chloroform on a Jasco instrument (P-1020-GT, JASCO Corporation, Tokyo, Japan) under ambient temperature. Reverse-phase HPLC separation of degraded fucoxanthin compounds was performed using a Waters HPLC system with a Mightysil RP-18 GP 250-20 column (20 × 250 mm, 5 μm; Kanto Chemical Co., Inc., Tokyo, Japan) and detected at 215 nm and 280 nm, simultaneously. For the HPLC conditions, solvent A was water and solvent B was ethanol. Initially, 75% of solvent A at a flow rate of 1 mL/min was used; subsequently, the flow rate was increased to 5 mL/min for 1 min, and this condition was maintained for 10 min. Then, a linear gradient of 25% to 100% of solvent B was applied for 60 min, and this condition was maintained. The HPLC analysis of compounds in medium for the antiproliferation assay was performed as follows: fucoxanthin (**1**), apo-13-fucoxanthinone (**2**), and apo-9′-fucoxanthinone (**3**) were analyzed by HPLC (using an LC-20AT pump, an SPD-M10A photodiode array detector, and a CTO-10AS column oven at a constant temperature of 25 °C; Shimadzu Corporation, Kyoto, Japan) on an ODS-80Ts column (2.0 × 150 mm) with an ODS-S1 precolumn (2.0 × 10 mm; Tosoh Corporation, Tokyo, Japan). An isocratic analysis was performed at a flow rate of 0.2 mL/min with acetonitrile/methanol/water (75:15:10, *v*/*v*/*v*) containing 0.1% ammonium acetate for fucoxanthin (**1**) or acetonitrile/methanol/water (45:9:46, *v*/*v*/*v*) containing 0.1% ammonium acetate for apo-13-fucoxanthinone (**2**) and apo-9′-fucoxanthinone (**3**), respectively. These compounds were quantified from the peak area at 450 nm for fucoxanthin (**1**), 288 nm for apo-13-fucoxanthinone (**2**), and 232 nm for apo-9′-fucoxanthinone (**3**), respectively, using an authentic standard calibration curve. Dry wakame from China was purchased from a local market in Tsukuba, Japan. Human colorectal carcinoma Caco-2 cells were obtained from the American Type Culture Collection (Rockville, MD, USA). Dulbecco's modified Eagle's medium (DMEM) was purchased from Nissui Pharmaceutical Co., Ltd. (Tokyo, Japan). The DMEM used a low glucose-type (1.0 g/L glucose), so we increased the glucose concentration to 4.5 g/L. Tetrahydrofuran (THF) was purchased from Nacalai Tesque, Inc. (Kyoto, Japan). THF was purified in a neutral alumina column just before use. All reagents and solvents used were reagent grade.

### 3.2. Synthesis of Two Six-Membered Ring Derivatives {apo-13-Fucoxanthinone (2) and apo-9′-Fucoxanthinone (3)}

Fucoxanthin was purified from wakame (*Undaria pinnatifida*). The extraction solvents and purification method have been referenced in previous experiments [30]. A mixed solvent (chloroform/methanol/water 5:4:1) was used for the extraction and subsequent purification of the residue on the silica gel (ethyl acetate/toluene 1:5) through chromatography, resulting in pure fucoxanthin (1.09 mg/g dry wakame). The $^1$H-NMR and $^{13}$C-NMR spectra were identical to those previously reported. The purified fucoxanthin (58 mg, 88 μmol) was dissolved in dichloromethane/methanol (39 mL/66 mL) and cooled to 0 °C. Ozone gas from a homemade ozone generator was bubbled though the cooled and stirred mixture for 2 h. Afterwards, the ozone gas was replaced with nitrogen gas, dimethyl sulfide (130 μL, 1.77 mmol) was added to the mixture, and the mixture was stirred from 0 °C to room temperature for a further 2 h. Then, silica gel (35 g) was added to the mixture, and the solvent was removed under reduced pressure to absorb the reaction product into the silica gel. Thereafter, the polar substances were removed by silica gel column chromatography (ethyl acetate/*n*-hexane 2:1). The obtained syrup was purified by reverse phase HPLC to obtain compound **2** (apo-13-fucoxanthinone, 1.87 mg, 6.9%) and compound **3** (apo-9′-fucoxanthinone, 7.85 mg, 33%). All experiments were done under dim light in order to minimize the isomerization and degradation of fucoxanthin derivatives due to light irradiation. Compound **2** (apo-13-fucoxanthinone): $[\alpha]_D^{25} = -13$ ($c$ = 0.06, chloroform); $^1$H-NMR (400 MHz, $CDCl_3$), $\delta$ = 0.94 (s, 3H, Me-15), 1.03 (s, 3H, Me-14), 1.21 (s, 3H, Me-16), 1.34 and 1.50 (each m, 2H, H-2a and H-2b), 1.79 and 2.31 (each m, 2H, H-4a and H-4b), 2.03 (s, 3H, Me-17), 2.35 (s, 3H, Me-18), 2.61 (d, 1H, $J_{gem}$ 18.6 Hz, H-7a), 3.63 (d, 1H, $J_{gem}$ 19.0 Hz, H-7b), 3.81 (m, 1H, H-3), 6.46 (d, 1H, $J_{11,12}$ 15.4 Hz, H-12), 7.05 (d, 1H, $J_{10,11}$ 11.2 Hz, H-10), 7.47 (dd, 1H, $J_{10,11}$ 11.3 Hz, $J_{11,12}$ 15.4 Hz, H-11); $^{13}$C-NMR (100 MHz, $CDCl_3$), $\delta$ = 12.5 (C-17), 21.0 (C-16), 24.9 (C-14), 28.1 (C-15), 28.2 (C-18), 35.1 (C-1), 41.4 (C-7), 41.6 (C-4), 47.2 (C-2), 64.2 (C-3), 66.1 (C-5), 66.7 (C-6), 134.5 (C-10), 135.6 (C-12), 136.7 (C-11), 143.1 (C-9), 197.7 (C-13), 198.1 (C-8); ESI-Orbitrap-MS, calcd. for $C_{18}H_{27}O_4{}^+$ [M + H]$^+$: 307.1904, found $m/z$: 307.1905. Compound **3** (apo-9′-fucoxanthinone): $[\alpha]_D^{25} = -19$ ($c$ = 0.08, chloroform); $^1$H-NMR (400 MHz, $CDCl_3$), $\delta$ = 1.16 (s, 3H, Me-10′), 1.43 (s, 6H, Me-11′ and Me-12′), 1.44 (1H, H-2′a), 1.53 (dd, 1H, $J$ 11.4 Hz, $J$ 12.9 Hz, H-4′a), 2.02 (dd, 1H, $J$ 2.2 Hz, $J$ 4.3 Hz, H-2′b), 2.05 (s, 3H, Ac), 2.19 (s, 3H, Me-13′), 2.33 (ddd, 1H, $J$ 2.2 Hz, $J$ 4.3 Hz, $J$ 12.9 Hz, H-4′b), 5.39 (tt, 1H, $J$ 4.2 Hz, $J$ 11.5 Hz, H-3′), 5.87 (s, 1H, H-8); $^{13}$C-NMR (100 MHz), $\delta$ = 21.3 (OCO$\underline{C}H_3$), 26.4 (C-13′), 28.9 (C-11′), 30.8 (C-12′), 31.6 (C-10′), 36.0 (C-1′), 45.00 (C-2′), 45.04 (C-4′), 67.4 (C-3′), 72.0 (C-5′), 100.9 (C-8′), 118.5 (C-6′), 170.4 (O$\underline{C}OCH_3$), 198.0 (C-9′), 209.5 (C-7′); ESI-Orbitrap-MS, calcd. for $C_{15}H_{23}O_4{}^+$ [M + H]$^+$: 267.1591, found $m/z$: 267.1595.

### 3.3. Antiproliferation Activity of Caco-2 Cells by apo-13-Fucoxanthinone (2) and apo-9′-Fucoxanthinone (3)

Caco-2 cells were cultured in DMEM supplemented with 0.1 mM nonessential amino acids, 10% heat-inactivated fetal bovine serum, 4 mM L-glutamine, and antibiotics (40 units/mL penicillin and 40 mg/mL streptomycin) [36]. The culture was carried out at 37 °C in a humidified atmosphere with 5% $CO_2$ in air. In order to evaluate the effects of these compounds on the viability of the cells, the cells were seeded at a density of $5 \times 10^3$ cells per well containing 100 µL of culture medium in 96-well plates for 24 h, and fresh medium was used for the treatment with the compounds described below. Media containing the compounds were prepared using a liquid-drying method, as in our previous study [37]. In brief, fucoxanthin and the degraded compounds dissolved in the purified tetrahydrofuran (THF) and were added to the culture medium. The control culture received only THF in the medium (vehicle alone). The THF in the medium was dried in a centrifugal evaporator. The medium was passed through a 0.2-µm filter to be sterilized, and was then used as fresh medium supplemented with the compounds. To determine the concentration of the compounds, one part of the fresh medium was diluted 41-fold with ethanol for apo-13-fucoxanthinone (**2**) and apo-9′-fucoxanthinone (**3**), or with dimethyl sulfoxide/methanol (2:7, $v/v$) for fucoxanthin (**1**), respectively, and subjected to HPLC analysis. After 72 h of cultivation, cell viability was evaluated by 3-(4,5-dimethylthiazol-2-yl)-2,5-diphenyl tetrazolium bromide (MTT) assay [31]. Data represent means ± standard deviations. The antiproliferation experiments were done under dim yellow light in order to minimize the isomerization and degradation of xanthophylls due to light irradiation. The results were analyzed by one-way ANOVA and the Tukey–Kramer test in order to identify significant differences between treatments, with $p$-values < 0.05 considered significant.

## 4. Conclusions

We succeeded in decomposing fucoxanthin under very mild and neutral conditions. The decomposition obtained here had the same structure as degraded fucoxanthin found in nature. By refining HPLC more precisely, there is a good possibility of obtaining a new degradation product. Currently, more detailed isolation and purification processes are being conducted. In addition, it was found that the two types of degradation product obtained here cause growth suppression in Caco-2 cells. In particular, the compound with an allene structure preferentially inhibited proliferation compared to that without an allene structure. From this result, we predict that an allene structure is important to inhibit the proliferation of Caco-2 cells.

**Supplementary Materials:** The following are available online at http://www.mdpi.com/1660-3397/16/8/275/s1, Figure S1: HSQC-NMR chart of apo-13-fucoxanthinone (**2**), Figure S2: HSQC-NMR chart of apo-9′-fucoxanthinone (**3**).

**Author Contributions:** S.K. and W.T. conceived and designed the experiments; S.K. and E.K.-N. performed the experiments; S.K. and E.K.-N. analyzed the data; W.T. contributed reagents/materials/analysis tools; S.K. wrote the paper.

**Funding:** This research received no external funding.

**Acknowledgments:** This research was conducted using NMR and MS equipment owned by the Advanced Analysis Center, NARO. We thank H. Ono (Advanced Analysis Center, NARO, JP) and his staff for the NMR, ESI-Orbitrap-MS measurements.

**Conflicts of Interest:** The authors declare no conflict of interest. The founding sponsors had no role in the design of the study; in the collection, analyses, or interpretation of data; in the writing of the manuscript, or in the decision to publish the results.

## References

1. Kim, S.M.; Jung, Y.J.; Kwon, O.N.; Cha, K.H.; Um, B.H.; Chung, D.; Pan, C.H. A potential commercial source of fucoxanthin extracted from the microalga *Phaeodactylum tricornutum*. *Appl. Biochem. Biotechnol.* **2012**, *166*, 1843–1855. [CrossRef] [PubMed]
2. Kanazawa, K.; Ozaki, Y.; Hashimoto, T.; Das, S.K.; Matsushita, S.; Hirano, M.; Okada, T.; Komoto, A.; Mori, N.; Nakatsuka, M. Commercial-scale preparation of biofunctional fucoxanthin from waste parts of brown sea algae *Laminalia japonica*. *Food Sci. Technol. Res.* **2008**, *14*, 573–582. [CrossRef]
3. Xiao, X.H.; Si, X.X.; Yuan, Z.Q.; Xu, X.F.; Li, G.K. Isolation of fucoxanthin from edible brown algae by microwave-assisted extraction coupled with high-speed countercurrent chromatography. *J. Sep. Sci.* **2012**, *35*, 2313–2317. [CrossRef] [PubMed]
4. Mori, K.; Ooi, T.; Hiraoka, M.; Oka, N.; Hamada, H.; Tamura, M.; Kusumi, T. Fucoxanthin and its metabolites in edible brown algae cultivated in deep seawater. *Mar. Drugs* **2004**, *2*, 63–72. [CrossRef]
5. Jaswir, I.; Noviendri, D.; Salleh, H.M.; Miyashita, K. Fucoxanthin extractions of brown seaweeds and analysis of their lipid fraction in methanol. *Food Sci. Technol. Res.* **2012**, *18*, 251–257. [CrossRef]
6. Jaswir, I.; Noviendri, D.; Salleh, H.M.; Taher, M.; Miyashita, K.; Ramli, N. Analysis of fucoxanthin content and purification of all-trans-fucoxanthin from *Turbinaria turbinata* and *Sargassum plagyophyllum* by $SiO_2$ open column chromatography and reversed phase-HPLC. *J. Liq. Chromatogr. Relat. Technol.* **2013**, *36*, 1340–1354.
7. Kim, S.M.; Kang, S.W.; Kwon, O.N.; Chung, D.; Pan, C.H. Fucoxanthin as a major carotenoid in *isochrysis* aff. *galbana*: Characterization of extraction for commercial application. *J. Korean Soc. Appl. Biol. Chem.* **2012**, *55*, 477–483. [CrossRef]
8. Pasquet, V.; Cherouvrier, J.R.; Farhat, F.; Thiery, V.; Piot, J.M.; Berard, J.B.; Kaas, R.; Serive, B.; Patrice, T.; Cadoret, J.P.; et al. Study on the microalgal pigments extraction process: Performance of microwave assisted extraction. *Process Biochem.* **2011**, *46*, 59–67. [CrossRef]
9. Xia, S.; Wang, K.; Wan, L.L.; Li, A.F.; Hu, Q.; Zhang, C.W. Production, characterization, and antioxidant activity of fucoxanthin from the marine diatom *Odontella aurita*. *Mar. Drugs* **2013**, *11*, 2667–2681. [CrossRef] [PubMed]
10. Peng, J.; Yuan, J.P.; Wu, C.F.; Wang, J.H. Fucoxanthin, a marine carotenoid present in brown seaweeds and diatoms: Metabolism and bioactivities relevant to human health. *Mar. Drugs* **2011**, *9*, 1806–1828. [CrossRef] [PubMed]
11. Okuzumi, J.; Takahashi, T.; Yamane, T.; Kitao, Y.; Inagake, M.; Ohya, K.; Nishino, H.; Tanaka, Y. Inhibitory effects of fucoxanthin, a natural carotenoid, on *N*-ethyl-*N*′-nitro-*N*-nitrosoguanidine-induced mouse duodenal carcinogenesis. *Cancer Lett.* **1993**, *68*, 159–168. [CrossRef]
12. Das, S.K.; Hashimoto, T.; Baba, M.; Nishino, H.; Komoto, A.; Kanazawa, K. Japanese kelp (kombu) extract suppressed the formation of aberrant crypt foci in azoxymethane challenged mouse colon. *J. Clin. Biochem. Nutr.* **2006**, *38*, 119–125. [CrossRef]
13. Kotake-Nara, E.; Kushiro, M.; Zhang, H.; Sugawara, T.; Miyashita, K.; Nagao, A. Carotenoids affect proliferation of human prostate cancer cells. *J. Nutr.* **2001**, *131*, 3303–3306. [CrossRef] [PubMed]

14. Hosokawa, M.; Kudo, M.; Maeda, H.; Kohno, H.; Tanaka, T.; Miyashita, K. Fucoxanthin induces apoptosis and enhances the antiproliferative effect of the PPAR$\gamma$ ligand, troglitazone, on colon cancer cells. *Biochim. Biophys. Acta* **2004**, *1675*, 113–119. [CrossRef] [PubMed]
15. Das, S.K.; Hashimoto, T.; Shimizu, K.; Yoshida, T.; Sakai, T.; Sowa, Y.; Komoto, A.; Kanazawa, K. Fucoxanthin induces cell cycle arrest at $G_0/G_1$ phase in human colon carcinoma cells through up-regulation of $p21^{WAF1/Cip1}$. *Biochim. Biophys. Acta* **2005**, *1726*, 328–335. [CrossRef] [PubMed]
16. Funahashi, H.; Imai, T.; Mase, T.; Sekiya, M.; Yokoi, K.; Hayashi, H.; Shibata, A.; Hayashi, T.; Nishikawa, M.; Suda, N.; et al. Seaweed prevents breast cancer? *Jpn. J. Cancer Res.* **2001**, *92*, 483–487. [CrossRef] [PubMed]
17. Funahashi, H.; Imai, T.; Tanaka, Y.; Tsukamura, K.; Hayakawa, Y.; Kikumori, T.; Mase, T.; Itoh, T.; Nishikawa, M.; Hayashi, H.; et al. Wakame seaweed suppresses the proliferation of 7,12-dimethylbenz(*a*)anthracene-induced mammary tumors in rats. *Jpn. J. Cancer Res.* **1999**, *90*, 922–927. [CrossRef] [PubMed]
18. Okai, Y.; Higashiokai, K.; Nakamura, S. Identification of heterogenous antimutagenic activities in the extract of edible brown seaweeds, *Laminaria-japonica* (Makonbu) and *Undaria-pinnatifida* (Wakame) by the *umu* gene expression system in *Salmonella typhimurium* (TA1535/pSK1002). *Mutat. Res.* **1993**, *303*, 63–70. [CrossRef]
19. Teas, J.; Vena, S.; Cone, D.L.; Irhimeh, M. The consumption of seaweed as a protective factor in the etiology of breast cancer: Proof of principle. *J. Appl. Phycol.* **2013**, *25*, 771–779. [CrossRef] [PubMed]
20. Moghadamtousi, S.Z.; Karimian, H.; Khanabdali, R.; Razavi, M.; Firoozinia, M.; Zandi, K.; Kadir, H.A. Anticancer and antitumor potential of fucoidan and fucoxanthin, two main metabolites isolated from brown algae. *Sci. World J.* **2014**, *2014*, 768323. [CrossRef]
21. Kumar, S.R.; Hosokawa, M.; Miyashita, K. Fucoxanthin: A marine carotenoid exerting anti-cancer effects by affecting multiple mechanisms. *Mar. Drugs* **2013**, *11*, 5130–5147. [CrossRef] [PubMed]
22. Satomi, Y. Antitumor and cancer-preventative function of fucoxanthin: A marine carotenoid. *Anticancer Res.* **2017**, *37*, 1557–1562. [CrossRef] [PubMed]
23. Maeda, H.; Hosokawa, M.; Sashima, T.; Takahashi, N.; Kawada, T.; Miyashita, K. Fucoxanthin and its metabolite, fucoxanthinol, suppress adipocyte differentiation in 3T3-L1 cells. *Int. J. Mol. Med.* **2006**, *18*, 147–152. [CrossRef] [PubMed]
24. Maeda, H.; Hosokawa, M.; Sashima, T.; Funayama, K.; Miyashita, K. Fucoxanthin from edible seaweed, *Undaria pinnatifida*, shows antiobesity effect through UCP1 expression in white adipose tissues. *Biochem. Biophys. Res. Commun.* **2005**, *332*, 392–397. [CrossRef] [PubMed]
25. Miyashita, K.; Nishikawa, S.; Beppu, F.; Tsukui, T.; Abe, M.; Hosokawa, A. The allenic carotenoid fucoxanthin, a novel marine nutraceutical from brown seaweeds. *J. Sci. Food Agric.* **2011**, *91*, 1166–1174. [CrossRef] [PubMed]
26. Heo, S.J.; Yoon, W.J.; Kim, K.N.; Oh, C.; Choi, Y.U.; Yoon, K.T.; Kang, D.H.; Qian, Z.J.; Choi, I.W.; Jung, W.K. Anti-inflammatory effect of fucoxanthin derivatives isolated from *Sargassum siliquastrum* in lipopolysaccharide-stimulated RAW 264.7 macrophage. *Food Chem. Toxicol.* **2012**, *50*, 3336–3342. [CrossRef] [PubMed]
27. Sugawara, T.; Matsubara, K.; Akagi, R.; Mori, M.; Hirata, T. Antiangiogenic activity of brown algae fucoxanthin and its deacetylated product, fucoxanthinol. *J. Agric. Food. Chem.* **2006**, *54*, 9805–9810. [CrossRef] [PubMed]
28. Fung, A.; Hamid, N.; Lu, J. Fucoxanthin content and antioxidant properties of *Undaria pinnatifida*. *Food Chem.* **2013**, *136*, 1055–1062. [CrossRef] [PubMed]
29. Kotake-Nara, E.; Sugawara, T.; Nagao, A. Antiproliferative effect of neoxanthin and fucoxanthin on cultured cells. *Fish. Sci.* **2005**, *71*, 459–461. [CrossRef]
30. Komba, S.; Kotake-Nara, E.; Machida, S. Fucoxanthin derivatives: Synthesis and their chemical properties. *J. Oleo Sci.* **2015**, *64*, 1009–1018. [CrossRef] [PubMed]
31. Mosmann, T. Rapid colorimetric assay for cellular growth and survival: Application to proliferation and cytotoxicity assays. *J. Immunol. Methods* **1983**, *65*, 55–63. [CrossRef]
32. Kim, E.A.; Kim, S.Y.; Ye, B.R.; Kim, J.; Ko, S.C.; Lee, W.W.; Kim, K.N.; Choi, I.W.; Jung, W.K.; Heo, S.J. Anti-inflammatory effect of Apo-9′-fucoxanthinone via inhibition of MAPKs and NF-κB signaling pathway in LPS-stimulated RAW 264.7 macrophages and zebrafish model. *Int. Immunopharmacol.* **2018**, *59*, 339–346. [CrossRef] [PubMed]
33. Chae, D.; Manzoor, Z.; Kim, S.C.; Kim, S.; Oh, T.H.; Yoo, E.S.; Kang, H.K.; Hyun, J.W.; Lee, N.H.; Ko, M.H.; et al. Apo-9′-fucoxanthinone, isolated from *Sargassum muticum*, inhibits CpG-induced inflammatory response by attenuating the mitogen-activated protein kinase pathway. *Mar. Drugs* **2013**, *11*, 3272–3287. [CrossRef] [PubMed]

34. Yang, E.J.; Ham, Y.M.; Lee, W.J.; Lee, N.H.; Hyun, C.G. Anti-inflammatory effects of apo-9′-fucoxanthinone from the brown alga, *Sargassum muticum*. *DARU J. Pharm. Sci.* **2013**, *21*, 62. [CrossRef] [PubMed]
35. Kang, J.I.; Yoo, E.S.; Hyun, J.W.; Koh, Y.S.; Lee, N.H.; Ko, M.H.; Ko, C.S.; Kang, H.K. Promotion effect of apo-9′-fucoxanthinone from *Sargassum muticum* on hair growth via the activation of Wnt/β-catenin and VEGF-R2. *Biol. Pharm. Bull.* **2016**, *39*, 1273–1283. [CrossRef] [PubMed]
36. Sugawara, T.; Kushiro, M.; Zhang, H.; Nara, E.; Ono, H.; Nagao, A. Lysophosphatidylcholine enhances carotenoid uptake from mixed micelles by Caco-2 human intestinal cells. *J. Nutr.* **2001**, *131*, 2921–2927. [CrossRef] [PubMed]
37. Kotake-Nara, E.; Hase, M.; Kobayashi, M.; Nagao, A. 3′-Hydroxy-ε,ε-caroten-3-one inhibits the differentiation of 3T3-L1 cells to adipocytes. *Biosci. Biotechnol. Biochem.* **2016**, *80*, 518–523. [CrossRef] [PubMed]

*marine drugs*

*Article*

# A Rapid Method for the Determination of Fucoxanthin in Diatom

**Li-Juan Wang [1,2], Yong Fan [2,*], Ronald L. Parsons [3], Guang-Rong Hu [2], Pei-Yu Zhang [1,*] and Fu-Li Li [2]**

1 College of Environmental Science and Engineering, Qingdao University, Qingdao 266071, China; w873644513@163.com
2 Key Laboratory of Biofuels, Shandong Provincial Key Laboratory of Synthetic Biology, Qingdao Engineering Laboratory of Single Cell Oil, Qingdao Institute of Bioenergy and Bioprocess Technology, Chinese Academy of Sciences, Qingdao 266101, China; hugr@qibebt.ac.cn (G.-R.H.); lifl@qibebt.ac.cn (F.-L.L.)
3 Solix Algredients Inc., 120 Commerce Dr., Ste 4, Fort Collins, CO 80524, USA; ron.parsons@solixalgredients.com
* Correspondence: fanyong@qibebt.ac.cn (Y.F.); envbio@163.com (P.-Y.Z.); Tel.: +86-0532-80662656 (Y.F.); +86-0532-83780155 (P.-Y.Z.)

Received: 4 December 2017; Accepted: 6 January 2018; Published: 22 January 2018

**Abstract:** Fucoxanthin is a natural pigment found in microalgae, especially diatoms and *Chrysophyta*. Recently, it has been shown to have anti-inflammatory, anti-tumor, and anti-obesityactivity in humans. *Phaeodactylum tricornutum* is a diatom with high economic potential due to its high content of fucoxanthin and eicosapentaenoic acid. In order to improve fucoxanthin production, physical and chemical mutagenesis could be applied to generate mutants. An accurate and rapid method to assess the fucoxanthin content is a prerequisite for a high-throughput screen of mutants. In this work, the content of fucoxanthin in *P. tricornutum* was determined using spectrophotometry instead of high performance liquid chromatography (HPLC). This spectrophotometric method is easier and faster than liquid chromatography and the standard error was less than 5% when compared to the HPLC results. Also, this method can be applied to other diatoms, with standard errors of 3–14.6%. It provides a high throughput screening method for microalgae strains producing fucoxanthin.

**Keywords:** *Phaeodactylum tricornutum*; fucoxanthin; spectrophotometry; high through-put screening

## 1. Introduction

Fucoxanthin is a carotenoid belonging to the xanthophyll class of carotenoids [1]. In recent years, much work has focused on studying the effect of dietary fucoxanthin and on demonstrating that fucoxanthin can be used as a safe and effective dietary supplement. It has anti-inflammatory, anti-tumor, anti-obesity, anti-diabetes, anti-malarial, and other physiological activities [2–4]. Clinical studies have shown that taking fucoxanthin can speed up the metabolism, but will not stimulate the central nervous system [5]. Fucoxanthin is widely found in brown algae and diatoms [6]. In macroalgae, the content of fucoxanthin is about 0.1–1 $mg{\cdot}g^{-1}$ (dry cell weight). Health products containing fucoxanthin derived from brown algae are already sold on markets. However, the production price is too high to meet the expectations of markets, due to the long growth cycle of macroalgae and the low extraction efficiency [7,8]. In microalgae, especially diatoms, fucoxanthin is one of the main pigments in cells, and accounts for about 1–2.5% of the dry cell weight, which is several fold higher than in macroalgae [9,10]. Fucoxanthin is a major component in the Fucoxanthin-Chlorophyll Protein (FCP) complex [11]. The FCP has the function of trapping light energy and light protection, and plays an important role in photosynthesis in diatoms. In the extraction procedures, the high content of lipid is beneficial to the extraction of fucoxanthin. At present, all over the global market, the pharmaceutical and food from the microalgae are mainly produced by spirulina, *Chlorella*, *Dunaliella*, and *Haematococcus pluvialis*. Due to diverse compounds found in microalgae, there are great potential to develop new products derived from microalgae, such as fucoxanthin, nervonic acid, etc. [12–14].

*Phaeodactylum tricornutum* is a model species of diatoms. In the early part of 21st century, this species of microalgae has been widely investigated as a potential source for biofuel and/or eicosapentaenoic acid (EPA) with a number of reports discussing the growth of *P. tricornutum* under laboratory or pilot scale [15,16]. *P. tricornutum* grows rapidly in the laboratory and at scale, moreover, the genome and a genetic transformation system have also been published [17–19]. Fucoxanthin in the *P. tricornutum* ranges from 15.42–16.51 $mg \cdot g^{-1}$ [20], which is a suitable level for fucoxanthin production. In order to improve the yield of fucoxanthin, physical and chemical mutagenesis could be applied to generate mutants. However, a rapid method to screen for mutants with higher fucoxanthin content is needed to accelerate the discovery of high-content strains.

At present, the concentration of pigments is mainly determined by high performance liquid chromatography (HPLC) [21,22], which limits the throughput because of extraction time, column time for each run, and requires skilled operators to maintain the equipment. For a high-throughput method, it is essential to simplify the extraction and detection methods, while maintaining accuracy. In plants, plankton and green algae, the Chl *a*, *b*, and total carotenoid contents can be determined by using spectrophotometric methods [23–26]. However, the reported methods either have complicated extraction methods or are biased towards pigment complexes found in Chlorophytes. In this study, we show that the concentration of fucoxanthin can be assessed using a spectrophotometer measuring values at 663 nm, 445 nm, and 750 nm. This work will not only improve screening efficiency of *P. tricornutum* mutants, but also increase the monitoring efficiency of fucoxanthin content in the cultivation process.

## 2. Results

When the total pigments of *P. tricornutum* at different culture stages were quantified by HPLC, there are mainly five product peaks: chlorophyll *c*, fucoxanthin, diatoxanthin, chlorophyll *a*, and β-carotene. Chl *c* levels varied greatly at different culture stages, but Chl *a* and fucoxanthin were relatively stable over the culture (Figure 1). The visible spectrum scan of the total pigment extract contained two main peaks. A broad peak centered at 445 nm and a sharper peak centered at 663 nm. But the purified fucoxanthin exhibits little absorption at 663 nm. However, at 445 nm, the absorption peak has contributions from both fucoxanthin and Chl *a* (Figure 2). Based on the Lambert-Beer law, the content of Chl *a* could be calculated by the absorbance values at 445 nm and 663 nm, respectively. Chl *a* of *P. tricornutum*, *Nannochloropsis oceanica*, *Mychonastes afer*, and tobacco leaf were obtained by thin-layer chromatography (TLC). The Chl *a* concentration was calculated by the absorbance at 663 nm based on Arnon's method ($\varepsilon_{663\ nm} = 82.04$; Formula (2)) [27]. In addition, the light absorption of the samples was measured at 445 nm and a calibration curve was established between the results for the two wavelengths (Figure 3). The extinction coefficient of Chl *a* is 66.8 at 445 nm with the concentration range from 1 to 7 $mg \cdot L^{-1}$.

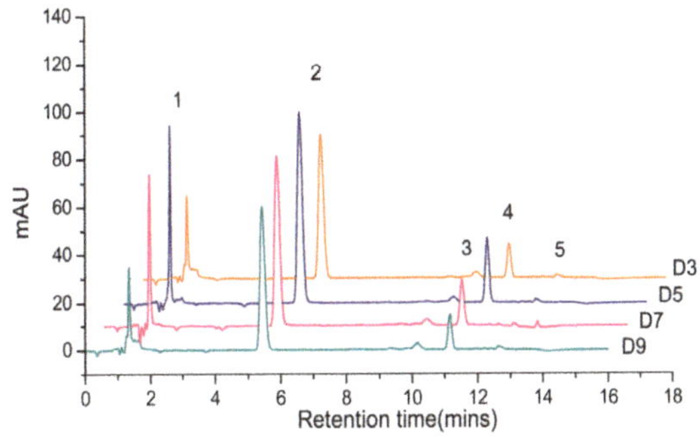


**Figure 1.** The high performance liquid chromatography (HPLC) chromatogram of total pigments in *P. tricornutum*. Different colored lines represent different culture days. 1: Chl *c*; 2: Fucoxanthin; 3: Diatoxanthin; 4: Chl *a*; 5: β-carotene.

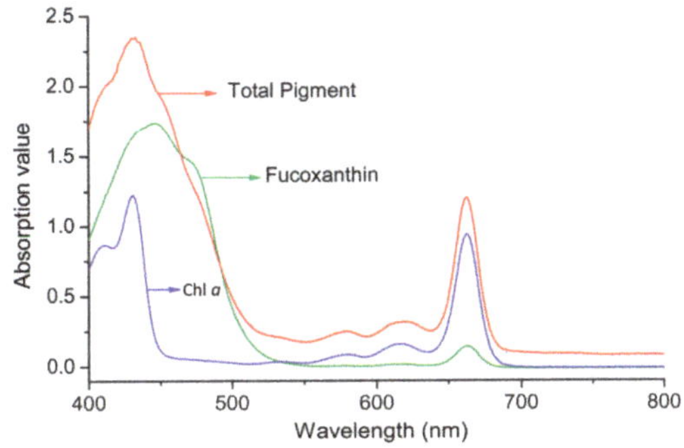


**Figure 2.** The visible spectrum scan of pigments in *P. tricornutum*. Green line: fucoxanthin; red line: total pigment; blue line: Chl *a*.

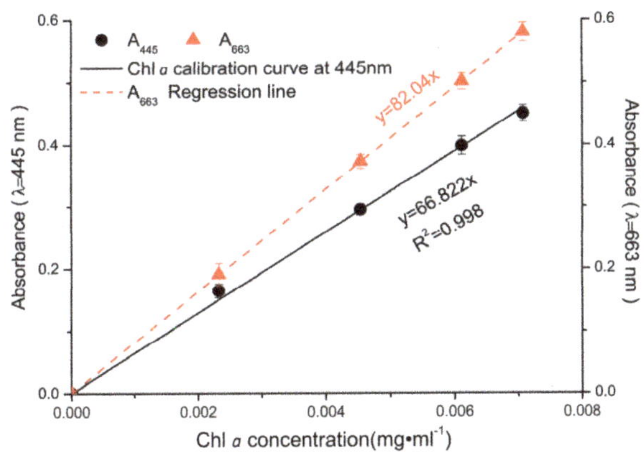


**Figure 3.** The absorbance values of Chl *a* at 445 nm and 663 nm. Red dots: $A_{663}$ of Chl *a* from tobacco leaf, microalgae *N. oceanica*, *M. afer*, and *P. tricornutum*, respectively; red line: the line of best fit for the $A_{663}$ values. Black dots and line show the data for the samples at $A_{445}$.

According to the literature [28,29], the extinction coefficient of fucoxanthin at 449 nm is 1600, with a concentration of 1% (*w*/*v*, g/100 mL). The extinction coefficient at 445 nm was corrected to 156.54 (with a concentration of 1 g·L$^{-1}$) using the purified standard as a sample over the concentration range from 0.005 to 0.05 mg·L$^{-1}$.

According to the additivity law of absorbance, the $A_{445}$ of the pigment extract is the absorbance of fucoxanthin and Chl *a* at 445 nm (Formula (1); $a_1$ is the extinction coefficient for fucoxanthin; and, $a_2$ is the extinction coefficient for Chl *a*), because they are the main absorbing components at 445 nm in *P. tricornutum*.

$$A_{445} = a_1 \times C_{fuc} + a_2 \times C_{\mathrm{Chl}\ a} \tag{1}$$

The concentration of Chl *a* is the absorption value at 663 nm divided by the extinction coefficient. The Chl *a* absorption value at 445 nm is the extinction coefficient at 445 nm multiplied by the Chl *a* concentration (Formula (2)). Because we desire to calculate the content of fucoxanthin, Formula (2) was rearranged into the form of Formula (3). When the numerical values for the extinction coefficients were placed into Formula (3), we could further simplify the formula to Formula (4). In the process, we transformed $C_{fuc}$ to $C_{fuc}'$ so the results of the equation would be in mg·L$^{-1}$.

$$A_{445} = a_1 \times C_{fuc} + \frac{a_2 \times A_{663}}{82.04} \tag{2}$$

$$C_{fuc} = \frac{A_{445}}{a_1} - \frac{\frac{a_2 \times A_{663}}{82.04}}{a_1} \tag{3}$$

$$C_{fuc}' = 6.39 \times A_{445} - 5.18 \times A_{663} \tag{4}$$

Using Formula (4), we calculated the fucoxanthin content of the culture using culture extracted by ethanol and cell debris removed by centrifugation prior to reading the samples in the spectrophotometer. However, this process was not accurate enough (data not shown), so we needed to optimize the formula further.

We tried to calculate the fucoxanthin concentration in whole cells using the absorbance value of cell culture at 445 nm and 663 nm. However, fucoxanthin is insoluble in the culture medium, and its extinction coefficients in water and ethanol are different. The $A_{445}$ and $A_{663}$ of algal cells suspension in f/2 medium or in ethanol showed no strong correlation coefficients (Figure 4). The absorbance values of algae suspension in culture medium (*ASC*) and algae suspension in ethanol (*ASE*) were collected. The coefficient of determination of the absorbance ($R^2$) at 750 nm is 0.366 between *ASC* and *ASE*, they also showed significant difference ($p = 0.004$; Figure 4c). On the other hand, as mentioned before, the $A_{445}$ and $A_{663}$ of the *ASC* also exhibited significant differences from *ASE* ($R^2_{445\ \mathrm{nm}} = 0.618$; $p_{445\ \mathrm{nm}} = 0.003$; $R^2_{663\ \mathrm{nm}} = 0.733$; $p_{663\ \mathrm{nm}} = 0.041$; Figure 4a,b). Therefore, if we simplify this method by measuring the data in one situation, we are unable to obtain a correct result.

The previous results suggested that we had to determine the absorbance of algal cells in ethanol at 445 nm and 663 nm with some correction for the interference from other pigments and/or cell debris. Therefore, we needed to remove the "background noise" present in the whole-cell suspensions at 445 nm and 663 nm, which were represented by *n1* and *n2* in the Formula (5), respectively.

$$C_{fuc}' = 6.39 \times (A_{445} - n1) - 5.18 \times (A_{663} - n2) \tag{5}$$

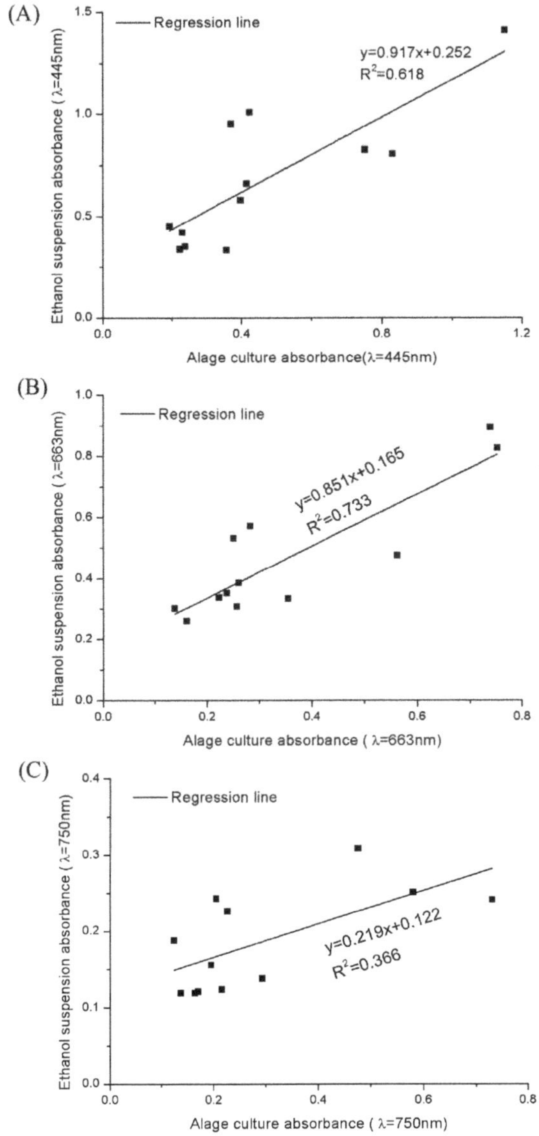


**Figure 4.** The regression lines of the absorbance values in different wavelength between algae suspension in culture medium (*ASC*) and algae suspension in ethanol (*ASE*). (**A**–**C**) represent the absorbance data at 445 nm, 663 nm and 750 nm, respectively.

We took an experimental approach to determine the values of $n1$ and $n2$ in formula 5. In theory, the "background noise" would be a function of the number of cells used to make the measurement. We tested this by determining whether there was a correlation between the number of cells (using $A_{750}$ as a proxy) and cell debris with all of the pigments, except fucoxanthin and Chl *a*. We extracted the pigments from cells, saving the cell debris. After Chl *a* and fucoxanthin were separated out, all of the other pigments (including Chl *c*, diatoxanthin, and β-carotene) and the saved cell debris were resuspended with ethanol and measured at 445 nm and 663 nm. With the absorption value at 750 nm as ordinate, and the absorption value of other pigments and cell debris at 445 nm and 663 nm as abscissa, respectively, the regression curves shown in Figure 5 were obtained.

$$n1(A_{445}{}') = 0.14 \times A_{750} + 1,\ R^2 = 0.9968$$

$$n2(A_{663}{}') = 0.233 \times A_{750} + 0.217,\ R^2 = 0.9962$$

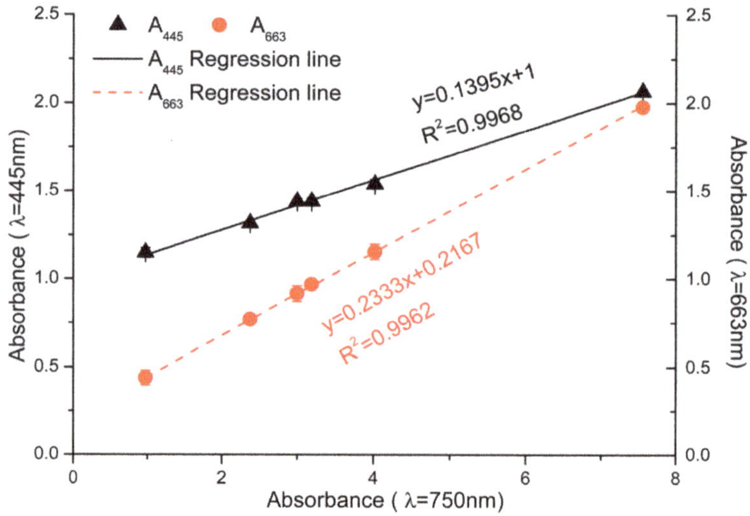


**Figure 5.** The absorbance value of the residue pigments mixture and cell debris at 445 nm and 663 nm. Black triangles: $A_{445}$ of the mixture; red dots: $A_{663}$ of the mixture. The abscissa showed the absorbance of cell culture at 750 nm. Black and red lines represented the fit curve of the data at different wavelength.

The equations for the regression lines shown in Figure 5 represent the "background noise" interfering at 445 nm and 663 nm. We can substitute the equations into Formula (5), which results in Formula (6). This can be further simplified to Formula (7). Using Formula (7), the concentration of fucoxanthin in algal cells could be determined through measuring the absorbance of cell culture at 750 nm and algal cell suspension in ethanol at 445 nm and 663 nm (Figure 6A).

$$C_{fuc}{}' = 6.39 \times [A_{445} - (0.14 \times A_{750} + 1)] - 5.18 \times [A_{663} - (0.233 \times A_{750} + 0.217)] \quad (6)$$

$$C_{fuc}{}' = 6.39 \times A_{445} - 5.18 \times A_{663} + 0.312 \times A_{750} - 5.27 \quad (7)$$

In order to verify the correctness of Formula (7), *P. tricornutum* cultures at a cell density of $2 \times 10^7$–$1 \times 10^8$ cells·mL$^{-1}$ at different growth stages were selected for fucoxanthin assay by HPLC and spectrophotometry. Regression analysis and paired *t*-test of these two results were carried out. The statistical analysis of these data derived from above two methods showed in early stage the standard errors were 0.80–4.56%, in middle stage the standard errors were 1.38–4.73%, in late stage the standard errors were 0.23–4.70% (Figure 6B). Therefore, the spectroscopic method has enough accuracy to calculate the fucoxanthin concentration of *P. tricornutum* culture throughout its growth cycle.

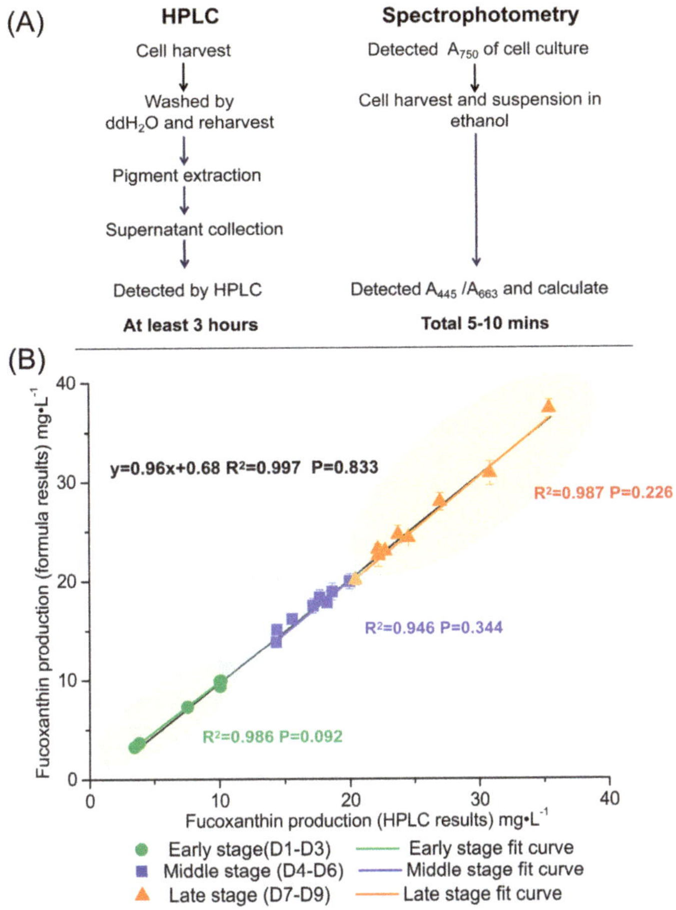


**Figure 6.** The contrast between HPLC and spectrophotometry. (**A**) The flow-process diagram and time cost of these two methods. (**B**) Spectrophotometric determination is verified at different culture stages. The *x*-axis is fucoxanthin production determined by HPLC, the *y*-axis is fucoxanthin production determined by spectrophotometry. Different colored squares represent different growth stages of the culture. The green area represents the early stage (from 1st day to 3rd day); the blue area represents the middle stage (from 4th day to 6th day); and the orange area represents the late stage (from 7th day to 9th day). The black, green, blue and orange lines represent the curve fit for the total data set, the early stage data set, middle stage data set and late stage data set, respectively. $R^2$ is the coefficient of determination. The *p* value is calculated by paired *t*-test. All of the experiments were repeated three times and expressed as mean ± standard deviation (SD).

In order to further improve the efficiency of testing fucoxanthin with our spectroscopic method, a multi-mode microplate reader (Synergy™ HT, BioTek, Winooski, VT, USA) was used, in which 96 samples could be analyzed in 10 min for one batch. It was noted that the volume of samples in wells of 96-well plate was 200 μL and the light path was about 0.5 cm. The absorption should be revised by the calibration coefficient of 0.504, 0.497, and 0.444 at $A_{445}$, $A_{663}$, and $A_{750}$, respectively (Figure 7).

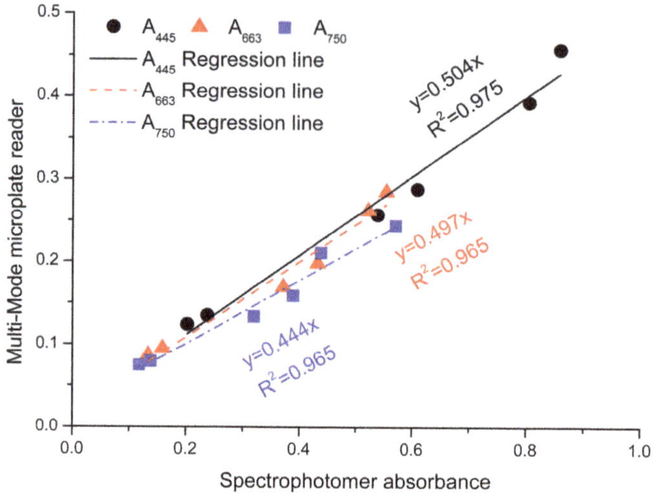


**Figure 7.** The calibration curves of Multi-Mode microplate reader at different wavelengths. The $x$-axis has the results from the spectrophotometer and the $y$-axis has the results from the Multi-Mode microplate reader. Black dots: absorbance values of *ASE* at 445 nm; red dots: absorbance values of *ASE* at 663 nm; blue dots: absorbance values of *ASC* at 750 nm.

## 3. Discussion

Diatoms are a potentially important resource for the production of fucoxanthin. In order to screen for high-fucoxanthin producing strains, a high throughput method for measuring the concentration of fucoxanthin is required. Based on the properties of pigments and statistical analysis, Formula (7) was obtained, which can be efficiently used to calculate the content of fucoxanthin in cultures of *P. tricornutum*.

In developing this method, we wanted to use ethanol to extract the pigment, but the extinction coefficient used for calculating Chl *a* is based on 80% acetone [27,30]. In order to ensure that the extinction coefficient determined in 80% acetone could be used with our method, we compared the absorbance value (at 663 nm) of Chl *a* isolated from different plant or algal material in either 80% acetone or ethanol. The results for the samples in different solvents were analyzed by paired $t$-test, and all of the samples show no significant differences ($p = 0.063$). Based on that result, we were convinced that we could use the classic extinction coefficient in our formulas.

When detecting the content of pigment using the HPLC, we found that the extraction efficiency was dependent on whether the cells are washed or not, as well as the wash solution used. We tested three methods to wash the cells after centrifugation prior to extracting the pigments: the first consisted in dd$H_2O$, the second in f/2 culture medium, and the third was a control group that was not washed (simply resuspended in the supernatant in the tube after centrifugation). Our results show that washing with dd$H_2O$ yielded the highest extraction efficiency (Figure 8). This is probably because washing with dd$H_2O$ started lysing the cells, improving pigment extraction. Based on these results, we used dd$H_2O$ washed cells to compare the spectrophotometric method against the HPLC method.

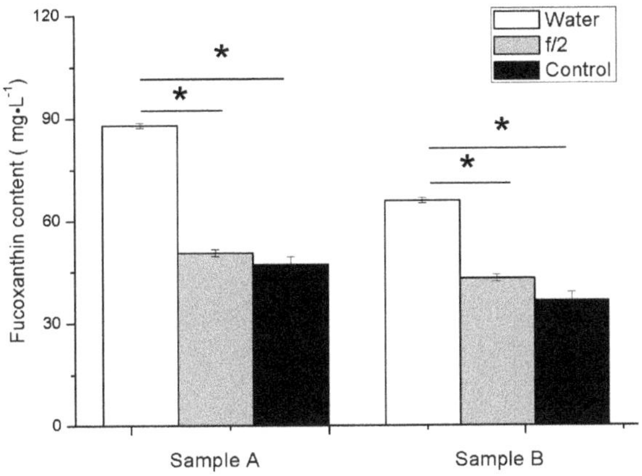


**Figure 8.** The fucoxanthin concentration of different samples detected by HPLC. Sample A is cells on the 9th day, sample B is cells on the 7th day. Statistically significant differences of fucoxanthin content among different methods are indicated with asterisks above the columns ($p = 0.05$).

Fucoxanthin is widely found in diatoms so we tested our method with two other diatoms: *C. muelleri* and *T. pseudonana* (both centric diatoms). We did a direct test without modifying Formula (7). The results demonstrate that the formula can measure the concentration of fucoxanthin in these two diatom strains with good confidence (Table 1). The standard error for the measurement is larger than what we saw for *P. tricornutum*, presumably due to a lower fucoxanthin content. Therefore, in a rough comparison, this spectroscopic method could be applied to screen other diatom species, as well as being used to quickly detect the fucoxanthin content in production.

**Table 1.** Comparison of the spectrophotometric methods with HPLC during different culture stages of *Chaetoceros muelleri* and *Thalassiosira pseudonana*.

| Culture Stages | *Chaetoceros muelleri* Fucoxanthin Production (mg·L$^{-1}$) | | | *Thalassiosira pseudonana* Fucoxanthin Production (mg·L$^{-1}$) | | |
|---|---|---|---|---|---|---|
| | Spectrophotometry | HPLC | Error (%) | Spectrophotometry | HPLC | Error (%) |
| Early stage | 2.20 ± 0.067 | 2.5 | 12.03 ± 2.68 | 2.27 ± 0.04 | 2.55 | 10.98 ± 1.57 |
| Middle stage | 2.30 ± 0.043 | 2.55 | 5.89 ± 1.69 | 3.05 ± 0.01 | 3.38 | 9.76 ± 0.39 |
| Late stage | 9.24 ± 0.042 | 8.93 | 3.47 ± 0.47 | 3.67 ± 0.01 | 3.51 | 4.7 ± 0.14 |

The values are mean ± SD; SD: standard deviation. All experiments were repeated three times.

Due to the differences in pigment distribution, cell size, and other cell components at different grow stages (as well as between pennate and centric diatoms), we expected that there would be significant differences within the growth cycle as well as between different species of diatoms that would impact the accuracy of Formula (7). However, we did not see a significant difference among different growth stages of *P. tricornutum* (Figure 6). We did see an increase in the error associated with the centric diatoms. It is possible that developing species-specific measurements of the "background noise" and substituting them in for the *P. tricornutum* values could improve the measurements. In comparison with the HPLC method of fucoxanthin detection, the spectrophotometric method described here simplified the operation process and saved time without markedly affecting accuracy.

## 4. Materials and Methods

### 4.1. Microalgae and Cultivation

Algal strains *P. tricornutum*, *Mychonastes afer*, and *Nannochloropsis oceanica* IMET1 were provided by the center of BioEnergy Culture Collection (CBECC) in Qingdao Institute of Bioenergy and Bioprocess Technology. *Chaetoceros muelleri* (CCMP 1316) and *Thalassiosira pseudonana* (CCMP 1335) were purchased from National Center for Marine Algae and Microbiota (NCMA, East Boothbay, ME, USA). *P. tricornutum* and *N. oceanica* IMET1 were grown photoautotrophically in modified f/2 medium [31], with increased concentration of sodium nitrate (1 $g \cdot L^{-1}$). *M. afer* was culture in BG-11 medium. Cells were cultured at 23 °C in 100 mL bubble columns (30 cm in height, 4 cm in diameter, 100 mL medium) at a photon flux density of 80 μmol photons$\cdot m^{-2} \cdot s^{-1}$ with a 12:12-h light/dark photoperiod.

### 4.2. Extraction and Purification of the Fucoxanthin

The *P. tricornutum* cells in different culture stages were centrifuged at 4000× *g*, then rinsed with $ddH_2O$, and recollected by centrifugation. Pellets were suspended in ethanol for pigment extraction (ethanol:algae culture volume = 1:1; *v*/*v*). In agreement with others, we found ethanol to be the most effective solvent in the extraction of fucoxanthin, with the extraction yield being ethanol > acetone > ethyl acetate [20]. The extraction system was incubated at 45 °C for 2 h, and mixed by vortex mixer every half an hour. Finally, the pigment solution was separated by centrifugation at 4000× *g*. The visible spectrum of the pigment solution was obtained by scanning from 400 to 800 nm with a spectrometer (Perkin Elmer UV-VIS Spectrometer Lambda 25, Waltham, MA, USA).

Purification of the fucoxanthin was carried out by using solid phase extraction (SPE) columns (Agilent Bond Elut HF Mega BE-SI, 5 mg 20 mL, Santa Clara, CA, USA). The pigment extracts were dried under nitrogen and resuspended in the mobile phase (n-hexane:acetone = 6:4) [6]. The total pigments were loaded on the SPE columns, then eluted by the mobile phase. Chl *a* eluted first, followed by fucoxanthin.

Pigments purity was checked using silica plates (Merck TLC Silica gel 60, Darmstadt, Germany) using hexane:acetone = 6:4 as the mobile phase. The pigment spots were detected visually, scraped from the plate, and then resuspended in ethanol for spectrophotometric analysis.

Quantification of fucoxanthin by HPLC was accomplished using a HITACHI Primaide HPLC system (HITACHI, Tokyo, Japan) with a $C_{18}$ reverse phase column (2.7 μm particle size, 100 × 4.6 mm). The mobile phase consisted of acetonitrile and water with a flow rate of 1 $mL \cdot min^{-1}$. After loading the column with the fucoxanthin extract in ethanol, the mobile phase was an acetonitrile:water solution with the ratio increasing from 80:20 to 100:0 over 8 min, maintained at 100:0 for 3 min, and then decreased back to 80:20 over 5 min. The chromatogram was recorded at 445 nm. Fucoxanthin standards (ChromaDex, fucoxanthin (P), ASB-00006296-010, Irvine, CA, USA) were used for the construction of standard curve in the concentration range of 0.01–1 $mg \cdot mL^{-1}$.

### 4.3. Spectrophotometric Assay

The extinction coefficient of the Chl *a* at 445 nm was calculated using a published extinction coefficient at 663 nm ($\varepsilon_{663\ nm}$ = 82.04) [30] using Chl *a* prepared from various sources, including tobacco leaf, *N. oceanica*, and *P. tricornutum*. The absorbance values at 445 nm and 663 nm were measured and a calibration curve was established. This allowed for us to estimate the extinction coefficient of Chl *a* at 445 nm.

For detecting the concentration of fucoxanthin in *P. tricornutum* using a spectrophotometry, the cultures were diluted with culture medium, and the absorbance measured at 750 nm ($A_{750}$ ranges from 0.1 to 0.8). In parallel, a volume of culture was centrifuged and the cells resuspended in an equal volume of ethanol, then the $A_{445}$ and $A_{663}$-values were detected after dilution with ethanol ($A_{445}$ & $A_{663}$ range from 0.2 to 1). Samples were protected from light exposure as much as possible using foil.

The cells were suspended in ethanol and analyzed at $A_{445}$ and $A_{663}$ within 5 min. With this data, the concentration of fucoxanthin could be calculated using our formula.

A multi-mode microplate reader (Synergy$^{TM}$ HT, BioTek, Winooski, VT, USA) with 96-well plates was used to study the feasibility of high-throughput analysis following the method described above with the following modifications. The volume of samples in each well was 200 µL. We corrected to a 1 cm path-length using the software provided with the plate reader.

### 4.4. Statistical Analysis

Statistical significance of the values obtained from each experiment was evaluated by regression analysis and paired *t*-test using the software SPSS (version 19.0, IBM, Chicago, IL, USA). All of the experiments were repeated three times. Unless otherwise stated, all data were expressed as mean ± standard deviation (SD). The *p* values of less than 0.05 were considered statistically significant.

## 5. Conclusions

Fucoxanthin is a bioactive substances from marine with high economic value. Using microalgae for the fucoxanthin production enjoy great development and market potential. In this study, we developed an accurate and convenient method to test the concentration of fucoxanthin in diatoms by spectrophotometer. This method not only improve screening efficiency of diatom mutants, but also increase the monitoring efficiency of fucoxanthin content in the cultivation process.

**Acknowledgments:** This work is supported by the grants from the National Science Foundation of China (31602154); the National Key Technology R&D Program of China (2015BAD15B00) and Solix Algredients, Inc.

**Author Contributions:** Conceived and designed the experiments: Li-Juan Wang, Yong Fan. Performed the experiments: Li-Juan Wang, Yong Fan. Analyzed the data and discussed the results: Ronald L. Parsons, Guang-Rong Hu, Fu-Li Li, Pei-Yu Zhang. Wrote the paper: Li-Juan Wang, Yong Fan.

**Conflicts of Interest:** The founding sponsors had no role in the design of the study; in the collection, analyses, or interpretation of data; in the writing of the manuscript, and in the decision to publish the results.

## References

1. Gammone, M.A.; D'Orazio, N. Anti-obesity activity of the marine carotenoid fucoxanthin. *Mar. Drugs* **2015**, *13*, 2196–2214. [CrossRef] [PubMed]
2. Peng, J.; Yuan, J.P.; Wu, C.F.; Wang, J.H. Fucoxanthin, a marine carotenoid present in brown seaweeds and diatoms: Metabolism and bioactivities relevant to human health. *Mar. Drugs* **2011**, *9*, 1806–1828. [CrossRef] [PubMed]
3. Woo, M.N.; Jeon, S.M.; Kim, H.J.; Lee, M.K.; Shin, S.K.; Shin, Y.C.; Park, Y.B.; Choi, M.S. Fucoxanthin supplementation improves plasma and hepatic lipid metabolism and blood glucose concentration in high-fat fed C57BL/6N mice. *Chem. Biol. Interact.* **2010**, *186*, 316–322. [CrossRef] [PubMed]
4. Hu, X.; Li, Y.; Li, C.; Fu, Y.; Cai, F.; Chen, Q.; Li, D. Combination of fucoxanthin and conjugated linoleic acid attenuates body weight gain and improves lipid metabolism in high-fat diet-induced obese rats. *Arch. Biochem. Biophys.* **2012**, *519*, 59–65. [CrossRef] [PubMed]
5. Maeda, H.; Hosokawa, M.; Sashima, T.; Miyashita, K. Dietary combination of fucoxanthin and fish oil attenuates the weight gain of white adipose tissue and decreases blood glucose in obese/diabetic KK-Ay mice. *J. Agric. Food Chem.* **2007**, *55*, 7701–7706. [CrossRef] [PubMed]
6. Xia, S.; Wang, K.; Wan, L.; Li, A.; Hu, Q.; Zhang, C. Production, characterization, and antioxidant activity of fucoxanthin from the marine diatom *Odontella aurita*. *Mar. Drugs* **2013**, *11*, 2667–2681. [CrossRef] [PubMed]
7. Mori, K.; Ooi, T.; Hiraoka, M.; Oka, N.; Hamada, H.; Tamura, M.; Kusumi, T. Fucoxanthin and its metabolites in edible brown algae cultivated in deep seawater. *Mar. Drugs* **2004**, *2*, 63–72. [CrossRef]
8. Kim, S.M.; Shang, Y.F.; Um, B.H. A preparative method for isolation of fucoxanthin from *Eisenia bicyclis* by centrifugal partition chromatography. *Phytochem. Anal.* **2010**, *22*, 322–329. [CrossRef] [PubMed]
9. Guo, B.; Liu, B.; Yang, B.; Sun, P.; Lu, X.; Liu, J.; Chen, F. Screening of diatom strains and characterization of *Cyclotella cryptica* as a potential fucoxanthin producer. *Mar. Drugs* **2016**, *14*, 125. [CrossRef] [PubMed]

10. Yi, Z.; Xu, M.; Magnusdottir, M.; Zhang, Y.; Brynjolfsson, S.; Fu, W. Photo-oxidative stress-driven mutagenesis and adaptive evolution on the marine diatom *Phaeodactylum tricornutum* for enhanced carotenoid accumulation. *Mar. Drugs* **2015**, *13*, 6138–6151. [CrossRef] [PubMed]
11. Owens, T.G. Light-harvesting function in the diatom *Phaeodactylum tricornutum*: II. Distribution of excitation energy between the photosystems. *Plant Physiol.* **1986**, *80*, 732–738. [CrossRef] [PubMed]
12. Yuan, C.; Zheng, Y.L.; Zhang, W.L.; He, R.; Fan, Y.; Hu, G.R.; Li, F.L. Lipid accumulation and anti-rotifer robustness of microalgal strains isolated from eastern china. *J. Appl. Phycol.* **2017**, *29*, 2789–2800. [CrossRef]
13. Yuan, C.; Liu, J.H.; Fan, Y.; Ren, X.H.; Hu, G.R.; Li, F.L. *Mychonastes afer*HSO-3-1 as a potential new source of biodiesel. *Biotechnol. Biofuels* **2011**, *4*, 47. [CrossRef] [PubMed]
14. Yuan, C.; Xu, K.; Sun, J.; Hu, G.R.; Li, F.L. Ammonium, Nitrate, and Urea Play Different Roles for Lipid Accumulation in the Nervonic Acid - Producing Microalgae *Mychonastes afer*HSO-3-1. *J. Appl. Phycol.* **2017**, 1–9. [CrossRef]
15. Dambek, M.; Eilers, U.; Breitenbach, J.; Steiger, S.; Büchel, C.; Sandmann, G. Biosynthesis of fucoxanthin and diadinoxanthin and function of initial pathway genes in *Phaeodactylum tricornutum*. *J. Exp. Bot.* **2012**, *63*, 5607–5612. [CrossRef] [PubMed]
16. Fernández Sevilla, J.M.; Cerón García, M.C.; Sánchez, M.A.; Belarbi, E.H.; García, C.F.; Molina, G.E. Pilot-plant-scale outdoor mixotrophic cultures of *Phaeodactylum tricornutum* using glycerol in vertical bubble column and airlift photobioreactors: Studies in fed-batch mode. *Biotechnol. Prog.* **2010**, *20*, 728–736. [CrossRef] [PubMed]
17. Bowler, C.; Allen, A.E.; Badger, J.H.; Grimwood, J.; Jabbari, K.; Kuo, A.; Maheswari, U.; Martens, C.; Maumus, F.; Otillar, R.P. The *Phaeodactylum* genome reveals the evolutionary history of diatom genomes. *Nature* **2008**, *456*, 239–244. [CrossRef] [PubMed]
18. Hamilton, M.L.; Haslam, R.P.; Napier, J.A.; Sayanova, O. Metabolic engineering of *Phaeodactylum tricornutum* for the enhanced accumulation of omega-3 long chain polyunsaturated fatty acids. *Metab. Eng.* **2014**, *22*, 3–9. [CrossRef] [PubMed]
19. Zaslavskaia, L.A.; Lippmeier, J.C.; Kroth, P.G.; Grossman, A.R.; Apt, K.E. Transformation of the diatom *Phaeodactylum tricornutum* (bacillariophyceae) with a variety of selectable marker and reporter genes. *J. Phycol.* **2000**, *36*, 379–386. [CrossRef]
20. Kim, S.M.; Jung, Y.J.; Kwon, O.N.; Cha, K.H.; Um, B.H.; Chung, D.; Pan, C.H. A potential commercial source of fucoxanthin extracted from the microalga *Phaeodactylum tricornutum*. *Appl. Biochem. Biotechnol.* **2012**, *166*, 1843–1855. [CrossRef] [PubMed]
21. Schwartz, S.J.; Woo, S.L.; Elbe, J.H.V. High-performance liquid chromatography of chlorophylls and their derivatives in fresh and processed spinach. *J. Agric. Food Chem.* **1981**, *29*, 533–535. [CrossRef]
22. Braumann, T.; Grimme, L.H. Reversed-phase high-performance liquid chromatography of chlorophylls and carotenoids. *Biochim. Biophys. Acta* **1981**, *637*, 8–17. [CrossRef]
23. Carreto, J.I.; Catoggio, J.A. An indirect method for the rapid estimation of carotenoid contents in *Phaeodactylum tricornutum*: Possible application to other marine algae. *Mar. Biol.* **1977**, *40*, 109–116. [CrossRef]
24. Carreto, J.I.; Catoggio, J.A. Variations in pigment contents of the diatom *Phaeodactylum tricornutum* during growth. *Mar. Biol.* **1976**, *36*, 105–112. [CrossRef]
25. Parsons, T.R.; Maita, Y.; Lalli, C.M. Determination of chlorophylls and total carotenoids: Spectrophotometric method. In *A Manual of Chemical & Biological Methods for Seawater Analysis*; Pergamon Press: Oxford, UK, 1984; pp. 101–104.
26. Dere, S.; Gunes, T.; Sivaci, R. Spectrophotometric determination of chlorophyll-A, B and total carotenoid contents of some algae species using different solvents. *Turk. J. Bot.* **1998**, *22*, 13–17. [CrossRef]
27. Bruuinsma, J. The quantitative analysis of chlorophylls *a* and *b* in plant extracts. *Photochem. Photobiol.* **1963**, *2*, 241–249. [CrossRef]
28. Jeffrey, S. Chlorophyll and carotenoid extinction coefficients. In *Phytoplankton Pigments in Oceanography: Guidelines to Modern Methods*; UNESCO: Paris, France, 1997; pp. 595–596.

29. Davies, B. Carotenoids. In *Chemistry and Biochemistry of Plant Pigments*; Academic Press: London, UK, 1976; Volume 2, pp. 38–165.
30. Arnon, D.I. Copper enzymes in isolated chloroplasts. Polyphenoloxidase in beta vulgaris. *Plant Physiol.* **1949**, *24*, 1–15. [CrossRef] [PubMed]
31. Guillard, R.R.; Ryther, J.H. Studies of marine planktonic diatoms.1. Cyclotella nana hustedt, and detonula confervacea (cleve) gran. *Can. J. Microbiol.* **1962**, *8*, 229–239. [CrossRef] [PubMed]

*marine drugs*

MDPI

*Article*

# Protective Effects of Fucoxanthin against Alcoholic Liver Injury by Activation of Nrf2-Mediated Antioxidant Defense and Inhibition of TLR4-Mediated Inflammation

**Jiawen Zheng [1], Xiaoxiao Tian [1], Wen Zhang [1], Pingan Zheng [2], Fangfang Huang [1,*], Guofang Ding [1] and Zuisu Yang [1,*]**

1 Zhejiang Provincial Engineering Technology Research Center of Marine Biomedical Products, School of Food and Pharmacy, Zhejiang Ocean University, Zhoushan 316022, China; jwzheng1996@163.com (J.Z.); TIANXIAOXIAO0208@163.com (X.T.); zhangwen1225z@163.com (W.Z.); dinggf2007@163.com (G.D.)

2 Zhejiang Hailisheng Group Co., Ltd., Zhoushan 316021, China; zhengpingan@hailisheng.com

* Correspondence: gracegang@126.com (F.H.); abc1967@126.com (Z.Y.); Tel.: +86-0580-226-0600 (Z.Y.); Fax: +86-0580-254-781 (Z.Y.)

Received: 11 September 2019; Accepted: 27 September 2019; Published: 27 September 2019

**Abstract:** Fucoxanthin (Fx) is a natural extract from marine seaweed that has strong antioxidant activity and a variety of other bioactive effects. This study elucidated the protective mechanism of Fx on alcoholic liver injury. Administration of Fx was associated with lower pathological effects in liver tissue and lower serum marker concentrations for liver damage induced by alcohol. Fx also alleviated oxidative stress, and lowered the level of oxides and inflammation in liver tissue. Results indicate that Fx attenuated alcohol-induced oxidative lesions and inflammatory responses by activating the nuclear factor erythrocyte-2-related factor 2 (Nrf2)-mediated signaling pathway and down-regulating the expression of the toll-like receptor 4 (TLR4)-mediated nuclear factor-kappa B (NF-κB) signaling pathway, respectively. Our findings suggest that Fx can be developed as a potential nutraceutical for preventing alcohol-induced liver injury in the future.

**Keywords:** alcoholic liver injury; fucoxanthin; oxidative stress; Nrf2; TLR4

## 1. Introduction

The liver is the main organ used for alcohol metabolism; alcohol damages liver cells, which can lead to alcoholic liver disease (ALD) [1]. Acute ALD refers to a disease caused by liver damage associated with heavy drinking over a short period [2,3]. ALD is a worldwide public health problem; its prevalence and morbidity have increased each year as a consequence of increased alcohol abuse rates, which damages human physical health [4,5]. Excessive drinking can cause liver damage to varying degrees, especially in the short term; a large amount of alcohol abuse, for instance, can cause severe liver damage [1,6]. Protective methods are one way of counteracting the rising rates of ALD, which may be possible by utilizing the properties of natural active substances, rather than traditional drugs.

Large amounts of alcohol in the body can be dehydrogenated into acetaldehyde and acetate under the catalysis of alcohol dehydrogenase (ADH) and acetaldehyde dehydrogenase (ALDH) [7]. Acetaldehyde can damage mitochondria and inhibit the tricarboxylic acid cycle; it also reacts with intracellular macromolecules such as proteins and lipids to form complexes. Acetaldehyde production and the release of adrenaline as a result of ethanol can also cause hepatic vasoconstriction, elevated intrahepatic sinus pressure, or hypoxia of liver tissue, which can lead to vacuolar degeneration of hepatocytes [8–10]. Alcohol metabolism also produces a large number of harmful free radicals, affecting the activities of antioxidant enzymes, especially glutathione levels in cells. The body's antioxidant capacity is insufficient to cope with the accumulation of free radicals, which can lead to the accumulation of lipid peroxides, causing damage to liver cells [11–13]. Furthermore, mass alcohol consumption stimulates the production of endotoxin (LPS) in the body, activates liver macrophages, namely Kupffer cells, and leads to high expression of the toll-like receptor 4 (TLR4), which in turn releases a large concentration of reactive oxygen species (ROS) and tumor necrosis factor (TNF-α) or other inflammatory factors that can accelerate downstream inflammation and oxidative damage [14–17]. When LPS enters the liver this directly damages hepatocytes and promotes the inflammatory cells to produce a large number of inflammatory mediators, which can induce inflammatory cell infiltration and hepatocyte necrosis [18]. Studies have shown that LPS and TNF-α can directly or indirectly exert toxic effects on liver cells [19]. At present, the exact etiopathogenesis of alcoholic liver damage has not been fully clarified which is, therefore, one of the hotspots of current research.

The ocean accounts for 70% of the Earth's surface area and is an important marine life support system as part of the biological world [20]. The multiplicity of the ocean environment has led to marine organisms that are diverse and widely distributed. Since the marine environment is completely different to terrestrial environments, many marine organisms produce active substances with specific functions and special structures: many of which have antibacterial, antiviral and antitumor activities [21]. In recent years, in-depth research has been conducted on marine resources, which has given a new outlook for drug development. Recent studies have reported the effects of natural extracts on the prevention or improvement of alcoholic liver damage, for example, astaxanthin [22] and aplysin [23] have shown protective effects on alcoholic liver damage in vivo. Fucoxanthin (Fx) is a red-orange carotenoid extracted from natural seaweed [24]. The chemical structure of Fx is shown in Figure 1A. Fx has been reported to have strong antioxidative effects, and has other biological activities, such as anti-obesity, anti-inflammatory and anti-cancer properties [25–27]. In addition, Fx was shown to improve glycolipid metabolism in type 2 diabetic mice and improve ventricular rhythm and muscle function models in aging mice [28]. Nevertheless, the role of Fx in alcohol-induced liver injury has not been reported. Our current research aims to investigate the prophylactic function of Fx on acute ALD by monitoring oxidative stress and inflammation. The potential mechanism of Fx was explored by observing oxidative stress signaling pathways and inflammatory signaling pathways in mice. These studies will help to elucidate the latent protective mechanisms of Fx for acute alcoholic liver injury.

## 2. Results and Discussion

### 2.1. Effects of Fucoxanthin (Fx) on Bodyweight and Relative Liver Weight

Mice were weighed each day during the experimental period; changes in the weight are shown in Figure 1C. After 7 days of treatment, mice given alcohol had significantly lower bodyweights than control mice. The bodyweight of H-Fx group mice was much greater than that of Model group mice. Regarding the relative liver weight (Figure 1D), except for the L-Fx group, there was no obvious difference between the Positive group, Fx groups and Control group, but the liver weight of these groups was significantly lower than those of the Model group, indicating that Fx may have a protective effect on liver tissue.

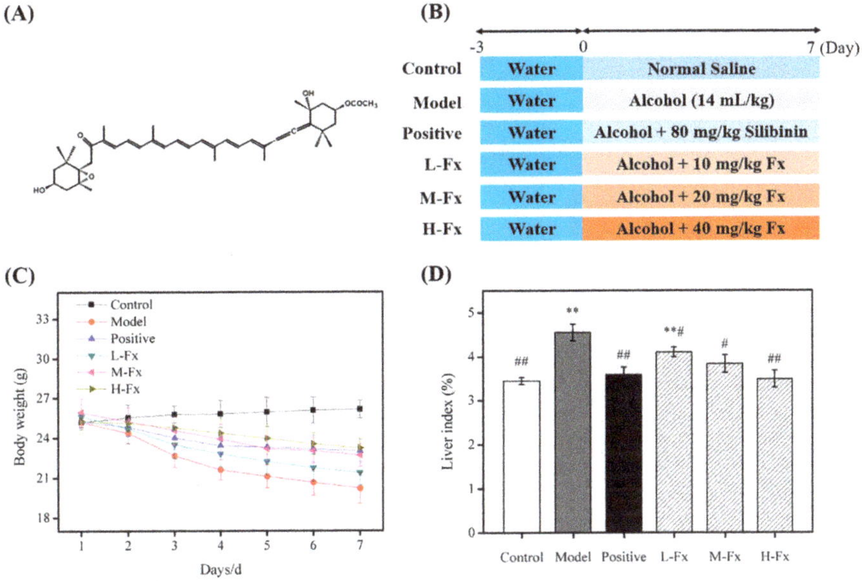


**Figure 1.** Effect of fucoxanthin (Fx) on bodyweight and liver index in mice with alcoholic liver injury. (**A**) The chemical structure of Fx. (**B**) The experimental design in 7 days. (**C**) The bodyweight of mice. (**D**) The liver index of mice. Data are given as mean ± standard deviation (SD) ($n = 8$). ** $p < 0.01$ vs. Control group, # $p < 0.05$, ## $p < 0.01$ vs. Model group.

### 2.2. Effects of Fx on Alcohol-Induced Liver and Stomach Injury

Changes in liver and stomach tissue were evaluated by histopathological observation and hematoxylin and eosin (H&E) staining to assess the protective effects of Fx on these tissues. Each group of liver specimens is shown in Figure 2A. Macro-observation showed that the Model group had an enlarged liver with a rough surface and plaque degeneration. However, administration of Fx and Silibinin modified the degree of swelling and degeneration of liver induced by alcohol.

As shown in Figure 2B, H&E staining further confirmed that Fx inhibits alcohol-induced liver injury. The hepatic lobule was structurally intact, the hepatic cord was arranged radially around the central vein, and the hepatic sinus was clear in the Control group. In contrast, in the Model group, the hepatic lobule structure was disordered and incomplete, the hepatic cord arrangement was disordered, and some hepatocytes were necrotic. In contrast to the Model group, the Fx groups had obvious signs of repair; the low, medium, and high doses were associated with a trend of gradual repair. In the H-Fx group, hepatic lobules were structurally intact and hepatic cords were neatly arranged; the hepatic sinus gap was clear and no vacuoles were observed. The signs of repair in the H-Fx group was more obvious than those in the Positive group. The results were consistent with results from Liu et al. [29], who found that astaxanthin can prevent alcoholic fatty liver disease by modulating mouse gut microbiota, indicating that Fx could possess a protective effect on alcohol-induced liver injury in vivo.

H&E staining results for stomach tissue are shown in Figure 2C. The epithelial and lamina propria cells were arranged closely and neatly in the Control group, and the main cells and parietal cells were normal. However, large areas of epithelium were exfoliated in the Model group, the structures of main cells and parietal cells were destroyed, their arrangement was disordered, and local necrosis was observed. There more signs of greater cellular health in Positive and Fx groups than in the Model group, such as closer arrangement of cells and clearer cell structures. This indicates that Fx has a protective effect on the gastric mucosa.

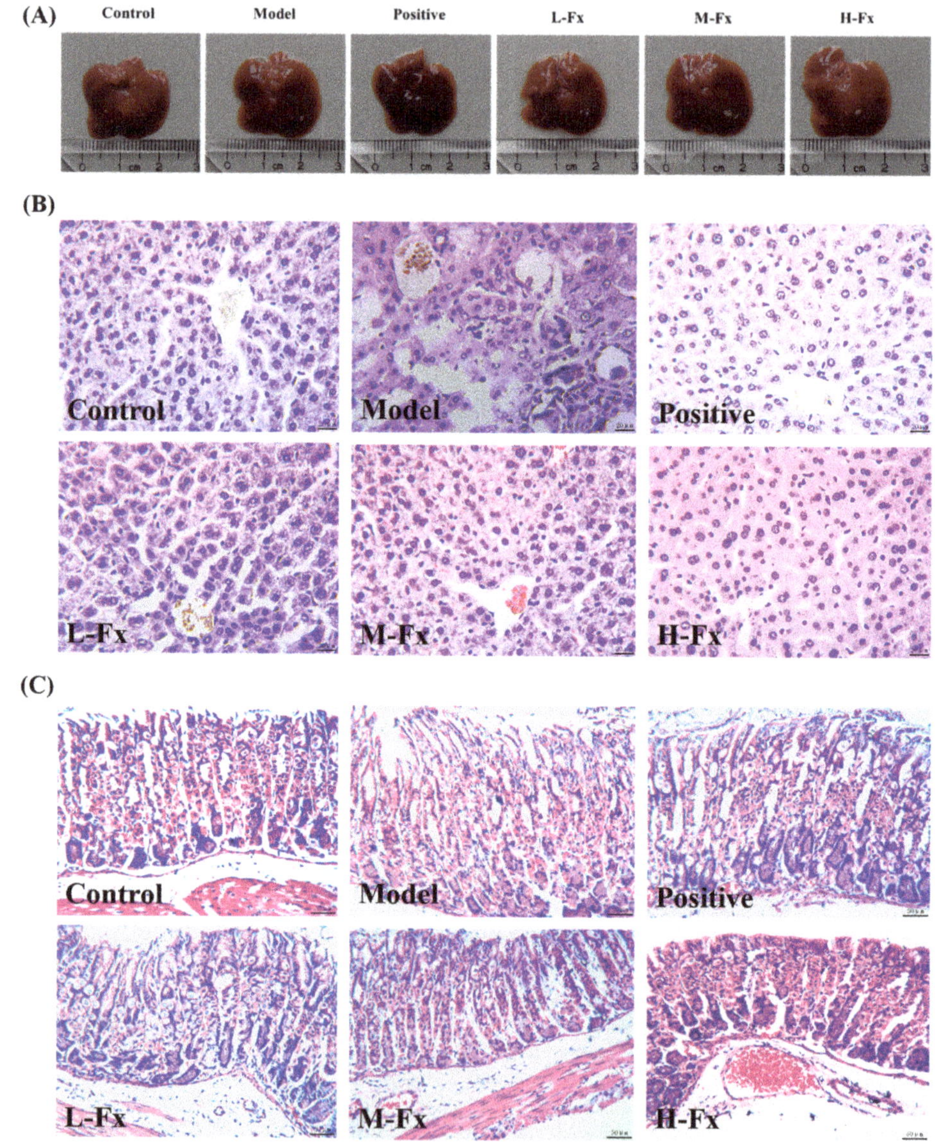


**Figure 2.** Effect of Fx on alcohol-induced liver and stomach injury. (**A**) Macroscopic picture of livers; (**B**) hematoxylin and eosin (H&E) stained liver tissues. (400× magnification); (**C**) H&E stained stomach tissues. (200× magnification).

### 2.3. *Effects of Fx on Serum Aspartate Transaminase (AST) and Alanine Transaminase (ALT) Activities*

Aspartate transaminase (AST) and alanine transaminase (ALT) are soluble enzymes present in hepatocyte cytoplasm [15]. After hepatocyte injury, the permeability of the cell membrane increases, causing ALT and AST to enter the bloodstream. Therefore, serum ALT and AST activities can reflect the degree of hepatocyte damage. Serum ALT and AST activities are conventionally used as sensitive markers to evaluate liver damage [30,31]. The ALT and AST activities recorded from Model group mice were significantly greater than those from Control group mice, indicating that Model group mice were successfully established as an alcoholic liver injury model (Figure 3). ALT and AST activities of Positive group and Fx group mice were significantly less than those of the Model group and gradually tended to Control group (Figure 3). These results indicate that Fx could control ALT and AST activities, thereby reducing alcohol-induced liver injury; this hypothesis is also supported by findings by Han et al. [32].

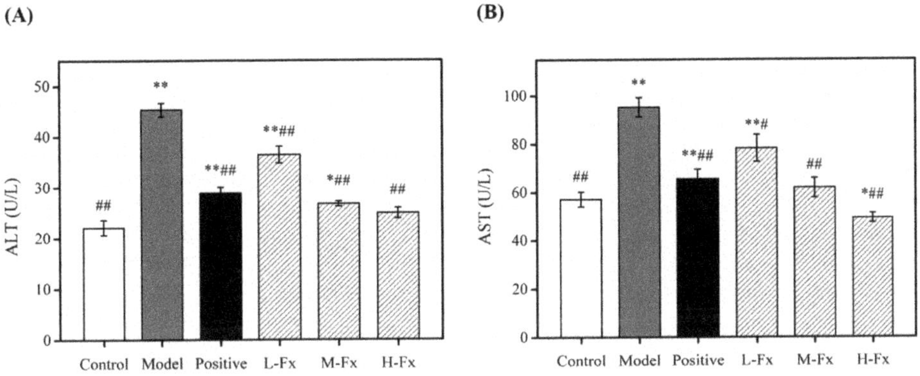


**Figure 3.** Effect of Fx on serum level of ALT (**A**) and AST (**B**) activities in mice with alcoholic liver injury. Data are given as mean ± SD ($n = 8$). * $p < 0.05$, ** $p < 0.01$ vs. Control group, # $p < 0.05$, ## $p < 0.01$ vs. Model group.

## 2.4. Effects of Fx on Serum and Hepatic Level of Triglyceride (TG)

Heavy drinking can lead to accumulation of lipids in the liver. During metabolic processing of alcohol, large amounts of lipase are released, which promotes the synthesis of triglyceride (TG) and causes fat to deposit in liver cells. If a fatty lesion develops in liver tissue, blood lipid levels can rise [33]. To further illustrate the effect of Fx pretreatment on lipid metabolism in alcohol-induced mice, serum TG levels (A) hepatic TG levels (B) were measured, as shown in Figure 4. TG levels in the serum and liver of Model group mice increased significantly compared to the Control group, indicating that alcohol caused lipid accumulation, resulting in the disorder of lipid metabolism and liver damage. The TG levels in serum and liver were significantly less in Positive and Fx groups than Model group. These results indicate that Fx can effectively alleviate alcohol-induced liver fat deposition.

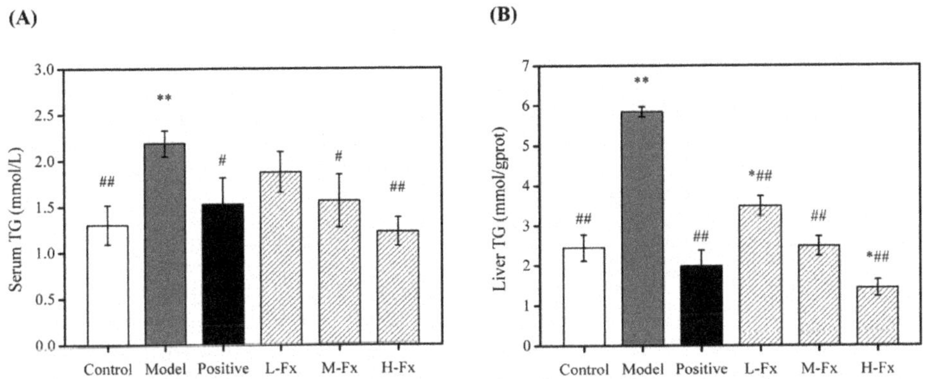


**Figure 4.** Effects of Fx on serum level of triglyceride (TG) (**A**) and hepatic level of TG (**B**) in mice with alcoholic liver injury. Data are given as mean ± SD ($n = 8$). * $p < 0.05$, ** $p < 0.01$ vs. Control group, # $p < 0.05$, ## $p < 0.01$ vs. Model group.

### 2.5. *Effects of Fx on Hepatic Levels of Malonaldehyde (MDA) and Antioxidant Enzymes*

Alcohol-induced oxidative stress is a vital element in the nosogenesis of acute ALD [34]. Metabolism of alcohol can generate high levels of ROS, which can affect the body's antioxidant systems, increasing the formation of products that cause oxidative stress; this can increase the level of oxidative stress in many tissues, especially those of the liver [35,36]. Antioxidants protect cells from ROS, but antioxidant activity is easily quenched by excess lipid peroxides, which can lead to liver damage [37,38]. Table 1 shows the effect of Fx on liver total antioxidant capacity (T-AOC), and malonaldehyde (MDA), glutathione peroxidase (GSH-Px), superoxide dismutase (SOD) and catalase (CAT) content in mice. The activities of T-AOC, GSH-Px, SOD, and CAT in the Model group were significantly lower than those of the Control group, but the MDA content was significantly higher. The activities of T-AOC, GSH-Px, SOD and CAT measured for Positive and the Fx groups were significantly higher than those of the Model group, and the MDA content was significantly lower. GSH-Px and SOD have a crucial effect on the balance between oxidation and antioxidant capacity in vivo, suggesting that alcohol significantly increased the peroxide levels in the liver and decreased the liver's antioxidant capacity [39]. Therefore, MDA may be a product of free radicals acting on lipid peroxidation. The degree of lipid peroxidation in the body can be reflected by MDA content, which is a common indicator of membrane lipid peroxidation [40]. This indicates that Fx may effectively prevent the excessive oxidation of mouse hepatocytes caused by alcohol intake, and could enhance the body's antioxidant capacity, greatly protecting against alcoholic liver injury. Qu et al. [41] reported that ginsenoside Rk3 protected against alcohol-induced liver injury in mice, which is consistent with the results of this study.

**Table 1.** Effects of Fx on Hepatic levels of MDA and Antioxidant Enzymes in mice.

| Group | T-AOC (U/mg prot) | MDA (nmol/mg prot) | SOD (U/mg prot) | GSH-Px (U/mg prot) | CAT (U/mg prot) |
|---|---|---|---|---|---|
| Control | $5.72 \pm 0.05$ ## | $3.07 \pm 0.21$ ## | $168.52 \pm 8.73$ ## | $129.57 \pm 5.03$ ## | $89.61 \pm 2.13$ ## |
| Model | $1.32 \pm 0.12$ ** | $5.90 \pm 0.46$ ** | $88.05 \pm 14.71$ ** | $69.60 \pm 4.41$ ** | $64.11 \pm 2.54$ ** |
| Positive | $9.34 \pm 0.52$ ** ## | $3.20 \pm 0.22$ ## | $156.57 \pm 3.81$ ## | $129.28 \pm 6.87$ ## | $96.93 \pm 2.14$ * ## |
| L-Fx | $3.39 \pm 0.17$ ** ## | $4.14 \pm 0.22$ ** ## | $121.36 \pm 6.23$ ** # | $86.69 \pm 3.43$ ** ## | $85.60 \pm 2.09$ ## |
| M-Fx | $5.22 \pm 0.23$ * ## | $3.79 \pm 0.30$ ## | $152.34 \pm 8.24$ ## | $100.75 \pm 6.89$ ** ## | $105.41 \pm 1.29$ ** ## |
| H-Fx | $9.64 \pm 0.14$ ** ## | $2.91 \pm 0.26$ ## | $175.74 \pm 12.68$ ## | $128.79 \pm 7.73$ ## | $134.00 \pm 1.36$ ** ## |

Data are given as mean ± SD ($n = 8$). * $p < 0.05$, ** $p < 0.01$ vs. Control group, # $p < 0.05$, ## $p < 0.01$ vs. Model group.

### 2.6. *Effects of Fx on Hepatic Alcohol Dehydrogenase (ADH) and Acetaldehyde Dehydrogenase (ALDH) Activities*

The liver is the main organ used in alcohol metabolism, ADH and ALDH are important enzymes for this process [42]. When the body ingests a large amount of alcohol, it is dehydrogenated through to form acetaldehyde via ADH, then converted to acetic acid by the action of ALDH; finally, the acetic acid metabolized to water and $CO_2$ via the tricarboxylic acid cycle. ADH and ALDH can metabolize 80% of liver ethanol, which has important anti-alcoholic activity. By detecting the viability of ADH and ALDH, it is possible to reflect the anti-alcoholic capacity of hepatocytes, and further evaluate the protection of Fx on the liver [8,43]. As shown in Figure 5, after alcohol was administered to mice, hepatic ADH (A) and ALDH (B) activities in the Model group increased slightly, but these activities were not significantly different to those of Control group. However, ADH and ALDH activities of groups that had Fx and Silibinin administered were significantly greater than those of Control group. This indicates that Fx has a certain anti-alcoholic effect, thereby preventing alcohol-induced liver damage.

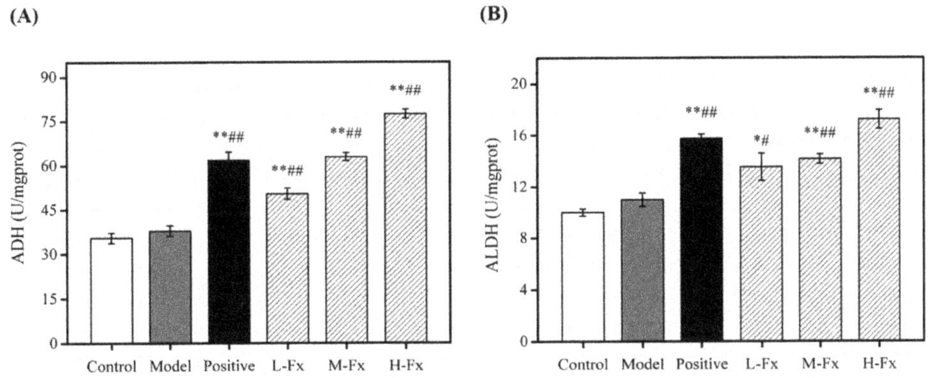


**Figure 5.** Effects of Fx on hepatic alcohol dehydrogenase (ADH) (**A**) and acetaldehyde dehydrogenase (ALDH) (**B**) activities in mice with alcoholic liver injury. Data are given as mean ± SD ($n = 8$). * $p < 0.05$, ** $p < 0.01$ vs. Control group, # $p < 0.05$, ## $p < 0.01$ vs. Model group.

### 2.7. *Effects of Fx on the Levels of Pro-Inflammatory Cytokines in Liver Tissue*

After long-term or heavy drinking, the excessive accumulation of lipids caused by alcohol intake may cause direct or indirect damage to liver cells, produce inflammatory responses, increase the expression of pro-inflammatory factors, and lead to alcoholic liver inflammation. The decomposition of a large amount of alcohol also stimulates the immune functioning of liver cells and activates Kupffer cells [44,45]. Activated Kupffer cells rapidly activate transcriptional regulators such as nuclear factor-kappa B (NF-κB), producing large amounts of inflammatory factors such as TNF-α, interleukin-1β (IL-1β), interleukin-6 (IL-6), interferon-γ (IFN-γ), and inflammatory mediators; this can aggravate the inflammatory infiltration and injury of liver tissue, eventually leading to liver inflammation and hepatocyte necrosis and apoptosis [46,47]. To investigate whether Fx can reduce the inflammatory response caused by alcohol-induced liver injury, we tested the expression of related inflammatory factors. After alcohol induction, levels of 4 cytokines released in liver tissue had significantly increased (Figure 6), indicating that these 4 pro-inflammatory factors participated in the process of alcohol-induced liver injury in mice. The expression of various factors in liver tissue was inhibited by different degrees in mice administered with silibinin and Fx. This indicates that Fx can effectively control the secretion of pro-inflammatory factors, thereby reducing the inflammatory response caused by the alcohol-induced liver injury.

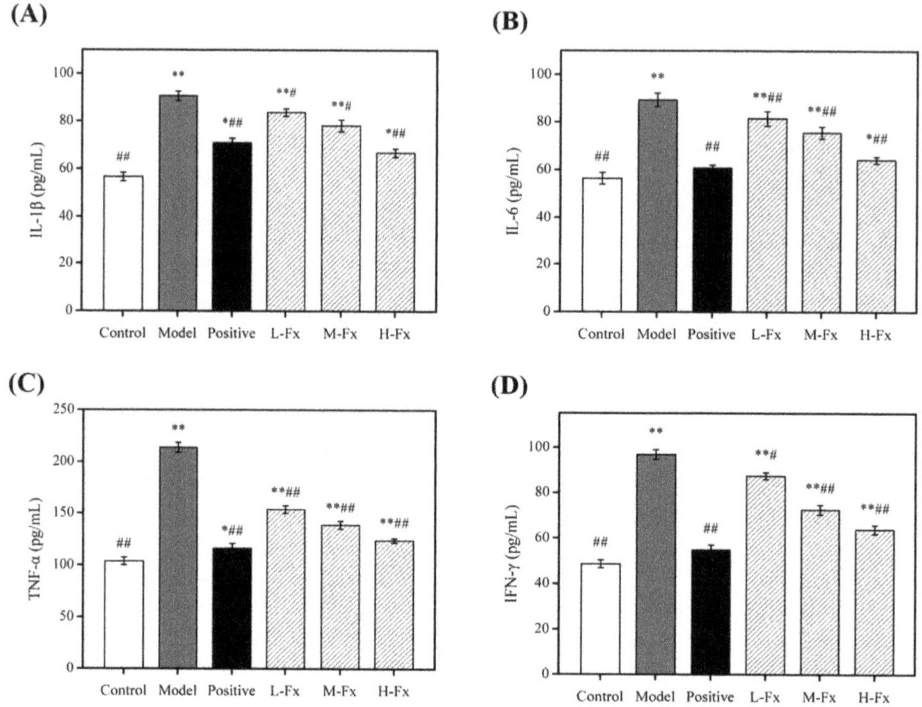


**Figure 6.** Effects of Fx on the levels of pro-inflammatory cytokines in liver. The release of interleukin-1β (IL-1β) (**A**), interleukin-6 (IL-6) (**B**), tumor necrosis factor (TNF-α) (**C**) and interferon-γ (IFN-γ) (**D**). Data are given as mean ± SD ($n$ = 8). * $p < 0.05$, ** $p < 0.01$ vs. Control group, # $p < 0.05$, ## $p < 0.01$ vs. Model group.

### 2.8. Effects of Fx on Nuclear Factor Erythrocyte-2-Related Factor 2 (Nrf2)-Mediated Antioxidant Response

To ulteriorly investigate the antioxidant mechanism of Fx on alcohol-induced liver injury, the nuclear factor erythrocyte-2-related factor 2 (Nrf2) content of cells and its downstream proteins were detected by Western blot. Nrf2 is a crucial regulatory transcription factor that acts upstream of the antioxidant defense system and plays a major part in regulating redox equilibrium [48]. Phosphorylation occurs when Nrf2 is stimulated by free radicals or nucleophiles; Nrf2 and cytoskeleton-related proteins then dissociate and Nrf2 enters the nucleus to initiate the expression of antioxidant enzymes such as nicotinamide quinone oxidoreductase 1 (NQO1), heme oxygenase-1 (HO-1), and glutamate-cysteine ligase modifier subunit (GCLM). *NQO1*, *HO-1* and *GCLM* are downstream target genes of Nrf2 that function as antioxidant molecules [49–51]. Studies have shown that a large amount of alcohol can inhibit the normal activation of Nrf2 expression, leading to an increase in alcohol-induced oxidative stress [52]. As shown in Figure 7, alcohol-induced liver injury can significantly reduce the levels of Nrf2 protein, further decreasing the content of downstream antioxidant protein regulated by Nrf2, and the expression of downstream proteins NQO1, HO-1 and GCLM was significantly reduced, indicating that alcoholic liver injury can significantly inhibit the Nrf2 signaling pathway. However, Positive and Fx groups had significantly greater levels of these proteins than Model groups. Fx was associated with an increase in the levels of Nrf2 protein and its downstream target proteins in a dose-dependent manner. The results indicate that Fx minimizes the damage caused by alcohol to the liver of mice by activating Nrf2-mediated antioxidant responses.

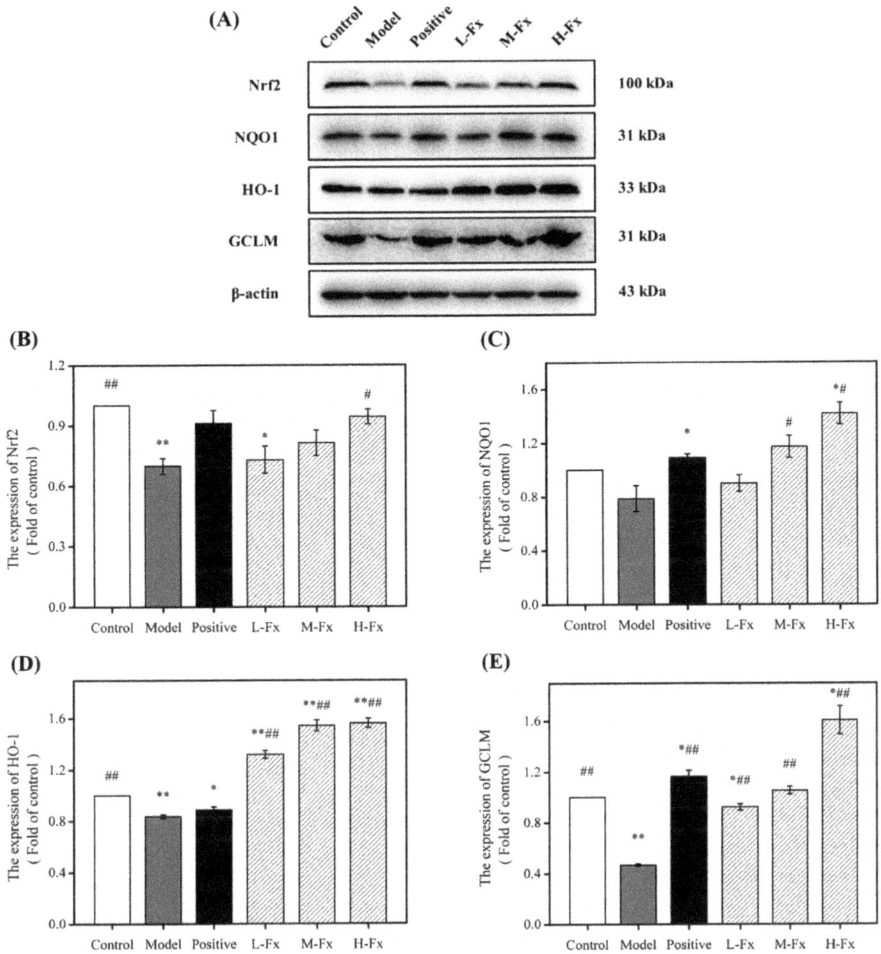


**Figure 7.** Effect of Fx on the nuclear factor erythrocyte-2-related factor 2 (Nrf2)-mediated antioxidant response. (**A**) Western blot analysis of Nrf2, NQO1, HO-1 and GCLM in livers; (**B**) quantitative analysis for Nrf2; (**C**) Quantitative analysis for NQO1; (**D**) quantitative analysis for HO-1; (**E**) quantitative analysis for GCLM. Data are given as mean ± SD ($n$ = 8). * $p < 0.05$, ** $p < 0.01$ vs. Control group, ## $p < 0.01$ vs. Model group.

### 2.9. Effect of Fx on Toll-Like Receptor 4 (TLR4)-Induced Inflammatory Response

TLR-induced signaling pathways are the main pathway leading to inflammatory responses in alcohol-induced liver injury [46]. TLR4 is a specific exogenous receptor of LPS; after ingesting a large amount of alcohol, TLR4 converts to form LPS activated TLR4, creating downstream signals, and activating transcription factors through intracellular signaling pathways to promote inflammation [53,54]. Myeloid differentiation factor 88 (MyD88) is an important adaptor protein molecule of TLR4, that can be activated by TLR4. Activated MyD88 can activate downstream NF-κB and mediate the release of inflammatory factors [53]. NF-κB as the most important transcriptional regulator in the LPS/TLR4 inflammatory signal transduction pathway, and plays a key regulatory role in the transcriptional synthesis of the pathway's inflammatory mediators [55,56]. Under normal conditions, NF-κB forms a complex with its inhibitor IκB in an inactive state, which causes the phosphorylation and degradation of IκB. NF-κB is then activated and enters the nucleus, which then leads to the transcription and release of inflammatory cytokines. To investigate the anti-inflammatory mechanism of Fx on the protective effects of alcohol-induced liver injury, Western blots were used to analyze the concentrations of TLR4-induced signaling pathway-related proteins. As shown in Figure 8, after alcohol induction, the concentrations of TLR4 were significantly increased, and a downstream signal cascade was initiated, leading to significant up-regulation of MyD88, p-IκBα and p-NF-κBp65. However, Fx pretreatment groups had significantly greater concentrations of downstream proteins than the Control group after alcohol administration. Overall, results indicate that Fx can attenuate alcohol-induced hepatic inflammatory responses by inhibiting TLR4-induced signaling pathways.

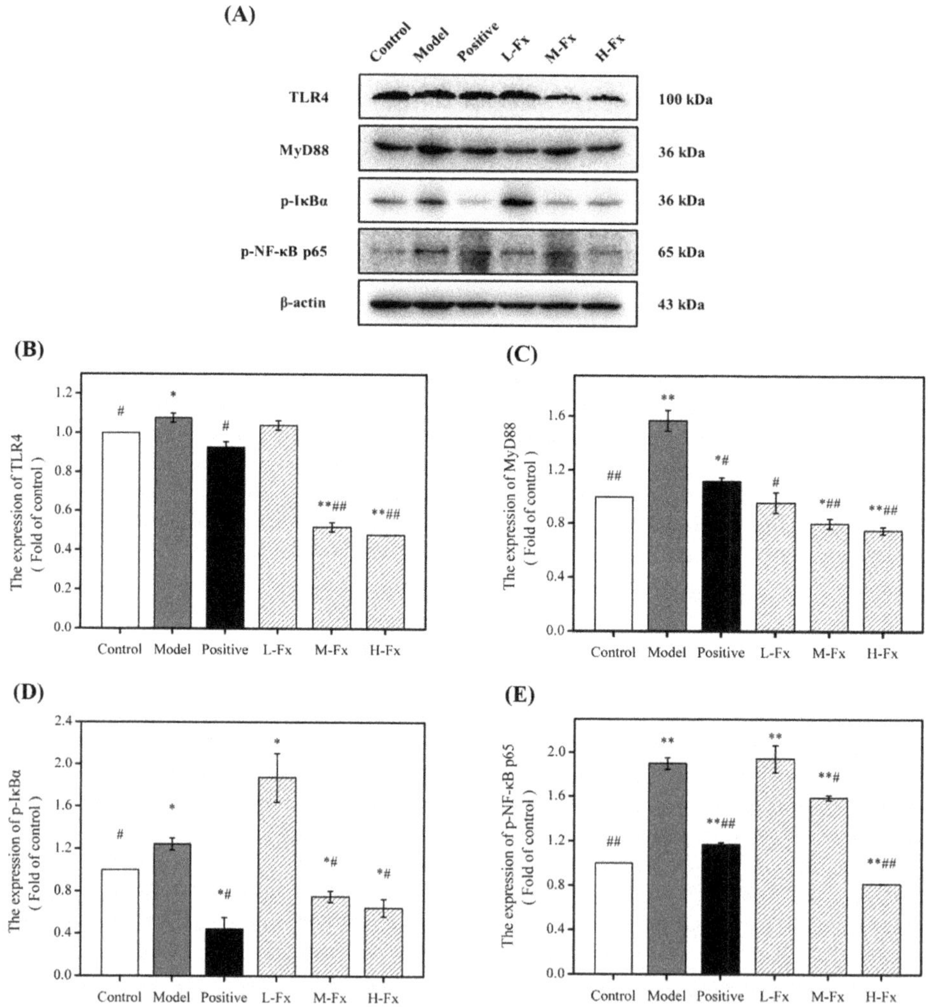


**Figure 8.** Effect of Fx on toll-like receptor 4 (TLR4)-induced inflammatory response. (**A**) Western blot analysis of TLR4, myeloid differentiation factor 88 (MyD88), p-IκBα and p-NF-κB p65; (**B**) the protein levels of TLR4; (**C**) the protein levels of MyD88; (**D**) the protein levels of p-IκBα; (**E**) the protein levels of p-NF-κB p65. Data are given as mean ± SD ($n = 8$). * $p < 0.05$, ** $p < 0.01$ vs. Control group, # $p < 0.05$, ## $p < 0.01$ vs. Model group.

## 3. Materials and Methods

### 3.1. Materials and Chemicals

Fucoxanthin was purchased from Shandong Jiejing Group Corporation (Rizhao, Shandong, China). Silibinin (powdered capsule) was purchased from Tianshili Shengte Pharmaceutical Co., Ltd. (Tianjin, China). Enzyme-linked immunosorbent assay (ELISA) kits and bicinchoninic acid (BCA) protein assay kit were purchased from BOSTER (Wuhan, China). AST, ALT, TG, T-AOC, MDA, GSH-Px, SOD, CAT, ADH and ALDH kits were all obtained from the Jiancheng Bioengineering Institute (Nanjing, China). Antibodies against Nrf-2, NQO1, TLR4 were obtained from Affinity Biosciences, Inc. (Cincinnati, OH, USA). Antibodies against MyD88 were purchased from BOSTER (Wuhan, China). Antibodies against HO-1, p-IκBα, GCLM and p-NF-κB p65 were obtained from Proteintech Group, Inc. (Princeton, NJ, USA). All reagents were of analytical grade.

### 3.2. Animals and Treatment

Male ICR mice (6 weeks old) were purchased from the Experiment Animal Center of Zhejiang Province (Hangzhou, China). The protocol was approved by the Experimental Animal Ethics Committee of Zhejiang Ocean University (Zhoushan, China), and our animal certificate was No. SCXK (ZHE 2014-0001). Mice were housed in a constant condition of temperature (22 ± 2 °C) and humidity (55 ± 5%) on a 12-h light/dark cycle. After 7 days of adaptive feeding, 48 mice were stochastically allocated into 6 groups: control group mice were given normal saline (Control), model group mice were given alcohol by gavage (56% *w/v*, total 10 g/kg b.w.) twice a day, separated by half an hour (Model), positive group mice were given alcohol by gavage 80 mg/kg b.w. Silibinin was used on mice in the positive control group (Positive) [57], and for the different Fx groups, mice were given alcohol supplemented with Fx by oral administration of 10 (L-Fx), 20 (M-Fx), and 40 (H-Fx) mg/kg b.w. Positive group and Fx group mice were given alcohol after taking the drug for half an hour. The experimental procedure is shown in Figure 1B and the whole experiment lasted for 7 days. All mice were given controlled food allowances but they could drink water at will. Mouse weights were recorded each day. At day 7, mice were fasted after 12 hours and euthanized by cervical dislocation; serum and tissues were stored at −80 °C.

### 3.3. Histopathological Analysis

Histopathological alteration of the liver and stomach tissues were measured by H&E staining using a standard procedure [58]. For H&E staining, tissue from the isolated stomach and the left lobe of the liver were fixed in 4% neutral-buffered formalin solution between 24 h to 48 h, dehydrated stepwise with gradient alcohol, clarified using xylene, then embedded in paraffin. The tissue was then cut into 4 μm thick pieces using a microtome (Leica RM2135, Leica Instruments GmbH, Wetzlar, Germany) for H&E staining. Photomicrographs were observed under an optical microscope (Biological microscope CX31, Olympus, Tokyo, Japan) and photographed at 200× or 400× magnification.

### 3.4. Serum Biochemical Analysis

The retro-orbital blood samples were centrifuged (10,000 rpm) at 4 °C for 5 min, then the collected supernatant was stored at 4 °C for further experiments. Serum AST, ALT, TG levels were used for serum biochemical analysis after administration of alcohol, Fx or Silibinin. The levels of ALT, AST, and TG in serum or liver tissue were determined according to the following commercial kit protocols obtained from Nanjing Jiancheng Bioengineering Institute (Nanjing, China).

### 3.5. Determination of Hepatic MDA and Antioxidant Enzymes

When preparing liver homogenate, 1 g of liver tissue was mixed on ice with 9 mL of normal saline and centrifuged (4000 rpm, 10 min). The protein concentration in the supernatant of liver homogenates was quantified using a BCA total protein assay kit. The hepatic level of MDA and activities of T-AOC, GSH-Px, SOD and CAT were measured by following commercial kit protocols.

### 3.6. Determination of Hepatic ADH and ALDH

A 10% liver homogenate was obtained as described in Section 3.5. The activities of hepatic ADH and ALDH were measured according to the commercial assay kits.

### 3.7. Measurement of Pro-Inflammatory Cytokines in Liver

A 10% liver homogenate was obtained as described in Section 3.5, using phosphate-buffered saline (PBS) instead of normal saline. The levels of IL-1β, IL-6, TNF-α and IFN-γ were quantified using enzyme-linked immunosorbent assay (ELISA) kits according to the manufacturer's instructions.

### 3.8. Western Blot Analysis

Liver tissue stored at −80 °C was quickly placed in a mortar and liquid nitrogen was added, followed by pulverization of the material to obtain tissue powder. The powder was collected in a microcentrifuge tube for use in immunoblot analysis. The procedure for Western blotting was consistent with the method described by Tang et al. [59]. Experiments were performed using liver tissue samples with a protein concentration of 50 μg. Immunoreactive bands were colored with enhanced chemiluminescence (ECL, TransGen Biotech, Beijing, China), imaged using a FluorChem FC3 system (ProteinSimple, Waltham, MA, USA), and protein expression were quantified using Image Lab software. β-actin probed with antibody (Cincinnati, OH, USA) was used as a control.

### 3.9. Statistical Analysis

All experiments were repeated at least three times. Data are expressed as the mean ± standard deviation (SD) ($n = 8$) and were analyzed by analysis of variance (ANOVA) using SPSS 19.0 software (IBM SPSS Statistics, Ehningen, Germany). The statistical difference was considered to be significant at $p < 0.05$.

## 4. Conclusions

In summary, our research indicates that Fx is an effective substance to prevent alcoholic liver injury. This study demonstrates that Fx attenuates alcohol-induced oxidative stress by lowering the concentration of oxidative products and up-regulating Nrf2-mediated antioxidant responses. In addition, Fx also prevents liver inflammation by inhibiting TLR4-induced signaling pathways. These findings indicate that Fx has broad prospects for the development of health foods that protect against alcoholic liver injury.

**Author Contributions:** F.H. and Z.Y. conceived and designed the experiments. J.Z., X.T., W.Z., P.Z., and G.D. performed the statistical analysis of the data. J.Z. wrote the manuscript.

**Funding:** This work was financially supported by the National Natural Science Foundation of China (grant No. 81773629).

**Conflicts of Interest:** The authors declare no conflict of interest.

## References

1. Altamirano, J.; Bataller, R. Alcoholic liver disease: Pathogenesis and new targets for therapy. *Nat. Rev. Gastroenterol. Hepatol.* **2011**, *8*, 491–501. [CrossRef] [PubMed]
2. Gao, B.; Bataller, R. Alcoholic liver disease: Pathogenesis and new therapeutic targets. *Gastroenterology* **2011**, *141*, 1572–1585. [CrossRef] [PubMed]
3. Wruck, W.; Adjaye, J. Meta-analysis reveals up-regulation of cholesterol processes in non-alcoholic and down-regulation in alcoholic fatty liver disease. *World J. Hepatol.* **2017**, *9*, 443–454. [CrossRef] [PubMed]

4. Basra, S.; Anand, B.S. Definition, epidemiology and magnitude of alcoholic hepatitis. *World J. Hepatol.* **2011**, *3*, 108–113. [CrossRef] [PubMed]
5. Dugum, M.; McCullough, A. Diagnosis and management of alcoholic liver disease. *J. Dig. Dis.* **2015**, *82*, 226–236.
6. Pavlov, C.S.; Casazza, G.; Semenistaia, M.; Nikolova, D.; Tsochatzis, E.; Liusina, E.; Ivashkin, V.T.; Gluud, C. Ultrasonography for diagnosis of alcoholic cirrhosis in people with alcoholic liver disease. *Cochrane Database Syst. Rev.* **2016**, *3*, CD011602. [CrossRef]
7. Beier, J.I.; Arteel, G.E.; McClain, C.J. Advances in alcoholic liver disease. *Curr. Gastroenterol. Rep.* **2011**, *13*, 56–64. [CrossRef]
8. Beier, J.I.; McClain, C.J. Mechanisms and cell signaling in alcoholic liver disease. *Biol. Chem.* **2010**, *391*, 1249–1264. [CrossRef]
9. Dunn, W.; Shah, V.H. Pathogenesis of alcoholic liver disease. *Clin. Liver Dis.* **2016**, *20*, 445–456. [CrossRef]
10. Kim, S.J.; Kim, D.G.; Lee, M.D. Effects of branched-chain amino acid infusions on liver regeneration and plasma amino acid patterns in partially hepatectomized rats. *Hepatogastroenterology* **2011**, *58*, 1280–1285. [CrossRef]
11. Leung, T.M.; Nieto, N. CYP2E1 and oxidant stress in alcoholic and non-alcoholic fatty liver disease. *J. Hepatol.* **2013**, *58*, 395–398. [CrossRef] [PubMed]
12. Lu, Y.; Cederbaum, A.I. CYP2E1 and oxidative liver injury by alcohol. *Free Radic. Biol. Med.* **2008**, *44*, 723–738. [CrossRef] [PubMed]
13. Tang, C.C.; Lin, W.L.; Lee, Y.J.; Tang, Y.C.; Wang, C.J. Polyphenol-rich extract of Nelumbo nucifera leaves inhibits alcohol-induced steatohepatitis via reducing hepatic lipid accumulation and anti-inflammation in C57BL/6J mice. *Food Funct.* **2014**, *5*, 678–687. [CrossRef] [PubMed]
14. Sarumathi, A.; Sethupathy, S.; Saravanan, N. The protective efficacy of spirulina against bacterial endotoxin potentiated alcoholic liver disease. *J. Funct. Foods* **2014**, *9*, 254–263. [CrossRef]
15. Sun, H.; Mu, T.; Liu, X.; Zhang, M.; Chen, J. Purple sweet potato (Ipomoea batatas L.) anthocyanins: Preventive effect on acute and subacute alcoholic liver damage and dealcoholic effect. *J. Agric. Food Chem.* **2014**, *62*, 2364–2373. [CrossRef] [PubMed]
16. Wang, M.; Zhang, X.; Liu, F.; Hu, Y.; He, C.; Li, P.; Su, H.; Wan, J. Saponins isolated from the leaves of Panax notoginseng protect against alcoholic liver injury via inhibiting ethanol-induced oxidative stress and gut-derived endotoxin-mediated inflammation. *J. Funct. Foods* **2015**, *19*, 214–224. [CrossRef]
17. Shin, S.M.; Yang, J.H.; Ki, S.H. Role of the Nrf2-ARE pathway in liver diseases. *Oxid. Med. Cell. Longev.* **2013**, *2013*, 1–9. [CrossRef] [PubMed]
18. Kim, G.J.; Song, D.; Yoo, H.; Chung, K.H.; Lee, K.; An, J. Hederagenin supplementation alleviates the pro-inflammatory and apoptotic response to alcohol in rats. *Nutrients* **2017**, *9*, 41. [CrossRef]
19. Ambade, A.; Mandrekar, P. Oxidative stress and inflammation: Essential partners in alcoholic liver disease. *Int. J. Hepatol.* **2012**, *2012*, 853175. [CrossRef]
20. Kiuru, P.; Auria, M.; Muller, C. Exploring marine resources for bioactive compounds. *Planta Med.* **2014**, *80*, 1234–1246. [CrossRef]
21. Arumugam, V.; Venkatesan, M.; Ramachandran, S.; Sundaresan, U. Bioactive Peptides from Marine Ascidians and Future Drug Development–A Review. *Int. J. Pept. Res. Ther.* **2017**, *24*, 13–18. [CrossRef]
22. Liu, H.; Liu, H.; Zhu, L.; Zhang, Z.; Zheng, X.; Liu, J. Comparative Transcriptome Analyses Provide Potential Insights into the Molecular Mechanisms of Astaxanthin in the Protection against Alcoholic Liver Disease in Mice. *Mar. Drugs* **2019**, *17*, 181. [CrossRef] [PubMed]
23. Ge, N.; Liang, H.; Zhao, Y.; Liu, Y.; Gong, A.; Zhang, W. Aplysin Protects Against Alcohol-Induced Liver Injury Via Alleviating Oxidative Damage and Modulating Endogenous Apoptosis-Related Genes Expression in Rats. *J. Food Sci.* **2018**, *83*, 2612–2621. [CrossRef] [PubMed]
24. D'Orazio, N.; Gammone, M.A.; Gemello, E.; Girolamo, M.D.; Cusenza, S.; Riccioni, G. Marine Bioactives: Pharmacological Properties and Potential Applications against Inflammatory Diseases. *Mar. Drugs* **2012**, *10*, 812–833. [CrossRef] [PubMed]
25. Peng, J.; Yuan, J.P.; Wu, C.F.; Wang, J.H. Fucoxanthin, a marine carotenoid present in brown seaweeds and diatoms: Metabolism and bioactivities relevant to human health. *Mar. Drugs* **2011**, *9*, 1806–1828. [CrossRef] [PubMed]

26. Gammone, M.; Riccioni, G.; D'Orazio, N. Marine carotenoids against oxidative stress: Effects on human health. *Mar. Drugs* **2015**, *13*, 6226–6246. [CrossRef] [PubMed]
27. Nishino, H.; Murakoshi, M.; Tokuda, H.; Satomi, Y. Cancer prevention by carotenoids. *Arch. Biochem. Biophys.* **2009**, *483*, 165–168. [CrossRef] [PubMed]
28. Lin, H.T.; Tsou, Y.C.; Chen, Y.T.; Lu, W.J.; Hwang, P.A. Effects of low-molecular-weight fucoidan and high stability fucoxanthin on glucose homeostasis, lipid metabolism, and liver function in a mouse model of type II diabetes. *Mar. Drugs* **2017**, *15*, 113. [CrossRef]
29. Liu, H.; Liu, M.; Fu, X.; Zhang, Z.; Zhu, L.; Zheng, X.; Liu, J. Astaxanthin Prevents Alcoholic Fatty Liver Disease by Modulating Mouse Gut Microbiota. *Nutrients* **2018**, *10*, 1298. [CrossRef]
30. Li, X.; Jin, Q.; Zhang, Y.; Wu, Y.L.; Jin, C.M.; Cui, B.W.; Li, Y.; Jin, M.J.; Shang, Y.; Jiang, M. Inhibition of P2 × 7R–NLRP3 Inflammasome Activation by Pleurotus citrinopileatus: A Possible Protective Role in Alcoholic Hepatosteatosis. *J. Agric. Food Chem.* **2018**, *66*, 13183–13190. [CrossRef]
31. Huang, Q.; Zhang, S.; Zheng, L.; Liao, M.; He, M.; Huang, R.; Zhuo, L.; Lin, X. Protective effect of isoorientin-2″-O-α-L-arabinopyranosyl isolated from Gypsophila elegans on alcohol induced hepatic fibrosis in rats. *Food Chem. Toxicol.* **2012**, *50*, 1992–2001. [CrossRef] [PubMed]
32. Han, J.; Ju, J.; Lee, Y.; Park, J.; Yeo, I.; Park, M.; Roh, Y.; Han, S.; Hong, J. Astaxanthin alleviated ethanol-induced liver injury by inhibition of oxidative stress and inflammatory responses via blocking of STAT3 activity. *Sci. Rep.* **2018**, *8*, 14090–14099. [CrossRef] [PubMed]
33. Zhang, M.; Ye, P. Research Progress of Alcohol Intake and Dyslipidemia. *Adv. Cardiovasc. Dis.* **2010**, *31*, 370–373.
34. Birben, E.; Sahiner, U.M.; Sackesen, C.; Erzurum, S.; Kalayci, O. Oxidative stress and antioxidant defense. *World Allergy Organ. J.* **2012**, *5*, 9–19. [CrossRef] [PubMed]
35. Jeon, S.M. Regulation and function of AMPK in physiology and diseases. *Exp. Mol. Med.* **2016**, *48*, e245. [CrossRef] [PubMed]
36. Tang, X.; Wei, R.; Deng, A.; Lei, T. Protective Effects of Ethanolic Extracts from Artichoke, an Edible Herbal Medicine, against Acute Alcohol-Induced Liver Injury in Mice. *Nutrients* **2017**, *9*, 1000. [CrossRef]
37. Yuan, R.; Tao, X.; Liang, S.; Pan, Y.; He, L.; Sun, J.; Wenbo, J.; Li, X.; Chen, J.; Wang, C. Protective effect of acidic polysaccharide from Schisandra chinensis on acute ethanol-induced liver injury through reducing CYP2E1-dependent oxidative stress. *Biomed. Pharmacother.* **2018**, *99*, 537–542. [CrossRef] [PubMed]
38. Fu, J.; Qi, L.; Hu, M.; Liu, Y.; Yu, K.; Liu, Q.; Liu, X. Salmonella proteomics under oxidative stress reveals coordinated regulation of antioxidant defense with iron metabolism and bacterial virulence. *J. Proteom.* **2017**, *157*, 52–58. [CrossRef] [PubMed]
39. Wang, M.; Zhu, P.; Jiang, C.; Ma, L.; Zhang, Z.; Zeng, X. Preliminary characterization, antioxidant activity in vitro and hepatoprotective effect on acute alcohol-induced liver injury in mice of polysaccharides from the peduncles of Hovenia dulcis. *Food Chem. Toxicol.* **2012**, *50*, 2964–2970. [CrossRef]
40. Ding, R.; Tian, K.; Cao, Y.; Bao, J.; Wang, M.; He, C.; Hu, Y.; Su, H.; Wan, J. Protective Effect of Panax notoginseng Saponins on Acute Ethanol-Induced Liver Injury Is Associated with Ameliorating Hepatic Lipid Accumulation and Reducing Ethanol-Mediated Oxidative Stress. *J. Agric. Food Chem.* **2015**, *63*, 2413–2422. [CrossRef]
41. Qu, L.; Zhu, Y.; Liu, Y.; Yang, H.; Zhu, C.; Ma, P.; Deng, J.; Fan, D. Protective effects of ginsenoside Rk3 against chronic alcohol-induced liver injury in mice through inhibition of inflammation, oxidative stress, and apoptosis. *Food Chem. Toxicol.* **2019**, *126*, 277–284. [CrossRef] [PubMed]
42. Becker, U.; Grønbaek, M.; Johansen, D.; Sørensen, T.I. Lower risk for alcohol-induced cirrhosis in wine drinkers. *Hepatology* **2002**, *35*, 868–875. [CrossRef] [PubMed]
43. Liu, W.; Gao, F.; Li, Q.; Lv, J.; Wang, Y.; Hu, P.; Xiang, Q.; Wei, L. Protective Effect of Astragalus polysaccharides on Liver Injury Induced by Several Different Chemotherapeutics in Mice. *Asian Pac. J. Cancer Prev.* **2014**, *15*, 10413–10420. [CrossRef]
44. Tang, C.; Huang, H.; Lee, Y.; Tang, Y.; Wang, C. Hepatoprotective effect of mulberry water extracts on ethanol-induced liver injury via anti-inflammation and inhibition of lipogenesis in C57BL/6J mice. *Food Chem. Toxicol.* **2013**, *62*, 786–796. [CrossRef] [PubMed]
45. Shu, L.; Lei, T.; Chai, G.; Bo, W.; Wang, B. Targeting heme oxygenase-1 by quercetin ameliorates alcohol-induced acute liver injury via inhibiting NLRP3 inflammasome activation. *Food Funct.* **2018**, *9*, 4184–4193.

46. Wang, M.; Shen, G.; Xu, L.; Liu, X.; Brown, J.; Feng, D.; Ross, R.; Gao, B.; Liangpunsakul, S.; Ju, C. IL-1 receptor like 1 protects against alcoholic liver injury by limiting NF-κB activation in hepatic macrophages. *J. Hepatol.* **2018**, *68*, S0168827817322638. [CrossRef] [PubMed]
47. Neyrinck, A.M.; Etxeberria, U.; Taminiau, B.; Daube, G.; Hul, M.V.; Everard, A.; Cani, P.D.; Bindels, L.B.; Delzenne, N.M. Rhubarb extract prevents hepatic inflammation induced by acute alcohol intake, an effect related to the modulation of the gut microbiota. *Mol. Nutr. Food Res.* **2017**, *61*, 1500899. [CrossRef]
48. Nguyen, T.; Nioi, P.; Pickett, C.B. The Nrf2-Antioxidant Response Element Signaling Pathway and Its Activation by Oxidative Stress. *J. Biol. Chem.* **2009**, *284*, 13291–13295. [CrossRef]
49. Wang, Z.; Dou, X.; Li, S.; Zhang, X.; Sun, X.; Zhou, Z.; Song, Z. Nrf2 activation-induced hepatic VLDLR overexpression in response to oxidative stress contributes to alcoholic liver disease in mice. *Hepatology* **2014**, *59*, 1381–1392. [CrossRef]
50. Klaassen, C.D.; Reisman, S.A. Nrf2 the rescue: Effects of the antioxidative/electrophilic response on the liver. *Toxicol. Appl. Pharmacol.* **2010**, *244*, 57–65. [CrossRef]
51. Nioi, P.; Mcmahon, M.; Itoh, K.; Yamamoto, M.; Hayes, J.D. Identification of a novel Nrf2-regulated antioxidant response element (ARE) in the mouse NAD(P)H:quinone oxidoreductase 1 gene: Reassessment of the ARE consensus sequence. *Biochem. J.* **2003**, *374*, 337–348. [CrossRef] [PubMed]
52. Xia, T.; Yao, J.; Jin, Z.; Yu, Z.; Jia, S.; Min, W. Protective effects of Shanxi aged vinegar against hydrogen peroxide-induced oxidative damage in LO2 cells through Nrf2-mediated antioxidant responses. *RSC Adv.* **2017**, *7*, 17377–17386. [CrossRef]
53. Mandrekar, P.; Szabo, G. Signalling pathways in alcohol-induced liver inflammation. *J. Hepatol.* **2009**, *50*, 1258–1266. [CrossRef] [PubMed]
54. Hritz, I.; Velayudham, A.; Dolganiuc, A.; Kodys, K.; Mandrekar, P.; Kurt-Jones, E.; Szabo, G. Bone marrow-derived immune cells mediate sensitization to liver injury in a myeloid differentiation factor 88-dependent fashion. *Hepatology* **2008**, *48*, 1342–1347. [CrossRef] [PubMed]
55. Petrasek, J.; Mandrekar, P.; Szabo, G. Toll-Like Receptors in the Pathogenesis of Alcoholic Liver Disease. *Gastroenterol. Res. Pract.* **2010**, *2010*, 70–80. [CrossRef] [PubMed]
56. Yang, S.; Zhuang, T.; Si, Y.; Qi, K.; Zhao, J. Coriolus versicolor mushroom polysaccharides exert immunoregulatory effects on mouse B cells via membrane Ig and TLR-4 to activate the MAPK and NF-κB signaling pathways. *Mol. Immunol.* **2015**, *64*, 144–151. [CrossRef] [PubMed]
57. Lin, Y.; Tai, S.; Chen, J.; Chou, C.; Fu, S.; Chen, Y. Ameliorative effects of pepsin-digested chicken liver hydrolysates on development of alcoholic fatty livers in mice. *Food Funct.* **2017**, *8*, 1763–1771. [CrossRef] [PubMed]
58. Hong, K.S.; Yun, S.M.; Cho, J.M.; Lee, D.Y.; Ji, S.D.; Son, J.-G.; Kim, E.H. Silkworm (Bombyx mori) powder supplementation alleviates alcoholic fatty liver disease in rats. *J. Funct. Foods* **2018**, *43*, 29–36. [CrossRef]
59. Tang, Y.; Yu, F.; Zhang, G.; Yang, Z.; Huang, F.; Ding, G. A purified serine protease from Nereis virens and its impaction of apoptosis on human lung cancer cells. *Molecules* **2017**, *22*, 1123. [CrossRef]

marine drugs

Article

# Anti-Obesity Effect of Standardized Extract of Microalga *Phaeodactylum tricornutum* Containing Fucoxanthin

**Song Yi Koo [1,5], Ji-Hyun Hwang [1], Seung-Hoon Yang [3], Jae-In Um [4], Kwang Won Hong [4], Kyungsu Kang [1,6], Cheol-Ho Pan [1,6], Keum Taek Hwang [5] and Sang Min Kim [2,6,*]**

1 Natural Product Informatics Center, KIST Gangneung Institute of Natural Products, Gangneung 25451, Korea; ninesong2@kist.re.kr (S.Y.K.); hwangjh@kist.re.kr (J.-H.H.); kskang@kist.re.kr (K.K.); panc@kist.re.kr (C.-H.P.)
2 Smart Farm Research Center, KIST Gangneung Institute of Natural Products, Gangneung 25451, Korea
3 Department of Medical Biotechnology, College of Life Science and Biotechnology, Dongguk University, Seoul 04620, Korea; shyang@dongguk.edu
4 R&D Department, AlgaeTech Co. Ltd., Gangneung 25457, Korea; korea4633@naver.com (J.-I.U.); hongtol2@gmail.com (K.W.H.)
5 Department of Food and Nutrition, and Research Institute of Human Ecology, Seoul National University, Seoul 08826, Korea; keum@snu.ac.kr
6 Division of Bio-Medical Science & Technology, University of Science & Technology, Daejeon 34113, Korea
* Correspondence: kimsm@kist.re.kr; Tel.: +82-33-650-3640

Received: 10 April 2019; Accepted: 24 May 2019; Published: 27 May 2019

**Abstract:** Fucoxanthin (FX), a marine carotenoid found in macroalgae and microalgae, exhibits several beneficial effects to health. The anti-obesity activity of FX is well documented, but FX has not been mass-produced or applied extensively or commercially because of limited availability of raw materials and complex extraction techniques. In this study, we investigated the anti-obesity effect of standardized FX powder (*Phaeodactylum* extract (PE)) developed from microalga *Phaeodactylum tricornutum* as a commercial functional food. The effects of PE on adipogenesis inhibition in 3T3-L1 adipocytes and anti-obesity in high-fat diet (HFD)-fed C57BL/6J mice were evaluated. PE and FX dose-dependently decreased intracellular lipid contents in adipocytes without cytotoxicity. In HFD-fed obese mice, PE supplementation for six weeks decreased body weight, organ weight, and adipocyte size. In the serum parameter analysis, the PE-treated groups showed attenuation of lipid metabolism dysfunction and liver damage induced by HFD. In the liver, uncoupling protein-1 (UCP1) upregulation and peroxisome proliferator activated receptor $\gamma$ (PPAR$\gamma$) downregulation were detected in the PE-treated groups. Additionally, micro computed tomography revealed lower fat accumulation in PE-treated groups compared to that in the HFD group. These results indicate that PE exerts anti-obesity effects by inhibiting adipocytic lipogenesis, inducing fat mass reduction and decreasing intracellular lipid content, adipocyte size, and adipose weight.

**Keywords:** anti-obesity effect; *Phaeodactylum tricornutum*; fucoxanthin; *Phaeodactylum* extract; microalgae

## 1. Introduction

Fucoxanthin (FX), a marine xanthophyll carotenoid, is abundantly in macroalgae, such as *Laminaria japonica* and *Undaria pinnatifida* [1]. Recently, some microalgae including *Phaeodactylum tricornutum*, *Odontella aurita*, and *Isochrysis galbana* were reported as new sources for FX production [2–4]. FX has multiple health-promoting effects, such as antioxidant, anticancer, anti-inflammatory, and anti-obesity effects [5–7]. The anti-obesity effect is the most remarkable property and has been supported by numerous in vitro and in vivo studies [8,9]. FX exerts its anti-obesity effects through several mechanisms, which are influenced by numerous factors, including nutritional, hormonal, and genetic elements [10]. FX significantly reduces triglyceride concentrations in the liver and adipose tissue and beneficially influences cholesterol-regulating enzymes [11]. FX also affects gene expression associated with lipid metabolism, such as hepatic acetyl-CoA carboxylase, sterol regulatory element-binding protein, fatty acid synthase, stearoyl-coenzyme A desaturase-1, CCAAT/enhancer-binding protein $\alpha$ (C/EBP$\alpha$), and peroxisome proliferator activated receptor (PPAR) $\alpha$ and $\gamma$ [12,13]. Recently, the anti-obesity effect of FX was found to involve stimulation of the uncoupling protein-1 (UCP1) expression in white adipose tissue (WAT) [14–16]. Increased expression of mitochondrial UCP1 leads to increased energy expenditure. UCP1 is found in brown adipose tissue and represents a crucial factor in thermogenesis, which is heat production contributing to the reduction of WAT and physiological defense against obesity [17–19]. Additionally, the hormone, leptin, is mainly expressed in differentiated adipocytes of white tissue to maintain homeostatic regulation of the adipose tissue and body weight by controlling food intake and energy expenditure [20]. Many previous studies suggested that FX alters plasma leptin levels to achieve anti-obesity effects [21]. Many studies reported that FX supplementation derived from brown seaweed (macroalgae) clearly reduced WAT in rats and mice and improved weight-loss effects as well as plasma and hepatic lipid metabolism [22,23]. Numerous in vivo studies have been conducted to evaluate seaweed-derived FX, while few in vivo tests have used microalgae-derived FX.

*Phaeodactylum tricornutum* (PT) is known as a potential source of FX production [2]. In PT, FX production is at least 10-fold higher than that in macroalgae. Therefore, PT may be valuable for treating and preventing obesity. PT biomass itself has been studied for its anti-obesity effect, but no studies have examined standardized materials containing FX extracted from marine microalgae as a health functional food, and no studies have evaluated the anti-obesity effects of these materials [24].

The aim of the present study was to evaluate the anti-obesity effect of the standardized extract of PT (*Phaeodactylum* extract, PE) containing 3.5~6% FX (*w/w*) on lipid accumulation in 3T3-L1 adipocytes and high-fat diet-induced obese mice. From in vitro and in vivo studies, we confirmed that PE exerted a strong anti-obesity effect by decreasing body weight, organ weight, and adipocyte size. In addition, UCP1 upregulation and PPAR$\alpha$ downregulation were also observed in both studies, indicating the mechanism through which PE regulates lipid metabolism. Therefore, we concluded that PE, a new FX source produced from microalga PT, might be a potent anti-obesity material applicable to commercial functional foods.

## 2. Results

### 2.1. PE Reduced Lipid Accumulation in Adipocytes

PE is the standardized FX powder, bright red in color, which was produced from the microalga PT for commercial purposes. AlgaeTech Co. Ltd. (Gangneung, Korea) supplied FX content information of 3.5~6% (*w/w*) in PE. In order to quantify FX in PE, HPLC analysis was performed, as described in our previous report, with slight modification [2]. The result demonstrated that the FX peak in PE showed the same UV-VIS absorption spectrum with the FX standard (Figure S1A) and the average amount of FX was 47.74 ± 0.04 mg/g in three independent PE batches (Figure S1B). This indicated that the FX contents in the PE batches used in this study falls within the standardization range.

Prior to evaluating the effects of FX and PE in 3T3-L1 cells, we performed a cytotoxicity assay to determine the proper concentration of FX and PE for further analysis. We indicated that FX was not toxic to cells below 80 μM, and PE showed cytotoxicity at 1000 μg/mL (Figure S2). Therefore, we conducted experiments by using non-toxic concentrations of FX and PE.

To determine the effect of FX and PE on lipid accumulation, the cells were cultured during differentiation (for six days) with FX, PE, and curcumin (CCM; reference). Staining with Oil-red O (ORO), which is commonly used to detect intracytoplasmic lipid accumulation, demonstrated that FX causes a dose-dependent reduction in lipid accumulation. FX at 40 μM inhibited lipid accumulation by 30% compared to the control during differentiation. PE decreased lipid accumulation at 250 μg/mL during adipocyte differentiation (Figure 1A,B). Thus, FX and PE alleviated the effect of lipid accumulation on insulin-stimulated 3T3-L1 cells.

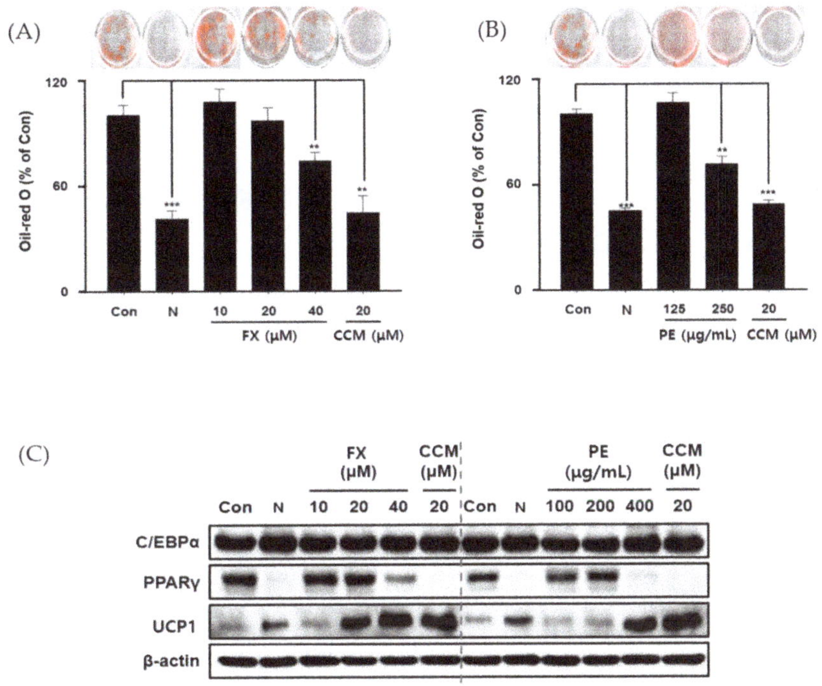


**Figure 1.** Effect of fucoxanthin (FX) and *Phaeodactylum* extract (PE) on lipid accumulation in 3T3-L1 cells during adipogenesis. Cells were cultured during differentiation (for six days) with FX, PE, or curcumin (CCM; reference). Accumulated lipids were stained with Oil-red O reagent and quantified by measuring the absorbance at 500 nm. (**A**) FX or CCM suppressed lipid accumulation; (**B**) PE and CCM inhibited lipid accumulation; (**C**) expression of proteins related to lipid accumulation (CCAAT/enhancer-binding protein α (C/EBPα), peroxisome proliferator activated receptor γ (PPARγ), uncoupling protein-1 (UCP1)). The experiment was performed in triplicate. ***/**/* indicate significant differences at $p < 0.001$/ $p < 0.01$/$p < 0.05$ compared to the control (differentiation). N indicates no differentiation.

Adipogenic differentiation is related to various adipogenic factors, such as C/EBPα, PPARγ, and UCP1, as mentioned above [12–16]. To determine whether these factors are controlled by FX and PE, we analyzed protein levels in the absence or presence of FX or PE during insulin-induced differentiation. At high levels, FX and PE clearly decreased PPARγ levels and dose-dependently increased the UCP1 level (Figure 1C). However, neither FX nor PE affected the level of C/EBPα. FX was reported to inhibit the intercellular lipid accumulation by reducing the expression of PPARγ and C/EBPα. In the case of C/EBPα, there have been some conflicting results about the effect of FX on the expression of this enzyme during the differentiation period [10,25,26]. In this study, we conclude that FX and PE treatment induced downregulation of PPARγ and upregulation of UCP1 in two key major enzymes (PPARγ and UCP1) in 3T3-L1 cells. We have also shown that FX, a major component of PE, can be one of the main ingredients for the anti-obesity effect of PE. Thus, it was concluded that PE has an anti-obesity effect by controlling lipid metabolism through PPARγ and UCP1.

## 2.2. Effect of PE on Body Weight, Liver, and Inguinal Fat Weights

Before treatment, the average initial body weight of the mice was 24.38 g, with no significant differences among the 7 groups. As shown in Figure 2A,B, the high-fat diet (HFD) group showed a large increase in body weight during the experimental period. After 6 weeks, the area under the curve (AUC) of body weights was higher in the HFD group than in the normal diet (ND) group and other groups ($p < 0.001$). Mice in the HFD plus PE treatment group showed a lower AUC level than the HFD group, while mice in the PE plus conjugated linoleic acid (CLA) group showed the lowest AUC level. For body weight gain, all PE-treated groups showed weight loss in a dose-dependent manner compared to the HFD group, with greater effects than in the positive control (FX). As shown in Figure 2C, we weighed the livers and inguinal adipose tissue. As a result, of liver weight measurement, PE treatment groups (PE-L, PE-M, PE-H) and the PE-M plus CLA group showed significantly lower weight values (1.12%~7.45% at $p < 0.001$ or $p < 0.01$) than in the HFD group. Inguinal adipose tissue weight was significantly higher in the HFD group than the ND group on both the right and left. However, the PE-treated groups exhibited lower values than the HFD group in inguinal adipose tissue weight over 40% ($p < 0.01$ or $p < 0.05$). Based on this result, we concluded that PE exerts anti-obesity effects in vivo in mice.

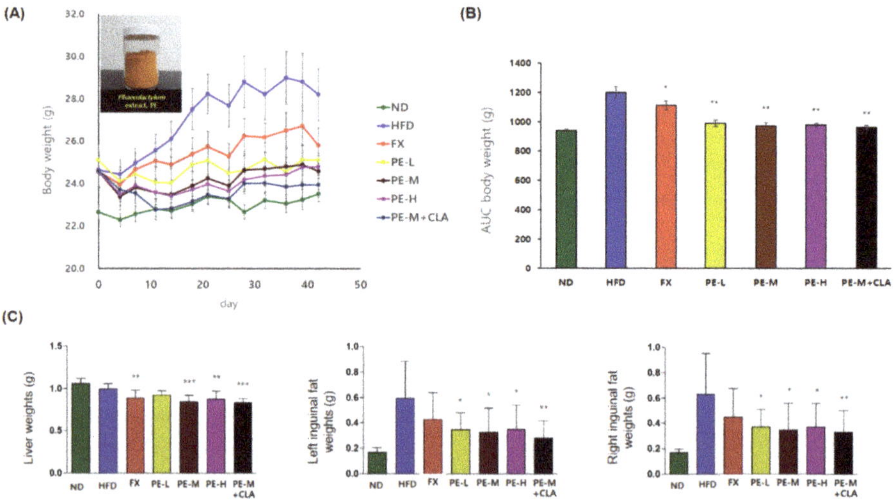


**Figure 2.** (**A**) Changes in body weight of C57BL/6J mice fed PE, (**B**) area under the curve data for body weight, (**C**) effects of PE on liver and inguinal fat weights in high-fat fed C57BL/6J mice. The groups are abbreviated as: Normal diet (ND); high-fat diet (HFD); fucoxanthin (FX), HFD + FX 0.1 mg/kg/day; PE-L, HFD + PE 0.81 mg/kg/day; PE-M, HFD + PE 1.62 mg/kg/day; PE-H, HFD + PE 3.25 mg/kg/day; PE-M + conjugated linoleic acid (CLA), HFD + PE 1.62 mg/kg/day plus CLA 410 mg/kg/day. ***/**/* indicate significant differences at $p < 0.001/p < 0.01/p < 0.05$ level compared to HFD. Values are shown as the mean ± SD ($n = 10$ per group).

## 2.3. Effect of PE on Serum Lipid Parameters

Next, we measured serum parameters to evaluate the effect of PE on lipid metabolism as well as liver damage. As shown in Table 1, the HFD group showed a slight increase in the level of triacylglyceride (TG) and a significant increase in total cholesterol (TC) and low-density lipoprotein (LDL) compare with the ND group. These data demonstrated that HFD consumption induces lipid dysmetabolism. Administration of PE-M and PE-M + CLA for 6 weeks significantly decreased the TG level compared with that in the HFD group, while groups treated with FX, PE-L, and PE-H showed a slight decrease in the TG level compared with that of the HFD group. Neither FX nor PE affected the HFD-induced increase in TC and high-density lipoprotein (HDL) levels. FX treatment did not affect the levels of LDL, which is generally considered as a pathological marker. In contrast, PE administration decreased the LDL levels in a dose-dependent manner, especially PE-H, which significantly decreased the LDL levels. In addition, the HFD group also showed an abnormal increase in alanine aminotransferase (ALT) levels and a significant increase in aspartate aminotransferase (AST) levels compared with those in the ND group, indicating that HFD also induced liver injury [27]. The FX, PE, and PE-M + CLA groups showed a slight decrease in AST levels compared with the HFD group, although the difference was not statistically significant. FX and PE-M + CLA treatment significantly decreased ALT levels, whereas PE treatment slightly decreased ALT levels, compared with those in the HFD group. Based on these data, we concluded that PE-M, PE-H, and PE-M + CLA could attenuate lipid metabolism dysfunction and liver damage induced by HFD.

**Table 1.** Effects of PE on serum parameters in the plasma.

| | ND | HFD | FX | PE-L | PE-M | PE-H | PE-M + CLA |
|---|---|---|---|---|---|---|---|
| **Plasma** | | | | | | | |
| TG (mg/dL) | 45.3 ± 10.3 | 50.6 ± 9.2 | 44.7 ± 17.2 | 47.6 ± 10.2 | 34.8 ± 13.5 # | 35.7 ± 18.2 | 31.3 ± 6.7 # |
| TC (mg/dL) | 79.5 ± 5.8 | 130.5 ± 6.1 *** | 128.2 ± 30.9 *** | 126.3 ± 20.3 *** | 137.2 ± 9.2 *** | 123.7 ± 18.4 *** | 133.1 ± 14.3 *** |
| HDL (mg/dL) | 48.1 ± 6.0 | 71.2 ± 4.0 *** | 63.8 ± 15.7 ** | 63.3 ± 11.9 ** | 68.2 ± 5.6 *** | 60.9 ± 10.8 * | 64.3 ± 7.8 ** |
| LDL (mg/dL) | 6.8 ± 1.0 | 17.2 ± 1.6 *** | 17.3 ± 5.3 *** | 15.7 ± 1.7 *** | 14.9 ± 1.9 *** | 14.1 ± 1.7 ***,# | 14.4 ± 1.8 *** |
| AST (U/L) | 30.8 ± 4.2 | 31.5 ± 10.6 * | 23.1 ± 4.8 | 23.4 ± 6.1 | 25.0 ± 3.2 | 26.8 ± 9.1 | 22.3 ± 8.0 |
| ALT (U/L) | 52.7 ± 6.6 | 64.7 ± 10.9 | 61.4 ± 12.8 # | 55.1 ± 7.1 | 56.1 ± 4.5 | 62.5 ± 10.5 | 55.8 ± 10.7 *,# |

Note: Data are expressed as the mean ± SD ($n$ = 10 per group). Triacylglyceride (TG); total cholesterol (TC); high-density lipoprotein (HDL); low-density lipoprotein (LDL);aspartate transaminase (AST); alanine transaminase (ALT); normal diet (ND); high fat diet (HFD); fucoxanthin (FX), HFD + FX 0.1 mg/kg/day; PE-L, HFD + PE 0.81 mg/kg/day; PE-M, HFD + PE 1.62 mg/kg/day; PE-H, HFD + PE 3.25 mg/kg/day; PE-M + CLA, HFD + PE 1.62 mg/kg/day plus CLA 410 mg/kg/day. ***/**/* indicate a significant difference at $p < 0.001/p < 0.01/p < 0.05$ levels compared to the ND. # indicates a significant difference at $p < 0.05$ level compared to the HFD.

## *2.4. Effect of PE on Fat Accumulation*

### 2.4.1. Histological Analysis of the Liver

For histological analysis, liver tissues were stained with ORO and hematoxylin and eosin (H&E) to observe the nucleus and cytoplasm of the cell [28]. As shown in Figure 3A, there was no significant difference in the cell staining pattern between the HFD and PE treatment groups in the H&E staining, indicating that the cell states of all groups were normal. However, ORO staining revealed that the livers of the HFD groups contained more lipid droplets than those of the ND group and the FX- and PE-treated groups with the increased fat globule size (Figure 3A). After treatment with FX or PE, the lipid droplet numbers decreased (Figure 3A) and the fat globule sizes reduced strongly to 75.66% of that in the HFD group in a dose-dependent manner (Figure 3B). In addition, the PE-M + CLA group showed the biggest reduction in fat globule size (72.5% of that in the HFD group, $p < 0.001$) indicating the synergistic effect of CLA on liver fat metabolism. In Figure 3C, the level of liver in TG in the PE-treated group reduced significantly and dose-dependently. These results clearly demonstrated the effect of PE on hepatic steatosis.

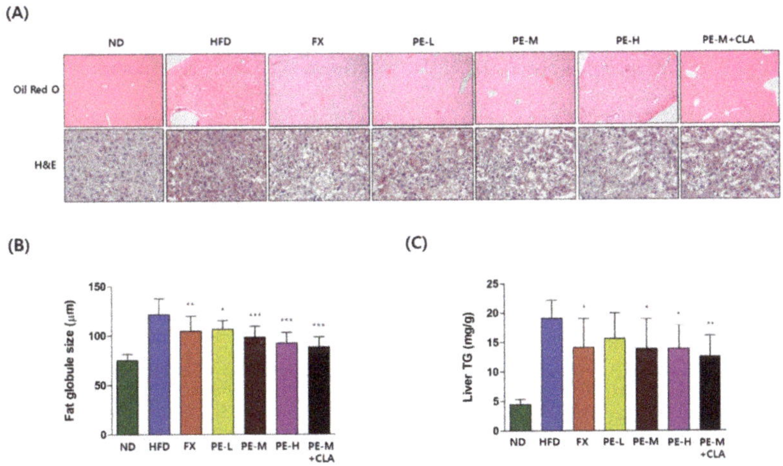


**Figure 3.** (**A**) Histological analysis of liver tissues. Liver tissue section stained with Oil Red O and hematoxylin and eosin (H&E) (magnification, ×100 (Oil Red O), ×400 (H&E)), (**B**) fat globule size, (**C**) liver TG. The groups are abbreviated as: Normal diet (ND); high-fat diet (HFD); fucoxanthin (FX), HFD + FX 0.1 mg/kg/day; PE-L, HFD + PE 0.81 mg/kg/day; PE-M, HFD + PE 1.62 mg/kg/day; PE-H, HFD + PE 3.25 mg/kg/day; PE-M + CLA, HFD + PE 1.62 mg/kg/day plus CLA 410 mg/kg/day. Lipid droplet numbers and fat globule size in the Oil Red O staining were dose-dependently decreased by PE treatment (**A**,**B**) and there was no significant difference in the cell staining pattern between the HFD and PE treatment groups in the H&E staining (**A**). ***/**/* indicate significant differences at $p < 0.001/p < 0.01/p < 0.05$ level compared to the HFD group. Values are shown as the mean ± SD ($n = 10$ per group).

### 2.4.2. Micro Computed Tomography Analysis

The abdominal and subcutaneous fat in each mouse was assessed by micro computed tomography (CT) analysis. As shown in Figure 4A, fat accumulation was greater in the HFD group than in any other groups. PE-treated groups (PE-L, PE-M, PE-H, and PE-M + CLA) showed less distributed fat accumulation than the HFD group and less fat than in the positive control, FX. As shown in Figure 4B, total, abdominal, and subcutaneous fat volumes were evaluated and quantified by micro CT analysis. The results showed that all groups (HFD, FX, PE-L, PE-M, PE-H, and PE-M + CLA) were significantly different compared to the ND group (statistical data not shown). Additionally, the PE-treated groups (PE-L, PE-M, PE-H, and PE-M + CLA) showed decreased fat contents compared to the HFD group ($p < 0.01$ to $< 0.05$), and the fat content was the lowest in the PE-M + CLA combination group ($p < 0.01$). The total fat volumes in the PE-treated groups were 53.30~65.66% of that in the HFD group. For abdominal fat, the values in the PE-treated groups were 43.34~57.96% of that in the HFD, whereas for subcutaneous fat, the values in the PE-treated groups were 67.68~76.78% of that of the HFD group. These results suggest that PE inhibits adipocytic lipogenesis and reduces fat levels. These are consistent with those of weight loss.

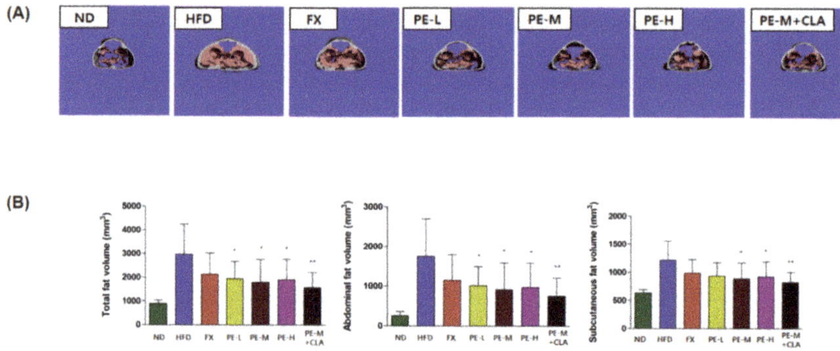


**Figure 4.** (**A**) Micro computed tomography (CT) analysis, (**B**) total, abdominal, and subcutaneous fat volumes. The groups are abbreviated as: Normal diet (ND); high-fat diet (HFD); fucoxanthin (FX), HFD + FX 0.1 mg/kg/day; PE-L, HFD + PE 0.81 mg/kg/day; PE-M, HFD + PE 1.62 mg/kg/day; PE-H, HFD + PE 3.25 mg/kg/day; PE-M + CLA, HFD + PE 1.62 mg/kg/day plus CLA 410 mg/kg/day. ***/**/* indicate significant differences at $p < 0.001/p < 0.01/p < 0.05$ levels compared to the HFD group. Values are shown as the mean ± SD ($n = 10$ per group).

### *2.5. PE Regulated the Expression Proteins Related to Lipid Metabolism in Adipose Tissue*

We assessed the expression of UCP1, PPARγ, and C/EBPα, which are related to lipid metabolism in adipose tissue. As shown in Figure 5, UCP1 expression was increased in the PE-treated groups. Particularly, compared to in the HFD group, UCP1 expression levels in the PE-M, PE-H, and PE-M + CLA groups were 2.02-, 2.34-, and 2.35-fold higher, respectively, indicating significant and dose-dependent effects. PPARγ expression levels were significantly lower in the PE-L, PE-M, PE-H, and PE-M + CLA groups than in the HFD group by 0.62-, 0.54-, 0.51-, and 0.45- fold, respectively, with high significance ($p < 0.001$ or $p < 0.01$). In addition, C/EBPα expression levels were lower in the PE-treated groups. Significance was low ($p < 0.05$), which is similar to the in vitro result in Figure 1. Overall, these results suggest that PE can upregulate UCP1, which is related to energy expenditure, and downregulate PPARγ, which is related to adipocyte differentiation.

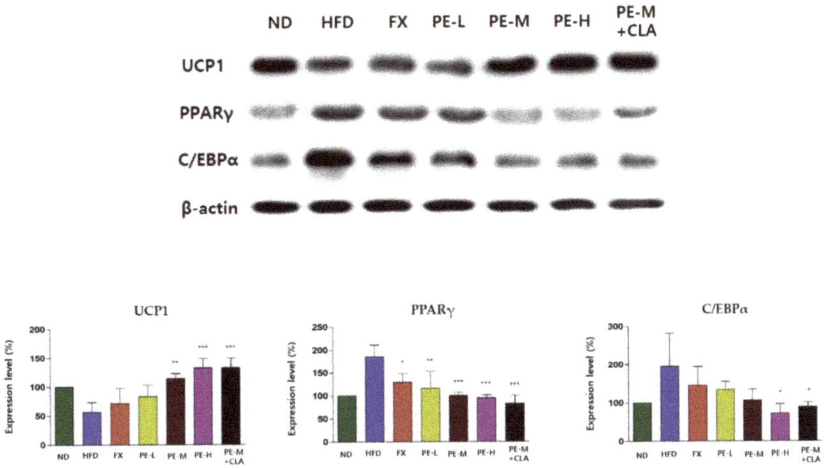


**Figure 5.** Effect of PE on the relative expression of lipid metabolism-related proteins in the liver. All expression levels (%) of each group were relatively compared with that of ND group (100%). The groups are abbreviated as: Normal diet (ND); high-fat diet (HFD); fucoxanthin (FX), HFD + FX 0.1 mg/kg/day; PE-L, HFD + PE 0.81 mg/kg/day; PE-M, HFD + PE 1.62 mg/kg/day; PE-H, HFD + PE 3.25 mg/kg/day; PE-M + CLA, HFD + PE 1.62 mg/kg/day plus CLA 410 mg/kg/day. ***/**/* indicate significant differences at $p < 0.001/p < 0.01/p < 0.05$ levels compared with HFD. Values are shown as the mean ± SD ($n = 3$ per group).

## 3. Discussion

Application of microalgae has extended from biomass production for biodiesel to value-added products [29,30], and FX is one of the most prominent value-added products obtained from microalgae. However, to date, FX has been produced mostly by macroalgae, such as *L. japonica*, *U. pinnatifida*, and *Eisenia bicyclis* [1]. Numerous commercial products containing FX show anti-obesity activity; xanthigen is among the most representative products and contains 3 mg of FX for daily intake. However, FX has not been mass-produced or applied extensively in the global market. The first reason is the difficulty of extracting highly concentrated FX from the materials. Macroalgae contains polysaccharides, such as fucoidan and chlorophylls. These components make it difficult to purify FX after extraction, using solvents such as ethanol and acetone. Although several studies have attempted specific FX extraction or have used discarded parts of macroalgae [31,32], no economical method is available for producing FX with high purity. Additionally, macroalgae contains a low amount of FX. As reported in our previous study, the FX content is typically below 0.5% in fresh material [2,3], making solvent extraction difficult and reducing the economic feasibility of FX purification using simple solvents. Thus, solvent-extracted seaweed is used as an FX source in many commercial FX products available on the market but is not applied extensively in other fields, such as in food, cosmetic, and pharmaceutical products. Finally, seaweed has an unpleasant odor when extracted using simple solvents. Seaweed extract contains high concentration of chlorophylls and sticky materials. These components give seaweed a specific odor and dark color, making it difficult to apply in other industrial fields.

In this study, we used the standardized FX powder (PE) from microalga PT. The amount of FX in PE ranged from 3.5% to 6% (*w/w*), and the sample was bright red in color and contained trace amounts of chlorophylls. This material can be utilized as a global source of FX like the global antioxidant carotenoid astaxanthin powder from the microalga *Haematococcus pluvialis*, which contains generally over 5% (*w/w*) astaxanthin [33]. Here, we evaluated the anti-obesity activity of this new standardized FX powder in vitro as well as in vivo using 3T3-L1 cells and C57BL/6J mice. The results showed that PE decreased lipid accumulation in 3T3-L1 cells in a dose-dependent manner. We confirmed that FX and

PE alleviated lipid accumulation in adipocytes. Additionally, regarding adipogenetic factors, FX and PE downregulated PPARγ protein levels at high concentrations and upregulated the UCP1 protein level in a dose-dependent manner, but did not influence the C/EBPα protein level. In obese mice fed an HFD, six weeks of supplementation with PE decreased the body weight, organ weight, fat volume, and adipocyte size without affecting food intake (Table S1). Simultaneously, serum parameters, such as TC, TG, HDL, LDL, AST, and ALT were decreased, indicating attenuation of lipid metabolism dysfunction and liver damage induced by HFD.

Hepatic steatosis, defined as induced excessive lipid accumulation in the liver, is highly related to obesity and other diseases [34]. In this study, PE effectively reduced the fat globule size compared to the HFD group, indicating that the enhanced anti-obesity effect of PE on hepatic steatosis was achieved mainly by lowering excessive fat accumulation (Figure 3). Additionally, the level of liver TG was significantly reduced in a dose-dependent manner, indicating decreased excessive hepatic lipid accumulation. Simultaneously, serum AST and ALT levels in the PE-treated groups were reduced and liver protective effects were observed.

Some mechanisms associated with the anti-obesity effect of FX have been reported. The mechanisms of the anti-obesity effect are influenced by numerous factors, including nutritional, hormonal, and genetic elements [35]. The most well-known mechanism of the anti-obesity effect of FX is related to thermogenesis and lipolysis [36]. In our study, the anti-obesity effect of PE was evaluated in thermogenesis by examining UCP1 induction and lipolysis-related genes. UCP1 is mainly distributed in brown adipose tissue and acts in thermogenesis, controls energy expenditure, and protects against oxidative stress [37]. Additionally, UCP1 is activated by free fatty acids and is involved in the transport of hydrogen ions. This increase in UCP1 expression increases energy expenditure and thus is helpful for improving obesity. Additionally, the expression of genes related to WAT was evaluated to determine the effect of PE containing FX on adipogenesis. PPARγ and C/EBPα promote the proliferation and differentiation of mature adipocytes and are the most important transcription factors in adipogenesis [38]. Our results suggest that upregulation of UCP1 in WAT by PE containing FX significantly reduced abdominal and subcutaneous fat accumulation; the PE-M + CLA combination groups also showed increased UCP1 expression compared to the HFD group. The expression of PPARγ and C/EBPα was downregulated after administration of PE, contributing to the suppression of adipogenesis in WAT. Therefore, we concluded that the standardized PE derived from microalga PT exerted anti-obesity effects, with FX being the main functional compound.

PE in combination with CLA appeared to have synergistic anti-obesity effects. The body weight- and body fat-lowering effects of CLA are controversial. CLA was reported to be effective for reducing weight and body fat in rats [39,40]. However, in hamsters, body fat was reduced but without weight loss [41]. Our results clearly demonstrated the increased anti-obesity effect of PE-M + CLA with respect to reduction in body weight, liver weight, adipose tissue weight, fat globules, fat volume, and the expression of proteins related to adipogenesis. Because FX is very hydrophobic, combining FX with other oils can further improve its transport through inter- and intracellular barriers, increasing its bioavailability to exert its effects. A previous study reported that the anti-obesity effects of FX were improved by combining it with medium-chain triacylglycerols; the adipose tissue weight gain was obviously lower and metabolic thermogenesis in the WAT was markedly increased in diabetic/obese KK-$A^y$ mice fed a mixture of FX and medium-chain triacylglycerol oil [42]. Xanthigen is a well-known supplement used to control weight. This agent is a promising combination of FX from seaweed extract and pomegranate seed oil (an omega-5 long-chain polyunsaturated fatty acid). Xanthigen was shown to improve lipid metabolism in experimental and clinical studies [43]. In our study, a combination of PE and CLA synergistically enhanced the anti-obesity effect of FX, as well as its bioavailability and transport in the PE, suggesting the necessity to process or formulate PE with proper oils to increase the bioavailability of FX.

## 4. Materials and Methods

### 4.1. Materials and Chemicals

*Phaeodactylum* extract (PE) used in this study was supplied by AlgaeTech Co., Ltd. (Gangneung, Republic of Korea). PE was in powder form and bright red in color and the amount of FX in PE was standardized to a range of 3.5~6% (*w/w*). The FX standard compound was purchased from Sigma-Aldrich (St. Louis, MO, USA). Dulbecco's modified Eagle's medium (DMEM), bovine calf serum, fetal bovine serum (FBS), penicillin/streptomycin, phosphate-buffered saline, insulin, and trypsin–EDTA were purchased from Gibco (Grand Island, NY, USA). Dexamethasone (DEX), 3-isobutyl-1-methylxanthine (IBMX), insulin, and Oil Red O were purchased from Sigma. All other chemicals were purchased from Sigma.

### 4.2. HPLC Analysis

The HPLC analysis was performed with an Agilent 1260 HPLC system (Agilent Technologies, Santa Clara, CA, USA) with a CAPCELL PAK C18 MG II (5 μm particle size, 250 × 4.6 mm I.D.). The mobile phase consisted of acetonitrile and water with a flow rate of 1 mL/min. After loading the column with the PE in ethanol, the mobile phase was an acetonitrile: Water solution with the ratio increasing from 90:10 to 100:0 over 8 min, maintained at 100:0 for 3 min, and then decreased to 80:20 over 5 min. The absorption spectrum was obtained from 210 to 600 nm and the chromatogram was recorded at 450 nm. The FX standard was used for the construction of the calibration curve in the concentration range of 1~200 μg/mL.

### 4.3. Cell Cultures

3T3-L1 preadipocytes were cultured in DMEM containing 10% (*v/v*) bovine calf serum and 1% penicillin/streptomycin as antibiotics in a humidified atmosphere of 5% $CO_2$ at 37 °C. The medium was changed every few days. When the cells were over 70% confluent, they were harvested via trypsinization and re-seeded into 25 $cm^2$ culture flasks at a density of $3 \times 10^3$ cells per $cm^2$ in a fresh medium.

### 4.4. Cell Toxicity and Proliferation Assay

To evaluate cell cytotoxicity, 3T3-L1 cells ($5 \times 10^3$ cells/well) were cultured in a 96-well plate overnight. PE, FX, and curcumin were dissolved in dimethyl sulfoxide (DMSO) for cellular treatments. Cells were exposed or not exposed to the FX or PT extracts for 48 h. A cell toxicity assay was conducted using an EZ-Cytox cell viability assay kit (Daeil Lab Service, Seoul, Korea).

### 4.5. Cell Differentiation

Cells were seeded into a 6-well plate in DMEM containing 10% calf serum at a density of $3 \times 10^4$ cells per well. The medium was changed every other day. Two-day post-confluent cells were stimulated for 2 days with a differentiation medium (DMEM containing 10% FBS, 0.5 mM IBMX, 1.0 μM DEX, and 1.0 μg/mL insulin). After 48 h, the medium was replaced with the adipocyte maintenance medium (DMEM containing 10% FBS and 10 μg/mL insulin) for 2 days. The cells were incubated with DMEM containing 10% FBS until day 8, when the cells should be fully differentiated. 3T3-L1 preadipocytes were treated with or without curcumin, FX, and PE starting on day 0 and maintained in the medium during the experiment.

### 4.6. Oil Red O Staining and Intracellular Triacylglycerol Level Measurements

The area of intracellular lipid accumulation was determined on day 6 by ORO staining. Briefly, the cells were fixed in 10% formaldehyde in phosphate-buffered saline for 1 h, and then washed with 60% isopropanol. The cells were stained with 0.5% ORO solution in 60:40 (*v/v*) isopropanol:distilled water (DW) for 20 min at room temperature, washed four times with water, and dried. Differentiation was quantified by dissolving the cells in isopropanol and measuring the optical density at 500 nm.

### 4.7. Animal Treatment

A total of 70e female C57BL/6J mice (4 weeks old) were purchased from Central Lab. Animal, Inc. (Seoul, Korea). All mice were housed in a room with controlled temperature (21 ± 2 °C), humidity (55 ± 15%), and lighting (12-h light/dark cycle). Mice had free access to water throughout the experiment. This study was approved by the Institutional Animal Care and Use Committee of Knotus (No. IACUC-18-KE-096).

After a 1-week adaptation period, the mice were randomly divided into the following seven groups ($n$ = 10 per group): Normal diet (ND), high-fat diet (HFD), HFD plus FX 0.1 mg/kg/day, HFD plus PE 0.81 mg/kg/day (PE-L), HFD plus PE 1.62 mg/kg/day (PE-M), HFD plus PE 3.25 mg/kg/day (PE-H), and HFD plus PE 1.62 mg/kg/day groups in combination with a CLA 410 mg/kg/day (PE-M + CLA) group. Animals were fed with either a ND or HFD for 6 weeks ad libitum. The mice were given free access to food and water. For 6 weeks of feeding, FX and PE were suspended in distilled water by vortex mixing and sonication just before administration. Each suspension was administered orally at the volume of 10 mL/kg/day.

Body weight was measured twice a week for 6 weeks and an area under curve (AUC) was calculated using Graphpad Prism 5.03 software (San Diego, CA, USA) by one-way ANOVA. All data were expressed as mean ± standard error or mean ± standard deviation. At the end of the experiment period, the mice were fasted for 12 h and sacrificed. Blood was collected and organs (adipose tissues, liver) were rapidly removed, rinsed with physiological saline solution, weighed, and stored at −80 °C.

### 4.8. Serum Biochemical Measurements

The collected blood was incubated for 2 h at room temperature. Serum was prepared by centrifugation at 3500× $g$ for 10 min at 4 °C and stored at −80 °C until use. The concentrations of triglyceride (TG), total cholesterol (TC), high-density lipoprotein (HDL), low-density lipoprotein (LDL), aspartate transaminase (AST), and alanine transaminase (ALT) were analyzed using a HITACHI 7180 chemistry analyzer (Tokyo, Japan) and enzyme-linked immunosorbent assay kit (R&D system, Inc., Minneapolis, MN, USA).

### 4.9. Histopathological Analysis

Adipose and liver pathological states were observed, and adipocyte size was quantified with a light microscope (Olympus BX53, Tokyo, Japan). Images were obtained using an image analyzer (Zen 2.3 blue edition, Carl Zeiss, Oberkochen, Germany).

### 4.10. Micro CT Analysis of Abdominal and Subcutaneous Fat Volume

The abdominal and subcutaneous fat in each mouse was scanned at an isotropic voxel size of 76 μm (45 kV, 177 μA, 200 ms integration time) with a viva CT 80 scanner (Scanco Medical, Brüttisellen, Switzerland). Two-dimensional gray-scale image slices were reconstructed by three-dimensional tomography. Density values for soft tissue were calibrated from a 5-point linear fit line with mixtures in various ratios of two liquids, ranging from 0.78 mg/mL (100% ethanol; Sigma) to 1.26 mg/mL (100% glycerol; J.T. Baker, Phillipsburg, NJ, USA). The measured body fat area was the abdominal fat and subcutaneous fat in the space from the second lumbar spine to the fifth lumbar vertebra. All animals were imaged using the standard micro-CT imaging protocol (220 views, 16 ms X-ray exposure time, and 70 kV/32 mA penetration energy). Reconstruction of micro-CT images was performed using MicroView (GE Healthcare, Little Chalfont, UK) software packages.

### 4.11. Western Blot Analysis

The cells or tissues were lysed in a cold RIPA buffer (pH 7.4) containing 10 μM phenylmethanesulfonyl fluoride and a 1% protease inhibitor cocktail. Cells and tissues lysates were centrifuged at 13,000 rpm for 10 min at 4 °C. The protein contents were analyzed via the BCA (bicinchoninic acid) protein assay kit (Vazyme Biotech Co., Ltd., Nanjing, China). Equal quantities of protein were electrophoresed by sodium dodecyl sulfate-polyacrylamide gel electrophoresis and transferred to polyvinylidene fluoride membranes (Millipore, Billerica, MA, USA). Membranes were incubated with antibodies including those for UCP1, PPARγ, and C/EBPα at 4 °C overnight. β-Actin was used as a control with a β-actin antibody (Santa Cruz Biotechnology, Dallas, TX, USA). The protein markers were visualized via chemiluminescence (Western Lighting Plus-ECL, PerkinElmer, Waltham, MA) and with a Chemidoc image analyzer (Bio-Rad, Hercules, CA, USA).

### 4.12. Statistical Analysis

Data were analyzed using the GraphPad Prism 5.03 software (San Diego, CA, USA). All data were expressed as the mean ± standard error or mean ± standard deviation. Statistical significances between groups were determined by one-way analysis of variance, followed by Dunnett's multiple tests. A value of $p < 0.05$ represents a significant difference.

## 5. Conclusions

In conclusion, PE significantly reduced body weight gain and adipose tissue weight by activating UCP1 and inactivating PPARγ and C/EBPα. Moreover, PE remarkably reduced fat mass in HFD-induced mice. The efficacy of PE was equal to that of the positive control substance, FX, and was potentiated when combined with CLA. Our results indicate that PE containing FX exerts anti-obesity effects by promoting lipolysis and inhibiting lipogenesis and is a good candidate for the development of anti-obesity foods and health functional foods derived from new marine microalgae.

**Supplementary Materials:** The following are available online at http://www.mdpi.com/1660-3397/17/5/311/s1, Figure S1: HPLC chromatogram of FX standard and PE recorded at 450 nm, Figure S2: Cytotoxicity of fucoxanthin, PE and curcumin toward 3T3-L1 cells, Table S1: The average food intake (g/day) contents of C57BL/6J mice.

**Author Contributions:** Conceived and designed the experiments: S.M.K., C.-H.P., and K.T.H. performed the experiments: S.Y.K., J.-I.U. and J.-H.H. Analyzed the data and discussed the results: S.-Y.K., J.-H.H., K.K., and K.W.H. Wrote the paper: S.-Y.K., K.K., S.-H.Y., and S.M.K. All authors have read and approved the manuscript.

**Funding:** This research was funded by the grants from Marine Biotechnology Program (2MP0630) funded by Ministry of Oceans and Fisheries, Korea and an intramural grant (2Z05620, 2Z05630) from KIST Gangneung Institute of Natural Products.

**Conflicts of Interest:** The authors declare no conflict of interest.

## References

1. D'Orazio, N.; Gemello, E.; Gammone, M.A.; de Girolamo, M.; Ficoneri, C.; Riccioni, G. Fucoxanthin: A treasure from the sea. *Mar. Drugs* **2012**, *10*, 604–616. [CrossRef]
2. Kim, S.M.; Jung, Y.J.; Kwon, O.N.; Cha, K.H.; Um, B.H.; Chung, D.H.; Pan, C.H. A potential commercial source of fucoxanthin extracted from the microalga *Phaeodactylum tricornutum*. *Appl. Biochem. Biotechnol.* **2012**, *166*, 1843–1855. [CrossRef]
3. Kim, S.M.; Kang, S.W.; Kwon, O.N.; Chung, D.W.; Pan, C.H. Fucoxanthin as a major carotenoid in *Isochrysis aff. galbana*: Characterization of extraction for commercial application. *J. Korean Soc. Appl. Biol. Chem.* **2012**, *55*, 477–483. [CrossRef]
4. Xia, S.; Wang, K.; Wan, L.; Li, A.; Hu, Q.; Zhang, C. Production, characterization, and antioxidant activity of fucoxanthin from the marine diatom *Odontella aurita*. *Mar. Drugs.* **2013**, *11*, 2667–2681. [CrossRef] [PubMed]
5. Lee, S.H.; Min, K.H.; Han, J.S.; Lee, D.H.; Park, D.B.; Jung, W.K.; Park, P.J.; Jeon, B.T.; Kim, S.K.; Jeon, Y.J. Effects of Brown Algae, *Ecklonia cava* on Glucose and Lipid Metabolism in C57BL/KsJ-db/db Mice, a Model of Type 2 Diabetes mellitus. *Food Chem. Toxicol.* **2012**, *50*, 575–582. [CrossRef] [PubMed]

6. Okada, T.; Mizuno, Y.; Sibayama, S.; Hosokawa, M.; Miyashita, K. Antiobesity Effects of *Undaria* Lipid Capsules Prepared with Scallop Phospholipids. *J. Food Sci.* **2011**, *76*, H2–H6. [CrossRef]
7. Fung, A.; Hamid, N.; Lu, J. Fucoxanthin Content and Antioxidant Properties of *Undaria pinnatifida*. *Food Chem.* **2013**, *136*, 1055–1062. [CrossRef]
8. Jeon, S.M.; Kim, H.J.; Woo, M.N.; Lee, M.K.; Shin, Y.C.; Park, Y.B.; Choi, M.S. Fucoxanthin-rich seaweed extract suppresses body weight gain and improves lipid metabolism in high-fat-fed C57BL/6J mice. *Biotechnol. J.* **2010**, *5*, 961–969. [CrossRef]
9. Woo, M.N.; Jeon, S.M.; Kim, H.J.; Lee, M.K.; Shin, S.K.; Shin, Y.C.; Park, Y.B.; Choi, M.S. Fucoxanthin supplementation improves plasma and hepatic lipid metabolism and blood glucose concentration in high-fat fed C57BL/6N mice. *Chem. Biol. Interact.* **2010**, *186*, 316–322. [CrossRef]
10. Gammone, M.A.; D'Orazio, N. Anti-Obesity Activity of the Marine Carotenoid Fucoxanthin. *Mar. Drugs* **2015**, *13*, 2196–2214. [CrossRef]
11. Beppu, F.; Hosokawa, M.; Niwano, Y.; Miyashita, K. Effects of dietary fucoxanthin on cholesterol metabolism in diabetic/obese KK-A(y) mice. *Lipids Health Dis.* **2012**, *11*, 112. [CrossRef] [PubMed]
12. Rangwala, S.M.; Lazar, M.A. Transcriptional control of adipogenesis. *Annu. Rev. Nutr.* **2000**, *20*, 535–559. [CrossRef] [PubMed]
13. Tong, L. Acetyl-coenzyme A carboxylase: Crucial metabolic enzyme and attractive target for drug delivery. *Cell. Mol. Life Sci.* **2005**, *62*, 1784–1803. [CrossRef]
14. Maeda, H.; Hosokawa, M.; Sashima, T.; Murakami-Funayama, K.; Miyashita, K. Anti-obesity and anti-diabetic effects of fucoxanthin on diet-induced obesity conditions in a murine model. *Mol. Med. Rep.* **2009**, *2*, 897–902. [CrossRef] [PubMed]
15. Woo, M.N.; Jeon, S.M.; Shin, Y.C.; Lee, M.K.; Kang, M.A.; Choi, M.S. Anti-obese property of fucoxanthin is partly mediated by altering lipid-regulating enzymes and uncoupling proteins of visceral adipose tissue in mice. *Mol. Nutr. Food Res.* **2009**, *53*, 1603–1611. [CrossRef] [PubMed]
16. Rebello, C.J.; Greenway, F.L.; Johnson, W.D.; Ribnicky, D.; Poulev, A.; Stadler, K.; Coulter, A.A. Fucoxanthin and its metabolite fucoxanthinol do not induce browning in human adipocytes. *J. Agric. Food Chem.* **2017**, *65*, 10915–10924. [CrossRef] [PubMed]
17. Hu, X.; Tao, N.; Wang, X.; Xiao, J.; Wang, M. Marine-derived bioactive compounds with anti-obesity effect: A review. *J. Func. Foods* **2016**, *21*, 372–387. [CrossRef]
18. Miyashita, K.; Hosokawa, M. Fucoxanthin in the management of obesity and its related disorders. *J. Func. Foods* **2017**, *36*, 195–202. [CrossRef]
19. Muradian, K.H.; Vaiserman, A.; Min, K.J.; Frafield, V.E. Fucoxanthin and lipid metabolism: A mini review. *Nutr. Metab. Cardiovasc. Dis.* **2015**, *25*, 891–897. [CrossRef]
20. Gautron, L.; Elmquist, J.K. Sixteen years and counting: An update on leptin in energy balance. *J. Clin. Invest.* **2011**, *121*, 2087–2093. [CrossRef]
21. Roujeau, C.; Jockers, R.; Dam, J. New pharmacological perspectives for the leptin receptor in the treatment of obesity. *Front. Endocrinol.* **2014**, *13*, 5–167. [CrossRef]
22. Heilbronn, L.K.; Noakes, M.; Clifton, M.P. Energy restriction and weight loss on very-low-fat diets reduce C-reactive protein concentrations in obese, healthy women. *Atheroscler. Thromb. Vasc. Biol.* **2001**, *21*, 968–970. [CrossRef]
23. Maeda, H.; Tsukui, T.; Sashima, T.; Hosokawa, M.; Miyashita, K. Seaweed carotenoid, fucoxanthin, as a multi-functional nutrient. *Asia Pac. J. Clin. Nutr.* **2008**, *17*, 196–199.
24. Kim, J.H.; Kim, S.M.; Cha, K.H.; Mok, I.K.; Koo, S.I.; Pan, C.H.; Lee, J.K. Evaluation of the anti-obesity effect of the microalga *Pheodactylum tricornutum*. *Appl. Biol. Chem.* **2016**, *59*, 283–290. [CrossRef]
25. Kang, S.I.; Ko, H.C.; Shin, H.S.; Kim, H.M.; Hong, Y.S.; Lee, N.H.; Kim, S.J. Fucoxanthin exerts differing effects on 3T3-L1 cells according to differentiation stage and inhibits glucose uptake in mature adipocytes. *Biochem. Biophys. Res. Commun.* **2011**, *409*, 769–774. [CrossRef] [PubMed]
26. Wu, Z.; Rosen, E.D.; Brun, R.; Hauser, S.; Adelmant, G.; Troy, A.E.; McKeon, C.; Darlington, G.J.; Spiegelman, B.M. Cross-Regulation of C/EBPα and PPARγ Controls the Transcriptional Pathway of Adipogenesis and Insulin Sensitivity. *Mol. Cell.* **1999**, *3*, 151–158. [CrossRef]
27. Kim, E.H.; Bae, J.S.; Hahm, K.B.; Cha, J.Y. Endogenously synthesized n-3 polyunsaturated fatty acids in *fat-1* mice ameliorate high-fat diet-induced non-alcoholic fatty liver disease. *Biochem. Pharmacol.* **2012**, *84*, 1359–1365. [CrossRef] [PubMed]

28. Fuster, J.J.; Castillo, A.I.; Zaragoza, C.; Ibáñez, B.; Andrés, V. Animal models of atherosclerosis. *Prog. Mol. Biol. Transl. Sci.* **2012**, *105*, 1–23. [PubMed]
29. Park, S.; Kim, K.; Han, S.I.; Kim, E.J.; Choi, Y.E. Organic solvent-free lipid extraction from wet *Aurantiochytrium* sp. Biomass for co-production of biodiesel and value-added products. *Appl. Biol. Chem.* **2017**, *60*, 101–108. [CrossRef]
30. Kumar, V.; Kumar, R.; Rawat, D.; Nanda, M. Synergistic dynamics of light, photoperiod and chemical stimulants influences biomass and lipid productivity in *Chlorella singularis* (UUIND5) for biodiesel production. *Appl. Biol. Chem.* **2018**, *61*, 7–13. [CrossRef]
31. Conde, E.; Moure, A.; Domíngues, H. Supercritial $CO_2$ extraction of fatty acids, phenolics and fucoxanthin from free-dried *Sargassum muticum*. *J. Appl. Phycol.* **2015**, *27*, 957–964. [CrossRef]
32. Kanazawa, K.; Ozaki, Y.; Hashimoto, T.; Das, S.K.; Mastushita, S.; Hirono, M.; Okada, T.; Komoto, A.; Mori, N.; Kakatsuka, M. Commrecial-scale preparation of biofunctional fucoxanthin from waste parts of brown sea algae *Laminalia japonica*. *Food Sci. Technol. Res.* **2008**, *14*, 573–582. [CrossRef]
33. Ambati, R.; Phang, S.M.; Ravi, S.; Aswathanarayana, R. Astaxanthin: Source, extraction, stability, biological activities and its commercial applications-a review. *Mar. Drugs* **2014**, *12*, 128–152. [CrossRef]
34. Renella, M.E. Nonalcoholic fatty liver disease: A systematic review. *JAMA* **2015**, *313*, 2263–2273. [CrossRef]
35. Pan, H.; Fu, C.; Huang, L.; Jiang, Y.; Deng, X.; Guo, J.; Su, Z. Anti-obesity effect of chitosan oligosaccharide capsules (COSCs) in obese rats by ameliorating leptin resistance and adipogenesis. *Mar. Drugs* **2018**, *16*, 198. [CrossRef]
36. Maeda, H.; Hosokawa, M.; Sashima, T.; Funayama, K.; Miyashita, K. Fucoxanthin from edible seaweed *Undaria Pinnatifida*, shows anti-obesity effect through UCP1 expression in white adipose tissues. *Biochem. Biophys. Res. Commun.* **2005**, *332*, 392–397. [CrossRef] [PubMed]
37. Maeda, H. Nutraceutical effects of fucoxanthin for obesity and diabetes therapy: A review. *J. Oleo Sci.* **2015**, *64*, 125–132. [CrossRef] [PubMed]
38. Ali, A.T.; Hochfeld, W.E.; Myburgh, R.; Pepper, M.S. Adipocyte and adipogenesis. *Eur. J. Cell. Biol.* **2013**, *92*, 229–236. [CrossRef]
39. Hu, X.; Li, Y.; Li, C.; Fu, Y.; Cai, F.; Chen, Q.; Li, D. Combination of fucoxanthin and conjugated linoleic acid attenuates body weight gain and improves lipid metabolism in high-fat diet-induced obese rats. *Arch. Biochem. Biophys.* **2012**, *519*, 59–65. [CrossRef] [PubMed]
40. Banni, S.; Carta, G.; Angioni, E.; Murru, E.; Scanu, P.; Melis, M.P.; Bauman, D.E.; Fischer, S.M.; Ip, C. Distribution of conjugated linoleic acid and metabolites in different lipid fractions in the rat liver. *J. Lipid Res.* **2001**, *42*, 1056–1061.
41. Navarro, V.; Zabala, A.; Macarulla, M.T.; Fernández-Quintela, A.; Rodríguez, V.M.; Simón, E.; Portillo, M.P. Effects of conjugated linoleic acid on body fat accumulation and serum lipids in hamsters fed an astherogenic diet. *J. Physiol. Biochem.* **2003**, *59*, 193–200. [CrossRef] [PubMed]
42. Maeda, H.; Hosokawa, M.; Sashima, T.; Miyashita, K. Dietary combination of fucoxanthin and fish oil attenuates the weight gain of white adipose tissue and decrease blood glucose in obese/diabetic KK-$A^y$ mice. *J. Agric. Food Chem.* **2007**, *55*, 7701–7706. [CrossRef] [PubMed]
43. Abidov, M.; Ramazanov, Z.; Seifulla, R.; Grachev, S. The effect of Xanthigen in the weight management of obese premenopausal women with non-alcoholic fatty liver disease and normal liver fat. *Diabetes Obes. Metab.* **2010**, *12*, 72–81. [CrossRef] [PubMed]

*marine drugs*

*Review*

# Siphonaxanthin, a Green Algal Carotenoid, as a Novel Functional Compound

**Tatsuya Sugawara [1,*], Ponesakki Ganesan [1,2], Zhuosi Li [1], Yuki Manabe [1] and Takashi Hirata [1,3]**

1 Division of Applied Biosciences, Graduate School of Agriculture, Kyoto University, Kitashirakawaoiwakecho, Sakyo-ku, Kyoto 606-8502, Japan; E-Mails: ganesan381980@yahoo.com (P.G.); lizhuosi@kais.kyoto-u.ac.jp (Z.L.); yuukim@kais.kyoto-u.ac.jp (Y.M.); hiratan@kais.kyoto-u.ac.jp (T.H.)

2 SRM Research Institute, SRM University, Kattankulathur, Tamilnadu 603 203, India

3 Department of Rehabilitation, Shijonawate Gakuen University, 5-11-10 Hojo, Daito, Osaka 574-0011, Japan

* Author to whom correspondence should be addressed; E-Mail: sugawara@kais.kyoto-u.ac.jp; Tel./Fax: +81-75-753-6212.

*Received: 10 March 2014; in revised form: 14 April 2014 / Accepted: 17 April 2014 / Published: 17 June 2014*

**Abstract:** Siphonaxanthin is a specific keto-carotenoid in green algae whose bio-functional properties are yet to be identified. This review focuses on siphonaxanthin as a bioactive compound and outlines the evidence associated with functionality. Siphonaxanthin has been reported to potently inhibit the viability of human leukemia HL-60 cells via induction of apoptosis. In comparison with fucoxanthin, siphonaxanthin markedly reduced cell viability as early as 6 h after treatment. The cellular uptake of siphonaxanthin was 2-fold higher than fucoxanthin. It has been proposed that siphonaxanthin possesses significant anti-angiogenic activity in studies using human umbilical vein endothelial cells and rat aortic ring. The results of these studies suggested that the anti-angiogenic effect of siphonaxanthin is due to the down-regulation of signal transduction by fibroblast growth factor receptor-1 in vascular endothelial cells. Siphonaxanthin also exhibited inhibitory effects on antigen-induced degranulation of mast cells. These findings open up new avenues for future research on siphonaxanthin as a bioactive compound, and additional investigation, especially *in vivo* studies, are required to validate these findings. In addition, further studies are needed to determine its bioavailability and metabolic fate.

**Keywords:** angiogenesis; apoptosis; carotenoid; inflammation; green algae; siphonaxanthin

## 1. Introduction

Marine algae are a potential renewable resource in the marine environment that are used as a culinary item in East Asia and they have been reported to confer several beneficial effects in human health. They are an excellent source of nutritional and bioactive compounds, such as carotenoids, dietary fibers, amino acids, essential fatty acids, vitamins, and minerals [1,2]. Algal carotenoids have received much attention, as they are structurally different from those found in terrestrial plants. It is important to characterize the novel bio-functional activities of carotenoids from marine algae.

One of the major carotenoids in marine algae is fucoxanthin, and it is found mainly in brown macroalgae and in some classes of microalgae [3–5]. The chemical structure of fucoxanthin includes an allenic bond and oxygenic functional groups, such as hydroxyl, epoxy, carbonyl, and acetyl groups, in addition to its polyene chain (Figure 1A). Recently, fucoxanthin has been reported to possess several beneficial effects related to its anti-cancerous, anti-oxidative, and anti-obesity properties [6–12]. Moreover, fucoxanthin induces uncoupling protein 1 in the mitochondria of abdominal white adipose tissue, leading to the oxidation of fatty acids and heat production [13]. On the other hand, we found that fucoxanthin and its deacetylated product, fucoxanthinol, effectively suppress angiogenesis [14]. This implies that fucoxanthin, with its anti-angiogenic activity, would be useful in preventing angiogenesis-related diseases such as cancer and diabetic retinopathy. Consequently, brown algal fucoxanthin seems to be a useful bioactive and nutraceutical compound for human health.

**Figure 1.** Structures of fucoxanthin (**A**); siphonaxanthin (**B**) and siphonein (**C**).

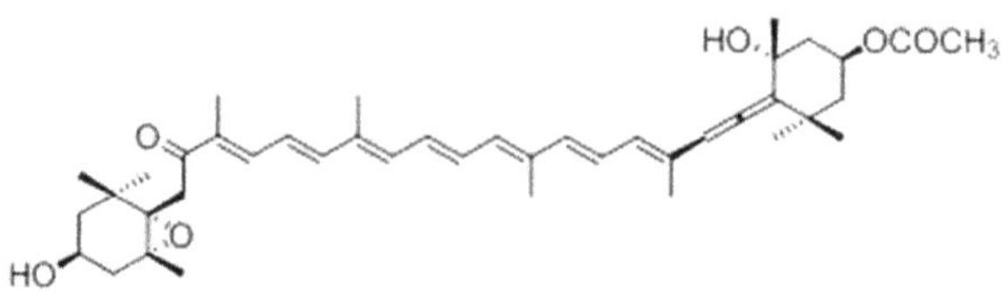


(**A**)

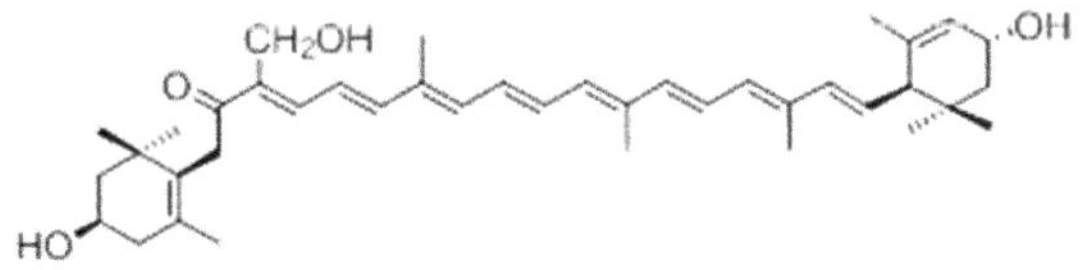

(B)

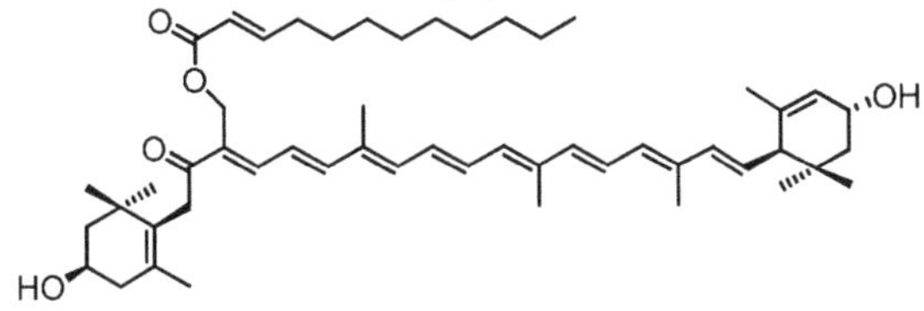

(C)

Scientific evidence on the bio-functional properties of algal carotenoids is still very limited. Siphonaxanthin (Figure 1B) is a specific keto-carotenoid of green algae, and its bio-functional properties are yet to be clarified. We focus on the potential of siphonaxanthin as a bioactive compound and outline the evidence associated with functionality functionality. The objective of this review is to summarize the novel bio-functional activities of green algal siphonaxanthin that have been highlighted in our studies.

## 2. Siphonaxanthin in Green Algae

Siphonaxanthin is a specific keto-carotenoid of siphonaceous green algae, which helps in absorbing available green and blue green light under water [15,16]. Brown algal fucoxanthin has epoxide and an allenic bond in its structure, whereas siphonaxanthin does not contain either of those functional groups. However, it possesses an additional hydroxyl group on the 19th carbon atom (Figure 1B). In edible green algae such as *Codium fragile, Caulerpa lentillifera,* and *Umbraulva japonica,* siphonaxanthin content is approximately 0.03%–0.1% of the dry weight. Siphonein, identified as siphonaxanthin 19-(*trans*-Δ2-dodecenoate), is present at the same level as siphonaxanthin in these algae. *C. fragile* is consumed as a part of the staple diet in ancient Japanese food culture.

Generally, carotenoids in plants play important roles in the photosynthetic process, such as light-absorption and quenching of excess energy [17]. β-Carotene is present in most of the divisions of the reaction-center complexes (RC) and the light-harvesting complexes (LHC) of photosystem I (PSI). On the other hand, in the peripheral LHC of photosystem II (PSII), the bound carotenoids are heterogenous, depending on the classes [5]. In green algae, the major carotenoid in the peripheral LHC of PSII is siphonaxanthin, which exhibited an additional absorption band (the 535 nm band) in the blue to green region. β-Carotene in both RC and LHC-PSI might have protective functions, and carotenoids in the peripheral LHC of PSII might have primarily light-harvesting functions. Employing femtosecond time-resolved fluorescence spectroscopy to analyze purified carotenoids in organic solvents and the LHC in solution, this keto-carotenoid was found to facilitate highly efficient energy transfer of carotenoids to chlorophylls [18]. Siphonaxanthin might have a largely light-harvesting function in the green light-rich underwater habitat [19].

## 3. Apoptosis-Inducing Effect

There is a wealth of information pertaining to apoptosis in anti-cancer research, and the
apoptosis-inducing properties of carotenoids may be an important approach to chemo-prevention and/or chemotherapy. Carotenoids from terrestrial plants, such as β-carotene and lycopene, have been extensively studied and implicated as cancer preventive agents. Carotenoids from marine sources have received much attention, as they are structurally different from those found in terrestrial sources. Recent studies have found that carotenoids from algae, including fucoxanthin and peridinin, induce apoptosis in cancer cells [6–9,20,21].

We investigated whether various carotenoids present in marine sources are potentially involved in cancer-preventing action in human leukemia HL-60 cells. We found that the 11 carotenoids investigated, siphonaxanthin most potently at inhibiting the viability of HL-60 cells [22]. Compared to fucoxanthin, siphonaxanthin, at a concentration of 20 μM, markedly reduced cell viability as early as 6 h after treatment. The cellular uptake of siphonaxanthin was 2-fold higher than fucoxanthin, demonstrating a positive correlation between cellular uptake and cell viability. The apoptotic activity of siphonaxanthin was characterized by increases in TUNEL-positive cells, and by increased chromatin condensation in the cells. This induction of apoptosis was accompanied by decreased expression of Bcl-2, and subsequent increase in the activation of caspase-3. These observations indicate that siphonaxanthin may be a more potent growth-inhibitor in cancer cells than fucoxanthin, possibly due to the differences in the structure of their different functional groups. The presence of hydroxyl group on the 19th carbon atom appeared to contribute to the strong apoptosis-inducing effect. This is further supported by the fact that siphonein (Figure 1C), an esterified form of siphonaxanthin that does not contain an additional hydroxyl group, had a reduced inhibitory effect on cell viability.

Interestingly, siphonaxanthin has been reported to up-regulate the expression of death receptor 5 (DR5) [22]. Tumor necrosis factor-related apoptosis-inducing ligand (TRAIL) promotes apoptosis selectively in tumor cells without much effect on normal cells by binding to the transmembrane receptors TRAIL-R1/DR4 and TRAIL-R2/DR5 [23]. Thus, any chemotherapeutic agent that activates TRAIL-induced apoptosis in cancer cells may be an attractive strategy in anti-cancer research.
Yoshida *et al.* reported that halocynthiaxanthin and peridinin sensitize cancer cells

to TRAIL-induced apoptosis by up-regulating the expression of DR5 [21], which indicates that some carotenoids can induce apoptosis through this pathway. In our study, siphonaxanthin potently up-regulated the expression of DR5, but fucoxanthin did not [22]. Siphonaxanthin could be potentially useful as a chemopreventive and/or chemotherapeutic agent.

## 4. Anti-Angiogenic Effect

Angiogenesis is characterized by the growth and remodeling process of the primitive network of blood vessels into a complex network. In this process, endothelial cells secrete proteases, migrate through the extracellular matrix, proliferate, and differentiate [24]. Angiogenesis is involved in many physiological and pathological situations. In normal adults, most vasculature is quiescent, with only 0.01% of endothelial cells undergoing division [25]. However, angiogenesis is an essential process in the female reproductive cycle, along with the remodeling and regeneration of tissues [26]. Pathological angiogenesis is implicated in the pathogenesis of many diseases, including cancer, atherosclerosis, diabetic retinopathy, and rheumatoid arthritis [27]. The newly-formed blood vessels promote cancer growth by supplying nutrients and oxygen and by removing waste products. Metastasis also depends on angiogenesis, as tumor cells are shed from a primary tumor and grow in their target organs [28]. Angiogenesis is activated under other pathological conditions, such as ocular and inflammatory disorders [29]. Hence, prevention of angiogenesis under pathological conditions is a promising approach in the prevention of cancer and other pro-angiogenic diseases. It has been reported that some natural products, such as vitamin $B_6$ [30], algal polysaccharides [31], and nasunin [32], suppress angiogenesis. Given this background, studies on natural bioactive molecules from marine algae have reported that fucoidans, polysaccharides from marine brown algae, and fucoxanthin and its deacetylated product, fucoxanthinol, exert anti-angiogenic properties [14,33].

To evaluate the anti-angiogenic effect of siphonaxanthin from green algae, we examined its anti-angiogenic effect in cell culture model systems and by employing *ex vivo* approaches using human umbilical vein endothelial cells (HUVECs) and rat aortic ring [34,35], respectively. Siphonaxanthin significantly suppressed HUVEC proliferation at relatively lower concentration of 2.5 μM, while its effect on chemotaxis was not significant [36]. In addition, siphonaxanthin exhibited a strong inhibitory effect on HUVEC tube formation in an *in vitro* angiogenesis model. Siphonaxanthin suppressed the tube length at a concentration of 10 μM, while no tube formation was observed at 25 μM, suggesting that this could be due to the suppression of angiogenic mediators. The *ex vivo* angiogenesis assay using rat aortic ring indicated reduced microvessel outgrowth in a dose-dependent manner, and the reduction was significant at concentrations of more than 2.5 μM.

To elucidate the molecular mechanism underlying the anti-angiogenic activity of siphonaxanthin compared to fucoxanthin, we focused on the vascular specific pro-angiogenic factor, vascular endothelial growth factor (VEGF), and the other angiogenic factors, such as fibroblast growth factors (FGFs) [37,38]. It has been shown that both siphonaxanthin and fucoxanthin suppress the mRNA expression of fibroblast growth factor 2 (FGF-2) and its receptor (FGFR-1), as well as their *trans*-activation factor, EGR-1 [39]. These suppressive effects of siphonaxanthin were more effective than fucoxanthin. However, the mRNA expression of VEGFR-2, a potent signal transducer involved in the
VEGF-mediated signaling pathway, was not significantly affected by these carotenoids. Furthermore, these two marine algal carotenoids can down-regulate the phosphorylation of FGF-2-mediated intracellular signaling proteins such as ERK1/2 and Akt. Inhibition of FGF-2-mediated intracellular signaling proteins by these carotenoids represses the migration of endothelial cells, as well as their differentiation into tube-like structures on Matrigel. These results demonstrate, for the first time, the possible molecular mechanism underlying the anti-angiogenic effect of these two algal carotenoids and suggest that their anti-angiogenic effect is due to the down-regulation of signal transduction by
FGFR-1 in vascular endothelial cells. Siphonaxanthin exhibited its anti-angiogenic effect at lower concentrations than fucoxanthin, indicating its potential as a strong angiogenesis inhibitor.
These findings provide new insights into the novel bio-functional property of marine algal carotenoids, especially that of siphonaxanthin, which has the potential to enhance current anti-angiogenic therapies in the treatment of cancer and other pro-angiogenic diseases.

## 5. Anti-Inflammatory Effect

Currently, one of the most common social problems in the world is an increasing number of patients with type I allergy. Mast cells play pivotal roles in localized inflammation and immediate type allergic reactions by secreting biologically active substances including histamine, eicosanoids, proteolytic enzymes, cytokines, and chemokines after antigen-induced degranulation. The antigen-induced aggregation of the high affinity IgE receptor (FcεRI) expressed on the cell surface triggers the degranulation of mast cells [40]. We have previously reported that astaxanthin, β-carotene, fucoxanthin, and zeaxanthin significantly inhibit antigen-induced degranulation of rat basophilic leukemia RBL-2H3 cells, which were used as a mast cell model, and bone marrow-derived mast cells [41]. Interestingly, these carotenoids inhibited antigen-induced translocation of FcεRI to lipid rafts, which are known as platforms of the aggregation of FcεRI [42]. Furthermore, oral administration of these four carotenoids for a week significantly inhibited dinitrofluorobenzene (DNFB)-induced ear swelling and the increased content of histamine in the DNFB-treated mice [43]. These results suggested that dietary carotenoids exert an anti-inflammatory effect by suppressing mast cell degranulation *in vivo*. However, information about the anti-degranulation effect of the other carotenoids is limited.

In this context, we evaluated the effects of eleven additional carotenoids using the RBL-2H3 cells. Results from our screening showed that nine carotenoids, including siphonaxanthin, had inhibitory effects on antigen-induced degranulation of mast cells [44]. The inhibitory activity of carotenoids was not related to their cellular uptake. It is speculated that carotenoids, including siphonaxanthin, may modify the functions of lipid rafts by localizing in the cell membrane and inhibiting the translocation of FcεRI to lipid rafts. Nevertheless, it is important to verify whether carotenoids affect other signaling pathways involved in lipid rafts in order to understand the biological relationship between carotenoids and lipid rafts.

## 6. Conclusions and Future Perspectives

To clarify the biological action of siphonaxanthin, further studies are required to determine its bioavailability and biological metabolism. Our previous studies, using cultured cells and mice, demonstrated that orally administered fucoxanthin is metabolized to fucoxanthinol and amarouciaxanthin A in the intestinal tract and liver, respectively [45,46]. However, information on the bioavailability and metabolic conversion of siphonaxanthin *in vivo* are limited. Siphonein, which is identified as siphonaxanthin 19-(trans-Δ2-dodecenoate) is present in similar levels as that of siphonaxanthin in siphonaceous green algae. Based on our data using cultured cells and *ex vivo* studies, the biological functions of siphonein were found to be weaker than those of siphonaxanthin.
If siphonein could be hydrolyzed in the intestine by digestive enzymes and absorbed in a manner similar to siphonaxanthin, oral administration of siphonein could be expected to exert the same biological activities as siphonaxanthin.

Siphonaxanthin isolated from marine green algae, such as *C. fragile*, remarkably suppresses cell viability, induces apoptosis in cancer cells, and possesses more potent anti-angiogenic activity than fucoxnathin. Although several reports have indicated that fucoxanthin could be immensely beneficial to human health, our studies indicate that siphonaxanthin is more effective than fucoxanthin. Nevertheless, these findings uncover new avenues for future research on siphonaxanthin as a bioactive compound, and additional investigation, especially *in vivo* studies, are needed to validate the reported findings.

## Acknowledgments

This research was partly supported by JSPS KAKENHI Grant Number 23380124 and the Kieikai Research Foundation.

## Author Contributions

All the authors contributed to the writing of the manuscript. They jointly developed the structure and arguments for the paper. All the authors reviewed and approved the final manuscript.

## Conflicts of Interest

The authors declare no conflict of interest.

## References

1. Mabeau, S.; Fleurence, J. Seaweed in food products: Biochemical and nutritional aspects. *Trends Food Sci. Technol.* **1993**, *4*, 103–107.
2. Fleurence, J. Seaweed proteins: Biochemical, nutritional aspects and potential uses. *Trends Food Sci. Technol.* **1999**, *10*, 25–28.
3. Peng, J.; Yuan, J.P.; Wu, C.F.; Wang, J.H. Fucoxanthin, a marine carotenoid present in brown seaweeds and diatoms: Metabolism and bioactivities relevant to human health. *Mar. Drugs* **2011**, *9*, 1806–1828.
4. Kumar, S.R.; Hosokawa, H.; Miyashita, K. Fucoxanthin: A marine carotenoid exerting anti-cancer effects by affecting multiple mechanisms. *Mar. Drugs* **2013**, *11*, 5130–5147.
5. Takaichi, S. Carotenoids in algae: Distributions, biosynthesis and functions. *Mar. Drugs* **2011**, *9*, 1101–1118.
6. Kotake-Nara, E.; Kushiro, M.; Zhang, H.; Sugawara, T.; Miyashita, K.; Nagao, A. Carotenoids affect proliferation of human prostate cancer cells. *J. Nutr.* **2001**, *131*, 3303–3306.
7. Kotake-Nara, E.; Sugawara, T.; Nagao, A. Antiproliferative effect of neoxanthin and fucoxanthin on cultured cells. *Fish. Sci.* **2005**, *71*, 459–461.
8. Hosokawa, M.; Kudo, M.; Maeda, H.; Kohno, H.; Tanaka, T.; Miyashita, K. Fucoxanthin induces apoptosis and enhances the antiproliferative effect of the PPARγ ligand, troglitazone, on colon cancer cells. *Biochim. Biophys. Acta* **2004**, *1675*, 113–119.
9. Das, S.K.; Hashimoto, T.; Shimizu, K.; Yoshida, T.; Sakai, T.; Sowa, Y.; Komoto, A.; Kanazawa, K. Fucoxanthin induces cell cycle arrest at G0/G1phase in human colon carcinoma cells through up-regulation of p21WAF1/Cip1. *Biochim. Biophys. Acta* **2005**, *1726*, 328–335.
10. Sachindra, N.M.; Sato, E.; Maeda, H.; Hosokawa, M.; Niwano, Y.; Kohno, M.; Miyashita, K. Radical scavenging and singlet oxygen quenching activity of marine carotenoid fucoxanthin and its metabolites. *J. Agric. Food Chem.* **2007**, *55*, 8516–8522.
11. Maeda, H.; Hosokawa, M.; Sashima, T.; Funayama, K.; Miyashita, K. Fucoxanthin from edible seaweed, *Undaria pinnatifida*, shows antiobesity effect

through UCP1 expression in white adipose tissues. *Biochem. Biophys. Res. Commun.* **2005,** *332,* 392–397.
12. Maeda, H.; Hosokawa, M.; Sashima, T.; Takahashi, N.; Kawada, T.; Miyashita, K. Fucoxanthin and its metabolite, fucoxanthinol, suppress adipocyte differentiation in 3T3-L1 cells. *Int. J. Mol. Med.* **2006,** *18,* 147–152.
13. Miyashita, K.; Nishikawa, S.; Beppu, A.; Tsukui, A.; Abe, M.; Hosokawa, M. The allenic carotenoid fucoxanthin, a novel marine nutraceutical from brown seaweeds. *J. Sci. Food Agric.* **2011,** *91,* 1166–1174.
14. Sugawara, T.; Matsubara, K.; Akagi, R.; Mori, M.; Hirata, T. Antiangiogenic activity of brown algae fucoxanthin and its deacetylated product, fucoxanthinol. *J. Agric. Food Chem.* **2006,** *54,* 9805–9810.
15. Akimoto, S.; Tomo, T.; Naitoh, Y.; Otomo, A.; Murakami, A.; Mimuro, M. Identification of a new excited state responsible for the *in vivo* unique absorption band of siphonaxanthin in the green alga *Codium fragile. J. Phys. Chem.* **2007,** *111,* 9179–9181.
16. Chen, G.; Niu, X.; Chen, X.; Li, L.; Kuang, T.; Li, S. Characterization of chlorophyll protein complexes isolated from a siphonous green alga, *Bryopsis corticulans. Photosynth. Res.* **2008,** *96,* 75–81.
17. Mimuro, M.; Katoh, T. Carotenoids in photosynthesis: Absorption, transfer and dissipation of light energy. *Pure Appl. Chem.* **1991,** *63,* 123–130.
18. Akimoto, S.; Yokono, M.; Higuchi, M.; Tomo, T.; Takaichi, S.; Murakami, A.; Mimuro, M. Solvent effects on excitation relaxation dynamics of a keto-carotenoid, siphonaxanthin. *Photochem. Photobiol. Sci.* **2008,** *7,* 1206–1209.
19. Wang, W.; Qin, X.; Sang, M.; Chen, D.; Wang, K.; Lin, R.; Lu, C.; Shen, J.; Kuang, T. Spectral and functional studies on siphonaxanthin-type light-harvesting complex of photosystem II from *Bryopsis corticulans. Photosynth. Res.* **2013,** *117,* 267–279.
20. Sugawara, T.; Yamashita, K.; Sakai, S.; Asai, A.; Nagao, A.; Shiraishi, T.; Imai, I.; Hirata, T. Induction of apoptosis in DLD-1 human colon cancer cells by peridinin isolated from the dinoflagellate, *Heterocapsa triguetra. Biosci. Biotechnol. Biochem.* **2007,** *71,* 1069–1072.
21. Yoshida, T.; Maoka, T.; Das, S.K.; Kanazawa, K.; Horinaka, M.; Wakada, M.; Satomi, Y.; Nishino, H.; Sakai, T. Halocynthiaxanthin and peridinin sensitize colon cancer cell lines to tumor necrosis factor-related apoptosis-inducing ligand. *Mol. Cancer Res.* **2007,** *5,* 615–625.
22. Ganesan, P.; Noda, K.; Manabe, Y.; Ohkubo, T.; Tanaka, Y.; Maoka, T.; Sugawara, T.; Hirata, T. Siphonaxanthin, a marine algal carotenoids from green algae, effectively induces apoptosis in human leukemia (HL-60) cells. *Biochim. Biophys. Acta* **2011,** *1810,* 497–503.
23. Srivastava, R.K. TRAIL/Apo-2L: Mechanisms and clinical applications in cancer. *Neoplasia* **2001,** *3,* 535–546.
24. Folkman, J.; Shing, Y. Angiogenesis. *J. Biol. Chem.* **1992,** *267,* 10931–10934.

25. Carmeliet, P.; Jain, R.K. Angiogenesis in cancer and other diseases. *Nature* **2000**, *407*, 249–257.
26. Folkman, J. Angiogenesis in cancer, vascular, rheumatoid and other diseases. *Nat. Med.* **1995**, *1*, 27–31.
27. Virmani, R.; Kolodgie, F.D.; Burke, A.P.; Finn, A.V.; Gold, H.K.; Tulenko, T.N.; Wrenn, S.P.; Narula, J. Atherosclerotic plaque progression and vulnerability to rupture: Angiogenesis as a source of intraplaque hemorrhage. *Atheroscler. Thromb. Vasc. Biol.* **2005**, *25*, 2054–2061.
28. Folkman, J. Influence of geometry on growth of normal and malignant cells. *Adv. Pathobiol.* **1976**, *4*, 12–28.
29. Carmeliet, P. Angiogenesis in life, disease and medicine. *Nature* **2005**, *438*, 932–936.
30. Matsubara, K.; Mori, M.; Matsuura, Y.; Kato, N. Pyridoxal 5′-phosphate and pyridoxal inhibit angiogenesis in serum-free rat aortic ring assay. *Int. J. Mol. Med.* **2001**, *8*, 505–508.
31. Matsubara, K.; Mori, M.; Matsumoto, H.; Hori, K.; Miyazawa, K. Antiangiogenic properties of a sulfated galactan isolated from a marine alga, *Codium cylindricum*. *J. Appl. Phycol.* **2003**, *15*, 87–90.
32. Matsubara, K.; Kaneyuki, T.; Miyake, T.; Mori, M. Antiangiogenic activity of nasunin, an antioxidant anthocyanin, in eggplant peels. *J. Agric. Food Chem.* **2005**, *53*, 6272–6275.
33. Matsubara, K.; Xue, C.; Zhao, X.; Mori, M.; Sugawara, T.; Hirata, T. Effects of middle molecular weight fucoidans on *in vitro* and *ex vivo* angiogenesis of endothelial cells. *Int. J. Mol. Med.* **2005**, *15*, 695–699.
34. Mori, M.; Sadahira, Y.; Kawasaki, S.; Hayashi, T.; Notohara, K.; Awai, M. Capillary growth from reversed rat aortic segments cultured in collagen gel. *Acta Pathol. Jpn.* **1988**, *38*, 1503–1512.
35. Kawasaki, S.; Mori, M.; Awai, M. Capillary growth of rat aortic segments cultured in collagen gel without serum. *Acta Pathol. Jpn.* **1989**, *39*, 712–718.
36. Ganesan, P.; Matsubara, K.; Ohkubo, T.; Tanaka, Y.; Noda, K; Sugawara, T.; Hirata, T. Anti-angiogenic effect of siphonaxanthin from green alga, *Codium fragile*. *Phytomedicine* **2010**, *17*, 1140–1144.
37. Hicklin, D.J.; Ellis, L.M. Role of vascular endothelial growth factor pathway in tumor growth and angiogenesis. *J. Clin. Oncol.* **2005**, *23*, 1011–1027.
38. Fahmy, R.G.; Dass, C.R.; Sun, L.; Chesterman, C.N.; Khachigian, L.M. Transcription factor Egr-1 supports FGF-dependent angiogenesis during neovascularization and tumor growth. *Nat. Med.* **2003**, *9*, 1026–1032.
39. Ganesan, P.; Matsubara, K.; Sugawara, T.; Hirata, T. Marine algal carotenoids inhibit angiogenesis by down-regulating FGF-2-mediated intracellular signals in vascular endothelial cells. *Mol. Cell. Biochem.* **2013**, *380*, 1–9.
40. Gilfillan, A.M.; Tkaczyk, C. Integrated signaling pathways for mast-cell activation. *Nat. Rev. Immunol.* **2006**, *6*, 218–230.

41. Sakai, S.; Sugawara, T.; Matsubara, K.; Hirata, T. Inhibitory effect of carotenoids on the degranulation of mast cells via suppression of antigen-induced aggregation of high affinity IgE receptors. *J. Biol. Chem.* **2009**, *284*, 28172–28179.
42. Field, K.A.; Holowka, D.; Baird, B. Compartmentalized activation of the high affinity immunoglobulin E receptor within membrane domains. *J. Biol. Chem.* **1997**, *272*, 4276–4280.
43. Sakai, S.; Sugawara, T.; Hirata, T. Inhibitory effect of dietary carotenoids on dinitrofluorobenzene-induced contact hypersensitivity in mice. *Biosci. Biotechnol. Biochem.* **2011**, *75*, 1013–1015.
44. Manabe, Y.; Hirata, T.; Sugawara, T. Suppressive effects of carotenoids on the antigen-induced degranulation in RBL-2H3 rat basophilic leukemia cells. *J. Oleo Sci.* **2014**, *63*, 291–294.
45. Sugawara, T.; Baskaran, V.; Tsuzuki, W.; Nagao, A. Brown algae fucoxanthin is hydrolyzed to fucoxanthinol during absorption by Caco-2 human intestinal cells and mice. *J. Nutr.* **2002**, *132*, 946–951.
46. Asai, A.; Sugawara, T.; Ono, H.; Nagao, A. Biotransformation of fucoxanthinol to amarouciaxanthin A in mice and HepG2 cells: Formation and cytotoxicity of fucoxanthin metabolites. *Drug Metab. Dispos.* **2004**, *32*, 205–211.

*marine drugs*

MDPI

*Article*

# Absorption and Tissue Distribution of Siphonaxanthin from Green Algae

**Zhuosi Li [1], Jiawen Zheng [1], Xiaolin Luo [1], Yuki Manabe [1], Takashi Hirata [1,2] and Tatsuya Sugawara [1,*]**

1 Division of Applied Biosciences, Graduate School of Agriculture, Kyoto University, Kyoto 6068502, Japan; lizhuosi624@gmail.com (Z.L.); feitianmao0715@gmail.com (J.Z.); shelyluo@gmail.com (X.L.); manabe.yuki.8c@kyoto-u.ac.jp (Y.M.); hiratan@mbox.kyoto-inet.or.jp (T.H.)

2 Department of Rehabilitation, Shijonawate Gakuen University, Osaka 5740011, Japan

* Correspondence: sugawara@kais.kyoto-u.ac.jp; Tel.: +81-75-753-6212

Received: 10 April 2020; Accepted: 24 May 2020; Published: 01 June 2020

**Abstract:** Siphonaxanthin has been known to possess inhibitory effects against obesity, inflammation, and angiogenesis. However, little information on its in vivo bioavailability and biotransformation is available. To assess the bioavailability and metabolism of siphonaxanthin, its absorption and accumulation were evaluated using intestinal Caco-2 cells and Institute of Cancer Research (ICR) mice. Siphonaxanthin was absorbed and exhibited non-uniform accumulation and distribution patterns in tissues of ICR mice. Notably, in addition to siphonaxanthin, three main compounds were detected following dietary administration of siphonaxanthin. Because the compounds showed changes on mass spectra compared with that of siphonaxanthin, they were presumed to be metabolites of siphonaxanthin in ICR mice. Siphonaxanthin mainly accumulated in stomach and small intestine, while putative metabolites of siphonaxanthin mainly accumulated in liver and adipose tissues. Furthermore, siphonaxanthin and its putative metabolites selectively accumulated in white adipose tissue (WAT), especially mesenteric WAT. These results provide useful evidence regarding the in vivo bioactivity of siphonaxanthin. In particular, the results regarding the specific accumulation of siphonaxanthin and its metabolites in WAT have important implications for understanding their anti-obesity effects and regulatory roles in lipid metabolism.

**Keywords:** siphonaxanthin; dehydro-metabolite; white adipose tissue; metabolic pathway in vivo

## 1. Introduction

Carotenoids are structurally and functionally an diverse group of natural pigments of the polyene type. Carotenoids with a structure of conjugated double bonds are known to be extremely potent natural antioxidants because of their ability to physically and chemically quench singlet oxygen and scavenge other reactive oxygen species [1,2]. The antioxidant potential of carotenoids is of particular significance to human health, owing to their ameliorative effects on oxidative stress, an essential contributor to the pathogenic processes of many diseases [3,4]. For example, many carotenoids with great antioxidant properties displayed a risk reduction in chronic inflammatory diseases such as preventing inflammation and insulin resistance, providing protection against UV light damage and age-related diseases, and promoting the immune response in the liver, kidneys, and eyes [5].

Several factors influence the bioavailability, absorption, transport, metabolism, and accumulation of dietary carotenoids. Current research has been focused on exploring the potential of carotenoids in human health and elucidating important aspects regarding the digestion, absorption, and metabolism of carotenoids. The degree of food matrix disruption and other dietary components could affect the bioavailability of carotenoids [6]. For instance, the efficiency of carotenoid absorption in the digestive tract can be enhanced by food processing and additional oil and fat, while the coexistence of dietary fibers has been known to suppress their absorption [7]. Other factors, such as genetic factors, gender, age, and nutritional status, also affect the bioavailability of carotenoids. It has been suggested that the absorption of most carotenoids involves several steps, including the release of carotenoids from the food matrix by digestion, dispersion into lipid emulsion particles, incorporation into mixed micelles, uptake by intestinal cells, and secretion into the lymphatic system as lipoprotein particles [3,7,8]. Next, carotenoids are exclusively transported in plasma by lipoproteins and are further distributed in tissues [9].

Polar xanthophylls, such as astaxanthin, fucoxanthin, and siphonaxanthin, which are generally present in aquatic organisms, have been shown to possess beneficial bioactivity in animal models or humans [5]. Siphonaxanthin is a marine carotenoid and an oxidative metabolite of lutein, possessing a structure similar to lutein, except for one keto group located at C-8 and an extra hydroxyl group at C-19. Siphonaxanthin is a special xanthophyll found in green algae, such as *Caulerpa lentillifera*, *Codium fragile*, and *Codium cylindricum* [10].

Siphonaxanthin (Figure 1) has been shown to possesses several potential bioactivities [11]. In vitro siphonaxanthin potently inhibits angiogenesis in vascular endothelial cells and induces apoptosis in human leukemia (HL-60) cells [12,13]. Additionally, siphonaxanthin can modulate inflammatory responses and suppress advanced glycation end product-induced inflammatory responses in vitro [14,15]. In particular, we have focused on the regulatory effect of siphonaxanthin on lipid metabolism. In vitro, we have observed that siphonaxanthin powerfully inhibits lipogenesis both in 3T3-L1 cells [16] and hepatocytes [17]. In vivo, siphonaxanthin has shown inhibitory effects on lipid accumulation in obese mice [16,18]. Furthermore, siphonaxanthin protects Ob/Ob mice fed a high-fat diet against lipotoxicity by ameliorating somatic stress and restoring the anti-oxidative capacity [19]. Thus, we hypothesized that the bioavailability and biotransformation of siphonaxanthin in vivo would allow its use for a dietary supplementation.

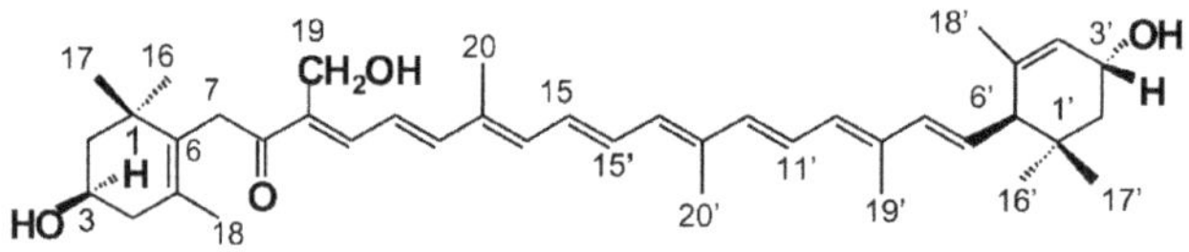

**Figure 1.** Chemical structure of siphonaxanthin (3,19,3′-Trihydroxy-7,8-dihydro-β,ε-caroten-8-one).

In this study, we first evaluated the absorption and biotransformation of siphonaxanthin using intestinal Caco-2 cells, which are widely used as a model to study and predict intestinal absorption and the transport of compounds at an early stage of drug or supplement development [20]. To understand the in vivo metabolism of siphonaxanthin, we investigated the tissue distribution, metabolic transformation, and accumulation of dietary siphonaxanthin using ICR mice [21]. This study will be useful for developing the applications of siphonaxanthin as a functional food. In particular, this study provides important reference information to further improve our knowledge regarding the anti-obesity effect of siphonaxanthin.

## 2. Results

### 2.1. Uptake of Siphonaxanthin by Caco-2 Cells

First, the uptake of siphonaxanthin was evaluated in Caco-2 cells as a model for intestinal epithelial absorption. Caco-2 cells were incubated with 1 nmol/well of siphonaxanthin solubilized micelles for 1, 3, 6, and 24 h. The concentration of siphonaxanthin in the cells and medium was quantified (Figure 2A). The cellular concentration of siphonaxanthin rapidly and linearly increased until 3 h after incubation, and then gradually increased until 24 h. The concentration of siphonaxanthin in the medium decreased with increasing incubation time (Figure 2A).

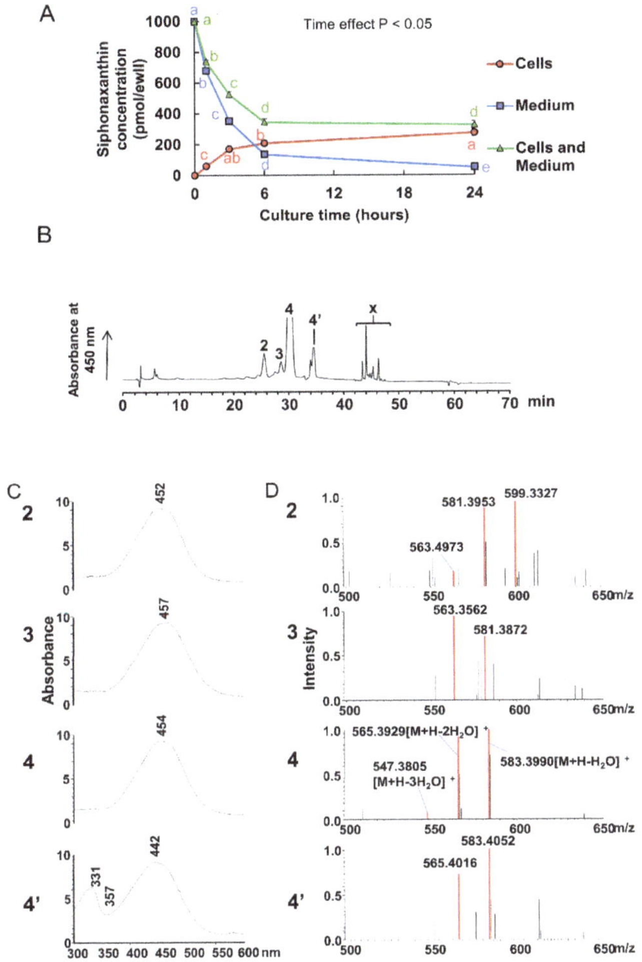


**Figure 2.** Uptake and metabolite analysis of siphonaxanthin in Caco-2 cells (**A**) Changes in siphonaxanthin concentration in Caco-2 cells and medium. Values are means ± SEM, n = 3. Data were analyzed by 1-factor ANOVA with repeated measures. "Time effect $p < 0.05$" indicates that there are differences in siphonaxanthin concentration variables at each time point within every group (cells, medium and cell+medium groups). Mean values without the same letter label indicate that they are significantly different within cells (red), medium (blue), or medium and cells (green), respectively, $p < 0.05$. (**B**) Representative HPLC chromatograms of the extract from Caco-2 cells after treatment with siphonaxanthin-containing micelles for 24 h. HPLC analysis was performed as described in the experimental methods. Peaks: 2, 3, and x, unknown metabolites; 4, siphonaxanthin; 4′, cis isomer of siphonaxanthin. Retention time: peak 2 at 25 min, peak 3 at 28 min, peak 4 at 30 min and peak x at 43–48 min. In the Caco-2 cells, we did not detect the peak corresponding to the peak 1 in the Figure 4 (mice data). Thus, in order to compare the corresponding peak at the same retention time compared with the Figure 4, here, the label "1" were not used. The detection wavelength was 450 nm. UV-vis spectra (**C**) and APCI-MS spectra (**D**) of peaks 2, 3, 4, and 4′. LC-MS with APCI analysis was performed as described in the experimental methods. HPLC, High-performance liquid chromatography; APCI, Atmospheric pressure chemical ionization; LC, Liquid chromatography; MS, Mass spectrometry.

The representative high-performance liquid chromatography (HPLC) chromatogram of an extract of Caco-2 cells after 24 h of incubation with micelles containing siphonaxanthin is shown in Figure 2B. The most predominant peak at approximately 30 min (peak 4) was completely fitted with the peak corresponding to the siphonaxanthin standard. Peak 4 was identified as siphonaxanthin based on its UV-vis spectrum (Figure 2C) and characteristic ions at a charge ratio (*m/z*) of 583.3990 $[M + H\text{-}H_2O]^+$, 565.3929 $[M + H\text{-}2H_2O]^+$, and 547.3805 $[M + H\text{-}3H_2O]^+$ by atmospheric pressure chemical ionization (APCI) (Figure 2D) and 623.4 $[M + Na]^+$ by electrospray ionization (ESI). Peak 4′ was identified as *cis*-siphonaxanthin because its APCI-mass spectrometry (MS) spectrum showed the same ion peaks as siphonaxanthin and its UV-vis spectrum presented a *cis* peak at 331 nm (Figure 2C). Except for peaks 4 and 4′, no additional peaks were detected during incubation for 1–6 h (data not shown). However, after incubation for 24 h with siphonaxanthin, peaks 2, 3, and x appeared in the HPLC chromatogram of the cell extract (Figure 2B). The UV-vis spectrum of peak 2 with maximum absorbance at 452 nm was almost consistent with that of siphonaxanthin. Peak 2 showed ion peaks at *m/z* 599.3327, 581.3953, and 563.4973 in the APCI-MS spectrum and *m/z* 621.4 in the ESI-MS spectrum. Peak 3 at 28 min was detected in the cell extract (Figure 2B). The maximum absorbance of peak 3 was not significantly shifted when compared with that of siphonaxanthin (Figure 2C). The APCI-MS spectrum of peak 3 showed two fragment ions at *m/z* 581.3872 and 563.3562; however, the ion peak at *m/z* 599.4 was not observed. The *m/z* values of two compounds (peak 2 and 3) reduced 2 mass compared with these of siphoanxanthin.

### 2.2. Absorption and Tissue Distribution of Siphonaxanthin in Mice

The body weight of siphonaxanthin fed mice was not altered when compared to the body weight of mice fed the control diet (Figure 3A). During the 16-day experimental period, food intake was not altered between the control and siphonaxanthin groups (Figure 3B). Siphonaxanthin supplementation significantly decreased the weight of the spleen when compared to the control diet (Figure 3C). The weight of the other tissues of siphonaxanthin fed mice did not significantly decrease when compared to that of mice fed the control diet (Figure 3C).

The representative HPLC chromatograms of the extract from each tissue of mouse fed a diet containing siphonaxanthin for 16 days are shown in Figure 4. Peaks 1, 2, 3, 4, and 5 appeared in the HPLC chromatogram of extracts obtained from most mice tissues; however, these peaks were not detected in control mice. Peak 4 at about 30 min was identified as siphonaxanthin based on its UV-vis and MS spectra (Figure 5). Siphonaxanthin was widely distributed in blood and tissues, except in the bladder. The maximum absorbance of peak 1 at 459 nm shifted 5 nm when compared with that of siphonaxanthin (Figure 5A). Peak 1 showed ion peaks at *m/z* 597.3820 and 579.3779 in the APCI-MS spectrum (Figure 5B) and *m/z* 619.4 in the ESI-MS spectrum. The *m/z* values of peak 1 compound reduced 4 mass compared with these of siphoanxanthin. Peak 2 showed ion peaks at *m/z* 599.4002, 581.3764, and 563.3693 in the APCI-MS spectrum and *m/z* 621.4 in the ESI-MS spectrum. The maximum absorbance of peak 3 was not significantly shifted when compared with that of siphonaxanthin (Figure 5). The APCI-MS spectrum of peak 3 showed two fragment ions at *m/z* 581.3953 and 563.3899; however, the ion peak at *m/z* 599 was not observed. The *m/z* values of peak 2 and 3 reduced 2 mass compared with these of siphoanxanthin. The UV-vis spectrum of peak 5 with maximum absorbance at 455 nm was almost consistent with that of siphonaxanthin (Figure 5A). As the APCI-MS spectrum showed ion peaks at *m/z* 581.8720 and 563.3562 and its retention time was longer than 46 min (Figure 5B).

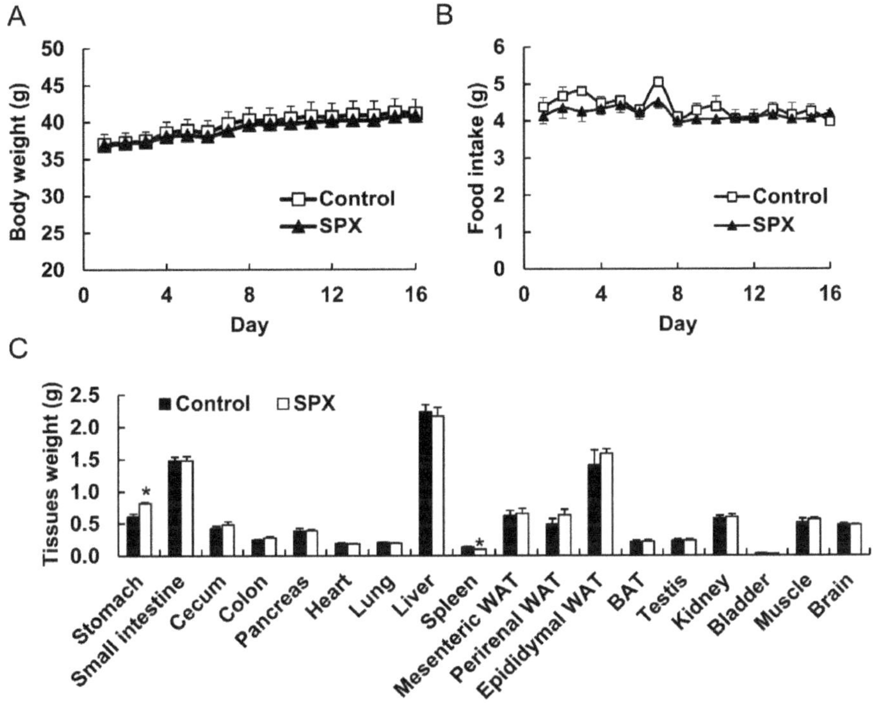


**Figure 3.** Body weight (**A**), food intake (**B**), and tissue weight (**C**) of ICR mice. ICR mice were fed a control or siphonaxanthin supplementation diet for 16 days. Body weight and food intake were measured daily. After the 16-day feeding period, the mice were killed and their weight was measured. Values are means ± SEM, n = 4. The difference between the control and siphonaxanthin groups was analyzed using the Student's t-test. * Different from the control group, $p < 0.05$. BAT, brown adipose tissue; WAT, white adipose tissue; SPX, siphonaxanthin.

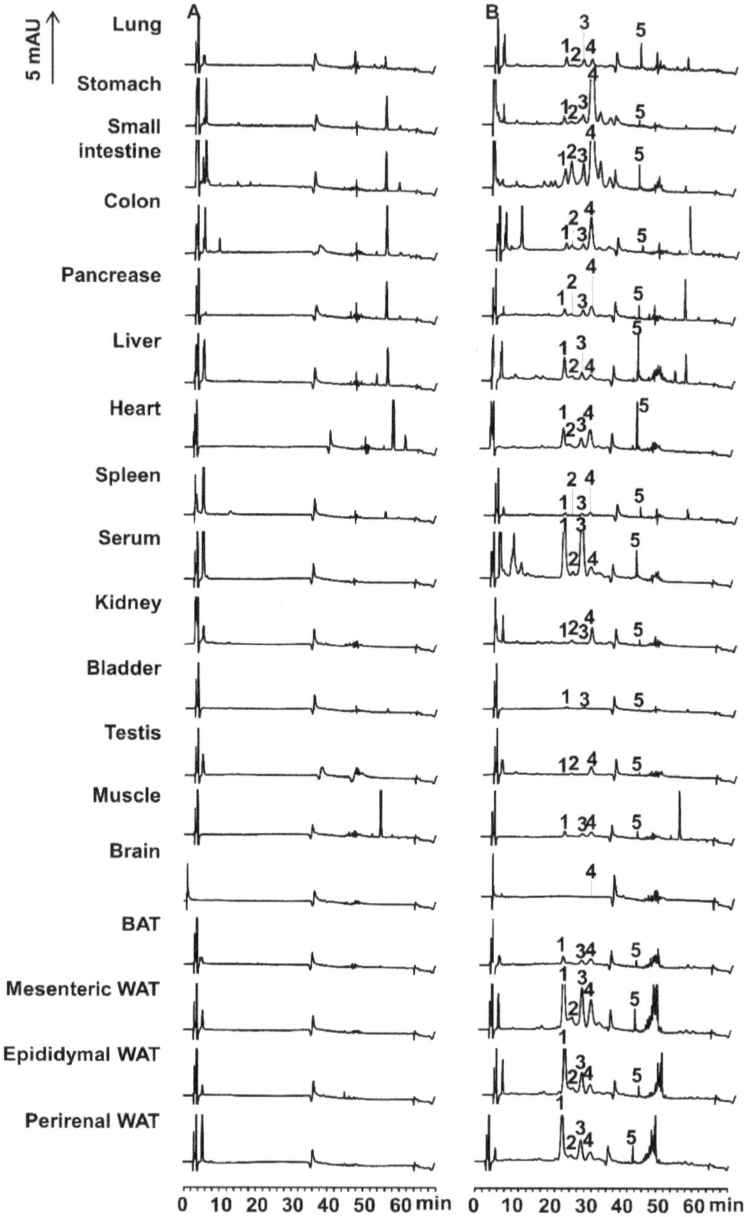


**Figure 4.** Representative HPLC chromatograms of the extracts from each tissue of mice fed a diet without (**A**) or with (**B**) siphonaxanthin for 16 days. HPLC analysis was performed as described in the experimental methods. The detection wavelength was 450 nm. Peaks with the same number in different chromatograms show similar UV-vis and MS spectra, as shown in Figure 5A,B. 1, 2, 3, and 5, unknown metabolites; 4, siphonaxanthin; Retention time: peak 1 at 22 min, peak 2 at 25 min, peak 3 at 28 min, peak 4 at 30 min and peak 5 at 43 min.

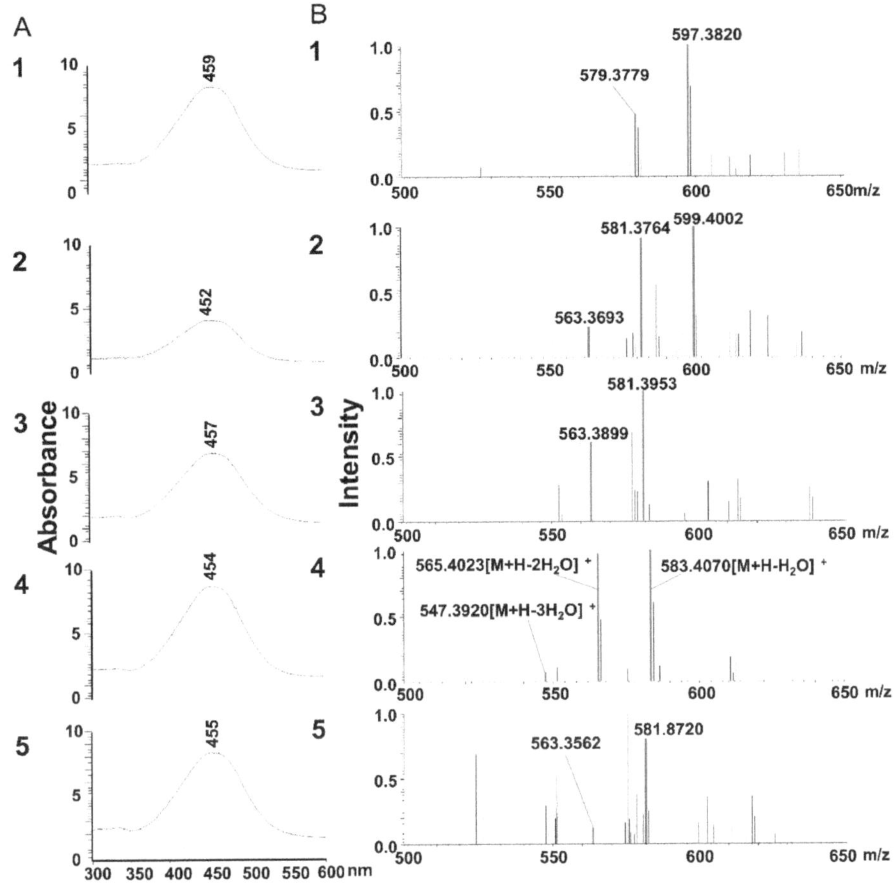


**Figure 5.** UV-vis spectra (**A**) and APCI-MS spectra (**B**) of peaks 1, 2, 3, 4, and 5 are shown in Figure 4. LC-MS (APCI) analysis was performed as described in the experimental methods.

## 2.3. Tissue Distribution and Accumulation of Siphonaxanthin and Its Metabolites in Mice

At the end of this feeding study, a higher concentration of intact siphonaxanthin was detected in the small intestine, stomach, and colon than in the plasma and other tissues (Table 1). Notably, the concentration of siphonaxanthin in the mesenteric white adipose tissue was higher than that in the epididymal and perirenal WAT (Table 1). Furthermore, siphonaxanthin was detectable in the mouse brain, although the concentration was extremely low (Table 1).

**Table 1.** Concentration of siphonaxanthin and compounds corresponding to peak 1, 2, and 3 in the plasma and tissues of mice at the end of the 16-day dietary supplementation with siphonaxanthin [1].

| Tissue | Peak 1 | Peak 2 | Peak 3 | Siphonaxanthin | Total |
|---|---|---|---|---|---|
| Stomach, pmol/g (%) | 26.6 ± 13.5 (1.2) | 19.0 ± 3.3 (0.9) | 86.1 ± 11.0 (3.9) | 2054 ± 636 (94.0) | 2186 ± 612 |
| Small intestine, pmol/g (%) | 172 ± 114 (4.0) | 287 ± 132 (6.7) | 423 ± 273 (9.8) | 3428 ± 1389 (79.5) | 4310 ± 1837 |
| Pancreas, pmol/g (%) | 24.7 ± 14.3 (24.3) | 6.4 ± 3.8 (6.3) | 24.4 ± 14.1 (24.0) | 46.1 ± 12.5 (45.4) | 102 ± 44.2 |
| Spleen, pmol/g (%) | 10.5 ± 5.2 (23.2) | 2.9 ± 1.0 (6.4) | 11.1 ± 4.9 (24.7) | 20.6 ± 3.4 (45.7) | 45.1 ± 12.5 |
| Colon, pmol/g (%) | 12.3 ± 7.5 (5.1) | 8.7 ± 0.4 (3.6) | 20.9 ± 11.4 (8.7) | 198 ± 24.5 (82.5) | 239 ± 11.3 |
| Lung, pmol/g (%) | 18.3 ± 10.8 (30.9) | 0.6 ± 0.4 (1.0) | 17.3 ± 10.0 (29.3) | 22.9 ± 8.2 (38.8) | 59.0 ± 29.3 |
| Heart, pmol/g (%) | 35.0 ± 20.2 (27.2) | 4.6 ± 2.6 (3.5) | 21.9 ± 12.7 (17.0) | 67.1 ± 25.2 (52.2) | 129 ± 60.6 |
| Liver, pmol/g (%) | 110 ± 63.4 (55.3) | 11.3 ± 4.6 (5.7) | 40.3 ± 21.2 (20.3) | 37.1 ± 14.4 (18.7) | 198 ± 102 |
| Bladder, pmol/g (%) | 14.6 ± 8.5 (58.5) | N.D. | 10.4 ± 6.1 (41.5) | N.D. | 25.0 ± 14.6 |
| Muscle, pmol/g (%) | 9.9 ± 5.7 (36.5) | 0.7 ± 0.7 (2.6) | 2.2 ± 2.2 (8.3) | 14.2 ± 7.5 (52.6) | 27.1 ± 15.1 |
| Plasma, pmol/mL (%) | 162 ± 93.4 (48.8) | 3.5 ± 2.0 (1.0) | 147.4 ± 78.6 (44.6) | 18.3 ± 4.0 (5.5) | 331 ± 177 |
| BAT, pmol/g (%) | 25.3 ± 15.4 (34.6) | 0.6 ± 0.6 (0.8) | 16.7 ± 9.5 (22.9) | 30.5 ± 11.4 (41.7) | 73.1 ± 36.0 |
| Mesenteric WAT, pmol/g (%) | 188 ± 106 (41.8) | 43.6 ± 21.8 (9.7) | 108.1 ± 58.7 (24.0) | 110 ± 50.8 (24.5) | 449 ± 236 |
| Perirenal WAT, pmol/g (%) | 95.5 ± 62.2 (57.4) | 4.3 ± 2.5 (2.6) | 44.4 ± 29.4 (26.7) | 22.1 ± 12.0 (13.3) | 166 ± 105 |
| Epididymal WAT, pmol/g (%) | 112 ± 69.0 (58.9) | N.D. | 50.6 ± 30.3 (26.5) | 27.8 ± 14.3 (14.6) | 190 ± 113 |
| Kidney, pmol/g (%) | 0.5 ± 0.5 (1.0) | 6.6 ± 2.9 (14.6) | 0.1 ± 0.1 (0.3) | 38.2 ± 15.7 (84.1) | 45.4 ± 18.9 |
| Testis, pmol/g (%) | 1.2 ± 0.8 (4.1) | 3.0 ± 1.8 (10.2) | 0.4 ± 0.4 (1.3) | 25.0 ± 10.0 (84.3) | 29.6 ± 12.6 |
| Brain, pmol/g (%) | N.D. | N.D. | N.D. | 0.4 ± 0.3 (100) | 0.4 ± 0.3 |

[1] The quantification of siphonaxanthin and metabolites was based on the peaks (shown in Figure 4) in the HPLC chromatogram. Siphonaxanthin and metabolites were quantified using the standard curve of siphonaxanthin. The values in parenthesis for siphonaxanthin (peak 4) and metabolites (peak 1–3) are also expressed as % of the total carotenoids. Values are means ± SEM, n = 4. BAT, interscapular brown adipose tissue; WAT, white adipose tissue; N.D., not detected.

In addition to siphonaxanthin, peaks 1, 2, and 3 were quantified using the standard curve of siphonaxanthin (Table 1). Following dietary supplementation with siphonaxanthin, the concentration of the compound corresponding to peak 2 was relatively higher in the small intestine than in the mesenteric WAT. The compound corresponding to peak 3 preferentially accumulated in the small intestine followed by the plasma, WATs, and liver. The compound corresponding to peak 1 mainly accumulated in mesenteric WAT followed by the small intestine, plasma, liver, and perirenal and epididymal WATs. However, the compounds of peaks 1, 2 and 3 failed to accumulate in the mouse brain. The proportion of peak 1 in the liver and WATs was 40–60% of total carotenoids, while the proportion was less than 10% in the small intestine, stomach, and colon, and less than 5% in the kidney and testis (Table 1). In all tissues, the proportion peak 2 was limited and less than 15% of the total carotenoids (Table 1). The proportion of peak 3 in the liver and WATs was more than 20–27% of total carotenoids, while the proportion was less than 10% in the small intestine, stomach and colon, and less than 2% in the kidney and testis (Table 1).

## 3. Discussion

To enhance understanding of the potential health benefits of dietary carotenoids, studies regarding their bioavailability and biotransformation are crucial. In this study, the absorption, distribution, and accumulation of dietary siphonaxanthin were investigated in differentiated Caco-2 cells and ICR mice.

At incubation with siphonaxanthin for 24 h in the Caco-2 cells, some unknown peaks in HPLC chromatography were appeared. The *m/z* values of two compounds (peaks 2 and 3) reduced 2 mass compared with these of siphoanxanthin. Thus, we speculated that siphonaxanthin might be oxidized into two types of didehydro-metabolites in differentiated Caco-2 cells. It has been suggested that siphonaxanthin is derived from lutein via an oxidative pathway in green algae [10]. Reportedly, metabolites of lutein, keto-carotenoids, were detected in the plasma and tissues of mice and humans [22]. Owing to the similar chemical structures of lutein and siphonaxanthin, it was assumed that the possible structures of the two siphonaxanthin metabolites found in Caco-2 cells were the dehydrogenization products of siphonaxanthin, in which a hydroxyl group could be oxidized into a keto group. Siphonaxanthin has three hydroxyl groups at positions 3, 19, and 3′, all of which may be oxidized into the keto group. Previously, the oxidation of lutein at the 3 and 3′ position [22], the oxidation of capsanthin at the 3′ position to capsanthone [23], and the dehydrogenation of fucoxanthinol at position 3 to amarouciaxanthin A in mice liver [24] has been reported after ingestion of these carotenoids in vivo. These reports indicate that a hydroxyl group at the end group would be liable to be oxidized. Thus, the two dehydrogenated metabolites of siphonaxanthin were presumed to be the 3′-didehydro-metabolite and 3-didehydro-metabolite of siphonaxanthin. In the future, we plan to identify the differences between these two metabolites.

Two compounds of peaks 2 and 3 were present in both Caco-2 cells and the small intestine, siphonaxanthin might be converted to the above didehydro-metabolites of siphonaxanthin by some enzymes associated with oxidation reactions in the small intestine of mice. These two compounds were observed in the plasma and other tissues. Additionally, in most tissues, the compound of peak 3 (Figure 4) was more abundant than peak 2 (Figure 4), especially in the plasma and WATs (Table 1), indicating that the structure of the compound of peak 3 might be more stable in vivo. Lutein has 3-hydroxy β-end and 3′-hydroxy ε-end groups. The former is preferentially oxidized into 3-didehydro ε-end [25]. It has been reported that the 3-didehydro-metabolite of lutein (3′-hydroxyl-ε,ε-caroten-3-one), found to be largely accumulated in mouse tissues, is more stable than the 3′-didehydro-metabolite of lutein (3-hydroxyl-β,ε-caroten-3′-one), detected in the human serum [22]. Furthermore, the 3-hydroxy β-end group in lutein is converted to a 3-didehydro ε-end group via the 3-didehydro β-end group, when incubated in a mixture of lutein with a post-mitochondrial fraction of mouse liver [25]. The intermediates containing the 3-didehydro β-end group, produced by dehydrogenation, demonstrate unstable conformations, thus resulting in structural isomerization into an ε-end group by double-bond migration [25]. However, the intermediates containing the 3-didehydro β-end group are not detected after the intake of lutein [25]. Therefore, we presumed that the structure of the putative final 3-didehydro-metabolites of siphonaxanthin might include the 3-dehydroxy ε-end group but not the β-end group because the β-end group is unstable. Based on the above discussion, peak 2, which demonstrated minimal accumulation in most tissues except for testis and kidney, might also be an unstable 3-didehydro-metabolite of siphonaxanthin, containing a 3-didehydroxy β-end group or a 3′-didehydro-metabolite of siphonaxanthin. In contrast, peak 3, with abundant accumulation, was presumed to be a 3-didehydro-metabolite of siphonaxanthin.

The most abundant metabolite eluted at 22 min (peak 1) accounted for 40–60% of siphonaxanthin and its metabolites in plasma, liver and WATs. Based on the MS spectra, this metabolite was also considered as a dehydrogenization product of siphonaxanthin. It is speculated that two hydroxyl groups were transformed into two keto groups in the structure of the compound. Reportedly, lutein is oxidized into ε,ε-carotene-3,3′-dione in mice [22], and 4,4′-dimethoxy-β-carotene is oxidized into 4,4′-diketo-β-carotene in human subjects [26]. Thus, the metabolite eluted at 22 min might be identified as a 3,3′-tetradehydro-metabolite of siphonaxanthin. And it was not possible to directly analyze the small amounts of crude metabolites of siponaxanthin found in each tissue by NMR. In the future, we will isolate these compounds and reveal the structure of the oxidized products of siphonaxanthin, especially the location of the carbonyl groups.

Based on the comparison with the other carotenoids which have similar structures with siphonaxanthin, the change of mass, and also not significantly shifted the maximum absorbance of UV-vis spectra, we speculated the compounds of peaks 1, 2 and 3 to be the dehydro-metabolites of siphonaxanthin. And thus, we speculated the formulas of peaks 1, 2 and 3 (Supplemental Table S1). Based on these predicted formulas, the predicted ion and the observed ion are shown in the Supplemental Table S1. However, the error between the predicted ion and the observed ion was always large (2–39 ppm). Peak 4 were identified as siphonaxanthin compared with the standard, however, the error of peak 4 was also large. Thus, we considered that the error was caused by the factors of environment, temperature, and the instrument itself. However, another possibility is that real formulas of peaks 1, 2 and 3 might not be our predicted formulas. In the future, we will verify this using the internal calibration method.

Furthermore, the carotenoids eluted at approximately 43–52 min in cells and mice were assumed to be the esters of siphonaxanthin. In differentiated Caco-2 cells, peridinin is converted to peridiniol and its fatty acid esters [27]. No information is available regarding the fatty acid esters of carotenoids in mice. At present, due to the coexistence of large amounts of triacylglycerols in the extract from cells and mice, the identification of the metabolites is challenging and needs to be addressed in future studies. The metabolic pathways of lutein and zeaxanthin have been suggested in many reports [26,28,29].

The metabolites of fucoxanthin, such as fucoxanthinol and amarouciaxanthin A, mainly accumulate in adipose tissues [30]. Additionally, it has been reported that lutein [22] and β-carotene [31,32] mainly accumulate in the liver instead of adipose tissues. In the present study, siphonaxanthin was not completely converted to metabolites in the gastrointestinal tract, and an unchanged form of siphonaxanthin was also detected in the plasma and most tissues, except the bladder, after the 16-day feeding period, along with its metabolites. Canthaxanthin, lycopene, and β-carotene are known to accumulate to a small extent in the colon, small intestine, stomach, and large intestine [33–35]. Siphonaxanthin might merely stick to the intestinal mucosa, rather than accumulate in the intestine. Notably, siphonaxanthin accumulated more easily in the mesenteric adipose tissue than epididymal WAT, perirenal WAT, and BAT, which might be attributed to the fact that the mesenteric WAT is closer to the gastrointestinal tract. These results are consistent with our previous findings that siphonaxanthin mainly accumulates in the mesenteric WAT and significantly inhibits the lipid accumulation of mesenteric WAT in KK-Ay mice administrated siphonaxanthin for 6 weeks. In our previous study, siphonaxanthin and its metabolites were not separated, and thus the total accumulation of siphonaxanthin and its metabolites in the mesenteric WAT was 649 pmol/g [16]. In this study, although the experimental conditions differed, the total accumulation of siphonaxanthin and its presumed metabolites reached approximately 70% of the above stated 649 pmol/g. However, the accumulation level did not change the weight of mesenteric WAT in ICR mice. Thus, we speculated that siphonaxanthin might have low toxicity and fewer side effects, such as excessive weight loss, in normal mice. To evaluate the effect of siphonaxanthin and its metabolites on normal subjects, a longer feeding period should be undertaken in the future.

The distribution of each possible metabolite depended on the tissues. The compounds corresponding to peak 3 and peak 1 were more abundant than siphonaxanthin in the liver, serum, and WATs, whereas the proportion of these two compounds was inversed in other evaluated tissues. The compound corresponding to peak 2 mainly accumulated in the small intestine, with small amounts observed in the other tissues. The tissue-dependent distribution of presumed siphonaxanthin metabolites might be associated with the tissue-specific distribution of enzymes or the rate of metabolism and transport of each metabolite.

In general, the bioavailability of functional food ingredients in diets can affect their biological effectiveness. The bioavailability of dietary carotenoids depends on several steps, including absorption in the intestinal epithelia and metabolism. In this study, siphonaxanthin was absorbed and metabolized by mice. In contrast with mice, humans tend to accumulate carotenoids selectively, possibly through discriminative uptake and/or re-excretion by intestinal cells and metabolism in the body [28]. When siphonaxanthin, solubilized in the mixed micelles compatible with those formed in the intestine, were incubated with human intestinal Caco-2 cells, the accumulation of siphonaxanthin increased. The results demonstrated that siphonaxanthin might be absorbed and metabolized by humans. Our previous study shows that the content of siphonaxanthin in *C. cylindricum* harvested in autumn is approximately 230 μg/g dry weight [18]. In this study, 65.79 nmol siphonaxanthin/g of diet (0.004%) is equivalent to a daily dosage of approximately 18 g dried *C. cylindricum* powder/kg body weight. Thus, purified siphonaxanthin is useful as a dietary supplement. In the future, it is necessary to investigate the behavior of siphonaxanthin in human subjects.

In addition to intestinal absorption, tissue distribution and metabolism also affect the bioavailability of carotenoids. The present study indicates that dietary siphonaxanthin might be partly dehydrogenated after absorption, and both intact and metabolized forms of siphonaxanthin could accumulate in the body. To understand the bioavailability of siphonaxanthin, further efforts are needed to clarify the metabolic pathway, metabolic rate, and the enzymes involved in the oxidative transformation of siphonaxanthin. In some tissues and plasma, the accumulation of the oxidative metabolites was much greater than that of intact siphonaxanthin; therefore, to elucidate the role of siphonaxanthin in human health, the bioactivity of siphonaxanthin metabolites is worth evaluating.

In summary, siphonaxanthin was absorbed accumulated in the mice. Three possible metabolites of siphonaxanthin were also found in ICR mice. The distribution of siphonaxanthin and its presumed metabolites depended on the tissue. Although further studies are needed to elucidate the metabolic mechanisms of siphonaxanthin and identify the structure of metabolites of siphonaxanthin, this study has provided useful information for developing siphonaxanthin applications beneficial to human health. In particular, this work provides important reference information to understand the bioavailability and tissue accumulation of other carotenoids with similar structures.

## 4. Materials and Methods

### 4.1. Preparation of Siphonaxanthin

Siphonaxanthin was extracted from green alga (*Codium cylindricum*) as previously described [16]. The extracted carotenoid was purified by HPLC (LC-6, Shimadzu, Kyoto, Japan) and the purified siphonaxanthin (purity > 98%) was used. Siphonaxanthin was stored at −80 °C until further use.

### 4.2. Cell Culture

Caco-2 cells obtained from the RIKEN Gene Bank (Tsukuba, Japan) were cultured in Dulbecco's Modified Eagle's Medium (DMEM) containing 10% fetal bovine serum (FBS), 1% penicillin-streptomycin (PS), and 1% non-essential amino acids in a humidified atmosphere of 95% air and 5% $CO_2$ at 37 °C. For differentiation, cells were seeded in 12-well plates at a density of $2.0 \times 10^5$ cells/well in 1 mL DMEM medium described above and allowed to differentiate until day 22 from seeding. The medium was regularly changed three times a week. Experiments were performed at day 22 post-seeding on 12-well plates.

### 4.3. Treatment of Caco-2 Cells with Micellar Siphonaxanthin

Siphonaxanthin solubilized in micelles was added to the medium for treatment, and micelle formation was performed as previously described [36]. Briefly, sodium taurocholate, monoolein, oleic acid, lysophosphatidylcholine, and siphonaxanthin dissolved in dichloromethane or methanol were mixed using a vortex mixer, and the organic solvent was evaporated using nitrogen gas. The residue was then dissolved in serum-free DMEM. The final concentration of each component in the medium was adjusted to 2 mmol/L sodium taurocholate, 100 μmol/L monoolein, 33.3 μmol/L oleic acid, 50 μmol/L lysophosphatidylcholine, and 1.0 μmol/L siphonaxanthin. The resultant solution should be optically clear. This medium was filtered using a 0.22 μm filter before supplementation to the culture cell. The concentration of micellar carotenoid was determined as 1.0 ± 0.05 μmol/L by HPLC before use in the following experiment.

The cell monolayers on 12-well plates were rinsed with the serum-free medium and then incubated in 1 mL of medium containing micellar siphonaxanthin. After incubation for 0, 1, 3, 6, and 24 h, the medium was collected, and the cells were washed twice with ice-cold phosphate-buffered saline (PBS) containing 10 mmol/L sodium taurocholate to remove surface-bonded carotenoids, followed by an additional washing with PBS. The cells were scraped into PBS and then centrifuged at 1000× *g* at 4 °C for 5 min. The cell pellets were resuspended in 0.5 mL PBS and homogenized with a sonicator (Qsonica Q55). To extract siphonaxanthin, 0.4 mL of the cell homogenate was mixed with 1.5 mL of dichloromethane/methanol (1:2, v/v), and vigorously vortexed. Then, 0.75 mL of hexane was added to the mixture, followed by strong agitation, and centrifugation at 1690× *g* at 4 °C for 10 min. The upper organic phase was transferred to a fresh test tube, the sample was extracted again by adding 0.5 mL of dichloromethane, and then 0.75 mL of hexane. This extraction was repeated three times. All organic phases were pooled together and evaporated gently under a nitrogen stream. The residue was dissolved in methanol and subjected to quantitative and qualitative analysis. Because there were no other carotenoids except for siphonaxanthin were observed at 6, 12, and 18 h, the quantitative analysis of the cell extraction at all time points was performed using a photodiode array detector (SPD-M20A, Shimadzu, Kyoto, Japan) connected to the HPLC system (LC-6, Shimadzu, Kyoto, Japan). The qualitative analysis of the cell extraction at 24 h was performed using LC-MS system as described below. The concentration of siphonaxanthin in the medium was also analyzed. An aliquot of the medium was mixed with 4-fold methanol and subjected to the above HPLC analysis for quantify.

### 4.4. Animal Studies

All experimental animal protocols were approved by the Animal Experimentation Committee of Kyoto University, Japan, for the care and use of experimental animals (Approval No. 26–35). Male ICR mice (7 weeks of age) were purchased from Japan SLC, Inc. (Hamamatsu, Japan). All mice were housed individually and maintained on an alternating 12-h light/dark cycle at a temperature of 23 ± 1 °C and free access to drinking water and chow (Oriental Yeast Co., Ltd., Tokyo, Japan). After an acclimatization period of 1 week, the mice were randomly divided into control and siphonaxanthin groups (n = 4 per group). Mice in the control group were fed an AIN-93G diet. The siphonaxanthin mouse group was fed the AIN-93G diet with siphonaxanthin supplementation, 65.79 nmol/g of diet (0.004%). The total food intake and body weight were recorded daily. After dietary supplementation for 16 days, the mice were anesthetized with isoflurane. Blood was collected from the caudal vena cava. Plasma was prepared by centrifuging at 400× *g* for 15 min at 4 °C. The tissues were rapidly removed, weighed, and immediately frozen in liquid nitrogen, and stored at −80 °C until use.

Carotenoids were then extracted from the plasma and tissues. Briefly, aliquots of tissue samples (0.2 g) were homogenized in a 9-fold volume of 0.9% NaCl saline with a homogenizer dispenser (T10 basic ULTRA-TURRAX, IKA). Plasma samples were diluted with a 3-fold volume of Milli-Q water. The resultant tissue homogenates (0.9 mL) or diluent plasma samples (0.9 mL) were mixed with 3 mL of dichloromethane/methanol (2:1, v/v) to extract carotenoids. The samples were extracted three times, and the dichloromethane layer was collected after centrifugation at 1690× *g* at 4 °C for 10 min. After evaporation under nitrogen, the residue was resuspended in 20 μL of dichloromethane and methanol (1:1, v/v) for HPLC analysis.

### 4.5. HPLC and MS Analysis

HPLC analysis was performed using a Prominence LC system (Shimadzu, Kyoto, Japan) connected to a photodiode array detector (SPD-M20A, Shimadzu, Kyoto, Japan), followed by an ion trap-time of flight mass spectrometer (LCMS-IT-TOF, Shimadzu, Kyoto, Japan) equipped with an atmospheric pressure chemical ionization (APCI) source or electrospray ionization (ESI) source. Siphonaxanthin was separated on a TSK gel ODS-80Ts QA column (2.0 × 250 mm, 5 μm, Tosoh, Tokyo, Japan). The binary gradient mobile phase was methanol/water (83:17, v/v) containing 0.1% ammonium acetate as mobile phase A and ethyl acetate/methanol (30:70, v/v) containing 0.1% ammonium acetate as mobile phase B. The column was eluted at a flow rate of 0.2 mL/min using the following gradients: 0–30 min, 0% B; 30–45 min, 0–100% B; 45–60 min, 100% B; 60–65 min, 100–0% B; 65–70 min, 0% B. Siphonaxanthin was detected at 450 nm. The APCI source was heated at 200 °C, and the probe was maintained at 400 °C. Nitrogen was used as sheath gas at 2.0 L/min, and drying gas was used at 25 kPa. Mass spectra were recorded in the positive ion mode. For the ESI source, sheath gas was set at 1.5 L/min, and drying gas was used at 120 kPa. A spray voltage of 4.5 kV was used for positive ionization. The peak identities of siphonaxanthin and its possible metabolites were further confirmed from their characteristic UV-vis spectra and their positive ions. Siphonaxanthin was quantified from its peak area at 450 nm using an external standard calibration curve with purified siphonaxanthin. Due to the unavailability of standards, possible metabolites of siphonaxanthin were estimated from the siphonaxanthin standard curve [22].

### 4.6. Statistical Analysis

Data analyses were performed using the statistical program SPSS 16.0 for Windows (SPSS Inc., Chicago, IL, USA). Changes in the concentration of siphonaxanthin in Caco-2 cells were analyzed by 1-factor ANOVA with repeated measures. Data are represented as mean ± SEM. Statistical significance was defined as $p < 0.05$.

**Supplementary Materials:** The following are available online at http://www.mdpi.com/1660-3397/18/6/291/s1, Table S1: Maximum absorbance of UV-vis spectra and LC–MS data in the positive ion mode of compounds identified in small intestine samples of mice at the end of the 16-day dietary supplementation with siphonaxanthin.

**Author Contributions:** Z.L. and T.S. designed the experiment; Z.L., J.Z. and X.L. conducted the research and analyzed the data; Y.M. provided the methods for critical experiments and provided useful suggestions to this research; Z.L. wrote the manuscript; T.H. provided instruments and reagents for analyses; T.S. had primary responsibility for the final content. All authors have read and agreed to the published version of the manuscript.

**Funding:** This study was partly supported by JSPS KAKENHI Grant Numbers JP23380124 (T.H.), 16K14926 (T.S.), 18K14406 (Y.M.), and Kieikai Research Foundation (Y.M.).

**Conflicts of Interest:** The authors declare no conflicts of interest.

## References

1. Foote, C.S.; Chang, Y.C.; Denny, R.W. Chemistry of singlet oxygen. XI. Cis-trans isomerization of carotenoids by singlet oxygen and a probable quenching mechanism. *J. Am. Chem. Soc.* **1970**, *92*, 5218–5219. [CrossRef] [PubMed]
2. Martin, H.D.; Ruck, C.; Schmidt, M.; Sell, S.; Beutner, S.; Mayer, B.; Walsh, R. Chemistry of carotenoid oxidation and free radical reactions. *Pure Appl. Chem.* **1999**, *71*, 2253–2262. [CrossRef]
3. Fiedor, J.; Burda, K. Potential role of carotenoids as antioxidants in human health and disease. *Nutrients* **2014**, *6*, 466–488. [CrossRef] [PubMed]
4. Riccioni, G. Marine carotenoids and oxidative stress. *Mar. Drugs* **2012**, *10*, 116–118. [CrossRef] [PubMed]
5. Gammone, M.A.; Riccioni, G.; D'Orazio, N. Marine carotenoids against oxidative stress: Effects on human health. *Mar. Drugs* **2015**, *13*, 6226–6246. [CrossRef]
6. Hof, K.H.V.; West, C.E.; Weststrate, J.A.; Hautvast, J.G.A.J. Dietary factors that affect the bioavailability of carotenoids. *J. Nutr.* **2000**, *130*, 503–506.
7. Yonekura, L.; Nagao, A. Intestinal absorption of dietary carotenoids. *Mol. Nutr. Food Res.* **2007**, *51*, 107–115. [CrossRef]
8. Furr, H.C.; Clark, R.M. Intestinal absorption and tissue distribution of carotenoids. *J. Nutr. Biochem.* **1997**, *8*, 364–377. [CrossRef]
9. Parker, R.S. Carotenoids in human blood and tissues. *J. Nutr.* **1989**, *119*, 101–104. [CrossRef]
10. Takaichi, S. Carotenoids in algae: Distributions, biosyntheses and functions. *Mar. Drugs* **2011**, *9*, 1101–1118. [CrossRef]
11. Sugawara, T.; Ganesan, P.; Li, Z.; Manabe, Y.; Hirata, T. Siphonaxanthin, a green algal carotenoid, as a novel functional compound. *Mar. Drugs* **2014**, *12*, 3660–3668. [CrossRef] [PubMed]
12. Ganesan, P.; Matsubara, K.; Sugawara, T.; Hirata, T. Marine algal carotenoids inhibit angiogenesis by down-regulating FGF-2-mediated intracellular signals in vascular endothelial cells. *Mol. Cell. Biochem.* **2013**, *380*, 1–9. [CrossRef] [PubMed]
13. Ganesan, P.; Noda, K.; Manabe, Y.; Ohkubo, T.; Tanaka, Y.; Maoka, T.; Sugawara, T.; Hirata, T. Siphonaxanthin, a marine carotenoid from green algae, effectively induces apoptosis in human leukemia (HL-60) cells. *Biochim. Biophys. Acta* **2011**, *1810*, 497–503. [CrossRef] [PubMed]
14. Manabe, Y.; Takii, Y.; Sugawara, T. Siphonaxanthin, a carotenoid from green algae, suppresses advanced glycation end product-induced inflammatory responses. *J. Nat. Med.* **2020**, *74*, 127–134. [CrossRef]
15. Manabe, Y.; Hirata, T.; Sugawara, T. Inhibitory effect of carotenoids on ligand-induced lipid raft translocation of immunoreceptors. *J. Oleo Sci.* **2019**, *68*, 149–158. [CrossRef] [PubMed]
16. Li, Z.S.; Noda, K.; Fujita, E.; Manabe, Y.; Hirata, T.; Sugawara, T. The green algal carotenoid siphonaxanthin inhibits adipogenesis in 3T3-L1 preadipocytes and the accumulation of lipids in white adipose tissue of KK-Ay mice. *J. Nutr.* **2015**, *145*, 490–498. [CrossRef]

17. Zheng, J.; Li, Z.; Manabe, Y.; Kim, M.; Goto, T.; Kawada, T.; Sugawara, T. Siphonaxanthin, a carotenoid from green algae, inhibits lipogenesis in hepatocytes via the suppression of liver X receptor alpha activity. *Lipids* **2018**, *53*, 41–52. [CrossRef]
18. Li, Z.S.; Zheng, J.W.; Manabe, Y.; Hirata, T.; Sugawara, T. Anti-obesity properties of the dietary green alga, Codium cylindricum, in high-fat diet-Induced obese mice. *J. Nutr. Sci. Vitaminol. (Tokyo)* **2018**, *64*, 347–356. [CrossRef]
19. Zheng, J.; Manabe, Y.; Sugawara, T. Siphonaxanthin, a carotenoid from green algae Codium cylindricum, protects Ob/Ob mice fed on a high-fat diet against lipotoxicity by ameliorating somatic stresses and restoring anti-oxidative capacity. *Nutr. Res.* **2020**, *77*, 29–42. [CrossRef]
20. Biganzoli, E.; Cavenaghi, L.A.; Rossi, R.; Brunati, M.C.; Nolli, M.L. Use of a Caco-2 cell culture model for the characterization of intestinal absorption of antibiotics. *Farmaco* **1999**, *54*, 594–599. [CrossRef]
21. Kim, J.E.; Nam, J.H.; Cho, J.Y.; Kim, K.S.; Hwang, D.Y. Annual tendency of research papers used ICR mice as experimental animals in biomedical research fields. *Lab. Anim. Res.* **2017**, *33*, 171–174. [CrossRef]
22. Yonekura, L.; Kobayashi, M.; Terasaki, M.; Nagao, A. Keto-carotenoids are the major metabolites of dietary lutein and fucoxanthin in mouse tissues. *J. Nutr.* **2010**, *140*, 1824–1831. [CrossRef] [PubMed]
23. Etoh, H.; Utsunomiya, Y.; Komori, A.; Murakami, Y.; Oshima, S.; Inakuma, T. Carotenoids in human blood plasma after ingesting paprika juice. *Biosci. Biotechnol. Biochem.* **2000**, *64*, 1096–1098. [CrossRef]
24. Asai, A.; Sugawara, T.; Ono, H.; Nagao, A. Biotransformation of fucoxanthinol into amarouciaxanthin A in mice and HepG2 cells: Formation and cytotoxicity of fucoxanthin metabolites. *Drug Metab. Dispos.* **2004**, *32*, 205–211. [CrossRef] [PubMed]
25. Nagao, A.; Maoka, T.; Ono, H.; Kotake-Nara, E.; Kobayashi, M.; Tomita, M. A 3-hydroxy β-end group in xanthophylls is preferentially oxidized to a 3-oxo ε-end group in mammals. *J. Lipid Res.* **2015**, *56*, 449–462. [CrossRef] [PubMed]
26. Khachik, F.; de Moura, F.F.; Chew, E.Y.; Douglass, L.W.; Ferris, F.L.; Kim, J.; Thompson, D.J. The effect of lutein and zeaxanthin supplementation on metabolites of these carotenoids in the serum of persons aged 60 or older. *Invest. Ophthalmol. Vis. Sci.* **2006**, *47*, 5234–5242. [CrossRef]
27. Sugawara, T.; Yamashita, K.; Asai, A.; Nagao, A.; Shiraishi, T.; Imai, I.; Hirata, T. Esterification of xanthophylls by human intestinal Caco-2 cells. *Arch. Biochem. Biophys.* **2009**, *483*, 205–212. [CrossRef]
28. Nagao, A. Bioavailability of dietary carotenoids: Intestinal absorption and metabolism. *Jpn. Agric. Res. Q.* **2014**, *48*, 385–391. [CrossRef]
29. Khachik, F. Distribution and metabolism of dietary carotenoids in humans as a criterion for development of nutritional supplements. *Pure Appl. Chem.* **2006**, *78*, 1551–1557. [CrossRef]
30. Hashimoto, T.; Ozaki, Y.; Taminato, M.; Das, S.K.; Mizuno, M.; Yoshimura, K.; Maoka, T.; Kanazawa, K. The distribution and accumulation of fucoxanthin and its metabolites after oral administration in mice. *Br. J. Nutr.* **2009**, *102*, 242–248. [CrossRef]
31. Yamanushi, T.; Igarashi, O. The mobilization and tissue distribution of beta-carotene in the rat by the venous injection method. *J. Nutr. Sci. Vitaminol. (Tokyo)* **1995**, *41*, 169–177. [CrossRef] [PubMed]
32. Schweigert, F.; Rosival, I.; Rambeck, W.; Gropp, J. Plasma transport and tissue distribution of [14C] beta-carotene and [3H] retinol administered orally to pigs. *Int. J. Vitam. Nutr. Res.* **1995**, *65*, 95–100.
33. Wang, X.D. Lycopene metabolism and its biological significance. *Am. J. Clin. Nutr.* **2012**, *96*, 1214S–1222S. [CrossRef] [PubMed]
34. Tang, G.; Blanco, M.C.; Fox, J.G.; Russell, R.M. Supplementing ferrets with canthaxanthin affects the tissue distributions of canthaxanthin, other carotenoids, vitamin A and vitamin E. *J. Nutr.* **1995**, *125*, 1945–1951. [CrossRef]
35. Van Lieshout, E.M.; Peters, W.H.; JJansen, J.B. Effect of oltipraz, alpha-tocopherol, beta-carotene and phenethylisothiocyanate on rat oesophageal, gastric, colonic and hepatic glutathione, glutathione S-transferase and peroxidase. *Carcinogenesis* **1996**, *17*, 439–1445. [CrossRef] [PubMed]
36. Sugawara, T.; Kushiro, M.; Zhang, H.; Nara, E.; Ono, H.; Nagao, A. Lysophosphatidylcholine enhances carotenoid uptake from mixed micelles by Caco-2 human intestinal cells. *J. Nutr.* **2001**, *131*, 2921–2927. [CrossRef]

*marine drugs*

MDPI

*Article*

# Anti-Oxidative Activity of Mytiloxanthin, a Metabolite of Fucoxanthin in Shellfish and Tunicates

**Takashi Maoka [1,*], Azusa Nishino [2], Hiroyuki Yasui [3], Yumiko Yamano [4] and Akimori Wada [4]**

1 Research Institute for Production Development, 15 Shimogamo, Morimoto Cho, Sakyoku, Kyoto 606-0805, Japan

2 Institute of Health Sciences, Ezaki Glico Co., Ltd., 4-6-5 Utajima, Nishiyodogawa-ku, Osaka 555-8502, Japan; nishino-azusa@gf.glico.co.jp

3 Department of Analytical and Bioinorganic Chemistry, Division of Analytical and Physical Chemistry, Kyoto Pharmaceutical University, 5 Nakauchi-cho, Misasagi, Yamashina-ku, Kyoto 607-8414, Japan; yasui@mb.kyoto-phu.ac.jp

4 Department of Organic Chemistry for Life Science, Kobe Pharmaceutical University, Motoyamakita-machi, Higashinada-ku, Kobe 658-8558, Japan; y-yamano@kobepharma-u.ac.jp (Y.Y.); a-wada@kobepharma-u.ac.jp (A.W.)

* Correspondence: maoka@mbox.kyoto-inet.or.jp; Tel.: +81-75-781-1107; Fax: +81-75-791-7659

Academic Editor: Peer Jacobson

Received: 8 April 2016; Accepted: 4 May 2016; Published: 11 May 2016

**Abstract:** Anti-oxidative activities of mytiloxanthin, a metabolite of fucoxanthin in shellfish and tunicates, were investigated. Mytiloxanthin showed almost the same activities for quenching singlet oxygen and the inhibition of lipid peroxidation as those of astaxanthin, which is a well-known singlet oxygen quencher. Furthermore, mytiloxanthin showed excellent scavenging activity for hydroxyl radicals and this activity was markedly higher than that of astaxanthin.

**Keywords:** mytiloxanthin; anti-oxidative activity; singlet oxygen; hydroxyl radical; lipid peroxidation

---

## 1. Introduction

Mytiloxanthin (**1**) is a carotenoid possessing a unique cyclopentyl enolic β-diketone group (Figure 1). It is distributed in several marine invertebrates such as shellfish and tunicates [1,2]. It was first isolated from the sea mussel *Mytilus californianus* by Sheer [3]. Its structure was determined to be 3,3′,8′-trihydroxy-7,8-didehydro-β,κ-caroten-6′-one through chemical and spectroscopic studies by Khare *et al.*, in 1973 [4]. Subsequently, Chopra *et al.* synthesized 9*Z*-(3*R*,3′*S*,5′*R*)-mytiloxanthin [5]. Its absolute configuration was determined to be (3*R*,3′*S*,5′*R*) using a modified Mosher's method by Maoka and Fujiwara [6]. Total synthesis of all-*E*-(3*R*,3′*S*,5′*R*)-mytiloxanthin was achieved for the first time by Tode *et al.* [7], and it was recently improved by Yamano *et al.* [8].

It was reported that mytiloxanthin was converted from fucoxanthin (**2**) through a pinacol-like rearrangement [1,2,4]. Namely, dietal fucoxanthin (**2**) from diatoms was a metabolite for mytiloxanthin (**1**) via fucoxanthinol and halocynthiaxanthin in shellfish and tunicates, as shown in Figure 2 [1,2,9,10].

It is well known that marine carotenoids such as astaxanthin show excellent anti-oxidative activity [11,12]. However, because of limited availability from natural sources, the anti-oxidative activity of mytiloxanthin has not yet to be reported. Therefore, we synthesized mytiloxanthin and studied its quenching effect on singlet oxygen, its scavenging effect on hydroxyl radicals and its inhibitory effect on lipid peroxidation of mytiloxanthin. In the present paper, we present these experimental results.

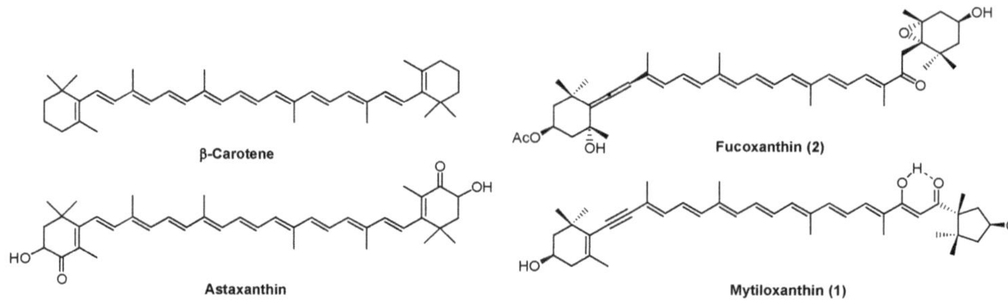


**Figure 1.** Structure of carotenoids used in this study.

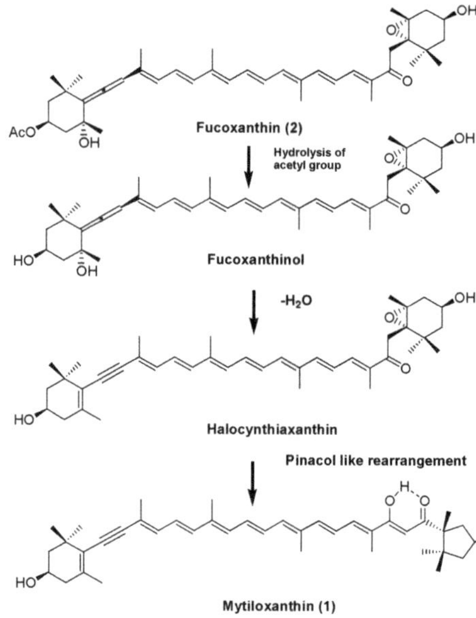


**Figure 2.** Metabolic conversion of fucoxanthin to mytiloxanthin in shellfish and tunicates.

## 2. Results and Discussion

In order to investigate the anti-oxidative activities of mytiloxanthin, it was synthesized using a recently reported method [8], as described in the Section 3 [8]. Furthermore, β-carotene, astaxanthin, and fucoxanthin were used as positive controls of the anti-oxidative activity.

Figure 3A,B show the scavenging effect (% of control group) on singlet oxygen and hydroxyl radicals, respectively, by carotenoids. Mytiloxanthin showed almost the same quenching activity (61.6%) for singlet oxygen as that of astaxanthin (61.0%), which is a well-known and excellent singlet oxygen quencher [11,12]. On the other hand, fucoxanthin, a precursor of mytiloxanthin, hardly showed singlet oxygen-quenching activity (99.5%) in this experimental system. It was reported that the singlet oxygen-quenching activity of carotenoids depends on the number of conjugated double bonds, polyene chain structures, and functional groups [12,13]. Fucoxanthin contains nine conjugated double bonds including one carbonyl group and one allenic group in its molecule. On the other hand, mytiloxanthin has an 11-conjugated-double-bond polyene system, including one acetylenic and one carbonyl group in its molecule. Therefore, it was suggested that the strong quenching activity for singlet oxygen shown by mytiloxanthin was due to this long conjugated polyene system.

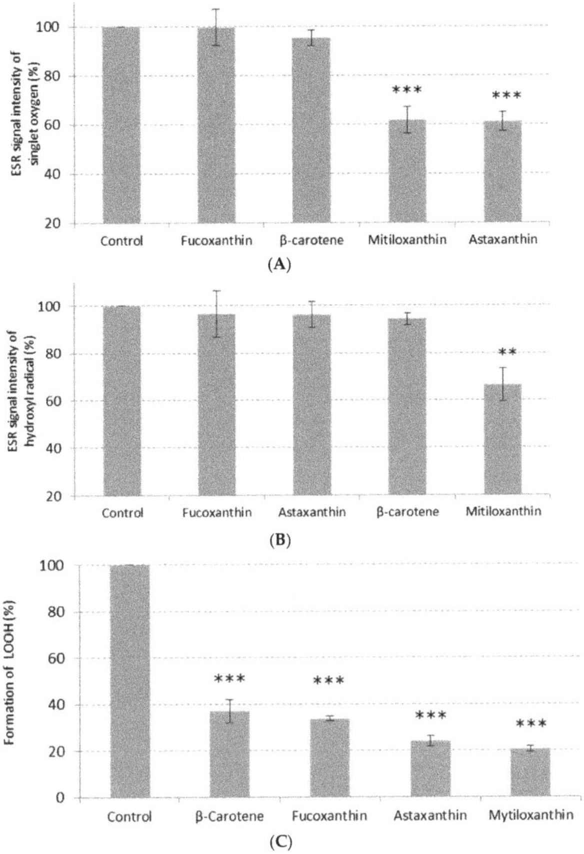


**Figure 3.** Scavenging activities (% of control group) for singlet oxygen (**A**) and hydroxyl radicals (**B**) and inhibitory effect (% of control group) on the lipid peroxidation (**C**) of carotenoids. Significance compared with the control group: * $p < 0.05$, ** $p < 0.01$, *** $p < 0.01$.

It has been consistent that carotenoids do not directly scavenge superoxide anions or hydroxyl radicals [14]. Recently, Hama *et al.* reported that astaxanthin could scavenge hydroxyl radicals in a liposome system [15]. In the present study, we also found that mytiloxanthin could scavenge hydroxyl radicals (66.4%) and that this activity was markedly higher than that of astaxanthin (96.1%) in this experimental system. Fucoxanthin hardly showed scavenging activity for hydroxyl radicals as shown in Figure 3B.

Inhibitory activities of mytiloxanthin and fucoxanthin on lipid peroxidation were monitored by measuring the accumulation of methyl linolate hydroperoxides during the incubation of methyl linolate with 2,2′-Azobis(2,4-dimethylvaleronitrile) (AMVN) as a radical initiator, with astaxanthin and β-carotene as positive controls. Figure 3C shows the results of inhibitory activity on lipid peroxidation by carotenoids at a concentration of 2 mM (final concentration of 167 μM). Mytiloxanthin showed slightly stronger activity than astaxanthin, which is a well-known antioxidant [11,12], and it also showed higher activity than fucoxanthin and β-carotene. Several investigators have reported that an increased number of conjugated double bonds in carotenoids, and the presence of functional groups such as carbonyl and hydroxyl groups in carotenoids, enhance their anti-oxidant effects [11–13]. Therefore, it was suggested that the strong inhibitory activity on lipid peroxidation shown by mytiloxanthin was due to the presence of the long conjugated polyene system described above. Mytiloxanthin has an enolic β-diketone group. Along with the 3-hydroxy-4-keto-β-end group in astaxanthin, this enolic β-diketone group may contribute to anti-oxidative activity.

Many crustaceans oxidatively convert dietary β-carotene to astaxanthin [1,2]. The scallop and sea angel also oxidatively convert dietary diatoxanthin to pectenolone [2,16]. By these oxidative metabolic conversions, the anti-oxidative activities of dietary carotenoids are increased. Therefore, these marine animals metabolize dietary carotenoids to a more active anti-oxidative form and accumulate them in their bodies and gonads.

Similarly, shellfish and tunicates accumulate fucoxanthin from dietary algae and convert it to mytiloxanthin. By this conversion, the carotenoid changes color from orange to red and shows increased anti-oxidative activities, as described above. Mytiloxanthin in shellfish and tunicates may contribute to protection against oxidative stress and promote reproduction, similarly to astaxanthin in crustaceans and pectenolonein in the scallop and sea angel.

## 3. Experimental Section

### 3.1. Reagents

Hematoporphyrin, riboflavin, and hydrogen peroxide were purchased from Wako Pure Chemicals (Osaka, Japan); 2,2,6,6-Tetramethyl-4-piperidone (TMPD) was purchased from Aldrich (Milwaukee, WI, USA); 5,5-Dimethyl-1-pyrroline-*N*-oxide (DMPO) was purchased from Labotec (Tokyo, Japan); 2,2′-Azobis(2,4-dimethylvaleronitrile) (AMVN) as purchased from Wako Pure Chemicals (Osaka, Japan). Methyl linolate, β-carotene, and astaxanthin were purchased from Aldrich (Milwaukee, WI, USA). Fucoxanthin was prepared from brown algae. Mytiloxanthin was synthesized as described below. Figure 1 shows structures of carotenoids used in this study.

### 3.2. ESR Spin-Trapping Analysis

ESR spectra were recorded at room temperature on a JEOL JES-FR30 spectrometer (JEOL, Tokyo, Japan) using an aqueous quartz flat cell (Labotec, Tokyo, Japan). TMPD was used as a singlet oxygen-trapping agent, and DMPO was used as superoxide anion and hydroxyl radicals, respectively. Superoxide anion radical ($O_2^-$) and OH were generated by addition of both the 100 μL of 25 μM of riboflavin, or 100 μL of 8 mM $H_2O_2$ solution and 10 μL of 250 mM DMPO to 100 μL of 8.8 μg/mL carotenoid $CH_3CN$ solution by UV-A irradiation. In a similar manner described above, $^1O_2$ was generated by the addition of both the 100 μL of 0.25 mM hematoporphyrin and 10 μL of 500 mM TMPD to 100 μL of 8.8 μg/mL of carotenoid $CH_3CN$ solution by UV-A irradiation. ESR spectra were started simultaneously to measure after UV-A irradiation. The all spin-trapped ESR spectra were monitored between the third and fourth signals from the low magnetic field due to the external standard, Mn(II)-doped MnO.

### 3.3. Inhibition of Lipid Peroxidation

Carotenoids were dissolved in EtOH at a concentration of 2 mM (final concentration of 167 μM in the reaction mixture). The sample solution, 100 μL, was added to 1 mL of 100 mM methyl linolate solution [*n*-hexane/2-propanol (1:1, *v*/*v*)], and the solution was incubated at 37 °C for 5 min. As a control, EtOH alone was used instead of the sample solution. The oxidation reaction was then performed by adding 100 μL of 100 mM n-hexane solution of AMVN and the mixture was incubated with air at 37°C. At regular intervals, the oxidation reaction products, methyl linolate hydroperoxides, were quantified by high performance liquid chromatography (HPLC). HPLC was performed with a Hitachi L-6000 intelligent pump and an L-4250 UV-VIS detector. The following HPLC conditions were employed for the quantitative analysis of methyl linolate hydroperoxides: column Lichrosorb Si 100 (5 μm particle size) (4.6 × 250 mm) (Merck, Damstraat, Germany); solvent system: 2-propanol/n-hexane (1:99, *v*/*v*); flow rate: 1 mL/min; and detection: 235 nm.

### 3.4. Synthesis of Mytiloxanthin

Mytiloxanthin (**1**) was synthesized by a recently reported method [8] as shown in Scheme 1. The anti(α)-epoxy alcohol **4**, stereoselectively prepared from (–)-actinol (**3**) [17,18], was treated with $BF_3 \cdot OEt_2$ to provide the cyclopentyl ketone **5** in 60% yield, by opening of C-6-oxygen bond of the oxirane ring and subsequent ring contraction. The yield of **5** from **4** was improved by a stepwise route through the di-*tert*-butyldimethylsilyl (TBS) ether **6** (86% for 3 steps). The compound **5** was converted into the methyl ketone **9** in four steps and this was condensed with the separately prepared conjugated ester **10** to give the desired β-diketone **11**. After hydrolysis of acetal moiety of **11** and subsequent desilylation, the resulting apocarotenal **12** [5] was condensed with the acetylenic phosphonium salt **13** [19] and then desilylated to preferentially provide all-*E*-mytiloxanthin (**1**) in a good yield. The total yield of **1** from epoxy alcohol **4** was 30% over 13 steps (21% from (–)-actinol over 18 steps).

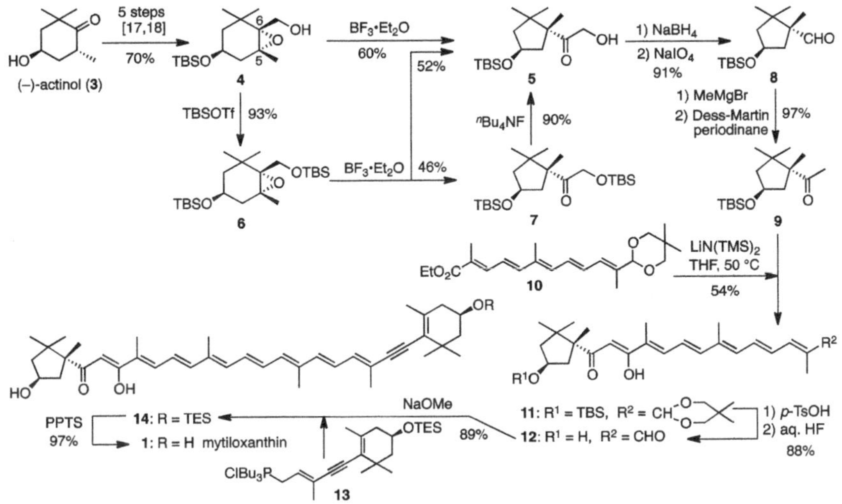


**Scheme 1.** Synthesis of mytiloxanthin (**1**).

## 4. Conclusions

The anti-oxidative activities of mytiloxanthin, a metabolite of fucoxanthin in shellfish and tunicates, were investigated. Mytiloxanthin showed excellent anti-oxidative activities for quenching singlet oxygen, scavenging hydroxyl radicals, and inhibiting lipid peroxidation. These activities were higher than those of fucoxanthin as a precursor of mytiloxanthin. Therefore, it was suggested that marine animals accumulate dietary fucoxanthin and convert it to a more anti-oxidative active form, mytiloxanthin.

**Author Contributions:** Basic idea of the research was proposed by all authors collaboratively. Synthesis of mytiloxanthin was performed by Y. Yamano and A. Wada. Anti-oxidative activities of carotenoids were studied by T. Maoka, A. Nishino, and H. Yasui.

**Conflicts of Interest:** The authors declare no conflict interest.

## References

1. Liaaen-Jensen, S. Carotenoids in Food Chain. In *Carotenoids Volume 3: Biosynthesis and Metabolism*; Britton, G., Liaaen-Jensen, S., Pfander, H., Eds.; Birkhäuser: Basel, Switzerland, 1998; pp. 359–371.
2. Maoka, T. Carotenoids in marine animals. *Mar. Drugs* **2011**, *9*, 278–293. [CrossRef] [PubMed]

3. Scheer, J.B.T. Some features of the metabolism of the carotenoid pigments in the California sea mussel (*Mytilis californicus*). *J. Biol. Chem.* **1940**, *136*, 275–299.
4. Khare, A.; Moss, G.P.; Weedon, B.C.L. Mytiloxanthin and isomytiloxanthin, two novel acetylenic carotenoids. *Tetrahedron Lett.* **1973**, 3921–3924. [CrossRef]
5. Chopra, A.K.; Khare, A.; Moss, G.P.; Weedon, B.C.L. Carotenoids and related compounds. Part 41. Structure of mytiloxanthin and synthesis of a cis isomer. *J. Chem. Soc. Perkin Trans. 1* **1988**, 1383–1388. [CrossRef]
6. Maoka, T.; Fujiwara, Y. Absolute configuration of mytiloxanthin and 9-*E*-mytiloxanthin. *J. Jpn. Oil Chem. Soc.* **1996**, *45*, 667–670. [CrossRef]
7. Tode, C.; Yamano, Y.; Ito, M. Carotenoids and related polyenes, Part 8. Total synthesis of optically active mytiloxanthin applying the stereoselective rearrangements of tetrasubstituted epoxide. *J. Chem. Soc. Perkin Trans. 1* **2002**, 1581–1587. [CrossRef]
8. Yamano, Y.; Ohata, M.; Wada, A. Stereocontrolled total synthesis of mytiloxanthin. *Carotenoid Sci.* **2015**, *20*, 35–39.
9. Partali, V.; Tangen, Y.K.; Liaaen-Jensen, S. Carotenoids in food chain studies-III. Reapportion and metabolic transformation of carotenoids in *Mytilus. edulis* (edible mussel). *Comp. Biochem. Physiol.* **1989**, *92B*, 239–246.
10. Matsuno, T. New structures of carotenoids in marine animals. *Pure Apple. Chem.* **1985**, *57*, 659–666. [CrossRef]
11. Miki, W. Biological functions and activities of animal carotenoids. *Pure Appl. Chem.* **1991**, *63*, 141–146. [CrossRef]
12. Shimidzu, N.; Goto, M.; Miki, W. Carotenoids as singlet oxygen quenchers in marine organism. *Fish. Sci.* **1996**, *62*, 134–137.
13. Hirayama, O.; Nakamura, K.; Hamada, S.; Kobayashi, K. Singlet oxygen quenching ability of naturally occurring carotenoids. *Lipids* **1994**, *29*, 149–150. [CrossRef] [PubMed]
14. Trevithick-Sutton, C.C.; Foote, C.S.; Collinus, M.; Trevithick, J.R. The retinal carotenoids zeaxanthin and lutein scavenge superoxide and hydroxy radicals: A. chemiluminesence and ESR study. *Mol. Vis.* **2006**, *12*, 1127–1135. [PubMed]
15. Hama, S.; Takahashi, K.; Inai, Y.; Shirota, K.; Sakamato, R.; Yamada, A.; Tsuchidya, H.; Kanamura, K.; Yamashita, E.; Kogure, K. Protective effects of topical application of apoorly soluble antioxidant astaxanthin liposomat formation on ultraviolet-induced skin damage. *J. Pharm Sci.* **2012**, *101*, 2909–2916. [CrossRef] [PubMed]
16. Maoka, T.; Kuwahara, T.; Narita, M. Carotenoids of sea angels *Clione limacina* and *Paedoclione doliiformis*, from the perspective of food chain. *Mar. Drugs* **2014**, *12*, 1460–1470. [CrossRef] [PubMed]
17. Yamano, Y.; Tode, C.; Ito, M. Carotenoids and related polyenes, Part 5. Lewis acid-promoted stereoselective rearrangement of 5,6-epoxy carotenoid model compounds. *J. Chem. Soc. Perkin Trans. 1* **1998**, 2569–2581. [CrossRef]
18. Furuichi, N.; Hara, H.; Osaki, T.; Nakano, M.; Mori, H.; Katsumura, S. Stereocontrolled total synthesis of a polyfunctional carotenoid, peridinin. *J. Org. Chem.* **2004**, *69*, 7949–7959. [CrossRef] [PubMed]
19. Yamano, Y.; Chary, V.M.; Wada, A. Stereoselective total synthesis of the acetylenic carotenoids alloxanthin and triophaxanthin. *Org. Biomol. Chem.* **2012**, *10*, 4103–4108. [CrossRef] [PubMed]


**Samples Availability:** Available from the authors.

*marine drugs*

*Article*

# Synthesis of (3*S*,3′*S*)- and *meso*-Stereoisomers of Alloxanthin and Determination of Absolute Configuration of Alloxanthin Isolated from Aquatic Animals

**Yumiko Yamano [1,*], Takashi Maoka [2] and Akimori Wada [1]**

[1] Kobe Pharmaceutical University, Motoyamakita-machi, Higashinada-ku, Kobe 658-8558, Japan;
E-Mail: a-wada@kobepharma-u.ac.jp

[2] Research Institute for Production Development, 15 Shimogamo-morimoto-cho, Sakyo-ku,
Kyoto 606-0805, Japan; E-Mail: maoka@mbox.kyoto-inet.or.jp

* Author to whom correspondence should be addressed; E-Mail: y-yamano@kobepharma-u.ac.jp;
Tel./Fax.: +81-78-441-7562.

*Received: 20 March 2014; in revised form: 15 April 2014 / Accepted: 15 April 2014 / Published: 8 May 2014*

**Abstract:** In order to determine the absolute configuration of naturally occurring alloxanthin, a HPLC analytical method for three stereoisomers **1a–c** was established by using a chiral column. Two authentic samples, (3*S*,3′*S*)- and *meso*-stereoisomers **1b** and **1c**, were chemically synthesized according to the method previously developed for (3*R*,3′*R*)-alloxanthin (**1a**). Application of this method to various alloxanthin specimens of aquatic animals demonstrated that those isolated from shellfishes, tunicates, and crucian carp are identical with (3*R*,3′*R*)-stereoisomer **1a**, and unexpectedly those from lake shrimp, catfish, biwa goby, and biwa trout are mixtures of three stereoisomers of **1a–c**.

**Keywords:** carotenoid; alloxanthin; synthesis; chiral HPLC separation; absolute configuration

## 1. Introduction

Alloxanthin (**1**) (Figure 1) was first isolated from *Cryptomonas* algae [1] and its structure was determined to be 7,8,7′,8′-tetreradehydro-β,β-carotene-3,3′-diol by MS, IR and [1]H-NMR spectroscopies [2]. Additionally, cynthiaxanthin [3] from the tunicate *Cynthia rorezi* (*Halocynthia rorezi*) and pectenoxanthin [4] from giant scallop *Pecten maximus* were isolated by Japanese scientists. In 1967, Campbel *et al.* demonstrated that these two carotenoids were identical with alloxanthin [5]. Therefore, cynthiaxanthin and pectenoxanthin were synonyms of alloxanthin. The absolute configuration of alloxanthin isolated form algae was deduced to be 3*R*,3′*R* by X-ray analysis of degradation product of fucoxanthin and in view of biogenetic grounds [6]. Bartlett *et al.* reported that the ORD spectra of alloxanthin specimens from *Cryptomonas* algae and tunicate showed an identical shape each other and that both specimens are assumed to have an identical absolute configuration [7].

**Figure 1.** Structures of stereoisomers of alloxanthin (**1a–c**) and other related carotenoids.

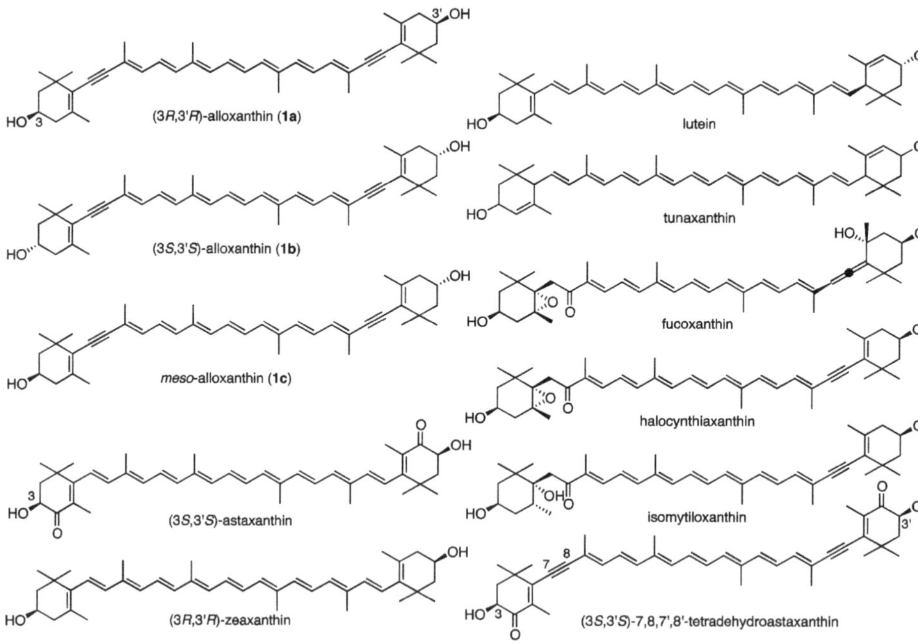


Since then, alloxanthin was isolated from several aquatic animals, such as shellfishes [8,9], starfishes [10], tunicates [11,12] and freshwater fishes [13,14], *etc.* These alloxanthin specimens showed similar non-conservative CD with weak negative Cotton effects.

Carotenoids such as astaxanthin, zeaxanthin, lutein, and tunaxanthin in animals are known to exist as a mixture of stereoisomers. Namely, astaxanthin in crustaceans and marine fishes exists as a mixture of three stereoisomers at C3 and C3′-positions [15,16]. Zeaxanthin [17], lutein [18], and
tunaxanthin [19] in marine fishes also consist of these stereoisomers. Their absolute configurations were determined by CD spectra and chiral HPLC analyses. Due to its non-conservative CD, absolute configurations of alloxanthin in several origins could not be determined exactly by CD spectra.

In order to determine the absolute configuration of naturally occurring alloxanthin, we synthesized stereoisomers of alloxanthin (**1a**–**c**) and established a HPLC analytical method using a chiral column. Applying this method, the absolute configurations of alloxanthin specimens isolated from shellfishes, tunicates and fishes were investigated. Here, we describe these results.

## 2. Results and Discussion

### 2.1. *Synthesis of (3S,3′S)-Alloxanthin (**1b**) and meso-Alloxanthin (**1c**)*

We previously reported [20] stereoselective total synthesis of (3*R*,3′*R*)-alloxanthin (**1a**) by use of $C_{15}$-acetylenic tri-*n*-butylphosphonium salt **5a** (Scheme 1) as a versatile synthon for syntheses of acetylenic carotenoids. This time, (3*S*,3′*S*)-alloxanthin (**1b**) and its *meso*-stereoisomer **1c** were newly synthesized using (3*S*)-phosphonium salt **5b**, which was prepared from 3-epi-actinol **6** [21] in the same procedure [20] as preparation of (3*R*)-one **5a**.

**Scheme 1.** Synthesis of $C_{15}$-acetylenic tri-*n*-butylphosphonium salts **5a** and **5b**.

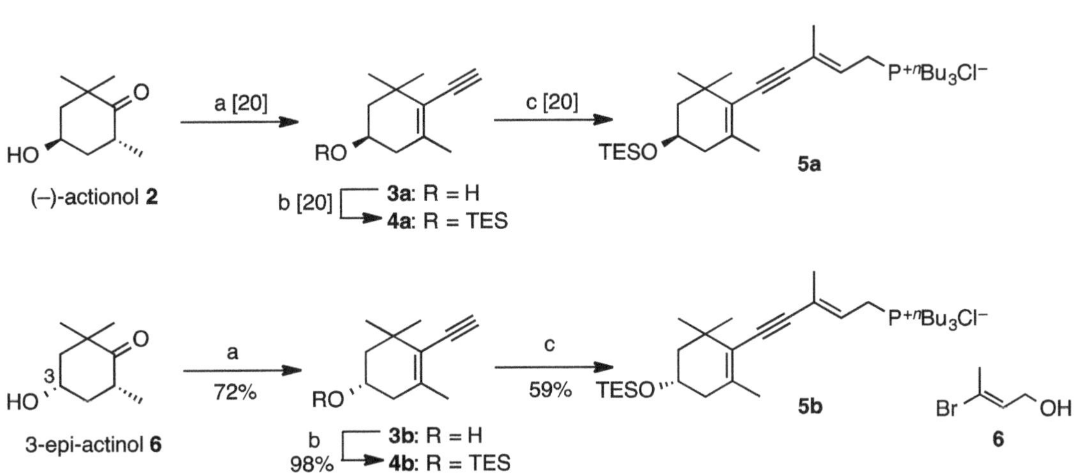


Reagents: (**a**) (i) TMSCl, $Et_3N$, DMAP, (ii) TMSC≡CH, $^n$BuLi, then aq. KOH, (iii) $Ac_2O$, pyridine, (iv) $CuSO_4$, xylene, reflux (Dean-Stark), (v) $LiAlH_4$; (**b**) TESCl, $Et_3N$, DMAP, (**c**) (i) vinyl bromide **6**, $Pd(PPh_3)_4$, CuI, BHT, $^iPr_2NH$, (ii) MsCl, LiCl, γ-collidine, (iii) $P^nBu_3$, $Et_3N$.

Compound **6** was converted into terminal alkyne **3b** via the addition of lithium acetylide in 72% yield over six steps. The high enantiomeric purity of **3b** (99% ee) was confirmed by HPLC analysis [CHIRALPAK AY-H; Daicel, 2-PrOH–*n*-hexane (5:95)]. Compound **3b** was then transformed into the phoshonium salt **5b** via Sonogashira cross-coupling of the triethylsilyl (TES)-protected terminal alkyne **4b** with vinylbromide **6** in 59% over four steps.

Wittig condensation of $C_{10}$-dialdehyde **7** with excess amount of (3*S*)-phosphonium salt **5b** in the presence of sodium methoxide in dichloromethane at room temperature and subsequent desilylation stereoselectively provided (3*S*,3′*S*)-alloxanthin (**1b**) (Scheme 2). On the other hand, *meso*-alloxanthin (**1c**) was synthesized via condensation between (3*S*)-phosphonium salt **5b** and (3*R*)-$C_{25}$-acetylenic apocarotenal **8**, which was prepared by Wittig reaction of $C_{10}$-dialdehyde **7** with (3*R*)-phosphonium salt **5a** in the presence of sodium methoxide in dichloromethane at 0 °C.

**Scheme 2.** Synthesis of three stereoisomers of alloxanthin (**1a**–**c**).

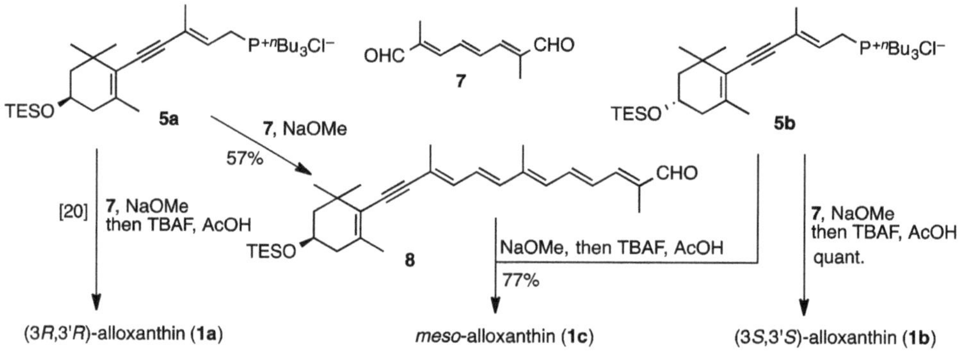


CD spectrum of (3*S*,3′*S*)-alloxanthin (**1b**) showed an antisymetrical curve having week Cotton effects to that of previously synthesized [20] (3*R*,3′*R*)-alloxanthin (**1a**) as shown in Figure 2.

**Figure 2.** CD spectra in $Et_2O$–isopentane–EtOH (5:5:2) of synthesized (3*R*,3′*R*)-alloxanthin (**1a**) and (3*S*,3′*S*)-alloxanthin (**1b**).

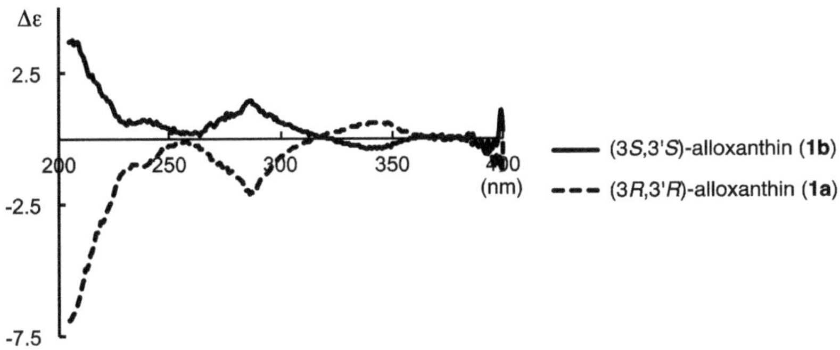


## 2.2. Determination of Absolute Configuration of Alloxanthin Isolated from Aquatic Animals by HPLC

In order to determine the absolute configuration of naturally occurring alloxanthin, a HPLC analytical method for three stereoisomers **1a**–**c** was investigated. As a result, three synthetic stereoisomers of alloxanthin can be separated using a chiral column (CHIRALPAK AD-H; Daicel) as shown in Figure 3.

Next, alloxanthin specimens isolated from scallop *Mizuhopecten yessoensis*, oyster *Crassostrea gigas*, pacific pearl oyster *Pinctada margaritifera*, freshwater bivalve *Unio douglasiae*, tunicate *Halocynthia roretzi*, and crucian carp *Carassius auratus grandoculis* were subjected to the HPLC method to find that these consist of only (3*R*,3′*R*)-stereoisomer **1a**. On the other hand, alloxanthin specimens isolated from lake shrimp *Palaemon paucidens*, catfish *Silurus asotus*, biwa goby *Gymnogobius isaza*, and biwa trout *Oncorhynchus masou rhodurus* consisted of three stereoisomers **1a**–**c** (Table 1).

**Figure 3.** HPLC elution profile of a mixture of three stereoisomers of alloxanthin (**1a**–**c**).

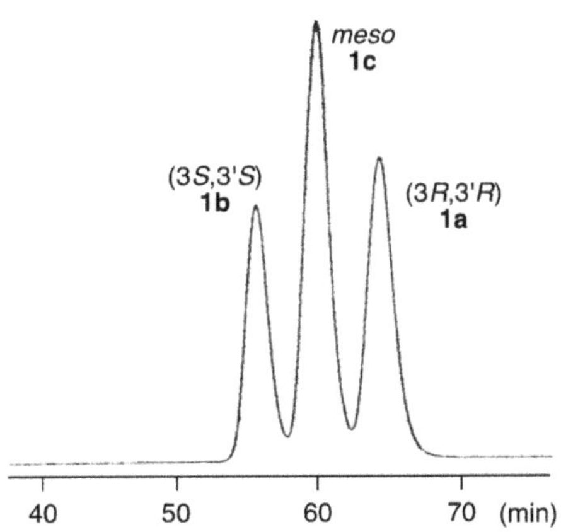

Column: CHIRALPAK AD-H 0.46 × 25 cm (Daicel, Tokyo, Japan); eluent: 2-PrOH–*n*-hexane (4:96); flow rate: 0.6 mL/min; temperature: 23 °C; detection: 450 nm.

**Table 1.** Occurrence and percentage composition of alloxanthin stereoisomers in aquatic animals.

| | Species | 3*R*,3′*R* **1a** | 3*S*,3′*S* **1b** | *meso* **1c** |
|---|---|---|---|---|
| Shellfish | | | | |
| Scallop | *Mizuhopecten yessoensis* | 100 | n.d. | n.d. |
| Oyster | *Crassostrea gigas* | 100 | n.d. | n.d. |
| Pacific pearl oyster | *Pinctada margaritifera* | 100 | n.d. | n.d. |
| Freshwater bivalves | *Unio douglasiae* | 100 | n.d. | n.d. |
| Tunicate | | | | |
| Sea squirt | *Halocynthia roretzi* | 100 | n.d. | n.d. |
| Crustacean | | | | |
| Lake shrimp | *Palaemon paucidens* | 53.7 | 9.6 | 36.7 |
| Fish | | | | |
| Crucian carp | *Carassius auratus grandoculis* | 100 | n.d. | n.d. |
| Biwa goby | *Gymnogobius isaza* | 91.4 | 0.9 | 7.7 |
| Biwa trout | *Oncorhynchus masou rhodurus* | >99.9 | trace | trace |
| Catfish | *Silurus asotus* | 82.9 | 1.5 | 15.6 |

n.d.: not detected.

Previously, one of the authors reported that zeaxanthin in plants, shellfishes, and tunicates consisted of only (3*R*,3′*R*)-stereoisomer, whereas zeaxanthin in fishes consisted of three stereoisomers [17]. Similar results were obtained in the case of alloxanthin in aquatic animals. Alloxanthin is *de novo* synthesized in *Chryptophyceae* and *Euglenophyceae* micro algae [22]. However, origin of alloxanthin in aquatic animals was remained uncertain. Patrali *et al.* (1989) [22] and Liaaen-Jensen (1998) [23] reported that alloxanthin in *Mytilus edulis* might be a terminal metabolite of fucoxanthin through intermediates, halocynthiaxanthin and isomytiloxanthin, based on observation in feeding experiment. However, conversion of isomytiloxanthin into alloxanthin is too complex and there were no direct evidences for the conversion, especially in aquatic animals. In our experience, isomytiloxanthin has not been isolated from these animals [24].

Shellfishes (bivalves) and tunicates are filter-feeders, which accumulate carotenoids from micro algae. Therefore, alloxanthin in these animals is assumed to originate from *Chryptophyceae* and *Euglenophyceae* micro algae, *etc.* Thus, these alloxanthin specimes consist of only (3*R*,3′*R*)-stereoisomer. Crucian carp is omnivorous and feeds not only animal planktons belonging to Cladocera but also micro algae. Therefore, alloxanthin in crucian carp is also assumed to originate from micro algae. On the other hand, alloxanthin in lake shrimp, catfish, biwa goby, and biwa trout exist as a mixture of three stereoisomres. These crustacean and fishes are carnivorous. Especially, lake shrimp contains a large amount of (3*S*,3′*S*)- and *meso*-alloxanthin (Table 1). Lake shrimp is a one of the major food of catfish and biwa trout. Therefore, (3*S*,3′*S*)- and *meso*-alloxanthin in these fishes might be originated from lake shrimp. However, origin of (3*S*,3′*S*)- and *meso*-alloxanthin in lake shrimp is uncertain.

Catfish is a top predator in Japanese freshwater ecosystems. Catfish ingests astaxanthin from crustaceans whose astaxanthin exists as a mixture of three stereoisomers. Catfish can convert astaxanthin into zeaxanthin [24]. Therefore, zeaxanthin in catfish exists as a mixture of three stereoisomers. Although the origin of stereoisomers of alloxanthin in catfish is uncertain, it might be naturally formed by epimerization of 7,8,7′,8′-tetradehydroastaxanthin originated from crustacean at C3 and C3′-positions and subsequent reduction at C4 and C4′-positions. Further studies are need to reveal the origin of (3*S*,3′*S*)- and *meso*-alloxanthin in crustaceans and fishes.

This is the first report of the occurrence of (3*S*,3′*S*) and *meso*-alloxanthin in nature.

## 3. Experimental Section

### 3.1. General

IR spectrum was measured on a Perkin-Elmer FT-IR spectrometer (Perkin-Elmer, Yokohama, Japan), spectrum 100. $^1$H and $^{13}$C NMR spectra were determined on a Varian Gemini-300 superconducting FT-NMR spectrometer (Agilent Technologies, Santa Clara, CA, USA) and the chemical shifts were referenced to tetramethylsilane. Mass spectrum was taken on a Thermo Fisher Scientific Exactive spectrometer (Thermo Fisher Scientific, Bremen, Germany). CD spectra were measured on a Shimadzu-AVIN 62A DS circular dichroism spectrometer (Shimadzu, Kyoto, Japan).

HPLC analyses were performed on Simadzu-LC-20AT instrument (Shimadzu, Kyoto, Japan) with a photodiode array detector (Waters 996, Tokyo, Japan) and column oven (GL Sciences Model 552, Tokyo, Japan).

NMR assignments are given using the carotenoid nmbering system.

### 3.2. Synthesis of (3S,3′S)-Alloxanthin (1b) and meso-Alloxanthin (1c)

In the same procedure [20] as preparation of (3*R*)-phosphonium salt **5a** and (3*R*,3′*R*)-alloxanthin (**1a**), (3*S*)-**5b** and (3*S*,3′*S*)-alloxanthin (**1b**) were prepared. Spectral data except for optical data of compounds **1b**, **3b**, **4b** and **5b** were identical with the corresponding previous reported [20] enantiomers **1a**, **3a**, **4a** and **5a**.

(3*S*,3′*S*)-Alloxanthin (**1b**): HRMS (ESI) *m*/*z* calcd for $C_{40}H_{53}O_2$ $[M + H]^+$ 565.4040, found 565.4038.

Compound **3b**: $[\alpha]_D^{26}$ 102.9 (*c* 1.03, MeOH); HRMS (ESI) *m*/*z* calcd for $C_{11}H_{17}O$ $[M + H]^+$ 165.1274, found 165.1277.

Compound **4b**: $[\alpha]_D^{23}$ 68.1 (*c* 1.00, MeOH); HRMS (ESI) *m*/*z* calcd for $C_{17}H_{31}OSi$ [M + H]$^+$ 279.2139, found 279.2139.

Compound **5b**: HRMS (ESI) *m*/*z* calcd for $C_{33}H_{62}OPSi$ [M − Cl]$^+$ 533.4302, found 533.4293.

*meso*-Alloxanthin (**1c**) was synthesized via condensation between **5b** and (3*R*)-$C_{25}$-acetylenic apocarotenal **8**, which was prepared by Wittig reaction of $C_{10}$-dialdehyde **7** with **5a** as follows.

(2*E*,4*E*,6*E*,8*E*,10*E*)-2,7,11-trimethyl-13-[(*R*)-2,6,6-trimethyl-4-triethylsilyloxycyclohex-1-en-1-yl] trideca-2,4,6,8,10-pentaen-12-ynal (**8**). NaOMe (1 M in MeOH; 1.2 mL, 1.2 mmol) was added to a solution of the (3*R*)-phosphonium salt **5a** (409 mg, 0.73 mmol) and $C_{10}$-dialdehyde **7** (100 mg, 0.61 mmol) in $CH_2Cl_2$ (10 mL) at 0 °C. After being stirred at 0 °C for 15 min, the mixture was poured into saturated aq. $NH_4Cl$ and extracted with AcOEt. The extracts were washed with brine, dried over $Na_2SO_4$ and evaporated to afford a residue, which was purified by flash column chromatography (AcOEt–*n*-hexane, 1:4) to give the (3*R*)-$C_{25}$-acetylenic apocarotenal **8** (165 mg, 57%) as an orange viscous oil: UV-VIS $\lambda_{max}$ (EtOH)/nm 420; IR $\nu_{max}$ ($CHCl_3$)/cm$^{-1}$ 2170 (C≡C), 1663 (conj. CHO), 1610 and 1599 (split) (C=C), 1552 (C=C); $^1$H-NMR ($CDCl_3$, 300 MHz) δ 0.61 (6H, q, *J* = 8 Hz, $SiCH_2$ × 3), 0.97 (9H, t, *J* = 8 Hz, $CH_2CH_3$ × 3), 1.14 and 1.18 (each 3H, s, 1-gem-Me), 1.49 (1H, t, *J* = 12 Hz, 2-$H_\beta$), 1.74 (1H, ddd, *J* = 12, 3.5, 2 Hz, 2-$H_\alpha$), 1.89 (3H), 1.91 (3H) and 2.03 (6H) (each s, 5-Me, 9-Me, 13-Me and 13′-Me), 2.11 (1H, br dd, *J* = 17.5, 9.5 Hz, 4-$H_\beta$), 2.30 (1H, br dd, *J* = 17.5, 5.5 Hz, 4-$H_\alpha$), 3.94 (1H, m, 3-H), 6.32 (1H, br d, *J* = 12 Hz, 14-H), 6.37 (1H, d, *J* = 15 Hz, 12-H), 6.46 (1H, br d, *J* = 11.5 Hz, 10-H), 6.66 (1H, dd, *J* = 15, 11.5 Hz, 11-H), 6.70 (1H, dd, *J* = 14.5, 11.5 Hz, 15′-H), 6.96 (1H, br d, *J* = 11.5 Hz, 14′-H), 7.03 (1H, dd, *J* = 14.5, 12 Hz, 15-H), 9.46 (1H, s, CHO); $^{13}$C-NMR ($CDCl_3$, 75 MHz) δ 4.82 (C × 3), 6.83 (C × 3), 9.59, 12.96, 18.17, 22.53, 28.61, 30.45, 36.53, 42.11, 47.04, 65.01, 90.10, 98.16, 121.15, 123.84, 126.60, 127.73, 131.75, 134.51, 137.02, 137.07, 137.47, 138.70, 141.26, 148.75, 194.45; HRMS (ESI) *m*/*z* calcd for $C_{31}H_{47}O_2Si$ (MH)$^+$ 479.3340, found 479.3347.

Preparation of *meso*-alloxanthin (**1c**). NaOMe (1 M in MeOH; 0.24 mL, 0.24 mmol) was added to a solution of the (3*S*)-phosphonium salt 5b (113 mg, 0.20 mmol) and (3*R*)-$C_{25}$-acetylenic apocarotenal 8 (59 mg, 0.12 mmol) in $CH_2Cl_2$ (10 mL) at room temperature. After being stirred for further 15 min, the mixture was poured into saturated aq. $NH_4Cl$ and extracted with AcOEt. The extracts were washed with brine, dried over $Na_2SO_4$ and evaporated to afford a residue, which was purified by flash column chromatography (AcOEt–*n*-hexane, 1:4) to give the TES-protected condensed product. Subsequently, to a solution of this condensed product in dry THF (5 mL) were added AcOH (1 M in THF; 0.20 mL, 0.20 mmol) and then tetrabutylammonium fluoride (TBAF) (1 M in THF, 0.40 mL, 0.40 mmol). After being stirred at room temperature for 2 h, the mixture was concentrated to give a residue, which was purified by flash column chromatography (AcOEt–*n*-hexane–MeOH, 50:45:5) to provide *meso*-alloxanthin (**1c**) (70 mg, quant.) as red solids. Its spectral data were identical with those of (3*R*,3′*R*)-alloxanthin (**1a**) [20]. HRMS (ESI) *m*/*z* calcd for $C_{40}H_{53}O_2$ $[M + H]^+$ 565.4040, found 565.4033.

### 3.3. Configurational Analysis of Natural Alloxanthin

#### 3.3.1. Animal Materials

Scallop *Mizuhopecten yessoensis* was provided from Hokkaido Research Organization, Abashiri Fisheries Research Institute, Hokkaido, Japan. Oyster *Crassostrea gigas*, and sea squirt *Halocynthia roretzi* were purchased from fisheries market at Kyoto city. Pacific pearl oyster *Pinctada margaritifera* was provided from a pearl aquaculture industry, Ishigaki city, Okinawa Prefecture. Freshwater bivalve *Unio douglasiae*, crucian carp *Carassius auratus grandoculis*, and catfish *Silurus asotus* were purchased from Katata fisheries cooperative, Shiga Prefecture. Biwa trout *Oncorhynchus masou rhodurus* was purchased from Nango Fisheries Center, Shiga Prefecture. Biwa goby *Gymnogobius isaza* and lake shrimp *Palaemon paucidens* were purchased from fisheries market at Maibara city.

#### 3.3.2. Isolation of Alloxanthin from Aquatic Animals

According to our routine methods, carotenoid was extracted with acetone from animal tissue. The extract was partitioned between $Et_2O$–*n*-hexane (1:1) and water in separating funnel. The organic phase was evaporated and saponified with 5% KOH/MeOH at room temperature for 2 h. Then, unsaponifiable compounds were extracted with $Et_2O$–*n*-hexane (1:1, v/v) from the reaction mixture by addition of water. The organic layer was dried over $Na_2SO_4$ and evaporated. The residue was subjected to silica gel column chromatography increasing percentage of $Et_2O$ in *n*-hexane. The fraction eluted with $Et_2O$ was subjected to HPLC on silica gel with acetone–*n*-hexane (3:7) to afford alloxanthin. Purity of alloxanthin was checked by UV-Vis, $^1$H-NMR, and MS spectral data. Then alloxanthin obtained from aquatic animals was subject to configurational analysis using a chiral column described above.

## 4. Conclusions

In conclusion, we synthesized stereoisomers of alloxanthin (**1a**–**c**) and established a HPLC analytical method using a chiral column to identify them for naturally occurring alloxanthin. Application of this method to various alloxanthin specimens of aquatic animals demonstrated that those isolated from shellfishes, tunicates, and crucian carp are identical with (3*R*,3′*R*)-stereoisomer **1a**, and unexpectedly those from lake shrimp, catfish, biwa goby, and biwa trout are mixtures of three stereoisomers of **1a–c**. This is the first report of the occurrence of (3*S*,3′*S*) and *meso*-alloxanthin in nature. The analytical method can be a powerful tool to identify stereoisomers of alloxanthin in nature in a straightforward manner.

## Acknowledgments

We thank M. Kurimoto and M Shoji for technical assistance.

## Author Contributions

Basic idea of the research was proposed by three authors collaboratively. The synthetic and analytical experiments were designed and performed by Y. Yamano. The isolation of natural products was designed and carried out by T. Maoka.

## Conflicts of Interest

The authors declare no conflict of interest.

## References

1. Haxo, F.T.; Fork, D.C. Photosynthetically active accessory pigments of cryptomonads. *Nature* **1959**, *184*, 1051–1052.
2. Mallams, A.K.; Waight, E.S.; Weedon, B.C.L.; Chapman, D.J.; Haxo, F.T.; Goodwin, T.W.; Thomas, B.M. A new class of carotenoids. *Chem. Commun.* **1967**, 301–302; doi:10.1039/c19670000301.
3. Tsuchiya, Y.; Suzuki, Y. Biochemical studies of the ascidian, Cynthia rorezi variety drasche. IV Carotenoids in test. *Tohoku J. Agric Res.* **1959**, *10*, 397–407.
4. Nishibori, K. Pigments of marine animlas—VIII. Carotenoids of some shellfish. *Publ. Seto Mar. Biol. Lab.* **1960**, *8*, 317–326.
5. Campbell, S.A.; Mallams, A.K.; Waight, E.S.; Weedon, B.C.L. Pectenoxanthin, cynthiaxanthin, and a new acetylenic carotenoid, pectenolone. *Chem. Commun.* **1967**, 941–942; doi:10.1039/c19670000941.
6. DeVille, T.E.; Hursthouse, M.B.; Russell, S.W.; Weedon, B.C.L. Absolute configuration of carotenoids. *Chem. Commun.* **1969**, 1311–1312; doi:10.1039/c29690001311.
7. Bartlett, L.; Klyne, W.; Mose, W.P.; Scopes, P.M.; Galasko, G.; Mallams, A.K.; Weedon, B.C.L.; Szabolcs, J.; Toth G. Optical rotatory dispersion of carotenoids. *J. Chem. Soc. Perkin Trans. 1* **1969**, 2527–2544; doi:10.1039/j39590002527.
8. Hertzberg, S.; Partali, V.; Liaaen-Jensen, S. Animal carotenoids. 32. Carotenoids of *Mytilus edulis* (edible mussel). *Acta Chem. Scand.* **1988**, *B42*, 495–503.
9. Maoka, T.; Matsuno, T. Carotenoids of shellfishes—IX. Isolation and structural elucidation of three new acetylenic carotenoids from the Japanese sea mussel *Mytilus coruscus*. *Nippon Suisan Gakkaishi* **1988**, *54*, 1443–1447.

10. Maoka, T.; Tsushima, M.; Matsuno, T. New acetylenic carotenoids from the starfishes *Asterina pectinifera* and *Asterias amurensis*. *Comp. Biochem. Physiol.* **1989**, *93B*, 829–834.
11. Matsuno, T.; Ookubo, M.; Nishizawa, T.; Shimizu, I. Carotenoids of sea squirts. I. New marine carotenoids, halocynthiaxanthin and mytiloxanthinone from *Halocynthia roretzi*. *Chem. Pharm. Bull.* **1984**, *32*, 4309–4315.
12. Ookubo, M.; Matsuno, T. Carotenoids of sea squirts—II. Comparative biochemical studies of carotenoids in sea squirts. *Comp. Biochem. Physiol.* **1985**, *81B*, 137–141.
13. Matsuno, T.; Maoka, T.; Ikuno, Y. Comparative biochemical studies of carotenoids in fish—XXVII. Carotenoids in the eggs of three species of *Cyprinidae*. *Comp. Biochem. Physiol.* **1986**, *83B*, 335–337.
14. Maoka, T.; Akiomoto, N. Structures of minor carotenoids from the Japanese common catfish, *Silurus asotus*. *Chem. Phram. Bull.* **2011**, *59*, 140–145.
15. Ronneberg, H.; Renstrom, B.; Aareskjold, K.; Liaaen-Jensen, S.; Vecchi, M.; Leuenberger, F.J.; Müller, R.K.; Mayer, H. Naturally occurrence of enantiomeric and *meso*-astaxanthin 1. Ex lobster eggs (*Homarus gammarus*). *Helv. Chim. Acta* **1980**, *63*, 711–715.
16. Matsuno, T.; Maoka, T.; Katsuyama, M.; Ookubo, M.; Katagiri, K.; Jimura, H. The occurrence of enantiomeric and *meso*-astaxanthin in aquatic animals. *Nippon Suisan Gakkaishi* **1984**, *50*, 1589–1592.
17. Maoka, T.; Arai, A.; Shimizu, M.; Matsuno, T. The first isolation of enantiomeric and *meso*-zeaxanthin in nature. *Comp. Biochem. Physiol.* **1986**, *83B*, 121–124.
18. Matsuno, T.; Maoka, T.; Katsuyama, M.; Hirono, T.; Ikuno, Y.; Shimizu, M.; Komori, T. Comparative biochemical studies of carotenoids in fishes—XXIX. Isolation of new luteins, lutein F and lutein G from marine fishes. *Comp. Biochem. Physiol.* **1986**, *85B*, 77–80.
19. Ikuno, Y.; Shimizu, M.; Koshino, Y.; Maoka, T.; Matsuno, T. Comparative biochemical studies of carotenoids in fishes—XXVII. Stereochemical investigation of carotenoids from yellow-tail rockfish *Sebastes flavidus*. *Nippon Suisan Gakkaishi* **1985**, *51*, 2033–2035.
20. Yamano, Y.; Chary, V.M.; Wada, A. Stereoselective total synthesis of the acetylenic carotenoids alloxanthin and triophaxanthin. *Org. Biomol. Chem.* **2012**, *10*, 4103–4108.
21. Leuenberger, H.G.W.; Boguth, W.; Widmer, E.; Zell, R. Synthesis of optically active natural carotenoids and structurally related compounds. I. Synthesis of the chiral key compound (4*R*,6*R*)-4-hydroxy-2,2,6-trimethylcyclohexanone. *Helv. Chim. Acta* **1976**, *59*, 1832–1849.
22. Partali, V.; Tangen, K.; Liaaen-Jensen, S. Carotenoids in food chain studies. III. Resorption and metabolic transformation of carotenoids in *Mytilus edulis* (edible mussel). *Comp. Biochem. Physiol.* **1989**, *92B*, 239–246.

23. Liaaen-Jensen, S. Carotenoids in food chain. In *Carotenoids Volume 3: Biosynthesis and Metabolism*; Britton, G., Liaaen-Jensen, S., Pfander, H., Eds.; Birkhäuser: Basel, Switzerland, 1998; pp. 359–371.
24. Maoka, T. Carotenoids in marine animals. *Mar. Drugs* **2011**, *9*, 278–293.

*marine drugs*

MDPI

*Article*

# Peridinin from the Marine Symbiotic Dinoflagellate, *Symbiodinium* sp., Regulates Eosinophilia in Mice

**Ken-ichi Onodera [1,*], Yuko Konishi [2], Takahiro Taguchi [3], Sumio Kiyoto [1] and Akira Tominaga [3]**

1 Oceanography Section, Science Research Center, Kochi University, Okoh, Nankoku,
Kochi 783-8505, Japan; E-Mail: jm-sumio.kiyoto@kochi-u.ac.jp

2 Medical Research Center, Kochi University, Okoh, Nankoku, Kochi 783-8505, Japan;
E-Mail: jm-konishiy@kochi-u.ac.jp

3 Laboratory of Human Health and Medical Science, Graduate School of Kuroshio Science, Kochi University, Okoh, Nankoku, Kochi 783-8505, Japan; E-Mails: ttaguchi@kochi-u.ac.jp (T.T.); tominaga@kochi-u.ac.jp (A.T.)

* Author to whom correspondence should be addressed; E-Mail: onoderak@kochi-u.ac.jp;
Tel.: +81-88-880-2184; Fax: +81-88-880-2185.

*Received: 10 December 2013; in revised form: 25 February 2014 / Accepted: 28 February 2014 /
Published: 25 March 2014*

**Abstract:** Peridinin and fucoxanthin, which are natural carotenoids isolated from a symbiotic dinoflagellate, *Symbiodinium* sp., and a brown alga, *Petalonia fascia*, respectively, were compared for inhibitory effects on delayed-type hypersensitivity in mice. The number of eosinophils at the site of inflammation and in peripheral blood was compared for the administration of peridinin and fucoxanthin applied by painting and intraperitoneally. Peridinin, but not the structurally-related fucoxanthin, significantly suppressed the number of eosinophils in both the ear lobe and peripheral blood. Furthermore, peridinin applied topically, but not administered intraperitoneally, suppressed the level of eotaxin in the ears of sensitized mice. Fucoxanthin weakly suppressed the concentration of eotaxin in ears only by intraperitoneal administration. Although both carotenoids inhibited the migration of eosinophils toward eotaxin, the inhibitory effect of peridinin was higher than that of fucoxanthin. Peridinin may be a potential agent for suppressing allergic inflammatory responses, such as atopic dermatitis, in which eosinophils play a major role in the increase of inflammation.

**Keywords:** peridinin; fucoxanthin; delayed-type hypersensitivity; eosinophils; eotaxin; *Symbiodinium*; dinoflagellate

## 1. Introduction

Dinoflagellates are unicellular phytoplankton and are known to produce various bioactive secondary metabolites [1–6]. The genus, *Symbiodinium*, which belongs to the zooxanthellae, is a representative symbiont found in many marine invertebrates. They also produce unique and complex bioactive secondary metabolites [1,2,4] together with a peridinin. Peridinin is one of the carotenoids that is synthesized in dinoflagellates.

Carotenoids have numerous bioactivities. In particular, the marine carotenoid, fucoxanthin, has multiple functions and has been reported to have considerable potential for applications for improving human health [7–12]. For example, fucoxanthin induces the uncoupling of protein 1 expression in white adipose tissue mitochondria [7]. It also improves insulin resistance and decreases blood glucose levels [7]. Kim *et al.* [9] reported that fucoxanthin reduces the levels of pro-inflammatory mediators, including NO, $PGE_2$, IL-1β, TNF-α and IL-6, via the inhibition of NF-κB activation and the suppression of MAPK phosphorylation in leukemic monocyte RAW264 cells. Sakai *et al.* [8] reported that fucoxanthin, astaxanthin, zeaxanthin and β-carotene significantly inhibit the antigen-induced release of β-hexosaminidase in basophilic leukemia 2H3 and mast cells and that these carotenoids also inhibit antigen-induced aggregation of the high affinity IgE receptor on mast cells. Further, Sakai *et al.* [10] reported that these carotenoids inhibit dinitrofluorobenzene-induced contact hypersensitivity in the ears by reducing the TNF-α and histamine levels in the ear.

In contrast, the bioactivities of peridinin, which has a structure similar to that of fucoxanthin (Figure 1), have not been well studied. Tsushima *et al.* [13] reported that peridinin has a strong inhibitory effect on Epstein–Barr virus early antigen activation in Raji cells at a lower concentration, but shows cytotoxicity to Raji cells at a higher concentration. Sugawara *et al.* [14] reported that peridinin induced apoptosis of human colorectal cancer cells by activating both caspase-8 and caspase-9.

**Figure 1.** Chemical structures of peridinin and fucoxanthin.

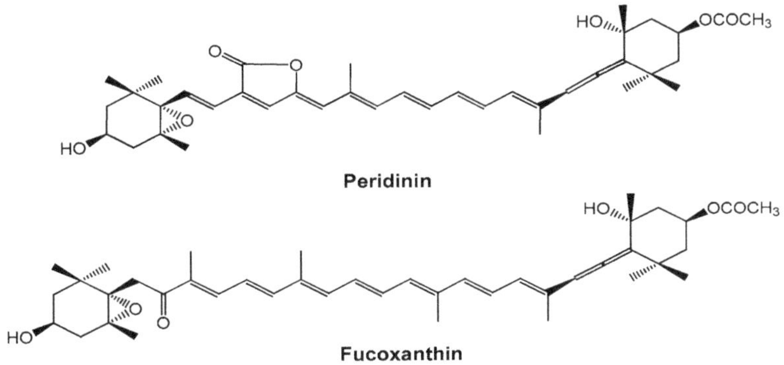


Here, we examine the bioactivity of peridinin as a functional material for human health. Recently, the number of allergy sufferers has increased worldwide. In this paper, we examined the inhibitory effect of peridinin on delayed-type hypersensitivity in mice and compared it with that of fucoxanthin.

## 2. Results and Discussion

### 2.1. *Effect of Peridinin on DTH in BALB/cAJc1 Mice*

Eosinophils are well-known granulocytes that increase at the sites of inflammation as part of the delayed-type hypersensitivity (DTH) response. Eosinophils are reported to increase in response to picryl chloride (PCl; 2,4,6-trinitrochlorobenzene) in BALB/cAJc1 mice [15]. In that study, treatment with cyclophosphamide 2 days before sensitization resulted in marked blood and tissue eosinophilia. We applied the same experimental protocol to elicit the DTH response.

Hydrocortisone strongly suppressed the DTH response at 24 and 48 h (22.4% and 18.2% suppression, respectively; Figure 2) after the antigen challenge. Peridinin also suppressed the DTH response in BALB/cAJc1 mice, both at 24 or 48 h after the antigen challenge by both routes of administration, painted onto ears (paint) (8.9% and 9.2% suppression, respectively; Figure 2) or administered intraperitoneally (i.p.) (12.7% and 11.7% suppression, respectively; Figure 2). In contrast, fucoxanthin did not suppress the DTH response at either 24 or 48 h after the antigen challenge (Figure 2).

Increased ear thickness is caused by the accumulation of lymphocytes, macrophages, neutrophils and eosinophils. Furthermore, macrophages stimulate the proliferation of fibroblasts. The accumulation of these cells to the site of antigen challenge results in the increase of ear thickness. It is suggested that peridinin reduced at least one of these factors. We focused on eosinophils, because they are typical white blood cells that increase in allergic reactions.

### 2.2. Induction of Tissue Eosinophilia in Peridinin-Treated Mice Sensitized with PCl

We examined whether the migration of eosinophils to the site of inflammation was suppressed by administering peridinin or fucoxanthin. We measured the number of eosinophils in the ear section of BALB/cAJc1 mice and compared it to the number in the negative control group (sensitized, but not challenged with PCl) and the hydrocortisone group (68.5% suppression). Peridinin by either route of administration, paint (79.9% suppression) or i.p. (60.3% suppression), significantly decreased the number of eosinophils at the site of inflammation (ear section) at 48 h after antigen challenge. In contrast, fucoxanthin did not inhibit the number of eosinophils that migrated to the site of inflammation at 48 h after the antigen challenge (Figure 3A,B). Eosinophils were not observed in the negative control group without the challenge with PCl. Eosinophils were more abundant after the challenge of sensitized mice with PCl. In the group treated with hydrocortisone by painting at 3 h before the challenge, the number of eosinophils was lower than that of the positive control. Groups treated with peridinin either by painting or i.p. showed lower number of eosinophils than the positive control. There were no significant difference between the hydrocortisone-treated group and peridinin-treated group. In contrast, the fucoxanthin-treated group, either by painting or i.p., did not have a lower number of eosinophils compared with the positive control group.

**Figure 2.** Suppression of delayed-type hypersensitivity (DTH) responses by peridinin and fucoxanthin in BALB/cAJc1 mice. Mice were sensitized with picryl chloride (PCl) after pretreatment with cyclophosphamide. Two weeks after 10 μg of peridinin or fucoxanthin or hydrocortisone in 10 μL of olive oil was painted onto the ears (per earlobe), 50 μg of peridinin or fucoxanthin in 100 μL of olive oil was administered i.p. (intraperitoneally) to these mice 3 h before the antigen challenge. Ear thickness was measured with a dial thickness gauge before (0 h) and after the antigen challenge (4, 24 and 48 h). Each value for each treatment group is expressed as the mean ± SD ($n$ = 10). Pos cont, positive control (sensitized and challenged); Hydroc, hydrocortisone; Peri, peridinin; Fuco, fucoxanthin; Peri i.p., peridinin administered intraperitoneally; Fuco i.p., fucoxanthin administered intraperitoneally; Neg cont, negative control (sensitized, but not challenged). * An asterisk indicates a significant difference between the treatment group and the positive control ($p < 0.01$, Tukey–Kramer's *post hoc* test). Asterisks are only shown for groups showing suppression compared to the positive control at 24 or 48 h after the antigen challenge.

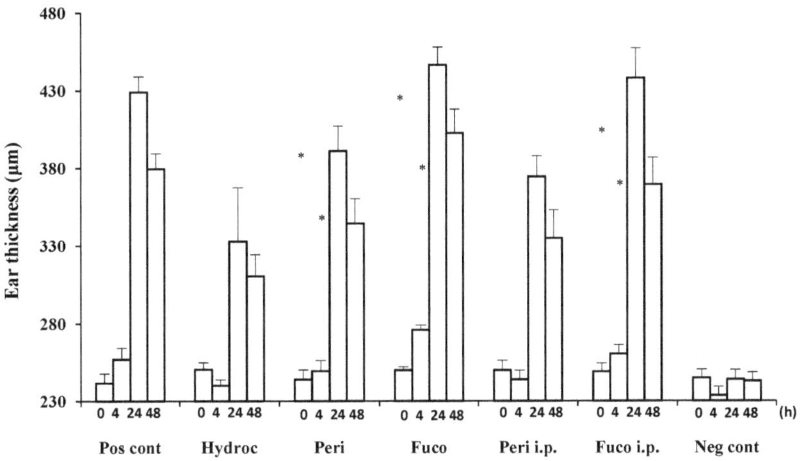


### *2.3. Suppression of the Percentage of Eosinophils in Peripheral Blood in Peridinin-Treated Mice Sensitized with PCl*

We measured and compared the percentage of eosinophils among white blood cells in the peripheral blood of individual mice between the non-challenged and hydrocortisone-treated groups and found the effect of treatment to result in 83.1% suppression. Peridinin administered by either route, paint (82.2% suppression) or i.p. (78.4% suppression), significantly suppressed the number of eosinophils in peripheral blood at 48 h after the antigen challenge. In contrast, fucoxanthin did not suppress the number of eosinophils in peripheral blood at 48 h after the antigen challenge (Figure 4). These results suggest that peridinin, but not fucoxanthin, suppressed the proliferation of eosinophils. In this DTH model, serum levels of IFN-γ and IL-5 were increased 48 h after the antigen challenges, and peridinin did not suppress the serum level of either cytokine significantly in either route of administration (IFN-γ, negative control, 37.8 ± 14; positive control, 104.4 ± 28; peridinin painted, 94 ± 32; peridinin i.p., 72.6 ± 31. IL-5: negative control, 12.3 ± 8.5; positive control, 18.6 ± 3.5; peridinin painted, 25.0 ± 0.1; peridinin i.p., 23.2 ± 12. Results are expressed as the mean (picogram per millimeter) ± SD). The serum level of IL-17 was not increased in this experiment [16]. Thus, we hypothesized that the production of eotaxin, the most potent chemo-attractant of the eosinophils, may be suppressed at the site of inflammation.

**Figure 3.** The effects of peridinin or fucoxanthin on the number of eosinophils in ear sections. Mice were sensitized with PCl, as described in the Experimental Section. Two weeks after the sensitization, 10 μg of peridinin, fucoxanthin or hydrocortisone was painted onto the ears and 50 μg of peridinin or fucoxanthin was administered i.p. to these mice 3 h before the antigen challenge. At 48 h after the antigen challenge, tissue specimens were fixed and stained with hematoxylin/eosin solution to count the number of eosinophils. (**A**) The number of eosinophils in each section from each group is shown. (**B**) Representative section photomicrographs from each group of mice are shown. Each value is expressed as the mean ± SD ($n$ = 10). Pos cont, positive control; Hydroc, hydrocortisone; Peri, peridinin; Fuco, fucoxanthin; Peri i.p., peridinin intraperitoneally; Fuco i.p., fucoxanthin intraperitoneally; Neg cont, negative control. * An asterisk indicates a significant difference between the treatment group and the positive control ($p < 0.01$, Tukey–Kramer's *post hoc* test).

**A**

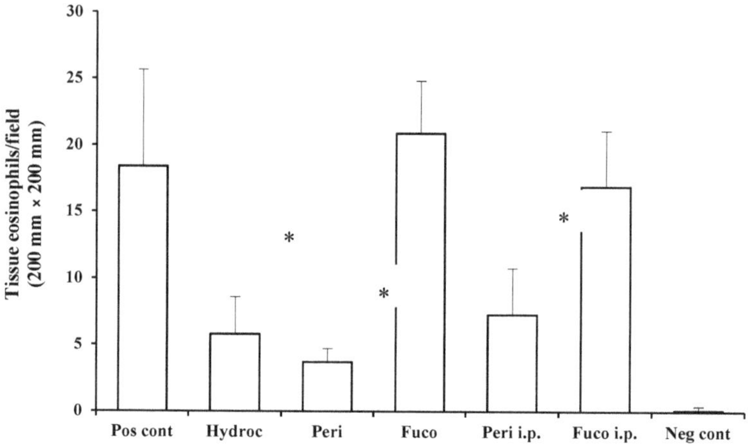

**Figure 3.** *Cont.*

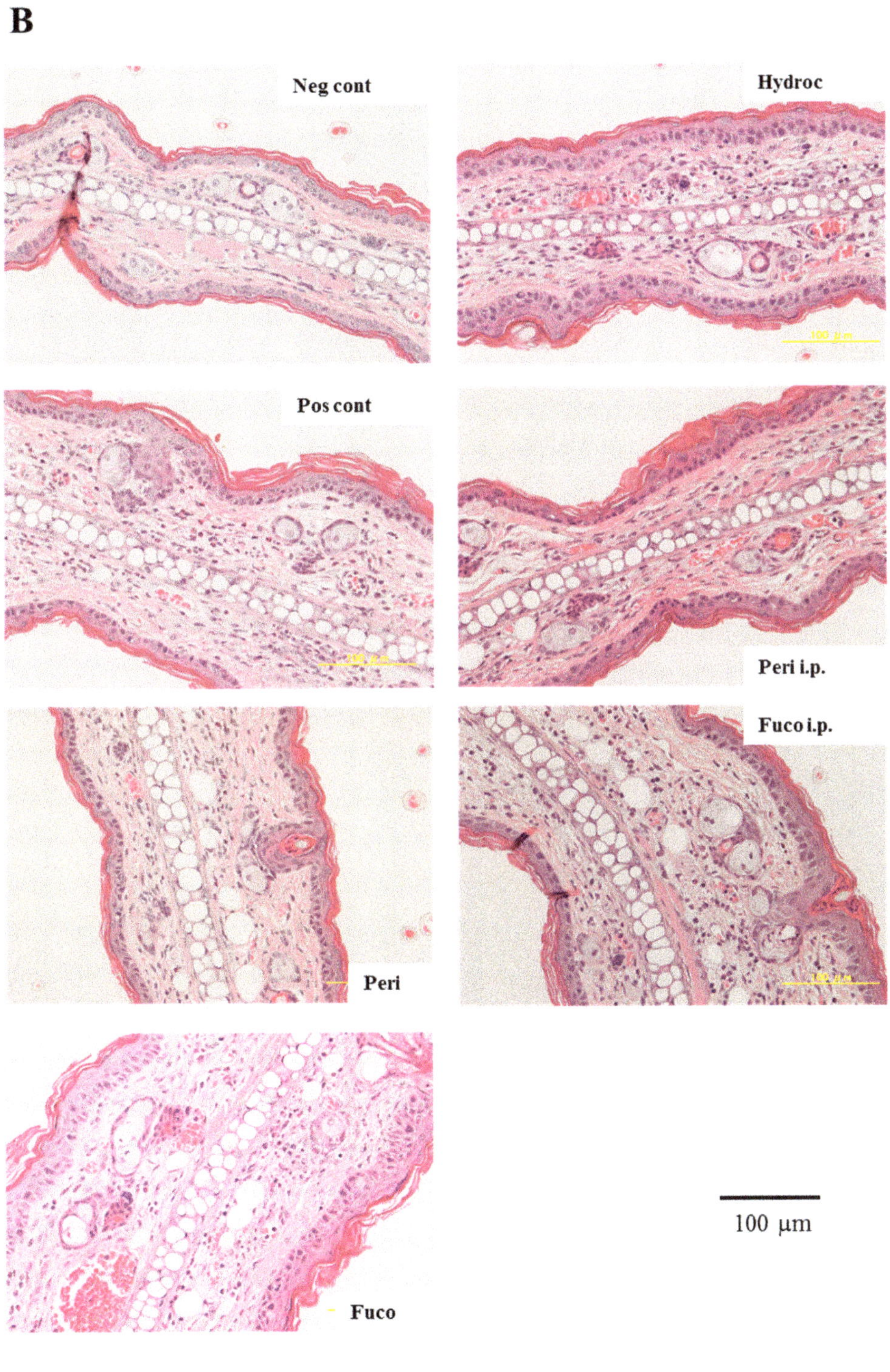

**Figure 4.** Effects of peridinin or fucoxanthin on the number of eosinophils in peripheral blood. Mice were sensitized with PCl, as described in the Experimental Section. Two weeks after the sensitization, 10 μg of peridinin, fucoxanthin or hydrocortisone was painted onto the ears and 50 μg of peridinin or fucoxanthin was administered i.p. to these mice 3 h before the antigen challenge. At 48 h after the antigen challenge, mice were bled from the retro-orbital plexus. The percentage of eosinophils among white blood cells was determined by making a smear on a slide glass for each mouse. Slide glasses were stained with Giemsa solution, and the percentage of eosinophils among white blood cells was estimated by counting at least 200 white blood cells from each sample. There was no significant difference in the total numbers of leukocytes among groups. Each value is expressed as the mean ± SD ($n$ = 10). Pos cont, positive control; Hydroc, hydrocortisone; Peri, peridinin; Fuco, fucoxanthin; Peri i.p., peridinin intraperitoneally; Fuco i.p., fucoxanthin intraperitoneally; Neg cont, negative control. * An asterisk indicates a significant difference between the treatment group and the positive control ($p < 0.01$, Tukey–Kramer's *post hoc* test).

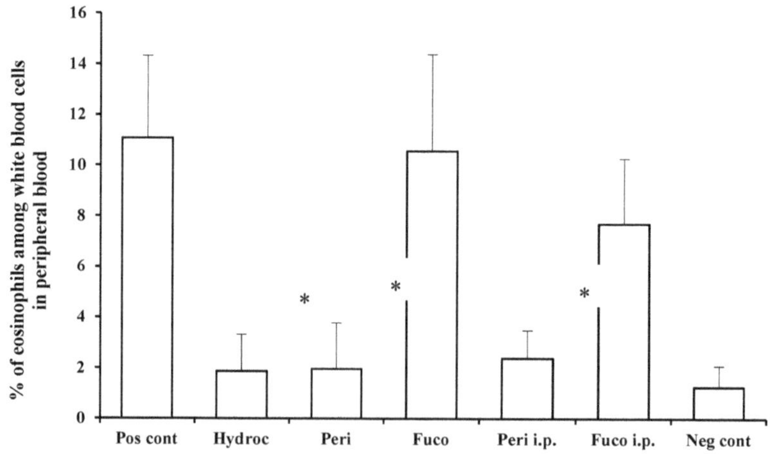


## 2.4. *Effects of Peridinin on Eotaxin Production*

Figure 5 shows that eotaxin production was stimulated by the challenge with PCl, and this was inhibited by peridinin following painting onto the ears (30.6% inhibition), but not by i.p administration. This inhibitory effect was comparable to that of hydrocortisone (42.1% inhibition). On the other hand, fucoxanthin weakly inhibited the production of eotaxin in mice administered i.p (19.4% inhibition), whereas it had no effect in the case of painting. These results suggest that metabolites of peridinin and fucoxanthin administered i.p. may have different effects on eotaxin-producing cells, such as endothelial cells, epithelial cells and macrophages. It is also possible that there is a difference between these two compounds in terms of penetration through the skin. These differences between peridinin and fucoxanthin remain to be clarified.

**Figure 5.** Peridinin inhibits eotaxin production in challenged skin. Mice were sensitized as described in the Experimental Section. At 48 h after the antigen challenge, ear lobes were removed, and the concentration of eotaxin in each ear lobe was measured as described in the Experimental Section. Each value is expressed as the mean ± SD ($n$ = 3). Pos cont, positive control; Hydroc, hydrocortisone; Peri, peridinin; Fuco, fucoxanthin; Peri i.p., peridinin intraperitoneally; Fuco i.p., fucoxanthin intraperitoneally; Neg cont, negative control. * An asterisk indicates a significant difference between the treatment group and the positive control ($p < 0.01$, Tukey–Kramer's *post hoc* test). ** There is a significant difference between Fuco i.p. and the positive control ($p < 0.05$, Tukey–Kramer's *post hoc* test). Asterisks are shown only for treatment groups showing suppression.

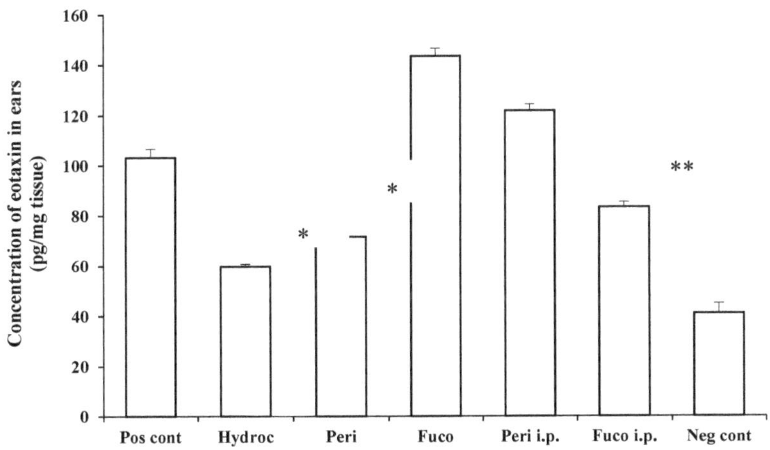


*2.5. Inhibitory Effects of Peridinin and Fucoxanthin on the Chemotaxis of Eosinophils toward Eotaxin*

Next, we asked whether peridinin inhibits the migration of eosinophils using murine eosinophils isolated from IL-5 transgenic mice. Peridinin and fucoxanthin were tested in an eosinophil chemotaxis assay *in vitro*. Chemotaxis of eosinophils toward eotaxin (20 ng/mL) was suppressed by one and
3 μg/mL of peridinin (57.4% and 72.8% suppression, respectively; Figure 6) and by one and 3 μg/mL of fucoxanthin (24.2% and 61.7% suppression, respectively; Figure 6). Peridinin showed higher suppression of the migration of eosinophils toward eotaxin than did fucoxanthin. More than 95% of eosinophils were viable after one

hour of incubation for the chemotaxis toward eotaxin. The viability of eosinophils was checked by trypan blue exclusion.

Although the painting of peridinin on earlobes decreased the level of eotaxin in the ears, the i.p. administration of peridinin did not have any effect on the production of eotaxin in the ears. In the peripheral blood, however, the number of eosinophils was suppressed significantly in either route of administration, suggesting that peridinin suppressed the number of eosinophils in the ears by reducing the number of eosinophils in the peripheral blood. Therefore, the level of eotaxin in the ears does not have a primary role in case of the i.p. administration of peridinin. Since peridinin did not change the serum level of IL-5 in either route of administration as described above, the decreased number of eosinophils in peripheral blood may be regulated by other factors.

In contrast, the number of eosinophils in peripheral blood was not influenced by fucoxanthin in either route of administration. Although fucoxanthin suppressed the migration of eosinophils *in vitro*, this effect seems to be marginal *in vivo*, because there is an increased level of eosinophils in the peripheral blood and a higher level of eotaxin is produced in the ears.

**Figure 6.** The effect of peridinin or fucoxanthin on the migration of eosinophils toward eotaxin in chemotaxis assays. Eosinophils prepared as described in the Experimental Section were suspended in Roswell Park Memorial Institute (RPMI) 1640 medium containing 0.1% bovine serum albumin (BSA) and placed in the Transwells of a 24-well chemotaxis chamber. The lower chamber contained RPMI 1640 medium containing 0.1% BSA and mouse eotaxin (final concentration: 20 ng/mL). Peridinin or fucoxanthin at a final concentration of 1 or 3 μg was added to each Transwell. Assay plates were incubated for 1 h at 37 °C in 5% $CO_2$. After 1 h incubation, the cells that migrated across the filter to the lower chamber were counted. Each value is expressed as the mean ± SD ($n = 18$). Pos cont, positive control; Peri 1, 1 μg of peridinin; Peri 3, 3 μg of peridinin; Fuco 1, 1 μg of fucoxanthin; Fuco 3, 3 μg of fucoxanthin; Neg cont, negative control. * An asterisk indicates a significant difference between the treatment group and the positive control ($p < 0.01$, Tukey–Kramer's *post hoc* test).

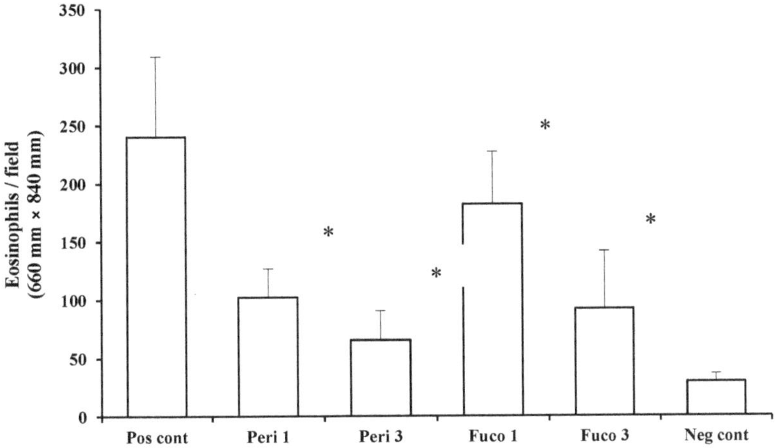


## 3. Experimental Section

### 3.1. Materials

Picryl chloride (PCl; 2,4,6-trinitrochlorobenzene) was purchased from Nacalai Tesque (Kyoto, Japan). Cyclophosphamide was purchased from Shionogi Pharmaceutical Co. (Osaka, Japan). Hydrocortisone was purchased from Sigma-Aldrich (St. Louis, MO, USA). Recombinant mouse eotaxin (CCL11) was purchased from ProSpec-Tany TechnoGene Ltd. (Ness Ziona, Israel). Peridinin and fucoxanthin were isolated from a dinoflagellate, *Symbiodinium* sp. (OTCL2A strain), and brown alga, *Petalonia fascia,* in culture, respectively.

### 3.2. Preparation of Peridinin

Peridinin was isolated from a cultured dinoflagellate, *Symbiodinium* sp. (OTCL2A strain). This *Symbiodinium* sp. was isolated from the giant clam, *Tridacna crocea,* by the author and cultured by a method reported previously [2]. The frozen cells (82.0 g from 132 L of culture) were homogenized in 70% EtOH (150 mL) with an Ultra-Turrax T25 homogenizer (Janke & Kunkel GmbH & Co. KG IKA-Labortechnik, Staufen, Germany), soaked for 3 days at 4 °C and centrifuged. The supernatant was collected; the precipitate was extracted twice with 70% EtOH (150 mL each). The ethanol extracts from the centrifugation and extractions were combined and evaporated *in vacuo*. The residue was suspended in water (100 mL) and extracted with EtOAc (3 × 200 mL). The combined EtOAc extracts were concentrated under reduced pressure and yielded an EtOAc-soluble fraction (903 mg). A portion (676 mg) of the EtOAc-soluble fraction was chromatographed on silica gel (60 mL of Silica Gel 60, Nacalai Tesque) and eluted with 120 mL of $CH_2Cl_2$:MeOH in fractions on a gradient of 99:1, 98:2, 96:4 and 92:8. A portion (199.2 mg) eluted from the last two extractions with $CH_2Cl_2$:MeOH (96:4 and 92:8; 264.8 mg) was then separated by chromatography on ODS (19 mL of Cosmosil 75C18-OPN, Nacalai Tesque) followed by eluting steps with 40 mL of 80% MeOH and 80 mL of 85% MeOH. The fraction eluted with 85% MeOH had a crude yield of 98.8 mg and, following separation by HPLC (COSMOSIL 5C18-AR-II, 20 mmφ × 250 mm, Nacalai Tesque) with 85% MeCN at a flow rate of .0 mL $min^{-1}$, yielded peridinin (24.7 mg).

### 3.3. Preparation of Fucoxanthin

Fucoxanthin was extracted from the brown alga, *Petalonia fascia,* cultured as described [17]. A sample (54.5 g) of frozen *Petalonia fascia* was extracted four times with 500 mL of MeOH. The combined methanol extracts were evaporated *in vacuo*. The residue was dissolved in 100 mL of 90% MeOH and then partitioned with hexane (2 × 100 mL). The 90% MeOH layer was removed to a different vessel and evaporated. The residue was resuspended in 100 mL of $H_2O$ and then extracted with EtOAc (2 × 100 mL). The EtOAc-soluble fraction (602.1 mg) was applied to a silica gel column (12 mL of Silica Gel 60, Nacalai Tesque), and the column was eluted with 170 mL of 1:1 hexane:EtOAc to give seven fractions. The fifth fraction (32.4 mg) was then separated by chromatography on ODS (10 mL of Cosmosil 75C18-OPN, Nacalai Tesque), and elution was performed with 40 mL of 85% MeOH and 30 mL of 90% MeOH. Fucoxanthin with a yield of 12.8 mg was eluted in the fraction of 90% MeOH.

### 3.4. Animals

BALB/cAJc1 mice were purchased from CLEA Japan, Inc. (Osaka, Japan). IL-5 transgenic mice (C3H/HeN-TgN(IL-5)-Imeg) were developed by our group [18] and were maintained in our animal facility under specific pathogen-free conditions. Mice used in the experiments were all female, and BALB/cAJc1 mice were 8 to 10 weeks of age and IL-5 mice were 8 to 15 weeks of age at the time of the experiments. All experiments were performed under the ethical guidelines of Kochi University.

### 3.5. Sensitization and Challenge

Two days before sensitization with PCl, cyclophosphamide was injected subcutaneously (150 mg/kg in distilled water) to remove proliferating immunosuppressive cells [15]. After removing coat hair, the mice were immunized by painting their abdominal skin with 0.05 mL of 7% PCl in ethanol:acetone (3:1). Two weeks after sensitization, 10 μg of peridinin, or fucoxanthin or hydrocortisone was painted onto the ears. 50 μg of peridinin or fucoxanthin was administered i.p. to these mice 3 h before the antigen challenge, and 0.02 mL of 1% PCl in acetone:olive oil (1:4) was painted on each ear lobe to challenge the mice. Ear thickness was measured with a dial thickness gauge (Peacock G-1M, Ozaki Mfg. Co. Ltd., Tokyo, Japan) before and after the challenge and was expressed as the mean thickness of each ear in micrometers.

### 3.6. Tissue Eosinophil Counts

At 48 h after being challenged, mice were sacrificed, and the ears were removed and fixed with 4% paraformaldehyde in a 0.1 M Phosphate buffer, pH 7.2. Then, the ears were embedded in paraffin, and sections of ears were stained with hematoxylin/eosin solution. The number of eosinophils infiltrated into the dermis was examined at a magnification of 400×. At least 10 fields were examined for each ear lobe. The number of eosinophils was counted and expressed as the number of cells in a field
(200 μm × 200 μm).

### 3.7. Percentage of Eosinophils in Peripheral Blood

At 48 h after the antigen challenge, mice were bled from the retro-orbital plexus. The percentage of eosinophils among the white blood cells was determined by making a smear on a slide glass for each mouse. Slide glasses were stained with Giemsa solution, and the percentage of eosinophils among the white blood cells was estimated by counting eosinophils and all white blood cells in the field until at least 200 white blood cells were counted in each sample. At a minimum, 10 slides were counted and the mean of the counts were taken for each group.

### 3.8. Cytokine Measurements by ELISA

Capture antibodies (50 μL of 1 mg/mL in 0.05 M sodium carbonate, pH 9.6) were added to coat each well of Immuno 96-microwell plates (9018, Corning Costar, Ithaca, NY, USA) and incubated overnight at 4 °C. After washing wells twice with 0.15 M sodium phosphate buffer, phosphate buffered saline (PBS), pH 7.2, containing 0.05% Tween 20 (PBS-T), blocking buffer (PBS containing 0.05% Tween 20 and 0.5% bovine serum albumin (BSA)) was added and incubated for 30 min at room temperature. After removing blocking buffer, diluted sera (5 times with blocking buffer) were added and incubated for 2 h at room temperature or at 37 °C. After washing wells with PBS-T three times, 50 μL of 0.5 μg/mL biotin-coupled antibodies (detection antibodies) against each cytokine were added to each well and incubated at room temperature for 45 min. After washing with PBS-T three times, streptavidin-horseradish peroxidase (Invitrogen, Camarillo, CA, USA) was added and incubated at room temperature for 45 min, according to the manufacturer's protocol. After washing wells with PBS-T three times, the substrate solution containing freshly prepared 0.7 mg/mL of *o*-phenylenediamine dihydrochloride (WAKO Pure Chemical Industries Ltd., Osaka, Japan) in citric acid buffer, pH 5.0, containing 0.006% hydrogen peroxide was added and incubated for 10 min. Reactions were stopped by adding 50 μL of 10% sulfuric acid to each well. Cytokine concentration was estimated by measuring the optical density (OD) at 490 nm using a micro plate reader (Model 680, Bio-Rad, Hercules, CA, USA) and a standard curve. Each cytokine level is expressed as the mean (pg/mL) ± SD.

### 3.9. Antibodies Used in ELISA

Capture antibodies included: rat anti-mouse IFN-γ antibodies (1 mg/mL), clone R46A2 (Becton Dickinson, Mountain View, CA, USA); rat anti-IL-5 antibodies (1 mg/mL), clone TRFK5 (prepared by Shiro Ono, Osaka Ohtani University, Tondabayashi, Japan); and rat anti-IL17 antibodies (2 mg/mL), clone TC11-18H10.1 (Becton Dickinson). Detection antibodies included: rat anti-mouse IFN-γ antibodies, clone XMG1.2 (Becton Dickinson); rat anti-IL-5 antibodies, clone TRFK4 (Becton Dickinson); and rat anti-IL17 antibodies, clone TC11-8H4.1 (Becton Dickinson).

### 3.10. Measurement of Eotaxin in Skin

At 48 h after being challenged, mice were sacrificed, and the ears were removed. The ear from each mouse in each experimental group was homogenized with PBS containing 0.1% Tween 20 (100 μL/10 mg tissue), using a mixer mill (MM 300, Retsch, Haan, Germany) with zirconia beads (5 mm) for 2 min at 30 Hz. The homogenates were centrifuged at 12,000× *g* for 10 min. The supernatants were kept at −80 °C until being assayed. The concentration of eotaxin in the supernatants was measured by a Mouse Eotaxin ELISA kit (Ray Biotech, Inc., Norcross, GA, USA). The sensitivity of the eotaxin assay was >0.01 pg/mg tissue.

### 3.11. Eosinophil Preparation

Enriched preparations of eosinophils were obtained from the peripheral blood of IL-5-transgenic mice. Eosinophil-enriched cells were obtained by the Percoll density gradient separation method described previously [19], with modification. Briefly, isotonic Percoll was prepared using a 10× solution of Krebs Ringer PBS (KRP; 10 mM sodium phosphate buffer, pH 7.5, containing 154 mM NaCl, 6 mM KCl and 1 mM $MgCl_2$) and diluted with KRP to achieve concentrations of 60%, 70% and 80%. In 15-mL conical polypropylene tubes (BD Falcon 352096, Becton, Dickinson and Company, Franklin Lakes, NJ, USA), 2 mL of cell suspensions of 20 to 50 × $10^6$ cells in KRP were placed, followed by the careful layering of aliquots (2.5 mL) of each concentration of Percoll solution starting with the lowest concentration at the bottom. The tubes were spun at 1000× *g* for 20 min at room temperature. Eosinophils were extracted from the 70% to 80% Percoll fractions by removing B and T lymphocytes using anti-B220- and anti-Thy1.2-coupled Dynabeads (DYNAL A.S., Oslo, Norway). Briefly, lymphocytes bound to these beads were removed using a permanent magnet. Unbound cells were used as an eosinophil-enriched fraction. To identify eosinophils, aliquots were removed and assessed using eosinophil peroxidase staining as described previously [19]. More than 93% of the cells prepared by this method were eosinophils.

### 3.12. Chemotaxis Assay toward Eotaxin

Eosinophils ($1.3 \times 10^6$ cells/mL) were suspended in RPMI 1640 medium containing 0.1% BSA and were placed in the upper well (Transwell) of a 24-well chemotaxis chamber (KURABO Co., Osaka, Japan). Transwells with 5 μm pores were inserted into each well, and $4 \times 10^5$ cells in 300 μL of RPMI 1640 medium containing 0.1% BSA were added to the upper chamber. The lower chamber was set with 800 μL of RPMI 1640 medium containing 0.1% BSA and mouse eotaxin (20 ng/mL). Then, 3 or 1 μg of peridinin or fucoxanthin ($n = 6$) was added to each Transwell. Assay plates were incubated for 1 h at 37 °C in 5% $CO_2$. Cells that migrated across the filter to the lower chamber were counted, and the results were expressed as the number of cells in a field of 660 μm × 840 μm. For each group, eosinophils in three fields for each well were counted, and results were reported as the mean of 18 fields (cell number ± SD).

### 3.13. Statistical Analysis

For all experiments, ANOVA was performed, and the Tukey–Kramer *post hoc* test was used to identify significantly different means. A $p$-value <0.01 was considered statistically significant.

## 4. Conclusions

In conclusion, peridinin suppressed DTH responses in mice. Peridinin also suppressed the numbers of eosinophils in ear tissues and peripheral blood. When painted on the ears, peridinin inhibited both the migration of eosinophils toward eotaxin and the production of eotaxin in ears. However, the suppressive effect of peridinin on the production of eotaxin was not observed when administered i.p. A structurally related carotenoid, fucoxanthin, inhibited the migration of eosinophils toward eotaxin only *in vitro* and did not suppress the DTH response. The major structural difference between peridinin and fucoxanthin is the presence of a butenolide moiety in peridinin. The butenolide moiety of peridinin may be important for suppressing these effects on eosinophils and for the production of eotaxin. Comparison of the inhibitory effects of peridinin and other carotenoids with the butenolide moiety remains to be clarified.

As described above, peridinin may ameliorate the allergic responses in which eosinophils play a major role in inflammation responses, such as asthma or atopic dermatitis.

## Acknowledgments

We appreciate Shiro Ono and Hiromi Okuyama (Osaka Ohtani University) for their help in the measurement of the serum levels of cytokines. This work was supported by the Program to Disseminate the Tenure Tracking System of the Ministry of Education, Culture, Sports, Science and Technology, the Japanese Government, and by the Adaptable and Seamless Technology Transfer Program through target-driven R & D, Japan Science and Technology Agency (AS242Z03463Q).

## Author Contributions

Ken-ichi Onodera and Akira Tominaga planned the experiments and wrote the manuscript. Yuko Konishi, Takahiro Taguchi and Sumio Kiyoto are engaged in the experiments of this research.

## Conflicts of Interest

The authors declare no conflict of interest.

## References

1. Nakamura, H.; Asari, T.; Murai, A.; Kan, Y.; Kondo, T.; Yoshida, K.; Ohizumi, Y. Zooxanthellatoxin-A, a potent vasoconstrictive 62-membered lactone from a symbiotic dinoflagellate. *J. Am. Chem. Soc.* **1995**, *117*, 550–551.
2. Onodera, K.; Nakamura, H.; Oba, Y.; Ohizumi, Y.; Ojika, M. Zooxanthellamide Cs: Vasoconstrictive polyhydroxylated macrolides with the largest lactone ring size from a marine dinoflagellate of *Symbiodinium* sp. *J. Am. Chem. Soc.* **2005**, *127*, 10406–10411.
3. Yasumoto, T. Chemistry, etiology, and food chain dynamics of marine toxins. *Proc. Jpn. Acad. Ser. B* **2005**, *81*, 43–51.
4. Kita, M.; Ohno, O.; Han, C.; Uemura, D. Bioactive secondary metabolites from symbiotic marine dinoflagellates: Symbiodinolide and durinskiols. *Chem. Rec.* **2010**, *10*, 57–69.
5. Ianora, A.; Bentley, M.G.; Caldwell, G.S.; Casotti, R.; Cembella, A.D.; Engström-Öst, J.; Halsband, C.; Sonnenschein, E.; Legrand, C.; Llewellyn, C.A.; *et al.* The relevance of marine chemical ecology to plankton and ecosystem function: An emerging field. *Mar. Drugs* **2011**, *9*, 1625–1648.
6. Ciminiello, P.; Dell'Aversano, C.; Dello Iacovo, E.; Fattorusso, E.; Forino, M.; Grauso, L.; Tartaglione, L.; Guerrini, F.; Pezzolesi, L.; Pistocchi, R.; *et al.* Isolation and structure elucidation of Ovatoxin-a, the major toxin produced by *Ostreopsis ovate*. *J. Am. Chem. Soc.* **2012**, *134*,
1869–1875.

7. Miyashita, K. The carotenoid fucoxanthin from brown seaweed affects obesity. *Lipid Technol.* **2009**, *21*, 186–190.
8. Sakai, S.; Sugawara, T.; Matsubara, K.; Hirata, T. Inhibitory effect of carotenoids on the degranulation of mast cells via suppression of antigen-induced aggregation of high affinity IgE receptors. *J. Biol. Chem.* **2009**, *284*, 28172–28179.
9. Kim, K.N.; Heo, S.J.; Yoon, W.J.; Kang, S.M.; Ahn, G.; Yi, T.H.; Jeon, Y.J. Fucoxanthin inhibits the inflammatory response by suppressing the activation of NF-κB and MAPKs in lipopolysaccharide-induced RAW 264.7 macrophages. *Eur. J. Pharmacol.* **2010**, *649*, 369–375.
10. Sakai, S.; Sugawara, T.; Hirata, T. Inhibitory effect of dietary carotenoids on dinitrofluorobenzene-induced contact hypersensitivity in mice. *Biosci. Biotechnol. Biochem.* **2011**, *75*, 1013–1015.
11. Peng, J.; Yuan, J.P.; Wu, C.F.; Wang, J.H. Fucoxanthin, a marine carotenoid present in brown seaweeds and diatoms: Metabolism and bioactivities relevant to human health. *Mar. Drugs* **2011**, *9*, 1806–1828.
12. D'Orazio, N.; Gemello, E.; Gammone, M.A.; de Girolamo, M.; Ficoneri, C.; Riccioni, G. Fucoxantin: A treasure from the sea. *Mar. Drugs* **2012**, *10*, 604–616.
13. Tsushima, M.; Maoka, T.; Katsuyama, M.; Kozuka, M.; Matsuno, T.; Tokuda, H.; Nishino, H.; Iwashima, A. Inhibitory effect of natural carotenoids on Epstein-Barr virus activation activity of a tumor promoter in Raji cells. A screening study for anti-tumor promoters. *Biol. Pharm. Bull.* **1995**, *18*, 227–233.
14. Sugawara, T.; Yamashita, K.; Sakai, S.; Asai, A.; Nagao, A.; Shiraishi, T.; Imai, I.; Hirata, T. Induction of apoptosis in DLD-1 human colon cancer cells by peridinin isolated from the dinoflagellate, *Heterocapsa triquetra*. *Biosci. Biotechnol. Biochem.* **2007**, *71*, 1069–1072.
15. Satoh, T.; Chen, Q.-J.; Sasaki, G.; Yokozeki, H.; Katayama, I.; Nishioka, K. Cyclophosphamide-induced eosinophilia in contact sensitivity: Mechanism of hapten-induced eosinophil recruitment into the skin. *Eur. J. Immunol.* **1997**, *27*, 85–91.
16. Ono, S.; Okayama, H. Osaka Ohtani University, Tondabayashi, Japan. 2013, Unpublished work.
17. Kimiya, T.; Ohtani, K.; Satoh, S.; Abe, Y.; Ogita, Y.; Kawakita, H.; Hamada, H.; Konishi, Y.; Kubota, S.; Tominaga, A. Inhibitory effects of edible marine algae extracts on degranulation of RBL-2H3 cells and mouse eosinophils. *Fish. Sci.* **2008**, *74*, 1157–1165.
18. Tominaga, A.; Takaki, S.; Koyama, N.; Katoh, S.; Matsumoto, R.; Migita, M.; Hitoshi, Y.; Hosoya, Y.; Yamauchi, S.; Kani, Y.; *et al.* Transgenic mice expressing a B cell growth and differentiation factor gene (interleukin5) develop eosinophilia and autoantibody production. *J. Exp. Med.* **1991**, *173*, 429–437.

19. Watanabe, Y.; Hashizume, M.; Kataoka, S.; Hamaguchi, E.; Morimoto, N.; Tsuru, S.; Katoh, S.; Miyake, K.; Matsushima, K., Tominaga, M.; *et al.* Differentiation of eosinophils characterized by hyaluronic acid binding via CD44 and responsiveness to stimuli. *DNA Cell Biol.* **2001**, *20*, 189–202.

*marine drugs*

*Review*

# Carotenoids in Algae: Distributions, Biosyntheses and Functions

**Shinichi Takaichi**

Department of Biology, Nippon Medical School, Kosugi-cho, Nakahara, Kawasaki 211-0063, Japan; E-Mail: takaichi@nms.ac.jp; Tel.: +81-44-733-3584; Fax: +81-44-733-3584

*Received: 2 May 2011; in revised form: 31 May 2011 / Accepted: 8 June 2011 / Published: 15 June 2011*

**Abstract:** For photosynthesis, phototrophic organisms necessarily synthesize not only chlorophylls but also carotenoids. Many kinds of carotenoids are found in algae and, recently, taxonomic studies of algae have been developed. In this review, the relationship between the distribution of carotenoids and the phylogeny of oxygenic phototrophs in sea and fresh water, including cyanobacteria, red algae, brown algae and green algae, is summarized. These phototrophs contain division- or class-specific carotenoids, such as fucoxanthin, peridinin and siphonaxanthin. The distribution of $\alpha$-carotene and its derivatives, such as lutein, loroxanthin and siphonaxanthin, are limited to divisions of Rhodophyta (macrophytic type), Cryptophyta, Euglenophyta, Chlorarachniophyta and Chlorophyta. In addition, carotenogenesis pathways are discussed based on the chemical structures of carotenoids and known characteristics of carotenogenesis enzymes in other organisms; genes and enzymes for carotenogenesis in algae are not yet known. Most carotenoids bind to membrane-bound pigment-protein complexes, such as reaction center, light-harvesting and cytochrome $b_6f$ complexes. Water-soluble peridinin-chlorophyll *a*-protein (PCP) and orange carotenoid protein (OCP) are also established. Some functions of carotenoids in photosynthesis are also briefly summarized.

**Keywords:** algal phylogeny; biosynthesis of carotenoids; distribution of carotenoids; function of carotenoids; pigment-protein complex

---

## 1. Introduction

Algae are classified throughout many divisions of the Kingdom Plantae. Their sizes range from single cells of picophytoplankton—the smallest of which are less than 1 μm—to seaweeds, the largest of which are more than 50 m. Attempts have been made to cultivate single-cell algae for a long time, but numbers were limited. With the recent development of culture techniques, some single-cell species can be cultured, and their characteristics, including pigments, can be studied. With the development of taxonomic technology, including DNA base sequences of 16S or 18S rRNA and some genes, algae phylogenetics has been developed.

More than 750 structurally defined carotenoids are reported from nature; land plants, algae, bacteria including cyanobacteria and photosynthetic bacteria, archaea, fungus and animals [1]. Except for animals, these organisms can synthesize many kinds of carotenoids, which are synthesized from diverse carotenogenesis pathways. These carotenoids and carotenogenesis pathways can be used as chemotaxonomic markers [2–7]. In addition, characteristics of carotenogenesis enzymes and genes are investigated. Some carotenogenesis genes have high similarity from bacteria to land plants, but some have low similarity. Some homologous genes have been proposed [8,9], but some carotenogenesis enzymes and genes, especially algae-specific ones, are not found.

In this review, the term algae refers to an oxygenic phototroph found in both seawater and fresh water, including cyanobacteria but excluding land plants. Distribution of carotenoids, carotenogenesis enzymes and pathways, and function of carotenoids in photosynthesis in algae are summarized.

## 2. Distribution of Carotenoids

Many different kinds of carotenoids were found from the algal species studied. Structures of some important carotenoids in algae are illustrated in Figure 1. Among them, approximately 30 types may have functions in photosynthesis, and others may be intermediates of carotenogenesis or accumulated carotenoids. Some carotenoids are found only in some algal divisions or classes; therefore, these carotenoids and also chlorophylls can be used as chemotaxonomic markers, and their distribution in algae is summarized in Table 1 [2–6].

Allene (C=C=C) is a unique structure in natural products, and is found mainly in carotenoids [10]; fucoxanthin in brown algae and diatoms, 19′-acyloxyfucoxanthin in Haptophyta and Dinophyta, peridinin only in dinoflagellates, and 9′-*cis* neoxanthin in green algae and land plants. Acetylene (C≡C) is also a unique structure, and acetylenic carotenoids are found only in algae; alloxanthin, crocoxanthin and monadoxanthin in Cryptophyta, and diadinoxanthin and diatoxanthin in Heterokontophyta, Haptophyta, Dinophyta and Euglenophyta. Acetylated carotenoids (-O-CO-$CH_3$), such as fucoxanthin, peridinin and dinoxanthin, are also mainly found in algae, such as Heterokontophyta, Haptophyta and Dinophyta. These carotenoids are specific to certain algal divisions and classes, and they are summarized in Table 1 based on our results [11–14] and some references [1–6].

Many cyanobacteria contain β-carotene, zeaxanthin, echinenone and myxol pentosides (myxoxanthophyll), while some species lack part of these and some contain additional carotenoids, such as nostoxanthin, canthaxanthin and oscillol dipentoside (Table 1, Figure 1) [13]. In addition, the carotenoid compositions of cyanobacteria are very different from those of chloroplasts in algae; consequently, during symbiosis of cyanobacteria to eukaryotic cells, carotenoids might be considerably restructured [13]. Note that since the name of myxoxanthophyll cannot specify the glycoside moieties, we have proposed the name of myxol glycosides to specify the glycosides, such as myxol 2′-α-L-fucoside, 4-ketomyxol 2′-rhamnoside and oscillol dichinovoside [13,15].

Rhodophyta (red algae) can be divided into two groups based on carotenoid composition; the unicellular type contains only β-carotene and zeaxanthin, and the macrophytic type contains additional α-carotene and lutein (Table 1, Figure 1) [16]. The relationship between phylogenetics of red algae and carotenoid composition is not clear [14]. Cryptophyta also contains α-carotene and its acetylenic derivatives, crocoxanthin and monadoxanthin, which are only found in this division.

**Figure 1.** Structures of some carotenoids.

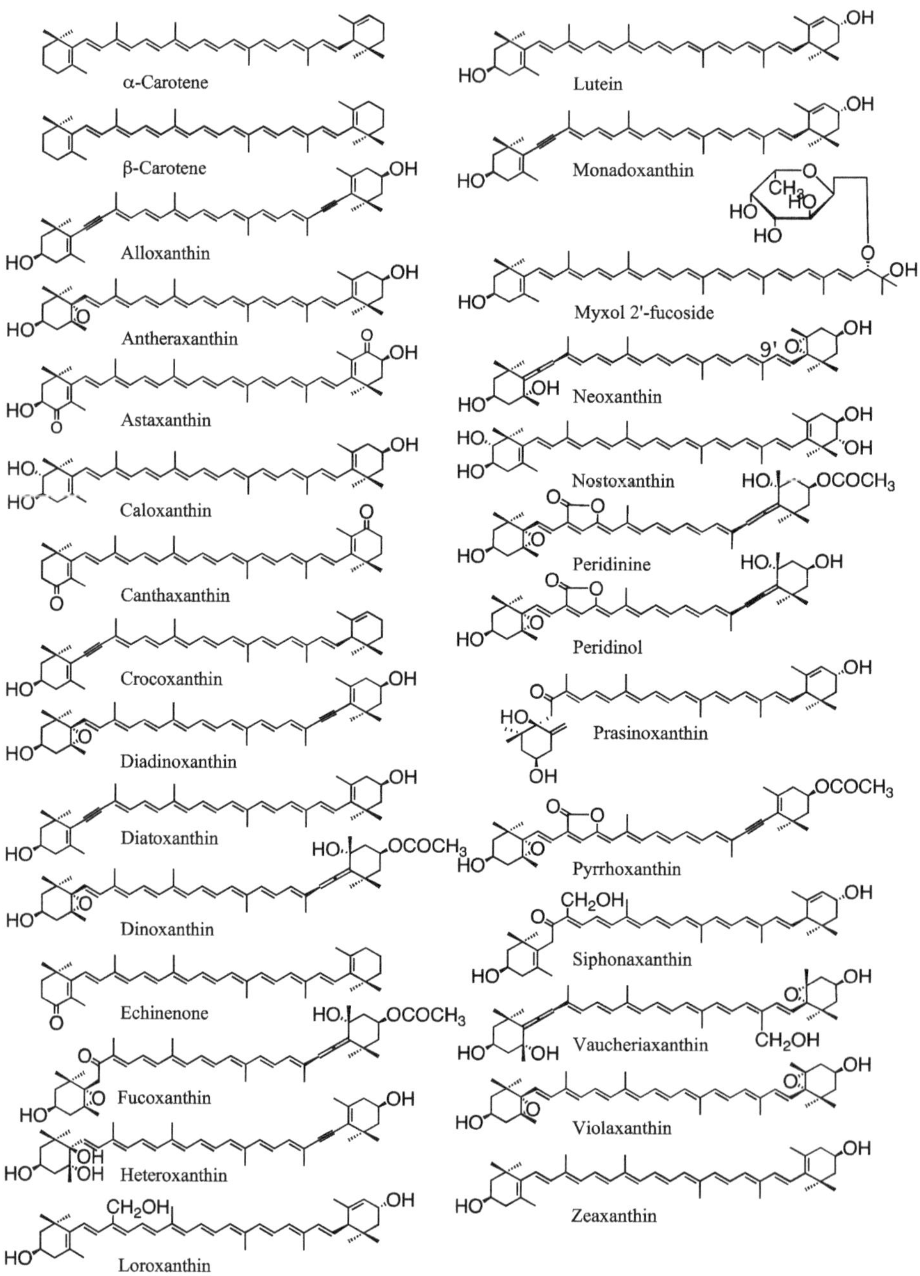

**Table 1.** Distribution of carotenoids in algae.

| Division Class | Carotene β | Carotene α | Xanthophyll Ze | Vi | Ne | Da | Dd | Fx | Va | Lu | Lo | Sx | Other xanthophyll(s) | Chlorophyll *a* | *b* | *c* |
|---|---|---|---|---|---|---|---|---|---|---|---|---|---|---|---|---|
| Cyanophyta | H | L | H | | | | | | | | | | No, L; Ec, H; My, H | H | L | |
| Glaucophyta | H | | H | | | | | | | | | | | H | | |
| Rhodophyta | | | | | | | | | | | | | | | | |
| Unicellular type | H | | H | | | | | | | | | | | H | | |
| Macrophytic type | L | L | H | L | | | | L | | H | | | | H | | |
| Cryptophyta | | H | L | | | | | | | | | | Al, L; Cr, L; Mo, L | H | | H |
| Heterokontophyta | | | | | | | | | | | | | | | | |
| Chrysophyceae | H | | L | | | L | L | H | L | | | | | H | | H |
| Raphidophyceae | H | | H | L | | L | L | L | | | | | | H | | H |
| Bacillariophyceae | H | | L | | | L | L | H | | | | | | H | | H |
| Phaeophyceae | H | | H | H | | L | L | H | | | | | | H | | H |
| Xanthophyceae | H | | L | | | H | H | | | | | | Va-FA, L | H | | H |
| Eustigmatophyceae | H | | | H | | | | | L | | | | | H | | |
| Haptophyta | H | | L | | | L | H | H | | | | | Fx-FA, L | H | | H |
| Dinophyta | L | | L | | | L | H | L | | | | | Pe, H | H | | H |
| Euglenophyta | H | | L | | L | L | H | | | | L | L | | H | H | |
| Chlorarachniophyta | H | | L | L | L | | | | | L | L | | Lo-FA, L | H | H | |
| Chlorophyta | | | | | | | | | | | | | | | | |
| Prasinophyceae | H | L | L | H | H | | | | | L | L | H | Pr, L; Lo-FA, L; Sx-FA, H | H | H | |
| Chlorophyceae | H | H | L | H | H | | | | | H | L | L | Sx-FA, L | H | H | |
| Ulvophyceae | H | L | L | H | H | | | | | L | L | L | Sx-FA, H | H | H | |
| Trebouxiophyceae | H | | L | H | H | | | | | H | | | | H | H | |
| Charophyceae | H | | L | H | H | | | | | H | | | | H | H | |

| Land Plants | H | L | L | H | H | H | H | H |
|---|---|---|---|---|---|---|---|---|

H, Major carotenoid in most species of the class; L, Low content in most species or major carotenoid in some species. α, α-carotene; β, β-carotene; Al, alloxanthin; Cr, crocoxanthin; Da, diatoxanthin; Dd, diadinoxanthin; Ec, echinenone; -FA, fatty acid ester; Fx, fucoxanthin; Lo, loroxanthin; Lu, lutein; Mo, monadoxanthin; My, myxol glycosides and oscillol glycosides; Ne, neoxanthin; No, nostoxanthin; Pe, peridinin; Pr, prasinoxanthin; Sx, siphonaxanthin; Va, vaucheriaxanthin; Vi, violaxanthin; Ze, zeaxanthin. Red, α-carotene and its derivatives

Heterokontophyta, Haptophyta and Dinophyta contain β-carotene and its derivatives as well as chlorophyll *c* (Table 1, Figure 1). These divisions, except for Eustigmatophyceae, which lacks chlorophylls *c*, contain unique acetylenic carotenoids of diadinoxanthin and diatoxanthin. Fucoxanthin and its derivatives are found in only four classes of Heterokontophyta (Chrysophyceae, Raphidophyceae, Bacillariophyceae and Phaeophyceae), Haptophyta and Dinophyta. Peridinin and its derivatives are found only in Dinophyta. Fucoxanthin and peridinin have unique structures (Figure 1) and are class-specific carotenoids (Table 1).

Euglenophyta, Chlorarachniophyta and Chlorophyta contain the same carotenoids, such as β-carotene, violaxanthin, 9′-*cis* neoxanthin [11] and lutein, as well as chlorophyll *a* and *b* with land plants (Table 1, Figure 1). Some classes contain additional carotenoids, such as loroxanthin, siphonaxanthin and prasinoxanthin, which are derivatives of lutein, and are class specific.

Note that identifications of some carotenoids were lacking because of insufficient analysis, and that some algae names were changed because of new developments in taxonomic technology and phylogenetic classification.

## 3. Carotenogenesis Pathways, Enzymes and Genes

Carotenogenesis pathways and their enzymes are mainly investigated in cyanobacteria [13] and land plants among oxygenic phototrophs [17]. Especially in land plants, carotenogenesis pathways and characteristics of enzymes are studied in detail (Figure 2). On the other hand, algae have common pathways with land plants and also additional algae-specific pathways, which are solely proposed based on the chemical structures of carotenoids (Figure 2). Some common carotenogenesis genes in algae are suggested from homology of the known genes [8,9], but most genes and enzymes for algae-specific pathways are still unknown (Figure 2). In cyanobacteria, since carotenoid compositions are different from those in land plants and algae, the pathways and enzymes are also different from those in Figure 2, and they are shown in Figure 3. In addition, carotenogenesis enzymes and genes, whose functions are confirmed in algae, including cyanobacteria, are summarized in Table 2. Unfortunately, these enzymes are mostly from cyanobacteria and green algae (Table 2).

### *3.1. Lycopene Synthesis*

#### 3.1.1. Isopentenyl Pyrophosphate to Phytoene Synthesis

Isopentenyl pyrophosphate (IPP), a $C_5$-compound, is the source of isoprenoids, terpenes, quinones, sterols, phytol of chlorophylls, and carotenoids. There are two known independent pathways of IPP synthesis: the classical mevalonate (MVA) pathway and the alternative, non-mevalonate, 1-deoxy-D-xylulose-5-phosphate (DOXP) pathway [18,19]. In the MVA pathway, acetyl-Coenzyme A is converted to IPP through mevalonate, and the enzymes and genes are well studied [20]. The pathway is found in plant cytoplasm, animals and some bacteria [18,20]. The DOXP pathway was found in the 1990s, and in this pathway, pyruvate and glycelaldehyde are converted to IPP. The DOXP pathway is found in cyanobacteria, the plastids of algae and land plants, and some bacteria [18]. Carotenoids are synthesized in plastids. Exceptionally among oxygenic phototrophs, Euglenophyceae has only the MVA pathway, and Chlorophyceae has only the DOXP pathway [18].

**Figure 2.** Carotenogenesis pathways and enzymes, whose functions are confirmed, in oxygenic phototrophs.

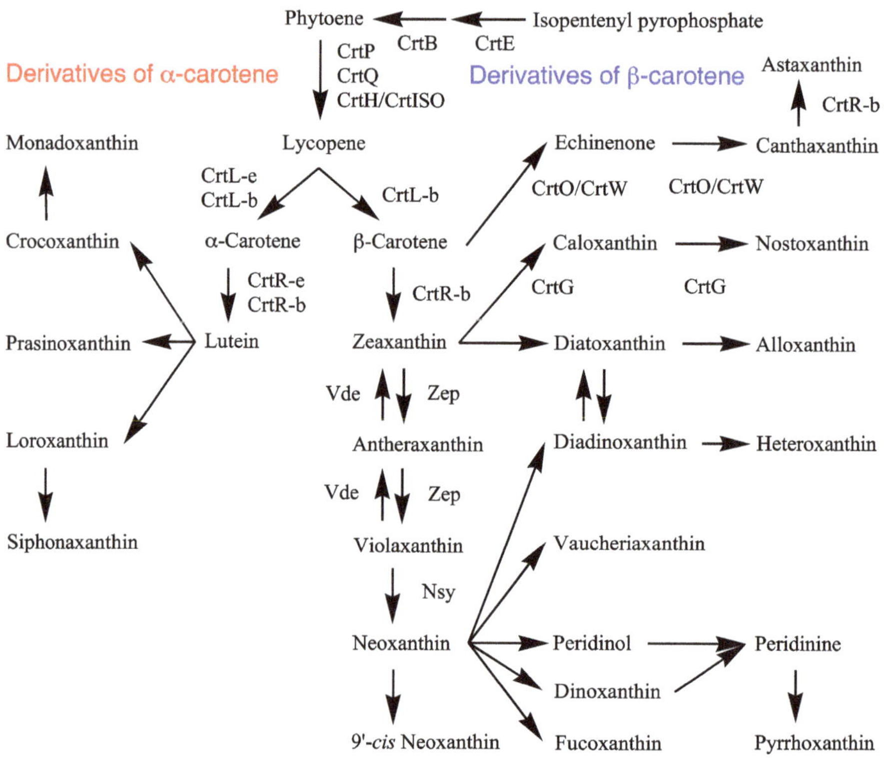

**Figure 3.** Carotenogenesis pathways and enzymes in cyanobacteria.

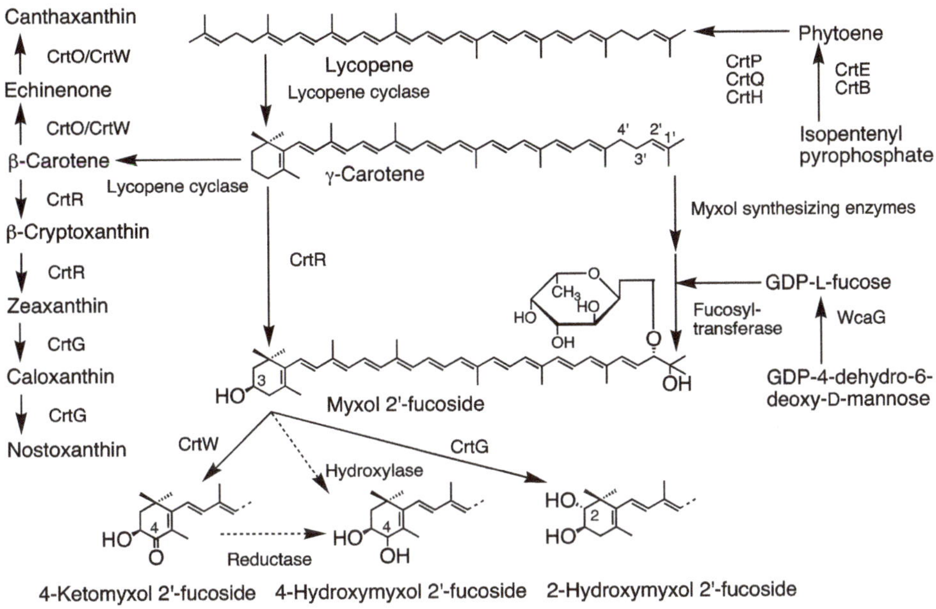


**Table 2.** Carotenogenesis genes and enzymes, whose functions are confirmed, in algae.

| Gene | Enzyme | Species | References |
|---|---|---|---|
| *crtE, ggps* | Geranylgeranyl pyrophosphate synthase | *Thermosynechococcus elongates* BP-1 | [21] |
| *crtB, pys, psy* | Phytoene synthase | *Gloeobacter violaceus* PCC 7421 | [22] |
| | | *Synechococcus elongatus* PCC 7942 | [23] |
| | | *Synechocystis* sp. PCC 6803 | [24] |
| | | *Chlamydomonas reinhardtii* | [25] |
| | | *Haematococcus pluvialis* NIES-144 | [26] |
| *crtI* | Phytoene desaturase (bacterial type) | *Gloeobacter violaceus* PCC 7421 | [22,27] |
| *crtP, pds* | Phytoene desaturase (plant type) | *Synechococcus elongatus* PCC 7942 | [23] |
| | | *Synechocystis* sp. PCC 6803 | [28] |
| | | *Chlamydomonas reinhardtii* | [29] |
| | | *Chlorella zofingiensis* ATCC 30412 | [30,31] |
| *crtQ, zds* | ζ-Carotene desaturase | *Anabaena* sp. PCC 7120 | [32] |
| | | *Synechocystis* sp. PCC 6803 | [33] |
| *crtH, crtISO* | Carotene isomerase | *Synechocystis* sp. PCC 6803 | [34,35] |
| *crtL, crtL-b, lcy-b* | Lycopene β-cyclase | *Synechococcus elongatus* PCC 7942 | [36] |
| | | *Prochlorococcus marinus* MED4 | [37] |
| | | *Cyanidioschyzon merolae* NIES-1332 | [38] |
| | | *Dunaliella salina* CCAP 19/30 | [39] |
| | | *Haematococcus pluvialis* NIES-144 | [40] |
| *crtL-e, lcy-e* | Lycopene ε-cyclase | *Prochlorococcus marinus* MED4 | [37] |
| *crtR* | β-Carotene hydroxylase | *Anabaena* sp. PCC 7120 | [41,42] |
| | | *Anabaena variabilis* ATCC 29413 | [42] |

| | | | |
|---|---|---|---|
| | | *Synechocystis* sp. PCC 6803 | [42–45] |
| | | *Haematococcus pluvialis* NIES-144 | [46] |
| *crtG* | β-Carotene 2-hydroxylase | *Thermosynechococcus elongates* BP-1 | [47] |
| *zep, npq* | Zeaxanthin epoxidase | *Chlamydomonas reinhardtii* CC-125 | [48] |
| *vde* | Violaxanthin de-epoxidase | *Mantonilla squamata* | [49] |
| *crtO* | β-Carotene ketolase | *Anabaena* sp. PCC 7120 | [50] |
| | | *Gloeobacter violaceus* PCC 7421 | [22] |
| | | *Synechocystis* sp. PCC 6803 | [42,45,51] |
| *crtW, bkt* | β-Carotene ketolase | *Anabaena* sp. PCC 7120 | [42,50] |
| | | *Gloeobacter violaceus* PCC 7421 | [22,27,42] |
| | | *Nostoc punctiforme* PCC 73102 | [42,52] |
| | | *Chlorella zofingiensis* ATCC 30412 | [53] |
| | | *Haematococcus pluvialis* NIES-144 | [54,55] |
| | | *Haematococcus pluvialis* strain 34/7 | [56] |

Red, genes and enzymes related to α-carotene.

Most carotenoids consist of eight IPP units. Farnesyl pyrophosphate ($C_{15}$) is synthesized from three IPPs, after which one IPP is added to farnesyl pyrophosphate by geranylgeranyl pyrophosphate synthase (CrtE, GGPS) to yield geranylgeranyl pyrophosphate ($C_{20}$). In a head-to-head condensation of the two $C_{20}$ compounds, the first carotene, phytoene ($C_{40}$), is formed by phytoene synthase (CrtB, Pys, Psy) using ATP [57,58]. This pathway has been confirmed by cloning genes from two species of *Rhodobacter* (purple bacteria) and two species of *Pantoea* (previously *Erwinia*) [57–59]. Among oxygenic phototrophs, the functions of CrtE of *Thermosynechococcus elongatus* BP-1 [21], and CrtB of three species of cyanobacteria [22–24] and two species of green algae [25,26] have also been confirmed (Table 2). The *crtE* and *crtB* genes have high sequence similarity from bacteria to land plants, respectively.

### 3.1.2. Phytoene to Lycopene Synthesis

Four desaturation steps are needed in the conversion from phytoene to lycopene. Oxygenic phototrophs require three enzymes: phytoene desaturase (CrtP, Pds), ζ-carotene desaturase (CrtQ, Zds) and *cis*-carotene isomerase (CrtH, CrtISO) (Figure 2). CrtP catalyzes the first two desaturation steps, from phytoene to ζ-carotene through phytofluene, and CrtQ catalyzes two additional desaturation steps, from ζ-carotene to lycopene through neurosporene. During desaturation by CrtQ, neurosporene and lycopene are isomerized to poly-*cis* forms, and then CrtH isomerizes to all-*trans* forms. Light is also effective for their photoisomerization to all-*trans* forms [34]. The functions of these enzymes have been mainly confirmed in cyanobacteria, green algae and land plants (Table 2): CrtP from *Synechocystis* sp. PCC 6803 [28], *Synechococcus elongatus* PCC 7942 [23], *Chlamydomonas reinhardtii* [29] and *Chlorella zofingiensis* [30,31], CrtQ from *Anabaena* sp. PCC 7120 (CrtQa,
*crtI*-like sequence) [32] and *Synechocystis* sp. PCC 6803 (CrtQb, plant *crtQ*-like) [33], and CrtH from *Synechocystis* sp. PCC 6803 [34,35]. The CrtP of *S. elongatus* PCC 7942 is stimulated by NAD(P) and oxygen as a possible final electron acceptor [60]. CrtQa has sequence homology with bacterial phytoene desaturase (CrtI) and CrtH, while CrtQb has sequence homology with

CrtP. In addition, genes homologous to *crtQa* are not found in cyanobacteria; therefore, among oxygenic phototrophs, *Anabaena* sp. PCC 7120 is the only species to have functional CrtQa.

In contrast, the bacterial type uses only one enzyme, phytoene desaturase (CrtI), to convert from phytoene to lycopene, and the primitive cyanobacterium of *Gloeobacter violaceus* PCC 7421 uses this type of CrtI, and the homologous genes of *crtP*, *crtQ* and *crtH* are not found in the genome [22,27]; therefore, *G. violaceus* is the first oxygenic phototroph that has been shown to use this type (Table 2). These observations suggest the following evolutionary scheme for this step in the reaction: the desaturation of phytoene was initially carried out by CrtI in ancestral cyanobacteria, *crtP* and related desaturase genes were acquired, and ultimately, there was replacement of *crtI* by *crtP* [27]. Among anoxygenic phototrophs, purple bacteria, green filamentous bacteria and heliobacteria use CrtI, whereas green sulfur bacteria use CrtP, CrtQ and CrtH [61].

### 3.2. *β-Carotene and α-Carotene Synthesis by Lycopene Cyclases*

All carotenoids in oxygenic phototrophs are dicyclic carotenoids; β-carotene, α-carotene and their derivatives, are derived from lycopene (Figures 1 and 2). Exceptionally, myxol glycosides and oscillol diglycosides in cyanobacteria are monocyclic and acyclic carotenoids, respectively.

Lycopene is cyclized into either β-carotene through γ-carotene, or α-carotene through γ-carotene or δ-carotene. Three distinct families of lycopene cyclases have been identified in carotenogenetic organisms [13,62,63]. One large family contains CrtY in some bacteria except cyanobacteria, and CrtL (CrtL-b, Lcy-b) in some cyanobacteria and land plants. Lycopene ε-cyclases (CrtL-e, Lcy-e) from land plants and lycopene β-monocyclases (CrtYm, CrtLm) from bacteria are also included. Their amino acid sequences exhibit a significant five conserved regions [39,62,64], and have an NAD(P)/
FAD-binding motif [65]. Note that Maresca *et al.* [63] divide this family into two CrtY and CrtL families. Three enzymes from Rhodophyta, *Cyanidioschyzon merolae* [38], and Chlorophyceae, *Dunaliella salina* [39] and *Haematococcus pluvialis* [40], are functionally confirmed (Table 2).

Some cyanobacteria also contain these enzymes (Table 2). *Synechococcus elongatus* PCC 7942 contains a functional CrtL [36]. *Prochlorococcus marinus* MED4 contains two lycopene cyclases (Table 2), which have sequence homology to CrtL. CrtL-b exhibits lycopene β-cyclase activity, while CrtL-e is a bifunctional enzyme having both lycopene ε-cyclase and lycopene β-cyclase activities [37]. The combination of these two cyclases allows the production of β-carotene, α-carotene and ε-carotene. Both enzymes might have originated from the duplication of a single gene. The characteristics of this CrtL-e are somewhat different from those in land plants [66]. In addition, the β-end groups of both
β-carotene and α-carotene (left half) might be hydroxylated by CrtR to zeaxanthin through β-cryptoxanthin and 3-hydroxy-α-carotene, respectively, in *P. marinus*. *Acaryochloris marina* MBIC 11017, which produces α-carotene, contains only one *crtL*-like gene from genome sequence [14].

The second family of lycopene cyclases, heterodimer (*crtYc* and *crtYd*) or monomer (*crtYc-Yd*), has been found in some bacteria, archaea and fungi [62,67], but not in phototrophs.

Recently, a new family of functional lycopene cyclase, CruA, has been found in *Chlorobaculum* (previously *Chlorobium*) *tepidum* (green sulfur bacterium), and the main product is γ-carotene in *Escherichia coli*, which produces lycopene [68]. Homologous genes, *cruA* and *cruP*, have been found in the genome of *Synechococcus* sp. PCC 7002, and their main products are γ-carotene, in *E. coli*, which produces lycopene [63]. In addition, their homologous genes are widely distributed in cyanobacteria, such as *Synechocystis* sp. PCC 6803 and *Anabaena* sp. PCC 7120; however, these *cruA*- and *cruP*-like genes from both *Synechocystis* sp. PCC 6803 and *Anabaena* sp. PCC 7120 did not show the lycopene dicyclase or monocyclase activities [14]. *S. elongatus* PCC 6301 and PCC 7942, and *A. marina* MBIC 11017 contain *crtL-*, *cruA-* and *cruP-*like genes; consequently, distributions of functional lycopene cyclases (CrtL-, CruA- and CruP-like) in cyanobacteria are unknown.

Since *Synechocystis* sp. PCC 6803 and *Anabaena* sp. PCC 7120 lack *crtL*-like genes and contain non-functional *cruA*-like genes, there is a possibility to present a fourth new family of lycopene cyclases in these cyanobacteria. Further studies of distributions of functional lycopene cyclases (CrtL- and CruA-like, or others) in cyanobacteria are needed.

Distribution of α-carotene, that is, CrtL-e, is limited in some algae classes (Table 1). Genes and enzymes of CrtL-e are not found in algae. In some species of land plants, the characteristics of CrtL-e were investigated [66], and were shown to have sequence homology with *crtL-b*. Lycopene is first converted to δ-carotene by CrtL-e, and then to α-carotene by CrtL-b. γ-Carotene produced by CrtL-b is not a suitable substrate for CrtL-e.

### 3.3. *β-Carotene Derivatives and Their Synthesis*

#### 3.3.1. Cyanobacteria

Some cyanobacteria produce zeaxanthin, and some produce both zeaxanthin and nostoxanthin (Figure 3). First, the C-3 and C-3′ hydroxyl groups of zeaxanthin are introduced to β-carotene by β-carotene hydroxylase (CrtR) through β-cryptoxanthin. Then, the C-2 and C-2′ hydroxyl groups of nostoxanthin are introduced by 2,2′-β-hydroxylase (CrtG) through caloxanthin (Table 2) [13,41–43,47]. The same enzymes, CrtR and CrtG, can also introduce hydroxyl groups to deoxymyxol and myxol to produce myxol and 2-hydroxymyxol, respectively [13,44,47]; consequently, the same enzymes are used in two pathways.

Cyanobacteria contain two ketocarotenoids, namely, canthaxanthin and 4-ketomyxol. Two distinct β-carotene ketolases, CrtO and CrtW, are known, and only seven enzymes are functionally confirmed in four species of cyanobacteria (Table 2) [13]. CrtO catalyzes β-carotene to echinenone, and the final product is canthaxanthin [22,42,45,50,51]. CrtW can introduce a keto group into β-carotene, zeaxanthin and myxol to produce canthaxanthin, astaxanthin and 4-ketomyxol, respectively (Figure 3) [22,27,42,50,52]; therefore, these ketolases are properly used in two pathways, β-carotene and myxol, depending on the species [13].

The pathway and the enzymes to produce the right half of myxol 2′-pentoside are still unknown (Figure 3) [13].

#### 3.3.2. Land Plants

In land plants, most of the carotenogenesis pathways and the functionally confirmed enzymes are known (Figure 2). Hydroxyl groups are introduced into β-carotene to produce zeaxanthin by β-carotene hydroxylase (CrtR, CrtR-b, BCH). Epoxy groups are introduced into zeaxanthin by zeaxanthin epoxidase (Zep, NPQ) to produce violaxanthin through antheraxanthin. Under high light conditions, violaxanthin is changed into zeaxanthin by violaxanthin de-epoxidase (Vde) for dispersion of excess energy from excited chlorophylls. One end group of violaxanthin is changed to an allene group of neoxanthin by neoxanthin synthase (Nsy). Because all neoxanthin in chloroplasts has the 9′-*cis* form, unknown 9′-isomerase for all *trans* neoxanthin to 9′-*cis* neoxanthin should be present [11].

#### 3.3.3. Algae

Little is known for the carotenogenesis pathways among algae, but some are proposed based on the chemical structures of carotenoids (Figure 2). Functionally confirmed enzymes are mainly reported in Chlorophyceae including *Chlorella, Chlamydomonas, Dunaliella* and *Haematococcus* for CrtB, CrtP, CrtL-b, CrtR-b [46], Zep [48], Vde [49], and CrtW (Table 2).

In the cell-free preparation of *Amphidinium carterae* (Dinophyta), $^{14}$C-labellled zeaxanthin was incorporated into allenic carotenoid of neoxanthin, and then into acetylenic diadinoxanthin and $C_{37}$ peridinin (Figure 2). In addition, the three carbon atoms of C-13′,14′,20′ of peridinin were eliminated from neoxanthin (C-13,14,20) [69,70]. In organic chemistry, the C-7,8 double bond of zeaxanthin can be oxidized to the triple bond (acetylene group) of diatoxanthin [17].

Allenic carotenoids are very limited in algae. From their chemical structures, all *trans* neoxanthin might be changed to fucoxanthin, dinoxanthin, peridinin, vaucheriaxanthin and diadinoxanthin, but the pathways and enzymes are still unknown (Figures 1 and 2).

Under a stressful environment, such as high light, UV irradiation and nutrition stress, some Chlorophyceae, such as *Haematococcus, Chlorella* and *Scenedesmus*, accumulate ketocarotenoids, canthaxanthin and astaxanthin, which are synthesized by combining CrtR-b and β-carotene ketolase (CrtW, BKT) (Table 2) [53–56,71]. Note that although β-carotene ketolase of *Haematococcus* and *Chlorella* were named CrtO at first [53,56], they are CrtW-type not CrtO-type from amino acid sequences (Table 2).

### 3.4. *α-Carotene Derivatives and Their Synthesis*

In *Arabidopsis thaliana*, β-carotene is hydroxylated mainly by the non-heme di-iron enzymes, BCH1 and BCH2 (CrtR-b), to produce zeaxanthin, while α-carotene is mainly hydroxylated by the cytochrome P450 enzymes, CYP97A3 for the β-end group and CYP97C1 for the β- and ε-end groups, to produce lutein [72].

Lutein and its derivatives are found only in Rhodophyta (macrophytic type), Cryptophyta, Euglenophyta, Chlorarachniophyta and Chlorophyta (Table 1), but nothing is known for hydroxylation of α-carotene. From the chemical structures of siphonaxanthin [12], loroxanthin, prasinoxanthin and monadoxanthin, it could be considered that they are derived from lutein, but the pathways and enzymes are still unknown (Figures 1 and 2).

## 4. Function of Carotenoids

For photosynthesis, both carotenoids and chlorophylls are necessarily bound to peptides to form pigment-protein complexes in the thylakoid membrane. Five main kinds of the complexes described below are isolated from some algae, and the pigment compositions are investigated [73–75]. Exceptionally in cyanobacteria, myxol glycosides and some carotenoids are located in the cytoplasmic membrane for protection from high-light [76,77].

β-Carotene is presented in the most divisions of the reaction-center complexes (RC) and the light-harvesting complexes (LHC) of photosystem I (PSI) as well as the RC and the core LHC of photosystem II (PSII); exceptionally zeaxanthin is presented in some red algae of the LHC of PSI.

On the other hand, in the peripheral LHC of PSII, the bound carotenoids are heterogenous depending on the classes. Major carotenoids are alloxanthin (Cryptophyta); fucoxanthin (Chrysophyceae, Raphidophyceae, Bacillariophyceae, Phaeophyceae and Haptophyta); diadinoxanthin and vaucheriaxanthin (Xanthophyceae); violaxanthin and vaucheriaxanthin (Eustigmatophyceae);

peridinin (Dinophyta); diadinoxanthin (Euglenophyta); siphonaxanthin (Chlorophyceae and Ulvophyceae); and lutein, violaxanthin and 9′-*cis* neoxanthin (land plants) (Figure 1) [73–75]. β-Carotene in both RC might have protective functions, and carotenoids in the peripheral LHC of PSII mainly might have light-harvesting functions.

The dimeric cytochrome $b_6f$ complexes of the cyanobacterium *Mastigocladus laminosus* [78] and the green alga *Chlamydomonas reinhardtii* [79] contain two β-carotene and two chlorophyll *a* molecules, while that of the cyanobacterium *Synechocystis* sp. PCC 6803 contains two echinenone and two chlorophyll *a* molecules [80]. These carotenoids might have protective functions.

The water-soluble peripheral LHC of peridinin-chlorophyll-protein (PCP) isolated from *Amphidinium carterae* (Dinophyta) has a trimeric structure, and the monomer contains eight peridinin and two chlorophyll *a* molecules [81]. The water-soluble orange carotenoid protein (OCP) isolated from the cyanobacterium *Arthrospira maima* forms a homodimer with two 3′-hydroxyechinenone molecules [82]. OCP is also found in some cyanobacteria, and its function might regulate energy dissipation from phycobilisomes to PSII [83].

The keto groups at C-8 of fucoxanthin [84], siphonaxanthin [85,86] and prasinoxanthin [87], which are found only in algae, are the single-bond *trans*-conformation for the conjugated double bond

(Figure 1). From the femtosecond time-resolved fluorescence spectroscopy of the purified

carotenoids in organic solvents and the LHC in solution, these keto-carotenoids and peridinin have been found to have highly efficient energy transfer from the $S_1$ state, not the $S_2$ state, of carotenoids to chlorophylls. From the comparison of other structural carotenoids, these keto groups are essential for high
efficiency [88,89]. These keto-carotenoids mainly might have light-harvesting functions.

The xanthophyll cycle, also known as the violaxanthin cycle, is the cyclical interconversion of violaxanthin, antheraxanthin and zeaxanthin in green algae and land plants (Figure 2) [90]. Zep catalyzes zeaxanthin to violaxanthin through antheraxanthin during biosynthesis. Violaxanthin is found in the peripheral LHC of PSII. Under high light conditions, Vde is activated and catalyzes de-epoxidation of violaxanthin to zeaxanthin through antheraxanthin. Zeaxanthin is used for the dissipation of excess energy from excited chlorophylls. Zep from Chlorophyceae *Chlamydomonas reinhardtii* [48] and Vde from Pracinophyceae *Mantonilla squamata* [49] are functionally confirmed (Table 2). Similarly, the diadinoxanthin cycle occurs in Heterokontophyta, Haptophyta and Dinophyta, which contain diadinoxanthin and diatoxanthin (Figure 2). The enzymes of diadinoxanthin de-epoxidase and diatoxanthin epoxidase have not yet been found [9,91], but the characteristics of partially purified diadinoxanthin de-epoxidase from the diatom *Cyclotella meneghinaina* are reported [92].

## References


1. Britton, G.; Liaaen-Jensen, S.; Pfander, H. *Carotenoids Handbook*; Birkhäuser: Basel, Switzerland, 2004.
2. Rowan, K.S. *Photosynthetic Pigments of Algae*; Cambridge University Press: Cambridge, UK, 1989.
3. Bjørnland, T.; Liaaen-Jensen, S. Distribution patterns of carotenoids in relation to chromophyte phylogeny and systematics. In *The Chromophyte Algae: Problems and Perspectives*; Green, J.C., Leadbeater, B.S.C., Diver, W.I., Eds.; Clarendon Press: Oxford, UK, 1989; pp. 37–60.
4. Liaaen-Jensen, S. Marine carotenoids. *New J. Chem.* **1990**, *14*, 747–759.
5. Mackey, M.D.; Mackey, D.J.; Higgins, H.W.; Wright, S.W. CHEMTAX-a program for estimating class abundances from chemical markers: Application to HPLC measurements of phytoplankton. *Mar. Ecol. Prog. Ser.* **1996**, *144*, 265–283.
6. Jeffrey, S.W.; Vesk, M. Introduction to marine phytoplankton and their pigment signatures. In *Phytoplankton Pigments in Oceanography: Guidelines to Modern Methods*; Jeffrey, S.W., Mantoura, R.F.C., Wright, S.W., Eds.; UNESCO Publishing: Paris, France, 1997; pp. 37–84.
7. Liaaen-Jensen, S. Carotenoids in chemosystematics. In *Carotenoids: Biosynthesis and Metabolism*; Britton, G., Liaaen-Jensen, S., Pfander, H., Eds.; Birkhäuser: Basel, Switzerland, 1998; Volume 3, pp. 217–247.
8. Frommolt, R.; Werner, S.; Paulsen, H.; Goss, R.; Wilhelm, C.; Zauner, S.; Maier, U.G.; Grossman, A.R.; Bhattacharya, D.; Lohr, M. Ancient recruitment by chromists of green algal genes encoding enzymes for carotenoid biosynthesis. *Mol. Biol. Evol.* **2008**, *25*, 2653–2667.

9. Bertrand, M. Carotenoid biosynthesis in diatoms. *Photosynth. Res.* **2010**, *106*, 89–102.
10. Dembitsky, V.M.; Maoka, T. Allenic and cumulenic lipids. *Prog. Lipid Res.* **2007**, *46*, 328–375.
11. Takaichi, S.; Mimuro, M. Distribution and geometric isomerism of neoxanthin in oxygenic phototrophs: 9′-*cis*, a sole molecular form. *Plant Cell Physiol.* **1998**, *39*, 968–977.
12. Yoshii, Y.; Takaichi, S.; Maoka, T.; Suda, S.; Sekiguchi, H.; Nakayama, T.; Inouye, I. Variation of siphonaxanthin series among the genus *Nephroselmis* (Prasinophyceae, Chlorophyta), including a novel primary methoxy carotenoid. *J. Phycol.* **2005**, *41*, 827–834.
13. Takaichi, S.; Mochimaru, M. Carotenoids and carotenogenesis in cyanobacteria: Unique ketocarotenoids and carotenoid glycosides. *Cell. Mol. Life Sci.* **2007**, *64*, 2607–2619.
14. Takaichi, S. Nippon Medical School, Kawasaki, Japan. Unpublished works, 2011.
15. Takaichi, S.; Maoka, T.; Masamoto, K. Myxoxanthophyll in *Synechocystis* sp. PCC 6803 is myxol 2′-dimethyl-fucoside, (3*R*,2′*S*)-myxol 2′-(2,4-di-*O*-methyl-α-L-fucoside), not rhamnoside. *Plant Cell Physiol.* **2001**, *42*, 756–762.
16. Schubert, N.; García-Mendoza, E. Photoinhibition in red algal species with different carotenoid profiles. *J. Phycol.* **2008**, *44*, 1437–1446.
17. Britton, G. Overview of carotenoid biosynthesis. In *Carotenoids: Biosynthesis and Metabolism*; Britton, G., Liaaen-Jensen, S., Pfander, H., Eds.; Birkhäuser: Basel, Switzerland, 1998; Volume 3, pp. 13–147.
18. Lichtenthaler, H.K. The 1-deoxy-D-xylulose-5-phosphate pathway of isoprenoid biosynthesis in plants. *Annu. Rev. Plant Physiol. Plant Mol. Biol.* **1999**, *50*, 47–65.
19. Eisenreich, W.; Bacher, A.; Arigoni, D.; Rohdich, F. Biosynthesis of isoprenoids via the non-mevalonate pathway. *Cell. Mol. Life Sci.* **2004**, *61*, 1401–1426.
20. Miziorko, H.M. Enzymes of the mevalonate pathway of isoprenoid biosynthesis. *Arch. Biochem. Biophys.* **2011**, *505*, 131–143.
21. Ohto, C.; Ishida, C.; Nakane, H.; Muramatsu, M.; Nishino, T.; Obata, S. A thermophilic cyanobacterium *Synechococcus elongatus* has three different Class I prenyltransferase genes. *Plant Mol. Biol.* **1999**, *40*, 307–321.
22. Steiger, S.; Jackisch, Y.; Sandmann, G. Carotenoid biosynthesis in *Gloeobacter violaceus* PCC4721 involves a single crtI-type phytoene desaturase instead of typical cyanobacterial enzymes. *Arch. Microbiol.* **2005**, *184*, 207–214.
23. Chamovitz, D.; Misawa, N.; Sandmann, G.; Hirschberg, J. Molecular cloning and expression in *Escherichia coli* of a cyanobacterial gene coding for phytoene synthase, a carotenoid biosynthesis enzyme. *FEBS Lett.* **1992**, *296*, 305–310.
24. Martínez-Férez, I.; Fernández-González, B.; Sandmann, G.; Vioque, A. Cloning and expression in *Escherichia coli* of the gene coding for phytoene synthase from the cyanobacterium *Synechocystis* sp. PCC6803. *Biochim. Biophys. Acta* **1994**, *1218*, 145–152.
25. McCarthy, S.S.; Kobayashi, M.C.; Niyogi, K.K. White mutants of *Chlamydomonas reinhardtii* are defective in phytoene synthase. *Genetics* **2004**, *168*, 1249–1257.
26. Steinbrenner, J.; Linden, H. Regulation of two carotenoid biosynthesis genes coding for phytoene synthase and carotenoid hydroxylase during stress-induced astaxanthin formation in the green alga *Haematococcus pluvialis*. *Plant Physiol.* **2001**, *125*, 810–817.

27. Tsuchiya, T.; Takaichi, S.; Misawa, N.; Maoka, T.; Miyashita, H.; Mimuro, M. The cyanobacterium *Gloeobacter violaceus* PCC 7421 uses bacterial-type phytoene desaturase in carotenoid biosynthesis. *FEBS Lett.* **2005**, *579*, 2125–2129.
28. Martínez-Férez, I.M.; Vioque, A. Nucleotide sequence of the phytoene desaturase gene from *Synechocystis* sp. PCC 6803 and characterization of a new mutation which confers resistance to the herbicide norflurazon. *Plant Mol. Biol.* **1992**, *18*, 981–983.
29. Vila, M.; Couso, I.; León, R. Carotenoid content in mutants of the chlorophyte *Chlamydomonas reinhardtii* with low expression levels of phytoene desaturase. *Process Biochem.* **2008**, *43*, 1147–1152.
30. Huang, J.; Liu, J.; Li, Y.; Chen, F. Isolation and characterization of the phytoene desaturase gene as a potential selective marker for genetic engineering of the astaxanthin-producing green alga *Chlorella zofingiensis* (Chlorophyta). *J. Phycol.* **2008**, *44*, 684–690.
31. Liu, J.; Zhong, Y.; Sun, Z.; Huang, J.; Sandmann, G.; Chen, F. One amino acid substitution in phytoene desaturase makes *Chlorella zofingiensis* resistant to norflurazon and enhances the biosynthesis of astaxanthin. *Planta* **2010**, *232*, 61–67.
32. Linden, H.; Vioque, A.; Sandmann, G. Isolation of a carotenoid biosynthesis gene coding for ζ-carotene desaturase from *Anabaena* PCC 7120 by heterologous complementation. *FEMS Microbiol. Lett.* **1993**, *106*, 99–104.
33. Breitenbach, J.; Fernández-González, B.; Vioque, A.; Sandmann, G. A higher-plant type ζ-carotene desaturase in the cyanobacterium *Synechocystis* PCC6803. *Plant Mol. Biol.* **1998**, *36*, 725–732.
34. Masamoto, K.; Wada, H.; Kaneko, T.; Takaichi, S. Identification of a gene required for *cis*-to-*trans* carotene isomerization in carotenogenesis of the cyanobacterium *Synechocystis* sp. PCC 6803. *Plant Cell Physiol.* **2001**, *42*, 1398–1402.
35. Breitenbach, J.; Vioque, A.; Sandmann, G. Gene *sll0033* from *Synechocystis* 6803 encodes a carotene isomerase involved in the biosynthesis of all-*E* lycopene. *Z. Naturforsch.* **2001**, *56c*, 915–917.
36. Cunningham, F.X., Jr.; Sun, Z.; Chamovitz, D.; Hirschberg, J.; Gantt, E. Molecular structure and enzymatic function of lycopene cyclase from the cyanobacterium *Synechococcus* sp. strain PCC7942. *Plant Cell* **1994**, *6*, 1107–1121.
37. Stickforth, P.; Steiger, S.; Hess, W.R.; Sandmann, G. A novel type of lycopene ε-cyclase in the marine cyanobacterium *Prochlorococcus marinus* MED4. *Arch. Microbiol.* **2003**, *179*, 409–415.
38. Cunningham, F.X., Jr.; Lee, H.; Gantt, E. Carotenoid biosynthesis in the primitive red alga *Cyanidioschyzon merolae*. *Eukaryot. Cell* **2007**, *6*, 533–545.
39. Ramos, A.; Coesel, S.; Marques, A.; Rodrigues, M.; Baumgartner, A.; Noronha, J.; Rauter, A.; Brenig, B.; Varela, J. Isolation and characterization of a stress-inducible *Dunaliella salina* *Lyc-β* gene encoding a functional lycopene β-cyclase. *Appl. Microbiol. Biotechnol.* **2008**, *79*, 819–828.

40. Steinbrenner, J.; Linden, H. Light induction of carotenoid biosynthesis genes in the green alga *Haematococcus pluvialis*: Regulation by photosynthetic redox control. *Plant Mol. Biol.* **2003**, *52*, 343–356.
41. Mochimaru, M.; Msukawa, H.; Maoka, T.; Mohamed, H.E.; Vermaas, W.F.J.; Takaichi, S. Substrate specificities and availability of fucosyltransferase and β-carotene hydroxylase for myxol 2′-fucoside synthesis in *Anabaena* sp. strain PCC 7120 compared with *Synechocystis* sp. strain PCC 6803. *J. Bacteriol.* **2008**, *190*, 6726–6733.
42. Makino, T.; Harada, H.; Ikenaga, H.; Matsuda, S.; Takaichi, S.; Shindo, K.; Sandmann, G.; Ogata, T.; Misawa, N. Characterization of cyanobacterial carotenoid ketolase CrtW and hydroxylase CrtR by complementation analysis in *Escherichia coli*. *Plant Cell Physiol.* **2008**, *49*, 1867–1878.
43. Masamoto, K.; Misawa, N.; Kaneko, T.; Kikuno, R.; Toh, H. β-Carotene hydroxylase gene from the cyanobacterium *Synechocystis* sp. PCC6803. *Plant Cell Physiol.* **1998**, *39*, 560–564.
44. Lagarde, D.; Vermaas, W. The zeaxanthin biosynthesis enzyme β-carotene hydroxylase is involved in myxoxanthophyll synthesis in *Synechocystis* sp. PCC 6803. *FEBS Lett.* **1999**, *454*, 247–251.
45. Lagarde, D.; Beuf, L.; Vermaas, W. Increased production of zeaxanthin and other pigments by application of genetic engineering techniques to *Synechocystis* sp. strain PCC 6803. *Appl. Environ. Microbiol.* **2000**, *66*, 64–72.
46. Linden, H. Carotenoid hydroxylase from *Haematococcus pluvialis*: cDNA sequence, regulation and functional complementation. *Biochim. Biophys. Acta* **1999**, *1446*, 203–212.
47. Iwai, M.; Maoka, T.; Ikeuchi, M.; Takaichi, S. 2,2′-β-Hydroxylase (CrtG) is involved in carotenogenesis of both nostoxanthin and 2-hydroxymyxol 2′-fucoside in *Thermosynechococcus elongatus* strain BP-1. *Plant Cell Physiol.* **2008**, *49*, 1678–1687.
48. Baroli, I.; Do, A.D.; Yamane, T.; Niyogi, K.K. Zeaxanthin accumulation in the absence of a functional xanthophyll cycle protects *Chlomydomonas reinhardtii* from photooxidative stress. *Plant Cell* **2003**, *15*, 992–1008.
49. Goss, R. Substrate specificity of the violaxanthin de-epoxidase of the primitive green alga *Mantoniella squamata* (Prasinophyceae). *Planta* **2003**, *217*, 801–812.
50. Mochimaru, M.; Msukawa, H.; Takaichi, S. The cyanobacterium *Anabaena* sp. PCC 7120 has two distinct β-carotene ketolase: CrtO for echinenone and CrtW for ketomyxol synthesis. *FEBS Lett.* **2005**, *579*, 6111–6114.
51. Fernández-González, B.; Sandmann, G.; Vioque, A. A new type of asymmetrically acting β-carotene ketolase is required for the synthesis of echinenone in the cyanobacterium *Synechocystis* sp. PCC 6803. *J. Biol. Chem.* **1997**, *272*, 9728–9733.
52. Steiger, S.; Sandmann, G. Cloning of two carotenoid ketolase genes from *Nostoc punctiforme* for the heterologous production of canthaxanthin and astaxanthin. *Biotechnol. Lett.* **2004**, *26*, 813–817.
53. Huang, J.-C.; Wang, Y.; Sandmann, G.; Chen, F. Isolation and characterization of a carotenoid oxygenase gene from *Chlorella zofingiensis* (Chlorophyta). *Appl. Microbiol. Biotechnol.* **2006**, *71*, 473–479.

54. Kajiwara, S.; Kakizono, T.; Saito, T.; Kondo, K.; Ohtani, T.; Nishio, N.; Nagai, S.; Misawa, N. Isolation and functional identification of a novel cDNA from astaxanthin biosynthesis from *Haematococcus pluvialis*, and astaxanthin synthesis in *Escherichia coli*. *Plant Mol. Biol.* **1995**, *29*, 343–352.
55. Huang, J.-C.; Chen, F.; Sandmann, G. Stress-related differential expression of multiple β-carotene ketolase genes in the unicellular green alga *Haematococcus pluvialis*. *J. Biotechnol.* **2006**, *122*, 176–185.
56. Lotan, T.; Hirschberg, J. Cloning and expression in *Escherichia coli* of the gene encoding β-C-4-oxygenase, that converts β-carotene to the ketocarotenoid canthaxanthin in *Haematococcus pluvialis*. *FEBS Lett.* **1995**, *364*, 125–128.
57. Sandmann, G. Carotenoid biosynthesis in microorganisms and plants. *Eur. J. Biochem.* **1994**, *223*, 7–24.
58. Armstrong, G.A. Genetics of eubacterial carotenoid biosynthesis: A colorful tale. *Annu. Rev. Microbiol.* **1997**, *51*, 629–659.
59. Misawa, N.; Nakagawa, M.; Kobayashi, K.; Yamano, S.; Izawa, Y.; Nakamura, K.; Harashima, K. Elucidation of the *Erwinia uredovora* carotenoid biosynthetic pathway by functional analysis of gene products expressed in *Escherichia coli*. *J. Bacteriol.* **1990**, *172*, 6704–6712.
60. Schneider, C.; Böger, P.; Sandmann, G. Phytoene desaturase: Heterologous expression in an active state, purification, and biochemical properties. *Protein Expr. Purif.* **1997**, *10*, 175–179.
61. Takaichi, S. Distribution and biosynthesis of carotenoids. In *The Purple Phototrophic Bacteria*; Hunter, C.N., Daldal, F., Thurnauer, M.C., Beatty, J.T., Eds.; Springer: Dordrecht, The Netherlands, 2009; pp. 97–117.
62. Krubasik, P.; Sandmann, G. Molecular evolution of lycopene cyclases involved in the formation of carotenoids with ionone end groups. *Biochem. Soc. Trans.* **2000**, *28*, 806–810.
63. Maresca, J.A.; Graham, J.E.; Wu, M.; Eisen, J.A.; Bryant, D.A. Identification of a fourth family of lycopene cyclases in photosynthetic bacteria. *Proc. Natl. Acad. Sci. USA* **2007**, *104*, 11784–11789.
64. Sandmann, G. Molecular evolution of carotenoid biosynthesis from bacteria to plants. *Physiol. Plant.* **2002**, *116*, 431–440.
65. Harker, M.; Hirschberg, J. Molecular biology of carotenoid biosynthesis in photosynthetic organisms. *Methods Enzymol.* **1998**, *297*, 244–263.
66. Cunningham, F.X., Jr.; Gantt, E. One ring or two? Determination of ring number in carotenoids by lycopene ε-cyclases. *Proc. Natl. Acad. Sci. USA* **2001**, *98*, 2905–2910.
67. Hemmi, H.; Ikejiri, S.; Nakayama, T.; Nishino, T. Fusion-type lycopene β-cyclase from a thermoacidophilic archaeon *Sulfolobus solfataricus*. *Biochem. Biophys. Res. Commun.* **2003**, *305*, 586–591.
68. Maresca, J.A.; Frigaard, N.-U.; Bryant, D.A. Identification of a novel class of lycopene cyclases in photosynthetic organisms. In *Photosynthesis: Fundamental Aspects to Global Perspectives*;
van der Est, A., Bruce, D., Eds.; Allen Press: Lawrence, KS, USA, 2005; pp. 884–886.

69. Swift, I.E.; Milborrow, B.V.; Jeffrey, S.W. Formation of neoxanthin, diadinoxanthin and peridinin from [14C]zeaxanthin by a cell-free system from *Amphidinium carterae*. *Phytochemistry* **1982**, *21*, 2859–2864.
70. Swift, I.E.; Milborrow, B.V. Stereochemistry of allene biosynthesis and the formation of the acetylenic carotenoid diadinoxanthin and peridinin ($C_{37}$) from neoxanthin. *Biochem. J.* **1981**, *199*, 69–74.
71. Lemoine, Y.; Schoefs, B. Secondary ketocarotenoid astaxanthin biosynthesis in algae: A multifunctional response to stress. *Photosynth. Res.* **2010**, *106*, 155–177.
72. Kim, J.; Smith, J.J.; Tian, L.; DellaPenna, D. The evolution and function of carotenoid hydroxylases in Arabidopsis. *Plant Cell Physiol.* **2009**, *50*, 463–479.
73. Durnford, D.G. Structure and regulation of algal light-harvesting complex genes. In *Photosynthesis in Algae*; Larkum, A.W.D., Douglas, S.E., Raven, J.A., Eds.; Kluwer: Dordrecht, The Netherlands, 2003; pp. 63–82.
74. Macpherson, A.N.; Hiller, R.G. Light-harvesting systems in chlorophyll *c*-containing algae. In *Light-Harvesting Antennas in Photosynthesis*; Green, B.R., Parson, W.W., Eds.; Kluwer: Dordrecht, The Netherlands, 2003; pp. 323–352.
75. Neilson, J.A.D.; Durnford, D.G. Structural and functional diversification of the light-harvesting complexes in photosynthetic eukaryotes. *Photosynth. Res.* **2010**, *106*, 57–71.
76. Kana, T.M.; Glibert, P.M.; Goericke, R.; Welschmeyer, N.A. Zeaxanthin and β-carotene in *Synechococcus* WH7803 respond differently to irradiance. *Limnol. Oceanogr.* **1998**, *33*, 1623–1627.
77. Masamoto, K.; Zsiros, O.; Gombos, Z. Accumulation of zeaxanthin in cytoplasmic membranes of the cyanobacteirum *Synechococcus* sp. Strain PCC 7942 grown under high light condition. *J. Plant Physiol.* **1999**, *155*, 136–138.
78. Kurisu, G.; Zhang, H.; Smith, J.L.; Cramer, W.A. Structure of the cytochrome $b_6f$ complex of oxygenic photosynthesis: Tuning the cavity. *Science* **2003**, *302*, 1009–1014.
79. Stroebel, D.; Choquet, Y.; Popot, J.-L.; Picot, D. An atypical haem in the cytochrome $b_6f$ complex. *Nature* **2003**, *426*, 413–418.
80. Boronowsky, U.; Wenk, S.-O.; Schneider, D.; Jäger, C.; Rögner, M. Isolation of membrane protein subunits in their native state: Evidence for selective binding of chlorophyll and carotenoid to the $b_6$ subunit of the cytochrome $b_6f$ complex. *Biochim. Biophys. Acta* **2001**, *1506*, 55–66.
81. Hofmann, E.; Wrench, P.M.; Sharples, F.P.; Hiller, R.G.; Welte, W.; Diederichs, K. Structural basis of light harvesting by carotenoids: Peridinin-chlorophyll-protein from *Amphidinium carterae*. *Science* **1996**, *272*, 1788–1791.
82. Kerfeld, C.A.; Sawaya, M.R.; Brahmandam, V.; Cascio, D.; Ho, K.K.; Trevithick-Sutton, C.C.; Krogmann, D.W.; Yeates, T.O. The crystal structure of a cyanobacterial water-soluble carotenoid binding protein. *Structure* **2003**, *11*, 55–65.

83. Wilson, A.; Ajlani, G.; Verbavatz, J.-M.; Vass, I.; Kerfeld, C.A.; Kirilovsky, D. A soluble carotenoid protein involved in phycobilisome-related energy dissipation in cyanobacteria. *Plant Cell* **2006**, *18*, 992–1007.
84. Englert, G.; Bjørnland, T.; Liaaen-Jensen, S. 1D and 2D NMR study of some allenic carotenoids of the fucoxanthin series. *Magn. Reson. Chem.* **1990**, *28*, 519–528.
85. Egeland, E.S.; Guillard, R.R.L.; Liaaen-Jensen, S. Additional carotenoid prototype representatives and a general chemosystematic evaluation of carotenoids in Prasinophyceae (Chlorophyta). *Phytochemistry* **1997**, *44*, 1087–1097.
86. Yoshii, Y.; Takaichi, S.; Maoka, T.; Hanada, S.; Inouye, I. Characterization of two unique carotenoid fatty acid esters from *Pterosperma cristatum* (Prasinophyceae, Chlorophyta). *J. Phycol.* **2002**, *38*, 297–303.
87. Egeland, E.S.; Liaaen-Jensen, S. Ten minor carotenoids from Prasinophyceae (Chlorophyta). *Phytochemistry* **1995**, *40*, 515–520.
88. Mimuro, M.; Nagashima, U.; Takaichi, S.; Nishimura, Y.; Yamazaki, I.; Katoh, T. Molecular structure and optical properties of carotenoids for the *in vivo* energy transfer function in the algal photosynthetic pigment system. *Biochim. Biphys. Acta* **1992**, *1098*, 271–274.
89. Akimoto, S.; Yokono, M.; Higuchi, M.; Tomo, T.; Takaichi, S.; Murakami, A.; Mimuro, M. Solvent effects on excitation relaxation dynamics of a keto-carotenoid, siphonaxanthin. *Photochem. Photobiol. Sci.* **2008**, *7*, 1206–1209.
90. Yamamoto, H.Y.; Bugos, R.C.; Hieber, A.D. Biochemistry and molecular biology of the xanthophyll cycle. In *The Phytochemistry of Carotenoids*; Frank, H.A., Young, A.J., Britton, G., Cogdell, R.J., Eds.; Kluwer: Dordrecht, The Netherlands, 1999; pp. 293–303.
91. Goss, R.; Jakob, T. Regulation and function of xanthophyll cycle-dependent photoprotection in algae. *Photosynth. Res.* **2010**, *106*, 103–122.
92. Grouneva, I.; Jakob, T.; Wilhelm, C.; Goss, R. Influence of ascorbate and pH on the activity of the diatom xanthophyll cycle-enzyme diadinoxanthin de-epoxidase. *Physiol. Plant.* **2006**, *126*, 205–211.

*marine drugs*

MDPI

*Review*

# Microalgae as Sources of Carotenoids

**Ana Catarina Guedes [1], Helena M. Amaro [1] and Francisco Xavier Malcata [2,3,*]**

[1] CIMAR/CIIMAR—Centro Interdisciplinar de Investigação Marinha e Ambiental, Universidade do Porto, Rua dos Bragas 177, P-4050-123 Porto, Portugal; E-Mails: acatarinaguedes@gmail.com (A.C.G.); lena.amaro@gmail.com (H.M.A.)

[2] ISMAI—Instituto Superior da Maia, Avenida Carlos Oliveira Campos, P-4475-690 Avioso S. Pedro, Portugal

[3] Instituto de Tecnologia Química e Biológica, Universidade Nova de Lisboa, Avenida da República, P-2780-157 Oeiras, Portugal

* Author to whom correspondence should be addressed; E-Mail: fmalcata@ismai.pt;
Tel.: +351-968-017-411; Fax: +351-229-825-331.

*Received: 14 March 2011 / Accepted: 14 April 2011 / Published: 20 April 2011*

**Abstract:** Marine microalgae constitute a natural source of a variety of drugs for pharmaceutical, food and cosmetic applications—which encompass carotenoids, among others. A growing body of experimental evidence has confirmed that these compounds can play important roles in prevention (and even treatment) of human diseases and health conditions, e.g., cancer, cardiovascular problems, atherosclerosis, rheumatoid arthritis, muscular dystrophy, cataracts and some neurological disorders. The underlying features that may account for such favorable biological activities are their intrinsic antioxidant, anti-inflammatory and antitumoral features. In this invited review, the most important issues regarding synthesis of carotenoids by microalgae are described and discussed—from both physiological and processing points of view. Current gaps of knowledge, as well as technological opportunities in the near future relating to this growing field of interest, are also put forward in a critical manner.

**Keywords:** lutein; astaxanthin; β-carotene; bioproduction; extraction

## 1. Introduction

Microalgae occupy the bottom of the food chain in aquatic ecosystems; they possess the intrinsic ability to take up $H_2O$ and $CO_2$—which, with the aid of sunlight, are converted to complex organic compounds that are subsequently kept inside or released from the cell. Those microorganisms have a worldwide distribution, and are well-adapted to survive under a large spectrum of environmental stresses—including (but not limited to) heat, cold, drought, salinity, photo-oxidation, anaerobiosis, osmotic pressure and UV exposure [1].

Microalgae combine, in a balanced way, a few properties typical of higher plants (*viz.* efficient oxygenic photosynthesis and simple nutritional requirements) with biotechnological attributes proper of microorganisms (*viz.* fast growth rates, and ability to accumulate or secrete primary and secondary metabolites). This rather useful combination has led to selection of such microorganisms for applied processes, and represents the basic rationale for the usefulness of microalgal biotechnology. Besides being currently used as feed for aquatic and terrestrial animals, the nutritional value of microalgal biomass goes well beyond—and includes use as colorant in aquaculture, and high-protein or polyunsaturated fatty acid supplement in human diets. The food, pharmaceutical and cosmetic markets have accordingly benefited from a growing array of microalgal products [2,3].

Furthermore, the large number of existing species of microalgae constitutes a unique reservoir of biodiversity, which supports potential commercial exploitation of many novel products besides vitamins, pigments and polyunsaturated fatty acids [4–6]. The key factor for their eventual economic feasibility is the possibility of operating large photobioreactors, able to handle biomass and metabolites to sufficiently high levels [7,8].

This review covers the most relevant features of a family of specialty products originated in microalgae that have already reached commercial expression—by presenting bioprocess considerations and reviewing practical applications, mainly in the food and health industries.

## 2. Cellular Location and Function

Carotenoids constitute a class of terpenoid pigments, derived from a 40-carbon polyene
chain, which can be envisaged as their molecular backbone—indeed it provides carotenoids with distinctive molecular structures, and the associated chemical properties including light-absorption features that are essential for photosynthesis and, in general, for life in the presence of oxygen [9]. The aforementioned backbone may be complemented by cyclic groups (rings) and oxygen-containing functional groups. Hence, hydrocarbon carotenoids are denoted as carotenes as a whole, but oxygenated derivatives are known specifically as xanthophylls—with oxygen being

present as –OH groups (e.g., lutein), as oxi-groups (e.g., cantaxanthin) or as a combination of both (e.g., astaxanthin) [9]. All xanthophylls synthesized by higher plants—e.g., violaxanthin, antheraxanthin, zeaxanthin, neoxanthin and lutein, can also be synthesized by green microalgae; however, these possess additional xanthophylls, e.g., loroxanthin, astaxanthin and canthaxanthin. Diatoxanthin, diadinoxanthin and fucoxanthin can also be produced by brown algae or diatoms [10].

A distinction is usually made between primary and secondary carotenoids: primary ones (*i.e.*, xanthophylls) are structural and functional components of the cellular photosynthetic apparatus, so they are essential for survival [10]; whereas secondary ones encompass those produced by microalgae to large levels, but only after exposure to specific environmental stimuli (via carotenogenesis).

Xanthophylls are relatively hydrophobic molecules, so they are typically associated with membranes and/or involved in non-covalent binding to specific proteins; they are usually localized in the thylakoid membrane, whereas secondary carotenoids are found in lipid vesicles—in either the plastid stroma or the cytosol. Most xanthophylls in cyanobacteria and oxygenic photosynthetic bacteria are associated with chlorophyll-binding polypeptides of the photosynthetic apparatus [11]; however, in most green microalgae, carotenes and xanthophylls are synthesized within plastids, and accumulate therein only. Conversely, secondary xanthophylls in some green microalgae—e.g., astaxanthin in *Haematococcus* sp., accumulate in the cytoplasm; this realization raises the possibility of an extra-plastidic site of carotenoid biosynthesis in that genus. Alternatively, xanthophylls synthesized in the chloroplast may be exported, and consequently accumulate in the cytoplasm [10,12,13]—so they may be found in essentially all cellular compartments.

Carotenoids perform several functions in microalgae: they are involved in light harvesting, but also contribute to stabilize the structure and aid in the function of photosynthetic complexes—besides quenching chlorophyll triplet states, scavenging reactive oxygen species and dissipating excess energy [14]. The intrinsic antioxidant activity of carotenoids constitutes the basis for their protective action against oxidative stress; however, not all biological activities claimed for carotenoids relate to their ability to inactivate free radicals and reactive oxygen species.

## 3. Practical Applications

Several researchers have actively focused on carotenoids from microalgal sources; the major areas, in terms of actual or potential industrial applications, are food and health—and the antioxidant properties exhibited by that class of compounds constitutes at present its core interest. Pigments of microalgal origin are indeed experiencing a strong market demand: the price of microalgal β-carotene easily attains 700 €/kg, whereas its synthetic counterpart cannot reach more than half that figure. Natural β-carotene is preferred by the health market because it is a mixture of *trans* and *cis* isomers—the latter of which possess anticancer features; such a mixture can hardly be obtained via chemical synthesis [14].

### *3.1. Uses for Food and Feed Formulation*

Manufacture of carotenoids via microbiological routes has undergone a greater and greater scientific and commercial importance within the alimentary and aquaculture fields [15], especially in view of environmental and health awareness by consumers at large.

Recall that most oxidation reactions in foods are deleterious—e.g., degradation of vitamins, pigments and lipids, with consequent loss of nutritional value and development of off-flavors [16,17]. Antioxidants—which are adventitious in, or deliberately added to foods, can inhibit oxidation or slow down initiation by free alkyl radicals, as well as interrupt propagation of such free radical chains. The threshold of synthetic food additives legally permitted has been steadily decreasing, due to their suspected role as promoters of carcinogenesis, besides claims of liver and renal toxicities [18]; hence, substitution thereof by natural pigments has become common practice. One good example is the application of *Dunaliella* spp. for mass production of carotenoids aimed at a preservation role [19,20]. Another advantage of carotenoids is that they are not affected by the presence of ascorbic acid, often used as acidulant to constrain unwanted microbial growth, nor by heating/freezing cycles employed in foods with a similar goal.

On the other hand, carotenoids are particularly strong dyes, even at levels of parts per million. Specifically, canthaxanthin, astaxanthin and lutein from *Chlorella* have been in regular use as pigments, and have accordingly been included as ingredients of feed for salmonid fish and trout, as well as poultry—to enhance the reddish color of said fish or the yellowish color of egg yolk [4,21–23]. Furthermore, β-carotene has experienced an increasing demand as pro-vitamin A (retinol) in multivitamin preparations; it is usually included in the formulation of healthy foods, although only under antioxidant claims [24–26].

### *3.2. Uses for Health and Well-Being Promotion*

In the human being, oxidation reactions driven by reactive oxygen species can lead to protein damage and DNA decay or mutation; these may in turn lead to several syndromes, *viz.* cardiovascular diseases, some kinds of cancer and degenerative diseases, and ageing at large [17,27]. As potent biological antioxidants, carotenoids are able to absorb the excitation energy of singlet oxygen radicals into their complex ringed chain—thus promoting energy dissipation, while protecting tissues from chemical damage. They can also delay propagation of such chain reactions as those initiated by degradation of polyunsaturated fatty acids—which are known to dramatically contribute to the decay of lipid membranes, thus seriously hampering cell integrity [21].

One illustrative example is the decline of cognitive ability accompanying Alzheimer's disease, which is apparently caused by persistent oxidative stress in the brain [28]. Using transgenic mice fed with extracts from *Chlorella* sp. containing β-carotene and lutein, Nakashima *et al.* [29] claimed significant prevention of cognitive impairment. Wu *et al.* [30] used also *Chlorella* extracts containing 2–4 mg/$g_{DW}$ of lutein, and reported reduction in the incidence of cancer, as well as prevention of macular degeneration [31]. Likewise, carotenoids extracted specifically from *Chlorella ellipsoidea* and *Chlorella vulgaris* inhibited colon cancer development [23]. Furthermore, astaxanthin obtained from *Haematococcus pluvialis* decreased expression of cyclin D1, but increased that of p53 and some cyclin kinase inhibitors of colon cancer cell lines [32].

Carotenoids have also the ability to stimulate the immune-system, thus being potentially involved in more than 60 life-threatening diseases—including various form of cancer, coronary heart diseases, premature ageing and arthritis [33]; this is specifically the case of canthaxanthin and astaxanthin, and other nonprovitamin A carotenoids from *Chlorella* but to a lesser degree [23]. A few epidemiological studies encompassing β-carotene from *Dunalliela* sp.—which contains readily bioavailable 9-*cis* and all-*trans* stereoisomers (*ca.* 40% and 50%, respectively), have indeed provided evidence of a lower incidence of several types of cancer and degenerative diseases [34]. Finally, carotenoids exhibited hyperlipidemic and hypercholesterolemic effects [19].

## 4. Industrial Production

The worldwide demand for carotenoids was *ca.* 640 M€ in 2004, but it has been rising ever since at an average yearly rate of 2.2% [9]; β-carotene has specifically risen from *ca.* 175 M€ in 2004 to *ca.* 183 M€ in 2009 [35]. A growing fraction has been accounted for by carotenoids from biotechnological sources; and β-carotene, as well as such xanthophylls as astaxanthin, cantaxanthin and lutein have consequently been in higher and higher demand [9]. The most famous source microalgae are *Chlorella, Chlamydomonas, Dunaliella, Muriellopsis* and *Haematococcus* spp.—all of which belong to the Chlorophyceae family [2]. They tend to accumulate carotenoids as an intrinsic part of their biomass, thus offering economical alternatives to chemical synthesis [36].

Among all natural sources studied to date, *Dunaliella* possesses the highest content of 9-*cis* β-carotene [20,34]—reaching levels up to 100 g/$kg_{DW}$, [19,37,38]; β-carotene-rich *Dunaliella* powder has been commercially exploited in many countries since the 1980s. Although many microalgae can produce xanthophylls, *H. pluvialis* is the one that accumulates them to the highest levels (e.g., asthaxanthin [10]), so it is now cultivated at large scale by several companies using distinct approaches [39].

On the other hand, *Muriellopsis* sp. holds a high lutein content (up to 35 mg $L^{-1}$), coupled with a high growth rate; hence, it has been exploited for commercial production of lutein [10]. Finally, *C. ellipsoidea* was reported to produce violaxanthin, together with two other minor xanthophylls, *viz.* antheraxanthin and zeaxanthin—whereas the main carotenoid in *C. vulgaris* was lutein [23]. Further pieces of related information are gathered in Table 1.

**Table 1.** Carotenoids produced by selected microalgae.

| Microalga source | Active compound | Reference |
| --- | --- | --- |
| *Dunaliella salina* | β-carotene | [13,14] |
| *Haematococcus pluvialis* | Astaxanthin, cantaxanthin, lutein | [14,18] |
| *Chlorella vulgaris* | Cantaxanthin, astaxanthin | [14,19] |
| *Coelastrella striolata* var. *multistriata* | Canthaxanthin, astaxanthin, β-carotene | [40] |
| *Scenedesmus almeriensis* | Lutein, β-carotene | [41] |

## 5. Biotechnological Processes

Despite a few useful features already referred to above, microalgae are in general expensive to produce, so concerted efforts have been on the way toward more cost-efficient modes of mass cultivation.

With regard to open systems, the best choice seems to be the open shallow pond—made of leveled raceways, 2–10 m wide and 15–30 cm deep, which run as simple loops or meandering pathways; each unit may cover an area of several hundred to a few thousand $m^2$. However, this configuration poses several problems—which restrict its use to strains that, in view of their weed-like behavior (e.g., *Chlorella*) or their ability to withstand adverse growing conditions (e.g., *Spirulina* or *Dunaliella*), can outgrow other microorganisms.

**Table 2.** Optimal conditions of production of carotenoids by selected microalgae.

| Carotenoid | Microalga source | Processing conditions | Reactor configuration | Productivity | Ref |
|---|---|---|---|---|---|
| β-carotene | *Dunaliella salina* | **T**: 25 °C; **pH**: 7.5 ± 0.5; **LI**: 281 ± 89 $\mu mol_{photon}$ $m^{-2}$ $s^{-1}$; **SR**: 38 $cms^{-1}$ $s^{-1}$ | Semi-continuous outdoor, closed tubular (55 L) | Biomass: 2 g $m^{-2}$ $d^{-1}$ Total carotenoids: 102.5 ± 33.1 mg $m^{-2}$ $d^{-1}$ (β-carotene: 10% of biomass) | [42] |
| | | **T**: 30 °C; **pH**: 7.5; **LI**: 200–1200 $\mu mol_{photon}$ $m^{-2}$ $s^{-1}$; **SR:** 0.6 L $min^{-1}$ ($N_2$) | Continuous turbidostat, flat-panel (2.5 L) | β-Carotene: 13.5 mg $L^{-1}$ $d^{-1}$ (15.0 pg $cell^{-1}$) | [43] |
| | | **T**: 30 °C; **pH**: 7.5; **LI**: 200–1200 $\mu mol_{photon}$ $m^{-2}$ $s^{-1}$; **SR:** 0.286 $L_{solvent}$ $L^{-1}$ $min^{-1}$ (organic solvent) | Continuous turbidostat, flat-panel (1.9 L) with *in situ* extraction | β-Carotene: 0.7 mg $L^{-1}$ $d^{-1}$ β-Carotene: 8.3 mg $L^{-1}$ $d^{-1}$ (8.9 pg $cell^{-1}$) | |
| Lutein | *Muriellopsis* sp. | **T**: 28 °C; **pH** 6.5; **LI**: 460 $\mu mol_{photon}$ $m^{-2}$ $s^{-1}$ | Batch (0.2 L, 4–7 d) | Lutein content: 5.5 mg $g^{-1}$ $L^{-1}$ $d^{-1}$ Lutein: 0.8–1.4 mg $L^{-1}$ $d^{-1}$ | [44] |
| | | **T**: 28 °C; **pH**: 7; **LI**: continuous 200 $\mu mol_{photon}$ $m^{-2}$ $s^{-1}$; **AF**: 50–100 $L^{-1}$ $h^{-1}$ (1 %, v/v $CO_2$) | Continuous outdoor, tubular (55 L) | Biomass: 7.2 mg $L^{-1}$ $d^{-1}$ Lutein: 5.5 mg $g^{-1}$ $L^{-1}$ $d^{-1}$ | [44] |
| | | - | Semicontinuous outdoor, open tank (100 L) | Biomass: 100 mg $m^{-2}$ $d^{-1}$ Lutein: 100 mg $g^{-1}$ $L^{-1}$ $d^{-1}$ | [42] |
| | *Scenedesmus almeriensis* | **T**: 30 °C; **pH**: 8.0; **$LI_{max}$**: 1700 μE $m^{-2}$ $s^{-1}$; **AF**: 0.5 (v/v)/$min^{-2}$ $s^{-1}$; **LDC**: solar cycle | Continuous (2 L) | Lutein: 4.9 mg $L^{-1}$ $d^{-1}$ | [8] |
| | | **T**: 35 °C; **LI**: 1900 μE $m^{-2}$ $s^{-1}$ | Continuous outdoor, tubular | Lutein: 5.31 mg $m^{-2}$ $d^{-1}$ | [45] |
| | *Chlorella protothecoides* | **T**: 28 °C; **pH**: 6.5; **LI**: absence of light; **MM**: heterotrophic | Batch (16 L) | Lutein: 10 mg $L^{-1}$ $d^{-1}$ | [46] |
| | *Chlorella zofingiensis* | **T**: 28 °C; **pH**: 7; **LI**: 200 $\mu mol_{photon}$ $m^{-2}$ $s^{-1}$; **AF**: 50–100 $L^{-1}$ $h^{-1}$ (1%, v/v $CO_2$) **LDC**: continuous light | Batch (0.2 L) | Lutein: 3.4 mg $L^{-1}$ $d^{-1}$ | [44] |
| | *Chlorococcum citriforme* | | | Lutein: 1.05 mg $L^{-1}$ $h^{-1}$ | |
| | *Neospongiococcus gelatinosum* | | | Lutein: 0.70 mg $L^{-1}$ $h^{-1}$ | |

**Table 2.** *Cont.*

| Carotenoid | Microalga source | Processing conditions | Reactor configuration | Productivity | Ref |
|---|---|---|---|---|---|
| Astaxanthin | *C. zofingiensis* | **T**: 30 °C; **pH**: 6.5; **LI**: darkness; **SR**: 130 rpm; **MM**: heterothrophic | Batch (250 mL) | Astaxanthin:10.3 mg $L^{-1}$ | [47] |
| | *Haematococcu s pluvialis* | **LI**: day light cycle | Continuous chemostat, tubular (50 L) | Biomass: 0.7 g $L^{-1}d^{-1}$ Astaxanthin: 8.0 mg $L^{-1}d^{-1}$ | [48] |
| | | **T**: 28 °C; **LI**: 345 $\mu mol_{photon}$ $m^{-2}$ $s^{-1}$ | Batch (1 L) | Astaxanthin content: 98 mg $g^{-1}_{biomass}$ | [49] |
| | | **T**: 15–25 °C; **LI**$_{max}$: 2000 $\mu mol_{photon}$ $m^{-2}$ $s^{-1}$ | Enclosed outdoor (25,000 L) | Biomass: 90 g $m^{-2}$ Astaxanthin: 13 g $m^{-2}$ $d^{-1}$ | [39] |

**AF**: air flow; **LDC**: light/dark cycle; **LI**: light irradiance; **MM**: metabolic mode; **SR**: stirring rate; **T**: temperature.

More advanced technologies have meanwhile been made available pertaining to closed systems; these provide better options for growth of most microalgal strains, by protecting the culture from contamination by unwanted (and often ill-defined) microorganisms, and allowing comprehensive and integrated control of processing conditions. Such photobioreactors are either flat or tubular, and may adopt a variety of designs and operation modes. They lead to higher volumetric productivities and an overall better quality for the biomass (or product) generated—but they are also more expensive to build and operate than their open counterparts [9].

Some microalgae exhibit unique productivity and plasticity features: when grown under distinct sets of operating conditions, they may accumulate different products to high levels; hence, careful design and control of medium composition, temperature, pH, aeration, stirring and irradiance are recommended. A few examples of optimum conditions of operation of microalgal reactors—using productivity of carotenoids as objective function, are listed in Table 2.

During microalgal cultivation, a few processing parameters can be specifically manipulated for maximum synthesis of carotenoids; the better studied cases are lutein, astaxanthin and β-carotene—which will be discussed below at some length.

### 5.1. Lutein

The most important factors that affect lutein content in microalgae are temperature, irradiance, pH, availability and source of nitrogen, salinity (or ionic strength) and presence of oxidizing substances (or redox potential); however, specific growth rate also plays a crucial role.

High temperature favors accumulation of lutein, as happens with other carotenoids (e.g., β-carotene) in *Dunaliella* sp. [42], close to the limit of environmental stress;

further temperature increases would thus be harmful, and eventually reduce biomass productivity.

A high irradiance level appears beneficial—but its effect depends on whether indoor or outdoor cultivation is considered; *in vitro* mimicking of all parameters that characterize outdoor operation, e.g., solar cycle and temperature fluctuation, is indeed difficult. Furthermore, the concentration of molecular oxygen outdoors cannot be manipulated, despite its interacting with illumination and temperature. Both irradiance and temperature influence the rate of lutein production, yet cultures of *Murielopsis* sp. and *Scenedesmus almeriensis* produced contradictory results; hence, these two factors should be considered in a combined, rather than independent fashion [8].

Likewise, the reported effects of pH are not consistent between batch and continuous cultivations. In the former, lutein content increased at extreme pH values, whereas the best results under continuous operation were observed at the optimum pH for growth rate. It is worth noting that pH is particularly relevant in microalgal cultures because it interferes with $CO_2$ availability (which is essential for photosynthesis); hence, continuous supply of $CO_2$, as a fraction of the aeration stream, and pH-controlled injection lead to different results. In general, the maximum lutein productivity is achieved at the optimum pH for biomass productivity [45].

The concentration of nitrogen in the culture medium (in the form of nitrate) does not apparently cause a significant effect upon the lutein content of biomass; however, N-limitation reduces biomass productivity, and consequently leads to poor overall lutein synthesis. Hence, nitrate should be supplied to a moderate excess—so that growth rate is not hampered, while avoiding saline stress that dramatically affects culture performance [8].

Lutein synthesis is enhanced via addition of such chemicals as $H_2O_2$ and NaClO, which behave as inducers: in the presence of $Fe^{2+}$, they affect the redox state and generate stress-inducing chemical species. This induction of oxidative stress is expected because lutein holds a protection role conveyed by its antioxidant features—particularly under heterotrophic growth, where spontaneous oxidative stress is normally absent (unlike happens with phototrophic cultures) [45].

Finally, the specific growth rate affects both continuous and semicontinuous cultures: lutein tends to accumulate at low dilution rates, but not to levels sufficient to balance the decrease in biomass productivity under such circumstances. Therefore, the maximum lutein productivity is again typically attained at the optimal dilution rate for biomass production [45].

### 5.2. Astaxanthin

Commercial production of astaxanthin by *Haematococcus* sp. has been implemented by more than one microalga company (e.g., Cyanotech and Aquasearch); they resorted to a two-stage system, consisting of a first step to produce green biomass under optimal growth conditions ("green" stage), followed by a second stage when the microalga is exposed to adverse environmental conditions to induce accumulation of astaxanthin ("red" stage) [50]. Astaxanthin productivities in large scale facilities are typically *ca.* 2.2 mg $L^{-1}$ [39]—even though maximum astaxanthin productivities of 11.5 mg $L^{-1}$ $d^{-1}$ can be attained at bench scale [51].

Micro Gaia, a marine biotech firm engaged in production of microalgae rich in astaxanthin, proposed a single-step, continuous manufacture process using moderate nitrogen limitation [52,53]: the biomass and astaxanthin productivities obtained were 8.0 and 0.7 mg $L^{-1}d^{-1}$, respectively [54]. The feasibility of the latter approach for production of astaxanthin by *H. pluvialis* was tested continuous-wise in outdoor apparatuses [48]: Aquasearch Growth Modules (AGM)—*i.e.*, 25,000 L enclosed, computerized photobioreactors, were combined up to three units to obtain large amounts of clean, fast growing *H. pluvialis*; they were transferred daily to a pond culture system, where carotenogenesis and astaxanthin accumulation were induced. After 5 days of synthesis, cells were harvested by gravitational settling—with a typical content of 2.5% ($w/w_{DW}$) astaxanthin; a high pressure homogenizer was used to disrupt the cells, and then drying was carried out to less than 5% (w/w) moisture. The performance of AGM could be improved 2-fold within the first 9 mo of operation; and the biomass concentration increased from 50 to 90 g $m^{-2}$, with associated productivities increasing from 9 to 13 g $m^{-2}$ $d^{-1}$ within the same period [39].

However, the production capacity of *H. pluvialis* was constrained by its intrinsic slow growth, low cell yield, ease of contamination by bacteria and protozoa, and susceptibility to adverse weather conditions [5]. Moreover, *H. pluvialis* cannot be efficiently cultivated in dark heterotrophic mode—so production of astaxanthin should adopt the photosynthetic mode, and thus resort to levels of irradiance (e.g., 950 μmol $m^{-2}$ $s^{-1}$) well beyond what would be economically reasonable [39]. Owing to its ease of culturing and high tolerance to environmental fluctuations, *C. zofingiensis* (another green microalga) has been put forward as an alternative for astaxanthin production: it grows quite fast (*ca.* three times faster than *H. pluvialis*), and accumulates significant amounts of secondary carotenoids in the dark, thus facilitating large-scale cultivation of denser biomass [47,55].

Oxidative stress induced by intense illumination has been found to play a crucial role upon astaxanthin synthesis [56]; active oxygen molecules, generated by excess photooxidation caused by high light irradiance, do apparently trigger synthesis of carotenoids as part of a cellular strategy aimed at cell protection against oxidative damage [47]. In particular, flashing light increased the rate of astaxanthin production per photon in *H. pluvialis* by at least 4-fold relative to that under continuous light sources [57]—thus proving that light quality is more important than quantity [58].

The effect of irradiance depends also on such operating variables as culture density, cell maturity (flagellates are much more sensitive than palmelloids), medium nutrient profile and light path [59]. The predominant role of light stress and nitrogen deprivation towards induction and enhancement of biosynthesis in the aplanospores of *H. pluvialis* was originally suggested in the 1950s [60]; astaxanthin accumulation comes along with growth halting, as happens in most cases of stress imposed upon microalgae [59,61]. Imamoglu *et al.* [54] compared the effect of various stress media, under high light intensities, upon astaxanthin accumulation; those authors concluded that addition of $CO_2$ in an N-free medium, under 546 $\mu mol_{photon}\ m^{-2}\ s^{-1}$, were the best conditions for accumulation of astaxanthin—which attained *ca.* 30 $mg\ g^{-1}$.

Astaxanthin may thus be efficiently produced outdoors in continuous mode, if accurate nitrate dosage is provided [48]; besides N, such oligoelements as iron play a role. This essential oligoelement takes part in assimilation of nitrate and nitrite, deoxidation of sulphate, fixation of N, and synthesis of chlorophyll [62–65]. Iron deficiency was reported to constrain microalga growth, even in rich nutrient media [64]; whereas its addition enhanced astaxanthin synthesis [66–69]. Cai *et al.* [67] further tested how iron electrovalencies and counter ions affect cell growth and accumulation of astaxanthin; 18 $\mu mol\ L^{-1}$ $Fe^{2+}$-EDTA stimulated synthesis of astaxanthin more effectively, up to contents of 30.7 $mg\ g^{-1}$; and despite the lower cell density attained ($2.3 \times 10^5$ cell $mL^{-1}$), a higher concentration (36 $\mu mol\ L^{-1}$) of $FeC_6H_5O_7$ yielded cell density and astaxanthin production levels that were 2- and 7-fold those reached under iron-limitation.

In the "red stage" of growth, *Haematococcus* cells require only carbon as major nutrient—which this is usually supplied via directly injecting $CO_2$ into the photobioreactor during daylight [61]. Furthermore, high irradiance provides more energy for photosynthetic fixation of C, which leads to a higher rate of astaxanthin synthesis [68]; this may be further enhanced by raising the C/N ratio [69].

Finally, Chen *et al.* [70] experimented with heterotrophic conditions—using pyruvate, citrate and malate as substrates, towards synthesis of astaxanthin by *C. zofingiensis* in the absence of light. Presence of any of the aforementioned substrates above 10 mM stimulated biosynthesis of astaxanthin (and other secondary carotenoids); *ca.* 100 mM pyruvate led to yields of 8.4–10.7 $mg\ L^{-1}$ astaxanthin, which correspond to a 28%-increase.

### 5.3. *β-Carotene*

Semicontinuous cultivation of *D. salina* at 25 °C produced 80 $g\ m^{-3}\ d^{-1}$ biomass [42]—from which 1.25 $mg\ L^{-1}$ of β-carotene was recovered [71]; however, this figure could be improved up to 2.45 $mg\ m^{-3}\ d^{-1}$ in continuous biphasic bioreactors [72]. When cultivated photoheterotrophically, a significant increase of cellular β-carotene content was experimentally observed: the maximum score was 70 pg cell$^{-1}$, in a culture enriched with 67.5 mM acetate and 450 μM $FeSO_4$ [33].

As with astaxanthin, $Fe^{2+}$ plays an important role in β-carotene accumulation in *D. salina*; by inducing oxidative stress, those cations stimulate said synthesis, especially in the presence of a carbon source. UV-A radiation (320–400 nm) added to the photosynthetically active radiation (PAR, *i.e.*, 400–700 nm) can be regarded as another stress factor during growth of, and carotenoid accumulation by *Dunalliela bardawil;* compared with cultures exposed to PAR only, addition of 8.7 W m$^{-2}$ UV-A radiation to 250 W m$^{-2}$ PAR stimulated long-term growth of that microalga, and led to a 2-fold enhancement in β-carotene accumulation by 24 d [38].

## 6. Extraction and Purification

Although microalga-mediated synthesis of carotenoids is crucial in biotechnological production thereof, a major portion (if not most) of their cost actually lies on downstream separation—e.g., biomass drying and disruption, followed by solvent extraction and purification. Hence, these issues are addressed below, in view of their importance toward commercial scale processes.

### *6.1. Cell Disruption*

A major practical problem in using such microalgae as *Murielopsis* sp. or *S. almeriensis* is the need for cell wall disruption. This can be accomplished through a variety of ways, e.g., milling, ultrasound, microwave, freezing, thawing or chemical attack [45].

The mortar-and-pestle procedure described by Mínguez-Mosquera *et al.* [73] provides full recovery, but it cannot be scaled up to industrial practice; sonication and ball milling produce results similar to that procedure, as long as alumina is employed as disaggregating agent [45]. Ceron *et al.* [74] complemented the alumina-based cell disruption with alkaline treatment using 4% (w/v) aqueous KOH (40 °C); disaggregation and lipid expression were both facilitated.

### *6.2. Biomass Extraction*

Microalgal biomass is usually processed via solvent extraction, to render carotenoid extracts—with typical contents of 25% [45]; this can be used directly in the formulation of supplements, or undergo further multistep purification—encompassing hydrolysis to release hydroxylated carotenoids from the accompanying fatty acids, and final recrystallization to polish the product.

Obtaining a carotenoid-rich oleoresin from microalgae—dried or in wet paste form, is a more straightforward task; such extracts may then be subjected to classical processes to obtain purer lutein [45,74] that may successfully compete with that extracted from marigold.

### 6.2.1. Organic Solvent-Mediated Extraction

Solvent extraction usually resorts to hexane—and has advantages over alkaline treatment because all lutein and zeaxanthin are converted to their free forms, while carboxylic acids and chlorophylls remain in the aqueous phase [45]; this method has been optimized for *S. almeriensis* [74]. Extraction was maximized with a 1:1 (v/v) ratio of hexane to sample, and the optimal number of extraction steps was typically six—which led to 95% recovery of lutein. Less conventional solvents—e.g., ethyl lactate, have been recently proposed [76] for plant matter at large, but can in principle be applied also to microalgae.

A significant improvement would be to eliminate the drying step of microalgal biomass prior to extraction; Fernández-Sevilla *et al.* [77] have accordingly proposed a modification of a previous approach [74] that can handle wet biomass paste (*ca.* 20% DW), based on an extraction phase composed by hexane/ethanol/water and KOH—which simultaneously effects an alkaline treatment to saponify susceptible lipids and extract the intended carotenoids.

Another enhancement is the accelerated solvent extraction methodology, which uses a special type of contactor to circulate solvent at high pressure through a tightly packed bed of biomass. However, high temperatures are required (over 60 °C, and usually as high as 170 °C) to lower the viscosity of the solvent, which leads to formation of pheophorbide from the microalgal chlorophylls that are of a major toxicological concern. In any case, extraction with hexane or ethanol allows easy solvent removal afterwards, as well as high-content lutein extracts [45].

For selective extraction of free astaxanthin from red encysted *Haematococcus* sp., an alternative procedure has been designed that resorts to dodecane and methanol [75]; it consists of dodecane-mediated extraction of the crude mixture, followed by extraction with methanol. The first stage did not require previous cell harvesting, and separation of the dodecane-rich phase from the culture medium containing cell debris proceeded rapidly via plain settling. In the second stage, the free astaxanthin in the former extract was selectively solubilized in methanol along with saponification of astaxanthin esters—thus leading to a total recovery of astaxanthin above 85%.

### 6.2.2. Green Solvent-Mediated Extraction

An environment-friendly downstream process using common vegetable oils was proposed by Kang *et al.* [79] for direct extraction of astaxanthin from *Haematococcus* sp. As said crude microalgal astaxanthin consists of *ca.* 70% monoesters, 25% diesters and 5% free forms, a rather lipophilic nature results, so vigorous stirring is required to gradually disrupt the cells; the oily extracts are then simply separated from the culture medium containing cell debris by gravity settling. When using olive oil, recoveries of up to 93.9% were possible [79]. Apparently, a similar method had been proposed long before by Nonomura [80], who then claimed up to 7.5% yield of lutein.

### 6.2.3. Supercritical Fluid-Mediated Extraction

Classical extraction with organic solvents has attained purity degrees sufficient to meet commercial specifications for large-scale production of lutein; however, selective precipitation with supercritical $CO_2$ constitutes a promising alternative. Note that conventional liquid extraction of carotenoids from microalgal matrices is time-consuming—as multiple extraction steps are typically required; and large relative ratios of organic solvents have to be used, which are often expensive and potentially harmful. Supercritical fluid extraction (SFE) using modified $CO_2$ permits more straightforward purification and shorter extraction times [81].

In general, SFE is relatively rapid and efficient because of the low viscosities and high diffusivities that characterize supercritical fluids. Furthermore, extraction can be made selective by controlling solvent density; the material extracted will be recovered afterwards by simply depressurizing, thus allowing the supercritical fluid to return to its gaseous form and leaving no (or little) residual solvent in the precipitate thus originated [82]. Supercritical $CO_2$ has so far been the most employed supercritical fluid—because it is non-flammable, non-toxic, inexpensive and relatively inert from a chemical point of view.

Previous studies demonstrated the feasibility of extracting pigments from plants using supercritical $CO_2$—e.g., carotenoids from carrots [83] and cabbages [84]; Mendes *et al.* [85], Careri *et al.* [86] and Macías-Sánchez *et al.* [87–89] have meanwhile extended such a technique to extraction of carotenoids from *C. vulgaris, Spirulina platensis, Nannochloropsis gaditana, Synechococcus* sp. and *S. almeriensis*, respectively—and satisfactory results were consistently reported, as emphasized in Table 3.

**Table 3.** SFE yields of total carotenoids (including lutein), and of lutein specifically, by selected microalgae.

| Microalga source | Operating conditions (pressure/temperature /time) | Total carotenoids (mg/g DW biomass) | Lutein (mg/g DW biomass) | Total carotenoids/ Chlorophyll *a* ratio | Reference |
|---|---|---|---|---|---|
| *Nannochloropsis gaditana* | 200 bar/40 °C/180 min | 0.152 | - | 0.524 | [87] |
| | 200 bar/50 °C/180 min | 0.152 | - | 0.410 | |
| | 200 bar/60 °C/180 min | 0.125 | - | 1.389 | |
| | 300 bar/40 °C/180 min | 0.208 | - | 0.258 | |
| | 300 bar/50 °C/180 min | 0.248 | - | 0.230 | |
| | 300 bar/60 °C/180 min | 0.250 | - | 0.179 | |
| *Chlorella vulgaris* | 200 bar/40 °C/198 min | 0.011 | - | - | [85] |
| | 200 bar/55 °C/180 min | 0.008 | - | - | |
| | 350 bar/55 °C/252 min | 0.080 | - | - | |

| Microalga source | Operating conditions (pressure/temperature /time) | Total carotenoids (mg/g DW biomass) | Lutein (mg/g DW biomass) | Total carotenoids/ Chlorophyll *a* ratio | Reference |
|---|---|---|---|---|---|
| *Synechococcus* sp. | 200 bar/40 °C/180 min | 0.386 | - | 193.000 | [88] |
| | 200 bar/50 °C/180 min | 1.225 | - | 23.113 | |
| | 200 bar/60 °C/180 min | 0.405 | - | 101.25 | |
| | 300 bar/40 °C/180 min | 0.748 | - | 32.522 | |
| | 300 bar/50 °C/180 min | 1.511 | - | 19.372 | |
| | 300 bar/60 °C/180 min | 0.808 | - | 46.316 | |

**Table 3.** *Cont.*

| Microalga source | Operating conditions (pressure/temperature /time) | Total carotenoids (mg/g DW biomass) | Lutein (mg/g DW biomass) | Total carotenoids/ Chlorophyll *a* ratio | Reference |
|---|---|---|---|---|---|
| *Scenedesmus almeriensis* | 200 bar/32 °C/300 min | - | 0.0013 | - | [89] |
| | 200 bar/46 °C/300 min | - | 0.0000 | - | |
| | 200 bar/60 °C/300 min | - | 0.0109 | - | |
| | 300 bar/39 °C/300 min | - | 0.0236 | - | |
| | 300 bar/53 °C/300 min | - | 0.0090 | - | |

However, this mode of extraction tends to recover chlorophylls more efficiently than carotenoids, thus producing extracts with relatively poor specifications [90]. Furthermore, the cost of supercritical fluids and associated equipment make it difficult for SFE to compete with classical solvent extraction—especially because the former requires dry biomass.

The selective adsorption of lutein might constitute an alternative in terms of separation/purification, especially if specific solid phases can be used [91], coupled with contacting conveyed by expanded beds [92]; this allows raw extracts to be processed, and tolerates the presence of cell debris or other particulate matter that causes major problems in conventional preparative chromatography. Selective precipitation was also described by Miguel *et al.* [93], who proposed use of supercritical $CO_2$ after organic solvent extraction; the first solvent (containing carotenoids) was accordingly mixed with supercritical $CO_2$, and the conditions of pressure and temperature were duly adjusted to promote preferential precipitation of lutein. However, simple standard mixtures—rather than complex microalgal extracts have been considered, so a long way of improvement is still anticipated prior to practical use.

### 6.2.4. *In Situ* Extraction

*In situ* extraction of β-carotene from *Dunaliella salina* was recently reported by Kleinegris *et al.* [44], using a flat-panel photobioreactor operated as a turbidostat—where the numbers of stressed cells were kept essentially constant via a continuous, well-defined level of irradiation. This two-stage system comprised an organic phase of dodecane, sparged at a rate of 286 $L_{dodecane}\ L_{reactor}^{-1}\ min^{-1}$ that promoted formation of an emulsion in the aqueous phase; β-carotene was then continuously extracted from the aqueous to the organic phase, at a rate of *ca.* 2.75 $mg_{\beta\text{-carotene}}\ L_{dodecane}^{-1}\ d^{-1}$ (equivalent to 0.7 $mg_{\beta\text{-carotene}}\ L_{reactor}^{-1}\ d^{-1}$). However, this process exhibited a poor efficiency—as the yield of β-carotene extracted by the solvent was a mere one-tenth of that removed from the reactor via biomass overflow.

If the aforementioned carotenoid-rich biomass was extracted as well, then the overall volumetric productivity of the system would reach 8.3 $mg_{\beta\text{-carotene}}\ L_{reactor}^{-1}\ d^{-1}$; this is still below the yield attained if downstream rather than *in situ* extraction was promoted (*ca.* 13.5 $mg_{\beta\text{-carotene}}\ L_{reactor}^{-1}\ d^{-1}$) [44], so in this system simultaneous biosynthesis and extraction cannot be justified relative to the classical sequential approach.

## 7. Final Considerations

Carotenoid production appears to be one of the most successful case studies of blue biotechnology. The rising market demand for pigments from natural sources has promoted large-scale cultivation of microalgae for synthesis of such compounds, so significant decreases in production costs are expected in coming years.

The recognized therapeutic value of some carotenoids (especially lutein) in prevention and treatment of degenerative diseases has indeed opened new avenues for development of mass production systems. Advances in knowledge of the underlying physiology, biochemistry and molecular genetics of carotenoid-producing microalgae are now urged—which would have a major impact upon development and optimization of this (and alternative) microalga-based technologies. In this regard, the genes encoding enzymes that are directly involved in specific carotenoid syntheses need in particular to be investigated—so that further development of transformation techniques will permit considerable increase of carotenoid cellular contents, and accordingly contribute to increase the volumetric productivities of the associated processes.

## Acknowledgements

A PhD fellowship (ref. SFRH/BD/62121/2009), supervised by author F.X.M., was granted to author H.M.A., under the auspices of ESF (III Quadro Comunitário de Apoio) and the Portuguese State. A postdoctoral fellowship (ref. SFRH/BPD/72777/2010), also supervised by author F.X.M., was granted to author A.C.G., under the auspices of ESF (III Quadro Comunitário de Apoio) and the Portuguese State. This work received partial financial support via projects OPTIC-ALGAE (PTDC/BIO/71710/2006) and MICROPHYTE (PTDC/EBB-EBI/102728/2008), both coordinated by author F.X.M., also under the auspices of ESF (III Quadro Comunitário de Apoio) and the Portuguese State.

## References

1. Tandeau-de-Marsac, N.; Houmard, J. Adaptation of cyanobacteria to environmental stimuli: new steps towards molecular mechanisms. *FEMS Microb. Rev.* **1993**, *104*, 119–190.
2. Pulz, O.; Gross, W. Valuable products from biotechnology of microalgae. *Appl. Microb. Biotechnol.* **2004**, *65*, 635–648.
3. Richmond, A. *Handbook of Microalgal Culture, Biotechnology and Applied Phycology*; Blackwell Science: Oxford, UK, 2004.
4. Lorenz, T.R.; Cysewski, G.R. Commercial potential for *Haematococcus* microalgae as a natural source of astaxanthin. *Trends Biotechnol.* **2000**, *18*, 160–167.
5. Ip, P.F.; Wong, K.H.; Chen, F. Enhanced production of astaxanthin by the green microalga *Chlorella zofingiensis* in mixotrophic culture. *Proc. Biochem.* **2004**, *39*, 1761–1766.
6. León, R.; Martín, M.; Vigara, J.; Vilchez, C.; Vega, J. Microalgae-mediated photoproduction of β-carotene in aqueous organic two phase systems. *Biomol. Eng.* **2003**, *20*, 177–182.
7. Kim, M.K.; Park, J.W.; Park, C.S.; Kim, S.J.; Jeune, K.H.; Chang, M.U.; Acreman, J. Enhanced production of *Scenedesmus* spp. (green microalgae) using a new medium containing fermented swine wastewater. *Biores. Technol.* **2007**, *98*, 2220–2228.
8. Sánchez, J.F.; Fernández, J.M.; Acién, F.G.; Rueda, A; Pérez-Parra, J.; Molina, E. Influence of culture conditions on the productivity and lutein content of the new strain *Scenedesmus almeriensis*. *Proc. Biochem.* **2008**, *43*, 398–405.
9. del Campo, A.J.; García-González, M.; Guerrero, M.G. Outdoor cultivation of microalgae for carotenoid production: current state and perspectives. *Appl. Microb. Biotechnol.* **2007**, *74*, 1163–1174.
10. Eonseon, J.; Polle, J.E.W.; Lee, H.K.; Hyund, S.M.; Chang, M. Xanthophylls in microalgae: from biosynthesis to biotechnological mass production and application. *Microb. Biotechnol.* **2003**, *13*, 165–174.
11. Grossman, A.R.; Bhaya, D.; Apt, K.E.; Kehoe, D.M. Light-harvesting complexes in oxygenic photosynthesis: diversity, control, and evolution. *Annu. Rev. Genet.* **1995**, *29*, 231–288.
12. Tardy, F.; Havaux, M. Photosynthesis, chlorophyll fluorescence, light-harvesting system and photoinhibition resistance of a zeaxanthin-accumulating mutant of *Arabidopsis thaliana*. *J. Photochem. Photobiol. A* **1996**, *34*, 87–94.
13. Rabbani, S.; Beyer, P.; von Lintig, J.; Hugueney, P.; Kleinig, H. Induced β-carotene synthesis driven by triacylglycerol deposition in the unicellular alga *Dunaliella bardawil*. *Plant Physiol.* **1998**, *116*, 1239–1248.

14. Demming-Adams, B.; Adams, W.W. Antioxidants in photosynthesis and human nutrition. *Science* **2002**, *298*, 2149–2153.
15. Lamers, P.P.; Janssen, M.; de Vos, R.C.H.; Bino, R.J.; Wijffels, R.H. Exploring and exploiting carotenoid accumulation in *Dunaliella salina* for cell-factory applications. *Trends Biotechnol.* **2008**, *26*, 631–638.
16. Fennema, O.R. *Food Chemistry*; Marcel Dekker: New York, NY, USA, 1999; pp. 780–782.
17. Halliwell, B.; Gutteridge, J.M.C. *Free Radicals in Biology and Medicine*; Oxford University Press: New York, NY, USA, 2007; pp. 79–350.
18. El-Baky, H.H.A.; El-Baz, F.K.; El-Baroty, G.S. *Spirulina* species as a source of carotenoids and $\alpha$-tocopherol and its anticarcinoma factors. *Biotechnology* **2003**, *2*, 22–240.
19. El-Baky, H.H.A.; El-Baroty, G.S. Enhancement of carotenoids in *Dunaliella salina* for use as dietary supplements and in the preservation of foods. *Food Chem. Toxicol.* **2011**, in press.
20. Hsu, Y.W.; Tsai, C.F.; Chang, W.H.; Ho, Y.C.; Chen, W.K.; Lu, F.J. Protective effects of *Dunaliella salina*—a carotenoid-rich alga, against carbon tetrachloride-induced hepatoxicity in mice. *Food Chem. Toxicol.* **2008**, *46*, 3311–3317.
21. Guerin, M.; Huntley, M.E.; Olaizola, M. *Haematococcus* astaxanthin: applications for human health and nutrition. *Trends Biotechnol.* **2003**, *21*, 210–215.
22. Cysewski, G.R.; Lorenz, R.T. Industrial production of microalgal cell-mass and secondary products—species of high potential: *Haematococcus*. In *Handbook of Microalgal Culture, Biotechnology and Applied Phycology*; Richmond, A., Ed.; Blackwell Science: Oxford, UK, 2004; pp. 281–288.
23. Plaza, M.; Herrero, M.; Cifuentes, A.; Ibáñez, E. Innovative natural functional ingredients from microalgae. *J. Agric. Food Chem.* **2009**, *57*, 7159–7170.
24. Krinsky, N.I.; Johnson, E.J. Carotenoid actions and their relation to health and disease. *Mol. Aspects Med.* **2005**, *26*, 459–516.
25. Murthy, K.N.C.; Vanitha, A.; Rajesha, J.; Swamy, M.M.; Sowmya, P.R.; Ravishanka G.A.
*In vivo* antioxidant activity of carotenoids from *Dunaliella salina*—a green microalga. *Life Sci.* **2005**, *76*, 1381–1390.
26. Spolaore, P.; Joannis-Cassan, C.; Duran, E.; Isambert, A. Commercial applications of microalgae. *J. Biosci. Bioeng.* **2006**, *101*, 87–96.
27. Kohen, R.; Nyska, A. Oxidation of biological systems: Oxidative stress phenomena, antioxidants, redox reactions, and methods for their quantification. *Toxicol. Pathol.* **2002**, *30*, 620–650.
28. Mattson, M.P. Pathways towards and away from Alzheimer's disease. *Nature* **2004**, *430*,
631–639.

29. Nakashima, Y.; Ohsawa, I.; Konishi, F.; Hasegawa, T.; Kumamoto, S.; Suzuki, Y.; Ohta, S. Preventive effects of *Chlorella* on cognitive decline in age-dependent dementia model mice. *Neur. Lett.* **2009**, *464*, 193–198.
30. Wu, Z.; Wu, S.; Shi, X. Supercritical fluid extraction and determination of lutein in heterotrophically cultivated *Chlorella pyrenoidosa*. *J. Food Proc. Eng.* **2007**, *30*, 174–185.
31. Zhao, L.; Sweet, B.V. Lutein and zeaxanthin for macular degeneration. *Am. J. Health-Syst. Pharm.* **2008**, *65*, 1232–1238.
32. Palozza, P.; Torelli, C.; Boninsegna, A.; Simone, R.; Catalano, A.; Mele, M.C.; Picci, N. Growth-inhibitory effects of the astaxanthin-rich alga *Haematococcus pluvialis* in human colon cancer cells. *Cancer Lett.* **2009**, *283*, 108–117.
33. Mojaat, M.; Pruvost, J.; Foucault, A.; Legrand, J. Effect of organic carbon sources and $Fe^{2+}$ ions on growth and β-carotene accumulation by *Dunaliella salina*. *Biochem. Eng. J.* **2008**, *39*, 177–184.
34. Ben-Amotz, A. *Dunaliella* β-carotene: From science to commerce. In *Enigmatic Microorganisms and Life in Extreme Environments*; Seckbach, J., Ed.; Kluwer: Deventer, The Netherlands, 1999; pp. 401–410.
35. Ye, Z.-W.; Jiang, J.-G.; Wu, G.-H. Biosynthesis and regulation of carotenoids in *Dunaliella*: progresses and prospects. *Biotechnol. Adv.* **2008**, *26*, 352–360.
36. Bhosale, P.; Bernstein, P.S. Microbial xanthophylls. *Appl. Microbiol. Biotechnol.* **2005**, *68*, 445–455.
37. Coesel, S.N.; Baumgartner, A.C.; Teles, L.M.; Ramos, A.A.; Henriques, N.M.; Cancela, L.; Varela, J.C.S. Nutrient limitation is the main regulatory factor for carotenoid accumulation and for *Psy* and *Pds* steady state transcript levels in *Dunaliella salina* (Chlorophyta) exposed to high light and salt stress. *Mar. Biotechnol.* **2008**, *10*, 602–611.
38. Mogedas, B.; Casal, C.; Forján, E.; Vílchez, C. β-Carotene production enhancement by UV-A radiation in *Dunaliella bardawil* cultivated in laboratory reactors. *J. Biosci. Bioeng.* **2009**, *108*, 47–51.
39. Olaizola, M. Commercial production of astaxanthin from *Haematococcus pluvialis* using 25,000-liter outdoor photo-bioreactors. *J. Appl. Phycol.* **2000**, *12*, 499–506.
40. Abe, K.; Hattor, H.; Hiran, M. Accumulation and antioxidant activity of secondary carotenoids in the aerial microalga *Coelastrella striolata* var. *multistriata*. *Food Chem.* **2005**, *100*, 656–661.
41. Macías-Sánchez, M.D.; Fernandez-Sevilla, M.; Acién-Fernández, F.G.; Cerón-Garcia, M.C.; Molina-Grima, E. Supercritical fluid extraction of carotenoids from *Scenedesmus almeriensis*. *Food Chem.* **2010**, *123*, 928–935.

42. García-González, M.; Moreno, J.; Manzano, J.C.; Florêncio, F.J.; Guerrero, M.G. Production of *Dunaliella salina* biomass rich in 9-*cis*-β-carotene and lutein in a closed tubular photobioreactor. *J. Biotechnol.* **2005**, *115*, 81–90.
43. del Campo, J.A.; Moreno, J.; Rodriguez, H.; Vargas, M.A.; Rivas, J.; Guerrero, M.G. Lutein production by *Muriellopsis* sp. in an outdoor tubular photobioreactor. *J. Biotechnol.* **2001**, *85*, 289–295.
44. Kleinegris, D.M.M.; Janssen, M.; Brandenburg, W.A.; Wijffels, R.H. Continuous production of carotenoids from *Dunaliella salina*. *Enzyme Microb. Technol.* **2011**, *48*, 253–259.
45. Fernández-Sevilla, J.M.; Acién-Fernández, F.G.; Molina-Grima E. Biotechnological production of lutein and its applications. *Appl. Microbiol. Biotechnol.* **2010**, *86*, 27–40.
46. Wei, D.; Chen, F.; Chen, G.; Zhang, X.W.; Liu, L.J.; Zhang, H. Enhanced production of lutein in heterotrophic *Chlorella protothecoides* by oxidative stress. *Sci. China Ser. C Life Sci.* **2008**, *51*, 1088–1093.
47. Ip, P.-F.; Chen, F. Production of astaxanthin by the green microalga *Chlorella zofingiensis* in the dark. *Process Biochem.* **2005**, *40*, 733–738.
48. García-Malea, M.C.; Acién, F.G.; del Río, E.; Fernández, J.M.; Cerón, M.C.; Guerrero, M.G.; Molina-Grima, E. Production of astaxanthin by *Haematococcus pluvialis*: taking the one-step system outdoors. *Biotechnol. Bioeng.* **2009**, *102*, 651–657.
49. Domínguez-Bocanegra, A.R.; Guerrero, L.I.; Jerónimo, F.M.; Campocosio, A.T. Influence of environmental and nutritional factors in the production of astaxanthin from *Haematococcus pluvialis*. *Biores. Technol.* **2004**, *92*, 209–214.
50. Guerin, M.; Huntley, M.E.; Olaizola, M. *Haematococcus* astaxanthin: applications for human health and nutrition. *Trends Biotechnol.* **2003**, *21*, 210–216.
51. Aflalo, C.; Meshulam, Y.; Zarka, A.; Boussiba, S. On the relative efficiency of two *vs.* one-stage production of astaxanthin by the green alga *Haematococcus pluvialis*. *Biotechnol. Bioeng.* **2007**, *98*, 300–305.
52. del Rio, E.; Acién, F.G.; García-Malea, M.C.; Rivas, J.; Molina-Grima, E.; Guerrero, M.G. Efficient one-step production of astaxanthin by the microalgae *Haematococcus pluvialis* in continuous culture. *Biotechnol. Bioeng.* **2005**, *91*, 808–815.
53. del Rio, E.; Acién, F.G.; Rivas, J.; Molina-Grima, E.; Guerrero, M.G. Efficiency assessment of the one-step production of astaxanthin by the microalga *Haematococcus pluvialis*. *Biotechnol. Bioeng.* **2008**, *100*, 397–402.
54. Imamoglu, E.; Dalay, M.C.; Sukan, F.V. Influences of different stress media and high light intensities on accumulation of astaxanthin in the green alga *Haematococcus pluvialis*. *New Biotechnol.* **2009**, *26*, 199–204.

55. del Campo, J.A; Rodriguez, H.; Moreno, J.; Vargas, M.A.; Rivas, J.; Guerrero, M.G. Accumulation of astaxanthin and lutein in *Chlorella zofingiensis* (Chlorophyta). *Appl. Microbiol. Biotechnol.* **2004**, *64*, 848–854.
56. Salguero, A.; de la Morena, B.; Vigara, J.; Veja, J.M.; Vilchez, C.; Leon, R. Carotenoids as protective response against oxidative damage in *Dunaliella bardawil. Biomol. Eng.* **2003**, *20*, 249–253.
57. Fábregas, J.; Otero, A.; Maseda, A.; Dominguez, A. Two-stage cultures for the production of astaxanthin from *Haematococcus pluvialis. J. Biotechnol.* **2001**, *89*, 65–71.
58. Kim, Z.-H.; Kim, S.-H.; Lee, H.-S.; Lee, C.-G. Enhanced production of astaxanthin by flashing light using *Haematococcus pluvialis. Enzyme Microb.Technol.* **2006**, *39*, 414–419.
59. Wang, B.; Zarka, A.; Trebest, A.; Boussiba, S. Astaxanthin accumulation in *Haematococcus pluvialis* (Chlorophyceae) as an active photoprotective process under high irradiance. *J. Phycol.* **2003**, *39*, 1116–1124.
60. Droop, M.R. Conditions governing haematochrome formation and loss in the algae *Haematococcus pluvialis. Arch. Microbiol.* **1954**, *20*, 391–397.
61. Boussiba, S. Carotenogenesis in the green alga *Haematococcus pluvialis*: Cellular physiology and stress response. *Physiol. Plant.* **2000**, *108*, 111–117.
62. Zhu, M.Y.; Mu, X.Y.; Li, R.X.; Lü, R.H. The effects of iron on the growth, the photosynthesis, and the biochemical composition of *Phaeodactylum triconutum. Acta Oceanol. Sinica* **2000**, *22*, 110–116.
63. Liu, C.Y.; Zhang, Z.B.; Chen, X.R. Mutual effects of nitric oxide and iron on the growth of marine algae. *Acta Oceanol. Sinica* **2005**, *24*, 100–109.
64. Naito, K.; Matsui, M.; Imai, I. Ability of marine eukaryotic red tide microalgae to utilize insoluble iron. *Harmful Algae* **2005**, *4*, 1021–1032.
65. Harker, M.; Tsavalos, A.J.; Young, A.J. Autotrophic growth and carotenoid production of *Haematococcus pluvialis* in a 30 liter air-lift photobioreactor. *J. Ferm. Bioeng.* **1996**, *82*, 113–118.
66. Choi, Y.E.; Yun, Y.S.; Park, J.M. Evaluation of factors promoting astaxanthin production by a unicellular green alga, *Haematococcus pluvialis,* with fractional factorial design. *Biotechnol. Prog.* **2002**, *18*, 1170–1175.
67. Cai, M.; Li, Z.; Qi, A. Effects of iron electrovalence and species on growth and astaxanthin production of *Haematococcus pluvialis. Chin. J. Oceanol. Limnol.* **2009**, *27*, 370–375.
68. Fábregas, J.; Dominguez, A.; Maseda, A.; Otero, A. Interactions between irradiance and nutrient availability during astaxanthin accumulation and degradation in *Haematococcus pluvialis. Appl. Microbiol. Biotechnol.* **2003**, *10*, 253–261.

69. Kang, C.D.; An, J.Y.; Park, T.H.; Sim, S.J. Astaxanthin biosynthesis from simultaneous N and P uptake by the green alga *Haematococcus pluvialis* in primary-treated wastewater. *Biochem. Eng. J.* **2006**, *31*, 234–238.
70. Chen, W.; Zhang, C.; Song, L.; Sommerfeld, M.; Hu, Q. A high throughput Nile red method for quantitative measurement of neutral lipids in microalgae. *J. Microbiol. Meth.* **2009**, *77*, 41–47.
71. Pruvost, J.; Cornet, J.F.; Legrand, J. Hydrodynamics influence on light conversion in photobioreactors: an energetically consistent analysis. *Chem. Eng. Sci.* **2008**, *3*, 679–694.
72. Hejazi, M.A.; Holwerda, E.; Wijffels, R.H. Milking microalga *Dunaliella salina* for β-carotene production in two-phase bioreactors. *Biotechnol. Bioeng.* **2004**, *85*, 475–481.
73. Mínguez-Mosquera, I.; Gandul-Rojas, M.; Lourdes, B.; Gallardo-Guerrero, M. Rapid method of quantification of chlorophylls and carotenoids in virgin olive oil by high-performance liquid chromatography. *J. Agric. Food Chem.* **1992**, *40*, 60–63.
74. Ceron, M.C.; Campos, I.; Sánchez, J.F.; Acien, F.G; Molina, E.; Fernandez-Sevilla, J.M. Recovery of lutein from microalgae biomass: development of a process for *Scenedesmus almeriensis*. *J. Agric. Food Chem.* **2008**, *56*, 11761–11766.
75. Kang, C.D.; Sim, S.J.; Selective extraction of free astaxanthin from *Haematococcus* culture using a tandem organic solvent system. *Biotechnol. Prog.* **2007**, *23*, 866–871.
76. Farrow, W.M.; Tabenkin, K. Process for the preparation of lutein. *US Patent 3,280,502*,
25 October 1966.
77. Ishida, B.K.; Chapman, M.H. Carotenoid extraction from plants using a novel, environmentally friendly solvent. *J. Agric. Food Chem.* **2009**, *57*, 1051–1059.
78. Fernández-Sevilla, J.M.; Acién-Fernández, F.G.; Pérez-Parra, J.; Magán-Cañadas, J.J.;
Granado-Lorencio, F.; Olmedilla, B. Large-scale production of high-content lutein extracts from *S. almeriensis*. In *Proceedings of 11th International Conference on Applied Phycology*, Galway, Ireland, 21–26 June 2008.
79. Kang, C.D.; Sim S.J. Direct extraction of astaxanthin from *Haematococcus* culture using vegetable oils. *Biotechnol. Lett.* **2008**, *30*, 441–444.
80. Nonomura, A.M. Process for producing a naturally-derived carotene/oil composition by direct extraction from algae. *US Patent 4,680,314*, 14 July 1987.
81. Zougagh, M.; Valcarcel, M.; Ríos, A. Supercritical fluid extraction: a critical review of its analytical usefulness. *Trends Anal. Chem.* **2004**, *23*, 399–405.
82. Bravi, E.; Perretti, G.; Motanari, L.; Favati, F.; Fantozzi, P. Supercritical fluid extraction for quality control in beer industry. *J. Supercrit. Fluids* **2007**, *42*, 342–346.

83. Bath, M.M.; Zhou, C.; Kute, K.M.; Rosenthal, G.A. Determination of optimum conditions for supercritical fluid extraction of carotenoids from carrot (*Daucus carota* L.) tissue. *J. Agric. Food Chem.* **1995**, *43*, 2876–2878.
84. Albino, A.M.S.; Penteado, M.D.V.C.; Lanças, F.M.; Vilegas, J.H.Y. Carotenoids extraction using $CO_2$ in supercritical state, from kale (*Brassica oleracea*, Lin. var. *acephala*). In *Proceedings of International Congress on Pigments in Food Technology*, Sevilla, Spain, 24–26 March 1999; pp. 65–69.
85. Mendes, R.L.; Fernandes, H.L.; Coelho, J.P.; Reis, E.C.; Cabral, J.M.S.; Novais, J.M.; Palabra, A.F. Supercritical $CO_2$ extraction of carotenoids and other lipids from *Chlorella vulgaris*. *Food Chem.* **1995**, *53*, 99–103.
86. Careri, M.; Furlattini, L.; Mangia, A.; Musci, M.; Anklam, E.; Theobald, A.; von Holst, C. Supercritical fluid extraction for liquid chromatographic determination of carotenoids in *Spirulina pacifica* algae: a chemometric approach. *J. Chromat. A* **2001**, *912*, 61–71.
87. Macías-Sánchez, M.D.; Mantell, C.; Rodríguez, M.; Martínez de la Ossa, E.; Lubián, L.M.; Montero, O. Supercritical fluid extraction of carotenoids and chlorophyll *a* from *Nannochloropsis gaditana*. *J. Food Eng.* **2005**, *66*, 245–251.
88. Macías-Sánchez, M.D.; Mantell, C.; Rodríguez, M.; Martínez de la Ossa, E.; Lubián, L.M.; Montero, O. Supercritical fluid extraction of carotenoids and chlorophyll *a* from *Synechococcus* sp. *J. Supercrit. Fluid.* **2007**, *39*, 323–329.
89. Macías-Sánchez, M.D.; Mantell, C.; Rodríguez, M.; Martínez de la Ossa, E.; Lubián, L.M.; Montero, O. Comparison of supercritical fluid and ultrasound-assisted extraction of carotenoids and chlorophyll a from *Dunaliella salina*. *Talanta* **2009**, *77*, 948–952.
90. Kitada, K.; Machmudah, S.; Sasaki, M.; Goto, M.; Nakashima, Y.; Kumamoto, S.; Hasegawa, T. Supercritical $CO_2$ extraction of pigment components with pharmaceutical importance from *Chlorella vulgaris*. *J. Chem. Technol. Biotechnol.* **2009**, *84*, 657–661.
91. Shen, Y.; Hu, Y.; Huang, K.; Yin, S.; Chen, B.; Yao, S. Solid-phase extraction of carotenoids. *J. Chromat.* **2009**, *1216*, 5763–5768.
92. Bermejo, R.; Ruiz, E.; Acién, F.G. Recovery of B-phycoerythrin using expanded bed adsorption chromatography: Scale-up of the process. *Enzyme Microb. Technol.* **2007**, *40*, 927–933.
93. Miguel, F.; Martin, A.; Mattea, F.; Cocero, M.J. Precipitation of lutein and co-precipitation of lutein and poly-lactic acid with the supercritical anti-solvent process. *Chem. Eng. Proc.* **2008**, *47*, 1594–1602.

*marine drugs*

MDPI

*Review*

# Synthetic Biology and Metabolic Engineering for Marine Carotenoids: New Opportunities and Future Prospects

**Chonglong Wang, Jung-Hun Kim and Seon-Won Kim ***

Division of Applied Life Science (BK21 Plus), PMBBRC, Gyeongsang National University,
Jinju 660-701, Korea; E-Mails: wangchonglong@gmail.com (C.W.); cremoris2000@hotmail.com (J.-H.K.)

* Author to whom correspondence should be addressed; E-Mail: swkim@gnu.ac.kr;
Tel.: +82-55-772-1362; Fax: +82-55-759-9363.

*Received: 14 June 2014; in revised form: 29 August 2014 / Accepted: 1 September 2014 / Published: 15 September 2014*

**Abstract:** Carotenoids are a class of diverse pigments with important biological roles such as light capture and antioxidative activities. Many novel carotenoids have been isolated
from marine organisms to date and have shown various utilizations as nutraceuticals and pharmaceuticals. In this review, we summarize the pathways and enzymes of carotenoid synthesis and discuss various modifications of marine carotenoids. The advances in metabolic engineering and synthetic biology for carotenoid production are also reviewed, in hopes that this review will promote the exploration of marine carotenoid for their utilizations.

**Keywords:** marine carotenoids; carotenoid synthesis; carotenoid modification; metabolic engineering; synthetic biology; protein engineering

## 1. Introduction

Carotenoids are a class of naturally occurring pigments originated in the chloroplasts and chromoplasts of plants, algae and some photosynthetic microorganisms [1–4]. As of 2004, over 750 known carotenoids, which can be divided into xanthophylls (containing oxygen) and carotenes (pure hydrocarbons), have been isolated from natural sources [5]. These structurally diverse pigments play important biological roles in light capture, protection of cells from the damaging effects of free radicals, and synthesis of many hormones as a precursor [6–10]. Carotenoids are traditionally used as food colorants, animal feed supplements, and, very recently, as nutraceuticals and pharmaceuticals [11,12]. Over the past few decades, researches have supported that the ability of carotenoids to reduce the risk of certain cancers, cardiovascular diseases, and degenerative pathogenesis (e.g., Alzheimer and Parkinson) due to their antioxidative properties [13,14]. According to "Carotenoids: A Global Strategic Business Report" from Global Industry Analysts (GIA), the global market for carotenoids was estimated at approximately $1.07 billion in 2010 and is projected to top $1.2 billion by 2015 [15]. Therefore, many efforts have been made to improve the production of these natural compounds for ever-increasing demands [12,16,17].

The ocean is a complex aquatic ecosystem covering about 71% of the Earth's surface, which is around 300 times larger than the habitable volume of the terrestrial habitats on Earth. A large proportion of all life on Earth lives in the ocean. Ecologically distinct from the terrestrial ecosystem, the ocean constitutes a unique reservoir of marine biodiversity and provides a vast resource of foodstuffs, medicines, and other useful materials. As such, more than 250 novel carotenoids have originated from marine species [10], many of which show great potential in commercial applications [18]. With the advent of synthetic biology and metabolic engineering, many engineering tools including vectors, genetic controllers, and enzyme designing, have been developed for heterologous production of valuable chemicals. These tools create new opportunities for exploring marine carotenoids for food and health industries. In this review, we describe diverse and novel carotenoids from marine resources and summarize recent progresses in synthetic biology and metabolic engineering which provide great application potential for marine carotenoids.

## 2. Diversity of Marine Carotenoids

Many carotenoids have been reported from a wide range of marine species. The advances in current technologies facilitate the elucidation of the carotenoid biosynthetic pathways and relevant enzymes from marine species, which would enable the production of important carotenoids from marine organisms.

### 2.1. Pathways and Diverse Enzymes for Biosynthesis of Carotenoids

Biosynthetic routes to carotenoids begin with the basic building blocks isopentenyl diphosphate (IPP) and its isomer dimethylallyl diphosphate (DMAPP), although carotenoids are very diverse in chemical structure. Two distinct pathways, the 2-C-methyl-D-erythritol 4-phospahte (MEP) pathway and the mevalonic acid

(MVA) pathway, are responsible for the synthesis of IPP and DMAPP. These two pathways have been reviewed in detail elsewhere [19–22]. IPP and DMAPP are head-to-tail condensed to generate farnesyl diphosphate (FPP) and geranylgeranyl diphosphate (GGPP) by isoprenyl diphosphate synthases (e.g., IspA of *Escherichia coli* and CrtE of *Pantoea agglomerans*) [23,24]. As shown in Figure 1, FPP and GGPP are further head-to-head condensed to produce symmetric hydrosqualene (C30) and phytoene (C40), which are dehydrogenated in a stepwise manner by desaturating enzymes representing an important branch point for pathway diversification [25,26].

**Figure 1.** Synthesis pathway of phytoene-based C40 carotenoid backbones. Most C40 marine carotenoids are modified from the backbones of α/β/γ-carotenes or isorenieratene. Carotenoid structures are presented with two symmetric or asymmetric halves (Figure 2), for example, lycopene is shown as $H_oH'_o$ in this review. Conjugated double bonds are shown in red.

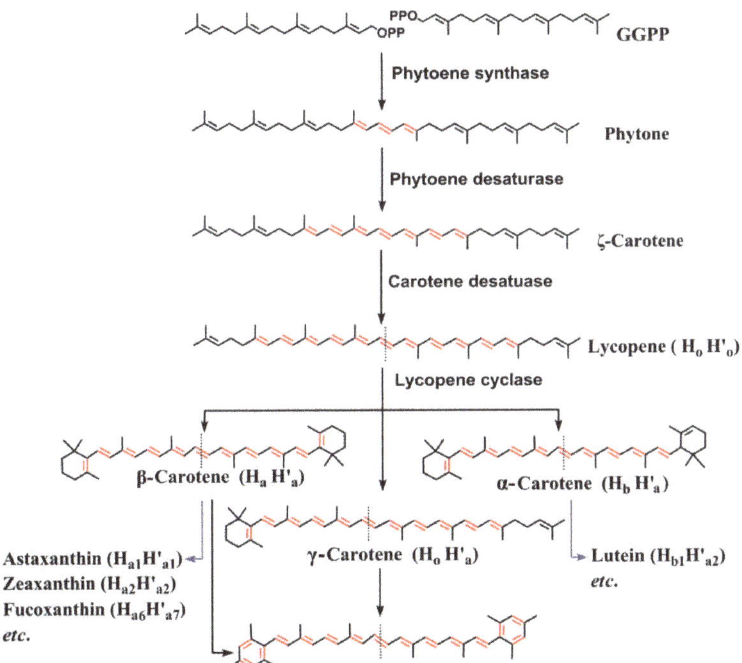


Enzymes involved in the biosynthesis of carotenoids have been mainly investigated in carotenogenic cyanobacteria and land plants [27,28]. They are mostly associated with cytoplasmic and organelle membranes where the hydrophobic substrates of carotenogenic enzymes are located [29]. So far, very few crystal structures of carotenogenic enzymes have been elucidated because of their association with the membranes [30,31]. More than 95% of carotenoids have been characterized in nature to be phytoene-based [32], which will be extensively discussed in this review.

Phytoene synthase is positioned early in the carotenoid synthesis pathway and is responsible as a pathway gatekeeper to discriminate GGPP substrate from enormous isoprenyl diphosphates [29]. Phylogenic analysis of 20 phytoene synthases from marine organisms supports the endosymbiotic theory that plastids evolve from a cyanobacterium, which is engulfed and retained by a unicellular protist [33,34]. Cyanobacteria *Acaryochloris marina* and *Prochlorococcus marinus* are clustered with green algae and land plant tomato (Figure 2A). However, phytoene synthases still display a significant diversification by evolution. A consensus position of 24.5% (identity of 0.5%) is remained among phytoene synthases from marine algae, bacteria, Achaea and land plants. There is only a similarity of 31.9% even between the two proteobacteria phyla $\alpha$-proteobacteria and $\gamma$-proteobacteria.

The photochemical properties of a carotenoid depend on the size of the chromophore formed by conjugated double bonds, and a C40 backbone can accumulate up to 15 conjugated double bonds [35]. Thus, six sequential desaturation steps are required to dehydrogenate colorless phytoene, which has three conjugated double bonds in the center [26]. Lycopene containing a chromophore with eleven conjugated double bonds is the direct precursor of $\alpha/\beta/\gamma$-carotenes or isorenieratene, the phytoene-based C40 carotenoid backbone (Figure 1). In general, oxygenic phototrophs require three enzymes, phytoene desaturase, $\zeta$-carotene desaturase and *cis*-carotene isomerase to generate lycopene [6]. However, most bacterial phytoene desaturases are able to catalyze all three reactions [30]. There are also some organisms that disobey this general rule. Primitive cyanobacteirum *Gloeobacter violaceus* PCC 7421 uses bacterial type phytoene desaturase, and no homolog of $\zeta$-carotene desaturase or *cis*-carotene isomerase is found in its genome [36,37]. Among anoxygenic phototrophs, green sulfur bacteria use three enzymes to catalyze desaturation, whereas purple bacteria, green filamentous bacteria, and heliobacteria use only one enzyme [38,39]. Phytoene desaturases also exhibit significant diversities among different organisms (Figure 2B). Just as phytoene synthase, green algae are clustered with tomato but they are distinguished from cyanobacteria. There is just a similarity of 23.2% among 21 proteins. It is suggested that phytoene desaturase exhibits a much faster evolution from the ancestral blueprint and higher diversities among species than phytoene synthases, which may correspond to promiscuous activities of phytoene desaturase.

**Figure 2.** Phylogenic trees of (**A**) phytoene synthases and (**B**) phytoene desaturases. Trees were built using MEGA6.0 software by Neighbor-Joining method [40]. Protein sequences were obtained from *National Center for Biotechnology Information* (NCBI).

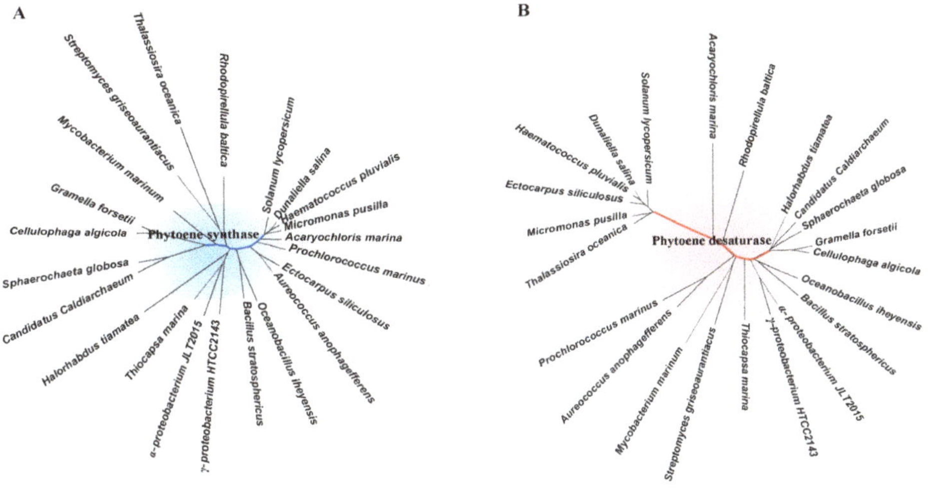


## 2.2. Diversity of Marine Carotenoids

Carotenogenic organisms in ocean are algae and bacteria, which possess all the genes for *de novo* synthesis of carotenoids [2–4]. Unicellular microalge *Dunaliella salina* and *Dunaliella bardawil* are rich in the orange pigment β-carotene ($H_aH'_a$, Figure 1) [41,42]. Two rings of β-carotene are often oxidized to form astaxanthin ($H_{a1}H'_{a1}$, Figure 3) in some microalgae by β-carotene hydroxylase and ketolase [43], which can individually catalyze the modification of β-carotene to generate zeaxanthin ($H_{a2}H'_{a2}$, Figure 3) in *Spriulina platensis* and *Spriulina maxima* [44], and canthaxanthin ($H_{a3}H'_{a3}$, Figure 3) in *Haematococcus pluvialis, Clorella vulgaris* and *Colastrella striolata* [44–46]. The modifications can just occur in one ring to generate asymmetric intermediates such as β-cryptoxanthin ($H_aH'_{a2}$, Figure 3) and echinenone ($H_aH'_{a3}$, Figure 3). Chlorophyta *Scenedesmus almeriensis* and *Muriellopsis sp.* accumulate a large amount of lutein ($H_{b1}H'_{a2}$, Figure 3), which is derived from α-carotene ($H_bH'_a$, Figure 1) [47]. Cryptophyta also synthesize α-carotene as well as acetylenic derivatives crocoxanthin ($H_bH'_{a4}$, Figure 3) and monadoxanthin ($H_{b1}H'_{a4}$, Figure 3) [28]. Acetylenic groups are also found in β-carotene derivatives alloxanthin ($H_{a4}H'_{a4}$, Figure 3) in Cryptophyta [48], and diatoxanthin ($H_{a2}H'_{a4}$, Figure 3) and epoxy oxidized diadinoxanthin ($H_{a4}H'_{a5}$, Figure 3) in Heterokontophyta, Haptophyta, Dinophyta, and Euglenophyta [28,49,50]. The unique acetylenic carotenoids are only found in algae. In brown algae and diatoms, acetylated and unique allenic modifications produce dinoxanthin ($H'_{a5}H_{a6}$, Figure 3) and chain-oxidized fucoxanthin ($H_{a6}H'_{a7}$, Figure 3) [2,51]. Some Chlorophyta species modify the methyl group of lutein to generate loroxanthin ($H_{b1}H'_{a8}$, Figure 3) in *Scenedesmus obliquus* and *Chlorella vulgaris* [52], and siphonaxanthin ($H_{b1}H'_{a9}$, Figure 3) in *Codium fragile* [53]. Aromatic isorenieratene ($H_cH'_c$, Figure 1) is a usual biomarker compound, which is synthesized from β-carotene in actinobacteria or γ-carotene ($H'_oH_a$, Figure 1) in green and purple sulfur bacteria [54,55]. γ-Carotene can also be converted to chlorobactene ($H_cH'_o$, Figure 1) and OH-chlorobactene ($H_cH'_{o1}$, Figure 3). Glycoside modifications generate OH-chlorobactene glucoside ($H_cH'_{o2}$, Figure 3) in green sulfur bacteria and myxol 2'-fucoside ($H_{o3}H'_{a2}$, Figure 3) in Cyanophyta [54,56]. Dinophyta can synthesize C37-skeletal carotenoids such as peridinin ($H_{d1}H'_{a6}$, Figure 3) [57]. Animals do not have pathways for *de novo* synthesis of carotenoids, but they obtain carotenoids from food and further modify carotenoids by oxidation, reduction, translocation of double bonds, cleavage of double bonds, *etc.* Peridinin-originated carotenoids such as peridininol ($H_{e1}H'_{d1}$, Figure 3) and cyclopyrrhoxanthin ($H_{e2}H'_{d2}$, Figure 3) have been isolated from bivalves *Crassostrea gigas*, *Paphia amabillis,* and *Corbicula japonica* [58–60]. Two unique nor-carotenoids, 2-nor-astaxanthin ($H_{f1}H'_{a1}$, Figure 3) and actinoerythrin ($H_{f2}H'_{f2}$, Figure 3), have been found in sea anemones *Actinia equine* and *Tealia feline* [61]. The carotenoid diversity in marine animals has been well summarized in detail elsewhere [61]. It is also worthy to note that some carotenoids are present in different stereo configurations among organisms (not covered in this review), which also greatly contributes to the diversification of carotenoids.

**Figure 3.** Diverse modifications of carotenoids. The structure shows the modification of a half carotenoid backbone. The stereo configurations are not shown in the structures. Glycosyl moieties of fucose and glucose are represented by Fuc and Glu, respectively.

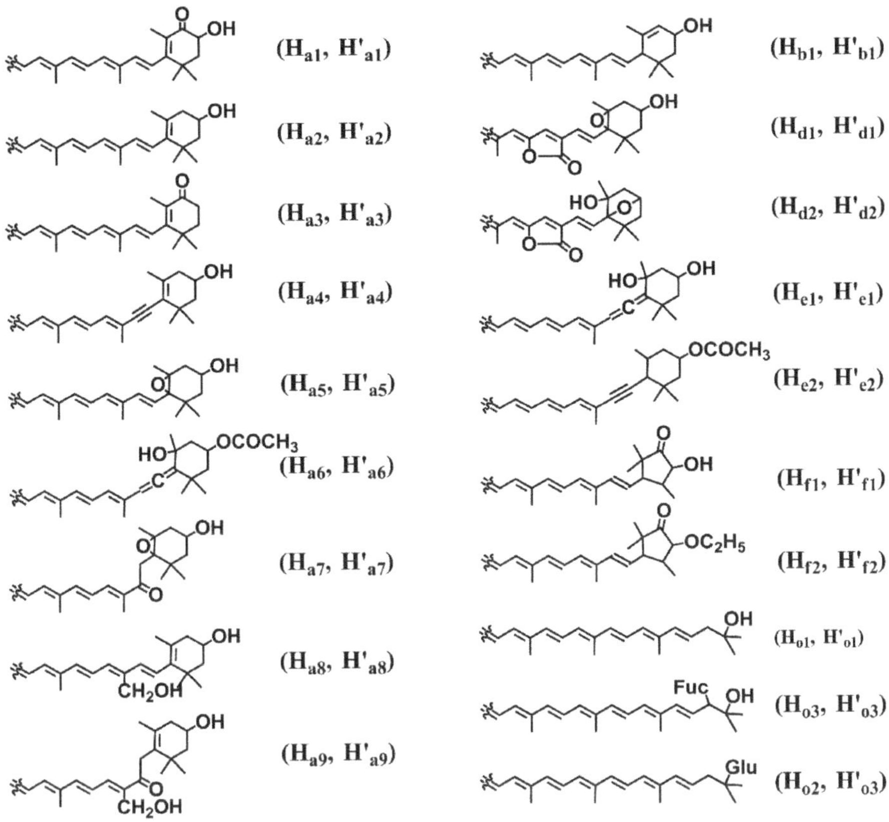


### *2.3. Synthesis of Some Important Marine Carotenoids and Enzymes*

β-Carotene as well as xanthophylls astaxanthin, zeaxanthin, lutein, and fucoxanthin are some representative marine carotenoids due to their abundance in marine organisms and their inherent antioxidant properties. β-Carotene is synthesized from the cyclization of lycopene, a key step in generating carotenoid diversity by lycopene cyclases, which can also lead to α/γ-carotene formation (Figure 1). The β-cyclase catalyzes the symmetrical formation of two identical β-ionone rings of β-carotene. On the other hand, α-carotene contains two different ring structures (ε and β) formed by the action of additional ε-cyclase with β-cyclase. Four distinct families of lycopene cyclases, CrtY-type β-cyclases in proteobacteria, CrtL β/ε-cyclases in some cyanobacteria, the heterodimeric cyclases in some Gram-positive bacteria and FixC dehydrogenase superfamily lycopene cyclases in *Chlorobium tepidum* and *Synechococcus sp.* PCC 7002, have been identified to date [62]. Further decorations occur via a variety of ketolation (oxidation), hydroxylation (Figure 4), which are the major causes for the diversity among carotenoids [29]. β-Carotene ketolase (CrtW or CrtO) adds the keto groups at the 4,4′-position of the ring and β-carotene hydroxylase (CrtZ) adds the hydroxyl group at the 3,3′-position [63]. Both enzymes are responsible for the formation of astaxanthin via zeaxanthin or canthaxanthin routes in some cyanobacteria and algae (Figure 3). Lutein formation is ascribed to the hydroxylation of $\alpha$-carotene by cytochrome P450 enzymes in *Arabidopsis thaliana* [64], while the pathway and enzymes remain to be elucidated from marine organisms. Fucoxanthin with a unique allenic and epoxide structure is derived from zeaxanthin in brown seaweeds, diatoms and dinoflagellates. Genome analysis indicates that zeaxanthin epoxidases epoxidize zeaxanthin to form antheraxanthin ($H_{a2}H'_{a5}$, Figure 3) and violaxanthin ($H_{a5}H'_{a5}$, Figure 3) [65]. Two possible routes have been proposed for the synthesis of fucoxanthin from violaxanthin via neoxanthin ($H_{e1}H'_{a5}$, Figure 3) or diadinoxanthin [66]. Very recently, a cytochrome P450-type carotene hydroxylase (PuCHY1) has been isolated from red alga *Porphyra umbilicalis*. The compensatory expression of PuCHY1 results in the formation of violaxanthin, neoxanthin, and lutein in *A. thaliana* by the β/ε-hydroxylation activities [67]. Some of the carotenogenic enzymes characterized from marine organisms have been summarized in the literature [28].

**Figure 4.** Synthetic pathway of astaxanthin, lutein, and fucoxanthin from lycopene. Arrows indicate each catalysis reaction, and enzymes are shown in blue. Dashed arrows indicate hypothesized reactions. Reaction intermediates are shown in gray.

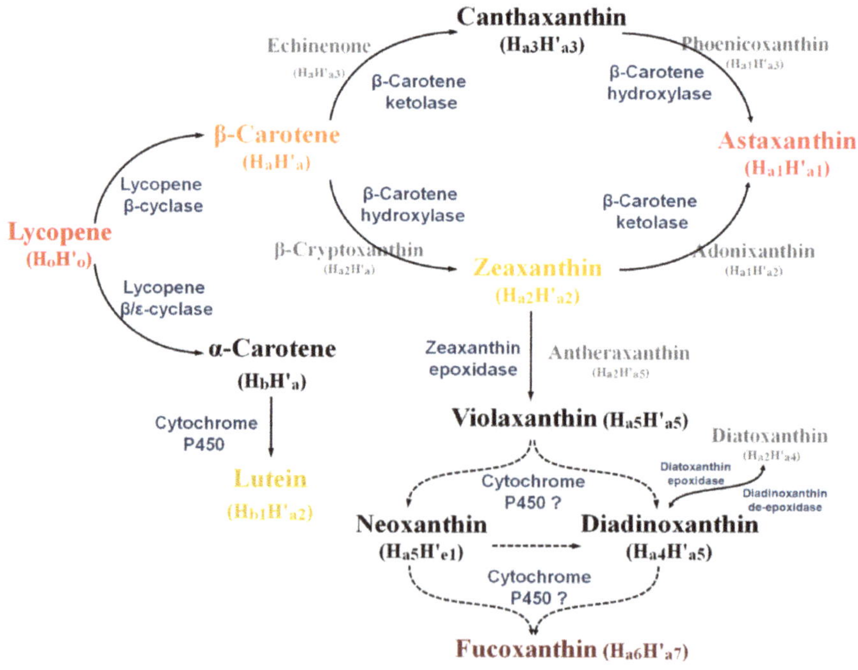


## 3. Technology Developments for Production of Carotenoids

Over the decades, many researches have been done for the production of carotenoids. Carotenogenic pathways have been identified and manipulated in several organisms, and advances in metabolic engineering and synthetic biology have resulted in significant improved production of carotenoids including astaxanthin, zeaxanthin, and lutein.

### *3.1. Easy Colorimetric Screening of Production of Carotenoids*

Carotenoids contain chromophores absorbing visible light and appear as being yellow (e.g., β-carotene) to red (e.g., lycopene), which benefits carotenogenic gene mining and engineering upon carotenoid synthesis pathway. To date, many carotenoid biosynthetic genes have been cloned from plants, bacteria, and fungi based on their abilities to render different colors to the host [68–70]. This merit has been vigorously implemented for random mutagenesis, directed evolution, and proof-of-principle experiments in synthetic biology. Moreover, cellular carotenoids can be easily extracted into an organic solvent and differentiated in a sensitivity of submilligrams per liter with a linear correlation between carotenoid contents and color intensity [71,72]. This provides an easy and high-throughput way to evaluate the performance of newly built synthetic circuits or methodologies for improved biosynthesis of carotenoid (Figure 5A).

**Figure 5.** Technologies and efforts to improve carotenoid production. (**A**) Colorimetric screening of desired producer; (**B**) Pathway engineering approaches for production improvement; (**C**) Genetic modifications for host strain development; (**D**) Protein engineering for enzyme and pathway improvement.

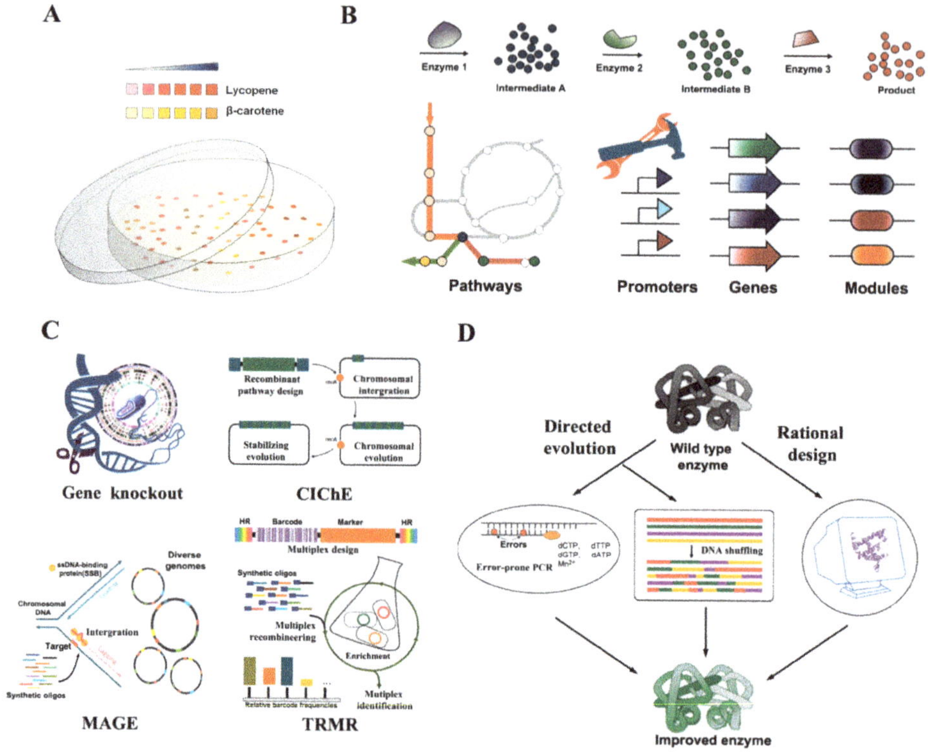


### 3.2. *Pathway Engineering for Production of Carotenoids*

Carotenoid biosynthesis emerges from the central isoprenoid pathway, either the MEP pathway or the MVA pathway, existing in all organisms [19,22]. The expression of carotenogenic genes can yield carotenoids of interest in a heterologous organism [16,73–75]. The early attempts led to the production of lycopene, β-carotene, and astaxanthin in *Saccharomyces cerevisiae* and *Candida utilis* by the expression of carotenogenic enzymes from *Pantoea ananatis* [74,76]. *Corynebacterium glutamicum* is a native producer of decaprenoxanthin and its glucosides, and it has been engineered to synthesize C50 carotenoids C.P.450 and sarcinaxanthin [77]. To date, there have been many exemplary illuminations to achieve high carotenoid titers from non-native producers. Carotenogenic enzymes from different sources exhibit different capacities in carotenoid biosynthesis. A two-fold higher lycopene production is obtained in *E. coli* by the expression of carotenogenic enzymes from *P. agglomerans* (27 mg/L) than from *P. ananatis* (12 mg/L) [78]. Metabolic engineering approaches allow the assembly of genes from different organisms for production purposes or for building new carotenoids [32,79,80]. β-Carotene production has been improved by hybrid expression of carotenogenic genes from *P. agglomeras* and *P. ananatis* in *E. coli* [81]. In another example, expression of β-end ketolase from *Agrobacterium aurantiacum* extends the zeaxanthin β-D-diglucoside pathway from *P. ananatis*, and synthesizes novel astaxanthin β-D-diglucoside and adonixanthin β-

D-diglucoside [29]. Generally, a sufficient precursor supply is a prerequisite for high-yield production of carotenoids. Overexpression of the rate-limiting enzymes 1-deoxy-D-xylulose-5-phosphate synthase and reductoisomerase led to a 3.6-fold increase in lycopene production in *E. coli* when compared with the native MEP pathway for IPP and DMAPP supply [71]. Overexpression of the rate-limiting enzyme 3-hdroxy-3-methyl-glutaryl-coenzyme A (HNG-CoA) reductase of the MVA pathway from *Xanthophyllomyces dendrorhous* significantly increased β-carotene production in *S. cerevisiae* [82]. A great effort in metabolic engineering of the central carotenoid building block pathway is the introduction of a hybrid MVA pathway of *Streptococcus pneumonia* and *Enterococcus faecalis* into *E. coli*, which enables the recombinant host to produce 465 mg/L of β-carotene [83]. With more available genetic tools, microbial organisms such as *Pseudomonas putida* and *Bacillus subtilis* have also been developed as platform hosts for carotenoid production [84,85].

Carotenoids synthesis involves multiple enzymes [2]. The expression level of all the components of a multigene circuit should be orchestrated to optimize metabolic flux and to gain a high yield (Figure 5B) [86]. A random approach is screening of the best orchestra from numerous combinatorial assemblies of required genes and control elements. BioBrick™ paradigm is capable of rapidly assembling a biosynthetic pathway in a variety of gene orders from different promoters in plasmids with different copy numbers [87]. It is possible to build a hybrid carotenoid pathway wherein each enzyme possesses a right turnover number, however, BioBrick™ assembly is still not in a high throughput to create vast combinatorial expression constructs for the best combination of carotenogenic genes. Recently, several advanced assembly methods using homologous recombination, such as sequence and ligation-independent cloning (SLIC), Gibson DNA assembly and reiterative recombination, have been applied to construct multigene circuits [88–90]. These advances promise to randomize all genetic components, including genes, promoters, ribosome binding sites, and other control modules to build a large number of individual genetic circuits for screening purposes. A so-called "randomized BioBrick assembly" approach has been applied to the optimization of the lycopene synthesis pathway wherein the expression construct was designed to independently express each enzyme from its own promoter, which resulted in an increase by 30% in lycopene production [91]. A longer and more complicated pathway can be modularized into subsets, which contain pathway enzymes with similar turnover numbers. Modulating these subsets would be more convenient and efficient than regulating all components of the entire pathway for improved production [92]. By using this multivariate modular metabolic engineering (MMME) approach, recent work achieved a 15,000-fold increase in production of taxadiene, a precursor of the anti-cancer drug taxol [93]. There are also a variety of promising approaches, such as tunable promoters, tunable intergenic regions, and ribosome binding site design, which can be applied to fine tuning the expression of modules [94–96]. In the other approaches, a multi-genic operon is transcribed into a single polycistronic mRNA, and then the large transcript can be spliced to small monocistronic transcripts through post-transcriptional RNA processing such as ribozyme cleavage and *clustered regularly interspaced short palindromic repeats* (CRISPR) editing. Thus, the stability of the monocistronic transcripts can be independently modulated to differentiate the expression level of each enzyme even in a multi-gene operon. These RNA processing tools have been developed as insulating elements between operonic genes to reduce the context dependence of the genes in a polycistronic transcription unit [97]. The diffusion of pathway intermediates can decrease the effective concentrations of intermediates for following enzyme reactions and some intermediates may serve for competing pathways. By learning from Mother Nature, synthetic biologists spatially organize enzymes of the MVA pathway by protein scaffolds in *E. coli* to minimize diffusion limitation and achieve a 77-fold increase in mevalonic acid production [98]. The propanediol utilization machinery of *Citrobacter freundii* has been heterologously recasted in *E. coli* [99]. Some intermediates of carotenoid synthesis such as isoprenyl diphosphates are toxic when they accumulate over the concentration threshold [100]. To avoid the accumulation of toxic intermediates, genetic sensors can potentially be coupled with gene expression cassettes to regulate the intermediate flux in a dynamic manner. The native *E. coli* promoters that respond to the toxic FPP have been successfully used to dynamically regulate the amorphadiene synthesis pathway and improve the production by two-fold over common inducible promoters and constitutive promoters [101]. The Ntr regulon has been engineered to control lycopene synthesis in response to glycolytic flux dynamics, resulting in an 18-fold increase in lycopene production [102].

### *3.3. Genome Engineering for Strain Development*

For the most efficient carotenoid production, the biological system of the host organism also needs to be optimized, by, for example, redirecting cellular carbon flux to the carotenoid synthesis pathway. The *de novo* synthesis of carotenoids is initiated from acetyl-coA by the MVA pathway or glycolytic metabolites pyruvate and glycraldehyde-3-phosphate (G3P) by the MEP pathway. The direct efforts are focused on the modification of associated genes to these pathways. Deletion of pyruvate kinases PykFA can balance the availability of pyruvate and G3P for the MEP pathway, and increase lycopene production by 2.8-fold in *E. coli* [103]. The deletion of glucose-6-phospahte (G6P) dehydrogenase Zwf, which branches G6P to pentose phosphate pathway results in an increase by 30% in lycopene production [104]. Deletion of carbohydrate phosphotransferase system yields a seven-fold increase in lycopene production in another study [105]. Replacement of native promoters of the rate-limiting genes of the MEP pathway with the T5 promoter has been carried out for enhancement of the targeted pathway flux, which results in a 4.5-fold increase in β-carotene production [106].

A heterologous pathway is not just an independent entity. It communicates with the native cellular metabolism and is therefore governed by the global regulation of the host organisms. Adaptive laboratory evolution is a traditional route for strain engineering to achieve desirable industrially relevant phenotypes. Owing to the antioxidant properties of carotenoids, adaptive evolution has been successfully applied to an engineered *S. cerevisiae* with periodic hydrogen peroxide shocking, resulting in a three-fold increasee of β-carotene production. Subsequent transcriptome analysis indicates that some genes related with lipid biosynthesis and MVA pathways are up-regulated in the adopted strains [107]. It also suggests that carotenoid production can be improved by modifications (knock-out or overexpression) of distant genes, which are responsible for the overall regulation of the metabolic network or the physiological fitness of the host (Figure 5C). In a genome-wide screening of yeast deletion collection, 24 deletions exhibit significant higher carotenoid levels than the wild type. The triple deletion of *ROX1, YJL064W,* and *YJL062W* shows an almost four-fold increase in total carotenoid production [108]. Gene deletions of *hnr, yjfP,* and *yjiD* related to the improvement of lycopene production have been identified from a global transposon *E. coli* mutant library [109]. Other gene deletions such as *gdhA, cyoA, ppc, gpmA, gpmB, eno, glyA,* aceE, *talB,* and *fdhF* have been *in silico* identified using a stoichiometric model [110]. The triple mutation of *gdhA, aceE* and *fdhF* was validated to increase lycopene production by nearly 40% in *E. coli* over the engineered parental strain. A similar set of gene deletions *dhA, cyoA, gpmA, gpmB, icdA,* and *eno* have been also *in silico* identified using different metabolic network models [111]. Overexpression of some genes encoding global regulatory proteins AppY, Crl, RpoS, and ElbAB, oxidoreductases TorC, YdgK, and YeiA, and hypothetical proteins YedR and YhbL, result in a significant increase in lycopene production in *E. coli* [112]. With a profound understanding of the landscape of genome manipulation, all these knocked-out and overexpressed alleles have been combined and optimized to generate high-fitness host strains for lycopene production [113,114]. ATP and NADPH are also important cofactors for the production of carotenoids. Using engineering ATP synthesis, pentose phosphate and TCA modules, recent work has shown the highest β-carotene production of 2.1 g/L by a fed-batch fermentation process in *E. coli* [115]. The advances in synthetic biology greatly boost genome manipulation on a large scale. Multiplex automated genome engineering (MAGE) simultaneously targets many locations on the chromosome for modification in a single cell or across a population of cells by directing ssDNA to the lagging strand of the replication fork during DNA replication [116]. The modifications can cover gene inactivation, expression regulations, and so on. Aforementioned twenty genes related to lycopene production have been targeted to tune their expression using a complex pool of synthetic DNAs, and lycopene production is increased more than five-fold. A complementary method called trackable multiplex recombineering (TRMR) has been developed to simultaneously map genome modifications that affect a trait of interest, which combines parallel DNA synthesis, recombineering and molecular barcode technology to enable rapid modification of all *E. coli* genes in an *a priori* knowledge-independent way [117].

Metabolic engineering for the production of valuable compounds often heavily relies on plasmid-based expression of the synthesis pathway in a heterologous host. Although plasmids are easily manipulated and allow strong expression of targeted enzymes, the plasmid-based systems suffer from genetic instability such as plasmid loss, an additional antibiotic cost, and a potential risk of antibiotic marker spreading to other organisms [118]. Accordingly, chromosomal integration of the production pathway promises the host to achieve stable overproduction of the desirable chemicals including carotenoids. By λ-Red homologous recombination, plasmid-free engineered *E. coli* strain has been developed to produce lycopene and astaxanthin [119]. The expression cassettes can be integrated into different loci to increase the number of gene copies. P1 transduction usually plays a role in transfering the different alleles between host strains. Recently, an intelligent strategy called chemically inducible chromosomal evolution (ClChE) has been developed to reduce the daunting repeated one-at-a-time tasks in the chromosomal integration of target genes [88]. ClChE allows the host to acquire a high gene copy (up to 40 copies) expression of integrated pathways with increasing concentration of selective chemicals, and the increased copy number is stabilized by the removal of the *recA* gene. With this approach, lycopene production has been increased by 60% from single copy integrated strain. The ClChE strategy has been further modified to eliminate antibiotic marker for environmental safety and health issue after the evolution of the recombinant host strain [120].

### *3.4. Protein Engineering for Improvement of Carotenoid Production Enzymes*

Pathway engineering for efficient production of desired chemicals is often challenged by limitations associated with the pathway enzymes themselves, such as low turnover numbers and promiscuities generating unwanted by-products [121]. Protein engineering provides a powerful solution to improve specific activity and substrate specificity of enzymes, and even to create new activity. Methods of protein engineering include directed evolution and computer-assisted rational design (Figure 5D) [122,123]. Directed evolution is an iterative process that imitates Darwinian evolution in the laboratory to select or screen a desired phenotype from mutagenesis. Typically, error-prone polymerase chain reaction (PCR) is used to generate mutant libraries, and DNA shuffling is carried out to recombine existing mutations. It can be performed in a blind manner with limited information on target enzymes, such as structures and reaction mechanisms, but it relies on an effective screening strategy. It is practical for the evolution of carotenogenic enzymes due to the innate traits of carotenoid pigments. Six mutants ((H96L, R203W, A205V, A208V, F213L and A215T) have been isolated to improve the catalytic activity of β-carotene ketolase from *Sphingomonas sp.* [124]. Three mutations (L175M, M99V, and M99I) of ketolase from *Paracoccus sp.* result in the improvement if its specificity of to synthesize astaxanthin [125]. *Staphylococcus aureus* dehydrosqualene (C30) synthase has evolved to synthesize lycopene by mutation F26L or F26S [126]. DNA shuffling of phytoene desaturases from *P. agglomerans* and *P. ananatis* results in the isolation of a variant favoring the production of fully conjugated tetradehydrolycopene [127]. Rational design of proteins is based on the *in silico* simulation and the prediction using *a priori* enzyme information, which greatly liberates biologists from onerous screening task. This strategy requires adequate information to predict specific targeted amino acid mutations, which can confer desired enzyme traits [128]. Unfortunately, the limited information on carotenogenic enzymes leads to few achievements using such a method.

As aforementioned, carotenoids are derived from the central isoprenoid pathway, which is also employed to synthesize several essential and secondary metabolites in nature. The carotenoid-based colorimetric screening has been developed for evolution of other isoprenoid pathway enzymes. Mutations of GGPP synthase are hypothesized to affect the binding efficiency of the magnesium ions needed for substrate anchoring and improve its catalysis. An error-prone PCR library of *Tsuga canadensis* GPPS has been screened using the lycopene synthesis pathway as a colorimetric reporter. The GPPS variant (S239C and G295D) is created to increase levopimaradiene production with a 1.7-fold increase over the wild type in *E. coli* [129]. Augmentation of one pathway can tamper with other pathways, which utilize the same substrate in one organism. Based on this fact, mutagenesis libraries of terpene synthases have been screened by depigmentation of colonies due to the competition between terpene synthases and carotenoid synthases for isoprenyl diphosphates, since the weakened carotenoid color intensity indicates an improvement of terpene synthase activity [130].

### 3.5. Development of Microalgae for Carotenoid Production

Algae are a diverse group of aquatic, photosynthetic organisms, generally categorized as macroalgae (*i.e.*, seaweed) and unicellular microalgae. Microalgae have recently garnered interest for production of valuable chemicals including carotenoids [41,131], because they are generally regarded as safe (GRAS) for human consumption and possess the renewable-energy capturing ability of photosynthesis. Moreover, these organisms can be used for genetic manipulation and high-throughput analysis [132]. Some microalgae are also native carotenoid producers (*i.e.*, *D. salina* for β-carotene and *H. pluvialis* for astaxanthin). The carotenoid production from microalgae is closely related to culture conditions such as illumination, pH, temperature, nitrogen availability and source, salinity, the oxidant substances, and growth rate [12,133,134]. *D. salina* is a model species of green microalgae which is widely cultivated outdoors for β-carotene production [131]. A systematic evaluation has been done to decipher the relationship between abiotic stresses (Nitrate concentration, salinity and light quality) and lutein synthesis in *D. salina* [135]. The abiotic stresses can also be applied to adaptive evolution of microalgae [136], in a similar manner to strain evolution in yeast for β-carotene production [107]. The freshwater microalga *Chlamydomonas reinhardtii* is the first and the best studied transformed Chlorophyte, and the nuclear genetic manipulation is easy and well established. It has been engineered with β-carotene ketolase from *H. pluvialis* to synthesize ketolutein ($H_{b1}H'_{a1}$, Figure 3) and adonixanthin ($H_{a1}H'_{a2}$, Figure 3) [137]. It is possible to produce diverse valuable carotenoids from marine microalgae with the development of more available genetic tools and technologies.

## 4. Opportunities and Challenges

The vast and mysterious ocean breeds diverse marine lives and provides unexhausted foodstuffs, nutriment, and drugs for humans. Diverse carotenoids are found from marine species and show broad utilities as colorant fragrance cosmetics and pharmaceuticals. The synthetic pathway of several carotenoids has been illuminated from marine species, which could benefit engineering processes in several host organisms for the production of carotenoids such as β-carotene, astaxanthin, and lutein. On the other hand, carotenoids such as β-carotene often undergo a series of modifications in the miraculous marine ecosphere. And indeed, several novel carotenoids have been isolated during the exploration of the marine ecosphere, while their pharmaceutical potentials remain to be examined due to the limited amount of extracts. Metabolic engineering and synthetic biology allow the assembly of such a chimeric pathway in a tractable organism for the mass production of rare carotenoids and also exhibit the potential to extend the catalogs of carotenoids to non-natural carotenoids, which could accelerate the exploration of novel carotenoids. It is noted that decoded carotenoid pathways and enzymes are still limited to a few marine organisms, although the J. Craig Venter Institute with worldwide collaboration had sequenced and annotated the genomes of 177 marine microbes up until 2010. However, we believe that the developed and developing technologies will allow us to search for novel marine carotenoid pathways in the future.

## Acknowledgments

This work was supported by a grant (NRF-2013R1A1A2008289) from the National Research Foundation, the Intelligent Synthetic Biology Center of Global Frontier Project funded by the MSIP (2011-0031964), and a grant from the Next-Generation BioGreen 21 Program (SSAC, grant#: PJ00952003), Rural Development Administration (RDA), Korea. J.K. is supported by scholarships from the BK21 Plus Program, Ministry of Education, Science & Technology (MEST), Korea.

## Author Contributions

S.K. conceived the idea and held and corrected the manuscript. C.W. and J.K collected the literature, analyzed the data, and wrote the manuscript. C.W. and J.K contributed to this manuscript equally.

## Conflicts of Interest

The authors declare no conflict of interest.

## References

1. Cazzonelli, C. Carotenoids in nature: Insights from plants and beyond. *Funct. Plant Biol.* **2011**, *38*, 833–847.
2. Bertrand, M. Carotenoid biosynthesis in diatoms. *Photosynth. Res.* **2010**, *106*, 89–102.
3. Guedes, A.C.; Amaro, H.M.; Malcata, F.X. Microalgae as sources of carotenoids. *Mar. Drugs* **2011**, *9*, 625–644.
4. Shindo, K.; Misawa, N. New and rare carotenoids isolated from marine bacteria and their antioxidant activities. *Mar. Drugs* **2014**, *12*, 1690–1698.
5. Britton, G.; Liaaen-Jensen, S.; Pfander, H. *Carotenoids Handbook*; Birkhäuser: Basel, Switzerland, 2004.
6. Armstrong, G.A.; Hearst, J.E. Carotenoids 2: Genetics and molecular biology of carotenoid pigment biosynthesis. *FASEB J. Off. Publ. Fed. Am. Soc. Exp. Biol.* **1996**, *10*, 228–237.
7. Mathews, M.M.; Sistrom, W.R. The function of the carotenoid pigments of *Sarcina lutea*. *Archiv. Mikrobiol.* **1960**, *35*, 139–146.
8. Liu, J.; Novero, M.; Charnikhova, T.; Ferrandino, A.; Schubert, A.; Ruyter-Spira, C.;
Bonfante, P.; Lovisolo, C.; Bouwmeester, H.J.; Cardinale, F. Carotenoid cleavage dioxygenase 7 modulates plant growth, reproduction, senescence, and determinate nodulation in the model legume *Lotus japonicus*. *J. Exp. Bot.* **2013**, *64*, 1967–1981.
9. Bolhassani, A.; Khavari, A.; Bathaie, S.Z. Saffron and natural carotenoids: Biochemical activities and anti-tumor effects. *Biochim. Biophys. Acta* **2014**, *1845*, 20–30.
10. Rodrigues, E.; Mariutti, L.R.; Mercadante, A.Z. Scavenging capacity of marine carotenoids against reactive oxygen and nitrogen species in a membrane-mimicking system. *Mar. Drugs* **2012**, *10*, 1784–1798.
11. Van Den Berg, H.; Faulks, R.; Fernando Granado, H.; Hirschberg, J.; Olmedilla, B.; Sandmann, G.; Southon, S.; Stahl, W. The potential for the improvement of carotenoid levels in foods and the likely systemic effects. *J. Sci. Food Agric.* **2000**, *80*, 880–912.

12. Fernandez-Sevilla, J.M.; Acien Fernandez, F.G.; Molina Grima, E. Biotechnological production of lutein and its applications. *Appl. Microbiol. Biotechnol.* **2010**, *86*, 27–40.
13. Kirsh, V.A.; Mayne, S.T.; Peters, U.; Chatterjee, N.; Leitzmann, M.F.; Dixon, L.B.; Urban, D.A.; Crawford, E.D.; Hayes, R.B. A prospective study of lycopene and tomato product intake and risk of prostate cancer. *Cancer Epidemiol. Biomark. Prev. Publ. Am. Assoc. Cancer Res. Cosponsored Am. Soc. Prev. Oncol.* **2006**, *15*, 92–98.
14. Wang, W.; Shinto, L.; Connor, W.E.; Quinn, J.F. Nutritional biomarkers in alzheimer's disease: The association between carotenoids, n-3 fatty acids, and dementia severity. *J. Alzheimer's Dis. JAD* **2008**, *13*, 31–38.
15. Cosgrove, J. The carotenoid market: Beyond beta-carotene. *Nutraceuticals World*, 13 December 2010. Available online: http://www.nutraceuticalsworld.com/contents/view_online-exclusives/2010-12-13/the-carotenoid-market-beyond-beta-carotene/ (accessed on 5 September 2014).
16. Rodriguez-Saiz, M.; de la Fuente, J.L.; Barredo, J.L. Xanthophyllomyces dendrorhous for the industrial production of astaxanthin. *Appl. Microbiol. Biotechnol.* **2010**, *88*, 645–658.
17. Kim, Y.S.; Lee, J.H.; Kim, N.H.; Yeom, S.J.; Kim, S.W.; Oh, D.K. Increase of lycopene production by supplementing auxiliary carbon sources in metabolically engineered *Escherichia coli*. *Appl. Microbiol. Biotechnol.* **2011**, *90*, 489–497.
18. Vilchez, C.; Forjan, E.; Cuaresma, M.; Bedmar, F.; Garbayo, I.; Vega, J.M. Marine carotenoids: Biological functions and commercial applications. *Mar. Drugs* **2011**, *9*, 319–333.
19. Miziorko, H.M. Enzymes of the mevalonate pathway of isoprenoid biosynthesis. *Arch. Biochem. Biophys.* **2011**, *505*, 131–143.
20. Hunter, W.N. The non-mevalonate pathway of isoprenoid precursor biosynthesis. *J. Biol. Chem.* **2007**, *282*, 21573–21577.
21. Wang, C.; Kim, J.Y.; Choi, E.S.; Kim, S.W. Microbial production of farnesol (FOH): Current states and beyond. *Process. Biochem.* **2011**, *46*, 1221–1229.
22. Rohdich, F.; Kis, K.; Bacher, A.; Eisenreich, W. The non-mevalonate pathway of isoprenoids: Genes, enzymes and intermediates. *Curr. Opin. Chem. Biol.* **2001**, *5*, 535–540.
23. Fujisaki, S.; Hara, H.; Nishimura, Y.; Horiuchi, K.; Nishino, T. Cloning and nucleotide sequence of the *ispA* gene responsible for farnesyl diphosphate synthase activity in *Escherichia coli*. *J. Biochem.* **1990**, *108*, 995–1000.
24. Math, S.K.; Hearst, J.E.; Poulter, C.D. The *crtE* gene in *Erwinia herbicola* encodes geranylgeranyl diphosphate synthase. *Proc. Natl. Acad. Sci. USA* **1992**, *89*, 6761–6764.

25. Raisig, A.; Bartley, G.; Scolnik, P.; Sandmann, G. Purification in an active state and properties of the 3-step phytoene desaturase from *Rhodobacter capsulatus* overexpressed in *Escherichia coli*. *J. Biochem.* **1996**, *119*, 559–564.
26. Garcia-Asuaa, G.; Langa, H.; Cogdellb, R.; Hunter, C.N. Carotenoid diversity: A modular role for the phytoene desaturase step. *Trends Plant Sci.* **1998**, *3*, 445–449.
27. Takaichi, S.; Mochimaru, M. Carotenoids and carotenogenesis in cyanobacteria: Unique ketocarotenoids and carotenoid glycosides. *Cell. Mol. Life Sci. CMLS* **2007**, *64*, 2607–2619.
28. Takaichi, S. Carotenoids in algae: Distributions, biosyntheses and functions. *Mar. Drugs* **2011**, *9*, 1101–1118.
29. Umeno, D.; Tobias, A.V.; Arnold, F.H. Diversifying carotenoid biosynthetic pathways by directed evolution. *Microbiol. Mol. Biol. Rev. MMBR* **2005**, *69*, 51–78.
30. Schaub, P.; Yu, Q.; Gemmecker, S.; Poussin-Courmontagne, P.; Mailliot, J.; McEwen, A.G.; Ghisla, S.; Al-Babili, S.; Cavarelli, J.; Beyer, P. On the structure and function of the phytoene desaturase CrtI from *Pantoea ananatis*, a membrane-peripheral and FAD-dependent oxidase/isomerase. *PLoS One* **2012**, *7*, e39550.
31. Liu, C.I.; Liu, G.Y.; Song, Y.; Yin, F.; Hensler, M.E.; Jeng, W.Y.; Nizet, V.; Wang, A.H.; Oldfield, E. A cholesterol biosynthesis inhibitor blocks *Staphylococcus aureus* virulence. *Science* **2008**, *319*, 1391–1394.
32. Tobias, A.V.; Arnold, F.H. Biosynthesis of novel carotenoid families based on unnatural carbon backbones: A model for diversification of natural product pathways. *Biochim. Biophys. Acta* **2006**, *1761*, 235–246.
33. Martin, W.; Rujan, T.; Richly, E.; Hansen, A.; Cornelsen, S.; Lins, T.; Leister, D.; Stoebe, B.; Hasegawa, M.; Penny, D. Evolutionary analysis of *Arabidopsis*, cyanobacterial, and chloroplast genomes reveals plastid phylogeny and thousands of cyanobacterial genes in the nucleus. *Proc. Natl. Acad. Sci. USA* **2002**, *99*, 12246–12251.
34. Reyes-Prieto, A.; Hackett, J.D.; Soares, M.B.; Bonaldo, M.F.; Bhattacharya, D. Cyanobacterial contribution to algal nuclear genomes is primarily limited to plastid functions. *Curr. Biol. CB* **2006**, *16*, 2320–2325.
35. Sandmann, G. Evolution of carotene desaturation: The complication of a simple pathway. *Arch. Biochem. Biophys.* **2009**, *483*, 169–174.
36. Steiger, S.; Jackisch, Y.; Sandmann, G. Carotenoid biosynthesis in *Gloeobacter violaceus* PCC4721 involves a single CrtI-type phytoene desaturase instead of typical cyanobacterial enzymes. *Arch. Microbiol.* **2005**, *184*, 207–214.

37. Tsuchiya, T.; Takaichi, S.; Misawa, N.; Maoka, T.; Miyashita, H.; Mimuro, M. The cyanobacterium *Gloeobacter violaceus* PCC4721 uses bacterial-type phytoene desaturase in carotenoid biosynthesis. *FEBS Lett.* **2005**, *579*, 2125–2129.
38. Harada, J.; Nagashima, K.V.; Takaichi, S.; Misawa, N.; Matsuura, K.; Shimada, K. Phytoene desaturase, CrtI, of the purple photosynthetic bacterium, *Rubrivivax gelatinosus*, produces both neurosporene and lycopene. *Plant Cell Physiol.* **2001**, *42*, 1112–1118.
39. Frigaard, N.U.; Maresca, J.A.; Yunker, C.E.; Jones, A.D.; Bryant, D.A. Genetic manipulation of carotenoid biosynthesis in the green sulfur bacterium *Chlorobium tepidum*. *J. Bacterial.* **2004**, *186*, 5210–5220.
40. Tamura, K.; Stecher, G.; Peterson, D.; Filipski, A.; Kumar, S. MEGA6: Molecular evolutionary genetics analysis version 6.0. *Mol. Biol. Evol.* **2013**, *30*, 2725–2729.
41. Lamers, P.P.; Janssen, M.; de Vos, R.C.; Bino, R.J.; Wijffels, R.H. Exploring and exploiting carotenoid accumulation in *Dunaliella salina* for cell-factory applications. *Trends Biotechnol.* **2008**, *26*, 631–638.
42. Rabbani, S.; Beyer, P.; Lintig, J.; Hugueney, P.; Kleinig, H. Induced β-carotene synthesis driven by triacylglycerol deposition in the unicellular alga *Dunaliella bardawil*. *Plant Physiol.* **1998**, *116*, 1239–1248.
43. Cui, H.; Yu, X.; Wang, Y.; Cui, Y.; Li, X.; Liu, Z.; Qin, S. Evolutionary origins, molecular cloning and expression of carotenoid hydroxylases in eukaryotic photosynthetic algae. *BMC Genomics* **2013**, *14*, 457.
44. El-Baky, H.A.; BaZ, F.E.; El-Baroty, G. *Spirulina* species as a source of carotenoids and α-tocopherol and its anticarcinoma factors. *Biotechnology* **2003**, *2*, 222–240.
45. Choubert, G.; Heinrich, O. Carotenoid pigments of the green alga *Haematococcus pluvialis*: Assay on rainbow trout, *Oncorhynchus mykiss*, pigmentation in comparison with synthetic astaxanthin and canthaxanthin. *Aquaculture* **1993**, *112*, 217–226.
46. Abe, K.; Hattori, H.; Hirano, M. Accumulation and antioxidant activity of secondary carotenoids in the aerial microalga *Coelastrella striolata* var. *Multistriata*. *Food Chem.* **2007**, *100*, 656–661.
47. Sanchez, J.F.; Fernandez-Sevilla, J.M.; Acien, F.G.; Ceron, M.C.; Perez-Parra, J.; Molina-Grima, E. Biomass and lutein productivity of *Scenedesmus almeriensis*: Influence of irradiance, dilution rate and temperature. *Appl. Microbiol. Biotechnol.* **2008**, *79*, 719–729.
48. Fietz, S.; Nicklisch, A. Acclimation of the diatom *Stephanodiscus neoastraea* and the cyanobacterium *Planktothrix agardhii* to simulated natural light fluctuations. *Photosynth. Res.* **2002**, *72*, 95–106.
49. Takaichi, S. Distributions, biosyntheses and functions of carotenoids in algae. *Agro FOOD Industry Hi Tech.* **2013**, *24*, 55–58.
50. Kupper, H.; Seibert, S.; Parameswaran, A. Fast, sensitive, and inexpensive alternative to analytical pigment HPLC: Quantification of chlorophylls and

carotenoids in crude extracts by fitting with Gauss peak spectra. *Anal. Chem.* **2007**, *79*, 7611–7627.
51. Miyashita, K.; Nishikawa, S.; Beppu, F.; Tsukui, T.; Abe, M.; Hosokawa, M. The allenic carotenoid fucoxanthin, a novel marine nutraceutical from brown seaweeds. *J. Sci. Food Agric.* **2011**, *91*, 1166–1174.
52. Aitzetmüller, K.; Strain, H.H.; Svec, W.A.; Grandolfo, M.; Katz, J.J. Loroxanthin, a unique xanthophyll from *Scenedesmus obliquus* and *Chlorella vulgaris*. *Phytochemistry* **1969**, 1761–1770.
53. Ganesan, P.; Matsubara, K.; Ohkubo, T.; Tanaka, Y.; Noda, K.; Sugawara, T.; Hirata, T. Anti-angiogenic effect of siphonaxanthin from green alga, *Codium fragile*. *Phytomed. Int. J. Phytother. Phytopharmacol.* **2010**, *17*, 1140–1144.
54. Maresca, J.A.; Romberger, S.P.; Bryant, D.A. Isorenieratene biosynthesis in green sulfur bacteria requires the cooperative actions of two carotenoid cyclases. *J. Bacterial.* **2008**, *190*, 6384–6391.
55. Brocks, J.J.; Love, G.D.; Summons, R.E.; Knoll, A.H.; Logan, G.A.; Bowden, S.A. Biomarker evidence for green and purple sulphur bacteria in a stratified Palaeoproterozoic sea. *Nature* **2005**, *437*, 866–870.
56. Graham, J.E.; Bryant, D.A. The biosynthetic pathway for myxol-2' fucoside (myxoxanthophyll) in the cyanobacterium *Synechococcus* sp. strain PCC 7002. *J. Bacterial.* **2009**, *191*, 3292–3300.
57. Song, P.S.; Koka, P.; Prezelin, B.B.; Haxo, F.T. Molecular topology of the photosynthetic light-harvesting pigment complex, peridinin-chlorophyll A-protein, from marine dinoflagellates. *Biochemistry* **1976**, *15*, 4422–4427.
58. Maoka, T.; Hashimoto, K.; Akimoto, N.; Fujiwara, Y. Structures of five new carotenoids from the oyster *Crassostrea gigas*. *J. Nat. Prod.* **2001**, *64*, 578–581.
59. Maoka, T.; Akimoto, N.; Yim, M.J.; Hosokawa, M.; Miyashita, K. New C37 skeletal carotenoid from the clam, *Paphia amabillis*. *J. Agric. Food Chem.* **2008**, *56*, 12069–12072.
60. Maoka, T.; Fujiwara, Y.; Hashimoto, K.; Akimoto, N. Structure of new carotenoids from corbicula clam *Corbicula japonica*. *J. Nat. Prod.* **2005**, *68*, 1341–1344.
61. Maoka, T. Carotenoids in marine animals. *Mar. Drugs* **2011**, *9*, 278–293.
62. Maresca, J.A.; Graham, J.E.; Wu, M.; Eisen, J.A.; Bryant, D.A. Identification of a fourth family of lycopene cyclases in photosynthetic bacteria. *Proc. Natl. Acad. Sci. USA* **2007**, *104*, 11784–11789.
63. Misawa, N. Carotenoid beta-ring hydroxylase and ketolase from marine bacteria-promiscuous enzymes for synthesizing functional xanthophylls. *Mar. Drugs* **2011**, *9*, 757–771.

64. Kim, J.; DellaPenna, D. Defining the primary route for lutein synthesis in plants: The role of *Arabidopsis* carotenoid β-ring hydroxylase CYP97A3. *Proc. Natl. Acad. Sci. USA* **2006**, *103*, 3474–3479.
65. Coesel, S.; Obornik, M.; Varela, J.; Falciatore, A.; Bowler, C. Evolutionary origins and functions of the carotenoid biosynthetic pathway in marine diatoms. *PLoS One* **2008**, *3*, e2896.
66. Mikami, K.; Hosokawa, M. Biosynthetic pathway and health benefits of fucoxanthin, an
algae-specific xanthophyll in brown seaweeds. *Int. J. Mol. Sci.* **2013**, *14*, 13763–13781.
67. Yang, L.E.; Huang, X.Q.; Hang, Y.; Deng, Y.Y.; Lu, Q.Q.; Lu, S. The P450-type carotene hydroxylase PuCHY1 from *Porphyra* suggested the evolution of carotenoid metabolism in red algae. *J. Integr. Plant Boil.* **2014**, in press.
68. Misawa, N.; Satomi, Y.; Kondo, K.; Yokoyama, A.; Kajiwara, S.; Saito, T.; Ohtani, T.; Miki, W. Structure and functional analysis of a marine bacterial carotenoid biosynthesis gene cluster and astaxanthin biosynthetic pathway proposed at the gene level. *J. Bacteriol.* **1995**, *177*, 6575–6584.
69. Netzer, R.; Stafsnes, M.H.; Andreassen, T.; Goksoyr, A.; Bruheim, P.; Brautaset, T. Biosynthetic pathway for γ-cyclic sarcinaxanthin in *Micrococcus luteus*: Heterologous expression and evidence for diverse and multiple catalytic functions of $C_{50}$ carotenoid cyclases. *J. Bacteriol.* **2010**, *192*, 5688–5699.
70. Zhang, W.; Hu, X.; Wang, L.; Wang, X. Reconstruction of the carotenoid biosynthetic pathway of *Cronobacter sakazakii* BAA894 in *Escherichia coli*. *PLoS One* **2014**, *9*, e86739.
71. Kim, S.W.; Keasling, J.D. Metabolic engineering of the nonmevalonate isopentenyl diphosphate synthesis pathway in *Escherichia coli* enhances lycopene production. *Biotechnol. Bioeng.* **2001**, *72*, 408–415.
72. Harker, M.; Bramley, P.M. Expression of prokaryotic 1-deoxy-D-xylulose-5-phosphatases in *Escherichia coli* increases carotenoid and ubiquinone biosynthesis. *FEBS Lett.* **1999**, *448*,
115–119.
73. Wang, G.S.; Grammel, H.; Abou-Aisha, K.; Sagesser, R.; Ghosh, R. High-Level production of the industrial product lycopene by the photosynthetic bacterium *Rhodospirillum rubrum*. *Appl. Environ. Microbiol.* **2012**, *78*, 7205–7215.
74. Miura, Y.; Kondo, K.; Saito, T.; Shimada, H.; Fraser, P.D.; Misawa, N. Production of the carotenoids lycopene, β-carotene, and astaxanthin in the food yeast *Candida utilis*. *Appl. Environ. Microbiol.* **1998**, *64*, 1226–1229.
75. Harada, H.; Misawa, N. Novel approaches and achievements in biosynthesis of functional isoprenoids in *Escherichia coli*. *Appl. Microbiol. Biotechnol.* **2009**, *84*, 1021–1031.

76. Yamano, S.; Ishii, T.; Nakagawa, M.; Ikenaga, H.; Misawa, N. Metabolic engineering for production of β-carotene and lycopene in *Saccharomyces cerevisiae*. *Biosci. Biotechnol. Biochem.* **1994**, *58*, 1112–1114.
77. Heider, S.A.; Peters-Wendisch, P.; Netzer, R.; Stafnes, M.; Brautaset, T.; Wendisch, V.F. Production and glucosylation of $C_{50}$ and $C_{40}$ carotenoids by metabolically engineered *Corynebacterium glutamicum*. *Appl. Microbiol. Biotechnol.* **2014**, *98*, 1223–1235.
78. Yoon, S.H.; Kim, J.E.; Lee, S.H.; Park, H.M.; Choi, M.S.; Kim, J.Y.; Shin, Y.C.; Keasling, J.D.; Kim, S.W. Engineering the lycopene synthetic pathway in *E. coli* by comparison of the carotenoid genes of *Pantoea agglomerans* and *Pantoea ananatis*. *Appl. Microbiol. Biotechnol.* **2007**, *74*, 131–139.
79. Sandmann, G.; Albrecht, M.; Schnurr, G.; Knorzer, O.; Boger, P. The biotechnological potential and design of novel carotenoids by gene combination in *Escherichia coli*. *Trends Biotechnol.* **1999**, *17*, 233–237.
80. Misawa, N. Pathway engineering for functional isoprenoids. *Curr. Opin. Biotechnol.* **2011**, *22*, 627–633.
81. Yoon, S.H.; Park, H.M.; Kim, J.E.; Lee, S.H.; Choi, M.S.; Kim, J.Y.; Oh, D.K.; Keasling, J.D.; Kim, S.W. Increased β-carotene production in recombinant *Escherichia coli* harboring an engineered isoprenoid precursor pathway with mevalonate addition. *Biotechnol. Prog.* **2007**, *23*, 599–605.
82. Verwaal, R.; Wang, J.; Meijnen, J.P.; Visser, H.; Sandmann, G.; van den Berg, J.A.; van Ooyen, A.J. High-level production of beta-carotene in *Saccharomyces cerevisiae* by successive transformation with carotenogenic genes from *Xanthophyllomyces dendrorhous*. *Appl. Environ. Microbiol.* **2007**, *73*, 4342–4350.
83. Yoon, S.H.; Lee, S.H.; Das, A.; Ryu, H.K.; Jang, H.J.; Kim, J.Y.; Oh, D.K.; Keasling, J.D.;
Kim, S.W. Combinatorial expression of bacterial whole mevalonate pathway for the production of β-carotene in *E. coli*. *J. Biotechnol.* **2009**, *140*, 218–226.
84. Yoshida, K.; Ueda, S.; Maeda, I. Carotenoid production in *Bacillus subtilis* achieved by metabolic engineering. *Biotechnol. Lett.* **2009**, *31*, 1789–1793.
85. Beuttler, H.; Hoffmann, J.; Jeske, M.; Hauer, B.; Schmid, R.D.; Altenbuchner, J.; Urlacher, V.B. Biosynthesis of zeaxanthin in recombinant *Pseudomonas putida*. *Appl. Microbiol. Biotechnol.* **2011**, *89*, 1137–1147.
86. Ye, V.M.; Bhatia, S.K. Pathway engineering strategies for production of beneficial carotenoids in microbial hosts. *Biotechnol. Lett.* **2012**, *34*, 1405–1414.
87. Vick, J.E.; Johnson, E.T.; Choudhary, S.; Bloch, S.E.; Lopez-Gallego, F.; Srivastava, P.;
Tikh, I.B.; Wawrzyn, G.T.; Schmidt-Dannert, C. Optimized compatible set of Biobrick vectors for metabolic pathway engineering. *Appl. Microbiol. Biotechnol.* **2011**, *92*, 1275–1286.
88. Li, M.Z.; Elledge, S.J. Harnessing homologous recombination *in vitro* to generate recombinant DNA via SLIC. *Nat. Methods* **2007**, *4*, 251–256.

89. Gibson, D.G.; Young, L.; Chuang, R.Y.; Venter, J.C.; Hutchison, C.A. 3rd; Smith, H.O. Enzymatic assembly of DNA molecules up to several hundred kilobases. *Nat. Methods* **2009**, *6*, 343–345.
90. Wingler, L.M.; Cornish, V.W. Reiterative recombination for the *in vivo* assembly of libraries of multigene pathways. *Proc. Natl. Acad. Sci. USA* **2011**, *108*, 15135–15140.
91. Sleight, S.C.; Sauro, H.M. Randomized Biobrick assembly: A novel DNA assembly method for randomizing and optimizing genetic circuits and metabolic pathways. *ACS Synth. Boil.* **2013**, *2*, 506–518.
92. Yadav, V.G.; de Mey, M.; Lim, C.G.; Ajikumar, P.K.; Stephanopoulos, G. The future of metabolic engineering and synthetic biology: Towards a systematic practice. *Metab. Eng.* **2012**, *14*, 233–241.
93. Ajikumar, P.K.; Xiao, W.H.; Tyo, K.E.; Wang, Y.; Simeon, F.; Leonard, E.; Mucha, O.;
Phon, T.H.; Pfeifer, B.; Stephanopoulos, G. Isoprenoid pathway optimization for taxol precursor overproduction in *Escherichia coli*. *Science* **2010**, *330*, 70–74.
94. Temme, K.; Hill, R.; Segall-Shapiro, T.H.; Moser, F.; Voigt, C.A. Modular control of multiple pathways using engineered orthogonal T7 polymerases. *Nucleic Acids Res.* **2012**, *40*, 8773–8781.
95. Salis, H.M.; Mirsky, E.A.; Voigt, C.A. Automated design of synthetic ribosome binding sites to control protein expression. *Nat. Biotechnol.* **2009**, *27*, 946–950.
96. Pfleger, B.F.; Pitera, D.J.; Smolke, C.D.; Keasling, J.D. Combinatorial engineering of intergenic regions in operons tunes expression of multiple genes. *Nat. Biotechnol.* **2006**, *24*, 1027–1032.
97. Qi, L.; Haurwitz, R.E.; Shao, W.; Doudna, J.A.; Arkin, A.P. RNA processing enables predictable programming of gene expression. *Nat. Biotechnol.* **2012**, *30*, 1002–1006.
98. Dueber, J.E.; Wu, G.C.; Malmirchegini, G.R.; Moon, T.S.; Petzold, C.J.; Ullal, A.V.;
Prather, K.L.; Keasling, J.D. Synthetic protein scaffolds provide modular control over metabolic flux. *Nat. Biotechnol.* **2009**, *27*, 753–759.
99. Parsons, J.B.; Dinesh, S.D.; Deery, E.; Leech, H.K.; Brindley, A.A.; Heldt, D.; Frank, S.;
Smales, C.M.; Lunsdorf, H.; Rambach, A.; *et al.* Biochemical and structural insights into bacterial organelle form and biogenesis. *J. Biol. Chem.* **2008**, *283*, 14366–14375.
100. Martin, V.J.; Pitera, D.J.; Withers, S.T.; Newman, J.D.; Keasling, J.D. Engineering a mevalonate pathway in *Escherichia coli* for production of terpenoids. *Nat. Biotechnol.* **2003**, *21*, 796–802.
101. Zhang, F.; Carothers, J.M.; Keasling, J.D. Design of a dynamic sensor-regulator system for production of chemicals and fuels derived from fatty acids. *Nat. Biotechnol.* **2012**, *30*, 354–359.

102. Farmer, W.R.; Liao, J.C. Improving lycopene production in *Escherichia coli* by engineering metabolic control. *Nat. Biotechnol.* **2000**, *18*, 533–537.
103. Farmer, W.R.; Liao, J.C. Precursor balancing for metabolic engineering of lycopene production in *Escherichia coli*. *Biotechnol. Prog.* **2001**, *17*, 57–61.
104. Zhou, Y.; Nambou, K.; Wei, L.; Cao, J.; Imanaka, T.; Hua, Q. Lycopene production in recombinant strains of *Escherichia coli* is improved by knockout of the central carbon metabolism gene coding for glucose-6-phosphate dehydrogenase. *Biotechnol. Lett.* **2013**, *35*, 2137–2145.
105. Zhang, C.; Chen, X.; Zou, R.; Zhou, K.; Stephanopoulos, G.; Too, H.P. Combining genotype improvement and statistical media optimization for isoprenoid production in *E. coli*. *PLoS One* **2013**, *8*, e75164.
106. Suh, W. High isoprenoid flux *Escherichia coli* as a host for carotenoids production. *Methods Mol. Boil.* **2012**, *834*, 49–62.
107. Reyes, L.H.; Gomez, J.M.; Kao, K.C. Improving carotenoids production in yeast via adaptive laboratory evolution. *Metab. Eng.* **2014**, *21*, 26–33.
108. Ozaydin, B.; Burd, H.; Lee, T.S.; Keasling, J.D. Carotenoid-Based phenotypic screen of the yeast deletion collection reveals new genes with roles in isoprenoid production. *Metab. Eng.* **2013**, *15*, 174–183.
109. Alper, H.; Miyaoku, K.; Stephanopoulos, G. Construction of lycopene-overproducing *E. coli* strains by combining systematic and combinatorial gene knockout targets. *Nat. Biotechnol.* **2005**, *23*, 612–616.
110. Alper, H.; Jin, Y.S.; Moxley, J.F.; Stephanopoulos, G. Identifying gene targets for the metabolic engineering of lycopene biosynthesis in *Escherichia coli*. *Metab. Eng.* **2005**, *7*, 155–164.
111. Choi, H.S.; Lee, S.Y.; Kim, T.Y.; Woo, H.M. *In silico* identification of gene amplification targets for improvement of lycopene production. *Appl. Environ. Microbiol.* **2010**, *76*, 3097–3105.
112. Kang, M.J.; Lee, Y.M.; Yoon, S.H.; Kim, J.H.; Ock, S.W.; Jung, K.H.; Shin, Y.C.; Keasling, J.D.; Kim, S.W. Identification of genes affecting lycopene accumulation in *Escherichia coli* using a shot-gun method. *Biotechnol. Bioeng.* **2005**, *91*, 636–642.
113. Alper, H.; Stephanopoulos, G. Uncovering the gene knockout landscape for improved lycopene production in *E. coli*. *Appl. Microbiol. Biotechnol.* **2008**, *78*, 801–810.
114. Jin, Y.S.; Stephanopoulos, G. Multi-dimensional gene target search for improving lycopene biosynthesis in *Escherichia coli*. *Metab. Eng.* **2007**, *9*, 337–347.
115. Zhao, J.; Li, Q.; Sun, T.; Zhu, X.; Xu, H.; Tang, J.; Zhang, X.; Ma, Y. Engineering central metabolic modules of *Escherichia coli* for improving β-carotene production. *Metab. Eng.* **2013**, *17*, 42–50.

116. Wang, H.H.; Isaacs, F.J.; Carr, P.A.; Sun, Z.Z.; Xu, G.; Forest, C.R.; Church, G.M. Programming cells by multiplex genome engineering and accelerated evolution. *Nature* **2009**, *460*, 894–898.
117. Warner, J.R.; Reeder, P.J.; Karimpour-Fard, A.; Woodruff, L.B.; Gill, R.T. Rapid profiling of a microbial genome using mixtures of barcoded oligonucleotides. *Nat. Biotechnol.* **2010**, *28*, 856–862.
118. Kachroo, A.H.; Jayaram, M.; Rowley, P.A. Metabolic engineering without plasmids. *Nat. Biotechnol.* **2009**, *27*, 729–731.
119. Chiang, C.J.; Chen, P.T.; Chao, Y.P. Replicon-free and markerless methods for genomic insertion of dnas in phage attachment sites and controlled expression of chromosomal genes in *Escherichia coli*. *Biotechnol. Bioeng.* **2008**, *101*, 985–995.
120. Chen, Y.Y.; Shen, H.J.; Cui, Y.Y.; Chen, S.G.; Weng, Z.M.; Zhao, M.; Liu, J.Z. Chromosomal evolution of *Escherichia coli* for the efficient production of lycopene. *BMC Biotechnol.* **2013**, *13*, 6.
121. Shin, J.H.; Kim, H.U.; Kim, D.I.; Lee, S.Y. Production of bulk chemicals via novel metabolic pathways in microorganisms. *Biotechnol. Adv.* **2013**, *31*, 925–935.
122. Zanghellini, A. *De novo* computational enzyme design. *Curr. Opin. Biotechnol.* **2014**, *29C*, 132–138.
123. Damborsky, J.; Brezovsky, J. Computational tools for designing and engineering enzymes. *Curr. Opin. Chem. Biol.* **2014**, *19C*, 8–16.
124. Tao, L.; Wilczek, J.; Odom, J.M.; Cheng, Q. Engineering a β-carotene ketolase for astaxanthin production. *Metab. Eng.* **2006**, *8*, 523–531.
125. Ye, R.W.; Stead, K.J.; Yao, H.; He, H. Mutational and functional analysis of the β-carotene ketolase involved in the production of canthaxanthin and astaxanthin. *Appl. Environ. Microbiol.* **2006**, *72*, 5829–5837.
126. Umeno, D.; Tobias, A.V.; Arnold, F.H. Evolution of the $C_{30}$ carotenoid synthase CrtM for function in a $C_{40}$ pathway. *J. Bacteriol.* **2002**, *184*, 6690–6699.
127. Schmidt-Dannert, C.; Umeno, D.; Arnold, F.H. Molecular breeding of carotenoid biosynthetic pathways. *Nat. Biotechnol.* **2000**, *18*, 750–753.
128. Fuxreiter, M.; Mones, L. The role of reorganization energy in rational enzyme design. *Curr. Opin. Chem. Biol.* **2014**, *21C*, 34–41.
129. Leonard, E.; Ajikumar, P.K.; Thayer, K.; Xiao, W.H.; Mo, J.D.; Tidor, B.; Stephanopoulos, G.; Prather, K.L. Combining metabolic and protein engineering of a terpenoid biosynthetic pathway for overproduction and selectivity control. *Proc. Natl. Acad. Sci. USA* **2010**, *107*, 13654–13659.
130. Furubayashi, M.; Ikezumi, M.; Kajiwara, J.; Iwasaki, M.; Fujii, A.; Li, L.; Saito, K.; Umeno, D. A high-throughput colorimetric screening assay for terpene synthase activity based on substrate consumption. *PLoS One* **2014**, *9*, e93317.

131. Del Campo, J.A.; Garcia-Gonzalez, M.; Guerrero, M.G. Outdoor cultivation of microalgae for carotenoid production: Current state and perspectives. *Appl. Microbiol. Biotechnol.* **2007**, *74*, 1163–1174.
132. Rosenberg, J.N.; Oyler, G.A.; Wilkinson, L.; Betenbaugh, M.J. A green light for engineered algae: Redirecting metabolism to fuel a biotechnology revolution. *Curr. Opin. Biotechnol.* **2008**, *19*, 430–436.
133. Binti Ibnu Rasid, E.N.; Mohamad, S.E.; Jamaluddin, H.; Salleh, M.M. Screening factors influencing the production of astaxanthin from freshwater and marine microalgae. *Appl. Biochem. Biotechnol.* **2014**, *172*, 2160–2174.
134. Casal, C.; Cuaresma, M.; Vega, J.M.; Vilchez, C. Enhanced productivity of a lutein-enriched novel acidophile microalga grown on urea. *Mar. Drugs* **2011**, *9*, 29–42.
135. Fu, W.; Paglia, G.; Magnusdottir, M.; Steinarsdottir, E.A.; Gudmundsson, S.; Palsson, B.O.; Andresson, O.S.; Brynjolfsson, S. Effects of abiotic stressors on lutein production in the green microalga *Dunaliella salina*. *Microb. Cell Factories* **2014**, *13*, 3.
136. Fu, W.; Guethmundsson, O.; Paglia, G.; Herjolfsson, G.; Andresson, O.S.; Palsson, B.O.; Brynjolfsson, S. Enhancement of carotenoid biosynthesis in the green microalga *Dunaliella salina* with light-emitting diodes and adaptive laboratory evolution. *Appl. Microbiol. Biotechnol.* **2013**, *97*, 2395–2403.
137. Leon, R.; Couso, I.; Fernandez, E. Metabolic engineering of ketocarotenoids biosynthesis in the unicelullar microalga *Chlamydomonas reinhardtii*. *J. Biotechnol.* **2007**, *130*, 143–152.

*marine drugs*

*Review*

# Innovative Alternative Technologies to Extract Carotenoids from Microalgae and Seaweeds

**Mahesha M. Poojary [1,2], Francisco J. Barba [3,*], Bahar Aliakbarian [4], Francesco Donsì [5,6], Gianpiero Pataro [5,6], Daniel A. Dias [1] and Pablo Juliano [7]**

1 Discipline of Laboratory Medicine, School of Health and Biomedical Sciences, RMIT University, 3083 Bundoora, Australia; mahesha.poojary@unicam.it (M.M.P.); daniel.dias@rmit.edu.au (D.A.D.)
2 Chemistry Section, School of Science and Technology, University of Camerino, via S. Agostino 1, 62032 Camerino, Italy
3 Nutrition and Food Science Area, Preventive Medicine and Public Health, Food Sciences, Toxicology and Forensic Medicine Department, Faculty of Pharmacy, Universitat de València, Avda. Vicent Andrés Estellés, s/n, 46100 Burjassot, València, Spain
4 Department of Civil, Chemical and Environmental Engineering, Pole of Chemical Engineering, University of Genoa, via Opera Pia 15, 16145 Genoa, Italy; Bahar.Aliakbarian@unige.it
5 Department of Industrial Engineering, University of Salerno, via Giovanni Paolo II 132, 84084 Fisciano, Italy; fdonsi@unisa.it (F.D.); gpataro@unisa.it (G.P.)
6 ProdAl Scarl, via Ponte don Melillo, 84084 Fisciano, SA, Italy
7 CSIRO Agriculture and Food, 671 Sneydes Road, 3030 Werribee, VIC, Australia; Pablo.Juliano@csiro.au
* Correspondence: francisco.barba@uv.es; Tel.: +34-963-544972; Fax: +34-963-544954

Academic Editors: Tatsuya Sugawara and Takashi Maoka
Received: 29 September 2016; Accepted: 11 November 2016; Published: 22 November 2016

**Abstract:** Marine microalgae and seaweeds (microalgae) represent a sustainable source of various bioactive natural carotenoids, including β-carotene, lutein, astaxanthin, zeaxanthin, violaxanthin and fucoxanthin. Recently, the large-scale production of carotenoids from algal sources has gained significant interest with respect to commercial and industrial applications for health, nutrition, and cosmetic applications. Although conventional processing technologies, based on solvent extraction, offer a simple approach to isolating carotenoids, they suffer several, inherent limitations, including low efficiency (extraction yield), selectivity (purity), high solvent consumption, and long treatment times, which have led to advancements in the search for innovative extraction technologies. This comprehensive review summarizes the recent trends in the extraction of carotenoids from microalgae and seaweeds through the assistance of different innovative techniques, such as pulsed electric fields, liquid pressurization, supercritical fluids, subcritical fluids, microwaves, ultrasounds, and high-pressure homogenization. In particular, the review critically analyzes technologies, characteristics, advantages, and shortcomings of the different innovative processes, highlighting the differences in terms of yield, selectivity, and economic and environmental sustainability.

**Keywords:** marine microalgae; seaweeds; carotenoids; nonconventional extraction; electrotechnologies; pulsed electric field-assisted extraction; supercritical fluid extraction; green processing; microwave-assisted extraction; marine drugs

## 1. Introduction

Carotenoids are a class of terpenoid pigments with a tetraterpenes ($C_{40}$) backbone, responsible for a range of colors, such as brilliant yellow, orange and red in fruits, vegetables, and aquatic creatures [1]. They contain highly conjugated polyene chromophoric chains which give rise to distinct colors and functions [2], constituting two major classes of molecules: (i) carotenes, which are strictly hydrocarbons (e.g., α-carotene, β-carotene, and lycopene) and (ii) xanthophylls, which are similar to carotenes but contain oxygen (e.g., lutein, zeaxanthin, neoxanthin, violaxanthin, flavoxanthin, and fucoxanthin). Carotenoids are predominantly found in plants; however, they are also present in many algae, bacteria and some fungi and play a key role in light harvesting and photo protection in photosynthetic organisms [3]. To date, over 600 unique carotenoids have been identified [4].

In recent decades, there has been a considerable amount of evidence supporting the role of carotenoids as food colorants and antioxidants with beneficial effects on human health, especially with regards to the prevention of chronic diseases, particularly certain cancers, cardiovascular and eye diseases [5–7]. They have been widely used as nutraceuticals, cosmeceuticals and feed supplements in aquaculture sectors [8,9]. For this reason, the global demand for carotenoids is growing remarkably with the worldwide carotenoids market estimated at USD 1.24 billion in 2016 and is expected to reach USD 1.53 billion by 2021 [10]. Currently, the major commercial carotenoids are produced by chemical synthesis [11]; however, in recent year, the increasing consumers' concerns for public health, safety and environmental burden have driven the growth of the market demand for natural carotenoids-based products.

Marine microalgae and seaweeds serve as a unique, sustainable and alternative source of carotenoids [9,12]. Therefore, in recent years, the industrial interest towards the production of natural carotenoids using algae has considerably increased, as they offer cost, scale, time and yield advantages over terrestrial plants.

In microalgae, carotenoids can be classified into two groups, primary and secondary carotenoids, based on their metabolism and function. Primary carotenoids are structural and functional components in the photosynthetic apparatus, which take direct part in photosynthesis. Secondary carotenoids refer to extra-plastidic pigments produced in large quantities, through carotenogenesis, after exposure to specific environmental stimuli [13,14]. Primary and secondary carotenoids are of considerable interest as natural colorants as well as their potential in human health. Specifically, they possess a wide range of distinctive biological activities, including antioxidant, cardiovascular protection, anticancer, antidiabetic, and anti-obesity, which have been recently reviewed [15]. The primary microalgae carotenoids include α-carotene, β-carotene, lutein, fucoxanthin, violaxanthin, zeaxanthin, and neoxanthin, among others. Examples of secondary carotenoids, include: astaxanthin, canthaxanthin, and echinenone [16]. Interestingly, among these compounds, astaxanthin, zeaxanthin, β-carotenes, fucoxanthin and lutein are commercially important carotenoids, which are widely found in marine microalgae.

Seaweeds (macroalgae) also serve as an important source of carotenoids [17], such as fucoxanthin, lutein, β-carotene and siphonaxanthin. In particular, fucoxanthin is a characteristic orange xanthophyll, which is abundant in several brown seaweeds including *Undaria pinnatifida* [18–23], *Hijikia fusiformis* [24], *Laminaria japonica* [21,25,26], *Sargassum* sp. [27–32], and *Fucus* sp. [33]. It is one of the most abundant carotenoids in seaweeds, accounting for more than 10% of the estimated total natural production of carotenoids [34], with remarkable biological properties, including anticancer, anti-inflammatory, antiobesity and neuroprotective bioactivities [34–38].

Carotenoids produced by algae are generally localized in the chloroplast or accumulated in vesicles, cytoplasmic matrix or bound to membranes and other macromolecules in the intracellular space. The cell wall and plasma membrane surrounding the cell, as well the chloroplast membranes, act as a barrier which greatly limits the rate of mass transfer of carotenoids and other intracellular compounds during conventional extraction processes. Figure 1 illustrates a marine microalga, highlighting that the extraction of carotenoid yield requires the disruption or permeabilization of the cell wall, of the plasma membrane and, depending on the biological features (e.g., organelle localization), eventually of the chloroplast membrane. Algae display complex cell envelope structures and their composition varies from species to species [39]. Therefore, it is essential to develop and optimize efficient methods for the selective extraction of these compounds, which takes into account the biological diversity as well as the localization of carotenoids within specific organelles.

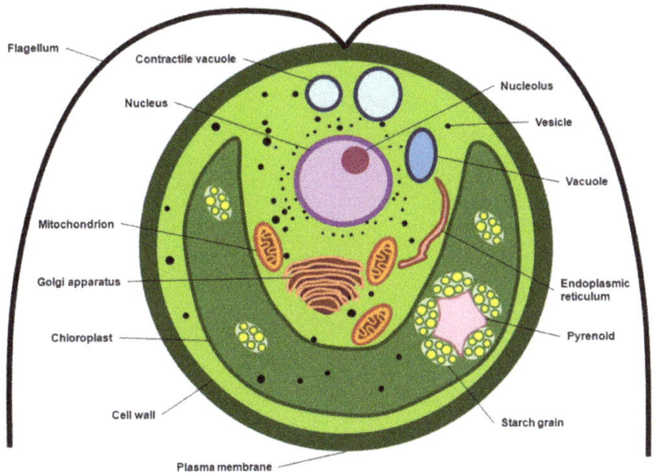


**Figure 1.** Illustration detailing organelles present in a typical marine unicellular microalgae.

Generally, carotenoids are recovered from microalgae and seaweeds by means of conventional solvent extraction (e.g., Soxhlet extraction) using organic solvents [40]. However, these methods are time-consuming, and often require the usage of relatively large amounts of solvents, which is expensive and not environmental friendly. The use of innovative non-conventional techniques, based on the physical membrane permeabilization or lysis, to selectively or non-selectively increase the rate of mass transfer of carotenoids from the intracellular space of microalgae and seaweeds, has gained growing interest in recent years. In particular, this review extensively details the recent advances in the use of novel technologies to recover carotenoids from microalgae and seaweeds, including electrotechnologies-assisted extraction, such as pulsed electric field (PEF), moderate electric field (MEF), high-voltage electric discharges (HVED), as well as supercritical fluid extraction (SFE), subcritical fluid extraction, pressurized liquid extraction (PLE), microwave-assisted extraction (MAE), ultrasound-assisted extraction (UAE) and high pressure homogenization (HPH) treatment. These alternative technologies have several advantages, including rapid extraction (e.g., PLE, MAE, UAE, PEF, MEF, HVED and HPH), low solvent consumption (e.g., PEF, PLE, MAE and UAE), use of "green" environmentally friendly solvents (e.g., SFE), superior recovery (e.g., MAE, subcritical fluid extraction, HVED, UAE and HPH) and higher selectivity (e.g., PEF and SFE).

## 2. Extraction Technologies for Carotenoids

### *Conventional Extraction Methods*

Conventional extraction of algae intracellular products is typically conducted from dry biomass and is based on maceration and thermal extraction using organic or aqueous solvents, depending on the polarity of the target compounds to be extracted. Carotenoids exhibit varying polarities, solubilities and chemical stabilities. Therefore, a suitable solvent system must be selected on the basis of the target carotenoids, which could selectively and efficiently extract carotenoids with greater purity. Since most carotenoids possess a high degree of hydrophobicity, their effective extractions requires the use of non-polar solvents, for example *n*-hexane, dichloromethane, dimethyl ether, diethyl ether etc. [23,41]. Acetone, octane, biphasic mixtures of several organic solvents have also been studied for the selective extraction of carotenoids [42–44]. Recently, the use of several green solvents such as ethanol, limonene and biphasic mixtures of water and organic solvents have been investigated for recovery of carotenoids from microalgae [45,46]. However, extraction efficiency, selectivity and high solvent consumption still remain a limiting factor in the conventional solvent extraction process. One avenue to overcome this problem can be based on the usage of a multi-stage extraction procedure eventually assisted by different physical and chemical methodologies which may selectively target the desired intracellular carotenoids.

## 3. Nonconventional Extraction of Carotenoids

### 3.1. Electrotechnologies

In recent years, there has been an emerging interest in the use of electrotechnologies, such as PEF, MEF, and HVED as a non-thermal, green extraction techniques for targeting intracellular compounds from plant or biosuspensions [47–50]. As illustrated in Figure 2, although each of these electrotechnologies has its own mode of treatment and mechanism of delivering electrical current through the processed biomaterial, they all induce a certain degree of cell disintegration allowing for the selective extraction of intracellular compounds. The principles and recent advances in application of electrotechnologies to extract carotenoids from microalgae and seaweeds are described in the following subsections.

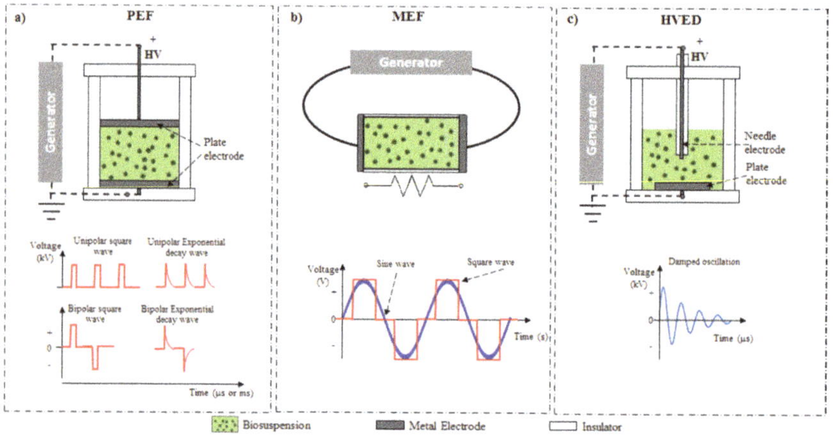


**Figure 2.** Schematics of typical experimental set-up and pulse protocols for (**a**) pulsed electric field (PEF)-assisted extraction; (**b**) moderate electric field (MEF)-assisted extraction; and (**c**) high voltage electric discharge (HVED)-assisted extraction.

#### 3.1.1. Pulsed Electric Field (PEF)-Assisted Extraction

PEF processing is a non-thermal technique, which has received increasing attention in recent years [51]. In PEF-assisted extraction, the sample matrix is placed between two electrodes in a batch or a continuous flow treatment chamber and exposed to repetitive electric frequencies (Hz–kHz) with an intense (0.1–80 kV/cm) electric field for very short periods (from several nanoseconds to several milliseconds). The pulses commonly used in PEF treatments are unipolar or bipolar, with either exponential or square-wave shaped frequencies (Figure 2a). The application of electric pulses causes the formation of reversible or irreversible pores in the cell membranes, defined as electroporation or electropermeabilization, which consequently aids the rapid diffusion of the solvents and the enhancement of the mass transfer of intracellular compounds [51]. The selective extraction of analytes can be achieved by controlling the pore formation, which is dependent on the intensity of the treatment applied (electric field strength, single pulse duration, treatment time or total specific energy input) and the cell characteristics (i.e., size, shape, orientation in the electric field) [52].

Recently, a few authors have investigated the ability of PEF to enhance the extractability of carotenoids from microalgae obtaining controversial results (Table 1). A PEF pretreatment at 15 kV/cm and 100 kJ/kg increased the extraction of carotenoids from marine microalgae *Chlorella vulgaris* and *Spirulina platensis* up to 525 and 150%, respectively, compared to the conventional ball milling homogenization process alone [53]. Subsequently, a PEF treatment at 20 kV/cm electric field strength with an energy density of 13.3–53.1 kJ/kg for 1–4 ms did not increase the carotenoids yield extracted from marine microalga *Nannochloropsis* sp. [54]. It is likely that the use of a polar solvent, such as water, together with the thick cell wall structure of *Nannochloropsis* sp., with the presence of secondary structures, complex polysaccharide networks and an outer algaenan layer [55] could be the reasons for the observed scarce efficiency of PEF treatments. Furthermore, extraction of carotenoids which are bound to chloroplasts requires more intense treatments, as the chloroplast membranes has to be electroporated along with the plasma membrane to improve the mass transfer of extractable compounds. Considering that the external electric field threshold required to trigger electroporation is inversely related to cell size [56,57], higher electric field strengths are required for permeabilization of smaller internal organelles like chloroplasts [58]. Specifically, nanosecond pulses with electric field strengths around 100 kV/cm have been reported to cause the electroporation of intracellular organelles [59]. However, more recently, it has been demonstrated that also longer pulses of lower intensity, currently used for the electroporation of plasma membranes, can also cause non-thermal intracellular effects, including organelles electroporation [58].

In a recent study, it has been found that PEF treatments in the microsecond range (3 μs pulse duration) with a field strength in the range of 20–25 kV/cm can cause significant, irreversible electroporation of *C. vulgaris*, resulting in improved carotenoid yield. Lower electric field strengths (10 kV/cm) in the microsecond range resulted only in the reversible electroporation, leading to lower extraction yields [60]. However, the lower degree of electroporation at lower electric field strengths can be compensated by increasing the treatment duration to the milliseconds range [61], at the expense of a specific energy higher than that of treatment at higher electric field strengths in the microsecond range [61]. Interestingly, when the extraction of carotenoids was performed after 1 h of PEF treatment, the authors noted that the yield increased up to 1.58 mg/L of culture for the PEF treated samples in the microsecond range, while no further increase was observed for the sample treated in the millisecond range [61]. Nevertheless, the authors did not find an increase in the degree of permeabilization in the PEF treated cells during the incubation period. The increase in the yield was, therefore, attributed to subsequent plasmolysis of the chloroplast during the incubation time due to osmolytic disequilibrium in the cytoplasmatic space, which facilitated the diffusion of both the solvent into the chloroplast and the carotenoid pigments towards the cytoplasm [61].

The application of multi-step extraction procedures, based on the combination of PEF and solvent extractions at various pH and the usage of biphasic mixture of organic solvents, can also assist in the recovery of low water solubility carotenoids with optimum yields [62,63]. Using this approach, Parniakov et al. [63] observed a noticeable increase in the concentrations of carotenoids in the aqueous extracts from *Nannochloropsis* spp. after PEF was applied at pH 8.5, followed by extraction at pH 11. In a further study, the authors efficiently recovered carotenoids and other pigments from *Nannochloropsis* spp. with the application of biphasic mixtures of organic solvents [(i.e., dimethyl sulfoxide (DMSO) and ethanol (EtOH)] and water. A two stage extraction procedure involving PEF treatment (20 kV/cm) of microalgae suspension and extraction in water at the first step, followed by the conventional extraction using biphasic mixtures in the second step, allowed for the efficient extraction of carotenoids in less concentrated mixtures of organic solvents with water [62].

**Table 1.** Summary of reported applications of electrotechnology (PEF, MEF, and HVED)-assisted extraction of carotenoids from microalgae. PEF: pulsed electric field; MEF: moderate electric field; HVED: high-voltage electric discharges.

| Microalgae | Biomass Concentration | Electrical Treatment | Extraction Conditions | Carotenoid Yield | Notes | Reference |
|---|---|---|---|---|---|---|
| MEF | | | | | | |
| *Heterochlorella luteoviridis* | 4 g dry biomass/100 mL 25% ethanol solution | 0–180 V, 60 Hz, 10 min, <35 °C | 25%–75% ethanol, 50 min, 30 °C | Total carotenoids 1.21 mg/g dw | MEF induced a reversible electroporation improving the extraction efficiency. Xanthophylls all-*trans*-lutein and all-*trans*-zeaxanthin were the major carotenoids extracted. | [64] |
| PEF | | | | | | |
| *Chlorella vulgaris* | ~3% dw | 15 kV/cm, 100 kJ/kg | N/A [a] | Total carotenoids 525% recovery compared with the conventional ball milling homogenization process | Antioxidant activity of the extract was increased by almost 100%. | [53] |
| *Chlorella vulgaris* | $10^9$ CFU/mL in McIlvaine buffer (pH 7) | 10–25 kV/cm 0.6–93 kJ/L of culture | 96% ethanol, 20 °C, 0–1 h | Total carotenoids: ~0.82 mg/g dw and 1.04 mg/g dw after, respectively, 0 h and 1 h of incubation after PEF treatment | Extraction yield significantly increased after 1 h of the application of PEF, likely caused by the plasmolysis of the chloroplast during the incubation time. | [60] |
| *Chlorella vulgaris* | $2 \times 10^8$ CFU/mL in McIlvaine buffer (pH 7) | Millisecond range: 1–40 ms pulses, 3.5–5 kV/cm 9–150 kJ/L of culture Microsecond range: 3 μs pulses 10–25 kV/cm 1.5–93 kJ/L | 96% ethanol 20 °C, 0–1 h | Total carotenoids: ~1.06 mg/L after 0 h and 1 h of incubation in the ms range; 1.09 mg/L and 1.58 mg/L after, respectively, 0 h and 1 h of incubation in the μs range | PEF in the ms range at a lower electric field strength created irreversible alterations, while in the μs range the defects were a dynamic structure along the post-pulse time. Higher energy efficiency of treatment in the μs range than in the ms range. | [61] |
| *Chlorella vulgaris* | $10^9$ CFU/mL in McIlvaine buffer (pH 7) | 10–40 °C 10–25 kV/cm 1.5–93 kJ/L of culture | 96% ethanol, 20 °C, 0–1 h | Lutein up to 0.753 mg/g dw | Increasing temperature increased the sensitivity of microalgae cells to irreversible electroporation, and decreased the total specific energy required to obtain a given extraction yield. PEF treatment did not cause pigment degradation. | [56] |
| *Spirulina platensis* | ~3% dw | 15 kV/cm, 100 kJ/kg | N/A [a] | Total carotenoids 150% recovery compared with the conventional ball milling homogenization process | Antioxidant activity of the extract was increased by almost 100% | [53] |
| *Nannochloropsis* sp. | 1% (*w*/*w*) in distilled water | 20 kV/cm, 1–4 ms, 13.3–53.1 kJ/kg | N/A [b] | N/A [c] | PEF allowed selective extraction of water-soluble ionic components and water-soluble proteins, but was ineffective for extraction of pigments. | [54] |

**Table 1.** *Cont.*

| Microalgae | Biomass Concentration | Electrical Treatment | Extraction Conditions | Carotenoid Yield | Notes | Reference |
|---|---|---|---|---|---|---|
| *Nannochloropsis* sp. | 1% (*w*/*w*) in distilled water | 20 kV/cm, 0.01–6 ms, 13.3–53.1 kJ/kg | Distilled water, up to 3 h, 50 °C, pH = 8.5–11 | Total carotenoids: ~0.04 mg/g dw after PEF (pH 8.5); ~0.2 mg/g dw after PEF (pH 8.5) + extraction at pH 11 | Extraction efficiency after PEF (pH 8.5) was comparable with that of the aqueous extraction at pH 11. PEF (pH 8.5) treatment was more efficient than PEF (pH 11) treatment. Supplementary extraction at pH = 11 allowed a noticeable increase of the concentrations yield. PEF extracts showed high purity. | [63] |
| *Nannochloropsis* sp. | 1% (*w*/*w*) in distilled water | 20 kV/cm, 0.01–6 ms, 13.3–53.1 kJ/kg | Aqueous DMSO, ethanol solutions: 0%, 30%, 50%, and 100%; 20 °C; 240 min | $K_{PEF}$ [d] ≈3.0 at 50% DMSO<br>$K_{PEF}$ [d] ≈ 2.4 at 30% EtOH | High levels of extracted proteins at the first step with water, and noticeable enhancement of extraction of pigments at the second step with binary mixtures. The two-stage PEF-assisted procedure allowed effective extraction using less concentrated mixtures of organic solvents with water. | [62] |
| HVED | | | | | | |
| *Nannochloropsis* sp. | 1% (*w*/*w*) in distilled water | 40 kV/cm, 1–4 ms, 13.3–53.1 kJ/kg | N/A [a] | N/A [c] | Noticeably agglomeration of microalgae cells in the HVED-treated suspensions. Higher pigment recovery than PEF, but less than UAE and HPH. | [54] |

[a] N/A: not available; [b] Extract analyzed immediately after electrical treatment; [c] Results provided as UV absorption spectra and absorption peaks at 415 nm; [d] PEF efficiency coefficient defined as the ratio of concentration values of the extracts obtained for two-stage (PEF/water extraction + extraction with binary mixture) and one-stage (extraction with binary mixture); dw—dry weight.

A combination of PEF and moderate thermal treatment could also assist to achieve the required permeabilization effect of rigid algal membrane structures with less severe processing conditions, or to achieve higher efficacy at the same treatment conditions [65]. It was found that a mild thermal treatment enhanced electroporation efficiency of PEF treatment in plant tissues [66,67]. In the case of extraction of luteinfrom *C. vulgaris*, PEF pretreated samples (25 kV/cm for 75 μs) resulted in a 4.5-fold higher concentration of lutein (753 μg/g dw of *C. vulgaris* culture), with respect to to untreated samples, when carried out at 40 °C, whereas the yield increased only of 2.3 and 3.2 fold when carried out at 10 and 25 °C [56]. The increase in yield at higher temperatures was positively correlated with the increase in membrane permeabilization. Furthermore, the yield also increased with the applied electric field strength, which was correlated to the irreversible electroporation of algal membranes. Interestingly, temperature enhanced electroporation under the PEF treatment and decreased the treatment time to achieve the desired yields, consequently reducing the total specific energy required for the treatment [56].

3.1.2. Moderate Electric Field (MEF)-Assisted Extraction

MEF-assisted extraction could also be an attractive alternative method to extract carotenoids from microalgae. MEF can promote cell membrane permeabilization and consequently assists in the diffusion of intracellular compounds from the intracellular matrix (for a detailed principle and mechanism MEF processing, see ref. [68]). The MEF-assisted extraction process involves the application of relatively low electric fields (arbitrarily defined between 1 and 1000 V/cm) in the range of Hz up to tens of kHz, with or without heating, to biomaterials placed between two electrodes [64] (Figure 2b). MEF can cause a wide variety of effects on biological samples depending on the electrical and thermal conditions used in the treatment. Interestingly, in spite of the relatively low field strength applied, it has been shown that MEF treatment can promote at least reversible electroporation of the cell membranes, increasing their permeability [69].

In comparison with PEF, there are only limited data available about the use of MEF in the recovery of intracellular compounds from microalgae (Table 1), although interesting studies have been reported on the extraction of valuable compounds from plant material [49,50]. In a recent study, Nezammahalleh et al. [70] showed that up to 73% of carotenoids (1.21 mg lutein Eq./g sample dw) can be recovered from the *Heterochlorella luteoviridis* microalga biomass using MEF combined with ethanol as solvent (180 V, 60 Hz, 75 mL/100 mL of ethanol solution). In this case, carotenoid extraction yield increased with electrical field strength and ethanol concentration. HPLC-UV-Vis analysis identified the presence of all-*trans*–lutein (856 μg/g), all-*trans*-zeaxanthin (244 μg/g) and all-*trans*-β-carotene (185 μg/g) in major quantities. Besides, all-*trans*-α-carotene, 9-13-15-*cis*-β-carotene, *cis*-violaxanthin, all-*trans*-violaxanthin and 13-13′-*cis*-lutein were detected in minor quantities [70]. In another study Jaeschke et al. [64] used a two-stage approach to evaluate the effect of MEF pretreatment (0–180 V, 10 min) in the presence of 25 mL/100 mL of ethanol/water solution, followed by the subsequent extraction with ethanol at varying concentrations (25–75 mL/100 mL, 50 min) for the extraction of carotenoids from the microalga *Heterochlorella luteoviridis*. It was observed that the extraction of carotenoids increased as the electrical field strength and ethanol concentration increased, with the highest extraction yields (73%) measured at the maximum values of the two variables (180 V and 75 mL/100 mL of ethanol concentration). The carotenoid profile of the extract revealed that the xanthophylls all-*trans*-lutein and all-*trans*-zeaxanthin were the major carotenoids extracted, owing to their polarity. Interestingly, the use of ethanol alone (75 mL/100 mL) was found to be insufficient for the extraction of carotenoids. It was anticipated that MEF was supposed to act on the cell membranes promoting their permeabilization. However, micrographic images of biomass samples revealed no visible damage caused by MEF to the cell structure, suggesting that a reversible electroporation occurred.

### 3.1.3. High Voltage Electric Discharges (HVED)-Assisted Extraction

HVED is a cell disintegration technique based on the phenomenon of electrical breakdown of water. As illustrated in Figure 2c, during HVED treatment, the biomaterial of interest is placed in a treatment chamber with a high voltage needle electrode and a plated grounded electrode exposed to pulsed shockwaves (typically, 40–60 kV/cm, 2–5 μs) [50]. To date, the mechanisms of HVED, due to their complexity, are not well understood. However, the combination of electrical breakdown with a number of secondary phenomena (high-amplitude pressure shock waves, bubbles cavitation, creation of liquid turbulence, etc.) occurring during HVED treatment have been reported to cause cell structural damages, including cell wall disruption, which accelerates the extraction of intracellular compounds. The attempts to study the effect of HVED on the extraction of intracellular compounds from microalgae have shown that this technology is effective in achieving the extraction of water-soluble, as well as high molecular weight intracellular compounds. However, as reported in Table 1, HVED does not appear to be very effective for the extraction of pigments (e.g., chlorophylls or carotenoids), which instead requires the use of organic solvents or the application of harsher, mechanical homogenization techniques, such as US or HPH [54].

Based on the available literature data, it can be concluded that electrotechnologies (PEF, MEF and HVED) offer a considerable potential for improving the extraction of carotenoids from microalgae. However, greater research is required in order to deeply understand the mechanisms regulating the electrically induced disintegration of cell wall, plasma and organelles (chloroplast) membranes, as well as the subsequent mass transfer of the target intracellular compounds. Moreover, the efficacy of the electrotechnologies on extraction improvement requires careful optimization of process parameters, depending on the compounds of interest, as well as algal cell size, shape, and envelope structures. It is likely that the potential of electrotechnologies could be exploited by using them as a first disintegration step in a multi-stage approach. In the first stage, water-soluble compounds could be extracted; in subsequent stages, either more powerful cell homogenization techniques or "green" solvents could be applied to achieve higher extraction yields of pigments or other hydrophobic compounds.

### 3.2. *Pressurized Liquid Extraction*

PLE, also known as accelerated solvent extraction, has been acknowledged as a green alternative technique for the extraction of compounds from biological matrices. It was first described by Richter et al. [71] in 1996. PLE involves the extraction using liquid solvents at elevated temperature and pressure (always below their critical points), normally in the ranges of 50–200 °C and 35–200 bar, respectively [72,73]. The use of solvents at temperatures above their atmospheric boiling point reduces their viscosity and surface tension significantly and enhances solubility and mass transfer of analytes. The main advantage of using PLE is that it allows for rapid extraction and reduces solvent consumption [74–77]. PLE allows for the efficient usage of green solvents such as water and ethanol for the extraction of a wide variety of compounds by changing their dielectric constants (polarity) to values similar to those of organic solvents [72]. Although water is the most widely used polar solvent for PLE (also referred as pressurized water extraction, subcritical water extraction, superheated water extraction, pressurized hot-water extraction), alternatively, bio-ethanol [78], methanol [79–82], *n*-hexane [83], propane, dichloromethane [84], acetone [85], ethyl acetate [86], ionic liquids [87], surfactants [88] can also be applied.

To date, PLE has been extensively investigated for the recovery of commercially and industrially valuable compounds from varying plant sources (reviewed in [89,90]). Nevertheless, the use of PLE in the recovery of carotenoids from microalgae and seaweeds is relatively limited. In an early study, Denery et al. [91] demonstrated the efficiency of PLE as an alternative technology for the extraction of oxygen and light-sensitive carotenoids from two green microalgae, namely *Dunaliella salina* and *Hematococcus pluvialis*. They confirmed that PLE required a lower amount of solvent and shorter extraction times compared to traditional extraction methods. They extracted equal amount of astaxanthin, β-carotene, lutein, and total pigments from *D. salina* and *H. pluvialis* compared to the traditional method.

In a subsequent study, Herrero et al. [92] detected all-*trans*-β-carotene and its isomers along with several minor carotenoids in the extract of the microalga *Dunaliella salina* while extracting antioxidant compounds using PLE [92]. The *n*-hexane extract obtained at 160 °C for 17.5 min showed the highest levels of β-carotene isomers (25.07 mg/100 g) and total carotenoids (29.50 mg/100g). This amount was more than seven times higher than the one obtained for ethanol extracts, with the antioxidant activity displaying double the activity to that of the ethanol extract. The authors selected ethanol as the most suitable solvent for PLE for these antioxidant compounds considering the total extraction yield and reduction in environmental impact using *n*-hexane. Interestingly, the β-carotene recovery increased with the extraction temperature and the best yield was obtained at a temperature of 160 °C, indicating PLE at high temperatures was not detrimental to the extraction of carotenoids, provided a short extraction time was applied.

In another study, PLE revealed the presence of several antioxidative carotenoids in the extracts of seaweed, *Himanthalia elongata* (commonly known as sea spaghetti) and from the microalgae *Synechocystis* sp. [93]. Fucoxanthin (0.82 mg/g) and zeaxanthin (0.13 mg/g) were the major carotenoids found in the *H. elongata*, while β-carotene (2.04 mg/g) zeaxanthin (1.64 mg/g), myxoxanthophyll (0.58 mg/g) and echinenone (0.24 mg/g) were abundant in the *Synechocystis* sp. extract. Overall, the *Synechocystis* sp. ethanolic extracts obtained at 100 °C showed a higher carotenoid yield than the those obtained at different temperatures (50, 150 and 200 °C), and using other solvents (*n*-hexane and water) [93].

Similarly, PLE was found to be a suitable technique for the extraction of bioactive carotenoids from *C. vulgaris* [94]. PLE followed by HPLC-DAD analysis revealed the presence of lutein, *cis*-lutein, and β-carotene in the extract and their presence was positively correlated to their antioxidant activity. In general, PLE resulted in a greater carotenoids yield when compared to conventional maceration and UAE [94]. In a subsequent study, the PLE process was optimized and its efficiency was compared with maceration, Soxhlet extraction and UAE [95]. Ethanol at 90% was found to be a suitable solvent for the PLE of carotenoids from *C. vulgaris* when compared to acetone, *n*-hexane and water. A temperature of 116.8 °C and an extraction time of 25.1 min were found to be optimal for the extraction of β-carotene, while 48.2 °C and 34.6 min yielded an optimum quantity of lutein. At these conditions the yields of β-carotene and lutein was 0.67 and 3.70 mg/g of sample, respectively. In general, β-carotene and lutein were more effectively extracted by PLE than conventional Soxhlet extraction and maceration techniques. The efficiencies of PLE and UAE for lutein extraction were similar, however, PLE found to be less time consuming [95].

In another study, extraction of antioxidant carotenoids from the microalga *Haematococcus pluvialis* was investigated by PLE using *n*-hexane and ethanol as the extraction solvents [96]. Lutein followed by neoxanthin and β-carotene were the main carotenoids extracted from the green phase, *H. pluvialis* vegetative cells, whereas astaxanthin derivatives were observed in abundance from the red phase encysted cells formed under the stress condition. Overall, ethanol was found to be the most suitable solvent for the recovery of total carotenoids, although the *n*-hexane extract showed a higher astaxanthin content (35.1 mg/g dw) in red phase cells [96]. Plaza et al. [97] demonstrated that acetone was the most suitable solvent to extract *C. vulgaris* carotenoids (α-carotene, β-carotene, neoxanthin and violaxanthin) with respect to ethanol and water in PLE. The authors also showed that PLE provided a higher yield in carotenoid compared to UAE. In a similar study, Kim et al. [98] showed that ethanol was a suitable solvent for the extraction of fucoxanthin from the microalga *Phaeodactylum tricornutum*. PLE was performed at 100 °C for 30 min at 103 bar yielding 16.51 mg/g dw of fucoxanthin. Although the obtained yield was similar to conventional maceration methods, PLE enabled to reduce solvent use and extraction time [98].

Recently, Shang et al. attempted to optimize the PLE protocols for the efficient recovery of carotenoids, using statistical experimental designs for the extraction of fucoxanthin from the edible seaweed, *Eisenia bicyclis* [99]. Results revealed that temperature and ethanol concentration significantly influenced the extraction efficiency. Fucoxanthin was found to be relatively stable at 80 °C when extracted for 1 h, however, slight degradation was observed at 100 °C when extracted for 1 h. Optimized conditions obtained by Response Surface Methodology (RSM) revealed that at 110 °C using 90% ethanol resulted in 0.39 mg/g of fucoxanthin. In a similar study, Koo et al. [100] optimized the pressurized liquid extraction of zeaxanthin from *C. ellipsoidea* using central composite design. According to their results, the highest recovery of zeaxanthin was obtained using ethanol when compared to *n*-hexane and isopropanol. The extraction temperature showed the strongest influence on the extraction of zeaxanthin. The optimum extraction temperature and time for zeaxanthin found to be 115.4 °C and 23.3 min, respectively and the maximum yield obtained under these conditions was 4.28 mg/g. Similarly, Castro-Puyana et al. [45] attempted to optimize the extraction conditions for the recovery of carotenoids from *Neochloris oleoabundans* using PLE with food grade solvents such as ethanol and limonene. A three-level factorial design was employed to optimize the extraction conditions; at a temperature of 112 °C and 100% ethanol (0% limonene) as the extraction solvent provided optimum yields of carotenoids. Under these conditions approximately 97–98 mg/g of the extract of carotenoids were detected with lutein the major carotenoid identified in the extract. Several other secondary carotenoids including canthaxanthin, echinenone, and astaxanthin monoesters and diesters were also detected [45].

In another recent study, Taucher et al. [101] showed that use of dichloromethane as an extraction solvent in PLE yielded significantly higher levels of carotenoids from *H. pluvialis* when compared to acetone, ethanol, ethyl acetate and *n*-hexane. The extraction temperature up to 60 °C demonstrated a positive effect on the recovery of carotenoids whilst higher temperatures resulted in the degradation of carotenoids. At optimized conditions (PLE at 60 °C for 10 min, in 1 cycle, using dichloromethane), 3.69 μg/mg dw astaxanthin and 4.78 μg/mg dw total carotenoids were recovered from *H. pluvialis*. Furthermore, 1.48 μg/mg dw lutein and 1.29 μg/mg dw astaxanthin were recovered from *Chromochloris zofingiensis* and 2.08 μg/mg dw lutein was obtained from *C. sorokiniana* under these conditions. The recovery was also dependent on mechanical cell-disruption techniques used (high pressure homogenization and ball mill disruption) [101].

Overall, PLE has been demonstrated to provide an alternative for the extraction of carotenoids from microalgae and seaweeds. It is evident that ethanol and in some cases acetone [97], dichloromethane [101] are the most suitable solvents for the PLE of bioactive carotenoids rather than organic solvents such as *n*-hexane [92,93,95,96,100,102]. Ethanol noted as a "green" solvent minimizes cost and environmental impact. Although, for the extraction of carotenoids, several researchers claim that PLE reduced the solvent consumption, a comparative study is elusive. A detailed investigation including optimization data is required for the potential usage of PLE in commercial and industrial applications.

### 3.3. Supercritical Fluid Extraction (SFE)

The current literature identifies that SFE is the most extensively studied non-conventional extraction technique for the recovery of carotenoids from algae and microalgae. SFE has been considered as a sustainable "green" technology for the selective isolation of compounds. SFE uses supercritical fluids i.e., fluids at a temperature and pressure above its critical limit as the extraction solvent. Since supercritical fluids possess low viscosity and high diffusivity, they provide better solvating and transport properties than liquids. As an important advantage, the solvating power (polarity) of supercritical fluid can be adjusted by manipulating the temperature and pressure of the fluid, allowing for the selective extraction of a wide range of compounds [103]. Nowadays, many laboratories and industries are replacing conventional extraction techniques with SFE in order to minimize organic solvent consumption and increase high throughput [104]. Currently, carbon dioxide is the preferred solvent (referred as supercritical $CO_2$ extraction) as it can easily attain supercritical conditions and has several advantages including low toxicity, flammability and cost, and high purity when compared to other fluids [104]. Supercritical carbon dioxide provides a nonpolar environment and its polarity can be occasionally modified by using co-solvents, such as ethanol, to extract relatively polar xanthophylls, such as lutein and astaxanthin. In some studies, ethane and ethylene were also used as SFE solvents for the extraction of carotenoids [105].

SFE technology for extraction of carotenoids has been employed from laboratory to the commercial scale. Reported applications of SFE to extract a wide range of carotenoids from microalgae and seaweeds are summarized in Table 2. In many cases, SFE has been found to be a superior technique for the extraction of heat sensitive carotenoids. A number of investigations are available to describe several SFE issues, such as the effect of temperature, pressure, co-solvents, solvent flow rate [19,106] and pretreatment, the extraction of carotenoids, selectivity [107], kinetics [29,107,108], and the modelling of extraction [26,109].

In the supercritical $CO_2$ extraction of carotenoids, in general, the extraction efficiency increases with $CO_2$ pressure and temperature up to a optimal level [110–114], nevertheless, this trend is dependent on the combined effect of pressure and temperature [19,112,115–117]. In some instances, high $CO_2$ pressure (>400 bar) has caused lower recovery of carotenoids [108,112], and on the other hand, some researchers have observed a reduction in carotenoid yields at low $CO_2$ pressure, the latter dependent on the temperature used [107,113]. Pressure has contrasting effects on the extraction yield; increasing pressure (at a constant temperature) increases the density of $CO_2$, and consequently, the solvation power of the fluids, which in turn increases the solubility of the compounds and extraction yield. However, high pressure can obstruct the diffusivity of supercritical fluid into the matrix, therefore decreasing the extraction yield [112]. Similarly, an increase in temperature at a constant pressure increases the vapour pressure resulting in improved solubility of pigments. An increase in temperature results in a decrease in the fluid density, which in turn results in lower solubility of pigments. Therefore, the recovery of carotenoids is highly dependent on the complex interaction of temperature and pressure, which greatly affects density, viscosity and vapour pressure in the system. The predominance of one or other effects is responsible for the extraction efficiency [112]. For instance, Macías-Sánchez et al. [116] obtained the highest carotenoid yield from the marine microalgae *Synechococcus* sp. using supercritical $CO_2$ at a temperature of 50 °C when the operating pressure was 200 and 300 bar, while the yield decreased when the pressure was increased to 400 and 500 bar. At these pressures, the maximum extraction yield was obtained when the temperature was 60 °C. The observed variation in the yield with respect to pressure and temperature was correlated to the dominating effects of density or vapour pressure at these conditions. Similar observations were made in the authors subsequent study on SFE of carotenoids from *Scenedesmus almeriensis* [112]. While studying the effects of pressure (200–600 bar ) and temperature (32–60 °C), the maximum yield of lutein was recovered at intermediate pressures, except for the extraction at 46 °C where the maximum yield was obtained at 600 bar [112]. Furthermore, a number of studies have shown similar effects of pressure and temperature on the recovery of several carotenoids [107,108,110,111,114–121].

Several researchers have used co-solvents such as ethanol [26,108,109,111,113,114,118,120–127], acetone [113], vegetable oil [26,106,128] as polarity modifiers for the efficient recovery of carotenoids such as β-carotene [113,118], astaxanthin [106,109,114,120–122,125,129], lutein [111,113] and zeaxanthin [118]. Since supercritical $CO_2$ is non-polar, addition of a small amount of co-solvent increases the ability of supercritical $CO_2$ to dissolve relatively polar carotenoids. Addition of co-solvents can cause swelling [130] of algal cells, facilitating the rapid mass transfer of analytes from the matrix [109,114]. Some co-solvents such as ethanol can enhance mass transfer by creating hydrogen bonding with analytes [114,131]. In a study on the supercritical $CO_2$ extraction of lutein from *Scenedesmus* sp., ethanol was found to be the superior co-solvent compared to methanol, propanol, butanol and acetone [111]. In another study, presence of the co-solvent ethanol improved the total carotenoid recovery from the microalga *H. pluvialis* by up to 25%. Similarly, it increased the recovery of fucoxanthin from the seaweeds *U. pinnatifida* and *Sargassum muticum* by up to 90 [29] and 10 times [23], respectively. Interestingly, Krichnavaruk et al. [106] showed that vegetable oils such as soybean oil or olive oil can be used as a co-solvent to enhance astaxanthin recovery from *H. pluvialis* in supercritical $CO_2$ extraction. The presence of 10% olive oil increased the recovery up to 51% at 70 °C and 400 bar, which was equivalent to that obtained using ethanol as a co-solvent. Recently, Saravana et al. [26] showed that sunflower oil as a co-solvent with supercritical-$CO_2$ increased the recovery of carotenoids and fucoxanthin from the brown seaweed *Saccharina japonica* and its efficiency was greater than canola oil, soybean oil, and ethanol.

Although the use of co-solvents improves the extraction yield, in some cases, the presence of co-solvents can decrease the selectivity of the extraction [113,118]. Cardoso et al. [118] showed that although the obtained yield of β-carotene increased on using $CO_2$ and 5% ethanol as a co-solvent, under these conditions zeaxanthin was co-extracted. In any case, the selectivity was also dependent on SFE parameters such as pressure and temperature [118], including extraction time [113]. Thus, the use of a co-solvent might tend to compromise product purity and should be considered.

The initial pretreatment of algae with physical or mechanical cell disintegration techniques such as crushing, sonication, ball-milling has been found to enhance the extraction efficiency in SFE [110,114,119,122]. The yield of total carotenoids obtained using SFE (with $CO_2$ and ethanol co-solvent) was significantly higher when the most homogenized form of the microalga *Synechococcus* sp., was used for the extraction when compared to uncrushed cells (91.8% recovery against 58.7%, respectively) [114]. Valderrama et al. [122] found that the astaxanthin yield increased with the degree of crushing microalga *H. pluvialis* during SFE performed at 60 °C at 300 bar with $CO_2$. Similar results were observed while extracting total carotenoids from *C. vulgaris* [132]. Crushing enhances the accessibility of supercritical fluids to the carotenoid bound to the cell organelles, thereby, increases extraction efficiency.

Several SFE parameters (pressure, temperature, flow rate, time, co-solvents etc.) significantly influence the extraction efficiency as well as selectivity of target compounds for extraction. Therefore, these parameters must be carefully considered and optimized for an efficient and selective recovery of target analytes. RSM could be a good statistical tool to design experiments, optimize experimental parameters and to determine the effect of these parameters on carotenoid yield. In a study Thana et al. [133] employed RSM with central composite design to investigate the effect of operating temperatures (40–80 °C), operating pressures (300–500 bar) and extraction times (1–4 h) on astaxanthin yields in supercritical $CO_2$ extraction. The optimal conditions for extraction of astaxanthin were found to be at 70 °C temperature, 500 bar pressure, and 4 h time. Under these conditions, the predicted astaxanthin extraction yield was 23.04 mg/g (83.78% recovery). Recently, Saravana et al. [26] showed that ~50 °C temperature, 300 bar pressure, and 2% of sunflower oil co-solvent are the most suited conditions for the extraction of total carotenoids and fucoxanthin from the brown seaweed, *Saccharina japonica* using supercritical $CO_2$. The authors attained 2.391 mg/g total carotenoids and 1.421 mg/g of fucoxanthin under these conditions. The optimum pressures and temperatures required for the extraction of carotenoids from microalgae such as *Nannochloropsis gaditana* [115,123], *Scenedesmus almeriensis* [112], *Dunaliella salina* [123], *Chlorella vulgaris* [113], *Scenedesmus* sp. [111], *Synechococcus* sp. [116], *Undaria pinnatifida* [19,20] has also been reported and shown in Table 2. Overall, SFE has been shown to be an excellent technique for the selective extraction of carotenoids from a wide range of algae and microalgae. In general, using supercritical $CO_2$ alone enhances selectivity, while, efficiency can be enhanced by using co-solvents such as ethanol, however, selectivity must be compromised.

**Table 2.** Applications of supercritical fluid extraction (SFE) for recovery of carotenoids from algae and seaweeds.

| Microalga/ Seaweed | Extraction Condition | | | | | Carotenoid Yield | Notes | Reference |
|---|---|---|---|---|---|---|---|---|
| | Solvent [a] | Pretreatment | P [b] (bar) | T [c] (°C) | T [d] (h) | | | |
| Microalgae | | | | | | | | |
| *Haematococcus pluvialis* | $CO_2$ and 9.4% ethanol | Crushing and then grinding in dry ice | 300 | 60 | – | Astaxanthin >97% recovery | Co-solvent enhanced the recovery slightly | [122] |
| *Synechococcus* sp. | $CO_2$ | Freeze drying | 500 | 60 | 4 | Total carotenoids 2.76 mg/g dw [e] | Optimal extraction conditions for β-carotene was 50 °C, 358 bar; for β-cryptoxanthin was 59 °C, 454 bar; and for zeaxanthin was 60 °C, 500 bar. | [134] |
| *Haematococcus pluvialis* | $CO_2$ | Drying (powder form) | 70 | 500 | 4 | Astaxanthin 23.04 mg/g dw [e] | Pressure, extraction time, and the interaction between temperature and pressure had significant effect on astaxanthin yield. | [133] |
| *Dunaliella salina* | $CO_2$ | Homogenization | 60 | 400 | 3 | Total carotenoids 12.17 mg/g algae dw | SFE was more selective than the UAE. | [117] |
| *Chlorella vulgaris* | $CO_2$ and 7.5% ethanol | – | 80 | 500 | 3 | Lutein ~≥1.8 mg/g algae; β-carotene ~≥0.2 mg/g | Supercritical $CO_2$ has high selectivity for lutein extraction, however the yield was lower than Soxhlet extraction; ethanol was better co-solvent than acetone. | [113] |
| *Chlorococcum littorale* | $CO_2$ and 10 mol % ethanol | Freeze drying | 60 | 300 | 1–3 | Total carotenoids 0.094%–0.21% dw | Co-solvent enhanced the recovery slightly. | [127] |
| *Scenedesmus* sp. | $CO_2$ and ethanol (30 mol %) | Freeze drying and milling | 70 | 400 | 1 | Lutein 2.210 mg/g algae [e] | Higher temperature lead to increased impurity. | [111] |
| *Nannochloropsis gaditana* | $CO_2$ | Freeze drying (powder form) | 60 | 400 | 3 | Total carotenoids 0.343 mg/g algae dw [e] | Higher temperature lead to degradation. | [115] |
| *Haematococcus pluvialis* | $CO_2$ and 5% ethanol | – | 70 | 550 | 4 | Astaxanthin 77.9% recovery with respect to 34.3 mg/g dw total content found in the sample using Soxhlet extraction | Astaxanthin yield increased with increasing cosolvent concentration up to 5% (*v*/*v*) ethanol. | [120] |
| *Haematococcus pluvialis* | $CO_2$ and 10% ethanol | Freeze drying and ball milling | 60 | 300 | – | Carotenoid recovery 92%; esterified astaxanthin ~75%; lutein >90%; astaxanthin >90%; β-carotene >90%; and canthaxanthin ~85% | Crushing improved the recovery significantly. | [114] |
| *Chlorella vulgaris* | $CO_2$ and 5% ethanol | Crushing | 40 | 300 | – | Total carotenoids up to 0.299% | Crushing increased pigments recovery. | [132] |

**Table 2.** *Cont.*

| Microalga/ Seaweed | Extraction Condition | | | | | Carotenoid Yield | Notes | Reference |
|---|---|---|---|---|---|---|---|---|
| | Solvent [a] | Pretreatment | P [b] (bar) | T [c] (°C) | T [d] (h) | | | |
| *Dunaliella salina* | $CO_2$ | Freeze drying | 9.8 | 443 | 1.6 | Total carotenoids 6.72% (predicted) | Higher yields were obtained at high pressures and low temperatures. | [135] |
| *Nannochloropsis* sp. | $CO_2$ and 20% ethanol | Ball milling | 40 | 300 | >1 | – | Co-solvent increased the yield. | [136] |
| *Scenedesmus almeriensis* | $CO_2$ | Freeze drying (powder form) | 60 | 400 | 5 | Lutein 0.0466 mg/g dw [e] β-carotene 1.5 mg/g dw [e] | Recovery was lower compared with conventional acetone extraction. | [112] |
| *Synechococcus* sp. | $CO_2$ | Freeze drying | 50 | 300 | 3 | Total carotenoids 1.511 mg/g algae dw [e] | The highest carotenoids/chlorophylls selectivity was obtained at 200 bar and 60 °C. | [116] |
| *Nannochloropsis oculata* | $CO_2$ and 16.7 wt % ethanol | – | 50 | 350 | – | Total carotenoids 7.61 mg/g dw | Anti-solvent precipitation of carotenoids allowed pure Zeaxanthin. | [124] |
| *Nannochloropsis oculata* | $CO_2$ and ethanol | Grinding and freeze drying | 50 | 350 | – | Zeaxanthin 13.17 mg/g | Ethanol as a co solvent increased the yield, and was efficient than dichloromethane, toluene and soybean oil | [128] |
| *Monoraphidium* sp. | $CO_2$ and ethanol | Freeze drying | 60 | 200 | 1 | Astaxanthin 2.02 mg/g dw | Ethanol as a co-solvent improved astaxanthin yield. | [125] |
| *Chlorella vulgaris* | $CO_2$ and ethanol | Pretreatment process using alcohol as elution solvent | 40 | 400 | 0.75 | Lutein 1.78% recovery based on 7.9 mg/g obtained in Soxhlet extraction | Ethanol as an elution solvent removed chlorophyll *a*, *b* and β-carotene and improved selectivity of lutein | [107] |
| *Haematococcus pluvialis* | $CO_2$ and 10% olive oil | Drying | 70 | 400 | 5 | Asthaxanthin 51% recovery | Olive oil co-solvent lead to a recovery comparable to ethanol as a co-solvent. | [106] |
| *Nannochloropsis gaditana* | $CO_2$ and 5% ethanol | Freeze drying | 40–60 | 100–500 | 3 | Carotenoid yield up to 0.3% | Extraction kinetics was studies. | [108] |
| *Synechococcus* sp. | | | | | | Carotenoid yield up to 0.12% | | |
| *Dunaliella salina* | | | | | | Carotenoid yield up to 1.3% | | |
| *Nannochloropsis gaditana* | $CO_2$ and 5% ethanol | Freeze drying | 60 | 500 | 3 | Total carotenoids 2.893 mg/g algae dw [e] | Supercritical extraction process with co-solvent was more selective than conventional methanol extraction. | [123] |
| *Synechococcus* sp. | | | 50 | 300 | | Total carotenoids 1.860 mg/g algae dw [e] | | |
| *Dunaliella salina* | | | | | | Total carotenoids 9.629 mg/g algae dw [e] | | |

**Table 2.** *Cont.*

| Microalga/Seaweed | Extraction Condition | | | | | Carotenoid Yield | Notes | Reference |
|---|---|---|---|---|---|---|---|---|
| | Solvent [a] | Pretreatment | P [b] (bar) | T [c] (°C) | T [d] (h) | | | |
| *Haematococcus pluvialis* | $CO_2$ and 2.3 mL/g sample ethanol | Freeze drying (powder form) | 65 | 435 | 3.5 | Astaxanthin recovery of 87.42% from sample containing 2.26% astaxanthin. | Increasing co-solvent amount resulted in an improving astaxanthin yield. | [121] |
| *Synechococcus* sp. | $CO_2$ and ethanol | – | 40 and 60 | 400 and 200 | 3 | β-carotene 0.70 mg/g algae dw at 40 °C 400 bar [e] Zeaxanthin 0.70 mg/g algae dw at 60 °C 200 bar | $CO_2$ with ethanol simultaneously extracted β-carotene and zeaxanthin. | [118] |
| *Arthrospira platensis* [f] | $CO_2$ and 26.7% ethanol | Air drying and milling | 60 | 150 | 0.83 | Total carotenoids 283 mg/g algae [e] | MAE resulted in better extraction yield than SFE. | [126] |
| Seaweeds | | | | | | | | |
| *Undaria pinnatifida* | $CO_2$ and ethanol | Freeze drying | 50 | 200 | 1 | Fucoxanthin 7.53 mg/g dw | Yield was dependent on pressure and temperature combination. | [19] |
| *Undaria pinnatifida* | $CO_2$ | Milling and microwave-assisted cell disruption | 40 | 400 | 3 | Fucoxanthin 38.5 mg/g [e] | MW pretreatment increased fucoxanthin yield. | [20] |
| *Sargassum muticum* | $CO_2$ and ethanol | Freeze drying and comminutating | 50 | 200 | 1 | Fucoxanthin ~0.12 mg/g algae dw | Use of co-solvent increased fucoxanthin yield by 90 times. | [29] |
| *Undaria pinnatifida* | $CO_2$ and 3.23% ethanol | Drying | 60 | 400 | 3 | Fucoxanthin 0.9945 mg/g dw [e] | Use of co-solvent increased fucoxanthin yield by 10 times. | [23] |
| *Undaria pinnatifida* | $CO_2$ | Drying | 60 | 400 | 2.5 | Fucoxanthin ~0.058 mg/g dw [e] | Pressure, temperature and extraction time affected fucoxanthin recovery. | [129] |
| *Saccharina japonica* | $CO_2$ and ethanol | Freeze drying and grinding | 45 | 250 | 2 | Fucoxanthin 0.41 mg/g dw [e] | SFE process extracted a similar content of fucoxanthin as when acetone–methanol conventional extraction was used. | [137] |
| *Sargassum horneri* | | | | | | Fucoxanthin 0.77 mg/g dw [e] | | |
| *Saccharina japonica* | $CO_2$ and 2% sunflower oil | Freeze drying | 50.62 | 200 | 2 | Total carotenoids 2.391 mg/g dw [e]; fucoxanthin 1.421 mg/g dw [e] | Sunflower oil as a co-solvent found to be the most effective, than soybean oil, canola oil, ethanol, and water. | [26] |

[a] Ethanol/vegetable oils mentioned in the column served as a co-solvent in the extraction; [b] Operating temperature; [c] Operating pressure; [d] Extraction time; [e] Maximal yield obtained at optimum conditions; [f] Considered as cyanobacteria.

### 3.4. Subcritical Fluid Extraction

Subcritical fluid extraction is a technique, similar to SFE, where subcritical (liquefied) fluids are used as extraction solvent. Compared to SFE, only a limited number of reports are available describing subcritical fluid extraction of carotenoids from microalgae and seaweeds. Subcritical fluid extraction operates at relatively lower temperature and pressure than SFE [18,23,129,138,139]. In recent studies, subcritical $CO_2$, 1,1,1,2-tetrafluoroethane and dimethyl ether (DME) have shown potential to extract carotenoids from microalgae and seaweeds [18,23,129,138,139].

Subcritical $CO_2$ extraction (SCCE) uses liquid $CO_2$ as the extraction solvent. It operates at relatively lower temperature (lower than critical temperature of $CO_2$, 31.06 °C), and therefore is effective in extracting thermally labile compounds. In this case, the operating pressure is maintained (sometimes higher) to the critical pressure of $CO_2$ (73.8 bar). Recently, Fan et al. [138] reported the extraction of lutein from *C. pyrenoidosa* using ultrasound-enhanced subcritical $CO_2$ extraction (UCCE) using ethanol as the co-solvent. The authors achieved excellent recovery of lutein (124 mg/100 g) using USCCE when compared to Soxhlet extraction, subcritical water extraction and supercritical $CO_2$ extraction. Table 3 compares the efficiency of UCCE with other conventional and non-conventional extraction techniques in terms of process conditions and lutein yield.

**Table 3.** Comparison of different extraction techniques for extraction of lutein from *Chlorella pyrenoidosa* (reproduced with permission from [138]).

| Extraction Method | Temperature (°C) | Pressure (MPa) | Ultrasound Power (W) | Time (h) | Lutein Yield (μg/g) |
|---|---|---|---|---|---|
| SE | 43 | 0.1 | 0 | 18 | 546.4 |
| SWE | 150 | 5 | 0 | 1/3 | 0 |
| SCE | 50 | 25 | 0 | 4 | 393.3 |
| SCCE | 27 | 21 | 0 | 4 | 422.9 |
| SCCE with pretreatment | 27 | 21 | 0 | 4 (+3 h pretreatment) | 921.5 |
| USCCE with pretreatment | 27 | 21 | 1000 | 4 (+3 h pretreatment) | 1240.1 |

SE—Soxhlet extraction; SWE—subcritical water extraction; SCE—supercritical $CO_2$ extraction; SCCE—subcritical $CO_2$ extraction; USCCE—ultrasound-enhanced subcritical $CO_2$ extraction; pretreatment includes enzymatic treatment with cellulose prior to extraction.

Recently, subcritical (liquefied) dimethyl ether (DME) was also used as an extraction solvent replacing $CO_2$ [18,23,129]. DME below its critical temperature and pressure (critical temperature, 126.85 °C; critical pressure, 53.7 bar) can dissolve a wide range of polar and nonpolar compounds [129]. DME can enhance mass transfer by forming hydrogen bonds with extractable compounds. DME is considered a non-toxic [140] extraction solvent [129], and unlike supercritical $CO_2$ extraction, raw samples can be used for the recovery of carotenoids without drying the biomass, reducing process time and cost. As an additional advantage, the liquefied DME can be evaporated as a gas under low-pressure, which is a highly effective and energy efficient method for solvent recovery [141]. Recently, Billakanti et al. [18], Goto et al. [129] and Kanda et al. [23] reported the extraction of fucoxanthin from *U. pinnatifida* using subcritical DME. At 25 °C, 5.9 bar pressure and 0.72 h extraction time, the amount of fucoxanthin recovered was approximately 390 μg/g dw [23,129], which was significantly higher than that attained by conventional Soxhlet extraction using ethanol (50 μg/g dw) [23]. However, the yield was lower than that obtained in supercritical $CO_2$ extraction [23]. The recovery of fucoxanthin using conventional extraction, subcritical DME extraction and supercritical $CO_2$ is compared in Table 4 [adapted from ref. [23]]. In a study, Billakanti et al. [18] found that enzyme pretreatment prior to subcritical DME extraction has no significant positive effects on the recovery of fucoxanthin from wet or dry *U. pinnatifida* biomass, however, the addition of ethanol as a co-solvent slightly enhanced the relative recovery from the wet biomass [18].

**Table 4.** Comparison of conventional and pressurized extraction techniques for recovery of fucoxanthin from *Undaria pinnatifida* (reproduced with permission from ref. [23]).

| Extraction Techniques | Time (h) | Temperature (°C) | Pressure (MPa) | Fucoxanthin Yield (μg/g) |
|---|---|---|---|---|
| Ethanol (Soxhlet) | 12 | 78 | ND | 50 |
| Liquefied DME | 0.72 | 25 | ND | 390 |
| Supercritical $CO_2$ | 3 | 60 | 40 | 60.12 |
| | 3 | 70 | 40 | 59.51 |
| Supercritical $CO_2$ with entrainer (3.23%) | 3 | 60 | 40 | 994.53 |

ND—Not determined.

In a recent subcritical fluid extraction study, carotenoids of marine seaweed *Laminaria japonica* were extracted using ethanol-modified subcritical 1,1,1,2-tetrafluoroethane (R134a) [142]. Response surface methodology (RSM) combined with a Box–Behnken design was applied to investigate the effects of pressure (50–170 bar), temperature (30–50 °C) and the amount of co-solvent (2%–6% R134a, *w*/*w*) on the recovery of carotenoids. An extraction temperature of 51 °C, extraction pressure 170 bar and a co-solvent amount of 4.73% yielded optimum quantity of carotenoids (0.233 g/kg), however, the yield was lower than that obtained using UAE with methanol solvent (0.336 g/kg) [142].

### 3.5. Microwave-Assisted Extraction

Microwaves are non-ionizing electromagnetic radiations with a frequency ranging from 300 MHz to 300 GHz. Microwave radiation can transfer heat to the system by means of dipole rotation of molecules and ionic conduction in the medium. This principle has been the basis for the development of microwave-assisted extraction (MAE), where the extraction is facilitated by microwave radiation that transfers heat in the extraction medium and aids in the dissolution and mass transfer of analytes. Heat transfer resulted by microwave irradiation can also cause evaporation of moisture inside the cell, developing significant pressure inside the biological matrix. This pressure change can rupture cell membranes and increase the cell porosity, which in turn accelerates the penetration of solvent and the release of intracellular compounds. Microwave radiation can also cause the disruption of hydrogen bonds and migration of dissolved ions, which further enhances the extraction of analytes (for an extensive explanation on theory and principles, see [143]). MAE can be performed in open or closed reaction vessels. Open vessels are used for low temperature extraction at atmospheric pressure whereas closed vessel systems are used for high temperature extractions.

Recently, several researchers have shown that MAE has the potential for the recovery of carotenoids from microalgae and seaweeds. Existing reports suggest that the efficiency of MAE is mainly dependent on the extraction condition and algal cell structures. In some cases, researchers have observed the selective degradation of carotenoids such as astaxanthin during intense microwave treatments, but not the subsequent degradation of other carotenoids such as fucoxanthin. Mild microwave treatment is sufficient to extract compounds from algae, however, species with complex exopolysaccharide envelopes require slightly intense microwave treatment. In a study, fucoxanthin was extracted from a frustulated diatom, *Cylindrotheca closterium* in acetone using MAE technique by Pasquet et al. [144]. MAE at 50 W resulted in the total extraction of fucoxanthin in 3–5 min with an extraction yield of 4.24 μg/mg. The yield obtained by MAE was comparable to that obtained by conventional cold and hot soaking extractions performed for 60 min (4.68 and 5.23 μg/mg, respectively), nevertheless, MAE significantly reduced the extraction time. MAE assisted the disruption of frustule structure associated with diatoms helping in the rapid extraction of analytes.

In their study, increasing microwave power and irradiation times did not show any effect on the extraction yield, indicating a higher stability of fucoxanthin under microwave radiation. In the same study, authors did not observe a significant improvement in the extraction yield for β-carotene isolated from *Dunaliella tertiolecta* when they compared MAE with other conventional methods. This was mainly due to a simple cell-wall structure associated with *D. tertiolecta*, which may not require additional energy for cell disruption and mass transfer.

MAE has also been successfully applied for the extraction of fucoxanthin from seaweeds. In one study Xiao et al. [21] showed the optimized process conditions for MAE of fucoxanthin from edible seaweeds and brown algae. *U. pinnatifida* was used as model matrix for the optimization process, and the solvent/sample ratio, irradiation time was significant on the recovery of fucoxanthin, whereas, the microwave power had insignificant influence. The use of ethanol and acetone as extraction solvent resulted in similar extraction yields, whilst a 50% *n*-hexane in ethanol caused lower recovery of fucoxanthin. Applying ethanol as the extraction solvent, with a solvent to sample ratio of 15:1 mL/g, an extraction temperature of 60 °C, a time of 10 min and microwave power of 300 W resulted in the optimum recovery of fucoxanthin. Under these conditions, the maximal yield of fucoxanthin from fresh *L. japonica*, dry *U. pinnatifida*, and dry *S. fusiforme* was 5.13, 109.3, and 2.12 mg/100 g, respectively. Based on the studies of Pasquet et al. [144] and Xiao et al. [21] it can be concluded that microwave energy optimal level does not affect the stability of fucoxanthin, indicating fucoxanthin is relatively stable carotenoid under microwave irradiation.

MAE offers great potential for the extraction of astaxanthin from microalgae. In one study, a closed system MAE resulted in the highest astaxanthin recovery from marine alga *H. pluvialis* in a shorter duration (5 min) when compared to conventional solvent extractions and ultrasound assisted extractions (UAE), at a time of 60 min for the recovery of astaxanthin [145]. Acetone was found to be a suitable solvent for the recovery of astaxanthin when compared to methanol, ethanol and acetonitrile. MAE at a temperature of 75 °C resulted in 74% recovery of astaxanthin. A temperature above 75 °C in the microwave system caused a rapid loss in the recovery of astaxanthin [145]. Most of the carotenoids are temperature sensitive due to their structure and chemical bonding and can undergo isomerization and/or degradation at elevated temperature [146,147]. Therefore, it is important to optimize the process parameters to recover these molecules in an acceptable yield and purity. Optimizing this process also aids in the reduction of solvent and overall energy consumptions. In this approach, Zhao et al. [148] attempted to optimize MAE of astaxanthin from *H. pluvialis* in ethanol and ethyl acetate (2:1, *v*/*v*) medium using RSM [148]. The authors found that the extraction parameters such as microwave power, extraction time, solvent volume, the number of extraction, and their interaction effects had significant influence on the recovery of astaxanthin. A microwave power of 141 W, extraction time of 83 s, solvent volume of 9.8 mL and four consecutive extractions was found to be optimum for the recovery of astaxanthin provided yields of about 5.94 μg/mg dw. The response surface plots showed that an increase in microwave power beyond 141 W decreased the recovery of astaxanthin. Higher microwave power can lead to increase in the temperature of the extraction medium, which can disrupt the structure of astaxanthin, leading to its lower recovery [148].

Recently, Esquivel-Hernández et al. [126] showed that MAE is an excellent technique for the recovery of total carotenoids from microalgae/cyanobacteria *Arthrospira platensis* (Spirulina). MAE extraction performed using a mixture of methanol/ethyl acetate/light petroleum (1:1:1 *v*/*v*) at 400 W power, 50 °C temperature and 1 bar pressure, for 15 min yielded 629 μg/g of total carotenoids, which was significantly higher than the yield obtained by SFE (283 μg/g) (ref). In general, it can be concluded that MAE is a promising technology for the rapid extraction of carotenoids. However, in this case microwave power and extraction temperature must be accurately adjusted; as it could lead to the subsequent degradation of selected, valuable carotenoids.

### 3.6. Ultrasound-Assisted Extraction

Ultrasound is composed by sound pressure waves ranging from 20 kHz to 10 MHz with intensities greater than 1 W/cm$^2$ which can be disruptive to matter, depending on the frequency utilized [149]. Most applications in extraction have dealt with low frequency ultrasound, defined between 18 and 200 kHz and recent literature defined the application of high frequency ultrasound standing waves between 400 kHz and 2 MHz to enhance separation [150].

The propagation of ultrasonic waves through liquid medium results in alternating compression and rarefaction cycles. During these rarefaction cycles, small bubbles filled with vapors are created, and these bubbles are able to grow to a certain size and shrink periodically. The formation of small bubbles in a liquid is defined as cavitation. Low frequency ultrasound produces large bubbles and bubble size decreases with frequency [151]. Bubbles formed at lower frequencies (between 18 and 200 kHz) typically attain a critical diameter up to several microns and collapse during the compression cycle, releasing large amounts of heat and shockwaves, creating localized temperatures around 5000 K and pressure jets from strong bubble implosions due to unstable cavitation [149]. Conversely, sound waves produced at frequencies in the megasonic range (>1MHz) produce more stable cavitation, producing tiny bubbles that open and close, creating localized microstreaming effects. Cavitation in general enhances diffusion through cell membranes; furthermore, the high temperature and pressure generated due to low frequency unstable cavitation can also destroy cell structures releasing intracellular components into the medium [152]. Therefore, low frequency ultrasound, also known as ultrasound-assisted extraction (UAE), has been more largely explored for the extraction of components from biological matter (Table 5). UAE is an alternative extraction technique that can be performed using four types of equipment, (i) ultrasonic bath; (ii) ultrasonic probe; (iii) ultrasound plates; and (iv) tubular devices populated with small transducer ceramics. In UAE, a number of parameters such as ultrasonic power, frequency, intensity, shape and size of the ultrasonic reactor, solvent type, temperature, presence of dissolved gases and an external pressure greatly influences the extraction efficiency (recently reviewed by [153]). Extraction temperatures can be controlled during UAE using heat-exchange systems, which are helpful in extracting thermally labile compounds in particular carotenoids.

In recent years, UAE has been employed to extract fucoxanthin, lutein, β-carotene, and astaxanthin from microalgae and seaweeds (Table 5). In a study Macías-Sánchez et al. [117] investigated the efficiency of UAE for the recovery of total carotenoids from the microalga *D. salina*. UAE was performed using lyophilized samples using methanol and *N*,*N*-dimethylformamide (DMF) as extraction solvents. According to their results, UAE performed using DMF recovered up to 27.7 μg/mg dw carotenoids and was significantly higher than the yield obtained by SFE (up to 14.92 μg/mg dw). On the other hand, the UAE had lower selectivity for carotenoids when compared to SFE [117].

In another study, Pasquet et al. [144] compared the extraction efficiency of UAE, MAE and conventional, cold and hot soaking methods for the extraction of fucoxanthin from *Cylindrotheca closterium* and β-carotene from *Dunaliella tertiolecta*. Although UAE performed at a power level of 4.3–12.2 W which appeared to be a rapid extraction technique compared to conventional soaking extractions and did not improve pigment yields at tested conditions. Similar results were observed in a subsequent study performed using the microalga, *Phaeodactylum tricornutum* [98], where the fucoxanthin yield attained through UAE was 15.96 mg/g dw was similar to conventional Soxhlet extraction and maceration; however, UAE reduced the extraction time significantly. Controlling process parameters or adjusting the power level could enhance the recovery of these pigments from these microalgae.

**Table 5.** Summary of ultrasound applications to enhance carotenoids from microalgae.

| Microalgae | Extraction Condition | | | | | | | | Carotenoid Yield/Recovery | Notes | Reference |
|---|---|---|---|---|---|---|---|---|---|---|---|
| | Solvent [a] | Cell Concentration (g Cells Dry Weight/L) | Pretreatment | $f$ [b] (kHz) | P [c] (W) | t [d] (min) | E [e] (kJ/kg) | T [f] (°C) | | | |
| *Chlamydomonas reinhardtii* | Water | 1.5 | Frozen cells with glycerol, thawing and suspension in artificial seawater | 20 | 2200 | 0.17 or 0.5 min at various amplitudes | 0–450 | N/A | Carotenoids—0.3 carotenoids/mg cells | 91%–95% disruption; 80 kJ/kg regardless of cell concentration | [154] |
| *Chlorella pyrenoidosa* | Subcritical $CO_2$ at 5–35 MPa | N/A | no treatment, ethanol soaking or enzymatic pretreatment | 20–24 | 0–19 W/cm$^2$; 0–1000 W 15–45 kg/h, time 0–6 h, | | | 15–33 | Lutein—87–124 mg lutein/100 g Chlorella | Ultrasound-enhanced subcritical $CO_2$ extraction | [138] |
| *Haematococcus pluvialis* | Ethanol and ethyl acetate | 50 | From dried algae | 40 | 200 | 10–20 | 120–240 | 30–50 | Astaxanthin—27 mg/g | US led to higher astaxanthin compared with conventional treatment | [155] |
| *Chlorella vulgaris* | Ethanol (90%) | N/A (31 mL solid/g solvent) | With or without enzymatic pre-treatment, 50 °C | 35 | 56 W/cm$^2$ | 60–240 | N/A | 37 | Lutein—3.16–3.36 mg/g wet weight | Highest ultrasound-based extraction was with enzymatic pre-treatment | [156] |
| *Cylindrotheca closterium (bacillariophyte)* | Acetone | N/A, 30 mL | Freeze dried | N/A | 4.3–12.2 | 3–15 | 25–350 | 8.5 | Fucoxanthin 3.5–4.5 g/mg | - | [144] |
| *Dunaliella salina* | Water | N/A | None | 20, 580, 864 and 1146 | 32.3, 3, 20, 60 | 30 | 5.4 | 15–20 | Carotenoids (yield not reported) | Inactivation efficiency 20 < 580 = 864 < 1146 kHz | [157] |
| *Dunaliella tertiolecta (chlorophyte)* | Water | 30 mL | None | N/A | 4.3–12.2 | 3–15 | 25–350 | 8.5 | β-carotene—5 mg/g | | [144] |
| *Haematococcus pluvialis* | Methanol, ethanol, acetonitrile, acetone | 0.1 g/30 mL | None | 38.5 | 18.4 | 0–90 | 2000 | 30–60 | Astaxanthin—73% recovery | 55%–60% yield increase of astaxanthin after US | [145] |
| *Spirulina platensis* | *n*-heptane, diethyl ether and hexane | 10–60 g/L solvent | Spray dried spirulina mixed with methanol and kept fat various times | 20 | 50–165 W (167 W/cm$^2$) | 8 min with cycling | 220 kWh/m$^3$ | 10–50 | β-carotene—0.8–1.0 mg/g | Extraction had variable increase with acoustic intensity. | [158] |

[a] 1–2 mL solvent/g; [b] Frequency; [c] Power; [d] Time; [e] Specific energy; [f] Temperature.

UAE was found to be a suitable technique also for the of extraction of violaxanthin, neoxanthin, β-carotene and lutein from *C. vulgaris* [97]. Superior extraction efficiency was obtained with acetone when compared to ethanol or *n*-hexane. However, UAE resulted in significantly lower carotenoids when compared to PLE [97]. Similar observations were also deduced by Cha et al. [95], who reported that a lower recovery of β-carotene was obtained in UAE when compared to PLE. On the other hand, lutein was recovered in similar levels both in UAE and PLE (3.83 and 3.78 mg/g, respectively [95]. Optimization of the UAE process parameters is crucial to increase pigment recovery. For example, Dey et al. [158] studied the effects of various UAE parameters such as extraction time, solvents, solid to solvent ratio, temperature, intensity, probe immersion length, duty cycles and pretreatment on the extraction of β-carotene from *Spirulina platensis*. Authors found that *n*-heptane was a better solvent than *n*-hexane and diethyl ether for ultrasound-assisted recovery of β-carotene. The extraction rate increased up to 4 min, and then slowed down, reaching saturation at 8 min. While investigating the effect of electrical acoustic intensity (in the range of 64–210 W/cm$^2$), the extraction yield increased up to 167 W/cm$^2$, then decreased at 185 W/cm$^2$ and increased again at 210 W/cm$^2$. The higher extraction yield observed up to 167 W/cm$^2$ was attributed to the increase in cavitation, whereas the lower yield at 185 W/cm$^2$ was ascribed to the formation of excessive bubbles due to higher acoustic intensity, which hindered the propagation of waves resulting in lower recovery. The authors suspected that a higher yield obtained from 185 to 210 W/cm$^2$ was due to the positive influence of thermal effects generated at higher electrical acoustic intensities [158]. Since the authors have performed experiments utilizing a one-factor-at-a-time approach, the effect of uncontrolled parameters or interaction effects of controlled parameters on the extraction process could be a plausible reason for this ambiguous observation. On a comprehensive level, 1.5 g sample (2 min pre-soaked in methanol) in 50 mL *n*-heptane at 30 °C temperature, 167 W/cm$^2$ electrical acoustic intensity and 61.5% duty cycle for 8 min with probe dip length of 0.5 cm resulted in the optimum recovery of β-carotene (up to 47%) [158]. Interestingly, a pretreatment of samples in methanol for 2 min dramatically enhanced the extraction yield of 12 times [158].

Similarly, Deenu et al. [156] optimized UAE and UAE combined with enzymatic pretreatment for the recovery of lutein from the green microalga *C. vulgaris* using RSM. UAE was performed at a frequency of 35 kHz and intensity of 56.58 W/cm$^2$ using 90% ethanol as an extraction solvent. Ultrasonic treatment at 37.7 °C with solvent to solid ratio of 31 mL/g for 300 min resulted in an optimal lutein yield of 3.16 mg/g (wet basis). Enzymatic pretreatment for 2 h using viscozyme [1.23% (*v*/*w*)] reduced the extraction time from 300 min to 162 min with a slight increase in lutein yield (3.36 mg/g). Enzymatic pretreatment using cell-wall degrading enzymes aid cell disruption techniques [159], facilitated the recovery of analytes in subsequent UAE.

A recent optimization study by Zou et al. [155] revealed that ultrasound irradiation power of 200 W, frequency 40 kHz, a solvent composition 48.0% ethanol in ethyl acetate, liquid-to-solid ratio 20:1 (mL/g), extraction time 16.0 min and a temperature 41.1 °C resulted in the optimum recovery of astaxanthin (27.58 mg/g) from marine the microalgae, *Haematococcus pluvialis*. In this case, a higher extraction temperature and longer irradiation time also resulted in the degradation of astaxanthin with a similar result observed by Ruen-ngam et al. [145]. In another study, the effect of the ultrasonic treatment on microalgal cell disruption was investigated by evaluating the release of intracellular carotenoids from the green microalga, *Chlamydomonas reinhardtii* [154]. Although this research work was not aimed at extracting carotenoids, cells were sonicated under cold conditions for 10 or 30 s at amplitudes of 160, 128, 96, 64, and 10 μm and energy levels of ≥800 J/10 mL, with results showing 0.3 μg carotenoids/mg of cells.

Recently, UAE was successfully employed to the extraction of carotenoids from the microalgae, *Phormidium autumnale* cultivated from agro-industrial wastes [160]. The carotenoids were extracted from dried microalgae in acetone for 20 min at 20 °C using an ultrasonic probe instrument. The amplitude and frequency applied for the extraction was ~61 μm and 20 kHz, respectively and the ultrasound probe depth was 25 mm inside the sample. HPLC-PDA-MS/MS analysis identified the presence of twenty carotenoids. The all-*trans*-β-carotene (70.22 μg/g), all-*trans*-zeaxanthin (26.25 μg/g), all-*trans*-utein (21.92 μg/g), all-*trans*-echinenone (19.87 μg/g), *cis*-echinenone (15.70 μg/g) and 9-*cis*-β-carotene (15.70 μg/g) was found to be the major carotenoids present in *P. autumnale* [160]. Minor carotenoids identified (amount >6 μg/g) included all-*trans*-neoxanthin, 9-*cis*-neoxanthin, 9-*cis*-violaxanthin, 13-*cis*-lutein, 13′-*cis*-lutein, 13-*cis*-zeaxanthin, 9-*cis*-lutein, 9-*cis*-zeaxanthin, all-*trans*-canthaxanthin, *cis*-carotenoid, all-*trans*-myxoxantophyll, all-*trans*-zeinoxanthin and α-carotene. The total carotenoid content extracted was 183.03 μg/g [160].

Yamamoto et al. [157] have compared the efficiency for disruption of *Dunaliella salina*, a high producer of carotenoids, using both low and high frequencies. They have shown that even though low frequencies were most effective, higher frequencies between 580 and 1146 kHz also had high impact on cell disruption. High frequency transducer plates are now used in industrial applications for oil separation by promotion of droplet coalescence and through microstreaming "cleaning" mechanisms in vegetable matter and these systems could be amenable for integration of carotenoid production processes [150,161]. Standing wave high frequency (also known as megasonic) units have also shown the possibility for algal agglomeration and pre-separation during harvest, which may facilitate extraction process pretreatments [162].

UAE can be used as a pretreatment step or complementary technique with other extraction methods. In a study Fan et al. [138] showed that UAE combined with subcritical $CO_2$ extraction is an efficient technique for the extraction of lutein from the microalga, Chlorella pyrenoidosa [138]. The combination of UAE and enzyme-assisted extraction for the extraction of lutein was previously described by Deenu et al. [156]. Recently, Parniakov et al. [63] found that the sonication pretreatment (60–600 s at power of 400 W and frequency of 24 kHz), followed by pH assisted extraction was efficient in the extraction of carotenoids from *Nannochloropsis* sp. However, ultrasonic pretreatment resulted in lower selectivity and required supplementary purification of the final products. Additional, sonication consumed greater power (≈250 kJ/kg, 200 W, 600 s) when compared to PEF (≈100 kJ/kg, 20 kV/cm, 6 ms).

### 3.7. High Pressure Homogenization (HPH)

A physical or mechanical pretreatment prior to extraction can be exploited to disrupt the cell wall of microalgae, promoting the recovery of carotenoids. In fact, to achieve high product yields, efficient cell disruption and extraction steps are required [101]. High pressure homogenization (HPH) is a wet milling process, where particle or cell disruption is achieved by applying high intensity fluid-mechanical stresses, as a consequence of the flow of the process fluid under high pressures (50–400 MPa) through a specifically designed homogenization valve chamber [163]. Schematic of a HPH system is given in Figure 3. Despite each manufacturer offers specific proprietary disruption valve chambers, the designs can be broadly classified as piston valves, where the valve gap is adjusted to set the operating pressure, and orifice valves, with a fixed opening, and operating pressure being adjusted by controlling the flow rate at the pressure intensifier [164]. In comparison with other physical comminution processes, such as ball or colloid milling, and ultrasounds, it offers significant advantages in terms of ease of operation, industrial scalability, reproducibility and high throughput [164–166].

HPH is a promising technique, particularly applied to micro- and macroalgae, as it is effective with respect to aqueous and/or fresh samples up to 25% *w*/*w* solids [167], omitting the energy intensive drying steps, and can be easily scaled up production purposes [168–170]. In HPH, the extraction process is facilitated by the mechanical disruption of the cell wall and cell membranes, enabling the non-selective release of the intracellular compounds. Several factors associated with HPH contribute to cell wall disruption, which includes the development of high pressure gradients, turbulence, cavitation, collision with hard surfaces, viscous and high pressure shear, pressure drop, as well as temperature increases due to the inherent heating associated to the rapid reduction in pressure [163,165,171,172]. In particular, while some of them depend on selected operating conditions, others are affected by suspension properties (density and viscosity) and by the feed concentration. Previously reported data have shown no significant impact on process performance for feed concentrations up to 9% $w/w$ for *Desmodesmus* sp. [173] and up to 25% $w/w$ for *Nannochloropsis* sp. [167].

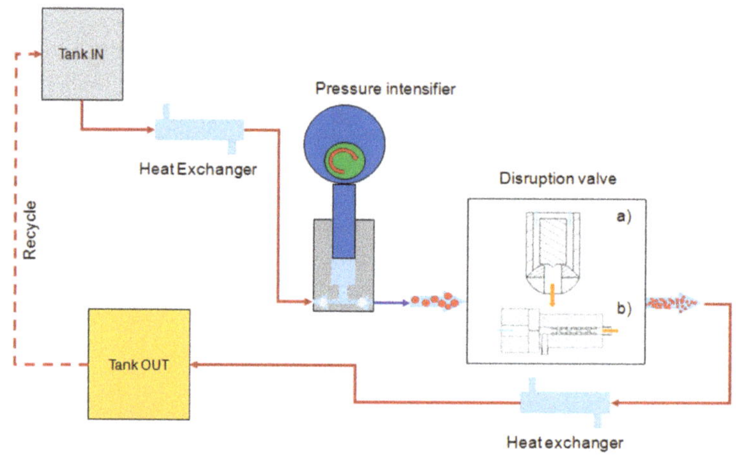


**Figure 3.** Schematics of a high pressure homogenization system, equipped with a (**a**) piston; or (**b**) orifice disruption valve.

Recent studies have suggested that the HPH systems exhibit higher microalgal disruption efficiencies than other methods, such as those based on electrotechnologies (PEF, and HVED) [54], ultrasounds (US) [54,101,174,175], microwaves (MW) [175], ball milling, colloidal milling, freeze drying or thawing [101]. Moreover, when used as a pretreatment to enhance extraction, HPH resulted in significantly higher yields of intracellular compounds, such as lipids [174–178] or carotenoids [101,173]. During HPH, due to the balance establishing between cell wall resistance and tearing forces transmitted to cell walls by the fluid-mechanical stresses generated in the fluid upon its passage through the homogenization valve, a certain variability in the degree of cell disruption efficiency can be observed. Previous comparative studies of different valve geometries on HPH-induced cell disruption showed that piston valves in general ensure a higher homogenization efficiency than orifice ones [165]. Moreover, available reports suggest that pressure is a more important variable than the number of passes [54,101,167,170,174–180] (which translates in higher investment costs but lower operating costs) and that, at least up to 25% $w/w$, biomass concentration does not significantly affects HPH performance, with significant advantages in terms of process intensification.

Although, the HPH technique has been successfully applied to the extraction of lipids, pigments, proteins, sugars etc. from algae and microalgae [174,181–183], to date, only a limited number of research has investigated the efficacy of HPH on the recovery and improvement of carotenoids from algae and microalgae. In an early study, HPH was used (700 bar, 1 pass) for commercial large scale production of astaxanthin from *Haematococcus pluvialis* [170]. In another study, HPH (1500 bar, 1–10 passes, 150–1500 kJ/kg) was found to be effective to extract carotenoids from the microalgae, *Nannochloropsis* sp. [54]. Recently, Taucher et al. [101] found that HPH (1000 bar, 3 passes) pretreatment resulted in the highest total carotenoid extraction yield [4.21 μg/mg (dry weight)] from *H. pluvialis* when compared to other cell disruption methods used such as ball mill [3.56 μg/mg (dry weight)], Ultra Turrax [3.42 μg/mg (dry weight)] and repeated freeze-and-thaw-cycles [0.02 μg/mg (dry weight)]. The HPH pretreatment also assisted in enhanced recovery of lutein and astaxanthin from *C. zofingensis* [101].

Although HPH has shown to be a useful technique for cell disruption and subsequent recovery of analytes, its main disadvantage lies in the non-selective extraction of compounds. Generally,

HPH targets desired products along with high amounts of cell debris/other intracellular matrices, which complicates downstream separation and purification processes. Moreover, the usage of very high pressures in the process can result in increases in temperatures (in the range of 15–20 °C/1000 bar), which can cause negative effects on the recovery of thermally labile compounds, such as carotenoids, if adequate and rapid cooling is not applied [164]. Nevertheless, the use of lower pressures and multi-step treatments, together with process optimization, could be useful in selective extraction of intracellular components [183]. To date and to our knowledge, no results have been reported about the use of HPH in the treatment of seaweeds, whose processing would require preliminary comminution steps to transform them into a slurry which can be pumped through the HPH systems (average size <500 μm).

## 4. Conclusions and Future Perspectives

It is evident that microalgae and seaweeds are sources of commercially and industrially valuable, health-promoting and biologically active carotenoids. This comprehensive review has focused on a number of emerging and innovative alternative extraction technologies, including PEF-, MEF-, and HVED-assisted extractions, PLE, SFE, subcritical fluid extraction, MAE, UAE, and HPH, used for the extraction of carotenoids from microalgae and seaweeds. Growing evidence suggest that these non-conventional techniques offer superior efficiency, selectivity, and a reduction in treatment time or solvent consumption. Nevertheless, available reports clearly indicate that the algal cell structure has a marked influence on extraction efficiency.

In general, MAE and UAE are suitable for rapid extraction, whilst PLE reduces solvent consumption. However, temperatures associated with these techniques can cause degradation of thermolabile carotenoids. On the other hand, PEF was found to be an excellent, non-thermal process for the recovery of thermally labile carotenoids. As previously mentioned, PEF was also found to be inefficient when the alga of interest had a complex cell-wall structure. SFE using $CO_2$ as a solvent was documented to be a superior "green" technique for the selective extraction of carotenoids, and the extraction efficiency was drastically improved by using co-solvents such as ethanol; however, the disadvantage was that it compromised the selectivity of carotenoids. In addition to this, SFE requires careful optimization of various factors to improve extraction efficiency often requiring dried samples, which may increase time and the cost of extraction. HPH was found to be one of the most effective techniques for completely unlocking algal, intracellular compounds, enabling the rapid extraction of carotenoids; however, its lack of selectivity has been documented as a major limitation, suggesting its use only in combination with other technologies, in the form of a biorefinery concept. More recently, subcritical fluid extraction and MEF-assisted extraction has also shown a potential to extract carotenoids.

Finally, it is evident that existing and emerging, non-conventional techniques could be a sustainable alternative in comparison to traditional extraction methods. Future developments should be directed towards overcoming the limitations associated with these techniques, the hyphenating and optimizing processes for improving yield and selectivity, and reducing the multitude of instrumentation, energy costs, and scaling-up of these processes from both commercial and industrial applications. The energy and cost efficiency of these emerging technologies, particularly electrotechnologies, are still under debate, as most of them have been tested only at lab or pilot scale, and are still far from commercial readiness. A detailed scientific investigation of these technologies at the industrial scale is required to understand their commercial viability. These developments could provide an innovative avenue to increase the production and target of selected carotenoids to use as nutraceuticals, pharmaceuticals, cosmeceuticals, food, and feed supplements.

**Acknowledgments:** Mahesha M. Poojary is grateful to the Australian Government for being awarded an Endeavour Research Fellowship under the supervision of Daniel A. Dias, RMIT University, Bundoora, Australia, and Regione Marche and The University of Camerino for the Eureka Ph.D. Fellowship.

**Conflicts of Interest:** The authors declare no conflict of interest.

## References


1. Lorenz, R.T.; Cysewski, G.R. Commercial potential for *Haematococcus* microalgae as a natural source of astaxanthin. *Trends Biotechnol.* **2000**, *18*, 160–167. [CrossRef]
2. Liaaen-Jensen, S. Basic carotenoid chemistry. In *Carotenoids in Health and Disease*; Krinsky, N.I., Mayne, S.T., Sies, H., Eds.; Marcel Dekker: New York, NY, USA, 2004; pp. 1–30.
3. Ritz, T.; Damjanović, A.; Schulten, K. Light-harvesting and photoprotection by carotenoids: Structure-based calculations for photosynthetic antenna systems. In *Photosynthesis: Mechanisms and Effects*; Springer: Dordrecht, The Netherlands, 1998; pp. 487–490.
4. Hammond, B.R.; Renzi, L.M. Carotenoids. *Adv. Nutr. Int. Rev. J.* **2013**, *4*, 474–476. [CrossRef] [PubMed]
5. Krinsky, N.I.; Mayne, S.T.; Sies, H. *Carotenoids in Health and Disease*; Marcel Dekker: New York, NY, USA, 2004.
6. Fiedor, J.; Burda, K. Potential role of carotenoids as antioxidants in human health and disease. *Nutrients* **2014**, *6*, 466–488. [CrossRef] [PubMed]
7. Johnson, E.J. The role of carotenoids in human health. *Nutr. Clin. Care* **2002**, *5*, 56–65. [CrossRef] [PubMed]
8. Vílchez, C.; Forján, E.; Cuaresma, M.; Bédmar, F.; Garbayo, I.; Vega, J.M. Marine carotenoids: Biological functions and commercial applications. *Mar. Drugs* **2011**, *9*, 319–333. [CrossRef] [PubMed]
9. Guedes, A.C.; Amaro, H.M.; Malcata, F.X. Microalgae as Sources of Carotenoids. *Mar. Drugs* **2011**, *9*, 625–644. [CrossRef] [PubMed]
10. MarketsandMarkets. Carotenoids Market by Type (Astaxanthin, Beta-Carotene, Canthaxanthin, Lutein, Lycopene, & Zeaxanthin), Source (Synthetic and Natural), Application (Supplements, Food, Feed, and Cosmetics), by Region—Global Trends and Forecasts to 2021. Available online: http://www.marketsandmarkets.com/Market-Reports/carotenoid-market-158421566.html (accessed on 24 September 2016).
11. Rodriguez-Amaya, D.B. *Food Carotenoids: Chemistry, Biology and Technology*; Wiley-Blackwell: West Sussex, UK, 2016.
12. Safafar, H.; Van Wagenen, J.; Møller, P.; Jacobsen, C. Carotenoids, phenolic compounds and tocopherols contribute to the antioxidative properties of some microalgae species grown on industrial wastewater. *Mar. Drugs* **2015**, *13*, 7339–7356. [CrossRef] [PubMed]
13. Grung, M.; D'Souza, F.M.L.; Borowitzka, M.; Liaaen-Jensen, S. Algal Carotenoids 51. Secondary carotenoids 2. *Haematococcus pluvialis* aplanospores as a source of (3S, 3′S)-astaxanthin esters. *J. Appl. Phycol.* **1992**, *4*, 165–171. [CrossRef]
14. Minhas, A.K.; Hodgson, P.; Barrow, C.J.; Adholeya, A. A review on the assessment of stress conditions for simultaneous production of microalgal lipids and carotenoids. *Front. Microbiol.* **2016**, *7*, 1–19. [CrossRef] [PubMed]
15. Zhang, J.; Sun, Z.; Sun, P.; Chen, T.; Chen, F. Microalgal carotenoids: Beneficial effects and potential in human health. *Food Funct.* **2014**, *5*, 413–425. [CrossRef] [PubMed]
16. Grama, B.S.; Delhaye, A.; Agathos, S.N.; Jeffryes, C. Industrial biotechnology of vitamins, biopigments, and antioxidants. In *Industrial Biotechnology of Vitamins, Biopigments, and Antioxidants*; Vandamme, E.J., Revuelta, J.L., Eds.; Wiley-VCH Verlag GmbH & Co. KGaA: Weinheim, Germany, 2016; pp. 265–286.
17. Boominathan, M.; Mahesh, A. Seaweed carotenoids for cancer therapeutics. In *Handbook of Anticancer Drugs from Marine Origin*; Kim, S.-K., Ed.; Springer: Cham, Switzerland, 2015; pp. 185–203.
18. Billakanti, J.M.; Catchpole, O.J.; Fenton, T.A.; Mitchell, K.A.; MacKenzie, A.D. Enzyme-assisted extraction of fucoxanthin and lipids containing polyunsaturated fatty acids from *Undaria pinnatifida* using dimethyl ether and ethanol. *Process Biochem.* **2013**, *48*, 1999–2008. [CrossRef]
19. Roh, M.K.; Uddin, M.S.; Chun, B.S. Extraction of fucoxanthin and polyphenol from *Undaria pinnatifida* using supercritical carbon dioxide with co-solvent. *Biotechnol. Bioprocess Eng.* **2008**, *13*, 724–729. [CrossRef]
20. Quitain, A.T.; Kai, T.; Sasaki, M.; Goto, M. Supercritical carbon dioxide extraction of fucoxanthin from *Undaria pinnatifida*. *J. Agric. Food Chem.* **2013**, *61*, 5792–5797. [CrossRef] [PubMed]
21. Xiao, X.; Si, X.; Yuan, Z.; Xu, X.; Li, G. Isolation of fucoxanthin from edible brown algae by microwave-assisted extraction coupled with high-speed countercurrent chromatography. *J. Sep. Sci.* **2012**, *35*, 2313–2317. [CrossRef] [PubMed]
22. Piovan, A.; Seraglia, R.; Bresin, B.; Caniato, R.; Filippini, R. Fucoxanthin from *Undaria pinnatifida*: Photostability and coextractive effects. *Molecules* **2013**, *18*, 6298–6310. [CrossRef] [PubMed]

23. Kanda, H.; Kamo, Y.; Machmudah, S.; Wahyudiono, E.Y.; Goto, M. Extraction of fucoxanthin from raw macroalgae excluding drying and cell wall disruption by liquefied dimethyl ether. *Mar. Drugs* **2014**, *12*, 2383–2396. [CrossRef] [PubMed]
24. Yan, X.; Chuda, Y.; Suzuki, M.; Nagata, T. Fucoxanthin as the major antioxidant in *Hijikia fusiformis*, a common edible seaweed. *Biosci. Biotechnol. Biochem.* **1999**, *63*, 605–607. [CrossRef] [PubMed]
25. Wang, W.-J.; Wang, G.-C.; Zhang, M.; Tseng, C.K. Isolation of fucoxanthin from the rhizoid of *Laminaria japonica* Aresch. *J. Integr. Plant Biol.* **2005**, *47*, 1009–1015. [CrossRef]
26. Saravana, P.S.; Getachew, A.T.; Cho, Y.-J.; Choi, J.H.; Park, Y.B.; Woo, H.C.; Chun, B.S. Influence of co-solvents on fucoxanthin and phlorotannin recovery from brown seaweed using supercritical $CO_2$. *J. Supercrit. Fluids* **2016**, in press. [CrossRef]
27. Heo, S.-J.; Jeon, Y.-J. Protective effect of fucoxanthin isolated from *Sargassum siliquastrum* on UV-B induced cell damage. *J. Photochem. Photobiol. B* **2009**, *95*, 101–107. [CrossRef] [PubMed]
28. Noviendri, D.; Jaswir, I.; Salleh, H.M.; Taher, M.; Miyashita, K.; Ramli, N. Fucoxanthin extraction and fatty acid analysis of *Sargassum binderi* and *S. duplicatum*. *J. Med. Plants Res.* **2011**, *5*, 2405–2412.
29. Conde, E.; Moure, A.; Domínguez, H. Supercritical $CO_2$ extraction of fatty acids, phenolics and fucoxanthin from freeze-dried *Sargassum muticum*. *J. Appl. Phycol.* **2014**, *2*, 957–964. [CrossRef]
30. Nomura, M.; Kamogawa, H.; Susanto, E.; Kawagoe, C.; Yasui, H.; Saga, N.; Hosokawa, M.; Miyashita, K. Seasonal variations of total lipids, fatty acid composition, and fucoxanthin contents of *Sargassum horneri* (Turner) and *Cystoseira hakodatensis* (Yendo) from the northern seashore of Japan. *J. Appl. Phycol.* **2013**, *25*, 1159–1169. [CrossRef]
31. Afolayan, A.F.; Bolton, J.J.; Lategan, C.A.; Smith, P.J.; Beukes, D.R. Fucoxanthin, tetraprenylated toluquinone and toluhydroquinone metabolites from *Sargassum heterophyllum* inhibit the in vitro growth of the malaria parasite *Plasmodium falciparum*. *Z. Naturforschung C J. Biosci.* **2008**, *63*, 848–852. [CrossRef]
32. Heo, S.-J.; Ko, S.-C.; Kang, S.-M.; Kang, H.-S.; Kim, J.-P.; Kim, S.-H.; Lee, K.-W.; Cho, M.-G.; Jeon, Y.-J. Cytoprotective effect of fucoxanthin isolated from brown algae *Sargassum siliquastrum* against $H_2O_2$-induced cell damage. *Eur. Food Res. Technol.* **2008**, *228*, 145–151. [CrossRef]
33. Zaragozá, M.C.; López, D.; P Sáiz, M.; Poquet, M.; Pérez, J.; Puig-Parellada, P.; Màrmol, F.; Simonetti, P.; Gardana, C.; Lerat, Y.; et al. Toxicity and antioxidant activity in vitro and in vivo of two *Fucus vesiculosus* extracts. *J. Agric. Food Chem.* **2008**, *56*, 7773–7780. [CrossRef] [PubMed]
34. Peng, J.; Yuan, J.-P.; Wu, C.-F.; Wang, J.-H. Fucoxanthin, a marine carotenoid present in brown seaweeds and diatoms: Metabolism and bioactivities relevant to human health. *Mar. Drugs* **2011**, *9*, 1806–1828. [CrossRef] [PubMed]
35. Kim, S.-K.; Pangestuti, R. Biological activities and potential health benefits of fucoxanthin derived from marine brown algae. *Adv. Food Nutr. Res.* **2011**, *64*, 111–128. [PubMed]
36. Zhang, H.; Tang, Y.; Zhang, Y.; Zhang, S.; Qu, J.; Wang, X.; Kong, R.; Han, C.; Liu, Z.; Zhang, H.; et al. Fucoxanthin: A promising medicinal and nutritional ingredient. *Evid. Based Complement. Altern. Med.* **2015**, *2015*, 1–10. [CrossRef] [PubMed]
37. Gammone, M.A.; D'Orazio, N. Anti-obesity activity of the marine carotenoid fucoxanthin. *Mar. Drugs* **2015**, *13*, 2196–2214. [CrossRef] [PubMed]
38. Pangestuti, R.; Kim, S.-K. Carotenoids, bioactive metabolites derived from seaweeds. In *Springer Handbook of Marine Biotechnology*; Kim, S.-K., Ed.; Springer: Berlin/Heidelberg, Germany, 2015; pp. 816–821.
39. Siegel, B.Z.; Siegel, S.M. The chemical composition of algal cell walls. *CRC Crit. Rev. Microbiol.* **1973**, *3*, 1–26. [CrossRef] [PubMed]
40. Fernández-Sevilla, J.M.; Acién Fernández, F.G.; Molina Grima, E. Biotechnological production of lutein and its applications. *Appl. Microbiol. Biotechnol.* **2010**, *86*, 27–40. [CrossRef] [PubMed]
41. Sarkar, C.R.; Das, L.; Bhagawati, B.; Goswami, B.C. A comparative study of carotenoid extraction from algae in different solvent systems. *Asian J. Plant Sci. Res.* **2012**, *2*, 546–549.
42. Mendes-Pinto, M.M.; Raposo, M.F.J.; Bowen, J.; Young, A.J.; Morais, R. Evaluation of different cell disruption processes on encysted cells of *Haematococcus pluvialis*: Effects on astaxanthin recovery and implications for bio-availability. *J. Appl. Phycol.* **2001**, *13*, 19–24. [CrossRef]
43. Mojaat, M.; Foucault, A.; Pruvost, J.; Legrand, J. Optimal selection of organic solvents for biocompatible extraction of beta-carotene from *Dunaliella salina*. *J. Biotechnol.* **2008**, *133*, 433–441. [CrossRef] [PubMed]

44. Hejazi, M.A.; de Lamarliere, C.; Rocha, J.M.S.; Vermuë, M.; Tramper, J.; Wijffels, R.H. Selective extraction of carotenoids from the microalga *Dunaliella salina* with retention of viability. *Biotechnol. Bioeng.* **2002**, *79*, 29–36. [CrossRef] [PubMed]
45. Castro-Puyana, M.; Herrero, M.; Urreta, I.; Mendiola, J.A.; Cifuentes, A.; Ibáñez, E.; Suárez-Alvarez, S. Optimization of clean extraction methods to isolate carotenoids from the microalga *Neochloris oleoabundans* and subsequent chemical characterization using liquid chromatography tandem mass spectrometry. *Anal. Bioanal. Chem.* **2013**, *405*, 4607–4616. [CrossRef] [PubMed]
46. Castro-Puyana, M.; Herrero, M.; Mendiola, J.A.; Suárez-Alvarez, S.; Cifuentes, A.; Ibáñez, E. Extraction of new bioactives from neochloris oleoabundans using pressurized technologies and food grade solvents. In Proceedings of the III Iberoamerican Conference on Supercritical Fluids Cartagena de Indias (Combodia), Cartagena, Colombia, 1–5 April 2013.
47. Golberg, A.; Sack, M.; Teissie, J.; Pataro, G.; Pliquett, U.; Saulis, G.; Stefan, T.; Miklavcic, D.; Vorobiev, E.; Frey, W. Energy-efficient biomass processing with pulsed electric fields for bioeconomy and sustainable development. *Biotechnol. Biofuels* **2016**, *9*, 1–22. [CrossRef] [PubMed]
48. Donsi, F.; Ferrari, G.; Pataro, G. Applications of pulsed electric field treatments for the enhancement of mass transfer from vegetable tissue. *Food Eng. Rev.* **2010**, *2*, 109–130. [CrossRef]
49. Kulshrestha, S.; Sastry, S. Frequency and voltage effects on enhanced diffusion during moderate electric field (MEF) treatment. *Innov. Food Sci. Emerg. Technol.* **2003**, *4*, 189–194. [CrossRef]
50. Vorobiev, E.; Lebovka, N. Enhanced extraction from solid foods and biosuspensions by pulsed electrical energy. *Food Eng. Rev.* **2010**, *2*, 95–108. [CrossRef]
51. Poojary, M.M.; Roohinejad, S.; Barba, F.J.; Koubaa, M.; Puértolas, E.; Jambrak, A.R.; Greiner, R.; Oey, I. Application of pulsed electric field treatment for food waste recovery operations. In *Handbook of Electroporation*; Miklavcic, D., Ed.; Springer: Cham, Switzerland, 2017; in press.
52. Raso, J.; Frey, W.; Ferrari, G.; Pataro, G.; Knorr, D.; Teissie, J.; Miklavčič, D. Recommendations guidelines on the key information to be reported in studies of application of PEF technology in food and biotechnological processes. *Innov. Food Sci. Emerg. Technol.* **2016**, *37*, 312–321. [CrossRef]
53. Töpfl, S. Pulsed Electric Fields (PEF) for Permeabilization of Cell Membranes in Food- and Bioprocessing. Applications, Process and Equipment Design and Cost Analysis. Ph.D. Thesis, Technische Universität Berlin, Berlin, Germany, 2006.
54. Grimi, N.; Dubois, A.; Marchal, L.; Jubeau, S.; Lebovka, N.I.; Vorobiev, E. Selective extraction from microalgae *Nannochloropsis* sp. using different methods of cell disruption. *Bioresour. Technol.* **2014**, *153*, 254–259. [CrossRef] [PubMed]
55. Scholz, M.J.; Weiss, T.L.; Jinkerson, R.E.; Jing, J.; Roth, R.; Goodenough, U.; Posewitz, M.C.; Gerken, H.G. Ultrastructure and composition of the *Nannochloropsis gaditana* cell wall. *Eukaryot. Cell* **2014**, *13*, 1450–1464. [CrossRef] [PubMed]
56. Luengo, E.; Martínez, J.M.; Bordetas, A.; Álvarez, I.; Raso, J. Influence of the treatment medium temperature on lutein extraction assisted by pulsed electric fields from *Chlorella vulgaris*. *Innov. Food Sci. Emerg. Technol.* **2015**, *29*, 15–22. [CrossRef]
57. Kotnik, T.; Kramar, P.; Pucihar, G.; Miklavčič, D.; Tarek, M. Cell membrane electroporation—Part 1: The phenomenon. *IEEE Electr. Insul. Mag.* **2012**, *28*, 14–23. [CrossRef]
58. Esser, A.T.; Smith, K.C.; Gowrishankar, T.R.; Vasilkoski, Z.; Weaver, J.C. Mechanisms for the intracellular manipulation of organelles by conventional electroporation. *Biophys. J.* **2010**, *98*, 2506–2514. [CrossRef] [PubMed]
59. Schoenbach, K.H.; Beebe, S.J.; Buescher, E.S. Intracellular effect of ultrashort electrical pulses. *Bioelectromagnetics* **2001**, *22*, 440–448. [CrossRef] [PubMed]
60. Luengo, E.; Condón-Abanto, S.; Álvarez, I.; Raso, J. Effect of pulsed electric field treatments on permeabilization and extraction of pigments from *Chlorella vulgaris*. *J. Membr. Biol.* **2014**, *247*, 1269–1277. [CrossRef] [PubMed]
61. Luengo, E.; Martínez, J.M.; Coustets, M.; Álvarez, I.; Teissié, J.; Rols, M.-P.; Raso, J. A comparative study on the effects of millisecond- and microsecond-pulsed electric field treatments on the permeabilization and extraction of pigments from *Chlorella vulgaris*. *J. Membr. Biol.* **2015**, *248*, 883–891. [CrossRef] [PubMed]

62. Parniakov, O.; Barba, F.J.; Grimi, N.; Marchal, L.; Jubeau, S.; Lebovka, N.; Vorobiev, E. Pulsed electric field assisted extraction of nutritionally valuable compounds from microalgae *Nannochloropsis* spp. using the binary mixture of organic solvents and water. *Innov. Food Sci. Emerg. Technol.* **2015**, *27*, 79–85. [CrossRef]
63. Parniakov, O.; Barba, F.J.; Grimi, N.; Marchal, L.; Jubeau, S.; Lebovka, N.; Vorobiev, E. Pulsed electric field and pH assisted selective extraction of intracellular components from microalgae *Nannochloropsis*. *Algal Res.* **2015**, *8*, 128–134. [CrossRef]
64. Jaeschke, D.P.; Menegol, T.; Rech, R.; Mercali, G.D.; Marczak, L.D.F. Carotenoid and lipid extraction from *Heterochlorella luteoviridis* using moderate electric field and ethanol. *Process Biochem.* **2016**, *51*, 1636–1643. [CrossRef]
65. Postma, P.R.; Pataro, G.; Capitoli, M.; Barbosa, M.J.; Wijffels, R.H.; Eppink, M.H.M.; Olivieri, G.; Ferrari, G. Selective extraction of intracellular components from the microalga *Chlorella vulgaris* by combined pulsed electric field-temperature treatment. *Bioresour. Technol.* **2016**, *203*, 80–88. [CrossRef] [PubMed]
66. Lebovka, N.I.; Praporscic, I.; Vorobiev, E. Effect of moderate thermal and pulsed electric field treatments on textural properties of carrots, potatoes and apples. *Innov. Food Sci. Emerg. Technol.* **2004**, *5*, 9–16. [CrossRef]
67. Lebovka, N.I.; Praporscic, I.; Ghnimi, S.; Vorobiev, E. Temperature enhanced electroporation under the pulsed electric field treatment of food tissue. *J. Food Eng.* **2005**, *69*, 177–184. [CrossRef]
68. Kulshrestha, S.; Sarang, S.; Loghavi, L.; Sastry, S. Moderate electrothermal treatments of cellular tissues. In *Electrotechnologies for Extraction from Food Plants and Biomaterials*; Vorobiev, E., Lebovka, N., Eds.; Springer: New York, NY, USA, 2009; pp. 83–94.
69. Kusnadi, C.; Sastry, S.K. Effect of moderate electric fields on salt diffusion into vegetable tissue. *J. Food Eng.* **2012**, *110*, 329–336. [CrossRef]
70. Nezammahalleh, H.; Ghanati, F.; Adams, T.A.; Nosrati, M.; Shojaosadati, S.A. Effect of moderate static electric field on the growth and metabolism of *Chlorella vulgaris*. *Bioresour. Technol.* **2016**, *218*, 700–711. [CrossRef] [PubMed]
71. Richter, B.E.; Jones, B.A.; Ezzell, J.L.; Porter, N.L. Accelerated solvent extraction: A technique for sample preparation. *Anal. Chem.* **1996**, *68*, 1033–1039. [CrossRef]
72. Plaza, M.; Turner, C. Pressurized hot water extraction of bioactives. *TrAC Trends Anal. Chem.* **2015**, *71*, 39–54. [CrossRef]
73. Osorio-Tobón, J.F.; Meireles, M.A.A.; Osorio-Tobón, J.F.; Meireles, M.A.A. Recent applications of pressurized fluid extraction: Curcuminoids extraction with pressurized liquids. *Food Public Health* **2013**, *3*, 289–303. [CrossRef]
74. Santos, D.T.; Veggi, P.C.; Meireles, M.A.A. Optimization and economic evaluation of pressurized liquid extraction of phenolic compounds from jabuticaba skins. *J. Food Eng.* **2012**, *108*, 444–452. [CrossRef]
75. Gao, Q.; Haglund, P.; Pommer, L.; Jansson, S. Evaluation of solvent for pressurized liquid extraction of PCDD, PCDF, PCN, PCBz, PCPh and PAH in torrefied woody biomass. *Fuel* **2015**, *154*, 52–58. [CrossRef]
76. Li, P.; Li, S.P.; Lao, S.C.; Fu, C.M.; Kan, K.K.W.; Wang, Y.T. Optimization of pressurized liquid extraction for Z-ligustilide, Z-butylidenephthalide and ferulic acid in Angelica sinensis. *J. Pharm. Biomed. Anal.* **2006**, *40*, 1073–1079. [CrossRef] [PubMed]
77. Iqbal, J.; Theegala, C. Optimizing a continuous flow lipid extraction system (CFLES) used for extracting microalgal lipids. *GCB Bioenergy* **2013**, *5*, 327–337. [CrossRef]
78. Hu, J.; Guo, Z.; Glasius, M.; Kristensen, K.; Xiao, L.; Xu, X. Pressurized liquid extraction of ginger (*Zingiber officinale* Roscoe) with bioethanol: An efficient and sustainable approach. *J. Chromatogr. A* **2011**, *1218*, 5765–5773. [CrossRef] [PubMed]
79. Aliakbarian, B.; Casazza, A.A.; Perego, P. Valorization of olive oil solid waste using high pressure–high temperature reactor. *Food Chem.* **2011**, *128*, 704–710. [CrossRef]
80. Ben Hamissa, A.M.; Seffen, M.; Aliakbarian, B.; Casazza, A.A.; Perego, P.; Converti, A. Phenolics extraction from *Agave americana* (L.) leaves using high-temperature, high-pressure reactor. *Food Bioprod. Process.* **2012**, *90*, 17–21. [CrossRef]
81. Casazza, A.A.; Aliakbarian, B.; Sannita, E.; Perego, P. High-pressure high-temperature extraction of phenolic compounds from grape skins. *Int. J. Food Sci. Technol.* **2012**, *47*, 399–405. [CrossRef]
82. Latoui, M.; Aliakbarian, B.; Casazza, A.A.; Seffen, M.; Converti, A.; Perego, P. Extraction of phenolic compounds from *Vitex agnus-castus* L. *Food Bioprod. Process.* **2012**, *90*, 748–754. [CrossRef]

83. Lopresto, C.G.; Petrillo, F.; Casazza, A.A.; Aliakbarian, B.; Perego, P.; Calabrò, V. A non-conventional method to extract D-limonene from waste lemon peels and comparison with traditional Soxhlet extraction. *Sep. Purif. Technol.* **2014**, *137*, 13–20. [CrossRef]
84. Moreau, R.A.; Powell, M.J.; Singh, V. Pressurized liquid extraction of polar and nonpolar lipids in corn and oats with hexane, methylene chloride, isopropanol, and ethanol. *J. Am. Oil Chem. Soc.* **2003**, *80*, 1063–1067. [CrossRef]
85. Jeannotte, R.; Hamel, C.; Jabaji, S.; Whalen, J.K. Comparison of solvent mixtures for pressurized solvent extraction of soil fatty acid biomarkers. *Talanta* **2008**, *77*, 195–199. [CrossRef] [PubMed]
86. Bermejo, D.V.; Luna, P.; Manic, M.S.; Najdanovic-Visak, V.; Reglero, G.; Fornari, T. Extraction of caffeine from natural matter using a bio-renewable agrochemical solvent. *Food Bioprod. Process.* **2013**, *91*, 303–309. [CrossRef]
87. Wu, H.; Chen, M.; Fan, Y.; Elsebaei, F.; Zhu, Y. Determination of rutin and quercetin in Chinese herbal medicine by ionic liquid-based pressurized liquid extraction–liquid chromatography–chemiluminescence detection. *Talanta* **2012**, *88*, 222–229. [CrossRef] [PubMed]
88. Choi, M.P.K.; Chan, K.K.C.; Leung, H.W.; Huie, C.W. Pressurized liquid extraction of active ingredients (ginsenosides) from medicinal plants using non-ionic surfactant solutions. *J. Chromatogr. A* **2003**, *983*, 153–162. [CrossRef]
89. Mustafa, A.; Turner, C. Pressurized liquid extraction as a green approach in food and herbal plants extraction: A review. *Anal. Chim. Acta* **2011**, *703*, 8–18. [CrossRef] [PubMed]
90. Ong, E.S.; Cheong, J.S. H.; Goh, D. Pressurized hot water extraction of bioactive or marker compounds in botanicals and medicinal plant materials. *J. Chromatogr. A* **2006**, *1112*, 92–102. [CrossRef] [PubMed]
91. Denery, J.R.; Dragull, K.; Tang, C.; Li, Q.X. Pressurized fluid extraction of carotenoids from *Haematococcus pluvialis* and *Dunaliella salina* and kavalactones from *Piper methysticum*. *Anal. Chim. Acta* **2004**, *501*, 175–181. [CrossRef]
92. Herrero, M.; Jaime, L.; Martin-Alvarez, P.J.; Cifuentes, A.; Ibanez, E. Optimization of the extraction of antioxidants from *Dunaliella salina* microalga by pressurized liquids. *J. Agric. Food Chem.* **2006**, *54*, 5597–5603. [CrossRef] [PubMed]
93. Plaza, M.; Santoyo, S.; Jaime, L.; García-Blairsy Reina, G.; Herrero, M.; Señoráns, F.J.; Ibáñez, E. Screening for bioactive compounds from algae. *J. Pharm. Biomed. Anal.* **2010**, *51*, 450–455. [CrossRef] [PubMed]
94. Cha, K.H.; Kang, S.W.; Kim, C.Y.; Um, B.H.; Na, Y.R.; Pan, C.H. Effect of pressurized liquids on extraction of antioxidants from *Chlorella vulgaris*. *J. Agric. Food Chem.* **2010**, *58*, 4756–4761. [CrossRef] [PubMed]
95. Cha, K.H.; Lee, H.J.; Koo, S.Y.; Song, D.G.; Lee, D.U.; Pan, C.H. Optimization of pressurized liquid extraction of carotenoids and chlorophylls from chlorella vulgaris. *J. Agric. Food Chem.* **2010**, *58*, 793–797. [CrossRef] [PubMed]
96. Jaime, L.; Rodríguez-Meizoso, I.; Cifuentes, A.; Santoyo, S.; Suarez, S.; Ibáñez, E.; Señorans, F.J. Pressurized liquids as an alternative process to antioxidant carotenoids' extraction from *Haematococcus pluvialis* microalgae. *LWT—Food Sci. Technol.* **2010**, *43*, 105–112. [CrossRef]
97. Plaza, M.; Santoyo, S.; Jaime, L.; Avalo, B.; Cifuentes, A.; Reglero, G.; García-Blairsy Reina, G.; Señorans, F.J.; Ibáñez, E. Comprehensive characterization of the functional activities of pressurized liquid and ultrasound-assisted extracts from *Chlorella vulgaris*. *LWT—Food Sci. Technol.* **2012**, *46*, 245–253. [CrossRef]
98. Kim, S.M.; Jung, Y.J.; Kwon, O.N.; Cha, K.H.; Um, B.H.; Chung, D.; Pan, C.H. A potential commercial source of fucoxanthin extracted from the microalga *Phaeodactylum tricornutum*. *Appl. Biochem. Biotechnol.* **2012**, *166*, 1843–1855. [CrossRef] [PubMed]
99. Shang, Y.F.; Kim, S.M.; Lee, W.J.; Um, B.H. Pressurized liquid method for fucoxanthin extraction from *Eisenia bicyclis* (Kjellman) Setchell. *J. Biosci. Bioeng.* **2011**, *111*, 237–241. [CrossRef] [PubMed]
100. Koo, S.Y.; Cha, K.H.; Song, D.G.; Chung, D.; Pan, C.H. Optimization of pressurized liquid extraction of zeaxanthin from *Chlorella ellipsoidea*. *J. Appl. Phycol.* **2012**, *24*, 725–730. [CrossRef]
101. Taucher, J.; Baer, S.; Schwerna, P.; Hofmann, D.; Hümmer, M.; Buchholz, R.; Becker, A. Cell disruption and pressurized liquid extraction of carotenoids from microalgae. *Thermodyn. Catal.* **2016**, *7*, 1–7. [CrossRef]
102. Jeon, Y.; Wijesinghe, W.A.J.P.; Kim, S. Enzyme-assisted extraction and recovery of bioactive components from seaweeds. In *Handbook of Marinemacroalgae: Biotechnology and Applied Phycology*, 1st ed.; Kim, S.-K., Ed.; John Wiley & Sons, Ltd.: New York, NY, USA, 2012; pp. 221–228.

103. Lang, Q.; Wai, C.M. Supercritical fluid extraction in herbal and natural product studies—A practical review. *Talanta* **2001**, *53*, 771–782. [CrossRef]
104. Zougagh, M.; Valcárcel, M.; Rios, A. Supercritical fluid extraction: A critical review of its analytical usefulness. *TrAC Trends Anal. Chem.* **2004**, *23*, 399–405. [CrossRef]
105. Talisic, G.C.; Yumang, A.N.; Salta, M.T.S. Supercritical Fluid Extraction of β-carotene from *D. Salina* algae using $C_2H_6$ and $C_2H_2$. In Proceedings of the 2012 International Conference on Geological and Environmental Sciences, Jeju Island, Korea, 29–30 June 2012; pp. 30–34.
106. Krichnavaruk, S.; Shotipruk, A.; Goto, M.; Pavasant, P. Supercritical carbon dioxide extraction of astaxanthin from *Haematococcus pluvialis* with vegetable oils as co-solvent. *Bioresour. Technol.* **2008**, *99*, 5556–5560. [CrossRef] [PubMed]
107. Ruen-ngam, D.; Shotipruk, A.; Pavasant, P.; Machmudah, S.; Goto, M. Selective extraction of lutein from alcohol treated *Chlorella vulgaris* by supercritical $CO_2$. *Chem. Eng. Technol.* **2012**, *35*, 255–260. [CrossRef]
108. Macías-Sánchez, M.D.; Serrano, C.M.; Rodríguez, M.R.; Martínez de la Ossa, E. Kinetics of the supercritical fluid extraction of carotenoids from microalgae with $CO_2$ and ethanol as cosolvent. *Chem. Eng. J.* **2009**, *150*, 104–113. [CrossRef]
109. Bustamante, A.; Roberts, P.; Aravena, R.; Del Valle, J.M. Supercritical extraction of astaxanthin from *H. pluvialis* using ethanol-modified $CO_2$. Experiments and modeling. In Proceedings of the 11th International Conference of Eng Food, Athens, Greece, 22–26 May 2011.
110. Aravena, R.I.; del Valle, J.M. Effect of microalgae preconditioning on supercritical $CO_2$ extraction of astaxanthin from *Haematococcus pluvialis*. In Proceedings of the 10th International Symposium of Supercritical Fluids, San Francisco, CA, USA, 13–16 May 2012.
111. Yen, H.W.; Chiang, W.C.; Sun, C.H. Supercritical fluid extraction of lutein from *Scenedesmus cultured* in an autotrophical photobioreactor. *J. Taiwan Inst. Chem. Eng.* **2012**, *43*, 53–57. [CrossRef]
112. Macías-Sánchez, M.D.; Fernandez-Sevilla, J.M.; Fernández, F.G.A.; García, M.C.C.; Grima, E.M. Supercritical fluid extraction of carotenoids from *Scenedesmus almeriensis*. *Food Chem.* **2010**, *123*, 928–935. [CrossRef]
113. Kitada, K.; Machmudah, S.; Sasaki, M.; Goto, M.; Nakashima, Y.; Kumamoto, S.; Hasegawa, T. Supercritical $CO_2$ extraction of pigment components with pharmaceutical importance from *Chlorella vulgaris*. *J. Chem. Technol. Biotechnol.* **2009**, *84*, 657–661. [CrossRef]
114. Nobre, B.; Marcelo, F.; Passos, R.; Beirão, L.; Palavra, A.; Gouveia, L.; Mendes, R. Supercritical carbon dioxide extraction of astaxanthin and other carotenoids from the microalga *Haematococcus pluvialis*. *Eur. Food Res. Technol.* **2006**, *223*, 787–790. [CrossRef]
115. Macias-Sánchez, M.D.; Mantell, C.; Rodriguez, M.; de la Ossa, E.J.M.; Lubián, L.M.; Montero, O. Supercritical fluid extraction of carotenoids and chlorophyll a from *Nannochloropsis gaditana*. *J. Food Eng.* **2005**, *66*, 245–251. [CrossRef]
116. Macías-Sánchez, M.D.; Mantell, C.; Rodríguez, M.; de la Ossa, E.J.M.; Lubián, L.M.; Montero, O. Supercritical fluid extraction of carotenoids and chlorophyll a from *Synechococcus* sp. *J. Supercrit. Fluids* **2007**, *39*, 323–329. [CrossRef]
117. Macías-Sánchez, M.D.; Mantell, C.; Rodríguez, M.; de la Ossa, E.J.M.; Lubián, L.M.; Montero, O. Comparison of supercritical fluid and ultrasound-assisted extraction of carotenoids and chlorophyll a from *Dunaliella salina*. *Talanta* **2009**, *77*, 948–952. [CrossRef] [PubMed]
118. Cardoso, L.C.; Serrano, C.M.; Rodríguez, M.R.; de la Ossa, E.J.M.; Lubián, L.M. Extraction of carotenoids and fatty acids from microalgae using supercritical technology. *Am. J. Anal. Chem.* **2012**, *3*, 877–883. [CrossRef]
119. Mendes, R.L.; Nobre, B.P.; Cardoso, M.T.; Pereira, A.P.; Palavra, A.F. Supercritical carbon dioxide extraction of compounds with pharmaceutical importance from microalgae. *Inorg. Chim. Acta* **2003**, *356*, 328–334. [CrossRef]
120. Machmudah, S.; Shotipruk, A.; Goto, M.; Sasaki, M.; Hirose, T. Extraction of astaxanthin from *Haematococcus p luvialis* using supercritical $CO_2$ and ethanol as entrainer. *Ind. Eng. Chem. Res.* **2006**, *45*, 3652–3657. [CrossRef]
121. Wang, L.; Yang, B.; Yan, B.; Yao, X. Supercritical fluid extraction of astaxanthin from *Haematococcus pluvialis* and its antioxidant potential in sunflower oil. *Innov. Food Sci. Emerg. Technol.* **2012**, *13*, 120–127. [CrossRef]
122. Valderrama, J.O.; Perrut, M.; Majewski, W. Extraction of astaxantine and phycocyanine from microalgae with supercritical carbon dioxide. *J. Chem. Eng. Data* **2003**, *48*, 827–830. [CrossRef]

123. Macías-Sánchez, M.D.; Mantell Serrano, C.; Rodríguez Rodríguez, M.; de la Ossa, E.J.M.; Lubián, L.M.; Montero, O. Extraction of carotenoids and chlorophyll from microalgae with supercritical carbon dioxide and ethanol as cosolvent. *J. Sep. Sci.* **2008**, *31*, 1352–1362. [CrossRef] [PubMed]
124. Liau, B.C.; Shen, C.T.; Liang, F.P.; Hong, S.E.; Hsu, S.L.; Jong, T.T.; Chang, C.M.J. Supercritical fluids extraction and anti-solvent purification of carotenoids from microalgae and associated bioactivity. *J. Supercrit. Fluids* **2010**, *55*, 169–175. [CrossRef]
125. Fujii, K. Process integration of supercritical carbon dioxide extraction and acid treatment for astaxanthin extraction from a vegetative microalga. *Food Bioprod. Process.* **2012**, *90*, 762–766. [CrossRef]
126. Esquivel-Hernández, D.; López, V.; Rodríguez-Rodríguez, J.; Alemán-Nava, G.; Cuéllar-Bermúdez, S.; Rostro-Alanis, M.; Parra-Saldívar, R. Supercritical carbon dioxide and microwave-assisted extraction of functional lipophilic compounds from *Arthrospira platensis*. *Int. J. Mol. Sci.* **2016**, *17*, 658. [CrossRef] [PubMed]
127. Ota, M.; Watanabe, H.; Kato, Y.; Watanabe, M.; Sato, Y.; Smith, R.L.; Inomata, H. Carotenoid production from *Chlorococcum littorale* in photoautotrophic cultures with downstream supercritical fluid processing. *J. Sep. Sci.* **2009**, *32*, 2327–2335. [CrossRef] [PubMed]
128. Liau, B.-C.; Hong, S.-E.; Chang, L.-P.; Shen, C.-T.; Li, Y.-C.; Wu, Y.-P.; Jong, T.-T.; Shieh, C.-J.; Hsu, S.-L.; Chang, C.-M.J. Separation of sight-protecting zeaxanthin from *Nannochloropsis oculata* by using supercritical fluids extraction coupled with elution chromatography. *Sep. Purif. Technol.* **2011**, *78*, 1–8. [CrossRef]
129. Goto, M.; Kanda, H.; Wahyudiono; Machmudah, S. Extraction of carotenoids and lipids from algae by supercritical $CO_2$ and subcritical dimethyl ether. *J. Supercrit. Fluids* **2015**, *96*, 245–251. [CrossRef]
130. Fahmy, T.M.; Paulaitis, M.E.; Johnson, D.M.; McNally, M.E.P. Modifier effects in the supercritical fluid extraction of solutes from clay, soil, and plant materials. *Anal. Chem.* **1993**, *65*, 1462–1469. [CrossRef]
131. Ritter, D.C.; Campbell, A.G. Supercritical carbon dioxide extraction of southern pine and ponderosa pine. *Wood Fiber Sci.* **1991**, *23*, 98–113.
132. Gouveia, L.; Nobre, B.P.; Marcelo, F.M.; Mrejen, S.; Cardoso, M.T.; Palavra, A.F.; Mendes, R.L. Functional food oil coloured by pigments extracted from microalgae with supercritical $CO_2$. *Food Chem.* **2007**, *101*, 717–723. [CrossRef]
133. Thana, P.; Machmudah, S.; Goto, M.; Sasaki, M.; Pavasant, P.; Shotipruk, A. Response surface methodology to supercritical carbon dioxide extraction of astaxanthin from *Haematococcus pluvialis*. *Bioresour. Technol.* **2008**, *99*, 3110–3115. [CrossRef] [PubMed]
134. Montero, O.; Macías-Sánchez, M.D.; Lama, C.M.; Lubián, L.M.; Mantell, C.; Rodríguez, M.; de la Ossa, E.J.M. Supercritical $CO_2$ extraction of β-carotene from a marine strain of the Cyanobacterium *Synechococcus* species. *J. Agric. Food Chem.* **2005**, *53*, 9701–9707. [CrossRef] [PubMed]
135. Jaime, L.; Mendiola, J.A.; Ibáñez, E.; Martin-Álvarez, P.J.; Cifuentes, A.; Reglero, G.; Señoráns, F.J. β-carotene isomer composition of sub- and supercritical carbon dioxide extracts. Antioxidant activity measurement. *J. Agric. Food Chem.* **2007**, *55*, 10585–10590. [CrossRef] [PubMed]
136. Nobre, B.P.; Villalobos, F.; Barragán, B.E.; Oliveira, A.C.; Batista, A.P.; Marques, P.A.S.S.; Mendes, R.L.; Sovová, H.; Palavra, A.F.; Gouveia, L. A biorefinery from *Nannochloropsis* sp. microalga—Extraction of oils and pigments. Production of biohydrogen from the leftover biomass. *Bioresour. Technol.* **2013**, *135*, 128–136. [CrossRef] [PubMed]
137. Sivagnanam, S.P.; Yin, S.; Choi, J.H.; Park, Y.B.; Woo, H.C.; Chun, B.S. Biological properties of fucoxanthin in oil recovered from two brown seaweeds using supercritical $CO_2$ extraction. *Mar. Drugs* **2015**, *13*, 3422–3442. [CrossRef] [PubMed]
138. Fan, X.D.; Hou, Y.; Huang, X.X.; Qiu, T.Q.; Jiang, J.G. Ultrasound-enhanced subcritical $CO_2$ extraction of lutein from *Chlorella pyrenoidosa*. *J. Agric. Food Chem.* **2015**, *63*, 4597–4605. [CrossRef] [PubMed]
139. King, J.W.; Srinivas, K.; Zhang, D. Advances in Critical Fluid Processing. In *Alternatives to Conventional Food Processing*; Proctor, A., Ed.; The Royal Society of Chemistry: Cambridge, UK, 2010; pp. 93–144.
140. Semelsberger, T.A.; Borup, R.L.; Greene, H.L. Dimethyl ether (DME) as an alternative fuel. *J. Power Sources* **2006**, *156*, 497–511. [CrossRef]
141. Kanda, H.; Li, P. Simple extraction method of green crude from natural blue-green microalgae by dimethyl ether. *Fuel* **2011**, *90*, 1264–1266. [CrossRef]
142. Lu, J.; Feng, X.; Han, Y.; Xue, C. Optimization of subcritical fluid extraction of carotenoids and chlorophyll a from *Laminaria japonica* Aresch by response surface methodology. *J. Sci. Food Agric.* **2014**, *94*, 139–145. [CrossRef] [PubMed]

143. Chemat, F.; Cravotto, G. (Eds.) *Microwave-Assisted Extraction for Bioactive Compounds*; Food Engineering Series; Springer: Boston, MA, USA, 2013.
144. Pasquet, V.; Chérouvrier, J.-R.; Farhat, F.; Thiéry, V.; Piot, J.-M.; Bérard, J.-B.; Kaas, R.; Serive, B.; Patrice, T.; Cadoret, J.-P.; et al. Study on the microalgal pigments extraction process: Performance of microwave assisted extraction. *Process Biochem.* **2011**, *46*, 59–67. [CrossRef]
145. Ruen-ngam, D.; Shotipruk, A.; Pavasant, P. Comparison of extraction methods for recovery of astaxanthin from *Haematococcus pluvialis*. *Sep. Sci. Technol.* **2011**, *46*, 64–70. [CrossRef]
146. Poojary, M.M.; Passamonti, P. Optimization of extraction of high purity all-trans-lycopene from tomato pulp waste. *Food Chem.* **2015**, *188*, 84–91. [CrossRef] [PubMed]
147. Poojary, M.M.; Passamonti, P. Extraction of lycopene from tomato processing waste: Kinetics and modelling. *Food Chem.* **2015**, *173*, 943–950. [CrossRef] [PubMed]
148. Zhao, L.; Chen, G.; Zhao, G.; Hu, X. Optimization of microwave-assisted extraction of astaxanthin from *Haematococcus pluvialis* by response surface methodology and antioxidant activities of the extracts. *Sep. Sci. Technol.* **2009**, *44*, 243–262. [CrossRef]
149. Meullemiestre, A.; Breil, C.; Abert-Vian, M.; Chemat, F. Innovative techniques and alternative solvents for extraction of microbial oils. In *Modern Techniques and Solvents for the Extraction of Microbial Oils*; Springer: New York, NY, USA, 2015; pp. 19–42.
150. Juliano, P.; Augustin, M.A.; Xu, X.-Q.; Mawson, R.; Knoerzer, K. Advances in high frequency ultrasound separation of particulates from biomass. *Ultrason. Sonochem.* **2016**, in press. [CrossRef] [PubMed]
151. Leong, T.; Johansson, L.; Juliano, P.; McArthur, S.L.; Manasseh, R. Ultrasonic separation of particulate fluids in small and large scale systems: A Review. *Ind. Eng. Chem. Res.* **2013**, *52*, 16555–16576. [CrossRef]
152. Dolatowski, Z.J.; Stasiak, D.M. Ultrasonically Assisted Diffusion Processes. In *Enhancing Extraction Processes in the Food Industry*; Lebovka, F., Vorobiev, N., Chemat, E., Eds.; CRC Press: Boca Raton, FL, USA, 2012; pp. 123–144.
153. Chemat, F.; Rombaut, N.; Sicaire, A.-G.; Meullemiestre, A.; Fabiano-Tixier, A.-S.; Abert-Vian, M. Ultrasound assisted extraction of food and natural products. Mechanisms, techniques, combinations, protocols and applications. A review. *Ultrason. Sonochem.* **2017**, *34*, 540–560. [CrossRef] [PubMed]
154. Gerde, J.A.; Montalbo-Lomboy, M.; Yao, L.; Grewell, D.; Wang, T. Evaluation of microalgae cell disruption by ultrasonic treatment. *Bioresour. Technol.* **2012**, *125*, 175–181. [CrossRef] [PubMed]
155. Zou, T.B.; Jia, Q.; Li, H.W.; Wang, C.X.; Wu, H.F. Response surface methodology for ultrasound-assisted extraction of astaxanthin from *Haematococcus pluvialis*. *Mar. Drugs* **2013**, *11*, 1644–1655. [CrossRef] [PubMed]
156. Deenu, A.; Naruenartwongsakul, S.; Kim, S.M. Optimization and economic evaluation of ultrasound extraction of lutein from *Chlorella vulgaris*. *Biotechnol. Bioprocess Eng.* **2013**, *18*, 1151–1162. [CrossRef]
157. Yamamoto, K.; King, P.M.; Wu, X.; Mason, T.J.; Joyce, E.M. Effect of ultrasonic frequency and power on the disruption of algal cells. *Ultrason. Sonochem.* **2015**, *24*, 165–171. [CrossRef] [PubMed]
158. Dey, S.; Rathod, V.K. Ultrasound assisted extraction of β-carotene from *Spirulina platensis*. *Ultrason. Sonochem.* **2013**, *20*, 271–276. [CrossRef] [PubMed]
159. Gerken, H.G.; Donohoe, B.; Knoshaug, E.P. Enzymatic cell wall degradation of *Chlorella vulgaris* and other microalgae for biofuels production. *Planta* **2013**, *237*, 239–253. [CrossRef] [PubMed]
160. Rodrigues, D.B.; Flores, É.M.M.; Barin, J.S.; Mercadante, A.Z.; Jacob-Lopes, E.; Zepka, L.Q. Production of carotenoids from microalgae cultivated using agroindustrial wastes. *Food Res. Int.* **2014**, *65*, 144–148. [CrossRef]
161. Leong, T.; Knoerzer, K.; Trujillo, F.J.; Johansson, L.; Manasseh, R.; Barbosa-Cánovas, G.V.; Juliano, P. Megasonic separation of food droplets and particles: Design considerations. *Food Eng. Rev.* **2015**, *7*, 298–320. [CrossRef]
162. Bosma, R.; van Spronsen, W.A.; Tramper, J.; Wijffels, R.H. Ultrasound, a new separation technique to harvest microalgae. *J. Appl. Phycol.* **2003**, *15*, 143–153. [CrossRef]
163. Donsì, F.; Ferrari, G.; Lenza, E.; Maresca, P. Main factors regulating microbial inactivation by high-pressure homogenization: Operating parameters and scale of operation. *Chem. Eng. Sci.* **2009**, *64*, 520–532. [CrossRef]
164. Donsì, F.; Ferrari, G.; Maresca, P. High-Pressure Homogenization for Food Sanitization. In *Global Issues in Food Science and Technology*; Barbosa-Cánovas, G., Mortimer, A., Lineback, D., Spiess, W., Buckle, K., Colonna, P., Eds.; Elesvier: Cambridge, MA, USA, 2009; pp. 309–352.

165. Donsì, F.; Annunziata, M.; Ferrari, G. Microbial inactivation by high pressure homogenization: Effect of the disruption valve geometry. *J. Food Eng.* **2013**, *115*, 362–370. [CrossRef]
166. Donsì, F.; Sessa, M.; Ferrari, G. Nanometric-size delivery systems for bioactive compounds for the nutraceutical and food industries. In *Bio-Nanotechnology: A Revolution in Food, Biomedical and Health Sciences;* Bagchi, D., Bagchi, M., Moriyama, H., Shahidi, F., Eds.; Blackwell Publishing Ltd.: Oxford, UK, 2013; pp. 619–666.
167. Yap, B.H.J.; Dumsday, G.J.; Scales, P.J.; Martin, G.J.O. Energy evaluation of algal cell disruption by high pressure homogenisation. *Bioresour. Technol.* **2015**, *184*, 280–285. [CrossRef] [PubMed]
168. Samarasinghe, N.; Fernando, S.; Lacey, R.; Faulkner, W.B. Algal cell rupture using high pressure homogenization as a prelude to oil extraction. *Renew. Energy* **2012**, *48*, 300–308. [CrossRef]
169. Samarasinghe, N.; Fernando, S.; Faulkner, B. Effect of high pressure homogenization on aqueous phase solvent extraction of lipids from *Nannochloris oculata* microalgae. *J. Energy Nat. Resour.* **2012**, *1*, 1–7. [CrossRef]
170. Olaizola, M. Commercial production of astaxanthin from *Haematococcus pluvialis* using 25,000-liter outdoor photobioreactors. *J. Appl. Phycol.* **2000**, *12*, 499–506. [CrossRef]
171. Clarke, A.; Prescott, T.; Khan, A.; Olabi, A.G. Causes of breakage and disruption in a homogeniser. *Appl. Energy* **2010**, *87*, 3680–3690. [CrossRef]
172. Lee, A.K.; Lewis, D.M.; Ashman, P.J. Disruption of microalgal cells for the extraction of lipids for biofuels: Processes and specific energy requirements. *Biomass Bioenergy* **2012**, *46*, 89–101. [CrossRef]
173. Xie, Y.; Ho, S.H.; Chen, C.N.N.; Chen, C.Y.; Jing, K.; Ng, I.S.; Chen, J.; Chang, J.S.; Lu, Y. Disruption of thermo-tolerant *Desmodesmus* sp. F51 in high pressure homogenization as a prelude to carotenoids extraction. *Biochem. Eng. J.* **2016**, *109*, 243–251. [CrossRef]
174. Mulchandani, K.; Kar, J.R.; Singhal, R.S. Extraction of lipids from *Chlorella saccharophila* using high-pressure homogenization followed by three phase partitioning. *Appl. Biochem. Biotechnol.* **2015**, *176*, 1613–1626. [CrossRef] [PubMed]
175. Lee, C.G.; Kang, D.H.; Lee, D.B.; Lee, H.Y. Pretreatment for simultaneous production of total lipids and fermentable sugars from marine alga, *Chlorella* sp. *Appl. Biochem. Biotechnol.* **2013**, *171*, 1143–1158. [CrossRef] [PubMed]
176. Olmstead, I.L.D.; Kentish, S.E.; Scales, P.J.; Martin, G.J.O. Low solvent, low temperature method for extracting biodiesel lipids from concentrated microalgal biomass. *Bioresour. Technol.* **2013**, *148*, 615–619. [CrossRef] [PubMed]
177. Choi, W.Y.; Lee, H.Y. Effective production of bioenergy from marine *Chlorella* sp. by high-pressure homogenization. *Biotechnol. Biotechnol. Equip.* **2016**, *30*, 81–89. [CrossRef]
178. Halim, R.; Rupasinghe, T.W.T.; Tull, D.L.; Webley, P.A. Mechanical cell disruption for lipid extraction from microalgal biomass. *Bioresour. Technol.* **2013**, *140*, 53–63. [CrossRef] [PubMed]
179. Spiden, E.M.; Yap, B.H.J.; Hill, D.R.A.; Kentish, S.E.; Scales, P.J.; Martin, G.J.O. Quantitative evaluation of the ease of rupture of industrially promising microalgae by high pressure homogenization. *Bioresour. Technol.* **2013**, *140*, 165–171. [CrossRef] [PubMed]
180. Wang, D.; Li, Y.; Hu, X.; Su, W.; Zhong, M. Combined enzymatic and mechanical cell disruption and lipid extraction of green alga *Neochloris oleoabundans*. *Int. J. Mol. Sci.* **2015**, *16*, 7707–7722. [CrossRef] [PubMed]
181. Shene, C.; Monsalve, M.T.; Vergara, D.; Lienqueo, M.E.; Rubilar, M. High pressure homogenization of *Nannochloropsis oculata* for the extraction of intracellular components: Effect of process conditions and culture age. *Eur. J. Lipid Sci. Technol.* **2016**, *118*, 631–639. [CrossRef]
182. Cho, S.-C.; Choi, W.-Y.; Oh, S.-H.; Lee, C.-G.; Seo, Y.-C.; Kim, J.-S.; Song, C.-H.; Kim, G.-V.; Lee, S.-Y.; Kang, D.-H.; et al. Enhancement of lipid extraction from marine microalga, *Scenedesmus* associated with high-pressure homogenization process. *J. Biomed. Biotechnol.* **2012**, *2012*, 1–6. [CrossRef] [PubMed]
183. Jubeau, S.; Marchal, L.; Pruvost, J.; Jaouen, P.; Legrand, J.; Fleurence, J. High pressure disruption: A two-step treatment for selective extraction of intracellular components from the microalga *Porphyridium cruentum*. *J. Appl. Phycol.* **2013**, *25*, 983–989. [CrossRef]

marine drugs

MDPI

*Review*

# Exploring the Valuable Carotenoids for the Large-Scale Production by Marine Microorganisms

**Javier Torregrosa-Crespo [1,†], Zaida Montero [2,†], Juan Luis Fuentes [2,†], Manuel Reig García-Galbis [3], Inés Garbayo [2], Carlos Vílchez [2] and Rosa María Martínez-Espinosa [1,*]**

1 Department of Agrochemistry and Biochemistry, Biochemistry and Molecular Biology division, Faculty of Science, University of Alicante, Ap. 99, E-03080 Alicante, Spain; javitorregrosa@ua.es
2 Algal Biotechnology Group, University of Huelva, CIDERTA and Faculty of Science, Marine International Campus of Excellence (CEIMAR), Parque Huelva Empresarial S/N, 21007 Huelva, Spain; mariazaida.montero@hotmail.com (Z.M.); juanlufc@gmail.com (J.L.F.); garbayo@dqcm.uhu.es (I.G.); bital.uhu@gmail.com (C.V.)
3 Department of Nutrition and Dietetics, Faculty of Health Sciences, University of Atacama, Copayapu 2862, CP 1530000 Copiapó, Chile; manuel.reig@uda.cl
* Correspondence: rosa.martinez@ua.es; Tel.: +34-965-901-258
† These authors have contributed equally to this work.

Received: 11 April 2018; Accepted: 5 June 2018; Published: 8 June 2018

**Abstract:** Carotenoids are among the most abundant natural pigments available in nature. These pigments have received considerable attention because of their biotechnological applications and, more importantly, due to their potential beneficial uses in human healthcare, food processing, pharmaceuticals and cosmetics. These bioactive compounds are in high demand throughout the world; Europe and the USA are the markets where the demand for carotenoids is the highest. The in vitro synthesis of carotenoids has sustained their large-scale production so far. However, the emerging modern standards for a healthy lifestyle and environment-friendly practices have given rise to a search for natural biocompounds as alternatives to synthetic ones. Therefore, nowadays, biomass (vegetables, fruits, yeast and microorganisms) is being used to obtain naturally-available carotenoids with high antioxidant capacity and strong color, on a large scale. This is an alternative to the in vitro synthesis of carotenoids, which is expensive and generates a large number of residues, and the compounds synthesized are sometimes not active biologically. In this context, marine biomass has recently emerged as a natural source for both common and uncommon valuable carotenoids. Besides, the cultivation of marine microorganisms, as well as the downstream processes, which are used to isolate the carotenoids from these microorganisms, offer several advantages over the other approaches that have been explored previously. This review summarizes the general properties of the most-abundant carotenoids produced by marine microorganisms, focusing on the genuine/rare carotenoids that exhibit interesting features useful for potential applications in biotechnology, pharmaceuticals, cosmetics and medicine.

**Keywords:** carotenoids; antioxidants; bioactive compounds; blue biotechnology; marine microorganisms

## 1. Introduction

Carotenoids are a class of pigments distributed ubiquitously in nature [1]. These pigments have received considerable attention because of their biotechnological applications and, more importantly, due to their potential beneficial applications in the fields of human healthcare, food processing, pharmaceuticals and cosmetics [2,3]. Chemically, these compounds are mainly $C_{40}$ lipophilic isoprenoids that range from colorless to yellow, orange and red [1,2,4]. These pigments may be categorized into two groups based on the presence or absence of oxygen in their structures: carotenes (do not contain oxygen) and xanthophylls (contain oxygen). The production of these pigments from plants, fungi and certain microorganisms, on a medium-scale level, and even on a large-scale level, has been described previously [5–11]. In fact, the studies focused on the regulation of carotenogenesis, the biological roles performed by the carotenoids, the general characterization of the carotenoids and the analytical procedures utilized in order to describe their structure are abundantly available in the literature [1,12,13].

Carotenoids have received much attention because of the variety of important biological roles they perform in all living systems [1,14–16]. In the majority of organisms, the most relevant biological functions performed by the carotenoids are associated with their antioxidant properties, which are a direct outcome of their molecular structure [10]. Xanthophylls, for instance, perform the role of free-radical scavengers, potent quenchers of ROS (reactive oxygen species) and RNS (reactive nitrogen species) and chain-breaking antioxidants. Therefore, astaxanthin and canthaxanthin (which are xanthophylls) are better antioxidants and scavengers of free radicals compared to β-carotene. In recent years, understanding the ROS-induced oxidative stress mechanisms and the search for suitable strategies in order to fight oxidative stress have become the major goals of medical research efforts [10]. Furthermore, carotenoids are the natural compounds that are responsible for conferring color to animals, plants and microorganisms [17].

Animals, as well as humans are not capable of synthesizing the carotenoids de novo; the pigments are, therefore, acquired through diet. However, they are capable of converting these pigments into vitamin A and the retinoid compounds, which are required for morphogenesis and embryonic development [18,19]. Vitamin A is a well-recognized factor of great importance in child health and survival; its deficiency causes disturbances in vision and also lung, trachea and oral cavity pathologies. Consuming carotenoids through diet is the only way to carry out retinol synthesis in animals and humans; fruits, vegetables and microalgae being the major sources of carotenoids that exhibit pro-vitamin A activity [10,19]. Other biological roles and functions performed by the carotenoids in animals and humans include absorption of light energy, oxygen transport [20], enhancing in vitro antibody production and antitumor activity [21,22] and antioxidant and anti-inflammatory activities [23].

In birds and fishes, carotenoids are an important indication of a satisfactory nutritional condition and are used in ornamental displays as a sign of fitness and to increase sexual attractiveness [16,24,25]. In algae and higher plants, carotenoids serve as the regulators of plant growth and development, the accessory pigments for photosynthesis and as photoprotectors. Thus, they contribute to light harvesting, maintenance of the structures and functions of the photosynthetic complexes, quenching of the chlorophyll triplet states, scavenging of reactive oxygen species and dissipation of excess energy [26]. Besides, carotenoids also serve as precursors for certain hormones such as abscisic acid (ABA) and strigolactones, as well as attractants for other organisms, such as insect pollinators and seed-dispersing herbivores [19]. Apart from the above-mentioned important roles, carotenoids also serve as important floral pigments, due to their striking and rich color, to attract pollinators and seed dispersers. Microorganisms are a great source of a variety of carotenoids. In microorganisms, carotenoids oversee photoprotection, provide color to microbial cells and regulate the mechanisms against oxidative stress. It is important to highlight that certain carotenoids, such as salinixanthin or bacterioruberin (produced by halophiles), are produced solely by certain extremophilic microorganisms [8,27,28].

For the last 30 years, researchers and the R&D companies have paid much attention to the microorganisms that are capable of producing significant concentrations of carotenoids, because these biocompounds obtained from these natural sources are in high demand these days. This huge demand is a result of consumers' preference for natural as opposed to synthetic products. Such demand has encouraged the production of scientific literature on novel carotenoid-producing microorganisms and has prompted major efforts to enhance the isolation of carotenoids from their biological sources instead of their chemical synthesis. This fact, coupled with fresh insights into the molecular biology techniques and downstream processes, epitomizes the carotenoid-producing microorganisms as suitable natural sources for the large-scale synthesis of carotenoids, as an alternative to in vitro carotenoid production. Among all the natural sources of carotenoids, marine microorganisms have emerged as the natural sources from which the production of these pigments may be relatively easy.

This review is an attempt to compile the latest studies and research works in this field. It summarizes the main findings that have been described to date about the marine microbial carotenoids, highlighting their potential beneficial effects on human health and their relevance in the natural compounds market.

## 2. Marine Microorganisms as a Source of Bioactive Compounds

Bioactive compounds are the compounds that produce specific effects within a living organism, tissue or cell. In the field of nutrition, bioactive compounds have been distinguished from essential nutrients; essential nutrients are necessary for the sustainability of life, while bioactive compounds are not. However, bioactive compounds may have an influence (usually positive) on metabolism and on health in general. Bioactive compounds are usually the secondary metabolites produced by microorganisms, yeast, algae and fungi [29].

Marine life forms have long adapted their physiology and metabolism to the extreme ambient conditions present in the seas, oceans and other closely-related environments, such as marshes, coastal salt ponds, etc. Consequently, they have evolved within themselves protective mechanisms that include the accumulation of bioactive compounds [30,31]. Recent research on marine ecosystems has revealed that the marine aquatic biomass (free-living cells or symbiotic) and its bioactive compounds have many potential applications in various fields; for example, ensuring future food and nutrition security and supplying raw materials for other high added-value chains and products, such as bioenergy, pharmaceuticals and cosmetics, while factoring in the environmental and climate change risks.

Therefore, international institutions, such as the European Commission, are promoting investigations in the field of biotechnology in general, particularly in blue biotechnology, a field that is concerned with the exploration and exploitation of the resulting diversity of marine organisms in order to develop novel products [32]. Besides, a significant number of research groups are working in a coordinated manner to improve the knowledge about carotenoids and the novel natural sources for their production, as is evident from the presence of networks such as Eurocaroten [33], CaRed (the Spanish Carotenoid Network; [34], the International Carotenoid Society [35] and IBERCAROT (the Ibero-American network for the study of carotenoids as functional food ingredients, [36]).

In line with this fact, the microorganisms that produce a significant amount of bioactive compounds (lipids, carotenoids, vitamins, etc.) are abundant in the marine environment; therefore, they are considered suitable targets for exploring their potential as the natural biosources from which carotenoids may be obtained at a large scale [37–39]. Besides, modern approaches such as genomics, transcriptomics, proteomics, etc., have been recently used as powerful tools to investigate the production of bioactive compounds from marine organisms [40–42]. Therefore, marine plants (kelp, for instance), phytoplanktons, marine algae and microorganisms such as marine bacteria and Haloarchaea, are perceived as attractive sources for the production and isolation of common, as well as novel and rare carotenoids (Table 1).

Marine microorganisms, in addition to their capacity to synthesize unique bioactive compounds, offer certain advantages specific to a large-scale production of carotenoids; for example, the risk of contamination with other microorganisms is reduced due to the high-salinity conditions used in their culture media [43]. This feature becomes more significant when the extreme halophilic marine microorganisms such as Haloarchaea are used as the natural source of bioactive compounds [8,44]. However, the concern is to produce the bioactive compounds at a price that is competitive; which is difficult because the production costs of the microbial biomass are high. Therefore, further investigation is required to achieve a more profitable production of the bioactive compounds.

**Table 1.** Microorganisms that produce carotenoids in the marine environment and biological properties with potential benefits for human health.

| Marine Microorganism | Carotenoid | Biological Properties | References |
|---|---|---|---|
| **Abundant Carotenoids in the Marine Environment** | | | |
| • Microalgae<br>• Bacteria<br>• Cyanobacteria | β-Carotene | Antioxidant immune response<br>Anti-inflammatory<br>Benefits for cognitive function and atherogenesis<br>Antidiabetic activity<br>Antitumor activity | [10,45–47] |
| • Haloarchaea<br>• Bacteria<br>• Cyanobacteria<br>• Microalgae | Astaxanthin | Immune response anti-inflammatory<br>Antioxidant activity<br>Antitumor activity<br>Ocular protective effect<br>Antidiabetic activity<br>Against:<br>○ benign prostatic hyperplasia<br>○ cancer<br>○ asthma<br>○ rheumatoid arthritis<br>○ metabolic syndrome<br>○ diabetic nephropathy<br>○ cardiovascular diseases<br>○ neurodegenerative diseases | [8,48–55] |
| • Microalgae | Fucoxanthin | Reduction of cardiovascular risk factors<br>Electron donor<br>Involved in lipid metabolism increasing production of energy<br>Antioxidant activity<br>Anti-inflammatory effect<br>Anticancer activity<br>Anti-obese effect<br>Antidiabetic activity<br>Hepatoprotective effect<br>Skin-Protective effect<br>Antiangiogenic effect<br>Cerebrovascular protective effect<br>Bone-protective effect<br>Ocular protective effect<br>Antimalarial effect | [15,51,56–60] |
| • Microalgae<br>• Bacteria<br>• Cyanobacteria | Zeaxanthin | Reduction of cardiovascular risk factors<br>Prevention of coronary syndromes<br>Helps in maintaining visual function<br>Antitumor activity (breast cancer)<br>Anti-cardiovascular diseases<br>Antioxidative, anti-inflammatory and structural actions in neural tissue | [15,61–65] |
| • Cyanobacteria | β-Cryptoxanthin | Antioxidant<br>Immune response anti-inflammatory<br>Improves respiratory function and lowers lung cancer rates<br>Stimulation of bone formation<br>Reduces the rate of oral and pharyngeal cancer<br>Modulation response to phytosterols in post-menopausal women<br>Protection of leukocyte telomeres' length<br>Decreases risk of some cancers and degenerative diseases.<br>Bone-protective effect | [15,66–71] |
| **Less abundant carotenoids in the marine environment** | | | |
| • Haloarchaea | Bacterioruberin | Antioxidant<br>Anticancer activity | [8] |
| • Bacteria | Saproxanthin | Antioxidant<br>Apoptosis-inducing effect | [15,72] |

**Table 1.** *Cont.*

| Abundant Carotenoids in the Marine Environment | | | |
|---|---|---|---|
| **Marine Microorganism** | **Carotenoid** | **Biological Properties** | **References** |
| • Cyanobacteria<br>• Bacteria | Myxol | Antioxidant<br>Anticancer activity<br>Against cardiovascular pathologies | [15,73] |
| • Actinomycetes | Sioxanthin | Antioxidant | [74] |
| • Microalgae<br>• Haloarchaea | Lutein | Antioxidant<br>Anti-macular eyes degradation<br>Prevention coronary syndromes and stroke<br>Prevention of cataract<br>Prevention of retinitis<br>Ocular protective effect<br>Dopaminergic neurons protection against MPTP-induced apoptotic death<br>Anti-atherosclerosis | [8,15,62,74–77] |
| • Microalgae<br>• Cyanobacteria<br>• Haloarchaea<br>• Bacteria | Canthaxanthin | Antioxidant<br>Antitumoral activity<br>Provitamin A activity | [8,46,51,78–80] |
| • Microalgae<br>• Cyanobacteria | Echinenone | Antioxidant | [51,74,81] |
| • Microalgae | Violaxanthin | Food additive E161e (not approved in the EU and USA)<br>Anti-inflammatory effects in macrophages | [74,82] |
| • Microalgae<br>• Haloarchaea<br>• Bacteria | Phytoene | Antitumoral activity | [8,78,83] |
| • Microalgae<br>• Bacteria | Lycopene | Reduction risk of atherosclerosis and coronary heart disease<br>Antioxidant activity<br>Antiulcer activity<br>Gene regulation<br>Gap-junction communication activity<br>Immune modulation<br>Antitumor activity | [8,84–86] |
| • Microalgae<br>• Haloarchaea | Salinixanthin | Anticancer activity (human liver cancer cell lines showed dose-dependent cytotoxicity of the carotenoids) | [8] |

## 3. Properties of the Most Demanded Carotenoids Isolated from Marine Microorganisms

This section summarizes the general aspects of the carotenoids that are most demanded in the pharmaceutical, cosmetics and biotechnological markets. These are known as the common carotenoids. These carotenoids are present in marine microorganisms, as well as in other organisms, such as terrestrial plants, aquatic plants, fungi, yeast, etc. Although these carotenoids have been obtained mainly from plants, fungi, yeast and algae, several recent works on microorganisms (mainly from the marine environments) have revealed that certain microbial genera are the producers of significantly large amounts of several carotenoids with potential applications in biomedicine and biotechnology.

Table 2 enlists the main features of the structure of each of the following carotenoids.

**Table 2.** IUPAC name, molecular formula and chemical structure of the most marketed carotenoids.

| Common Name | IUPAC Name | Molecular Formula | Chemical Structure | Reference |
|---|---|---|---|---|
| Astaxanthin | 3,3′-Dihydroxy-β,β-carotene-4,4′-dione | $C_{40}H_{52}O_4$ | 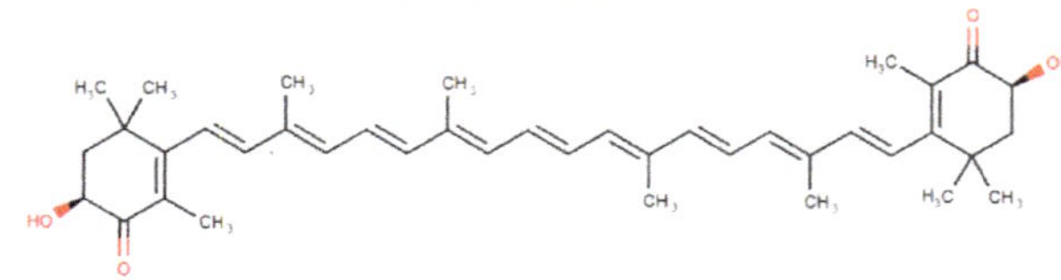 (3*S*,3′*S*)-3,3′-dihydroxy-β,β-carotene-4,4′-dione | [1] |
| β-Carotene | β,β-Carotene | $C_{40}H_{56}$ | 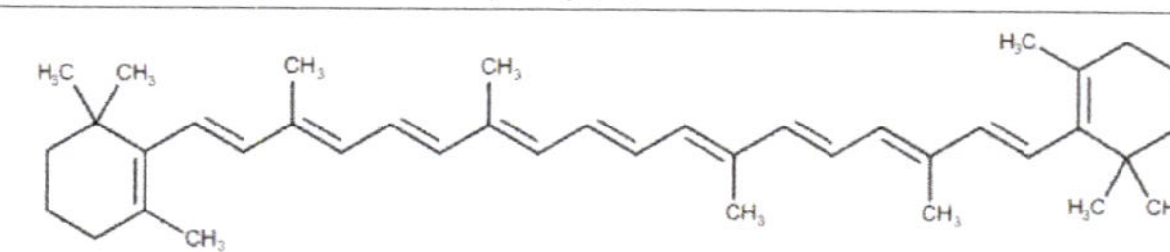 *all-trans*-β-carotene | [2] |
| Canthaxanthin | β,β-Carotene-4,4′-dione | $C_{40}H_{52}O_2$ | 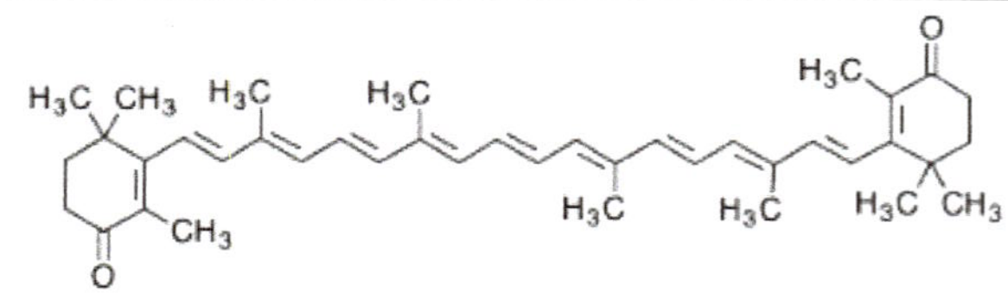 *trans*-β-carotene-4,4′-dione | [3] |
| β-Cryptoxanthin | β,β-Caroten-3-ol | $C_{40}H_{56}O$ | 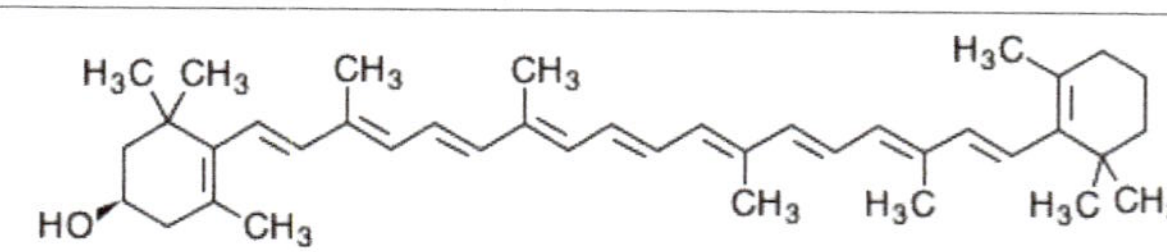 (3*R*)-β,β-Caroten-3-ol | [4] |

**Table 2.** *Cont.*

| Common Name | IUPAC Name | Molecular Formula | Chemical Structure | Reference |
| --- | --- | --- | --- | --- |
| Fucoxanthin | 3,5′-Dihydroxy-8-oxo-6′,7′-didehydro-5,6-epoxy-5,6,7,8,5′,6′-hexahydro-β,β-caroten-3′-yl acetate | $C_{42}H_{58}O_6$ | 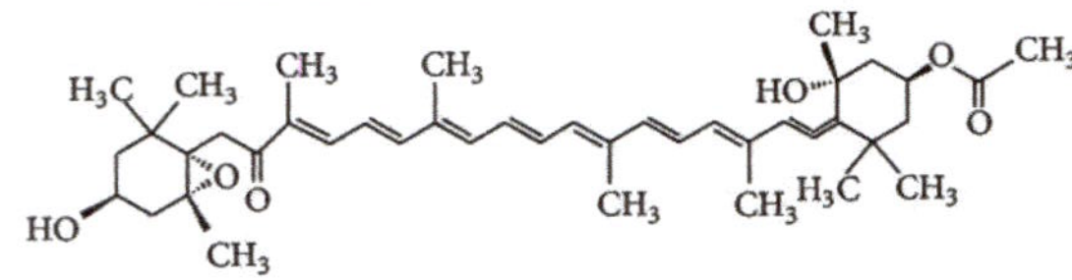 (3*S*,5*R*,6*S*,3′*S*,5′*R*,6′*R*)-3,5′-dihydroxy-8-oxo-6′,7′-didehydro-5,6-epoxy-5,6,7,8,5′,6′-hexahydro-β,β-caroten-3′-yl acetate | [5] |
| Lycopene | ψ,ψ-Carotene | $C_{40}H_{56}$ | 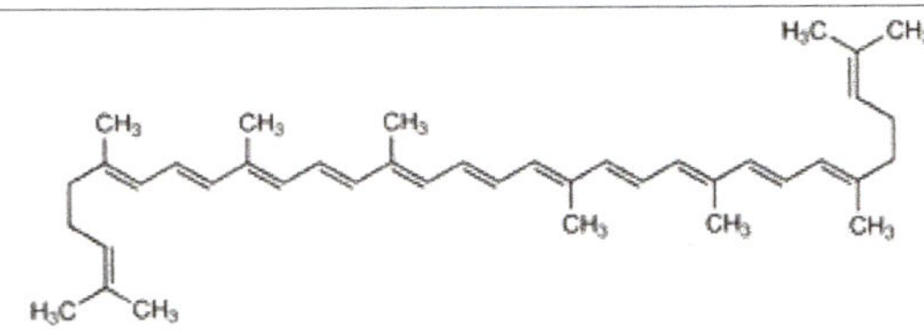 *all-trans*-lycopene | [6] |
| Lutein | β-ϵ-Carotene-3,3′-diol | $C_{40}H_{56}O_2$ | 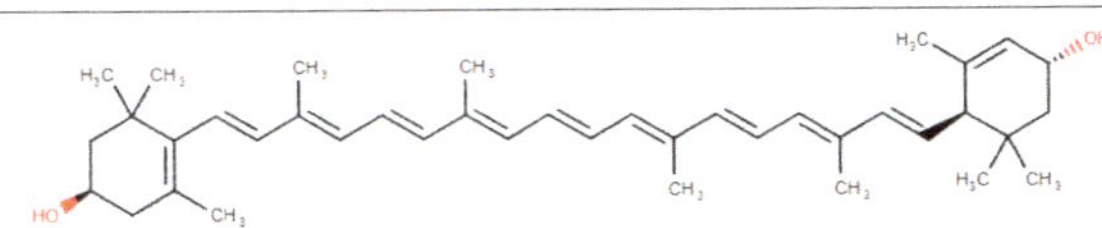 (3*R*,3′*R*,6′*R*)-β,ε-carotene-3,3′-diol | [7] |
| Zeaxanthin | β,β-Carotene-3,3′-diol | $C_{40}H_{56}O_2$ | 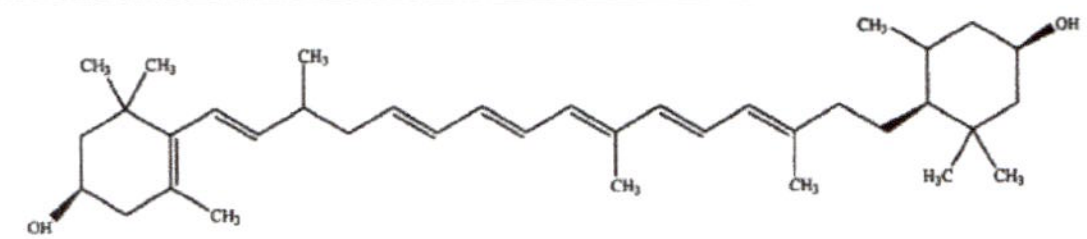 (3*R*,3′*R*)-β,β-carotene-3,3′-diol | [7] |
| Violaxanthin | 5,5′,6,6′-Tetrahydro-5,6:5′,6′-diepoxy-β,β-carotene-3,3′-diol | $C_{40}H_{56}O_4$ | 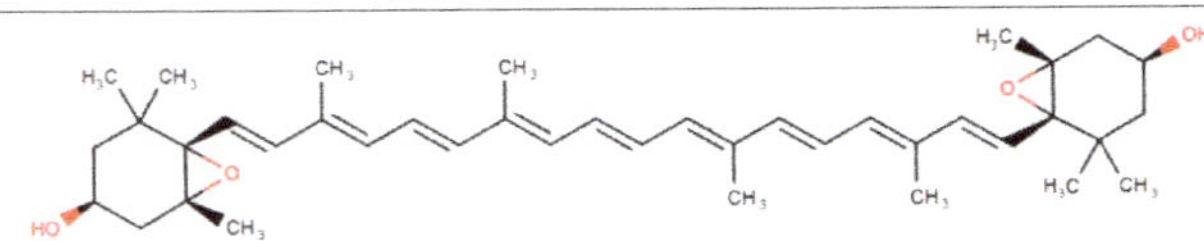 (3*S*,3′*S*,5*R*,5′*R*,6*S*,6′*S*)-5,5′,6,6′-tetrahydro-5,6:5′,6′-diepoxy-β,β-carotene-3,3′-diol | [8] |

### 3.1. Astaxanthin

Astaxanthin (3,3′-dihydroxy-β,β′-carotene-4,4′-dione) is a red-pink-colored xanthophyll carotenoid, which contains two additional oxygenated groups on each ring compared to the other carotenoids, which results in enhanced antioxidant properties of this compound [56]. It is a β-carotene derivative, which is eleven-times more potent as a quencher of singlet oxygen than β-carotene and 550-times more potent than α-tocopherol [54]. Astaxanthin exhibits a higher biological activity than the other antioxidants, because of its ability to link across the entire cell membrane from the inside to the outside [87]. This compound occurs naturally in a wide variety of living organisms, including microalgae, fungi, plants, marine organisms and certain birds such as flamingos; it confers salmon, shrimp and lobster their distinctive color [54]. Astaxanthin can neither be synthesized de novo by animals, nor converted into vitamin A; therefore, it must be consumed in the diet [56]. The high potency, as well as the polar property of astaxanthin makes it an attractive antioxidant nutraceutical suitable for further investigation in the field of biomedicine. Several works have demonstrated that astaxanthin, when used as a nutritional supplement (it is associated with the E number E161j), has potential beneficial effects on human health. The biotechnological production of astaxanthin from various sources has been studied in depth [88] in order to achieve its production on a large scale for several commercial applications [87]. The astaxanthin products are used for commercial applications in dosage forms, such as tablets, capsules, syrups, oils, soft gels, creams, biomass and granulated powders. Astaxanthin-related patent applications are available in the areas of food, feed and nutraceuticals and are currently the major market driver for the pigment.

### 3.2. β-Carotene

β-Carotene is an intensely-colored orange pigment, abundant in green leafy plants (parsley, spinach, broccoli), certain fruits (mandarin, peach) and several vegetables (carrot, pumpkin) [26]. Its distinguishing characteristic is the beta rings present at both ends of the molecule. β-Carotene occurs as several isomers, two of which (9-*cis* and all-*trans*) constitute approximately 80% of the total β-carotene present in the microalga *Dunaliella bardawil* [89]. β-Carotene is used as a food coloring agent with the E number E160 [90]. In nature, β-carotene is a precursor (inactive form) of vitamin A, which is synthesized from carotenoids via the action of the enzyme β-carotene 15,15′-monooxygenase. Following its synthesis, vitamin A is assimilated or further converted into retinoids so that it does not cause hypervitaminosis A. The isolation of β-carotene from fruits is commonly performed by using column chromatography [91,92]. β-Carotene is deeply colored and a highly-conjugated carotenoid, which lacks functional groups, causing it to be highly lipophilic. Overconsumption of β-carotene may cause carotenosis, a benign condition under which the skin turns orange. Chronic intake of high doses of β-carotene supplementation has been correlated directly with the increase in the probability of lung cancer in cigarette smokers [93].

### 3.3. Canthaxanthin

Canthaxanthin (β,β-carotene-4,4′-dione) is a diketo carotenoid pigment, which is orange-red in color. It was first isolated from edible mushrooms. In several green algae and also in blue-green algae, this pigment is produced as a secondary carotenoid at the end of the growth phase, either in place of, or in addition to, the primary carotenoids. It has also been found in bacteria, crustaceans, birds (even in the yolk of bird eggs) [94] and in various species of fish, including the common carp (*Cyprinus carpio*), golden mullet (*Mugil auratus*), annular seabream (*Diplodus annularis*) and thrush wrasse (*Crenilabrus tinca*) [57]. It is used as a food coloring agent, associated with the E number E161g, in different countries, including the United States and the EU member states. As canthaxanthin has a high commercial value, its biosynthesis has been studied extensively. Canthaxanthin is biosynthesized from β-carotene, through the action of a single enzyme, known as β-carotene ketolase, which adds carbonyl groups to the carbon atoms at the 4 and 4′ positions in the β-carotene molecule. Although functionally identical, several distinct β-carotene ketolase proteins are known, which differ (from an evolutionary perspective) in their primary protein/amino acid sequence. In order to improve the large-scale canthaxanthin production using microorganisms, the regulation of its biosynthesis has been studied recently in *Haematococcus pluvialis* [95] and in *Dietzia natronolimnaea* [96,97].

### 3.4. *β-Cryptoxanthin*

β-Cryptoxanthin (hydroxy-β-carotene) occurs in a variety of sources, including petals of flowers and several fruits such as papayas, satsuma mandarins and apples [57]. It is present together with α-carotene, β-carotene, lycopene, lutein and zeaxanthin. It is also present in egg yolk [98] and sweet oranges [99]. β-Cryptoxanthin is a good source of vitamin A, and therefore, it is considered a pro-vitamin A [100]. This carotenoid is oxidized and isomerized in the presence of light [101]. β-Cryptoxanthin is used as a coloring agent to color food products in certain countries and is associated with the E number E161c. β-Cryptoxanthin obtained from its common food sources, exhibits relatively high bioavailability to the extent that certain β-cryptoxanthin-rich foods might be considered equivalent to β-carotene-rich foods as sources of retinol [102].

### 3.5. *Fucoxanthin*

Fucoxanthin is one of the most abundant carotenoids, and it is present as an accessory pigment in the chloroplasts of brown algae, phytoplankton, brown seaweeds and diatoms, giving them a brown or olive-green color [58]. It shares more than 10% of the estimated total production of carotenoids in nature, especially in the marine environments [57]. Fucoxanthin is a xanthophyll with a unique structure that includes an unusual allenic bond, an epoxide group and a conjugated carbonyl group in the polyene chain, conferring it the antioxidant properties. However, the difference is that fucoxanthin exhibits antioxidant properties even under anoxic conditions, while the other carotenoids demonstrate practically no quenching ability under those conditions. Most of the tissues have a low-oxygen status under physiological conditions. Consequently, fucoxanthin may be performing key roles under anoxic conditions. The unique molecular structure of fucoxanthin confers remarkable biological properties to it, similar to neoxanthin, dinoxanthin and peridinin. Fucoxanthin does not exhibit toxicity and mutagenicity under experimental conditions, and it may possess the ability to increase the levels of circulating cholesterol in rodents as a common feature [57]. Fucoxanthinol is the deacetylated derivative of fucoxanthin. In fact, it has been reported that several of the bioactive properties of fucoxanthin are due to fucoxanthinol [58,103–106]. Fucoxanthinol exhibits suppressive effects on lipid accumulation during adipocyte differentiation, and it also demonstrates anti-inflammatory and antioxidative properties.

### 3.6. *Lycopene*

Lycopene (ψ,ψ-carotene) is responsible for the red color in several fruits and vegetables, such as tomatoes. Unlike certain other carotenoids, lycopene lacks the terminal β-ionone ring in its structure, and therefore, provitamin A activity is not present in this carotenoid. Lycopene has a highly unsaturated, hydrocarbon chain, consisting of eleven conjugated and two unconjugated double bonds, which confers its antioxidant activity. Because of the presence of the double bonds in the structure of lycopene, it can exist in both *cis* and *trans* isomeric forms. In nature, lycopene is present primarily in the all-*trans* isomeric form. However, it may undergo mono- or poly-isomerization in the presence of light, thermal energy and temperature, which can convert it to the *cis* isomer. Lycopene is highly stable under the conditions of thermal processing and storage. It has been reported that 5-*cis* lycopene is the most stable isomer of lycopene, followed by the all-*trans*, 9-*cis*, 13-*cis*, 15-*cis*, 7-*cis* and 11-*cis* isomeric forms. The 5-*cis* isomer of lycopene has been demonstrated to exhibit the highest antioxidant activity, followed by the 9-*cis*, 7-*cis*, 13-*cis*, 11-*cis* and all-*trans* isomers [85]. Lycopene is associated with the E number E160d when used as a coloring agent for food products.

### 3.7. Lutein

Lutein (β,ε-carotene-3,3′-diol) is one of the two major components of the macular pigment of the retina and is present at a high concentration in the human serum [107]. It is synthesized only by plants, and similar to the other xanthophylls, it is present in high quantities in green leafy vegetables, such as spinach and kale, and yellow carrot. In green plants, lutein modulates light energy and serves as a non-photochemical quenching agent to deal with triplet chlorophyll (an excited form of chlorophyll), which is overproduced at high light intensity during photosynthesis. Lutein contains only 10 conjugated double bonds, which causes it more yellowish-green color compared to the other carotenoids. Lutein acts as a powerful antioxidant and is able to filter high-energy blue light [108]. When used as a food colorant, it is associated with the E number E161b. Recently, novel methodologies have been developed and optimized in order to isolate this pigment from vegetables, for example from spinach [109] or carrot [92].

### 3.8. Zeaxanthin

Zeaxanthin (β,β-carotene-3,3′-diol) is one of the most common carotenoid alcohols present in nature, which performs an important role in the xanthophyll cycle. It is synthesized by plants and certain microorganisms. Zeaxanthin is the pigment that confers paprika (produced from bell peppers), corn, saffron, wolfberries and several other plants and microorganisms (such as marine bacteria) their characteristic color [110]. It is also one of the two major components of the macular pigment of the retina. Zeaxanthin is isomeric with lutein, and the two carotenoids differ from each other only in the location of a single double bond; in zeaxanthin, all the double bonds are conjugated. Zeaxanthin does not exhibit vitamin A activity. This pigment and its close relative lutein perform a critical role in the prevention of AMD (age-related macular degeneration), one of the leading causes of blindness [57]. Like lutein, zeaxanthin has been found in significant concentrations in human milk [111]. Because of its high value and demand in the nutraceutical market, several methods have been developed to produce zeaxanthin on a large scale [112]. As a food colorant, it is associated with the E number E161 h.

### 3.9. Violaxanthin

Violaxanthin (5,6:5′,6′-diepoxy-5,5′,6,6′-tetrahydro-β-carotene-3,3′-diol) is a natural xanthophyll pigment, which is orange in color. It is present in a variety of brown algae and in plants including pansies. This pigment is biosynthesized from zeaxanthin through the process of epoxidation. As a food additive, it is used under the E number E161e; however, this use is approved only in Australia and New Zealand, where it is listed under the INS number 161e. The interconversions of violaxanthin and zeaxanthin are caused mainly by the action of the violaxanthin de-epoxidase enzyme. Violaxanthin de-epoxidation and the violaxanthin cycle were first studied in the late 1960s to the early 1970s [113–115]. Violaxanthin de-epoxidases are susceptible to DTT [116,117], and their thermostability is due to the disulfide bridges present in their structures [118]. These interconversion mechanisms have been observed to correlate directly to the dissipation of excess excitation energy in leaves in 2% $O_2$, 0% $CO_2$ [119]. So far, the major biological roles of violaxanthin, as well as its role as a precursor of abscisic acid, have been described using leaves [120] and fruits [121] as the sources of this carotenoid.

## 4. The Rare Carotenoids

The rare carotenoids include certain carotenoids recently found in only a few marine organisms, at the time of writing this review. Although there are probably several rare carotenoids to be described, only a few of them have been reported so far in the previous studies. The production of the biomass of marine organisms that contain the rare carotenoids is the first step toward the biotechnological production of these carotenoids. All the rare carotenoids that have been described so far were isolated mainly from microalgae or Haloarchaea (the halophilic microorganisms belonging to domain Archaea, which inhabit salty environments). So far, there are no examples of a large-scale production of rare carotenoids from microalgae, and only one indexed paper has reported the cultivation of *Nostoc* for this purpose [122]. Several wild-types, as well as genetically-engineered species of *Nostoc*, have been used to investigate the production of exopolysaccharides, such as polyhydroxyalkanoates [123], and the production of hydrogen [124]. Moreover, several species of *Anabaena* have also been reported to serve as attractive sources for the production of exopolysaccharides [125] and carotenoids as feedstock for biodiesel [126]. However, the production of rare carotenoids from *Nostoc* and *Anabaena*, as well as the processes for the large-scale production of their biomass have not yet developed completely.

Other rare carotenoids, such as bacterioruberin (and its derivatives) or salinixanthin, are produced mostly from Haloarchaea [8]. The recent works reported by Shindo and by Gamone and co-workers have referred to myxol, saproxanthin, sioxanthin and siphonaxanthin as rare carotenoids [15,127]. The main features describing each of the above-mentioned rare carotenoids are summarized in the following sections, as well as in Table 3.

### 4.1. Bacterioruberin

Bacterioruberin is the main carotenoid responsible for the color of the red halophilic Archaea belonging to the families Halobacteriaceae and Haloferacaceae. To date, the production of this rare carotenoid has been reported only from the Haloarchaea and *Micrococcus roseus* [8,128]. Bacterioruberin is a $C_{50}$ carotenoid pigment, exhibiting a unique molecular structure. It consists of a primary conjugated isoprenoid chain with 13 C=C units and no subsidiary conjugation arising from the terminal groups, which contain only four hydroxyl groups [8]. This pigment protects the microbial cells against the damage caused by high intensities of light in the visible and the ultraviolet range of the spectrum, and it also aids in photoreactivation [129]. The pigment is also involved in the reinforcement of the microbial cell membrane. This pigment was first described from the cells of a *Halobacterium* species [130]. Recent works on the production of bacterioruberin from Haloarchaea report that the production of this rare carotenoid may be easily enhanced by modifications of the temperature, salt concentrations, pH, oxygen availability or light incidence [8,44].

### 4.2. Myxol

Myxol is derived from γ-carotene. It is present in various forms in nature; however, as free myxol, it is present mainly in the marine environments. The marine bacterium MBIC 03313 was reported as the first microorganism to produce free myxol. *Anabaena variabilis* ATCC 29413 also produces this pigment, along with 4-hydroxymyxol [131]. *Robiginitalea myxolifaciens* sp. nov. has been reported to be capable of producing myxol in sea water [132]. Myxol, in its glyoxylated form, has been identified in both marine and non-marine aquatic microorganisms such as *Nostoc commune* [133]. The freshwater alga *Oscillatoria limosa* produces myxol-glyoxylate as myxol-2′-O-methyl-methylpentoside and 4-keto-myxol-2′-methylpentoside [73]. The potential applications of myxol in the area of human health are related to its strong antioxidant activity. Myxol and saproxanthin have demonstrated better antioxidant activity against lipid peroxidation and better neuroprotective effects against L-glutamate toxicity than β-carotene or zeaxanthin, in a rat brain homogenate model [134]. The antioxidant activity of myxol has also been associated with its capacity to decrease the oxygen permeability in lipid membranes [15].

### 4.3. Salinixanthin

This carotenoid is produced mainly by the halophilic bacterium Salinibacter ruber and by certain halophilic Archaea species belonging to the Halobacteriaceae and Haloferacaceae families [8,28,135]. The structure of salinixanthin ((all-E,2′S)-2′-hydroxy-1′-[6-O-(13-methyltetradecanoyl)-β-D-glycopyranosyloxy]-3′,4′-didehydro-1′,2′-dihydro-β,Ψ-caroten-4-one) was first determined using spectroscopic techniques [136]. In nature, this carotenoid functions as a light-harvesting antenna for supplying the additional excitation energy for retinal isomerization and proton transport [137].

### 4.4. Saproxanthin

Saproxanthin is a xanthophyll carotenoid. It was first reported and described by Aasen and Jensen in Saprospira grandis. A selected strain of Saprospira sp., SS90–1, which was identified as Saprospira toviformis, synthesizes saproxanthin as a major pigment [138]. The marine bacterial strain 04OKA–13–27 was reported as the second species to produce saproxanthin [127,134]. More recently, it was reported that alkaline conditions improved the synthesis of this rare carotenoid in the bacterial strains belonging to Jejuia pallidilutea [139].

### 4.5. Sioxanthin

Sioxanthin has been reported to be produced so far only by a marine actinomycete belonging to the genus Salinispora. It was described that this pigment has a novel C40 carotenoid structure, which is glycosylated at one end of the molecule, and contains an aryl moiety at the other end. Glycosylation is unusual among the actinomycete carotenoids, and therefore, sioxanthin forms a part of the rare group of compounds that possess polar, as well as non-polar head groups [140].

### 4.6. Siphonaxanthin

Siphonaxanthin is a keto-carotenoid, a xanthophyll pigment that is present in the species belonging to the Siphonales order [15,141]. Siphonales are the green algae present in both marine and freshwater environments and are quite common in deep, as well as shallow waters [142,143]. To date, the species that belong to this order are the only ones identified as the producers of siphonaxanthin. The characterization of this rare carotenoid remains poorly explored; nevertheless, several recent works have stated that siphonaxanthin could be considered a bioactive compound with potential beneficial effects on human health [15,144].

**Table 3.** IUPAC name, molecular formula and chemical structure of the rare carotenoids.

| Common Name | IUPAC Name | Molecular Formula | Chemical Structure | Reference |
|---|---|---|---|---|
| Bacterioruberin | (2S,2′S)-2,2′-Bis(3-hydroxy-3-methylbutyl)-3,4,3′,4′-tetradehydro-1,2,1′,2′-tetrahydro-y,y-carotene-1,1′-diol | $C_{50}H_{76}O_4$ | 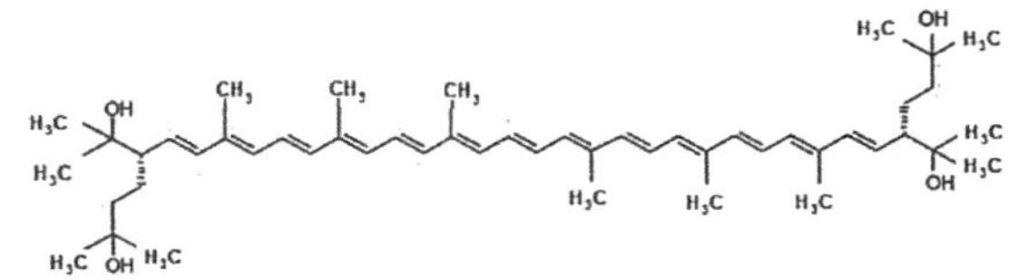 | [8] |
| Myxol | (3R,3′E)-3′,4′-Didehydro-1′,2′-dihydro-β,ψ-carotene-1′,2′,3-triol | $C_{40}H_{56}O_3$ | 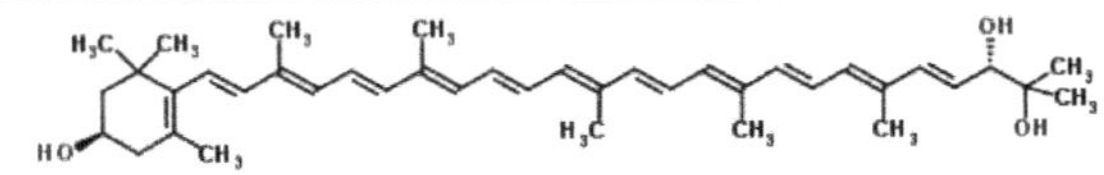 | [132] |
| Salinixanthin | (3′E)-2′-Hydroxy-4-oxo-3′,4′-didehydro-1′,2′-dihydro-β,ψ-caroten-1′-yl 6-O-(13-methyltetradecanoyl)-β- D-glucopyranoside | $C_{61}H_{92}O_9$ | 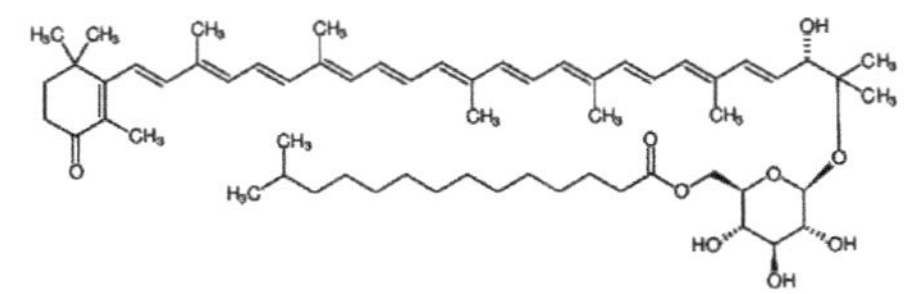 | [137] |
| Saproxanthin | (3′Z)-3′,4′-Didehydro-1′,2′-dihydro-β,ψ-carotene-1′,3-diol | $C_{40}H_{56}O_2$ | 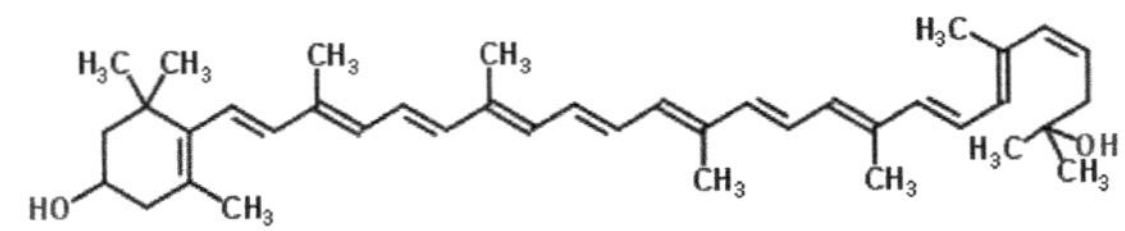 | [134] |
| Sioxanthin | (2′S)-1′-(β-D-Glucopyranosyloxy)-3′,4′-didehydro-1′,2′-dihydro-Φ,Ψ-caroten-2′-ol | $C_{46}H_{62}O_7$ | 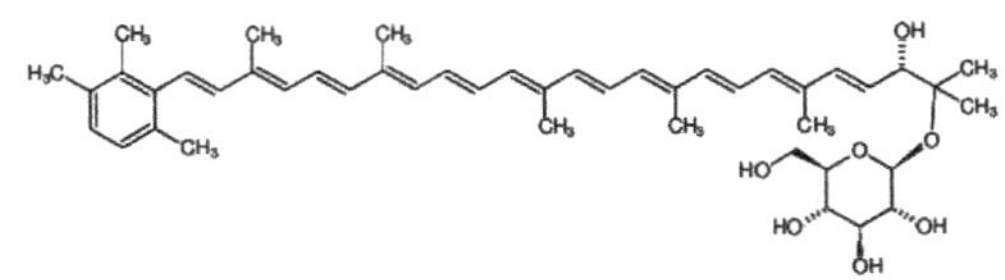 | [140] |
| Siphonaxanthin | (3R,3′R,6′R)-3,3′,19-Trihydroxy-4′,5′-didehydro-5′,6′,7,8-tetrahydro-β,β-caroten-8-one | $C_{40}H_{56}O_4$ | 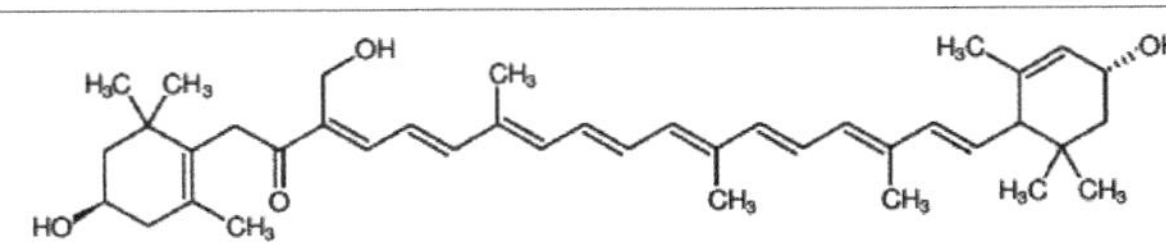 | [141] |

## 5. Carotenoids and Human Health

To date, more than 6000 works on the effects of carotenoids on human health have been published in the indexed scientific journals (PubMed: 6500; Scopus 3000; and WOS: 5978; "carotenoids" and "human health" were used as keywords to perform the search). Most of the studies have reported the general characterizations of carotenoids using biochemical techniques and the conclusions regarding their potential benefits [17]. In certain other works, the potential effects of carotenoids on cell lines or animal models have been described. Basically, only one-third of the research currently published in this area of study has been conducted in humans. Therefore, so far, the studies demonstrating a direct correlation between the carotenoid uptake and the beneficial effects on human health are scarce. Another important aspect that needs to be highlighted here is that the assays that are used to analyze the carotenoids in the human body must be optimized, as several of the studies continue to use empirical prediction models instead of real-time measurements in human fluids or tissues [145]. Consequently, further research is required in both basic and applied aspects.

The few connections that have been established directly between carotenoid consumption and the human health reveal that: (i) these pigments are powerful antioxidants, and consequently, they could possess antitumor activity; (ii) certain carotenoids exhibit pro-vitamin A activity; and (iii) certain carotenoids may be able to demonstrate regulatory activities at different levels in several tissues, and because of this regulatory role, they may exhibit a protective effect against the development of degenerative diseases, such as macular degeneration, cancer and heart diseases [15], or they may be able to prevent metabolic diseases, such as type 2 diabetes or dyslipidemia [146,147]. Table 1 summarizes the major potential beneficial effects of carotenoids (or their derivatives) on human health, as described by various studies, to date.

Nowadays, consumers are aware of the association among diet, health and the prevention of diseases [147]. As a result, bioactive compounds, such as antioxidants, peptides, carbohydrates and lipids, present in food have become important for human nutrition and the reason for the development of functional foods and nutraceuticals in the food industry. The terms such as "health-promoting foods" or "functional foods" refer to food that is rich in bioactive compounds and substances that are effective against diseases [148]. The outer appearance of these functional foods is similar to conventional foods; however, the difference is that functional foods offer advantages beyond the basic nutritional functions; they exhibit physiological benefits and are able to reduce the risk of chronic diseases [149]. The term "nutraceutical" was coined by combining two terms, "nutrition" and "pharmaceutical", in 1989 [150]; nutraceuticals are defined as "a food (or a part of food) that provides medical or health benefits, including the prevention and/or treatment of a disease" [150]. When the functional food aids in the prevention and/or treatment of a disease(s) and/or a disorder(s) other than anemia, it is referred to as a "nutraceutical" [151]. It is important to note that functional foods may not be a universal panacea for poor health habits [152]. In brief, health benefits and novel markets in the food industry may be obtained by incorporating carotenoids into foods that do not contain high amounts of these natural pigments; also, when added to food, the bioavailability of carotenoids is improved compared to when they are consumed from their natural sources [153].

The interest in marine-derived functional foods has increased, especially the functional foods from marine algae, which are considered an interesting source of bioactive compounds with biological properties to be used as functional food ingredients. Taking into consideration all of these aspects, carotenoid-rich foods, such as the Mediterranean diet, are being regarded as the best diet models for supporting healthy living standards and to promote the positive effects on human health, for instance cardiovascular health [15]. Host-related factors may also affect the capability to absorb, convert and metabolize dietary carotenoids. Factors such as gender, body fat and genetic variation may also play important roles in these processes.

It is noteworthy that several recent works have stated that the intake of food supplements without professional supervision is associated with certain problems, such as side effects caused when the supplements interact with classical synthetic drugs [154]. The excessive intake of carotenoids causes carotenemia in humans, which is benign. Moreover, unusual diets and the consumption of preformed vitamin A present in the food supplements may lead to vitamin A toxicity [155,156]. In summary, a significant number of previous studies support the use of food supplements, nutraceuticals and carotenoid-rich foods; however, in all cases, the consumption of these supplements must be accompanied by professional/medical supervision.

## 6. Marine Microalgae Biomass and Their Valuable Molecules in the Food Market

Microalgae have been used as food, feed and fertilizers for centuries. Their cultivation on a large scale may be extended to areas that are unsuitable for agricultural purposes, with productivities higher than those obtained with traditional crops [9]. Marine microalgae are rich in PUFAs (poly-unsaturated fatty acids), polysaccharides, sterols and pigments such as carotenoids, chlorophylls and phycobiliproteins [157,158] that could be used to increase the nutritional value of foods [159]. Antioxidants have become a major focus of interest, and several studies have associated antioxidant activity with carotenoid content in algae [160], which has resulted in an increased demand for algae in order to obtain carotenoids to be added to functional foods.

The global market for carotenoids is expected to reach U.S. $1.7 billion by the year 2022. The market size of functional foods derived from microalgae has increased five-fold since the beginning of the century, and its development has relatively matured now [161]. Microalgae are currently used both as dried whole algae and as the source for the extraction of high-valued food supplements, such as carotenoids and omega-3 fatty acids. A majority of microalgae used belong to the genera *Chlorella*, *Dunaliella*, *Spirulina*, *Haematococcus*, *Isochrysis* and *Schizochytrium* [162]. *Chlorella* and *Spirulina* are the most-marketed algae worldwide. The production of *Chlorella* is centered in Asia, while *Spirulina* is produced in Asia, as well as the USA; although both of these microalgae are also produced in a few other countries with warm climates [163]. Both *Chlorella* and *Spirulina* rank first in the microalgal biomass production rate worldwide (5000 and 2000 tons of dry weight/year, respectively), with estimated production values of about $40 million/year for each [164]. In food, carotenoids serve as the precursors to aroma compounds and also as natural antioxidants that may help prolong the shelf life of food products [165]. Therefore, Portugal [9,166,167] proposed to utilize microalgal biomass as pigments and functional ingredients in food products. The biomass of *C. vulgaris* and *H. pluvialis* has been incorporated into pea protein-stabilized emulsions, achieving a considerable compromise between the stability and the sensory properties. The *C. vulgaris* biomass, with organoleptic characteristics that are acceptable to consumers, was used as a source of natural green pigment to color Christmas cookies.

A fishy taste and a powdery consistency are a few inconveniences caused by the incorporation of microalgal biomass in traditional food products. It may be necessary to address the issues concerning these organoleptic characteristics, for the food products to be accepted by many consumers [147]. In the traditional cuisines of Asian countries, algae have been a common ingredient. As a result, the addition of microalgae to food is not perceived as a strong change and is appreciated by the consumers in these countries [168]. In Western nations, however, microalgae are considered a novel ingredient, and their addition to food has not been accepted yet.

Instead of using the whole biomass, Gantar and Svircev [169] suggested employing the microalgal biomass as a source of biomolecules of interest, so that it would be accepted by consumers [162]. Microalgae-based high-valued molecules have been produced on a smaller scale, with a larger market potential, mainly in Asia, the USA and Australia. Although the production costs are high, the quality of biomolecules produced is better than that from the alternative methods, such as chemical synthesis or the extraction from plants; this is mainly because the molecules produced from the microalgal biomass is more effective for food applications compared to their synthetic variants [170]. The most important molecules from the microalgal biomass that are currently on the market are carotenoids and fatty acids used as dietary supplements. Astaxanthin from dried *H. pluvialis*, which is used as a food additive or as a dietary supplement, is the most-developed product from this source [171].

As a result of the development of novel eating habits among consumers who are now demanding more sophisticated products, a few innovative food products that are enriched with microalgae have been developed and positioned in this emerging functional food market. One good example is *Spirulina*, an ingredient in beverages and shakes [162], which is used in combination with other ingredients. It is available in New Zealand as Charlie's Honest Superfood Spirulina and Fruit Smoothie. Coca-Cola's brand Odwalla manufactures a fruit juice drink with *Spirulina* as an ingredient [162]. Another example is PepsiCo, 100% Juice Smoothie with *Spirulina* [172]. Nestle Rowntree reintroduced the blue Smarties into the market after having identified the blue-green cyanobacteria as a natural source of the blue color [173].

Besides the health interests, it is important to assess the toxicity of the natural compounds isolated from microalgae (pigments, lipids, etc.), as certain compounds may accumulate in the human body, and this could affect the safe usage of microalgae in food products for human consumption [174]. It is undeniable that the interest in using microalgae or the other marine microorganisms as natural sources for functional food ingredients is growing, and certain reasons that support this fact are as follows: the number of marine species available and the prospect of discovering novel ones is significant [175], and the management of growth conditions is able to assist the marine microorganisms in accumulating bioactive compounds, which is helpful in economically-competitive processes [8,44,160].

*Regulation*

The regulations for the production of marine microbial biomass (or their bioactive compounds) are poorly established, increasing the chaos worldwide in this regard. Besides, the laws and regulations associated with food or food ingredients vary with different countries [176]. There are two EU regulations for the production and marketing of microalgae-based products for food and feed: the Food Safety Regulation (EC 178/2002) and the Novel Food Regulation (EC 258/97) (MB). The latter is particularly relevant, as it provides the authorization procedures for all novel food and feed products. In Europe, the first step to commercialize carotenoids or the microorganisms as food or food ingredients is to identify whether they have been consumed to a significant degree before 15 May 1997 (the date on which the regulation (EC) No. 258/97 entered into force), in at least one of the member states. If this is the case, the food or the food ingredient should be considered the same as a conventional food, if it demonstrates the same characteristics and composition as those of the conventional food. This simplifies the http://www.linguee.es/ingles-espanol/traduccion/commercialization+process.html commercialization process, and the food product may be positioned in the market within the EU after notification to the European Commission [177]. Prior to commercializing any type of novel food product (including the ones from microalgae), scientific evidence must be provided in order to confirm that the novel food products are equivalent to the traditional ones (EC Regulation, 1997) [178].

In another scenario, if the food or the food ingredient is identified as a novel food, the further procedure becomes complex. In order to introduce a novel food or food ingredient into the EU market, prior authorization is required, which includes the manufacturing processes, as well as a rigorous assessment of the toxicological, nutritional, compositional and other relevant data by the competent authorities of the respective EU member state. In the event of commercialization of the microorganism as a functional food, it must be demonstrated to affect one or more target functions in the body beneficially. The functional foods are not allowed to be in the form of pills or capsules; they must remain in the form of food. Furthermore, they must be demonstrated to achieve their effects when used in an amount that would normally be expected to be consumed in the diet (EAS 2008) [164].

Regulations in the USA assess the food product, while those in Europe are focused on the technology used to obtain the final food product. The competition from outside Europe (China and the USA) is growing fast, and this is the right time for Europe to explore the opportunities to increase the possibility of successful large-scale applications of microalgae and other marine microorganisms in food [149]. The insufficient domestic demand for marine microalgal food products in Europe and the difficulties in achieving commercial authorization due to the strict Novel Food regulation remain a challenge. In this regard, experts consider cost reductions, technical breakthroughs and achieving better cooperation between academia and industries as the most important challenges in the near future [164].

## 7. Conclusions

In the last few decades, the research on carotenoids, and their implementation in markets, have advanced considerably. However, further advancement of this development is required, especially in the processes used to obtain carotenoids; for example, chemical synthesis must be replaced by biological production. In this way, the reduction in the costs of the production of carotenoids and in the generation of non-active waste would be achieved.

Marine microorganisms offer numerous advantages as natural sources of carotenoids: they usually have low nutritional requirements during their growth and cultivation, which reduces the production costs; the culture medium in which they grow contains moderate or high salt concentrations, which prevents contamination with other microorganisms, reducing the costs and facilitating the downstream processes; these microorganisms are also a source of rare carotenoids, the properties of which are yet to be explored, for example bacterioruberin from Haloarchaea.

In a more applied sense, it should be noted that an extensive bibliography is available on the biochemical characteristics of carotenoids and their potential beneficial effects on human health. However, further studies on the direct and real-time effects of carotenoids on human populations are required, in order to corroborate their antioxidant, antitumor, anti-aging and various other roles. On the nutritional level, the use of the carotenoids obtained from marine biomass is well studied; also, their use is common in the diets of people in several countries, especially in Asia. The industry must continue to take steps in the direction of improving the organoleptic properties, which are altered by the addition of marine biomass to the conventional foods; especially in Europe, where such functional foods are relatively recent and not yet completely accepted.

**Acknowledgments:** This work was funded by a research grant from the MINECO Spain (CTM2013-43147-R).

**Conflicts of Interest:** The authors declare no conflict of interest.

## References

1. Nisar, N.; Li, L.; Lu, S.; Khin, N.C.; Pogson, B.J. Carotenoid metabolism in plants. *Mol. Plant* **2015**, *8*, 68–82. [CrossRef] [PubMed]
2. Zhang, J.; Sun, Z.; Sun, P.; Chen, T.; Chen, F. Microalgal carotenoids: Beneficial effects and potential in human health. *Food Funct.* **2014**, *5*, 413–425. [CrossRef] [PubMed]
3. Fiedor, J.; Burda, K. Potential role of carotenoids as antioxidants in human health and disease. *Nutrients* **2014**, *6*, 466–468. [CrossRef] [PubMed]
4. Yatsunami, R.; Ando, A.; Yang, Y.; Takaichi, S.; Kohno, M.; Matsumura, Y.; Ikeda, H.; Fukui, T.; Nakasone, K.; Fujita, N.; et al. Identification of carotenoids from the extremely halophilic archaeon *Haloarcula japonica*. *Front. Microbiol.* **2014**, *5*, 100–105. [CrossRef] [PubMed]
5. Goodwin, T.W.; Britton, G. Distribution and analysis of carotenoids. In *Plant Pigments*, 2nd ed.; Goodwin, T.W., Ed.; Academic Press: London, UK, 1980; Volume 1, pp. 61–132.
6. Cunningham, F.X.; Gantt, E. Genes and enzymes of carotenoid biosynthesis in plants. *Annu. Rev. Plant Physiol. Plant Mol. Biol.* **1998**, *49*, 557–583. [CrossRef] [PubMed]
7. Mata-Gómez, L.C.; Montañez, J.C.; Méndez-Zavala, A.; Aguilar, C.N. Biotechnological production of carotenoids by yeasts: An overview. *Microb. Cell Fact.* **2014**, *13*, 12. [CrossRef] [PubMed]
8. Rodrigo-Baños, M.; Garbayo, I.; Vílchez, C.; Bonete, M.J.; Martínez-Espinosa, R.M. Carotenoids from Haloarchaea and their potential in biotechnology. *Mar. Drugs* **2015**, *13*, 5508–5532. [CrossRef] [PubMed]
9. Gouveia, L.; Coutinho, C.; Mendonça, E.; Batista, A.P.; Sousa, I.; Bandarra, N.M.; Raymundo, A. Functional biscuits with PUFA-ω3 from *Isochrysis galbana*. *J. Sci. Food Agric.* **2008**, *88*, 891–896. [CrossRef]
10. Vílchez, C.; Forján, E.; Cuaresma, M.; Bédmar, F.; Garbayo, I.; Vega, J.M. Marine carotenoids: Biological functions and commercial applications. *Mar. Drugs* **2011**, *9*, 319–333. [CrossRef] [PubMed]

11. Forján, E.; Vílchez Lobato, C.; Vega Piqueres, J.M. *Biotecnología de Microalgas*, 1st ed.; Cepsa: Huelva, Spain, 2014; pp. 1–318, ISBN 978-84-617-2314-0.
12. Othman, R.; Mohd Zaifuddin, F.A.; Hassan, N.M. Carotenoid biosynthesis regulatory mechanisms in plants. *J. Oleo Sci.* **2014**, *63*, 753–760. [CrossRef] [PubMed]
13. Giuliano, G. Plant carotenoids: Genomics meets multi-gene engineering. *Curr. Opin. Plant Biol.* **2014**, *19*, 111–117. [CrossRef] [PubMed]
14. Burton, G.W.; Foster, D.O.; Perly, B.; Slater, T.F.; Smith, I.C.; Ingold, K.U. Biological antioxidants. *Philos. Trans. R. Soc. Lond. B Biol. Sci.* **1985**, *311*, 565–578. [CrossRef] [PubMed]
15. Gammone, M.A.; Riccioni, G.; D'Orazio, N. Marine carotenoids against oxidative stress: Effects on human health. *Mar. Drugs* **2015**, *13*, 6226–6246. [CrossRef] [PubMed]
16. LaFountain, A.M.; Prum, R.O.; Frank, H.A. Diversity, physiology, and evolution of avian plumage carotenoids and the role of carotenoid-protein interactions in plumage color appearance. *Arch. Biochem. Biophys.* **2015**, *572*, 201–212. [CrossRef] [PubMed]
17. Namitha, K.K.; Negi, P.S. Chemistry and biotechnology of carotenoids. *Crit. Rev. Food Sci. Nutr.* **2010**, *50*, 728–760. [CrossRef] [PubMed]
18. Zile, M.H. Vitamin A and embryonic development: An overview. *J. Nutr.* **1998**, *128*, 455S–458S. [CrossRef] [PubMed]
19. Rivera, S.M.; Canela-Garayoa, R. Analytical tools for the analysis of carotenoids in diverse materials. *J. Chromatogr. A* **2012**, *1224*, 1–10. [CrossRef] [PubMed]
20. Kaulmann, A.; Bohn, T. Carotenoids, inflammation, and oxidative stress—Implications of cellular signaling pathways and relation to chronic disease prevention. *Nutr. Res.* **2014**, *34*, 907–929. [CrossRef] [PubMed]
21. Gupta, C.; Prakash, D. Phytonutrients as therapeutic agents. *J. Complement. Integr. Med.* **2014**, *11*, 151–169. [CrossRef] [PubMed]
22. Ascenso, A.; Ribeiro, H.; Marques, H.C.; Oliveira, H.; Santos, C.; Simões, S. Chemoprevention of photocarcinogenesis by lycopene. *Exp. Dermatol.* **2014**, *23*, 874–878. [CrossRef] [PubMed]
23. Woodside, J.V.; McGrath, A.J.; Lyner, N.; McKinley, M.C. Carotenoids and health in older people. *Maturitas* **2015**, *80*, 63–68. [CrossRef] [PubMed]
24. Freeman, H.D.; Valuska, A.J.; Taylor, R.R.; Ferrie, G.M.; Grand, A.P.; Leighty, K.A. Plumage variation and social partner choice in the greater flamingo (*Phoenicopterus roseus*). *Zoo Biol.* **2016**, *35*, 409–414. [CrossRef] [PubMed]
25. Duarte, R.C.; Flores, A.A.V.; Stevens, M. Camouflage through colour change: Mechanisms, adaptive value and ecological significance. *Philos. Trans. R. Soc. Lond. B Biol. Sci.* **2017**, *372*, pii: 20160342. [CrossRef]
26. Del Campo, J.A.; García-González, M.; Guerrero, M.G. Outdoor cultivation of microalgae for carotenoid production: Current state and perspectives. *Appl. Microbiol. Biotechnol.* **2007**, *74*, 1163–1174. [CrossRef] [PubMed]
27. Mandelli, F.; Miranda, V.S.; Rodrigues, E.; Mercadante, A.Z. Identification of carotenoids with high antioxidant capacity produced by extremophile microorganisms. *World J. Microbiol. Biotechnol.* **2012**, *28*, 1781–1790. [CrossRef] [PubMed]
28. Jehlička, J.; Edwards, H.G.; Oren, A. Bacterioruberin and salinixanthin carotenoids of extremely halophilic Archaea and Bacteria: A Raman spectroscopic study. *Spectrochim. Acta Part A Mol. Biomol. Spectrosc.* **2013**, *106*, 99–103. [CrossRef] [PubMed]
29. Davies, J.; Ryan, K.S. Introducing the parvome: Bioactive compounds in the microbial world. *ACS Chem. Biol.* **2012**, *7*, 252–259. [CrossRef] [PubMed]
30. Gupta, S.; Abu-Ghannam, N. Bioactive potential and possible health effects of edible brown seaweeds. *Trends Food Sci. Technol.* **2011**, *22*, 315–326. [CrossRef]
31. Stengel, D.B.; Connan, S.; Popper, Z.A. Algal chemodiversity and bioactivity: Sources of natural variability and implications for commercial application. *Biotechnol. Adv.* **2011**, *29*, 483–501. [CrossRef] [PubMed]
32. Research and Innovation Website of the European Commission. Available online: http://ec.europa.eu/research (accessed on 20 February 2018).
33. European Network to Advance Carotenoid Research and Applications in Agro-Food and Health. Available online: http://www.cost.eu/COST_Actions/ca/CA15136 (accessed on 20 February 2018).
34. Spanish Network of Carotenoids. Available online: http://www.ictan.csic.es/1693/cared-red-espanola-de-carotenoides-desde-los-microbios-y-plantas-a-los-alimentos-y-la-salud/ (accessed on 20 February 2018).

35. International Carotenoid Society. Available online: http://www.carotenoidsociety.org (accessed on 20 February 2018).
36. Ibero-American Program of Science and Technology for Development—Carotenoids in Agro-Food and Health Section. Available online: http://www.cyted.org/es/ibercarot (accessed on 20 February 2018).
37. Zhang, L.; An, R.; Wang, J.; Sun, N.; Zhang, S.; Hu, J.; Kuai, J. Exploring novel bioactive compounds from marine microbes. *Curr. Opin. Microbiol.* **2005**, *8*, 276–281. [CrossRef] [PubMed]
38. Bhatnagar, I.; Kim, S.K. Immense essence of excellence: Marine microbial bioactive compounds. *Mar. Drugs* **2010**, *8*, 2673–2701. [CrossRef] [PubMed]
39. Habbu, P.; Warad, V.; Shastri, R.; Madagundi, S.; Kulkarni, V.H. Antimicrobial metabolites from marine microorganisms. *Chin. J. Nat. Med.* **2016**, *14*, 101–116. [CrossRef]
40. Felczykowska, A.; Bloch, S.K.; Nejman-Faleńczyk, B.; Barańska, S. Metagenomic approach in the investigation of new bioactive compounds in the marine environment. *Acta Biochim. Pol.* **2012**, *59*, 501–505. [PubMed]
41. Gurgui, C.; Piel, J. Metagenomic approaches to identify and isolate bioactive natural products from microbiota of marine sponges. *Methods Mol. Biol.* **2010**, *668*, 247–264. [CrossRef] [PubMed]
42. Schweder, T.; Lindequist, U.; Lalk, M. Screening for new metabolites from marine microorganisms. *Adv. Biochem. Eng. Biotechnol.* **2005**, *96*, 1–48. [PubMed]
43. Markou, G.; Nerantzis, E. Microalgae for high-value compounds and biofuels production: A review with focus on cultivation under stress conditions. *Biotechnol. Adv.* **2013**, *31*, 1532–1542. [CrossRef] [PubMed]
44. Calegari-Santos, R.; Diogo, R.A.; Fontana, J.D.; Bonfim, T.M. Carotenoid production by halophilic Archaea under different culture conditions. *Curr. Microbiol.* **2016**, *72*, 641–651. [CrossRef] [PubMed]
45. García-González, M.; Moreno, J.; Manzano, J.C.; Florencio, F.J.; Guerrero, M.G. Production of *Dunaliella salina* biomass rich in 9-cis-β-carotene and lutein in a closed tubular photobioreactor. *J. Biotechnol.* **2005**, *115*, 81–90. [CrossRef] [PubMed]
46. Misawa, N. Carotenoid β-ring hydroxylase and ketolase from marine bacteria- promiscuous enzymes for synthesizing functional xantophylls. *Mar. Drugs* **2011**, *9*, 757–771. [CrossRef] [PubMed]
47. Marshall, C.P.; Leuko, S.; Coyle, C.M.; Walter, M.R.; Burns, B.P.; Neilan, B.A. Carotenoid analysis of halophilic archaea by resonance Raman spectroscopy. *Astrobiology* **2007**, *7*, 631–643. [CrossRef] [PubMed]
48. Relevy, N.Z.; Harats, D.; Harari, A.; Ben-Amotz, A.; Bitzur, R.; Rühl, R.; Shaish, A. Vitamin A-deficient diet accelerated atherogenesis in apolipoprotein E(-/-) Mice and dietary β-carotene prevents this consequence. *Biomed Res. Int.* **2015**, *2015*, 758723. [CrossRef] [PubMed]
49. Sluijs, I.; Cadier, E.; Beulens, J.W.; van der A, D.L.; Spijkerman, A.M.; van der Schouw, Y.T. Dietary intake of carotenoids and risk of type 2 diabetes. *Nutr. Metab. Cadrivasc. Dis.* **2015**, *25*, 376–381. [CrossRef] [PubMed]
50. Boeke, C.E.; Tamimi, R.M.; Berkey, C.S.; Colditz, G.A.; Eliassen, A.H.; Malspeis, S.; Willett, W.C.; Frazier, A.L. Adolescent carotenoid intake and benign breast disease. *Pediatrics* **2014**, *133*, e1292–e1298. [CrossRef] [PubMed]
51. Takaichi, S. Carotenoids in algae: Distributions, biosynthesis and functions. *Mar. Drugs* **2011**, *9*, 1101–1118. [CrossRef] [PubMed]
52. Yokoyama, A.; Izumida, H.; Miki, W. Production of astaxanthin and 4-ketozeaxanthin by the marine bacterium *Agrobacterium aurantiacum. Biosci. Biotechnol. Biochem.* **1994**, *58*, 1842–1844. [CrossRef]
53. Dhankar, J.; Kadian, S.S.; Sharma, A. Astaxanthin: A potential carotenoid. *Int. J. Pharm. Sci. Res.* **2012**, *3*, 1246–1259. [CrossRef]
54. Fassett, R.G.; Coombes, J.S. Astaxanthin in cardiovascular health and disease. *Molecules* **2012**, *17*, 2030–2048. [CrossRef] [PubMed]
55. Higuera-Ciapara, I.; Félix-Valenzuela, L.; Goycoolea, F.M. Astaxanthin: A review of its chemistry and applications. *Crit. Rev. Food Sci. Nutr.* **2006**, *46*, 185–196. [CrossRef] [PubMed]
56. Riccioni, G.; D'Orazio, N.; Franceschelli, S.; Speranza, L. Marine carotenoids and cardiovascular risk markers. *Mar. Drugs* **2011**, *9*, 1166–1175. [CrossRef] [PubMed]
57. Tanaka, T.; Shnimizu, M.; Moriwaki, H. Cancer chemoprevention by carotenoids. *Molecules* **2012**, *17*, 3202–3242. [CrossRef] [PubMed]
58. Peng, J.; Yuan, J.P.; Wu, C.F.; Wang, J.H. Fucoxanthin, a marine carotenoid present in brown seaweeds and diatoms: Metabolism and bioactivities relevant to human health. *Mar. Drugs* **2011**, *9*, 1806–1828. [CrossRef] [PubMed]

59. Afolayan, A.F.; Bolton, J.J.; Lategan, C.A.; Smith, P.J.; Beukes, D.R. Fucoxanthin, tetraprenylatedtoluquinone and toluhydroquinone metabolites from *Sargassum heterophyllum* inhibit the in vitro growth of the malaria parasite *Plasmodium falciparum*. *Z. Naturforsch. C* **2008**, *63*, 848–852. [CrossRef] [PubMed]
60. Woo, M.N.; Jeon, S.M.; Shin, Y.C.; Lee, M.K.; Kang, M.A.; Choi, M.S. Anti-obese property of fucoxanthin is partly mediated by altering lipid-regulating enzymes and uncoupling proteins of visceral adipose tissue in mice. *Mol. Nutr. Food Res.* **2009**, *53*, 1603–1611. [CrossRef] [PubMed]
61. Shaina, M.; Hameed, A.; Lin, S.Y.; Lee, R.J.; Lee, M.R.; Young, C.C. *Gramella planctonica* sp. Nov., a zeaxanthin-producing bacterium isolated from surface seawater and emended descriptions of *Gramella estuarii* and *Gramella echinicola*. *Antonie Leeuwenhoek* **2014**, *105*, 771–779. [CrossRef] [PubMed]
62. Huang, Y.M.; Dou, H.L.; Huang, F.F.; Xu, X.R.; Zou, Z.Y.; Lin, X.M. Effect of supplemental lutein and zeaxanthin on serum, macular pigmentation, and visual performance in patients with early age-related macular degeneration. *Biomed Res. Int.* **2015**, *2015*, 564738. [CrossRef] [PubMed]
63. Bernstein, P.S.; Li, B.; Vachali, P.P.; Gorusupudi, A.; Shyam, R.; Henriksen, B.S.; Nolan, J.M. Lutein, zeaxanthin, and meso-zeaxanthin: The basic and clinical science underlying carotenoid-based nutritional interventions against ocular disease. *Prog. Retinal Eye Res.* **2016**, *50*, 34–66. [CrossRef] [PubMed]
64. Lu, M.S.; Fang, Y.J.; Chen, Y.M.; Luo, W.P.; Pan, Z.Z.; Zhong, X.; Zhang, C.X. Higher intake of carotenoid is associated with a lower risk of colorectal cancer in Chinese adults: A case-control study. *Eur. J. Nutr.* **2015**, *54*, 619–628. [CrossRef] [PubMed]
65. Wang, L.; Li, B.; Pan, M.X.; Mo, X.F.; Chen, Y.M.; Zhang, C.X. Specific carotenoid intake is inversely associated with the risk of breast cancer among Chinese women. *Br. J. Nutr.* **2014**, *111*, 1686–1695. [CrossRef] [PubMed]
66. Ghodratizadeh, S.; Kanbak, G.; Beyramzadeh, M.; Dikmen, Z.G.; Memarzadeh, S.; Habibian, R. Effect of carotenoid β-cryptoxanthin on cellular and humoral immune response in rabbit. *Vet. Res. Commun.* **2014**, *38*, 59–62. [CrossRef] [PubMed]
67. Granado-Lorencio, F.; de Las Heras, L.; Millán, C.S.; García-López, F.J.; Blanco-Navarro, I.; Pérez-Sacristán, B.; Domínguez, G. β-Cryptoxanthin modulates the response to phytosterols in post-menopausal women carrying NPC1L1 L272L and ABCG8 A632 V polymorphisms: An exploratory study. *Genes Nutr.* **2014**, *9*, 428. [CrossRef] [PubMed]
68. Chisté, R.C.; Freitas, M.; Mercadante, A.Z.; Fernandes, E. Carotenoids are effective inhibitors of in vitro hemolysis of human erythrocytes, as determined by a practical and optimized cellular antioxidant assay. *J. Food Sci.* **2014**, *79*, H1841–H1847. [CrossRef] [PubMed]
69. Min, K.B.; Min, J.Y. Association between leukocyte telomere length and serum carotenoid in US adults. *Eur. J. Nutr.* **2016**, *56*, 1045–1052. [CrossRef] [PubMed]
70. Liu, X.R.; Wang, Y.Y.; Dan, X.G.; Kumar, A.; Ye, T.Z.; Yu, Y.Y.; Yang, L.G. Anti-inflammatory potential of β-cryptoxanthin against LPS-induced inflammation in mouse Sertoli cells. *Reprod. Toxicol.* **2015**, *60*, 148–155. [CrossRef] [PubMed]
71. Ozaki, K.; Okamoto, M.; Fukasawa, K.; Iezaki, T.; Onishi, Y.; Yoneda, Y.; Sugiura, M.; Hinoi, E. Daily intake of β-cryptoxanthin prevents bone loss by preferential disturbance of osteoclastic activation in ovariectomized mice. *J. Pharmacol. Sci.* **2015**, *129*, 72–77. [CrossRef] [PubMed]
72. Aasen, A.J.; Liaaen-Jensen, S. The carotenoids of flexibacteria: II. A new xanthophyll from Saprospira grandis. *Acta Chem. Scand.* **1996**, *20*, 811–819. [CrossRef]
73. Francis, G.W.; Hertzberg, S.; Andersen, K.; Liaaen-Jensen, S. New carotenoid glycosides from *Oscillatoria limosa*. *Phytochemistry* **1970**, *9*, 629–635. [CrossRef]
74. Raposo, M.F.; De Morais, A.M.; De Morais, R.M. Carotenoids from marine microalgae: A valuable natural source for the prevention of chronic diseases. *Mar. Drugs* **2015**, *13*, 5128–5155. [CrossRef] [PubMed]
75. Sulich, A.; Hamułka, J.; Nogal, D. Dietary sources of lutein in adults suffering eye disease (AMD/cataracts). *Rocz. Państwowego Zakl. Hig.* **2015**, *66*, 55–60.
76. Nataraj, J.; Manivasagam, T.; Thenmozhi, A.J.; Essa, M.M. Lutein protects dopaminergic neurons against MPTP-induced apoptotic death and motor dysfunction by ameliorating mitochondrial disruption and oxidative stress. *Nutr. Neurosci.* **2016**, *19*, 237–246. [CrossRef] [PubMed]
77. Han, H.; Cui, W.; Wang, L.; Xiong, Y.; Liu, L.; Sun, X.; Hao, L. Lutein prevents high fat diet-induced atherosclerosis in ApoE-deficient mice by inhibiting NADPH oxidase and increasing PPAR expression. *Lipids* **2015**, *50*, 261–273. [CrossRef] [PubMed]

78. Mathews-Roth, M.M. Antitumor activity of β-carotene, canthaxanthin and phytoene. *Oncology* **1982**, *39*, 33–37. [CrossRef] [PubMed]
79. Mordi, R.C.; Walton, J.C. Identification of products from canthaxanthin oxidation. *Food Chem.* **2016**, *197*, 836–840. [CrossRef] [PubMed]
80. Surai, P.F. The antioxidant properties of canthaxanthin and its potential effects in the poultry eggs and on embryonic development of the chick, Part 1. *World's Poult. Sci. J.* **2012**, *68*, 465–476. [CrossRef]
81. Kent, M.; Welladsen, H.M.; Mangott, A.; Li, Y. Nutritional evaluation of Australian microalgae as potential human health supplements. *PLoS ONE* **2015**, *10*, e0118985. [CrossRef] [PubMed]
82. Soontornchaiboon, W.; Joo, S.S.; Kim, S.M. Anti-inflammatory effects of violaxanthin isolated from microalga *Chlorella ellipsoidea* in RAW 264.7 macrophages. *Biol. Pharm. Bull.* **2012**, *35*, 1137–1144. [CrossRef] [PubMed]
83. Issouf, M.; Mearns, S.A.; Fraser, K.; Hodgson, R. Biological Production of Zeaxanthin and Carotenoid Biosynthesis Control. World Intellectual Property Organization (WIPO) Patent Application WO2006120400 A1, 16 November 2006.
84. Viuda-Martos, M.; Sánchez-Zapata, E.; Sayas-Barberá, E.; Sendra, E.; Pérez-Álvarez, J.A.; Fernández-López, J. Tomato and tomato byproducts. Human health benefits of lycopene and its application to meat products: A review. *Crit. Rev. Food Sci. Nutr.* **2014**, *54*, 1032–1049. [CrossRef] [PubMed]
85. Rao, A.V.; Rao, L.G. Carotenoids and human health. *Pharmacol. Res.* **2007**, *55*, 207–216. [CrossRef] [PubMed]
86. Igielska-Kalwat, J.; Gościańska, J.; Nowak, I. Carotenoids as natural antioxidants. *Postepy Hig. Med. Dosw. (Online)* **2015**, *69*, 418–428. [CrossRef] [PubMed]
87. Ambati, R.R.; Phang, S.M.; Ravi, S.; Aswathanarayana, R.G. Astaxanthin: Sources, extraction, stability, biological activities and its commercial applications—A review. *Mar. Drugs* **2014**, *12*, 128–152. [CrossRef] [PubMed]
88. Schmidt, I.; Schewe, H.; Gassel, S.; Jin, C.; Buckingham, J.; Hümbelin, M.; Sandmann, G.; Schrader, J. Biotechnological production of astaxanthin with *Phaffia rhodozyma/Xanthophyllomyces dendrorhous*. *Appl. Microbiol. Biotechnol.* **2011**, *89*, 555–571. [CrossRef] [PubMed]
89. Mogedas, B.; Casal, C.; Forján, E.; Vílchez, C. Beta-carotene production enhancement by UV-A radiation in *Dunaliella bardawil* cultivated in laboratory reactors. *J. Biosci. Bioeng.* **2009**, *108*, 47–51. [CrossRef] [PubMed]
90. Milne, G.W.A. *Gardner's Commercially Important Chemicals: Synonyms, Trade Names, and Properties*, 1st ed.; Wiley-Interscience: New York, NY, USA, 2005; ISBN 978-0-471-73518-3.
91. Stutz, H.; Bresgen, N.; Eckl, P.M. Analytical tools for the analysis of β-carotene and its degradation products. *Free Radic. Res.* **2015**, *49*, 650–680. [CrossRef] [PubMed]
92. Englert, M.; Hammann, S.; Vetter, W. Isolation of β-carotene, α-carotene and lutein from carrots by countercurrent chromatography with the solvent system modifier benzotrifluoride. *J. Chromatogr. A* **2015**, *1388*, 119–125. [CrossRef] [PubMed]
93. Tanvetyanon, T.; Bepler, G. Beta-carotene in multivitamins and the possible risk of lung cancer among smokers versus former smokers: A meta-analysis and evaluation of national brands. *Cancer* **2008**, *113*, 150–157. [CrossRef] [PubMed]
94. Brulc, L.; Simonovska, B.; Vovk, I.; Glavnik, V. Determination of egg yolk xanthophylls by isocratic high-performance liquid chromatography. *J. Chromatogr. A* **2013**, *1318*, 134–141. [CrossRef] [PubMed]
95. Kathiresan, S.; Chandrashekar, A.; Ravishankar, G.A.; Sarada, R. Regulation of astaxanthin and its intermediates through cloning and genetic transformation of β-carotene ketolase in *Haematococcus pluvialis*. *J. Biotechnol* **2015**, *196–197*, 33–41. [CrossRef] [PubMed]
96. Rostami, F.; Razavi, S.H.; Sepahi, A.A.; Gharibzahedi, S.M. Canthaxanthin biosynthesis by *Dietziana tronolimnaea* HS-1: Effects of inoculation and aeration rate. *Braz. J. Microbiol.* **2014**, *45*, 447–456. [CrossRef] [PubMed]
97. Hojjati, M.; Razavi, S.H.; Rezaei, K.; Gilani, K. Stabilization of canthaxanthin produced by *Dietziana tronolimnaea* HS-1 with spray drying microencapsulation. *J. Food Sci. Technol.* **2014**, *51*, 2134–2140. [CrossRef] [PubMed]
98. Heying, E.K.; Tanumihardjo, J.P.; Vasic, V.; Cook, M.; Palacios-Rojas, N.; Tanumihardjo, S.A. Biofortified orange maize enhances β-cryptoxanthin concentrations in egg yolks of laying hens better than tangerine peel fortificant. *J. Agric. Food Chem.* **2014**, *62*, 11892–11900. [CrossRef] [PubMed]

99. Wei, X.; Chen, C.; Yu, Q.; Gady, A.; Yu, Y.; Liang, G.; Gmitter, F.G., Jr. Comparison of carotenoid accumulation and biosynthetic gene expression between Valencia and Rohde Red Valencia sweet oranges. *Plant Sci.* **2014**, *227*, 28–36. [CrossRef] [PubMed]
100. Burri, B.J. Beta-cryptoxanthin as a source of vitamin A. *J. Sci. Food Agric.* **2015**, *95*, 1786–1794. [CrossRef] [PubMed]
101. Li, D.; Xiao, Y.; Zhang, Z.; Liu, C. Light-induced oxidation and isomerization of all-trans-β-cryptoxanthin in a model system. *J. Photochem. Photobiol. B Biol.* **2015**, *142*, 51–58. [CrossRef] [PubMed]
102. Burri, B.J.; La Frano, M.R.; Zhu, C. Absorption, metabolism, and functions of β-cryptoxanthin. *Nutr. Rev.* **2016**, *74*, 69–82. [CrossRef] [PubMed]
103. Kim, S.M.; Jung, Y.J.; Kwon, O.N.; Cha, K.H.; Um, B.H.; Chung, D.; Pan, C.H. A potential commercial source of fucoxanthin extracted from the microalga *Phaeodactylum tricornutum*. *Appl. Biochem. Biotechnol.* **2012**, *166*, 1843–1855. [CrossRef] [PubMed]
104. Crupi, P.; Toci, A.T.; Mangini, S.; Wrubl, F.; Rodolfi, L.; Tredici, M.R.; Coletta, A.; Antonacci, D. Determination of fucoxanthin isomers in microalgae (*Isochrysis* sp.) by high-performance liquid chromatography coupled with diode-array detector multistage mass spectrometry coupled with positive electrospray ionization. *Rapid Commun. Mass Spectrom.* **2013**, *27*, 1027–1035. [CrossRef] [PubMed]
105. Martin, L.J. Fucoxanthin and its metabolite fucoxanthinol in cancer prevention and treatment. *Mar. Drugs* **2015**, *13*, 4784–4798. [CrossRef] [PubMed]
106. Satomi, Y. Antitumor and Cancer-preventative Function of Fucoxanthin: A Marine Carotenoid. *Anticancer Res.* **2017**, *37*, 1557–1562. [CrossRef] [PubMed]
107. Naziri, D.; Hamidi, M.; Hassanzadeh, S.; Tarhriz, V.; Maleki Zanjani, B.; Nazemyieh, H.; Hejazi, M.A.; Hejazi, M.S. Analysis of carotenoid production by *Halorubrum* sp. TBZ126; an extremely halophilic archeon from Urmia Lake. *Adv. Pharm. Bull.* **2014**, *4*, 61–67. [CrossRef] [PubMed]
108. Cuaresma, M.; Casal, C.; Forján, E.; Vílchez, C. Productivity and selective accumulation of carotenoids of the novel extremophile microalga *Chlamydomonas acidophila* grown with different carbon sources in batch systems. *J. Ind. Microbiol. Biotechnol.* **2011**, *38*, 167–177. [CrossRef] [PubMed]
109. Altemimi, A.; Lightfoot, D.A.; Kinsel, M.; Watson, D.G. Employing response surface methodology for the optimization of ultrasound assisted extraction of lutein and β-carotene from spinach. *Molecules* **2015**, *20*, 6611–6625. [CrossRef] [PubMed]
110. Prabhu, S.; Rekha, P.D.; Arun, A.B. Zeaxanthin biosynthesis by members of the genus *Muricauda*. *Pol. J. Microbiol.* **2014**, *63*, 115–119. [PubMed]
111. Costa, S.; Giannantonio, C.; Romagnoli, C.; Barone, G.; Gervasoni, J.; Perri, A.; Zecca, E. Lutein and zeaxanthin concentrations in formula and human milk samples from italian mothers. *Eur. J. Clin. Nutr.* **2015**, *69*, 531–532. [CrossRef] [PubMed]
112. Li, X.R.; Tian, G.Q.; Shen, H.J.; Liu, J.Z. Metabolic engineering of *Escherichia coli* to produce zeaxanthin. *J. Ind. Microbiol. Biotechnol.* **2015**, *42*, 627–636. [CrossRef] [PubMed]
113. Yamamoto, H.Y.; Chang, J.L.; Aihara, M.S. Light-induced interconversion of violaxanthin and zeaxanthin in New Zealand spinach-leaf segments. *Biochim. Biophys. Acta Gen. Subj.* **1967**, *141*, 342–347. [CrossRef]
114. Yamamoto, H.Y.; Kamite, L.; Wang, Y.Y. An ascorbate-induced absorbance change in chloroplasts from violaxanthin de-epoxidation. *Plant Physiol.* **1972**, *49*, 224–228. [CrossRef] [PubMed]
115. Sapozhnikov, D.I. Investigation on the violaxanthin cycle. *Pure Appl. Chem.* **1973**, *35*, 47–61. [CrossRef] [PubMed]
116. Pfündel, E.; Strasser, R.J. Violaxanthin de-epoxidase in etiolated leaves. *Photosynth. Res.* **1988**, *15*, 67–73. [CrossRef] [PubMed]
117. Winter, K.; Königer, M. Dithiothreitol, an inhibitor of violaxanthin de-epoxidation, increases the susceptibility of leaves of *Nerium oleander* L. to photoinhibition of photosynthesis. *Planta* **1989**, *180*, 24–31. [PubMed]
118. Hallin, E.I.; Guo, K.; Åkerlund, H.E. Violaxanthin de-epoxidase disulphides and their role in activity and thermal stability. *Photosynth. Res.* **2015**, *124*, 191–198. [CrossRef] [PubMed]
119. Demmig-Adams, B.; Winter, K.; Krüger, A.; Czygan, F.C. Zeaxanthin and the induction and relaxation kinetics of the dissipation of excess excitation energy in leaves in 2% $O_2$, 0% $CO_2$. *Plant Physiol.* **1989**, *90*, 887–893. [CrossRef] [PubMed]
120. Li, Y.; Walton, D.C. Violaxanthin is an abscisic acid precursor in water-stressed dark-grown bean leaves. *Plant Physiol.* **1990**, *92*, 551–559. [CrossRef] [PubMed]

121. Neuman, H.; Galpaz, N.; Cunningham, F.X., Jr.; Zamir, D.; Hirschberg, J. The tomato mutation nxd1 reveals a gene necessary for neoxanthin biosynthesis and demonstrates that violaxanthin is a sufficient precursor for abscisic acid biosynthesis. *Plant J.* **2014**, *78*, 80–93. [CrossRef] [PubMed]
122. Kang, B.; Yoon, H.-S. The application of two-step linear temperature program to thermal analysis for monitoring the lipid induction of *Nostoc* sp. KNUA003 in large scale cultivation. *Enzyme Microb. Technol.* **2015**, *69*, 54–61. [CrossRef] [PubMed]
123. Bhati, R.; Mallick, N. Carbon dioxide and poultry waste utilization for production of polyhydroxyalkanoate biopolymers by *Nostoc muscorum* Agardh: A sustainable approach. *J. Appl. Phycol.* **2016**, *28*, 161–168. [CrossRef]
124. Mona, S.; Kaushik, A.; Kaushik, C.P. Prolonged hydrogen production by *Nostoc* in photobioreactor and multi-stage use of the biological waste for column biosorption of some dyes and metals. *Biomass Bioenergy* **2013**, *54*, 27–35. [CrossRef]
125. Singh, S.; Verma, E.; Niveshika; Tiwari, B.; Mishra, A.K. Exopolysaccharide production in *Anabaena* sp. PCC 7120 under different $CaCl_2$ regimes. *Physiol. Mol. Biol. Plants* **2016**, *22*, 557–566. [CrossRef] [PubMed]
126. Ten, L.N.; Chae, S.M.; Yoo, S.A. Production of poly-3-hydroxybutyrate by cyanobacterium *Anabaena sp.* BD47. *Chem. Nat. Compd.* **2015**, *51*, 350–351. [CrossRef]
127. Shindo, K.; Misawa, N. New and rare carotenoids isolated from marine bacteria and their antioxidant activities. *Mar. Drugs* **2014**, *12*, 1690–1698. [CrossRef] [PubMed]
128. Strand, A.; Shivaji, S.; Liaaen-Jensen, S. Bacterial carotenoids 55. $C_{50}$-carotenoids 25. Revised structures of carotenoids associated with membranes in psychrotrophic *Micrococcus roseus*. *Biochem. Syst. Ecol.* **1997**, *25*, 547–552. [CrossRef]
129. Shahmohammadi, H.R.; Asgarani, E.; Terato, H.; Saito, T.; Ohyama, Y.; Gekko, K.; Yamamoto, O.; Ide, H. Protective roles of bacterioruberin and intracellular KCl in the resistance of *Halobacterium salinarium* against DNA-damaging agents. *J. Radiat. Res.* **1998**, *39*, 251–262. [CrossRef] [PubMed]
130. Kelly, M.; Norgard, S.; Liaaen-Jensen, S. Bacterial carotenoids. 31. C50-carotenoids 5. Carotenoids of *Halobacterium salinarium*, especially bacterioruberin. *Acta Chem. Scand.* **1970**, *24*, 2169–2182. [CrossRef] [PubMed]
131. Takaichi, S.; Mochimaru, M.; Maoka, T. Presence of free myxol and 4-hydroxymyxol and absence of myxol glycosides in *Anabaena variabilis* ATCC 29413, and proposal of a biosynthetic pathway of carotenoids. *Plant Cell Physiol.* **2006**, *47*, 211–216. [CrossRef] [PubMed]
132. Manh, H.D.; Matsuo, Y.; Katsuta, A.; Matsuda, S.; Shizuri, Y.; Kasai, H. *Robiginitalea myxolifaciens* sp. nov., a novel myxol-producing bacterium isolated from marine sediment, and emended description of the genus *Robiginitalea*. *Int. J. Syst. Evol. Microbiol.* **2008**, *58*, 1660–1664. [CrossRef] [PubMed]
133. Arima, H.; Horiguchi, N.; Takaichi, S.; Kofuji, R.; Ishida, K.I.; Wada, K.; Sakamoto, T. Molecular genetic and chemotaxonomic characterization of the terrestrial cyanobacterium *Nostoc commune* and its neighboring species. *FEMS Microbiol. Ecol.* **2012**, *79*, 34–45. [CrossRef] [PubMed]
134. Shindo, K.; Kikuta, K.; Suzuki, A.; Katsuta, A.; Kasai, H.; Yasumoto-Hirose, M.; Takaichi, S. Rare carotenoids, (3R)-saproxanthin and (3R, 2′ S)-myxol, isolated from novel marine bacteria (*Flavobacteriaceae*) and their antioxidative activities. *Appl. Microbiol. Biotechnol.* **2007**, *74*, 1350–1357. [CrossRef] [PubMed]
135. De Lourdes Moreno, M.; Sánchez-Porro, C.; García, M.T.; Mellado, E. Carotenoids' production from halophilic bacteria. *Methods Mol. Biol.* **2012**, *892*, 207–217. [CrossRef] [PubMed]
136. Lutnaes, B.F.; Oren, A.; Liaaen-Jensen, S. New C(40)-carotenoid acyl glycoside as principal carotenoid in *Salinibacter ruber*, an extremely halophilic eubacterium. *J. Nat. Prod.* **2002**, *65*, 1340–1343. [CrossRef] [PubMed]
137. Balashov, S.P.; Imasheva, E.S.; Lanyi, J.K. Induced chirality of the light-harvesting carotenoid salinixanthin and its interaction with the retinal of xanthorhodopsin. *Biochemistry* **2006**, *45*, 10998–11004. [CrossRef] [PubMed]
138. Sakata, T.; Yasumoto, H. Colony formation by algicidal *Saprospira* sp. on marine agar plates. *Nippon Suisan Gakkaishi* **1991**, *57*, 2139–2143. [CrossRef]
139. Takatani, N.; Nishida, K.; Sawabe, T.; Maoka, T.; Miyashita, K.; Hosokawa, M. Identification of a novel carotenoid, 2′-isopentenylsaproxanthin, by *Jejuia pallidilutea* strain 11shimoA1 and its increased production under alkaline condition. *Appl. Microbiol. Biotechnol.* **2014**, *98*, 6633–6640. [CrossRef] [PubMed]

140. Richter, T.K.; Hughes, C.C.; Moore, B.S. Sioxanthin, a novel glycosylated carotenoid, reveals an unusual subclustered biosynthetic pathway. *Environ. Microbiol.* **2015**, *17*, 2158–2171. [CrossRef] [PubMed]
141. Walton, T.J.; Britton, G.; Goodwin, T.W.; Diner, B.; Moshier, S. The structure of siphonaxanthin. *Phytochemistry* **1970**, *9*, 2545–2552. [CrossRef]
142. Kageyamam, A.; Yokohama, Y.; Shimura, S.; Ikawa, T. An efficient excitation energy transfer from a carotenoid, siphonaxanthin to chlorophyll a observed in a deep-water species of chlorophycean seaweed. *Plant Cell Physiol.* **1977**, *18*, 477–480. [CrossRef]
143. Yokohama, Y.; Kageyama, A.; Ikawa, T.; Shimura, S. A carotenoid characteristic of chlorophycean seaweeds living in deep coastal waters. *Bot. Mar.* **1977**, *20*, 433–436. [CrossRef]
144. Sugawara, T.; Ganesan, P.; Li, Z.; Manabe, Y.; Hirata, T. Siphonaxanthin, a green algal carotenoid, as a novel functional compound. *Mar. Drugs* **2014**, *12*, 3660–3668. [CrossRef] [PubMed]
145. Hendrickson, S.J.; Willett, W.C.; Rosner, B.A.; Eliassen, A.H. Food predictors of plasma carotenoids. *Nutrients* **2013**, *5*, 4051–4066. [CrossRef] [PubMed]
146. EFSA (European Food Safety Authority). Scientific Opinion of the Panel on Dietetic Products, Nutrition and Allergies on a request from the EC on Food-Based Dietary Guidelines. *EFSA J.* **2008**, *644*, 1–44.
147. Freitas, A.C.; Rodrigues, D.; Rocha-Santos, T.A.; Gomes, A.M.; Duarte, A.C. Marine biotechnology advances towards applications in new functional foods. *Biotechnol. Adv.* **2012**, *30*, 1506–1515. [CrossRef] [PubMed]
148. El-Sohaimy, S.A. Functional foods and nutraceuticals-modern approach to food science. *World Appl. Sci. J.* **2012**, *20*, 691–708. [CrossRef]
149. FAO: Food and Agriculture Organization of the United Nations. Report on Functional Foods, Food Quality and Standards Service (AGNS). Available online: http://www.fao.org/ag/agn/agns/files/Functional_Foods_Report_Nov2007.pdf (accessed on 25 February 2010).
150. Brower, V. Nutraceuticals: Poised for a healthy slice of the healthcare market? *Nat. Biotechnol.* **1998**, *16*, 728–731. [CrossRef] [PubMed]
151. Trottier, G.; Bostrom, P.J.; Lawrentschuk, N.; Fleshner, N.E. Nutraceuticals and prostate cancer prevention: A current review. *Nat. Rev. Urol.* **2010**, *7*, 21–30. [CrossRef] [PubMed]
152. Gaffe, R. The Current and Future Regulation of Dietary Supplements. 2010. Available online: http://www.richardjaffeesq.com/jaffe/dietarysupplements.asp (accessed on 25 February 2010).
153. Boon, C.S.; Mc Clements, D.J.; Weiss, J.; Decker, E.A. Factors influencing the chemical stability of carotenoids in foods. *Crit. Rev. Food Sci. Nutr.* **2010**, *50*, 515–532. [CrossRef] [PubMed]
154. Margină, D.; Ilie, M.; Grădinaru, D.; Androutsopoulos, V.P.; Kouretas, D.; Tsatsakis, A.M. Natural products-friends or foes? *Toxicol. Lett.* **2015**, *236*, 154–167. [CrossRef] [PubMed]
155. Gangakhedkar, A.; Somerville, R.; Jelleyman, T. Carotenemia and hepatomegaly in an atopic child on an exclusion diet for a food allergy. *Australas J. Dermatol.* **2017**, *58*, 42–44. [CrossRef] [PubMed]
156. Safi, K.H.; Filbrun, A.G.; Nasr, S.Z. Hypervitaminosis A causing hypercalcemia in cystic fibrosis. Case report and focused review. *Ann. Am. Thorac. Soc.* **2014**, *11*, 1244–1247. [CrossRef] [PubMed]
157. Gamal, A.A. Biological importance of marine algae. *Saudi Pharm. J.* **2010**, *18*, 1–25. [CrossRef] [PubMed]
158. Bule, M.H.; Ahmed, I.; Maqbool, F.; Bilal, M.; Iqbal, H.M.N. Microalgae as a source of high-value bioactive compounds. *Front. Biosci. (Sch. Ed.)* **2018**, *10*, 197–216. [CrossRef]
159. Kadam, S.U.; Prabhasankar, P. Marine foods as functional ingredients in bakery and pasta products. *Food Res. Int.* **2010**, *43*, 1975–1980. [CrossRef]
160. Plaza, M.; Herrero, M.; Cifuentes, A.; Ibáñez, E. Innovative natural functional ingredients from microalgae. *J. Agric. Food Chem.* **2009**, *57*, 7159–7170. [CrossRef] [PubMed]
161. Vicentini, A.; Liberatore, L.; Mastrocola, D. Functional foods: Trends and development of the global market. *Ital. J. Food Sci.* **2016**, *28*, 338–351. [CrossRef]
162. Chacon-Lee, T.L.; González-Mariño, G.E. Microalgae for "healthy" foods-possibilities and challenges. *Compr. Rev. Food Sci. Food Saf.* **2010**, *9*, 655–675. [CrossRef]
163. Milledge, J.J. Microalgae-Commercial Potential for Fuel, Food and Feed. *Food Sci. Technol.* **2012**, *26*. Available online: http://eprints.soton.ac.uk/336243/ (accessed on 20 February 2018).
164. Enzing, C.; Ploeg, M.; Barbosa, M.; Sijtsma, L. *Microalgae-Based Products for the Food and Feed Sector: An Outlook for EUROPE*, 1st ed.; Joint Research Centre Institute for Prospective Technological Studies of European Commission: Sevilla, Spain, 2014; pp. 1–75, ISBN 978-92-79-34037-6.

165. Rodríguez-Amaya, D.B. Status of carotenoid analytical methods and in vitro assays for the assessment of food quality and health effects. *Curr. Opin. Food Sci.* **2015**, *1*, 56–63. [CrossRef]
166. Gouveia, L.; Batista, A.P.; Raymundo, A.; Sousa, I.; Empis, J. *Chlorella vulgaris* and *Haematococcus pluvialis* biomass as colouring and antioxidant in food emulsions. *Eur. Food Res. Technol.* **2006**, *222*, 362–367. [CrossRef]
167. Gouveia, L.; Batista, A.P.; Miranda, A.; Empis, J.; Raymundo, A. *Chlorella vulgaris* biomass used as colouring source in traditional butter cookies. *Innov. Food Sci. Emerg. Technol.* **2007**, *8*, 433–436. [CrossRef]
168. Liang, S.; Liu, X.; Chen, F.; Chen, Z. Current microalgal health food R&D activities in China. *Hydrobiologica* **2004**, *512*, 45–48. [CrossRef]
169. Gantar, M.; Svircev, Z. Microalgae and cyanobacteria: Food for thought. *J. Phycol.* **2008**, *44*, 260–268. [CrossRef]
170. Wijesinghe, W.A.J.P.; Jeon, Y.J. Biological activities and potential cosmeceutical applications of bioactive components from brown seaweeds: A review. *Phytochem. Rev.* **2011**, *10*, 431–443. [CrossRef]
171. Shah, M.M.; Liang, Y.; Cheng, J.J.; Daroch, M. Astaxanthin-producing green microalga *Haematococcus pluvialis:* From single cell to high value commercial products. *Front. Plant Sci.* **2016**, *7*, 531. [CrossRef] [PubMed]
172. Hudson, E. Trend Watch: Spirulina: Healthy, Green, Versatile. 2008. Available online: http://www.portal.euromonitor.com.simsrad.net.ocs.mq.edu (accessed on 5 March 2009).
173. Spiralyn: Promoting Good Health. Available online: http://www.foodnavigator.com/Financial-Industry/Blue-Smarties-are-back-thanks-to-Spirulina (accessed on 9 March 2018).
174. Pooja, S. Algae used as medicine and food-a short review. *J. Pharm. Sci. Res.* **2014**, *6*, 33–35.
175. Raja, R.; Hemaiswarya, S.; Kumar, N.A.; Sridhar, S.; Rengasamy, R. A perspective on the biotechnological potential of microalgae. *Crit. Rev. Microbiol.* **2008**, *34*, 77–88. [CrossRef] [PubMed]
176. Borowitzka, M.A. High-value products from microalgae, their development and commercialisation. *J. Appl. Phycol.* **2013**, *25*, 743–756. [CrossRef]
177. Holdt, S.L.; Kraan, S. Bioactive compounds in seaweed: Functional food applications and legislation. *J. Appl. Phycol.* **2011**, *23*, 543–597. [CrossRef]
178. Sweetman, J.O. *Developing Food Products for Consumers with Specific Dietary Needs*, 1st ed.; Elsevier: Duxford, UK, 2016; pp. 63–77, ISBN 978-0-08-100329-9.

marine drugs

Review

# A Review of the Biological Activities of Microalgal Carotenoids and Their Potential Use in Healthcare and Cosmetic Industries

**Ramaraj Sathasivam and Jang-Seu Ki** *

Department of Biotechnology, Sangmyung University, Seoul 03016, Korea; ramarajbiotech@gmail.com
* Correspondence: kijs@smu.ac.kr; Tel.: +82-2-2287-5449; Fax: +82-2-2287-0070

Received: 17 November 2017; Accepted: 8 January 2018; Published: 12 January 2018

**Abstract:** Carotenoids are natural pigments that play pivotal roles in many physiological functions. The characteristics of carotenoids, their effects on health, and the cosmetic benefits of their usage have been under investigation for a long time; however, most reviews on this subject focus on carotenoids obtained from several microalgae, vegetables, fruits, and higher plants. Recently, microalgae have received much attention due to their abilities in producing novel bioactive metabolites, including a wide range of different carotenoids that can provide for health and cosmetic benefits. The main objectives of this review are to provide an updated view of recent work on the health and cosmetic benefits associated with carotenoid use, as well as to provide a list of microalgae that produce different types of carotenoids. This review could provide new insights to researchers on the potential role of carotenoids in improving human health.

**Keywords:** carotenoids; microalgae; anti-angiogenic; cardioprotective activity; anti-cancer; anti-diabetic; anti-inflammatory; anti-obesity; anti-oxidant; beauty

## 1. Introduction

Carotenoids have a wide range of applications in the healthcare and nutraceuticals industry. The growing importance of carotenoid use in improving food quality has also led to an increase in the demand for carotenoids in the global market. The global carotenoid market was estimated to be ~1.24 billion USD in 2016, and is projected to increase to ~1.53 billion USD by 2021, at a compound annual growth rate (CAGR) of 3.78% from 2016 to 2021 (www.bccresearch.com). Due to changes in life style and the rising health consciousness of the average population, the demand for nutrient-rich supplements with health benefits (such as immunity boosters and supplements rich in proteins and vitamins) has risen. Carotenoids have various medicinal properties and they are widely used as preventives against diseases such as cancer, diabetes, and cataract [1]. Carotenoids are also used in food supplements, cosmetics, and pharmaceuticals. Therefore, the use of carotenoids in supplements is likely to increase over the next few years. The major types of carotenoids that are used commercially in the global market are astaxanthin, β-carotene, lutein, canthaxanthin, lycopene, and zeaxanthin. Among these, β-carotene is the pigment carotene, which imparts a red, yellow, or orange color to plants, algae, and animals [2,3]. For this reason, production of carotenoids is considered as an important business opportunity for the healthcare and cosmetic industries in the future. As a result, many multinational companies have begun producing various types of carotenoids for use in different applications.

Currently, the bulk of industry-produced carotenoids are synthesized chemically, though a small portion of carotenoids is obtained naturally from plants or algae. Due to the adverse side effects commonly associated with drug therapy, public interest in recent times has focused on natural products with health-promoting properties as alternatives to conventional drugs. There is a growing demand for natural compounds because of an increasing global trend in the use of, and consumer preference for products made with natural ingredients. Since algae represent an alternative source of natural carotenoids, carotenoid extraction from cultivated algae may help in overcoming problems with balancing the supply of and demand for these products [4].

Microalgae are photosynthetic microorganisms, which may be widely used as a potential source for the production of several highly valuable bioproducts. There are ~13,000 species, of which ~8000 are described, and ~5000 species are yet to be described [5]. Microalgae are a rich source of bioactive compounds such as vitamins, proteins with essential amino acids, polysaccharides, fatty acids, minerals, photosynthetic pigments (carotenoids and chlorophylls), enzymes, and fiber [6–8]. Due to the high levels of biologically valuable components in microalgae, these organisms can represent a good source of supplements with various health benefits. Foods rich in chlorophyll are considered to be very nutritious for humans as they contain high levels of heme and can help in increasing the production of red blood cells [9]. In addition, consumption of microalgae has recently been shown to provide health benefits; microalgae can therefore function as nutraceutical agents with antioxidant, anti-inflammatory, anti-mutagenic, and anti-microbial properties [10].

Reviews on the benefits of carotenoids for human health, and the application-based uses of carotenoids ([11,12], respectively), unfortunately, do not provide much information on the uses of carotenoids in the healthcare and beauty industries. In addition to these reviews, a review by Raposo et al. [13] elaborates the use of marine microalgal carotenoids for the prevention of chronic diseases; another study by Di Pietro et al. [14] summarizes the use of carotenoids in cardiovascular disease prevention. Recently, there has been a rising interest in exploring the beneficial effects of carotenoids in the healthcare and cosmetics industries. Therefore, the main aim of this article is to critically review available data on the use of carotenoids, and their mechanisms of action in providing healthcare and cosmetic benefits. We will summarize the anti-angiogenic, cardioprotective activity anti-cancer, anti-diabetic, anti-inflammatory, anti-obesity, anti-oxidant, beauty-enhancing and other beneficial effects of carotenoids in this article. We hope this review will further improve our understanding of carotenoids and their therapeutic potential in healthcare and for cosmetic benefits.

## 2. Types of Carotenoids

The two main groups of carotenoids are carotenes and xanthophylls. Some familiar carotenes are β-carotene and lycopene—both these carotenoids are strict hydrocarbon carotenoids, and do not possess any substituent (or even oxygen) in their structures. Xanthophylls or oxycarotenoids, which belong to the second group, are oxygen-containing molecules. Lutein and zeaxanthin are two xanthophylls with –OH groups in their structures, whereas canthaxanthin and echinenone contain =O groups. Astaxanthin has both –OH and =O groups in its structure. Furthermore, some carotenoids such as violaxanthin and diadinoxanthin contain epoxy groups, and others such as dinoxanthin and fucoxanthin have acetyl groups in their structures. The two carotenoids with acetyl groups also contain the C=C=C (allene) group in their structures, which is unique to natural products [15]. In addition, some carotenoids such as allo-, diato-, diadino-, hetero-, croco-, pyro-, and monadoxanthin contain C≡C (acetylene) groups in their structures. The different types of carotenoids mentioned above are naturally produced by the microalgae showed in Table 1.

**Table 1.** Carotenoids produced by microalgae.

| Main Carotenoid | Microalgae | Other Carotenoids | Concentration | General Uses | References |
|---|---|---|---|---|---|
| Astaxanthin | *Haematococcus pluvialis* | β-carotene, Lutein, Canthaxanthin, Neoxanthin, Violaxanthin, Zeaxanthin, Echinenone | Up to 7% DW; 75% TC | In benign prostatic hyperplasia and prostate and liver tumors<br>Anti–inflammatory properties<br>Active against liver neoplasms<br>Strong anti-oxidant property<br>Cardiovascular health | [13,16–20] |
| | *Chlorella zofingiensis* | - | 3.7% DW | | [21] |
| β-carotene | *Chlorella zofingiensis* | Canthaxanthin (97% DW), Astaxanthin (0.7% DW) | 0.9% DW | Provitamin A function<br>In colorectal cancer<br>In the prevention of acute and chronic coronary syndromes<br>Photoprotection of skin against UV light<br>Prevent night blindness<br>Anti-oxidant property<br>Prevents liver fibrosis | [3,4,13,22–32] |
| | *Dunaliella salina* | Zeaxanthin, Lutein, α-carotene | 10–13% DW | | |
| | *Spirulina maxima* | Astaxanthin, Lutein, β-cryptoxanthin, Zeaxanthin, Echinenone, Oscillaxanthin, Myxoxanthophyll | 80% TC | | |
| Canthaxanthin | *Coelastrella striolata* var. *multistriata* | Astaxanthin (0.15% DW), β-carotene (0.7% DW) | 4.75% DW | Create a tan color<br>Anti-oxidant property | [33] |
| Canthaxanthin and Lutein | *Chlorella vulgaris* | Astaxanthin 12.5% TC Violaxanthin | 45% TC | Create a tan color<br>Anti-oxidant property<br>In the prevention of acute and chronic coronary syndromes and stroke<br>In the prevention of cataracts<br>To prevent macular degeneration associated with age<br>In the prevention of retinitis<br>To avoid gastric infection by *H. pylori* | [34,35] |
| Echinenone | *Botryococcus braunii* | Botryoxanthins A and B—0.03% DW<br>Braunixanthins 1 and 2—0.06% DW | 0.17% DW | - | [36] |
| Fucoxanthin | *Cyclotella* cf. *cryptica* | - | 0.7 mg $g^{-1}$ | Anti-obesity<br>Anti-oxidant property | [37] |
| | *Cyclotella meneghiniana* | - | 2.3 mg $g^{-1}$ | | [37] |
| | *Cylindrotheca closterium* | - | 0.52% DW | | [38] |
| | *Isochrysis* aff. *galbana* | - | 1.8% DW | | [39] |
| | *Mallomonas* sp. SBV13 | - | 26.6 mg $g^{-1}$ | | [37] |
| | *Nitzschia* cf. *carinospeciosa* | - | 5.5 mg $g^{-1}$ | | [37] |
| | *Odontella aurita* | Diadinoxanthin, β-carotene | up to 2.2% DW | | [40] |
| | *Paralia longispina* | - | 1.4 mg $g^{-1}$ | | [37] |
| | *Phaeodactylum tricornutum* | Diadinoxanthin, Zeaxanthin, Neoxanthin, Violaxanthin, β-carotene | 1.65% DW | | [39,41,42] |
| | | - | 10.2 mg $g^{-1}$ | | [37] |

**Table 1.** *Cont.*

| Main Carotenoid | Microalgae | Other Carotenoids | Concentration | General Uses | References |
|---|---|---|---|---|---|
| Lutein | *Auxenochlorella protothecoides* | Astaxanthin | 0.76 mg $g^{-1}$ | In the prevention of acute and chronic coronary syndromes and stroke<br>Helps to maintain a normal visual function<br>In the prevention of cataracts<br>To prevent macular degeneration associated with age<br>In the prevention of retinitis<br>To avoid gastric infection by *H. Pylori*<br>Anti-oxidant property<br>Anti-cancer activity | [43] |
| | *Chlorella protothecoides* | - | 5.4 mg $g^{-1}$ | | [44] |
| | *Chlorella pyrenoidosa* | Violaxanthin, Loroxanthin, α- and β-carotene | 0.2–0.4% DW | | [13,18,25,45–47] |
| | *Chlorella sorokiniana* | Astaxanthin | 5.90 mg $g^{-1}$ | | [43] |
| | *Chlorella* sp. | Astaxanthin | 2.26 mg $g^{-1}$ | | [43] |
| | *Coelastrella* sp. | Astaxanthin | 6.49 mg $g^{-1}$ | | [43] |
| | *Galdieria sulphuraria* | - | 0.4 mg $g^{-1}$ | | [48] |
| | *Parachlorella kessleri* | Astaxanthin | 0.28 mg $g^{-1}$ | | [43] |
| | *Scenedesmus almeriensis* | - | 0.54% DW | | [49] |
| | *Scenedesmus bijugus* | Astaxanthin | 2.9 mg $g^{-1}$ | | [43] |
| | *Scenedesmus* sp. | Astaxanthin | 1.8 mg $g^{-1}$ | | [43] |
| | *Vischeria stellata* | Astaxanthin | 1.50 mg $g^{-1}$ | | [43] |
| Violaxanthin | *Chlorella ellipsodea* | Antheraxanthin, Zeaxanthin | - | Anti-inflammatory activity | [35,50] |
| Zeaxanthin | *Porphyridium cruentum* | β-carotene | 97.4% TC | In the prevention of acute and chronic coronary syndromes<br>Helps to maintain a normal visual function<br>In the prevention of cataracts<br>To prevent macular degeneration associated with age | [13,19,25,46,47,51] |

DW—Dry weight; TC—Total carotenoids; '-'—No data available.

## 3. Functions in Algal Cells

In plants, carotenoids play a significant role in the photosynthetic process by forming pigment–protein complexes in the thylakoid membrane; in cyanobacteria, however, they may also be found in the plasma membrane, and function in protecting the cells during exposure to high light intensities [52,53]. In algae, carotenoids protect chlorophyll from the effects of excess light exposure (by scavenging reactive oxygen species (ROS) such as singlet oxygen molecules and free radicals [54]) and are also required for phototropism and phototaxis [55] as they are present in algal eyespots [56]. Previous studies have reported that β-carotene also plays a role in protecting cells from free radicals, and that other carotenoids are mainly involved in light-harvesting functions [57]. In addition, fucoxanthin in algae acts as a primary light-harvesting carotenoid that transfers energy to chlorophyll–protein complexes. The fucoxanthin molecule exhibits high energy transfer efficiency (~80%), a property that is attributed to the unique structure of this carotenoid [58]. Fucoxanthin also participates in photoprotection and has strong antioxidant activity [40,59]. In addition, under unfavorable conditions, certain microalgae defend themselves by producing secondary metabolites via the carotenogenesis pathway [20]. To date, little information is available in the public database about the genes and enzymes involved in the algal carotenogenesis pathway [60].

## 4. Molecular Details of the Carotenoid Pathway in Microalgae

Our current understanding of the carotenoid metabolic pathway and its regulation in microalgae is unclear, and is mainly inferred from our knowledge of the process in higher plants. In chloroplasts, several nucleus-encoded membrane proteins are involved in the synthesis of carotenoids. These proteins are produced in the cytoplasm as precursor polypeptides with amino-terminal extensions that target them to chloroplasts. There are many publications that comprehensively review the carotenoid biosynthetic pathway [61–64].

Carotenoids are synthesized from a C5 building block, a common precursor to all isoprenoids, which is the isopentenyl pyrophosphate (IPP) molecule, via the plastidial 2-C-methyl-D-erythriol 4-phosphate (MEP) pathway [65]. The first step in carotenoid biosynthesis is the condensation of two molecules of geranyl geranyl pyrophosphate (GGPP), catalyzed by the enzyme phytoene synthase (PSY), to yield phytoene, the first but uncolored carotenoid. Following this step, phytoene undergoes a series of sequential desaturations catalyzed by phytoene desaturase (PDS) and ζ-carotene desaturase (ZDS), which result in the formation of pro-lycopene. Pro-lycopene is then isomerized by a specific carotenoid isomerase (CRTISO) into all-trans lycopene. After this step, the pathway is divided into two branches. In one branch of the synthetic pathway, lycopene is cyclized at both ends by lycopene β-cyclase (LCYB), yielding β-carotene with two β–ionone end groups. These can be further hydroxylated by a non-heme di-iron hydroxylase, called β-carotene hydroxylase (CHYB), to yield zeaxanthin. In the other branch of the synthetic pathway, the combined actions of LCYB and lycopene ε-cyclases (LCYEs) result in the formation of α-carotene. The quantities of carotenoids produced by each branch of the pathway are determined by the absolute activities of LCYE and LCYB. The hydroxylation of α-carotene is catalyzed by two heme-containing cytochrome P450 monoxygenases (namely, carotene β-hydroxylase and carotene ε-hydroxylase), which leads to the formation of lutein. In another branch, zeaxanthin is converted to violaxanthin by zeaxanthin epoxidase (ZEP), which inserts two epoxy groups at positions C-5,6 and C-5′,6′ [62]. In another branch of the pathway, zeaxanthin is converted to the di-keto carotenoid canthaxanthin, and violaxanthin is converted to astaxanthin by the enzyme β-carotene ketolase (BKT) (Figure 1).

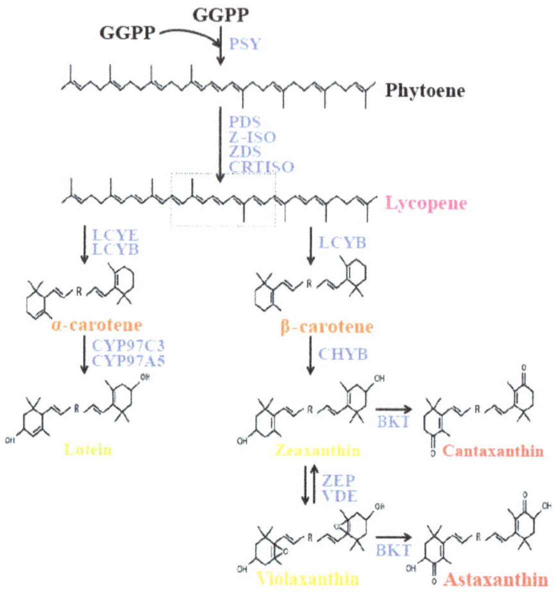


**Figure 1.** Carotenoid biosynthesis pathway in chlorophytes. GGPP—geranylgeranyl pyrophosphate; PSY—phytoene synthase; PDS—phytoene desaturase; Z-ISO—ζ-carotene isomerase; ZDS—ζ-carotene desaturase; CRTISO—carotene isomerase; LCYE—lycopene ε-cyclase; LCYB—lycopene β-cyclase; CYP97C3—cytochrome P450 ε-hydroxylase; CYP97A5—cytochrome P450 β-hydroxylase; CHYB—carotene β-hydroxylase; BKT—β-carotene oxygenase; ZEP—zeaxanthin epoxidase; VDE—violaxanthin de-epoxidase.

In most microalgae, the basic carotenoid synthesis pathways are highly conserved, although some species are able to accumulate unusual carotenoids via specific biosynthetic routes. In higher plants, most of the genes encoding enzymes for the carotenoid biosynthetic pathway have been isolated and characterized by functional complementation experiments [66]. In microalgae, most of the carotenoid biosynthetic pathway genes have also been isolated and characterized [67].

## 5. Carotenoids and Their Biological Activities

### 5.1. Anti-Angiogenic Activity

Angiogenesis is the process of formation of new blood vessels from pre-existing capillaries and involves a sequence of events that are fundamental to many physiological and pathological processes [68]. It occurs throughout life, during both healthy and diseased conditions, and is tightly regulated under normal physiological conditions such as during embryogenesis, ovary cycling, and wound healing. Chronic, unregulated angiogenesis can lead to several anomalous angiogenic conditions such as inflammatory diseases, rheumatoid arthritis, and tumor metastasis [69]. Tumor growth and metastasis are processes that are highly dependent on the formation of new blood vessels. Therefore, preventing angiogenesis under pathological conditions (such as cancer and other angiogenesis related diseases) is a promising approach for controlling or eradicating such diseases.

Studies involving in vivo and in vitro experiments on male C57BL/6 mice and B16F-10 cells have been used to evaluate the anti-angiogenic effects of β-carotene by Guruvayoorappan and Kuttan [70]. Their study found that treatment with β-carotene significantly reduces the number of tumor-directed capillaries (associated with altered serum cytokine levels) formed, and suppresses the proliferation, migration, and tube formation of endothelial cells. In addition, β-carotene treatment also inhibited the activation and nuclear translocation of p65, p50, and c-Rel sub-units of nuclear factor-κB (NF-κB), as well as other transcription factors such as c-fos, activated transcription factor-2, and cyclic adenosine monophosphate response element-binding protein in B16F-10 melanoma cells [70]. This study clearly showed that the anti-angiogenic effect of β-carotene occurs by affecting serum cytokine levels, and that β-carotene could inhibit the activation and nuclear translocation of transcription factors.

In another study, Sugawara et al. [71] reported that fucoxanthin (at concentrations of >10 μM) can significantly inhibit tube formation and proliferation in human umbilical vein endothelial cells (HUVECs). Fucoxanthin significantly suppressed the differentiation of endothelial progenitor cells into endothelial cells during the formation of new blood vessel. Fucoxanthin and its metabolite fucoxanthinol also suppressed the growth of microvessels during in vitro and ex vivo experiments using rat aortic rings [71]. In addition, a study in 2013 used HUVECs to understand the molecular mechanisms responsible for the anti-angiogenic activity of fucoxanthin [72]. The results of this study showed that fucoxanthin significantly reduced the genetic expression, and hence, mRNA levels of fibroblast growth factor 2 (FGF-2), its receptor (FGFR-1), as well as their trans-activation factor, EGR-1. However, the study found no significant changes in the mRNA levels of the vascular endothelial growth factor receptor-2 (VEGFR-2). Furthermore, fucoxanthin was found to down-regulate the phosphorylation of FGF-2-mediated intracellular signaling proteins such as extracellular signal-reduced kinase and protein kinase B (ERK1/2 and Akt). Matrigel invasion assays showed that fucoxanthin not only inhibited the migration of endothelial cells, but also inhibited their differentiation into tube-like structures by suppressing the phosphorylation of the FGF-2-mediated intracellular signaling proteins. However, there was no evidence to indicate that carotenoids activate the angiopoietins 1 and 2 (Ang1 and Ang2) pathways. The possible molecular mechanisms responsible for the anti-angiogenic effects of carotenoids are shown in Figure 2.

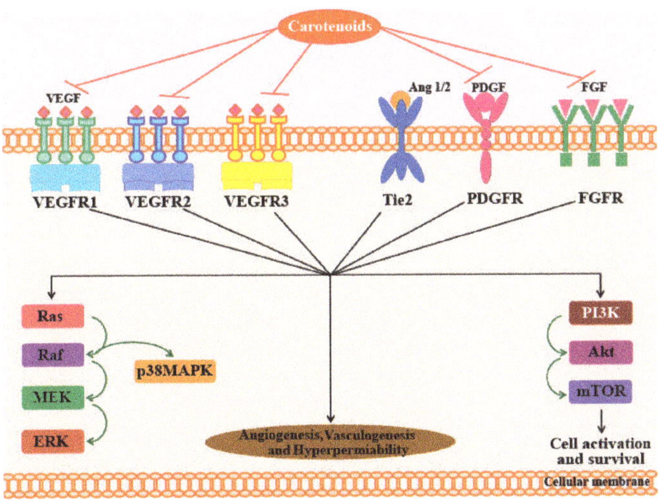


**Figure 2.** Diagrammatic representation of anti-angiogenic effect of carotenoid. Both VEGF and non-VEGF dependent pathways are noted. Akt—protein kinase B; ERK—extracellular signal-reduced kinase; FGF—fibroblast growth factor; FGFR—fibroblast growth factor receptor; MEK—mitogen-activated protein kinase/extracellular signal-reduced kinase kinase; mTOR—mechanistic target of rapamycin; p38 MAPK—p38 mitogen-activated protein kinase; PDGF—platelet-derived growth factor; PDGFR—platelet-derived growth factor receptor; PI3K—phosphoinositide 3-kinase; Tie2—tyrosine kinase with immunoglobulin-like and EGF-like domains 1; RAF—rapidly accelerated fibrosarcoma; RAS—rat sarcoma; VEGF—vascular endothelial growth factor; VEGFR—vascular endothelial growth factor receptor.

### 5.2. Cardioprotective Activity

A study in 2005 by Hussein et al. [73] reported the anti-hypertensive effects of astaxanthin in spontaneously hypertensive rats (SHRs). In their study, Hussein et al. [73] found that oral administration of astaxanthin (at a concentration of 50 mg/kg) for 14 days led to a significant decrease in blood pressure in the SHRs. In addition, long-term administration of astaxanthin (for 5 weeks) also considerably reduced blood pressure and postponed the occurrence of heart strokes in these rats. On the 4th day of treatment, 60% of the rats in the placebo group showed signs of heart stroke, whereas none of the rats in the astaxanthin-treated group showed any signs of heart stroke. In later studies, the same authors also reported the mechanism of how astaxanthin works to prevent heart strokes [74]. The authors found that SHRs treated with astaxanthin showed significantly higher levels of vasorelaxation in response to nitric oxide (NO), which enhanced thoracic aorta contractions, as compared to rats not treated with astaxanthin. These results suggest that the anti-hypertensive effect of astaxanthin is mediated by NO-related mechanisms. In addition, another study carried out on SHRs to explore the beneficial effects of astaxanthin on blood rheology found that the transit times of whole blood in astaxanthin-treated SHRs were significantly lower than those of placebo-treated SHRs. Histopathological measures, such as levels of vascular elastin in the aorta and arterial wall thickness were also improved in SHRs treated with astaxanthin [73,74]. A study by Preuss et al. [75] in Zucker fatty rats found that administration of 25 mg/kg of astaxanthin for one month significantly lowered systolic blood pressure. In addition, the astaxanthin treatment also decreased the activity of the renin-angiotensin system, which indicates that the lowering in blood pressure was dependent on changes in the renin-angiotensin, as well as the NO systems. Furthermore, in heat stress experiments, all rats fed with astaxanthin survived, whereas, >60% of the rats in the placebo group died.

In another study, female BALB/c mice treated with 800 mg/kg astaxanthin for eight weeks exhibited higher heart mitochondrial membrane potentials and contractility indices than mice in a placebo group [76]. An ex vivo study of 24 adult humans showed that astaxanthin has the potential to prevent atherosclerosis by delaying the prolonged oxidation of low-density lipoprotein (LDL)-cholesterol. In this study, volunteers consumed astaxanthin at doses of 1.8, 3.6, 14.4, or 21.6 mg/day for 14 days and the LDL lag times were longer (5.0%, 26.2%, 42.3%, and 30.7%, respectively) compared with the initial day [77]. Experiments by Miyawaki et al. [78] to determine the health benefits of astaxanthin extracted from *Haematococcus pluvialis* on human blood rheology were carried out on 20 adult men. After 10 days of astaxanthin (6 mg/day) administration, the whole blood transit time of the experimental group decreased from $52.8 \pm 4.9$ s to $47.6 \pm 4.2$ s, which is considerably lower than that of the control group ($54.2 \pm 6.7$ s) [78]. In addition, another study carried out on humans in an age group of 25–60 years [79] showed that 12 weeks of astaxanthin administration significantly decreased serum triglyceride levels, while significantly increasing high density lipoprotein (HDL)-cholesterol levels. However, LDL-cholesterol levels remained unchanged. Furthermore, astaxanthin intake increased serum adiponectin levels, which are positively correlated with changes in HDL-cholesterol levels independent of age and body mass index (BMI) [79]. Fucoxanthin and its derivative fucoxanthinol show cardioprotective activity; administration of these carotenoids in an in vivo study reduced triglyceride levels in blood (high triglyceride levels in blood are related to the development of atherosclerotic vascular disease) [80]. When rats were fed with 2 mg/kg of fucoxanthin or fucoxanthinol, they showed a significant reduction in triglyceride absorption in their jugular veins on being fed with non-pre-digested 10% soybean oil.

### 5.3. Anti-Cancer Activity

Numerous in vitro and in vivo studies have demonstrated the anti-cancer activities of carotenoids. The results of these studies indicate that carotenoids may prevent different types of cancers in humans, including bladder, breast, hepatic, intestinal, leukemic, lung, oral, and prostate cancer. The anti-cancer activity of carotenoids involves a variety of mechanisms, including induction of cell apoptosis and suppression of cell proliferation. In particular, one in vivo study showed that β-carotene, astaxanthin, canthaxanthin, and zeaxanthin help in reducing the sizes and numbers of liver neoplasias [51]. Another study also reports that dietary intake of carotenoids can reduce the risk of developing colon cancer [81, 82].

Many studies indicate that β-carotene shows great potential as an anti-tumor agent. In a study in China, administration of a combination of β-carotene, vitamin E, and selenium to humans was observed to decrease the incidence of mortality due to cancer [83]. Many other studies have also reported an inverse relationship between ingesting carotenoids and cancer prevalence [84,85]. Lycopene is one of the best studied carotenoids with respect to its potential health benefits [86,87]; this is because it exhibits much higher anti-cancer potential than most other carotenoids [51]. Several in vivo and in vitro studies using tumor cell lines indicate that lycopene can significantly reduce tumor cell growth [86,87]. Nishino et al. [51] have reported that the carotenoids α-carotene, lutein, zeaxanthin, lycopene, β-cryptoxanthin, fucoxanthin, astaxanthin, capsanthin, crocetin, and phytoene exhibit greater anti-carcinogenic activity than β-carotene.

The anti-proliferative and cancer-preventive activities of fucoxanthin and fucoxanthinol are dependent on different molecules and pathways involved in the processes of cell cycle arrest, apoptosis, and metastasis [88]. Furthermore, studies using human umbilical vein endothelial cells (HUVECs) have shown that fucoxanthin also has anti-angiogenic activity, which is helpful in preventing cancer. The detailed mechanisms of how fucoxanthin functions in this respect are explained in Section 5.1. Fucoxanthin can potentially inhibit the proliferation of cancer cells by increasing intercellular communication through gap junctions in human cancer cells, which increases intracellular signaling mechanisms that promote cell cycle arrest and apoptosis. Therefore, fucoxanthin and its metabolites show great potential as chemotherapy agents if administered in the initial stages of cancer [88]. In addition, fucoxanthin also lowers the viabilities of human leukemia (HL-60) cells. Fucoxanthin also shows anti-cancer activity against Caco-2, DLD-1, and HT-29, which are human colorectal adenocarcinoma cell lines. Although fucoxanthin treatment has been shown to reduce cell viability, the strength of the effect varies across cell types. After 72 h of fucoxanthin treatment (at a concentration of 15.2 mM), the viabilities of Caco-2, DLD-1, and HT-29 cells decreased to 14.8%, 29.4%, and 50.8%, respectively [89]. These remarkable reductions in cell viability levels were caused by a significant increase in cell apoptosis and DNA fragmentation [89]. Kim et al. [90] reported that astaxanthin, β-carotene, and fucoxanthin show potent anti-cancer activities when tested on HL-60 cancer cells at a concentration of 7.6 mM. At this concentration, fucoxanthin caused high levels of DNA fragmentation, whereas the other two carotenoids (astaxanthin and β-carotene) did not show any significant effects on DNA fragmentation. Kim et al. [90] stated that the mechanism of fucoxanthin-induced apoptosis in HL-60 cells involves the generation of ROS, which leads to cytotoxicity and apoptosis involving the cleavage of caspases-3 and -9 and poly-ADP-ribose polymerase (PARP), coupled with reductions in levels of Bcl-xL (Figure 3). Kotake-Nara et al. [91] investigated the effects of fucoxanthin (at concentrations of 5 and 10 mM) on the viabilities of six types of cancer cells. Incubation with fucoxanthin for 72 h showed that five of the cancer cell lines suffered significant reductions in cell viability. In addition, comparisons of the effects of fucoxanthin and lycopene on cancer cells indicate that at the same concentrations, fucoxanthin shows higher anti-cancer effects than lycopene. Fucoxanthin is a potential chemopreventive agent for urinary bladder cancers, as it inhibits the growth and causes apoptosis in EJ-1 cells (a urinary bladder cancer cell line). Treatment with fucoxanthin significantly reduced EJ-1 cell proliferation in a dose- and time-dependent manner. Treatment with 20 mM fucoxanthin for 72 h caused a high percentage of cells to undergo apoptosis (93%), which was evident by morphological changes, DNA fragmentation, increased percentages of hypodiploid cells, and caspase-3 activity [92].

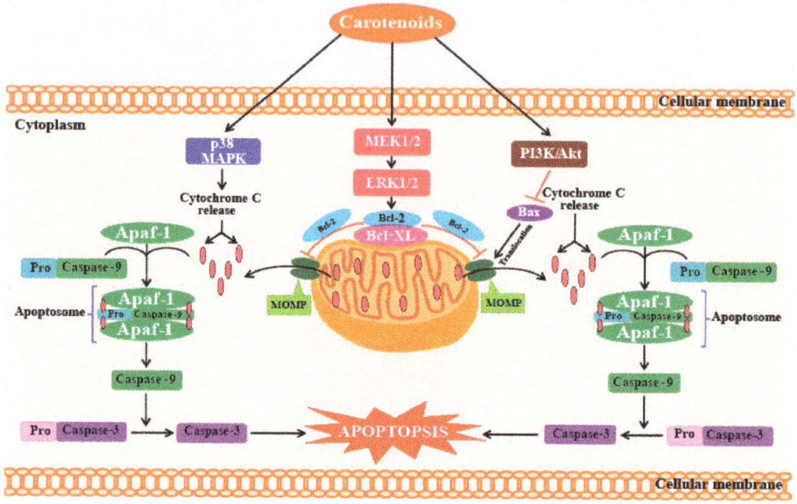


**Figure 3.** Factors (molecules and mechanisms) regulated by carotenoids, resulting in their anti-carcinogenic effects. Carotenoids induce the activation of PI3K/Akt survival pathway, trigger the phosphorylation-dependent inactivation of Bax (Bcl-2 associated X), which leads to apoptosis by decrease in caspase activity. Carotenoids also maintain the mitochondrial integrity by regulating the p38 MAPK signaling pathway, which leads to a decrease in cytochrome c release and inhibits caspase-dependent apoptotic cell death. APAF-1—Apoptotic protease activating factor-1; BAX—Bcl-2 associated X; Bcl-2—B-cell lymphoma 2; Bcl-XL—B-cell lymphoma-extra-large; ERK—extracellular signal-reduced kinase; MEK—mitogen-activated protein kinase/extracellular signal-reduced kinase kinase; MOMP—mitochondrial outer membrane permeabilization; p38 MAPK—p38 mitogen-activated protein kinase; PI3K—phosphoinositide 3-kinase.

The effect of different carotenoids such as β-cryptoxanthin, canthaxanthin, fucoxanthin, neoxanthin, phytoene, and zeaxanthin on the viabilities of three prostate cancer cell lines, namely, PC-3, DU 145, and LNCaP, has been investigated [93]. Among the carotenoids tested, fucoxanthin and neoxanthin exhibited the highest cell-growth inhibition rates; the other carotenoid showed no significant effects on cell viability. When compared to untreated control cells, the viabilities of cells treated with 20 mM fucoxanthin for 72 h were reduced to 14.9%, 5.0%, and 4.8% for PC-3, DU 145, and LNCaP cells, respectively [93]. In addition, another study investigating the anti-cancer activities of fucoxanthin and neoxanthin on PC-3 cells with an apoptosis assay [91] showed that treatment with 20 mM fucoxanthin for 48 h increased the percentage of apoptotic cells to >30%. This indicates that fucoxanthin induces apoptosis by activating caspase-3. Fucoxanthinol has also been found to induce apoptosis in PC-3 cells, and has a greater inhibitory effect on these cells as compared to fucoxanthin. The 50% inhibitory concentration (IC50) of fucoxanthin and fucoxanthinol on the proliferation of PC-3 cells was 3.0 and 2.0 mM, respectively [91]. A study to compare the effects of carotenoids such as β-carotene and astaxanthin, with those of xanthophyll carotenoids like fucoxanthin on human colon cancer cells [94] has showed that the xanthophyll carotenoid, fucoxanthin, has higher anti-cancer activity than the other carotenoids. Taken together, the study clearly showed that the fucoxanthin metabolites (halocynthiaxanthin and fucoxanthinol) have greater anti-cancer activities than fucoxanthin. However, the effects of these metabolites on cancer cells were highly variable depending on the types of cancer cells.

Canthaxanthin, which is another type of carotenoid with significant anti-cancer activity, has been reported to significantly inhibit the growth of JB/MS, B16F10 (a melanoma cell line), and PYB6 (a fibrosarcoma tumor cell line) cells at a concentration of 100 mM [95]. Treatment with canthaxanthin also induced apoptosis in WiDr (a human colon adenocarcinoma cell line) and SK-MEL-2 (a human melanoma cell line) cells by increasing the numbers of in situ nick-end labeled-positive nuclei [96]. Canthaxanthin induced apoptosis in both cell lines in a dose- and time-dependent manner. A study by Abdel-Fatth [97] found that treatment with 10 mM canthaxanthin for 48 h caused 18% and 20% of cells to undergo apoptosis in the WiDr and SK-MEL-2 cell lines, respectively. Furthermore, the growth of WiDr cells showed significantly higher inhibition than that of SK-MEL-2 cells. These results suggest that other pathways, such as stimulation of tumor necrosis factor-α (TNF-α) and other cytokines [97], or down-regulation of the epidermal growth factor receptor [98], are involved in the inhibitory effects of canthaxanthin on cancer cell lines. The effects of canthaxanthin on chemically induced mammary carcinogenesis in mice showed that dietary intake of canthaxanthin for three weeks prior to the induction of cancer with dimethylbenzanthracene could reduce the occurrence of cancer by 65% [99]. Another study reported that anti-cancer agents may have the ability to upregulate intercellular communication via gap junctions. Even at a low dosage of 1 mM, canthaxanthin increases the levels of connexin43 in C3H10T1/2 (mouse embryo) cells [100]. In addition to this, numerous studies on mice have investigated the effects of canthaxanthin on colon carcinogenesis [101], skin papillomas [102], and cervical cancer [103].

Several studies have reported that astaxanthin has significant anti-cancer effects on certain cancer types such as prostatic hyperplasias and prostatic cancers. Astaxanthin mainly inhibits the enzyme 5-α-reductase, which is involved in abnormal prostate growth [17,104]. The chemopreventive effect of these carotenoids against various cancer types has been extensively studied by Tanaka et al. [101]. In one study, the occurrence of colon cancer induced by azoxymethane in F344 rats was significantly lower in rats fed with 500 ppm astaxanthin or canthaxanthin for 34 weeks; furthermore, rats fed with these carotenoids had significantly lower multiplicity of neoplasms than those rats in the placebo group [101]. In addition, rats fed with carotenoids showed a significant reduction in cell proliferation activity and the development of aberrant crypt foci (ACF) in these rats was also observed to be inhibited [101]. The investigations of Tanaka et al. [105] on the effects of astaxanthin and canthaxanthin on mouse urinary bladder carcinogenesis revealed that lower incidences of pre-neoplastic lesions and neoplasms occurred in mice treated with 50 ppm astaxanthin or canthaxanthin for 20 weeks, as compared to the incidences of cancer in mice from the placebo group. In addition, it was also found that the number of silver-stained nucleolar organizer region proteins (AgNORs) in the transitional epithelium was reduced in the carotenoid-treated group. Furthermore, both carotenoids showed anti-proliferative effects on the cancer cells in the mouse urinary bladder, with astaxanthin showing higher levels of anti-proliferative activity as compared to canthaxanthin [105]. A study by Kozuki et al. [106] on the inhibitory activities of eight different carotenoids on AH109A cell-invasion showed that at concentrations $\geq$5 μM all carotenoids could significantly inhibit AH109A cell-invasion in a dose dependent manner. Among the carotenoids tested (which included canthaxanthin, astaxanthin, α-carotene, β-carotene, β-cryptoxanthin, lutein, lycopene, and zeaxanthin), canthaxanthin showed the highest effects on inhibiting the invasiveness of AH109A cells.

An investigation by Lyons and O'Brien [107] on the differential effects of algal extracts (containing 14% astaxanthin) and synthetic astaxanthin on cancer cells in culture showed that treatment with both, algal extracts and synthetic astaxanthin, can protect cells against UVA-induced DNA damage. In this study, it was found that 2 h of exposure to UVA could cause a significant increase in superoxide dismutase (SOD) activity, along with a marked decrease in glutathione (GSH) content in 1BR-3 cells. However, in cells pre-incubated with the algal extract (18 h prior to UVA exposure), there were no changes in the level of antioxidant enzymes even after UVA exposure. This result agrees with the result of another experiment, where intestinal cells treated with 10 mM astaxanthin were observed to maintain their GSH content, even after UVA exposure. In addition to these effects, astaxanthin has also been shown to inhibit prostate cancer cell proliferation in a dose-dependent manner by inducing androgen hormones [108]. Numerous in vivo studies have investigated the anti-cancer effects of astaxanthin. Jyonouchi et al. [109] observed that astaxanthin treatment can reduce the weights and sizes of tumors induced by transplantable methylcholanthrene-induced fibrosarcoma (Meth-A tumor) cells in mice. In addition, Kurihara et al. [110] reported that daily oral administration of astaxanthin to mice inhibited lipid peroxidation, which markedly attenuated the development of hepatic metastasis induced by restraint stress.

The anti-cancer properties of the carotenoid cryptoxanthin have not been investigated extensively. However, recent in vitro, in vivo, and human-intervention studies report that β-cryptoxanthin can differentially regulate the expression of P73 variants. In addition, these studies have found that this carotenoid has the ability to inhibit the proliferation of colon cancer cells and in conjunction with oxaliplatin, can induce apoptosis in cancer cells by negatively regulating ΔNP73 [111].

### 5.4. Anti-Diabetic Activity

Recent work on carotenoids suggests that these molecules may be more effective in treating and controlling diabetes than antioxidants. Studies have shown that levels of dietary carotenoids and concentrations of β-carotene in blood are inversely associated with fasting blood glucose levels and insulin resistance, respectively [112]. Numerous studies have reported that carotenoids reduce the risk of type 2 diabetes mellitus (T2DM) development in men and women [112,113]. It has also been observed that carotenoid intake is inversely related to HbA1c levels [114]. In addition, recent findings have confirmed that carotenoids such as lycopene, lutein, and zeaxanthin can protect against diabetic retinopathy [115].

Most studies on carotenoids and diabetes report the importance of carotenoids in dietary intake for the prevention and treatment of T2DM [116]. A recent study by Sugiura et al. [117] shows that in middle-aged and older Japanese patients, serum levels of α-carotene and β-cryptoxanthin are associated with lower incidences of T2DM. In addition, another study that investigated the interactions between serum concentrations of carotenoids and smoking with the incidence of diabetes mellitus over a time span of 15 years [118] showed that the incidence of T2DM is inversely associated with serum concentrations of carotenoids in nonsmokers. A similar result was also obtained by Ylonen et al. [112], who showed that serum concentrations of lutein, zeaxanthin, lycopene, α-carotene, and β-carotene were significantly lower in diabetic subjects. Most of these studies also report that there is an association between carotenoid intake and reductions in the risk of developing T2DM [113,118,119].

Astaxanthin, which is one of the best studied carotenoids, shows great potential in preventing and treating diabetes. Astaxanthin has higher antioxidant activity than other carotenoids such as lutein, β-carotene, and zeaxanthin [120], and can be consumed safely by humans [121]. In db/db mice (a well-known obesity model for T2DM), treatment with astaxanthin decreases glucose tolerance, enhances serum insulin levels, and attenuates blood glucose levels. These results indicate that astaxanthin has protective antioxidant effects that can help in the preservation of pancreatic β-cell function [122]. Bhuvaneswari et al. [123] have also reported similar anti-diabetic effects in high-fat, high-fructose diet HFFD mice. The effect of astaxanthin on metabolic syndrome has also been investigated in a rat experimental model. Astaxanthin was found to decrease blood glucose and triglyceride levels, as well as enhance serum levels of HDL-cholesterol and adiponectin [124]. Interestingly, recent studies have reported that astaxanthin primarily targets the peroxisome proliferator-activated receptor (PPARγ), which plays a pivotal role in carbohydrate metabolism. These studies also report that astaxanthin not only binds to PPARγ, but the carotenoid also affects the mRNA levels of this protein [125]. These results are consistent with another study that reports the anti-hyperglycemic effects of astaxanthin [126,127].

Another important carotenoid, β-carotene, has been investigated in detail regarding its usefulness in the treatment of diabetes. Hozumi et al. [128] reported a significantly inverse correlation between serum concentrations of β-carotene and serum levels of HbA1c in diabetic patients. Arnlov et al. [129] also reported that impaired insulin sensitivity is linked to low serum concentrations of β-carotene. In a study conducted by the European Prospective Investigation into Cancer and Nutrition–Netherlands

(EPIC), investigations on 37,846 men and women revealed an inverse association between dietary intake of β-carotene and the risk of T2DM development [113], a result similar to that obtained by Coyne et al. [130] in a population-based study in Queensland, Australia. Furthermore, serum levels of β-carotene are reported to be important determinants of metabolic syndrome outcome [131]. Although β-carotene is well-studied with respect to its usefulness in preventing or treating diabetes, other carotenoids such as lutein have not been well investigated. Katyal et al. [132] found that lutein can lower streptozotocin (STZ)-induced hyperglycemia and shows significant antioxidant effects in the kidneys of diabetic rats.

Another carotenoid, fucoxanthin, shows good potential as an anti-diabetic agent. A study on fucoxanthin reports that treatment with this carotenoid can restore blood glucose and insulin levels to normal in obese mice. The study reports that fucoxanthin upregulated the genetic expression and mRNA levels of the glucose transporter 4 (GLUT4) protein in skeletal muscle cells [133]. Nishikawa et al. [134] also report similar results (that fucoxanthin increases GLUT4 expression levels in skeletal muscle) and hypothesize that the induction of the PPARγ coactivator-1α mediates this process, which is accompanied by an upregulation in insulin receptor mRNA levels, along with increased phosphorylation levels of Akt, all of which play key roles in the regulation of GLUT4 translocation [134]. Maeda et al. [135] found that fucoxanthin significantly decreases the serum glucose and plasma insulin levels in diabetic/obese *KKAy* mice. Similarly, it has also been demonstrated that in *KKAy* mice, fucoxanthin can reduce hyperglycemia, although this carotenoid has no such effect on lean C57BL/6J mice [136]. Other studies have reported that the anti-diabetic activity of fucoxanthin involves several different mechanisms. For example, it has been demonstrated that fucoxanthin inhibits several enzymes such as aldose reductase in rat lens, human recombinant aldose reductase, protein tyrosine phosphatase 1B (PTP1B), and α-glucosidase, as well as processes such as advanced glycation end-product formation [137]. In addition, fucoxanthin has also been shown to increase the gene expression of PPARγ and GLUT4 proteins [135]. From these studies, it is clear that fucoxanthin manifests strong anti-diabetic effects through multiple mechanisms of action. Figure 4 illustrates the molecular targets involved in the anti-diabetic effects of carotenoids.

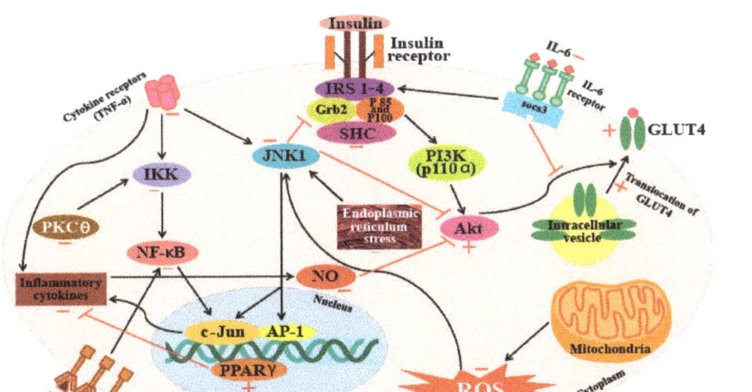


**Figure 4.** An overview on the targets of carotenoid with respect to their anti-diabetic effects. +, promote/activate; −, inactivate/inhibit. Akt—protein kinase B; AP1—activator protein 1 (c-jun and c-fos); GLUT 4—glucose transporter 4; Grb2—growth factor receptor-bound protein 2; IKK—IκB kinase; IL-6—interleukin-6; IRS 1–4—insulin receptor substrate 1–4; JNK 1—jun amino-terminal kinases 1; NF-κB—nuclear factor κB; NO—nitric acid; PI3K—phosphoinositide 3-kinase; PKCθ—protein kinase C θ; RAGE—receptor for advanced glycation end products; SHC—SH2-containing collagen-related proteins.

### 5.5. Anti-Inflammatory Activity

The first response of the immune system to infection or irritation is inflammation, which is also referred to as the innate cascade. However, some inflammatory reactions can have adverse effects on host cells or tissues; for example, chronic inflammation can cause many conditions such as arthritis, hepatitis, gastritis, periodontal disease, colitis, atherosclerosis, pneumonia, and neuro-inflammatory diseases [138]. Therefore, natural anti-inflammatory substances, especially carotenoids, are receiving much attention from researchers; carotenoids could potentially be used as drugs for preventing and controlling chronic inflammatory conditions due to their inhibitory effects on the production of NO, prostaglandin $E_2$ ($PGE_2$), and proinflammatory cytokines, as well as their inhibitory effects on enzymes such as inducible nitric oxide synthase (iNOS) and cyclooxygenase-2 (cox-2) [139,140].

Recently, astaxanthin has garnered much attention due to its potential as an anti-inflammatory agent. Both, in vitro and in vivo studies have been carried out in rat models to investigate the effects of astaxanthin on lipopolysaccharide (LPS)-induced inflammatory reactions [141,142]. The inhibitory effects of astaxanthin were compared with those of the common anti-inflammatory drug, prednisolone. The anti-inflammatory effect of astaxanthin (at a concentration of 100 mg/kg) was higher than that of 10 mg/kg of prednisolone [141]. LPS-fed mice treated with astaxanthin showed a dose-dependent anti-inflammatory effect; astaxanthin has been shown to function by suppressing the production of NO, $PGE_2$, TNF-$\alpha$, and interleukin-1$\beta$ (IL-1$\beta$), as well as by blocking the activity of NOS enzymes in RAW 264.7 cells. The results of this study agree with those of a previously conducted study, which showed that astaxanthin inhibited NO production, as well as the expression of iNOS and COX-2 in LPS-stimulated BV2 microglial cells [143].

Lee et al. [144] also reported that astaxanthin can inhibit the production of NO, $PGE_2$, as well as the expression of pro-inflammatory genes by suppressing the function of NF-κB. Furthermore, astaxanthin also suppressed the activity of the iNOS promoter by inhibiting IKK (IκB kinase) activity. A similar study found that LPS-stimulated mouse neutrophils treated with astaxanthin produce significantly lower levels of the proinflammatory cytokines TNF-$\alpha$ and IL-6. Macedo et al. [145] also report that treatment with 5 mM astaxanthin improved the phagocytic and microbicidal activity of neutrophils. In addition, oxidative damage to lipids and proteins in human neutrophils were significantly lower after astaxanthin treatment. The inhibitory activity of this xanthophyll carotenoid on the secretion of IL-1$\beta$, IL-6, and TNF-$\alpha$ has also been observed in U937 (a human lymphoma cell line) cells treated with $H_2O_2$. Cytokine levels in cells pre-incubated with 10 mM astaxanthin before $H_2O_2$ stimulation were significantly lower (less than half) of the levels seen in control cells (cells not pre-treated with astaxanthin). Furthermore, cells pre-incubated with astaxanthin also showed restoration of SHP-1 (a protein tyrosine phosphate) expression levels and reduced levels of NF-κB expression [146]. Bennedsen et al. [147] reported that *Helicobacter pylori*-infected mice fed with astaxanthin extracted from the microalga *H. pluvialis*, showed reduced levels of gastric inflammation. Mice fed with 200 mg/kg of algal extract for 10 days, showed significantly lower levels of inflammation and mucosa-bacterial loads in their stomachs than untreated mice. These results indicate a change in the T-lymphocyte response in mice; the response changes from a predominantly Th1-response to a mixed Th1/Th2-response. This shift was found to occur because of a block in IFN-$\gamma$ release that boosts IL-4 release in splenocytes in the infected mice pre-treated with astaxanthin. A clinical study to investigate the anti-oxidative and anti-inflammatory effects of astaxanthin was also undertaken on a cohort of healthy young women [148]. In this study, the women who ingested 2 mg of astaxanthin for 8 weeks had lowered blood levels of C-reactive protein, indicating that this compound has anti-inflammatory activity. In addition, the study also found that astaxanthin could reduce ROS production by down-regulating NF-κB and AP-1 transcription factors, as well as inflammatory cytokine production. From these results, it is clear that astaxanthin ingestion can decrease DNA damage, reduce acute phase protein levels, and enhance immune responses in healthy young women [148]. Overall, these studies indicate that astaxanthin inhibits inflammatory processes by blocking the expression of pro-inflammatory genes through suppression of NF-κB activation; a diagrammatic representation of this mechanism is represented in Figure 5.

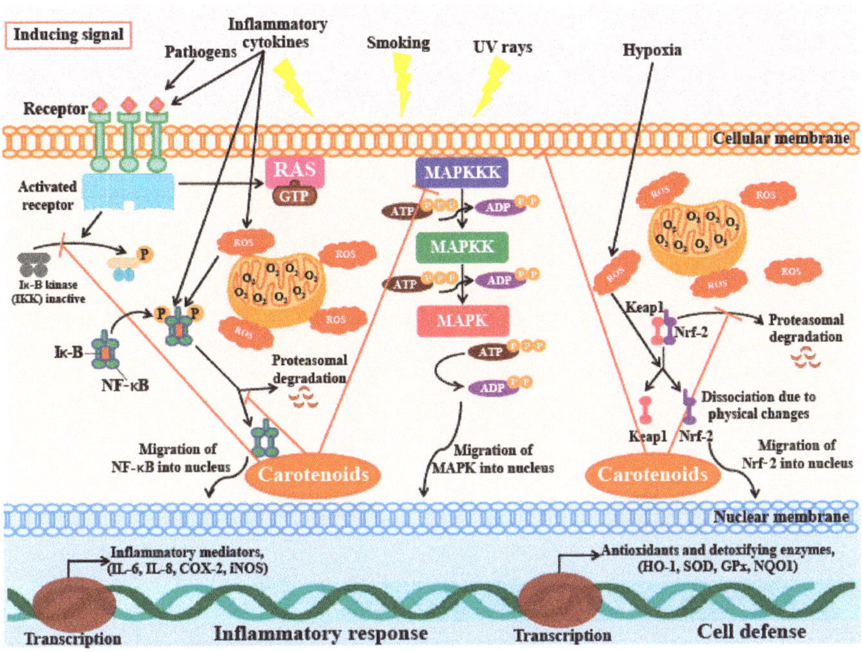


**Figure 5.** Schematic diagram of the interactions of anti-inflammatory signaling pathways and carotenoids. In the cytoplasm during the resting stage the NF-κB (nuclear factor-κB) is inactivated, which is bound to its inhibitory protein IκB (inhibitor of kappa B). During oxidative stress, inflammatory cytokines, or hypoxia, IκB protein is phosphorylated by the IKK (IκB kinase) complex, which leads to the ubiquitination and proteasomal degradation of IκB protein. At this time NF-κB was released which will migrate to the nucleus and the transcription of inflammatory mediators will start. It is assumed that carotenoids and their metabolites may interact with cysteine residues of the IKK and/or NF-κB subunits, which will inhibit the NF-κB pathway. In cytosol the Nrf2 (Nuclear factor (erythroid-derived 2)-like 2) is kept inactive by Keap1 (kelch-like ECH-associated protein 1) especially by poly-ubiquitination and rapid degradation through the proteasome. During redox imbalance, the Keap1-Nrf2 complex is disturbed, which leads to dissociation of Nrf2 from the complex. This Nrf2 migrates to the nucleus, which will induce the transcription of antioxidant and detoxifying enzymes, which promote cell protection. Carotenoids and their metabolites may interact with Keap1 by changing its physical properties. MAPK (mitogen-activated protein kinase) refers to a family of serine/threonine protein kinases. MAPK signaling cascades undergo consecutive and sequential step. MAPKs are phosphorylated and activated by MAPK-kinases (MAPKKs), which are further phosphorylated and activated by MAPKK-kinases (MAPKKKs). The MAPKKKs are in turn activated by interaction with small GTPases and/or other protein kinases family, which connect the MAPK module to cell surface receptors or external stimuli. However, it is still remains unclear how the carotenoids interact with the MAPK signaling pathway. COX—cyclooxygenase; GPx—glutathione peroxidase; GTP—guanine triphosphate; HO-1—heme oxygenase-1; IL-6—interleukin-6; iNOS—nitric oxide synthase 2; RAS—rat sarcoma; ROS—reactive oxygen species; SOD—superoxide dismutase; NQO1—NAD(P)H: quinone oxidoreductase 1.

### 5.6. Anti-Obesity Activity

Obesity is a condition where excessive accumulation and storage of fat in the body occurs, leading to inordinate increases in body weight [149]. Obesity leads to, and exacerbates several conditions, particularly those related to cardiovascular diseases, T2DM, obstructive sleep apnea, certain types of cancer, osteoarthritis, and depression [150]. Since the first half of this century, obesity has been one of the foremost issues of concern regarding public health. In developing countries, increased industrialization has increased the incidence of obesity in teenagers and senior citizens, causing a worrying health trend [151]. Therefore, the search for safe anti-obesity agents is now of great importance.

Wang et al. [152] reported that in obese individuals, there is an excessive accumulation of adipose tissue in organs that have large numbers of fat cells. Obesity is thought to result from adipocyte hypertrophy and the recruitment of new adipocytes from precursor cells. For this reason, the regulation of adipogenesis may be a potential strategy for the treatment of obesity. Okada et al. [153] reported that the chemical structures of carotenoids are important for suppression of adipocyte differentiation; investigations on 13 naturally occurring carotenoids have revealed that molecules with keto or epoxy groups, as well as epoxy-hydroxy carotenoids, hydroxyl-carotenoid, and keto-hydroxy carotenoid have no suppressive effects on adipocyte differentiation. The study found that only fucoxanthin and neoxanthin could significantly suppress adipocyte differentiation, suggesting that the presence of the allenic bond is an important factor for carotenoids to exhibit anti-obesity functions. From these results, it could be hypothesized that carotenoids containing an allenic group and an additional hydroxyl group may be effective in controlling adipocyte differentiation.

Maeda et al. [154] used a mouse model to show that oral intake of fucoxanthin could significantly decrease the amount of abdominal white adipose tissue (WAT) in obese mice. In addition, the study also found that this treatment had no such effects on normal mice kept on normal diets. This indicates that fucoxanthin specifically suppresses adiposity in obese mice. This study suggests that the anti-obesity effect of fucoxanthin is mediated by alterations in the functioning of lipid-regulating enzymes that could raise plasma adipokine levels and promote higher expression levels of uncoupling protein 1 (UCP1) and β3-adrenergic receptor (Adrb3) in abdominal fat tissues (Figure 6A). UCP1, which is abundant in the inner membrane of the mitochondria, is specifically expressed at high levels in brown adipocytes. UCP1 can dissipate energy by uncoupling the process of oxidative phosphorylation, which then produces heat instead of ATP (Figure 6B). It is well-known that brown adipose tissue (BAT) plays a vital role in the prevention and treatment of obesity [155]. The role of UCP1 in BAT is known to be a significant component of the regulatory system governing whole-body energy expenditure, and the protein is thought to be important in preventing the development of obesity [156]. Increasing UCP1 expression in BAT could be considered as a useful anti-obesity treatment option [157]. However, in humans, most of the body fat is stored in WAT [158]. Furthermore, WAT has now been recognized to function as an endocrine and active secretory organ as its produces biologically active mediators known as adipokines [159]. Fucoxanthin is likely to emerge as an important and attractive anti-obesity agent [94]. However, further studies are needed to clarify the various molecular mechanisms and intracellular signaling pathways that are involved in the anti-obesity activities of fucoxanthin. These studies indicate that natural pigments may play a vital role in the treatment and prevention of obesity, as these molecules may act as regulators of lipid metabolism in fat tissues. Natural pigments obtained from microalgae can be used in functional foods and pharmaceuticals, as these substances can be obtained at relatively low production costs, exhibit low cytotoxicity, and have gained wide acceptance as food supplements. Among the different types of carotenoids, fucoxanthin derived from marine algae may be considered a promising food supplement and weight-loss drug for the prevention and management of obesity.

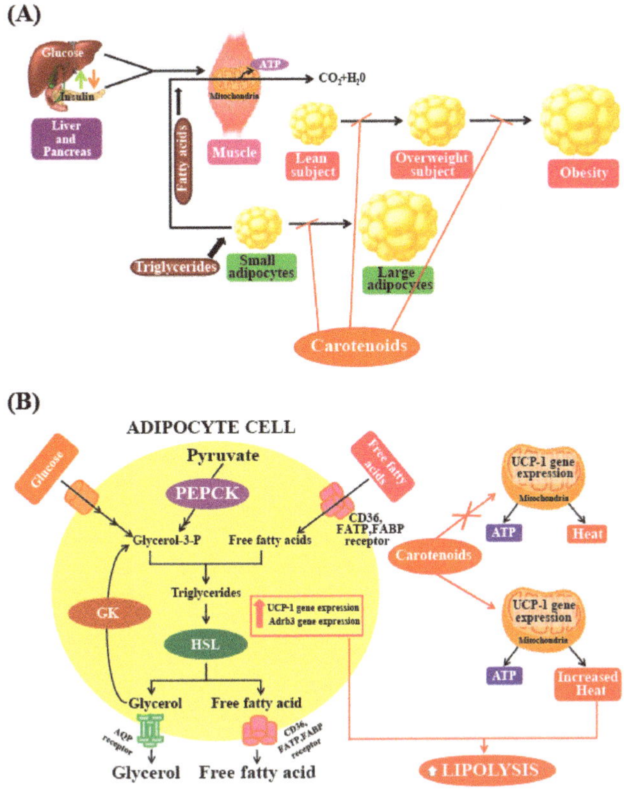


**Figure 6.** (**A**,**B**) Effects of carotenoid on thermogenesis and lipolysis: the muscle (**A**) and the adipose tissue (**B**). During excess caloric intake in the body, adipocytes take up free fatty acids (FFA), which are stored in the form of triglycerides (TG). For the synthesis of TG other metabolites are required, glycerol-3-phosphate, proceeds from three metabolic sources: (i) glucose; (ii) glycerol derived from lipolysis, which is phosphorylated by glycerol kinase (GK) and (iii) pyruvate, which is converted to glycerol by the activity of phosphoenolpyruvate carboxykinase (PEPCK). During fasting or exercise, TG are hydrolyzed to glycerol and FFA by the hormone-sensitive lipase (HSL) and released into the bloodstream. Several membrane proteins, including fatty acid binding protein (FABP), fatty acid translocase (FAT, CD36) or fatty acid transporter protein (FATP), facilitate the free fatty acid transport across the membrane. At this time, the uncoupling protein 1 (UCP1) and β3-adrenergic receptor (Adrb3) mRNA expression in the abdominal fat tissues and plasma adipokine levels was increased. Carotenoid plays an anti-obesity effect mainly by stimulating uncoupling protein-1 (UCP-1) expression in white adipose tissue (WAT). AQP—aquaporin; CD36—cluster of differentiation 36. 'X'—indicates that mechanism occur in the absence of carotenoid.

### *5.7. Anti-Oxidant Activity*

ROS and reactive nitrogen species (RNS) are generated during aerobic metabolism processes that occur in the cell; these include processes such as signal transduction, gene expression, and activation of cell signaling cascades [160]. ROS can damage biologically important molecules such as lipids, DNA, and proteins, which can in turn, negatively affect the integrity of cell membrane structures, enzyme functions, and gene expression; ROS are well-known to be involved in the patho-biochemistry of degenerative diseases [161]. The antioxidant defense systems in living organisms are complex networks that are comprised of several enzymatic and non-enzymatic antioxidants [162].

Carotenoids are known to play important roles in scavenging ROS such as singlet molecular oxygen ($^1O_2$) and peroxyl radicals, but there is little information regarding their roles in cellular defenses against RNS. Raposo et al. [13] have reported that the structural features of carotenoids play a significant role in their antioxidant activities. Fucoxanthin extracts from algae show great potential as antioxidants [163]. Fucoxanthin has strong radical-scavenging activity due to the presence of the unusual double allenic bonds at the C-7′ position of its structure, as demonstrated by Sachindra et al. [164]. Miyashita [165] has reported that fucoxanthin can significantly affect human health by altering the gene expression profiles of proteins involved in cell metabolism. Many studies have tested the antioxidant effects of fucoxanthin on different cell lines and animal models [166,167]. Another important carotenoid exhibiting strong antioxidant activity is astaxanthin, which shows higher levels of antioxidant activity than other carotenoids such as β-carotene, zeaxanthin, and canthaxanthin [168]. Rodrigues et al. [169] have reported that astaxanthin acts as a scavenger of various reactive species such as LOO•, HOCl, and $ONOO^-$. Several studies have reported that dietary intake of carotenoids can protect humans and animals from oxidative damage to lipophilic parts of cells; this is because carotenoids can limit lipid peroxidation events by scavenging the ROS formed during photo-oxidative processes [13]. To prevent oxidative damage, and disease conditions arising from such damage, a combination of carotenoids possessing different chemical characteristics can be used. Microalgae such as *Spirulina platensis*, *H. pluvialis*, and *Dunaliella salina* might be of great value in the production of various types of such carotenoids (Table 1).

### *5.8. Beauty-Enhancing Effect*

Skin has naturally occurring antioxidant agents which can block the effects of ROS and suppress cell disruption and damage [170]. However, when high levels of ROS are produced by ultraviolet (UV)-exposure, these defenses may not provide adequate protection. Apoptosis and necrosis are the two major modes of cell death that occur due to the accumulation of ROS in cells; excessive cell death can lead to wrinkling and dryness of skin. ROS accumulation also plays an important role in photo-aging conditions such as cutaneous inflammation, melanoma, and skin cancer [171]. Natural pigments can be used as therapeutic agents to overcome these problems. As many consumers prefer naturally derived compounds in their cosmetics, there is an increasing global demand for naturally derived carotenoids rather than those synthetized chemically. Due to this demand, the price of natural pigments isolated from algae is roughly double (~700 Euros/kg) of that of synthetic products [172].

Astaxanthin is an excellent antioxidant, exhibiting higher antioxidant activity than vitamins C and E; furthermore, this molecule aids in the preservation of proteins and essential lipids in human lymphocytes as it boosts superoxide dismutase and catalase enzyme activities [12,173]. Tominaga et al. [174] reported that both topical and oral use of astaxanthin can suppress skin hyper-pigmentation, inhibit synthesis of melanin, and improve the condition of all skin layers. Fucoxanthin has been reported to suppress tyrosinase activity in UVB-irradiated guinea pigs, and melanogenesis in UVB-irradiated mice. Studies have also found that oral administration of fucoxanthin decreases the mRNA levels of proteins linked to melanogenesis in skin cells. This indicates that fucoxanthin can negatively regulate melanogenesis factors at the transcriptional level [175]. In addition, fucoxanthin has the ability to counteract oxidative stresses caused by UV radiation, due to which it is currently used in cosmeceuticals [176]. Another important carotenoid exhibiting strong antioxidant activity is β-carotene, which helps in preventing the formation of free radicals that can cause premature aging in skin cells. In the epidermal and dermal layers of skin, the carotenoid lutein has been shown to protect against UV-induced oxidative damage, especially in combination with other antioxidant systems and immunoprotective substances [177].

A study conducted by Darvin et al. [178], which compared skin roughness with age in a cohort of women aged 40–50 years, indicated that there was no significant correlation between the two parameters. However, skin roughness was clearly correlated with the concentration of lycopene present in skin. Individuals with higher levels of antioxidants in their skin showed fewer furrows and wrinkles than those with lower levels of antioxidants [178]. Figure 7A shows that UV radiation from the sun is one of the major causes of premature skin aging. Figure 7B shows that UV radiation from sun rays destroys elastin and collagen fibers in the skin [179]. High concentrations of antioxidants such as carotenoids can efficiently neutralize free radicals before they can cause damage. These studies confirm the results of a study conducted by Heinrich et al. [180], which showed that a significant reduction in skin roughness could be achieved with supplements of antioxidant micronutrients such as lycopene.

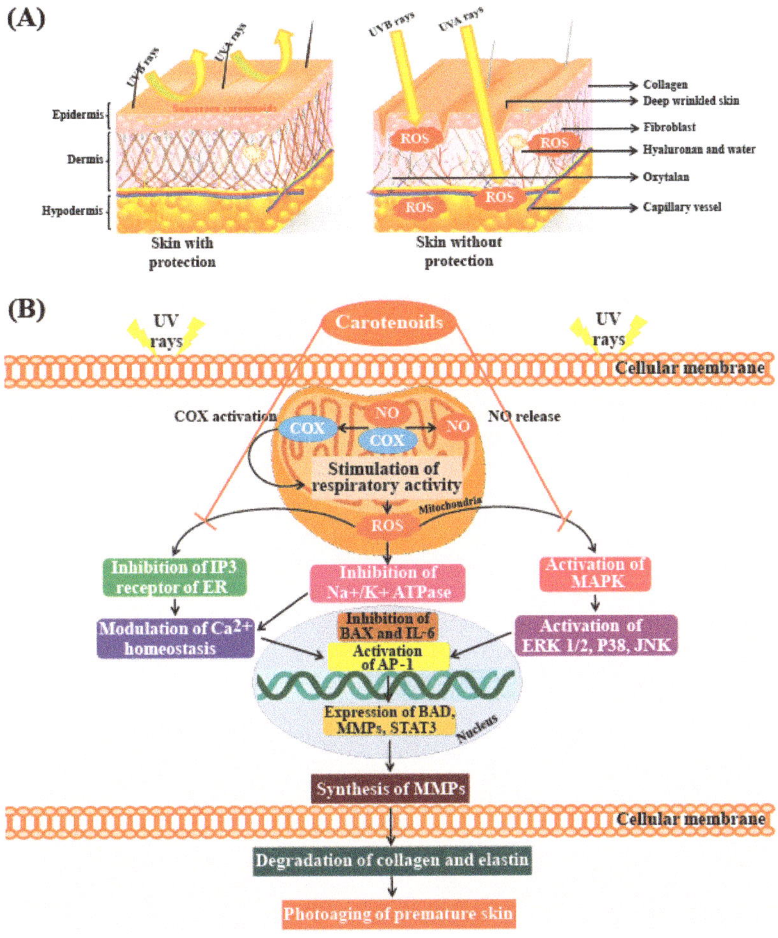


**Figure 7.** (**A**) Schematic diagram showing penetration of UV rays into the layers of the human skin. UVB rays do not penetrate the skin deeply because they are blocked by the epidermis, whereas UVA penetrate deep into the skin, which leads to damage of elastin and collagen fibers. Carotenoids act as a sunscreen and protect the skin from UV rays; (**B**) Inhibitory effects of UV rays-induced photodamage in premature skins. AP-1—activator protein-1; BAD—BCL2 associated agonist of cell death; BAX—Bcl-2 associated X; COX—cyclooxygenase; ER—endoplasmic reticulum; ERK—extracellular signal-reduced kinase; IL-6—interleukin-6; IP3—inositol 1,4,5-triphosphate; JNK—jun amino-terminal kinase; MAPK—mitogen-activated protein kinase; MMPs—matrix metalloproteinases; NO—nitric acid; ROS—reactive oxygen species; STAT3—signal transducer and activator of transcription 3.

## 6. Other Health Benefits

### 6.1. Age-Related Macular Degeneration

In older people, age-related macular degeneration (ARMD) is one of the leading causes of visual impairment. In the United States of America, approximately 2.07 million Americans were affected with advanced ARMD in 2010, and this number is expected to increase to 5.44 million in 2050. ARMD is a major cause of irreversible blindness among elderly people (>65 years of age) in western countries, and affects ~20% of the total population [181]. The macula lutea is an oval-shaped pigmented area near the center of the retina, and the area of maximal visual acuity. Recently, the focus of much research has been on investigating the protective effects of carotenoids against ARMD. As carotenoids absorb UV light and other forms of solar radiation that can damage our eyes, these molecules could help in preserving healthy cells in the eyes by reducing oxidative damage and vision loss [181].

An investigation using unilamellar liposomes as a model membrane showed that the filtering effects of lutein and zeaxanthin are higher than those of lycopene and β-carotene. Because of this, lutein and zeaxanthin are thought to be essential pigments in the lens and retina of the eye, and maintaining their levels could be critical in protecting vision in older people [182]. Although the major dietary carotenoids such as α-carotene and lycopene are efficient blue-light filters, these molecules are not found in the macula lutea [183]. The spectral properties of carotenoids, as well as their antioxidant activities can change with the environment. Epidemiological data indicate that macular pigments such as lutein play an important protective role in the eyes [184]. Another study found that the retinas of donors suffering from ARMD had lower levels of lutein and zeaxanthin than those of donors unaffected by ARMD [185]. Furthermore, several reports also indicate that dietary supplementation with lutein alone or lutein together with other nutrients can improve visual function in patients suffering from atrophic ARMD [186]. Stahl and Sies [116], who investigated the combined effects of administering high-doses of β-carotene with vitamin C, vitamin E, and zinc to ARMD patients, found that ARMD progression was slowed and visual acuity improved with this treatment [116].

### 6.2. Neuroprotective Activity

Neuroprotection strategies are defined as the mechanisms and strategies used to protect neuronal cells against injury, apoptosis, dysfunction and/or degeneration in the central nervous system (CNS) by limiting neuronal dysfunction or death after CNS injury [187]. As most synthetic neuroprotective agents have strong side effects, natural bioactive compounds that can act as neuroprotective agents have been under intensive investigation [138]. Many studies have focused on the neuroprotective properties of natural pigments obtained from algae. Okuzumi et al. [188] reported that fucoxanthin can inhibit N-myc expression and cell cycle progression in GOTO cells (a human neuroblastoma cell line). At a concentration of 10 μg/mL, fucoxanthin can significantly inhibit the growth rate of GOTO cells to just 38%, though its exact mode of action remains unclear. Furthermore, the study shows that fucoxanthin can protect cortical neurons from oxidative damage during hypoxia and oxygen reperfusion [189]. Much neuronal damage can occur during re-oxygenation after hypoxia, because re-oxygenation can lead to the generation of significant amounts of ROS. Fucoxanthin exhibits neuroprotective activity mainly because of its ability to scavenge ROS. Based on these reports, carotenoids could be considered as potential neuroprotective agents that can be used to treat or prevent neurodegenerative diseases. So far, most studies on the neuroprotective activities of carotenoids have been carried out in in vitro systems. Therefore, it is vital to conduct in vivo experimental studies to investigate the neuroprotective activities of carotenoids, especially in humans.

### 6.3. Osteo-Protective Activity

Osteoclasts are highly specialized bone cells that break down bone tissues. One of the most recent uses of fucoxanthin has been in the treatment of osteoclast diseases. Das et al. [190] reported that fucoxanthin significantly suppresses the differentiation of RAW264.7 cells. The study found that

2.5 μM of fucoxanthin can activate caspase-3 and induce apoptosis in osteoclast-like cells. These results suggest that fucoxanthin suppresses osteoclastogenesis by inhibiting osteoclast differentiation and by inducing apoptosis in osteoclasts. Furthermore, dietary supplements of fucoxanthin may be useful in preventing bone diseases such as osteoporosis and rheumatoid arthritis, which are known to be related to bone resorption.

### *6.4. Weight Loss*

Fucoxanthin is a well-known weight-loss agent that encourages the 'burning of fats' by enhancing thermogenin expression. As fucoxanthin has numerous health benefits, and the molecule is found in high concentrations in microalgae, industrial production of fucoxanthin from microalgae is on the rise [40]. In 2009, a double-blind, randomized, and placebo-controlled study on human volunteers showed that ingesting 2.4 mg of fucoxanthin can lead to significant weight loss. Women ingesting fucoxanthin also exhibited higher levels of resting energy expenditure, which helps in reducing fat and body weight. Furthermore, fucoxanthin consumption also led to significant reductions in blood pressure, and levels of liver fat, triglycerides, and C-reactive protein [191].

## 7. Conclusions and Future Direction of Research

Microalgae are rich sources of carotenoids, with great industrial potential and accessibility, and thus are likely to have a wide range of applications in the healthcare and cosmetic industries. Most studies on carotenoids have been focused mainly on the preventive and protective effects of these molecules in various chronic diseases such as diabetes mellitus, metabolic syndrome, cancer, and cardiovascular diseases. Recent studies, however, report that carotenoids might also play a significant role in the treatment of various other diseases. Although the mechanisms of antioxidant activity for some carotenoids have been well studied, most of the other effects of carotenoids, such as their pro-vitamin A activity, metabolic activity, effects on the immune and endocrine systems, as well as their effects on cell cycle regulation, apoptosis, and cell differentiation have not yet been studied in detail. Although there are a number of ongoing studies investigating the use of carotenoids to enhance healthcare and beauty, most of these studies have been carried out in animal models, with very few human clinical trials. Future areas of research will need to focus on human clinical trials. In addition, these studies must collect detailed data on subject selection, end point measurements and levels of carotenoids being tested. It is hoped that such studies, will help researchers understand the roles and potential uses of carotenoids in developing new strategies for the prevention, treatment, and management of diseases.

**Acknowledgments:** We thank S. Abassi for critical comments and English editing on the early version of manuscript. This work was supported by the National Research Foundation of Korea Grant funded by the Korean Government (2015M1A5A1041805 and 2016R1D1A1A09920198), and by a grant from the National Fisheries Research and Development (R2017047) funded to J.-S. Ki.

**Conflicts of Interest:** The authors declare no conflict of interest.

## References

1. Mein, J.R.; Lian, F.; Wang, X.D. Biological activity of lycopene metabolites: Implications for cancer prevention. *Nutr. Rev.* **2008**, *66*, 667–683. [CrossRef] [PubMed]
2. Becker, E.W. Microalgae in human and animal nutrition. In *Handbook of Microalgal Culture*; Richmond, A., Ed.; Blackwell: Oxford, UK, 2004; pp. 312–351.
3. Sathasivam, R.; Radhakrishnan, R.; Hashem, A.; Abd_Allahd, E.F. Microalgae metabolites: A rich source for food and medicine. *Saudi J. Biol. Sci.* **2018**, in press. [CrossRef]
4. Sathasivam, R.; Pongpadung, P.; Praiboon, J.; Chirapart, A.; Trakulnaleamsai, S.; Roytrakul, S.; Juntawong, N. Optimizing NaCl and $KNO_3$ concentrations for high β-carotene production in photobioreactor by *Dunaliella salina* KU11 isolated from saline soil sample. *Chiang Mai J. Sci.* **2018**, *45*, in press.
5. Guiry, M.D. How many species of algae are there? *J. Phycol.* **2012**, *48*, 1057–1063. [CrossRef] [PubMed]

6. Mimouni, V.; Ulmann, L.; Pasquet, V.; Mathieu, M.; Picot, L.; Bougaran, G.; Cadoret, J.-P.; Morant-Manceau, A.; Schoefs, B. The potential of microalgae for the production of bioactive molecules of pharmaceutical interest. *Curr. Pharm. Biotechnol.* **2012**, *13*, 2733–2750. [CrossRef] [PubMed]
7. Goh, L.P.; Loh, S.P.; Fatimah, M.Y.; Perumal, K. Bioaccessibility of carotenoids and tocopherols in marine microalgae, *Nannochloropsis* sp. and *Chaetoceros* sp. *Malays. J. Nutr.* **2009**, *15*, 77–86. [PubMed]
8. Matos, J.; Cardoso, C.; Bandarra, N.M.; Afonso, C. Microalgae as a healthy ingredient for functional food: A review. *Food Funct.* **2017**, *8*, 2672–2685. [CrossRef] [PubMed]
9. Solymosi, K.; Mysliwa-Kurdziel, B. Chlorophylls and their derivatives used in food industry and medicine. *Mini-Rev. Med. Chem.* **2017**, *13*, 1194–1222. [CrossRef] [PubMed]
10. Nuno, K.; Vilarruel-Lopez, A.; Puebla-Perez, A.M.; Romero-Velarde, E.; Puebla-Mora, A.G.; Ascencio, F. Effects of the marine microalgae *Isochrysis galbana* and *Nannochloropsis oculata* in diabetic rats. *J. Funct. Foods* **2013**, *5*, 106–115. [CrossRef]
11. Rao, A.V.; Rao, L.G. Carotenoids and human health. *Pharmacol. Res.* **2007**, *55*, 207–216. [CrossRef] [PubMed]
12. Vilchez, C.; Forjan, E.; Cuaresma, M.; Bedmar, F.; Garbayo, I.; Vega, J.M. Marine carotenoids: Biological functions and commercial applications. *Mar. Drugs* **2011**, *9*, 319–333. [CrossRef] [PubMed]
13. Raposo, M.F.D.J.; Morais, A.M.M.B.D.; Morais, R.M.S.C.D. Carotenoids from marine microalgae: A valuable natural source for the prevention of chronic diseases. *Mar. Drugs* **2015**, *13*, 5128–5155. [CrossRef] [PubMed]
14. Di Pietro, N.; Di Tomo, P.; Pandolfi, A. Carotenoids in cardiovascular disease prevention. *JSM Atheroscler.* **2016**, *1*, 1002.
15. Dembitsky, V.M.; Maoka, T. Allenic and cumulenic lipids. *Prog. Lipid Res.* **2007**, *46*, 328–375. [CrossRef] [PubMed]
16. Cerón, M.C.; García-Malea, M.C.; Rivas, J.; Acien, F.G.; Fernandez, J.M.; del Rio, E.; Guerrero, M.G.; Molina, E. Antioxidant activity of *Haematococcus pluvialis* cells grown in continuous culture as a function of their carotenoid and fatty acid content. *Appl. Microbiol. Biotechnol.* **2007**, *74*, 1112–1119. [CrossRef] [PubMed]
17. Guerin, M.; Huntley, M.E.; Olaizola, M. *Haematococcus* astaxanthin: Applications for human health and nutrition. *Trends Biotechnol.* **2003**, *21*, 210–216. [CrossRef]
18. Molnár, P.; Deli, J.; Tanaka, T.; Kann, Y.; Tani, S.; Gyémánt, N.; Molnár, J.; Kawases, M. Carotenoids with anti-*Helicobacter pylori* activity from *Golden delicious* apple. *Phytother. Res.* **2010**, *24*, 644–648. [PubMed]
19. Gradelet, S.; Le Bon, A.M.; Berges, R.; Suschetet, M.; Astorg, P. Dietary carotenoids inhibit aflatoxin B1-induced liver preneoplastic foci and DNA damage in the rat: Role of the modulation of aflatoxin B1 metabolism. *Carcinogenesis* **1998**, *19*, 403–411. [CrossRef] [PubMed]
20. Lemoine, Y.; Schoefs, B. Secondary ketocarotenoid astaxanthin biosynthesis in algae: A multifunctional response to stress. *Photosynth. Res.* **2010**, *106*, 155–177. [CrossRef] [PubMed]
21. Liu, J.; Sun, Z.; Gerken, H.; Liu, Z.; Jiang, Y.; Chen, F. *Chlorella zofingiensis* as an alternative microalgal producer of astaxanthin: Biology and industrial potential. *Mar. Drugs* **2014**, *12*, 3487–3515. [CrossRef] [PubMed]
22. Miranda, M.S.; Cintra, R.G.; Barros, S.B.; Mancini-Filho, J. Antioxidant activity of the microalga *Spirulina maxima*. *Braz. J. Med. Biol. Res.* **1998**, *31*, 1075–1079. [CrossRef] [PubMed]
23. El-Baky, H.H.A.; El-Baz, F.K.; El-Baroty, G.S. *Spirulina* species as a source of carotenoids and $\alpha$-tocopherol and its anticarcinoma factors. *Biotechnology* **2003**, *2*, 222–240.
24. Erhardt, J.G.; Meisner, C.; Bode, J.C.; Bode, C. Lycopene, $\beta$-carotene and colorectal adenomas. *Am. J. Clin. Nutr.* **2003**, *78*, 1219–1224. [PubMed]
25. Lidebjer, C.; Leanderson, P.; Ernerudh, J.; Jonasson, L. Low plasma levels of oxygenated carotenoids in patients with coronary artery disease. *Nutr. Metab. Cardiovasc. Dis.* **2007**, *17*, 448–456. [CrossRef] [PubMed]
26. Sies, H.; Stahl, W. Nutritional protection against skin damage from sunlight. *Annu. Rev. Nutr.* **2004**, *24*, 173–200. [CrossRef] [PubMed]
27. Aust, O.; Stahl, W.; Sies, H.; Tronnier, H.; Heinrich, U. Supplementation with tomato-based products increases lycopene, phytofluene, and phytoene levels in human serum and protects against UV-light-induced erythema. *Int. J. Vitam. Nutr. Res.* **2005**, *75*, 54–60. [CrossRef] [PubMed]
28. Ramaraj, S.; Hemaiswarya, S.; Raja, R.; Ganesan, V.; Anbazhagan, C.; Carvalho, I.S.; Juntawong, N. Microalgae as an attractive source for biofuel production. In *Environmental Sustainability: Role of Green Technologies*; Thangavel, P., Sridevi, G., Eds.; Springer: New Delhi, India, 2014; pp. 129–157.

29. Sathasivam, R.; Juntawong, N. Modified medium for enhanced growth of *Dunaliella* strains. *Int. J. Curr. Sci.* **2013**, *5*, 67–73.
30. Sathasivam, R.; Kermanee, P.; Roytrakul, S.; Juntawong, N. Isolation and molecular identification of β-carotene producing strains of *Dunaliella salina* and *Dunaliella bardawil* from salt soil samples by using species-specific primers and internal transcribed spacer (ITS) primers. *Afr. J. Biotechnol.* **2012**, *11*, 8425–8432.
31. Sathasivam, R.; Praiboon, J.; Chirapart, A.; Trakulnaleamsai, S.; Kermanee, P.; Roytrakul, S.; Juntawong, N. Screening, phenotypic and genotypic identification of β-carotene producing strains of *Dunaliella salina* from Thailand. *Indian J. Geo-Mar. Sci.* **2014**, *43*, 2198–2216.
32. Wu, Z.; Dejtisakdi, W.; Kermanee, P.; Ma, C.; Arirob, W.; Sathasivam, R.; Juntawong, R. Outdoor cultivation of *Dunaliella salina* KU 11 using brine and saline lake water with raceway ponds in northeastern Thailand. *Biotechnol. Appl. Biochem.* **2017**, *64*, 938–943. [CrossRef] [PubMed]
33. Abe, K.; Hattori, H.; Hirano, M. Accumulation and antioxidant activity of secondary carotenoids in the aerial microalga *Coelastrella striolata* var. *multistriata*. *Food Chem.* **2007**, *100*, 656–661. [CrossRef]
34. Mendes, R.L.; Fernandes, H.L.; Coelho, J.P.; Reis, E.C.; Cabral, J.M.S.; Novais, J.M.; Palavra, A.F. Supercritical $CO_2$ extraction of carotenoids and other lipids from *Chlorella vulgaris*. *Food Chem.* **1995**, *53*, 99–103. [CrossRef]
35. Cha, K.H.; Koo, S.Y.; Lee, D.U. Antiproliferative effects of carotenoids extracted from *Chlorella ellipsoidea* and *Chlorella vulgaris* on human colon cancer cells. *J. Agric. Food Chem.* **2008**, *56*, 10521–10526. [CrossRef] [PubMed]
36. Tonegawa, I.; Okada, S.; Murakami, M.; Yamagushi, K. Pigment composition of the green microalga *Botryococcus braunii* Kawagushi-1. *Fish. Sci.* **1998**, *64*, 305–308. [CrossRef]
37. Petrushkina, M.; Gusev, E.; Sorokin, B.; Zotko, N.; Mamaeva, A.; Filimonova, A.; Kulikovskiy, M.; Maltsev, Y.; Yampolsky, I.; Guglya, E.; et al. Fucoxanthin production by heterokont microalgae. *Algal Res.* **2017**, *24*, 387–393. [CrossRef]
38. Rijstenbil, J.W. Effects of UVB radiation and salt stress on growth, pigments and oxidative defence of the marine diatom *Cylindrotheca closterium*. *Mar. Ecol. Prog. Ser.* **2003**, *254*, 37–48. [CrossRef]
39. Kim, S.M.; Kang, S.W.; Kwon, O.N.; Chung, D.; Pan, C.H. Fucoxanthin as a major carotenoid in *Isochrysis* aff. *galbana*: Characterization of extraction for commercial application. *J. Korean Soc. Appl. Biol. Chem.* **2012**, *55*, 477–483. [CrossRef]
40. Xia, S.; Wang, K.; Wan, L.; Li, A.; Hu, Q.; Zhang, C. Production, characterization, and antioxidant activity of fucoxanthin from the marine diatom *Odontella aurita*. *Mar. Drugs* **2013**, *11*, 2667–2681. [CrossRef] [PubMed]
41. Ragni, M.; d'Alcalá, M.R. Circadian variability in the photobiology of *Phaeodactylum tricornutum*: Pigment content. *J. Plankton Res.* **2007**, *29*, 141–156. [CrossRef]
42. Dambek, M.; Eilers, U.; Breitenbach, J.; Steiger, S.; Büchel, C.; Sandmann, G. Biosynthesis of fucoxanthin and diadinoxanthin and function of initial pathway genes in *Phaeodactylum tricornutum*. *J. Exp. Bot.* **2012**, *63*, 5607–5612. [CrossRef] [PubMed]
43. Minhas, A.K.; Hodgson, P.; Barrow, C.J.; Sashidhar, B.; Adholeya, A. The isolation and identification of new microalgal strains producing oil and carotenoid simultaneously with biofuel potential. *Bioresour. Technol.* **2016**, *211*, 556–565. [CrossRef] [PubMed]
44. Shi, X.M.; Jiang, Y.; Chen, F. High-yield production of lutein by the green microalga *Chlorella protothecoides* in heterotrophic fed-batch culture. *Biotechnol. Prog.* **2002**, *18*, 723–727. [CrossRef] [PubMed]
45. Sujak, A.; Gabrielska, J.; Grudzinski, W.; Borc, R.; Mazurek, P.; Gruszecki, W.I. Lutein and zeaxanthin as protectors of lipid membranes against oxidative damage: The structural aspects. *Arch. Biochem. Biophys.* **1999**, *371*, 301–307. [CrossRef] [PubMed]
46. Le, M.; Xiao-Ming, L. Effects of lutein and zeaxanthin on aspects of eye health. *J. Sci. Food Agric.* **2010**, *90*, 2–12.
47. Olmedilla, B.; Granado, F.; Blanco, I.; Vaquero, M.; Cajigal, C. Lutein in patients with cataracts and age-related macular degeneration: A longterm supplementation study. *J. Sci. Food Agric.* **2001**, *81*, 904–909. [CrossRef]
48. Graziani, G.; Schiavo, S.; Nicolai, M.A.; Buono, S.; Fogliano, V.; Pionto, G.; Pollio, A. Microalgae as human food: Chemical and nutritional characteristics of the thermo-acidophilic microalga *Galdieria sulphuraria*. *Food Funct.* **2013**, *4*, 144–152. [CrossRef] [PubMed]
49. Sánchez, J.F.; Fernández-Sevilla, J.M.; Acién, F.G.; Cerón, M.C.; Pérez-Parra, J.; Molina-Grima, E. Biomass and Lutein Productivity of *Scenedesmus almeriensis*: Influence of irradiance, dilution rate and temperature. *Appl. Microbiol. Biotechnol.* **2008**, *79*, 719–729. [CrossRef] [PubMed]

50. Soontornchaiboon, W.; Joo, S.S.; Kim, S.M. Anti-inflammatory effects of violaxanthin isolated from microalga *Chlorella ellipsoidea* in RAW 264.7 macrophages. *Biol. Pharm. Bull.* **2012**, *35*, 1137–1144. [CrossRef] [PubMed]
51. Nishino, H.; Murakoshi, M.; Li, T.; Takemura, M.; Kuchide, M.; Kanazawa, M.; Mou, X.; Wada, S.; Masuda, M.; Ohsaka, Y.; et al. Carotenoids in cancer chemoprevention. *Cancer Metastasis Rev.* **2002**, *21*, 257–264. [CrossRef] [PubMed]
52. Kana, T.M.; Glibert, P.M.; Goericke, R.; Welschmeyer, N.A. Zeaxanthin and β-carotene in *Synechococcus* WH7803 respond differently to irradiance. *Limnol. Oceanogr.* **1998**, *33*, 1623–1627.
53. Masamoto, K.; Zsiros, O.; Gombos, Z. Accumulation of zeaxanthin in cytoplasmic membranes of the cyanobacterium *Synechococcus* sp. strain PCC7942 grown under high light condition. *J. Plant Physiol.* **1999**, *155*, 136–138. [CrossRef]
54. Esteban, R.; Martinez, B.; Fernandez-Marin, B.; Becerril, J.S.; Garcia-Plazaola, I. Carotenoid composition in Rhodophyta: Insights into xanthophyll regulation in *Corallina elongata*. *Eur. J. Phycol.* **2009**, *44*, 221–230. [CrossRef]
55. Borowitzka, M.A. Vitamins and fine chemicals from micro-algae. In *Micro-Algal Biotechnology;* Borowitzka, M.A., Borowitzka, L.J., Eds.; Cambridge University Press: Cambridge, UK, 1988; pp. 153–196.
56. Solymosi, K.; Keresztes, Á. Plastid structure, diversification and interconversions II. Land plants. *Curr. Chem. Biol.* **2012**, *6*, 187–204. [CrossRef]
57. Takaichi, S. Carotenoids in algae: Distributions, biosynthesis and functions. *Mar. Drugs* **2011**, *9*, 1101–1118. [CrossRef] [PubMed]
58. Kajikawa, T.; Okumura, S.; Iwashita, T.; Kosumi, D.; Hashimoto, H.; Katsumura, S. Stereocontrolled total synthesis of fucoxanthin and its polyene chain-modified derivative. *Org. Lett.* **2012**, *14*, 808–811. [CrossRef] [PubMed]
59. Egeland, E.S. Carotenoids. In *The Physiology of Microalgae:Developments in Applied Phycology;* Borowitzka, M.A., Beardall, J., Raven, J.A., Eds.; Springer: Cham, Switzerland, 2016; pp. 507–563.
60. Christaki, E.; Bonos, E.; Giannenas, I.; Florou-Paneri, P. Functional properties of carotenoids originating from algae. *J. Sci. Food Agric.* **2012**, *93*, 5–11. [CrossRef] [PubMed]
61. Farre, G.; Sanahuja, G.; Naqvi, S.; Bai, C.; Capell, T.; Zhu, C.; Christou, P. Travel advice on the road to carotenoids in plants. *Plant Sci.* **2010**, *179*, 28–48. [CrossRef]
62. Varela, J.C.; Pereira, H.; Vila, M.; León, R. Production of carotenoids by microalgae: Achievements and challenges. *Photosynth. Res.* **2015**, *125*, 423–436. [CrossRef] [PubMed]
63. Wani, S.H.; Sah, S.K.; Sági, L.; Solymosi, K. Transplastomic plants for innovations in agriculture. A review. *Agron. Sustain. Dev.* **2015**, *35*, 1391–1430. [CrossRef]
64. Gateau, H.; Solymosi, K.; Marchand, J.; Schoefs, B. Carotenoids of microalgae used in food industry and medicine. *Mini-Rev. Med. Chem.* **2017**, *12*, 1140–1172. [CrossRef] [PubMed]
65. Schwender, J.; Gemunden, C.; Lichtenthaler, H.K. Chlorophyta exclusively use the 1-deoxyxylulose 5-phosphate/2-C-methylerythritol 4-phosphate pathway for the biosynthesis of isoprenoids. *Planta* **2001**, *212*, 416–423. [CrossRef] [PubMed]
66. Cunningham, F.X., Jr.; Gantt, E. A portfolio of plasmids for identification and analysis of carotenoid pathway enzymes: *Adonis aestivalis* as a case study. *Photosynth. Res.* **2007**, *92*, 245–259. [CrossRef] [PubMed]
67. Couso, I.; Vila, M.; Vigara, J.; Cordero, B.F.; Vargas, M.Á.; Rodríguez, H.; León, R. Synthesis of carotenoids and regulation of the carotenoid biosynthesis pathway in response to high light stress in the unicellular microalga *Chlamydomonas reinhardtii*. *Eur. J. Phycol.* **2012**, *47*, 223–232. [CrossRef]
68. Carmeliet, P. Angiogenesis in health and disease. *Nat. Med.* **2003**, *9*, 653–660. [CrossRef] [PubMed]
69. Kirk, S.; Frank, J.A.; Karlik, S. Angiogenesis in multiple sclerosis: Is it good, bad or an epiphenomenon? *J. Neurol. Sci.* **2004**, *217*, 125–130. [CrossRef] [PubMed]
70. Guruvayoorappan, C.; Kuttan, G. β-Carotene inhibits tumor-specific angiogenesis by altering the cytokine profile and inhibits the nuclear translocation of transcription factors in B16F-10 melanoma cells. *Integr. Cancer Ther.* **2007**, *6*, 258–270. [CrossRef] [PubMed]
71. Sugawara, T.; Matsubara, K.; Akagi, R.; Mori, M.; Hirata, T. Antiangiogenic activity of brown algae fucoxanthin and its deacetylated product, fucoxanthinol. *J. Agric. Food Chem.* **2006**, *54*, 9805–9810. [CrossRef] [PubMed]

72. Ganesan, P.; Matsubara, K.; Sugawara, T.; Hirata, T. Marine algal carotenoids inhibit angiogenesis by down-regulating FGF-2-mediated intracellular signals in vascular endothelial cells. *Mol. Cell. Biochem.* **2013**, *380*, 1–9. [CrossRef] [PubMed]
73. Hussein, G.; Nakamura, M.; Zhao, Q.; Iguchi, T.; Goto, H.; Sankawa, U.; Watanabe, H. Antihypertensive and neuroprotective effects of astaxanthin in experimental animals. *Biol. Pharm. Bull.* **2005**, *28*, 47–52. [CrossRef] [PubMed]
74. Hussein, G.; Goto, H.; Oda, S.; Sankawa, U.; Matsumoto, K.; Watanabe, H. Antihypertensive potential and mechanism of action of astaxanthin: III. Antioxidant and histopathological effects in spontaneously hypertensive rats. *Biol. Pharm. Bull.* **2006**, *29*, 684–688. [CrossRef] [PubMed]
75. Preuss, H.G.; Echard, B.; Yamashita, E.; Perricone, N.V. High dose astaxanthin lowers blood pressure and increases insulin sensitivity in rats: Are these effects interdependent? *Int. J. Med. Sci.* **2011**, *8*, 126–138. [CrossRef] [PubMed]
76. Nakao, R.; Nelson, O.L.; Park, J.S.; Mathison, B.D.; Thompson, P.A.; Chew, B.P. Effect of astaxanthin supplementation on inflammation and cardiac function in BALB/c mice. *Anticancer Res.* **2010**, *30*, 2721–2725. [PubMed]
77. Iwamoto, T.; Hosoda, K.; Hirano, R.; Kurata, H.; Matsumoto, A.; Miki, W.; Kondo, K. Inhibition of low-density lipoprotein oxidation by astaxanthin. *J. Atheroscler. Thromb.* **2000**, *7*, 216–222. [CrossRef] [PubMed]
78. Miyawaki, H.; Takahashi, J.; Tsukahara, H.; Takehara, I. Effects of astaxanthin on human blood rheology. *J. Clin. Biochem. Nutr.* **2008**, *43*, 69–74. [CrossRef] [PubMed]
79. Yoshida, H.; Yanai, H.; Ito, K.; Tomono, Y.; Koikeda, T.; Tsukahara, H.; Tada, N. Administration of natural astaxanthin increases serum HDL-cholesterol and adiponectin in subjects with mild hyperlipidemia. *Atherosclerosis* **2010**, *209*, 520–523. [CrossRef] [PubMed]
80. Matsumoto, M.; Hosokawa, M.; Matsukawa, N.; Hagio, M.; Shinoki, A.; Nishimukai, M.; Hara, H. Suppressive effects of the marine carotenoids, fucoxanthin and fucoxanthinol on triglyceride absorption in lymph duct-cannulated rats. *Eur. J. Nutr.* **2010**, *49*, 243–249. [CrossRef] [PubMed]
81. Senesse, P.; Touvie, M.; Kesse, E.; Faivre, J.; Boutron-Ruault, M.C. Tobacco use and associations of β-carotene and vitamin intakes with colorectal adenoma risk. *J. Nutr.* **2005**, *135*, 2468–2472. [PubMed]
82. Ramadas, A.; Kandiah, M.; Jabbar, F.; Zarida, H. Dietary risk factors for colorectal adenomatous polyps: A mini review. *J. Sci. Technol.* **2010**, *18*, 321–349.
83. Blot, W.J.; Li, J.Y.; Taylor, P.R.; Guo, W.; Dawsey, S.; Wang, G.Q.; Yang, C.S.; Zheng, S.F.; Gail, M.; Yu, G.Y.L.Y.; et al. Nutrition intervention trials in Linxian, China: Supplementation with specific vitamin/mineral combinations, cancer incidence, and disease-specific mortality in the general population. *J. Natl. Cancer Inst.* **1993**, *85*, 1483–1492. [CrossRef] [PubMed]
84. Mayne, S.T. β-Carotene, carotenoids, and disease prevention in humans. *FASEB J.* **1996**, *10*, 690–701. [PubMed]
85. Rock, C.L. Carotenoids: Biology and treatment. *Pharmacol. Ther.* **1997**, *75*, 185–197. [CrossRef]
86. Graydon, R.; Gilchrist, S.; Young, I.; Obermüller-Jevic, U.; Hasselwander, U.; Woodside, J. Effect of lycopene supplementation on insulin-like growth factor-1 and insulin-like growth factor binding protein-3: A double-blind, placebo-controlled trial. *Eur. J. Clin. Nutr.* **2007**, *61*, 1196–1200. [CrossRef] [PubMed]
87. Schwarz, S.; Obermüller-Jevic, U.; Hellmis, E.; Koch, W.; Jacobi, G.; Biesalski, H.K. Lycopene inhibits disease progression in patients with benign prostate hyperplasia. *J. Nutr.* **2008**, *138*, 49–53. [PubMed]
88. Kumar, S.R.; Hosokawa, M.; Miyashita, K. Fucoxanthin: A marine carotenoid exerting anti-cancer effects by affecting multiple mechanisms. *Mar. Drugs* **2013**, *11*, 5130–5147. [CrossRef] [PubMed]
89. Hosokawa, M.; Kudo, M.; Maeda, H.; Kohno, H.; Tanaka, T.; Miyashita, K. Fucoxanthin induces apoptosis and enhances the antiproliferative effect of the PPAR c-ligand, troglitazone, on colon cancer cells. *BBA-Gen. Subj.* **2004**, *1675*, 113–119. [CrossRef] [PubMed]
90. Kim, K.N.; Heo, S.J.; Yoon, W.J.; Kang, S.M.; Ahn, G.; Yi, T.H.; Jeon, Y.J. Fucoxanthin inhibits the inflammatory response by suppressing the activation of NF-κB and MAPKs in lipopolysaccharide-induced RAW 264.7 macrophages. *Eur. J. Pharmacol.* **2010**, *649*, 369–375. [CrossRef] [PubMed]
91. Kotake-Nara, E.; Terasaki, M.; Nagao, A. Characterization of apoptosis induced by fucoxanthin in human promyelocytic leukemia cells. *Biosci. Biotechnol. Biochem.* **2005**, *69*, 224–227. [CrossRef] [PubMed]

92. Zhang, Z.; Zhang, P.; Hamada, M.; Takahashi, S.; Xing, G.; Liu, J.; Sugiura, N. Potential chemoprevention effect of dietary fucoxanthin on urinary bladder cancer EJ-1 cell line. *Oncol. Rep.* **2008**, *20*, 1099–1103. [CrossRef] [PubMed]
93. Kotake Nara, E.; Kushiro, M.; Zhang, H.; Sugawara, T.; Miyashita, K.; Nagao, A. Carotenoids affect proliferation of human prostate cancer cells. *J. Nutr.* **2001**, *131*, 3303–3306. [PubMed]
94. Miyashita, K.; Nishikawa, S.; Beppu, F.; Tsukui, T.; Abe, M.; Hosokawa, M. The allenic carotenoid fucoxanthin, a novel marine nutraceutical from brown seaweeds. *J. Sci. Food Agric.* **2011**, *91*, 1166–1174. [CrossRef] [PubMed]
95. Huang, D.S.; Odeleye, O.E.; Watson, R.R. Inhibitory effects of canthaxanthin on in vitro growth of murine tumor cells. *Cancer Lett.* **1992**, *65*, 209–213. [CrossRef]
96. Palozza, P.; Maggiano, N.; Calviello, G.; Lanza, P.; Piccioni, E.; Ranelletti, F.O.; Bartoli, G.M. Canthaxanthin induces apoptosis in human cancer cell lines. *Carcinogenesis* **1998**, *19*, 373–376. [CrossRef] [PubMed]
97. Abdel-Fatth, G.; Watzl, B.; Huang, D.; Watson, R.R. Beta-carotene *in vitro* stimulates tumor necrosis factor alpha and interleukin 1 alpha secretion by human peripheral blood mononuclear cells. *Nutr. Res.* **1993**, *13*, 863–871. [CrossRef]
98. Muto, Y.; Fujii, J.; Shidoji, Y.; Moriwaki, H.; Kawaguchi, T.; Noda, T. Growth retardation in human cervical dysplasia-derived cell lines by beta-carotene through down-regulation of epidermal growth factor receptor. *Am. J. Clin. Nutr.* **1995**, *62*, 1535S–1540S. [CrossRef] [PubMed]
99. Grubbs, C.J.; Eto, I.; Juliana, M.M.; Whitaker, L.M. Effect of canthaxanthin on chemically induced mammary carcinogenesis. *Oncology* **1991**, *48*, 239–245. [CrossRef] [PubMed]
100. Hanusch, M.; Stahl, W.; Schulz, W.A.; Sies, H. Induction of gap junctional communication by 4-oxoretinoic acid generated from its precursor canthaxanthin. *Arch. Biochem. Biophys.* **1995**, *317*, 423–428. [CrossRef] [PubMed]
101. Tanaka, T.; Kawamori, T.; Ohnishi, M.; Makita, H.; Mori, H.; Satoh, K.; Hara, A. Suppression of azoxymethane-induced rat colon carcinogenesis by dietary administration of naturally occurring xanthophylls astaxanthin and canthaxanthin during the postinitiation phase. *Carcinogenesis* **1995**, *16*, 2957–2963. [CrossRef] [PubMed]
102. Katsumura, N.; Okuno, M.; Onogi, N.; Moriwaki, H.; Muto, Y.; Kojima, S. Suppression of mouse skin papilloma by canthaxanthin and beta-carotene in vivo: Possibility of the regression of tumorigenesis by carotenoids without conversion to retinoic acid. *Nutr. Cancer* **1996**, *26*, 203–208. [CrossRef] [PubMed]
103. Palan, P.R.; Mikhail, M.S.; Goldberg, G.L.; Basu, J.; Runowicz, C.D.; Romney, S.L. Plasma levels of beta-carotene, lycopene, canthaxanthin, retinol, and alpha- and tau-tocopherol in cervical intraepithelial neoplasia and cancer. *Clin. Cancer Res.* **1996**, *2*, 181–185. [PubMed]
104. Anderson, M. Method of Inhibiting 5α-Reductase with Astaxanthin to Prevent and Treat Benign Prostate Hyperplasia (BPH) and Prostate Cancer in Human Males. U.S. Patent 6277417, 21 August 2001.
105. Tanaka, T.; Morishita, Y.; Suzui, M.; Kojima, T.; Okumura, A.; Mori, H. Chemoprevention of mouse urinary bladder carcinogenesis by the naturally occurring carotenoid astaxanthin. *Carcinogenesis* **1994**, *15*, 15–19. [CrossRef] [PubMed]
106. Kozuki, Y.; Miura, Y.; Yagasaki, K. Inhibitory effects of carotenoids on the invasion of rat ascites hepatoma cells in culture. *Cancer Lett.* **2000**, *151*, 111–115. [CrossRef]
107. Lyons, N.M.; O'Brien, N.M. Modulatory effects of an algal extract containing astaxanthin on UVA-irradiated cells in culture. *J. Dermatol. Sci.* **2002**, *30*, 73–84. [CrossRef]
108. Chuyen, H.V.; Eun, J.B. Marine carotenoids: Bioactivities and potential benefits to human health. *Crit. Rev. Food Sci. Nutr.* **2017**, *57*, 2600–2610. [CrossRef] [PubMed]
109. Jyonouchi, H.; Sun, S.; Iijima, K.; Gross, M.D. Antitumor activity of astaxanthin and its mode of action. *Nutr. Cancer* **2000**, *36*, 59–65. [CrossRef] [PubMed]
110. Kurihara, H.; Koda, H.; Asami, S.; Kiso, Y.; Tanaka, T. Contribution of the antioxidative property of astaxanthin to its protective effect on the promotion of cancer metastasis in mice treated with restraint stress. *Life Sci.* **2002**, *70*, 2509–2520. [CrossRef]
111. San Millan, C.S.; Soldevilla, B.; Martín, P.; Gil-Calderon, B.; Compte, M.; Perez-Sacristan, B.; Donoso, E.; Pena, C.; Romero, J.; Granado-Lorencio, F.; et al. β-Cryptoxanthin synergistically enhances the antitumoral activity of Oxaliplatin through ΔNP73 negative regulation in colon cancer. *Clin. Cancer Res.* **2015**, *21*, 4398–4409. [CrossRef] [PubMed]

112. Ylonen, K.; Alfthan, G.; Groop, L.; Saloranta, C.; Aro, A.; Virtanen, S.M. Dietary intakes and plasma concentrations of carotenoids and tocopherols in relation to glucose metabolism in subjects at high risk of type 2 diabetes: The Botnia dietary study. *Am. J. Clin. Nutr.* **2003**, *77*, 1434–1441. [PubMed]
113. Sluijs, I.; Cadier, E.; Beulens, J.W.; van der, A.D.; Spijkerman, A.M.; van der Schouw, Y.T. Dietary intake of carotenoids and risk of type 2 diabetes. *Nutr. Metab. Cardiovasc. Dis.* **2015**, *25*, 376–381. [CrossRef] [PubMed]
114. Suzuki, K.; Ito, Y.; Nakamura, S.; Ochiai, J.; Aoki, K. Relationship between serum carotenoids and hyperglycemia: A population-based cross-sectional study. *J. Epidemiol.* **2002**, *12*, 357–366. [CrossRef] [PubMed]
115. Brazionis, L.; Rowley, K.; Itsiopoulos, C.; O'Dea, K. Plasma carotenoids and diabetic retinopathy. *Br. J. Nutr.* **2009**, *101*, 270–277. [CrossRef] [PubMed]
116. Stahl, W.; Sies, H. Bioactivity and protective effects of natural carotenoids. *Biochim. Biophys. Acta* **2005**, *1740*, 101–107. [CrossRef] [PubMed]
117. Sugiura, M.; Nakamura, M.; Ogawa, K.; Ikoma, Y.; Yano, M. High-serum carotenoids associated with lower risk for developing type 2 diabetes among Japanese subjects: Mikkabi cohort study. *BMJ Open Diabetes Res. Care* **2015**, *3*, e000147. [CrossRef] [PubMed]
118. Hozawa, A.; Steffes, D.R., Jr.; Jacobs, M.W.; Gross, M.D.; Steffen, L.M.; Lee, D.H. Associations of serum carotenoid concentrations with the development of diabetes and with insulin concentration: Interaction with smoking: The coronary artery risk development in young adults (CARDIA) study. *Am. J. Epidemiol.* **2006**, *163*, 929–937. [CrossRef] [PubMed]
119. Akbaraly, T.N.; Fontbonne, A.; Favier, A.; Berr, C. Plasma carotenoids and onset of dysglycemia in an elderly population: Results of the epidemiology of vascular ageing study. *Diabetes Care* **2008**, *31*, 1355–1359. [CrossRef] [PubMed]
120. Yeh, P.T.; Huang, H.W.; Yang, C.M.; Yang, W.S.; Yang, C.H. Astaxanthin inhibits expression of retinal oxidative stress and inflammatory mediators in streptozotocin-induced diabetic rats. *PLoS ONE* **2016**, *11*, e0146438. [CrossRef] [PubMed]
121. Spiller, G.A.; Dewell, A. Safety of an astaxanthin-rich *Haematococcus pluvialis* algal extract: A randomized clinical trial. *J. Med. Food* **2003**, *6*, 51–56. [CrossRef] [PubMed]
122. Uchiyama, K.; Naito, Y.; Hasegawa, G.; Nakamura, N.; Takahashi, J.; Yoshikawa, T. Astaxanthin protects beta-cells against glucose toxicity in diabetic db/db mice. *Redox Rep.* **2002**, *7*, 290–293. [CrossRef] [PubMed]
123. Bhuvaneswari, S.; Yogalakshmi, B.; Sreeja, S.; Anuradha, C.V. Astaxanthin reduces hepatic endoplasmic reticulum stress and nuclear factor-κB-mediated inflammation in high fructose and high fat diet-fed mice. *Cell Stress Chaperones* **2014**, *19*, 183–191. [CrossRef] [PubMed]
124. Hussein, G.; Nakagawa, T.; Goto, H.; Shimada, Y.; Matsumoto, K.; Sankawa, U.; Watanabe, H. Astaxanthin ameliorates features of metabolic syndrome in SHR/NDmcr-cp. *Life Sci.* **2007**, *80*, 522–529. [CrossRef] [PubMed]
125. Inoue, M.; Tanabe, H.; Matsumoto, A.; Takagi, M.; Umegaki, K.; Amagaya, S.; Takahashi, J. Astaxanthin functions differently as a selective peroxisome proliferator-activated receptor gamma modulator in adipocytes and macrophages. *Biochem. Pharmacol.* **2012**, *84*, 692–700. [CrossRef] [PubMed]
126. Xu, L.; Zhu, J.; Yin, W.; Ding, X. Astaxanthin improves cognitive deficits from oxidative stress, nitric oxide synthase and inflammation through upregulation of PI3K/Akt in diabetes rat. *Int. J. Clin. Exp. Pathol.* **2015**, *8*, 6083–6094. [PubMed]
127. Sila, A.; Ghlissi, Z.; Kamoun, Z.; Makni, M.; Nasri, M.; Bougatef, A.; Sahnoun, Z. Astaxanthin from shrimp by-products ameliorates nephropathy in diabetic rats. *Eur. J. Nutr.* **2015**, *54*, 301–307. [CrossRef] [PubMed]
128. Hozumi, M.; Murata, T.; Morinobu, T.; Manago, M.; Kuno, T.; Tokuda, M.; Konishi, K.; Mingci, Z.; Tamai, H. Plasma beta-carotene retinol, and alpha-tocopherol levels in relation to glycemic control of children with insulin-dependent diabetes mellitus. *J. Nutr. Sci. Vitaminol.* **1998**, *44*, 1–9. [CrossRef] [PubMed]
129. Arnlov, J.; Zethelius, B.; Riserus, U.; Basu, S.; Berne, C.; Vessby, B.; Alfthan, G.; Helmersson, J. Serum and dietary β-carotene and α-tocopherol and incidence of type 2 diabetes mellitus in a community-based study of Swedish men: Report from the Uppsala Longitudinal Study of Adult Men (ULSAM) study. *Diabetologia* **2009**, *52*, 97–105. [CrossRef] [PubMed]
130. Coyne, T.; Ibiebele, T.I.; Baade, P.D.; Dobson, A.; McClintock, C.; Dunn, S.; Leonard, D.; Shaw, J. Diabetes mellitus and serum carotenoids: Findings of a population-based study in Queensland, Australia. *Am. J. Clin. Nutr.* **2005**, *82*, 685–693. [PubMed]

131. Suzuki, K.; Ito, Y.; Inoue, T.; Hamajima, N. Inverse association of serum carotenoids with prevalence of metabolic syndrome among Japanese. *Clin. Nutr.* **2011**, *30*, 369–375. [CrossRef] [PubMed]
132. Katyal, T.; Singh, G.; Budhiraja, R.; Sachdeva, M. Effect of lutein in development of experimental diabetic nephropathy in rats. *Afr. J. Pharm. Pharmacol.* **2013**, *7*, 2953–2959. [CrossRef]
133. Maeda, H.; Hosokawa, M.; Sashima, T.; Murakami-Funayama, K.; Miyashita, K. Anti-obesity and anti-diabetic effects of fucoxanthin on diet-induced obesity conditions in a murine model. *Mol. Med. Rep.* **2009**, *2*, 897–902. [CrossRef] [PubMed]
134. Nishikawa, S.; Hosokawa, M.; Miyashita, K. Fucoxanthin promotes translocation and induction of glucose transporter 4 in skeletal muscles of diabetic/obese KK-A(y) mice. *Phytomedicine* **2012**, *19*, 389–394. [CrossRef] [PubMed]
135. Maeda, H.; Hosokawa, M.; Sashima, T.; Takahashi, N.; Kawada, T.; Miyashita, K. Fucoxanthin and its metabolite, fucoxanthinol, suppress adipocyte differentiation in 3T3-L1 cells. *Int. J. Mol. Med.* **2006**, *18*, 147–152. [CrossRef] [PubMed]
136. Hosokawa, M.; Miyashita, T.; Nishikawa, S.; Emi, S.; Tsukui, T.; Beppu, F.; Okada, T.; Miyashita, K. Fucoxanthin regulates adipocytokine mRNA expression in white adipose tissue of diabetic/obese KK-Ay mice. *Arch. Biochem. Biophys.* **2010**, *504*, 17–25. [CrossRef] [PubMed]
137. Jung, H.A.; Islam, M.N.; Lee, C.M.; Jeong, H.O.; Chung, H.Y.; Woo, H.C.; Choi, J.S. Promising antidiabetic potential of fucoxanthin isolated from the edible brown algae *Eisenia bicyclis* and *Undaria pinnatifida*. *Fish. Sci.* **2012**, *78*, 1321–1329. [CrossRef]
138. Pangestuti, R.; Kim, S.K. Neuroprotective effects of marine algae. *Mar. Drugs* **2011**, *9*, 803–818. [CrossRef] [PubMed]
139. Rajapakse, N.; Kim, M.M.; Mendis, E.; Kim, S.K. Inhibition of inducible nitric oxide synthase and cyclooxygenase-2 in lipopolysaccharide-stimulated RAW264.7 cells by carboxybutyrylated glucosamine tales place via down-regulation of mitogen-activated protein kinase-mediated nuclear factor-κB signaling. *Immunology* **2008**, *123*, 348–357. [CrossRef] [PubMed]
140. Peerapornpisal, Y.; Amornlerdpison, D.; Jamjai, U.; Taesotikul, T.; Pongpaibul, Y.; Nualchareon, M.; Kanjanapothi, D. Antioxidant and anti-inflammatory activities of brown marine alga, *Padina minor* Yamada. *Chiang Mai J. Sci.* **2010**, *37*, 507–516.
141. Ohgami, K.; Shiratori, K.; Kotake, S.; Nishida, T.; Mizuki, N.; Yazawa, K.; Ohno, S. Effects of astaxanthin on lipopolysaccharide induced inflammation in vitro and in vivo. *Investig. Ophthalmol. Vis. Sci.* **2003**, *44*, 2694–2701. [CrossRef]
142. Shiratori, K.; Ohgami, K.; Ilieva, I.; Jin, X.H.; Koyama, Y.; Miyashita, K.; Ohno, S. Effects of fucoxanthin on lipopolysaccharide-induced inflammation in vitro and in vivo. *Exp. Eye Res.* **2005**, *81*, 422–428. [CrossRef] [PubMed]
143. Choi, Y.; Lee, M.K.; Lim, S.Y.; Sung, S.H.; Kim, Y.C. Inhibition of inducible NO synthase, cyclooxygenase-2 and interleukin-1β by torilin is mediated by mitogen-activated protein kinases in microglial BV2 cells. *Br. J. Pharmacol.* **2009**, *156*, 933–940. [CrossRef] [PubMed]
144. Lee, S.J.; Bai, S.K.; Lee, K.S.; Namkoong, S.; Na, H.J.; Ha, K.S.; Kim, Y.M. Astaxanthin inhibits nitric oxide production and inflammatory gene expression by suppressing I(kappa)B kinase-dependent NF-kappaB activation. *Mol. Cells* **2003**, *16*, 97–105. [PubMed]
145. Macedo, R.C.; Bolin, A.P.; Marin, D.P.; Otton, R. Astaxanthin addition improves human neutrophils function: In vitro study. *Eur. J. Nutr.* **2010**, *49*, 447–457. [CrossRef] [PubMed]
146. Speranza, L.; Pesce, M.; Patruno, A.; Franceschelli, S.; de Lutiis, M.A.; Grilli, A.; Felaco, M. Astaxanthin treatment reduced oxidative induced pro-inflammatory cytokines secretion in U937: SHP-1 as a novel biological target. *Mar. Drugs* **2012**, *10*, 890–899. [CrossRef] [PubMed]
147. Bennedsen, M.; Wang, X.; Willen, R.; Wadstrom, T.; Andersen, L.P. Treatment of *H. pylori* infected mice with antioxidant astaxanthin reduces gastric inflammation, bacterial load and modulates cytokine release by splenocytes. *Immunol. Lett.* **1999**, *70*, 185–189. [PubMed]
148. Park, J.S.; Chyun, J.H.; Kim, Y.K.; Line, L.L.; Chew, B.P. Astaxanthin decreased oxidative stress and inflammation and enhanced immune response in humans. *Nutr. Metab. (Lond.)* **2010**, *7*, 18. [CrossRef] [PubMed]

149. Jung, U.J.; Choi, M.S. Obesity and its metabolic complications: The role of adipokines and the relationship between obesity, inflammation, insulin resistance, dyslipidemia and nonalcoholic fatty liver disease. *Int. J. Mol. Sci.* **2014**, *15*, 6184–6223. [CrossRef] [PubMed]
150. Luppino, F.S.; De Wit, L.M.; Bouvy, P.F.; Stijnen, T.; Cuijpers, P.; Penninx, B.W.; Zitman, F.G. Overweight, obesity, and depression: A systematic review and meta-analysis of longitudinal studies. *Arch. Gen. Psychiatry* **2010**, *67*, 220–229. [CrossRef] [PubMed]
151. Kelishadi, R. Childhood overweight, obesity, and the metabolic syndrome in developing countries. *Epidemiol. Rev.* **2007**, *29*, 62–76. [CrossRef] [PubMed]
152. Wang, T.; Wang, Y.; Kontani, Y.; Kobayashi, Y.; Sato, Y.; Mori, N.; Yamashita, N. Evodiamine improves diet-induced obesity in a uncoupling protein-1-independent manner: Involvement of antiadipogenic mechanism and extracellularly regulated kinase/mitogen-activated protein kinase signaling. *Endocrinology* **2008**, *149*, 358–366. [CrossRef] [PubMed]
153. Okada, T.; Nakai, M.; Maeda, H.; Hosokawa, M.; Sashima, T.; Miyashita, K. Suppressive effect of neoxanthin on the differentiation of 3T3-L1 adipose cells. *J. Oleo Sci.* **2008**, *57*, 345–351. [CrossRef] [PubMed]
154. Maeda, H.; Hosokawa, M.; Sashima, T.; Miyashita, K. Antiobesity effect of fucoxanthin from edible seaweeds and its multibiological functions. In *Functional Food and Health*; ACS Publications: Washington, DC, USA, 2008; pp. 376–388.
155. Nicholls, D.G.; Locke, R.M. Thermogenic mechanisms in brown fat. *Physiol. Rev.* **1984**, *64*, 1–64. [CrossRef] [PubMed]
156. Lowell, B.B.; Susullc, V.S.; Hamann, A.; Lawitts, J.A.; Himms-Hagen, J.; Boyer, B.B.; Kozak, L.P.; Flier, J.S. Development of obesity in transgenic mice after genetic ablation of brown adipose tissue. *Nature* **1993**, *366*, 740–742. [CrossRef] [PubMed]
157. Cederberg, A.; Gronning, L.M.; Ahren, B.; Tasken, K.; Carlsson, P.; Enerback, S. *FOXC2* is a winged helix gene that counteracts obesity, hypertriglyceridemia, and diet-induced insulin resistance. *Cell* **2001**, *106*, 563–573. [CrossRef]
158. Trayhurn, P.; Wood, I.S. Signalling role of adipose tissue: Adipokines and inflammation in obesity. *Biochem. Soc. Trans.* **2005**, *33*, 1078–1081. [CrossRef] [PubMed]
159. Curat, C.; Wegner, V.; Sengenes, C.; Miranville, A.; Tonus, C.; Busse, R.; Bouloumie, A. Macrophages in human visceral adipose tissue: Increased accumulation in obesity and a source of resistin and visfatin. *Diabetologia* **2006**, *49*, 744–747. [CrossRef] [PubMed]
160. Hancock, J.T.; Desikan, R.; Neill, S.J. Role of reactive oxygen species in cell signalling pathways. *Biochem. Soc. Trans.* **2001**, *29*, 345–350. [CrossRef] [PubMed]
161. Halliwell, B. Antioxidants in human health and disease. *Annu. Rev. Nutr.* **1996**, *16*, 33–50. [CrossRef] [PubMed]
162. Sies, H. Strategies of antioxidant defense. *Eur. J. Biochem.* **1993**, *215*, 213–219. [CrossRef] [PubMed]
163. Heo, S.J.; Ko, S.C.; Kang, S.M.; Kang, H.S.; Kim, J.P.; Kim, S.H.; Lee, K.W.; Cho, M.G.; Jeon, Y.J. Cytoprotective effect of fucoxanthin isolated from brown algae *Sargassum siliquastrum* against $H_2O_2$-induced cell damage. *Eur. Food Res. Technol.* **2008**, *228*, 145–151. [CrossRef]
164. Sachindra, N.M.; Sato, E.; Maeda, H.; Hosokawa, M.; Niwano, Y.; Kohno, M.; Miyashita, K. Radical scavenging and singlet oxygen quenching activity of marine carotenoid fucoxanthin and its metabolites. *J. Agric. Food Chem.* **2007**, *55*, 8516–8522. [CrossRef] [PubMed]
165. Miyashita, K. Function of marine carotenoids. *Forum Nutr.* **2009**, *61*, 136–146. [PubMed]
166. Iio, K.; Okada, Y.; Ishikura, M. Single and 13-week oral toxicity study of fucoxanthin oil from microalgae in rats. *Shokuhin Eiseigaku Zasshi* **2011**, *52*, 183–189. [CrossRef] [PubMed]
167. Peng, J.; Yuan, J.P.; Wu, C.F.; Wang, J.H. Fucoxanthin, a marine carotenoid present in brown seaweeds and diatoms: Metabolism and bioactivities relevant to human health. *Mar. Drugs* **2011**, *9*, 1806–1828. [CrossRef] [PubMed]
168. Pashkow, F.J.; Watumull, D.G.; Campbell, C.L. Astaxanthin: A novel potential treatment for oxidative stress and inflammation in cardiovascular disease. *Am. J. Cardiol.* **2008**, *101*, 58–68. [CrossRef] [PubMed]

169. Rodrigues, E.; Mariutti, L.R.B.; Mercadante, A.Z. Scavenging capacity of marine carotenoids against reactive oxygen and nitrogen species in a membrane-mimicking system. *Mar. Drugs* **2012**, *10*, 1784–1798. [CrossRef] [PubMed]
170. Masaki, H. Role of antioxidants in the skin: Anti-aging effects. *J. Dermatol. Sci.* **2010**, *58*, 85–90. [CrossRef] [PubMed]
171. Pallela, R.; Na-Young, Y.; Kim, S.K. Anti-photoaging and photoprotective compounds derived from marine organisms. *Mar. Drugs* **2010**, *8*, 1189–1202. [CrossRef] [PubMed]
172. Guedes, A.C.; Amaro, H.M.; Malcata, F.X. Microalgae as sources of carotenoids. *Mar. Drugs* **2011**, *9*, 625–644. [CrossRef] [PubMed]
173. Pan, J.L.; Wang, H.M.; Chen, C.Y.; Chang, J.S. Extraction of astaxanthin from *Haematococcus pluvialis* by supercritical carbon dioxide fluid with ethanol modifier. *Eng. Life Sci.* **2012**, *12*, 638–647. [CrossRef]
174. Tominaga, K.; Hongo, N.; Karato, M.; Yamashita, E. Cosmetic benefits of astaxanthin on humans subjects. *Acta Biochim. Pol.* **2012**, *59*, 43–47. [PubMed]
175. Thomas, N.V.; Kim, S.K. Beneficial effects of marine algal compounds in cosmeceuticals. *Mar. Drugs* **2013**, *11*, 146–164. [CrossRef] [PubMed]
176. Wijesinghe, W.A.J.P.; Jeon, Y.J. Biological activities and potential cosmeceutical applications of bioactive components from brown seaweeds: A review. *Phytochem. Rev.* **2011**, *10*, 431–443. [CrossRef]
177. Lee, E.H.; Faulhaber, D.; Hanson, K.M.; Ding, W.; Peters, S.; Kodali, S.; Granstein, R.D. Dietary lutein reduces ultraviolet radiation-induced inflammation and immunosuppression. *J. Investig. Dermatol.* **2004**, *122*, 510–517. [CrossRef] [PubMed]
178. Darvin, M.; Patzelt, A.; Gehse, S.; Schanzer, S.; Benderoth, C.; Sterry, W.; Lademann, J. Cutaneous concentration of lycopene correlates significantly with the roughness of the skin. *Eur. J. Pharm. Biopharm.* **2008**, *6*, 943–947. [CrossRef] [PubMed]
179. Kawaguchi, Y.; Tanaka, H.; Okada, T.; Konishi, H.; Takahashi, M.; Ito, M.; Asai, J. Effect of reactive oxygen species on elastin mRNA expression in cultured human dermal fibroblast. *Free Radic. Biol. Med.* **1997**, *23*, 162–165. [CrossRef]
180. Heinrich, U.; Tronnier, H.; Stahl, W.; Béjot, M.; Maurette, J.M. Antioxidant supplements improve parameters related to skin structure in humans. *Skin Pharmacol. Physiol.* **2006**, *19*, 224–231. [CrossRef] [PubMed]
181. Krinsky, N.I.; Landrum, J.T.; Bone, R.A. Biologic mechanisms of the protective role of lutein and zeaxanthin in the eye. *Annu. Rev. Nutr.* **2003**, *23*, 171–201. [CrossRef] [PubMed]
182. Junghans, A.; Sies, H.; Stahl, W. Macular pigments lutein and zeaxanthin as blue light filters studied in liposomes. *Arch. Biochem. Biophys.* **2001**, *391*, 160–164. [CrossRef] [PubMed]
183. Khachik, F.; de Moura, F.F.; Zhao, D.Y.; Aebischer, C.P.; Bernstein, P.S. Transformations of selected carotenoids in plasma, liver, and ocular tissues of humans and in nonprimate animal models. *Investig. Ophthalmol. Vis. Sci.* **2002**, *43*, 3383–3392.
184. Schalch, W. Carotenoids in the retina—A review of their possible role in preventing or limiting damage caused by light and oxygen. In *Free Radicals and Aging*; Emerit, I., Chance, B., Eds.; EXS; Birkhauser: Basel, Switerland, 1992; Volume 62, pp. 280–298.
185. Bone, R.A.; Landrum, J.T.; Mayne, S.T.; Gomez, C.M.; Tibor, S.E.; Twaroska, E.E. Macular pigment in donor eyes with and without AMD: A case-control study. *Investig. Ophthalmol. Vis. Sci.* **2001**, *42*, 235–240.
186. Richer, S.; Stiles, W.; Statkute, L.; Pulido, J.; Frankowski, J.; Rudy, D.; Pei, K.; Tsipursky, M.; Nyland, J. Double-masked, placebo-controlled, randomized trial of lutein and antioxidant supplementation in the intervention of atrophic age-related macular degeneration: The Veterans LAST study (Lutein Antioxidant Supplementation Trial). *Optometry* **2004**, *75*, 216–230. [CrossRef]
187. Zarros, A. In which cases is neuroprotection useful? *Adv. Altern. Think. Neurosci.* **2009**, *1*, 3–4.
188. Okuzumi, J.; Nishino, H.; Murakoshi, M.; Iwashima, A.; Tanaka, Y.; Yamane, T.; Fujita, Y.; Takahashi, T. Inhibitory effects of fucoxanthin, a natural carotenoid, on N-*myc* expression and cell cycle progression in human malignant tumor cells. *Cancer Lett.* **1990**, *55*, 75–81. [CrossRef]
189. Khodosevich, K.; Monyer, H. Signaling involved in neurite outgrowth of postnatally born subventricular zone neurons in vitro. *BMC Neurosci.* **2010**, *11*, 18. [CrossRef] [PubMed]

190. Das, S.K.; Ren, R.; Hashimoto, T.; Kanazawa, K. Fucoxanthin induces apoptosis in osteoclast-like cells differentiated from RAW264.7 cells. *J. Agric. Food Chem.* **2010**, *58*, 6090–6095. [CrossRef] [PubMed]
191. Abidov, M.; Ramazanov, Z.; Seifulla, R.; Grachev, S. The effects of Xanthigen™ in the weight management of obese premenopausal women with non-alcoholic fatty liver disease and normal liver fat. *Diabetes Obes. Metab.* **2010**, *12*, 72–81. [CrossRef] [PubMed]

*marine drugs*

*Article*

# Enhanced Productivity of a Lutein-Enriched Novel Acidophile Microalga Grown on Urea

**Carlos Casal [1], Maria Cuaresma [2], Jose Maria Vega [3] and Carlos Vilchez [2,*]**

[1] CIDERTA, University of Huelva, Park Huelva Empresarial, 21007, Huelva, Spain;
E-Mail: carlos.bejarano@dqcm.uhu.es (C.C.)

[2] International Centre for Environmental Research (CIECEM), University of Huelva, Parque Dunar s/n, Matalascañas, Almonte, 21760, Huelva, Spain; E-Mail: maria.cuaresma@dqcm.uhu.es (M.C.)

[3] Faculty of Chemistry, Department of Plant Biochemistry and Molecular Biology, University of Seville, 41012, Seville, Spain; E-Mail: jmvega@us.es (J.M.V.)

* Author to whom correspondence should be addressed; E-Mail: bital.uhu@gmail.com;
Tel.: +34-959-21-99-47; Fax: +34-959-21-99-42.

*Received: 2 November 2010; in revised form: 18 December 2010 / Accepted: 23 December 2010 / Published: 24 December 2010*

**Abstract:** *Coccomyxa acidophila* is an extremophile eukaryotic microalga isolated from the Tinto River mining area in Huelva, Spain. *Coccomyxa acidophila* accumulates relevant amounts of β-carotene and lutein, well-known carotenoids with many biotechnological applications, especially in food and health-related industries. The acidic culture medium ($pH < 2.5$) that prevents outdoor cultivation from non-desired microorganism growth is one of the main advantages of acidophile microalgae production. Conversely, acidophile microalgae growth rates are usually very low compared to common microalgae growth rates. In this work, we show that mixotrophic cultivation on urea efficiently enhances growth and productivity of an acidophile microalga up to typical values for common microalgae, therefore approaching acidophile algal production towards suitable conditions for feasible outdoor production. Algal productivity and potential for carotenoid accumulation were analyzed as a function of the nitrogen source supplied. Several nitrogen conditions were assayed: nitrogen starvation, nitrate and/or nitrite, ammonia and urea. Among them, urea clearly led to the best cell growth ($\sim 4 \times 10^8$

cells/mL at the end of log phase). Ammonium led to the maximum chlorophyll and carotenoid content per volume unit (220 μg·mL 1 and 35 μg·mL 1, respectively). Interestingly, no significant differences in growth rates were found in cultures grown on urea as C and N source, with respect to those cultures grown on nitrate and $CO_2$ as nitrogen and carbon sources (control cultures). Lutein accumulated up to 3.55 mg·g in the mixotrophic cultures grown on urea. In addition, algal growth in a shaded culture revealed the first evidence for an active xanthophylls cycle operative in acidophile microalgae.

**Keywords:** urea; *Coccomyxa;* extremophile microorganisms; lutein; microalgae

## 1. Introduction

*Coccomyxa acidophila* is a novel microalgal specie isolated from Tinto River (Huelva, Spain), which is so-called the 'Red river' due to its high iron water concentration. This special feature causes the river bed to constitute an acidic environment where the pH value remains constantly between 2 and 3 along a stretch of 80 km [1]. Besides, this microalga is characterized by having important potential to accumulate high lutein concentrations, a carotenoid with powerful well-known antioxidant properties [2,3].

Nowadays, extremophile organisms are gaining increasing interest due to their faculty to be used as renewable source of different high value compounds including carotenoids, fatty acids (PUFAs), lipids, vitamins, toxins, enzymes, *etc.* [4–6]. Furthermore, the extremophile character of these microorganisms can be a benefit for getting axenic cultures with no interference from others microalgae. In general, apart from contamination risks, one of the main problems for microalgae cultivation is the relatively high costs, which is expected to be overcome by technological advances [7]. For that reason, since some time ago, efforts are being focused on reducing the cost of elements related to microalgae cultivation. One aspect that puts up the total price of the operation of production systems is the high $CO_2$ demand that photosynthetic microorganisms usually have. In any case, although there are currently various attempts for capturing carbon dioxide by means of algae cultures from industrial flue gases [8], one strategy aimed to reduce costs could be the replacement of the carbon source by another cheaper option.

A wide variety of nitrogen sources, such as ammonia, nitrate, nitrite and urea, can be used as nitrogen source for growing microalgae [9]. Urea ($CO(NH_2)_2$) is a small-molecular weight polar and relatively lipid-insoluble compound which is ubiquitous in nature. This organic compound can be considered as a combined source of nitrogen and carbon and it has diverse functions. In organisms containing the enzyme urease, a nickel-dependent metalloenzyme [10] present in bacteria, fungi and plants, urea is primarily used as a source of nitrogen necessary for growth. However, since urease metabolizes urea to $CO_2$ and ammonia, thus providing a ready source of base, metabolism of urea by urease can also enable microorganisms to respond to acid challenges [11]. On the other hand, in mammals, urea is the primary waste product of amino acid catabolism [12]. Urea is a versatile substance and its role largely depends on whether it is an end-product or can be further broken-down, and if so, the utilization of the break-down products also varies considerably, either for anabolic processes or for buffering under acidic conditions.

Previous works performed in our group with acidophile microalgae growing under mixotrophic conditions showed that urea can be a more than suitable alternative for cultivation of this microalga, showing good productivity and lutein accumulation results. Moreover, in the literature, several examples can be found where urea is shown to be an effective combined source of N and C for the production of *S. Platensis*, *Neochloris oleoabundans* and *Chlorella sp.* under different cultivation modes [7,9,13–16].

This work aimed at assessing the effect of different nitrogen sources on biomass productivity and carotenoid accumulation of *Coccomyxa acidophila,* paying special attention to the amount of accumulated lutein. In addition, the results will allow for assessing the use of nitrogen sources other than the conventional ones in growing acidophile microalgae.

## 2. Results and Discussion

### 2.1. *Coccomyxa acidophila enhanced growth on urea*

It was mentioned above that acidophile microalgae have so far never been used for massive production. Massive production requires fast growing microalgal strains. Most acidophile microalgae are slow growth strains, as reported in the literature [5,17]. However, the growth of acidophile microorganisms in acidic culture media becomes advantageous for biomass production, as under such conditions the growth of other microorganisms becomes difficult. Therefore, we attempted to find culture conditions under which *Coccomyxa acidophila* cultivation is enhanced, such that growth rates and productivity values approached those of common microalgae. In such a situation, the acidophile microalga should show fast-growth and could hopefully be grown in outdoor systems with limited risks for microbial contamination, in comparison to common microalgae.

One of the main growth conditions assayed was nitrogen source. In previous experiments, we first tested the effect of adding ammonium, nitrite or nitrate to photoautotrophically growing *Coccomyxa acidophila* cells. Growth on ammonium and nitrate resulted in the highest productivities. Unlike common microalgae, nitrite became toxic for *Coccomyxa acidophila*. In the experiments here, we also used urea as a combined source for C and N, with high $CO_2$ concentration (5% $v.v^{-1}$) as the main carbon source where indicated. Urea has been widely used instead of high $CO_2^-$ for sustaining microalgae growth and is also a cheap N source. As shown in Figure 1, urea promoted enhanced growth of *Coccomyxa acidophila*, both in terms of chlorophyll content (Figure 1A) and cell density (Figure 1B). This resulted in an increased growth rate with respect to control cultures (photoautotrophically grown on nitrate), as shown in Table 1. In addition, culture productivity was higher when the microalga was grown on urea (the highest productivity) and ammonium. Specifically, the highest productivity was reached in cultures grown on 0.67 $g{\cdot}L^{-1}$ urea ("control; air" in Figure 1). This urea concentration provided cultures with the same molar concentration of nitrogen than the nitrate added to control cultures. More interestingly, the best productivity values obtained from the microalgal growth on urea did not differ from those usually obtained for most of common microalgae (0.2–0.4 $g{\cdot}L^{-1}{\cdot}d^{-1}$). Unexpectedly, the simultaneous presence of urea and nitrate limited *Coccomyxa acidophila* growth. This will be discussed further.

**Figure 1.** Time-course of chlorophyll (**A**) and cell density (**B**) in *Coccomyxa acidophila* cultures grown on nitrate, urea or nitrate plus urea. Air alone or $CO_2$ in air (5% v/v) were used as carbon source, as indicated for each culture within the Figure legend.

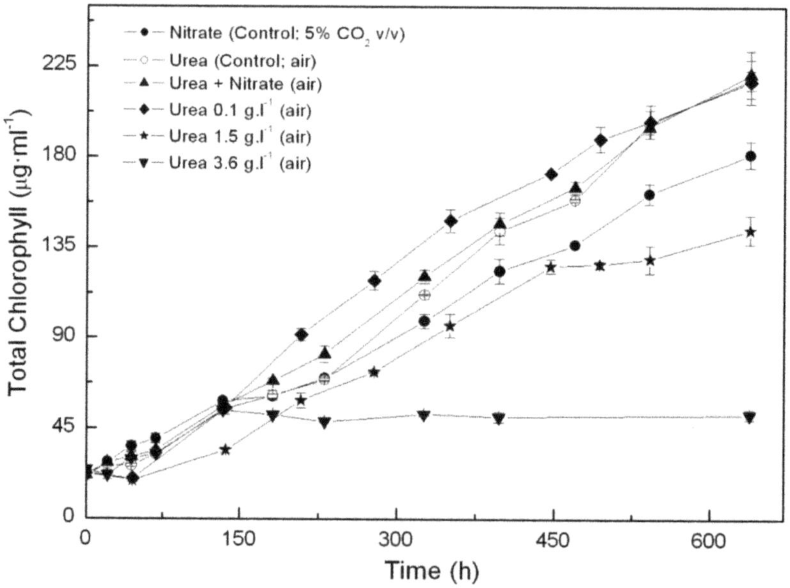

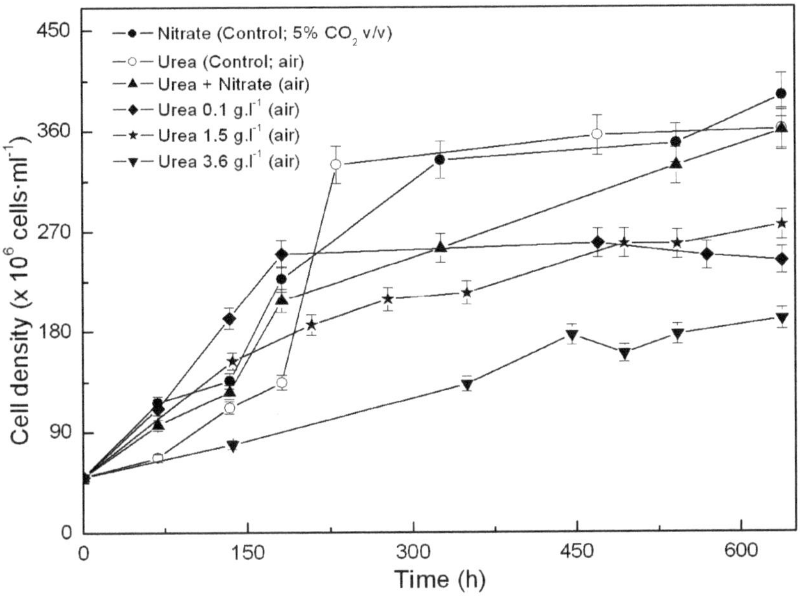


**Table 1.** Growth rates and productivity of *Coccomyxa acidophila* grown on different N-sources.

| **Nitrogen source** | **Growth rate (d $^{1}$)** | **Maximum productivity (g·L $^{1}$·d $^{1}$)** | **Maximum cellular carotenoids content (pg·cell $^{1}$)** |
|---|---|---|---|
| Nitrate | 0.27 | 0.23 | 0.084 |
| Nitrite | 0.02 | 0.13 | Non detectable |
| Ammonium | 0.31 | 0.24 | 0.055 |
| Urea | 0.34 | 0.25 | 0.104 |

The previous results were obtained by means of using mixotrophic or photoautotrophic cultures, *i.e.*, either urea or nitrate were added to culture media as N sources while high $CO_2$ concentration (5% v/v in air) was supplied as a carbon source (first set of experiments). It has also been discussed that the microalga cells could make use of urea as an additional carbon source [18,19], perhaps being one of the reasons behind the improved microalgal productivity of urea grown cultures. Low $CO_2$ solubility at acidic pHs makes carbon uptake more difficult than at pH 7 (standard pH for most common microalgal cultures). Therefore, the supply of additional carbon in a soluble form at low pH (e.g., urea, glucose) could help to increase microalgal productivity. This raises the question of whether addition of glucose as a carbon source to cultures of an acidophila microalga should also increased microalgal productivity. Such a question was investigated by our group in previous research [1] and the answer was "no". Urea should by far allow maximum productivity in *Coccomyxa acidophila* cultures when used as a carbon source.

From the results above, *Coccomyxa acidophila* apparently prefers urea to nitrate as nitrogen source. Therefore, another question we addressed was whether such consumption preference indeed occurred. For this purpose, nitrate and urea consumption were followed in time in photoautotrophic cultures to which nitrate (control culture) or urea and nitrate (with the same molar nitrogen concentration to that used in control cultures, 22.7 mM), were added. Results are shown in Figure 2. If urea and nitrate are added simultaneously, nitrate only started to be consumed at late exponential growth phase while urea was first consumed as the only nitrogen source. A decreasing time-course trend in urea concentration is observed from the beginning of the experiment, whereas the nitrate concentration time-course trend remains stable. Inhibition of nitrate consumption by the presence of urea has been reported to occur in microalgae, though not many references dealing with the subject have been published. Cochland and Harrison [20] reported about 30% inhibition of nitrate consumption by urea in the eukaryotic picoflagellate *Microsomas pusilla*. Following consumption, assimilatory reduction of nitrate also could be inhibited. One of the first classic references was published by Smith and Thompson [21] who observed 70% nitrate reductase inhibition by urea in *Chlorella*, evidencing nitrate assimilatory reduction down regulation to be behind nitrate consumption inhibition by urea.

As already mentioned, simultaneous addition of urea and nitrate as nitrogen sources slightly limits cell growth. Merigout *et al.* [22] evidenced in *Arabidopsis* plants that urea uptake was stimulated by urea but was reduced by the presence of nitrate in the growth medium. Such conclusions from their recent study on physiological and transcriptomic aspects of urea uptake and assimilation are in good concordance with the following observations from our results: (a) urea increased *Coccomyxa acidophila* growth and (b) simultaneous presence of urea and nitrate resulted in a decreased uptake of urea and culture productivity. These observations related to nitrogen uptake regulation in *Coccomyxa acidophila* are for the first time reported in acidophile microalgae and suggest that urea uptake and assimilation patterns in extreme acidophile microalgae (living in fully urea-free environments) and plants are similar. Further experiments in nitrogen assimilatory enzymes and gene expression are currently being developed in our group.

**Figure 2.** Time course of nitrogen consumption in *Coccomyxa acidophila* cultures grown on nitrate, urea or nitrate plus urea. Air alone or $CO_2$ in air (5% v/v) were used as carbon source, as indicated for each culture within the Figure legend. Dotted line with triangles corresponds to time-course of nitrate consumption of cultures incubated with nitrate plus urea.

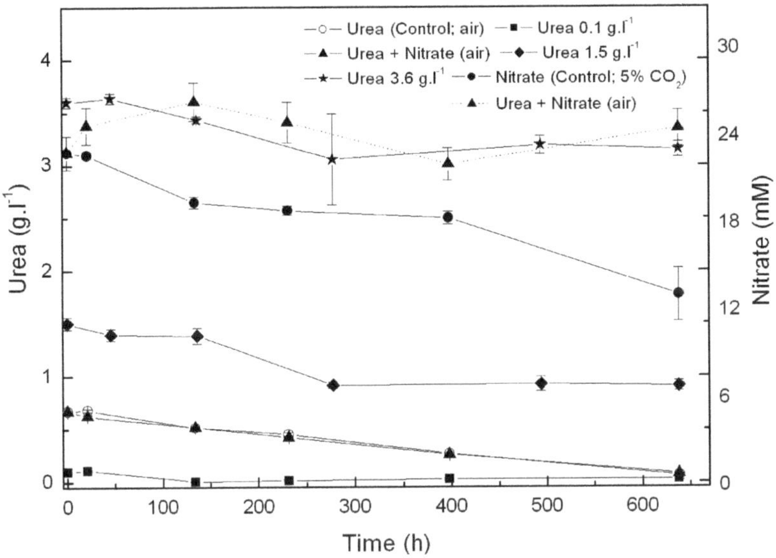


To determine whether simultaneous addition of urea and nitrate to the algal cultures has any impact on photosynthesis, relative electron transport rates were determined in each of the cultures (namely, control –nitrate ; urea; urea and nitrate, as nitrogen sources, respectively). Results are shown in Figure 3. Surprisingly, there was no nitrogen source-dependent impact on PS2 and on photosynthetic energy production chain, if the light intensity remained approximately below 150 $\mu E \cdot m^{-2} \cdot s^{-1}$. However, in urea grown cultures incubated under higher light intensities, photosynthetic energy production is shown to be clearly limited, up to the point that the electron transport chain becomes inhibited. This dramatically influences carbon assimilation and culture productivity.

According to our results, urea appears to be a suitable nitrogen source for *Coccomyxa acidophila* growth at relatively low light intensity; however, it has a dramatic impact on the photosynthetic energy production chain when exposed to high light intensity, which has never been reported for any other microalga. This is currently under study in our laboratories.

**Figure 3.** Light-dependent electron transport rates in *Coccomyxa acidophila* cultures grown on nitrate, urea or nitrate plus urea. Air alone or $CO_2$ in air (5% v/v) were used as carbon source, as indicated for each culture within the Figure legend.

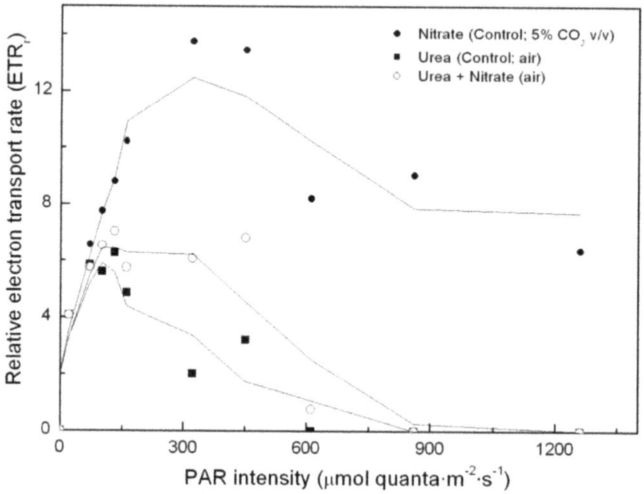


In addition, physiological responses of acidophile microalgae to urea and nitrate uptake processes anyhow differ, according to the observed pH changes in the culture media which tend to increase if nitrate (2.3 g·L $^{-1}$) is the only added nitrogen source and to decrease if urea (0.67 g·L $^{-1}$) is used (data not shown). So far, we have no evidence for antyport/symport mechanism details that help to elucidate the different physiological behavior.

## 2.2. *Carotenoid accumulation and xanthophylls cycle activity in urea grown* Coccomyxa acidophila *cells*

*Coccomyxa acidophila* accumulates commercial value carotenoids including lutein, β-carotene and zeaxanthin (Figure 4). Besides assessment of the best nitrogen sources for biomass production, carotenoid accumulation in urea and nitrate grown cultures was also studied (Figure 5). According to the best growth conditions inferred from Figure 1, for this experiment, the carotenoid content was followed in urea grown cultures (fluidized with air) and in nitrate grown cultures (fluidized with air supplemented with 5% v/v $CO_2$). In addition, carotenoid content was also followed in nitrogen-deprived cultures, as nitrogen depletion is a very well known carotenogenic condition for many microalgae species. In good agreement with the enhanced cell growth in urea grown cultures, total carotenoid content in the reactor also increased much more rapidly in urea grown cultures than in control cultures. Consequently, the carotenoid content of urea grown cultures ( g·mL $^{-1}$) became about two-fold that of the nitrate grown cultures (control cultures), until late exponential growth phase (Figure 5). This could be due to the increased biomass production in urea grown batch cultures, therefore higher carotenoid content in the reactor is not necessarily a consequence of faster carotenoid biosynthesis. However, simple calculations of the content of specific carotenoids per cell revealed a prompt carotenoid biosynthesis enhancement (namely -carotene, lutein, zeaxanthin) in urea grown *Coccomyxa acidophila* cells, as can be inferred from the carotenoid cell content data in Figure 6; lutein by far being the most abundant carotenoid. This means that urea clearly promotes increases in lutein and β-carotene cell content, at least up to late exponential growth phase, where lack of nutrients, shading effect and stress factors change the observed trend. Besides, it can be observed that cell content of violaxanthin inversely correlates with cell content of zeaxanthin over the time course. This is the first evidence of an active xanthophylls cycle in *Coccomyxa acidophila* that converts violaxanthin into zeaxanthin by means of violaxanthin de-epoxidase activity. Interestingly, nitrogen starvation did not promote carotenoid accumulation in *Coccomyxa acidophila* cultures, unlike other common microalgae including *Dunaliella, Haematococcus* and many others [23,24].

**Figure 4.** HPLC chromatogram showing the main carotenoids of *Coccomyxa acidophila*. AU: arbitrary units.

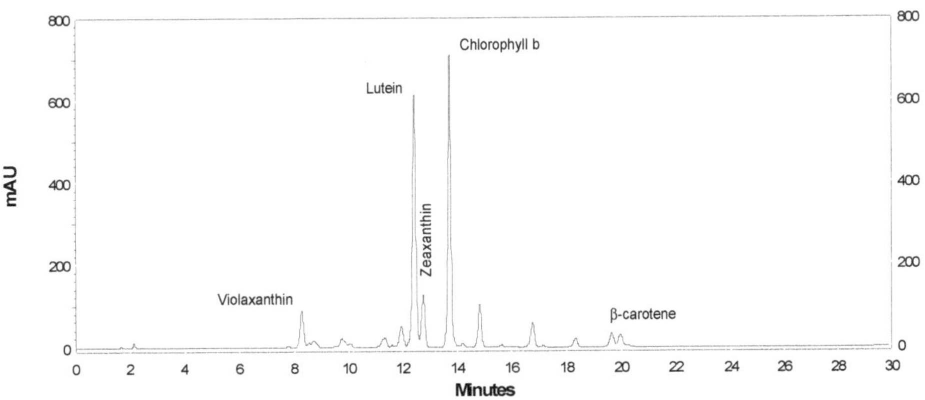

**Figure 5.** Total carotenoid content in *Coccomyxa acidophila* cultures grown on different nitrogen sources or under nitrogen starvation. Air alone or $CO_2$ in air (5% v/v) were used as carbon source, as indicated for each culture within the Figure legend.

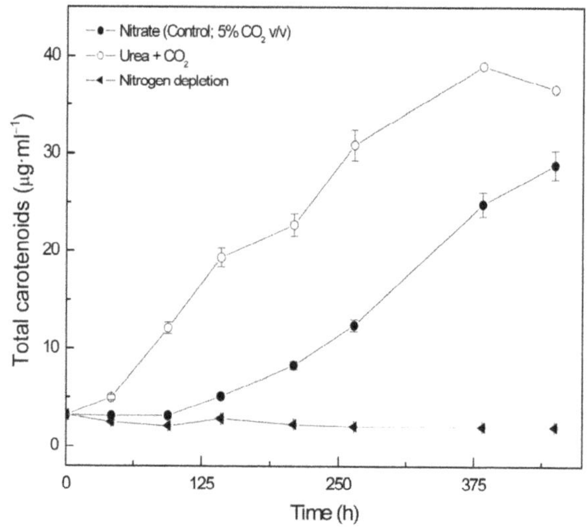


**Figure 6.** Time-course of the cell content of the indicated specific carotenoids in *Coccomyxa acidophila* cultures incubated in either nitrate or urea.

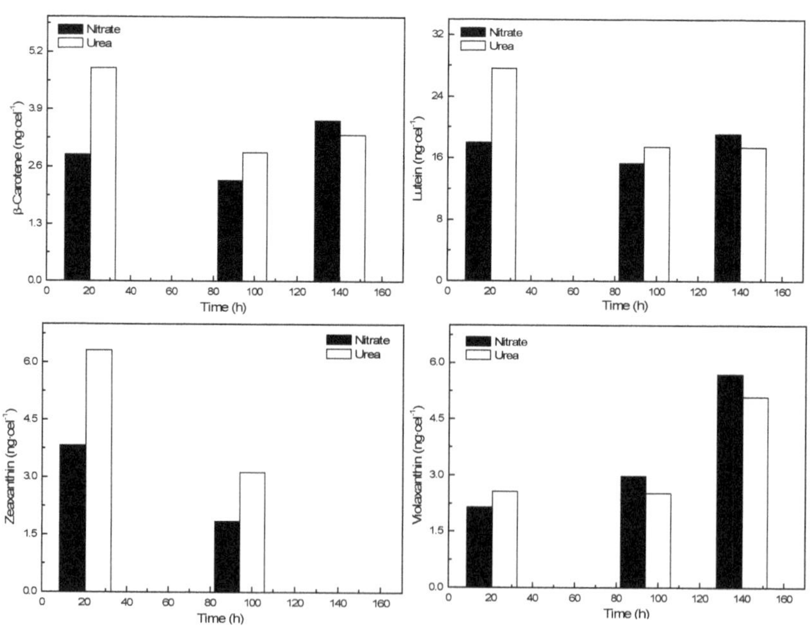

Fernández-Sevilla *et al.* [25] recently reviewed the biotechnological production of lutein. The paper includes an updated list of lutein production experiments performed on different scales using microalgae species. Table 2 shows the most relevant data of lutein productivity by microalgae and reactor type used for the production processes. Considering intracellular lutein cell content of each one of the promising species and lutein productivity in photobioreactors, *Scenedesmus almeriensis* [26], *Muriellopsis sp.* [27] and *Chlorella protothecoides* [28] emerge so far as the most efficient strains for the biotechnological production of lutein from microalgae. When incubated under standard culture conditions, *Coccomyxa acidophila* onubensis accumulates up to 6.1 mg·g $^{1}$ dry weight, which is within the upper range of lutein concentrations accumulated by the above mentioned microalgae. We are now running continuous cultures of *C. acidophila* in tubular laboratory photobioreactors in order to obtain lutein productivity data in long-term (weeks) production processes. Compared to continuous cultivation of other lutein producing species, *C. acidophila* has the practical advantage of growing well in an extremely selective culture medium at very low pH which preserves cultures from microbial contamination.

**Table 2.** Lutein productivity of lutein-enriched microalgae.

| **Microalga** | **Lutein (mg·g$^{-1}$)** | **Lutein productivity (mg·L $^{1}$·d $^{1}$)** | **Cultivation system** |
|---|---|---|---|
| *Scenedesmus almeriensis* | 5.5<br>4.5 | 4.9 mg·L ·d $^{1}$<br>290 mg·m ·d $^{1}$ | Laboratory, continuous culture, 2 L |
| *Muriellopsis sp* | 5.5<br>4.3 | 1.4 mg·L ·d $^{1}$<br>7.2 mg·L ·d $^{1}$ | Laboratory, batch, 0.2 L<br>Outdoor, tubular systems, 55 L |
| *Chlorella protothecoides* | 4.6 | 10 mg·L ·d $^{1}$ | Laboratory, batch, heterotrophic, 16 L |
| *Coccomyxa acidophila* | 6.1 | 2.0 mg·L ·d $^{1}$ | Laboratory, batch, 2 L |

## 3. Conclusions

The main conclusions of this manuscript are: (1) Mixotrophic cultivation on urea efficiently enhances growth and productivity of *C. acidophile*; signaling strategies towards suitable conditions definition for feasible outdoor production; (2) Urea clearly led to the fastest cell growth;

(3) Maximal lutein accumulation was found to occur in urea supplemented culture medium;

(4) In addition, algal growth in a shaded culture revealed the first evidence for an active xanthophylls cycle operative in acidophile microalgae.

## 4. Experimental Section

### *4.1. Microorganism and cultivation conditions*

*Coccomyxa acidophila,* the algal material used in this work, was isolated from the acidic water of the Tinto River's mining area, in Huelva (Spain).

Initially, an axenic culture of the microalga was obtained by streaking it on basal agar medium at

pH 2.5. After that, isolated colonies were transferred from the solid medium to a liquid medium modified by Silverman and Lundgren [29]. *Coccomyxa acidophila* mother cultures were maintained by periodic transfers in sterile medium adjusted to pH 2.5 with concentrated $H_2SO_4$. Unless otherwise indicated, standard cultivation conditions were batch cultures grown at 25 ºC into 1 L-Roux flasks, bubbled with air containing 5% (v/v) $CO_2$ and continuously illuminated with fluorescent lamps (Philips TLD, 30 W, 150 $\mu E \cdot m^{-2} \cdot s^{-1}$ at the surface of the flasks). In those cases where $CO_2$ was not supplied to the cultures, it was necessary to put a carbon dioxide trap with KOH 5 M buffer for removing it from the air mix. Every day, pH was controlled and adjusted at 2.5 ± 0.1 by adding diluted HCl or NaOH.

The irradiance was measured with a quantum/photometer Licor (mod. LI 250A).

### *4.2. Dry weight measurements*

Before filtering culture samples, filters of cellulose acetate with a 0.45 μm pore size, from Sartorius (Goettingen, Germany), were washed with distilled water and dried at 80 ºC in an oven for 24 h. After that, these were weighted and used to separate cells from the medium. Five milliliter culture samples were taken, vigorously homogenated, and filtered by means of a vacuum pump. Filters containing cells were dried and kept in an oven for 24 h, after which they were weighed.

### *4.3. Measurements of fluorescence*

Optimal chlorophyll fluorescence yield measurements (Fm/Fv) were performed with a pulse amplitude modulated fluorometer (Teaching-PAM from WALZ, Effeltrich, Germany). In order to make sure that there is no reduction of the PSII primary electron acceptor $Q_A$ and, therefore (consequently), all PSII reaction centers are open, cultures samples of 1 mL were previously adapted to dark conditions for 15 min [30]. After that period, a short saturating pulse of light (SP) was triggered. When necessary (e.g., low chlorophyll concentrations), the PAM modulated light (ML) had to be adjusted to higher values to obtain readings in the proper range.

### 4.4. Oxygen evolution

In addition of fluorescence measurements, the biological activity used to check cell viability was photosynthetic activity. For these determinations, 1 mL samples of *Coccomyxa acidophila* cultures were placed into a Clark-type electrode (Hansatech, U.K.) to measure $O_2$-evolution. Measurements were made at 25 ºC in the dark (endogenous respiration) or under saturating white light (1500 $\mu E \cdot m^{-2} \cdot s^{-1}$).

### 4.5. Analytical determinations

Total chlorophyll and carotenoid pigments were determined spectrophotometrically after centrifuging tubes containing samples for 6 min at 13000 rpm, heating them for 1 min, and extracting cell pellets with pure methanol. Sonication by ultrasound was also applied when necessary. After that, samples were spun down again for 5 min at 5000 rpm to eliminate cellular wastes. Calculations were done using equations according to [31].

For specific carotenoid analysis and quantification, separation was performed by liquid chromatography (HPLC; Merck Hitachi) using a RP-18 column with a flow rate of 1 mL/min. The applied gradient was the following (solvent A; ethyl acetate and solvent B; acetonitrile/agua, 9:1 v/v): 0–16 min, 0–60% solvent A; 16–30 min, 60% A; 30 – 35 min, 100% A. In order to quantify, pigment standards supplied by DHI-Water and Environment (Denmark) were injected.

Nitrate was determined following the method described by Cawse *et al.* [32]. Urea was determined according to the method from Wilcox [33].

### 4.6. Statistics

Unless otherwise indicated, all data included in figures and tables represent the average of triplicates.

### 4.7. Cell counting

Cellular density was determined by microscopy using an Olympus CX41 in a Neubauer chamber.

## Acknowledgements

This work has been supported by grant AGR-4337 (Proyecto de Excelencia, Junta de Andalucía) and grant Bioándalus (Junta de Andalucía, Estrategia de Impulso a la Biotecnología de Andalucía).

## References

1. Cuaresma, C.; Garbayo, I.; Vega, J.M.; Vílchez, C. Growth and photosynthetic utilization of inorganic carbon of the microalga Chlamydomonas acidophila isolated from Tinto river. *Enzyme Microb. Technol.* **2006**, *40*, 158–162.
2. Bartlett, H.E.; Eperjesi, F. Effect of lutein and antioxidant dietary supplementation on contrast sensitivity in age-related macular disease: a randomized controlled trial. *Eur. J. Clin. Nutr.* **2007**, *61*, 1121–1127.
3. Thurnham, D.I. Macular zeaxanthins and lutein a review of dietary sources and bioavailability and some relationships with macular pigment optical density and age-related macular disease. *Nut. Res. Rev.* **2007**, *20*, 163–179.
4. Schiraldi, C.; De Rosa, M. The production of biocatalysts and biomolecules from extremophiles. *Trends Biotechnol.* **2002**, *20*, 515–521.
5. Pulz, O.; Gross, W. Valuable products from biotechnology of microalgae. *Appl. Microbiol. Biotechnol.* **2004**, *65*, 635–648.
6. Spolaore, P.; Joannis-Cassan, C.; Duran, E.; Isambert, A. Commercial application of microalgae. *J. Biosci. Bioeng.* **2006**, *101*, 87–96.
7. Li, Y.; Horsman, M.; Wang, B.; Wu, N.; Lan, C.Q. Effect of nitrogen sources on cell growth and lipid accumulation of green alga *Neochloris oleoabundans*. *Appl. Microbiol. Biotechnol.* **2008**, *81*, 629–636.
8. Packer, M. Algal capture of carbon dioxide; biomass generation as a tool for greenhouse mitigation with reference to New Zealand energy strategy and policy. *Energ. Policy* **2009**, *37*, 3428–3437.
9. Becker, E.W. *Microalgae: Biotechnology and Microbiology*; Cambridge University Press: New York, NY, USA, 2004; pp. 18–24.
10. Mobley, H.L.T.; Hausinger, R.P. Microbial ureases: significance, regulation, and molecular characterization. *Microbiol. Rev.* **1989**, *53*, 85–108.
11. Sachs, G.; Kraut, J.A.; Wen, Y.; Feng, J.; Scott, D.R. Urea Transport in Bacteria: Acid Acclimation by Gastric *Helicobacter spp*. *J. Membr. Biol.* **2006**, *212*, 71–82.
12. Zawada, R.J.X.; Kwan, P.; Olszewski, K.L.; Llinas, M.; Huang, S.-G. Quantitative determination of urea concentrations in cell culture medium. *Biochem. Cell Biol.* **2009**, *87*, 541–544.
13. Rangel-Yagui, C.d.O.; Danesi, E.D.G.; Carvalho, J.C.M.; Sato, S. Chlorophyll production from *Spirulina platensis*: cultivation with urea addition by fed-batch process. *Bioresour. Technol.* **2004**, *92*, 133–141.

14. Sánchez-Luna, L.D.; Converti, A.; Tonini, G.C.; Sato, S.; Carvalho, J.C.M. Continuous and pulse feedings of urea as a nitrogen source in fed-batch cultivation of *Spirulina platensis*. *Aquacult. Eng.* **2004**, *31*, 237–245.
15. Soletto, D.; Binaghi, L.; Lodi, A.; Carvalho, J.C.M.; Converti, A. Batch and fed-batch cultivations of *Spirulina platensis* using ammonium sulphate and urea as nitrogen sources. *Aquaculture* **2005**, *243*, 217–224.
16. Hsieh, C.H.; Wu, W.T. Cultivation of microalgae for oil production with a cultivation strategy of urea limitation. *Bioresour. Technol.* **2009**, *100*, 3921–3026.
17. Tittel, J.; Bissinger, V.; Gaedke, U.; Kamjunke, N. Inorganic carbon limitation and mixotrophic growth in *Chlamydomonas* from an acidic mining lake. *Protist* **2005**, *156*, 63–75.
18. Rocha, J.M.S.; Garcia, J.E.C.; Henriques, M.H.F. Growth aspects of the marine microalga *Nannochloropsis gaditana*. *Biomol. Eng*. **2003**, *20*, 237 242.
19. Ellner, P.D.; Steers, E. Urea as a carbon source for *Chlorella* and *Scenedesmus*. *Arch. Biochem. Biophys*. **1955**, *59*, 534–535.
20. Cochlan, W.P.; Harrison, P.J. Inhibition of nitrate uptake by ammonium and urea in the eucaryotic picoflagellate *Micromonas pusilla* (Butcher) Manton et Parke. *J. Exp. Mar. Biol. Ecol.* **1991**, *153*, 143–152.
21. Smith, F.W.; Thompson, J.F. Regulation of Nitrate Reductase in *Chlorella vulgaris*. *Plant Physiol.* **1971**, *48*, 224–227.
22. Mérigout, P.; Lelandais, M.; Bitton, F.; Renou, J.-P.; Briand, X.; Meyer, C; Daniel-Vedele, F. Physiological and Transcriptomic Aspects of Urea Uptake and Assimilation in Arabidopsis Plants. *Plant Physiol*. **2008**, *147*, 1225–1238.
23. Boussiba, S. Carotenogenesis in the green alga *Haematococcus pluvialis*: Cellular physiology and stress response. *Physiol. Plant*. **2000**, *108*, 111–117.
24. Richmond, A. Biological Principles of Mass Cultivation. In *Handbook of Microalgal Culture*; Blackwell Press: Boca Raton, FL, USA, 2004; pp. 125–177
25. Fernández-Sevilla, J.M.; Acién-Fernández, F.; Molina-Grima, E. Biotechnological production of lutein and its applications. *Appl. Microbiol. Biotechnol*. **2010**, *86*, 27–40.
26. Sánchez, J.F.; Fernández-Sevilla, J.M.; Acién, F.G.; Cerón, M.C.; Pérez-Parra, J.; Molina-Grima, E. Biomass and lutein productivity of *Scenedesmus almeriensis*: influence of irradiance, dilution rate and temperature. *Appl. Microbiol. Biotechnol.* **2008**, *79*, 719–729.
27. Del Campo, J.A.; Rodríguez, H.; Moreno, J.; Vargas, M.A.; Rivas, J.; Guerrero, M.G. Lutein production by *Muriellopsis sp.* in an outdoor tubular photobioreactor. *J. Biotechnol.* **2001**, *85*, 289–295.
28. Wei, D.; Chen, F.; Chen, G.; Zhang, X.W.; Liu, L.J.; Zhang, H. Enhanced production of lutein in heterotrophic Chlorella protothecoides by oxidative stress. *Sci. China Ser. C Life Sci.* **2008**, *51*, 1088–1093.
29. Silverman, M.P.; Lundgren, D.G. Studies on the chemoautotrophic iron bacterium *Ferrobacillus ferrooxidans*. *J. Bacteriol.* **1959**, *77*, 642–647.

30. Schreiber, U.; Schliwa, U.; Bilger, W. Continuous recording of photochemical and non photochemical chlorophyll fluorescence quenching with a new type of modulation fluorometer. *Photosynth. Res.* **1986**, *10*, 51–62.
31. Harmut, A.; Lichtenthaler, K. Chlorophylls and carotenoids: pigments of photosynthetic membranes. *Method Enzymol.* **1987**, 148, 350–383.
32. Cawse, P.A. The determination of nitrate in soil solution by ultraviolet spectrophotometry. *Analyst* **1967**, *92*, 311–315.
33. Wilcox, A.A.; Carroll, W.E.; Sterling, R.E.; Davis, H.A.; Ware, A.G. Use of Berthelot reaction in automatic analysis of serum urea nitrogen. *Clin. Chem.* **1966**, *12*, 151–154.

*Samples Availability:* Available from the authors.

marine drugs

Article

# The New Carotenoid Pigment Moraxanthin Is Associated with Toxic Microalgae

**Olga Mangoni [1], Concetta Imperatore [2], Carmelo R. Tomas [3], Valeria Costantino [2], Vincenzo Saggiomo [4] and Alfonso Mangoni [2,*]**

1 Dipartimento delle Scienze Biologiche, Università di Napoli Federico II, Via Mezzocannone 8, 80134 Napoli, Italy; E-Mail: olga.mangoni@unina.it

2 Dipartimento di Chimica delle Sostanze Naturali, Università di Napoli Federico II, via D. Montesano 49, 80131 Napoli, Italy; E-Mails: cimperat@unina.it (C.I.); valeria.costantino@unina.it (V.C.)

3 Center for Marine Science, University of North Carolina at Wilmington, 5600 Marvin K. Moss Lane, Wilmington, NC 28409, USA; E-Mail: tomasc@uncw.edu

4 Stazione Zoologica "A. Dohrn", Villa Comunale I, 80121 Napoli, Italy; E-Mail: saggiomo@szn.it

* Author to whom correspondence should be addressed; E-Mail: alfonso.mangoni@unina.it;
Tel.: +39-081-678-532; Fax: +39-081-678-552.

*Received: 22 December 2010; in revised form: 25 January 2011 / Accepted: 4 February 2011 / Published: 10 February 2011*

**Abstract:** The new pigment "moraxanthin" was found in natural samples from a fish mortality site in the Inland Bays of Delaware, USA. Pure cultures of the species, tentatively named *Chattonella cf. verruculosa*, and natural samples contained this pigment as a dominant carotenoid. The pigment, obtained from a 10 L culture of *C. cf. verruculosa*, was isolated and harvested by HPLC and its structure determined from MS and 1D- and 2D-NMR. The data identified this pigment as a new acylated form of vaucheriaxanthin called moraxanthin after the berry like algal cell. Its presence in pure cultures and in natural bloom samples indicates that moraxanthin is specific to *C. cf. verruculosa* and can be used as a marker of its presence when HPLC is used to analyze natural blooms samples.

**Keywords:** *Chattonella cf. verruculosa*; Raphidophyceae; toxic algae; carotenoids; moraxanthin

## 1. Introduction

Phytoplankton, unicellular photosynthetic microorganisms, are ubiquitous in all aquatic environments. As primary producers, they are responsible for nearly half of the global primary production of organic carbon [1]. Photosynthesis, the process whereby energy is absorbed by pigments in algae and transformed into chemical energy, relies on the presence of energy trapping pigments. The main pigments, chlorophylls, carotenoids and phycobilins, absorb Photosynthetically Available Radiation (PAR) from 400–700 nm wavelengths [2]. However, pigments may also serve several functions including metabolic regulation, light harvesting (antenna pigments), electron donation or acceptance (in reaction centers), and photoprotection. The combination of different pigments and functions result in maximum efficiency and economy [3–6]. The kind of pigments produced and their relative proportions characterize the different phytoplankton groups.

In recent years, high performance liquid chromatography (HPLC) has been used to estimate phytoplankton composition by identifying photosynthetic pigments. Some pigments found exclusively in particular algal classes or genera may serve as useful taxonomic markers [7–13]. Such indicator pigments are termed 'finger prints'. Pigment analyses offer a valuable technique in oceanography for mapping phytoplankton populations and monitoring their abundance and composition [14–17].

Phytoplankton blooms occur naturally in coastal waters particularly during spring and summer seasons. However, a small number of microalgae are harmful, and although each individual is small, they may occur in huge numbers known as blooms [18–21]. Among the estimated phytoplankton species, about 7% (300 species) are known to produce red tides and of those, only 2% are actually harmful or toxic [22]. In marine and brackish water environments, most toxic species belong in the Dinophyceae, but also the Diatomophyceae, Haptophyceae, Raphidophyceae, and Cyanophyceae comprise toxic species [23–28]. The algal toxins may cause damage to other flora and fauna directly or they may accumulate through the food web in e.g., shellfish or finfish, thereby causing harm to their predators including humans [29–34]. Harmful algal blooms (HABs) are an ever more frequent phenomenon expanding in coastal regions on a world scale [35–38]. These have received much attention from researchers and local regulatory authorities due to their impact on the ecosystem and human health, influencing local economic issues [39].

Monitoring of coastal waters for harmful species is costly and labor-intensive and the possibility to recognize a potentially harmful algal species by means of chemical or biochemical analyses significantly reduces the time and costs of such monitoring. The one caveat is that the analysis, pigment or biochemical, involves a species specific marker for the HAB species in question [40]. Pigment signatures in the study of HABs have been very limited, particularly in monitoring programs [33,41,42].

During the summer period of 2000, ten fish mortality events occurred from unidentified causes in the Inland Bays of Delaware, USA. During the final fish-killing event of 28 August, 2000, over two million menhaden (*Brevoortia tyrranus*) perished when a bloom of an unidentified microalgal flagellate was observed [36]. This flagellate, accompanied by the presence of a potent neurotoxin, was tentatively called *Chattonella* cf. *verruculosa* since it resembled a fish killing species found in Japan thought to be of the class Raphidophyceae. Since none of the previously described Raphidophyceae completely agreed with the molecular features (18S rDNA; 16S rDNA) [43,44], further studies are underway to define its taxonomic position.

This work describes the isolation and structural elucidation of a new pigment (**1**) found in *C. cf. verruculosa* cultures and in natural samples where this species was dominant, which has been called moraxanthin after the berry like algal cell (Figure 1). Moraxanthin, which is a new acylated form of vaucheriaxanthin (**2**), is unique to *C. cf. verruculosa*, indicating that it can be used as a marker of its presence when HPLC analyses of natural blooms are performed.

**Figure 1.** Structures of moraxanthin (**1**) and vaucheriaxanthin (**2**).

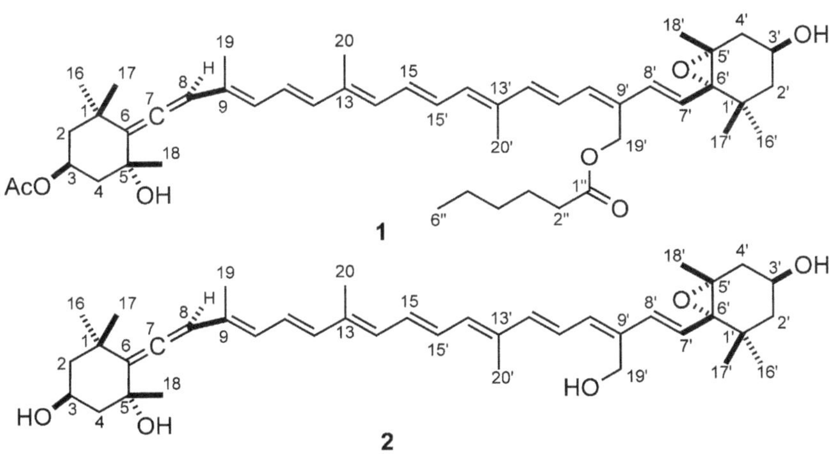

## 2. Results and Discussion

The chromatogram of the pigments of the *C. cf. verruculosa* culture showed a major peak (Figure 2, peak 4), whose retention time (Table 1) and UV spectrum (Table 1 and Figure 3) did not fit those of any known pigments, although the UV spectrum clearly showed the pigment to be a carotenoid.

**Figure 2.** HPLC absorbance chromatogram of the extract from cultured *C. cf. verruculosa.*

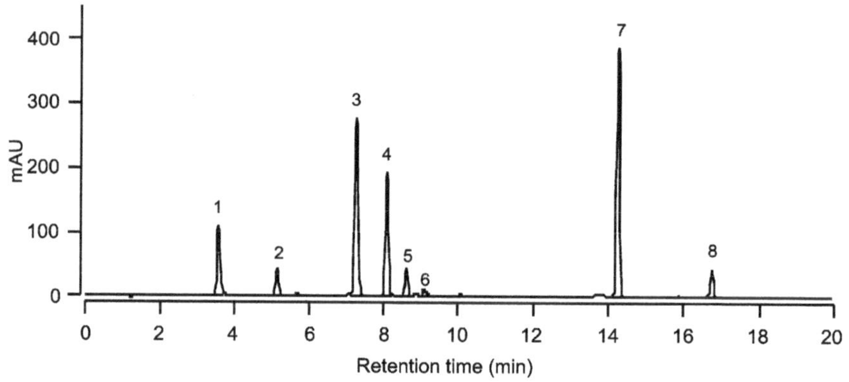


**Figure 3.** UV spectrum of moraxanthin **1** (EtOH).

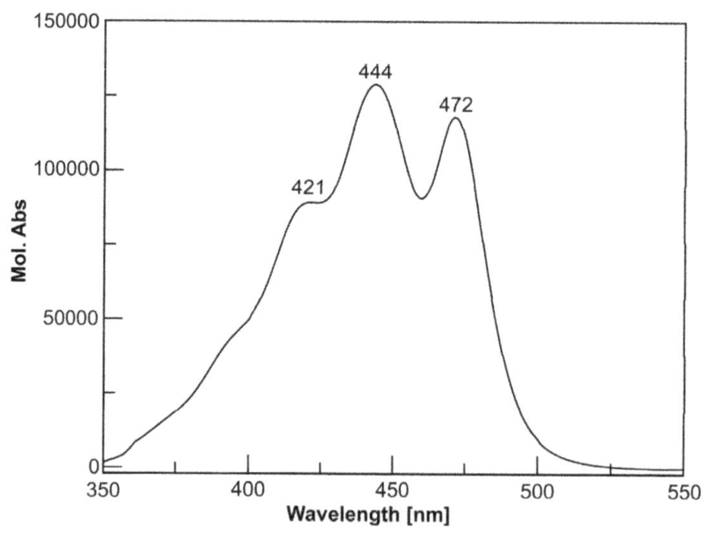


**Table 1.** Total pigments found in *C. cf. verruculosa* with relative retention times and specific absorption maxima.

| **Peak No.** | **Pigments** | **Retention time (min)** | **Absorption maxima (nm)** | | |
|---|---|---|---|---|---|
| 1 | Chlorophyll $c_1$ + $c_2$ (Chl$c_1$ + $c_2$) | 3.59 | 445 | 583 | 634 |
| 2 | Unknown (RT 5.15) | 5.15 | 423 | 446 | 476 |
| 3 | Diadinoxanthin (Dd) | 7.24 | 422 | 446 | 475 |
| 4 | **Moraxanthin** | **8.07** | **421** | **444** | **472** |

| | | | | | |
|---|---|---|---|---|---|
| 5 | Diatoxanthin (Dt) | 8.60 | 426 | 451 | 478 |
| 6 | Zeaxanthin (Zea) | 9.08 | | 449 | 477 |
| 7 | Chlorophyll *a* (Chl*a*) | 14.25 | 432 | 617 | 665 |
| 8 | -carotene ( -car) | 16.75 | | 450 | 478 |

The new pigment was then isolated to determine its structure by spectroscopic (MS and NMR) means. A large-scale (10 L) culture of *C. cf. verruculosa* was grown, and harvested by continuous flow centrifugation into an algal pellet and supernatant. The algal pellet (4 g) was extracted exhaustively with MeOH, and the extract was subjected to repeated HPLC separation, yielding 1.1 mg of the pure carotenoid moraxanthin (**1**). When re-injected in the same HPLC conditions as for the chromatogram in Figure 2, the isolated compound **1** showed retention time and UV spectrum identical to that of peak 4. Compound **1** showed a $[M + Na]^+$ pseudomolecolar ion peak in the ESI high-resolution mass spectrum at *m/z* 779.4879, in accordance with the formula $C_{48}H_{68}NaO_7$. Compared to the $C_{40}$ carotenoid skeleton, this formula contains eight additional carbon atoms. In addition, the ESI mass spectrum also contained a peak at *m/z* 663 ($C_{42}H_{56}NaO_5$, $[M + Na - C_6H_{12}O_2]^+$), which can be accounted for by the in-source loss of hexanoic acid.

Most of the information used for structure elucidation came from one- and two-dimensional NMR spectroscopy. The general features of the proton NMR spectrum ($C_6D_6$) resembled those of carotenoids, with several olefinic protons between δ 6 and δ 7, and 10 methyl singlets between δ 1.87 and 1.08. However, one of the ten methyl signals (δ 1.74) was part of an acetyl group, as shown by its correlation peak with the carbonyl carbon atom at δ 169.2 in the HMBC spectrum. Other notable features of the proton NMR spectrum were (i) the AB system at δ 5.13 and 5.07 ($H_2$-19') of an oxymethylene group, (ii) two oxymethine protons at δ 5.69 (H-3) and δ 3.78 (H-3'), a methyl triplet at δ 0.78 ($H_3$-6'') and a methylene triplet at δ 2.14 ($H_3$-2''), indicative of an acyl chain and (iv) an olefinic proton singlet at δ 6.03 (H-8), which showed correlation peaks in the HMBC spectrum with two non-protonated carbon atoms at δ 202.1 (C-7) and 117.8 (C-6), and was therefore part of an allene system. The observed structural features were suggestive of a structure similar to vaucheriaxanthin (**2**), but a direct comparison of spectral data was hindered by the presence of the two additional acyl groups. A detailed analysis of the correlation peaks observed in the COSY, HSQC, and HMBC (Figure 4) spectra demonstrated that the planar structure of moraxanthin is indeed the same as that of vaucheriaxanthin, except for the presence of an acetyl group at position 3 and a hexanoyl group at position 19'.

**Figure 4.** $^1$H-$^{13}$C long range couplings evidenced by the HMBC spectrum of moraxanthin (**1**).

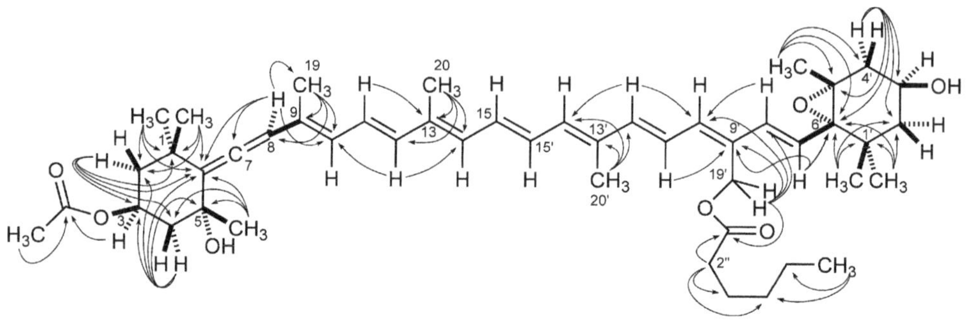

In addition to all the expected geminal and vicinal couplings, the COSY spectrum revealed several proton-proton long-range couplings. Among them, the quite large W couplings of H-2$\alpha$ with H-4$\alpha$ (2.2 Hz) and of H-2'$\alpha$ with H-4'$\alpha$ (1.7 Hz) indicated the 1,3-diequatorial relationship of these two pairs of protons. Furthermore, the methyl protons on the polyene system showed weak correlation peaks with the olefinic protons, arising from the usual allylic $^4J_{HH}$ couplings, but also from $^6J_{HH}$ couplings ($H_3$-19/H-12, $H_3$-20/H-15', $H_3$-20'/H-15) and even one remarkable $^8J_{HH}$ coupling ($H_3$-19/H-14). To the best of our knowledge, this is the first report of a $^8J_{HH}$ coupling in a carotenoid.

The *E* configuration of double bonds at positions 11, 15, 7', and 11' was evident from the large *trans* coupling constant values of the relevant protons (see Table 2). The *E* configuration of double bonds at positions 9, 13, and 13' and the *Z* configuration of the double bond at position 9' were determined from the ROESY spectrum, displaying correlation peaks of $H_3$-19 with H-11, $H_3$-20 with H-15, $H_3$-20' with H-15', and $H_2$-9' with H-11'.

**Table 2.** [1]H (700 MHz) and [13]C (175 MHz) NMR data of moraxanthin **1** in $C_6D_6$.

| Pos. | | $\delta_H$ (*J* in Hz) | $\delta_C$, mult | Pos. | | $\delta_H$ (*J* in Hz) | $\delta_C$, mult |
|---|---|---|---|---|---|---|---|
| 1 | | - | 35.8, C | 1' | | - | 35.2, C |
| 2 | α | 2.05, ddd (12.4, 4.2, 2.2) | 45.8, $CH_2$ | 2' | α | 1.48, ddd (12.8, 3.3, 1.7) | 47.1, $CH_2$ |
| | β | 1.39, dd (12.4, 11.5) | | | β | 1.11, dd (12.8, 10.2) | |
| 3 | | 5 .69, dddd (11.5, 11.5, 4.2, 4.2) | 67.7, CH | 3' | | 3.78, ddddd (10.2, 8.6, 5.2, 4.3, 3.3) | 63.7, CH |
| | | | | 3'-OH | | 0.57, d (4.3) | - |
| 4 | α | 2.30, ddd (12.8, 4.2, 2.2) | 45.5, $CH_2$ | 4' | α | 2.24, ddd (14.2, 5.2, 1.7) | 41.0, $CH_2$ |
| | β | 1.41, ddd (12.8, 11.5, 1.7) | | | β | 1.48, dd (14.2, 8.6) | |
| 5 | | - | 72.0, C | 5' | | - | 66.6, C |
| 5-OH | | 0.77, d (1.7) | - | | | - | - |
| 6 | | - | 117.8, C | 6' | | - | 69.9, C |
| 7 | | - | 202.1, C | 7' | | 6.22, d (15.7) | 125.9, CH |
| 8 | | 6.03, s | 103.3, CH | 8' | | 6.63, d (15.7) | 134.9, CH |
| 9 | | - | 131.9, C | 9' | | - | 132.4, C |
| 10 | | 6.20, d (11.4) | 129.0, CH | 10' | | 6.27, d (11.6) | 136.3, CH |
| 11 | | 6.70, dd (15.0, 11.4) | 125.3, CH | 11' | | 6.95, dd (14.9, 11.6) | 124.1, CH |
| 12 | | 6.46, d (15.0) | 137.5, CH | 12' | | 6.39, d (14.9) | 141.0, CH |
| 13 | | - | 136.8, C | 13' | | - | 136.4, C |
| 14 | | 6.28 , d (11.5) | 133.0, CH | 14' | | 6.27, d (11.5) | 134.5, $CH_2$ |
| 15 | | 6.66, dd (14.3, 11.5) | 131.1, CH | 15' | | 6.60, dd (14.3, 11.5) | 130.3, CH |
| 16 | | 1.43, s | 29.0, $CH_3$ | 16' | | 1.15, s | 25.2, $CH_3$ |
| 17 | | 1.08, s | 32.1, $CH_3$ | 17' | | 1.14, s | 29.4, $CH_3$ |
| 18 | | 1.17, s | 30.9, $CH_3$ | 18' | | 1.19, s | 20.0, $CH_3$ |
| 19 | | 1.78, s | 13.8, $CH_3$ | 19' | a | 5.07, d (12.3) | 58.1, $CH_2$ |
| | | | | | b | 5.13, d (12.3) | |
| 20 | | 1.87, s | 12.7, $CH_3$ | 20' | | 1.86, s | 12.6, $CH_3$ |
| 1'' | | - | 172.8, C | Ac | $CH_3$ | 1.74, s | 20.8, $CH_3$ |
| 2'' | | 2.14, t (7.5) | 34.2, $CH_2$ | | CO | - | 169.2, C |
| 3'' | | 1.55, quintet (7.5) | 24.8, $CH_2$ | | | | |
| 4'' | | 1.12, overlapped | 31.3, $CH_2$ | | | | |
| 5'' | | 1.14, overlapped | 22.4, $CH_2$ | | | | |
| 6'' | | 0.78, t (7.0) | 13.8, $CH_3$ | | | | |

The ROESY spectrum also provided information on the relative configuration of the two terminal six-membered rings (Figure 5). The allene terminus is in the chair conformation, with the two W-coupled H-2*α* and H-4*α* protons in the equatorial orientation. The large coupling constants of H-3 with the axial H-2β and H-4β (Table 2) showed the former proton to be axial, and therefore on the *α* face of the ring; as a consequence, the OAc group at C-3 must be β. The ROESY correlation of the methyl protons $H_3$-19 with H-2β and H-4β determined the axial chirality of the allene functionality as *R*. Finally, the ROESY correlation of $H_3$-19 with $H_3$-18 located C-18 on the β face of the ring, and therefore the OH group at C-5 on the *α* face.

**Figure 5.** ROESY correlations used to determine the relative configuration of moraxanthin (**1**).

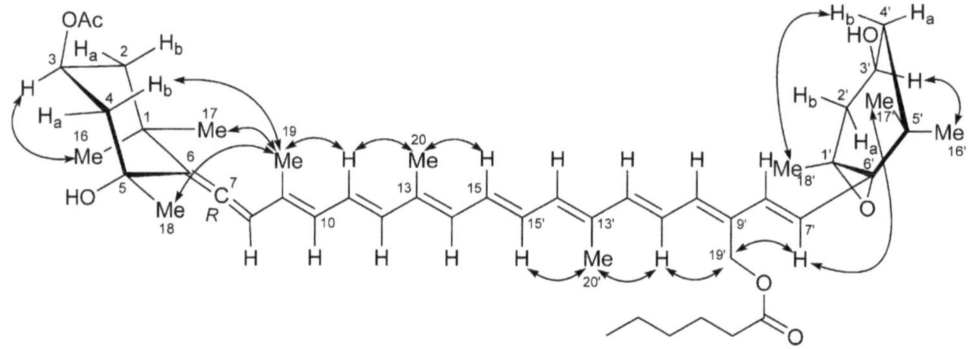

As for the other terminal ring, the W coupling (1.7 Hz) of the pseudoequatorial H-2'α and H-4'α suggests a half-chair conformation of this ring. The *trans* relationship between the epoxide ring and the hydroxyl group was established from the ROESY correlation peaks of the two geminal methyl groups $H_3$-16' and $H_3$-17' with, respectively, H-3' and H-7' (Figure 5), showing that H-3' and H-7' are on opposite faces of the six-membered ring. This was confirmed by the prominent peak between the psudoaxial H-4β and $H_3$-18 in the same spectrum. The relative configuration determined for moraxanthin matches that of vaucheriaxanthin (**2**), and it may be assumed that also the absolute configuration of moraxanthin is the same as in vaucheriaxanthin.

**Figure 6.** HPLC absorbance chromatogram of natural water sample collected at the
fish-kill site of Torque Canal, Delaware on 28 August 2000 during a *C. cf. verruculosa* bloom. The arrow indicates the moraxanthin peak.

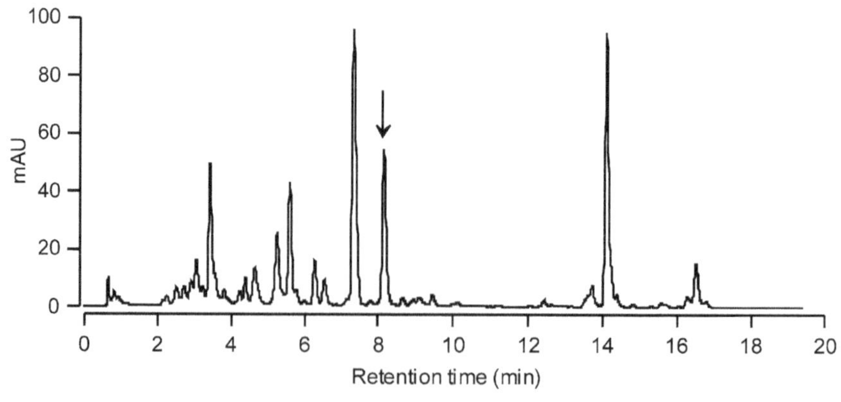

To investigate the utility of using moraxanthin as a marker for the toxic alga *C. cf. verruculosa*, natural bloom samples from the fish-kill site at Torque Canal, Delaware, collected on 28 August 2000 on Whatman GFF glass fiber filters and stored at −80 °C, were extracted in methanol and subjected to HPLC analysis. The HPLC chromatogram (Figure 6) definitely showed a peak for moraxanthin in the natural sample. The moraxanthin peak clearly separated from all the other pigment peaks having no overlap with other pigments. In addition, a peak with the same retention time and absorbance characteristics was present in the HPLC chromatogram from water samples collected in 2003–2007 at various sites in Delaware's Inland Bays where *C. cf. verruculosa* blooms occurred (data not shown). This shows that the HPLC analysis may provide a simple and rapid tool for detecting harmful blooms of *C. cf. verruculosa*.

## 3. Experimental Section

### *3.1. General experimental procedures*

ESI-MS experiments were performed on an Applied Biosystem API 2000 triple-quadrupole mass spectrometer. High Resolution ESI-MS spectra were performed on a Thermo Orbitrap XL mass spectrometer. All the mass spectra were recorded by infusion into the ESI source using MeOH as the solvent. CD spectra were recorded in MeOH solution on a Jasco J-710 spectrophotometer using a 1 cm cell. $^1H$ and $^{13}C$ NMR spectra were determined in $C_6D_6$ solution on a Varian UnityInova spectrometer at 700 and 175 MHz, respectively; chemical shifts were referenced to the residual solvent signal ($\delta_H$ 7.15, $\delta_C$ = 128.0). For an accurate measurement of the coupling constants, the one-dimensional $^1H$ NMR spectra were transformed at 64K points (digital resolution: 0.09 Hz). Homonuclear $^1H$ connectivities were determined by COSY experiments. Through-space $^1H$ connectivities were evidenced using a ROESY experiment with a mixing time of 500 ms. The reverse multiple-quantum heteronuclear correlation (HMQC) spectra was optimized for an average $^1J_{CH}$ of 142 Hz. The gradient-enhanced multiple-bond heteronuclear correlation (HMBC) experiment was optimized for a $^3J_{CH}$ of 8.3 Hz.

### *3.2. Plant material*

Clonal cultures of *C. cf. verruculosa* were established by single cell pipette isolation from a natural bloom sample taken at the time of a fish-kill in Torque Canal, Delaware. Individually isolated cells were grown in DYV medium [43] using sea water adjusted to a salinity of 20 to match that of the sample water. Successful isolates grown in 96 well microtiter plates were stepped up in volume eventually becoming stabilized cultures maintained in 150 mL volumes in erlenmeyer culture flasks. All cultures were maintained at 22 °C, with a fluence rate of 50 μ mol. quanta $m^{-2}$ $s^{-1}$ of cool white fluorescent light and a 12:12 h (LD) cycle. For this study, culture

CMS TAC1050 was used and is presently deposited in the Center for Marine Science Toxic Algal Collection housed at UNCW's marine facility. This collection of harmful species is under the direction of Dr. Carmelo Tomas, Professor of Biology and Marine Biology at the CMS location who was also the isolator of the original culture. Large volume cultivation consisting of 10 L batches were grown in Bellco stirred cell system under conditions mentioned above. After a growth period of 1 month, the 10 L culture was harvested using a Sorvall RCB-2 refrigerated centrifuge equipped with a KSB (Kendro) continuous centrifuge head. A 4 g (wet weight) pellet was harvested, transferred to 15 mL cryovials and kept frozen at −80 °C prior to analyses for pigments.

### 3.3. Pigment analysis

The algal pellet (1 g) from cultures of *C. cf. verruculosa* was extracted with MeOH (3 mL), and the extract was filtered through Whatman GFF (0.45 μm). A portion of extract (500 μL) was added to 250 μL of ion-pairing solution (1M ammonium acetate), and after 5 minutes injected to the HPLC system. Assessment of the pigment composition was performed using a Hewlett-Packard HPLC 1100 Series system, equipped with a quaternary pump system and diode array detector. Pigments were separated on a temperature-controlled (20 °C) Hypersil MOS C8 reverse phase column (Sigma-Aldrich, 3 μm, 100 × 4.6 mm) according to the HPLC method of Vidussi *et al.* [45]. The mobile phases were MeOH (eluent A) and MeOH/0.5 N ammonium acetate (7:3) (eluent B). The elution gradient was kept constant at 1.0 mL/min for 20 min. The ratio of eluent B was gradually increased from 25 to 100%, and then returned to the initial proportion at the end of the elution. Chlorophylls and carotenoids were detected at 440 nm and identified by a diode array detector ($\lambda$ = 350–750 nm, 1.2 nm spectral resolution). Standards of all the known pigments were provided by International Agency for $^{14}C$ Determination (VKI Water Quality Institute) and calibration was performed according to Mantoura and Repeta [46].

### 3.4. Analysis of algal bloom

Samples from blooms occurring in the Delaware Inland Bays were collected and returned to the laboratory or shipped by overnight courier to UNCW CMS Laboratory. Upon arrival, the sample was processed immediately. Pigment samples were taken as natural samples filtered on Whatman GFF filters and frozen immediately in liquid nitrogen and stored in a −80° C freezer until extraction and pigment analyses could be performed as described above. Species contents of the sample water were determined by direct observations using a Nikon Diaphot inverted microscope. Observations of the phytoplankton included species identification, at least to genus level of live cells, cell density estimates using standard inverted microscope techniques and extraction for lipid soluble toxins in chloroform. Preserved samples (Lugol's solution) were carefully mixed, placed into a 10 or 50 mL settling chamber and allowed to settle for 24 hours prior to observation. The dominant species were identified, enumerated and used to define the phytoplankton composition.

### 3.5. *Extraction and isolation of moraxanthin* (**1**)

The algal pellet (4 g) was extracted once with MeOH (40 mL). The extract was dried to give a dark green oil (104 mg) which was subjected to reversed-phase HPLC separation on a Varian Prostar 210 apparatus equipped with an Varian 325 UV detector [column: RP-18, 10 $\mu$m, 250 $\times$ 10 mm; eluent: MeOH/$H_2O$ (9:1), flow 5 mL/min, UV detector set at 430 nm] to give partially purified moraxanthin (3.4 mg). Further reversed phase HPLC [column: RP-18, 3 $\mu$m, 250 $\times$ 4.6 mm; eluent: MeOH/$H_2O$ (8:2), flow 1 mL/min, UV detector set at 430 nm] gave pure moraxanthin **1** (1.1 mg), whose identity was confirmed by HPLC analysis as described in Section 4.3.

### 3.6. *Moraxanthin* (**1**)

Dark yellow oil; CD (MeOH; *c* $3.06{\cdot}10^{-6}$ M): $\Delta\varepsilon_{471}$ +9.9, $\Delta\varepsilon_{446}$ +11.9, $\Delta\varepsilon_{422}$ +9.0; UV (EtOH): $\lambda_{max}$ nm ($\varepsilon$): 421 (89000), 444 (129000), 472 (118000); ESI MS *m/z* 779 $[M + Na]^+$; HRESIMS *m/z* 779.4879 $[M + Na]^+$ (calcd. for $C_{48}H_{68}NaO_7$, 779.4857). For $^1H$ and $^{13}C$ NMR spectroscopic data, see Table 2.

## 4. Conclusions

The newly proposed toxic species *C. cf. verruculosa* contains a new species-specific pigment, moraxanthin (**1**), whose structure was established as (3S,5*R*,7*R*,3'*S*,5'*R*,6'*S*)-3-acetoxy-5',6'-epoxy-19'-(hexanoyloxy)-6,7-didehydro-5,6,5',6'-tetrahydro-β,β-carotene-5,3'-diol, *i.e.*, 3-*O*-acetyl-19'-*O*-hexanoylvaucheriaxanthin. Two esterified forms of vaucheriaxanthin have been described, namely the 3-*O*-acetyl-19'-*O*-octanoate and the 3-*O*-acetyl-19'-*O*-decanoate derivatives [47]. However, none of them contains the hexanoyl residue present in moraxanthin, which therefore can be easily distinguished from these known compounds on the basis of the HPLC retention time.

New harmful species have been identified and the taxonomy of other species has been revised [48]. It is usually accepted that the routine identification of phytoplankton for monitoring studies in estuaries and coastal waters requires additional methods other than traditional microscopy, which can underestimate some taxonomic groups containing fragile or poorly differentiated small cells. In conjunction with microscopy, pigment separation using HPLC has become a more widely applied method for estimating and characterizing phytoplankton biomass and community structure [6–8].

However, algal pigments usually show complex overlapping patterns with different taxa, offering only a few unambiguous markers. In our case, *C. cf. verruculosa* may be readily identified in natural samples by means of HPLC chromatograms due to the distinct peak corresponding to the species-specific pigment moraxanthin described here.

## Acknowledgements

The research leading to these results has received funding from the European Union's Seventh Framework Programme FP7/2007–2013 under grant agreement n° 229893 (NatPharma) and from grant No. 01504-07 awarded by U.S. Center for Disease Control and Prevention and North Carolina Department of Health and Human Services to Carmelo Tomas. Mass and NMR spectra were recorded at the "Centro Interdipartimentale di Analisi Strumentale", Università di Napoli "Federico II". The assistance of the staff is gratefully acknowledged.

## References

1. Falkowski, P.G.; Raven, J.A. *Aquatic Photosynthesis*; Blackwell Science: Oxford, UK, 1997; p. 375.
2. Rowan, K.S. *Photosynthetic Pigments of Algae*; Cambridge University Press: Cambridge, UK, 1989.
3. Porra, R.J.; Pfündel, E.E.; Engel, N. Metabolism and function of photosynthetic pigments. In *Phytoplankton Pigments in Oceanography*; Jeffrey, S.W., Mantoura, R.F.C., Wright, S.W., Eds.; UNESCO Publishing: Paris, France, 1997; pp. 85–126.
4. Scheer, H. The pigments. In *Light-Harvesting Antennas in Photosynthesis. Advances in Photosynthesis and Respiration*; Green, B.R., Parsons, W.E., Eds.; Kluwer Academic Publishers: Dordrecht, The Netherlands, 2003; Volume 13, p. 29.
5. Rodríguez, F.; Chauton, M.; Johnsen, G.; Andresen, K.; Olsen, L.M.; Zapata, M. Photoacclimation in phytoplankton: implications for biomass estimates, pigment functionality and chemotaxonomy. *Mar. Biol.* **2006**, *148*, 963–971.
6. Mangoni, O.; Carrada, G.C.; Modigh, M.; Catalano, G.; Saggiomo, V. Photoacclimation in Antarctic bottom ice algae: an experimental approach. *Polar Biol.* **2009**, *32*, 325–335.
7. Mackey, M.D.; Mackey, D.J.; Higgins, H.W.; Wright, S.W. CHEMTAX- a program for estimating class abundances from chemical markers: application to HPLC measurements of phytoplankton. *Mar. Ecol. Progr. Ser.* **1996**, *144*, 265–283.
8. Jeffrey, S.W.; Vesk, M. Introduction to marine phytoplankton and their pigment signatures. In *Phytoplankton Pigments in Oceanography: Guidelines to Modern Methods*; Jeffrey S.W., Mantoura R.F.C., Wright, S.W., Eds.; UNESCO Publishing: Paris, France, 1997; pp. 37–84.

9. Zapata, M.; Jeffrey, S.W.; Wright, S.W.; Rodríguez, F.; Garrido, J.L.; Clementson, L. Photosynthetic pigments in 37 species (65 strains) of Haptophyta: implications for oceanographyand chemotaxonomy. *Mar. Ecol. Prog. Ser.* **2004**, *270*, 83–102.
10. Yoshii, Y.; Takaichi, S.; Maoka, T.; Suda, S.; Sekiguchi, H.; Nakayama, T.; Inouye, I. Variation of siphonaxanthin series among the genus *Nephroselmis* (Prasinophyceae, Chlorophyta), including a novel methoxyl carotenoid. *J. Phycol.* **2005**, *41*, 827–834.
11. Wright, S.W.; Jeffrey, S.W. Pigment markers for phytoplankton production. In *Marine Organic Matter: Biomarkers, Isotopes and DNA*; Volkman, J.K., Ed.; Springer-Verlag: Berlin, Germany, 2006; pp. 71–104.
12. Latasa, M. Improving estimations of phytoplankton class abundances using CHEMTAX. *Mar. Ecol. Prog. Ser.* **2007**, *329*, 13–21.
13. Jeffrey, S.W. Application of pigment methods to oceanography. In: *Phytoplankton Pigments in Oceanography: Guidelines to Modern Methods*; Jeffrey S.W., Mantoura R.F.C., Wright, S.W., Eds.; UNESCO Publishing: Paris, France, 1997; p. 127.
14. Laza-Martinez, A.; Seoane, S.; Zapata, M.; Orive, E. Phytoplankton pigment patterns in a temperate estuary: from unialgal cultures to natural assemblages. *J. Plankton Res.* **2007**, *29*, 913–929.
15. Mangoni, O.; Modigh, M.; Mozetic, P.; Bergamasco, A.; Rivaro, P.; Saggiomo, V. Structure and photosynthetic properties of phytoplankton assemblages in a highly dynamic system, the Northern Adriatic Sea. *Estuar. Coast. Shelf Sci.* **2008**, *77*: 633–644.
16. Ras, J.; Claustre, H.; Uitz, J. Spatial variability of phytoplankton pigment distributions in the Subtropical South Pacific Ocean: comparison between in situ and predicted data. *Biogeosciences* **2008**, *5*, 353–369.
17. Smith, W.O., Jr.; Tozzi, S.; DiTullio, G.R.; Dinnimand, M.; Mangoni, O.; Modigh, M.; Saggiomo, V. Phytoplankton photosynthetic pigments in the Ross Sea: patterns and relationships among functional groups. *J. Mar. Systems* **2010**, *82*: 177–185.
18. Anderson, D. M. Toxic algal blooms and red tides: a global perspective. In *Red Tides: Biology, Environmental Science and Tosicology*; Okaichi, T., Anderson, D.M., Nemoto, T., Eds.; Elsevier: New York, NY, USA, 1989; pp. 11–16.
19. Zingone, A.; Enevoldsen , H.O. The diversity of harmful algal blooms: a challenge for science and management. *Ocean Coastal Manage* **2000**, *43*, 725–748.
20. Masò, M.; Garcés, E. Harmful microalgae blooms (HAB); problematic and conditions that induce them. *Mar. Pollut. Bull.* **2006**, *53*, 620–630.
21. Zingone, A.; Siano, R.; D'Alelio, A.; Sarno, D. Potentially toxic and harmful microalgae from coastal waters of the Campania region (Tyrrhenian Sea, Mediterranean Sea). *Harmful Algae* **2006**, *5*, 321–337.

22. Smayda, T.J. Harmful algal blooms: Their ecophysiology and general relevance to phytoplankton blooms in the sea. *Limnol. Oceanogr.* **1997**, 1137–1153.
23. Smayda, T.J.; Reynolds, C.S. Community Assembly in Marine Phytoplankton: Application of Recent Models to Harmful Dinoflagellate Blooms. *J. Plankton Res.* **2001**, *23*, 447–461.
24. Hallegraeff, G.M., Anderson, D.M., Cembella, A.D., Eds.; *Manual on Harmful Marine Microalgae, Monographs on oceanographic methodology 11*; UNESCO, Paris, France, 2004.
25. Pearson, L.; Mihali, T.; Moffitt, M.; Kellmann, R.; Neilan, B. On the Chemistry, Toxicology and Genetics of the Cyanobacterial Toxins, Microcystin, Nodularin, Saxitoxin and Cylindrospermopsin. *Mar. Drugs* **2010**, *8*, 1650–1680.
26. Ma, H.Y.; Krock, B.; Tillmann, U.; Cembella, A. Preliminary Characterization of Extracellular Allelochemicals of the Toxic Marine Dinoflagellate Alexandrium tamarense Using a Rhodomonas salina Bioassay. *Mar. Drugs* **2009**, *7*, 497–522.
27. Lefebvre, K.A.; Bill, B.D.; Erickson, A.; Baugh, K.A.; O'Rourke, L.; Costa, P.R.; Nance, S.; Trainer, V.L. Characterization of intracellular and extracellular saxitoxin levels in both field and cultured Alexandrium spp. samples from Sequim Bay, Washington. *Mar. Drugs* **2008**, *6*, 103–116.
28. Wang, D.Z. Neurotoxins from marine dinoflagellates: A brief review. *Mar. Drugs* **2008**, *6*, 349–371.
29. Landsberg, J.H. The effects of harmful algal blooms on aquatic organisms. *Rev. Fish. Sci.* **2002**, 10, 113–390.
30. Backer, L.C.; Kirkpatrick, B.; Fleming, L.E.; Cheng ,Y.S.; Pierce, R.; Bean, J.A.; Clark, R.; Johnson, D.; Wanner, A.; Tamer, R.; Zhou, Y.; Baden, D.G. Occupational exposure to aerosolized brevetoxins during Florida red tide events: effects on a healthy worker population. *Environ. Health Persp.* **2005**, *113*, 644–649.
31. Pierce R.H.; Henry M.S. Harmful algal toxins of the Florida red tide (*Karenia brevis*): natural chemical stressors in South Florida coastal ecosystems. *Ecotoxicology* **2008**, *17*, 623–631.
32. Ramos, V.; Vasconcelos, V. Palytoxin and Analogs: Biological and Ecological Effects. *Mar. Drugs* **2010**, *8*, 2021–2037.
33. Richlen, M.L.; Morton, S.L.; Jamali, E.A.; Rajan A.; Anderson D.M. The catastrophic 2008–2009 red tide in the Arabian gulf region, with observations on the identification and phylogeny of the fish-killing dinoflagellate *Cochlodinium polykrikoides. Harmful Algae* **2010**, *9*, 163–172.
34. Wiese, M.; D'Agostino, Paul, M.; Mihali, T.K.; Moffitt, M.C.; Neilan, B.A. Neurotoxic Alkaloids: Saxitoxin and Its Analogs. *Mar. Drugs* **2010**, *8*, 2185–2211.
35. Hallegraeff, G.M. Harmful algal blooms: a global overview. In: *Manual on Harmful Marine Microalgae, Monographs on oceanographic methodology 11*; Hallegraeff, G.M., Anderson, D.M., Cembella, A.D., Eds.; UNESCO, Paris, France, 2004; pp. 25–50.

36. Bourdelais, A.; Tomas, C.R.; Narr, J.; Kubanek, J; Baden, D. A new fish-killing alga in coastal Delaware produces neurotoxins. *Environ. Health Perspect.* **2002,** *110*, 465–470.
37. Hallegraeff, G.M. A review of harmful algal blooms and their apparent global increase. *Phycologia* **1993**, *32*, 79–99.
38. Tomas, C.R.; Smayda, T.J. Red tide blooms of *Cochlodinium polykrikoides* in a coastal cove. *Harmful Algae* **2008**, *7*, 308–317.
39. Anderson, D.M. Harmful Algal Blooms and Ocean Observing Systems: Needs, Present Status and Future Potential. In *Fisheries for Global Welfare and Environment*; Memorial book of the 5th World Fisheries Congress, TERRAPUB, Tokyo, Japan, 2008; Tsukamoto, K., Kawamura, T., Takeuchi, T., Beard, T.D., Jr., Kaiser, M.J., Eds.; pp. 317–334.
40. Richardson, T.L.; Pinckney, J.L. Monitoring of the toxic dinoflagellate *Karenia brevis* using gyroxanthin-based detection methods. *J. Appl. Phycol.* **2004**, *16*, 315–328.
41. Gárate-Lizárraga, I.; Bustillos-Guzmán J.J.; Morquecho L.M.; Lechuga-Deveze C.H. First Outbreak of *Cochlodinium polykrikoides* in the Gulf of California. *Harmful Algae News* **2000**, *21*, 7.
42. Bustillos-Guzmán, J.; Gárate-Lizárraga, I.; López-Cortés, D.; Hernández-Sandoval, F. The use of pigment "fingerprints" in the study of harmful algal blooms. *Rev. Biol. Trop.* **2004**, *52* (Suppl. 1), 17–26.
43. Bowers, H.; Tengs, T.; Sayaka, G.; Tomas, C.R.; Ono, C.; Yoshimatsu, S.; Oldach, D. Development of real-time PCR assays for the detection of *Chattonella* species in culture and environmental samples. In *Harmful Algae*; Steidinger, K.A, Landsberg, J.H., Tomas, C.R., Vargo, G.A., Eds.; Florida Fish and Wildlife Conservation Commission, Florida Institute of Oceanography and Intergovernmental Oceanographic Commission: St. Petersburg, FL, USA, 2004; pp. 231–233.
44. Bowers, H.A.; Tomas, C.R.; Tengs, T.; Kempton, J. W.; Lewitus, A.J.; Oldach, D.W. 35 Raphidophyceae (Chadefaud ex Silva) systematics and rapid identification: sequence analysis and 36 real-time PCR analysis. *J. Phycol.* **2006,** *42*, 1333–1348.
45. Vidussi, F.; Claustre, H.; Bustillos-Guzman, J.; Cailliau, C.; Marty, J.-C. Determination of chlorophylls and carotenoids of marine phytoplankton: separation of chlorophyll a from divinyl-chlorophyll a and zeaxanthin from lutein. *J. Plank Res.* **1996**, *18*, 2377–2382.
46. Mantoura, R.F.C.; Repeta, D. In *Phytoplankton Pigments in Oceanography: Guidelines to Modern Methods*; Jeffrey, S.W., Mantoura, R.F.C., Wright, S.W., Eds.; UNESCO Publishing: Paris, France, 1997; pp. 407–428.
47. Britton, G.; Liaaen-Jensen, S.; Pfander, H. *Carotenoids Handbook*; Birkhäuser Verlag: Basel, Switzerland, 2004; p. 238.

48. Andersen, R. In *Algal Culturing Techniques*; Elsevier Academic Press: San Diego, CA, USA, 2005; p. 578.

*Samples Availability:* Available from the authors.

*marine drugs*

MDPI

*Article*

# Isolation and Analysis of the *Cppsy* Gene and Promoter from *Chlorella protothecoides* CS-41

**Meiya Li [1,2,†], Yan Cui [1,†], Zhibing Gan [1], Chunlei Shi [1,*] and Xianming Shi [1]**

1 MOST-USDA Joint Research Center for Food Safety, School of Agriculture and Biology, and State Key Lab of Microbial Metabolism, Shanghai Jiao Tong University, Shanghai 200240, China;
E-Mails: lmeiya@126.com (M.L.); cyan9028@163.com (Y.C.); forrestgzb@yahoo.cn (Z.G.); xmshi@sjtu.edu.cn (X.S.)

2 Analytical Testing Center, Zhejiang Chinese Medical University, Hangzhou 310053, China

† *These authors contributed equally to this work.*

* Author to whom correspondence should be addressed; E-Mail: clshi@sjtu.edu.cn;
Tel.: +86-21-3420-5755; Fax: +86-21-3420-6616.

Academic Editor: Graziano **Riccioni**

*Received: 25 June 2015 / Accepted: 9 September 2015 / Published:*

**Abstract:** Phytoene synthase (PSY) catalyzes the condensation of two molecules of geranylgeranyl pyrophosphate to form phytoene, the first colorless carotene in the carotenoid biosynthesis pathway. So it is regarded as the crucial enzyme for carotenoid production, and has unsurprisingly been involved in genetic engineering studies of carotenoid production. In this study, the *psy* gene from *Chlorella protothecoides* CS-41, designated *Cppsy*, was cloned using rapid amplification of cDNA ends. The full-length DNA was 2488 bp, and the corresponding cDNA was 1143 bp, which encoded 380 amino acids. Computational analysis suggested that this protein belongs to the Isoprenoid_Biosyn_C1 superfamily. It contained the consensus sequence, including three predicted substrate-$Mg^{2+}$ binding sites. The *Cppsy* gene promoter was also cloned and characterized. Analysis revealed several candidate motifs for the promoter, which exhibited light- and methyl jasmonate (MeJA)-responsive characteristics, as well as some typical domains universally discovered in promoter sequences, such as the TATA-box and CAAT-box. Light- and MeJA treatment showed that the

*Cppsy* expression level was significantly enhanced by light and MeJA. These results provide a basis for genetically modifying the carotenoid biosynthesis pathway in *C. protothecoides*.

**Keywords:** *Chlorella protothecoides* CS-41; phytoene synthase; *Cppsy*; promoter

---

## 1. Introduction

Lutein is one of more than 750 known naturally occurring carotenoids, and is synthesized by higher plants, bacteria, fungi, and algae. Based on its molecular structure (containing oxygen), it belongs to the xanthophyll family, one of the two major carotenoid families. In the plant kingdom, lutein provides photoprotection by scavenging singlet oxygen and peroxyl radicals [1], and its bright yellow color helps plants achieve effective cross pollination. Humans cannot synthesize lutein themselves, yet it is essential for the human body. Lutein is the predominant carotenoid in the infant brain [2], and is supplemented in newborn babies in the first hours of life. Lutein can increase biological antioxidant potential and reduce the plasma concentration of total hydroperoxides. It also reduces free radical-induced damage [3]. Lutein is the main carotenoid in the human retina; hence, it has been used as a therapeutic agent for the prevention of age-related macular degeneration [4,5]. Epidemiologic data suggest that lutein plays an active role in delaying chronic diseases [6], stimulating the immune response [7], and hampering the development of cataracts and atherosclerosis [8,9]. A recent study showed that a lutein-based dye used during chromovitrectomy in humans could improve the identification and removal of the vitreous, internal limiting membrane and the epiretinal membrane [10].

As lutein has many functions, it has become increasingly important to find and create more sources of lutein production. In recent years, algae have received a great deal of attention in the production of carotenoids and proteins. Previous studies in our laboratory showed that heterotrophically cultivated *Chlorella protothecoides* CS-41 can produce considerable amounts of lutein [11]. Furthermore, optimization of the cultivation conditions, medium composition, and extraction techniques can improve lutein yields [12,13]. However, to date, there are no reports of the enhancement of lutein production by this alga using genetic modification, although genetic engineering technologies have become increasingly popular in the field of carotenoid production. The first step is to determine the genes involved in lutein biosynthesis—information that is essential for genetic modification.

It has been found that phytoene synthase (PSY) is the rate-limiting enzyme in the carotenoid biosynthesis pathway in photosynthetic organisms [14–16]. In many cases, the rate of lutein formation through the carotenoid biosynthetic pathway appears to be controlled by PSY, which catalyzes the head-to-head condensation of two geranylgeranyl diphosphate molecules to yield phytoene—the first committed reaction in carotenogenesis. Since PSY plays a key role in the first step of carotenogenesis, it has unsurprisingly been chosen for genetic engineering studies of carotenoid production.

PSY has been extensively studied in bacteria and higher plants, but its study in algae is still in its infancy. For unicellular green algae, the *psy* gene has previously been investigated in *Chlamydomonas reinhardtii* [17], *Duanliella bardawil* [18], and *Haematococcus pluvialis* [19]. In *C. reinhardtii*, deletion of the *psy* gene resulted in a white phenotype [20]. For *Haematococcus*, *psy* was shown to be up-regulated under stress conditions of high light and low nutrient availability [21]. Overexpression of exogenous *psy* from *D. salina* [22] or *C. zofingiensis* [23] in *C. reinhardtii* has been shown to increase the lutein content to over 1.25 and 2.2-fold, respectively.

As an efficient lutein-production alga, *C. protothecoides* CS-41 has high potential for application in the commercial production of lutein; however, its lutein biosynthesis pathway has not been well studied. Our research group previously cloned other key enzyme genes in the lutein biosynthesis pathway of this alga, such as the phytoene desaturase (*pds*) (GenBank accession No. FJ968162) [24], zeta-carotene desaturase (*zds*) (GenBank accession No. GU269622) [25], and lycopene-ε-cyclase (*lyce*) (GenBank accession No. FJ752528) genes. The *psy* gene is essential for determining the complete lutein biosynthesis pathway in this alga. Therefore, in this study, the *psy* gene from the unicellular microalga *C. protothecoides* CS-41 and its promoter were isolated and analyzed. This study provides an important theoretical basis for the genetic modification of lutein biosynthesis in *C. protothecoides* CS-41, including gene sequences, expression promoter candidates, and possible regulatory environmental factors for gene expression.

## 2. Results and Discussion

### 2.1. Cloning and Characterization of the psy Gene from C. protothecoides

Touchdown PCR with primers YF and YR (Table 1) generated a predicted 373 bp fragment (Supplementary Figure S1, lane 1). BLAST analysis showed that the nucleotide sequence of this fragment shared about 74% and 73% identities with that of *C. reinhardtii* and *D. salina*, respectively, demonstrating that this fragment sequence is derived from a putative phytoene synthase.

**Table 1.** PCR primers and target fragments for *Cppsy*. F: forward; R: reverse; O: outer primer; I: inner primer.

| Aim | Primer Sequence 5′-3′ |
|---|---|
| **Partial *psy* fragment** | |
| YF | GCCATCTACGTGTGGTGCC |
| YR | CACGCAAGATGTTGGTCAGC |
| **5′-RACE-PCR** | |
| YFO1 | GACTTGTCCACGCCCATCAC |
| YRI1 | GGGGAAGCGGGAGATGGTGT |
| **3′-RACE-PCR** | |
| YFO2 | GATGCTGCCCTCACAGACAC |
| YRI2 | TGGATTTGGTCAAGTCACGC |
| **cDNA and DNA** | |
| YF1 | ATGAGCACGTTTCTGAGCACAGTG |
| YR1 | TCACATGCGCGCCCTCAG |
| **Probe** | |
| psy-F | GAAGTGACCAGCGAGTATGCC |
| psy-R | CTAAAGGGTTGGATGTGC |
| psyRTF | GAAGTGACCAGCGAGTATGCC |
| psyRTR | TCTCTAAAGGGTTGGATGTGC |
| **Promoter** | |
| PSYSP1 | CTGTGCATGCGAAGTCGGAGTGAGA |
| PSYSP2 | CGTCTTGGCATACTCGCTGGTCACTT |
| PSYSP3 | ACTCATGCTGGGGGCTAGGAAAG |
| PSYSP1′ | ATGGCGGGTGGCAGAGTCAATGTA C |
| PSYSP2′ | CCAGACACAATCACCTCGCAGCCCT T |
| PSYSP3′ | CGTTCACTCACCGCTCTCCATCACAA |

With the sequence information, specific primers were designed for 5′- and 3′-rapid amplification of cDNA ends (RACE) of the related gene. 5′-RACE generated a 598 bp fragment (Supplementary Figure S1, lane 2), and 3′-RACE produced an 816 bp fragment (Supplementary Figure S1, lane 3). They were displayed by sequencing as the 5′ and 3′ regions of the phytoene synthase gene of *C. protothecoides* (*Cppsy*). RT-PCR (Reverse Transcription) with a pair of primers YF1 and YR1 generated an 1143 bp fragment (Supplementary Figure S1, lane 4), which was identified as the full-length *Cppsy* cDNA (GenBank accession No. FJ968161).

The open reading frame of *Cppsy* cDNA encoded a protein of 380 amino acid residues with a calculated molecular mass of 43.035 kDa and an isoelectric point of 6.40 (http://cn.expasy.org/tools/protparam.html) and shared 81.7% identical sequence with *Chlorella* NC_64A.

To characterize the corresponding gene of *Cppsy* cDNA, genomic PCR was performed. A 2488 bp fragment (Supplementary Figure S1, lane5) (GenBank accession No. GU351883) was generated and sequenced. Analysis of the obtained nucleotide sequence revealed that the product was the corresponding *Cppsy* gene.

The Southern blot analysis results indicated that there is only one *Cppsy* gene copy in *C. protothecoides* CS-41 (Supplementary Figure S2), which is different to those in higher plants. *Psy* gene replication is common in dicot plants, such as tomato (SlPSY1 and SlPSY2), and in monocot plants, such as maize (ZmPSY1-3), rice (OsPSY1-3), and sorghum (SbPSY1-3) [16,26–28].

Analysis of the *Cppsy* gene structure (Figure 1) revealed that it is more complicated than those of dicot and monocot plants. It consists of ten exons and nine introns. *Chlorella* has a higher intron density than other algae and higher plants; in most of the higher plants, *psy* genes always have four or five introns, but this alga has nine introns. Compared with the structure of the *psy* gene from *C. reinhardtii* (*Crpsy*), it seems that there are two introns inserted into each of the first and second exons, and one intron inserted into the fourth exon, which makes the gene structure more complicated (Figure 1C).

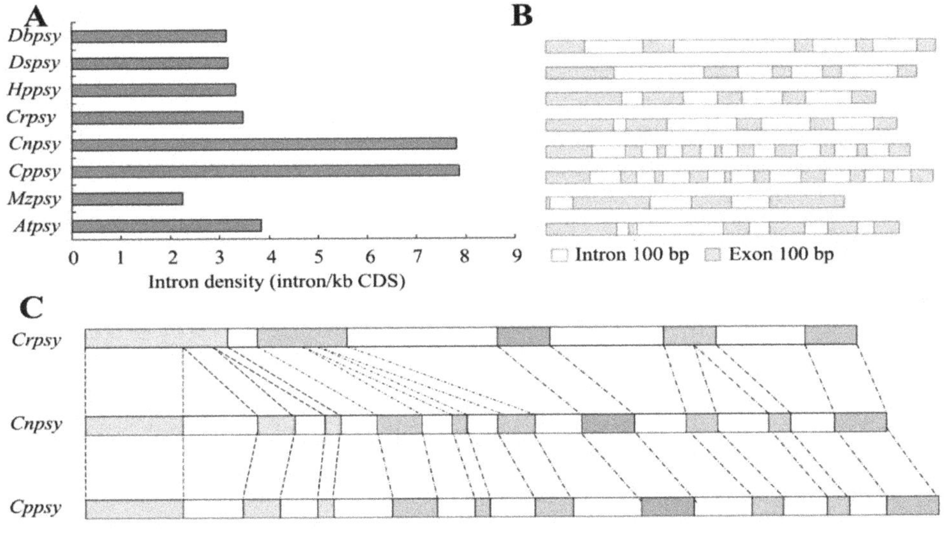


**Figure 1.** Exons and introns of the *Cppsy* gene in *C. protothecoides* CS-41.The ten exons are: (1) 1 bp to 280 bp; (2) 477 bp to 576 bp; (3) 691 bp to 743 bp; (4) 913 bp to 1030 bp; (5) 1139 bp to 1180 bp; (6) 1327 bp to 1427 bp; (7) 1635 bp to 1793 bp; (8) 1957 bp to 2045 bp; (9) 2171 bp to 2233 bp; (10) 2351 bp to 2488 bp. (**A**) Intron density; (**B**) DNA structure; (**C**) The relationship between introns and exons of *Cppsy*, *Crpsy*, and *Cnpsy* genes. *Dbpsy*, *Dspsy*, *Hppsy*, *Crpsy*, *Cnpsy*, *Cppsy*, *Mzpsy*, and *Atpsy* are the *psy* genes of *Duanliella bardawil*, *Duanliella salina*, *Haematococcus pluvialis*, *Chlamydomonas reinhardtii*, *Chlorella* NC_64A, *Chlorella protothecoides* CS-41, *Zea mays*, and *Arabidopsis thaliana*, respectively.

## *2.2. Sequence Alignment and Phylogenetic Reconstruction*

After the DNA and cDNA sequences of the *Cppsy* gene were determined, it was possible to investigate its evolutionary position among the various *psy* genes. Using MEGA 4.0 from Clustal W1.6 alignments, the phylogenetic tree of PSYs from different organisms was constructed based on their deduced amino acid sequences. It showed that *psy* was derived from an ancestor gene and later evolved into four subgroups, including higher plants, cyanobacteria, algae, and bacteria (Figure 2). According to the neighbor-joining (NJ) tree, *Cppsy* belongs to the algae group, and is more ancient than plant species (Figure 2).

The deduced amino acid sequence of *Cppsy* was submitted to NCBI for PSI-BLAST searches and the results showed that *Cppsy* has high homology with *psy* genes from other algal species, with 83% identity and 88% positives with *psy* from *Chlorella* NC_64A. *Cppsy* was also highly similar to *psy* from *C. reinhardtii* (67% identities, 79% positives), *H. pluvialis* (63% identities, 77% positives), *D. bardawil* (68% identities, 80% positives), and *D. salina* (68% identities, 79% positives), suggesting that *Cppsy* belongs to the algae *psy* family. In the algae family, CpPSY belongs to class I of PSY according to Tran's data [29]. BlastP analysis suggested that this protein has the essential characteristics of PSY. It belongs to the Isoprenoid_Biosyn_C1 superfamily, and contains the consensus sequence, including three predicted substrate-$Mg^{2+}$ binding sites (aspartate-rich regions) (DXXXD), 130-DELVD-134, 203-DELYD-207, and 256-DEGED-260 (Figure 3A). In other algae and higher plants, there are two (DELVD and DVGED) (Figure 3A); hence, CpPSY has one more DXXXD motif than other PSYs. The abundant 203-DELYD-207 site possibly plays an important role in the function of CpPSY, which should be studied further.

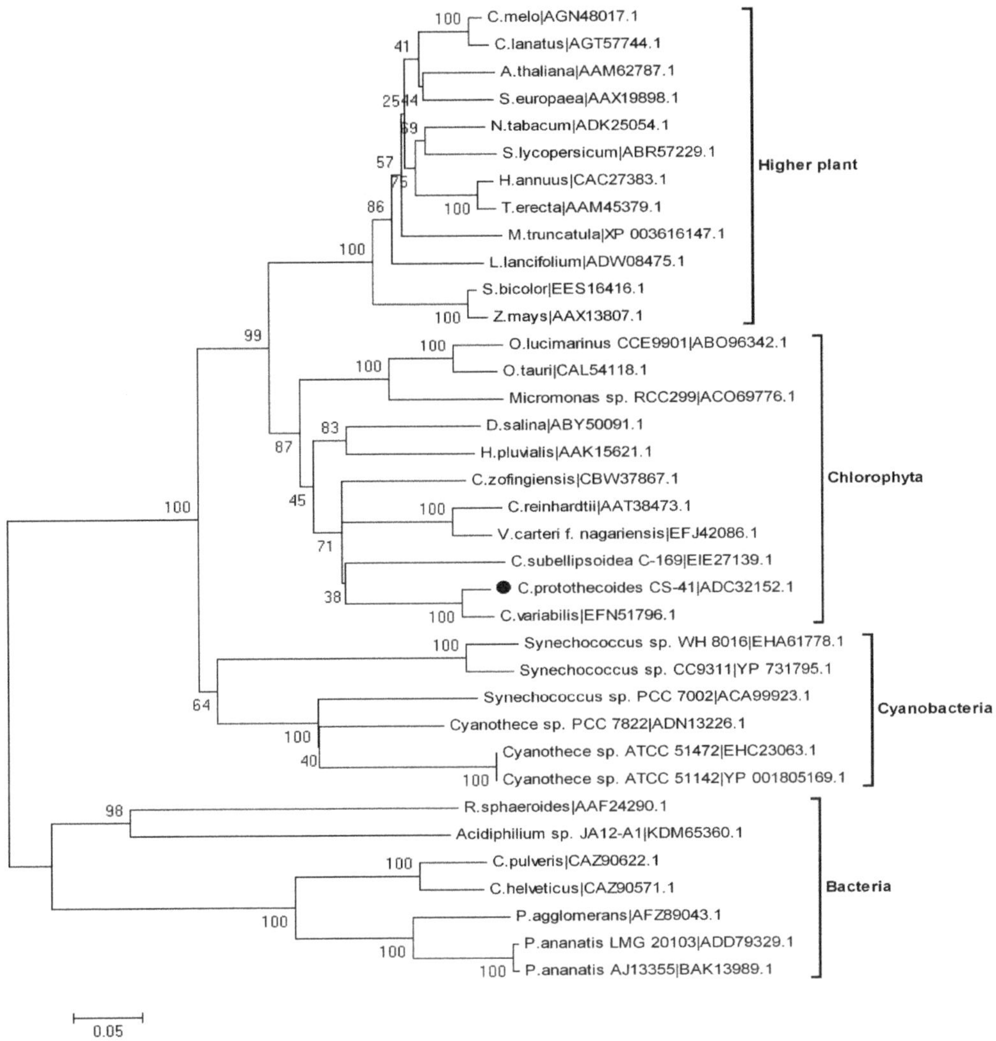


**Figure 2.** Phylogenetic tree of PSY sequences from various species. The phylogeny was derived using neighbor-joining analysis. The accession numbers of the amino acid sequences follow the taxon names. Horizontal branch lengths represent relative evolutionary distances, with the scale bar corresponding to 0.05 amino acid substitutions per site.

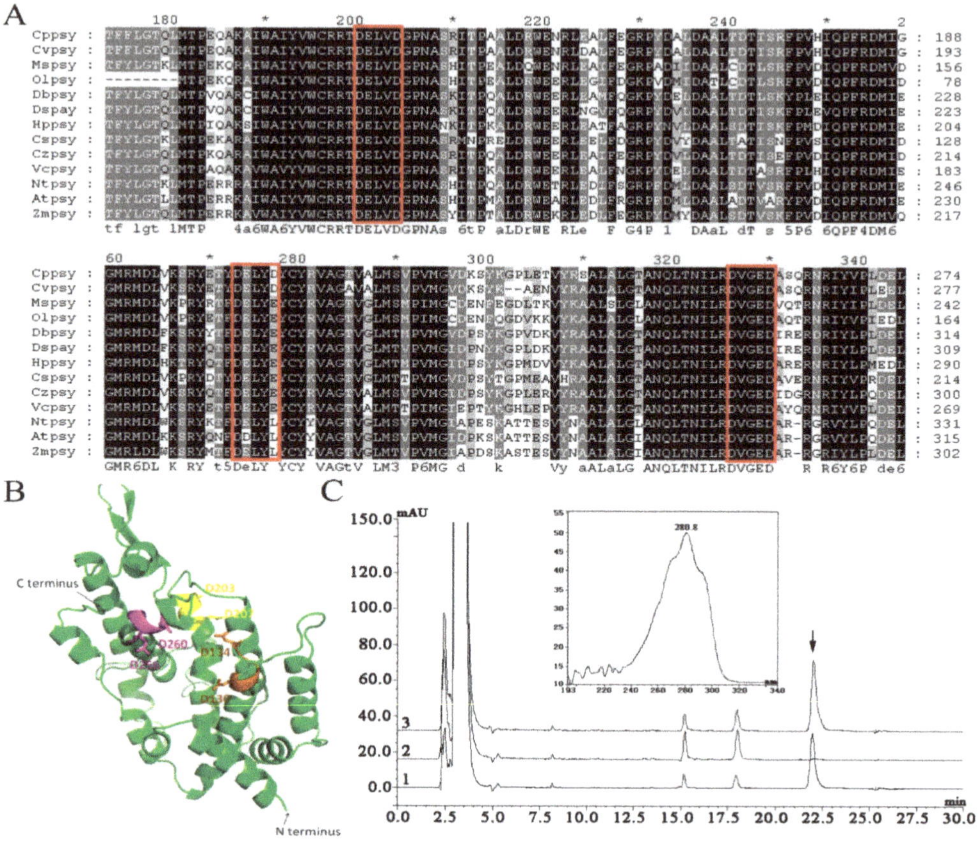

**Figure 3.** (**A**) Alignment of the selective PSY-deduced amino acid sequences from different algae produced with the GeneDoc program using Clustal W. The alignment indicates aspartate-rich regions/substrate-$Mg^{2+}$ binding sites (DXXXD). The three DXXXD motifs are shown by the red boxes. Cppsy, Cvpsy, Mspsy, Olpsy, Dbpsy, Dspsy, Hppsy, Cspsy, Czpsy, Vcpsy, Ntpsy, Atpsy, and Zmpsy are the PSY of *Chlorella protothecoides* CS-41, *Chlorella variabilis, Micromonas* sp. RCC299, *Ostreococcus lucimarinus, Duanliella bardawil, Duanliella salina, Haematococcus pluvialis, Coccomyxa subellipsodiea* C-169, *Chromochloris zofingiensis, Volvox carteri* f. *nagariensis, Nicotiana tabacum, Arabidopsis thaliana,* and *Zea mays,* respectively; (**B**) Three-dimensional model structure of CpPSY. Comparative modeling was performed using homology-based three-dimensional structural modeling. The three aspartate-rich motifs (DXXXD) are colored in orange (DELVD), yellow (DELYD), and magenta (DEGED); others are shown in green. The *N*-terminus and *C*-terminus are also shown; (**C**) High-performance liquid chromatography trace and UV spectrum of carotenoid pigments in the *E. coli* heterologous complementation system. Pigments extracted from *E. coli* cells transformed

with pACCRT-E and pUC-psy together (1), pUC-psy only (2), and pACCRT-EB only (3). Absorbance was recorded at 285 nm. The peak indicated by the arrow is phytoene.

The secondary structure prediction carried out at NPS@ (https://npsa-prabi.ibcp.fr/) showed that CpPSY consists of 58.68% alpha helix, 26.58% random coil, 10.79% extended strand, and 3.95% beta turn. The tertiary structure of CpPSY was constructed using homology-based modeling by Swiss-Model (Figure 3B). A total of 50 models were found. Squalene synthase (HpnC) was used as a template for molecular modeling, since the identity is the highest (30.42%). The modeled structure also showed that CpPSY consists mostly of alpha helices. The three conserved DXXXD motifs (orange DELVD, yellow DELYD, and magenta DEGED) were marked in the three-dimensional model structure (Figure 3B). It seems that the three DXXXD motifs form a circle-like structure, which could be important for enzyme activity.

All of the analysis results strongly suggest that PSY from *C. protothecoides* CS-41 is an algal phytoene synthase protein involved in the carotenoid biosynthesis pathway. Bacterial complementation assay further confirmed that this gene is functional. The expressed protein in pUC-psy could catalyze the GGPP produced by pACCRT-E (Figure 3C,1) to phytoene, similar to the function of the *crtB* gene in pACCRT-EB (Figure 3C,3).

## 2.3. Promoter Isolation and Analysis

The promoter region of the *Cppsy* gene was cloned from *C. protothecoides* CS-41 genomic DNA. The cloned *Cppsy* promoter region was determined to be 1980 bp in length, and the sequence is shown in Figure 4. Furthermore, the cloned *Cppsy* promoter region was analyzed using the PLACE and PlantCARE databases. Several core fragments were identified, which are homologous to the *cis*-acting elements of higher plants and of great importance for the promoter functions (Figure 4). Three types of elements, which have been found to be regulated by hormones in the upstream region of some plant genes, are present in the *Cppsy* promoter: the ABRE type (CCTGCGTGGC, CACGTG, and GCCTCGTGGC) involved in abscisic acid responsiveness; the TGACG-motif (TGACG), the *cis*-acting regulatory element involved in methyl jasmonate (MeJA) responsiveness; and the Sp1 (CCCCCGCCA and ACCCGCCATG), MNF (GTGCCCCATGCAGGTT) and Box I (TTTCAAA) types involved in light responsiveness.

The transcriptional start site (TSS) of the *Cppsy* promoter was determined by 5′-RACE using total RNA extracted from *C. protothecoides*. The TSS is an adenine (A) at 34 bp upstream of the translation initiation codon. The distance between the putative

TATA-box and TSS is approximately −24 to −28 bp, which is consistent with the distance of 32 bp ± 7 bp from previous data [30].

A previous study showed that the *psy* expression level is affected by light [31] and other biotic and abiotic stresses [16] in higher plants. Gene expression response to environmental stress is related to the regulation mechanism. To understand more about the regulation mechanism, we need to know more information about the gene, including the gene structure and regulatory domains. Here, many elements were found in the *Cppsy* promoter that belong to light-responsive elements, such as Sp1, MNF1, G-box, and chs-CMA2a. There were also some *cis*-acting elements involved in abscisic acid (ABRE) and MeJA (TGACG-motif) responsiveness. Light is one of the most important environmental factors for algae. To determine whether these motif sequences from the *Cppsy* promoter are involved in light, abscisic acid, and MeJA signaling, loss-of-function analysis needs to be carried out in future studies.

```
-1920   CCAGGCCATGCGGCATCACCGGGCGCTGCTCAAACAGCGCGGCTGAATCCGTGGCGAACC
-1860   CCACATCACTTTTGAAGGCCGTTGCCGTTGAATCTGCGGGCGCAAACGTGCGACCACGGC
-1800   TGATTCGCTCGCCGCCGCCGCTGCCGCTCGTCGACCCTTCCGCAGGCTGTGGCCGATGGC
-1740   GTCTTTTACGCCGAGCTGAACGAGCTGTTGATGCGGGAGCTGGGTGGCGAGGGCTACTCT
          TATA box
-1680   GGCGTGGAGGTGCGAACTGCTGTGGGCAGCTGGGCTGTGCATTGACCAACGGTTTGGGGC
-1620   GTTGGGTTGACCTTACTTTGCTGGGTTTGTCTTTTGTGGTCGCTCTGCAGGTGCGCGTGA
-1560   CCCCCATGCGCACTGAGGTCATCATTCGCGCCACTCGGACTCAGGCCGTGCTGGGTGAGC
-1500   GCCAGAACGATCTTGTTGCGCGGCGGTGGCTTTCAGCCTGTAGGCAGCGACCAGGGGCAT
-1440   GGACATGCAGCAGAGCTGGCTGCTGGCGCGGAGAGCGGGCTTGGCCTCCAGGCATCGTAG
-1380   CGAGCCTGCGTGGCATTGATGGCCGTGTGTGTGCGTGTGCGTGCAGGCGAGAAGGGACGC
             ABRE                                      rbcS-CMA7a
-1320   CGCATCCGTGAGCTGACGAGCGTGGTGCAGAAGCGCTTCAAGTTCCCCGAGGGGTCTGTG
                 TGACG-motif
-1260   GAGCTGTATGCTGAGAAGGTGGGCAGTCCCTGCAACGTACACGAGTCCACTGGCACTCAA
-1200   GCAAGGACAACCCTGGGTCCACAGCCACAGACAAGAGCCGCAGAGTCACAGAGCCACAGT
-1140   CAAGTTTGTCAGCCTTTTTGGGTGCACGTGCCGCTGGACAGCTCCGTGACACTTTTGTCC
                                  ABRE
-1080   TTTCCATGTGTGTGCCCCATGCAGGTTGCCAACCGCGGCCTGTGCGCTGTGGCCCAGGCT
                      MNFI
-1020   GAATCCATGCGCTACAAGCTGCTGGGTGGCCTGGCCGTGCGCCGTGCCTGCTACGGTGTC
-960    CTGCGCTTTGTGATGGAGAGCGGTGAGTGAACGCGTCCCGCCGCCGCTGCCGCTTCTGGC
-900    ACGGTTGCGGAGGCAGGCCTGTGGCGTGAAGGCGGCGCTGCCTGTGCCCCAAGGCGCGGA
-840    GCGCTGTGCCTCGTGGCGTCTGCCAAAGCCGACCGTTCCGTGTGTGTGTCGTGTCCCCCC
                ABRE
-780    GCAGGTGCCAAGGGCTGCGAGGTGATTGTGTCTGGCAAGCTGCGTGCGCAGCGTGCCAAG
-720    AGCATGAAGTTCAAGGACGGCTACATGGTGTCCAGCGGCAACCCGGTGAACGAGTACATT
-660    GACTCTGCCACCCGCCATGTCATGCTGCGCCAGGTGAGCGCAGCGCAGCCGTTTGAAGGC
                 Sp1
-600    CTTCCAACCCCAGTTGGGCGCAAAGGGGGATTGCTGGTTGGGGCGCGGTGTGTTAGTGCT
-540    GCTTTCTTGGGCTGGTGGTGTGGGCTGGTGGTGAGTGCGCTTCACTGGTGACGATCGTGC
                                                       TGACG-motif
-480    TGTGAGTGCATGCCTGGTTATTGCGTCAAGGCTTGGGCCCCTGTGCTGACTGTTGCCGGG
                             CGTCA-motif
-420    TTGCCTGCCCTCGTGCTTGTCTGCCTGCAGGGTGTGCTGGGCATCAAGGTGCGCATCATG
-360    AAGGGGTGGGACCCCAGCGGAAAGATCGGGCCCCGCACCCCCTCTGCCCGATGTGGTCAC
-300    TGTGCTGGCGCCCAAGGAGGAGGAGTATGTGTCGGACAAGCAGGAGAAGCTCAAGGCATA
-240    GGCGTGGTCCCCCCGCCATGCGATTCCCGATTGCCGAAACCGTCGCACACCCATC TGCCA
                 Sp1                      LTR
-180    CATCTTGCAAATGCCCACTCGTTTCAAATCAGCATTTGTAACGTGATAGGCTTTGAACGG
          CAAT box           Box I
-120    GCCAGAGGCATCAGAACACTCAGATGCAGGAGGGCAGCGCGTGAGAGAATGACAAATGCG
                                                             CAAT box
-60     GCGGCGGTGCGGTGCCATTTGTGCGCGTCGGCTTTTATCCGTCTTCGCTTGCTCCTCATC
        +1                                TATA box
+1      ACGCGCATTGTTTCTGTTATCACTTGATCGCCGGCATG
                            chs-CMA2a
```

**Figure 4.** Promoter sequence of *Cppsy* from *C. protothecoides* CS-41. Numbers indicate the positions relative to transcriptional start site (TSS). The TSS is indicated as +1 and in bold; important *cis*-elements are underlined.

### 2.4. Gene Expression Response to Light and MeJA

To investigate the effects of light and MeJA on the promoter inducibility at different time points,
we analyzed *Cppsy* mRNA expression levels after light and MeJA treatment. The results show that on treatment with light, *Cppsy* gene expression increased by up to 36 times, compared with the dark, which indicates that *Cppsy* gene expression is up-regulated in response to light (Figure 5A). When treated with MeJA, *Cppsy* gene expression peaked at 10 h after treatment (Figure 5B). Therefore, the gene expression of *Cppsy* is significantly induced by treatment with light and MeJA ($p < 0.01$). These results confirm that the *Cppsy* promoter is induced by light and MeJA, and can be used as a candidate promoter element for the genetic modification of carotenoid biosynthesis in *Chlorella*, other algae, or higher plants.

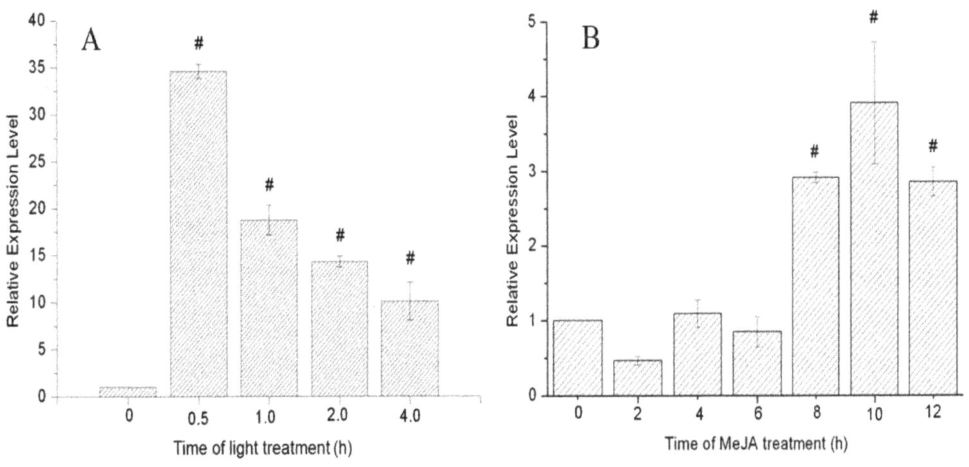


**Figure 5.** The expression of *Cppsy* gene induced by light (**A**) and MeJA (**B**) at different time points in *C. protothecoides* CS-41. Data (mean ± SEM) are combined from three independent experiments. # indicates that the gene expression levels were significantly different from that at 0 h ($p < 0.01$).

## 3. Experimental Section

### 3.1. Strains and Culture Conditions

The microalgal strain used in this study was *C. protothecoides* CS-41 obtained from CSIRO Marine Laboratories (Hobart, Australia). They were grown in modified basal medium [32] containing 10 g·L$^{-1}$ glucose at 28 °C and 180 rpm, and were collected at the log phase or late log phase.

### 3.2. Genomic DNA and RNA Isolation

Genomic DNA was extracted using a modified cetyltrimethylammonium bromide (CTAB) method [33]. The total RNA was isolated from *C. protothecoides* CS-41 cells at the late log phase (about 5-day incubation, with a cell density around $1 \times 10^8$ cells $mL^{-1}$) using TRIzol® reagent (Invitrogen, Carlsbad, CA, USA) according to the manufacturer's instructions.

### 3.3. Cloning of Full-Length Cppsy cDNA and Its Corresponding Gene

Degenerate primers (Table 1) were designed for the amplification of a partial *Cppsy* cDNA from *C. protothecoides* CS-41. The primers were derived from the highly conserved nucleotide and amino acid sequences reported for the *psy* genes from five kinds of algae (*C. reinhardtii, H. pluvialis, D. salina, D. bardawil,* and *Chlorella* NC_64A).

Sets of specific primers were synthesized based on the sequence of putative insert for 5′- and 3′-RACE [34]. YFO1 and YRI1 were used for 5′-RACE, and YFO2 and YRI2 (Table 1) were used for 3′-RACE. RACE was performed using the 5′-Full RACE Kit and 3′- Full RACE Core Set Version 2.0 (TaKaRa, Dalian, China) according to the manufacturer's protocol. The RACE products were gel purified and sequenced as previously described. One pair of specific primers, YF1 and YR1 (Table 1), was designed from the sequences of the 5′- and 3′-RACE fragments to amplify the full-length *Cppsy* cDNA and its corresponding gene.

### 3.4. Southern Blot Analysis

According to the *Cppsy* genomic DNA sequence, *Bam*HI, *Eco*RI, *Nco*I, *Sma*I, and *Xma*I, which showed no recognition sites in the probed region of the *Cppsy* gene, were chosen to digest the whole genomic DNA. The probe was prepared by amplifying genomic DNA with the primers psy-F and psy-R, resulting in a 552-bp fragment of the *Cppsy* gene. The digested DNA was transferred to a Hybond-N membrane (GE Healthcare, Little Chanfont, UK) by capillary transfer and hybridized with a $^{32}$P-labelled DNA probe at both low and high stringency overnight.

After hybridization, the radioactivity of the membrane was monitored using a Storm 840 Phosphor Imaging System (Molecular Dynamics, Sunnyvale, CA, USA).

### 3.5. Bioinformatics Analysis

Comparative and bioinformatic analyses of the nucleotide sequences and deduced amino acid sequences were carried out online at http://www.ncbi.nlm.nih.gov and http://cn.expasy.org. The nucleotide sequence, deduced amino acid sequence, and open reading frame (ORF) were analyzed, and sequence comparison was conducted through database searches using BLAST programs (http://www.ncbi.nlm.nih.gov/BLAST/) and GeneDoc software. The phylogenetic analysis of *psy* from other plant species was aligned with Clustal X program version 1.83 using default parameters [35] and manual adjustments where necessary. A phylogenetic tree was constructed using MEGA (molecular evolutionary genetics analysis) program, version 4.0 [36] from Clustal W1.6 alignments. The NJ [37] method was used to construct the tree. In the NJ method, the P distance was used to analyze the amino acid sequences. A total of 1000 repetitions were performed using the bootstrap method to determine the reliability of each node of the tree. The homology-based three-dimensional structural modeling of PSY was accomplished using Swiss-Model and WebLab Viewer Lite (http://swissmodel.expasy.org).

### 3.6. Functional Complementation Experiment in E. coli

*E. coli* JM109 (Table 2) was used as a host for complementation experiments by cotransformation of the plasmid pUC-psy with pACCRT-E (Table 2). *E. coli* JM109 harboring only plasmid pACCRT-EB (Table 2) was cultured as a positive control, and only plasmid pUC-*psy* was cultured as a negative control for PSY functional analysis. The different strains were cultivated in 100 mL LB medium containing 100 $\mu g \cdot mL^{-1}$ ampicillin and 50 $\mu g \cdot mL^{-1}$ chloramphenicol at 28 °C and 180 rpm. IPTG (1 mM) was added when the optical density at 600 nm ($OD_{600}$) reached 0.5, and the culture was kept at 28 °C for 2 days. The *E. coli* cells were collected by centrifugation at 12,000 rpm and used for high-performance liquid chromatography (HPLC) analysis.

**Table 2.** Strains and plasmids used in this study.

| Strains or Plasmids | Characteristics | Source |
|---|---|---|
| *E. coli* JM109 | Host for expression vectors | MOST-USDA Joint Research Center for Food Safety stock |
| pACCRT-E | pACYC184 containing *crtE* gene of *Erwinia uredovora* ($Cm^r$) (metabolite: GGPP) | Gift from Prof. Gerhard. Sandmann (J.W. Goethe University, Frankfurt, Germany) |
| pACCRT-EB | pACYC184 containing *crtE* and *ctrB* genes of *Erwinia uredovora* ($Cm^r$) (metabolite: Phytoene) | Gift from Prof. Gerhard. Sandmann (J.W. Goethe University, Frankfurt, Germany) |
| pUC19 | Expression vector ($Amp^r$) | MOST-USDA Joint Research Center for Food Safety stock |
| pUC-psy | pUC19 vector containing *Cppsy* gene ($Amp^r$) | This work |

Pigments in the bacteria were extracted according to procedures described by Breitenbach *et al.* [38]. *E. coli* JM109 cells harboring different plasmids were collected by centrifugation and freeze dried. Pigment extraction was carried out in acetone

(80%, v/v) using ultrasonication, and the solvent was removed by blowing with nitrogen gas. The carotenoids were then resuspended in acetone for subsequent HPLC analysis. All operations were carried out on ice under dim light to prevent photodegradation, isomerizations, and structural changes of the carotenoids. The samples were prepared for HPLC by dissolving the dried residues in 1 mL of acetone and filtered through a polycarbonate 0.22 μm filter (Millipore, Carrigtwohill, Ireland). The extracted pigments were separated on a Kromasil KR100-5$C_{18}$ analytical column (250 mm × 4.6 mm, 5 μm) using an UltiMate3000 HPLC system (Thermo Fisher Scientific, Waltham, MA, USA). The procedures described by Huang [39] were used with the following modifications: the mobile phase consisted of solvent A (acetonitrile/methanol/0.1 M Tris-HCl (pH 8.0), 84:2:14, v/v/v) and solvent B (methanol/ethyl acetate, 68:32, v/v). Pigments were eluted at a flow rate of 1 mL·min$^{-1}$ with a linear gradient from 100% solvent A to 100% solvent B over a 5 min period, followed by 25 min of solvent B. The column temperature was maintained at 30 °C and the sample volume was 20 μL. The pigments were monitored by diode array detector, and the targeted products were identified by their absorption spectra and typical retention times compared with the control.

### 3.7. Promoter Isolation and Analysis

The Genomic Walking Kit (TaKaRa, Dalian, China) was used to obtain promoter regions of the *Cppsy* gene from *C. protothecoides* CS-41. Based on the *Cppsy* genomic sequences, gene-specific primers were designed and are listed in Table 1. Primary and nested PCRs were performed with the *Cppsy* gene-specific primers and Genome Walking adapter primers (AP1) in the kit according to the manufacturer's instruction. The primary nested PCR products were diluted to 1:50 with distilled water for subsequent nested PCR. The nested PCR products were purified from 1.2% (w/v) agarose gel and sub-cloned into the pMD18T vector (TaKaRa, Dalian, China). The cloned vectors were then sequenced and the putative *cis*-regulatory elements were analyzed using the PLACE [40] and PlantCARE databases [41].

### 3.8. Gene Expression Response to Light and MeJA

To analyze the light regulation pattern of the *Cppsy* gene in *C. protothecoides*, algal cells in the late log phase were cultivated in the dark for more than 2 days, then collected by centrifugation at 5000 rpm for 15 min in the darkness. The pellet was resuspended in fresh medium without glucose, and then subjected to light treatment under a light intensity of 120 μmol m$^{-2}$·s$^{-1}$ for different induction times (0, 0.5, 1.0, 2.0, and 4.0 h). Each treatment was carried out with three parallel repetitions.

To analyze the MeJA regulation pattern of the *Cppsy* gene in *C. protothecoides*, algal cells in the log phase were treated with 100 μM MeJA (Sigma, St. Louis, MO, USA) diluted in dimethyl sulfoxide (DMSO) for 0, 2, 4, 6, 8 10, and 12 h. Control cultures were treated with DMSO only.

The effects of light and MeJA on the *Cppsy* gene transcripts in *C. protothecoides* were quantified by reverse transcriptase quantitative PCR (RT-qPCR). The RT-qPCR experiment was performed in two steps: the cDNA templates were synthesized from RNA samples using Prime Script™ Reverse Transcriptase Reagent according to the manufacturer's instructions (TaKaRa, Dalian, China) using oligo (dT) as the primer; then qPCR was conducted on an iQ Cycler (Bio-Rad, Watford, UK) using the specific primers (Table 1) and the SYBR ExScript RT-PCR kit (TaKaRa, Dalian, China). The specific primers for the corresponding genes included psyRT-F and psyRT-R for the *Cppsy* gene, and 16SRT-F and 16SRT-R for the 16S gene (Table 1). Before the qPCR expression analysis, we checked the amplification efficiency of each primer pairs, and all were well controlled between 99.83% and 101.25%.

Each qPCR measurement was carried out independently at least three times, and the mean value was used for quantification. The $2^{-\Delta\Delta CT}$ method was used to analyze the relative changes in gene expression, the expression of the 16S gene was used as a normalized control, and the expression of the untreated samples was used as a negative control.

## 4. Conclusions

Carotenoid pathways in plants have been described in great detail using genetic, biochemical, and molecular data, mainly from *Arabidopsis* and other higher plants; however, this is the first study in the unicellular microalga *C. protothecoides* CS-41.

We successfully isolated and analyzed the *Cppsy* gene, which encodes the functional phytoene synthase—a vital enzyme for carotenoid biosynthesis in *C. protothecoides* CS-41—as well as its promoter. Computational analysis suggested that this protein belongs to the Isoprenoid_Biosyn_C1 superfamily. It contains one more substrate-$Mg^{2+}$ binding site than other algae and higher plants. Analysis also demonstrated several candidate motifs for the promoter, which exhibited light- and MeJA-responsive characteristics. Light- and MeJA treatment showed that the *Cppsy* expression level was significantly enhanced by light and MeJA.

These achievements will be helpful to understand more about the regulatory mechanism of the carotenoid biosynthesis pathway in algae and the mechanisms for accumulation of lutein and other important carotenoids.

## Acknowledgments

This work was supported by the National 863 Program of China (2012AA101601), and the Agri-X Program of Shanghai Jiao Tong University (Agri-X2015005).

## Author Contributions

Chunlei Shi and Xianming Shi conceived and designed the experiments. Meiya Li, Yan Cui and Zhibing Gan performed the experiments. Chunlei Shi, Meiya Li and Yan Cui analyzed the data and prepared the manuscript.

## Conflict of Interest

The authors declare no conflict of interest.

## References

1. Demmig-Adams, B.; Adams, W.W. Antioxidants in photosynthesis and human nutrition. *Science* **2002**, *298*, 2149–2153.
2. Vishwanathan, R.; Kuchan, M.J.; Sen, S.; Johnson, E.J. Lutein and Preterm Infants with Decreased Concentrations of Brain Carotenoids. *J. Pediatr. Gastroenterol. Nutr.* **2014**, *59*, 659–665.
3. Perrone, S.; Tei, M.; Longini, M.; Santacroce, A.; Turrisi, G.; Proietti, F.; Felici, C.; Picardi, A.; Bazzini, F.; Vasarri, P.; *et al.* Lipid and protein oxidation in newborn infants after lutein administration. *Oxid. Med. Cell. Longev.* **2014**, *2014*, doi:10.1155/2014/781454.
4. Kalariya, N.M.; Ramana, K.V.; Vankuijk, F.J. Focus on molecules: Lutein. *Exp. Eye Res.* **2012**, *102*, 107–108.
5. Widomska, J.; Subczynski, W.K. Why has Nature Chosen Lutein and Zeaxanthin to Protect the Retina? *J. Clin. Exp. Ophthalmol.* **2014**, *5*, 326.
6. Mares-Perlman, J.A.; Millen, A.E.; Ficek, T.L.; Hankinson, S.E. The body of evidence to support a protective role for lutein and zeaxanthin in delaying chronic disease: Overview. *J. Nutr.* **2002**, *132*, 518S–524S.
7. Kim, H.W.; Chew, B.P.; Wong, T.S.; Park, J.S.; Weng, B.B.; Byrne, K.M.; Hayek, M.G.;
Reinhart, G.A. Dietary lutein stimulates immune response in the canine. *Vet. Immunol. Immunopathol.* **2000**, *74*, 315–327.
8. Krinsky, N.I.; Landrum, J.T.; Bone, R.A. Biologic mechanisms of the protective role of lutein and zeaxanthin in the eye. *Annu. Rev. Nutr.* **2003**, *23*, 171–201.
9. Granado, F.; Olmedilla, B.; Blanco, I. Nutritional and clinical relevance of lutein in human health. *Annu. Rev. Nutr.* **2003**, *90*, 487–502.
10. Maia, M.; Furlani, B.A.; Souza-Lima, A.A.; Martins, D.S.; Navarro, R.M.; Belfort, R.J. Lutein:
A new dye for chromovitrectomy. *Retina* **2014**, *34*, 262–272.
11. Shi, X.M.; Zhang, X.W.; Chen, F. Heterotrophic production of biomass and lutein by *Chlorella protothecoides* on various nitrogen sources. *Enzyme Microb. Technol.* **2000**, *27*, 312–318.

12. Shi, X.M.; Jiang, Y.; Chen, F. High-yield production of lutein by the green microalga
*Chlorella protothecoides* in heterotrophic fed-batch culture. *Biotechnol. Prog.* **2002**, *18*, 723–727.
13. Shi, X.M.; Wu, Z.Y.; Chen, F. Kinetic modeling of lutein production by heterotrophic *Chlorella* at various pH and temperatures. *Mol. Nutr. Food Res.* **2006**, *50*, 763–768.
14. Chen, Q.; Jiang, J.G.; Wang, F. Molecular phylogenies and evolution of *crt* genes in algae.
*Crit. Rev. Biotechnol.* **2007**, *27*, 77–91.
15. Salvini, M.; Bernini, A.; Fambrini, M.; Pugliesi, C. cDNA cloning and expression of the phytoene synthase gene in sunflower. *J. Plant Physiol.* **2005**, *162*, 479–484.
16. Li, F.; Vallabhaneni, R.; Yu, J.; Rocheford, T.; Wurtzel, E.T. The maize phytoene synthase gene family: Overlapping roles for carotenogenesis in endosperm, photomorphogenesis, and thermal stress tolerance. *Plant Physiol.* **2008**, *147*, 1334–1346.
17. Bohne, F.; Linden, H. Regulation of carotenoid biosynthesis genes in response to light in *Chlamydomonas reinhardtii*. *Biochim. Biophys. Acta* **2002**, *1579*, 26–34.
18. Lao, Y.M.; Xiao, L.; Ye, Z.W.; Jiang, J.G.; Zhou, S.S. *In silico* analysis of phytoene synthase and its promoter reveals hints for regulation mechanisms of carotenogenesis in *Duanliella bardawil*. *Bioinformatics* **2011**, *27*, 2201–2208.
19. Steinbrenner, J.; Linden, H. Regulation of two carotenoid biosynthesis genes coding for phytoene synthase and carotenoid hydroxylase during stress-induced astaxanthin formation in the green alga *Haematococcus pluvialis*. *Plant Physiol.* **2001**, *125*, 810–817.
20. McCarthy, S.S.; Kobayashi, M.C.; Niyogi, K.K. White mutants of *Chlamydomonas reinhardtii* are defective in phytoene synthase. *Genetics* **2004**, *168*, 1249–1257.
21. Steinbrenner, J.; Linden, H. Light induction of carotenoid biosynthesis genes in the green alga *Haematococcus pluvialis*: Regulation by photosynthetic redox control. *Plant Mol. Biol.* **2003**, *52*, 343–356.
22. Couso, I.; Vila, M.; Rodriguez, H.; Vargas, M.A.; Leon, R. Overexpression of an exogenous phytoene synthase gene in the unicellular alga *Chlamydomonas reinhardtii* leads to an increase in the content of carotenoids. *Biotechnol. Prog.* **2011**, *27*, 54–60.
23. Cordero, B.F.; Couso, I.; Leon, R.; Rodriguez, H.; Vargas, M.A. Enhancement of carotenoids biosynthesis in *Chlamydomonas reinhardtii* by nuclear transformation using a phytoene synthase gene isolated from *Chlorella zofingiensis*. *Appl. Microbiol. Biotechnol.* **2011**, *91*, 341–351.
24. Li, M.Y.; Gan, Z.B.; Cui, Y.; Shi, C.L.; Shi, X.M. Structure and function characterization of
the phytoene desaturase related to the lutein biosynthesis in *Chlorella*

*protothecoides* CS-41.
*Mol. Biol. Rep.* **2013**, *40*, 3351–3361.

25. Li, M.Y.; Gan, Z.B.; Cui, Y.; Shi, C.L.; Shi, X.M. Cloning and characterization of the
zeta-carotene desaturase gene from *Chlorella protothecoides* CS-41. *J. Biomed. Biotechnol.* **2011**, *2011*, doi:10.1155/2011/731542.
26. Gallagher, C.E.; Matthews, P.D.; Li, F.; Wurtzel, E.T. Gene duplication in the carotenoid biosynthetic pathway preceded evolution of the grasses. *Plant Physiol.* **2004**, *135*, 1776–1783.
27. Li, F.; Vallabhaneni, R.; Wurtzel, E. PSY3, a new member of the phytoene synthase gene family conserved in the Poaceae and regulator of abiotic stress-induced root carotenogenesis. *Plant Physiol.* **2008**, *146*, 1333–1345.
28. Welsch, R.; Wust, F.; Bar, C.; al-Babili, S.; Beyer, P. A third phytoene synthase is devoted to abiotic stress-induced abscisic acid formation in rice and defines functional diversification of phytoene synthase genes. *Plant Physiol.* **2008**, *147*, 367–380.
29. Tran, D.; Haven, J.; Qiu, W.G.; Polle, J.E. An update on carotenoid biosynthesis in algae: Phylogenetic evidence for the existence of two classes of phytoene synthase. *Planta* **2009**, *229*, 723–729.
30. Joshi, C.P. An inspection of the domain between putative TATA box and translation start site in
79 plant genes. *Nucleic Acids Res.* **1987**, *15*, 6643–6653.
31. Cazzonelli, C.I.; Pogson, B.J. Source to sink: Regulation of carotenoid biosynthesis in plants. *Trends Plant Sci.* **2010**, *15*, 266–274.
32. Shi, X.M.; Chen, F.; Yuan, J.-P.; Chen, H. Heterotrophic production of lutein by selected Chlorella strains. *J. Appl. Phycol.* **1997**, *9*, 445–450.
33. Stewart, C.J.; Via, L.E. A rapid CTAB DNA isolation technique useful for RAPD fingerprinting and other PCR applications. *BioTechniques* **1993**, *14*, 748–750.
34. Yeku, O.; Frohman, M.A. Rapid amplification of cDNA ends (RACE). In *RNA*; Nielsen, H., Ed.; Humana Press: Totowa, NJ, USA, 2011; pp. 107–122.
35. Thompson, J.D.; Gibson, T.J.; Plewniak, F.; Jeanmougin, F.; Higgins, D.G. The CLUSTAL_X windows interface: Flexible strategies for multiple sequence alignment aided by quality analysis tools. *Nucleic Acids Res.* **1997**, *25*, 4876–4882.
36. Tamura, K.; Dudley, J.; Nei, M.; Kumar, S. MEGA4: Molecular Evolutionary Genetics Analysis (MEGA) software version 4.0. *Mol. Biol. Evol.* **2007**, *24*, 1596–1599.
37. Saitou, N.; Nei, M. The neighbor-joining method: A new method for reconstructing phylogenetic trees. *Mol. Biol. Evol.* **1987**, *4*, 406–425.
38. Breitenbach, J.; Zhu, C.; Sandmann, G. Bleaching herbicide norflurazon inhibits phytoene desaturase by competition with the cofactors. *J. Agric. Food Chem.* **2001**, *49*, 5270–5272.

39. Huang, J.C.; Liu, J.; Li, Y.T.; Chen, F. Isolation and characterization of the phytoene desaturase gene as a potential selective marker for genetic engineering of the astaxanthin-producing green alga *Chlorella zofingiensis* (Chlorophyta). *J. Phycol.* **2008**, *44*, 684–690.
40. Higo, K.; Ugawa, Y.; Iwamoto, M.; Korenaga, T. Plant *cis*-acting regulatory DNA elements (PLACE) database: 1999. *Nucleic Acids Res.* **1999**, *27*, 297–300.
41. Lescot, M.; Dehais, P.; Thijs, G.; Marchal, K.; Moreau, Y.; van de Peer, Y.; Rouze, P.; Rombauts, S. PlantCARE, a database of plant *cis*-acting regulatory elements and a portal to tools for *in silico* analysis of promoter sequences. *Nucleic Acids Res.* **2002**, *30*, 325–327.

marine drugs

MDPI

Article

# Sequencing and Characterization of Novel PII Signaling Protein Gene in Microalga *Haematococcus pluvialis*

**Ruijuan Ma [1,2], Yan Li [2,*] and Yinghua Lu [1,*]**

1 Department of Chemical and Biochemical Engineering, College of Chemistry and Chemical Engineering, Xiamen University, Xiamen 361005, China; ruijuanma@hotmail.com

2 Algae Biotechnology Laboratory, School of Agriculture and Food Sciences, The University of Queensland, Brisbane 4072, Queensland, Australia

* Correspondence: y.li12@uq.edu.au (Y.Li); ylu@xmu.edu.cn (Y.Lu); Tel.: +61-7-336-58817 (Y.Li); +86-592-218-6038 (Y.Lu)

Received: 31 August 2017; Accepted: 30 September 2017; Published: 11 October 2017

**Abstract:** The PII signaling protein is a key protein for controlling nitrogen assimilatory reactions in most organisms, but little information is reported on PII proteins of green microalga *Haematococcus pluvialis*. Since *H. pluvialis* cells can produce a large amount of astaxanthin upon nitrogen starvation, its PII protein may represent an important factor on elevated production of *Haematococcus* astaxanthin. This study identified and isolated the coding gene (Hp*GLB1*) from this microalga. The full-length of Hp*GLB1* was 1222 bp, including 621 bp coding sequence (CDS), 103 bp 5′ untranslated region (5′ UTR), and 498 bp 3′ untranslated region (3′ UTR). The CDS could encode a protein with 206 amino acids (HpPII). Its calculated molecular weight (Mw) was 22.4 kDa and the theoretical isoelectric point was 9.53. When *H. pluvialis* cells were exposed to nitrogen starvation, the Hp*GLB1* expression was increased 2.46 times in 48 h, concomitant with the raise of astaxanthin content. This study also used phylogenetic analysis to prove that HpPII was homogeneous to the PII proteins of other green microalgae. The results formed a fundamental basis for the future study on HpPII, for its potential physiological function in *Haematococcus* astaxanthin biosysthesis.

**Keywords:** *Haematococcus pluvialis*; PII signaling protein; nitrogen starvation; gene cloning; mRNA expression

---

## 1. Introduction

Inorganic nitrogen acts as an important nutrient in autotrophic microalgae cultivation. It is a limiting factor for cell growth. While under a nitrogen-depleted condition, microalgae will transfer the energy from cell division/growth to produce more secondary metabolites like carotenoids [1,2] and lipids [3], which have widespread commercial value. Therefore, nitrogen regulation is an important approach for microalgae cultivation.

It is believed that nitrogen metabolism is regulated by multiple signal regulators and linked to carbon flux [4]. In most organisms, PII proteins act as the central signal-integrating molecules for controlling nitrogen and/or carbon metabolism, along with some effector molecules (e.g., ATP and 2-oxoglutarate) [5–7]. These molecules bind to the intercommunicating sites of the trimeric PII proteins, forming different PII conformations to control a variety of enzymes, transcription factors, and membrane transport proteins [8–10]. Although PII proteins have highly conserved structures, they interact with various target metabolisms in different organisms. In cyanobacteria and plants, for example, PII proteins can control the activity of *N*-acetyl-L-glutamate kinase (NAGK), and then regulate the metabolism of glutamate towards arginine and polyamines [5,11]. In plant chloroplast and bacteria, acetyl-CoA carboxylases, catalysing the committed and rate-limiting step in fatty acid biosynthesis, are regulated by PII proteins [12,13]. It is believed that PII proteins may also play some uncharacterized roles in plant cells [5,14].

Compared to the study on plants and bacteria, research of PII proteins in microalgae is still in the primary stage. To date, the PII proteins have only been reported in two green microalgae *Chlamydomonas reinhardtii* [15] and *Chlorella variabilis* [14]. It has been reported that PII proteins originate from cyanobacteria and conserve in the evolution of higher plants [5]. Since green algae are in the phylogenetic lineage between cyanobacterial ancestor and higher plants [14], phylogenetic analysis seems to be an efficient approach to verify the newly identified PII protein of the target microalgae, aligned with some reported PII proteins in plants and cyanobacteria. Given the importance of nitrogen metabolism in microalgae, it is essential to characterize more PII proteins and also verify relevant metabolic functions.

The green microalga *Haematococcus pluvialis* is well known due to its extreme capability of producing a large amount of powerful antioxidant-astaxanthin [16–18]. Driven by their nutrition condition, *H. pluvialis* cells have two different physiological traits: (1) in favorable conditions they are in a green motile stage; (2) under stress conditions (especially nitrogen depletion) the green cells will transform into a reddish non-motile resting stage, coupled with astaxanthin accumulation [19–21]. Several key genes related to astaxanthin biosynthesis and stress responding have been cloned and characterized, such as *pds* [22], *CYP97C* [23], *MnSOD* [24], and *TR1* [25]. Despite many reports on astaxanthin accumulation of *H. pluvialis* upon nitrogen depletion [26–28], there is no information of its PII protein and associated genetic transcription information on this unicellular microalga.

In this study, we cloned the full length of the PII signaling protein gene on *H. pluvialis*, analyzed the phylogenetic relationship and structure of this protein, and also investigated its time course-dependent transcriptional regulation. It is anticipated that the results will provide fundamental knowledge on the PII protein of *H. pluvialis*, and also highlight the importance of its potential regulation/interactions for astaxanthin biosynthesis.

## 2. Results

### 2.1. Cloning and Characterization of HpGLB1

A 148 bp cDNA fragment of Hp*GLB1* was obtained from *H. pluvialis* cells by RT-PCR in this study. This fragment was homologous to the *GLB1* of *Chlamydomonas reinhardtii* and *Chlorella variabilis*. The results showed that 5′-RACE PCR generated a 568 bp fragment, and 3′-RACE PCR generated a 702 bp fragment. Alignment assay indicated that the complete cDNA sequence of Hp*GLB1* was 1222 bp, including 621 bp coding sequence (CDS), 103 bp 5′ untranslated region (5′ UTR), and 498 bp 3′ untranslated region (3′ UTR). The length of the open reading frame of Hp*GLB1* in genomic DNA was 2123 bp, containing 7 exons and 6 introns (Figure 1). The length of introns varied from 66 bp to 437 bp. The sequence of Hp*GLB1* has been submitted to NCBI GenBank (Accession number: KT696441).

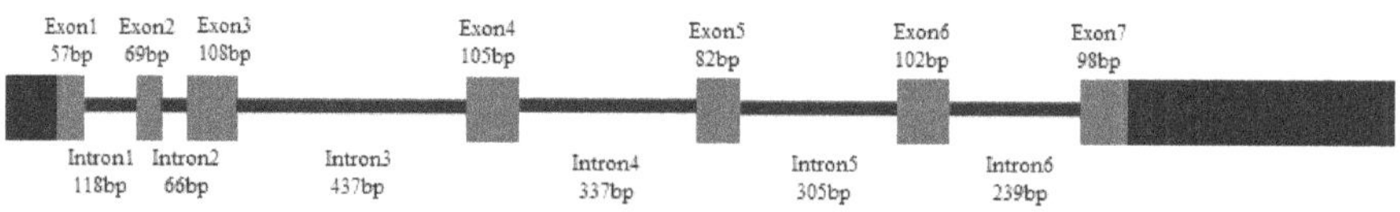


**Figure 1.** The gene structure of Hp*GLB1*. Grey boxes and black solid lines represent exons and introns, respectively. Black boxes represent 5′ and 3′ UTRs.

### 2.2. Characterization of HpPII Protein

Encoded by Hp*GLB1* cDNA, the deduced full-length HpPII protein consisted of 206 amino acid residues (Figure 2). According to the Computer pI/Mw Tool, the calculated values of pI and Mw were

9.53 and 22.4 kDa, respectively. However, the molecular mass was approximately 27 kDa after HpPII was expressed and purified in *E. coli* BL21 (DE3) (Figure 3). This was the sum of the calculated Mw of HpPII (22.4 kDa) and N-terminal tag (4 kDa), and the latter was from the plasmid pET28a used for purification.

```
1      ATGATTGGCACAGTGCTGCAGAAGGGCTACTGTAGGCCCTGCGTAGTGGCTAACAGGTCT
1       M  I  G  T  V  L  Q  K  G  Y  C  R  P  C  V  V  A  N  R  S

61     CAATCGCGACCCTGCGTGGCAAGGCTCTGCCTTCGCACAAGACGTTGCGTCCTTCTCGTA
21      Q  S  R  P  C  V  A  R  L  C  L  R  T  R  R  C  V  L  L  V

121    CGCGCTGGCGAGGACAATGGATCAGCTGCCACACAGCGAGCCACGTATGAGCAGTTGGAA
41      R  A  G  E  D  N  G  S  A  A  T  Q  R  A  T  Y  E  Q  L  E

181    AGCATCAAGGTCAACTTGGCAGCTTTCCCTGAAGTGCAGTTCTTCAGAGTTGAGGCCATC
61      S  I  K  V  N  L  A  A  F  P  E  V  Q  F  F  R  V  E  A  I

241    GTGCGACCATGGCGCCTGGCATTTGTCATTGAGCACCTCAGCAAGGAAGGCATCCGAGGC
81      V  R  P  W  R  L  A  F  V  I  E  H  L  S  K  E  G  I  R  G

301    CTTACTAACACACCCGTCAAGGGCGTAGGCATGCAAGGAGGCTTCCGGGAGCGCTATGCT
101     L  T  N  T  P  V  K  G  V  G  M  Q  G  G  F  R  E  R  Y  A

361    GGCACTGAGTTTGCACGCACCGACCTGGTGGTGAAGGAGAAGATCGACGTGGTGGTGTCA
121     G  T  E  F  A  R  T  D  L  V  V  K  E  K  I  D  V  V  V  S

421    CGCTCCCAGGTGGACATAGTGACGCGCATCATCGCGGCCGCTGCCTACACCGGGGAGATT
141     R  S  Q  V  D  I  V  T  R  I  I  A  A  A  A  Y  T  G  E  I

481    GGGGACGGCAAGATCTTTGTCCACCCAGTGGCAGATGTGGTGCGAGTGCGTACCGCGGAG
161     G  D  G  K  I  F  V  H  P  V  A  D  V  V  R  V  R  T  A  E

541    ACCGGTGCAGTGGCAGAGCGCATGCAGGGCGGCATGTCTGACCAGCAGAGCCTAGCAGGC
181     T  G  A  V  A  E  R  M  Q  G  G  M  S  D  Q  Q  S  L  A  G

601    AAGCCGGAGGCTAGTGGTTAG
201     K  P  E  A  S  G  *
```

**Figure 2.** Nucleotide and deduced amino acid sequence of Hp*GLB1*. The asterisk represents the stop codon.

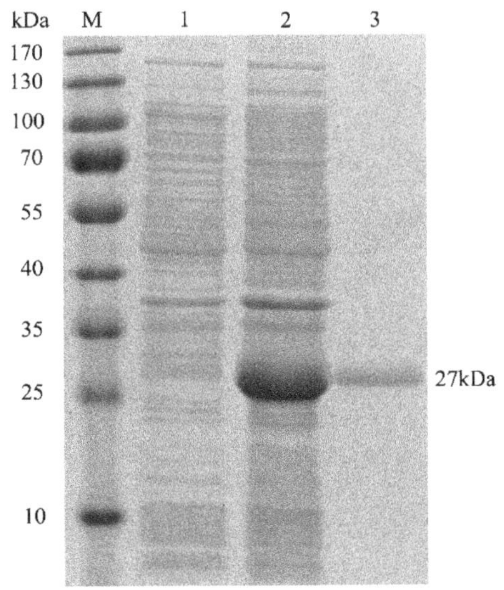


**Figure 3.** Sodium dodecyl sulfate-polyacrylamide gel electrophoresis (SDS-PAGE) analysis of HpPII in *E. coli* BL21 (DE3). M: protein marker; lane 1: total proteins extracted from uninduced pET28a-Hp*GLB1* (no HpPII expression, control); lane 2: total proteins extracted from induced pET28a-Hp*GLB1* (with HpPII expression); lane 3: purified HpPII protein.

### 2.3. Multiple Sequence Alignment and Structural Prediction

The derived HpPII protein was aligned with the sequences of representative PII protein from green algae, plants, cyanobacteria, and bacteria (Figure 4). Like the PII proteins of *C. reinhardtii* and *C. variabilis*, HpPII had N- and C-terminal extensions, which did not exist in prokaryotes. Furthermore, the sequence from residue 74 to 186 encompassed homology to the entire 112 residues of prokaryotic PII proteins. Within the homology region, the identities of HpPII were aligned up to 78% and 63% with microalgae *C. reinhardtii* and *C. variabilis*, respectively. HpPII also had a high similarity with the PII proteins of plants, cyanobacterial, bacterial and red algal, such as: *Arabidopsis thaliana* (55%), *Oryza sativa* (52%), *Solanum lycopersicum* (54%), *Synechocystis* sp. PCC 6803 (50%), *Synechococcus* sp. (47%), *Prochlorococcus marinus* (47%), *Escherichia coli* (45%), *Porphyra umbilicalis* (43%), and *Pyropia yezoensis* (41%).

Two signature patterns (I and II) of extremely high similarity have been proposed at the PROSITE (PS00496 and PS00638) (Figure 4). *Escherichia coli* of proteobacteria has tyrosyl-residue (Tyr-51) in signature pattern I, which is posttranslational modified by uridylylation. However, in HpPII and PII proteins of green algae *C. reinhardtii*, *C. variabilis*, and higher plants, the Tyr-51 residue is substituted by phenylalanyl-residue (Figure 4). Although PII proteins of *Synechococcus* comprise Tyr-51 residue, they are not uridylylated, due to being modified by phosphorylation at seryl-residue (Ser-49) [29]. Nevertheless, the corresponding Ser-49 residue is replaced by threoninyl-residue in HpPII and PII proteins of green algae *C. reinhardtii* and *C. variabilis* (Figure 4). Similar to the other PII proteins, HpPII has T loop, B loop, and C loop (Figure 4). In addition, HpPII also has a consensus sequence of Q loop (R/K, M, Q, G) structure in C-terminal extension (Figure 4), which is comparable to the PII proteins of *C. reinhardtii*, *C. variabilis*, and higher plants. They are known to constitute binding sites of metabolite effectors [8]. These binding residues are highlighted in Figure 4.

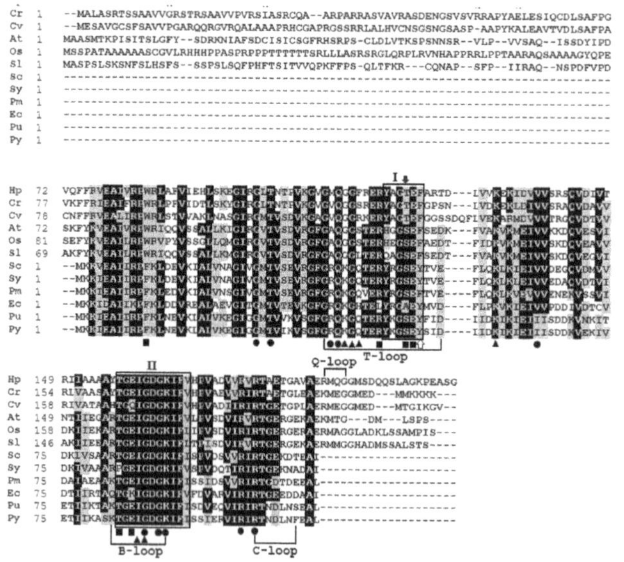


**Figure 4.** Alignment of the amino acid sequences of PII proteins among different organisms: *Haematococcus pluvialis* (Hp; AOO85416), *Chlamydomonas reinhardtii* (Cr; EDO96407.1), *Chlorella variabilis* (Cv; AHW46897.1), *Arabidopsis thaliana* (At; AAC78333.1), *Oryza sativa* (Os; NP_001054562.1), *Solanum lycopersicum* (Sl; NP_001234506.1), *Synechocystis* sp. PCC 6803 (Sc; WP_010873156.1), *Synechococcus* sp. (Sy; AAA27312.1), *Prochlorococcus marinus* (Pm; WP_036930683.1), *Escherichia coli* (Ec; CDZ21367.1), *Porphyra umbilicalis* (Pu; AFC39923.1), *Pyropia yezoensis* (Py; YP_536935.1). Residues in black represent

>60% identity of aligned PII proteins. Amino acids shaded with grey display similar residues. Box I and box II are PII signature patterns I and II. The white arrow indicates the residue of the corresponding uridylylated threonyl-residue in proteobacteria. The black arrow indicates the residue of corresponding phosphorylated serine-residue in cyanobacteria. Black dots, squares, and triangles show the ATP-, NAGK-, and 2KG-binding residues, respectively.

## 2.4. Phylogenetic Analysis

A Neighbor-Joining tree was generated by PII proteins of microalgae, plants, and bacteria via MEGA 6.06 software. There were five clusters: cyanobacteria, red algae, bacteria, green algae, and plants (Figure 5). HpPII protein was grouped with the PII proteins of Chlorophyta algae (*C. reinhardtii*, *C. variabilis*, *Micromonas pusilla*).

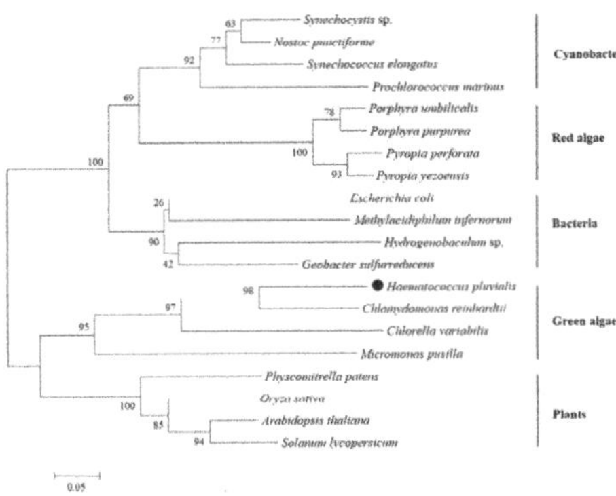


**Figure 5.** Phylogenetic tree of PII proteins in different organisms (the black dot represents the derived HpPII in this study). The phylogenetic tree was constructed by the Neighbor-joining method, and the numbers above the nodes represent the bootstrap values.

## 2.5. Transcription Analysis of HpGLB1 under Nitrogen Deprivation Condition

When *H. pluvialis* cells were exposed to nitrogen starvation, the expression level of Hp*GLB1* mRNA hardly changed from 0 to 24 h ($p > 0.05$) (Figure 6). However, its transcription level increased significantly at 48 h ($p < 0.05$).

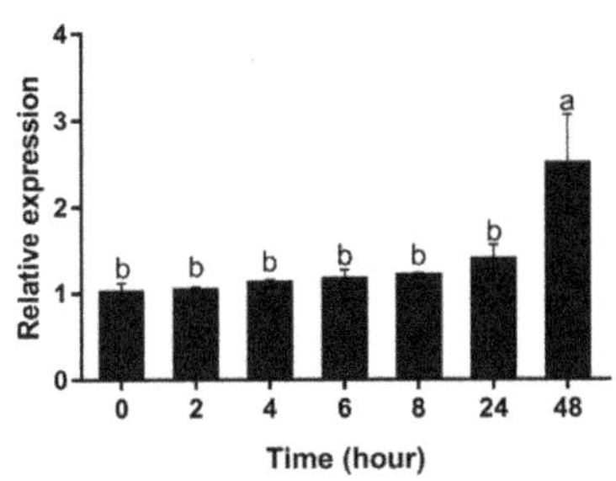


**Figure 6.** Transcription level of Hp*GLB1* under nitrogen deprivation condition. Data in this figure are the mean values of biological triplicates, and the error bar indicates their standard error (SE). Statistical significance differences are represented by different letters at $p < 0.05$. The *H. pluvialis* cells (of $5 \times 10^5$ cells $mL^{-1}$) were resuspended in BBM-N medium at 0 h (control). The transcription level of Hp*GLB1* were measured at 0, 2, 4, 6, 8, 24, 48 h by qRT-PCR. The 18S ribosomal RNA gene was used as the reference gene, and the values were normalized to the transcription level of the control.

### 2.6. Astaxanthin Accumulation Pattern under Nitrogen Deprivation Condition

The content of astaxanthin was maintained below 200 μg $g^{-1}$ in the first 24 h ($p > 0.05$). It started to increase at 48 h, and then continuously accelerated from 48 h (219.02 μg $g^{-1}$) to 96 h (303.25 μg $g^{-1}$) (Figure 7).

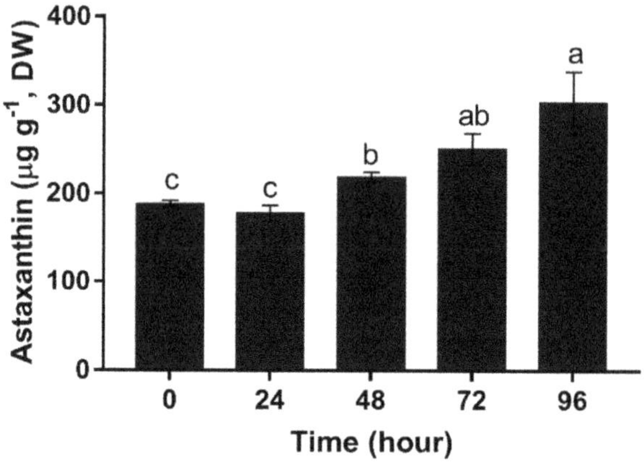


**Figure 7.** Total astaxanthin content of *H. pluvialis* upon nitrogen starvation. Data in this figure are the mean values of biological triplicates, and the error bar indicates their standard error (SE). Statistical significance differences are represented by different letters at $p < 0.05$.

## 3. Discussion

This study is the first to isolate and identify a novel gene of PII protein (Hp*GLB1*) on green algae *H. pluvialis*. The complete cDNA sequence of Hp*GLB1* was 1222 bp, including 621 bp of CDS, 103 bp 5′UTR, and 498 bp 3′UTR (Figure 1). The open reading frame of Hp*GLB1* in genomic DNA contained 7 exons and 6 introns, which is the same as for green microalga *C. variabilis* [14]. It is reported that PII proteins are chloroplast genome-encoded in red algae [6] but in green microalgae they are encoded by nuclear genome [14]. Since *H. pluvialis* is a green alga and has the same exons and introns distribution of *GLB1* as that of *C. variabilis*, the HpPII is deemed nuclear-encoded in *H. pluvialis* as well.

The coding sequence (CDS) of Hp*GLB1* gene encoded a protein of 207 amino acids with 22.4 kDa of calculated weight (Figure 2). When HpPII was expressed in *E. coli* BL21 (DE3), the purified protein Mw was about 27 kDa (Figure 3). Since there was 4 kDa of N-terminal tag, the remaining Mw of protein was similar to the calculated weight of HpPII. In this study, the length of Hp*GLB1* CDS and the Mw of HpPII were similar to the PII proteins of green microalgae *C. reinhardtii* (615 bp, 22 kDa) [15] and *C. variabilis* (630 bp, 22 kDa) [14]. These results indicate that these PII proteins are conserved in green microalgae.

HpPII had a high similarity in the segment of 75 and 185 amino acids to the eukaryotic PII proteins (Figure 4). This was consistent with the phylogenetic analysis of HpPII being in the same cluster with three other green algal PII proteins from *C. reinhardtii*, *C. varLiabilis*, and *M. pusilla*. Since the specific Tyr-51 (uridylylated site) was substituted with phenylalanyl-residue in the signature pattern I region of HpPII (Figure 4), it is indicated that HpPII is not modified by uridylylation. The absence of modification by uridylylation is similar to the literature reports on green algae *C. reinhardtii* and *C. variabilis* [14,15]. The Ser-49 was replaced by a threonyl residue in PII protein (Figure 4). This result is the same as PII protein of *C. reinhardtii*, which has been verified not being modified by phosphorylation [15]. Therefore, it is deduced that HpPII is also not be regulated by phosphorylation.

Different from prokaryotic PII proteins, HpPII had extra N- and C-terminal extensions. As highlighted in green algae and plants, they would represent some additional functions [15,30].

A consensus sequence of Q loop (R/K, M, Q, G) was found in the C-terminal extension of HpPII (Figure 4). Q loop was also observed in PII proteins of green microalgae *C. reinhardtii* and *C. variabilis* [14,15]. As reported in *C. reinhardtii*, Q loop was deemed to play a role in nitrogen assimilation, binding glutamine molecule and forming glutamine-dependent complex with *N*-acetyl-L-glutamate kinase (NAGK) [5]. Furthermore, the NAGK binding sites of HpPII were conserved with *C. reinhardtii* and *C. variabilis* (Figure 4). Hence, HpPII likely behaves as a glutamine sensor through the C-terminal Q loop extension to control the activity of NAGK, and further to regulate nitrogen metabolism. The residues involved in ATP and 2KG binding sites were conserved as well. These results demonstrate that HpPII possibly acts as a signaling protein in nitrogen metabolism. However, a future study is needed to further investigate the contribution of HpPII in nitrogen regulation, targeting the astaxanthin biosynthesis.

The expression of the PII-coding gene was species-specific between different phylogenetic clusters. For example, in cyanobacteria, the expression of PII-coding *glnB* was increased by nitrogen-poor conditions [31]. However, in the higher plant *Arabidopsis thaliana*, *GLB1* expression was not significantly affected by nitrogen [32]. It is worth noting that within the same phylogenetic cluster the *GLB1* expression to nitrogen depletion seems species-specific. The expression of *GLB1* in *H. pluvialis* (Hp*GLB1*) was enhanced under a nitrogen starvation condition. A similar result was also reported on the *GLB1* expression of green microalga *C. reinhardtii* [15]. The result in this study showed that Hp*GLB1* expression was stable in the first 24 h of nitrogen starvation, but increased dramatically at 48 h under nitrogen starvation condition. *C. reinhardtii GLB1* was induced from 0.5 h to 4 h upon nitrogen starvation [15]. In contrast, *C. variabilis' GLB1* expression was independent of nitrogen availability [14]. As these microalgae can produce different secondary metabolic bioproducts upon nitrogen depletion, it is deduced that the *GLB1* expression is associated with different metabolic activities in microalgae.

In this study, the astaxanthin content is correlated to the Hp*GLB1* expression. Similar to the Hp*GLB1* expression, the astaxanthin content was maintained at a low level within the first 24 h. The similar delayed astaxanthin increment was also observed previously in other *H. pluvialis* strains when exposed to stress condition [20]. In terms of the stable Hp*GLB1* expression, this may be associated with the initial physiological condition of cells, as these cells reaching exponential stage likely were exposed to N deficient condition prior to the N starvation test in this study. However, both of Hp*GLB1* expression and astaxanthin content started to increase at 48 h. Then, the astaxanthin content continuously increased after 48 h, even though it was under low light intensity (20 μmol $m^{-2}$ $s^{-1}$). It seems that the upregulation of Hp*GLB1* expression is likely regulated at a posttranslational level. Also, this is elucidated that there may be some physiological connection/interaction between nitrogen metabolism regulation and astaxanthin synthesis on *H. pluvialis* cells. A number of previous literatures have already confirmed that nitrogen condition plays an important role in *H. pluvialis* cells' transformation, growth, and lipid and astaxanthin sythesis; the derived sequence of Hp*GLB1* and the characterization of HpPII in this study will facilitate further understanding on these metabolic pathways at the molecular level. More importantly, our results provide solid and credible bases for HpPII protein study. These will also provide new insight into the HpPII protein and its potential functions on the nitrogen-driven metabolic mechanism for astaxanthin production on *H. pluvialis*.

## 4. Materials and Methods

### 4.1. Microalga Strain and Culture Conditions

A strain of microalga *H. pluvialis* HPH was obtained from the Algae Collection at the College of Ocean and Earth Sciences in Xiamen University, China. The algal cells were cultivated in a 1-L glass vessel (15.5 cm in length and 9.5 cm in diameter) with Bold's Basal Medium (BBM) [33] at 25 °C. The culture was continuously exposed to 20 μmol $m^{-2}$ $s^{-1}$ of light radiation, and 2.5% $CO_2$ of aeration at a flow rate of 0.2 vvm [34]. When reaching the late exponential growth phase (approx. $5 \times 10^5$ cells $mL^{-1}$), the *H. pluvialis* cells were centrifuge-collected at 8000× *g* for 5 min. The cells were rinsed twice with nitrogen-free medium (BBM-N) and resuspended in BBM-N at a density of $5 \times 10^5$ cells $mL^{-1}$. The culture was continued for the trial (with triplicates), under the same cultivation condition. The culture was sampled with 50 mL ($n$ = 3) at 0, 2, 4, 6, 8, 24, and 48 h, respectively. The sampled *H. pluvialis* cells ($3 \times 7$) were collected via centrifugation and used for the subsequent RNA isolation, cDNA synthesis, and transcription analysis.

### 4.2. RNA Isolation and cDNA Synthesis

The collected cells were washed with phosphate-buffered saline buffer (137 mM NaCl, 2.7 mM KCl, 10 mM $Na_2HPO_4$, 1.8 mM $KH_2PO_4$, pH 7.4), re-centrifuged, and immediately frozen in the liquid nitrogen. The frozen biomass was ground into a fine powder using mortar and pestle, which was cold-maintained by adding more liquid nitrogen. 50 mg of algal powder was used for total RNA extraction, which was achieved by a TaKaRa MiniBEST Universal RNA Extraction Kit (Takara Bio Inc., Kusatsu, Japan). The cDNA first-strand synthesis was carried out using High Capacity cDNA Transcription Kits (Applied Biosystems, Foster, CA, USA) according to the manufacturer's protocol.

### 4.3. Gene Cloning and Rapid Amplification of cDNA Ends (RACE)

Partial cDNA of the PII protein coding gene of *H. pluvialis* (Hp*GLB1*) was amplified by reverse transcription PCR (RT-PCR) using degenerate primers GLB1-de-F and GLB1-de-R (Table 1). The PCR conditions for amplification were set as a 5 min polymerase activation step at 94 °C followed by 30 cycles of denaturation at 94 °C for 30 s, annealing at 56 °C for 30 s, extension at 72 °C for 1 min, and a final 10 min of extension at 72 °C. The PCR product was purified with an E.Z.N.A® Gel Extraction Kit (Omega Bio-tek, Norcross, GA, USA), cloned into a pMD™19-T Vector (Takara Bio Inc., Kusatsu, Japan), and then sequenced at the Invitrogen Corporation (Waltham, MA, USA).

In order to generate the full length cDNA of the Hp*GLB1*, this study used the SMART™ RACE cDNA Amplification Kit (Clontech, Mountain View, CA, USA) to perform 3′ and 5′ rapid amplification of the cDNA ends (RACE). Based on the obtained PCR results, the primers GSP1 and GSP2 (Table 1) were designed for the 5′ RACE and 3′ RACE. Subsequently, genomic DNA of *H. pluvialis* was isolated using the cetyltrimethylammounium bromide (CTAB) method [35]. Primers GLB1-F and GLB1-R were used to amplify genomic DNA of Hp*GLB1*. As presented, all the PCR products were purified, cloned, and sequenced.

**Table 1.** Primer sequences used in this study.

| Name | Sequence (5′–3′) | Sequence Information |
|---|---|---|
| GLB1-de-F | AGCGNTACGCN GGCACNGAGTT | Homology cloning primer |
| GLB1-de-R | ATGTCCTCCATGCCNCCCTCCAT | Homology cloning primer |
| GSP1 | GATTACGCCAAGCTTACTATGTCCACCTGGGAGCGTGAC | 3′-RACE primer |
| GSP2 | GATTACGCCAAGCTTCGCTCCCAGGTGGACATAGTGAC | 5′-RACE primer |
| GLB1-F | ATGATTGGCACAGTGCTGCAGAA | Gene cloning primer |
| GLB1-R | CTAACCACTAGCCTCCGGCTTG | Gene cloning primer |
| GLB1-B-F | GGAATTCATGATTGGCACAGTGCTGCAGAA | Gene cloning primer |
| GLB1- B-R | CCCAAGCTTCTAACCACTAGCCTCCGGCTTG | Gene cloning primer |
| GLB1-Q-F | CGCCTGGCATTTGTCATTG | Real-time gene primer |
| GLB1-Q-R | AAACTCAGTGCCAGCATAGCG | Real-time gene primer |
| 18S-F | CAAAGCAAGCCTACGCTCT | Real-time gene primer |
| 18S-R | ATACGAATGCCCCCGACT | Real-time gene primer |

### 4.4. Bioinformatics Analysis

The open reading frame (ORF) was determined by BioEdit software. The deduced amino acid sequence was analyzed with Primer Premier 5. The homology searches for nucleotide and amino acid sequence similarities were conducted with Clustal W and Blast (http://www.ncbi.nlm.nih.gov/). The phylogenetic tree was constructed according to the amino acid sequences of the PII signaling protein by MEGA 6.06 software. Theoretical isoelectric point (pI) and molecular weight (Mw) were calculated with Compute pI/Mw tool (http://web.expasy.org/compute_pi/).

### *4.5. Plasmid Construction*

Based on the nucleotide sequences of start and stop codons of the Hp*GLB1* gene, the gene specific primers GLB1-B-F and GLB1-B-R were used to amplify the full length of the coding region. The PCR performance was programmed as 94 °C for 5 min followed by 30 cycles of denaturation at 94 °C for 30 s, annealing at 65 °C for 30 s, extension at 72 °C for 1 min, and a final 10 min extension at 72 °C. The 5′ and 3′ ends of amplified Hp*GLB1* coding region contain a Bam*HI* and an Eco*RI* restriction site, respectively. The amplified coding region was digested with Bam*HI* and Eco*RI* restriction endonucleases and inserted into the corresponding sites of pET28a for generating the recombinant plasmid pET28a-Hp*GLB1*, and then introduced into *Escherichia coli* BL21 (DE3) for expression.

### *4.6. Expression and Purification of HpPII*

HpPII protein expression and purification were performed according to the previous method [36], with some modifications. Briefly, the *E. coli* BL21 (DE3) harboring pET28a-Hp*GLB1* was cultured in 10 mL of Luria-Bertani (LB) broth with 50 μg $mL^{-1}$ of kanamycin (sigma-Aldrich, St. Louis, MO, USA) at 37 °C and retained on a shaker (200 rpm) overnight. Then, 0.5 mL of the resulting culture was inoculated to 50 mL of LB broth supplemented with 50 μg $mL^{-1}$ kanamycin and incubated under the same condition. When the bacterial concentration reached 0.6 of $OD_{600}$ value, the culture was added with 0.1 mM isopropyl β-D-1-thiogalactopyranoside (IPTG) and incubated for another 6 h. The cells were then harvested by centrifugation (4 °C, 5000× *g*, 10 min) and washed twice with sodium phosphate buffer (200 mM $Na_3PO_4$, 300 mM NaCl, pH 7.4). The pellets were suspended in the sodium phosphate buffer and lysed by sonication (100 W, 60 cycles of 5 s sonication), followed by 5 s incubation on ice. Finally, the soluble lysate was centrifuged (4 °C, 10,000× *g*, 10 min) for collection. The HpPII protein was purified with Ni-NTA His•Bind Resin. The molecular mass and purity, plus the total proteins, were checked by 12% SDS-PAGE.

### *4.7. Expression Analysis of HpGLB1 by qRT-PCR*

A total of 21 cDNA samples derived from the control and nitrogen depleted treatments (triplicates sampling at 0, 2, 4, 6, 8, 24, and 48 h) were used to determine mRNA expressions by real-time quantitative reverse transcriptase PCR (qRT-PCR). The initial samples (0 h) were treated as the control. qRT-PCR was performed on an ABI-7000 System (Applied Biosystems, Foster, CA, USA) using iTaq™ Universal SYBR® Green Supermix (Bio-Rad, Hercules, CA, USA). Primers GLB1-Q-F and GLB1-Q-R were designed to amplify the 120 bp sequence of HpPII for qRT-PCR. The 18S ribosomal RNA gene expression was used as the internal control. Each 20 μL PCR reaction consisted of 10 μL iTaq™ Universal SYBR® Green Supermix, 1 μL of forward and reverse primers (10 μM) each, and 8 μL of cDNA. The thermal cycles were set as stage 1: 95 °C for 10 min; stage 2: 40 cycles of 95 °C for 15 s, and 60 °C for 1 min; and stage 3: 1 cycle of 95 °C for 15 s, 60 °C for 1 min, and 95 °C for 15 s. The qRT-PCR data were analyzed by $2^{-\Delta\Delta CT}$ method based on the cycle threshold (Ct) values.

### *4.8. Astaxanthin Extraction and Quantification*

The freeze-dried cells (10 mg, taken at 0, 24, 48, 72, and 96 h with triplicated) were mixed with 0.5 g glass beads and vortexed for 2 min. Then 5 mL dichloromethane/methanol (1:1, *v*/*v*) was added and vortexed for 1 min. The mixture was centrifuged at 3000× *g* at 4 °C for 3 min, and the supernatant was collected. The pellet was extracted repeatedly with 3 ml dichloromethane/methanol (1:1, *v*/*v*) until it became colorless. The combined supernatant was dried in nitrogen gas, reconstituted with 1 mL of 0.025 M NaOH, and then saponified at 4 °C for 4 h. The pigment extract was diluted with 1 mL of the mix of methanol and acetonitrile/methanol (3:1), and used for astaxanthin quantification.

Separation and identification of astaxanthin was carried out on an Acquity UHPLC system (Waters) equipped with photodiode array (PDA) detector. Samples (5.0 μL) were quantitatively injected into an Acquity UPLC Shield C18 BEH column (2.1 × 100 mm, 1.7 μm particle size; Waters, Milford, WI, USA). The eluents were (A) 50% ($v/v$) acetonitrile in demineralized water; (B) acetonitrile; and (C) ethyl acetate, which all contained 0.10% ($v/v$) formic acid. The flow rate was maintained at 300 μL $min^{-1}$. The program was initiated from 25% A/75% B and then as follows: 0–10 min—linear gradient to 100% B, 10–15 min—isocratic at 100% B, 15–20 min—linear gradient to 87.5% B/12.5% C, 20–21 min—linear gradient to 70% B/30% C, 21–28 min—linear gradient to 100% C, and 28–29 min—isocratic at 100% C. After 29 min, the eluent composition reverted to its initial composition in 1 min followed by an equilibration phase of 12 min. Detection wavelength for UV–vis was adjusted to 460 nm. Since astaxanthin esters were saponified, the free astaxanthin peak represented total astaxanthin. The astaxanthin peak was identified by comparing retention time and spectra with astaxanthin standard (Sigma-Aldrich, St. Louis, MO, USA). The quantification of astaxanthin was determined by a calibration curve obtained with astaxanthin standard.

### 4.9. Statistical Analysis

The experiment was conducted with biological triplicates, and the data measurement was performed in triplicates. The data in the figure were showed as mean ± SE in this study. The statistical differences of transcription level between sampling times were detected by one-way ANOVA analysis using IBM SPSS Statistics 24, with nitrogen starvation time as variance and relative transcription level as dependent variables, followed by Duncan's test with a significant level of 0.05.

**Acknowledgments:** This work was supported by funds from the National Natural Science Foundation of China (No. 31071488), the National Marine Commonweal Research Program, China (No. 201205020-2), and the State Key Program of National Natural Science Foundation of China (No. 21336009).

**Author Contributions:** Ruijuan Ma and Yinghua Lu conceived and designed the experiments; Ruijuan Ma performed the experiments; Ruijuan Ma, Yan Li, and Yinghua Lu analyzed the data; Yinghua Lu contributed reagents/materials/analysis tools; Ruijuan Ma and Yan Li wrote the paper.

**Conflicts of Interest:** The authors declare no conflict of interest.

## References

1. Mulders, K.J.M.; Weesepoel, Y.; Bodenes, P.; Lamers, P.P.; Vincken, J.P.; Martens, D.E.; Gruppen, H.; Wijffels, R.H. Nitrogen-depleted *Chlorella zofingiensis* produces astaxanthin, ketolutein and their fatty acid esters: A carotenoid metabolism study. *J. Appl. Phycol.* **2015**, *27*, 125–140. [CrossRef]
2. Tran, N.P.; Park, J.K.; Lee, C.G. Proteomics analysis of proteins in green alga *Haematococcus lacustris* (chlorophyceae) expressed under combined stress of nitrogen starvation and high irradiance. *Enzyme Microb. Technol.* **2009**, *45*, 241–246. [CrossRef]
3. Longworth, J.; Wu, D.Y.; Huete-Ortega, M.; Wright, P.C.; Vaidyanathan, S. Proteome response of *Phaeodactylum tricornutum*, during lipid accumulation induced by nitrogen depletion. *Algal Res.* **2016**, *18*, 213–224. [CrossRef] [PubMed]
4. Commichau, F.M.; Forchhammer, K.; Stulke, J. Regulatory links between carbon and nitrogen metabolism. *Curr. Opin. Microbiol.* **2006**, *9*, 167–172. [CrossRef] [PubMed]
5. Chellamuthu, V.R.; Ermilova, E.; Lapina, T.; Luddecke, J.; Minaeva, E.; Herrmann, C.; Hartmann, M.D.; Forchhammer, K. A widespread glutamine-sensing mechanism in the plant kingdom. *Cell* **2014**, *159*, 1188–1199. [CrossRef] [PubMed]
6. Uhrig, R.G.; Ng, K.K.S.; Moorhead, G.B.G. PII in higher plants: A modern role for an ancient protein. *Trends Plant Sci.* **2009**, *14*, 505–511. [CrossRef] [PubMed]
7. Ninfa, A.J.; Jiang, P. PII signal transduction proteins: Sensors of alpha-ketoglutarate that regulate nitrogen metabolism. *Curr. Opin. Microbiol.* **2005**, *8*, 168–173. [CrossRef] [PubMed]
8. Huergo, L.F.; Chandra, G.; Merrick, M. PII signal transduction proteins: Nitrogen regulation and beyond. *FEMS Microbiol. Rev.* **2013**, *37*, 251–283. [CrossRef] [PubMed]

9. Helfmann, S.; Lu, W.; Litz, C.; Andrade, S.L.A. Cooperative binding of MgATP and MgADP in the trimeric PII protein GlnK2 from *Archaeoglobus fulgidus*. *J. Mol. Biol.* **2010**, *402*, 165–177. [CrossRef] [PubMed]
10. Gerhardt, E.C.; Araujo, L.M.; Ribeiro, R.R.; Chubatsu, L.S.; Scarduelli, M.; Rodrigues, T.E.; Monteiro, R.A.; Pedrosa, F.O.; Souza, E.M.; Huergo, L.F. Influence of the ADP/ATP ratio, 2-oxoglutarate and divalent ions on *Azospirillum brasilense* PII protein signalling. *Microbiology* **2012**, *158*, 1656–1663. [CrossRef] [PubMed]
11. Burillo, S.; Luque, I.; Fuentes, I.; Contreras, A. Interactions between the nitrogen signal transduction protein PII and *N*-acetyl glutamate kinase in organisms that perform oxygenic photosynthesis. *J. Bacteriol.* **2004**, *186*, 3346–3354. [CrossRef] [PubMed]
12. Bourrellier, A.B.F.; Valot, B.; Guillot, A.; Ambard-Bretteville, F.; Vidal, J.; Hodges, M. Chloroplast acetyl-CoA carboxylase activity is 2-oxoglutarate-regulated by interaction of PII with the biotin carboxyl carrier subunit. *Proc. Natl. Acad. Sci. USA* **2010**, *107*, 502–507. [CrossRef] [PubMed]
13. Gerhardt, E.C.M.; Rodrigues, T.E.; Muller-Santos, M.; Pedrosa, F.O.; Souza, E.M.; Forchhammer, K.; Huergo, L.F. The bacterial signal transduction protein GlnB regulates the committed step in fatty acid biosynthesis by acting as a dissociable regulatory subunit of acetyl-CoA carboxylase. *Mol. Microbiol.* **2015**, *95*, 1025–1035. [CrossRef] [PubMed]
14. Minaeva, E.; Ermilova, E. Sequencing and expression analysis of the gene encoding PII signal protein in *Chlorella variabilis* NC64A. *J. Plant Biochem. Biotechnol.* **2015**, *2*. [CrossRef]
15. Ermilova, E.; Lapina, T.; Zalutskaya, Z.; Minaeva, E.; Fokina, O.; Forchhammer, K. PII signal transduction protein in *Chlamydomonas reinhardtii*: Localization and expression pattern. *Protist* **2013**, *164*, 49–59. [CrossRef] [PubMed]
16. Gwak, Y.; Hwang, Y.S.; Wang, B.B.; Kim, M.; Jeong, J.; Lee, C.G.; Hu, Q.; Han, D.X.; Jin, E. Comparative analyses of lipidomes and transcriptomes reveal a concerted action of multiple defensive systems against photooxidative stress in *Haematococcus pluvialis*. *J. Exp. Bot.* **2014**, *65*, 4317–4334. [CrossRef] [PubMed]
17. Lorenz, R.T.; Cysewski, G.R. Commercial potential for *Haematococcus microalgae* as a natural source of astaxanthin. *Trends Biotechnol.* **2000**, *18*, 160–167. [CrossRef]
18. Regnier, P.; Bastias, J.; Rodriguez-Ruiz, V.; Caballero-Casero, N.; Caballo, C.; Sicilia, D.; Fuentes, A.; Maire, M.; Crepin, M.; Letourneur, D.; et al. Astaxanthin from *Haematococcus pluvialis* prevents oxidative stress on human endothelial cells without toxicity. *Mar. Drugs* **2015**, *13*, 2857–2874. [CrossRef] [PubMed]
19. Hong, M.E.; Hwang, S.K.; Chang, W.S.; Kim, B.W.; Lee, J.; Sim, S.J. Enhanced autotrophic astaxanthin production from *Haematococcus pluvialis* under high temperature via heat stress-driven Haber-Weiss reaction. *Appl. Microbiol. Biotechnol.* **2015**, *99*, 5203–5215. [CrossRef] [PubMed]
20. Gao, Z.Q.; Meng, C.X.; Chen, Y.C.; Ahmed, F.; Mangott, A.; Schenk, P.M.; Li, Y. Comparison of astaxanthin accumulation and biosynthesis gene expression of three *Haematococcus pluvialis* strains upon salinity stress. *J. Appl. Phycol.* **2015**, *27*, 1853–1860. [CrossRef]
21. Scibilia, L.; Girolomoni, L.; Berteotti, S.; Alboresi, A.; Ballottari, M. Photosynthetic response to nitrogen starvation and high light in *Haematococcus pluvialis*. *Algal Res.* **2015**, *12*, 170–181. [CrossRef]
22. Liang, C.W.; Zhao, F.Q.; Meng, C.X.; Tan, C.P.; Qin, S. Molecular cloning, characterization and evolutionary analysis of phytoene desaturase (*pds*) gene from *Haematococcus pluvialis*. *World J. Microbiol. Biotechnol.* **2006**, *22*, 59–64. [CrossRef]
23. Cui, H.L.; Yu, X.N.; Wang, Y.; Cui, Y.L.; Li, X.Q.; Liu, Z.P.; Qin, S. Gene cloning and expression profile of a novel carotenoid hydroxylase (CYP97C) from the green alga *Haematococcus pluvialis*. *J. Appl. Phycol.* **2014**, *26*, 91–103. [CrossRef]
24. Wang, J.X.; Sommerfeld, M.; Hu, Q. Cloning and expression of isoenzymes of superoxide dismutase in *Haematococcus pluvialis* (chlorophyceae) under oxidative stress. *J. Appl. Phycol.* **2011**, *23*, 995–1003. [CrossRef]
25. Zheng, Y.H.; Tao, M.; Li, Z.; Hu, Z.L. Cloning and characterization of selenoprotein thioredoxin reductase gene in *Haematococcus pluvialis*. *Algal Res.* **2016**, *17*, 97–104. [CrossRef]
26. Recht, L.; Topfer, N.; Batushansky, A.; Sikron, N.; Gibon, Y.; Fait, A.; Nikoloski, Z.; Boussiba, S.; Zarka, A. Metabolite profiling and integrative modeling reveal metabolic constraints for carbon partitioning under nitrogen starvation in the green algae *Haematococcus pluvialis*. *J. Biol. Chem.* **2014**, *289*, 30387–30403. [CrossRef] [PubMed]
27. Fabregas, J.; Dominguez, A.; Alvarez, D.G.; Lamela, T.; Otero, A. Induction of astaxanthin accumulation by nitrogen and magnesium deficiencies in *Haematococcus pluvialis*. *Biotechnol. Lett.* **1998**, *20*, 623–626. [CrossRef]

28. Boussiba, S.; Bing, W.; Yuan, J.P.; Zarka, A.; Chen, F. Changes in pigments profile in the green alga *Haeamtococcus pluvialis* exposed to environmental stresses. *Biotechnol. Lett.* **1999**, *21*, 601–604. [CrossRef]
29. Forchhammer, K.; Tandeau de Marsac, N. The PII protein in the cyanobacterium *Synechococcus* sp. strain PCC 7942 is modified by serine phosphorylation and signals the cellular N-status. *J. Bacteriol.* **1994**, *176*, 84–91. [CrossRef] [PubMed]
30. Mizuno, Y.; Moorhead, G.B.G.; Ng, K.K.S. Structural basis for the regulation of *N*-acetylglutamate kinase by PII in *Arabidopsis thaliana*. *J. Biol. Chem.* **2007**, *282*, 35733–35740. [CrossRef] [PubMed]
31. Forchhammer, K. Global carbon/nitrogen control by PII signal transduction in cyanobacteria: From signals to targets. *FEMS Microbiol. Rev.* **2004**, *28*, 319–333. [CrossRef] [PubMed]
32. Ferrario-Mery, S.; Bouvet, M.; Leleu, O.; Savino, G.; Hodges, M.; Meyer, C. Physiological characterisation of *Arabidopsis* mutants affected in the expression of the putative regulatory protein PII. *Planta* **2005**, *223*, 28–39. [CrossRef] [PubMed]
33. Olguin, E.J.; Dorantes, E.; Castillo, O.S.; Hernandez-Landa, V.J. Anaerobic digestates from vinasse promote growth and lipid enrichment in *Neochloris oleoabundans* cultures. *J. Appl. Phycol.* **2015**, *27*, 1813–1822. [CrossRef]
34. Xie, Y.P.; Ho, S.H.; Chen, C.N.N.; Chen, C.Y.; Ng, I.S.; Jing, K.J.; Chang, J.S.; Lu, Y.H. Phototrophic cultivation of a thermo-tolerant *Desmodesmus* sp. for lutein production: Effects of nitrate concentration, light intensity and fed-batch operation. *Bioresour. Technol.* **2013**, *144*, 435–444. [CrossRef] [PubMed]
35. Varela-Alvarez, E.; Andreakis, N.; Lago-Leston, A.; Pearson, G.A.; Serrao, E.A.; Procaccini, G.; Duarte, C.M.; Marba, N. Genomic DNA isolation from green and brown algae (*Caulerpales* and *Fucales*) for microsatellite library construction. *J. Phycol.* **2006**, *42*, 741–745. [CrossRef]
36. Ng, I.S.; Chi, X.Q.; Wu, X.M.; Bao, Z.W.; Lu, Y.H.; Chang, J.S.; Ling, X.P. Cloning and expression of Cel8A from *Klebsiella pneumoniae* in *Escherichia coli* and comparison to *cel* gene of *Cellulomonas uda*. *Biochem. Eng. J.* **2013**, *78*, 53–58. [CrossRef]

marine drugs

Communication

# Cloning and Functional Characterization of a Lycopene β-Cyclase from Macrophytic Red Alga *Bangia fuscopurpurea*

**Tian-Jun Cao [1], Xing-Qi Huang [1], Yuan-Yuan Qu [1], Zhong Zhuang [1], Yin-Yin Deng [2,*] and Shan Lu [1,*]**

1 State Key Laboratory of Pharmaceutical Biotechnology, School of Life Sciences, Nanjing University, Nanjing 210023, China; ctjsw@126.com (T.-J.C.); hl_david@msn.com (X.-Q.H.); someoneqyy@hotmail.com (Y.-Y.Q.); njcs2006@126.com (Z.Z.)
2 Jiangsu Institute of Oceanology and Marine Fisheries, Nantong 226007, China
* Correspondence: dengyinyin@yeah.net (Y.-Y.D.); shanlu@nju.edu.cn (S.L.); Tel.: +86-513-85228265 (Y.-Y.D); +86-25-8968-6217 (S.L.)

Academic Editor: Tatsuya Sugawara
Received: 10 November 2016; Accepted: 29 March 2017; Published: 11 April 2017

**Abstract:** Lycopene cyclases cyclize the open ends of acyclic lycopene (ψ,ψ-carotene) into β- or ε-ionone rings in the crucial bifurcation step of carotenoid biosynthesis. Among all carotenoid constituents, β-carotene (β,β-carotene) is found in all photosynthetic organisms, except for purple bacteria and heliobacteria, suggesting a ubiquitous distribution of lycopene β-cyclase activity in these organisms. In this work, we isolated a gene (*BfLCYB*) encoding a lycopene β-cyclase from *Bangia fuscopurpurea*, a red alga that is considered to be one of the primitive multicellular eukaryotic photosynthetic organisms and accumulates carotenoid constituents with both β- and ε-rings, including β-carotene, zeaxanthin, α-carotene (β,ε-carotene) and lutein. Functional complementation in *Escherichia coli* demonstrated that BfLCYB is able to catalyze cyclization of lycopene into monocyclic γ-carotene (β,ψ-carotene) and bicyclic β-carotene, and cyclization of the open end of monocyclic δ-carotene (ε,ψ-carotene) to produce α-carotene. No ε-cyclization activity was identified for BfLCYB. Sequence comparison showed that BfLCYB shares conserved domains with other functionally characterized lycopene cyclases from different organisms and belongs to a group of ancient lycopene cyclases. Although *B. fuscopurpurea* also synthesizes α-carotene and lutein, its enzyme-catalyzing ε-cyclization is still unknown.

**Keywords:** *Bangia fuscopurpurea*; red algae; lycopene cyclase; carotenoid; metabolism

## 1. Introduction

Carotenoids are one of the largest groups of natural pigments in land plants, algae, bacteria including cyanobacteria and photosynthetic bacteria, archaea, fungi and animals [1,2]. More than 750 carotenoids, not including their *cis*- and *trans*- isomers as distinct compounds, have been isolated and characterized from natural sources [3]. Carotenoids not only serve as essential pigments for photosynthesis and photoprotection in photosynthetic organisms, but also comprise many of the yellow, orange and red colors in flowers and fruits that attract pollinators and seed dispersers in higher plants [4]. Abscisic acid and strigolactones are also derivatives of carotenoids [5].

Higher plants generally have very similar leaf carotenoid profiles, with lutein, β-carotene (β,β-carotene), violaxanthin and neoxanthin being the major components. They may also accumulate specific carotenoids, such as capsanthin in pepper fruits [6], β-carotene in carrot tubers [7] and astaxanthin in *Adonis aestivalis* flowers [8]. In the past 20 years, most of the metabolic reactions in the biosynthesis of carotenoids have been deciphered, thanks to the cloning and characterization of genes encoding metabolic enzymes from different organisms [9]. In plants, the plastidic methylerythritol phosphate (MEP) pathway utilizes glyceraldehyde-3-phosphate and pyruvate to synthesize the 5-carbon ($C_5$) isopentenyl diphosphate (IPP) and its isomer dimethylallyl diphosphate (DMAPP). A geranylgeranyl diphosphate (GGPP, $C_{20}$) synthase (CrtE) catalyzes the condensation of these two substrates to produce GGPP as a precursor for the biosynthesis of carotenoids and other diterpenoids. Phytoene synthase (CrtB) utilizes two molecules of GGPP to produce phytoene and thus directs the metabolic flux from GGPP into carotenoid metabolic pathway. Phytoene is then converted to acyclic lycopene (ψ,ψ-carotene) by the combination of desaturases CrtP and CrtQ in eukaryotes, but only CrtI in prokaryotes [2,10].

Cyclization of lycopene by lycopene cyclases (LCY) is a crucial bifurcation step in the carotenogenic pathway (Figure 1). Each open end of lycopene can be cyclized by lycopene β-cyclase (LCYB) or ε-cyclase (LCYE) to produce β- or ε-ionone rings, respectively (Figure 1). The formation of two β-ionone rings results in the production of β-carotene and its derivatives (e.g., zeaxanthin and neoxanthin), whereas the combination of β- and ε-ionone rings forms α-carotene (β,ε-carotene) and its derivate lutein (Figure 1). Carotenoids with two ε-ionone rings are uncommon in most plants, with lactucaxanthin in lettuce (*Lactuca sativa*) being one of the exceptions [11]. Carotenes with at least one unsubstituted β-ionone ring can be utilized by the human body to produce vitamin A. Therefore, an increase in the activity of LCYB can direct the metabolic flux from lycopene to the production of this group of so-called provitamin A carotenes to increase the nutritional value of crops [12,13].

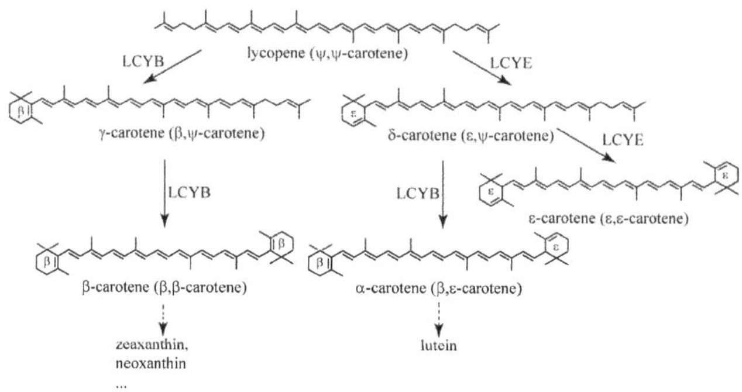


**Figure 1.** Cyclization of lycopene catalyzed by lycopene β-cyclase (LCYB) and ε-cyclase (LCYE).

Cumulative studies on carotenoids in red algae have uncovered broad compositional diversity [2,14–16]. Red algae can also be divided into three groups according to their specific carotenoid compositions [15,16]. Although carotenoid constituents with an ε-ionone ring are missing in some primitive species, β-carotene is found in all red algae, suggesting a ubiquitous distribution of the β-cyclization activity in carotenoid biosynthesis [16,17]. *LCYB* of *Cyanidioschyzon merolae* is also the only *LCY* gene from the entire red alga phylum that has been characterized [16,17].

The red algal genus *Bangia* Lyngb., although probably not monophyletic, is among the earliest multicellular eukaryotic photosynthetic organisms [18]. Carotenoids with both β- and ε-rings, including β-carotene, zeaxanthin, α-carotene and lutein, are accumulated in *Bangia* [15,16]. In this work, we isolated a gene encoding an LCY from *Bangia fuscopurpurea*, and functional characterization revealed that this enzyme has only β-cyclization activity.

## 2. Results

### 2.1. *BfLCYB* Is an Intron-Less *LCY* Gene

From the alignment of functionally characterized LCYs from different organisms, we designed two pairs of degenerate primers. Nested PCR using these primers resulted in amplicons of different sizes. After cloning and sequencing these fragments, a 262 base pair (bp) fragment was found to have high similarity to known LCYs (data not shown). Based on the sequence of this fragment, we successfully amplified a 2218 bp genomic DNA fragment and a 1932 bp complementary DNA (cDNA) encoding a putative protein with 643 amino acid residues, designated BfLCYB in this work (Figure 2A). Sequence alignment demonstrated that *BfLCYB* does not possess any introns. BfLCYB shares 38% sequence identity with the characterized LCYB from *C. merolae* [17]. The sequence of *BfLCYB* has been deposited in GenBank under the accession number KX943552.

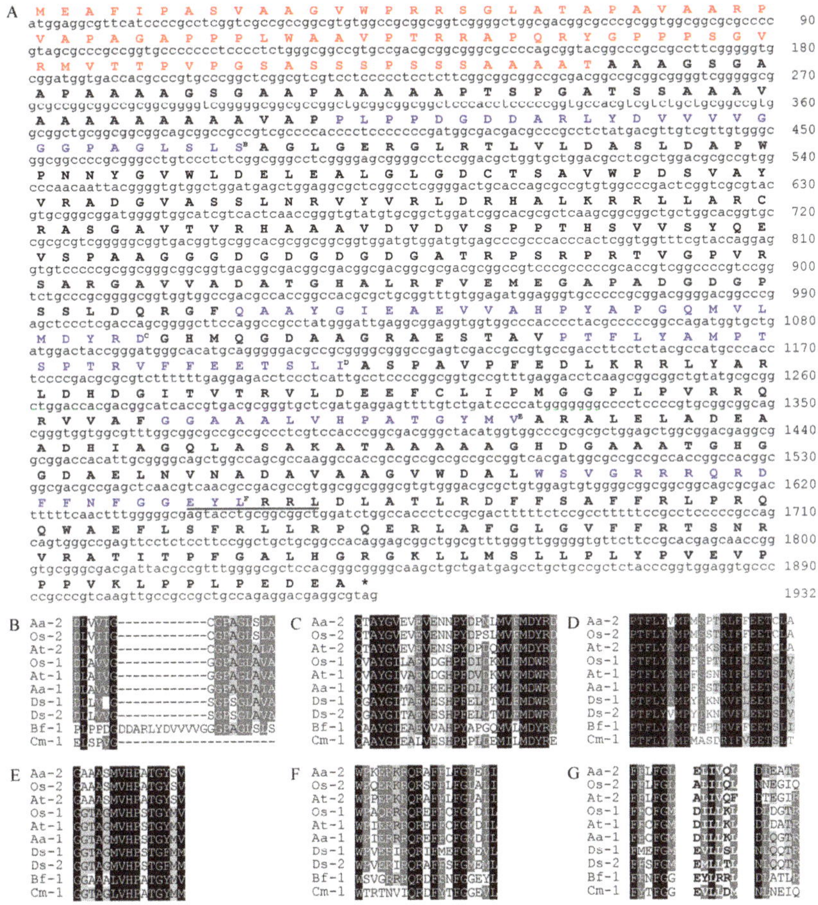

**Figure 2.** Sequence analysis of BfLCYB. (**A**) *BfLCYB* complementary DNA (cDNA) sequence and its deduced amino acid sequence. Sequence of the predicted N-terminal chloroplast transit peptide is in red. (**B–F**) Sequence comparison of the NAD(P)/FAD-binding domain (**B**) and other conserved domains (**C–F**). (**G**) Sequence comparison of the six-amino-acid region (in bold) that determines ring numbers of LCYE and flanking regions. The conserved domains in BfLCYB are in blue and indicated by superscript letters, and the ring number-determination region in BfLCYB is underlined in (**A**). Sequences of functionally characterized LCYEs from *Adonis aestivalis* (Aa-2, AAK07430.1), *Oryza sativa* (Os-2, XP_015622198.1) and *Arabidopsis thaliana* (At-2, NP_200513.1) and LCYBs from *A. aestivalis* (Aa-1, AAK07432.1), *O. sativa* (Os-1, XP_015627234.1), *A. thaliana* (At-1, NP_187634.1), *Dunaliella salina* (Ds-1, ACA34344.1; Ds-2, ANY98896.1), *Cyanidioschyzon merolae* (Cm-1, XP_005536481.1) and *Bangia fuscopurpurea* (Bf-1, KX943552) were compared.

Five conserved domains, including an NAD(P)/FAD-binding domain, were previously reported for functionally characterized LCYs [2,10]. These domains were also identified in BfLCYB (Figure 2). However, the NAD(P)/FAD-binding domain in BfLCYB is interrupted by an apparent insertion of 12 additional amino acid residues (Figure 2B). Cunningham and Gantt also identified a six-amino-acid region in LCYE that contributes to the determination of ring numbers in lycopene ε-cyclization products [11]. We found that this region is located immediately downstream from the fifth conserved domain in LCYs (Figure 2F,G). However, in BfLCYB, this region has much lower similarity to other LCYs than either its upstream or downstream flanking regions (Figure 2G).

### 2.2. *BfLCYB Localizes in Chloroplasts*

From online analysis using TargetP, ChloroP and YLoc, BfLCYB was predicted to have a chloroplast localization [19–21]. Its N-terminal chloroplast transit peptide predicted by TargetP and ChloroP is in square brackets in Figure 2A. To experimentally confirm its localization, the open reading frame (ORF) of *BfLCYB* was fused in frame to the 5′-end of the yellow fluorescent protein (*YFP*) in pA7-YFP. Due to the lack of a method to stably transform *Bangia*, we transiently expressed the BfLCYB-YFP fusion protein in *Arabidopsis thaliana* leaf protoplasts [22]. The YFP signal of the fusion protein was solely found in chloroplasts and merged nicely with chlorophyll autofluorescence, thus confirming its predicted chloroplast localization (Figure 3).

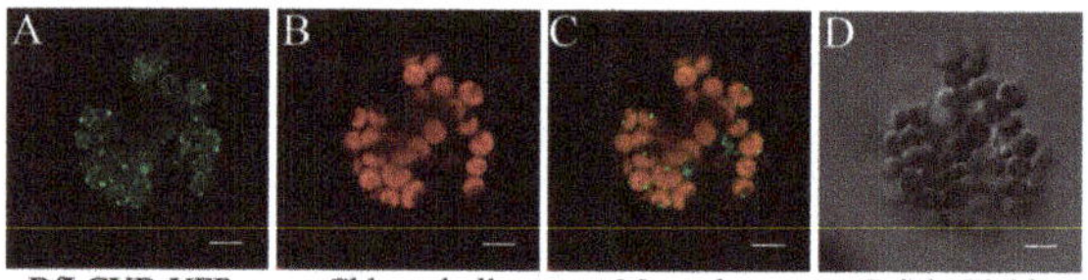


**Figure 3.** Subcellular localization of BfLCYB. The BfLCYB yellow fusion protein (BfLCYB-YFP) was transiently expressed in *Arabidopsis thaliana* protoplasts, and observed under a confocal microscope. Signals from the fusion protein (**A**) and chlorophyll autofluorescence (**B**) were merged (**C**) for a comparison with the protoplast observed under bright field (**D**). The transient expression experiment was repeated independently five times and only one representative protoplast is shown here. Scale bar, 5 μm.

### 2.3. *BfLCYB Is a Bicyclic Lycopene β-Cyclase*

To determine the enzymatic activity of BfLCYB, we used a bacterial pigment complementation system [23]. The plasmid pAC-LCY contains the genes from *Pantoea* for geranylgeranyl diphosphate synthase (*CrtE*), phytoene synthase (*CrtB*) and phytoene desaturase (*CrtI*). *Escherichia coli* transformed with pAC-LCY can utilize endogenous IPP and DMAPP to produce lycopene, which turns the cells pink (Figure 4A). When we co-transformed *E. coli* with both pMAL-BfLCYB and pAC-LCY, the positive transformants harboring both plasmids had an orange color (Figure 4B). High-performance liquid chromatography (HPLC) analysis revealed an accumulation of both the monocyclic γ-carotene (β,ψ-carotene) and the bicyclic β-carotene in these orange cells (Figure 4B). We did not find accumulation of either ε-carotene (ε,ε-carotene) or δ-carotene (ε,ψ-carotene) in the co-transformed bacteria. These results suggested that BfLCYB is able to cyclize one or both ends of lycopene into β-ionone rings but not ε-ionone rings.

We further confirmed the function of BfLCYB by co-transforming pMAL-BfLCYB with pAC-EPSILON. In addition to the genes carried by pAC-LCY, pAC-EPSILON carries the gene from lettuce for LCYE, which catalyzes the ε-cyclization at both ends of lycopene [11]. *E. coli* cells transformed with pAC-EPSILON accumulated ε-carotene, as expected (Figure 4C). When this plasmid was co-transformed with pMAL-BfLCYB, HPLC analysis revealed the accumulation of both α-carotene and β-carotene, together with a significantly reduced level of ε-carotene and a small amount of γ-carotene (Figure 4D). This demonstrated that BfLCYB can also β-cyclize the open end of monocyclic δ-carotene.

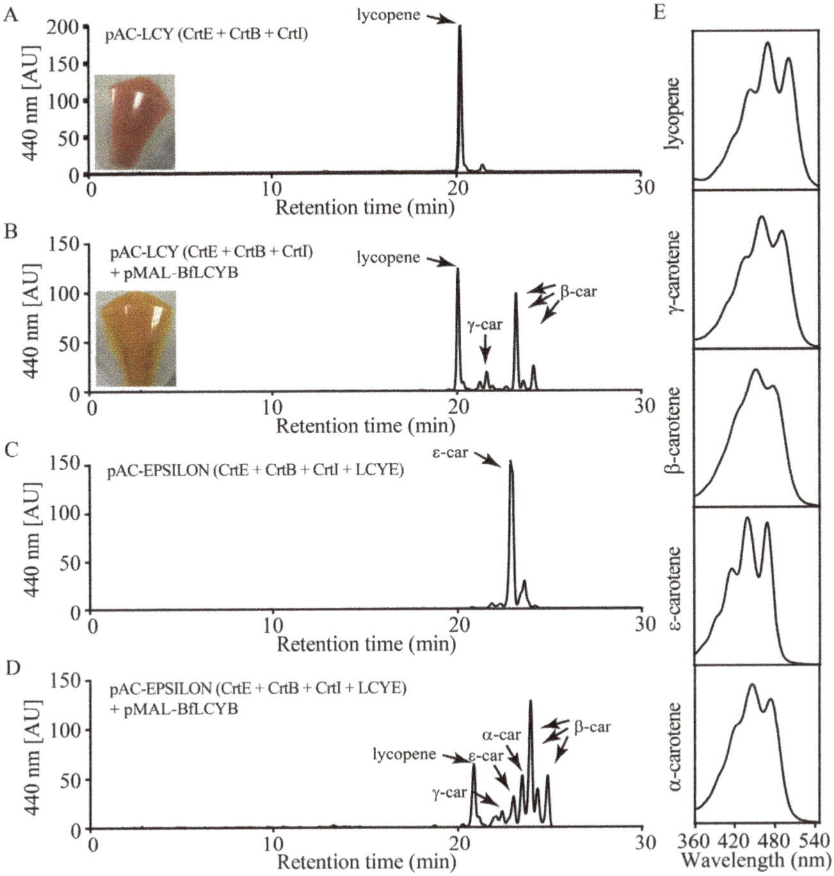


**Figure 4.** Functional characterization of BfLCYB. (**A**) Production of lycopene in *Escherichia coli* cells transformed with pAC-LCY. (**B**) Production of γ-carotene and β-carotene in *E. coli* cells transformed with both pAC-LCY and pMAL-BfLCY1. (**C**) Production of ε-carotene in *E. coli* cells transformed with pAC-EPSILON. (**D**) Production of α-, ε-, γ- and β-carotene in *E. coli* cells transformed with both pAC-EPSILON and pMAL-BfLCYB. (**E**) UV-visible absorption spectra of different carotenoid products. Insets in (**A**,**B**) showed different colors of the pelleted bacterial cells accumulating lycopene and β-carotene, respectively.

### 2.4. *BfLCYB Is an Ancient Type of Lycopene Cyclase in Plants*

Using sequences of functionally characterized LCYs, we identified LCY homologs in heterokonts and red algae of which full genomes have been sequenced. For heterokonts, in the genome of either *Ectocarpus siliculosus* (brown alga) or *Phaeodactylum tricornutum* (diatom), there is only a single gene encoding an LCY homolog (Figure 5A). However, for the multicellular red alga *Chondrus crispus*, our search identified two genes for putative LCYs in its genome (Figure 5A). Our phylogenetic analysis divided LCYs and their homologs into five major clades. Clade I comprises members from single-cell red algae, including the LCYB from *Cyanidioschyzon merolae* and a homolog from *Galdieria sulphuraria*. Clade II has members from multicellular red algae, including BfLCYB in this study and the two homologs from *Chondrus crispus*. Clade IV has LCY homologs from heterokonts. The other two clades, III and V, are LCYBs and LCYEs, respectively, from the green lineage (green algae and land plants) (Figure 5A).

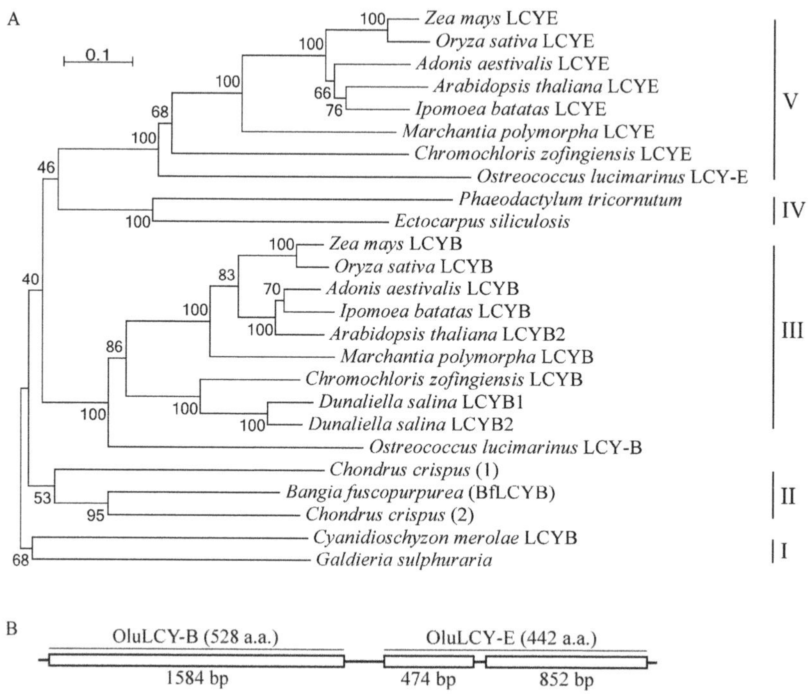


**Figure 5.** Phylogenetic analysis of plant LCYs. (**A**) Sequences of the functionally characterized LCYs from *Zea mays* (LCYB, NP_001169155.1; LCYE, NP_001146840.1), *Oryza sativa* (LCYB, XP_015627234.1; LCYE, XP_015622198.1), *Arabidopsis thaliana* (LCYB, NP_187634.1; LCYE, NP_200513.1), *Ipomoea batatas* (LCYB, AGL44392.1; LCYE, BAW34178.1), *Adonis aestivalis* (LCYB, AAK07430.1; LCYE, AAK07432.1), *Marchantia polymorpha* (LCYB, BAO27799.1; LCYE, BAO27800.1), *Chromochloris zofingiensis* (LCYB, CBH31263.1; LCYE, CCG06343.1), *Dunaliella salina* (LCYB1, ACA34344.1; LCYB2, ANY98896.1), *Ostreococcus lucimarinus* (LCYB, XP_001422489.1; LCYE, XP_001422490.1), *Bangia fuscopurpurea* (LCYB, KX943552) and *Cyanidioschyzon merolae* (XP_005536481.1), and their homologs from *Ectocarpus siliculosus* (CBN78004.1), *Phaeodactylum tricornutum* (XP_002176612.1), *Chondrus crispus* (1, XP_005713399.1; 2, XP_005715292.1) and *Galdieria sulphuraria* (XP_005708098.1) are analyzed using MEGA 6. Scale bar, 10% sequence divergence. (**B**) Structure of the *Ostreococcus lucimarinus LCY* gene that has two tandemly arranged regions for LCYB (OluLCY-B) and LCYE (OluLCY-E) activities.

A recently study by Blatt et al. [24] reported a gene (*OluLCY*) encoding a lycopene β-cyclase/lycopene ε-cyclase/light-harvesting complex-fusion protein from the green alga *Ostreococcus lucimarinus*. Our sequence analysis of this gene identified two coding regions (*OluLCY-B* and *OluLCY-E*) that are tandemly arranged, separated by a short intron (Figure 5B). In our phylogenetic tree, peptides encoded by these two regions are grouped with functionally characterized LCYBs and LCYEs, respectively (Figure 5B).

## 3. Discussion

In green lineage organisms (green algae and land plants), carotenoids with both β- and ε-rings are found, and separate LCYBs and LCYEs catalyzing the corresponding cyclization steps are also characterized from different species [1,25]. Our phylogenetic analysis clustered LCYBs and LCYEs from these organisms into separate clades (IV and V, Figure 5), showing a divergence of lycopene β- and ε-cyclase activities (Clades IV and V, Figure 5A) before the origination of the green lineage. This is also supported by the identification of the *OluLCY* gene, which has two tandemly arranged regions for peptides with LCYB and LCYE activities, respectively, suggesting probable gene duplication and divergence.

In red algae and heterokonts, although their carotenoid constituents have been studied, there is not sufficient information on their LCYs [2,26]. In both single-cell red algae and heterokonts, only β-carotene and its derivatives were found. Consistently, LCYs from *Cyanidioschyzon merolae* possess only β-cyclization activity, and studies in diatoms found no sequences showing similarity to *LCYE* genes either [16,26–28]. Therefore, it is reasonable to postulate that members of Clades I and IV in Figure 5A are all LCYBs.

For multicellular red algae, which accumulates carotenoids with both β- and ε-rings, studies on their LCYs are lacking. In this work, we studied lycopene cyclization in *Bangia fuscopurpurea*, which is considered to be one of the simplest macrophytic red algae. Our molecular cloning and functional characterization results demonstrated that *B. fuscopurpurea* has a functional LCYB (BfLCYB), which is capable of catalyzing β-cyclization at both ends of the acyclic lycopene or at either end of the monocyclic γ-carotene or δ-carotene. By scanning the genome sequence of another multicellular red alga, *Chondrus crispus*, we found two homolog genes for LCYs. These remind us of the possible existence of another as-yet-unidentified LCYE homolog responsible for ε-cyclization in *B. fuscopurpurea*. Cloning and functional characterization of this second LCY from *B. fuscopurpurea* should help to decipher carotenoid metabolism and its evolution in red algae.

## 4. Materials and Methods

### 4.1. Material and Growth Condition

Thalli of *Bangia fuscopurpurea* (#Bangia-MC1F, Jiangsu Institute of Oceanology and Marine Fisheries, Nantong, Jiangsu, China) were initially collected from cultivated populations in Putian, Fujian Province, China (24°59′ N, 118°48′ E) [29] and then subcultured in sterilized seawater in our laboratory. Ambient air filtered through a 0.45 μm filter was gently supplied to the culture. The growth conditions were 16 °C with a 12 h/12 h light/dark regime under 80 μmol $m^{-2}$ $s^{-1}$ irradiance. Seawater was changed every week.

### 4.2. Sequence Analysis

Sequences of functionally characterized LCYs from *Zea mays*, *Oryza sativa*, *Arabidopsis thaliana*, *Ipomoea batatas*, *Adonis aestivalis*, *Marchantia polymorpha*, *Chromochloris zofingiensis*, *Dunaliella salina*, *Ostreococcus lucimarinus* and *Cyanidioschyzon merolae* were downloaded from GenBank, and their homologs in *Ectocarpus siliculosus*, *Phaeodactylum tricornutum*, *Chondrus crispus* and *Galdieria sulphuraria* were searched in GenBank using the blastP algorithm with the sequences of known LCYs as queries. Sequence alignment was performed using ClustalW in MEGA 6 [30]. A bootstrap (1000 replicates) neighbor-joining phylogenetic tree was also generated using MEGA 6. 4.3. Molecular manipulation.

Total RNA was isolated from *Bangia fuscopurpurea* material as previously described [31] and treated with DNase (Promega, Madison, WI, USA) to eliminate any genomic DNA contamination. For rapid amplification of cDNA ends (RACE), the SMARTer RACE cDNA Amplification Kit (TaKaRa, Shiga, Japan) was used to synthesize a full-length cDNA pool.

According to the alignment of the functionally characterized LCYs, two pairs of degenerate primers (DF1/DR1, DF2/DR2, sequences of all primers used in this work are listed in Table 1) were designed from the conserved regions [32], and used for amplifying a corresponding fragment from the *B. fuscopurpurea* cDNA pool. After two rounds of nested PCR, amplified products were subcloned and sequenced by GenScript (Nanjing, China). The sequences of the cloned fragments were queried against GenBank using the tblastX algorithm to analyze their similarities to known or predicted *LCY* genes.

**Table 1.** Primers used in this work.

| Primer | Sequence (5′-3′) [1,2] |
|---|---|
| DF1 | MMNAAYTAYGGNKKNTGGBWNGAYGAR |
| DR1 | GGRTGNACSADNSHNGCNGYNSCNCCNAWNSC |
| DF2 | ATGBTNYTNATGGAYKDNMGNGA |
| DR2 | RNVNNCCNCCNAYNGGDATNWV |
| RR1 | GGCCCCCCCATGGGGATCAGACAAAACTC |
| RR2 | CAGACAAAACTCCTCATCGAGCACCCGC |
| RR3 | CACCCGCGTCACGGTGATGCCGTC |
| RF1 | GGTGCTGATGGACTACCGGGATGG |
| RF2 | GGGATGGGCACATGCAGGGGGAC |
| RF3 | GGACGCCGCGGGGCGGGCCGAGTC |
| HF | ATGGAGGCGTTCATCCCCGCCTC |
| ER | CTACGCCTCGTCCTCTGGCAGCGG |
| gHF | AATCAGTCAGTTTCGGTGATCTTTC |
| gER | CCGCCGAAACGGCCTACCCCT |
| SpeI-LCY-F | GACTAGTATGGAGGCGTTCATCCCCGCCTC |
| SpeI-LCY-R | GGGACTAGTCTACGCCTCGTCCTCTGGCAGCGG |
| BamHI-LCY-F | GGATCCATGGAGGCGTTCATCCCCGCCTC |
| BamHI-LCY-R | GGGGATCCCTACGCCTCGTCCTCTGGCAGCGG |

[1] Codes for degenerate nucleotides are M (A/C), R (A/G), W (A/T), S (C/G), Y (C/T), K (G/T), V (A/C/G), H (A/C/T), D (A/G/T), B (C/G/T) and N (any nucleotide); [2] Restriction sites are underlined.

Using the sequences that had high similarity to known *LCY*s, three pairs of primers (RF1/RR1, RF2/RR2 and RF3/RR3) were designed for 5′- and 3′-RACEs, respectively, using the SMARTer RACE kit. After three rounds of nested PCR, two overlapping fragments were amplified from the cDNA pool, subcloned and sequenced. The sequences were assembled into a putative cDNA sequence that contained both translation initiation and stop codons. We then designed an additional pair of primers (HF and ER) from the assembled sequence to amplify the full-length ORF from the cDNA pool for confirming the existence of this transcript (named *BfLCYB*).

To study the genomic sequence of *BfLCYB*, genomic DNA was extracted according to Yang et al. [31]. Primers HF and ER were used to amplify its coding region. A Genome Walking Kit (TaKaRa) was used to amplify the 5′ and 3′ flanking sequences. All amplicons were purified, subcloned, sequenced and assembled. Two primers (gHF and gER) were used to amplify from the genomic DNA for confirming the assembled genomic DNA sequence of *BfLCYB*.

### 4.3. Subcellular Localization

Prediction of the subcellular localization of BfLCYB was performed using TargetP [19], YLoc [20] and ChloroP [21].

For studying the subcellular localization of BfLCYB, protoplasts isolated from *Arabidopsis thaliana* mesophyll were transiently transformed [22]. The full-length ORF of *BfLCYB* was amplified with primers HF-SpeI and ER-SpeI to incorporate restriction sites. The amplicon was purified, digested with SpeI, and subcloned into pA7-YFP (kindly provided by Dr. Hongquan Yang). The transformed protoplasts were cultured in 24-well plates at room temperature for 16 h in the dark. Fluorescence signal from the BfLCYB-YFP fusion protein was observed using a confocal laser scanning microscope (Olympus FluoView FV1000, Tokyo, Japan). The transient expression experiments were repeated independently at least three times.

### 4.4. Functional Characterization

To assess the catalytic activity of BfLCYB, we used the pigment complementation system [23]. Two vectors, pAC-LCY and pAC-EPSILON were provided by Dr. Cunningham. The full-length ORF of *BfLCYB* was amplified with primers HF-BamHI and ER-BamHI to incorporate restriction sites.

The amplified product was subcloned into pMAL-C5X (New England BioLabs, Ipswich, MA, USA) to produce pMAL-BfLCYB. pMAL-BfLCYB was co-transformed with pAC-LCy, or pAC-EPSILON, into *E. coli* BL21(DE3) cells (New England BioLabs, Ipswich, MA, USA). After overnight growth at 37 °C on Luria-Bertani (LB), plants containing 34 μg $mL^{-1}$ chloramphenicol and 100 μg $mL^{-1}$ ampicillin, positive co-transformed single colonies were picked and used to inoculate 8 mL LB medium containing the same antibiotics. One mL of the overnight culture was transferred to 100 mL of the same medium and grown at 37 °C until reaching an OD600 of 0.30. Protein expression was induced by adding isopropyl β-thiogalactopyranoside (IPTG) to a final concentration of 50 μmol $L^{-1}$. After growing for 16 h at 18 °C, bacterial cells were harvested by centrifugation and resuspended in 80% (*v*/*v*) acetone to extract the pigments.

### 4.5. Carotenoid Analysis

Separation of carotenoid components was performed by reverse-phase HPLC [33]. A Waters 2695 separation module and 2998 photodiode array detector (PDA) were used with a Spherisorb ODS2 column (5 μm, 4.6 mm × 250 mm, Waters, Milford, MA, USA). A 45 min gradient of ethyl acetate (0% to 100%) in acetonitrile-water-triethylamine (9:1:0.01, *v/v/v*) at a flow rate of 1 mL $min^{-1}$ was used, and the absorbance of the eluent at 440 nm was monitored [34]. The identity of the carotenoid in each peak was further confirmed by their UV/visible spectra recorded by the PDA detector.

**Acknowledgments:** This study was supported by the State Key Basic Research Project of China (#2013CB127004) to S. Lu. We thank Elisabeth Gantt and Francis X. Cunningham Jr. (University of Maryland) for the bacterial complementation vectors, and Hongquan Yang (Shanghai Jiaotong University) for the pA7-YFP vector.

**Author Contributions:** T.-J.C., Y.-Y.D. and S.L. conceived and designed the experiments; T.-J.C., X.-Q.H., Y.-Y.D., Y.-Y.Q. and Z.Z. performed the experiments; T.-J.C., X.-Q.H. and S.L. analyzed the data; S.L. wrote the paper.

**Conflicts of Interest:** The authors declare no conflict of interest.

## References

1. Cazzonelli, C.I.; Pogson, B.J. Source to sink: Regulation of carotenoid biosynthesis in plants. *Trends Plant Sci.* **2010**, *15*, 266–274. [CrossRef] [PubMed]
2. Takaichi, S. Carotenoids in algae: Distributions, biosyntheses and functions. *Mar. Drugs* **2011**, *9*, 1101–1118. [CrossRef] [PubMed]
3. Britton, G.; Liaaen-Jensen, S.; Pfander, H. *Carotenoids: Handbook*; Birkhäuser: Basel, Switzerland, 2004.
4. Cazzonelli, C.I. Carotenoids in nature: Insights from plants and beyond. *Funct. Plant Biol.* **2011**, *38*, 833–847. [CrossRef]
5. Al-Babili, S.; Bouwmeester, H.J. Strigolactones, a novel carotenoid-derived plant hormone. *Annu. Rev. Plant Biol.* **2015**, *66*, 161–186. [CrossRef] [PubMed]
6. Tian, S.L.; Li, L.; Shah, S.N.M.; Gong, Z.H. The relationship between red fruit colour formation and key genes of capsanthin biosynthesis pathway in *Capsicum annuum*. *Biol. Plant.* **2015**, *59*, 507–513. [CrossRef]
7. Clotault, J.; Peltier, D.; Berruyer, R.; Thomas, M.; Briard, M.; Geoffriau, E. Expression of carotenoid biosynthesis genes during carrot root development. *J. Exp. Bot.* **2008**, *59*, 3563–3573. [CrossRef] [PubMed]
8. Cunningham, F.X., Jr.; Gantt, E. Elucidation of the pathway to astaxanthin in the flowers of *Adonis aestivalis*. *Plant Cell* **2011**, *23*, 3055–3069. [CrossRef] [PubMed]
9. Ruiz-Sola, M.Á.; Rodríguez-Concepción, M. Carotenoid biosynthesis in Arabidopsis: A colorful pathway. *Arabidopsis Book* **2012**, *10*, e0158. [CrossRef] [PubMed]
10. Sandmann, G. Molecular evolution of carotenoid biosynthesis from bacteria to plants. *Physiol. Plant.* **2002**, *116*, 431–440. [CrossRef]
11. Cunningham, F.X., Jr.; Gantt, E. One ring or two? Determination of ring number in carotenoids by lycopene ε-cyclases. *Proc. Natl. Acad. Sci. USA* **2001**, *98*, 2905–2910. [CrossRef] [PubMed]
12. Harjes, C.E.; Rocheford, T.R.; Bai, L.; Brutnell, T.P.; Kandianis, C.B.; Sowinski, S.G.; Stapleton, A.E.; Vallabhaneni, R.; Williams, M.; Wurtzel, E.T.; et al. Natural genetic variation in *Lycopene Epsilon Cyclase* tapped for maize biofortification. *Science* **2008**, *319*, 330–333. [CrossRef] [PubMed]

13. Moise, A.R.; Al-Babili, S.; Wurtzel, E.T. Mechanistic aspects of carotenoid biosynthesis. *Chem. Rev.* **2014**, *114*, 164–193. [CrossRef] [PubMed]
14. Römer, S. Carotenoids in higher plants and algae. In *The Chloroplast: From Molecular Biology to Biotechnology*; Argyroudi-Akoyunoglou, J.H., Senger, H., Eds.; Springer: Dordrecht, The Netherlands, 1999; pp. 217–223.
15. Schubert, N.; García-Mendoza, E.; Pacheco-Ruiz, I. Carotenoid composition of marine red algae. *J. Phycol.* **2006**, *42*, 1208–1216. [CrossRef]
16. Takaichi, S.; Yokoyama, A.; Uchida, H.; Murakami, A. Carotenogenesis diversification in phylogenetic lineages of Rhodophyta. *J. Phycol.* **2016**, *52*, 329–338. [CrossRef] [PubMed]
17. Cunningham, F.X., Jr.; Lee, H.; Gantt, E. Carotenoid biosynthesis in the primitive red alga *Cyanidioschyzon merolae*. *Eukaryot. Cell* **2007**, *6*, 533–545. [CrossRef] [PubMed]
18. Butterfield, N.J. *Bangiomorpha pubescens* n. gen., n. sp.: Implications for the evolution of sex, multicellularity, and the Mesoproterozoic/Neoproterozoic radiation of eukaryotes. *Paleobiology* **2000**, *26*, 386–404. [CrossRef]
19. Emanuelsson, O.; Nielsen, H.; Brunak, S.; von Heijne, G. Predicting subcellular localization of proteins based on their N-terminal amino acid sequence. *J. Mol. Biol.* **2000**, *300*, 1005–1016. [CrossRef] [PubMed]
20. Briesemeister, S.; Rahnenfuhrer, J.; Kohlbacher, O. YLoc—An interpretable web server for predicting subcellular localization. *Nucleic Acids Res.* **2010**, *38*, W497–W502. [CrossRef] [PubMed]
21. Emanuelsson, O.; Nielsen, H.; Von Heijne, G. ChloroP, a neural network-based method for predicting chloroplast transit peptides and their cleavage sites. *Protein Sci.* **1999**, *8*, 978–984. [CrossRef] [PubMed]
22. Yoo, S.-D.; Cho, Y.-H.; Sheen, J. *Arabidopsis* mesophyll protoplasts: A versatile cell system for transient gene expression analysis. *Nat. Protoc.* **2007**, *2*, 1565–1572. [CrossRef] [PubMed]
23. Cunningham, F.X., Jr.; Gantt, E. A portfolio of plasmids for identification and analysis of carotenoid pathway enzymes: *Adonis aestivalis* as a case study. *Photosynth. Res.* **2007**, *92*, 245–259. [CrossRef] [PubMed]
24. Blatt, A.; Bauch, M.E.; Porschke, Y.; Lohr, M. A lycopene β-cyclase/lycopene ε-cyclase/light-harvesting complex-fusion protein from the green alga *Ostreococcus lucimarinus* can be modified to produce α-carotene and β-carotene at different ratios. *Plant J.* **2015**, *82*, 582–595. [CrossRef] [PubMed]
25. Lohr, M.; Im, C.S.; Grossman, A.R. Genome-based examination of chlorophyll and carotenoid biosynthesis in *Chlamydomonas reinhardtii*. *Plant Physiol.* **2005**, *138*, 490–515. [CrossRef] [PubMed]
26. Bertrand, M. Carotenoid biosynthesis in diatoms. *Photosynth. Res.* **2010**, *106*, 89–102. [CrossRef] [PubMed]
27. Coesel, S.; Oborník, M.; Varela, J.; Falciatore, A.; Bowler, C. Evolutionary origins and functions of the carotenoid biosynthetic pathway in marine diatoms. *PLoS ONE* **2008**, *3*, e2896. [CrossRef] [PubMed]
28. Kuczynska, P.; Jemiola-Rzeminska, M.; Strzalka, K. Photosynthetic pigments in diatoms. *Mar. Drugs* **2015**, *13*, 5847–5881. [CrossRef] [PubMed]
29. Wang, W.-J.; Zhu, J.-Y.; Xu, P.; Xu, J.-R.; Lin, X.-Z.; Huang, C.-K.; Song, W.-L.; Peng, G.; Wang, G.-C. Characterization of the life history of *Bangia fuscopurpurea* (Bangiaceae, Rhodophyta) in connection with its cultivation in China. *Aquaculture* **2008**, *278*, 101–109. [CrossRef]
30. Tamura, K.; Stecher, G.; Peterson, D.; Filipski, A.; Kumar, S. MEGA6: Molecular Evolutionary Genetics Analysis version 6.0. *Mol. Biol. Evol.* **2013**, *30*, 2725–2729. [CrossRef] [PubMed]
31. Yang, L.-E.; Jin, Q.-P.; Xiao, Y.; Xu, P.; Lu, S. Improved methods for basic molecular manipulation of the red alga *Porphyra umbilicalis* (Rhodophyta: Bangiales). *J. Appl. Phycol.* **2012**, *25*, 245–252. [CrossRef]
32. Ramos, A.; Coesel, S.; Marques, A.; Rodrigues, M.; Baumgartner, A.; Noronha, J.; Rauter, A.; Brenig, B.; Varela, J. Isolation and characterization of a stress-inducible *Dunaliella salina Lcy-β* gene encoding a functional lycopene β-cyclase. *Appl. Microbiol. Biotechnol.* **2008**, *79*, 819–828. [CrossRef] [PubMed]
33. Yang, L.-E.; Huang, X.-Q.; Hang, Y.; Deng, Y.-Y.; Lu, Q.-Q.; Lu, S. The P450-type carotene hydroxylase PuCHY1 from *Porphyra* suggests the evolution of carotenoid metabolism in red algae. *J. Integr. Plant Biol.* **2014**, *56*, 902–915. [CrossRef] [PubMed]
34. Norris, S.R.; Barrette, T.R.; DellaPenna, D. Genetic dissection of carotenoid synthesis in Arabidopsis defines plastoquinone as an essential component of phytoene desaturation. *Plant Cell* **1995**, *7*, 2139–2149. [CrossRef] [PubMed]

*marine drugs*

MDPI

Article

# Carotenoid Profiling of a Red Seaweed *Pyropia yezoensis*: Insights into Biosynthetic Pathways in the Order Bangiales

**Jiro Koizumi [1], Naoki Takatani [1], Noritoki Kobayashi [1], Koji Mikami [2], Kazuo Miyashita [2], Yumiko Yamano [3], Akimori Wada [3], Takashi Maoka [4] and Masashi Hosokawa [2,*]**

1 Graduate School of Fisheries Sciences, Hokkaido University, Hakodate 041-8611, Japan; j_koizumi@eis.hokudai.ac.jp (J.K.); n-takatani@eis.hokudai.ac.jp (N.T.); air-norsquare.el-bor@ezweb.ne.jp (N.K.)

2 Faculty of Fisheries Sciences, Hokkaido University, Hakodate 041-8611, Japan; komikami@fish.hokudai.ac.jp (K.M.); kmiya@fish.hokudai.ac.jp (K.M.)

3 Laboratory of Organic Chemistry for Life Science, Kobe Pharmaceutical University, Kobe 658-8558, Japan; y-yamano@kobepharma-u.ac.jp (Y.Y.); a-wada@kobepharma-u.ac.jp (A.W.)

4 Research Institute for Production Development, 15 Shimogamo, Morimoto Cho, Sakyoku, Kyoto 606-0805, Japan; maoka@mbox.kyoto-inet.or.jp

* Correspondence: hoso@fish.hokudai.ac.jp; Tel.: +81-138-40-5530

Received: 17 October 2018; Accepted: 30 October 2018; Published: 1 November 2018

**Abstract:** Carotenoids are natural pigments that contribute to light harvesting and photo-protection in photosynthetic organisms. In this study, we analyzed the carotenoid profiles, including mono-hydroxy and epoxy-carotenoids, in the economically valuable red seaweed *Pyropia yezoensis*, to clarify the detailed biosynthetic and metabolic pathways in the order Bangiales. *P. yezoensis* contained lutein, zeaxanthin, α-carotene, and β-carotene, as major carotenoids in both the thallus and conchocelis stages. Monohydroxy intermediate carotenoids for the synthesis of lutein with an ε-ring from α-carotene, α-cryptoxanthin (β,ε-caroten-3′-ol), and zeinoxanthin (β,ε-caroten-3-ol) were identified. In addition, β-cryptoxanthin, an intermediate in zeaxanthin synthesis from β-carotene, was also detected. We also identified lutein-5,6-epoxide and antheraxanthin, which are metabolic products of epoxy conversion from lutein and zeaxanthin, respectively, by LC-MS and $^1$H-NMR. This is the first report of monohydroxy-carotenoids with an ε-ring and 5,6-epoxy-carotenoids in Bangiales. These results provide new insights into the biosynthetic and metabolic pathways of carotenoids in red seaweeds.

**Keywords:** *Pyropia yezoensis*; α-cryptoxanthin; zeinoxanthin; lutein-5,6-epoxide; antheraxanthin; carotenoid synthesis pathway; red seaweed

## 1. Introduction

Carotenoids are yellow-orange tetraterpenoid pigments generated from C4 isoprenoid units. To date, more than 700 carotenoids have been identified in nature. Carotenoids are synthesized in plants, algae (unicellular type), seaweeds (macrophytic type), bacteria, and fungi, but not animals. In photosynthetic organisms, carotenoids contribute to photosynthesis in light-harvesting antenna complexes in the thylakoid membrane of chloroplasts. They also play an important role in photo-protection from damage by reactive oxygen species under excess light energy [1]. In addition, carotenoids have various health benefits in humans [2]. For example, β-carotene, α-carotene, and β-cryptoxanthin have pro-vitamin A activity [3]. Lutein and zeaxanthin decrease the risk of age-related macular degeneration [4,5].

*Pyropia yezoensis*, belonging to Bangiales (Rhodophyta), is one of the most economically valuable marine foods in East Asia. In Japan, approximately 300,000–350,000 tons (wet weight) of thalli are produced every year [6]. Moreover, *P. yezoensis* is valuable for studies of seaweed biology owing to its heteromorphic life cycle, characterized by independent macroscopic leafy gametophyte (thallus) and microscopic, filamentous sporophyte (conchocelis) stages [7]. The genome sequence of *P. yezoensis* has been determined [8] and it is readily cultivated at a laboratory scale; accordingly, it is a useful model organism for studies of red seaweeds [9].

In most red seaweeds, major carotenoids include α-carotene, β-carotene, lutein, and zeaxanthin [10–12]. Lutein consists of two hydroxyl groups with α-carotene composed of a β-ring and ε-ring, and thus, has an asymmetrical structure. In terrestrial plants, lutein is produced by two biosynthetic pathways via α-cryptoxanthin (β,ε-caroten-3′-ol) and zeinoxanthin (β,ε-caroten-3-ol) from α-carotene by carotene hydroxylases, such as CYP97A, CYP97C, and BHY [13]. In the genome of *Porphyra umbilicalis*, which belongs to Bangiales, *CYP97A*, *CYP97C*, and *BHY* genes have not been found [14]. PuCHY$_1$ belonging to the CYP97B subfamily shows β-hydroxylation activity in the synthesis of zeaxanthin from β-carotene in *Po. umbilicalis* [15]. Moreover, intermediate monohydroxy carotenoids in lutein synthesis from α-carotene with an ε-ring are not well known (Figure 1). These results suggest that Bangiales have characteristic biosynthetic pathways different from those of land plants. Therefore, it is interesting to investigate the lutein synthesis pathway in *P. yezoensis*, including intermediate carotenoids.

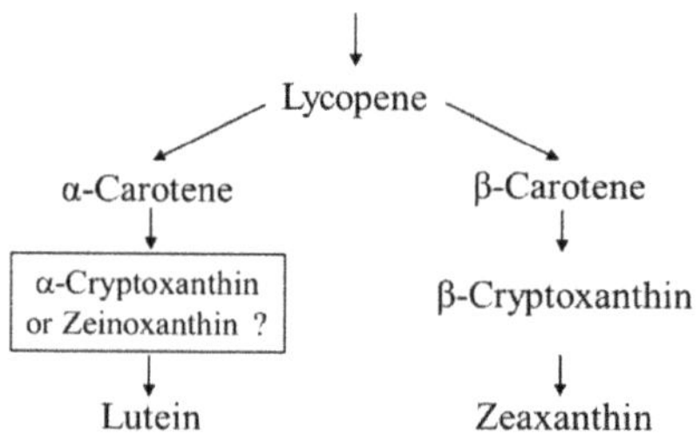


**Figure 1.** Occurrence of carotenoids in *Pyropia yezoensis*.

We recently found a gene candidate for zeaxanthin epoxidase (ZEP), which catalyzes the conversion from zeaxanthin to violaxanthin, in *P. yezoensis* [16]. Abscisic acid (ABA), a phytohormone, has also been detected in *P. yezoensis* [17]. Genes involved in ABA biosynthesis have also been found in the genomes of Bangiales. ABA attenuates oxidative stress under desiccation in seaweeds [18]. These findings suggest that intermediates in the biosynthetic pathway of ABA from zeaxanthin might be found in Bangiales. However, owing to the lack of information about epoxy-carotenoids, their identification is necessary to clarify the ABA synthetic pathway in this order.

In the present study, we analyzed intermediate carotenoids to clarify the carotenoid synthetic pathways in *P. yezoensis*. Based on the identification of monohydroxy-carotenoids and epoxy-carotenoids, we proposed novel carotenoid metabolic pathways in *P. yezoensis*.

## 2. Results

### 2.1. Lutein, Zeaxanthin, and α/β-Carotene Contents in *P. yezoensis*

Total lipids were extracted from *P. yezoensis* at the conchocelis stage, and major carotenoids were analyzed by high-performance liquid chromatography (HPLC) at 450 nm (Figure 2). Carotenoids were identified by comparing the HPLC retention times and absorption spectra with those of their authentic standards. Lutein, zeaxanthin, α-carotene, and β-carotene were detected as major carotenoids in the total lipids from *P. yezoensis*. Lutein was the most abundant carotenoid at 3.46 ± 0.57 mg/g dry weight and 1.44 ± 0.03 mg/g dry weight in the cultured conchocelis and the aquacultured thallus of

P. yezoensis, respectively. The contents of these carotenoids were greater in the conchocelis than in the thallus of P. yezoensis (Figure 3).

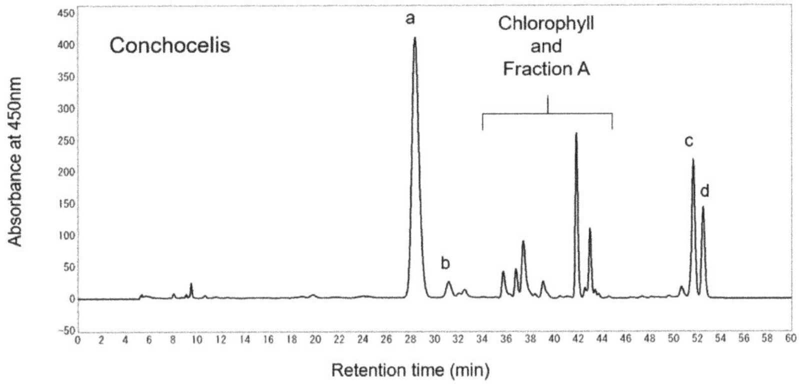


**Figure 2.** Chromatogram of total lipids extracted from the conchocelis of *Pyropia yezoensis*. Carotenoids were measured using a high-performance liquid chromatography (HPLC) system equipped with a diode array detector at 450 nm. Two Develosil C30-UG-5 columns were connected for the analysis. The mobile phase was methanol for 20 min, followed by dichloromethane, increased linearly from 0% to 50% over 20 min, and methanol: Dichloromethane (1:1, *v*/*v*) for an additional 20 min. The flow rate was 1.0 mL/min, and the sample injection volume was 50 μL. a, lutein; b, zeaxanthin; c, α-carotene; d, β-carotene.

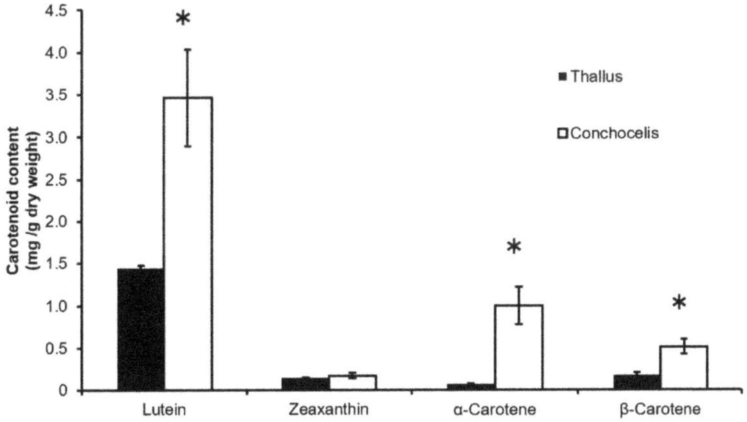


**Figure 3.** Carotenoid contents in the thallus and conchocelis of *Pyropia yezoensis*. Values are the means ± SD ($n$ = 3). Asterisks show significance by the Student's $t$-test (* $p < 0.05$).

## *2.2. Isolation and Identification of Monohydroxy-Carotenoids in P. yezoensis*

We observed a yellow band (Fraction A in Figure 4a) between α/β-carotene and lutein/zeaxanthin on the silica gel by thin layer chromatography (TLC) using total lipids from the *P. yezoensis* conchocelis. To remove chlorophyll in Fraction A eluted from the silica gel with acetone, silica gel TLC was performed using ethyl acetate: n-hexane (6:4, *v*/*v*) (Figure 4b). Peak 1, peak 2, and peak 3 were separated from Fraction A by silica gel-HPLC with n-hexane: Acetone (9:1, *v*/*v*) (Figure 4c).

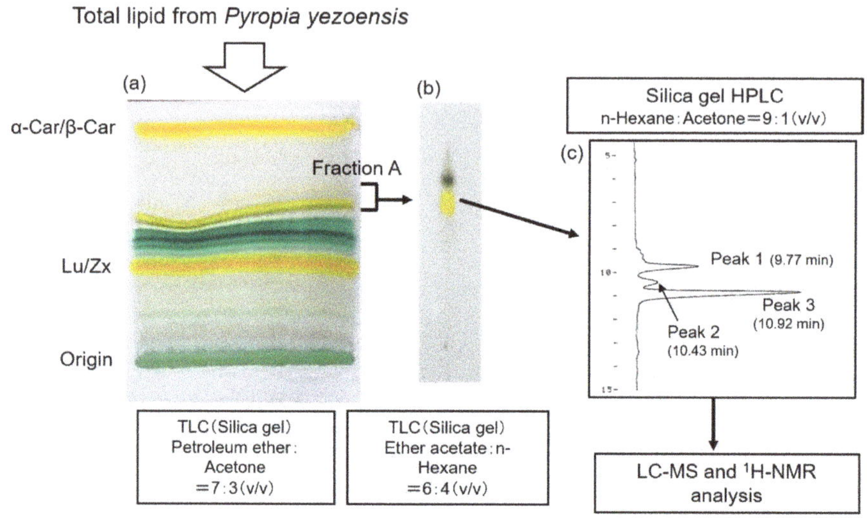


**Figure 4.** Scheme for the separation of monohydroxyl carotenoids in total lipids extracted from the conchocelis of *Pyropia yezoensis*. (**a**) Silica gel TLC of total lipids extracted from the conchocelis of *P. yezoensis*; (**b**) Silica gel TLC of Fraction A to remove the chlorophyll fraction; (**c**) Fractionation of peak 1, peak 2, and peak 3 in Fraction A by an HPLC system equipped with a Mightysil silica gel column (250 × 4.6 mm), with a flow rate of 1.0 mL/min with n-hexane: acetone (9:1, *v*/*v*) and detection at 450 nm. α-Car/β-Car: α-carotene and β-carotene; Lu/Zx: lutein and zeaxanthin.

Peak 3 was further separated into peak 3-1 and peak 3-2 using the LC-MS system (Shimadzu Triple Quadrupole Mass Spectrometer LCMS8040) with an ODS column and methanol as a solvent (Figure 5a). The retention times of peak 3-1 and peak 3-2 detected at 445 nm were the same as those of the zeinoxanthin and β-cryptoxanthin standards, respectively (Figure 5a–c). Mass chromatography at *m/z* 552.40 $[M]^+$ also indicated the same patterns as those of the zeinoxanthin and β-cryptoxanthin standards, respectively (Figure 5d–f)). Furthermore, product ions at *m/z* 535.40 $[MH\text{-}H_2O]^+$ were not detected in peak 3-1 and peak 3-2, consistent with zeinoxanthin and β-cryptoxanthin with an OH group on the β-ring, but not on the ε-ring. These data indicated that peak 3-1 and peak 3-2 were zeinoxanthin and β-cryptoxanthin, respectively.

Peak 1 (Figure 4c) corresponded to a minor carotenoid compared to peak 3, which included zeinoxanthin and β-cryptoxanthin. Therefore, we analyzed the peak by TOF-MS (Waters Acquity LC Xevo G2-S Q TOF Mass Spectrometer) with high sensitivity. An LC-MS analysis of peak 1 showed ions at *m/z* 552.4337 $[M]^+$ ($C_{40}H_{56}O$, calc. for 552.4331) and *m/z* 535.4299 $[MH\text{-}H_2O]^+$ as products (Figure 6). UV-VIS wavelengths were 420, 444, and 471 nm (methanol, Table 1). In particular, $[MH\text{-}H_2O]^+$ product ions showed higher intensities than those of $[M]^+$ molecular ions. OH group binding at the 3′ position of the ε-ring more easily produces $-H_2O$ product ions compared with binding to the hydroxylated β-ring in zeinoxanthin and β-cryptoxanthin. Furthermore, $^1$H-NMR signals indicate that peak 1 is α-cryptoxanthin with an OH group at the allylic position on the ε-end group [19] (Table 1).

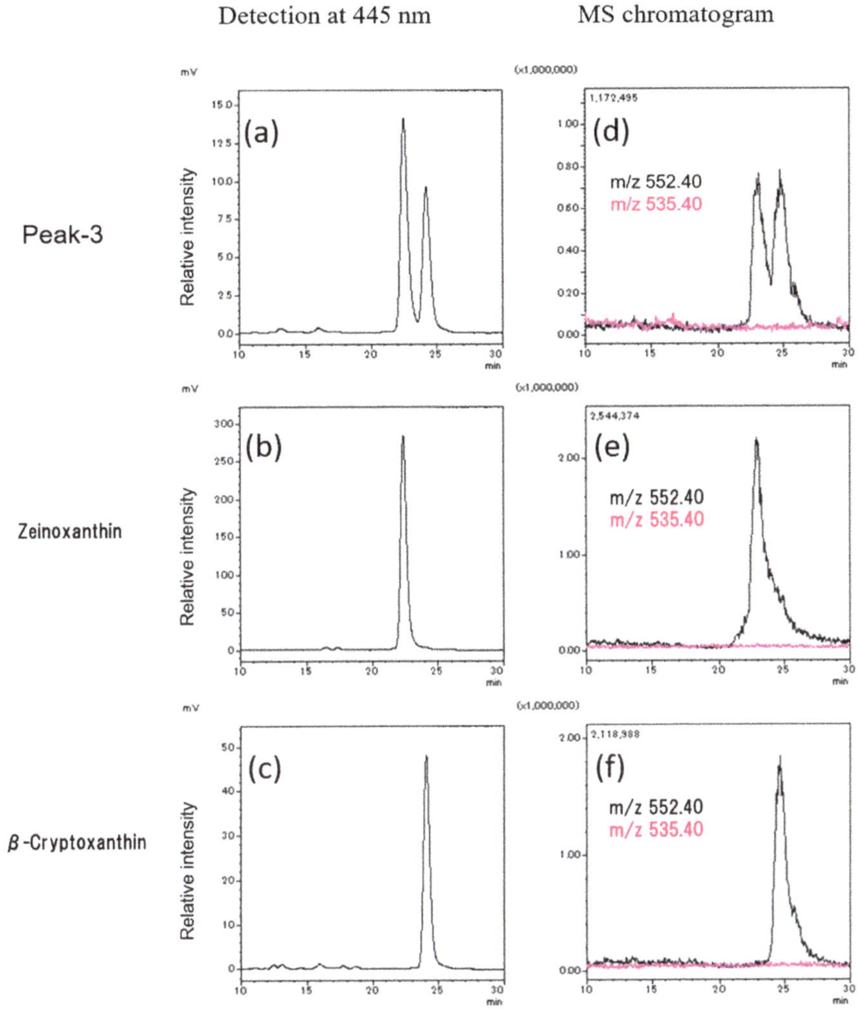


**Figure 5.** LC-MS analysis of peak 3 separated from the conchocelis of *Pyropia yezoensis*. Peak-3 separated from Fraction A was evaluated by LC-MS (Shimadzu LCMS8040) with an ODS-UG-3 column and methanol as a solvent at 0.1 mL/min. Detection at 445 nm of (**a**) peak 3 separated from the conchocelis of *Pyropia yezoensis*, (**b**) zeinoxanthin standard, (**c**) β-cryptoxanthin standard. MS chromatogram at *m/z* 552.40 (black) and *m/z* 535.40 (red) of (**d**) peak 3 separated from the conchocelis of *Pyropia yezoensis*, (**e**) zeinoxanthin standard, (**f**) β-cryptoxanthin standard.

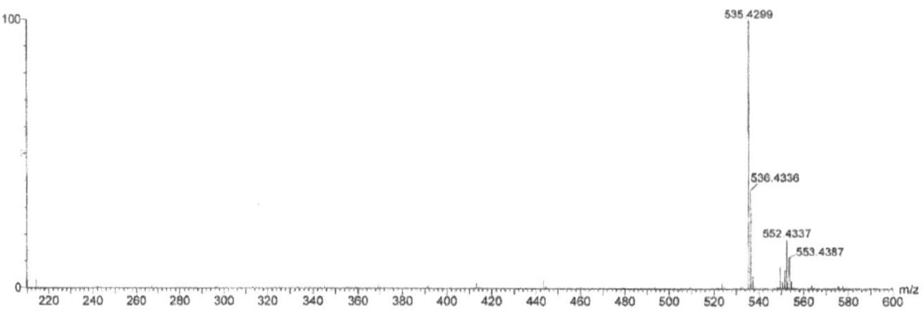


**Figure 6.** Mass spectrum of peak 1 separated from the conchocelis of *Pyropia yezoensis*.

**Table 1.** Spectrum data of peak 1 isolated from the conchocelis of *Pyropia yezoensis*.

| UV-VIS | 420, 444, 471 nm Methanol |
|---|---|
| $^1$H-NMR | δ ppm |
| 16/17 | 1.03 |
| 18 | 1.73 |
| 19 | 1.97 |
| 20 | 1.97 |
| 4′ | 5.55 |
| 7′ | 5.43 |
| 17′ | 0.85 |
| 18′ | 1.63 |
| 19′ | 1.91 |
| 20′ | 1.97 |

We tried to identify peak 2 in the total lipids isolated from *P. yezoensis*. However, it was difficult to determine the structure owing to the low content of the compound.

These results showed that *P. yezoensis* contains three monohydroxy-carotenoids, i.e., α-cryptoxanthin, β-cryptoxanthin, and zeinoxanthin, as intermediate carotenoids in the biosynthetic pathways from α-carotene to lutein or β-carotene to zeaxanthin.

### 2.3. Isolation and Identification of Epoxy-Carotenoids in P. yezoensis

We evaluated epoxy-carotenoids to reveal novel carotenoid metabolic pathways in *P. yezoensis*. We extracted total lipids from the aquacultured thallus of *P. yezoensis*, and the concentrated polar carotenoid fraction (Fraction B) was expected to contain epoxy-carotenoids according to silica gel open column chromatography. Furthermore, by HPLC separation using a C30 column (Develosil C30-UG-5, two columns were connected to enhance separation), peak 4 and peak 5 were isolated from Fraction B (Figure 7).

Peak 4 had the same HPLC retention time (19.80 min) and maximal absorption wavelength as those of the lutein-5,6-epoxide standard (Figure 8). An LC-MS (Shimadzu LCMS8040) analysis of peak 4 showed that the $[MH]^+$ ion and product ion $[MH-18]^+$ at *m/z* 585.40 and *m/z* 567.40 were identical to those of the standard. $^1$H-NMR data indicated that peak 4 is lutein-5,6-epoxide (Table 2). The HPLC retention time (22.18 min), maximal absorption wavelength, LC-MS, and $^1$H-NMR data showed that peak 5 is antheraxanthin (5,6epoxide derived from zeaxanthin) (Figure 9, Table 2). This is the first identification of lutein-5,6-epoxide and antheraxanthin in Bangiales.

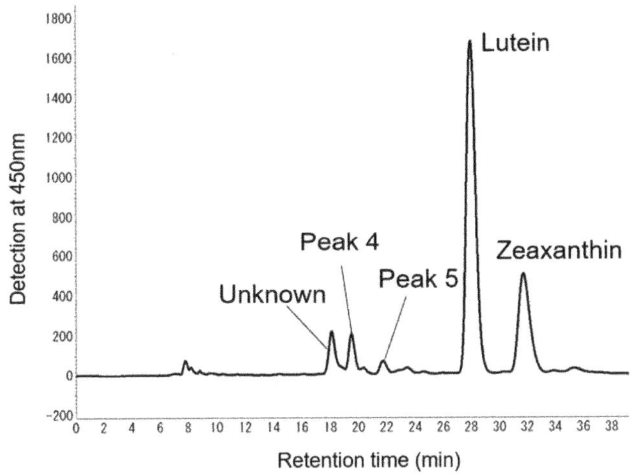


**Figure 7.** Purification of peak 4 and peak 5 in Fraction B separated from the thallus of *Pyropia yezoensis* by the HPLC system. Peak 4 and Peak 5 were isolated by the HPLC system equipped with a diode array detector at 450 nm. Two Develosil C30-UG-5 columns were connected and used for carotenoid isolation. The mobile phase was methanol. The flow rate was 1.0 mL/min.

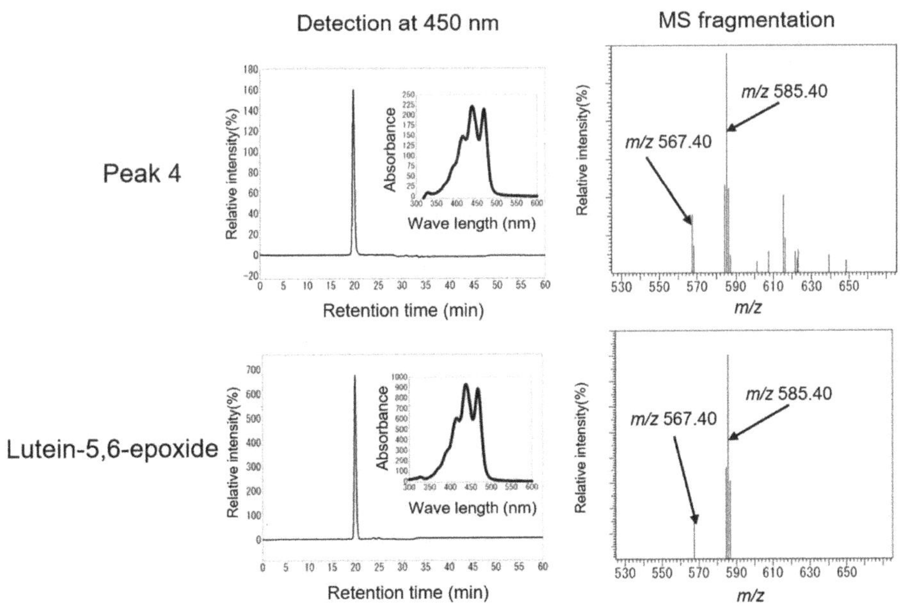


**Figure 8.** HPLC and LC-MS analyses of peak 4 and lutein-5,6-epoxide.

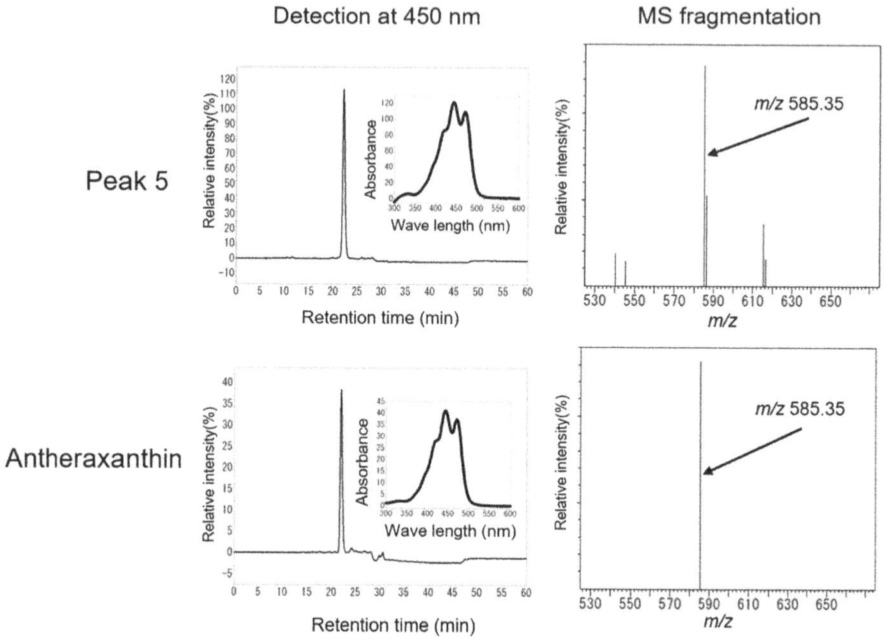


**Figure 9.** HPLC and LC-MS analyses of peak 5 and antheraxanthin.

**Table 2.** $^1$H-NMR spectral data of peak 4 and peak 5 isolated from the thallus of *Pyropia yezoensis*.

| | Peak 4 | | Peak 5 | |
|---|---|---|---|---|
| **Position** | **d** | **mult. J (Hz)** | **d** | **mult. J (Hz)** |
| 2 | 1.25 | dd (12, 7) | 1.25 | dd (12, 7) |
| 2 | 1.63 | ddd (12, 3, 1.5) | 1.63 | ddd (12, 3, 1.5) |
| 3 | 3.91 | m | 3.91 | m |
| 4 | 1.63 | dd (14, 9) | 1.63 | dd (14, 9) |
| 4 | 2.39 | ddd (14, 5, 1.5) | 2.39 | ddd (14, 5, 1.5) |
| 7 | 5.88 | d (16) | 5.88 | d (16) |
| 8 | 6.29 | d (16) | 6.29 | d (16) |
| 10 | 6.20 | d (11) | 6.20 | d (11) |
| 11 | 6.61 | dd (15, 11) | 6.61 | dd (15, 11) |
| 12 | 6.38 | d (15) | 6.38 | d (15) |
| 14 | 6.25 | m | 6.25 | m |
| 15 | 6.63 | m | 6.63 | m |
| 16 | 0.98 | s | 0.98 | s |
| 17 | 1.15 | s | 1.15 | s |
| 18 | 1.19 | s | 1.19 | s |
| 19 | 1.93 | s | 1.93 | s |
| 20 | 1.97 | s | 1.97 | s |
| 2′ | 1.37 | dd (13, 7) | 1.48 | |
| 2′ | 1.85 | dd (13, 6) | 1.77 | |
| 3′ | 4.25 | m | 4.00 | m |
| 4′ | 5.55 | br. S | 2.05 | dd (14, 9) |
| 4′ | | | 2.39 | ddd (14, 5.5, 1.5) |
| 6′ | 2.40 | d (9) | | |

**Table 2.** *Cont.*

| | Peak 4 | | Peak 5 | |
|---|---|---|---|---|
| **Position** | **d** | **mult. J (Hz)** | **d** | **mult. J (Hz)** |
| 7′ | 5.43 | dd (15.5, 9) | 6.10 | d (16) |
| 8′ | 6.14 | d (16) | 6.16 | d (16) |
| 10′ | 6.14 | d (10) | 6.16 | d (11) |
| 11′ | 6.62 | dd (15, 11) | 6.60 | dd (15, 11) |
| 12′ | 6.36 | d (15) | 6.35 | d (15) |
| 14′ | 6.26 | m | 6.25 | d (11) |
| 15′ | 6.63 | m | 6.63 | m |
| 16′ | 1.00 | s | 1.08 | s |
| 17′ | 0.85 | s | 1.08 | s |
| 18′ | 1.63 | s | 1.74 | s |
| 19′ | 1.91 | s | 1.93 | s |
| 20′ | 1.97 | s | 1.97 | s |

## 3. Discussion

Bangiales contains not only β-carotene and its derivatives, but also α-carotene and lutein with an ε-ring [11,15], and these carotenoids have important biological and nutraceutical functions in red seaweeds [20,21]. In the present study, we analyzed the carotenoid profile of one of the most popular edible red seaweeds, *P. yezoensis*.

Lutein was the predominant carotenoid in the conchocelis and thallus of *P. yezoensis*, similar to other red seaweeds (Figure 3). Lutein and zeaxanthin have established health benefits, such as protection against aged-related macular degeneration [4,5]. In addition, the improvement of cognitive function by dietary lutein and zeaxanthin has been reported in a clinical trial [22]. Thus, *P. yezoensis* containing lutein and zeaxanthin is a highly valuable food with respect to human health.

Minor intermediate carotenoids, which are important components of carotenoid biosynthetic pathways, have not been comprehensively identified [11,12]. We identified intermediate carotenoids in *P. yezoensis* and found that both α-cryptoxanthin and zeinoxanthin, which are monohydroxy-carotenoids, are produced as intermediates in the synthesis of lutein from α-carotene in *P. yezoensis* (Figure 10). α-Cryptoxanthin has been detected in the red seaweeds *Antithamnion plumula* (Ceramiales) and *Jania rubens* (Corallinales) [23,24]. However, α-cryptoxanthin has not been detected in Bangiales to date. To the best of our knowledge, this is the first report of both zeinoxanthin and α-cryptoxanthin in *P. yezoensis* of Bangiales, as evidenced by LC-MS and $^{1}$H-NMR analyses.

In higher plants, lutein is synthesized via α-cryptoxanthin and zeinoxanthin from α-carotene [25]. The hydroxylation of α-carotene is catalyzed by nonheme diiron β-hydroxylase (BHY) and heme-containing cytochrome P450-type carotene hydroxylase (CYP97A and C) [26–28]. CYP97C1 (the *Lut1* locus) catalyzes the ε-ring hydroxylation of both α-carotene and zeinoxanthin [28], and CYP97A3 catalyzes β-ring hydroxylation in *Arabidopsis thaliana* [13]. In the liverwort *Marchantia polymorpha*, CYP97C catalyzes the ε-ring hydroxylation of zeinoxanthin but not α-carotene, and BHY catalyzes β-ring hydroxylation [29]. The differences in lutein biosynthesis among taxa are likely to be associated with differences in the specificity of carotene hydroxylases for β- and ε-rings in the α-carotene molecules.

Recently, PuCHY1, which belongs to the CYP97B subfamily, was functionally characterized as a carotene hydroxylase in zeaxanthin synthesis from β-carotene in *Po. umbilicalis* [15]. It is noteworthy that there are no homologues of *CYP97A*, *CYP97C*, and *BHY* genes in the genomes of *Po. umbilicalis* [14] and *P. yezoensis* [unpublished observation]. Therefore, the identification of α-cryptoxanthin and zeinoxanthin together with genome information provides insight into the lutein biosynthetic pathway with unique enzymes in Bangiales.

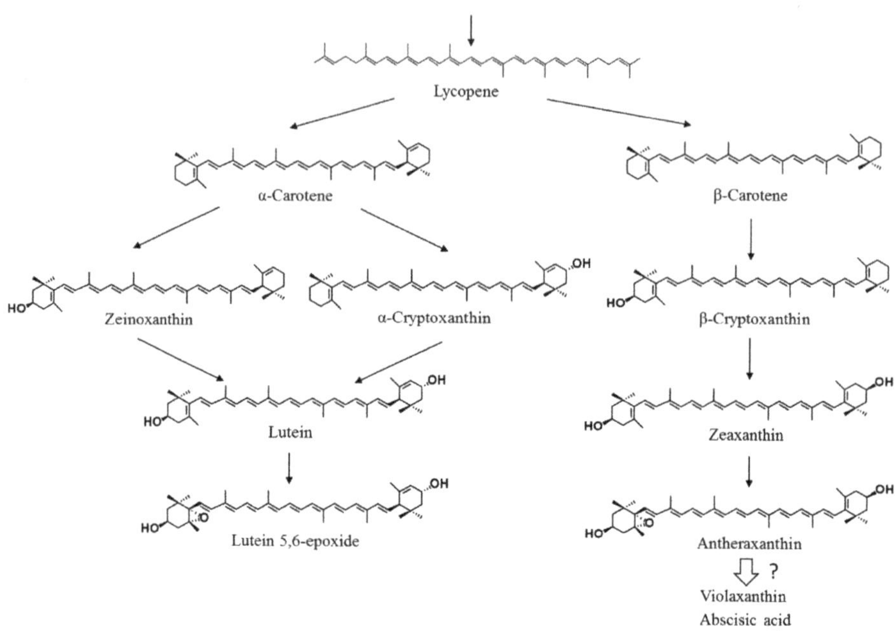


**Figure 10.** Summary of proposed pathways of carotenoid synthesis and metabolism in *Pyropia yezoensis*.

Antheraxanthin and violaxanthin are converted from zeaxanthin in land plants [30]. The reversible conversion among these carotenoids in response to light stress plays a pivotal role in photoprotection and photoabsorption, and is referred to as the xanthophyll cycle [31]. The xanthophyll cycle requires two enzymes, i.e., zeaxanthin epoxidase (ZEP), which converts zeaxanthin to violaxanthin via antheraxanthin, and violaxanthin de-epoxidase, which converts violaxanthin to zeaxanthin [30]. Antheraxanthin has been detected in red seaweeds in the orders Corallinales, Ceramiales, and Gracilariales, within the class Florideophyceae [11]. However, it has not been detected in Bangiales [11]. In the present study, we evaluated epoxy-carotenoids from zeaxanthin in *P. yezoensis*, because we previously found a ZEP-related sequence in the *P. yezoensis* genome, as well as ABA, which is a down-stream product of violaxanthin [17]. Since it was difficult to detect epoxy-carotenoids directly in total lipids extracted from *P. yezoensis*, the polar lipid fraction (Fraction B) containing epoxy-carotenoids was first separated from the total lipids by silica gel open column chromatography. Then, peak 4 and peak 5 (Figure 7) were isolated from the polar lipid fraction by HPLC and their structures were determined by LC-MS and $^{1}$H-NMR analyses. The analytical data demonstrated that peak 4 and peak 5 in *P. yezoensis* lipids were lutein-5,6-epoxide and antheraxanthin with an epoxide group, respectively. These results provide the first evidence of these 5,6-epoxy carotenoids in *P. yezoensis*, belonging to Bangiales, and provide insight into the epoxidation pathways for lutein and zeaxanthin (Figure 10), although its activity is weak as indicated by the low contents of epoxy-carotenoids.

A ZEP-related sequence has been detected in the genomes of Bangiales [17]. Dautermann and Lohr [32] reported ZEP enzyme activity in *Madagascaria erythrocladioides* (order Erythropeltidales), and the accumulation of antheraxanthin as a predominant carotenoid in a ZEP-deficient tobacco mutant. A distant relationship between the ZEP protein candidates from *P. yezoensis* and *M. erythrocladioides* was observed in an in silico analysis. In addition, the ZEP protein candidate in *P. yezoensis* is not involved in carotenoid metabolism based on its lack of a transit peptide and the lack of epoxy-carotenoids [32]. However, in the present study, we clearly detected antheraxanthin and lutein-5,6-epoxide in *P. yezoensis* (Figures 8 and 9). Since antheraxanthin and lutein-5,6-epoxide are derived from zeaxanthin and lutein by ZEP, our findings provide evidence for ZEP activity in *P. yezoensis*. Moreover, the production of antheraxanthin may be related to violaxanthin synthesis in *P. yezoensis* (Figure 10). The present results contribute to the elucidation of the biosynthesis of epoxy-carotenoids in the order Bangiales.

## 4. Materials and Methods

### *4.1. Chemicals*

α-Carotene, β-carotene, β-cryptoxanthin, lutein, zeaxanthin, antheraxanthin, and lutein-5,6-epoxide were purchased from Carote *Nature* GmbH (Münsingen, Switzerland). Zeinoxanthin used for standard was synthesized by Wittig condensation of $C_{25}$-apocarotenal derived from commercially available α-ionone, with previously prepared $C_{15}$-phosphonium salt possessing 3-hyxroxy-β-end group [33]. $^1$H NMR data of synthetic zeinoxanthin were identical with those reported in Reference [34].

### *4.2. Seaweed Materials*

Thalli and conchocelis filaments of *P. yezoensis* strain U-51 were used in the present study. Thalli maricultured at Shichigahama (Miyagi, Japan) were frozen and kept under −20 °C until extraction of total lipid. Conchocelis materials were prepared by laboratory culture in ESL medium [35] at 15 °C, under irradiation of 40 µmol photons $m^{-2}$ $s^{-1}$ provided by cool white fluorescent lamps with a photoperiod of 10 h light/14 h dark. The medium was bubbled continuously with filter-sterilized air and changed weekly. We used approximately 40–80 mg of conchocelis stage of *P. yezoensis* during cultivation for total lipid extraction.

### *4.3. Extraction of Total Lipids from P. yezoensis*

Total lipid was extracted from conchocelis and thallus of *P. yezoensis* with 20-fold methanol (*v*/*w*) for 24 h at room temperature under shading conditions. The extraction with methanol was conducted twice. Then, methanol was evaporated, and the extracts was solved in chloroform-methanol-water (10:5:3, *v*/*v*/*v*) to remove water soluble components. The total lipid fraction containing carotenoids was obtained from chloroform layer.

### *4.4. Lutein, Zeaxanthin, α/β-Carotene Contents in P. yezoensis*

Carotenoids content in *P. yezoensis* were measured by a high-performance liquid chromatography (HPLC) system (Hitachi, Tokyo, Japan) equipped with a diode array detector (Hitachi L2455, Tokyo, Japan). Two Develosil C30-UG-5 (Nomura Chemical Co., Aichi, Japan) columns were connected and used for carotenoid analysis. Detection was set at 450 nm. Mobile phase was methanol until 20 min, and thereafter, dichloromethane content increased linearly from 0% to 50% in 20 min, then was held at methanol:dichloromethane (1:1, *v*/*v*) for additional 20 min. The flow rate was maintained at 1.0 mL/min, and sample injection volume was 50 µL. α-Carotene, β-carotene, lutein, and zeaxanthin were identified by comparison of retention time and absorption spectra of HPLC analysis with their authentic standards. The contents of α-carotene, β-carotene, lutein, and zeaxanthin in *P. yezoensis* were calculated using a calibration curve prepared by each authentic standard.

### *4.5. Isolation of Monohydroxy Carotenoids*

The total lipid of *P. yezoensis* was developed on a silica gel TLC plate (Merk KGaA, Darnstadt, Germany) with petroleum ether:acetone (7:3, *v*/*v*) or n-hexane:acetone (7:3, *v*/*v*). Three pigment fractions with yellow-orange color were observed on the TLC plate. The yellow pigment fraction with relative front (Rf) value 0.61 between strong two yellow-orange fractions with Rf value 0.92 (α/β-carotene) and 0.42 (lutein/zeaxanthin) was scrapped off from the TLC plate developed with petroleum ether:acetone (7:3, *v*/*v*), and extracted with acetone. Separation of the yellow pigment fraction (Fraction A) was conducted by silica gel TLC plate with ethyl acetate:n-hexane (6:4, *v*/*v*) to remove remaining chlorophyll. Then, peak 1, peak 2, and peak 3 were separated from Fraction A by

HPLC system (Hitachi, Tokyo, Japan) equipped with Mightysil silica gel column (Kanto Chemical Co., Tokyo, Japan, 250 × 4.6 mm), flow rate: 1.0 mL/min with n-hexane:acetone (9:1, *v*/*v*), detection at 450 nm. Peak 3 was further fractionated and analyzed using an LC-MS system (Shimadzu Triple Quadrupole mass spectrometer LCMS8040, Shimadzu, Kyoto, Japan), equipped with ODS column (Develosil ODS-UG-3 150 × 2.0 mm, Nomura Chemical Co., Aichi, Japan). Methanol was used as a mobile phase at a flow rate of 0.1 mL/min.

### 4.6. Isolation of Epoxy-Carotenoids

The total lipid extracted from aquacultured thallus of *P. yezoensis* was separated silica gel column by n-hexane:acetone (7:3, *v*/*v*). Four fractions [strong yellow layer (α/β-carotene fraction), yellow-green layer (α/β-cryptoxanthin, zeinoxanthin and chrolophyll faction), strong orange layer (lutein/zeaxanthin fraction), and light-yellow layer (Fraction B)] were eluted in order. Then, peak 4 and peak 5 were isolated from Fraction B (light yellow layer) using an HPLC system (Hitachi, Tokyo, Japan) equipped with a Develosil C30-UG-5 column (250 × 4.6 mm, two columns were connected), flow rate: 1.0 mL/min with methanol, detection at 450 nm.

### 4.7. Identification of Monohydroxy- and Epoxy-Carotenoids in P. yezoensis

The chemical structures of peaks 1–5 were determined by LC-MS and $^1$H-NMR analyses. LC-MS analysis was carried out using a Shimadzu Triple Quadrupole mass spectrometer LCMS8040, as described in 2.6 or a Waters Xevo G2S Q TOF mass spectrometer (Waters Corporation, Milford, CT, USA), equipped with an Acquity UPLC system with Acquity 1.7 μm BEH UPLC C18 column (100 × 2.1 mm) (Waters Corporation, Milford, CT, USA), and methanol as a mobile phase at a flow rate of 0.4 mL/min. $^1$H-NMR (500 MHz) spectra in $CDCl_3$ were measured with an UNITY INOVA-500 system (Varian Corporation, Palo Alto, CA, USA).

### 4.8. Statistical Analysis

Analytical data of carotenoid content (Figure 3) were expressed as the mean ± standard deviation (SD). Statistical significance was determined between the two groups using the Student's *t*-test. A significant difference was defined at $p < 0.05$.

## 5. Conclusions

The results of the present study demonstrated that *P. yezoensis* synthesizes zeinoxanthin and α-cryptoxanthin, as well as β-cryptoxanthin, as intermediate monohydroxy-carotenoids. These results indicate that *P. yezoensis* has two lutein biosynthetic pathways via zeinoxanthin and α-cryptoxanthin from α-carotene (Figure 10). Furthermore, antheraxanthin and lutein-5,6-epoxide were found for the first time in Bangiales (Figure 10). In particular, antheraxanthin is considered an intermediate carotenoid in the conversion from zeaxanthin to violaxanthin. The carotenoid profile of *P. yezoensis* provides new insight into the biosynthetic and metabolic pathways of carotenoids in red seaweeds.

**Author Contributions:** M.H. conceived and designed the experiments and wrote the paper; K.M. (Koji Mikami) designed the experiments and reviewed the manuscript. J.K., N.T., and N.K. performed the experiments; T.M. analyzed NMR data for isolated carotenoids; Y.Y. and A.W. synthesized standard carotenoids and evaluated the data for carotenoid identification; K.M. (Kazuo Miyashita) reviewed the manuscript.

**Funding:** This research was funded by The Towa Foundation for Food Science & Research.

**Acknowledgments:** We thank the Marine Resources Research Center of Aichi Fisheries Research Institute (Gamagori-shi, Aichi, Japan) for providing the *P. yezoensis* strain U-51. We are grateful to the Shichigahama branch office of the Miyagi Prefecture Fisheries Cooperative Association for kindly providing raw maricultured *P. yezoensis* thalli.

**Conflicts of Interest:** The authors declare no conflict of interest.

## References


1. Nisar, N.; Li, L.; Lu, S.; Khin, N.C.; Pogson, B.J. Carotenoid metabolism in plants. *Mol. Plant* **2015**, *8*, 68–82. [CrossRef] [PubMed]
2. Eggersdorfer, M.; Wyss, A. Carotenoids in human nutrition and health. *Arch. Biochem. Biophys.* **2018**, *652*, 18–26. [CrossRef] [PubMed]
3. Castenmiller, J.J.; West, C.E. Bioavailability and bioconversion of carotenoids. *Annu. Rev. Nutr.* **1998**, *18*, 19–38. [CrossRef] [PubMed]
4. Johnson, E.J. The role of carotenoids in human health. *Nutr. Clin. Care* **2002**, *5*, 56–65. [CrossRef] [PubMed]
5. Fernández-Sevilla, J.M.; Acién Fernández, F.G.; Molina Grima, E. Biotechnological production of lutein and its applications. *Appl. Microbiol. Biotechnol.* **2010**, *86*, 27–40. [CrossRef] [PubMed]
6. E-Stat, Portal Site of Official Statistics of Japan. Available online: https://www.e-stat.go.jp/ (accessed on 1 November 2018).
7. Drew, K.M. Life-history of Porphyra. *Nature* **1954**, *173*, 1243–1244. [CrossRef]
8. Nakamura, Y.; Sasaki, N.; Kobayashi, M.; Ojima, N.; Yasuike, M.; Shigenobu, Y.; Satomi, M.; Fukuma, Y.; Shiwaku, K.; Tsujimoto, A.; et al. The first symbiont-free genome sequence of marine red alga, Susabi-nori (*Pyropia yezoensis*). *PLoS ONE* **2013**, *8*, e57122. [CrossRef] [PubMed]
9. Sahoo, D.; Tang, X.R.; Yarish, C. Porphyra—The economic seaweed as a new experimental system. *Curr. Sci.* **2002**, *83*, 1313–1316.
10. Takaichi, S. Carotenoids in algae: Distributions, biosyntheses and functions. *Mar. Drugs* **2011**, *9*, 1101–1118. [CrossRef] [PubMed]
11. Takaichi, S.; Yokoyama, A.; Mochimaru, M.; Uchida, H.; Murakami, A. Carotenogenesis diversification in phylogenetic lineages of Rhodophyta. *J. Phycol.* **2016**, *52*, 329–338. [CrossRef] [PubMed]
12. Schubert, N.; Garcia-Mendoza, E.; Pacheco-Ruiz, I. Carotenoid composition of marine red algae. *J. Phycol.* **2006**, *42*, 1208–1216. [CrossRef]
13. Kim, J.; Smith, J.J.; Tian, L.; DellaPenna, D. The evolution and function of carotenoid hydroxylases in Arabidopsis. *Plant Cell Physiol.* **2009**, *50*, 463–479. [CrossRef] [PubMed]
14. Brawley, S.H.; Blouin, N.A.; Ficko-Blean, E.; Wheeler, G.L.; Lohr, M.; Goodson, H.V.; Jenkins, J.W.; Blaby-Haas, C.E.; Helliwell, K.E.; Chan, C.X.; et al. Insights into the red algae and eukaryotic evolution from the genome of *Porphyra umbilicalis* (Bangiophyceae, Rhodophyta). *Proc. Natl. Acad. Sci. USA* **2017**, *114*, E6361–E6370. [CrossRef] [PubMed]
15. Yang, L.E.; Huang, X.Q.; Hang, Y.; Deng, Y.Y.; Lu, Q.Q.; Lu, S. The P450-type carotene hydroxylase PuCHY1 from *Porphyra* suggests the evolution of carotenoid metabolism in red algae. *J. Integr. Plant Biol.* **2014**, *56*, 902–915. [CrossRef] [PubMed]
16. Mikami, K.; Hosokawa, M. Biosynthetic pathway and health benefits of fucoxanthin, an algae-specific xanthophyll in brown seaweeds. *Int. J. Mol. Sci.* **2013**, *14*, 13763–13781. [CrossRef] [PubMed]
17. Mikami, K.; Mori, I.C.; Matsuura, T.; Ikeda, Y.; Kojima, M.; Sakakibara, H.; Hirayama, T. Comprehensive quantification and genome survey reveal the presence of novel phytohormone action modes in red seaweeds. *J. Appl. Phycol.* **2016**, *28*, 2539–2548. [CrossRef]
18. Guajardo, E.; Correa, J.A.; Contreras-Porcia, L. Role of abscisic acid (ABA) in activating antioxidant tolerance responses to desiccation stress in intertidal seaweed species. *Planta* **2016**, *243*, 767–781. [CrossRef] [PubMed]
19. Englert, G. NMR spectroscopy. In *Carotenoids*; Britton, G., Liaaen-Jensen, S., Pfander, H., Eds.; Birkhäuser: Basel, Switzerland, 1995; Volume 1B, pp. 147–260.
20. Chan, C.X.; Blouin, N.A.; Zhuang, Y.; Zäuner, S.; Prochnik, S.E.; Lindquist, E.; Lin, S.; Benning, C.; Lohr, M.; Yarish, C.; et al. *Porphyra* (Bangiophyceae) transcriptomes provide insights into red algal development and metabolism. *J. Phycol.* **2012**, *48*, 1328–1342. [CrossRef] [PubMed]
21. Christaki, E.; Bonos, E.; Giannenas, I.; Florou-Paneri, P. Functional properties of carotenoids originating from algae. *J. Sci. Food Agric.* **2013**, *93*, 5–11. [CrossRef] [PubMed]
22. Hammond, B.R., Jr.; Miller, L.S.; Bello, M.O.; Lindbergh, C.A.; Mewborn, C.; Renzi-Hammond, L.M. Effects of lutein/zeaxanthin supplementation on the cognitive function of community dwelling older adults: A randomized, double-masked, placebo-controlled trial. *Front. Aging Neurosci.* **2017**, *9*, 254. [CrossRef] [PubMed]

23. Bjørnland, T.; Borch, G.; Liaaen-Jensen, S. Configurational studies on red algae carotenoids. *Phytochemistry* **1984**, *23*, 1711–1715. [CrossRef]
24. Hegazi, M.M.; Pérez-Ruzafa, A.; Almela, L.; Candela, M.E. Separation and identification of chlorophylls and carotenoids from *Caulerpa prolifera*, *Jania rubens* and *Padina pavonica* by reversed-phase high-performance liquid chromatography. *J. Chromatogr.* **1998**, *829*, 153–159. [CrossRef]
25. Quinlan, R.F.; Shumskaya, M.; Bradbury, L.M.; Beltrán, J.; Ma, C.; Kennelly, E.J.; Wurtzel, E.T. Synergistic interactions between carotene ring hydroxylases drive lutein formation in plant carotenoid biosynthesis. *Plant Physiol.* **2012**, *160*, 204–214. [CrossRef] [PubMed]
26. Quinlan, R.F.; Jaradat, T.T.; Wurtzel, E.T. Escherichia coli as a platform for functional expression of plant P450 carotene hydroxylases. *Arch. Biochem. Biophys.* **2007**, *458*, 146–157. [CrossRef] [PubMed]
27. Li, Q.; Farre, G.; Naqvi, S.; Breitenbach, J.; Sanahuja, G.; Bai, C.; Sandmann, G.; Capell, T.; Christou, P.; Zhu, C. Cloning and functional characterization of the maize carotenoid isomerase and β-carotene hydroxylase genes and their regulation during endosperm maturation. *Transgenic Res.* **2010**, *19*, 1053–1568. [CrossRef] [PubMed]
28. Tian, L.; Musetti, V.; Kim, J.; Magallanes-Lundback, M.; DellaPenna, D. The Arabidopsis LUT1 locus encodes a member of the cytochrome p450 family that is required for carotenoid epsilon-ring hydroxylation activity. *Proc. Natl. Acad. Sci. USA* **2004**, *101*, 402–407. [CrossRef] [PubMed]
29. Takemura, M.; Maoka, T.; Misawa, N. Biosynthetic routes of hydroxylated carotenoids (xanthophylls) in Marchantia polymorpha, and production of novel and rare xanthophylls through pathway engineering in *Escherichia coli*. *Planta* **2015**, *241*, 699–710. [CrossRef] [PubMed]
30. Hieber, A.D.; Bugos, R.C.; Yamamoto, H.Y. Plant lipocalins: Violaxanthin de-epoxidase and zeaxanthin epoxidase. *Biochim. Biophys. Acta* **2000**, *1482*, 84–91. [CrossRef]
31. Niyogi, K.K.; Grossman, A.R.; Björkman, O. Arabidopsis mutants define a central role for the xanthophyll cycle in the regulation of photosynthetic energy conversion. *Plant Cell* **1988**, *10*, 1121–1134. [CrossRef]
32. Dautermann, O.; Lohr, M. A functional zeaxanthin epoxidase from red algae shedding light on the evolution of light-harvesting carotenoids and the xanthophyll cycle in photosynthetic eukaryotes. *Plant J.* **2017**, *92*, 879–891. [CrossRef] [PubMed]
33. Yamano, Y.; Ito, M. Total synthesis of capsanthin and capsorubin using Lewis acid-promoted regio- and stereoselective rearrangement of tetrasubsutituted epoxides. *Org. Biomol. Chem.* **2007**, *5*, 3207–3212. [CrossRef] [PubMed]
34. Khachik, F.; Chan, A.-N.; Gana, A.; Mazzola, E. Partial synthesis of (3*R*,6′*R*)-α-cryptoxanthin and (3*R*)-β-cryptoxanthin from (3*R*,3′*R*,6′*R*)-lutein. *J. Nat. Prod.* **2007**, *70*, 220–226. [CrossRef] [PubMed]
35. Kitade, Y.; Fukuda, S.; Nakajima, M.; Watanabe, T.; Saga, N. Isolation of a cDNA encoding a homologue of actin from *Porphyra yezoensis* (Rhodophyta). *J. Appl. Phycol.* **2002**, *14*, 135–141. [CrossRef]

*marine drugs*

MDPI

*Article*

# Nutrient Deprivation-Associated Changes in Green Microalga *Coelastrum* sp. TISTR 9501RE Enhanced Potent Antioxidant Carotenoids

**Monrawat Rauytanapanit [1,2], Kantima Janchot [1,3], Pokchut Kusolkumbot [4], Sophon Sirisattha [4], Rungaroon Waditee-Sirisattha [1,3] and Thanit Praneenararat [1,2,*]**

1 The Chemical Approaches for Food Applications Research Group, Faculty of Science, Chulalongkorn University, Phayathai Rd., Pathumwan, Bangkok 10330, Thailand; monrawat.mon@gmail.com (M.R.); kantima.janchot@gmail.com (K.J.); rungaroon.w@chula.ac.th (R.W.-S.)

2 Department of Chemistry, Faculty of Science, Chulalongkorn University, Phayathai Rd., Pathumwan, Bangkok 10330, Thailand

3 Department of Microbiology, Faculty of Science, Chulalongkorn University, Phayathai Rd., Pathumwan, Bangkok 10330, Thailand

4 Thailand Institute of Scientific and Technological Research (TISTR), Khlong Luang, Pathum Thani 12120, Thailand; pokchut@tistr.or.th (P.K.); sophon_si@tistr.or.th (S.S.)

* Correspondence: Thanit.P@chula.ac.th; Tel.: +66-2-218-7638

Received: 2 May 2019; Accepted: 28 May 2019; Published: 1 June 2019

**Abstract:** The utilization of microalgae as a source of carotenoid productions has gained increasing popularity due to its advantages, such as a relatively fast turnaround time. In this study, a newly discovered *Coelastrum* sp. TISTR 9501RE was characterized and investigated for its taxonomical identity and carotenoid profile. To the best of our knowledge, this report was the first to fully investigate the carotenoid profiles in a microalga of the genus *Coelastrum*. Upon use of limited nutrients as a stress condition, the strain was able to produce astaxanthin, canthaxanthin, and lutein, as the major carotenoid components. Additionally, the carotenoid esters were found to be all astaxanthin derivatives, and β-carotene was not significantly present under this stress condition. Importantly, we also demonstrated that this practical stress condition could be combined with simple growing factors, such as ambient sunlight and temperature, to achieve even more focused carotenoid profiles, i.e., increased overall amounts of the aforementioned carotenoids with fewer minor components and chlorophylls. In addition, this green microalga was capable of tolerating a wide range of salinity. Therefore, this study paved the way for more investigations and developments on this fascinating strain, which will be reported in due course.

**Keywords:** *Coelastrum*; microalgae; astaxanthin; canthaxanthin; lutein; carotenoids

---

## 1. Introduction

Carotenoids are an important class of natural tetraterpenes found in several plants, algae, fungi, and bacteria [1–5]. To date, there are several hundred characterized carotenoids found in various sources, some of which were found to play important roles in their respective organisms. Examples include being an integral part of the photosynthesis unit [6,7], or playing photoprotective roles [8–10]. Interestingly, the discovery of these aforementioned roles simultaneously inspired the investigation of these compounds for use in other applications. Consequently, several carotenoids, some of which are depicted in Figure 1, have found uses in a number of areas [11–14]. For instance, astaxanthin was found to be the strongest antioxidant carotenoid in nature, with power that was several-fold stronger than that of β-carotene [4,15,16]. Thus, it is currently in high demand for use in the nutraceutical, pharmaceutical, and cosmetic industries [17–21]. Another example is canthaxanthin, which was found to serve as a food colorant in animal feed for several food sources, such as poultry and aquatic animals [22–24]. Lastly, lutein has been demonstrated to exhibit crucial roles in human eye health [25–27]. Clearly, these attractive properties of carotenoids have led to an increasing demand for them, and thus the development of economic, high-yielding productions of carotenoids are continuously being studied [4,5,28–32].

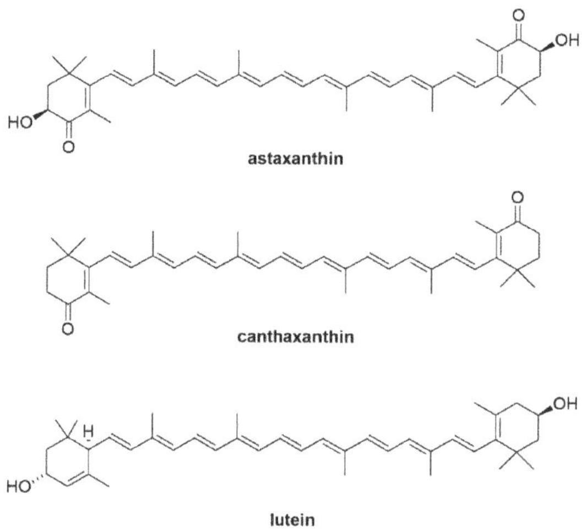


**Figure 1.** Chemical structures of important carotenoids with various applications.

Despite the fact that the chemical syntheses of some carotenoids are available, these methods have limited utilization in real applications owing to health concerns arising from the use of pure compounds instead of isomeric mixtures that typically exist in natural sources [4,33]. Hence, naturally obtained carotenoids are usually preferred, leading to the research and development of efficient carotenoid extraction methods from natural sources [4,5,31]. Nevertheless, except for some cases like lutein extraction from Marigold [34,35], microalgal sources have become more important than plants as a source of carotenoids [2,4,13,29]. This is due to the fact that microalgae possess a desirable balance between being autotrophs capable of producing a range of carotenoids, and having biotechnological-related advantages, such as fast growth and easier genetic manipulation. Prominent examples include astaxanthin from *Haematococcus pluvialis* and *Xanthophyllomyces dendrorhous*, and β-carotene from *Dunaliella salina*, both of which have been utilized in actual commercial production [36–38].

Notably, some enhancement cues are typically required for substantial production of the carotenoids of interest, which is usually in the form of stress conditions. Heat, light, high salt concentrations, and depleted nitrogen supply are common stress conditions used to induce carotenoid production in microalgae [30,39–43]. Interestingly, *Coelastrum* is a genus of green algae that was also found to be capable of producing carotenoids, although there have been few reports that investigated the carotenoid profiles of this genus. *Coelastrum* sp. HA-1, an isolated microalga from Bohai Bay, China, was cultivated under nitrogen-limiting conditions and it produced astaxanthin at a level of 6.36 mg/g dried cell weight [44]. The cultivation of *Coelastrum* cf. *pseudomicroporum* in urban wastewater under salt stress conditions produced carotenoids at a level of 33.4 ± 19.86 pg cell$^{-1}$ [45]. Lastly, Soares et al. reported that *Coelastrum sphaericum* provided a set of carotenoids with the prominent components being astaxanthin and lutein [46].

In this study, a strain of *Coelastrum* sp. was isolated and investigated for its ability to enhance carotenoid production under stress conditions. The induction condition was a drastic reduction of the nutrient contents, which could also be viewed as a practical advantage in a real production process (cost reduction). Notably, in contrast to the aforementioned studies [44,45], this study serves as the first example of this genus, where the carotenoid profile is characterized by both HPLC-PDA (photodiode array detection) and LC-MS. Molecular identification of this microalgal strain, its growth profiles, and the carotenoid identifications and quantifications are discussed herein.

## 2. Results and Discussion

### 2.1. Morphology and Genetic Identification of the Microalga Coelastrum sp. TISTR 9501RE

Isolated from a coastal ecosystem in the northern part of Thailand, this strain was discovered from the screening of strains that are capable of enhancing carotenoid accumulations upon the stress conditions of interest. In this regard, we employed a nutrient-depleted condition to induce the carotenoid biosynthesis. Notably, the condition used was an overall reduction of the required nutrient (BG11), such that it was only one-fourth of the normal formula. This provided the added benefit of drastically reducing the overall production costs for larger scale production. Apparently, this led to an obvious concern regarding whether this reduced nutrient had overly affected the growth of the microalga. Hence, we sought to observe the growth rates under both conditions. The results (Figure S1) showed that the growth, albeit being unsurprisingly suppressed, could still reach $OD_{730}$ at around 1.0. Compared to other studies [44], this level of cell mass was at an acceptable level. Prior to the induction condition, the algal colonies on the BG11 agar exhibited a dull-shiny texture, olive green color, and circular shape. Microscopic morphology observations showed that the algal cells were spherical vegetative cells varying between 8 and 15 μm in diameter (Figure 2A). The colonies were also spherical (Figure 2B). After 14 days of growth on one-fourth BG11 agar, there was an obvious accumulation of a bright orange color, which we hypothesized to be carotenoid-based pigments (Figure 2C). Furthermore, we also demonstrated that this microalga was capable of tolerating a wide range of salinities (up to the relevant concentration in the ocean at 500 mM NaCl) (Figure S2), thereby paving the way for more diverse applications. Given its spherical shape, this algal strain was preliminarily hypothesized to belong to the genus Coelastrum [47].

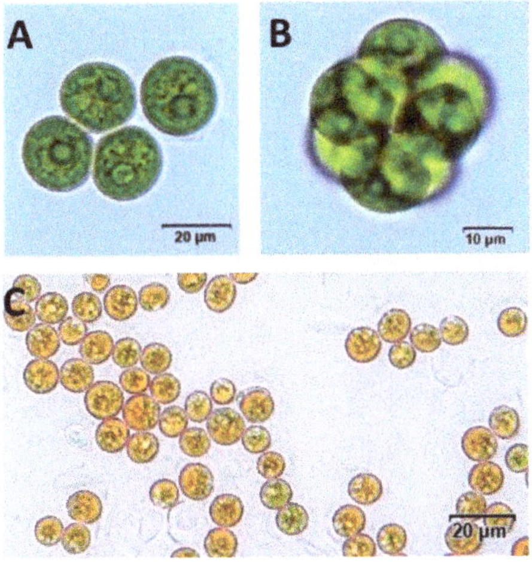


**Figure 2.** Light microscopic images of *Coelastrum* sp. TISTR 9501RE; (**A**) Green vegetative cells; (**B**) A spherical colony; (**C**) Carotenoids accumulating cells.

To better confirm its identity, standard genetic identification was accomplished using genomic sequencing of the internal transcribed spacer (ITS) 1 region of the 5.8S ribosomal RNA gene. To verify its taxonomical position, the ITS1 of the 5.8S rRNA gene was compared to sequences in public databases (e.g., GenBank). Similar sequences were used to construct independent molecular phylogenetic trees based on the ITS1-5.8s-ITS2 sequence. The reliability of the phylogenetic tree was evaluated using neighbor joining analysis. As shown in Figure 3, the ITS1-5.8s-ITS2 sequence comparison revealed that strain TISTR 9501RE was grouped together with other *Coelastrum* species in the same clade, thereby confirming that this strain was a species in the genus Coelastrum. Based on both the morphological and molecular evidence, this microalga was named *Coelastrum* sp. TISTR 9501RE, which is a member of the green algae (Chlorophyta).

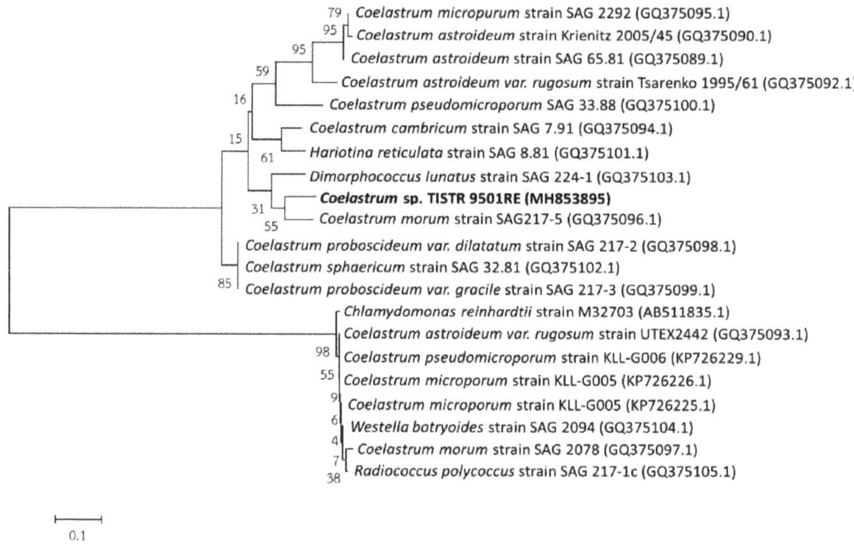


**Figure 3.** Maximum likelihood (ML) tree from the sequence-structure analysis of the ITS sequence data. The tree was generated by the neighbor joining method using the Molecular Evolutionary Genetic Analysis (MEGA6) software. The bootstrap value is expressed as a percentage of 100 replicates. The nucleotide sequence accession numbers are indicated in parentheses.

### *2.2. Carotenoid Profiles of Coelastrum sp. TISTR 9501RE*

To characterize the molecular components of the putative carotenoid mixture observed from the morphological change of the cell, an extraction procedure by bead-beating [4,31] was employed for both the control (normal nutrient strength) condition and the nutrient-depleted condition. Thereafter, the obtained pigment mixtures were subjected to total chlorophyll determination via UV-vis spectroscopy. This revealed that the nutrient-depleted condition resulted in a significant decrease in chlorophyll production in the *Coelastrum* sp. TISTR 9501RE. That is, the total chlorophyll content substantially decreased from 5.64 ± 0.12 mg of total chlorophylls per one gram of dried weight of the cells (mg/g DW) to 3.89 ± 0.03 mg/g DW when the nutrient amount was restricted. This could also be illustrated by calculation of the chlorophyll a and b contents, which clearly showed that the amounts of both compounds were significantly reduced (chlorophyll a: 5.08 ± 0.10 to 3.37 ± 0.02 mg/g DW; chlorophyll b: 1.09 ± 0.02 to 0.92 ± 0.01 mg/g DW). Nevertheless, these numbers merely served as a rough guideline, and the identifications of the specific types of compounds were deemed to be more important.

To gain more insight on the composition of the carotenoid mixtures, high performance liquid chromatography with photodiode array detection (HPLC-PDA) and liquid chromatography–mass spectrometry (LC-MS) experiments were conducted. These data, along with some comparisons from previous literature [48], resulted in the identification of the carotenoids of interest; most of which are shown in Figure 4, and Table 1; Table 2. Overall, three carotenoids of interest, namely astaxanthin (16.2

min), lutein (19.3 min, co-eluted with chlorophyll b), and canthaxanthin (23.5 min), were produced in appreciable amounts in this strain. Other prominent peaks included chlorophylls and carotenoid esters (putatively derived from astaxanthin), along with some unidentified steroid and glyceride species (as suggested by MS data not shown) scattered throughout the chromatograms. Compared to the control conditions, the nutrient-depleted microalgae showed discernible changes in the carotenoid biosynthesis. For instance, the syntheses of the unidentified steroid-related species around 10 min were suppressed, and the amounts of all the chlorophyll-related species, e.g., peak 6 and 9, were significantly decreased. On the other hand, the syntheses of all the aforementioned carotenoids were increased, as quantified by the LC-MS (see below). This suggested that the astaxanthin biosynthesis pathway could be legitimately up-regulated, as canthaxanthin was also produced in a significant amount. This was in-line with the pathway of *H. pluvialis* [49], where echinenone and canthaxanthin were direct precursors to astaxanthin. Furthermore, it seemed that fatty acid esters, which were commonly found in several organisms, were all derivatives of astaxanthin. Although many of them were present in minute amounts (there were in fact some additional astaxanthin esters in small but MS-detectable amounts that are not shown in Tables 1 and 2), the combined amounts of all the esters could be considered as a significant addition to the free form, thereby confirming the enhanced production of astaxanthin.

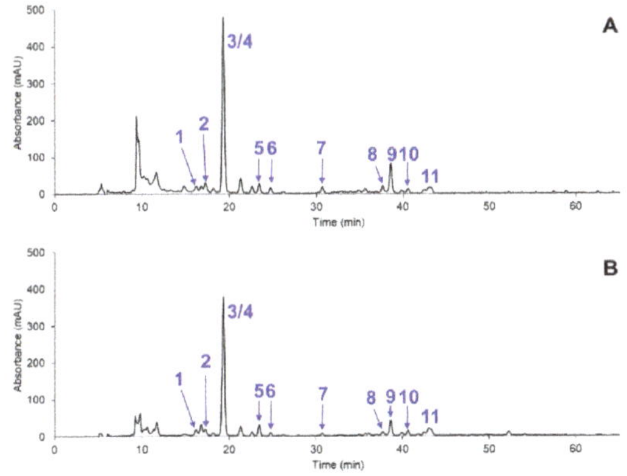


**Figure 4.** Chromatograms (representative data from three experiments at 450 nm) of the carotenoid extract from *Coelastrum* sp. TISTR 9501RE, with (**A**) the control condition, and (**B**) the nutrient depleted condition. Identities of the peaks can be found in Tables 1 and 2.

**Table 1.** Carotenoid profiles based on the LC-MS data of the extract of the control conditions from *Coelastrum* sp. TISTR 9501RE (a representative set of data from three experiments).

| Entry | Identity * | Retention Time ** | Proposed Formula | Precursor Mass | Found at Mass | Mass Error (ppm) |
|---|---|---|---|---|---|---|
| 1 | All-*trans*-astaxanthin | 16.24 | $C_{40}H_{52}O_4$ | 597.3938 | 597.3936 | −0.48 |
| 2 | Violaxanthin isomer | 17.28 | $C_{40}H_{56}O_4$ | 601.4251 | 601.4234 | −2.91 |
| 3 | All-*trans*-lutein | 19.32 | $C_{40}H_{56}O_2$ | 569.4353 | 569.4317 | −6.34 |
| 4 | Chlorophyll b | 19.32 | $C_{55}H_{70}N_4O_6Mg$ | 907.5219 | 907.5214 | −0.45 |
| 5 | All-*trans*-Canthaxanthin | 23.45 | $C_{40}H_{52}O_2$ | 565.4040 | 565.4030 | −1.75 |
| 6 | Chlorophyll a | 25.65 | $C_{55}H_{72}N_4O_5Mg$ | 893.5426 | 893.5393 | −3.66 |
| 7 | AME C18:4 isomer | 30.66 | $C_{58}H_{78}O_5$ | 855.5922 | 855.5877 | −5.27 |
| 8 | AME C18:1 isomer | 37.67 | $C_{58}H_{84}O_5$ | 861.6392 | 861.6346 | −5.29 |
| 9 | Chlorophyll b epimer | 38.62 | $C_{55}H_{70}N_4O_6Mg$ | 907.5219 | 907.5294 | 8.32 |
| 10 | AME C18:2 isomer | 41.76 | $C_{58}H_{82}O_5$ | 859.6235 | 859.6211 | −2.85 |
| 11 | Chlorophyll a epimer | 42.96 | $C_{55}H_{72}N_4O_5Mg$ | 893.5426 | 893.5474 | 5.43 |

* AME = Astaxanthin monoester. C18:n indicates an unsaturated fatty acyl part, with n being the number of double bonds in the molecule. ** Data from the HPLC-PDA run.

**Table 2.** Carotenoid profiles based on the LC-MS data of the extract of the nutrient-depleted conditions from *Coelastrum* sp. TISTR 9501RE (a representative set of data from three experiments).

| Entry | Identity * | Retention Time ** | Proposed Formula | Precursor Mass | Found at Mass | Mass Error (ppm) |
|---|---|---|---|---|---|---|
| 1 | All-*trans*-astaxanthin | 16.22 | $C_{40}H_{52}O_4$ | 597.3938 | 597.3935 | −0.53 |
| 2 | Violaxanthin isomer | 17.28 | $C_{40}H_{56}O_4$ | 601.4251 | 601.4251 | −0.04 |
| 3 | All-*trans*-lutein | 19.33 | $C_{40}H_{56}O_2$ | 569.4353 | 569.4327 | −4.54 |
| 4 | Chlorophyll b | 19.33 | $C_{55}H_{70}N_4O_6Mg$ | 907.5219 | 907.5218 | −0.03 |
| 5 | All-*trans*-Canthaxanthin | 23.46 | $C_{40}H_{52}O_2$ | 565.4040 | 565.4035 | −0.87 |
| 6 | Chlorophyll a | 25.65 | $C_{55}H_{72}N_4O_5Mg$ | 893.5426 | 893.5425 | −0.12 |
| 7 | AME C18:4 isomer | 30.67 | $C_{58}H_{78}O_5$ | 855.5922 | 855.5918 | −0.51 |
| 8 | AME C18:1 isomer | 37.69 | $C_{58}H_{84}O_5$ | 861.6392 | 861.6392 | 0.02 |
| 9 | Chlorophyll b epimer | 38.64 | $C_{55}H_{70}N_4O_6Mg$ | 907.5220 | 907.5329 | 12.2 |
| 10 | AME C18:2 isomer | 41.77 | $C_{58}H_{82}O_5$ | 859.6235 | 859.6231 | −0.49 |
| 11 | Chlorophyll a epimer | 43.00 | $C_{55}H_{72}N_4O_5Mg$ | 893.5426 | 893.5507 | 9.04 |

* AME = Astaxanthin monoester. C18:n indicates an unsaturated fatty acyl part, with n being the number of double bonds in the molecule. ** Data from the HPLC-PDA run.

Thereafter, some quantitative studies were conducted to allow for further comparisons with previous studies. In this regard, we prepared calibration plots (Figure S3) and quantified three compounds, namely astaxanthin, canthaxanthin, and lutein, using LC-MS experiments. This provided the benefit of not having to develop a new method, since certain compounds were co-eluted with other components, thereby causing some quantification errors if HPLC-PDA were to be used. As a result, we determined that the strain *Coelastrum* sp. TISTR 9501RE produced all the compounds in higher amounts when stressed with limited nutrients (Table 3). However, as alluded above, the total astaxanthin was higher owing to the conjugation with various fatty acids.

**Table 3.** Quantitative data from the LC-MS of three carotenoids (astaxanthin, lutein, and canthaxanthin) produced from *Coelastrum* sp. TISTR 9501RE under control and nutrient-depleted conditions. Data shown are averaged from three experiments.

| Compound | Amount (mg/g DW) | | |
|---|---|---|---|
| | Control Condition | Nutrient-Depleted Condition | Nutrient-Depleted Condition (20,000-L Pond) |
| All-*trans*-astaxanthin | 0.03 ± 0.001 | 0.11 ± 0.01 | 0.18 ± 0.004 |
| All-*trans*-lutein | 2.35 ± 0.05 | 4.18 ± 0.46 | 3.13 ± 0.07 |
| All-*trans*-Canthaxanthin | 0.27 ± 0.03 | 1.15 ± 0.10 | 1.37 ± 0.03 |

Interestingly, when compared to previous studies, the *Coelastrum* sp. TISTR 9501RE exhibited a unique carotenoid profile. For example, our strain yielded about one half of the amount of all-trans-astaxanthin as did *H. pluvialis* from Jin and coworkers (0.25 ± 0.04 mg/g DW vs. 0.11 ± 0.01 mg/g DW in our case) under that study's nitrogen deficiency condition [50]. However, none of their conditions, including conditions with higher astaxanthin contents, yielded comparable amounts of canthaxanthin and lutein as obtained in our case. On the other hand, a dark condition from their study did provide a significant amount of lutein, but at the expense of the total absence of astaxanthin and canthaxanthin. Significantly higher amounts of astaxanthin could be obtained, although with other stress conditions requiring high energy, such as 6000-lx cool white fluorescent light. Interestingly, whilst *H. pluvialis* is well known as the most efficient astaxanthin producer, it has some notable drawbacks, including slow growth at room temperature, ease of contamination by other microalgae, and a high light requirement [51]. Hence, it is relatively uncommon to cultivate this strain in more relaxed, but large-scale conditions, such as the one demonstrated herein (see below), which has prompted researchers to find alternative species [52]. For more similar organisms, a report by Liu et al. [44] showed that *Coelastrum* sp. HA-1 could produce astaxanthin at 6.36 mg/g DW, although it was not possible to compare it with other carotenoid profiles, since there was no identification of other carotenoids at all. Similarly, the work by Minhas et al. reported the production of 1.17 mg/L of astaxanthin and 0.64 mg/g DW of lutein for *Coelastrella* sp. (P63), but there was no information on other carotenoid species [53]. Importantly, the currently reported *Coelastrum* sp. TISTR 9501RE produced a relatively limited amount of β-carotene, and only under the control condition (around 62.5 min in Figure 4A), which was in contrast to some of the aforementioned works. Another example is from Hu and coworkers [54], who reported the presence of five components, including astaxanthin, canthaxanthin, lutein, β-carotene, and adonirubin, without the identification of other carotenoids.

Last but not least, since the stress condition used in this study provided the practical advantage of consuming less nutrients, we explored the possibility of growing this microalga on a much larger scale. To demonstrate its great applicability, we employed ambient sunlight, ambient temperature (ranging from 30 to 35 °C), groundwater as the water source, paddle wheels for stirring without active air feeding, and the same reduced nutrient (1/4 diluted BG11) to cultivate the *Coelastrum* sp. TISTR 9501RE in a 20,000-L open raceway pond. Interestingly, the HPLC and LC-MS analyses (Figure 5 and Table 4) of the extract showed a drastically simpler carotenoid profile, where all the major peaks were only astaxanthin, lutein, and canthaxanthin (excluding two chlorophyll species). Given that these carotenoids are attractive candidates for nutraceutical applications [21] due to their potent antioxidant activities, this condition thereby served as a prime example for more extensive application in large-scale productions in the future.

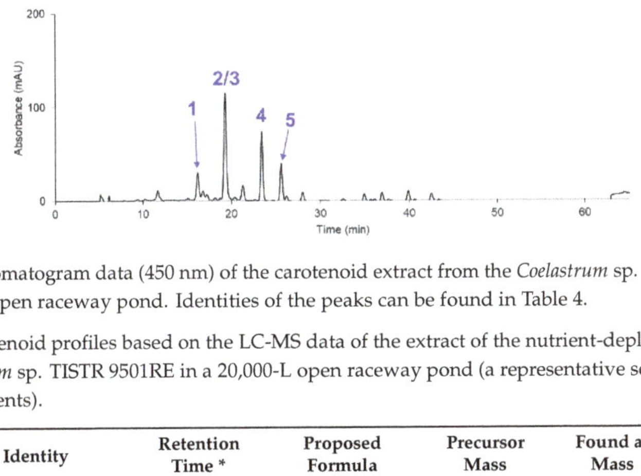


**Figure 5.** Chromatogram data (450 nm) of the carotenoid extract from the *Coelastrum* sp. TISTR 9501RE in a 20,000-L open raceway pond. Identities of the peaks can be found in Table 4.

**Table 4.** Carotenoid profiles based on the LC-MS data of the extract of the nutrient-depleted condition from *Coelastrum* sp. TISTR 9501RE in a 20,000-L open raceway pond (a representative set of data from three experiments).

| Entry | Identity | Retention Time * | Proposed Formula | Precursor Mass | Found at Mass | Mass Error (ppm) |
|---|---|---|---|---|---|---|
| 1 | All-*trans*-astaxanthin | 16.21 | $C_{40}H_{52}O_4$ | 597.3938 | 597.3935 | −0.60 |
| 2 | All-*trans*-lutein | 19.32 | $C_{40}H_{56}O_2$ | 569.4353 | 569.4351 | −0.39 |
| 3 | Chlorophyll b | 19.32 | $C_{55}H_{70}N_4O_6Mg$ | 907.5219 | 907.5227 | 0.88 |
| 4 | All-*trans*-Canthaxanthin | 23.44 | $C_{40}H_{52}O_2$ | 565.4040 | 565.4044 | 0.65 |
| 5 | Chlorophyll a | 25.64 | $C_{55}H_{72}N_4O_5Mg$ | 893.5426 | 893.5434 | 0.95 |

* Data from the HPLC-PDA run.

## 3. Materials and Methods

### 3.1. Microalgal Strain and Culture Conditions

The green microalga *Coelastrum* sp. was isolated from a coastal ecosystem in northern Thailand (obtained from the algae library of the Thailand Institute of Scientific and Technological Research (TISTR)). Cells were grown photoautotrophically (70 µmol $m^{-2}$ $s^{-1}$) in BG11 medium [55] or onto BG11 agar at 25 °C, unless otherwise stated. For carotenoids induction, the algal cells were inoculated at 3% (*v/v*) into 50 mL of BG11 and one-fourth strength BG11 medium (three replicates). All the flasks were shaken at 110 rpm under light 75–100 µmol $m^{-2}$ $s^{-1}$ for 14 days, with intermittent $OD_{730}$ measurements for growth study (which was repeated for three independent experiments). The cells were harvested using centrifugation (12,000 × $g$, 10 min, 4 °C), and then dehydrated using a Flexi-Dry MP freeze dryer (Kinetics, Stone Ridge, NY, USA). The freeze-dried cells were used for carotenoids extraction (see below), and the subsequent HPLC analyses. The effect of salinity on growth of the microalgal strain was performed onto BG11 agar supplemented with different doses of NaCl (0, 0.15, 0.30, 0.40, and 0.5 M).

For large-scale production, the algal cells (10% inoculum) were cultivated in a 20,000-L open raceway pond equipped with paddle wheels (to ensure thorough mixing) under natural sunlight in one-fourth strength BG11 medium for 14 days. Cells were harvested by precipitation and centrifugation, using a SSE80-06-077 centrifuge (at 6200 rpm as per the manufacturer's instructions, GEA Westfalia Separator Group GmbH, Oelde, Germany), followed by freeze drying and storage at −20 °C before analysis.

### *3.2. Molecular Identification and Polyphasic Taxonomy Approaches*

All molecular cloning methods were performed according to standard protocols described in Reference [56]. PCR amplification of the internal transcribed spacer (ITS) 1 of the 5.8S ribosomal RNA gene was performed with the following primers: ITS forward 1 5′-TCCGTAGGTGAACCTGCGG-3′ and ITS reverse4 5′-TCCTCCGCTTATTGATATGC-3′. DNA sequencing was performed using an ABI 310 Genetic Analyzer (Applied Biosystems, Foster City, CA, USA). The ITS1-5.8s-ITS2 sequence was deposited into the GenBank under accession number MH853895. The phylogenetic tree was constructed and analyzed by the neighbor-joining method using the Molecular Evolutionary Genetics Analysis (MEGA6) software (free of charge from http://www.megasoftware.net). The robustness of the tree was assessed using bootstrap analysis (100 replicates).

### *3.3. Carotenoids Extraction*

Thereafter, 70-mg freeze-dried cell powder was dissolved in 1-mL acetone and mixed with 0.5-mm silica glass beads (BeadBug™, Sigma-Aldrich, St. Louis, MO, USA). The resulting suspension was subjected to alternating cycles of vortexing and sonication as described below.

1. The suspension was vortexed for 5 min, followed by 2 min centrifugation at 10,000 rpm. The supernatant was collected, and 1 mL of fresh acetone was added to the crude precipitate. This process was repeated one more time.
2. The suspension from step 1 was sonicated for 30 min, followed by 2 min centrifugation at 10,000 rpm. Then, the supernatant was collected, and 1 mL of fresh acetone was added to the crude precipitate.
3. Step 1 was repeated exactly as shown above.
4. Step 2 was repeated exactly as shown above.
5. Step 1 was repeated exactly as shown above.

Then, all the supernatant fractions were combined. Thereafter, the solvent from the resulting solution was removed using a rotary evaporator. After that, the dried pigment mixture was ready for further analyses.

### *3.4. Determination of the Overall Chlorophyll Content*

The overall content of chlorophylls was determined by adapting the method from Ritchie [57]. Briefly, crude extract was added with acetone to create a solution at a concentration of 0.5 mg/mL. This solution was then subjected to absorption measurement at 630, 647, 664, and 691 nm using a Cary 100 Bio-UV visible spectrophotometer (Agilent, Santa Clara, CA, USA). The resulting absorbance data were used to calculate the content of chlorophylls and carotenoids based on the following formulae (as μg/mL) [57].

$$\text{Chlorophyll a} = -0.3319A_{630} - 1.7485A_{647} + 11.9442A_{664} - 1.4306A_{691}$$

$$\text{Chlorophyll b} = -1.2825A_{630} + 19.8839A_{647} - 4.8860A_{664} - 2.3416A_{691}$$

$$\text{Total Chlorophyll} = 21.3877A_{630} + 10.3739A_{647} + 5.3805A_{664} + 5.5309A_{691}$$

### 3.5. HPLC and LC-MS Analyses

A high performance liquid chromatograph (HPLC) with a photodiode array (PDA) detector was used to reveal the composition of the carotenoid mixtures. The following parameters were used in these experiments. Three separate experiments were performed for each sample, although one representative set was selected for illustration purposes in the results and discussion section.

Model: UltiMate 3000 HPLC ThermoFisher Scientific (Waltham, MA, USA); column: YMC30 reverse phase column (3 μm, ID 4.6 mm × 150 mm) (Kyoto, Japan); column temperature: 35 °C; injection volume: 10 μL; flow rate: 0.3 mL/min; mobile phase A: MeOH:MTBE (methyl *t*-butyl ether):$H_2O$ (81:15:4); mobile phase B: MeOH:MTBE:$H_2O$ (16:80.4:3.6). The time program can be found in Table S1 in the supplementary material.

Liquid chromatography–mass spectrometry (LC-MS) was performed to further characterize the carotenoids found in the mixture with the following parameters. Three separate experiments were performed for each sample, although one representative set was selected for illustration purposes in the results and discussion section.

Model: HPLC—ExionLC™ AD ultra-high performance liquid chromatograph (UHPLC); MS—SciEx X500R quadrupole time-of-flight MS (QTOF) (Framingham, MA, USA). MS parameters: mass range = 500–1300 *m/z*, positive mode; ion source gas 1 = 50 psi; ion source gas 2 = 50 psi; source temperature = 500 °C; spray voltage = 5500 V; declustering potential (DP) = 50 V; collision energy (CE) = 10 V.

The time programs for both the HPLC-PDA and LC-MS analyses were the same (Table S1). The concentrations of the mixtures used for the analysis were 10 and 2 mg/mL in acetone for the HPLC-PDA and LC-MS experiments, respectively. For the quantification experiments, standard solutions of the carotenoids of interest were prepared (500 ppm in acetonitrile for astaxanthin and canthaxanthin; 1000 ppm in acetone for lutein). Then each solution was serially diluted to different ranges that covered the relevant concentrations in the samples (0.5–8 ppm for astaxanthin, 1–20 ppm for canthaxanthin, and 10–150 ppm for lutein). All of these solutions were then analyzed and quantified by LC-MS in triplicates.

## 4. Conclusions

In conclusion, in this study, we reported the genetic identification and the study of carotenoid profiles of a new *Coelastrum* strain named *Coelastrum* sp. TISTR 9501RE. With the stress condition being the overall reduction in nutrients, the strain provided unique carotenoid compositions, with astaxanthin, canthaxanthin, and lutein being the major components. Interestingly, large-scale production could also be achieved under sustainable conditions such as ambient light, with even higher amounts of the desirable carotenoids. Further studies on a variety of stress conditions, especially the use of saltwater as a stress condition, and their effects on the biosynthesis pathways of carotenoids are an ongoing investigation in our group, so as to gain a better understanding and achieve more efficient carotenoid production.

**Supplementary Materials:** The following are available online at http://www.mdpi.com/1660-3397/17/6/328/s1, Figure S1: Growth curves of *Coelastrum* sp. TISTR 9501RE under (A) the control condition, and (B) the nutrient-depleted condition. Data shown are averaged from three experiments, Figure S2: Effect of salinity on the growth and cell morphology of *Coelastrum* sp. TISTR 9501RE. (A) Effect of salinity on the growth of *Coelastrum* sp. TISTR 9501RE was performed onto BG11 agar supplemented with different doses of NaCl (0, 0.15, 0.30, 0.40, and 0.50 M). For each concentration, 20 μL of microalgal culture in BG11 ($OD_{730}$ ~ 1) was streaked onto BG11 at an indicated NaCl concentration. Plate cultures were incubated photoautotrophically (70 μmol $m^{-2}$ $s^{-1}$) at 25 °C for 3 days. Salinity tolerance was scored by assessing the growth or lack of growth. (B) Light microscopic images of *Coelastrum* sp. TISTR 9501RE were grown in BG11 liquid medium supplemented with different concentrations of NaCl (0, 0.15, 0.30, 0.45 M). Microalgal culture in BG11 ($OD_{730}$ ~ 1) was inoculated into new BG11 liquid medium at each indicated NaCl concentration. Incubation condition was performed in the same way as indicated in (A). The morphology of the cells was observed under a light microscope (Olympus BX51, Japan), Figure S3: Calibration plots for astaxanthin (top), canthaxanthin (middle), and lutein (bottom) quantifications. Data shown are averaged from three experiments, Table S1: The time program for the HPLC-PDA and LC-MS experiments.

**Author Contributions:** Conceptualization, R.W.-S. and T.P.; methodology, R.W.-S. and T.P.; validation, M.R., K.J., and P.K.; formal analysis, M.R. and K.J.; investigation, M.R. and K.J.; resources, P.K., S.S., R.W.-S., and T.P.; writing—original draft preparation, R.W.S. and T.P.; writing—review and editing, R.W.-S. and T.P.; visualization, S.S., M.R., R.W.-S., and T.P.; supervision, R.W.-S. and T.P.; project administration, R.W.-S. and T.P.; funding acquisition, T.P.

**Funding:** This research was funded by the Thailand Research Fund (MRG6280238). A PhD scholarship was provided to MR from the Development and Promotion of Science and Technology Talented Project (DPST), Thailand.

**Conflicts of Interest:** The authors declare no conflict of interest.

## References

1. Maresca, J.A.; Graham, J.E.; Bryant, D.A. The biochemical basis for structural diversity in the carotenoids of chlorophototrophic bacteria. *Photosynth. Res.* **2008**, *97*, 121–140. [CrossRef] [PubMed]
2. Guedes, A.C.; Amaro, H.M.; Malcata, F.X. Microalgae as sources of carotenoids. *Mar. Drugs* **2011**, *9*, 625–644. [CrossRef]
3. Avalos, J.; Carmen Limón, M. Biological roles of fungal carotenoids. *Curr. Genet.* **2015**, *61*, 309–324. [CrossRef] [PubMed]
4. Gong, M.; Bassi, A. Carotenoids from microalgae: A review of recent developments. *Biotechnol. Adv.* **2016**, *34*, 1396–1412. [CrossRef] [PubMed]
5. Adadi, P.; Barakova, N.V.; Krivoshapkina, E.F. Selected methods of extracting carotenoids, characterization, and health concerns: A review. *J. Agric. Food Chem.* **2018**, *66*, 5925–5947. [CrossRef]
6. Polívka, T.; Frank, H.A. Molecular factors controlling photosynthetic light harvesting by carotenoids. *Acc. Chem. Res.* **2010**, *43*, 1125–1134. [CrossRef]
7. Christaki, E.; Bonos, E.; Giannenas, I.; Florou-Paneri, P. Functional properties of carotenoids originating from algae. *J. Sci. Food Agric.* **2013**, *93*, 5–11. [CrossRef]
8. Wang, B.; Zarka, A.; Trebst, A.; Boussiba, S. Astaxanthin accumulation in *Haematococcus pluvialis* (chlorophyceae) as an active photoprotective process under high irradiance. *J. Phycol.* **2003**, *39*, 1116–1124. [CrossRef]
9. Cazzonelli, C.I. Carotenoids in nature: Insights from plants and beyond. *Funct. Plant Biol.* **2011**, *38*, 833–847. [CrossRef]
10. Erickson, E.; Wakao, S.; Niyogi, K.K. Light stress and photoprotection in *Chlamydomonas reinhardtii*. *Plant J.* **2015**, *82*, 449–465. [CrossRef] [PubMed]
11. Breithaupt, D.E. Modern application of xanthophylls in animal feeding—A review. *Trends Food Sci. Technol.* **2007**, *18*, 501–506. [CrossRef]
12. Vílchez, C.; Forján, E.; Cuaresma, M.; Bédmar, F.; Garbayo, I.; Vega, J.M. Marine carotenoids: Biological functions and commercial applications. *Mar. Drugs* **2011**, *9*, 319–333. [CrossRef] [PubMed]
13. Borowitzka, M.A. High-value products from microalgae—Their development and commercialisation. *J. Appl. Phycol.* **2013**, *25*, 743–756. [CrossRef]
14. Sathasivam, R.; Ki, J.-S. A review of the biological activities of microalgal carotenoids and their potential use in healthcare and cosmetic industries. *Mar. Drugs* **2018**, *16*, 26. [CrossRef]
15. Naguib, Y.M.A. Antioxidant activities of astaxanthin and related carotenoids. *J. Agric. Food Chem.* **2000**, *48*, 1150–1154. [CrossRef]
16. Higuera-Ciapara, I.; Félix-Valenzuela, L.; Goycoolea, F.M. Astaxanthin: A review of its chemistry and applications. *Crit. Rev. Food Sci. Nutr.* **2006**, *46*, 185–196. [CrossRef]
17. Yuan, J.-P.; Peng, J.; Yin, K.; Wang, J.-H. Potential health-promoting effects of astaxanthin: A high-value carotenoid mostly from microalgae. *Mol. Nutr. Food Res.* **2011**, *55*, 150–165. [CrossRef] [PubMed]
18. Ambati, R.; Phang, S.-M.; Ravi, S.; Aswathanarayana, R. Astaxanthin: Sources, extraction, stability, biological activities and its commercial applications—A review. *Mar. Drugs* **2014**, *12*, 128. [CrossRef] [PubMed]
19. Visioli, F.; Artaria, C. Astaxanthin in cardiovascular health and disease: Mechanisms of action, therapeutic merits, and knowledge gaps. *Food Funct.* **2017**, *8*, 39–63. [CrossRef]

20. Davinelli, S.; Nielsen, M.; Scapagnini, G. Astaxanthin in skin health, repair, and disease: A comprehensive review. *Nutrients* **2018**, *10*, 522. [CrossRef]
21. Viera, I.; Pérez-Gálvez, A.; Roca, M. Bioaccessibility of marine carotenoids. *Mar. Drugs* **2018**, *16*, 26. [CrossRef] [PubMed]
22. Torrissen, O.J. Pigmentation of salmonids—Effect of carotenoids in eggs and start-feeding diet on survival and growth rate. *Aquaculture* **1984**, *43*, 185–193. [CrossRef]
23. Surai, A.P.; Surai, P.F.; Steinberg, W.; Wakeman, W.G.; Speake, B.K.; Sparks, N.H.C. Effect of canthaxanthin content of the maternal diet on the antioxidant system of the developing chick. *Br. Poult. Sci.* **2003**, *44*, 612–619. [CrossRef]
24. Esatbeyoglu, T.; Rimbach, G. Canthaxanthin: From molecule to function. *Mol. Nutr. Food Res.* **2017**, *61*, 1600469. [CrossRef]
25. Ma, L.; Lin, X.-M. Effects of lutein and zeaxanthin on aspects of eye health. *J. Sci. Food Agric.* **2010**, *90*, 2–12. [CrossRef]
26. Koushan, K.; Rusovici, R.; Li, W.; Ferguson, L.; Chalam, K. The role of lutein in eye-related disease. *Nutrients* **2013**, *5*, 1823–1829. [CrossRef] [PubMed]
27. Manayi, A.; Abdollahi, M.; Raman, T.; Nabavi, S.F.; Habtemariam, S.; Daglia, M.; Nabavi, S.M. Lutein and cataract: From bench to bedside. *Crit. Rev. Biotechnol.* **2016**, *36*, 829–839. [CrossRef]
28. Del Campo, J.A.; García-González, M.; Guerrero, M.G. Outdoor cultivation of microalgae for carotenoid production: Current state and perspectives. *Appl. Microbiol. Biotechnol.* **2007**, *74*, 1163–1174. [CrossRef]
29. Varela, J.C.; Pereira, H.; Vila, M.; León, R. Production of carotenoids by microalgae: Achievements and challenges. *Photosynth. Res.* **2015**, *125*, 423–436. [CrossRef] [PubMed]
30. Sun, X.-M.; Ren, L.-J.; Zhao, Q.-Y.; Ji, X.-J.; Huang, H. Microalgae for the production of lipid and carotenoids: A review with focus on stress regulation and adaptation. *Biotechnol. Biofuels* **2018**, *11*, 272. [CrossRef] [PubMed]
31. Saini, R.K.; Keum, Y.-S. Carotenoid extraction methods: A review of recent developments. *Food Chem.* **2018**, *240*, 90–103. [CrossRef] [PubMed]
32. Huang, W.; Lin, Y.; He, M.; Gong, Y.; Huang, J. Induced high-yield production of zeaxanthin, lutein, and β-carotene by a mutant of *Chlorella zofingiensis*. *J. Agric. Food Chem.* **2018**, *66*, 891–897. [CrossRef] [PubMed]
33. Patrick, L. Beta carotene: The controversy continues. *Altern. Med. Rev.* **2000**, *5*, 530–545. [PubMed]
34. Lin, J.-H.; Lee, D.-J.; Chang, J.-S. Lutein production from biomass: Marigold flowers versus microalgae. *Bioresour. Technol.* **2015**, *184*, 421–428. [CrossRef] [PubMed]
35. Sowbhagya, H.B.; Sampathu, S.R.; Krishnamurthy, N. Natural colorant from marigold-chemistry and technology. *Food Rev. Int.* **2004**, *20*, 33–50. [CrossRef]
36. Rodríguez-Sáiz, M.; de la Fuente, J.L.; Barredo, J.L. *Xanthophyllomyces dendrorhous* for the industrial production of astaxanthin. *Appl. Microbiol. Biotechnol.* **2010**, *88*, 645–658. [CrossRef] [PubMed]
37. Li, J.; Zhu, D.; Niu, J.; Shen, S.; Wang, G. An economic assessment of astaxanthin production by large scale cultivation of *Haematococcus pluvialis*. *Biotechnol. Adv.* **2011**, *29*, 568–574. [CrossRef]
38. Hejazi, M.A.; Holwerda, E.; Wijffels, R.H. Milking microalga *Dunaliella salina* for β-carotene production in two-phase bioreactors. *Biotechnol. Bioeng.* **2004**, *85*, 475–481. [CrossRef]
39. Kim, S.-H.; Liu, K.-H.; Lee, S.-Y.; Hong, S.-J.; Cho, B.-K.; Lee, H.; Lee, C.-G.; Choi, H.-K. Effects of light intensity and nitrogen starvation on glycerolipid, glycerophospholipid, and carotenoid composition in *Dunaliella tertiolecta* culture. *PLoS ONE* **2013**, *8*, e72415. [CrossRef] [PubMed]
40. Minyuk, G.; Chelebieva, E.; Chubchikova, I.; Dantsyuk, N.; Drobetskaya, I.; Sakhon, E.; Chekanov, K.; Solovchenko, A. Stress-induced secondary carotenogenesis in *Coelastrella rubescens* (Scenedesmaceae, Chlorophyta), a producer of value-added keto-carotenoids. *Algae* **2017**, *32*, 245–259. [CrossRef]
41. Zhang, P.; Li, Z.; Lu, L.; Xiao, Y.; Liu, J.; Guo, J.; Fang, F. Effects of stepwise nitrogen depletion on carotenoid content, fluorescence parameters and the cellular stoichiometry of *Chlorella vulgaris*. *Spectrochim. Acta Part A Mol. Biomol. Spectrosc.* **2017**, *181*, 30–38. [CrossRef] [PubMed]
42. Huang, J.J.; Lin, S.; Xu, W.; Cheung, P.C.K. Enhancement of the production of bioactive microalgal metabolites by ultraviolet radiation (UVA 365 nm). *J. Agric. Food Chem.* **2018**, *66*, 10215–10224. [CrossRef]

43. Janchot, K.; Rauytanapanit, M.; Honda, M.; Hibino, T.; Sirisattha, S.; Praneenararat, T.; Kageyama, H.; Waditee-Sirisattha, R. Effects of potassium chloride-induced stress on the carotenoids canthaxanthin, astaxanthin, and lipid accumulations in the green Chlorococcal microalga strain TISTR 9500. *J. Eukaryot. Microbiol.* **2019**. [CrossRef]
44. Liu, Z.; Liu, C.; Hou, Y.; Chen, S.; Xiao, D.; Zhang, J.; Chen, F. Isolation and characterization of a marine microalga for biofuel production with astaxanthin as a co-product. *Energies* **2013**, *6*, 2759–2772. [CrossRef]
45. Úbeda, B.; Gálvez, J.Á.; Michel, M.; Bartual, A. Microalgae cultivation in urban wastewater: *Coelastrum* cf. *pseudomicroporum* as a novel carotenoid source and a potential microalgae harvesting tool. *Bioresour. Technol.* **2017**, *228*, 210–217.
46. Soares, A.T.; da Costa, D.C.; Vieira, A.A.H.; Antoniosi Filho, N.R. Analysis of major carotenoids and fatty acid composition of freshwater microalgae. *Heliyon* **2019**, *5*, e01529. [CrossRef]
47. Štenclová, L.; Fučíková, K.; Kaštovský, J.; Pažoutová, M. Molecular and morphological delimitation and generic classification of the family Oocystaceae (Trebouxiophyceae, Chlorophyta). *J. Phycol.* **2017**, *53*, 1263–1282. [CrossRef]
48. Deli, J.; Gonda, S.; Nagy, L.Z.; Szabó, I.; Gulyás-Fekete, G.; Agócs, A.; Marton, K.; Vasas, G. Carotenoid composition of three bloom-forming algae species. *Food Res. Int.* **2014**, *65*, 215–223. [CrossRef]
49. Han, D.; Li, Y.; Hu, Q. Astaxanthin in microalgae: Pathways, functions and biotechnological implications. *Algae* **2013**, *28*, 131–147. [CrossRef]
50. Jin, H.; Lao, Y.M.; Zhou, J.; Zhang, H.J.; Cai, Z.H. Simultaneous determination of 13 carotenoids by a simple C18 column-based ultra-high-pressure liquid chromatography method for carotenoid profiling in the astaxanthin-accumulating *Haematococcus pluvialis*. *J. Chromatogr. A* **2017**, *1488*, 93–103. [CrossRef]
51. Masojídek, J.; Torzillo, G. Mass cultivation of freshwater microalgae. In *Encyclopedia of Ecology*; Jørgensen, S.E., Fath, B.D., Eds.; Academic Press: Oxford, UK, 2008; pp. 2226–2235.
52. Liu, J.; Sun, Z.; Gerken, H.; Liu, Z.; Jiang, Y.; Chen, F. *Chlorella zofingiensis* as an alternative microalgal producer of astaxanthin: Biology and industrial potential. *Mar. Drugs* **2014**, *12*, 3487–3515. [CrossRef] [PubMed]
53. Minhas, A.K.; Hodgson, P.; Barrow, C.J.; Sashidhar, B.; Adholeya, A. The isolation and identification of new microalgal strains producing oil and carotenoid simultaneously with biofuel potential. *Bioresour. Technol.* **2016**, *211*, 556–565. [CrossRef] [PubMed]
54. Hu, C.-W.; Chuang, L.-T.; Yu, P.-C.; Chen, C.-N.N. Pigment production by a new thermotolerant microalga *Coelastrella* sp. F50. *Food Chem.* **2013**, *138*, 2071–2078. [CrossRef] [PubMed]
55. Stanier, R.Y.; Kunisawa, R.; Mandel, M.; Cohen-Bazire, G. Purification and properties of unicellular blue-green algae (order Chroococcales). *Bacteriol. Rev.* **1971**, *35*, 171–205. [PubMed]
56. Sambrook, J.; Fritsch, E.F.; Maniatis, T. *Molecular Cloning: A Laboratory Manual*, 2nd ed.; Cold Spring Harbor Laboratory Press: New York, NY, USA, 1989.
57. Ritchie, R.J. Universal chlorophyll equations for estimating chlorophylls a, b, c, and d and total chlorophylls in natural assemblages of photosynthetic organisms using acetone, methanol, or ethanol solvents. *Photosynthetica* **2008**, *46*, 115–126. [CrossRef]

marine drugs

Article

# Phytoene Accumulation in the Novel Microalga *Chlorococcum* sp. Using the Pigment Synthesis Inhibitor Fluridone

**Kelly Laje [1], Mark Seger [2], Barry Dungan [1], Peter Cooke [3], Juergen Polle [4,5] and F. Omar Holguin [1,*]**

1 Department of Plant and Environmental Sciences, New Mexico State University, Las Cruces, NM 88003, USA; klaje@nmsu.edu (K.L.); bdungan@nmsu.edu (B.D.)
2 AzCATI, School of Sustainable Engineering and the Built Environment, Arizona State University, Mesa, AZ 85212, USA; mseger1@asu.edu
3 Core University Research Resources Laboratory, New Mexico State Univesrity, Las Cruces, NM 88003, USA; phcooke@nmsu.edu
4 Department of Biology, Brooklyn College of the City University of New York, Brooklyn, NY 11210, USA; JPolle@brooklyn.cuny.edu
5 The Graduate Center of the City University of New York, 365 Fifth Avenue, New York, NY 10016, USA
* Correspondence: frholgui@nmsu.edu; Tel.: +575-646-5913

Received: 26 February 2019; Accepted: 19 March 2019; Published: 22 March 2019

**Abstract:** Carotenoids are lipophilic pigments found in plants and algae, as well as some bacteria, archaea, and fungi that serve two functions—(1) as light harvesting molecules—primary carotenoids, and (2) as antioxidants, acting against reactive oxygen species–secondary carotenoids. Because of their strong antioxidant properties, they are also valuable for the development of anti-aging and photo-protective cosmetic applications. Of particular interest is the carotenoid phytoene, for its colorless and UV absorption characteristics. In this study, we targeted a reduction of phytoene desaturase (PDS) activity with the pigment-inhibiting herbicide 1-methyl-3-phenyl-5-[3-(trifluoromethyl)phenyl]pyridin-4-one (fluridone), which leads to the over-accumulation of phytoene in the recently characterized microalgal strain *Chlorococcum* sp. (UTEX B 3056). After post-incubation with fluridone, phytoene levels were measured at ~33 ug/mg cell tissue, as opposed to non-detectable levels in control cultures. Hence, the novel microalga *Chlorococcum* sp. is a viable candidate for the production of the high-value carotenoid phytoene and subsequent applications in cosmeceuticals, as well as more obvious nutraceutical and pharmaceutical applications.

**Keywords:** phytoene; carotenoids; antioxidants; fluridone; microalgae; cosmeceuticals

## 1. Introduction

Microalgae are known to be potential sources of natural products, abundant and versatile in their activity and applications. Of particular importance are the lipophilic pigments, carotenoids. Commonly used in the food and nutraceuticals industry as colorants and dietary supplements, carotenoids have received growing popularity in cosmetics in large part, due to their antioxidant properties [1–4]. Synthesized in chloroplasts, carotenoids are a part of the photosynthetic complex (primary carotenoids), absorbing light in the 400–500 nm range, and also acting as a defense system in the presence of high light intensity or oxidative stress (secondary carotenoids) [5–7]. Secondary carotenoids act to quench singlet oxygen species and trap peroxyl radicals, protecting the cell from lipid peroxidation in both plants and animals [8–12]. Studies have shown that carotenoids also possess anti-inflammatory and immunomodulatory effects in animal tissues [8,13,14]. These qualities have made secondary carotenoids the subject of intense research surrounding anti-cancer therapies and heart disease, among others [8,15,16].

Carotenoids are either pure hydrocarbon molecules (carotenes) or oxygenated derivatives of carotenes (xanthophylls), all of which are comprised of a 40 carbon atom chain. One conjugated double bond is added with every carotenoid produced downstream of phytoene, in the synthetic chain, having a direct impact on the antioxidant strength of the molecule [8,16,17]. Thus, carotenoids are of particular importance for their potential as a natural source of antioxidants. The first carotenoid in the terpenoid pathway is phytoene; a symmetric, linear branched carotenoid with nine conjugated double bonds, produced from two C20 molecules of geranylgeranyl pyrophosphate (GGPP), and catalyzed by the enzyme phytoene synthase (PSY) [17,18]. In plants and green algae, phytoene progresses to phytofluene and ζ-carotene via phytoene desaturase (PDS). Subsequently, the carotenoid biosynthesis pathway proceeds to the carotenes–lycopene, and by ring introduction, to α-carotene and β-carotene; and then further to the xanthophylls–lutein (from α-carotene) and zeaxanthin (from β-carotene), respectively [5,16,18]. Secondary carotenoids are synthesized and accumulated during unfavorable growth conditions, such as high irradiance and/or nutrient deprivation, in which carotenoids contribute to cell protection (e.g., light absorption at a photosynthetic range beyond the capacity of chlorophyll) [19,20]. Depending on the species of alga, these secondary carotenoids may accumulate in carotene globules within the chloroplast [21,22] or in oil bodies in the cytosol, as seen during astaxanthin production in *Haematococcus pluvialis* [23,24].

Phytoene absorbs light in the ultraviolet range, and is colorless in nature; qualities that add to its value in cosmetic formulation as a skin protectant [13,25]. Current sources of phytoene come from tomato extract [26,27] and the carotenogenic microalga *Dunaliella bardawil* [28–30]. However, phytoene is difficult to accumulate in large quantities because, as a precursor molecule, it is used in the downstream synthesis of other primary and secondary carotenoids [18]. Phytoene levels in tomato (ripe) and *D. bardawil* (stress-induced) range from ~2–9 μg/g dry weight [31–33], and 8% (80 mg/g) [28], respectively.

Previous studies successfully induced the over accumulation of phytoene through the use of pigment synthesis inhibiting herbicides [29,31–33]. These bleaching herbicides target the enzyme phytoene desaturase (PDS), responsible for the downstream production of carotenoids past the metabolic step of phytoene production [34]. The inability to synthesize carotenoids that are essential for structure and function of photosynthetic complexes results in chlorophyll degradation, and ultimately, plant cell death [10,35–37]. At non-lethal doses, effective inhibition of PDS leads to the over-accumulation of phytoene [23,29,31,32,35,38]. This has been demonstrated in the microalgae *D. bardawil* and *H. pluvialis*, in which phytoene accumulation increased sharply as a result of exposure to bleaching herbicides [29,31–33]. *Chlamydomonas reinhardtii*, *H. pluvialis*, and the cyanobacteria *Synechococcus* have been studied extensively for norflurazon (5-amino-4-chloro-2-[3-(trifluoromethyl)phenyl] pyridazin-3-one) and fluridone (1-methyl-3-phenyl-5-[3-(trifluoromethyl)phenyl]pyridine-4-one) resistance mechanisms and mutagenesis, as well as herbicide inhibition activity [33,34,38–41].

In this study, our objective was to over-accumulate the carotenoid phytoene in a novel strain of green microalga, *Chlorococcum* sp. (UTEX B 3056), a fresh-water algae that closely resembles *C. reinhardtii* [42–44]. *Chlorococcum* exists as a unicellular, spheroidal organism, in either a vegetative (non-motile) or a zoospore (bi-flagellate) state [42,43]. We chose to study this strain of *Chlorococcum* sp. because it is highly carotenogenic, fast-growing, produces large quantities of biomass, and can be cultivated outdoors in raceway-type ponds [42,45]. We optimized the concentration of fluridone to facilitate the accumulation of phytoene without inducing bleaching and cell death. Furthermore, we characterized the effects of phytoene accumulation on the carotenoid and fatty acid (FA) profiles of cell extracts.

## 2. Results

### 2.1. Strain Identification & Morphology

Briefly, sequencing of the 18S rDNA confirmed previous characterization of the ITS2 region by Neofotis, et al., linking this alga to *Chlorococcum* sp. (Supplementary Figure S1) [42]. Neofotis, et al. pointed out that query coverage is low with this species and that unambiguous identification of this group at the species level, even with use of the ITS2 marker, is not definitive due to a lack of sequence availability in the public databases [42]. Morphological characterization via bright field and scanning electron microscopy agreed with molecular taxonomy; these images are provided in supplementary materials (Supplementary Figure S2).

### 2.2. Microplate Bioassays

*Chlorococcum* sp. growth was analyzed in the presence of fluridone at serial concentrations via UV spectrophotometric readings at the following wavelengths: 750 nm (overall growth), 680 nm (chlorophyll content), 450 nm (carotenoid content) (Figure 1) [7]. Note that cultures were started at an OD of 0.1 (day zero), and growth monitoring began the following day (day 1) (Figure 1). The overall growth and chlorophyll/carotenoid content of the cultures was significantly impacted at all concentrations of fluridone; thus, there appears to be no difference between the OD at each wavelength amongst the trends (panels A–C, Figure 1) [7]. The graph representing 750/450 nm showed highest growth/lowest carotenoid content in the 152 µM concentration. Upon experimental scale-up, we chose to treat cultures with the two highest doses, 152 µM and 304 µM, to observe the effects of the optimal concentration (152 µM), as well as the effects of a stronger dose (304 µM), on culture growth and phytoene accumulation (panel D, Figure 1). Although 152 µM does not appear to be significantly different between early and later time points in the 750/450 nm ratio, this is likely due to cell death and pigment inhibition over the course of the treatment (panel D, Figure 1). A two way repeated measures ANOVA, using the Holm-Sidak method, was performed to measure the significance of growth period and concentration. Herbicidal effects were dosage dependent, with a statistically significant interaction between day and concentration ($P \leq 0.001$). Asterisks denote treatments in which significance was observed (panels A–C, Figure 1). However, it should be noted that 152 µM (panel A, Figure 1) and 38 µM (panels A–C, Figure 1) treatments have a $p$ value of 0.007 and ~0.02, respectively, on day seven. Significance is not noted in panel D (Figure 1), as there was no statistical significance observed between treatments within a given day, unlike for panels A–C.

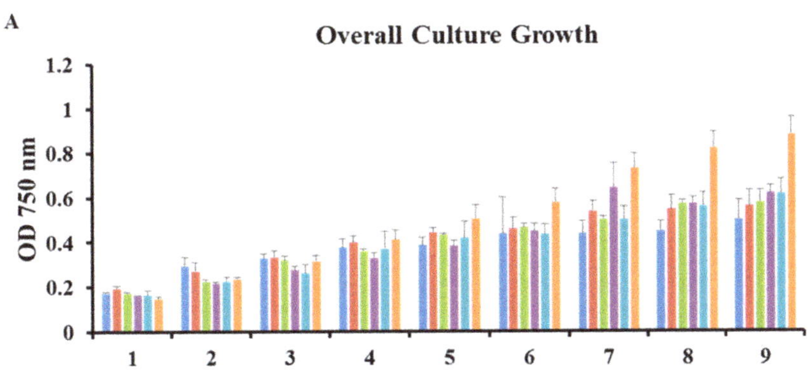


**Figure 1.** *Cont.*

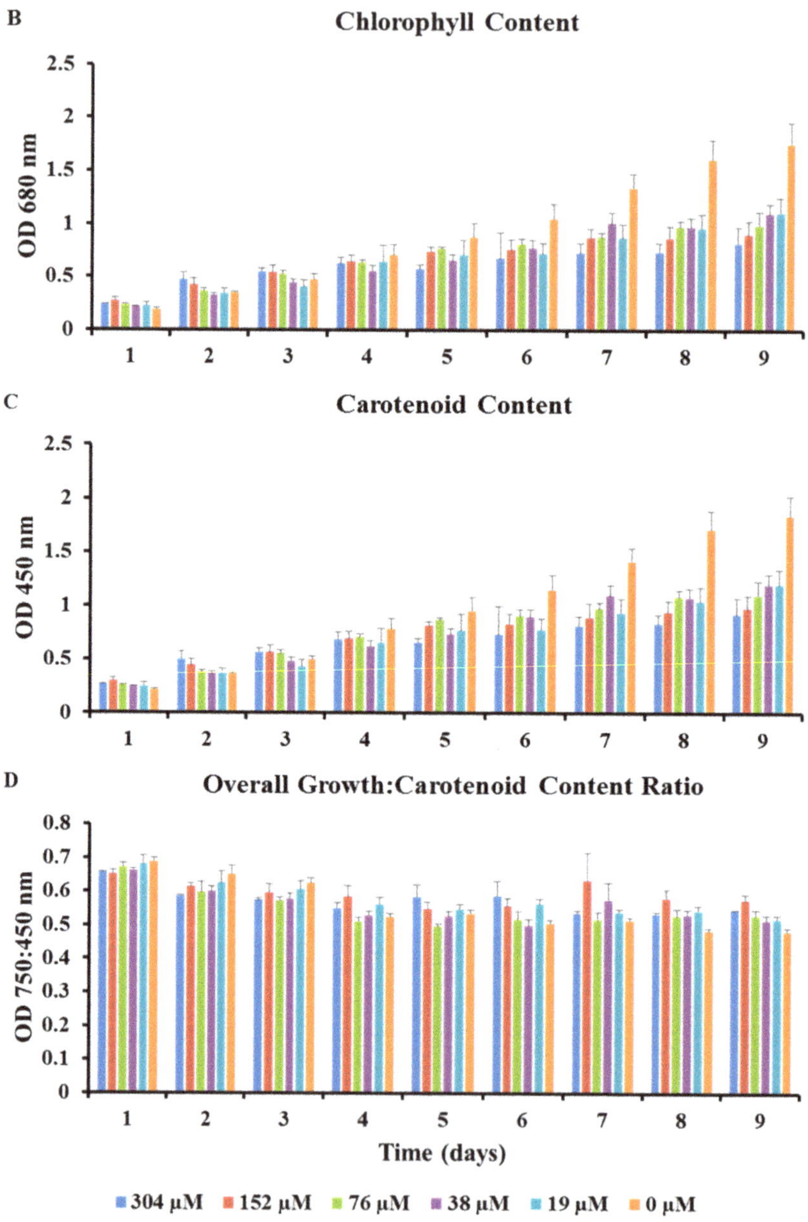


**Figure 1.** Micro-plate bioassay growth charts of *Chlorococcum* sp. over a series of fluridone treatments (**A**) Overall Culture Growth (750 nm); (**B**) Chlorophyll Content (680 nm); (**C**) Carotenoid Content (450 nm); (**D**) Overall Growth:Carotenoid Content Ratio used to determine optimal fluridone concentration (750:450 nm). * n = 4 for all samples; excluding day 6–304 µM, where n = 3. Asterisks indicate statistical significance in panels A–C. Significance in panel D not applicable.

## 2.3. Phytoene Quantification

Results in Panel B, Figure 1 (chlorophyll absorbance) indicate that algal growth begins to slow after day 4, and statistically significant differences in growth between treated and untreated cultures (panels A–C, Figure 1) are observed at day 7 and beyond. The statistical significance that occurs at days 7–9 indicates treated cultures were not growing as optimally as the control. Hence, we chose to harvest cell tissue for phytoene analysis when cultures were in optimal growth (day 4). Carotenoid extraction and subsequent HPLC analysis of 25 mL cultures *Chlorococcum* sp. incubated with 152 μM and 304 μM fluridone revealed the accumulation of phytoene at approximately 33 μg/mg of phytoene per dry cell weight when harvested on day four, as well as a reduction in downstream carotenoid production at both concentrations (Table 1). At the fourth-day harvest, there was no notable increase of phytoene accumulation when increasing the fluridone dose from 152 μM and 304 μM. (Table 1). Differences in phytoene levels became apparent when cell tissue was harvested after a nine-day incubation period. Phytoene quantification of this tissue revealed a reduction in the amount of phytoene accumulated in cultures treated with 304 μM as compared to 152 μM fluridone, at only 4.6 μg/mg, versus 14.6 μg/mg, respectively (Table 1). Carotenoid content at both harvest periods was reduced in fluridone-treated cultures by approximately half that seen in non-treated cultures ~40 μg/mg (treated cultures) vs. 70 μg/mg (controls), and ~70 μg/mg (treated cultures) vs. 145 μg/mg (controls), at the four-day and nine-day harvest, respectively (Table 1). Final carotenoid levels were within a standard deviation between concentrations.

Panel A, Figure 2 shows phytoene eluting at approximately twenty-seven minutes, absorbing at 284 nm in cultures that had been harvested on day 4 of treatment with 152 μM fluridone, and a relatively low amount of carotenoid production is observed. Chromatograms for controls (cultures without fluridone) contained no peak for phytoene (panel B, Figure 2) and exhibited downstream carotenoid products (i.e., lutein, zeaxanthin, and β-carotene).

**Table 1.** Total phytoene and carotenoids in *Chlorococcum* sp. with fluridone at 304 μM and 152 μM. n = 3, N.D. = not detected, SEM = standard error of the means.

| | Total Phytoene (μg/mg) | | | Total Carotenoids (μg/mg) | | |
|---|---|---|---|---|---|---|
| | | | **Day 4 Harvest** | | | |
| **Treatments (μM)** | **MEAN** | | **SEM** | **MEAN** | | **SEM** |
| 304 | 33.8 | ± | 1.7 | 40.4 | ± | 11.5 |
| 152 | 33 | ± | 0.3 | 38.6 | ± | 1.1 |
| 0 | N.D. | ± | N.D. | 70.1 | ± | 7.5 |
| | | | **Day 9 Harvest** | | | |
| **Treatments (μM)** | **MEAN** | | **SEM** | **MEAN** | | **SEM** |
| 304 | 4.6 | ± | 0.9 | 68.3 | ± | 4.9 |
| 152 | 14.6 | ± | 0.6 | 66.1 | ± | 4.8 |
| 0 | N.D. | ± | N.D. | 145.1 | ± | 7.2 |

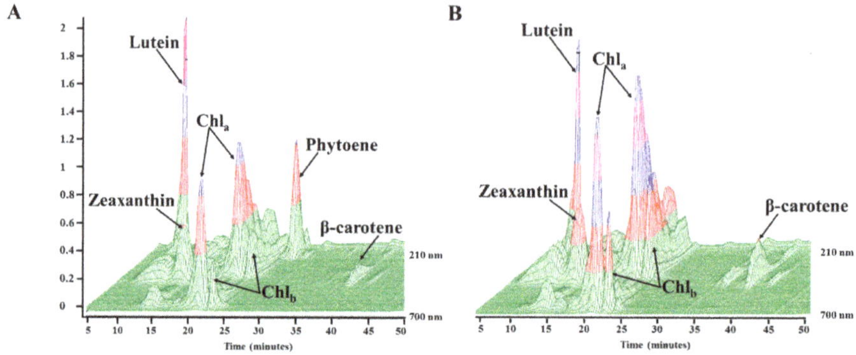


**Figure 2.** HPLC 3D chromatograms of cellular extracts from *Chlorococcum* sp., day 4 harvest (**A**) exposed to fluridone (152 μM). Note that phytoene was detected eluting at ~27 min. (**B**) the absence of phytoene without the addition of fluridone. Lutein, zeaxanthin, β-carotene and chlorophyll *a/b* are also denoted.

## 2.4. Fatty Acid Analysis

The fatty acid profile of cellular extracts obtained on day 4 from *Chlorococcum* sp. were analyzed for the observation of any potential downstream effects on fatty acid desaturase enzymes, in which previous studies have shown herbicides with this mode of action have exhibited inhibitory effects [23]. The FAs C16:0 and C16:3 remained relatively conserved within concentrations and controls, comparatively speaking. (panel A, Figure 3). However, the mono and poly-unsaturated FAs showed a slight increase in the presence of fluridone, from ~11 μg/mg (controls) to ~12 μg/mg (+ fluridone), and from ~8 μg/mg (controls) to ~11 μg/mg (+ fluridone), in C16:1 and C16:2, respectively (panels A & B, Figure 3). The increase in abundance of the aforementioned FAs was similar in both fluridone treatments (panels A & B, Figure 3). The abundance of the mono-unsaturated and poly-unsaturated FAs C18:1 cis/trans, C18:2 cis, and C18:3 in cultures incubated with 152 μM and 304 μM fluridone were not significantly different from that of the control cultures or between concentrations; <0.5 μg/mg difference (panel A, Figure 3). C18:0 concentration increased slightly in cultures incubated at 152 μM: from ~1.5 μg/mg (304 μM), to ~2.5 μg/mg (152 μM) (panel A, Figure 3). A two-way analysis of variance (ANOVA), using the Holm-Sidak method, was performed to determine any significance between FA levels, fluridone treatment, and treatment concentration (152 μM vs. 304 μM). Statistical significance has been noted for FA abundance between herbicide treatments and controls. However, statistical significance was not observed when comparing the two treatment concentrations. In other words, we did not see a significant change in the effect of 152 μM over 304 μM and the resulting FA abundance, overall. It should be noted, though, that C18:0 abundance was significantly different between the two concentrations, as an exception to the former statement. $P \leq 0.001$. Note: * n = 3 for all samples; excluding C16:2, where n = 2. Total FAME concentrations are also outlined in the supplementary material, Supplementary Table S2 (S3).

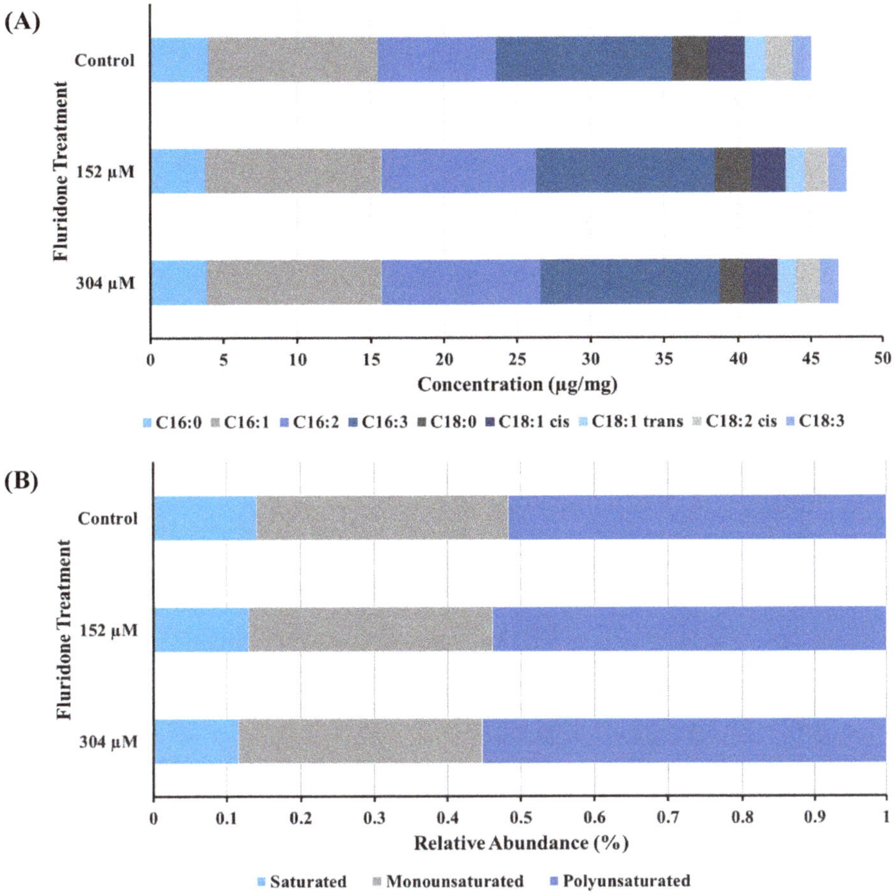


**Figure 3.** Fatty acid methyl ester (FAME) analysis of cellular extracts obtained on the day 4 harvest period from *Chlorococcum* sp. (**A**) total FA content, (**B**) relative abundance of saturated, monounsaturated, and polyunsaturated FAs.

### *2.5. Intracellular Oil Body Visualization*

Confocal fluorescence microscopy indicated non-uniformity/streaking of the chlorophyll (red fluorescence) in fluridone treated cultures, as opposed to control cultures, which showed fuller/more uniform chlorophyll fluorescence throughout the cell (panels C & E, Figure 4). This might indicate chloroplastic degradation in cultures incubated with fluridone. We also observed a minor increase in the number of oil bodies formed in cultures treated with both concentrations of fluridone, characterized by yellow fluorescent droplets within zoospores (smaller cells) and dormant aplanospores (larger cells) (panels C & E, Figure 4). Note that the dormant aplanospores are large cysts containing oil bodies that fluoresce yellow when observed microscopically [46]; whereas, the large cells that did not fluoresce yellow are simply cells undergoing multiple fission–a process whereby a mitotic cell gives rise to several daughter cells [47]. Dormant aplanospores and cells undergoing multiple fission are labeled in the differential interference contrast images (DIC)-panels B, D, & F, Figure 4, as the corresponding images to panels A, C, & E, Figure 4. DIC images were taken to better define intracellular bodies (panels B, D, & F, Figure 4). Further study is needed to elucidate the intracellular location of phytoene, and whether it is accumulated in oil bodies or elsewhere within the cells.

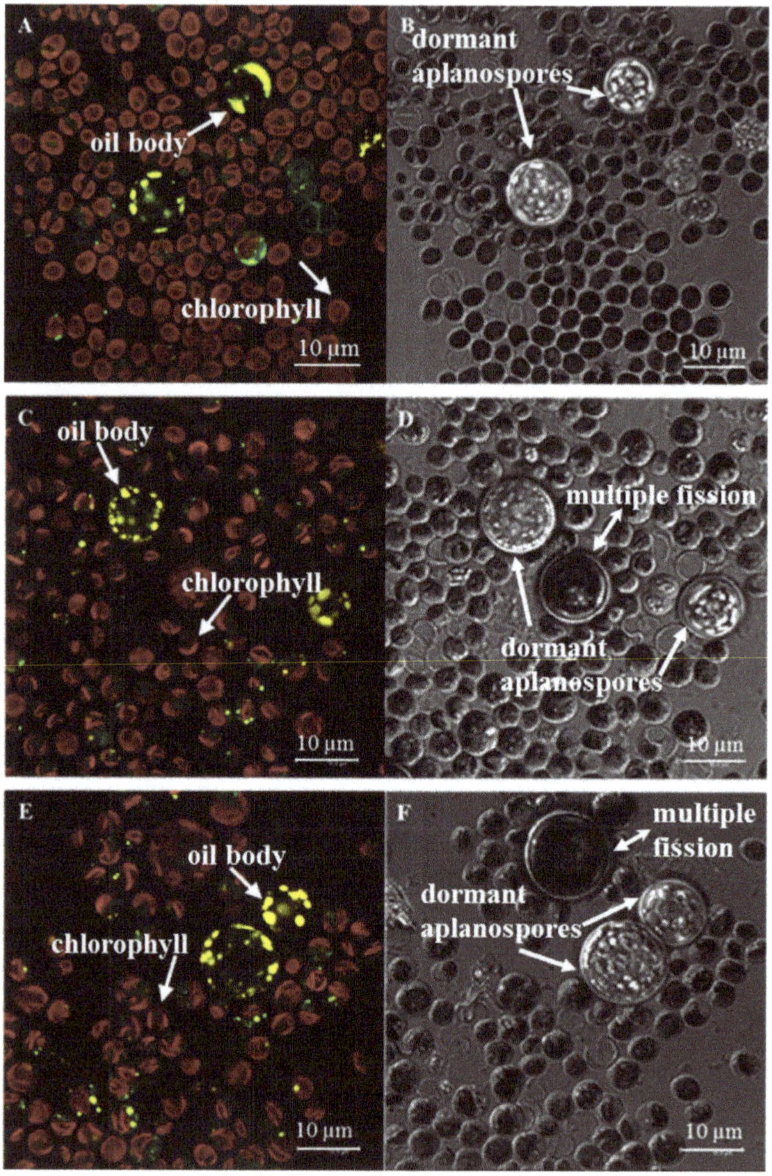


**Figure 4.** Confocal fluorescence images (left); corresponding DIC images (right) of *Chlorococcum* sp.—panels (**A**,**B**) without fluridone, (**C**,**D**) with 152 μM fluridone, (**E**,**F**) with 304 μM fluridone. Arrows in panels (**A**,**C**,**E**) point either to cells with oil bodies (yellow fluorescence) or chlorophyll (red fluorescence); arrows in panels (**B**,**D**,**F**) point to dormant aplanospores (containing oil bodies), and multiple fission cells (double-headed arrows).

## 3. Discussion

### 3.1. Strain Identification & Microplate Bioassays

*Chlorococcum* sp. identity was confirmed molecularly (DNA) and morphologically [41]. Microplate inhibition bioassays were used to determine appropriate herbicide concentration for optimal phytoene desaturase inhibition, adapted from Franz, et al. [48]. Phytoene absorbs at approximately 280 nm, whereas carotenoids downstream of phytoene absorb in the 400 nm to 500 nm range. Therefore, cultures that showed highest overall biomass accumulation as determined by measuring the optical density at 750 nm, coupled with lowest carotenoid development, measured at 450 nm, were indicative of the optimal herbicide concentration at which greatest PDS inhibition was achieved without cell death. As such, 152 μM fluridone was chosen as the optimal concentration to achieve carotenoid inhibition without severely limiting growth (panel D, Figure 1). Cultures were also treated with 304 μM fluridone upon experimental scale-up to observe any notable differences between the concentrations, of which no significant differences in the overall accumulation of phytoene were seen (Table 1).

Similar studies found that the pigment synthesis inhibitor norflurazon caused an 80% decrease of the secondary carotenoid β-carotene in the alga *D. bardawil* at a concentration of 0.1 μM, with concurrent accumulation of phytoene [29]. Other studies have found concentrations of norflurazon ranging from 0.02 μM to 0.3 μM and 100 μM to be effective concentrations for PDS inhibition in the algae *H. pluvialis* and *D. bardawil*, respectively, with substantial accumulation of phytoene in both species [23,31,32]. However, unlike similar studies where cultures were treated with pigment synthesis inhibitors during a carotenogenic state [28], we have chosen to treat during exponential growth phase for the purpose of achieving maximum biomass during phytoene accumulation. A study into the inhibitory effects of fluridone on *E. coli* expressed PDS from the cyanobacterium *Synechococcus* (PCC 7492), as well as purified *Synechococcus* PDS, revealed a concentration of 0.3 μM and 3.5 μM to cause 50% inhibition of carotenoid production, respectively [38]. Chalifour, et al. discovered that a range of temperatures influences the inhibitory effects of the herbicides norflurazon and fluridone in the model alga *C. reinhardtii* [35]. It was found that 1.25 μM fluridone had the greatest impact on secondary carotenoid formation at a temperature of 25 °C; whereas, secondary carotenoid formation was affected to a lesser extent at lower temperatures [35].

### 3.2. Phytoene Quantification

The insignificant increase of phytoene accumulation between 304 μM and 152 μM concentrations of fluridone at day four (Table 1) was likely due to the inhibition of downstream carotenoid synthesis, and therefore, the inability to maintain the photosynthetic complex at a fluridone concentration greater than 152 μM, resulting in increased cell death. This tentative conclusion is supported in previous studies where photosynthetic complexes I and II, particularly system II, are negatively impacted and experience some form of inhibition in the presence of pigment-synthesis inhibitors—fluridone and/or norflurazon [28,49,50]. Therefore, when carotenoid synthesis is inhibited, the photosynthetic complex degrades [10,35–37]. Phytoene levels were further reduced at the nine-day time point. Therefore, we suspect that the strongest concentration of fluridone applied for this study (304 μM), in conjunction with a longer incubation period (9 days), leads to increased cell death and an overall reduction in phytoene accumulation/carotenoid development. For future study, it would be wise to measure phytoene content, cell viability, and photosynthetic inhibition using Fv/Fm measurements, on a daily basis to draw better conclusions that may refute or support these statements.

Decreased carotenoid production coupled with a significant peak for phytoene, as seen in panel A, Figure 2, is a result of successful PDS inhibition by the herbicide fluridone. Results observed in Figure 2 are consistent with previous studies where inhibition of PDS by the pigment synthesis inhibitors fluridone and norflurazon resulted in the over-accumulation of phytoene. One study showed that phytoene constituted 60% of total carotenoid content in norflurazon treated *H. pluvialis* [23,32]. Large amounts of phytoene accumulation in the alga *D. bardawil* have been reported in two separate studies through the use of the inhibitor norflurazon [29,31]. Norflurazon has been a popular choice for PDS inhibition; thus, there is a need for further research into the inhibition capabilities of fluridone for the purpose of carotenoid regulation and potential phytoene accumulation.

### 3.3. *FAME Analysis & Confocal Fluorescence Microscopy*

We speculate that the increase in the unsaturated FAs C16:1 and C16:2 may be due to oil body formation as a response to induced stress (panel A, Figure 3). The literature describes inhibition of FA desaturase enzyme activity by pigment synthesis inhibitors, resulting in decreased levels of lipids, especially the mono-unsaturated and poly-unsaturated FAs [23,35]. The observed increase in C16:1 and C16:2 FAs in this study suggests that the applied concentrations of fluridone did not result in inhibition of FA desaturases, however was likely an effect of lipid remodeling during triacylglycerol synthesis and oil body formation. However, research has shown that, in the alga *C. reinhardtii*, temperature and inhibitor dosage play a large role in the amount of FA desaturase inhibition when exposed to a pigment synthesis inhibitor [35]. Zhekisheva et al. observed the simultaneous decrease in total FA and oleic FA content with increasing concentrations of norflurazon [23]. Notably, C18:0 abundance is markedly and significantly decreased in cultures treated with 304 μM. This same phenomenon was not observed in cultures treated with 152 μM, nor were there any statistically significant differences in the abundance of FAs downstream of C18:0 when compared to controls (no treatment), or in either treatment concentration. As previously mentioned, we suspect these observations are the result of cell death at higher concentrations of fluridone. As in the case of phytoene concentration, future studies should include daily FAME analysis and live cell counts to better understand the effects of various concentrations of fluridone on the metabolic profile and overall lifespan of *Chlorococcum* sp.

We further investigated oil body formation and phytoene accumulation via confocal fluorescence microscopy to determine a relationship between the two, if any. It has been reported that secondary carotenoid formation, specifically β-carotene and astaxanthin, and the accumulation/storage thereof, is directly related to overall FA content and oil body formation [23,29,35,51,52]. The increase in oil bodies within dormant aplanospores seen in fluorescence images, as well as the slight increase in C16:1 and C16:2 FAs in cultures treated with fluridone, may be explained as either a stress response to the herbicide, and/or an accumulation site for phytoene, as is seen in *H. pluvialis* for the storage of astaxanthin [23,24,46]. Therefore, FAME and fluorescence microscopy results should be considered together.

Fluorescence microscopy provided further insights into the effects of fluridone on the photosynthetic apparatus. We speculate that the chloroplastic bifurcation observed in panels C & E, Figure 4 occurs as a result of carotenoid inhibition. Chalifour et al. 2014 found a decrease in chlorophyll *a/b* content and photosynthetic capacity of *C. reinhardtii* when exposed to norflurazon and fluridone. This is not surprising, as carotenoid inhibition with bleaching herbicides results in a loss of the ability to maintain and protect the photosynthetic complex. Therefore, when carotenoid synthesis is inhibited, the photosynthetic complex degrades [10,35–37].

Although informative, the precise location of phytoene cannot be determined, conclusively, using the methods discussed above. Further investigation utilizing spatial and molecular signature tools, such as Raman spectroscopy, are needed to better understand the site and mechanism of phytoene accumulation.

## 4. Conclusions

The pigment synthesis inhibitor fluridone was effective in the over-accumulation of phytoene in the novel microalga *Chlorococcum* sp. Our observations indicate that higher concentrations of the inhibitor fluridone do not result in an increase of phytoene; therefore, lower concentrations of the inhibitor may be a more efficient and effective choice for producers utilizing this method. However, PDS mutagenesis for enhanced phytoene production may be even more effective than the use of pigment synthesis inhibitors. Thus, genomic sequencing of *Chlorococcum* sp., followed by bioinformatics research, is necessary to understand PDS expression in this strain, and how targeted mutagenesis may proceed from those findings. Based on these conclusions, *Chlorococcum* sp. should be considered a valuable candidate in the production of high-value carotenoids for cosmetics, and other biomedical studies for which carotenoids are relevant.

## 5. Materials and Methods

### 5.1. Cultivation

Cultivation of *Chlorococcum* sp. (UTEX B 3056) was performed in sterile BD Falcon™ Tissue Culture Flasks with vented caps from BD Biosciences (Erembodegem, BE), grown in BG11 media at 24 °C in an incubator with 1% $CO_2$ and atmosphere illuminated with cool white fluorescent lamps (22μE per $s^{-1}m^{-2}$). The composition of the liquid medium is as described by the UTEX Culture Collection of Algae (The University of Texas at Austin, Austin, TX, USA).

### 5.2. Strain Identification

Molecular characterization and identification were performed on the genomic DNA extracted from *Chlorococcum* sp. DOE 0101 using the PowerSoil DNA Isolation Kit (Mo Bio Laboratories; Carlsbad, CA, USA). Regions of the 18S rDNA and the RuBisCo Large subunit were amplified from genomic DNA by polymerase chain reaction (PCR) using universal primer sets 18S rDNA (Forward—GTCAGAGGTGAAATTCTTGGATTTA, Reverse—AAGGGCAGGGACGTAATCAACG) and the RuBisCo Large subunit (Forward—AACCTTTCATGCGTTGGAGAGA, Reverse—CCTG CATGAATACCACCAGAAGC) and the GoTaq® colorless master mix (Promega; Madison, WI, USA) according to the manufacturer's instructions. The PCR reactions were performed on a Mastercycler gradient machine (Eppendorf, Wesbury, NY, USA). The PCR program consisted of an initial denaturation/activation step at 95 °C (3 min), 35 cycles of amplification [DNA denaturation step at 95 °C (30 s), followed by an annealing step at 57 °C (30 s) and an elongation step at 72 °C (45 s)], and a final elongation step at 57 °C (10 min). Amplicons were checked for size verification and specificity by gel electrophoresis on a 1% agarose gel. The amplicons were purified from gels using an UltraClean GelSpin® DNA extraction kit (Mo Bio Laboratories; Carlsbad, CA, USA) for subsequent forward and reverse sequencing (Sanger; ABI 3730 DNA analyzer) at the Functional Biosciences laboratory (Madison, WI, USA). Sequence data was analyzed and assembled using Geneious® (V6.1.4; Biomatters Inc., Newark, NJ, USA). The consensus sequences were subjected to standard nucleotide similarity searches via BLASTn [53] against the NCBI non-redundant database using standard parameters to determine their identities and assess their similarities to those in NCBI GenBank.

### 5.3. Microscopy

Concentrated suspensions of three series of preparation: (1) fresh, (2) 2.5% glutaraldehdye-fixed (Electron Microscopy Sciences, Hatfield, PA, USA) and (3) fixed-Nile Red (Sigma-Aldrich Corp., St. Louis, MO, USA) (5 micrograms/mL) treated cells were deposited onto the coverslip areas in glass bottom microwell dishes (MatTek Corp., Ashland, MA, USA) and examined by confocal microscopy using a model TCS SP5 system coupled to a DMI 6000 inverted microscope equipped with a 100× objective lens (Leica Microsystems, Exton, PA, USA) in the x,y,z imaging mode and fluorescence scanning mode with excitation from the 488 nm line of an Argon laser. Images were collected in data sets of two channels (500–550 nm and 660–720 nm) for fresh- and glutaraldehyde-fixed cells or in three channels for Nile Red-treated cell suspensions (500–550 nm, 570–620 nm and 660–720 nm) and examined as maximum projections (8–12 micrometers deep) in separate and graphically overlaid image channels. Fluorescence emission scans were performed from 500–750 nm using a 15 nm detector window and frame averaging of selected focal planes. DIC images were taken using the same method for laser-scanning confocal microscopy; however, the transmitted light channel (non-confocal) was employed here.

### 5.4. Microplate Bioassays

A fluridone standard obtained from Sigma-Aldrich, Inc. (St. Louis, MO, USA) was dissolved in 100% ethanol and diluted accordingly: 304, 152, 76, 38, 19, 0 μM. *Chlorococcum* sp. culture was added to two individual 96 well microplates from BRAND® GmbH & Co. KG (Wertheim, DE) at an optical density (OD) of 0.1, 250 μL per well; adapted from Franz, et al., 2013 [48]. Fluridone was subsequently added to the microplates according to the above concentrations and incubated for ten days under the following conditions: light (22 μE per $s^{-1}m^{-2}$), $CO_2$ (1%), temperature (24 °C). Plates were read at wavelengths 750 nm (biomass), 680 nm (chlorophyll), and 450 nm (carotenoids) over a ten day period using a SPECTRAmax microplate spectrophotometer, Molecular Devices (Sunnyvale, CA, USA).

### 5.5. Scale-up Bioassays

*Chlorococcum* sp. was added to twelve sterile BD Falcon™ Tissue Culture Flasks with vented caps from BD Biosciences (Erembodegem, BE) at 25 mL per flask, OD 0.5. A total of six flasks were incubated with fluridone at 152 and 304 μM, respectively and subjected to the following two conditions: (1) light (22 μE per $s^{-1}m^{-2}$), $CO_2$ (1%), temperature (24 °C); (2) high light (2,000 μE per $s^{-1}m^{-2}$), $CO_2$ (1%), temperature (24 °C). Controls did not contain fluridone. Experiment was performed in biological replicates of three. UV spectrophotometric readings were taken daily on a SPECTRAmax microplate spectrophotometer, Molecular Devices (Sunnyvale, CA, USA), for a total of four days.

### 5.6. Carotenoid Analyses & Quantification

Phytoene Profile–Algae was collected by centrifugation at 10,000 rpm and lyophilized on a Labconco FreeZone 6 (Kansas City, MO, USA) for a minimum of 24h. Approximately 15 mg of algae was milled with 0.5 mm dia. zirconia/silica beads (BioSpec Products, Inc., Bartlesville, OK, USA) using a Mini Bead Beater™ (BioSpec Products, Inc., Bartlesville, OK, USA) for a total of 2 min to achieve cell lyses. HPLC grade acetone (Sigma-Aldrich Corp., St. Louis, MO, USA) was added to the samples in a 1:30, mg: μL, ratio and allowed to sit for 20 min., followed by centrifugation for 10 min at 13,000 rpm and the supernatant removed and collected in a separate vial. The extraction was repeated a second time to ensure complete pigment-tissue extraction, and the supernatants combined. Samples were analyzed on a Waters 2695 Alliance® HPLC (Waters Corp., Milford, MA, USA) with a 996 photodiode array detector and YMC America carotenoid column (YMC America, Inc., Allentown, PA, USA), using the YMC MTBE carotenoid method [54]. Phytoene quantification was performed using external calibration on a series of dilutions of a phytoene standard obtained from Sigma-Aldrich, product #78903 (Sigma-Aldrich Corp., St. Louis, MO, USA). Carotenoid content was quantified via UV-Vis using a series of dilutions of a β-carotene standard obtained from Sigma-Aldrich, product #1065480 (Sigma-Aldrich Corp., St. Louis, MO, USA), and read at 450 nm using a SPECTRAmax microplate spectrophotometer, Molecular Devices (Sunnyvale, CA, USA).

### 5.7. Lipid Analysis

Fatty Acid Methyl Ester (FAME) profiles were obtained for each treatment group via base catalyzed transesterification. 2 mL of KOH in methanol (2N) was applied to dried, ground tissue (~5 mg), vortexed and incubated at ~37 °C for 30 min. Samples were allowed to cool for approximately 15 min. 1 mL of acetic acid (1M) was added to samples to quench the reaction. Subsequently, 2 mL of HPLC grade hexane with C23:0 ISTD at 50 ppm was added to samples and vortexed thoroughly. All reagents for FAME extraction were obtained from (Sigma-Aldrich, St. Louis, MO, USA). 200 μL of the upper portion of the sample was removed and dispensed into GC vials, fitted with inserts, for analysis. Samples were then analyzed by GC/MS on a Varian 3800 Gas Chromatograph with a Varian 2000 Mass Spectrometer and a Varian 8200 Auto sampler (Agilent Technologies, Inc., Santa Clara, CA, USA). 2 μL were injected onto a 30 m × 0.25 mm diam. ×0.25 μm film DB-23 capillary column (Agilent Technologies, Inc., Santa Clara, CA, USA) with Helium carrier gas at 1 mL/min with a 5:1

split. The inlet and transfer line were held at 250 °C. The column temperature was held at 60 °C for 1 min. and then ramped at 30 °C $min^{-1}$ to 175 °C and maintained for 1 min., then ramped to 235 °C at 4 °C for a total run time of 21.83 min. The instrument was tuned with a standard auto tune method and a calibration curve prepared from a Supelco 37 Component FAME mix (10 mg $mL^{-1}$) in methylene chloride product # CRM47885 (Sigma-Aldrich, St. Louis, MO, USA). The mass spectrometer operated at 70 eV in electron ionization (EI) mode with 5 scans per second between the mass range 40 and 500.

### 5.8. Statistics

Statistics were performed using Sigma Plot V11.0 (Systat Software, Inc., San Jose, CA, USA). All data sets were run as a two-way analysis of variance (ANOVA), and using the Holm-Sidak method. Microplate bioassay data was run as a two-way ANOVA with repeated measures.

**Supplementary Materials:** The following are available online at http://www.mdpi.com/1660-3397/17/3/187/s1, Figure S1: PCR amplification, Figure S2: Scanning Electron Microscopic images of *Chlorococcum* sp. (UTEX B 3056).

**Author Contributions:** K.L. and F.O.H. conceived and designed the experiments; K.L. and B.D. performed the experiments; K.L., M.S., P.C., F.O.H., and J.P. analyzed the data; M.S., P.C., F.O.H., J.P. contributed reagents/materials/analysis tools; K.L., M.S., B.D., P.C., and F.O.H. wrote the paper.

**Funding:** We acknowledge NMSU RISE-MBRS (Supported by NIH NIGMS Grant # R25GM061222), and the National Science Foundation (award #IIA-1301346) for their funding of this research and the ability to publish this work open access.

**Conflicts of Interest:** The authors declare no conflict of interest, and the founding sponsors had no role in the design of the study; in the collection, analyses, or interpretation of data; in the writing of the manuscript, and in the decision to publish the results.

## Abbreviations

| | |
|---|---|
| DOE | Department of Energy |
| GGPP | Geranylgeranyl pyrophosphate |
| PDS | Phytoene desaturase |
| PCR | Polymerase chain reaction |
| HPLC | High pressure liquid chromatography |
| MTBE | Methyl tert-butyl ether |
| SEM | Standard error of the means |
| FAME | Fatty acid methyl ester |
| FA | Fatty acid |
| DIC | Differential interference contrast |
| SEM | Scanning electron microscopy |
| ANOVA | Analysis of variance |

## References

1. Del Campo, J.A.; García-González, M.; Guerrero, M.G. Outdoor cultivation of microalgae for carotenoid production: Current state and perspectives. *Appl. Microbiol. Biotechnol.* **2007**, *74*, 1163–1174. [CrossRef] [PubMed]
2. Kim, S.-K. *Marine Cosmeceuticals: Trends and Prospects*; CRC Press: Boca Raton, FL, USA, 2011.
3. Martins, A.; Vieira, H.; Gaspar, H.; Santos, S. Marketed marine natural products in the pharmaceutical and cosmeceutical industries: Tips for success. *Mar. Drugs* **2014**, *12*, 1066–1101. [CrossRef] [PubMed]
4. Spolaore, P.; Joannis-Cassan, C.; Duran, E.; Isambert, A. Commercial applications of microalgae. *J. Biosci. Bioeng.* **2006**, *101*, 87–96. [CrossRef] [PubMed]
5. Guedes, A.C.; Amaro, H.M.; Malcata, F.X. Microalgae as sources of carotenoids. *Mar. Drugs* **2011**, *9*, 625–644. [CrossRef] [PubMed]
6. Santos, M.F.; Mesquita, J. Ultrastructural Study of *Haematococcus lacustris* (Girod.) Rostafinski (Volvocales). *Cytologia* **1984**, *49*, 215–228. [CrossRef]

7. Lichtenthaler, H.K.; Buschmann, C. Chlorophylls and carotenoids: Measurement and characterization by UV-VIS spectroscopy. *Curr. Protoc. Food Anal. Chem.* **2001**, *1*. [CrossRef]
8. Paiva, S.A.; Russell, R.M. β-Carotene and other carotenoids as antioxidants. *J. Am. Coll. Nutr.* **1999**, *18*, 426–433. [CrossRef]
9. Stahl, W.; Sies, H. β-Carotene and other carotenoids in protection from sunlight. *Am. J. Clin. Nutr.* **2012**, *96*, 1179S–1184S. [CrossRef]
10. Salguero, A.; de la Morena, B.; Vigara, J.; Vega, J.M.; Vilchez, C.; León, R. Carotenoids as protective response against oxidative damage in Dunaliella bardawil. *Biomol. Eng.* **2003**, *20*, 249–253. [CrossRef]
11. Haegele, A.D.; Gillette, C.; O'Neill, C.; Wolfe, P.; Heimendinger, J.; Sedlacek, S.; Thompson, H.J. Plasma xanthophyll carotenoids correlate inversely with indices of oxidative DNA damage and lipid peroxidation. *Cancer Epidemiol. Prev. Biomark.* **2000**, *9*, 421–425.
12. Miki, W. Biological functions and activities of animal carotenoids. *Pure Appl. Chem.* **1991**, *63*, 141–146. [CrossRef]
13. Fuller, B.; Smith, D.; Howerton, A.; Kern, D. Anti-inflammatory effects of CoQ10 and colorless carotenoids. *J. Cosmet. Dermatol.* **2006**, *5*, 30–38. [CrossRef]
14. Guerin, M.; Huntley, M.E.; Olaizola, M. Haematococcus astaxanthin: Applications for human health and nutrition. *TRENDS Biotechnol.* **2003**, *21*, 210–216. [CrossRef]
15. Rao, A.V.; Rao, L.G. Carotenoids and human health. *Pharmacol. Res.* **2007**, *55*, 207–216. [CrossRef]
16. Bartley, G.E.; Scolnik, P.A. Plant carotenoids: Pigments for photoprotection, visual attraction, and human health. *Plant Cell* **1995**, *7*, 1027–1038. [CrossRef]
17. McGarvey, D.J.; Croteau, R. Terpenoid metabolism. *Plant Cell* **1995**, *7*, 1015. [CrossRef]
18. Meléndez-Martínez, A.J.; Mapelli-Brahm, P.; Benítez-González, A.; Stinco, C.M. A comprehensive review on the colorless carotenoids phytoene and phytofluene. *Arch. Biochem. Biophys.* **2015**, *572*, 188–200. [CrossRef]
19. Goodwin, T. Functions of carotenoids. In *The Biochemistry of the Carotenoids*; Springer: Berlin, Germany, 1980; pp. 77–95.
20. Grung, M.; Liaaen-Jensen, S. Algal carotenoids 52; secondary carotenoids of algae 3; carotenoids in a natural bloom of Euglena sanguinea. *Biochem. Syst. Ecol.* **1993**, *21*, 757–763. [CrossRef]
21. Katz, A.; Jimenez, C.; Pick, U. Isolation and characterization of a protein associated with carotene globules in the alga Dunaliella bardawil. *Plant Physiol.* **1995**, *108*, 1657–1664. [CrossRef]
22. Pick, U.; Zarka, A.; Boussiba, S.; Davidi, L. A hypothesis about the origin of carotenoid lipid droplets in the green algae Dunaliella and Haematococcus. *Planta* **2019**, *249*, 31–47. [CrossRef]
23. Zhekisheva, M.; Zarka, A.; Khozin-Goldberg, I.; Cohen, Z.; Boussiba, S.J.J. Inhibition of astaxanthin synthesis under high irradiance does not abolish triacylglycerol accumulation in the green alga haematococcus pluvialis (Chlorophyceae) 1. *J. Phycol.* **2005**, *41*, 819–826. [CrossRef]
24. Jin, E.-S.; Lee, C.-G.; Polle, J.E. Secondary carotenoid accumulation in Haematococcus (Chlorophyceae): Biosynthesis, regulation, and biotechnology. *J. Microbiol. Biotechnol.* **2006**, *16*, 821–831.
25. Von Oppen-Bezalel, L. Colorless carotenoids, phytoene and phytofluene for the skin: For prevention of aging/photo-aging from the inside and out. *SÖFW-Journal* **2007**, *133*, 38–40.
26. Von Oppen-Bezalel, L.; Fishbein, D.; Havas, F.; Ben-Chitrit, O.; Khaiat, A.J.G.D. The photoprotective effects of a food supplement tomato powder rich in phytoene and phytofluene, the colorless carotenoids, a preliminary study. *Glob. Dermatol.* **2015**, *2*, 178–182.
27. Engelmann, N.J.; Clinton, S.K.; Erdman, J.W., Jr. Nutritional aspects of phytoene and phytofluene, carotenoid precursors to lycopene. *Adv. Nutr.* **2011**, *2*, 51–61. [CrossRef]
28. Ben-Amotz, A.; Gressel, J.; Avron, M.J. Massive accumulation of phytoene induced by norflurazon in dunaliella bardawil (Chlorophyceae) prevents recovery from photoinhibition 1. *J. Phycol.* **1987**, *23*, 176–181. [CrossRef]
29. Ben-Amotz, A.; Lers, A.; Avron, M.J.P.P. Stereoisomers of β-carotene and phytoene in the alga Dunaliella bardawil. *Plant Physiol.* **1988**, *86*, 1286–1291. [CrossRef]
30. Soudant, E.; Bezalel, L.; Schickler, H.; Paltiel, J.; Ben-Amotz, A.; Shaish, A.; Perry, I. Carotenoid Preparation. U.S. Patent No. 6,383,474, 7 May 2002.
31. Shaish, A.; Avron, M.; Ben-Amotz, A. Effect of inhibitors on the formation of stereoisomers in the biosynthesis of β-carotene in Dunaliella bardawil. *Plant Cell Physiol.* **1990**, *31*, 689–696.

32. Harker, M.; Young, A.J. Inhibition of astaxanthin synthesis in the green alga, *Haematococcus pluvialis*. *Eur. J. Phycol.* **1995**, *30*, 179–187. [CrossRef]
33. Tjahjono, A.E.; Kakizono, T.; Hayama, Y.; Nishio, N.; Nagai, S. Isolation of resistant mutants against carotenoid biosynthesis inhibitors for a green alga *Haematococcus pluvialis*, and their hybrid formation by protoplast fusion for breeding of higher astaxanthin producers. *J. Ferment. Bioeng.* **1994**, *77*, 352–357. [CrossRef]
34. Chamovitz, D.; Sandmann, G.; Hirschberg, J. Molecular and biochemical characterization of herbicide-resistant mutants of cyanobacteria reveals that phytoene desaturation is a rate-limiting step in carotenoid biosynthesis. *J. Biol. Chem.* **1993**, *268*, 17348–17353.
35. Chalifour, A.; Arts, M.T.; Kainz, M.J.; Juneau, P. Combined effect of temperature and bleaching herbicides on photosynthesis, pigment and fatty acid composition of Chlamydomonas reinhardtii. *Eur. J. Phycol.* **2014**, *49*, 508–515. [CrossRef]
36. Takaichi, S. Carotenoids in algae: Distributions, biosyntheses and functions. *Mar. Drugs* **2011**, *9*, 1101–1118. [CrossRef]
37. Vílchez, C.; Forján, E.; Cuaresma, M.; Bédmar, F.; Garbayo, I.; Vega, J.M. Marine carotenoids: Biological functions and commercial applications. *Mar. Drugs* **2011**, *9*, 319–333. [CrossRef] [PubMed]
38. Sandmann, G.; Fraser, P.D. Differential inhibition of phytoene desaturases from diverse origins and analysis of resistant cyanobacterial mutants. *Z. Naturforsch. C* **1993**, *48*, 307–311. [CrossRef]
39. Sharon-Gojman, R.; Maimon, E.; Leu, S.; Zarka, A.; Boussiba, S. Advanced methods for genetic engineering of *Haematococcus pluvialis* (Chlorophyceae, Volvocales). *Algal Res.* **2015**, *10*, 8–15. [CrossRef]
40. Suarez, J.V.; Banks, S.; Thomas, P.G.; Day, A. A new F131V mutation in Chlamydomonas phytoene desaturase locates a cluster of norflurazon resistance mutations near the FAD-binding site in 3D protein models. *PLoS ONE* **2014**, *9*, e99894. [CrossRef] [PubMed]
41. Steinbrenner, J.; Sandmann, G. Transformation of the green alga *Haematococcus pluvialis* with a phytoene desaturase for accelerated astaxanthin biosynthesis. *Appl. Environ. Microbiol.* **2006**, *72*, 7477–7484. [CrossRef] [PubMed]
42. Neofotis, P.; Huang, A.; Sury, K.; Chang, W.; Joseph, F.; Gabr, A.; Twary, S.; Qiu, W.; Holguin, O.; Polle, J.E. Characterization and classification of highly productive microalgae strains discovered for biofuel and bioproduct generation. *Algal Res.* **2016**, *15*, 164–178. [CrossRef]
43. Watanabe, S.; Floyd, G.L. Variation in the ultrastructure of the biflagellate motile cells of six unicellular genera of the Chlamydomonadales and Chlorococcales (Chlorophyceae), with emphasis on the flagellar apparatus. *Am. J. Bot.* **1989**, *76*, 307–317. [CrossRef]
44. Miller, D.H. Cell wall chemistry and ultrastructure of chlorococcum oleofaciens (Chlorophyceae) 1, 2. *J. Phycol.* **1978**, *14*, 189–194. [CrossRef]
45. Lammers, P.J.; Huesemann, M.; Boeing, W.; Anderson, D.B.; Arnold, R.G.; Bai, X.; Bhole, M.; Brhanavan, Y.; Brown, L.; Brown, J. Review of the cultivation program within the National Alliance for Advanced Biofuels and Bioproducts. *Algal Res.* **2017**, *22*, 166–186. [CrossRef]
46. Chekanov, K.; Vasilieva, S.; Solovchenko, A.; Lobakova, E. Reduction of photosynthetic apparatus plays a key role in survival of the microalga *Haematococcus pluvialis* (Chlorophyceae) at freezing temperatures. *Photosynthetica* **2018**, *56*, 1268–1277. [CrossRef]
47. Bišová, K.; Zachleder, V. Cell-cycle regulation in green algae dividing by multiple fission. *J. Exp. Bot.* **2014**, *65*, 2585–2602. [CrossRef]
48. Franz, A.K.; Danielewicz, M.A.; Wong, D.M.; Anderson, L.A.; Boothe, J.R. Phenotypic screening with oleaginous microalgae reveals modulators of lipid productivity. *ACS Chem. Biol.* **2013**, *8*, 1053–1062. [CrossRef] [PubMed]
49. Dankov, K.; Busheva, M.; Stefanov, D.; Apostolova, E.L. Relationship between the degree of carotenoid depletion and function of the photosynthetic apparatus. *J. Photochem. Photobiol. B Biol.* **2009**, *96*, 49–56. [CrossRef] [PubMed]
50. Trebst, A.; Depka, B. Role of carotene in the rapid turnover and assembly of photosystem II in *Chlamydomonas reinhardtii*. *FEBS Lett.* **1997**, *400*, 359–362. [CrossRef]
51. Rise, M.; Cohen, E.; Vishkautsan, M.; Cojocaru, M.; Gottlieb, H.E.; Arad, S.M. Accumulation of secondary carotenoids in *Chlorella zofingiensis*. *J. Plant Physiol.* **1994**, *144*, 287–292. [CrossRef]

52. Vechtel, B.; Eichenberger, W.; Ruppel, G. Lipid bodies in Eremosphaera viridis De Bary (Chlorophyceae). *Plant Cell Physiol.* **1992**, *33*, 41–48.
53. Altschul, S.F.; Madden, T.L.; Schäffer, A.A.; Zhang, J.; Zhang, Z.; Miller, W.; Lipman, D.J. Gapped BLAST and PSI-BLAST: A new generation of protein database search programs. *Nucleic Acids Res.* **1997**, *25*, 3389–3402. [CrossRef] [PubMed]
54. Rodriguez-Uribe, L.; Guzman, I.; Rajapakse, W.; Richins, R.D.; O'connell, M.A. Carotenoid accumulation in orange-pigmented *Capsicum annuum* fruit, regulated at multiple levels. *J. Exp. Bot.* **2011**, *63*, 517–526. [CrossRef] [PubMed]

*marine drugs*

*Review*

# Carotenoids in Marine Animals

**Takashi Maoka**

Research Institute for Production Development, 15 Shimogamo-morimoto-cho, Sakyo-ku,
Kyoto 606-0805, Japan; E-Mail: maoka@mbox.kyoto-inet.or.jp; Tel.: +81-75-781-1107; Fax: +81-75-781-1118

*Received: 14 January 2011; in revised form: 16 February 2011 / Accepted: 21 February 2011 / Published: 22 February 2011*

**Abstract:** Marine animals contain various carotenoids that show structural diversity. These marine animals accumulate carotenoids from foods such as algae and other animals and modify them through metabolic reactions. Many of the carotenoids present in marine animals are metabolites of β-carotene, fucoxanthin, peridinin, diatoxanthin, alloxanthin, and astaxanthin, *etc.* Carotenoids found in these animals provide the food chain as well as metabolic pathways. In the present review, I will describe marine animal carotenoids from natural product chemistry, metabolism, food chain, and chemosystematic viewpoints, and also describe new structural carotenoids isolated from marine animals over the last decade.

**Keywords:** carotenoids; marine animals; metabolism; food chain; chemosystematic

## 1. Introduction

Since the first structural elucidation of β-carotene by Kuhn and Karrer in 1928–1930, about 750 naturally occurring carotenoids had been reported as of 2004 [1]. Improvements of analytical instruments such as NMR, MS, HPLC, *etc.*, have made it possible to perform the structural elucidation of very minor carotenoids in nature [2–4].

Marine animals contain various carotenoids that show structural diversity [3–9]. Among the 750 reported carotenoids found in nature, more than 250 are of marine origin. In particular, allenic carotenoids, except for neoxanthin and its derivatives, and all acetylenic carotenoids originate from marine algae and animals [1].

In general, animals do not synthesize carotenoids *de novo*, and so those found in animals are either directly accumulated from food or partly modified through metabolic reactions [5–9], as shown in Figure 1. The major metabolic conversions of carotenoids found in animals are oxidation, reduction, translation of double bonds, oxidative cleavage of double bonds, and cleavage of epoxy bonds.

Up until 2001, marine animal carotenoids were reviewed by Liaaen-Jensen [5,6], Matsuno [7,8], and Matsuno and Hirao [9]. Since then, there have been no reviews of carotenoids in marine animals. The present review describes progress in the field of carotenoids in marine animals over the last decade.

**Figure 1.** Accumulation and metabolism of carotenoids in marine animals through food chain.

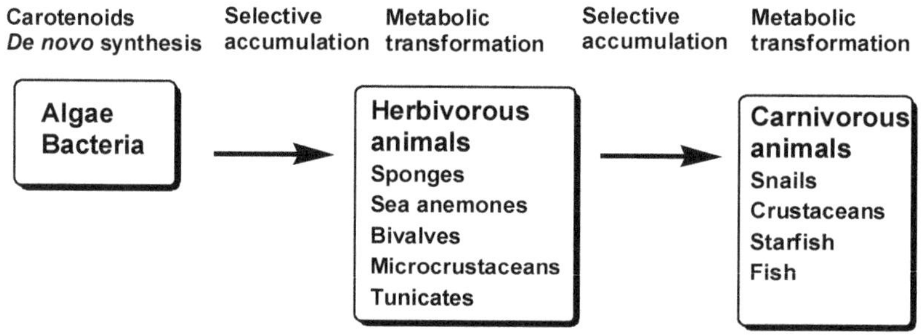


## 2. Porifera (Marine Sponges)

Characteristic carotenoids in marine sponges are shown in Figure 2. Many marine sponges are brilliantly colored due to the presence of carotenoids. Sponges are filter feeders and are frequently associated with symbionts such as microalgae or bacteria [6]. The characteristic carotenoids in sponges are aryl carotenoids such as isorenieratene (**1**), renieratene (**2**), and renierapurpurin (**3**) [6,7]. More than twenty aryl carotenoids have been reported in sponges [1]. Except for sea sponges, aryl carotenoids are found only in green sulfur bacteria [1,6]. Therefore, aryl carotenoids in sponges are assumed to originate from symbiotic bacteria [6,7]. Novel carotenoid sulfates having an acetylenic group, termed bastaxanthins (**4**), were isolated from the sea sponge *Ianthella basta* [1]. Recently, a new acetylenic carotenoid (**5**) was isolated from the marine sponge *Prianos osiros* [10]. Based on the structural similarity, bastaxanthins and compound **5** were assumed to be metabolites of fucoxanthin originating from microalgae.

**Figure 2.** Characteristic carotenoids in marine sponges.

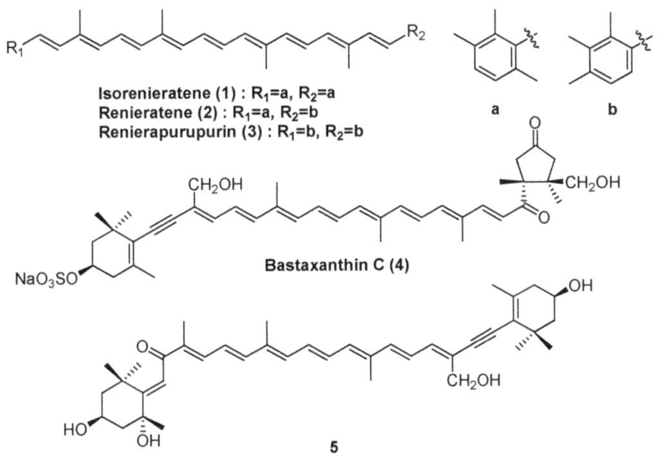


## 3. Coelenterata (Sea Anemones)

Astaxanthin, which originates from dietary zooplankton, was found in some jelly fish. Peridinin, pyrrhoxanthin, and diadinoxanthin were found in some corals [11]. They originate from symbiotic dinoflagellates. Unique nor carotenoids, 2-nor-astaxnthin (**6**) and actinioerythrin (**7**), have been reported in the sea anemones *Actinia equina* and *Tealia felina* [1] (Figure 3).

**Figure 3.** Characteristic carotenoids in sea anemones.

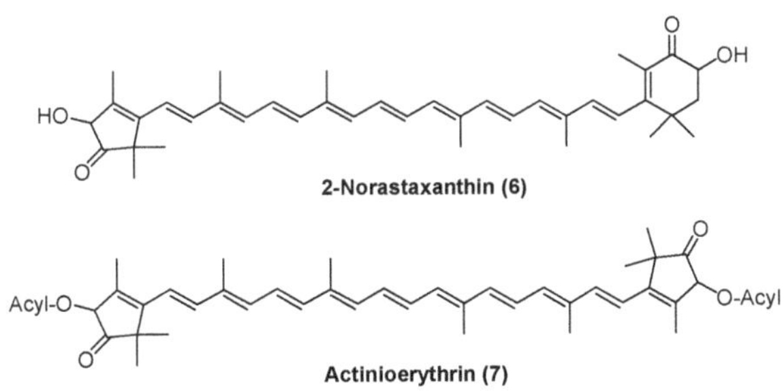


## 4. Mollusca (Mollusks)

Many chitons are herbivorous and feed on attached algae. Major carotenoids found in chitons are lutein, zeaxanthin, fucoxanthin, and their metabolites [12].

Abalone, *Haliotis discus discus*, and turban shell, *Turbo cornutus*, feed on brown and red algae. Carotenoids found in these shells are β-carotene, α-carotene, zeaxanthin, lutein, and fucoxanthin [11].

On the other hand, many sea snails are carnivores. The triton *Charonia sauliae* feeds on starfish. Therefore, astaxanthin (**8**), 7,8-didehydroastaxanthin (**9**), and

7,8,7′,8′-tetradehydroastaxanthin (**10**), characteristic carotenoids found in starfish, were isolated as major carotenoids in triton. Astaxanthin (**8**), originating from dietary microcrustaceans, was found to be a major carotenoid in the whelk *Buccinum bayani*. Alternatively, *Drupella fragum* preys upon corals. Thus, peridinin and diadinoxanthin are present as major carotenoids in this sea snail [11]. Carotenoids in sea snails well reflect their diet.

Canthaxanthin (**11**), (3*S*)-adonirubin (**12a**), and (3*S*,3′*S*)-astaxanthin (**8a**) were found to be major carotenoids in the spindle shell *Fushinus perplexus* [13]. Furthermore, a series of carotenoids with a 4-hydroxy-5,6-dihydro-β-end group and/or 3,4-dihydroxy-5,6-dihydro-β-end (**13**–**15**) were isolated from *Fushinus perplexus* [13] (Figure 4). They were assumed to correspond to reduction metabolites of canthaxanthin (**11**), (3*S*)-adonirubin (**12a**), and (3*S*,3′*S*)-astaxanthin (**8a**).

Sea slugs and sea hares also belong to Gastropoda. They are herbivorous and feed on brown and red algae. Several apocarotenoids have been reported in sea slugs and sea hares [1]. A series of 8′-apocarotenal and 8′-apocarotenols derived from β-carotene, lutein, and zeaxanthin were found in the sea hare *Aplysia kurodai* [14]. They are oxidative cleavage products of the polyene chain at C-8 in $C_{40}$ skeletal carotenoids [14].

Bivalves (oyster, clam, scallop, mussel, ark shell, *etc.*) contain various carotenoids that show structural diversity [3,6]. Bivalves accumulate carotenoids obtained from their dietary microalgae and modify them through metabolic reactions. Many of the carotenoids present in bivalves are metabolites of fucoxanthin, diatoxanthin, diadinoxanthin, and alloxanthin [3,6], which originate from microalgae.

**Figure 4.** Characteristic carotenoids in sea snails.

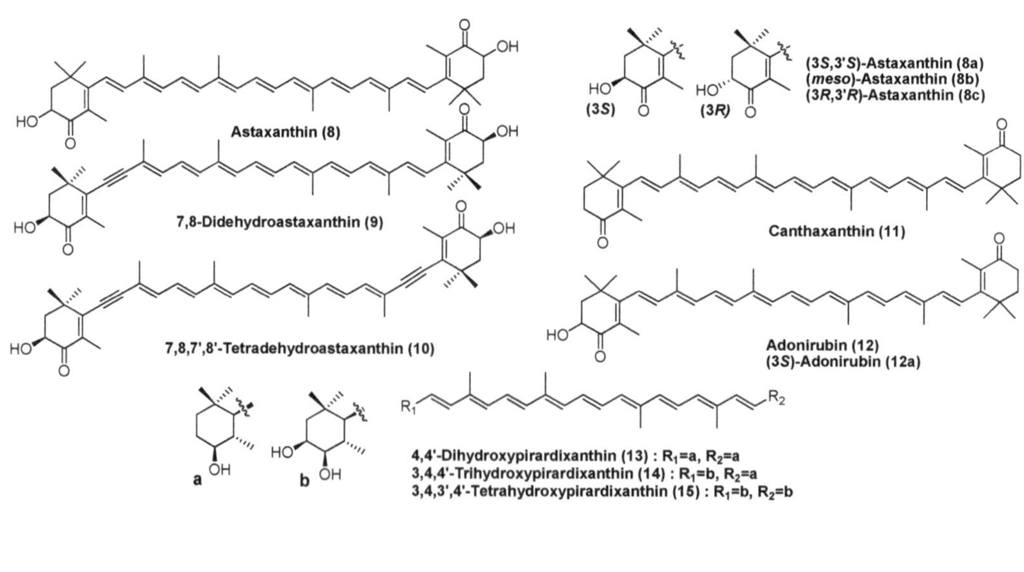

Oxidative metabolites of diatoxanthin (**16**) and alloxanthin (**17**), such as pectenol (**18**), pectenolone (**19**), 4-hydroxyalloxanthin (**20**), and 4-ketoalloxanthin (**21**), are distributed in scallops and ark shells [3,6,7]. 8′-Apoalloxanthinal (**22**), which is an oxidative cleavage product of alloxanthin, was also found in bivalves [15] (Figure 5).

A novel 3,6-epoxy derivative of diadinoxanthin (**23**), named cycloidadinoxanthin (**24**), was also isolated from the oyster [16] (Figure 5).

**Figure 5.** Metabolites of diatoxanthin, alloxanthin, and diadinoxanthin in bivalves.

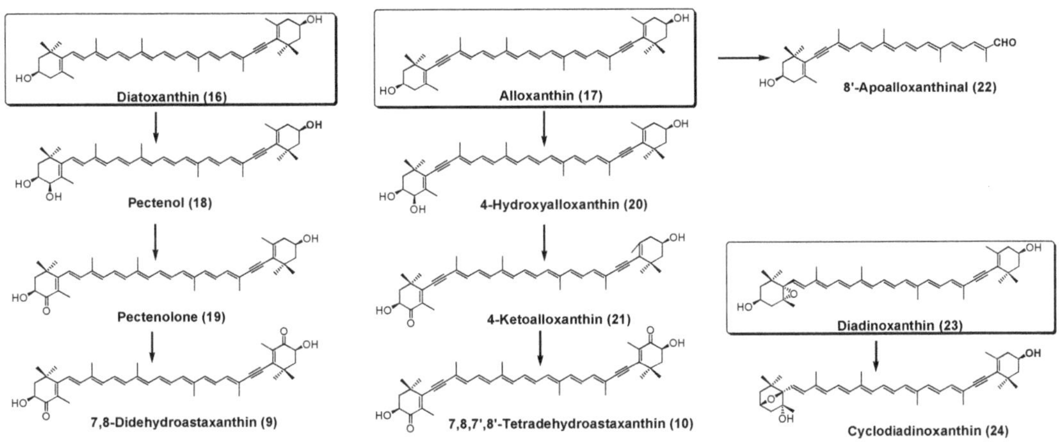


Fucoxanthin (**25**) and its metabolites fucoxanthinol (**26**) and halocynthiaxanthin (**27**) were found to be widely distributed in oysters and clams [3,6,7].

Mytiloxanthin (**28**), which has a unique enol hydroxy group at C-8′ in the polyene chain and a
3′-hydroxy-6′-oxo-κ-end group, is a characteristic carotenoid in marine mussels and oysters [6,7]. Furthermore, three mytiloxanthin analogues containing an allenic end group (**29**), a 3,6-epoxy-end group (**30**), and a 3,4-dihydroxy-β-end group (**31**) were isolated from the oyster [16,17]. Compound **29**, termed allenic mytiloxanthin, was assumed to be a metabolic intermediate from fucoxanthinol to mytiloxanthin.

Some edible clams have a bright orange or red color due to the presence of carotenoids. Fucoxanthin 3-ester (**32**) and fucoxanthinol 3-ester (**33**) were found to be major carotenoids in *Mactra chinensis* [18], *Ruditapes philippinarum*, and *Meretrix petechialis* [19]. Amarouciaxanthin A (**34**) and its ester were also identified as major carotenoids in *Paphia amabills* and *Paphia amabillis* [20].

Other metabolites of fucoxanthin, crasssostreaxanthin A (**35**) and crassostreaxanthin B (**36**), were isolated from the Japanese oyster *Crassostrea gigas* [21]. Tode *et al.* demonstrated that crassostreaxanthin B could be converted from halocynthiaxanthin by bio-mimetic chemical reactions [22,23]. Further studies of carotenoids in marine animals revealed that crassostreaxanthin A, crassostreaxanthin B, and their 3-acetates were widely distributed in marine bivalves [16,17]. Moreover, two crassostreaxanthin A analogues, **37** and **38**, were isolated from the oyster as minor components [16,17]. Metabolic pathways of fucoxanthin in bivalves are shown in Figure 6.

**Figure 6.** Metabolic pathways of fucoxanthin in bivalves.

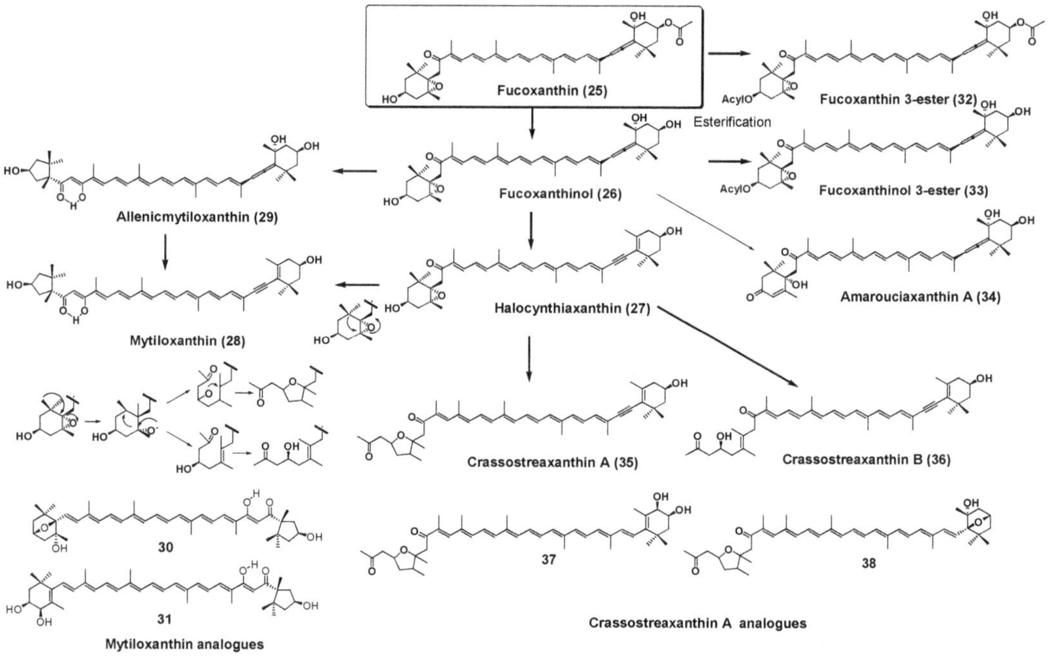


Bivalves also feed on dinoflagellates. Peridinin (**39**), a characteristic carotenoid in dinoflagellates with a $C_{37}$-skeletal structure, and its metabolites (**40–43**) were also found in some bivalves. Recently, four new $C_{37}$-skeletal carotenoids (**44–47**) were isolated from *Crassostrea gigas* [16,17], *Paphia amabillis* [20], and *Corbicula japonica* [24,25]. The metabolic pathways of peridinin in bivalves are shown in Figure 7. As well as fucoxanthin, the major metabolic conversions of peridinin in bivalves are hydrolysis of acetyl group, conversion of the allenic bond to an acetylenic bond, and hydrolysis cleavage of the epoxy ring, as shown in Figure 7.

**Figure 7.** Metabolic pathways of peridinin in bivalves.

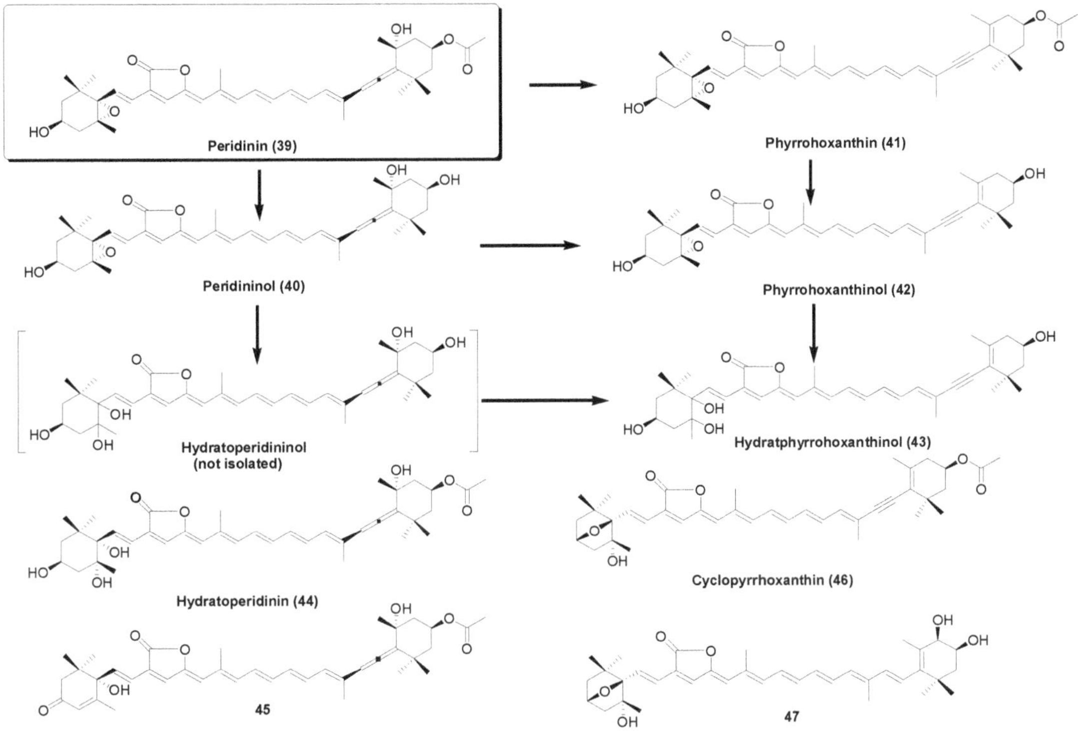


There are many reports on carotenoids in marine shellfish [6,7]. However, there are few reports on the carotenoids of shellfish inhabiting brackish or fresh water [24,25]. Four new carotenoids, corbiculaxanthin (**48**), corbiculaxanthin 3′-acetate (**49**), 6-epiheteroxanthin (**50**), and 7′,8′-didehydrodeepoxyneoxanthin (**51**), were isolated from the brackish clam *Corbicula japonica* and freshwater clam *Corbicula sandai* (Figure 8) [24,25]. 7′,8′-Didehydrodeepoxyneoxanthin (**51**) has an interesting structure, with both allenic and acetylenic moieties.

**Figure 8.** New carotenoids in corbicula clams.

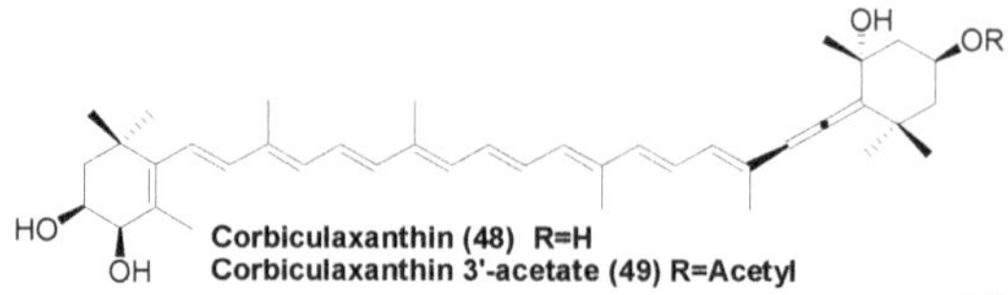

Corbiculaxanthin (48) R=H
Corbiculaxanthin 3'-acetate (49) R=Acetyl

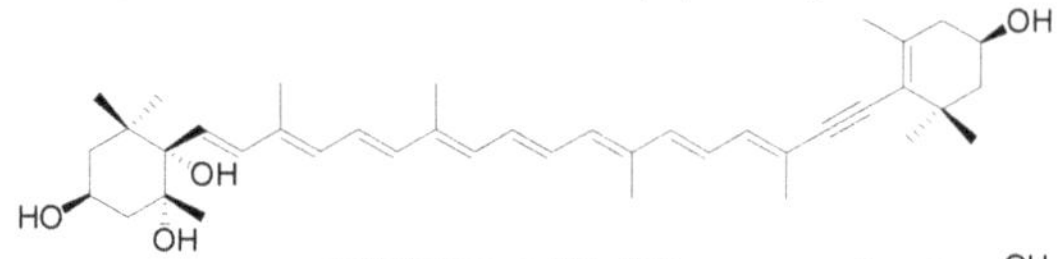

6-Epiheteroxanthin (50)

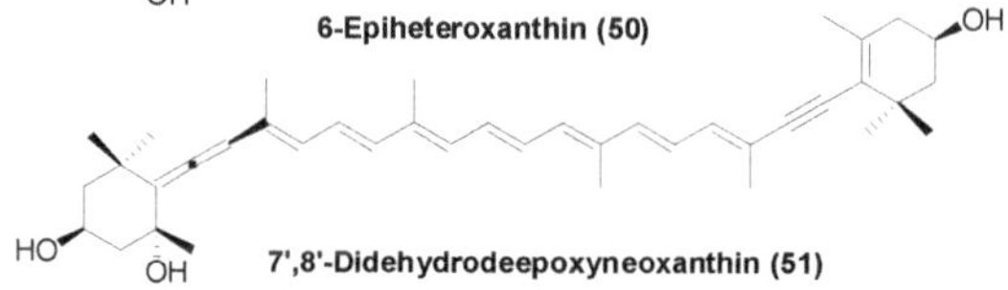

7',8'-Didehydrodeepoxyneoxanthin (51)

Carotenoids found in bivalves provide a key to the food chain as well as metabolic pathways.

Astaxanthin and its esters were found to be major carotenoids in species of octopus and cuttlefish. Their astaxanthins consisted of three optical isomers and originated from dietary zooplankton [26].

## 5. Arthropoda (Crustaceans)

Carotenoids in the carapace of crustaceans exist as both free and esterified forms. The principal carotenoid in crustaceans is astaxanthin [6,7]. In crustaceans, astaxanthin exists as carotenoproteins such as crustacyanin, and exhibits purple, blue, and yellow colors. Many crustaceans can synthesize astaxanthin (**8**) from β-carotene (**52**), ingested from dietary algae, via echinenone (**53**), 3-hydroxyechinenone (**54**), canthaxanthin (**11**), and adonirubin (**12**), as shown in Figure 9 [6,7].

In many crustaceans, hydroxylation at C-3 (C-3′) in the 4-oxo-β-end group is none-stereo-selective. Therefore, astaxanthin, adonixanthin, and 3-hydroxyechinenone, having a 3-hydroxy-4-oxo-β-end group, present in crustaceans, are comprised of a mixture of optical isomers [6,7].

**Figure 9.** Metabolism of β-carotene in crustaceans.

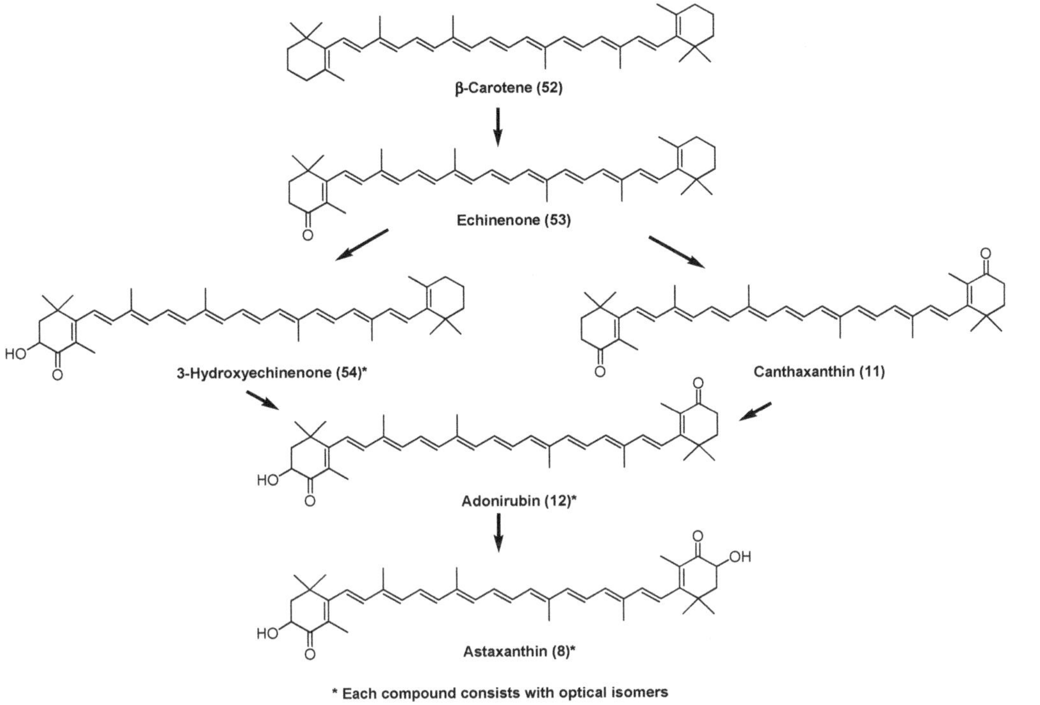


Some crustaceans can convert zeaxanthin to adonixanthin (**55**) and lutein to fritschiellaxanthin (**56**) and papyrioerythrinone (**57**) [1,6,7]. Crustaceans belonging to Isopoda can introduce a hydroxy group at C-2 in the β-end group. This hydroxylation is also none-stereo-selective. Therefore, β-caroten-2-ol (**58**) in the sea louse *Ligia exotica* exists as two optical isomers [1,6,7]. Recently, two new carotenoids, 2,3′-dihydroxycanthaxanthin (**59**) [27] and 2,3-dihydroserythrin (**60**) [28], were isolated from the hermit crab *Paralithodes brevipes* and crawfish *Procambarus clarkii*, respectively (Figure 10).

**Figure 10.** Characteristic carotenoids in crustaceans.

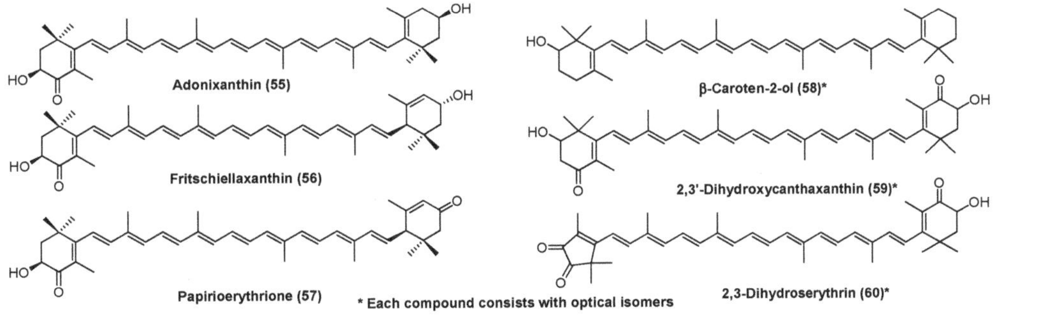


## 6. Echinodermata (Echinoderms)

Echinenone is a well-known major carotenoid in the gonads of sea urchins and is an oxidative metabolite of β-carotene [6,7]. Echinenone from the gonads of sea urchins was found to have a 9′Z configuration (**61**) [29].

Starfish are carnivorous and mainly feed on bivalves and small crustaceans. Principal carotenoids in starfish are astaxanthin (**8**), 7,8-didehydroastaxanthin (**9**), and 7,8,7′,8′-didehydroastaxanthin (**10**). They correspond to the oxidative metabolites of β-carotene, diatoxanthin, and alloxanthin, respectively. The crown-of-thorns starfish *Acanthaster planci* is a large, nocturnal sea star that preys upon coral polyps. Recently, four new carotenoids: 4-ketodeepoxyneoxanthin (**62**), 4-keto-4′-hydroxydiatoxanthin (**63**), 3′-epigobiusxanthin (**64**), and 7,8-dihydrodiadinoxanthin (**65**), were isolated from *A. planci* as minor components along with the major carotenoids 7,8-didehydroastaxanthin, peridininol, and astaxanthin, and several other minor carotenoids including 7,8,7′,8′-tetrahydroastaxanthin, diadinoxanthin, diatoxanthin, and alloxanthin [30].

3,4,3′,4′-Tetrahydroxypirardixanthin 4,4′-disulfate, named ophioxanthin (**66**), was reported in the brittle star *Ophioderma longicaudum* [31]. Canthaxanthin and astaxanthin were found in the gonads of sea cucumbers as major components. 5,6,5′,6′-Tetrahydro-β-carotene derivatives with 9Z, 9′Z configurations, termed cucumariaxanthin (**67**), were isolated from the sea cucumber *Cucumaria japonica* [32] (Figure 11).

Recently, zeaxanthin, astaxanthin, and lutein were identified from spiny sea-star *Marthasterias glacialis* by HPLC-PAD-atmospheric pressure chemical ionization-MS. These carotenoids showed strong cell proliferation inhibition activity against rat basophilic leukemia RBL-2H3 cancer cell line [33].

## 7. Protochordata (Tunicates)

As well as bivalves, tunicates are filter feeders. Carotenoids found in tunicates originate from phytoplankton such as diatoms, and are also metabolites of fucoxanthin, diatoxanthin, and alloxanthin [7,8].

Halocynthiaxanthin (**27**), an acetylenic analog of fucoxanthinol (**26**), and mytiloxanthinone (**68**), an oxidative metabolite of mytiloxanthin (**28**), were first isolated from the sea squirt *Halocynthia roretzi* [34]. They are widely distributed in various tunicates. Amarouciaxanthin A (**34**) and amarouciaxanthin B (**69**), having a unique 3-oxo-6-hydroxy-ε-end group, were first isolated from the tunicate *Amaroucium pliciferum* [35] (Figure 12). Peridinin and its metabolites are also found in tunicates.

**Figure 11.** Characteristic carotenoids in echinoderms.

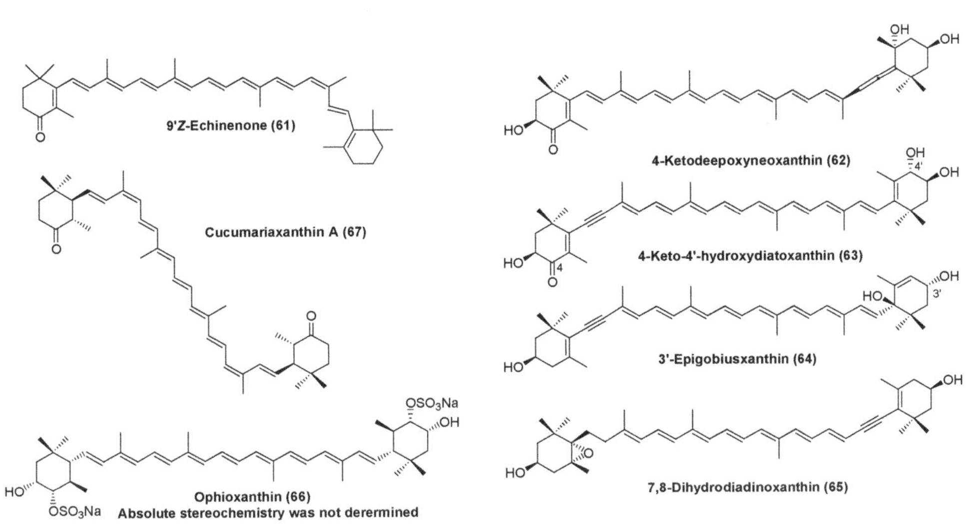


**Figure 12.** Metabolic pathways of fucoxanthin in tunicates.

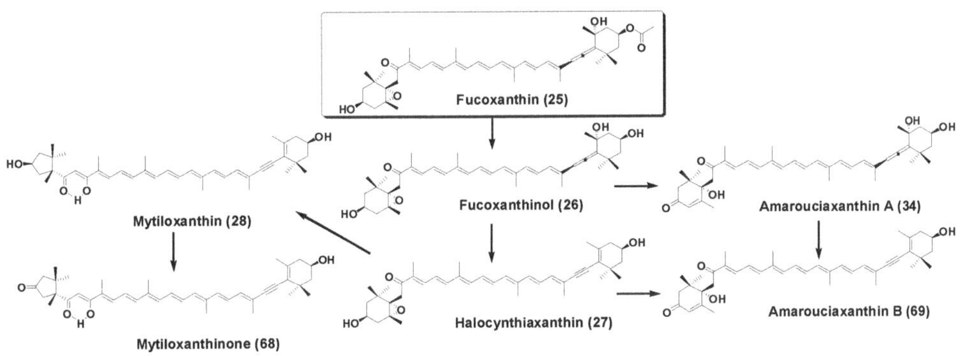


## 8. Pisces (Fish)

Many fish accumulate carotenoids in their integuments and gonads. On the other hand, Salmonidae fish peculiarly accumulate astaxanthin (**8**) in muscle. Except for catfish, carotenoids in the integuments of fish exist in an esterified form.

Astaxanthin (**8**) is widely distributed in both marine and fresh water fish. Cyprinidae fish, which inhabit fresh water, can synthesize (3S,3′S)-astaxanthin (**8a**) from zeaxanthin (**70**) by oxidative metabolic conversion (Figure 13). On the other hand, Perciformes and Salmonidae fish cannot synthesize astaxanthin from other carotenoids [6,7,36]. Therefore, astaxanthin present in these fish originates from dietary crustacean zooplankton. Astaxanthin in these marine fish comprises three optical isomers. Perciformes and Salmoidae fish can convert astaxanthin to zeaxanthin [36,37]. Therefore, zeaxanthin in these fish also exists as three optical isomers [38].

**Figure 13.** Metabolism of zeaxanthin in Cyprinidae and astaxanthin in Salmonidae fish.

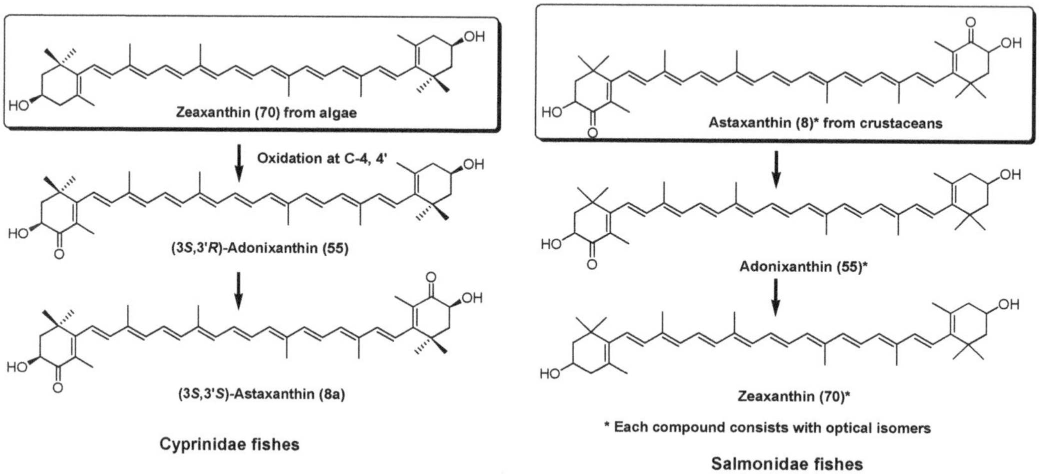


Tunaxanthin (**71**) is widely distributed in fish belonging to Perciformes. The bright yellow color in the fins and skin of marine fish is due to the presence of tunaxanthin. Feeding experiments involving red sea bream and yellow tail revealed that tunaxanthin (**71**) was metabolized from astaxanthin (**8**) via zeaxanthin, as shown in Figure 14 [7,36]. Carotenoids with a 3-oxo-ε-end group such as ε,ε-carotene-3,3′-dione (**72**) [37] are key intermediates in this metabolic conversion.

**Figure 14.** Metabolism of astaxanthin in Perciformes fish.

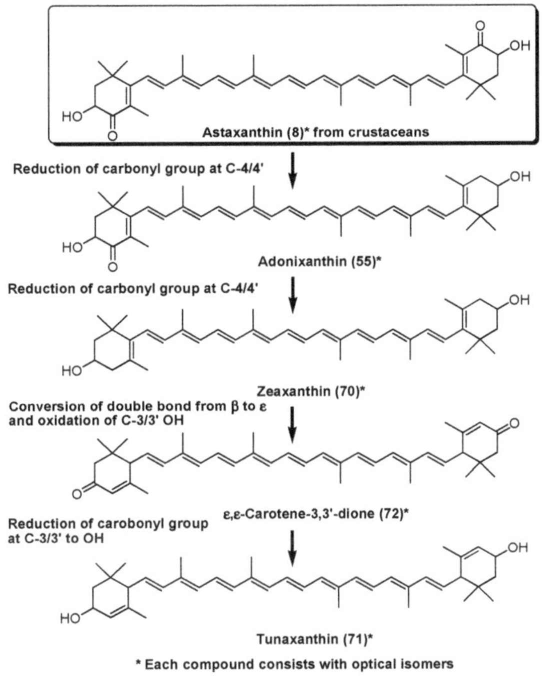


in Perciformes fishes

Unique apocarotenoids, micropteroxanthins (**73**–**76**), were reported from the integuments of the black bass *Micropterus salmoides* [39]. They were assumed to be corresponding oxidative cleavage products of tunaxanthin, lutein, and alloxanthin.

Since 2000, there are a few reports on new structures of carotenoids from fish (Figure 15). Carotenoids with a 3,6-dihydroxy-ε-end group, salmoxanthin (**77**), deepoxysalmoxanthin (**78**) (from the salmon *Oncorhynchus keta*) [40], and gobiusxanthin (**79**) (from the freshwater goby *Rhinogobius brunneus*) [41], were isolated. A series of carotenoids with a 7,8-dihydro- and/or 7,8,7′,8′-tetrahydro polyene chain were isolated from the integuments and eggs of the Japanese common catfish *Silurus asotus* [42]. Recently, new carotenoids, 7′,8′,9′,10′-tetrahydro-β-cryptoxanthin (**80**), 7′,8′-dihydrodiatoxanthin (**81**), and (3*S*,6*S*,6′*S*)-ε-cryptoxanthin (**82**), were isolated from the integuments and gonads of the Japanese common catfish as minor carotenoids [43].

**Figure 15.** New carotenoids from fish.

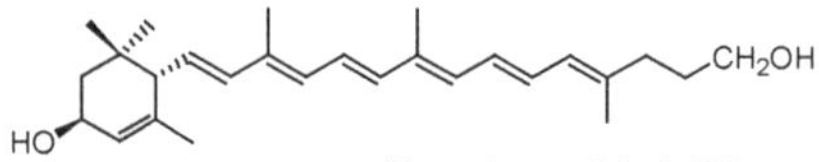

α-Micropteroxanthin A (73)

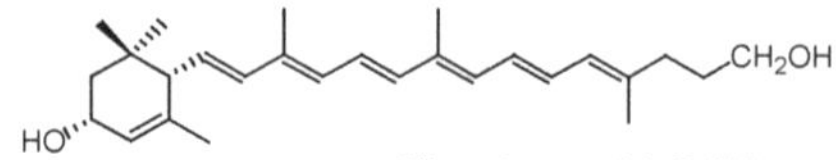

α-Micropteroxanthin B (74)

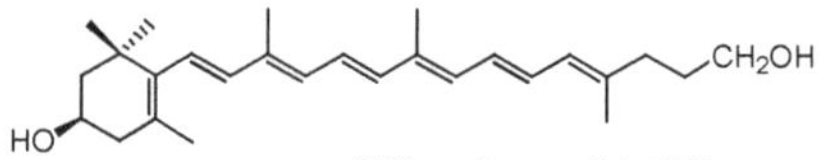

β-Micropteroxanthin (75)

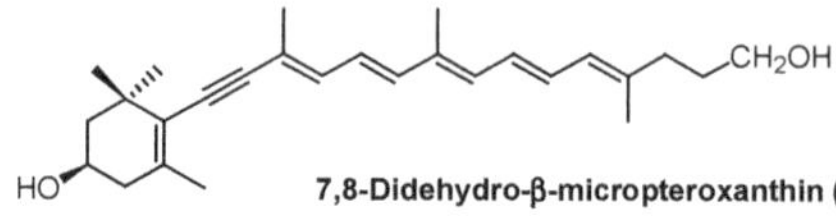

7,8-Didehydro-β-micropteroxanthin (76)

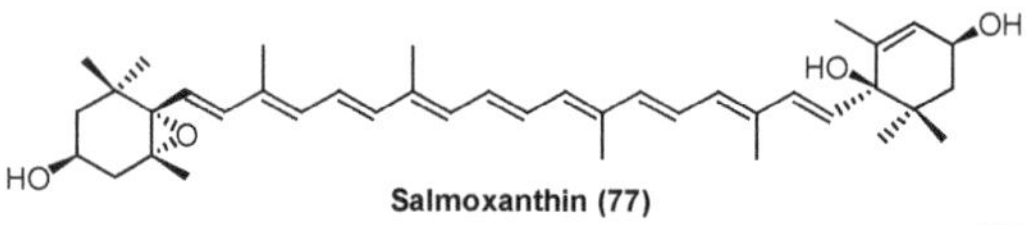

Salmoxanthin (77)

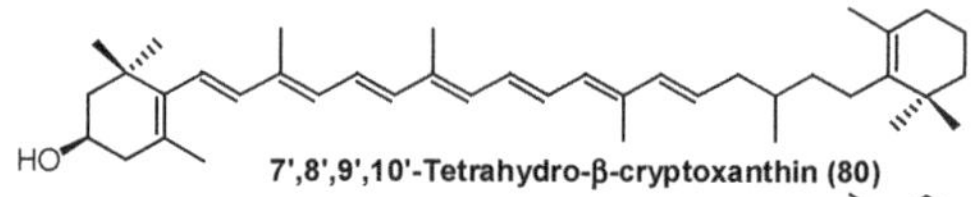

7',8',9',10'-Tetrahydro-β-cryptoxanthin (80)

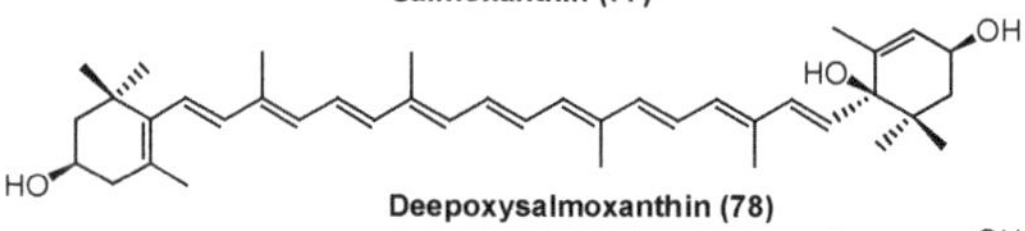

Deepoxysalmoxanthin (78)

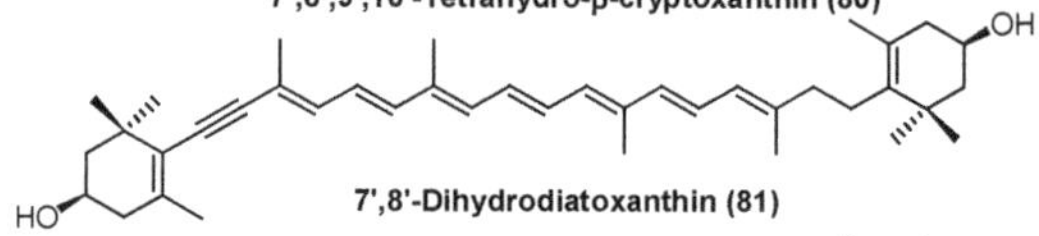

7',8'-Dihydrodiatoxanthin (81)

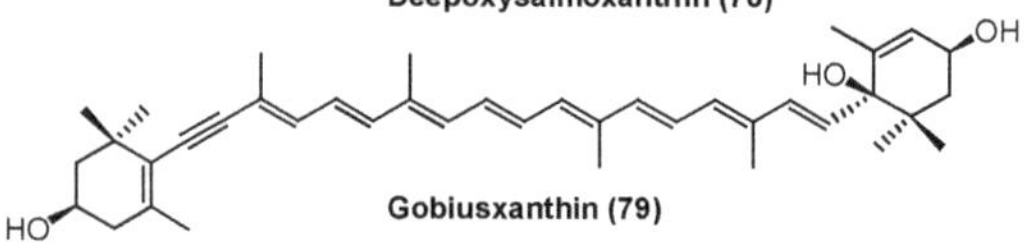

Gobiusxanthin (79)

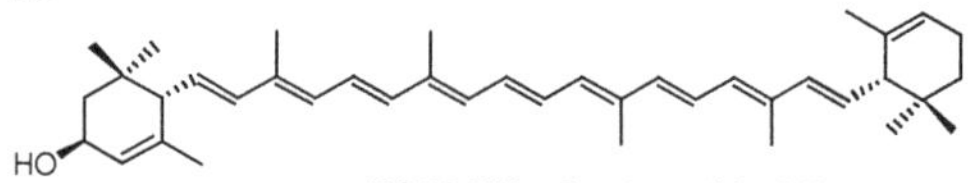

(3*S*,6*S*,6'*S*)-ε-Cryptoxanthin (82)

## 9. Mammalia (Mammals)

There are few reports available on carotenoids in marine mammals. Only, β-carotene and lutein were reported from the dolphin [44]. The whale is the biggest marine mammal. Whales feed on krill, which is an important dietary source of astaxanthin for marine animals. Therefore, whales might accumulate astaxanthin in the body.

Recently, absorption and metabolism of fucoxanthin (**25**) in mice was investigated. Dietary administrated fucoxanthin was converted to amarouciaxanthin A (**34**) via fucoxanthinol (**26**) in mice [45,46] (Figure 16). This metabolic conversion was also observed in human hepatoma cell HepG2 and required $NAD(P)^+$ as a cofactor [45].

**Figure 16.** Metabolism of fucoxanthin in mice.

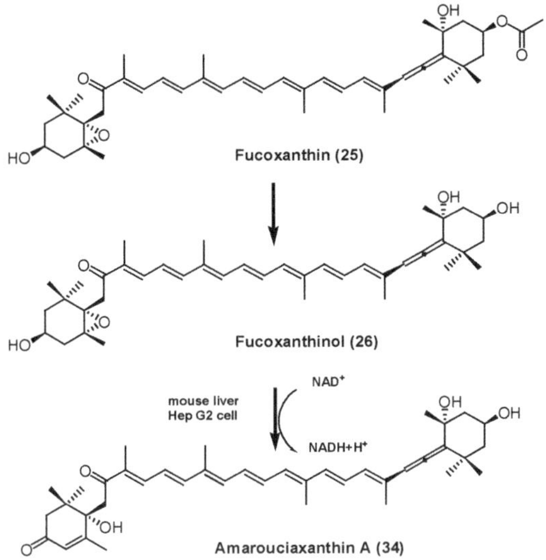


## 10. Role of Carotenoids in Marine Animals and Utilization of Carotenoids for Aquaculture

Carotenoids are not essential in the nutritional sense. However, they are beneficial for animal health. It is well-known that carotenoids have an unsubstituted β-end group, such as β-carotene, α-carotene, and the β-cryptoxanthin precursor of vitamin A in animals. Furthermore, canthaxanthin was also converted to retinol in Salmoidae fish. 3-Hydroxy carotenoids: lutein, zeaxanthin, and astaxanthin, were also reported to be precursors of 3,4-dehydroretinol (Vitamin A2) in some freshwater fish [36,47].

Many marine animals accumulate carotenoids in their integuments. Integumentary carotenoids may contribute to photoprotection, camouflage, and signaling such as breeding color.

Carotenoids have excellent antioxidative activities for quenching singlet oxygen and inhibiting lipid peroxidation. Astaxanthin supplementation in Salmonidae fish suppressed oxidative stress [48,49].

Marine animals also accumulate carotenoids in their gonads. Carotenoids are assumed to be essential for reproduction in marine animals. Astaxanthin supplementation in cultured salmon and red sea bream increased ovary development, fertilization, hatching, and larval growth [50]. In the case of the sea urchin, supplementation with β-carotene, which was metabolized to echinenone, also increased reproduction and the survival of larvae [51]. Carotenoids also enhance immune activity in marine animals [52,53].

Carotenoids are used for pigmentation in several aquaculture fish. Synthetic and natural astaxanthin from *Phaffia* yeast and *Haematococcus* algae is widely used for the pigmentation of salmon, trout, and red sea bream. Lutein from marigold is also used as a yellow coloration for cultured marine fish such as yellow tail and red sea bream. Zeaxanthin from spirulina is used as a red coloration for goldfish and ornamental carp.

## 11. Conclusions

In the present review, I have described marine animal carotenoids from natural product chemistry, metabolism, food chain, and chemosystematic viewpoints and also describe new structural carotenoids isolated from marine animals during the last decade.

In plants and photosynthetic bacteria, biosynthetic roots of carotenoids were identified at the enzymatic and gene level. On the other hand, neither enzymes nor genes for the metabolism of carotenoids in animals have been clarified. Therefore, chemical, biochemical, and analytical approaches are still important to clarify carotenoids in animals.

Interesting new structural carotenoids can still be found in marine animals. The structures of these new carotenoids provide information on the function and food chain, as well as metabolic pathways in marine animals.

## References and Notes

1. Britton, G.; Liaaen-Jensen, S.; Pfander, H. *Carotenoids Hand Book*; Birkhäuser: Basel, Switzerland, 2004.
2. Maoka, T.; Akimoto, N. Natural product chemistry in carotenoid, some experimental techniques for structural elucidation and analysis of natural carotenoids. *Carotenoid Sci.* **2008**, *13*, 10–17.
3. Maoka, T. Recent progress in structural studies of carotenoids in animals and plants. *Arch. Biochem. Biophys.* **2009**, *483*, 191–195.
4. Maoka, T. Structural studies of natural carotenoids by our research group during the three decade. *Carotenoid Sci.* **2009**, *14*, 26–36.
5. Liaaen-Jensen, S. Marine carotenoids—Selected topics. *New J. Chem.* **1990**, *14*, 747–759.
6. Liaaen-Jensen, S. Carotenoids in Food Chain. In *Carotenoids: Biosynthesis and Metabolism*; Britton, G., Liaaen-Jensen, S., Pfander, H., Eds.; Birkhäuser: Basel, Switzerland, 1998; Volume 3, pp. 359–371.
7. Matsuno, T. Aquatic animal carotenoids. *Fish. Sci.* **2001**, *67*, 771–789.
8. Matsuno, T. Animal carotenoids. In *Carotenoids Chemistry and Biology*; Krinsky, N.I., Mathews-Roth, M.M., Taylor, R.F., Eds.; Plenum Press: New York, NY, USA, 1989; pp. 59–74.
9. Matsuno, T.; Hirao, S. Marine carotenoids. In *Marine Biogenic Lipids, Fats, and Oils*; Ackman, R.G., Ed.; CRC Press: Boca Raton, FL, USA, 1989; Volume 1, pp. 251–388.
10. Rogers, E.W.; Molinski, T.F. A cytotoxic carotenoid from the marine sponge *Prianos osiro*. *J. Nat. Prod.* **2005**, *68*, 450–452. The IUPAC name (3*R*,3′*R*,5*S*)-3,3′,5,19′-tetrahydroxy-7′,8′-didehydro-γ,ε-carotene-8-one was given for

compound **5** in this literature. However, this IUPAC name is un-correct. Correct IUPAC name for **5** is (3*R*,3′*R*,5*S*)-3,3′,5,19′-tetrahydroxy-7′,8′-didehydro-5,8-dihydro-β,β-caroten-8-one.

11. Maoka, T.; Akimoto, N.; Tsushima, M.; Komemushi, S.; Harada, R.; Sameshima, N.; Iwase, F.; Sakagami, Y. Carotenoids in corals and sea snails. *Carotenoid Sci.* **2010**, submitted for publication.
12. Tsushima, M.; Maoka, T.; Matsuno, T. Comparative biochemical studies of carotenoids in marine invertebrates. The first positive identification of ε,ε-carotene derivatives and isolation of two new carotenoids from chitons. *Comp. Biochem. Physiol.* **1989**, *93B*, 665–671.
13. Tsushima, M.; Maoka, T.; Matsuno, T. Structure of carotenoids with 5,6-dihydro-β-end group from the spindle shell *Fusinus perplexus*. *J. Nat. Prod.* **2001**, *64*, 1139–1142.
14. Yamashita, E.; Matsuno, T. A new apocarotenoid from the sea hare *Aplysia kurodai*. *Comp. Biochem. Physiol.* **1990**, *96B*, 465–470.
15. Maoka, T. A new apocarotenoid from marine shellfish. *J. Nat. Prod.* **1997**, *60*, 616–617.
16. Maoka, T.; Hashimoto, K.; Akimoto, N.; Fujiwara, Y. Structures of five new carotenoids from the oyster *Crassostrea gigas*. *J. Nat. Prod.* **2001**, *64*, 578–581.
17. Maoka, T.; Fujiwara, Y.; Hashimoto, K.; Akimoto, N. Structures of new carotenoids with a 3,4-dihydroxy-β-end group from the oyster *Crassostrea gigas*. *Chem. Pharm. Bull.* **2005**, *53*, 1207–1209.
18. Maoka, T.; Fujiwara, Y.; Hashimoto, K.; Akimoto, N. Characterizatoin of fucoxanthin and fucoxanthinol esters in the chinese surf clam, *Mactra chinensis*. *J. Agric. Food Chem.* **2007**, *55*, 1563–1567.
19. Maoka, T.; Akimoto, N.; Murakoshi, M.; Sugiyama, K.; Nishino, H. Carotenoids in clams, *Ruditapes philippinarum* and *Meretrix petechialis*. *J. Agric. Food Chem.* **2010**, *58*, 5784–5788.
20. Maoka, T.; Akimoto, N.; Yim, M.-J.; Hosokawa, M.; Miyashita, K. A new $C_{37}$-skeletal carotenoid from the clam, *Paphia amabillis*. *J. Agric. Food Chem.* **2008**, *56*, 12069–12072.
21. Fujiwara, Y.; Maoka, T.; Ookubo, M.; Matsuno, T. Crassostreaxanthins A and B: novel marine carotenoids from the oyster *Crassostrea gigas*. *Tetrahedron Lett.* **1992**, *33*, 4941–4944.
22. Tode, C.; Yamano, Y.; Ito, M. First total synthesis of crassostreaxanthin B. *J. Chem. Soc., Perkin Trans. 1* **1999**, 1625–1626; doi: 10.1039/A901116A.
23. Tode, C.; Yamano, Y.; Ito, M. Carotenoids and related polyenes. Part 7. Total synthesis of crassostreaxanthin B Applying the stereoselective rearrangement of tetrasubstituted epoxides. *J. Chem. Soc., Perkin Trans. 1* **2001**, 3338–3345; doi: 10.1039/B108037G.

24. Maoka, T.; Fujiwara, Y.; Hashimoto, K.; Akimoto, N. Carotenoids in three species of corbicula clams, *Corbicula japonica*, *Corbicula sandai*, and *Corbicula* sp. (Chinese freshwater corbicula clam). *J. Agric. Food Chem.* **2005**, *53*, 8357–8364.
25. Maoka, T.; Fujiwara, Y.; Hashimoto, K.; Akimoto, N. Structure of new carotenoids from the corbicula clam *Corbicula japonic*. *J. Nat. Prod.* **2005**, *68*, 1341–1344.
26. Maoka, T.; Yokoi, S.; Matsuno, T. Comparative biochemical stidies of carotenoids in nine species of Cephalopoda. *Comp. Biochem. Physiol.* **1989**, *92B*, 247–250.
27. Maoka, T.; Akimoto, N. 2,3′-Dihydroxycanthaxanthin, a new carotenoid with a 2-hydroxy-4-oxo-β-end group from the hermit Crab, *Paralithodes brevipes*. *Chem. Pharm. Bull.* **2006**, *54*, 1462–1464.
28. Maoka, T.; Ando, S. Isolation of purple nor-carotenoid, 2,3-dihydroserythrin, from crawfish *Procambarus clarkii*. *Fish. Sci.* **2007**, *73*, 967–968.
29. Tsushima, M.; Matsuno, T. Occurrence of 9′Z-β-echinenone of the sea urchin *Pseudocentrotus depressus*. *Comp. Biochem. Physiol.* **1997**, *118B*, 921–925.
30. Maoka, T.; Akimoto, N.; Terada, Y.; Komemushi, S.; Harada, R.; Sameshima, N.; Sakagami, Y. Structure of minor carotenoids from crown-of-thorns starfish, *Acanthaster planci*. *J. Nat. Prod.* **2010**, *73*, 675–678.
31. D'auria, M.V.; Riccio, R.; Minale, L. Ophioxanthin, a new marine carotenoid sulphate from the ophiuroid *Ophioderma longicaudum*. *Tetrahedron Lett.* **1985**, *26*, 1871–1872.
32. Tsushima, M.; Fujiwara, Y.; Matsuno, T. Novel marine di-Z-carotenoids, cucumariaxanthins A, B and C from the sea cucumber *Cucumaria japonica*. *J. Nat. Prod.* **1996**, *59*, 30–34.
33. Ferreres, F.; Pereira, D.M.; Gil-Izquierdo, A.; Valentão, P.; Botelho, J.; Mouga, T.; Andrade, P.B. HPLC-PAD-atmospheric pressure chemical ionization-MS metabolite profiling of cytotoxic carotenoids from the echinoderm *Marthasterias glacialis* (spiny sea-star). *J. Sep. Sci.* **2010**, *33*, 2250–2257.
34. Matsuno, T.; Ookubo, M.; Nishizawa, T.; Shimizu, I. Carotenoids of sea suirts. I. New marine carotenids, halocynthiaxanthin and mytiloxanthinone from *Halocynthia roretzi*. *Chem. Pharm. Bull.* **1984**, *32*, 4309–4315.
35. Matsuno, T.; Ookubo, M.; Komori, T. Carotenoids of tunicates. III: The structural elucidation of two new marine carotenoids, amarouciaxanthin A and B. *J. Nat. Prod.* **1985**, *48*, 606–613.
36. Schiedt, K. Absorption and metabolism of carotenoids in birds, fish and crustaceans. In *Carotenoids Biosynthesis and Metabolism*; Britton, G., Liaaen-Jensen, S., Pfander, H., Eds.; Birkhäuser: Basel, Switzerland, 1998; Volume 3, pp. 285–358.

37. Matsuno, T.; Katsuyama, M.; Maoka, T.; Hirono, T.; Komori, T. Reductive metabolic pathways of carotenoids in fish (3S,3′S)-astaxanthin to tunaxanthin A, B and C. *Comp. Biochem. Physiol.* **1985**, *80B*, 779–789.
38. Maoka, T.; Arai, A.; Shimizu, M.; Matsuno, T. The first isolation of enantiomeric and
*meso*-zeaxanthin in nature. *Comp. Biochem. Physiol.* **1986**, *83B*, 121–124.
39. Yamashita, E.; Arai, S.; Matsuno, T. Metabolism of xanthophylls to vitamin A and new apocarotenoids in liver and skin of black bass, *Micropterus salmoides*. *Comp. Biochem. Physiol.* **1996**, *113B*, 485–489.
40. Matsuno, T.; Tsushima, M.; Maoka, T. Salmoxanthin, deepoxy-salmoxanthin and
7,8-didehydrodeepoxy-salmoxanthin from the salmon *Oncorhynchus keta*. *J. Nat. Prod.* **2001**, *64*, 507–510.
41. Tsushima, M.; Mune, E.; Maoka, T.; Matsuno, T. Isolation of stereoisomeric epoxy carotenoids and new acetylenic carotenoid from the common freshwater goby *Rhinogobius brunneus*.
*J. Nat. Prod.* **2000**, *63*, 960–964.
42. Tsushima, M.; Ikuno, Y.; Nagata, S.; Kodama, K.; Matsuno, T. Comparative biochemical studies of carotenoids in catfishes. *Comp. Biochem. Physiol.* **2002**, *133B*, 331–336.
43. Maoka, T.; Akiomoto, N. Structures of minor carotenoids from the Japanese common catfish, *Silurus asotus*. *Chem. Phram. Bull.* **2011**, *59*, 140–145.
44. Slifka, K.A.; Bowen, P.E.; Stacewicz-Sapuntzakis, M.; Susan D.; Crissey, S.D. A survey of serum and dietary carotenoids in captive wild animals. *J. Nutr.* **1999**, *129*, 380–390.
45. Asai, A.; Sugawara, H.; Ono, H.; Nagao, A. Biotransformation of fucoxanthinol into amarouciaxanthin A in mice and HepG2 cells: formation and cytotoxicity of fucoxanthin metabolites. *Drug Metab. Dispos.* **2004**, *32*, 205–211.
46. Hashimoto, T.; Ozaki, Y.; Taminato, M.; Das, S.K.; Mizuno, M.; Yoshimura, K.; Maoka, T.; Kanazawa, K. The distribution and accumulation of fucoxanthin and its metabolites after oral administration in mice. *Br. J. Nutr.* **2009**, *102*, 242–248.
47. Matsuno, T. Xanthophylls as precursor of retinoids. *Pure Appl. Chem.* **1991**, *63*, 81–88.
48. Nakano, T.; Tosa, M.; Takeuchi, M. Improvement of biochemical features in fish health by red yeast and synthetic astaxanthin. *J. Agric. Food Chem.* **1995**, *43*, 1570–1573.
49. Nakano, T.; Kanmuri, T.; Sato, M.; Takeuchi, M. Effect of astaxanthin rich red yeast (*Phaffia rhodozyma*) on oxidative stress in rainbow trout. *Biochim. Biophys. Acta* **1999**, *1426*, 119–125.

50. Torrissen, O.J.; Christiansen, R. Requirements for carotenoids in fish diets. *J. Appl. Ichthyol.* **1995**, *11*, 225–230.
51. Tsushima, M.; Kawakami, T.; Mine, M.; Matsuno, T. The role of carotenoids in the development of the sea urchin *Pseudocentrotus depressus*. *Invert. Reprod. Develop.* **1997**, *32*, 149–153.
52. Thompson, I.; Choubert, G.D.; Houlihan, F.; Secombes, C.J. The effect of dietary Vitamin A and Astaxanthin on the immunocompetence of rainbow trout. *Aquaculture* **1995**, *133*, 91–102.
53. Kawakami, T.; Tsushima, M.; Katabami, Y.; Mine, M.; Ishida, A.; Matsuno, T. Effect of β,β-carotene, β-echinenone, astaxanthin, fucoxanthin, vitamin A and vitamin E on the biological defense of the sea urchin *Pseudocentrotus depressus*. *J. Exp. Mar. Biol. Ecol.* **1998**, *226*, 165–174.

*marine drugs*

Article

# Carotenoids in Marine Invertebrates Living along the Kuroshio Current Coast

**Takashi Maoka [1], Naoshige Akimoto [2], Miyuki Tsushima [3], Sadao Komemushi [4], Takuma Mezaki [5], Fumihito Iwase [5], Yoshimitsu Takahashi [6], Naomi Sameshima [6], Miho Mori [6], and Yoshikazu Sakagami [6,*]**

1 Research Institute for Production Development, 15 Shimogamo-morimoto-cho, Sakyo-ku, Kyoto 606-0805, Japan; E-Mail: maoka@mbox.kyoto-inet.or.jp

2 *Graduate School of Pharmaceutical Sciences, Kyoto University, Yoshida-shimoadachi-cho, Sakyo-ku, Kyoto 606-8501, Japan; E-Mail: nmakimoto@leto.eonet.ne.jp*

3 *Kyoto Pharmaceutical University, Misasagi Yamashina-Ku, Kyoto 607-8412, Japan; E-Mail: tsushima@mb.kyoto-phu.ac.jp*

4 Osaka City Graduate School of Engineering and Faculty of Engineering, Osaka City University, 3-3-138 Sugimoto, Sumiyoshi-ku, Osaka 558-8585, Japan; E-Mail: volvo.s80.clsssic386@gmail.com

5 Kuroshio Biological Research Foundation, Nishidomari-560, Ootsuki-cho, Kochi 788-0333, Japan; E-Mails: mezaki@kuroshio.or.jp (T.M.); iwase@kuroshio.or.jp (F.I.)

6 *Faculty of Agriculture, Kinki University, Nakamachi 3327-204, Nara-shi 631-8505, Nara, Japan; E-Mails: srdkt24134@hera.eonet.ne.jp (Y.T.); ato10min@yahoo.co.jp (N.S.); mori@nara.kindai.ac.jp (M.M.)*

* Author to whom correspondence should be addressed; E-Mail: sakagami@nara.kindai.ac.jp; Tel.: +81-742-43-7154; Fax: +81-742-43-1593.

*Received: 30 June 2011; in revised form: 31 July 2011 / Accepted: 8 August 2011 / Published: 22 August 2011*

**Abstract:** Carotenoids of the corals *Acropora japonica*, *A. secale*, and *A. hyacinthus*, the tridacnid clam *Tridacna squamosa*, the crown-of-thorns starfish *Acanthaster planci*, and the small sea snail *Drupella fragum* were investigated. The corals and the tridacnid clam are filter feeders and are associated with symbiotic zooxanthellae. Peridinin and pyrrhoxanthin, which originated from symbiotic zooxanthellae, were found to be major carotenoids in corals and the tridacnid clam. The crown-of-thorns starfish and the sea snail *D. fragum* are carnivorous and mainly feed on corals. Peridinin-3-acyl esters were major carotenoids in the sea snail *D. fragum*. On the other hand, ketocarotenoids such as 7,8-didehydroastaxanthin and astaxanthin were major carotenoids in the crown-of-thorns starfish. Carotenoids found in these marine animals closely reflected not only their metabolism but also their food chains.

**Keywords:** carotenoid; marine invertebrates; food chain; metabolism

## 1. Introduction

Marine animals, especially marine invertebrates, contain various carotenoids, with structural diversity [1–4]. Interesting structural carotenoids are still being found in marine animals [4]. In general, animals do not synthesize carotenoids *de novo*, and so those found in animals are either directly accumulated from food or partly modified through metabolic reactions [2]. The major metabolic conversions of carotenoids found in marine animals are oxidation, reduction, transformation of double bonds, oxidative cleavage of double bonds, and cleavage of epoxy bonds [2,3]. Therefore, various structural varieties are found in carotenoids of marine animals [4].

We have studied carotenoids in several marine invertebrates from chemical and comparative biochemical points of view [4]. In the present study, we focused on carotenoids of the corals *Acropora japonica*, *A. secale*, and *A. hyacinthus*, the tridacnid clam (elongate giant clam) *Tridacna squamosa*, crown-of-thorns starfish *Acanthaster planci*, and small sea snail *Drupella fragum*, inhabiting the Kuroshio current coast. These animals are closely associated within the food chain. Corals and the tridacnid clam are filter feeders and are associated with symbiotic zooxanthellae (dinoflagellate algae). On the other hand, the crown-of-thorns starfish and small sea snail *D. fragum* are carnivorous and mainly prey upon corals. Therefore, carotenoids that originated from zooxanthellae are passed to starfish and small sea snails through this food chain. In the present paper, we describe the carotenoids of these marine invertebrates.

## 2. Results and Discussion

Structural formulae of carotenoids identified from *Acropora* corals, the tridacnid clam *T. squamosa*, starfish *A. planci*, and sea snail *D. fragum* are shown in Figure 1.

## 2.1. Carotenoids of Corals and the Tridacnid Clam

The carotenoids composition of the corals and the tridacnid clam were similar to each other (Table 1). β,β-Carotene, peridinin (including the 9′Z isomer), pyrrhoxanthin, diatoxanthin, and diadinoxanthin were found in these animals. These carotenoid patterns resembled those of symbiotic zooxanthellae [5,6]. The results indicate that corals and the tridacnid clam directly absorb carotenoids from symbiotic zooxanthellae and accumulate them without metabolic modification. In the eggs of corals, peridinin and pyrroxanthin were present as major carotenoids. It was assumed that peridinin and pyrroxanthin play important roles in reproduction in corals, as with astaxanthin in salmonid fishes [7].

Recently, Daigo *et al.* studied carotenoids of more than 20 species of coral inhabiting reefs in Okinawa [8]. They reported that carotenoids found in these corals were not only peridinin and diadinoxanthin, that originated from symbiotic zooxanthellae, but also zeaxanthin, lutein, and, fucoxanthin, that originated from cyanobacteria, green algae, and diatoms. Cyanobacteria, green algae, and diatoms were epizoic and/or endolithic algae that grew in association with the corals. Corals accumulated carotenoids from these epizoic and/or endolithic algae [8]. However, the present study found that carotenoids in *Acropora* corals, inhabiting the Kuroshio current coast of Kochi, only consisted of those that originated from zooxanthellae. These differences might reflect the constitution of associating algae with corals.

Peridinin and pyrrhoxanthin were found to be major carotenoids in the tridacnid clam. In general, major carotenoids found in clams are fucoxanthin and its metabolites originating from diatoms [9–11]. On the other hand, neither fucoxanthin nor its metabolites were found in the tridacnid clam. This indicates that the tridacnid clam only ingested carotenoids from dinoflagellate algae. Similar results were reported in carotenoids of the bivalves, *Modiolus modiolus* and *Pecten maximus* [12].

**Figure 1.** Carotenoids identified from *Acropora* corals, the tridacnid clam *T. squamosa*, starfish *A. planci*, and sea snail *D. fragum*.

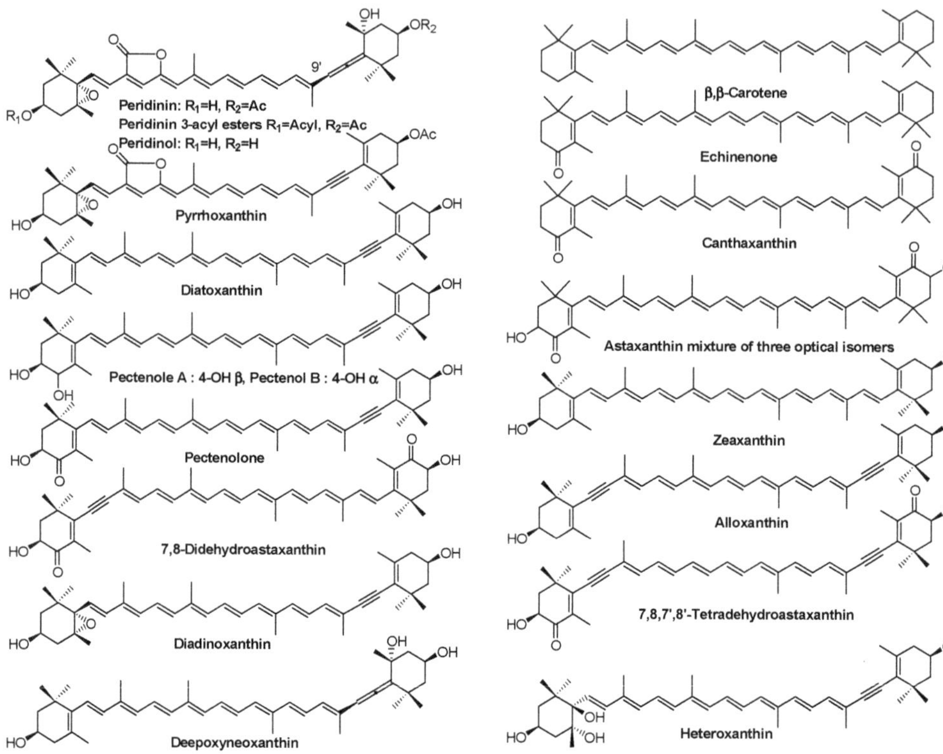
Peridinin: $R_1$=H, $R_2$=Ac
Peridinin 3-acyl esters $R_1$=Acyl, $R_2$=Ac
Peridinol: $R_1$=H, $R_2$=H
Pyrrhoxanthin
Diatoxanthin
Pectenole A : 4-OH β, Pectenol B : 4-OH α
Pectenolone
7,8-Didehydroastaxanthin
Diadinoxanthin
Deepoxyneoxanthin
β,β-Carotene
Echinenone
Canthaxanthin
Astaxanthin mixture of three optical isomers
Zeaxanthin
Alloxanthin
7,8,7',8'-Tetradehydroastaxanthin
Heteroxanthin

**Table 1.** Carotenoids of *Acropora* corals and the tridacnid clam *Tridacna squamosa*.

| | *Acropora japonica* | | *A. secale* | *A. hyacinthus* | *Tridacna squamosa* |
|---|---|---|---|---|---|
| | **Whole** | **Egg** | **Whole** | **Whole** | **Mantle and foot** |
| Carotenoid content | 3.3 (mg/100 g) | 0.02 | 2.4 | 2.9 | 10 |
| | composition (%) | % | % | % | % |
| β,β-Carotene | 15.5 | 5.0 | 12.0 | 13.4 | 5.1 |
| Diatoxanthin | 5.5 | 15.0 | 4.5 | 5.2 | 0.9 |
| Diadinoxanthin | 4.5 | 5.0 | 5.0 | 5.5 | 9.2 |
| Pyrrhoxanthin | 45.5 | 20.0 | 50.6 | 40.5 | 10.1 |
| Peridinin | 13.0 | 50.0 | 10.0 | 16.0 | 44.1 |
| 9′Z-Peridinin | 16.0 | 5.0 | 17.9 | 19.4 | 30.6 |

*2.2. Carotenoids of the Crown-of-Thorns Starfish*

The crown-of-thorns starfish, *A. planci,* is a large, nocturnal sea star that mainly preys upon coral polyps. Like other starfish [13], 7,8-didehydroastaxanthin and astaxanthin were found to be major carotenoids along with pectenolone, 7,8,7′,8′-tetradehydroastaxanthin, diatoxanthin, and alloxanthin (Table 2). In general, the starfish can introduce a hydroxy group at C-3 and carbonyl group at C-4 in the β-end group of carotenoids [6]. So, 7,8-didehydroastaxanthin and astaxanthin were oxidative metabolites of diatoxanthin and β-carotene, respectively, ingested from dietary corals. Echinenone and canthaxanthin were metabolic intermediates from β,β-carotene to astaxanthin. The acetylenic carotenoids, pectenolone, pectenol A, and pectenol B, were also metabolic intermediates from diatoxanthin to 7,8-didehydroastaxanthin. Peridinol, one of the major carotenoids in the crown-of-thorns starfish, was converted from peridinin, which originated from corals, by hydrolysis. Furthermore, four new carotenoids; 4-ketodeepoxyneoxanthin, 4-keto-4′-hydroxydiatoxanthin, 3′-epigobiusxanthin, and 7,8-dihydrodiadinoxanthin, were isolated [14]. Details of the structural elucidation of those compounds were described previously [14]. In the present paper, the biosynthetic origins of these compounds are discussed (Figure 2). 4-Keto-4′-hydroxydiatoxanthin was one of the metabolic intermediates from diatoxanthin to 7,8-didehydroastaxanthin. 4-Ketodeepoxyneoxanthin might be an oxidative metabolite of deepoxyneoxanthin derived from neoxanthin by deepoxydation.

3′-Epigobiusxanthin might be derived from diadinoxanthin. 7,8-Dihydrodiadinoxanthin, which has a unique single bond in the 7,8-saturated polyene chain, may be a reduction metabolite of diadinoxanthin. Therefore, it was concluded that carotenoids ingested from corals were oxidatively metabolized and accumulated in the crown-of-thorns starfish.

**Table 2.** Carotenoids of the crown-of-thorns starfish *Acanthaster planci.*

| | Whole | Gonad |
|---|---|---|
| | 0.46 mg/100 g | 6.6 mg/100 g |
| β,β-Carotene | 2.1 | 1.4 |
| Echinenone | 1.3 | 1.3 |
| Canthaxanthin | 1.6 | 1.6 |
| 7,8,7′,8′-Tetradehydroastaxanthin | 2.0 | 1.6 |
| 7,8-Didehydroastaxanthin | 35.6 | 35.3 |
| Astaxanthin | 9.8 | 5.8 |
| Pectenolone | 3.2 | 3.0 |
| Diatoxanthin | 3.2 | 15.8 |
| Alloxanthin | 3.2 | 11.8 |
| Diadinoxanthin | 3.0 | 8.6 |
| 7,8-Dihydrodiadinoxanthin | 4.0 | 1.0 |
| 3′-Epigobiusxanthin | 2.0 | 1.0 |
| Pectenol A | 2.0 | 1.6 |
| Pectenol B | 4.0 | 3.2 |
| 4-Keto-4′-hydroxydiatoxanthin | 5.5 | 1.3 |
| 4-Ketodeepoxyneoxanthin | 1.8 | 1.8 |
| Deepoxyneoxanthin | 1.0 | 0.3 |
| Heteroxanthin | 1.2 | 0.6 |
| Peridinol | 13.5 | 3.0 |

**Figure 2.** Possible bioformation roots of new carotenoids in crown-of-thorns starfish.

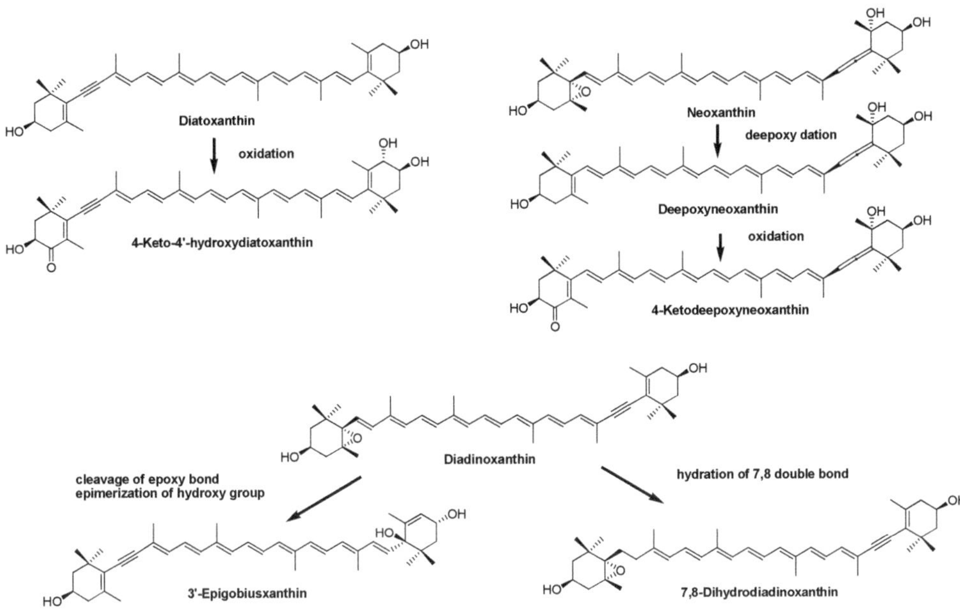


### 2.3. *Carotenoids of the Sea Snail* D. fragum

Like the crown-of-thorns starfish, the small sea snail *D. fragum* also feeds on corals. The carotenoid composition of this snail resembled that of the dietary corals (Table 3). This indicated that *D. fragum* also accumulated carotenoids from dietary corals without metabolic modification, except for the esterification of peridinin. In the present study, peridinin 3-acyl esters were fully characterized based on $^{1}$H-NMR and FAB MS spectral data. The $^{1}$H-NMR signal of H-3 (δ 4.95), which showed 1.04 ppm downfield shift relative to the corresponding signal in peridinin [15,16], indicated that the hydroxy group at C-3 was acylated. Fatty acids esterified with peridinin were assigned as palmitic acid, palmitoleic acid, and myristic acid based on FAB-MS data. Previously, peridinol fatty acid ester was characterized by Moaka *et al.* [10] and Sugawara *et al.* [17]. However, peridinin 3-acyl esters have not yet been reported. The origin of zeaxanthin in this snail was unclear. It might have originated from associated algae such as cyanobacteria [8].

**Table 3.** Carotenoids of the sea snail *Drupella fragum*.

| **Carotenoid content** | **4.03 mg/100 g** |
|---|---|
| | composition (%) |
| β,β-Carotene | 10.0 |
| Peridinin-3-acyl esters | 25.0 |
| Zeaxanthin | 15.0 |
| Diatoxanthin | 18.3 |

| | |
|---|---|
| Diadinoxanthin | 9.2 |
| Pyrrhoxanthin | 5.8 |
| Peridinin | 16.7 |

## 3. Experimental Section

### 3.1. General

The UV-Visible (UV-VIS) spectra were recorded with a Hitachi U-2001 in diethyl ether ($Et_2O$).
The positive ion FAB-MS spectra were recorded using a JEOL JMS-700 110A mass spectrometer with
*m*-nitrobenzyl alcohol as a matrix. The $^1$H-NMR (500 MHz) spectra were measured with a Varian UNITY INOVA 500 spectrometer in $CDCl_3$ with TMS as an internal standard. HPLC was performed on a Shimadzu LC-6AD with a Shimadzu SPD-6AV spectrophotometer set at 470 nm. The column used was a 250 × 10 mm i.d., 10 μm Cosmosil 5C18-II (Nacalai Tesque, Kyoto, Japan) with acetone:hexane (3:7, v/v) at a flow rate of 1.0 mL/min. The optical purity of astaxanthin was analyzed by chiral HPLC using a 300 × 8 mm i.d., 5 μm Sumichiral OA-2000 (Sumitomo Chemicals, Osaka, Japan) with *n*-hexane/$CHCl_3$-ethanol (48:16:0.8, v/v) at a flow rate of 1.0 mL/min [18].

### 3.2. Animal Material

The corals *A. japonica*, *A. secale*, and *A. hyacinthus*, the tridacnid clam *T. squamosa*, the crown-of-thorns starfish *A. planc*, and the sea snail *D. fragum* were collected along the Ootsuki coast, Kochi Prefecture, Japan from July to August 2009 and 2010.

### 3.3. Analysis of Carotenoids

The extraction and identification of carotenoids were carried out according to our routine methods [19]. Carotenoids were extracted from living or fresh animal specimens with acetone. The acetone extract was transferred to ether-hexane (1:1) layer after the addition of water. The total carotenoid contents were calculated employing an extinction coefficient of $E^{1\%}_{cm}$ = 2100 (astaxanthin) [20] for the starfish *A. planci* and $E^{1\%}_{cm}$ = 1350 (peridinin) [20] for *A. japonica*, *T. squamosa*, and *D. fragum* at λ max. The ether-hexane solution was evaporated. The residue was subjected to HPLC on silica gel. Carotenoid compositions were estimated by the peak area of the HPLC on silica gel with
acetone-hexane (3:7) monitored at 470 nm.

Individual carotenoids were identified by UV-VIS (ether), FAB MS, and partial $^1$H NMR (500 MHz, $CDCl_3$).

### *3.4. Identification of Carotenoids*

Identification of individual carotenoids were carried out on UV-VIS and FAB MS spectral data and compared with chromatographic property with authentic samples [19]. Optical isomer of astaxanthin in the crown-of-thorns starfish *Acanthaster planci* was analyzed by Chiral HPLC [18]. Astaxanthin fraction in *Acanthaster planci* was consisted of three optical isomers (3*R*,3′*R*):(3*R*,3′*S*):(3*S*,3′*S*) with a ratio of 32:14:54. Furthermore, peridiniol, peridinin and 9′Z-Peridinin were characterized by $^1$H NMR [15,16]. Structures of 7,8-ihydrodiadinoxanthin, 3′-epigobiusxanthin, 4-keto-4′-hydroxydiatoxanthin, 4-ketodeepoxyneoxanthin, and deepoxyneoxanthin were fully characterized by NMR [14].

### *3.5. Caracterization of Peridinin-3-acyl Esters*

Peridinin-3-acyl esters. FAB-MS: *m/z* 868.5860 $[M]^+$ (calcd. for $C_{55}H_{80}O_8$, 868.5856) peridinin 3-palmitate, *m/z* 866.5698 $[M]^+$ (calcd. for $C_{55}H_{79}O_8$, 866.5703) peridinin 3-palmitolate, *m/z* 840.5550 $[M]^+$ (calcd. for $C_{53}H_{76}O_8$, 840.5547) peridinin 3-myristate; UV-VIS 455, 475 nm; $^1$H NMR ($CDCl_3$), δ 0.88 (3H, t, *J* = 7.5 Hz, $CH_3$ in fatty acid moiety), 0.99 (3H, s, H-16), 1.07 (3H, s, H-17′), 1.20 (3H, s H-17), 1.21 (3H, s H-18), 1.25 (about 24H, s, -$CH_2$- in fatty acid moiety), 1.35 (3H, s, H-18′), 1.39 (3H, s, H-16′), 1.41 (1H, dd, *J* = 13, 7 Hz,H-2′β), 1.51 (1H, dd, *J* = 13, 12.5 Hz, H-4′β), 1.64 (1H, dd, *J* = 12.5, 12 Hz, H-2α eq), 1.79 (1H, dd, *J* = 12, 7 Hz, H-4β ax), 1.81 (3H, s, H-19′), 2.00 (H, ddd, *J* = 13, 4, 2 Hz, H-2′α), 2.04 (3H, s, $CH_3CO$-), 2.23 (3H, s, H-20), 2.29 (1H, overlapped, H-α), 2.28 (2H, t, *J* = 7.5 Hz, -$CH_2$-COOH in fatty acid moiety), 2.41 (1H, ddd, *J* = 14, 5, 1.5 Hz, H-4α), 4.95 (1H, m, H-3), 5.37 (1H, m, H-3′), 5.74 (1H, s, H-12), 6.06 (1H, s, H-8′), 6.11 (1H, d, *J* = 11 Hz, H-10′), 6.38 (1H, dg, *J* = 14, 11 Hz, H-11′), 6.40 (1H, d, *J* = 16 Hz, H-8), 6.46 (1H, d, *J* = 11 Hz, H-14′), 6.51 (1H, dd, *J* = 14, 11 Hz, H-15′), 6.61 (2H, dd, *J* = 14, 11 Hz, H-11′, 15′).

## 4. Conclusions

In conclusion, carotenoids found in the coral *A. japonica*, clam *T. squamosa*, starfish *A. planci*, and sea snail *D. fragum* well reflected not only their metabolism but also the food chain. The accumulation and metabolism of carotenoids that originate from zooxanthellae to the starfish through the food chain are summarized in Figure 3.

**Figure 3.** Accumulation and metabolism of carotenoids that originate from zooxanthellae to the starfish and sea snail through the food chain.

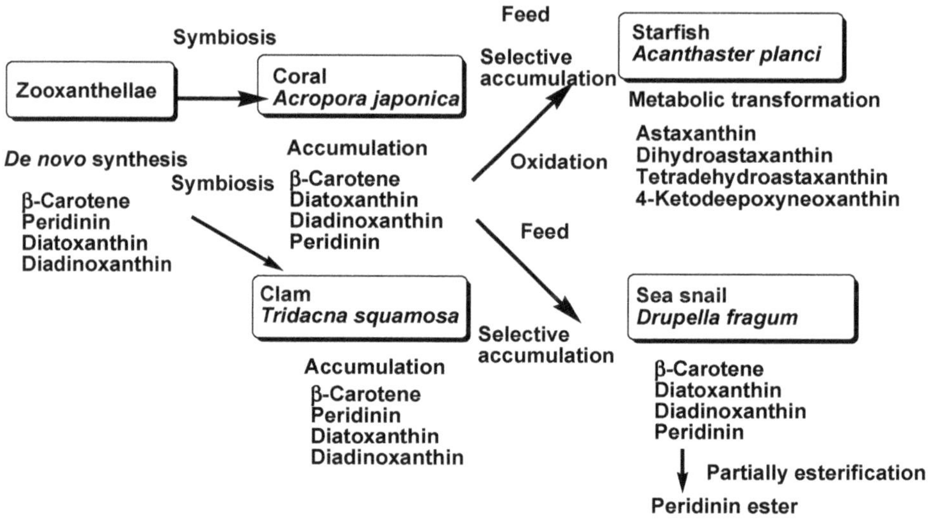



## Acknowledgements

We wish to thank Ryo Harada, a former student of Kinki University (now working in Torii Pharmaceutical Co., Ltd.) for his technical support.


## References


1. Britton, G.; Liaaen-Jensen, S.; Pfander, H. *Carotenoids Handbook*; Birkhäuser: Basel, Switzerland, 2004.
2. Liaaen-Jensen, S. Biosynthesis and Metabolism. In *Carotenoids in Food Chain*; Britton, G., Liaaen-Jensen, S., Pfander, H., Eds.; Birkhäuser: Basel, Switzerland, 1998; Volume 3,
pp. 359–371.
3. Matsuno, T. Aquatic animal carotenoids. *Fish. Sci.* **2001**, *67*, 771–789.
4. Maoka, T. Recent progress in structural studies of carotenoids in animals and plants.
*Arch. Biochem. Biophys.* **2009**, *483*, 191–195.
5. Skjenstad, T.; Haxo, F.T.; Liaaen-Jensen, S. Carotenoids of clam, coral and nudibranch zooxanthellae in aposymbiotic culture. *Biochem. Syst. Ecol.* **1984**, *12*, 149–153.
6. Liaaen-Jensen, S. Biosynthesis and Metabolism. In *Carotenoids in Chemosystematics*; Britton, G., Liaaen-Jensen, S., Pfander, H., Eds.; Birkhäuser: Basel, Switzerland, 1998; Volume 3, pp. 217–247.

7. Bjerkeng, B. Natural functions. In *Carotenoids in Aquaculture: Fish and Crustaceans*; Britton, G., Liaaen-Jensen, S., Pfander, H., Eds.; Birkhäuser: Basel, Switzerland, 2008; Volume 4, pp. 237–254.
8. Daigo, K.; Nakano, Y.; Casareto, B.E.; Suzuki, Y.; Shioi, Y. High-performance liquid chromatographic analysis of photosynthetic pigments in corals: An existence of a variety of epizoic, endozoic and endolithic algae. In *Proceedings of the 11th International Coral Reef Symposium*, Fort Lauderdale, FL, USA, 7–11 July 2008; pp. 123–127.
9. Maoka, T.; Fujiwara, Y.; Hashimoto, K.; Akimoto, N. Characterizatoin of fucoxanthin and fucoxanthinol esters in the chinese surf clam, *Mactra chinensis*. *J. Agric. Food Chem.* **2007**, *55*, 1563–1567.
10. Maoka, T.; Akimoto, N.; Yim, M.-J.; Hosokawa, M.; Miyashita, K. A new $C_{37}$-skeletal carotenoid from the clam, *Paphia amabillis*. *J. Agric. Food Chem.* **2008**, *56*, 12069–12072.
11. Maoka, T.; Akimoto, N.; Murakoshi, M.; Sugiyama, K.; Nishino, H. Carotenoids in clams, *Ruditapes philippinarum* and *Meretrix petechialis*. *J. Agric. Food Chem.* **2010**, *58*, 5784–5788.
12. Bjerkeng, B.; Hertberg, S.; Liaaen-Jensen, S. Carotenoids in food chain studies-V. Carotenoids of the bivalves *Modiolus modiolus* and *Pecten maximus*—structural, metabolic and food chain aspects. *Comp. Biochem. Physiol.* **1993**, *106B*, 243–250.
13. Maoka, T.; Tsushima, M.; Matsuno, T. New acetylenic carotenoids from the starfishes *Asterina pectinifera* and *Asterias amurensis*. *Comp. Biochem. Physiol.* **1989**, *93B*, 829–834.
14. Maoka, T.; Akimoto, N.; Terada, Y.; Komemushi, S.; Harada, R.; Sameshima, N.; Sakagami, Y. Structure of minor carotenoids from crown-of-thorns starfish, *Acanthaster planci*. *J. Nat. Prod.* **2010**, *73*, 675–678.
15. Englert, G. NMR Spectroscopy. In *Carotenoids*; Britton, G., Liaaen-Jensen, S., Pfander, H., Eds.; Birkhäuser Verlag: Basel, Switzerland, 1995; Volume 1B, pp. 147–260.
16. Haugan, J.A.; Englert, G.; Aakermann, T.; Glinz, E.; Liaaen-Jensen, S. Algal carotenoids 58. Isomerization studies of peridinin. *Acta Chem. Scand.* **1994**, *48*, 769–779.
17. Sugawara, T.; Yamashita, K.; Asai, A.; Nagao, A.; Shiraishi, T.; Imai, I.; Hirata, T. Esterification of xanthophylls by human intestinal Caco-2 cells. *Arch. Biochem. Biophys.* **2009**, *483*, 205–212.
18. Maoka, T.; Komori, T.; Matsuno, T. Direct resolution of diastereomeric carotenoid-I 3-oxo-β-end group. *J. Chromatogr. A* **1985**, *318*, 122–124.
19. Maoka, T.; Akimoto, N. Natural product chemistry in carotenoid, some experimental techniques for structural elucidation and analysis of natural carotenoids. *Carotenoid Sci.* **2008**, *13*, 10–17.

20. Britton, G. UV/Visible Spectroscopy. In *Carotenoids*; Britton, G., Liaaen-Jensen, S., Pfander, H., Eds.; Birkhäuser Verlag: Basel, Switzerland, 1995; Volume 1B, pp. 13–62.

*Samples Availability:* Available from the authors.

*marine drugs*

*Review*

# Carotenoid β-Ring Hydroxylase and Ketolase from Marine Bacteria—Promiscuous Enzymes for Synthesizing Functional Xanthophylls

**Norihiko Misawa**

Research Institute for Bioresources and Biotechnology, Ishikawa Prefectural University, Suematsu, Nonoichi-machi, Ishikawa 921-8836, Japan; E-Mail: n-misawa@ishikawa-pu.ac.jp;
Tel.: +81-76-227-7525; Fax: +81-76-227-7557

*Received: 21 March 2011; in revised form: 19 April 2011 / Accepted: 26 April 2011 / Published: 6 May 2011*

**Abstract:** Marine bacteria belonging to genera *Paracoccus* and *Brevundimonas* of the *α-Proteobacteria* class can produce $C_{40}$-type dicyclic carotenoids containing two β-end groups (β rings) that are modified with keto and hydroxyl groups. These bacteria produce astaxanthin, adonixanthin, and their derivatives, which are ketolated by carotenoid β-ring 4(4′)-ketolase (4(4′)-oxygenase; CrtW) and hydroxylated by carotenoid β-ring 3(3′)-hydroxylase (CrtZ). In addition, the genus *Brevundimonas* possesses a gene for carotenoid β-ring 2(2′)-hydroxylase (CrtG). This review focuses on these carotenoid β-ring-modifying enzymes that are promiscuous for carotenoid substrates, and pathway engineering for the production of xanthophylls (oxygen-containing carotenoids) in *Escherichia coli*, using these enzyme genes. Such pathway engineering researches are performed towards efficient production not only of commercially important xanthophylls such as astaxanthin, but also of xanthophylls minor in nature (e.g., β-ring(s)-2(2′)-hydroxylated carotenoids).

**Keywords:** *Paracoccus*; *Brevundimonas*; marine bacteria; ketocarotenoid; functional xanthophyll

## 1. Introduction

Many bacteria that have been isolated from marine environments can synthesize a variety of carotenoid pigments [1]. For example, acyclic $C_{30}$-type carotenoic acids were identified in some marine bacteria such as *Planococcus maritimus* [2] and *Rubritalea squalenifaciens* [3]. *Algoriphagus* sp. KK10202C of the *Flexibacteraceae* family, which was isolated from a marine sponge, was found to produce flexixanthin ((3*S*)-3,1′-dihydroxy-3′,4′-didehydro-1′2′-dihydro-β,ψ-caroten-4-one) and deoxyflexixanthin (1′-hydroxy-3′,4′-didehydro-1′2′-dihydro-β,ψ-caroten-4-one) [4], which are $C_{40}$-type monocyclic carotenoids containing one β-end group (β ring) (called monocyclic carotenoids in this review). Other marine bacteria including strain P99-3, which belong to the *Flavobacteriaceae* family, were shown to produce monocyclic carotenoids, myxol ((3*R*,2′*S*)-3′,4′-didehydro-1′,2′-dihydro-β,ψ-carotene-3,1′,2′-triol) and saproxanthin ((3*R*)-3′,4′-didehydro-1′,2′-dihydro-β,ψ-carotene-3,1′-diol), and zeaxanthin ((3*R*,3′*R*)-β,β-carotene-3,3′-diol) [5,6], which are a $C_{40}$-type dicyclic carotenoid containing two β-end groups (called dicyclic carotenids in this review). Marine bacteria belonging to genus *Paracoccus, Brevundimonas* or *Erythrobacter* in the *α-Proteobacteria* class have been revealed to synthesize dicyclic carotenoids that are ketolated at the 4(4′)-position(s) (called ketocarotenoids), e.g., astaxanthin ((3*S*,3′*S*)-3,3′-dihydroxy-β,β-carotene-4,4′-dione) and adonixanthin ((3*S*,3′*R*)-3,3′-dyhydroxy-β,β-caroten-4-one) (Figure 1) [7–9].

**Figure 1.** Chemical structures of ketocarotenoids produced in marine bacteria, *Paracoccus* sp. and *Brevundimonas* sp., and feasible functions of the carotenoid biosynthesis enzymes. These bacteria synthesize dicyclic carotenoids. *Paracoccus* sp. and *Brevundimonas* sp. are demonstrated to possess the unique genes *crtX* and *crtG*, respectively, in addition to the common genes, *crtE*, *crtB*, *crtI*, *crtY*, *crtZ*, and *crtW* [10,11].

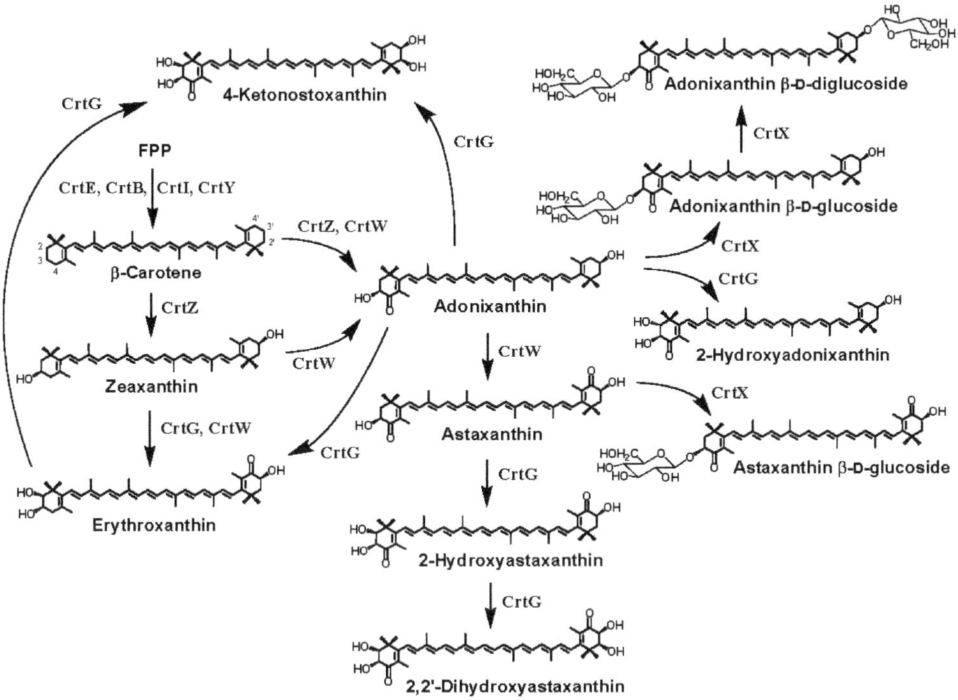


Among ketocarotenoids, astaxanthin and canthaxanthin (β,β-carotene-4,4′-dione) (specifically the former), are commercially important pigments as nutraceuticals and cosmetics that have anti-oxidation and anti-aging effects as well as colorants in aquaculture, while other ketocarotenoids are likely to have industrial potentials [12–16]. This review focuses on carotenoid β-ring 4(4′)-ketolase (4-oxygenase), carotenoid β-ring 3(3′)-hydroxylase, and carotenoid β-ring 2(2′)-hydroxylase, derived from the marine bacteria that belong to *α-Proteobacteria*, and pathway engineering for the production of functional xanthophylls via the incorporation of these β-ring-modifying enzyme genes.

## 2. Bacterial Strains Producing Ketocarotenoids

*Paracoccus* sp. strain N81106 (NBRC 101723), isolated from surface seawater near Aka island, Okinawa, Japan, was first shown to produce astaxanthin in bacteria [7,17]. This bacterium was also found to synthesize adonixanthin, adonixanthin β-D-glucoside, and astaxanthin β-D-glucoside [7,18]. *Paracoccus haeundaesis* BC74171$^T$, isolated from the Haeundae Coast, Korea, was shown to produce astaxanthin mainly [19]. *Paracoccus marinus* KKL-A5$^T$ (NBRC 100637$^T$), isolated from coastal seawater in Tokyo Bay, Japan, was found to produce adonixanthin diglucoside predominantly [20,21]. On the other hands, a marine bacterium *Brevundimonas* sp. strain SD212 (NBRC 101024) was revealed to synthesize not only astaxanthin and adonixanthin but also their 2(2′)-hydroxylated metabolites, that is, 2-hydroxyastaxanthin

((2*R*,3*S*,3′*S*)-2,3,3′-trihydroxy-β,β-carotene-4,4′-dione), 2-hydroxyadonixanthin ((2*R*,3*S*,3′*R*)-2,3,3′-trihydroxy-β,β-caroten-4-one), erythroxanthin ((3*S*,2′*R*,3′*R*)-3,2′,3′-trihydroxy-β,β-caroten-4-one), 4-ketonostoxanthin ((2*R*,3*S*,2′*R*,3′*R*)-2,3,2′,3′-tetrahydroxy-β,β-caroten-4-one) and 2,2′-dyhydroxyastaxanthin ((2*R*,3*S*,2′*R*,3′*S*)-2,3,2′,3′-tetrahydroxy-β,β-carotene-4,4′-dione) [9]. Figure 1 shows the structures of the ketocarotenoids shown in this section and their feasible biosynthetic pathway. Figure 2 shows phylogenetic tree of the marine bacteria that produce astaxanthin and other ketocarotenoids, which were isolated in Marine biotechnology Institute (Kamaishi, Japan), along with the type strains relative to these bacteria, many of which are not marine bacteria but soil bacteria. Interestingly, the phylogenetically closest strains to the marine bacteria *Paracoccus* sp. N81106 and *Brevundimonas* sp. SD212 were soil bacteria *Paracoccus marcusii* DSM 11574$^{T}$ [22] and *Brevundimonas aurantiaca* ATCC 15266$^{T}$, respectively (Figure 2).

## 3. Ketocarotenoid Biosynthesis Genes

Genes required for the biosynthesis of dicyclic carotenoids were first isolated from soil bacteria *Pantoea ananatis* (reclassified from *Erwinia uredovora*) [23] and *Pantoea agglomerans* (*Erwinia herbicola*) [24,25], which cannot produce ketocarotenoids and belong to the *Enterobacteriaceae* family of class γ-*Proteobacteria* (the same family to *Escherichia coli*). The *Pantoea* carotenoid biosynthesis genes composed a gene cluster for the synthesis of zeaxanthin β-D-diglucoside from farnesyl diphosphate (farnesyl pyrophosphate; FPP) [25–27], and comprised six genes that encode geranygeranyl diphosphate (GGPP) synthase (CrtE) [27,28], phytoene synthase (CrtB) [27,29], phytoene desaturase (CrtI) [23,30], lycopene (ψ,ψ-carotene) β-cyclase (CrtY) [23,31], β-carotene (β,β-carotene) 3-hydrocylase (CrtZ) [23], and zeaxanthin glucosyltransferase (CrtX) [23,32] (Figure 3).

**Figure 2.** Phylogenetic positions of *Paracoccus* sp., *Erythrobacer* sp., and *Brevundimonas* sp. strains deduced from their 16S rRNA sequences. ○ represents marine bacteria. Bacterial strains, whose carotenoid biosynthesis genes were elucidated, are shown in boldface, and the second accession numbers in the parentheses shows those of carotenoid biosynthesis genes. *Paracoccus* sp. strain N81106 (MBIC01143 = NBRC 101723) and *Paracoccus* sp. strain PC1 (MBIC03024 = NBRC 101025) were formerly classified as *Agrobacterium aurantiacum* [17] and *Alacaligenes* sp. PC-1 [33], respectively. The phylogenetic tree was constructed as described [10]. The scale bar indicates a genetic distance of 0.02 (*Knuc*).

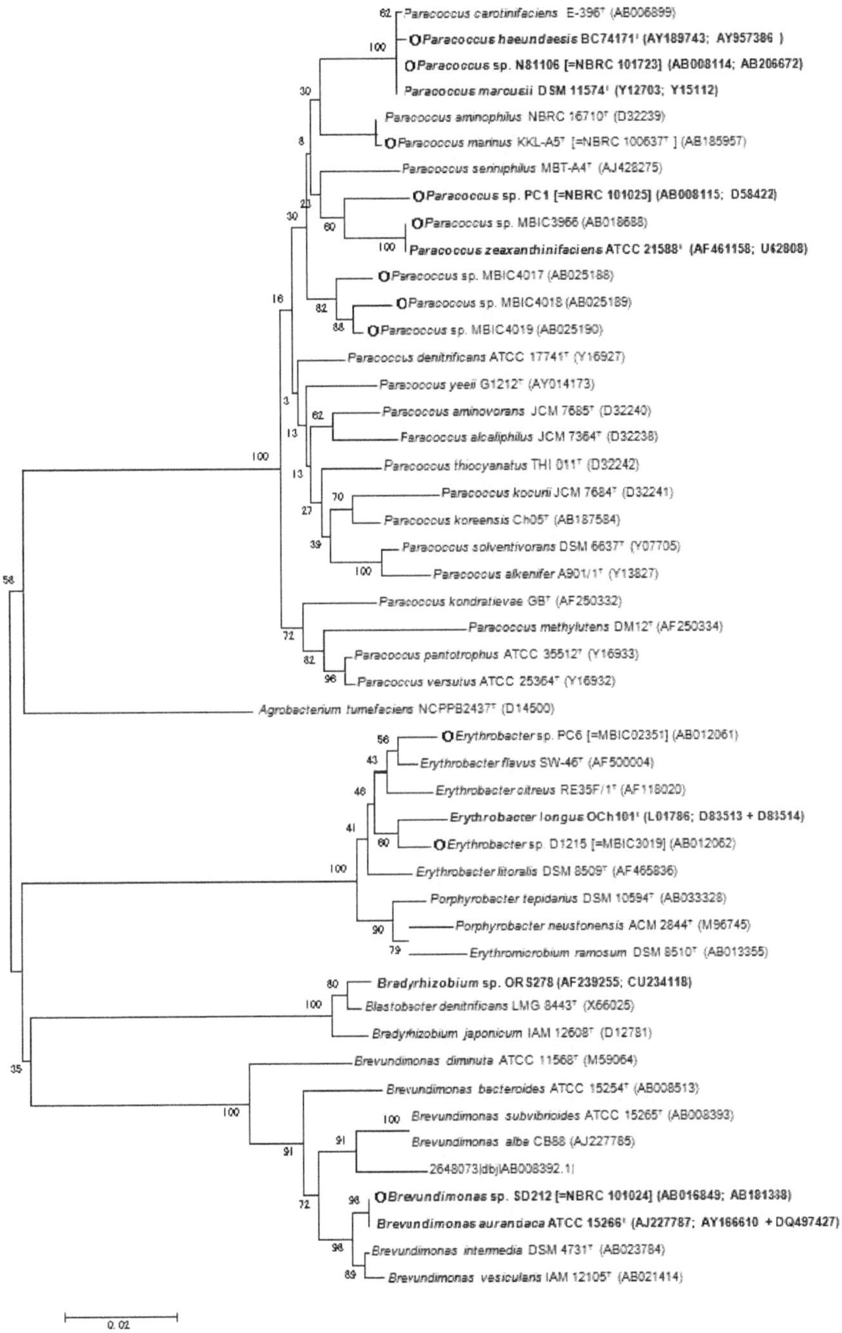
Paracoccus carotinifaciens E-396T (AB006899)
Paracoccus haeundaesis BC74171T (AY189743; AY957386)
Paracoccus sp. N81106 [=NBRC 101723] (AB008114; AB206672)
Paracoccus marcusii DSM 11574T (Y12703; Y15112)
Paracoccus aminophilus NBRC 16710T (D32239)
Paracoccus marinus KKL-A5T [=NBRC 100637T] (AB185957)
Paracoccus seriniphilus MBT-A4T (AJ428275)
Paracoccus sp. PC1 [=NBRC 101025] (AB008115; D58422)
Paracoccus sp. MBIC3966 (AB018688)
Paracoccus zeaxanthinifaciens ATCC 21588T (AF461158; U62808)
Paracoccus sp. MBIC4017 (AB025188)
Paracoccus sp. MBIC4018 (AB025189)
Paracoccus sp. MBIC4019 (AB025190)
Paracoccus denitrificans ATCC 17741T (Y16927)
Paracoccus yeeii G1212T (AY014173)
Paracoccus aminovorans JCM 7685T (D32240)
Paracoccus alcaliphilus JCM 7364T (D32238)
Paracoccus thiocyanatus THI 011T (D32242)
Paracoccus kocurii JCM 7684T (D32241)
Paracoccus koreensis Ch05T (AB187584)
Paracoccus solventivorans DSM 6637T (Y07705)
Paracoccus alkenifer A901/1T (Y13827)
Paracoccus kondratievae GBT (AF250332)
Paracoccus methylutens DM12T (AF250334)
Paracoccus pantotrophus ATCC 35512T (Y16933)
Paracoccus versutus ATCC 25364T (Y16932)
Agrobacterium tumefaciens NCPPB2437T (D14500)
Erythrobacter sp. PC6 [=MBIC02351] (AB012061)
Erythrobacter flavus SW-46T (AF500004)
Erythrobacter citreus RE35F/1T (AF118020)
Erythrobacter longus OCh101T (L01786; D83513 + D83514)
Erythrobacter sp. D1215 [=MBIC3019] (AB012062)
Erythrobacter litoralis DSM 8509T (AF465836)
Porphyrobacter tepidarius DSM 10594T (AB033328)
Porphyrobacter neustonensis ACM 2844T (M96745)
Erythromicrobium ramosum DSM 8510T (AB013355)
Bradyrhizobium sp. ORS278 (AF239255; CU234118)
Blastobacter denitrificans LMG 8443T (X66025)
Bradyrhizobium japonicum IAM 12608T (D12781)
Brevundimonas diminuta ATCC 11568T (M59064)
Brevundimonas bacteroides ATCC 15254T (AB008513)
Brevundimonas subvibrioides ATCC 15265T (AB008393)
Brevundimonas alba CB88 (AJ227785)
2648073|dbj|AB008392.1|
Brevundimonas sp. SD212 [=NBRC 101024] (AB016849; AB181338)
Brevundimonas aurantiaca ATCC 15266T (AJ227787; AY166610 + DQ497427)
Brevundimonas intermedia DSM 4731T (AB023784)
Brevundimonas vesicularis IAM 12105T (AB021414)
0.02

**Figure 3.** Pathway engineering for the production of functional xanthophylls using the carotenoid biosynthesis genes, *crtW, crtZ,* and/or *crtG,* which were isolated from the marine bacteria, *Paracoccus* sp. strain N81106 or *Brevundimonas* sp. strain SD212, in addition to the *crtE, crtB, crtI,* and *crtY* genes (and *crtX*) from *P. ananatis.*

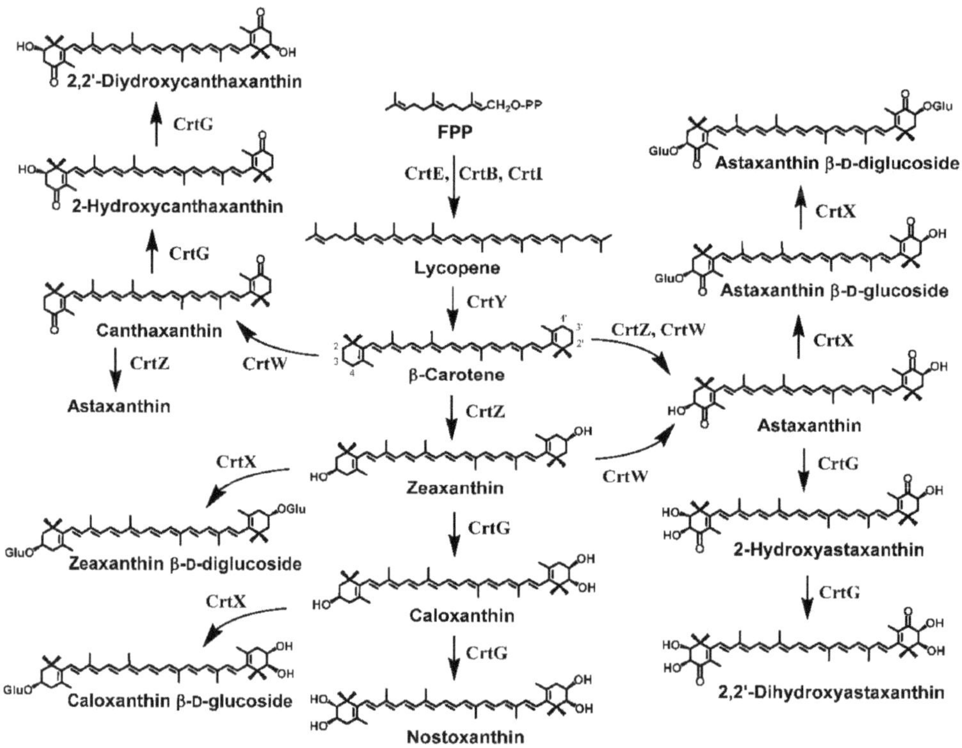


These *crt* genes have widely used for complementation analysis of carotenoid biosynthesis genes isolated from other organisms, since they are functionally expressed in *E. coli* with ease [11,34–37]. The *P. agglomerans* gene cluster contained a gene encoding isopentenyl diphosphate (IPP) isomerase (Idi; type 2) [38] in addition to the six *crt* genes [39]. These seven carotenogenic (carotenoid-biosynthetic) genes were also found to exist in a carotenoid biosynthesis gene cluster of *Paracoccus* sp. strain N81106 [10,39]. This cluster included an additional gene, designated CrtW, which was elucidated to code for an enzyme responsible for ketocarotenoid formation, that is, CrtW proved to catalyze the synthesis of canthaxanthin from β-carotene by complementation analysis using recombinant *E. coli* cells that contains the *P. ananatis crtE, crtB, crtI,* and *crtY* genes [33] (Figure 3). The hydropathy and transmembrane prediction analyses indicated that CrtW from *Paracoccus* sp. N81106 contains four transmembrane domains and two other hydrophobic regions, and its topology model is very similar to those for fatty acid desaturases [40]. It should be noted that it is recalcitrant to purify active CrtW and CrtZ proteins, which both are very likely iron-dependent integral membrane proteins, from the recombinant hosts as well as the native hosts, precluding their close enzymatic characterizations.

## 4. Carotenoid 4,4′-Ketolase

It has been revealed that only two enzymes, carotenoid 4,4′-ketolase (4,4′-oxygenase) (β-ring 4(4′)-ketolase; CrtW) and carotenoid 3,3′-hydroxylase (β-ring 3(3′)-hydroxylase; CrtZ), are sufficient to biosynthesize astaxanthin from β-carotene via eight intermediates including zeaxanthin, canthaxanthin and adonixanthin [35,40,41]. CrtW can convert not only the (un-substituted) β ring but also the 3-hydroxylated β ring into the respective 4-ketolated groups, and CrtZ can convert not only the (un-substituted) β ring but also the 4-ketolated β-ring into the respective 3-hydroxylated groups, as shown in Figure 4 [42–46]. An *in vitro* analysis with the crude enzymes of CrtW and CrtZ from the *E. coli* cells expressing the corresponding genes indicated that these enzymes are likely 2-oxoglutarate (α-ketoglutarate)-dependent dioxygenases [42].

**Figure 4.** Catalytic functions of carotenoid 4,4′-ketolases (oxygenases) and carotenoid 3′3′-hydroxylases. BKT means BKT1 or BKT2 from *H. pluvialis.*

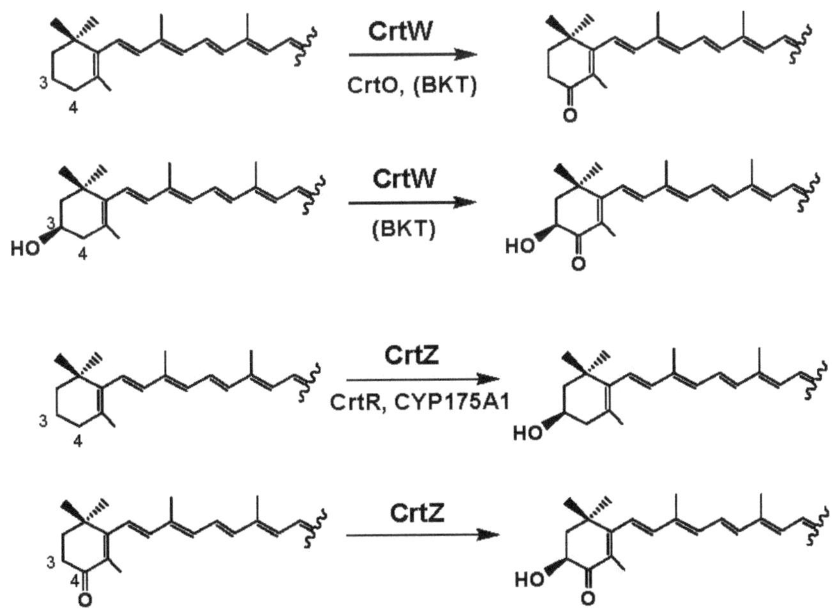


The *crtW* genes were present not only in the above-mentioned *α-Proteobacteria* (Figure 2) but also in the marine bacterium *Algoriphagus* sp. KK10202C [4] and cyanobacterial strains such as *Anabaena* (*Nostoc*) sp. PCC 7120 and *N. punctiforme* [47,48]. These cyanobacteria produced not astaxanthin but echinenone (β,β-caroten-

4-one), and 4-ketomyxol 2′-fucoside, a monocyclic carotenoid that includes the 4-ketolated β-ring [49]. Conversion efficiency to astaxanthin in several CrtWs was compared with recombinant *E. coli* cells that synthesize the carotenoid substrate zeaxanthin due to the presence of the *P. ananatis crtE, crtB, crtI, crtY,* and *crtZ* genes, in which each *crtW* gene from *Paracoccus* sp. N81106, *Paracoccus* sp. PC1, *Brevundimonas* sp. SD212, *Anabaena* sp. PCC7120, and *N. punctiforme* was expressed [44,46]. It was consequently shown that the *Brevundimonas* sp. SD212 CrtW, which exhibited the highest amino acid identity (96.3%) with that of the *B. aurantiaca* ATCC 15266 CrtW (accession no. AY166610), converted β-carotene to astaxanthin with the highest efficiency, along with the *P. ananatis* CrtZ [44,46]. In the case of the *Paracoccus* CrtWs, not only astaxanthin but also adonixanthin tended to accumulate, and this intermediate was difficult to be converted to astaxanthin [43,44]. The cyanobacterial CrtWs poorly converted zeaxanthin to astaxanthin via adonixanthin [46].

Two paralogous genes exhibiting significant homology to *crtW* were isolated from *H. pluvialis,* and designated *bkt* [50] or *crtO* [51]. These genes were renamed *bkt1* from *crtO* and *bkt2* from *bkt,* since "*crtO*" has been used for the other type of cyanobacterial β-ring 4(4′)-ketolase genes, as shown later [52]. The BKT1 and BKT2 enzymes are very likely to have catalytic function same to the *Paracoccus* (or *Brevundimonas*) CrtWs, considering results from the *in vitro* study on BKT2 with *E. coli* [42] and pathway engineering researches in higher plants as well as *E. coli* as the hosts [16,50,51,53].

A gene encoding a new type of β-ring 4(4′)-ketolase (named CrtO) that showed apparent homology not to CrtW-type ketolase but to CrtI-type phytoene desaturase was first found in cyanobacterium *Synechocystis* sp. strain PCC 6803 [54], which produced 3′-hydroxyechinenone (3′-hydroxy-β,β-caroten-4-one), zeaxanthin and myxol 2′-dimethyl-fucoside [55]. The *crtO* genes were also present in *Anabaena* sp. PCC 7120 [48], and an actinomycete *Rhodococcus erythropolis* and *Deinococcus radiodurans* R1 highly resistant to γ and UV radiation [56], which produced other monocyclic carotenoids, e.g., the latter strain produced deinoxanthin (2,1′-dihydroxy-3′,4′-didehydro-1′2′-dihydro-β,ψ-caroten-4-one) [1]. An *in vivo* analysis on *crtO* was performed with recombinant *E. coli* cells that synthesize the carotenoid substrate β-carotene or zeaxanthin, into which each *crtO* gene from *Synechocystis* sp. PCC 6803 and *R. erythropolis* was introduced and expressed there [57]. This result along with previous finding [48] suggested that the CrtO-type of β-ring 4(4′)-ketolases can accept only the (un-substituted) β ring(s) in β-carotene and probably in monocyclic carotenoids as the substrates (Figure 4).

## 5. Carotenoid 3,3′-Hydroxylase

The *crtZ* genes have been found not only in carotenogenic bacteria belonging to genera *Pantoea, Paracoccus* and *Brevundimonas,* but also in those belonging to the *Flavobacteriaceae* family [6,39]. Conversion efficiency to astaxanthin in several CrtZs was compared with recombinant *E. coli* cells that synthesize the carotenoid substrate canthaxanthin due to the presence of the *P. ananatis crtE, crtB, crtI* and *crtY* gens, and the *Paracoccus* N81106 *crtW* gene, into which each *crtZ* gene from *P. ananatis, Paracoccus* sp. N81106, *Paracoccus* sp. PC1, *Brevundimonas* sp. SD212, and marine bacterium strain P99-3 of the *Flavobacteriacea* family was introduced and expressed there [45]. It was consequently shown that the CrtZ enzymes from *Brevundimonas* sp. SD212 and the bacterial strain P99-3 converted β-carotene to astaxanthin with the highest and lowest efficiency, respectively, along with the *Paracoccus* N81106 CrtW [45].

On the other hands, no *crtZ* sequences have not been found in cyanobacteria, instead genes encoding a new type of β-ring 3(3′)-hydroxylases (named CrtR) that exhibited moderate homology to CrtW have been found there [58,59]. The *crtR* genes were isolated from *Synechocystis* sp. strain PCC 6803, *Anabaena* sp. PCC 7120, *Anabaena variabilis,* and *N. punctiforme* [46,58]. An *in vivo* analysis on *crtR* was performed with recombinant *E. coli* cells that synthesize the carotenoid substrate β-carotene or canthaxanthin, into which each *crtR* gene from *Synechocystis* sp. PCC 6803, *Anabaena* sp. PCC 7120, and *A. variabilis* was introduced and expressed there [46]. This result along with another result [60] indicated that the CrtR-type enzymes can hydroxylate the (un-substituted) β ring of monocyclic carotenoids such as deoxymyxol and deoxymyxol 2′-fucoside at the 3 position (Figure 4). Among them, only the *Synechocystis* sp. PCC 6803CrtR was able to convert β-carotene to zeaxanthin [46,58,60]. A thermophilic bacterium *Thermus thermophilus* HB27, which grows at temperatures above 75 °C, was found to possess another new type of β-ring 3(3′)-hydroxylase of the cytochrome P450 superfamily, named CYP175A1 [61]. The *in vivo* analysis with the gene strongly suggested that this thermostable P450 accepts only the (un-substituted) β ring of β-carotene as the substrate to form zeaxanthin [45,61].

## 6. Carotenoid 2,2′-Hydroxylase

Carotenoid 2,2′-hydroxylase (β-ring 2(2′)-hydroxylase) was first found in the marine bacterium *Brevundimonas* sp. strain SD212, and named CrtG [11]. An *in vivo* analysis on *crtG* was performed with recombinant *E. coli* cells that synthesize each carotenoid substrate (β-carotene, zeaxanthin, canthaxanthin, or astaxanthin), into which the *crtG* gene was introduced and expressed there [11]. The result indicated that the CrtG can hydroxylate the β rings substituted with 3-hydroxy and/or 4-keto groups in dicyclic carotenoids at the 2(2′)-positions (Figures 1 and 3) [11]. The *crtG* genes were also isolated from soil bacteria *Brevundimonas vesicularis* DC263 and *B. aurantiaca* ATCC 15266 [62]. The *in vivo* analysis with these genes indicated that the *B. aurantiaca* CrtG enzyme (accession no. DQ497427), which exhibited the highest amino acid identity (98.8%) to that of the *Brevundimonas* SD212 CrtG, accepted the (un-substituted) β rings of β-carotene in addition to the substituted β rings as the substrates [62]. A *crtG* gene sequence, whose encoded amino acid sequence was 41%

identical to the *Brevundimonas* sp. SD212 CrtG, was found in a thermophilic cyanobacterium *Thermosynechococcus elongatus*, which synthesized 2-hydroxylated carotenoids such as caloxanthin ((2*R*,3*R*,3′*R*)-β,β-carotene-2,3,3′-triol), nostoxanthin ((2*R*,3*R*,2′*R*,3′*R*)-β,β-carotene-2,3,2′,3′-tetrol) (Figure 3), and 2-hydroxymyxol 2′-fucoside [63].

## 7. Pathway Engineering for the Synthesis of Functional Xanthophylls via the Incorporation of *crtW*, *crtZ*, and/or *crtG* Genes

Figure 3 shows xanthophylls that were produced in recombinant *E. coli* cells via the incorporation of the marine bacterial *crtW*, *crtZ*, and/or *crtG* genes along with the *Pantoea crtE*, *crtB*, *crtI*, and *crtY* genes. The recombinant *E. coli* strain that expresses the four *Pantoea crt* genes can produce β-carotene predominantly (approximately 0.2–1 mg·g$^{-1}$ dry cell weight). The coexpression of the *crtW*, *crtZ*, and/or *crtG* genes in the β-carotene-synthesizing *E. coli* cells confer the ability to produce not only commercially important xanthophylls such as astaxanthin but also xanthophylls minor in nature (e.g., β-ring(s)-2(2′)-hydroxylated carotenoids), which are difficult to synthesize chemically. Particularly, the chemical synthesis of 2(2′)-hydroxycarotenoids are likely to be recalcitrant, due to high-density around the 1,2-positions of the β ring in these xanthophylls. We showed that the coexpression of the *Brevundimonas* sp. SD212 *crtW* gene and the *P. ananatis crtZ* gene in the β-carotene-synthesizing *E. coli* due to the presence of the four *crt* genes of *P. ananatis* resulted in predominant production of astaxanthin [44,46]. The *Paracoccus* sp. N81106 *crtW* gene was evolved by random mutagenesis to have improved activity [40]. It is also demonstrated that the coexpression of the *crtW* gene and the *crtG* gene from *Brevundimonas* sp. SD212 or from *B. aurantiaca* ATCC 15266 in the β-carotene-synthesizing *E. coli* resulted in dominant production of 2,2′-dihydroxycanthaxanthin and 2-hydroxycanthaxanthin, while the substrate canthaxanthin accumulated [11,62]. The coexpression of the *crtZ* gene and the *crtG* gene in the β-carotene-synthesizing *E. coli* resulted in predominant production of nostoxanthin along with small amounts of caloxanthin [11,62]. The coexpression of all the three genes (*crtW*, *crtZ*, and *crtG*) in the β-carotene-synthesizing *E. coli* resulted in dominant production of 2,2′-dihydroxyastaxanthin and 2-hydroxyastaxanthin [11]. When the *P. ananatis crtX* gene was coexpressed in addition to appropriate combinations of the above *crt* genes in *E. coli*, resultant *E. coli* cells were able to synthesize carotenoid-glycosides such as caloxanthin β-D-glucoside [64] and astaxanthin β-D-diglucoside [65], as shown in Figure 3.

The γ-ray-tolerant bacterium *D. radiodurans* R1 produces the monocyclic carotenoid including the 2-hydroxy-4-keto-β-ring, deinoxanthin [1]. 2,2′-Dihydroxycanthaxanthin was shown to have strong inhibitory effect against lipid peroxidation in a rat brain homogenate [11]. Such minor ketocarotenoids, which include the 2-hydroxy-4-keto-β-ring, may have beneficial effects on human health as well as anti-oxidation function, while few works are present examining their biological functions.

When carotenoid biosynthesis genes starting from the utilization of FPP are introduced in *E. coli*, as above-mentioned, amounts of carotenoids produced with the recombinant *E. coli* cells are far from the practical use, which was difficult to exceed 1 mg·g$^{-1}$ dry weight. In order to overcome this problem, many pathway engineering researches in *E. coli* have been performed for increasing intracellular concentration of FPP (e.g., recently reviewed [66,67]). For example, the coexpression of the *idi* (type 1) gene from *H. pluvialis*, *Xanthophyllomyces dendrorhous* (renamed from *Phaffia rhodozyma*), or *Saccharomyces cerevisiae*, as well as the *idi* (type 2) from *Streptomyces* sp. strain CL190, was shown to be effective to increase FPP content [68,69]. The introduction of heterologous mevalonate pathway genes in *E. coli* along with an *idi* (type 2) gene has been described to efficiently improve the productivity of carotenoids or sesquiterpenes that are synthesized from FPP [69–73]. For example, Yoon *et al.* [73] produced 22 mg·g$^{-1}$ dry cell weight of lycopene in 72 h using such mevalonate-pathway-engineered *E. coli* cells. On the other hand, production of lycopene reached high levels (near to 20 mg·g$^{-1}$ dry cell weight) in 24-h batch flask culture in pathway-engineered *E. coli*, which reflected results of multi-dimensional gene target search or gene-knockout analysis [74]. These finding should be applied to efficient production of the above-mentioned functional xanthophylls with *E. coli* cells.

Pathway engineering researches in higher plants have also been performed for efficient production of astaxanthin, which utilized the marine bacterial *crtW* genes from *Paracoccus* sp. N81106 or *Brevundimonas* sp. SD212, or the *H. pluvialis bkt1* or *bkt2* genes, as reviewed [16,39,53]. For example, the *Brevundimonas* sp. SD212 *crtW* and *crtZ* genes, whose nucleotide sequence is modified to codon usage of higher plants, were successfully overexpressed in the chloroplasts of tobacco plants (*Nicotiana tabacum*), and astaxanthin level produce there reached 5.44 mg·g$^{-1}$ dry weight (74% of total carotenoids) [75].

## 8. Conclusions

This review has focused on the carotenoid β-ring-modifying enzymes, CrtW, CrtZ and CrtG, derived from the marine bacteria of the *α-Proteobacteria* class, and pathway engineering for the production of xanthophylls in *E. coli*, using these enzyme genes. Such pathway engineering researches are performed towards efficient production not only of commercially important xanthophylls such as astaxanthin, but also of xanthophylls minor in nature, which are difficult to synthesize chemically, and expected to have beneficial effects on human health as well as anti-oxidation function.

## Acknowledgements

The author gratefully acknowledges the Marine Biotechnology Institute (MBI) that was closed on 30 June 2008, and Kirin Holdings Company, Limited (Kirin Brewery Co., Ltd.). This work was also supported by New Energy and Industrial Technology Development Organization (NEDO) of Japan.

## References

1. Britton, G.; Liaaen-Jensen, S.; Pfander, H. *Carotenoids Handbook*; Birkhauser Verlag: Basel, Switzerland, 2004.
2. Shindo, K.; Endo, M.; Miyake, Y.; Wakasugi, K.; Morritt, D.; Bramley, P.M.; Fraser, P.D.; Kasai, H.; Misawa, N. Methyl glucosyl-3,4-dehydro-apo-8′-lycopenoate, a novel antioxidative glyco-C30-carotenoic acid produced by a marine bacterium *Planococcus maritimus*. *J. Antibiot.* **2008**, *61*, 729–735.
3. Shindo, K.; Mikami, K.; Tamesada, E.; Takaichi, S.; Adachi, K.; Misawa, N.; Maoka, T. Diapolycopenedioc acid xylosyl ester, a novel glyco-$C_{30}$-carotenoic acid produced by a new marine bacterium *Rubritalea squalenifaciens*. *Tetrahedron Lett.* **2007**, *48*, 2725–2727.
4. Tao, L.; Yao, H.; Kasai, H.; Misawa, N.; Cheng, Q. A carotenoid synthesis gene cluster from *Algoriphagus* sp. KK10202C with a novel fusion-type lycopene-β-cyclase gene. *Mol. Genet. Genomics* **2006**, *276*, 79–86.
5. Shindo, K.; Kikuta, K.; Suzuki, A.; Katsuta, A.; Kasai, H.; Yasumoto-Hirose, M.; Matuo, Y.; Misawa, N.; Takaichi, S. Rare carotenoids, (3*R*)-Saproxanthin and (3*R*,2′*S*)-myxol, isolated from novel marine bacteria (*Flavobacteriaceae*) and their antioxidant activities. *Appl. Microbiol. Biotechnol.* **2007**, *74*, 1350–1357.
6. Teramoto, M.; Takaichi, S.; Inomata, Y.; Ikenaga, H.; Misawa, N. Structural and functional analysis of a lycopene β-monocyclase gene isolated from a unique marine bacterium that produces myxol. *FEBS Lett.* **2003**, *545*, 120–126.
7. Yokoyama, A.; Izumida, H.; Miki, W. Production of astaxanthin and 4-ketozeaxanthin by the marine bacterium, *Agrobacterium aurantiacum*. *Biosci. Biotechnol. Biochem.* **1994**, *58*, 1842–1844.
8. Yokoyama, A.; Izumida, H.; Shizuri, Y. New carotenoid sulfates isolated from a marine bacterium. *Biosci. Biotechnol. Biochem.* **1996**, *60*, 1877–1878.
9. Yokoyama, A.; Miki, W.; Izumida, H.; Shizuri, Y. New trihydroxy-keto-carotenoids isolated from an astaxanthin-producing marine bacterium. *Biosci. Biotechnol. Biochem.* **1996**, *60*, 200–203.
10. Maruyama, T.; Kasai, H.; Choi, S.K.; Ramasamy, A.K.; Inomata, Y.; Misawa, N. Structure of a complete carotenoid biosynthesis gene cluster of marine bacterium *Paracoccus* sp. strain N81106. *Carotenoid Sci.* **2007**, *11*, 50–55.
11. Nishida, Y.; Adachi, K.; Kasai, H.; Shizuri, Y.; Shindo, K.; Sawabe, A.; Komemushi, S.; Miki, W.; Misawa, N. Elucidation of a carotenoid biosynthesis

gene cluster encoding a novel enzyme, 2,2′-β-hydroxylase, from *Brevundimonas* sp. strain SD212 and combinatorial biosynthesis of new or rare xanthophylls. *Appl. Environ. Microbiol.* **2005**, *71*, 4286–4296.
12. Nishino, H.; Murakoshi, M.; Ii, T.; Takemura, M.; Kuchide, M.; Kanazawa, M.; Mou, X.Y.; Wada, S.; Masuda, M.; Ohsaka, Y.; *et al.* Carotenoids in cancer chemoprevention. *Cancer Metastasis Rev.* **2002**, *21*, 257–264.
13. Pashkow, F.J.; Watumull, D.G.; Campbell, C.L. Astaxanthin: A novel potential treatment for oxidative stress and inflammation in cardiovascular disease. *Am. J. Cardiol.* **2008**, *101*, 58D–68D.
14. Camera, E.; Matrofrancesco, A.; Fabbri, C.; Daubrawa, F.; Picardo, M.; Sies, H.; Stahl, W. Astaxanthin, canthaxanthin and β-carotene differently affect UVA-induced oxidative damage and expression of oxidative stress-responsive enzymes. *Exp. Dermatol.* **2009**, *18*, 222–231.
15. Jackson, H.; Braun, C.L.; Ernst, H. The chemistry of novel xanthophyll carotenoids. *Am. J. Cardiol.* **2008**, *101*, 50D–57D.
16. Misawa, N. Pathway engineering of plants toward astaxanthin production. *Plant Biotechnol.* **2009**, *26*, 93–99.
17. Izumida, H.; Adachi, K.; Nishizima, M.; Endo, M.; Miki, W. Akalone: A novel xanthine oxidase inhibitor produced by the marine bacterium, *Agrobacterium aurantiacum* sp. nov. *J. Mar. Biotechnol.* **1995**, 2, 115–118.
18. Yokoyama, A.; Adachi, K.; Shizuri, Y. New carotenoid glycosides, astaxanthin glucoside and adonixanthin glucoside, isolated from the astaxanthin-producing marine bacterium, *Agrobacterium aurantiacum*. *J. Nat. Prod.* **1995**, *58*, 1929–1933.
19. Lee, J.H.; Kim, Y.S.; Choi, T.J.; Lee, W.J.; Kim, Y.T. *Paracoccus haeundaensis* sp. nov., a Gram-negative, halophilic, astaxanthin-producing bacterium. *Int. J. Syst. Evol. Microbiol.* **2004**, *54*, 1699–1672.
20. Takaichi, S.; Maoka, T.; Akimoto, N.; Khan, S.T.; Harayama, S. Major carotenoid isolated from *Paracoccus schoinia* NBRC 100637$^T$ is adonixanthin diglucoside. *J. Nat. Prod.* **2006**, *69*, 1823–1825.
21. Khan, S.T.; Takaichi, S.; Harayama, S. *Paracoccus marinus* sp. nov., an adonixanthin diglucoside-producing bacterium isolated from coastal seawater in Tokyo Bay. *Int. J. Syst. Evol. Microbiol.* **2008**, *58*, 383–386.
22. Harker, M.; Hirschberg, J.; Oren, A. *Paracoccus marcusii* sp. nov., an orange gram-negative coccus. *Int. J. Syst. Bacteriol.* **1998**, *48*, 543–548.
23. Misawa, N.; Nakagawa, M.; Kobayashi, K.; Yamano, S.; Izawa, Y.; Nakamura, K.; Harashima, K. Elucidation of the *Erwinia uredovora* carotenoid biosynthetic pathway by functional analysis of gene products expressed in *Escherichia coli*. *J. Bacteriol.* **1990**, *172*, 6704–6712.
24. Armstrong, G.A.; Alberti, M.; Hearst, J.E. Conserved enzymes mediate the early reactions of carotenoid biosynthesis in nonphotosynthetic and photosynthetic prokaryotes. *Proc. Natl. Acad. Sci. USA* **1990**, *87*, 9975–9979.

25. Hundle, B.S.; Beyer, P.; Kleinig, H.; Englert, G.; Hearst, J.E. Carotenoids of *Erwinia herbicola* and an *Escherichia coli* HB101 strain carrying the *Erwinia herbicola* carotenoid gene cluster. *Photochem. Photobiol.* **1991**, *54*, 89–93.
26. Nakagawa, M.; Misawa, N. Analysis of carotenoid glycosides produced in gram-negative bacteria by introduction of the *Erwinia uredovora* carotenoid biosynthesis genes. *Agric. Biol. Chem.* **1991**, *55*, 2147–2148.
27. Sandmann, G.; Misawa, N. New functional assignment of the carotenogenic genes *crtB* and *crtE* with constructs of these genes from *Erwinia* species. *FEMS Microbiol. Lett.* **1992**, *90*, 253–258.
28. Math, S.K.; Hearst, J.E.; Poulter, C.D. The *crtE* gene in *Erwinia herbicola* encodes geranylgeranyl diphosphate synthase. *Proc. Natl. Acad. Sci. USA* **1992**, *89*, 6761–6764.
29. Neudert, U.; Martines-Ferez, I.M.; Fraser, P.D.; Sandmann, G. Expression of an active phytoene synthease from *Erwinia uredovora* and biochemical properties of the enzyme. *Biochim. Biophys. Acta* **1998**, *1392*, 51–58.
30. Fraser, P.D.; Misawa, N.; Linden, H.; Yamano, S.; Kobayashi, K.; Sandmann, G. Expression in *E. coli*, purification and reactivation of the recombinant *Erwinia uredovora* phytoene desturase. *J. Biol. Chem.* **1992**, *267*, 19891–19895.
31. Schnurr, G.; Misawa, N.; Sandmann, G. Expression, purification and properties of lycopene cyclase from *Erwinia uredovora*. *Biochem. J.* **1996**, *315*, 869–874.
32. Hundle, B.S.; O'Brien, D.A.; Alberti, M.; Beyer, P.; Hearst, J.E. Functional expression of zeaxanthin glucosyltransferase from *Erwinia herbicola* and a proposed uridine diphosphate binding site. *Proc. Natl. Acad. Sci. USA* **1992**, *89*, 9321–9325.
33. Misawa, N.; Kajiwara, S.; Kondo, K.; Yokoyama, A.; Satomi, Y.; Saito, T.; Miki, W.; Ohtani, T. Canthaxanthin biosynthesis by the conversion of methylene to keto groups in a hydrocarbon β-carotene by a single gene. *Biochem. Biophy. Res. Commun.* **1995**, *209*, 867–876.
34. Misawa, N.; Truesdale, M.R.; Sandmann, G.; Fraser, P.D.; Bird, C.; Schuch, W.; Bramley, P.M. Expression of a tomato cDNA coding for phytoene synthase in *Escherichia coli*, phytoene formation *in vivo* and *in vitro*, and functional analysis of the various truncated gene products. *J. Biochem.* **1994**, *116*, 980–985.
35. Misawa, N.; Satomi, Y.; Kondo, K.; Yokoyama, A.; Kajiwara, S.; Saito, T.; Ohtani, T.; Miki, W. Structure and functional analysis of a marine bacterial carotenoid biosynthesis gene cluster and astaxanthin biosynthetic pathway proposed at the gene level. *J. Bacteriol.* **1995**, *177*, 6575–6584.
36. Sun, Z.; Gantt, E.; Cunningham, F.X., Jr. Cloning and functional analysis of the β-carotene hydroxylase of *Arabidopsis thaliana*. *J. Biol. Chem.* **1996**, *271*, 24349–24352.
37. Hannibal, L.; Lorquin, J.; D'Ortoli, N.A.; Garcia, N.; Chaintreuil, C.; Masson-Boivin, C.; Dreyfus, B.; Giraud, E. Isolation and characterization of the

canthaxanthin biosynthesis genes from the photosynthetic bacterium *Bradyrhizobium* sp. strain ORS278. *J. Bacteriol.* **2000**, *182*, 3850–3853.

38. Kaneda, K.; Kuzuyama, T.; Takagi, M.; Hayakawa, Y.; Seto, H. An unusual isopentenyl diphosphate isomerase found in the mevalonate pathway gene cluster from *Streptomyces* sp. strain CL190. *Proc. Natl. Acad. Sci. USA* **2001**, *98*, 932–937.
39. Misawa, N. Carotenoids. In *Comprehensive Natural Products II Chemistry and Biology*; Mander, L., Lui, H.W., Eds.; Elsevier: Oxford, UK, 2010; Volume 1, pp. 733–753.
40. Ye, R.W.; Stead, K.J.; Yao, H.; He, H. Mutational and functional analysis of the β-carotene ketolase involved in the production of canthaxanthin and astaxanthin. *Appl. Environ. Microbiol.* **2006**, *72*, 5829–5837.
41. Sieiro, C.; Poza, M.; de Miguel, T.; Villa, T.G. Genetic basis of microbial carotenogenesis. *Int. Microbiol.* **2003**, *6*, 11–16.
42. Fraser, P.D.; Miura, Y.; Misawa, N. *In vitro* characterization of astaxanthin biosynthetic enzymes. *J. Biol. Chem.* **1997**, *272*, 6128–6135.
43. Fraser, P.D.; Shimada, H.; Misawa, N. Enzymic confirmation of reactions involved in routes to astaxanthin formation, elucidated using a direct substrate *in vitro* assay. *Eur. J. Biochem.* **1998**, *252*, 229–236.
44. Choi, S.K.; Nishida, Y.; Matsuda, S.; Adachi, K.; Kasai, H.; Peng, X.; Komemushi, S.; Miki, W.; Misawa, N. Characterization of β-carotene ketolases, CrtW, from marine bacteria by complementation analysis in *Escherichia coli*. *Mar. Biotechnol.* **2005**, *7*, 515–522.
45. Choi, S.K.; Matsuda, S.; Hoshino, T.; Peng, X.; Misawa, N. Characterization of bacterial β-carotene 3,3′-hydroxylases, CrtZ, and P450 in astaxanthin biosynthetic pathway and adonirubin production by gene combination in *Escherichia coli*. *Appl. Microbiol. Biotechnol.* **2006**, *72*, 1238–1246.
46. Makino, T.; Harada, H.; Ikenaga, H.; Matsuda, S.; Takaichi, S.; Shindo, K.; Sandmann, G.; Ogata, T.; Misawa, N. Characterization of cyanobacterial carotenoid ketolase CrtW and hydroxylase CrtR by complementation analysis in *Escherichia coli*. *Plant Cell Physiol.* **2008**, *49*, 1867–1878.
47. Steiger, S.; Sandmann, G. Cloning of two carotenid ketolase genes from *Nostoc punctiforme* for the heterologous production of canthaxanthin and astaxanthin. *Biotechnol. Lett.* **2004**, *26*, 813–817.
48. Mochimaru, M.; Masukawa, H.; Takaichi, S. The cyanobacterium *Anabaena* sp. PCC 7120 has two distinct β-carotene ketolases: CrtO for echinenone and CrtW for ketomyxol synthesis. *FEBS Lett.* **2005**, *579*, 6111–6114.
49. Takaichi, S.; Mochimaru, M.; Maoka, T.; Katoh, H. Myxol and 4-ketomyxol 2′-fucosides, not rhamnosides, from *Anabaena* sp. PCC 7120 and *Nostoc punctiforme* PCC 73102, and proposal for the biosynthetic pathway of carotenoids. *Plant Cell Physiol.* **2005**, *46*, 497–504.

50. Kajiwara, S.; Kakizono, T.; Saito, T.; Kondo, K.; Ohtani, T.; Nishio, N.; Nagai, S.; Misawa, N. Isolation and functional identification of a novel cDNA for astaxanthin biosynthesis from *Haematococcus pluvialis*, and astaxanthin synthesis in *Escherichia coli*. *Plant Mol. Biol.* **1995**, *29*, 343–352.
51. Lotan, T.; Hirschberg, J. Cloning and expression in *Escherichia coli* of the gene encoding β-C-4-oxygenase, that converts β-carotene to the ketocarotenoid canthaxanthin in *Haematococcus pluvialis*. *FEBS Lett.* **1995**, *364*, 125–128.
52. Huang, J.C.; Chen, F.; Sandmann, G. Stress-related differential expression of multiple β-carotene ketolase genes in the unicellular green alga *Haematococcus pluvialis*. *J. Biotechnol.* **2006**, *122*, 176–185.
53. Giuliano, G.; Tavazza, R.; Diretto, G.; Beyer, P.; Taylor, M.A. Metabolic engineering of carotenoid biosynthesis in plants. *Trends Biotechnol.* **2008**, *26*, 139–145.
54. Fernandez-Gonzalez, B.; Sandmann, G.; Vioque, A. A new type of asymmetrically acting β-carotene ketolase is required for the synthesis of echinenone in the cyanobacterium *Synechocystis* sp. PCC 6803. *J. Biol. Chem.* **1997**, *272*, 9728–9733.
55. Takaichi, S.; Maoka, T.; Masamoto, K. Myxoxanthophyll in *Synechocyctis* sp. PCC 6803 is myxol 2′-dimethyl-fucoside, (3*R*,2′*S*)-myxol 2′-(2,4-di-*O*-methyl-α-L-fucoside), not rhamnoside. *Plant Cell Physiol.* **2001**, *42*, 756–762.
56. Tao, L.; Cheng, Q. Novel β-carotene ketolases from non-photosynthetic bacteria for canthaxanthin synthesis. *Mol. Genet. Genomics* **2004**, *272*, 530–537.
57. Choi, S.K.; Harada, H.; Matsuda, S.; Misawa, N. Characterization of two β-carotene ketolases, CrtO and CrtW, by complementation analysis in *Escherichia coli*. *Appl. Microbiol. Biotechnol.* **2007**, *75*, 1335–1341.
58. Masamoto, K.; Misawa, N.; Kaneko, T.; Kikuno, R.; Toh, H. β-carotene hydroxylase gene from the cyanobacterium *Synechocystis* sp. strain PCC6803. *Plant Cell Physiol.* **1998**, *39*, 560–564.
59. Takaichi, S.; Mochimaru, M. Carotenoids and carotenogenesis in cyanobacteria: Unique ketocarotenoids and carotenoid glycosides. *Cell. Mol. Life Sci.* **2007**, *64*, 2607–2619.
60. Mochimaru, M.; Masukawa, H.; Maoka, T.; Mohamed, H.E.; Vermaas, W.L.; Takaichi, S. Substrate specificities and availability of fucosyltransferase and β-carotene hydroxylase for myxol 2′-fucoside synthesis in *Anabaena* sp. strain PCC 7120 compared with *Synechocystis* sp. strain PCC 6803. *J. Bacteriol.* **2008**, *190*, 6726–6733.
61. Blasco, F.; Kauffmann, I.; Schmid, R.D. CYP175A1 from *Thermus thermophilus* HB27, the first β-carotene hydroxylase of the P450 superfamily. *Appl. Microbiol. Biotechnol.* **2004**, *64*, 671–674.
62. Tao, L.; Rouvière, P.E.; Cheng, Q. A carotenoid synthesis gene cluster from a non-marine *Brevundimonas* that synthesizes hydroxylated astaxanthin. *Gene* **2006**, *379*, 101–408.

63. Iwai, M.; Maoka, T.; Ikeuchi, M.; Takaichi, S. 2,2′-β-Hydroxylase (CrtG) is involved in carotenogenesis of both nostoxanthin and 2-hydroxymyxol 2′-fucoside in *Thermosynechococcus elongatus* strain BP-1. *Plant Cell Physiol.* **2008**, *49*, 1678–1687.
64. Osawa, A.; Harada, H.; Choi, S.K.; Misawa, N.; Shindo, K. Production of caloxanthin 3′-β-D-glucoside, zeaxanthin 3,3′-β-D-diglucoside, and nostoxanthin in a recombinant *Escherichia coli* expressing system harboring seven carotenoid biosynthesis genes, including *crtX* and *crtG*. *Phytochemistry* **2011**, *72*, 711–716.
65. Yokoyama, A.; Shizuri, Y.; Misawa, N. Production of new carotenoids, astaxanthin glucosides, by *Escherichia coli* transformants carrying carotenoid biosynthetic genes. *Tetrahedron Lett.* **1998**, *39*, 3709–3712.
66. Harada, H.; Misawa, N. Novel approaches and achievements in biosynthesis of functional isoprenoids in *Escherichia coli*. *Appl. Microbiol. Biotechnol.* **2009**, *84*, 1021–1031.
67. Misawa, N. Pathway engineering for functional isoprenoids. *Curr. Opin. Biotechnol.* **2011**, doi:10.1016/j.copbio.2011.01.002.
68. Kajiwara, S.; Fraser, P.D.; Kondo, K.; Misawa, N. Expression of an exogenous isopentenyl diphosphate isomerase gene enhances isoprenoid biosynthesis in *Escherichia coli*. *Biochem. J.* **1997**, *324*, 421–426.
69. Harada, H.; Yu, F.; Okamoto, S.; Kuzuyama, T.; Utsumi, R.; Misawa, N. Efficient synthesis of functional isoprenoids from acetoacetate through metabolic pathway-engineered *Escherichia coli*. *Appl. Microbiol. Biotechnol.* **2009**, *81*, 915–925.
70. Kakinuma, K.; Dekishima, Y.; Matsushima, Y.; Eguchi, T.; Misawa, N.; Takagi, M.; Kuzuyama, T.; Seto, H. New approach to multiply deuterated isoprenoids using triply engineered *Escherichia coli* and its potential as a tool for mechanistic enzymology. *J. Am. Chem. Soc.* **2001**, *123*, 1238–1239.
71. Newman, J.D.; Marshall, J.; Chang, M.; Nowroozi, F.; Paradise, E.; Pitera, D.; Newman, K.L.; Keasling, J.D. High-level production of amorpha-4,11-diene in a two-phase partitioning bioreactor of metabolically engineered *Escherichia coli*. *Biotechnol. Bioeng.* **2006**, *95*, 684–691.
72. Vadali, R.V.; Fu, Y.; Bennett, G.N.; San, K.Y. Enhanced lycopene productivity by manipulation of carbon flow to isopentenyl diphosphate in *Escherichia coli*. *Biotechnol. Prog.* **2005**, *21*, 1558–1561.
73. Yoon, S.H.; Lee, Y.M.; Kim, J.E.; Lee, S.H.; Lee, J.H.; Kim, J.Y.; Jung, K.H.; Shin, Y.C.; Keasling, J.D.; Kim, S.W. Enhanced lycopene production in *Escherichia coli* engineered to synthesize isopentenyl diphosphate and dimethylallyl diphosphate from mevalonate. *Biotechnol. Bioeng.* **2006**, *94*, 1025–1032.
74. Alper, H.; Stephanopoulos, G. Uncovering the gene knockout landscape for improved lycopene production in *E. coli*. *Appl. Microbiol. Biotechnol.* **2008**, *78*, 801–810.

75. Hasunuma, T.; Miyazawa, S.; Yoshimura, S.; Shinzaki, Y.; Tomizawa, K.; Shindo, K.; Choi, S.K.; Misawa, N.; Miyake, C. Biosynthesis of astaxanthin in tobacco leaves by transplastomic engineering. *Plant J.* **2008**, *55*, 857–868.

*Samples Availability:* Available from the authors.

*marine drugs*

*Review*

# Marine Carotenoids: Biological Functions and Commercial Applications

**Carlos Vílchez [1,*], Eduardo Forján [1], María Cuaresma [1], Francisco Bédmar [2], Inés Garbayo [1] and José M. Vega [3]**

1. Algal Biotechnology Group, International Centre for Environmental Research (CIECEM), University of Huelva, 21760 Huelva, Spain; E-Mails: eduardo.forjan@dqcm.uhu.es (E.F.); maria.cuaresma@dqcm.uhu.es (M.C.); garbayo@uhu.es (I.G.)
2. Faculty of Business, University of Huelva, Plaza de la Merced 11, 21071 Huelva, Spain;
   E-Mail: f.bedmar@ono.com
3. Plant Biochemistry and Molecular Biology Department, Faculty of Chemistry, University of Seville, 41012 Seville, Spain; E-Mail: jmvega@us.es

* Author to whom correspondence should be addressed; E-Mail: cvilchez@uhu.es; Tel.: +34-959-219947; Fax: +34-959-219942.

*Received: 31 January 2011; in revised form: 15 February 2011 / Accepted: 17 February 2011 / Published: 3 March 2011*

**Abstract:** Carotenoids are the most common pigments in nature and are synthesized by all photosynthetic organisms and fungi. Carotenoids are considered key molecules for life. Light capture, photosynthesis photoprotection, excess light dissipation and quenching of singlet oxygen are among key biological functions of carotenoids relevant for life on earth. Biological properties of carotenoids allow for a wide range of commercial applications. Indeed, recent interest in the carotenoids has been mainly for their nutraceutical properties. A large number of scientific studies have confirmed the benefits of carotenoids to health and their use for this purpose is growing rapidly. In addition, carotenoids have traditionally been used in food and animal feed for their color properties. Carotenoids are also known to improve consumer perception of quality; an example is the addition of carotenoids to fish feed to impart color to farmed salmon.

**Keywords:** carotenoids; microalgae; applications; nutraceuticals; health benefits

## 1. Marine Carotenoids: Biological Functions and Benefits to Human Health

In photosynthetic organisms including plants and microalgae, carotenoids play various roles. Essentially, carotenoids may act as accessory pigments in light harvesting functions during the light phase of photosynthesis and are also able to photoprotect the photosynthetic machinery from excess light by scavenging reactive oxygen species (ROS) like singlet oxygen and other free radicals [1].

In humans, the most relevant biological functions of carotenoids are linked to their antioxidant properties, which directly emerge from their molecular structure. In recent years, the understanding of ROS-induced oxidative stress mechanisms and the search for suitable strategies to fight oxidative stress has become one the major goals of medical research efforts. A number of studies have been reported which implicate oxidative stress involvement in degenerative pathogenesis, e.g., Alzheimer and Parkinson [1,2]. In parallel, a carotenoid-enriched diet has been found to diminish the risk of suffering from degenerative diseases [2].

Moreover, far from being just a speculative hypothesis, the benefits of carotenoids (lutein, -carotene, lycopene) to human health have been shown based on the positive impacts of the antioxidant bioactivity of carotenoids in inmunoresponse modulation, in signaling transduction between cells and in anti-inflammatory response mechanisms [3–5]. These positive consequences are the result of either the direct chemical action of carotenoids on biological molecules and structures or through expression of different genes involved in antioxidant responses [2]. The main biological functions of carotenoids and benefits to health are listed in Table 1.

### 1.1. Provitamin A Activity

One of the most important functions of carotenoids in the human body is their ability to convert into retinol (provitamin A function), a faculty that about 10% of carotenoids identified in nature

possess [6]. Vitamin A is well recognized as a factor of great importance for child health and survival, its deficiency causes disturbances in vision and various related lung, trachea and oral cavity pathologies [7]. Animals and humans cannot synthesize carotenoids *de novo*, although they are able to convert them into vitamin A. Diet is the only source for these precursors for retinol synthesis, fruits, vegetables and microalgae being the major suppliers of provitamin A active carotenoids. As a reference value, a recommended daily intake of 6 mg of carotenoids has been proposed. This value is based on the contribution of compounds with provitamin A activity, specially β-carotene, which has been assigned a provitamin A activity of 100% [8].

### 1.2. Carotenoids and Cancer

In recent years, epidemiological evidence supporting a protective effect of carotenoids to the development of chronic and degenerative diseases has grown considerably. We must not forget that cancer and cardiovascular diseases are the leading causes of death in the world and that approximately 50% of all tumors are attributed to the diet [9].

From a nutritional point of view, an antioxidant can be defined as any substance present in foods that significantly reduces the adverse effects of reactive oxygen species in normal physiological conditions in humans [10]. Antioxidants, in particular carotenoids, are essential for cell health due to their protective action on cellular components against oxidative damage [2]. These activities have generated two lines of research related to the physiological functionality of carotenoids: on one hand, their activity as membrane antioxidants, therefore involved in the oxidative cell cycle [11] and, on the other hand, their involvement in control processes of cell differentiation and proliferation [12]. As an example of the first, a recent study showed that antioxidant enzymes including catalase, superoxide dismutase and peroxidase levels in plasma and liver of mice increased significantly when the animals were fed with microalgae biomass (*Haematococcus pluvialis*, *Scenedesmus platensis* or *Botryococcus braunii*), which reveals an increased antioxidant protection against free radicals [13].

It is well known that cellular proliferation is controlled by the communication established between the cells in a tissue. Cell communications reset or stimulation becomes essential if abnormal cell proliferation occurs. In that respect, it has been mentioned that carotenoids might stimulate expression of genes directly involved in regulation of cell communication processes. In more detail, carotenoids would directly act on DNA in order to regulate the production of RNA that is responsible for *gap-junctions* communications, which could successfully explain some anti-tumor activities of carotenoids [2,12]. Immune system cells also require intercellular communication to conduct their activity efficiently, so the previous action mechanism of carotenoids could also apply for supporting the immune system activity. As an example, high doses of β-carotene increase the CD4 to CD8 lymphocyte ratio, which is very low in patients suffering from HIV disease [14].

In the last decades, many laboratory and epidemiological studies have been conducted which suggest that intake of carotenoids and cancer prevalence are inversely related [4,15–17]. Among the carotenoids, lycopene has been one of the most extensively studied [4,18–20] probably due to the greater anticancer capacity shown with respect to other carotenoids [21]. Within the wide frame of research carried out by Giovannucci *et al.* [4], lycopene intake and prostate cancer were found to be inversely related. The inverse relation was based on *in vivo* and *in vitro* studies on the effect of lycopene in tumor cell lines that showed tumor cells growth inhibition by the action of lycopene [19,20]. Although the functional meaning of the lycopene distribution in the organism has not been fully elucidated, it is particularly interesting that this carotenoid predominates in testes and adrenal glands, with an abundance of about 60 to 80% of the total carotenoids [22]. It has also been inferred that astaxanthin could be effective against benign prostatic hyperplasia and against prostatic cancer through inhibition of the enzyme 5-a-reductase which is involved in abnormal prostate growth [2,23].

Antitumoral activity of carotenoids toward other type of cancer has also been observed. In particular, -carotene, astaxanthin, cantaxanthin and zeaxanthin have been shown to promote reduction in size and number of liver neoplasias *in vivo* [21,24]. Other studies have shown that inclusion of carotenoids in the diet and reduced risk of colon cancer might be directly related [25–27]. The antitumoral effect of b-carotene has also been associated to the nutritional situation of the studied population. As an example, -carotene implementation studies carried out at Linxian (China) in population that suffered from a diet deficient in vitamins and mineral salts, led to reduced incidence of total mortality from gastrointestinal cancer [28]. Interestingly, in population not affected by nutritional deficiency but included in cancer risk groups (e.g., smokers or asbestos-exposed groups) it has been shown that -carotene supplements even might increase cancer risk, probably due to generation of metabolites that increase the cell oxidative state and led to reduced control of cell differentiation and cell proliferation processes [29–32].

Although carotenoids including zeaxanthin, criptoxanthin and lutein antitumoral activities have still been scarcely studied, the strategy of using carotenoids as chemoprotecting agents is not yet endorsed by clinical trials. More on the contrary, in spite of using -carotene as pure drug for producing an intense punctual effect after any dosage intake, the derived positive action of carotenoids should be produced through continuous intake of usual quantities. This idea is in line with current dietary recommendations that suggest consumption of five fruit and vegetables portions a day, which will provide water, vitamins, fiber and phytochemical compounds including carotenoids in sufficient quantities to meet our body needs [10,16].

### *1.3. Carotenoids and Cardiovascular Diseases*

Cardiovascular diseases are the leading cause of death in developed countries, and have become the main health problem also in developing countries [33]. These include acute myocardial infarction and disorder of high morbidity and mortality [34]. Oxidative stress and inflammation are the main factors contributing to the pathophysiology of these disorders [35,36]. In particular, the oxidative stress induced by ROS can cause low density lipoproteins oxidation (LDL), an aspect that plays a key role in the pathogenesis of atherosclerosis [37,38].

Another major feature of carotenoids is protection of LDL against oxidation [39,40], which confers carotenoids antiatherogenic properties [2,36,41]. In addition, carotenoids have been shown to inhibit *in vivo* lipid peroxidation processes [42], by which the presence of carotenoids in cell membranes is essential to act as stabilizing elements of these structures [8,43]. In this sense, the antioxidant activity of some carotenoids during radical peroxide-induced cholesterol oxidation was investigated by Palozza *et al.* [44], showing that carotenoids exerted a significant antioxidant activity, in the decreasing activity order indicated: astaxanthin, cantaxanthin, lutein and β–carotene. Several authors have published that daily dietary -carotene supplementation in mammals led to decreased plasma levels of total lipids, cholesterol and triglycerides [45,46].

Numerous epidemiological studies suggest that diets rich in carotenoids could protect the human body from certain cardiovascular diseases due to the involvement of oxidizing substances and oxidative stress in the development and clinical expression of coronary heart disease [47]. In fact, high lycopene levels in plasma and tissues have been inversely linked to coronary heart disease [48], myocardial infarction [49] and risk to suffer from arteriosclerosis [50]. Low lutein levels in plasma have also been associated with an increased tendency to suffer from myocardial infarction [51], while a high intake of lutein has been inversely related with the risk of stroke [52].

Likewise, low α-carotene levels in serum have been shown to inversely correlate prevalence of coronary artery disease and formation of arterial plaque, by which α-carotene has been proposed as a potential marker for human atherosclerosis. In addition, carotenoids displaying high levels of provitamin A activity, including α-carotene, β-carotene and β-cryptoxanthin, have been associated with reduced risk of angina pectoris disease [53,54]. Other epidemiological studies have also found low levels of oxygenated carotenoids (namely xanthophylls: lutein, zeaxanthin, lycopene, β-cryptoxanthin, β-carotene and α-carotene) in plasma of patients with acute and chronic coronary syndromes [55,56]. Particularly, in the recent study by [38], high levels of β-cryptoxanthin and lutein in plasma have been shown to decrease risk for suffering from myocardial infarction, but no statistically significant associations with other carotenoids were found.

### *1.4. Carotenoids and Eye Health*

Many research studies showed that lutein and zeaxanthin are the main responsible pigments for both the yellowing and the maintenance of normal visual function of the human eye macula [57,58], while other major carotenoids in serum ( -carotene, β-carotene, lycopene and β-cryptoxanthin), are absent or are found in trace amounts in the human macula [59]. In the eye macula, lutein and zeaxanthin absorb blue light and also attenuate pernicious photooxidative effects caused by the excess blue light, while reducing eye chromatic aberration. Due to their antioxidant properties, carotenoids protect the eye macula from adverse photochemical reactions [60]. In people over the age of 64, visual sensitivity directly depends on lutein and zeaxanthin concentrations in retina [61].

Major prevalency of cataracts has also been linked to people with low levels of lutein and zeaxanthin [62]. Also macular degeneration, the main cause of irreversible loss of vision in people above 65 years in industrialized countries, has been associated with very low levels of lutein and zeaxanthin [63,64].

The spectra of lutein and zeaxanthin show a wide absorption band with a peak at 450 nm, which is thought to be involved in absorbing excess blue light before it comes to photoreceptors, therefore preventing the eye macula from being damaged by blue light [65]. Moreover, due to lutein's and zeaxanthin's biophysical and biochemical properties for ROS scavenging, these carotenoids might also preserve the membrane structure in the eye photoreceptors from lipid peroxidation processes [66], in contrast to non-polar carotenoids as lycopene and β-carotene [67]. Concentration of lutein and zeaxanthin in the retina can be increased on diet bases (spinach and maize) and on supplements of both pigments [60,68].

### *1.5. Other Physiological Functions of Carotenoids*

Carotenoids provide skin photoprotection against UV light [69–71]. Due to their scavenging action on ROS, carotenoids also possess anti-inflammatory properties [72–74]. In this sense, it has been recently described that astaxanthin raises anti-inflammatory effects while preserving essential lipids and proteins of human lymphocytes [74]. Astaxanthin would act by inducing superoxide dismutase and catalase enzyme activities [74]. Other studies have shown astaxanthin to protect from $CCl_4$-induced hepatic damage by inhibiting lipid peroxidation, stimulating the cellular antioxidant system and modulating the inflammatory process [73]. Table 1 resumes biological functions, benefits to health and applications of the main carotenoids, including their role in prevention of cataracts [75,76], macular degeneration [77–80], retinitis [81–83] and gastric infection [84].

Carotenoids have been used as preservatives in cosmetics and, combined with other antioxidants or algal bioactive substances, also in creams and lotions for sun protection [85]. The beneficial effect of carotenoids has also been shown in patients with psoriasis, skin inflammatory pathology. Lima and Kimball [86] found low levels of carotenoids in the skin correlate well with psoriasis prevalence.

Finally, it is interesting to note that, in recent years, carotenoids are being considered as important protective molecules in gastric disorders. It has been published that a high intake of carotenoids prevents the development of disorders caused by *Helicobacter pylori* [84,87,88], a Gram negative bacteria genus that colonizes the gastric mucosa of at least half of the human population [89].

**Table 1.** Biological functions, benefits to health and applications of the main carotenoids.

| Carotenoid | Functions and benefits to health | References |
|---|---|---|
| *Lycopene* | In prostatic hyperplasia and prostate cancer | [19–21] |
| | In the prevention of atherosclerosis and acute and chronic coronary syndromes | [48–50,55,56] |
| *β–carotene* | Provitamin A function | [6,8] |
| | In colorectal cancer | [25–27] |
| | In the prevention of acute and chronic coronary syndromes | [53–56] |
| | Photoprotection of skin against UV light | [69–71] |
| *Astaxanthin* | In benign prostatic hyperplasia and prostate and liver tumors | [2,21–24] |
| | Anti–inflammatory properties | [72–74] |
| *Zeaxanthin* | Active against liver neoplasms | [21,24] |
| | In the prevention of acute and chronic coronary syndromes | [55,56] |
| | Helps to maintain a normal visual function | [57,58] |
| | In the prevention of cataracts | [62,75,76] |
| | To prevent macular degeneration associated with age | [65,77–80] |
| *Lutein* | In the prevention of acute and chronic coronary syndromes and stroke | [38,55,56,66] |
| | Helps to maintain a normal visual function | [57,58] |
| | In the prevention of cataracts | [62,75,76] |
| | To prevent macular degeneration associated with age | [65,77–80] |
| | In the prevention of retinitis | [58,81–83] |
| | To avoid gastric infection by *H. Pylori* | [84] |

## 2. Marine Carotenoids: Applications

Carotenoids have been traditionally used in food and animal feed due to their color properties. The natural carotenoids are used to reinforce fish color, which increases consumers' perception of quality. An example is the addition of carotenoids to fish feed to impart color to farmed salmon. The nutraceutical properties of carotenoids also attracted attention of the food industry. Large numbers of scientific studies have confirmed the benefits of carotenoids to health and use for this purpose is growing rapidly. Besides, carotenoids have been proposed as added-value compounds that could contribute to make microalgal biofuel production economically feasible [90,91].

Among all existing natural carotenoids, five can be considered to be the most relevant ones in economical terms (Table 2). The main applications of carotenoids are currently as dietary supplements, fortified foods, food color, animal feed and pharmaceuticals and cosmetics.

**Table 2.** Main commercial carotenoids and origin.

| Carotenoid | Origin |
|---|---|
| -carotene | Synthetic and naturally extracted forms [6,22]. |
| Astaxanthin | Synthetic nature identical and naturally extracted forms [2,13] |
| Canthaxanthin | Comes in synthetic nature identical form [13]. |
| Lycopen | Only currently available in natural form [22] |
| Lutein | Only comes in natural form [6,13,85]. |

β-carotene, the most widely known of the carotenoids, is known to be a vitamin A precursor, likely several other carotenoids. Carotenoids have antioxidant properties and a large number of studies have confirmed their benefits to health. In particular, carotenoids are thought to reduce the risk of degenerative diseases and cancer especially in elderly people, as explained above [29,32,41].

The health industry uses carotenoids in over-the-counter (OTC) dietary supplements and fortified foods. This is one of the fastest growing segments of the industry but is still relatively small compared to the color segment. The pharmaceutical and cosmetics industries also use carotenoids mainly for their coloring properties, though their use by the pharmaceutical and cosmetics companies is growing rapidly due to their nutraceutical properties. An example of a new product from this segment is a 'beauty pill' containing the carotenoid lycopene. This product belongs to a new market segment known as 'cosmeceuticals', which aims to combine cosmetics and nutraceutical food ingredients to create products to improve skin and hair.

Chemically synthesized nature identical carotenoids dominate the market but naturally extracted carotenoids are growing in popularity due to increasing demand for natural products from consumers. Natural carotenoids can be extracted from plant material such as tomatoes, algae and fungi. Individual carotenoids are available in a variety of forms. The most common forms are cold water soluble powder, oil emulsion and beadlets. Concentrations range from 0.2 to 100%. The most common concentration is 10%. Blends or mixed carotenoids are also available containing two or more different carotenoids. Like the individual carotenoids, blends are available in a variety of forms including, water dispersible powder, oil suspension and beadlet forms. The concentration of blends ranges from 1 to 30%, with the most common concentration being 10% [91–93].

### 2.1. *Dietary Supplements and Food Color*

Carotenoids are widely used as color enhancers in natural foods including egg yolk, chicken meat or fish [90]. However, among more than 400 known carotenoids just few of them have been commercially used, including -carotene, lycopene, asthaxanthin and lutein [91]. One of the main advantages in the use of microalgae as a carotenoid carrier in the food industry is that many other antioxidant compounds present in the microalgal biomass have positive impact on human health, sometimes acting with carotenoids synergistically [92].

In addition, if carotenoids are disposed within the microalgal matrix (carotenoid enriched dry biomass) also a number of minerals whose presence is inherent to the algal biomass are provided in the formula. These mineral have positive effects to human health, especially in enhancing anabolic activities. Carotenoids have also been used as preservatives in cosmetics and solar protection products [85].

Because of the content of carotenoids, the commercial value of microalgae increased and their use extended widely into many applications of the food market. That includes the use of *Arthrospira, Chlorella, Dunaliella, Spirulina* and *Aphanizomenon* as functional foods which can be found in the market in the form of pills, tablets and capsules. These microalgae have also been integrated in nutritional formula of pasta, snacks, sweets, drinks and bubble gum [91,93].

Microalgae are also used in fish color quality improvement in aquaculture. Salmonids are supplied with astaxanthin-enriched microalgae species, in particular *Haematococcus pluvialis* [2].

### 2.2. *Environmental Applications: Carotenoids in Biorefining*

Microalgae have gained interest as promising feedstocks for biofuels. The productivity of these photosynthetic microorganisms in converting carbon dioxide into carbon-rich lipids greatly exceeds that of agricultural oleaginous crops, without competing for arable land. However, large scale production of lipid-enriched algal biomass is not yet economically feasible and still requires major efforts in developing suitable technology which allows for reducing biomass production costs at large scale by at least an order of magnitude. Recent advances in systems biology, genetic engineering and methods to profit from the fractions of the biomass residue open new scenarios to make biofuel production from microalgae economically suitable within a period of about 15 years. Production of biodiesel and other bio-products from microalgae can be more cost-effective and profitable if combined with processes such as wastewater and flue gas treatments [94,95]. Carotenoids are, indeed, one of the main bio-products whose production is required to make biofuel production economically feasible. The paradox, therefore, is that production of high-added value compounds as carotenoids should so far be the only way to approach economical production of a low value energy source as biofuel from microalgae.

## 2.3. *Commercial Value for Carotenoids*

In recent years, production of carotenoids has become one of the most successful activities in microalgal biotechnology. The demand for carotenoids obtained from natural sources is increasing. This has promoted major efforts to improve carotenoid production from biological sources instead of chemical synthesis [96]. According to the report published by Business Communications in March, 2008, the global market for all commercial carotenoids accounted for 766 million dollars, with expectations of rising to 919 million dollars in 2015. In particular, beta-carotene market volume in 2007 was 247 million dollars, with expectations of reaching 285 million dollars in 2015. Besides lycopene and -carotene, xanthophylls lutein, astaxanthin and cantaxanthin appear as the most demanded and valuable carotenoids. Astaxanthin market volume in aquaculture in 2009 was 260 million dollars and about 2500 \$ $kg^{-1}$. In addition, lutein market volume in 2010 accounted for about 190 million dollars, the carotenoid experiencing the most rapid growth in sales [97]. Therefore, carotenoid-containing microalgae find many applications in a wide range of commercial activities, the reason for which carotenoid-enriched microalgae production is steeply becoming an attractive business (Table 3).

**Table 3.** Main applications of microalgae due to their carotenoid content.

| Microalga | Application | Product formula | Price |
|---|---|---|---|
| *Chorella vulgaris* | Aquaculture, cosmetics, nutraceuticals, food ingredient | Dry powder, tablets | \$30–100 $kg^{-1}$ |
| *Isochrysis galbana* | Aquaculture, cosmetics, nutraceutical | Paste, dry powder | \$100–400 $kg^{-1}$ |
| *Nannochloropsis gaditana* | Aquaculture, cosmetics | Paste, dry powder | \$300 $kg^{-1}$ |
| *Pavlova lutheri* | Aquaculture | Paste, dry powder | >\$300 $kg^{-1}$ |
| *Phaeodactylum tricornutum* | Aquaculture, nutraceuticals | Paste, dry powder | >\$200 $kg^{-1}$ |
| *Tetraselmis* | Aquaculture | Paste, dry powder | \$600–800 $kg^{-1}$ |
| *Thalassiosira weissflogii* | Aquaculture | Paste, dry powder | >\$300 $kg^{-1}$ |
| *Arthrospira* | Cosmetics, nutraceuticals | Paste, dry powder | >\$200 $kg^{-1}$ |
| *Haematococcus pluvialis* | Aquaculture, nutraceuticals | Dry powder | >\$600 $kg^{-1}$ |
| *Dunaliella salina* | Nutraceuticals, food ingredients | Dry powder, tablets | \$100–400 $kg^{-1}$ |

**Acknowledgements**

This work has been supported by grant AGR-4337 (Proyecto de Excelencia, Junta de Andalucía) and grant Bioándalus (Junta de Andalucía, Estrategia de Impulso a la Biotecnología).

## References

1. Murthy, K.; Vanitha, A.; Rajesha, J.; Swamy, M.; Sowmya, P.; Ravishankar, G.A. *In vivo* antioxidant activity of carotenoids from *Dunaliella salina*, a green microalga. *Life Sci.* **2005**, *76*, 1381–1390.
2. Guerin, M.; Huntley, M.E.; Olaizola, M. *Haematococcus* astaxanthin: Applications for human health and nutrition. *Trends Biotech.* **2003**, *21*, 210–216.
3. Le Marchand, L.; Hankin, J.H.; Kolonel, L.N.; Beecher, G.R.; Wilkens, L.R.; Zhao, L.P. Intake of specific carotenoids and lung cancer risk. *Cancer Epidemiol. Biomarkers Prev.* **1993**, *2*, 183–187.
4. Giovannucci, E.; Ascherio, A.; Rimm, E.B.; Stampfer, M.J.; Colditz, G.A.; Willett, W.C. Intake of carotenoids and retinol in relation to risk of prostate cancer. *J. Nat. Cancer Inst.* **1995**, *87*, 1767–1776.
5. Biesalski, H. Evidence from Intervention Studies. In *Functions of Vitamins beyond Recommended Dietary Allowances*; Walter, P., Hornig, D., Moser, U., Eds.; Woodhead Publishing Limited: Cambridge, UK, 2001; pp. 92–134.
6. García-González, M.; Moreno, J.; Manzano, J.C.; Florencio, F.J.; Guerrero, M.G. Production of *Dunaliella salina* biomass rich in 9-*cis*-β-carotene and lutein in a closed tubular photobioreactor. *J. Biotechnol.* **2005**, *115*, 81–90.
7. Indicators for assessing vitamin A deficiency and their implications in monitoring and evaluating intervention programmes. WHO/NUT/96.10; World Health Organization: Geneva, Switzerland, 1998.
8. Graham, R.D.; Rosser, J.M. Carotenoids in staple foods: Their potential to improve human nutrition. *Food Nutr. Bull.* **2000**, *21*, 405–409.
9. Williams, G.M.; Williams, C.L.; Weisburger, J.H. Diet and cancer prevention: The fiber first diet. *Toxicol. Sci.* **1999**, *52*, 72–86.
10. Food and Nutrition Board. *Dietary Reference Intakes for Vitamin C, Vitamin E, Selenium and Carotenoids*; Institute of Medicine, National Academy Press: Washington, DC, USA, 2000.
11. McNulty, H.P.; Byun, J.; Lockwood, S.F.; Jacob, R.F.; Mason, R.P. Differential effects of carotenoids on lipid peroxidation due to membrane interactions: X-ray diffraction analysis. *Biochim. Biophys. Acta* **2007**, *1768*, 167–174.
12. Bertram, J.S. Carotenoids and gene regulation. *Nutr. Rev.* **1999**, *57*, 182–191.
13. Ranga Rao, A.; Raghunath, R.L.; Baskaran, V.; Sarada, R.; Ravishankar, G.A. Characterization of microalgal carotenoids by mass spectrometry and their bioavailability and antioxidant properties elucidated in rat model. *J. Agric. Food Chem.* **2010**, *58*, 8553–8559.
14. Olson, J.A. Carotenoids and human health. *Arch. Latinoamer. Nutr.* **1999**, *49* (1-S), 7–11.
15. Garewall, H. Antioxidants in oral cancer prevention. *Am. J. Clin. Nutr.* **1995**, *62* (Suppl.), 1403S–1417S.

16. Mayne, S.T. β-Carotene, carotenoids, and disease prevention in humans. *FASEB J.* **1996**, *10*, 690–701.
17. Rock, C.L. Carotenoids: Biology and treatment. *Pharmacol. Ther.* **1997**, *75*, 185–197.
18. Narisawa, T.; Fukaura, Y.; Hasebe, M.; Ito, M.; Aizawa, R.; Murakoshi, M.; Uemura, S.; Khachik, F.; Nishino, H. Inhibitory effects of natural carotenoids, alpha-carotene, beta-carotene, lycopene and lutein, on colonic aberrant crypt foci formation in rats. *Cancer Lett.* **1996**, *107*, 137–142.
19. Graydon, R.; Gilchrist, S.; Young, I.; Obermüller-Jevic, U.; Hasselwander, U.; Woodside, J. Effect of lycopene supplementation on insulin-like growth factor-1 and insulin-like growth factor binding protein-3: A double-blind, placebo-controlled trial. *Eur. J. Clin. Nutr.* **2007**, *61*, 1196–1200.
20. Schwarz, S.; Obermüller-Jevic, U.; Hellmis, E.; Koch, W.; Jacobi, G.; Biesalski, H.-K. Lycopene inhibits disease progression in patients with benign prostate hyperplasia. *J. Nutr.* **2008**, *138*, 49–53.
21. Nishino, H.; Murakoshi, M.; Li, T.; Takemura, M.; Kuchide, M.; Kanazawa, M.; Mou, X.; Wada, S.; Masuda, M.; Ohsaka, Y.; Yogosawa, S.; Satomi, Y.; Jinno, K. Carotenoids in cancer chemoprevention. *Cancer Metast. Rev.* **2002**, *21*, 257–264.
22. Stahl, W.; Sundquist, A.R.; Hanusch, M.; Schwarz, W.; Sies, H. Separation of β-carotene and lycopene geometrical isomers in biological samples. *Clin. Chem.* **1993**, *39*, 810–814.
23. Anderson, M. Method of inhibiting 5α-reductase with astaxanthin to prevent and treat benign prostate hyperplasia (BPH) and prostate cancer in human males. *US Patent No. 6277417*, 2001.
24. Gradelet, S.; Le Bon, A.M.; Berges, R.; Suschetet, M.; Astorg, P. Dietary carotenoids inhibit aflatoxin B1-induced liver preneoplastic foci and DNA damage in the rat: Role of the modulation of aflatoxin B1 metabolism. *Carcinogenesis* **1998**, *19*, 403–411.
25. Erhardt, J.G.; Meisner, C.; Bode, J.C.; Bode, C. Lycopene, β-carotene and colorectal adenomas. *Am. J. Clin. Nutr.* **2003**, *78*, 1219–1224.
26. Senesse, P.; Touvie, M.; Kesse, E.; Faivre, J.; Boutron-Ruault, M.-C. Tobacco use and associations of β-carotene and vitamin intakes with colorectal adenoma risk. *J. Nutr.* **2005**, *135*, 2468 –2472.
27. Ramadas, A.; Kandiah, M.; Jabbar, F.; Zarida, H. Dietary risk factors for colorectal adenomatous polyps: a mini review. *J. Sci. Technol.* **2010**, *18* (2), 321–349.
28. Blot, W.L.; Li, J.-Y.; Taylor, P.R.; Guo, W.; Dawsey, S.; Wang, G.; Yang, C.; Zheng, S.; Gail, M.; Li, G.; *et al.* Nutrition intervention trials in Linxian, China: Supplementation with specific vitamin/mineral combinations, cancer incidence, and disease-specific mortality in the general population. *J. Nat. Cancer Inst.* **1993**, *85*, 1483–1492.

29. Albanes, D.; Heinonen, O.P.; Taylor, P.R.; Virtamo, J.; Edwards, B.K.; Rautalahti, M.; Hartman, A.M.; Palmgren, J.; Freedman, L.S.; Haapakoski, J.; *et al.* α-Tocopherol and β-carotene supplements and lung cancer incidence in the α-tocopherol, β-carotene cancer prevention study: Effects of base-line characteristics and study compliance. *J. Nat. Cancer Inst.* **1996**, *88*, 1560–1570.
30. Hennekens, C.; Buring, J.; Manson, J.; Stampfer, J.; Rosner, B.; Cook, N.; Belanger, C.; LaMotte, F.; Gaziano, J.; Ridker, P.; Willet, W.; Peto, R. Lack of effect of long-term supplementation with beta carotene on the incidence of malignant neoplasm and cardiovascular disease. *N. Engl. J. Med.* **1996**, *334*, 1145–1149.
31. Rowe, P.M. β-Carotene takes a collective beating. *Lancet* **1996**, *347*, 249.
32. Lee, I.M.; Cook, N.R.; Manson, J.E.; Buring, J.E.; Hennekens, C.H. β-carotene supplementation and incidence of cancer and cardiovascular disease: the Women's Health Study. *J. Nat. Cancer Inst.* **1999**, *91*, 2102– 2106.
33. American Heart Association. *Heart Disease and Stroke Statistics—2008 Update*; American Heart Association: Dallas, TX, USA, 2008.
34. Lloyd-Jones, D.; Adams, R.; Carnethon, M.; De Simone, G.; Ferguson, T.B.; Flegal, K.; Ford, E.; Furie, K.; Go, A.; Greenlund, K.; *et al.* Heart disease and stroke statistics—2009 update: A report from the American Heart Association Statistics Committee and Stroke Statistics Subcommittee. *Circulation* **2009**, *119*, 480–486.
35. Heller, F.R.; Descamps, O.; Hondekijn, J.C. LDL oxidation: Therapeutic perspectives. *Atherosclerosis* **1998**, *137* (Suppl.), S25–S31.
36. Riccioni, G. Carotenoids and cardiovascular disease. *Curr. Atheroscl. Rep.* **2009**, *11*, 434–439.
37. Yla-Herttuala, S. Macrophages and oxidized low density lipoproteins in the pathogenesis of atherosclerosis. *Ann. Med.* **1991**, *23*, 561–567.
38. Koh, W.-P.; Yuan, J.; Wang, R.; Lee, Y.-P.; Lee, B.-L.; Yu, M.C.; Ong, C.N. Plasma carotenoids and risk of acute myocardial infarction in the Singapore Chinese Health Study. *Nutr. Metab. Cardiovasc. Dis.* **2011**, *21*, 1–6.
39. Dugas, T.R.; Morel, D.W.; Harrison, E.H. Dietary supplementation with β-carotene, but not with lycopene, inhibits endothelial cell-mediated oxidation of low-density lipoprotein. *Free Radic. Biol. Med.* **1999**, *26*, 1238–1244.
40. Bub, A.; Watzl, B.; Abrahamse, L.; Delincée, H.; Adam, S.; Wever, J.; Müller, H.; Rechkemmer, G. Moderate intervention with carotenoid-rich vegetable products reduces lipid peroxidation in men. *J. Nutr.* **2000**, *130*, 2200–2206.
41. Sies, H.; Stahl, W. Vitamins E and C, β-carotene and other carotenoids as antioxidants. *Amer. J. Clin. Nutr.* **1995**, *62*, 1315S–1321S.
42. Seppanen, S.M.; Csallany, A.S. The effect of paprika carotenoids on *in vivo* lipid peroxidation measured by urinary excrection of secondary oxidation products. *Nutr. Res.* **2002**, *22*, 1055–1065.

43. Wieslaw, I. Carotenoid orientation: Role in membrane stabilization. In *Carotenoids in Health and Disease*; Krinsky, N.I., Mayne, S.T., Sies, H., Eds.; CRC Press Inc.: Boca Raton, FL, USA, 2004; pp. 151–163.
44. Palozza, P.; Barone, E.; Mancuso, C.; Picci, N. The protective role of carotenoids against 7-keto-cholesterol formation in solution. *Mol. Cell. Biochem.* **2008**, *309*, 61–68.
45. Ben-Amotz, A.; Yatziv, S.; Sela, M.; Greenberg, S.; Rachmilevich, B.; Shawarzman, M.; Weshler, Z. Effect of natural beta-carotene supplementation in children exposed to radiation from the Chernobyl accident. *Rad. Environ. Biophys.* **1998**, *37*, 187–193.
46. El-Baky, H.; El-Baz, F.; El-Baroty, G.S. Production of carotenoids from marine microalgae and its evaluation as safe food colorante and lowering cholesterol agents. *Amer-Eur. J. Agric. Environ. Sci.* **2007**, *2* (6), 792–800.
47. Kohlmeier, L.; Hastings, S.B. Epidemiologic evidence of a role of carotenoids in cardiovascular disease prevention. *Amer. J. Clin. Nutr.* 1995, *62* (Suppl.), 1370S–1377S.
48. Kristenson, M.; Zieden, B.; Kucinskiene, Z.; Elinder, L.S.; Bergdahl, B.; Elwing, B.; Abaravicius, A.; Razinkoviene, L.; Calkauskas, H.; Olsson, A.G. Antioxidant state and mortality from coronary heart diases in Lithuanian and Swedish men: Concomitant cross sectional study on men aged 50. *BMJ* **1997**, *314*, 629–633.
49. Kohlmeier, L.; Kark, J.D.; Gomez-Gracia, E.; Martin, B.C.; Steck, S.E.; Kardinaal, A.; Ringstad, J.; Thamm, M.; Masaev, V.; Riemersma, R.; Martin-Moreno, J.M.; Huttunen, J.K.; Kok, F.J. Lycopene and myocardial infarction risk in the EURAMIC study. *Amer. J. Epidemiol.* **1997**, *146*, 618–626.
50. Klipstein-Grobush, K.; Launer, L.; Geleijnse, J.M.; Boeing, H.; Hofman, A.; Witteman J.C. Serum antioxidant and atherosclerosis. The rotterdam study. *Atherosclerosis* **2000**, *148*, 49–56.
51. Street, D.A.; Comstock, G.W.; Salkeld, R.M.; Schuep, W.; Klag, M. Serum antioxidants and myocardial infarction: Are low levels of carotenoids and $\alpha$-tocopherol risk factors for myocardial infarction? *Circulation* **1994**, *90*, 1154–1161.
52. Ascherio, A.; Rimm, E.B.; Hernán, M.; Giovannucci, E.; Kawachi, I.; Stampfer, M.; Willet, W. Relation of consumption of vitamin E, vitamin C and carotenoids to risk for stroke among men in the United States. *Ann. Intern. Med.* **1999**, *130*, 963–970.
53. Kritchewsky, S.B. β-carotene, carotenoids and the prevention of coronary heart disease. *J. Nutr.* **1999**, *129*, 5–8.
54. Ford, E.S.; Giles, W.H. Serum vitamins, carotenoids and angina pectoris: Findings from the National Health and Nutrition Examination Survey III. *Ann. Epidemiol.* **2000**, *10*, 106–116.

55. Sesso, H.D.; Buring, J.E.; Norkus, E.P.; Gaziano, J.M. Plasma lycopene, other carotenoids, retinol and the risk of cardiovascular disease in men. *Amer. J. Clin. Nutr.* **2005**, *81*, 990–997.
56. Lidebjer, C.; Leanderson, P.; Ernerudh, J.; Jonasson, L. Low plasma levels of oxygenated carotenoids in patients with coronary artery disease. *Nutr. Metab. Cardiovasc. Dis.* **2007**, *17*, 448–456.
57. Khachik, F.; De Moura, F.F.; Zhao, D.Y.; Aebischer, C.P.; Bernstein, P.S. Transformations of selected carotenoids in plasma, liver, and ocular tissues of humans and in nonprimate animalmodels. *Invest. Ophthalmol. Vis. Sci.* **2002**, *43*, 3383–3392.
58. Le, M.; Xiao-Ming, L. Effects of lutein and zeaxanthin on aspects of eye health. *J. Sci. Food Agr.* **2010**, *90*, 2–12.
59. Bates, C.J.; Chen, S.-J.; MacDonald, A.; Holden, R. Quantitation of vitamin E and a carotenoid pigment in cataractous human lenses and the effect of a dietary supplement. *Internat. J. Vit. Nutr. Res.* **1996**, *66*, 316–321.
60. Landrum, J.T.; Bohne, R. Lutein, zeaxanthin and the macular pigment. *Arch. Biochem. Biophys.* **2001**, *385*, 28–40.
61. Hammond, B.R.; Wooten, B.; Snodderly, D.M. Preservation of visual sensitivity of older individuals: Association with macular pigment density. *Invest. Ophthalmol. Vis. Sci.* **1998**, *39*, 397–406.
62. Olmedilla, B.; Granado, F.; Blanco, I.; Vaquero, M.; Cajigal, C. Lutein in patients with cataracts and age-related macular degeneration: A longterm supplementation study. *J. Sci. Food Agr.* **2001**, *81*, 904–909.
63. Snodderly, M.D. Evidence for protection against age-related macular degeneration by carotenoids and antioxidant vitamins. *Amer. J. Clin. Nutr.* **1995**, *62*, 1448–1461.
64. Friedman, D.S.; O'Colmain, B.J.; Muñoz, B.; Tomany, S.C.; McCarty, C.; De Jong, P.T.; Nemesure, B.; Mitchell, P.; Kempen, J.; Congdon, N. Prevalence of age-related macular degeneration in the United States. *Arch. Ophthalmol.* **2004**, *122*, 564–572.
65. Greenstein, V.C.; Chiosi, F.; Baker, P.; Seiple, W.; Holopigian, K.; Braunstein, R.E.; Sparrow, J.R. Scotopic sensitivity and color vision with a blue-lightabsorbing intraocular lens. *J. Cataract. Refract. Surg.* **2007**, *33*, 667–672.
66. Sujak, A.; Gabrielska, J.; Grudzinski, W.; Borc, R.; Mazurek, P.; Gruszecki, W.I. Lutein and zeaxanthin as protectors of lipid membranes against oxidative damage: The structural aspects. *Arch. Biochem. Biophys.* **1999**, *371*, 301–307.
67. Woodall, A.A.; Britton, G.; Jackson, M.J. Carotenoids and protection of phospholipids in solution or in liposomes against oxidation by peroxyl radicals: Relationship between carotenoid structure and protective ability. *Biochim. Biophys. Acta* **1997**, *1336*, 575–586.
68. Jonhson, E.; Hammond, B.R.; Yeum, K.-J.; Wang, X.D.; Castaneda, C.; Snodderly, D.M.; Russell, R.M. Relation among serum and tissue concentrations

of lutein and zeaxanthin and macular pigment density. *Amer. J. Clin. Nutr.* **2000**, *71*, 1555–1562.

69. Sies, H.; Stahl, W. Nutritional protection against skin damage from sunlight. *Annu. Rev. Nutr.* **2004**, *24*, 173–200.
70. Aust, O.; Stahl, W.; Sies, H.; Tronnier, H.; Heinrich, U. Supplementation with tomato-based products increases lycopene, phytofluene, and phytoene levels in human serum and protects against UV-light-induced erythema. *Int. J. Vitam. Nutr. Res.* **2005**, *75*, 54–60.
71. Wertz, K.; Hunziker-Buchwald, P.; Seifert, N.; Riss, G.; Neeb, M.; Steiner, G.; Goralczyk, R. β-carotene interferes with ultraviolet light A-induced gene expression by multiple pathways. *J. Invest. Dermatol.* **2005**, *124*, 428–434.
72. Akyon, Y. Effect of antioxidants on the immune response of *Helicobacter pylori*. *Clin. Microbiol. Infect.* **2002**, *8*, 438–441.
73. Kim, S.H.; Jean, D.; Lim, Y.P.; Lim, C.; An, G. Weight gain limitation and liver protection by long-term feeding of astaxanthin in murines. *J. K. Soc. Appl. Biol. Chem.* **2009**, *52*, 180–185.
74. Bolin, A.P.; Macedo, R.C.; Marin, D.P.; Barros, M.P.; Otton, R. Astaxanthin prevents *in vitro* auto-oxidative injury in human lymphocytes. *Cell Biol. Toxicol.* **2010**, *26*, 457–467.
75. Gale, C.R.; Hall, N.F.; Phillips, D.; Martyn, C.N. Plasma antioxidant vitamins and carotenoids and age-related cataract. *Ophthalmology* **2001**, *108*, 1992–1998.
76. Vu, H.T.; Robman, L.; Hodge, A.; McCarty, C.A.; Taylor H.R. Lutein and zeaxanthin and the risk of cataract: The melbourne visual impairment project. *Invest. Ophthalmol. Vis. Sci.* **2006**, *47*, 3783–3786.
77. Neelam, K.; O'Gorman, N.; Nolan, J.; O'Donovan, O.; Wong, H.B.; Eong, K.G.; Beatty, S. Measurement of macular pigment: Raman spectroscopy *versus* heterochromatic flicker photometry. *Invest. Opthalmol. Vis. Sci.* **2005**, *46*, 1023–1032.
78. Moeller, S.M.; Parekh, N.; Tinker, L.; Ritenbaugh, C.; Blodi, B.; Wallace, R.B.; Mares, J.A. Associations between intermediate age-related macular degeneration and lutein and zeaxanthin in the Carotenoids in Agerelated Eye Disease Study (CAREDS): Ancillary study of the Women's Health Initiative. *Arch. Ophthalmol.* **2006**, *124*, 1151–1162.
79. Christen, W.G.; Liu, S.; Glynn, R.J.; Gaziano, J.M.; Buring, J.E. Dietary carotenoids, vitamins C and E, and risk of cataract in women: A prospective study. *Arch. Ophthalmol.* **2008**, *126*, 1606–1607.
80. Tan, J.S.; Wang, J.J.; Flood, V.; Rochtchina, E.; Smith, W.; Mitchell, P. Dietary antioxidants and the long-term incidence of agerelated macular degeneration: The Blue Mountains Eye Study. *Ophthalmology* **2008**, *115*, 334–341.
81. Dagnelie, G.; Zorge, I.S.; McDonald, T.M. Lutein improves visual function in some patients with retinal degeneration: A pilot study via the Internet. *Optometry* **2000**, *71*, 147–164.

82. Aleman, T.S.; Duncan, J.L.; Bieber, M.L.; De Castro, E.; Marks, D.A.; Gardner, L.M.; Steinberg, J.D.; Cideciyan, A.V.; Maguire, M.G.; Jacobson, S.G. Macular pigment and lutein supplementation in retinitis pigmentosa and Usher syndrome. *Invest. Ophthalmol. Vis. Sci.* **2001**, *42*, 1873–1881.
83. Bahrami, H.; Melia, M.; Dagnelie, G. Lutein supplementation in retinitis pigmentosa: PC-based vision assessment in a randomized double-masked placebo-controlled clinical trial. *BMC Ophthalmol.* **2006**, *6*, 23:1–23:12.
84. Molnár, P.; Deli, J.; Tanaka, T.; Kann, Y.; Tani, S.; Gyémánt, N.; Molnár, J.; Kawases, M. Carotenoids with anti-*Helicobacter pylori* activity from *Golden Delicious* apple. *Phytother. Res.* **2010**, *24*, 644–648.
85. Del Campo, J.A.; Moreno, J.; Rodríguez, H.; Vargas, M.A.; Rivas, J.; Guerrero, M.G. Carotenoid content of chlorophycean microalgae: Factors determining lutein accumulation in *Muriellopsis sp.* (*Chlorophyta*). *J. Biotechnol.* **2000**, *76*, 51–59.
86. Lima, X.T.; Kimball, A.B. Skin carotenoid levels in adult patients with psoriasis. *J. Eur. Acad. Derm. Vener.* **2010**, 25, 11–16
87. Zhang, Z.W.; Patchett, S.E.; Perrett, D.; Domizio, P.; Farthing, M. Gastric *α-tocopherol* and *β-carotene* concentrations in association with *Helicobacter pylori* infection. *Eur J. Gastr. Hep.* **2000**, *12*, 497–503.
88. Kupcinskas, L.; Lafolie, P.; Lignell, A.; Kiudelis, G.; Jonaitis, L.; Adamonis, K.; Andersen, L.P.; Wadstrom, T. Efficacy of the natural antioxidant astaxanthin in the treatment of functional dyspepsia in patients with or without *Helicobacter pylori* infection: A prospective, randomized, double blind, and placebo-controlled study. *Phytomedicine* **2008**, *15*, 391–399.
89. Kusters, J.G.; Van Vliet, A.H.; Kuipers, E.J. Pathogenesis of *Helicobacter pylori* infection. *Clin. Microbiol. Rev.* **2006**, *19*, 449–490.
90. Becker, E.W. Microalgae in human and animal nutrition. In *Handbook of Microalgal Culture*; Richmond, A., Ed.; Blackwell: Oxford, UK, 2004; pp. 312–351.
91. Pulz, O.; Gross, W. Valuable products from biotechnology of microalgae. *Appl. Microbiol. Biotechnol.* **2004**, *65*, 635–648.
92. Belay, A. Current knowledge on potential health benefits of *Spirulina platensis*. *J. Appl. Phycol.* **1993**, *5*, 235–240.
93. Liang, S.; Wueming, L.; Chen, F.; Chen, Z. Current microalgal health food R&D activities in China. *Hydrobiologia* **2004**, *512*, 45–48.
94. Wijffels, R.; Barbosa, M. An outlook on microalgal biofuels. *Science* **2010**, *329*, 796–799.
95. Mata, T.; Martins, A.; Caetano, N. Microalgae for biodiesel production and other applications: A review. *Renew. Sustain. Energ. Rev.* **2011**, *14*, 217–232.
96. Del Campo, J.A.; García-González, M.; Guerrero, M.G. Outdoor cultivation on microalgae for carotenoid production: Current state and perspectives. *Appl. Microbiol. Biotechnol.* **2007**, *74*, 1163–1174.

97. Fernandez-Sevilla, J.M.; Acien Fernandez, F.G.; Molina Grima, E. Biotechnological production of lutein and its applications. *Appl. Microbiol. Biotechnol.* **2010**, *86*, 27–40.

*Samples Availability:* Available from the authors.

*marine drugs*

MDPI

*Review*

# New and Rare Carotenoids Isolated from Marine Bacteria and Their Antioxidant Activities

**Kazutoshi Shindo [1,*] and Norihiko Misawa [2]**

[1] Department of Food and Nutrition, Japan Women's University, 2-8-1 Mejirodai, Bunkyo-ku,
Tokyo 112-8681, Japan

[2] Research Institute for Bioresources and Biotechnology, Ishikawa Prefectural University,
1-308 Suematsu, Nonoichi-shi, Ishikawa 921-8836, Japan; E-Mail: n-misawa@ishikawa-pu.ac.jp

* Author to whom correspondence should be addressed; E-Mail: kshindo@fc.jwu.ac.jp;
Tel./Fax: +81-35-981-3433.

*Received: 10 February 2014; in revised form: 3 March 2014 / Accepted: 4 March 2014 / Published: 19 March 2014*

**Abstract:** Marine bacteria have not been examined as extensively as land bacteria. We screened carotenoids from orange or red pigments-producing marine bacteria belonging to rare or novel species. The new acyclic carotenoids with a $C_{30}$ aglycone, diapolycopenedioc acid xylosylesters A–C and methyl 5-glucosyl-5,6-dihydro-apo-4,4′-lycopenoate, were isolated from the novel Gram-negative bacterium *Rubritalea squalenifaciens*, which belongs to phylum Verrucomicrobia, as well as the low-GC Gram-positive bacterium *Planococcus maritimus* strain iso-3 belonging to the class Bacilli, phylum Firmicutes, respectively. The rare monocyclic $C_{40}$ carotenoids, (3*R*)-saproxanthin and (3*R*,2′*S*)-myxol, were isolated from novel species of Gram-negative bacteria belonging to the family Flavobacteriaceae, phylum Bacteroidetes. In this review, we report the structures and antioxidant activities of these carotenoids, and consider relationships between bacterial phyla and carotenoid structures.

**Keywords:** diapolycopenedioc acid xylosylesters A–C; methyl 5-glucosyl-5,6-dihydro-apo-4,4′-lycopenoate; (3*R*)-saproxanthin; (3*R*,2′*S*)-myxol; antioxidant activity

## 1. Introduction

Some species of bacteria, yeast, and fungi, as well as algae and higher plants, synthesize a large number of carotenoids with different molecular structures, and more than 750 carotenoids with different structures have been isolated from natural sources [1]. Many beneficial pharmaceutical effects of carotenoids have recently been reported. Therefore, evaluating the pharmaceutical potentials of various carotenoids may represent an interesting field in medical research. However, the number of carotenoid species that have been examined for this purpose has been limited, and has included $C_{40}$ carotenoids possessing skeletons composed of 40 carbon atoms, such as dicyclic carotenoids, e.g., β-carotene, α-carotene, β-cryptoxanthin, zeaxanthin, lutein, canthaxanthin, astaxanthin, and fucoxanthin, and the acyclic carotenoid lycopene [2–8]. Difficulties have been associated with identifying natural sources to supply sufficient amounts of new or rare carotenoids, with the exception of carotenoids that can be isolated from a species of higher plants or algae or chemically synthesized. It has therefore been desirable to find cultivable bacteria that produce new or rare carotenoids, since they can easily be reproduced.

Marine bacteria have not been examined as extensively as land bacteria. Thus, the Marine Biotechnology Institute Co., Ltd. (MBI, Kamaishi, Japan) was established in December, 1988, and continued to isolate novel or rare marine bacteria until March, 2008, the number of which reached more than ten thousand [9–12]. Many bacteria have been shown to produce dicyclic or monocyclic $C_{40}$ carotenoids, in addition to some acyclic $C_{30}$ carotenoids with a 30 carbon skeleton [1,13]. The MBI isolated new or rare dicyclic $C_{40}$ carotenoids with the β-carotene (β,β-carotene) skeleton from Gram-negative marine bacteria belonging to the class α-Proteobacteria, phylum Proteobacteria, e.g., astaxanthin glucoside from *Paracoccus* sp. strain N81106 (re-classified from *Agrobacterium aurantiacum*) [14], 2-hydroxyastaxanthin from *Brevundimonas* sp. strain SD212 [15], and 4-ketonostoxanthin 3′-sulfate from *Erythrobacter* sp. strain. PC6 (re-classified from *Flavobacterium* sp. PC-6; MBIC02351) [16]. These marine bacteria were also able to produce astaxanthin [17]. The carotenoid biosynthesis gene clusters of these marine bacteria have been elucidated in detail [17–19].

The generation of free radicals has been suggested to play a major role in the progression of a wide range of pathological disturbances, including myocardial and cerebral ischemia [20], atherosclerosis [21], renal failure [22], inflammation [23], and rheumatoid arthritis [24]. The subsequent peroxidative disintegration of cells and organelle membranes has also been implicated in various pathological processes [25]. Carotenoid pigments, which have been shown to possess strong antioxidant activities, have been attracting increasing attention due to their beneficial effects on human health, e.g., their potential to prevent diseases such as cancer and cardiovascular diseases [26].

We have attempted to identify novel or rare types of carotenoids from yellow or red pigment-producing marine bacteria that were classified to belong to rare or novel species by 16S rRNA analyses since 2002. The results of this screening led to the isolation of diapolycopenedioc acids xylosylesters A–C (new carotenoids) from *Rubritalea squalenifaciens* [27,28], methyl 5-glucosyl-5,6-dihydro-apo-4,4′-lycopenoate (a new carotenoid) from *Planococcus maritimus* [29], and (3*R*)-saproxanthin and (3*R*,2′*S*)-myxol (rare carotenoids) from a novel species belonging to the family Flavobacteriaceae [30]. In this review, we report the structures and antioxidant activities of these carotenoids, and consider relationships between bacterial phyla and carotenoid structures.

## 2. Results

### 2.1. Diapolycopenedioc Acid Xylosylesters A–C from Rubritalea Squalenifaciens [27,28]

A yellow pigment-producing bacterium (strain HOact23$^T$) that was found to produce squalene was isolated from the homogenate of the marine sponge *Halicondria okadai*, which had been collected from the Miura peninsula (Kanagawa, Japan), and was subsequently classified as a novel species in the genus *Rubritalea*, belonging to phylum Verrucomicrobia, based on 16S rRNA gene sequence data. The name proposed for the new taxon was *Rubritalea squalenifaciens* [31], with the type strain HOact23$^T$ (=MBIC08254$^T$ = DSM 18772$^T$).

*R. squalenifaciens* was cultured in 100 mL of medium (1.0% starch, 0.4% yeast extract, and 0.2% peptone in seawater) in a 500 mL Erlenmeyer flask at 30 °C on a rotary shaker at 120 rpm for 2 days, and the carotenoids produced were purified from the cells using chromatographic methods (EtOAc/$H_2O$ partition → silica gel column chromatography $CH_2Cl_2$–MeOH (20:1) → preparative silica gel HPLC $CH_2Cl_2$–MeOH (15:1) → preparative ODS HPLC (MeOH)). Three carotenoids were purified from cells in the 42-liter culture (diapolycopenedioc acids xylosylesters A (**1**) 10.2 mg, B (**2**) 3.0 mg, and C (**3**) 2.2 mg, respectively). The structures of compounds **1**–**3** were determined by HRESI-MS and spectroscopic (UV-Vis, NMR (1D and 2D investigations on $^1H$ and $^{13}C$ nuclei), and $[\alpha]_D$) analyses as shown in Figure 1. Compounds **1**–**3** were all new carotenoids.

Compounds **1–3** possessed diapolycopenedioc acid ($C_{30}$ carotenoid) [32,33] as their aglycone. Diapolycopenedioic acid glucosyl ester and diapolycopenedioic acid diglucosyl were previously shown to be carotenoids that possessed diapolycopenedioc acid as the aglycone [32]. Compounds **1–3** were the first carotenoids to include 2-acyl-D-xylose in their structures.

The antioxidant activity of compound **1** was evaluated using $^1O_2$ suppression activity. Its $IC_{50}$ was 5.1 μM (the $IC_{50}$ values of astaxanthin and β-carotene were 8.9 μM and >100 μM, respectively).

**Figure 1.** The structures of diapolycopenedioc acids A (**1**), B (**2**) and C (**3**).

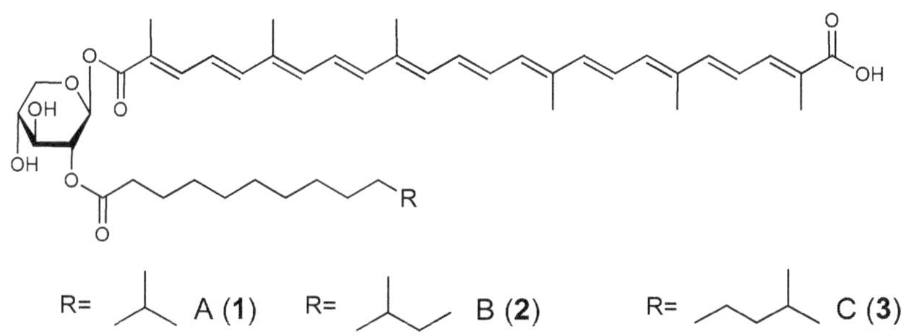


*2.2. Methyl 5-Glucosyl-5,6-Dihydro-Apo-4,4'-Lycopenoate from Planococcus Maritimus [29]*

A yellow pigment-producing bacterium (strain iso-3), which was found to be solvent-tolerant, was isolated as an orange-pigmented colony from the microbial analysis of a sample derived from an intertidal sediment from the Clyde estuary, UK. The 16S rRNA gene sequence of strain iso-3 was the most similar to that of type strain *Planococcus maritimus* (99.5 as a similarity score, and 96.4 as an s_ab score, from the Sequence match analysis of RDP), which belongs to the class Bacilli, phylum Firmicutes, and was identified as *Planococcus maritimus* strain iso-3.

Strain iso-3 was cultured in 100 mL of medium (Marine Broth 2216, Difco, Sparks, MD, USA) in a 500 mL Erlenmeyer flask at 30 °C on a rotary shaker at 120 rpm for 1 day, and the carotenoid produced was purified from the alkaline-digested cells using chromatographic methods (EtOAc/$H_2O$ partition → silica gel column chromatography ($CH_2Cl_2$–MeOH (10:1) → preparative silica gel HPLC $CH_2Cl_2$–MeOH (10:1) → preparative ODS HPLC (96% MeOH)). A total of 2.5 mg of pure methyl 5-glucosyl-5,6-dihydro-apo-4,4'-lycopenoate (**4**) was obtained from the cells in the 18-liter culture, and the structure of compound **4** was determined by HRESI-MS and spectroscopic (UV-Vis, NMR (1D and 2D investigations on $^1H$ and $^{13}C$ nuclei), and $[\alpha]_D$) analyses, as shown in Figure 2. Compound **4** was a new carotenoid. Compound **4** possessed 5,6-dihydro-5-hydroxy-

apo-4,
4′-lycopene-4′-oic acid ($C_{30}$ carotenoid) as its aglycone. Although 4,4′-diapocarotene-4-oic acid [32] was previously reported to be a related $C_{30}$ carotenoid aglycone, 5,6-dihydro and 5-hydroxy functions in the aglycone of compound **4** were demonstrated for the first time. The antioxidant activity
of compound **4** was evaluated using $^1O_2$ suppression activity, and its $IC_{50}$ value was 5.1 μM. We previously described the isolated carotenoid as methyl glucosyl-3,4-dihydro-apo-8′-lycopenoate [29], but confirmed its structure as methyl 5-glucosyl-5,6-dihydro-apo-4,4′-lycopenoate, as shown in this review. Corrigenda is currently being prepared for the previous study.

**Figure 2.** The structure of methyl 5-glucosyl-5,6-dihydro-apo-4,4′-lycopenoate (**4**).

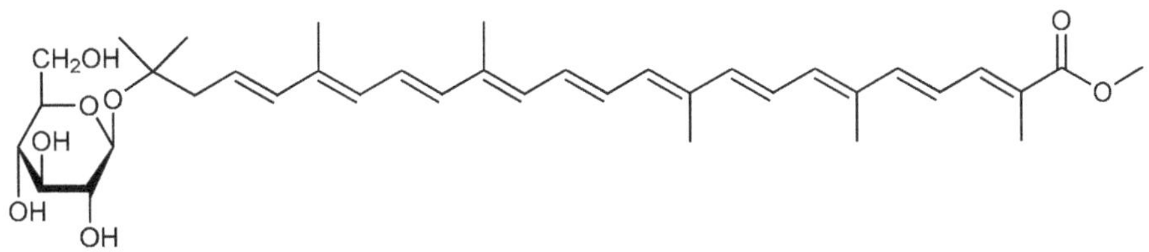


*2.3. (3R)-Saproxanthin and (3R,2′S)-Myxol [30]*

Strain 04OKA-13-27 (MBIC08261) was isolated from the dense mats of filamentous algae from within the territory of damselfish (*Stegastes nigricans*). Strain YM6-073 (MBIC06409) was isolated from a sediment sample collected 0.1 m below the surface of the sea by cultivating for 30 days on an HSV medium. The two marine bacteria, which had been collected off the coast of Okinawa Prefecture, were classified on the basis of this 16S rRNA gene sequences. A similarity search in the databases
of the DNA Data Bank of Japan (DDBJ) and RNA Database Project II (RDPII) showed the 16S rRNA gene sequences of the both strains (04OKA-13-27 and YM6-073) to be 96.5% (1408 bp/1459 bp) similar to *Stanierella latercula* ATCC 23177$^T$, 95.5% (1324 bp/1386 bp) similar to *Gaetbulimicrobium brevivitae* strain SMK-19$^T$, and 94.2% (1306 bp/1386 bp) similar to *Robiginitalea biformata* strain HTCC2501$^T$. The phylogenetic relationship between these strains was deduced with already known species in the family Flavobacteriaceae. The result obtained revealed that the two bacterial strains should be classified as novel species of the family Flavobacteriaceae.

Both 04OKA-13-27 and YM6-073 were cultured in 100 mL of medium (Marine Broth 2216, Difco) in a 500 mL Sakaguchi flask at 30 °C on a rotary shaker at 100 rpm for 1 day, and the carotenoids produced were each purified from the cells using chromatographic methods (EtOAc/$H_2O$ partition → silica gel column chromatography hexane–ethyl acetate (2:1) → preparative silica gel high performance thin layer chromatography (HPTLC; Merck, Darmstadt, Germany)

$CH_2Cl_2$–MeOH
(10:1) → preparative ODS HPLC (MeOH)). A total of 0.3 mg (04OKA-13-27) and 0.5 mg (YM6-073) of pure carotenoids were obtained from the cells of each 2 liter culture, and the carotenoids were identified as (3*R*)-saproxanthin (04OKA-13-27) (**5**) and (3*R*,2′*S*)-myxol (YM6-073) (**6**) by MS, $^1$H-NMR, and CD analyses, respectively (Figure 3).

The antioxidative activities of compounds **5** and **6** were examined using rat brain homogenate model. Compounds **5** and **6** showed potent antioxidant activities ($IC_{50}$ 2.1 μM (**5**) and 6.2 μM (**6**))
($IC_{50}$ 10.9 μM (β-carotene)).

**Figure 3.** The structures of (3*R*)-saproxanthin (**5**) and (3*R*,2′*S*)-myxol (**6**).

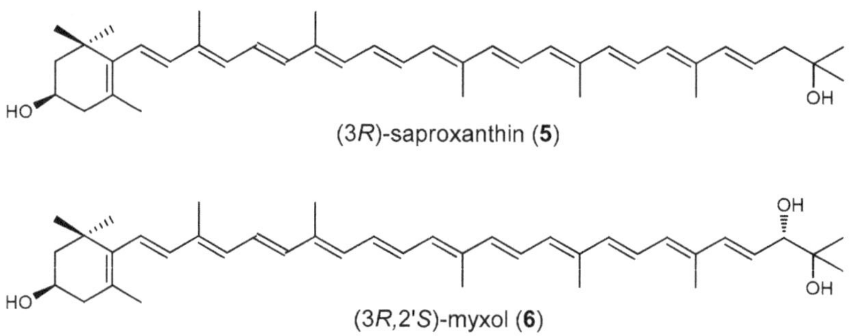


## 3. Discussion

The MBI has isolated approximately 1000 pigment-producing marine bacteria. We selected
10 strains, which were identified as rare or novel species by 16S rRNA, including strain HOact23$^T$ (*Rubritalea squalenifaciens* sp. nov., phylum Verrucomicrobia), strain iso-3 (*Planococcus maritimus*, the class Bacilli, phylum Firmicutes), strain 04OKA-13-27 (novel species of the family Flavobacteriaceae), and strain YM6-073 (novel specie of the family Flavobacteriaceae) from these isolated bacteria.

We found two-type new $C_{30}$ carotenoids diapolyconedioc acid xylosylesters (compound **1**–**3**) from HOact23$^T$ and methyl 5-glucosyl-5,6-dihydro-apo-4,4′-lycopenoate (compound **4**) from iso-3 through the isolation and structural analyses of carotenoids produced by these strains. Acyclic $C_{30}$ carotenoids were previously shown to be contained in land bacteria including *Staphylococcus aureus*, belonging to the class Bacilli, and the methanotrophs *Methylobacterium rhodium* (formerly *Pseudomonas rhodos*), belonging to the class α-Proteobacteria, and *Methylomonas* sp., belonging to the class
γ-Proteobacteria [17,34]. Thus, acyclic $C_{30}$ carotenoids are likely to widely exist in domain bacteria (prokaryotes), *i.e.*, they are present not only in some low-GC Gram-positive bacteria, but also in some phyla in Gram-negative bacteria. The strong

singlet-oxygen-quenching activities of our $C_{30}$ carotenoids also indicated that such $C_{30}$ carotenoids are promising as functional carotenoids, although these *in vivo* functional analyses have not yet been conducted.

We isolated two rare monocyclic $C_{40}$ carotenoids with one 3-hydroxy-β-ring ((3*R*)-saproxanthin (compound **5**) from 04OKA-13-27 and (3*R*,2′*S*)-myxol (compound **6**) from YM6-073), which belong to the family Flavobacteriaceae, phylum Bacteroidetes. (3*R*)-Saproxanthin has only previously been detected from *Saprospira grandis*, which belongs to the family Saprospiracea, phylum Bacteroidetes [35]. Hence, marine bacterial strain 04OKA-13-27 was the second species to produce saproxanthin. (3*R*,2′*S*)-Myxol has only previously been detected in marine bacterial strain P99-3 (MBIC03313), belonging to the family Flavobacteriaceae [15], and in the cyanobacterium *Anabaena variabilis* ATCC 29413, phylum Cyanobacteria [36]. Therefore, marine bacterial strain YM6-073 was the third species to produce myxol. Myxoxanthophyll (myxol 2′-fucoside), which is widely distributed in phylum Cyanobacteria, contains myxol as its aglycone. These findings indicated that such monocyclic $C_{40}$ carotenoids with one 3-hydroxy-β-ring exist in phylum Bacteroidetes as well as phylum Cyanobacteria.

The carotenoids produced by the six other strains isolated were all zeaxanthin, which is a common carotenoid in domain bacteria. Our study may be effective for identifying rare and new carotenoids based on its ratio (4/10). In addition, all the rare and new carotenoids (**1**–**6**) isolated possessed potent antioxidant activities.

## 4. Conclusions

Marine bacteria are likely to produce carotenoids to protect themselves from activated oxygen produced by sunlight (mainly $^1O_2$); therefore, their potent antioxidant activities were expected and reasonable. Therefore, the techniques performed in our study effectively identified new antioxidant carotenoids.

## Author Contributions

Kazutoshi Shindo performed the experiments and wrote the text; Norihiko Misawa supervised the project and corrected the manuscript.

## Conflicts of Interest

The authors declare no conflict of interest.

## References

1. Britton, G.; Liaaen-Jensen, S.; Pfander, H. *Carotenoids Handbook*; Birkhäuser Verlag: Basel, Switzerland, 2004.
2. Nishino, H.; Tokuda, H.; Murakoshi, M.; Satomi, Y.; Masuda, M.; Onozuka, M.; Yamaguchi, S.; Takayasu, J.; Tsuruta, J.; Okuda, M.; *et al.* Cancer prevention by natural carotenoids. *Biofactors* **2000**, *13*, 89–94.
3. Iwamoto, T.; Hosoda, K.; Hirano, R.; Kurata, H.; Matsumoto, A.; Miki, W.; Kamiyama, M.; Itakura, H.; Yamamoto, S.; Kondo, K. Inhibition of low-density lipoprotein oxidation
by astaxanthin. *J. Atheoscler. Thomb.* **2000**, *7*, 216–222.
4. Krinsky, N.I.; Landrum, J.T.; Bone, R.A. Biological mechanism of the protective role of lutein
and zeaxanthin in the eye. *Annu. Rev. Nutr.* **2003**, *23*, 171–201.
5. Naito, Y.; Uchiyama, K.; Aoi, W.; Hasegawa, G.; Nakamura, N.; Yoshida, N.; Maoka, T.; Takahashi, J.; Yoshikawa, T. Prevention of diabetic nephropathy by treatment with astaxanthin in diabetic db/db mice. *Biofactors* **2004**, *20*, 49–59.
6. Hosokawa, M.; Kudo, M.; Maeda, H.; Kohno, H.; Tanaka, T.; Miyashita, K. Fucoxanthin induces apoptosis and enhances the antiproliferative effect of PPARgamma ligand, troglitazone, on colon cancer cells. *Biochim. Biophys. Acta* **2004**, *1675*, 113–119.
7. Talegawkar, S.A.; Johnson, E.J.; Carithers, T.C.; Taylor, H.A., Jr.; Bogle, M.L.; Tucker, K.L. Carotenoid intakes, assessed by food-frequency quesionnaires (FFQs), are associated with serum carotenoid concentrations in the Jackson Heart Study: Validation of the Jackson Heart Study Delta NIRI Adult FFQs. *Public Health Nutr.* **2008**, *11*, 989–997.
8. Sugiura, M.; Nakamura, M.; Ogawa, K.; Ikoma, M. High serum carotenoids associated with lower risk for bone loss and osteoporosis in post-menopausal Japanese female subjects: Prospective cohort study. *PLoS One* **2012**, *7*, e52643.
9. Katsuta, A.; Adachi, K.; Matsuda, S.; Shizuri, Y.; Kasai, H. *Ferrimonas marina* sp. nov. *Int. J. Syst. Evol. Microbiol.* **2005**, *55*, 1851–1855.
10. Tao, L.; Yao, H.; Kasai, H.; Misawa, N.; Cheng, Q. A carotenoid synthesis gene cluster from *Algoriphagus* sp. KK10202C with a novel fusion-type lycopene-β-cyclase gene. *Mol. Genet. Genomics* **2006**, *276*, 79–86.
11. Peng, X.; Adachi, K.; Chen, C.; Kasai, H.; Kanoh, K.; Shizuri, Y.; Misawa, N. Discovery of a marine bacterium producing 4-hydroxybenzoate and its alkyl esters, parabens. *Appl. Environ. Microbiol.* **2006**, *72*, 5556–5561.
12. Matsuo, Y.; Katsuta, A.; Matsuda, S.; Shizuri, Y.; Yokota, A.; Kasai, H. *Mechercharimyces mesophilus* gen. nov., sp. nov. and *Mecherocharimyces asporophorigenens* sp. nov., antitumor substance-producing marine bacteria, and description of *Thermoactinomycetaceae* fam. nov. *Int. J. Syst. Evol. Microbiol.* **2006**, *56*, 2837–2842.
13. Goodwin, T.M. *The Biochemistry of Barotenoids*, 2nd ed.; Plant Chapman and Hall: London, UK, 1980; Volume 1, pp. 291–319.

14. Yokoyama, A.; Adachi, K.; Shizuri, Y. New carotenoid glucosides, astaxanthin glucoside and adonixanthin glucoside, isolated from the astaxanthin-producing marine bacterium, *Agrobacterium aurantiacum*. *J. Nat. Prod.* **1995**, *58*, 1929–1933.
15. Yokoyama, A.; Miki, W.; Izumida, H.; Shizuri, Y. New trihydroxy-keto-carotenoids isolated from an astaxanthin-producing marine bacterium. *Biosci. Biotechnol. Biochem.* **1996**, *60*, 200–203.
16. Yokoyama, A.; Izumida, H.; Shizuri, Y. New carotenoid sulfates isolated from a marine bacterium. *Biosci. Biotechnol. Biochem.* **1996**, *60*, 1877–1878.
17. Misawa, N. Carotenoids. In *Comprehensive Natural Products II Chemistry and Biology*; Mander, L., Liu, H.-W., Eds.; Elsevier: Oxford, UK, 2010; Volume 1, pp. 733–753.
18. Misawa, N.; Satomi, Y.; Kondo, K.; Yokoyama, A.; Kajiwara, S.; Saito, T.; Ohtani, T.; Miki, W. Structure and functional analysis of a marine bacterial carotenoid biosynthesis gene cluster and astaxanthin biosynthetic pathway proposed at the gene level. *J. Bacteriol.* **1995**, *177*, 6575–6584.
19. Nishida, Y.; Adachi, K.; Kasai, H.; Shizuri, Y.; Shindo, K.; Sawabe, A.; Komemushi, S.; Miki, W.; Misawa, N. Elucidation of a carotenoid biosynthesis gene cluster encoding a novel enzyme, 2,2′-β-hydroxylase, from *Brevundimonas* sp. strain SD212 and combinatorial biosynthesis of new or rare xanthophylls. *Appl. Environ. Microbiol.* **2005**, *71*, 4286–4296.
20. Traysman, R.J.; Kirsch, J.R.; Koehler, R.C. Oxygen radical mechanisms of brain injury follwing ischemia and reperfusion. *J. Appl. Physiol.* **1991**, *71*, 1185–1195.
21. Palinski, W.; Rosenfeld, M.E.; Yla, H.S.; Gurtner, G.C.; Socher, S.S.; Butler, S.W.; Parthasarathy, S.; Carew, T.E.; Steinberg, D.; Witztum, J.L. Low density lipoprotein undergoes oxidative modification *in vivo*. *Proc. Natl. Acad. Sci. USA* **1989**, *86*, 1372–1376.
22. Erdogan, C.; Unlucerci, Y.; Turkmen, A.; Kuru, A.; Cetin, O.; Bekpinar, S. The evaluation of oxidative stress in patients with chronic renal failure. *Clin. Chim. Acta* **2002**, *322*, 157–161.
23. Cheeseman, K.H.; Forni, L.G. An investigation of the novel anti-inflammatory agents ONO-3144 and MK-447. Studies on their potential antioxidant activity. *Biochem. Pharmacol.* **1988**, *37*, 4225–4233.
24. Bodamyali, T.; Kanczler, J.M.; Millar, T.M.; Stevens, C.R.; Blake, D.R. Free radicals in rheumatoid arthritis: Mediators and modulators. *Oxid. Stress Dis.* **2004**, *10*, 591–610.
25. Mylonas, C.; Kouretas, D. Lipid peroxidation and tissue damage. *In Vivo* **1999**, *13*, 295–309.
26. Van den Berg, H.; Faulks, R.; Granado, H.F.; Hirschberg, J.; Olmedilla, B.; Sandmann, G.; Southon, S.; Stahl, W. The potential for the improvement of

carotenoid levels in foods and the likely systemic effects. *J. Sci. Food Agric.* **2000**, *80*, 880–912.

27. Shindo, K.; Mikami, K.; Tamesada, E.; Takaichi, S.; Adachi, K.; Misawa, N.; Maoka, T. Diapolycopendioic acid xylosyl ester, a novel glycol-$C_{30}$-carotenoic acid produced by a new marine bacterium *Rubritalea squalenifaciens*. *Tetrahedron Lett.* **2007**, *48*, 2725–2727.
28. Shindo, K.; Asagi, E.; Sano, A.; Hotta, E.; Minemura, N.; Mikami, K.; Tamesada, E.; Misawa, N.; Maoka, T. Diapolycopenedioic acid xlosyl esters A, B, and C, novel antioxidative glycol-$C_{30}$-carotenoic acids produced by a new marine bacterium *Rubritalea squalenifaciens*. *J. Antibiot.* **2008**, *61*, 185–191.
29. Shindo, K.; Endo, M.; Miyake, Y.; Wakasugi, K.; Morritt, D.; Bramley, M.P.; Fraser, D.P.; Kasai, H.; Misawa, N. Methyl glucosyl-3,4-dehydro-apo-8′-lycopenoate, a novel antioxidative glycol-C30-carotenoic acid produced by a marine bacterium *Planococcus maritimus*. *J. Antibiot.* **2008**, *61*, 729–735.
30. Shindo, K.; Kikuta, K.; Suzuki, A.; Katsuta, A.; Kasai, H.; Yasumoto-Hirose, M.; Matsuo, Y.; Misawa, N.; Takaichi, S. Rare carotenoids, (3*R*)-saproxanthin and (3*R*,2′*S*)-myxol, isolated from novel marine bacteria (*Flavobacteriaceae*) and their antioxidative activities. *Appl. Microbiol. Biotechnol.* **2007**, *74*, 1350–1357.
31. Kasai, H.; Katsuta, A.; Sekiguchi, H.; Matsuda, S.; Adachi, K.; Shindo, K.; Yoon, J.; Yokota, A.; Shizuri, Y. *Rubritalea squalnifaciens* sp. nov., a squalene-producing marine bacterium belonging to subdivision 1 of the phylum "*Verrucomicrobia*". *Int. J. Syst. Evol. Microbiol.* **2007**, *57*, 1630–1634.
32. Kleing, H.; Schmitt, R.; Meister, W.; Englert, G.; Thommen, H. New carotenoic acid glucosyl esters from *Pseudomonas rhodos*. *Z. Naturforsch C* **1979**, *34*, 181–185.
33. Osawa, A.; Iki, K.; Sandmann, G.; Shindo, K. Isolation and identification of 4,4′-diapolycopene-4, 4′-dioic acids produced by *Bacillus firmus* GB1 and its singlet oxygen quenching activity. *J. Oleo Sci.* **2013**, *62*, 955–960.
34. Tao, L.; Schenzle, A.; Odom, J.M.; Cheng, Q. Novel carotenoid oxidase involved in biosynthesis of 4,4′-diapolycopene dialdehyde. *Appl. Environ. Microbiol.* **2005**, *71*, 3294–3301.
35. Aasen, A.J.; Liaaen-Jensen, S. The carotenoids of flexibacteria: II. A new xanthophyll from *Saprospira grandis*. *Acta Chem. Scand.* **1966**, *20*, 811–819.
36. Takaichi, S.; Mochimaru, M.; Maoka, T. Presence of free myxol and 4-hydroxymyxol and absence of myxol glycosides in *Anabaena variabilis* ATCC 29413, and proposal of biosynthetic pathway of carotenoids. *Plant Cell Physiol.* **2006**, *47*, 211–216.

*marine drugs*

*Article*

# Carotenoids of Sea Angels *Clione limacina* and *Paedoclione doliiformis* from the Perspective of the Food Chain

**Takashi Maoka [1,*], Takashi Kuwahara [2] and Masanao Narita [3]**

[1] Research Institute for Production Development, Shimogamo-Morimoto-cho 15, Sakyo-ku, Kyoto 606-0805, Japan

[2] Okhotsk Sea Ice Museum of Hokkaido, Motomombetsu, Monbetsu, Hokkaido 094-0023, Japan; E-Mail: kanri1@giza-ryuhyo.com

[3] Hokkaido Research Organization, Abashiri Fisheries Research Institute, Minatomachi, Monbetsu, Hokkaido 094-0011, Japan; E-Mail: narita-masanao@hro.or.jp

* Author to whom correspondence should be addressed; E-Mail: maoka@mbox.kyoto-inet.or.jp; Tel.: +81-75-781-1107; Fax: +81-75-791-7659.

*Received: 16 January 2014; in revised form: 19 February 2014 / Accepted: 3 March 2014 / Published: 10 March 2014*

**Abstract:** Sea angels, *Clione limacina* and *Paedoclione doliiformis*, are small, floating sea slugs belonging to Gastropoda, and their gonads are a bright orange-red color. Sea angels feed exclusively on a small herbivorous sea snail, *Limacina helicina*. Carotenoids in *C. limacina, P. doliiformis*, and *L. helicina* were investigated for comparative biochemical points of view. β-Carotene, zeaxanthin, and diatoxanthin were found to be major carotenoids in *L. helicina*. *L. helicina* accumulated dietary algal carotenoids without modification. On the other hand, keto-carotenoids, such as pectenolone, 7,8-didehydroastaxanthin, and adonixanthin were identified as major carotenoids in the sea angels *C. limacina* and *P. doliiformis*. Sea angels oxidatively metabolize dietary carotenoids and accumulate them in their gonads. Carotenoids in the gonads of sea angels might protect against oxidative stress and enhance reproduction.

**Keywords:** carotenoids; sea angels; food chain; metabolism

## 1. Introduction

*Clione limacina* is a small, floating sea slug (0.5~3 cm body length) belonging to the family Clionidae, which is a group of pelagic marine gastropods. *Paedoclione doliiformis* is a very small, floating sea slug (<0.5 cm body length) that also belongs to the family Clionidae. Their shells are lost during development and their body is gelatinous and transparent. On the other hand, their gonads and viscera are a bright orange-red color. They float by flapping their "wings". Their floating styles resemble angels and so they are called "sea angels" [1]. From spring to autumn, sea angels live at a depth of 200 m in the Sea of Okhotsk. In winter, they migrate to the coast of north Hokkaido with drift ice. The sea angels, *C. limacina* and *P. doliiformis,* are carnivorous and feed exclusively on *Limacina helicina,* which is a small, swimming predatory sea snail belonging to the family Limacinidae (Gastropoda) which feed on micro algae such as diatoms and dinoflagellates [2]. Chum salmon, *Oncorhynchus keta,* is one of the major predators of sea angels in the Okhotsk Sea of north Hokkaido [3,4].

Marine animals, especially marine invertebrates, contain various carotenoids, showing structural diversity [5–8]. New carotenoids are still being discovered in marine animals [9]. In general, animals do not synthesize carotenoids *de novo,* and so those found in animals are either directly accumulated from food or partly modified through metabolic reactions [6–8]. The major metabolic conversions of carotenoids found in marine animals are oxidation, reduction, the translation of double bonds, oxidative cleavage of double bonds, and cleavage of epoxy bonds. Therefore, structural diversity is found in carotenoids of marine animals [6–8].

We have studied carotenoids in several marine animals from chemical and comparative biochemical points of view [8–10]. We have been interested in the orange-red pigments, which were assumed to be carotenoids, of sea angels. Thus, we studied the carotenoids of the sea angels *C. limacina* and *P. doliiformis*. Furthermore, carotenoids in the small snail *L. helicina* and chum salmon *O. keta* were studied from the perspective of the food chain (Figure 1). In the present paper, we describe the carotenoids of these marine animals from the viewpoints of comparative biochemistry and the food chain.

**Figure 1.** Food chains from phytoplankton to salmon via sea angels in the Okhotsk Sea of north Hokkaido.

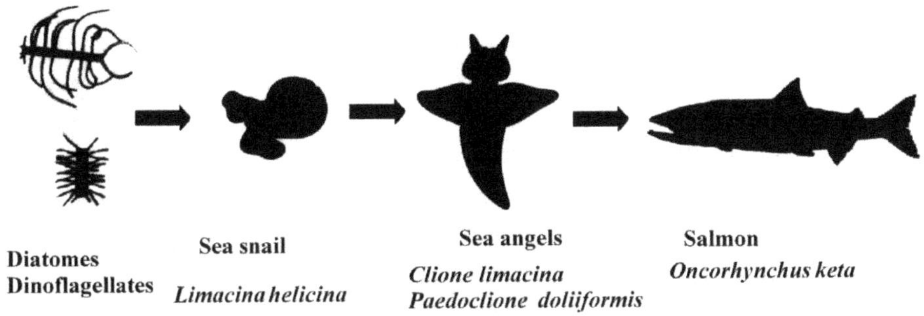


## 2. Results

Structural formulae of carotenoids identified from the sea angels *C. limacina* and *P. doliiformis* and the small herbivorous sea snail *L. helicina* are shown in Figure 2.

**Figure 2.** Structure of carotenoids found in *C. limacina, P. doliiformis,* and *L. helicina.*

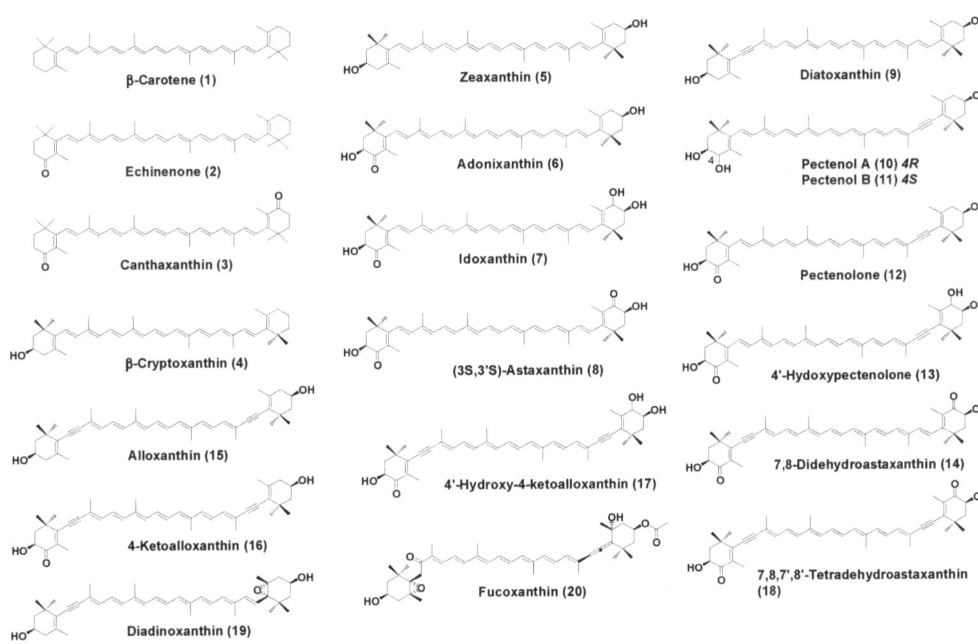


### 2.1. *Carotenoids of L. helicina*

The carotenoid content and composition of the small herbivorous sea snail *L. helicina* are shown in Table 1. The total carotenoid content of *L. helicina* was 21.0 μg/g wet weight. β-Carotene (32.2%), zeaxanthin (24.2%), diatoxanthin (11.1%), and β-cryptoxanthin (10.4%) were found to be major carotenoids. Characteristic algal carotenoids, fucoxanthin (5.2%) and diadinoxanthin (2.4%), were also found.

## 2.2. *Carotenoids of C. limacinea*

The carotenoid content and composition of the sea angel *C. limacina* are shown in Table 1. The total carotenoid content of *C. limacina* was 47.0 μg/g wet weight. Fifteen carotenoids were identified. β-Carotene (27.6%), β-cryptoxanthin (13.5%), and echinenone (9.2%) were found to be major components. Monoacetylenic carotenoids, such as diatoxanthin, 7,8-didehydroastaxanthin, pectenolone, pectenol A, pectenol B, and 4′-hydroxypectenolone, comprised 25.9% of the total carotenoids. Diacetylenic carotenoids, such as alloxanthin, 7,8,7′,8′-tetradehydroastaxanthin, 4-ketoalloxanthin, and 4′-hydroxy-4-ketoalloxanthin, comprised 13.3% of the total carotenoids.

## 2.3. *Carotenoids of P. doliiformis*

The carotenoid content and composition of the sea angel *P. doliiformis* are shown in Table 1. *P. doliiformis* contained 159.8 μg/g wet weight carotenoid in the body. This was about three times higher than that of *C. limacina.* It was uncertain why *P. doliiformis* accumulated carotenoids three times higher than *C. limacina. P. doliiformis* showed more bright red color than *C. limacina.* This might reflect difference of species. The carotenoid composition of *P. doliiformis* was similar to that of *C. limacina.* Pectenolone (30.5%) was found to be a major component, followed by β-cryptoxanthin (12.8%) and β-carotene (10.2%). The monoacetylenic carotenoid diatoxanthin and its oxidative metabolites, 7,8-didehydroastaxanthin, pectinolone, pectenol A, pectenol B, and 4′-hydroxypectenolone, comprised with 25.9% of the total carotenoids. Diacetylenic carotenoids, alloxanthin, 7,8,7′,8′-tetradehydroastaxanthin, 4-ketoalloxanthin, and 4′-hydroxy-4-ketoalloxanthin, comprised 13.3% of the total carotenoids.

**Table 1.** Carotenoids content and composition of *L. helicina, C. limacina,* and *P. doliiformis.*

| | ***L. helicina*** | ***C. limacina*** | ***P. doliiformis*** |
|---|---|---|---|
| **Carotenoid content (μg/g wet weight)** | 21.0 | 47.0 | 159.8 |
| **(μg/specimen)** | 0.70 | 0.75 | 2.68 |
| **Composition** | % | % | % |
| β-Carotene (1) | 32.2 | 27.1 | 10.2 |
| Echinenone (2) | | 9.2 | 6.4 |
| Canthaxanthin (3) | | | 3.3 |
| β-Cryptoxanthin (4) | 10.4 | 13.1 | 12.8 |

| | | | |
|---|---|---|---|
| Zeaxanthin (5) | 24.2 | 1.1 | 1.2 |
| Adonixanthin (6) | | 9.1 | 1.4 |
| Idoxanthin (7) | | | 2.5 |
| (3*S*,3′*S*)-Astaxanthin (8) | | 1.1 | 5.5 |
| Diatoxanthin (9) | 11.1 | 3.5 | 3.6 |
| Pectenol A (10) | | 1.2 | 2.2 |
| Pectenol B (11) | | 1.2 | 2.2 |
| Pectenolone (12) | | 9.2 | 30.5 |
| 4′-Hydroxypectenolone (13) | | 4.2 | 2.5 |
| 7,8-Didehydroastaxanthin (14) | | 4.5 | 6.4 |
| Alloxanthin (15) | 6.4 | 2.1 | 1.1 |
| 4-Ketoalloxanthin (16) | | 3.5 | 2.3 |
| 4′-Hydroxy-4-ketoalloxanthin (17) | | 3.2 | 2.5 |
| 7,8,7′,8′-Tetradehydroastaxanthin (18) | | 4.5 | 1.2 |
| Diadinoxanthin (19) | 2.4 | | |
| Fucoxanthin (20) | 5.2 | | |
| Others | 8.1 | 2.2 | 2.2 |

### 2.4. Carotenoids of the Chum Salmon O. keta

The carotenoid content and composition of flesh of the chum salmon *O. keta*, collected in Monbetsu bay, are shown in Table 2. Acetylenic carotenoids, pectenolone and 7,8-didehydroastaxanthin, were found in *O. keta* as minor carotenoids, along with astaxanthin.

**Table 2.** Carotenoids content and composition of flesh of the chum salmon *O. keta* collected in Monbetsu bay.

| **Carotenoids Content and Composition of Flesh of the Chum Salmon *O. keta*** | |
|---|---|
| **Carotenoid content (μg/g wet weight)** | 0.89 |
| **Composition** | % |
| Astaxanthin * | 83.5 |
| 9-*cis*-Astaxanthin * | 5.1 |
| 13-*cis*-Astaxanthin * | 2.5 |
| 7,8-Didehydroastaxanthin | 0.5 |
| Adonixanthin | 1.1 |
| Pectenolone | 2.5 |
| Others | 4.8 |

* Astaxanthin consisted of three optical isomers (3*R*,3′*R*),(3*R*,3′*S*), and (3*S*,3′*S*) at the ratio of 82:2:16.

## 3. Discussion

It has been reported that animals do not synthesize carotenoids *de novo*, and so those found in animals are either directly accumulated from food or partly modified through metabolic reactions [6–8]. *L. helicina* is a herbivorous animal that feeds on micro algae such as diatoms and dinoflagellates [2]. Sea angels, *C. limacina* and *P.*

*doliiformis* are carnivorous animals that exclusively feed on the small mollusk *L. helicina* [1]. Therefore, carotenoids produced by micro algae are made available to sea angels through *L. helicina* in the food chain. As shown in Table 1, β-carotene, zeaxanthin, diatoxanthin, and β-cryptoxanthin were found to be major carotenoids along with alloxanthin, fucoxanthin, and diadinoxanthin in *L. helicina*. They are characteristic carotenoids in diatoms and microalgae belonging to Cyanophyceae, Rhodophyceae, *etc.* [5,6]. The results indicate that *L. helicina* directly absorbs carotenoids from dietary algae and accumulates them without metabolic modification. On the other hand, keto-carotenoids such as pectenolone, 7,8-didehydroastaxanthin, 4-ketoalloxanthin, and echinenone were found to be major components in sea angels. The results clearly indicate that sea angels oxidatively metabolize ingested carotenoids from *L. helicina*. So, β-carotene was oxidatively converted to astaxanthin via echinenone and canthaxanthin. β-Cryptoxanthin was also metabolized to astaxanthin via asteroidenone and adonirubin, as shown in Figure 3. There are three optical isomers of astaxanthin in nature. However, sea angels contain only one (3*S*,3′*S*) isomer. This shows that hydroxylation at C-3 and/or C-3′ of 4-keto and/or 4′-keto β-end group of carotenoid in sea angels is stereo-selective to form (3*S*,3′*S*)-astaxanthin. This stereo-selective hydroxylation has also been reported in other snails: *Fushinus perplexus*, *F. perplexus ferrugineus*, *F. forceps* [11,12], *Cipangopaludina chinensis laeta*, *Semisulcospia libertina* [13], and *Pomacea canaliculata* [14].

**Figure 3.** Accumulation and metabolic pathways of carotenoids that originated from phytoplankton in the sea angels *C. limacina* and *P. doliiformis*.

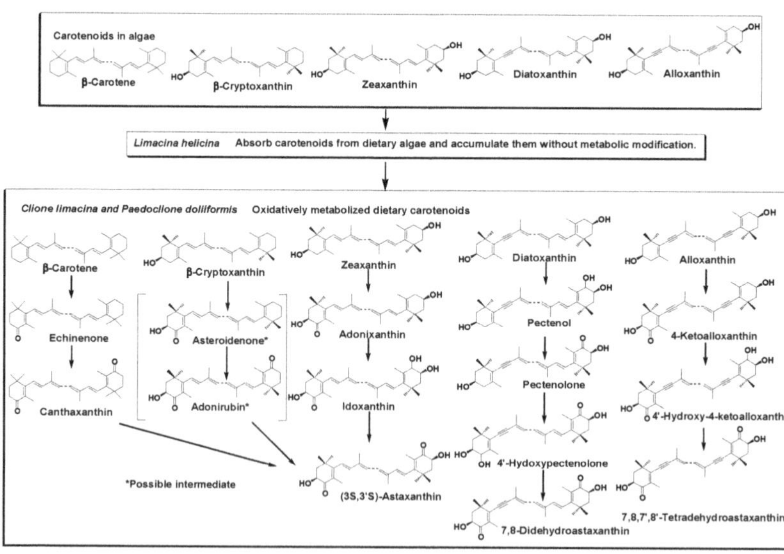

Sea angels also introduced a carbonyl group at C-4 and/or C-4′ in the 3-hydroxy- and/or 3′-hydroxy-β-end group. Namely, zeaxanthin was metabolized to astaxanthin via adonixanthin and idoxanthin. Similarly, an acetylenic carotenoid, diatoxanthin, was metabolized to 7,8-didehydroastaxanthin via pectenol, pectenolone, and 4′-hydroxypectenolone. Alloxanthin was also oxidatively metabolized to 7,8,7′,8′-tetradehydroastaxanthin via 4′-hydroxy-4-ketoalloxanthin, and 4-ketoalloxanthin, as shown in Figure 3. By introducing a carbonyl group at C-4 and/or C-4′ in the 3-hydroxy- and/or 3′-hydroxy-β-end group, carotenoids changed their color from yellow to red. Therefore, the red color of the gonads of sea angels is due to the presence of keto-carotenoids such as pectenolone, 7,8-didehydroastaxanthin, and 7,8,7′,8′-tetradehydroastaxanthin. Epoxy carotenoids, diadinoxanthin and fucoxanthin, which are present in *L. helicina,* were not found in sea angels. It is suggested that sea angels cannot absorb epoxy carotenoids.

Chum salmon, *O. keta,* feeds not only on micro crustaceans but also on sea angels [3–5]. Astaxanthin, which consists of three optical isomers, was found to be a major carotenoid, along with the acetylenic carotenoids pectenolone and 7,8-didehydroastaxanthin, in *O. keta*. It is well-known that astaxanthin in crustaceans such as krill also consists of three optical isomers [6–8,15]. Therefore it is clear that astaxanthin in salmon originates from crustaceans. On the other hand, the acetylenic carotenoids pectenolone and 7,8-didehydroastaxanthin were not found in these crustaceans [6–8,15]. So, they are suggested to originate from sea angels.

It has been reported that marine animals accumulate carotenoids in their gonads, such as astaxanthin in salmon, pectenolone in scallops, and echinenone in sea urchins and that carotenoids are essential for reproduction in marine animals [8]. For example, astaxanthin supplementation in cultured salmon and red sea bream increased ovary development, fertilization, hatching, and larval growth [16]. In the case of sea urchins, supplementation with β-carotene, which was metabolized to echinenone, also increased reproduction and the survival of larvae [17].

As described above, sea angels converted dietary carotenoids to corresponding keto-carotenoids by introducing a carobonyl group and accumulated these keto-carotenoids in their gonads. Several investigators have reported that introducing a carobonyl group at C-4 and/or C-4′ of the β-end group of carotenoids enhanced their antioxidant effects, such as the quenching of singlet oxygen ($^1O_2$), inhibiting lipid peroxidation, and protection from photo-oxidation [18–21]. As well as astaxanthin, pectenolone, an oxidative metabolite of diatoxanthin, showed excellent antioxidative activity by inhibiting lipid peroxidation [22] and quenching singlet oxygen ($^1O_2$). Therefore, keto-carotenoids such as pectenolone may contribute to protection against oxidative stress and promote the reproduction of sea angels through antioxidative activity.

## 4. Experimental Section

### *4.1. General*

The UV-visible (UV-VIS) spectra were recorded with a Hitachi U-2001 (Hitachi High-Technologies Corporation, Tokyo, Japan) in diethyl ether ($Et_2O$). The positive ion electro spray ionization time of flight mass (ESI-TOF MS) spectra were recorded using a Waters Xevo G2S Q TOF mass spectrometer (Waters Corporation, Milford, CT, USA). The $^1$H-NMR (500 MHz) spectra were measured with a Varian UNITY INOVA 500 spectrometer (Agilent Technologies, Santa Clara, CA, USA) in $CDCl_3$ with TMS as an internal standard. HPLC was performed on a Shimadzu LC-6AD with a Shimadzu SPD-6AV spectrophotometer (Shimadzu Corporation, Kyoto, Japan) set at 470 nm. The column used was a 250 × 10 mm i.d., 10 μm Cosmosil 5C18-II (Nacalai Tesque, Kyoto, Japan) with acetone:hexane (3:7, v/v) at a flow rate of 1.0 mL/min, run time of 60 min. The optical purity of astaxanthin was analyzed by chiral HPLC using a 300 × 8 mm i.d., 5 μm Sumichiral OA-2000 (Sumitomo Chemicals, Osaka, Japan) with *n*-hexane/$CHCl_3$/ethanol (48:16:0.8, v/v) at a flow rate of 1.0 mL/min [23].

### *4.2. Animal Specimens*

The sea angel *C. limacina* (30 specimens, 464 mg wet weight) was collected at Monbetsu bay, Monbetsu City, Hokkaido, Japan in December 2011. Another sea angel, *P. doliiformis* (60 specimens, 1041 mg wet weight), was also collected at Monbetsu bay in April 2013. The small sea snail *L. helicina* (6 specimens, 200 mg wet weight) was collected at Monbetsu bay in May 2013. Chum salmon, *O. keta* (3 specimens, five to six years of age), was collected at Monbetsu in September 2013.

### *4.3. Analysis of Carotenoids*

The extraction and identification of carotenoids were carried out according to our routine methods [24]. Carotenoids were extracted from living or fresh animal specimens with acetone. The acetone extract was translated to an ether-hexane (1:1) layer by the addition of water. The total carotenoid contents were calculated employing an extinction coefficient of $E^{1\%}_{cm}$ = 2100 [25] at $\lambda$ max. The ether-hexane solution was evaporated. The residue was subjected to HPLC on silica gel. Carotenoid compositions were estimated by the peak area of the HPLC on silica gel with acetone–hexane (2:8)–(4:6) monitored at 450 nm.

Individual carotenoids were identified by retention time in HPLC, UV-vis (ether), ESI-TOF MS, and $^1$H NMR (500 MHz, $CDCl_3$) in the case of pecetenolone.

### *4.4. Identification of Carotenoids*

β-Carotene (**1**). ESI-TOF MS: *m/z* 536.4372 [M]$^+$ (calcd for $C_{40}H_{56}$, 536.4382); UV-VIS: 425, 449, 475 nm.

Echinenone (**2**). ESI-TOF MS: *m/z* 551.4271 [M + H]$^+$ (calcd for $C_{40}H_{53}O$, 551.4253); UV-VIS: 460 nm.

Canthaxanthin (**3**). ESI-TOF MS: *m/z* 565.4044 [M + H]$^+$ (calcd for $C_{40}H_{53}O_2$, 565.4046); UV-VIS 470 nm.

β-Cryptoxanthin (**4**). ESI-TOF MS: *m/z* 553.4511 [M + H]$^+$ (calcd for $C_{40}H_{53}O$, 553.4409); UV-VIS: (425), 450, 475 nm.

Zeaxanthin (**5**). ESI-TOF MS: *m/z* 569.4353 [M + H]$^+$ (calcd for $C_{40}H_{57}O_2$,569.4359); UV-VIS: (425) 450, 475 nm.

Adonixanthin (**6**). ESI-TOF MS: *m/z* 583.4139 [M + H]$^+$ (calcd for $C_{40}H_{55}O_3$, 583.4151); UV-VIS 460 nm.

Idoxanthin (**7**). ESI-TOF MS: *m/z* 599.4090 [M + H]$^+$ (calcd for $C_{40}H_{55}O_4$, 599.4100); UV-VIS 460 nm.

Astaxanthin (**8**). ESI-TOF MS: *m/z* 597.3942 [M + H]$^+$ (calcd for $C_{40}H_{53}O_4$, 597.3944); UV-VIS 472 nm, Chiral HPLC [13] revealed that astaxanthin fraction in sea angels was consisted of only (3*S*,3′*S*) optical isomers.

Diatoxanthin (**9**). ESI-TOF MS: *m/z* 567.4225 [M + H]$^+$ (calcd for $C_{40}H_{55}O_2$, 567.4202); UV-VIS: (426), 451, 478 nm.

Pectenol A (**10**). ESI-TOF MS: *m/z* 583.4173 [M + H]$^+$ (calcd for $C_{40}H_{55}O_3$, 583.4152); UV-VIS: (426), 451 478 nm.

Pectenol B (**11**). ESI-TOF MS: *m/z* 583.4170 [M + H]$^+$ (calcd for $C_{40}H_{55}O_3$, 583.4152); UV-VIS: (426), 451, 478 nm.

Pectenolone (**12**). ESI-TOF MS: *m/z* 581.3983 [M + H]$^+$ (calcd for $C_{40}H_{53}O_3$, 581.3995); UV-VIS: 460 nm; $^1$H-NMR ($CDCl_3$, 500 MHz) δ 1.15 ($H_3$-16′, s), 1.20 ($H_3$-17′, s), 1.21 ($H_3$-17, s), 1.32 ($H_3$-16, s), 1.45 (H-2′β, dd, *J* = 12, 11), 1.82 (H-2β, d, *J* = 13, 13), 1.84 (H-2′α, ddd, *J* = 12, 4, 1.5), 1.92 ($H_3$-19′, s), 1.95 ($H_3$-19, s), 2.07 (H-2′β, dd, *J* = 18, 10), 2.15 (H-2α, dd, *J* = 13, 6), 2.43 (H-4′α, ddd, *J* = 18, 6, 1.5), 3.68 (OH-3, d, *J* = 2), 3.99 (H-3′, m), 4.32 (H-3, ddd, *J* = 13, 6, 2), 6.22 (H-7, d, *J* = 16), 6,28 (H-14′, d, *J* = 11), 6.30 (H-10, d, *J* = 11), 6.30 (H-14, d, *J* = 11), 6.36 (H-12′, d, *J* = 15), 6.43 (H-8, d, *J* = 16), 6.45 (H-12, d, *J* = 15), 6.45 (H-10′, d, *J* = 11), 6.53 (H-11′, dd, *J* = 15, 11), 6.63 (H-15 and H-15′, m), 6.65 (H-11, dd, *J* = 15, 11).

4′-Hydroxypectenolone (**13**). ESI-TOF MS: *m/z* 597.3942 [M + H]$^+$ (calcd for $C_{40}H_{53}O_4$, 597.3944); UV-VIS: 460 nm.

7,8-Didehydroastaxanthin (**14**). ESI-TOF MS: *m/z* 595.3789 [M + H]+ (calcd for $C_{40}H_{51}O_4$, 595.3787); UV-VIS: 474 nm.

Alloxanthin (**15**). ESI-TOF MS: *m/z* 565.4028 [M + H]+ (calcd for $C_{40}H_{53}O_2$, 565.4046); UV-VIS: (426), 451 478 nm.

4-Ketoalloxanthin (**16**). ESI-TOF MS: *m/z* 579.3851 [M + H]+ (calcd for $C_{40}H_{51}O_3$, 579.3838); UV-VIS: 460 nm.

4′-Hydroxy-4-Ketoalloxanthin (**17**). ESI-TOF MS: *m/z* 595.3801 [M + H]+ (calcd for $C_{40}H_{51}O_4$,595.3787); UV-VIS: 469 nm.

7,8,7′,8′-Tetradehydroastaxanthin (**18**). ESI-TOF MS: *m/z* 593.3649 [M + H]+ (calcd for $C_{40}H_{49}O_4$,593.3631); UV-VIS: 476 nm.

Diadinoxanthin (**19**). ESI-TOF MS: *m/z* 583.4173 [M + H]+ (calcd for $C_{40}H_{55}O_3$, 583.4151); UV-VIS: 420, 433, 470 nm.

Fucoxanthin (**20**). ESI-TOF MS: *m/z* 659.4333 [M + H]+ (calcd for $C_{42}H_{59}O_6$,659.4312); UV-VIS: 445, 470 nm.

### 4.5. $^1O_2$ Quenching Activity of Carotenoids

Quenching activity of $^1O_2$ was measured according to the method described in the literature [26]. $^1O_2$ quenching activities ($IC_{50}$ values) of pectenolone and astaxanthin were 7.9 and 6.5 μM, respectively.

## 5. Conclusions

Carotenoids originating from phytoplankton are accumulated in the sea angels, *C. limacina* and *P. doliiformis,* through eating the herbivorous sea snail, *L. helicina,* in the food chain. In sea angels, dietary carotenoids were oxidatively metabolized, as shown in Figure 3. Sea angels mainly accumulate carotenoids in their gonads. Carotenoids in the gonads of sea angels might protect against oxidative stress and enhance reproduction. Furthermore, carotenoids in sea angels can then be found in salmon through the food chain.

## Acknowledgements

We wish to thank Kazutoshi Shindo and Ayako Osawa; Department of Food and Nutrition; Japan Women's University for measurement of the $^1O_2$ quenching activities of pectenolone and astaxanthin.

## Authors Contributions

T.M. analyzed carotenoids of marine animals. T.K. and M.N. collected marine animals and studied their ecology.

## Conflicts of Interest

The authors declare no conflict of interest.

## References

1. Carol, M.; Ronald, L.; Gilmer, W. *Pelagic Snails*; Stanford University Press: Stanford, CA, USA, 1989.
2. Gilmer, R.W.; Harbison, G.R. Diet of *Limacina helicina* (Gastropoda: Thecosomata) in arctic waters in midsummer. *Mar. Ecol. Prog. Ser.* **1991**, *77*, 125–134.
3. Azuma, T. Biological mechanisms enabling sympatry between salmoids with special reference to sockeye and chum salmon in oceanic water. *Fish. Res.* **1995**, *24*, 291–300.
4. Davis, N.D.; Volkov, A.V.; Efimkin, A.Y.; Kuznetsova, N.A.; Armstrong, J.L.; Sakai, O. Review of basis salmon food habits studies. *N. Pac. Anadr. Fish Comm. Bull.* **2009**, *5*, 197–208.
5. Britton, G.; Liaaen-Jensen, S.; Pfander, H. *Carotenoids Hand Book*; Birkhäuser: Basel, Switzerland, 2004.
6. Liaaen-Jensen, S. Carotenoids in Food Chain. In *Carotenoids Volume 3: Biosynthesis and Metabolism*; Britton, G., Liaaen-Jensen, S., Pfander, H., Eds.; Birkhäuser: Basel, Switzerland, 1998; pp. 359–371.
7. Matsuno, T. Aquatic animal carotenoids. *Fish. Sci.* **2001**, *67*, 771–789.
8. Maoka, T. Carotenoids in marine animals. *Mar. Drugs* **2011**, *9*, 278–293.
9. Maoka, T. Recent progress in structural studies of carotenoids in animals and plants. *Arch. Biochem. Biophys.* **2009**, *483*, 191–195.
10. Maoka, T.; Akimoto, N.; Tsushima, M.; Komemushi, S.; Mezaki, T.; Iwase, F.; Takahashi, Y.; Sameshima, N.; Mori, M.; Sakagami, Y. Carotenoids in marine invertebrates living along the Kuroshio current coast. *Mar. Drugs* **2011**, *9*, 1419–1427.
11. Matsuno, T.; Katagiri, K.; Maoka, T.; Komori, T. Novel reductive metabolic pathways of 4-oxo-β-end group in carotenoids of the spindle shell *Fushinus perplexus*. *Comp. Biochem. Physiol.* **1985**, *81B*, 905–908.
12. Katagiri, K.; Maoka, T.; Matsuno, T. Carotenoids of shellfishes VIII. Comparative biochemical studies of carotenoids in three species of spindle shell, *Fushinus perplexus, F. perplexus ferrugineus* and *F. forceps*. *Comp. Biochem. Physiol.* **1986**, *84B*, 473–476.

13. Maoka, T.; Ochi, J.; Mori, M.; Sakagami, Y. Identification of Carotenoids in the Freshwater Shellfish *Unio douglasiae nipponensis, Anodonta lauta, Cipangopaludina chinensis laeta,* and *Semisulcospira libertina*. *J. Oleo Sci.* **2012**, *61*, 69–74.
14. Tsushima, M.; Katsuyama, M.; Matsuno, T. Metabolism of carotenoids in the apple snail *Pomacea canaliculata*. *Comp. Biochem. Physiol.* **1997**, *118B*, 431–436.
15. Maoka, T.; Katsuyama, M.; Kaneko, N.; Matsuno, T. Stereochemical investigation of carotenoids in the antarctic krill *Euphausia superba*. *Bull. Jpn. Soc. Sci. Fish* **1985**, *51*, 1671–1673.
16. Torrissen, O.J.; Christiansen, R. Requirements for carotenoids in fish diets. *J. Appl. Ichthyol.* **1995**, *11*, 225–230.
17. Tsushima, M.; Kawakami, T.; Mine, M.; Matsuno, T. The role of carotenoids in the development of the sea urchin *Pseudocentrotus depressus*. *Invertebr. Reprod. Dev.* **1997**, *32*, 149–153.
18. Terao, J. Antioxidative activity of beta-carotene-related carotenoids in solution. *Lipids* **1989**, *24*, 659–661.
19. Shimidzu, N.; Goto, M.; Miki, W. Carotenoids as singlet oxygen quenchers in marine organism. *Fish. Sci.* **1996**, *62*, 134–137.
20. Miki, W. Biological functions and activities of animal carotenoids. *Pure Appl. Chem.* **1991**, *63*, 141–146.
21. Maoka, T.; Goto, Y.; Isobe, K.; Fujiwara, Y.; Hashimoto, K.; Mochida, K. Antioxidative activity of capsorubin and related compounds from paprika (*Capsicum annuum*). *J. Oleo Sci.* **2001**, *50*, 663–665.
22. Narita, M.; Maoka, T.; Ebitani, K.; Nishino, H. Characteristics of chemical constituents and red pigment of orange adductor muscle of scallop *Mizuhopecten yessoensis* in the Okhotsk Sea and anti-oxidative activity of the pigment. *Nippon Suisan Gakkaishi* **2013**, *79*, 48–54.
23. Maoka, T.; Komori, T.; Matsuno, T. Direct resolution of diastereomeric carotenoid-I. 3-oxo-β-end group. *J. Chromatogr.* **1985**, *318*, 122–124.
24. Maoka, T.; Akimoto, N. Natural product chemistry in carotenoid, some experimental techniques for structural elucidation and analysis of natural carotenoids. *Carotenoid Sci.* **2008**, *13*, 10–17.
25. Britton, G. UV/Visible Spectroscopy. In *Carotenoids*; Britton, G., Liaaen-Jensen, S., Pfander, H., Eds.; Birkhäuser Verlag: Basel, Switzerland, 1995; pp. 13–62.
26. Maoka, T.; Yasui, H.; Ohmori, A.; Tokuda, H.; Suzuki, N.; Osawa, A.; Shindo, K.; Ishibashi, T. Anti-oxidative, anti-tumor-promoting, and anti-carcinogenic activities of adonirubin and adonixanthin. *J. Oleo Sci.* **2013**, *62*, 181–186.

MDPI
St. Alban-Anlage 66
4052 Basel
Switzerland
Tel. +41 61 683 77 34
Fax +41 61 302 89 18
www.mdpi.com

*Marine Drugs* Editorial Office
E-mail: marinedrugs@mdpi.com
www.mdpi.com/journal/marinedrugs

www.ingramcontent.com/pod-product-compliance
Lightning Source LLC
Chambersburg PA
CBHW061445220926
51850CB00004B/21
* 9 7 8 3 0 3 9 4 3 1 9 0 8 *